Mathematics

From the Birth of Numbers

Mathematics

From the Birth of Numbers

Jan Gullberg

Technical Illustrations

Pär Gullberg

W · W · NORTON & COMPANY

New York · London

Printed in the United States of America

Art and cartoons: *AnnaGreta Nordvall*

Decorations and cartoons: *Ann Gullberg, Kamen, and Kalin*

Camera-ready copy for this book was produced entirely by the author, utilizing a combination of modern desktop-publishing and traditional paste-up methods. The text was written on an Apple Macintosh® Plus computer with Microsoft® Word 3.0 and 4.0; fonts used included New Century Schoolbook, Symbol, ITC Zapf Cancery®, and ITC Zapf Dingbats®. Technical illustrations were generated with a Macintosh® computer. Software used for the illustrations included Adobe Illustrator®, Aldus FreeHand®, CA-Cricket® Draw™, Claris® CAD, and Mathematica®. The three-dimensional, opaque fishnet graphics were rendered with software designed by Pär Gullberg. The main text of the manuscript was produced on a Hewlett-Packard LaserJet 4 MP 600 DPI printer; for technical illustrations an Apple LaserWriter Plus 300 DPI printer and a LaserJet 4 MP 600 DPI printer were used.

The nontechnical illustrations are a mixed bag of work by professionals and nonprofessionals: original works include two cartoons by Leif Ekerling and various illustrations by AnnaGreta Nordvall; reproductions of copyrighted cartoons were obtained from sources specified in the text; the greatest number of illustrations were made by the home-based artist Ann Gullberg; some illustrations are works by Kamen and Kalin, who during their participation in the project aged from nine to ten.

Library of Congress Cataloging-in-Publication Data

Gullberg, Jan.
Mathematics: from the birth of numbers / Jan Gullberg
p. cm.
Includes bibliographical references and index.
ISBN 0-393-04002-X
1. Mathematics—History. I. Title.
QA21.G78 1996
510' .9—dc20

96-13428
CIP

W. W. Norton & Company, Inc., 500 Fifth Avenue, New York, N.Y. 10110
http://web.wwnorton.com

W. W. Norton & Company Ltd., 10 Coptic Street, London WC1A 1PU

7 8 9 0

To Per-Ola Asplund, a friend in deed

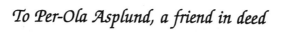

Contents in Brief

Preface

My goal during the ten years that I have been writing this text has been to bring together the history of mathematics with a broad treatment of its foundations.

Three concurrent events led me to this pursuit:

- I had started preparing a translation and revision of a text I had written on units and measurements that I thought could benefit from a 10- to 12-page introductory mathematics text.
- My son studied mathematics as part of an engineering degree, and mathematics became the common subject of our conversations, making the intended introductory text snowball.
- As a physician, I had become disillusioned with an ever-increasing number of "studies" in which medical and other life science matters were proved and disproved and then again, in a cyclic manner, proved, disproved, proved ... with "irrefutable mathematical support"; this use of mathematics strengthened my desire to enjoy mathematics where it was not brutalized.

In the beginning, I thought I would be able to gain something useful for my units and measurements text (now relegated to a locked drawer), but I mainly wrote for myself – as a means of exploring mathematics and its history more completely than I ever had before.

After focusing on numbers and their symbols, my quest continued from the four fundamental rules of arithmetic to calculus and its culmination in differential equations. Along the way, I delved into algebra, the theory of functions, geometry, trigonometry, hyperbolic functions, and analytic geometry, and also followed trails to number theory, symbolic logic, set theory, Boolean algebra, transfinite numbers, topology, fractals, vector analysis, and probability theory.

My draft text was later polished with the help of distinguished professional mathematicians, linguists, and historians. In its final form, I hope this format of mathematics and its history will appeal to readers who have, or would like to develop, a fondness for mathematics.

Among those who have given me constructive criticism, advice, suggestions, and contributions to the text, Johan C. Martensen – bookseller, specialist in technical and scientific publications – has been available throughout the project; his never-ending enthusiasm, encouragement, and wealth of ideas were an enormous asset. I thank Johan and all other individuals wholeheartedly for their help, and ask pardon if I have not always been fully receptive to their wise counsel and erudition.

My thanks are extended to librarians in Sweden, Norway, and the United States for their kind help and efforts in procuring originals and facsimiles of old texts.

I am delighted that Peter Hilton adorns my work with a foreword: "Mathematics in Our Culture".

The knowledge and persistence of my editor, Joseph Wisnovsky, have aided the project enormously. Not only did he marshal the resources of W. W. Norton, but also went through the entire text himself.

I began writing this text in Angola, Indiana, in November 1986 and continued in Forks, Sequim, and Moses Lake, Washington; in Karlskoga, Sweden; in Rjukan, Norway; in Moab, Utah. The final design was completed in June 1996 in Mosjøen, Norway, and I checked the galley proofs in Morton, Washington, in the autumn of 1996. My peripatetic lifestyle has been crucial to the gathering of data and development of ideas.

Exploring Mathematics …

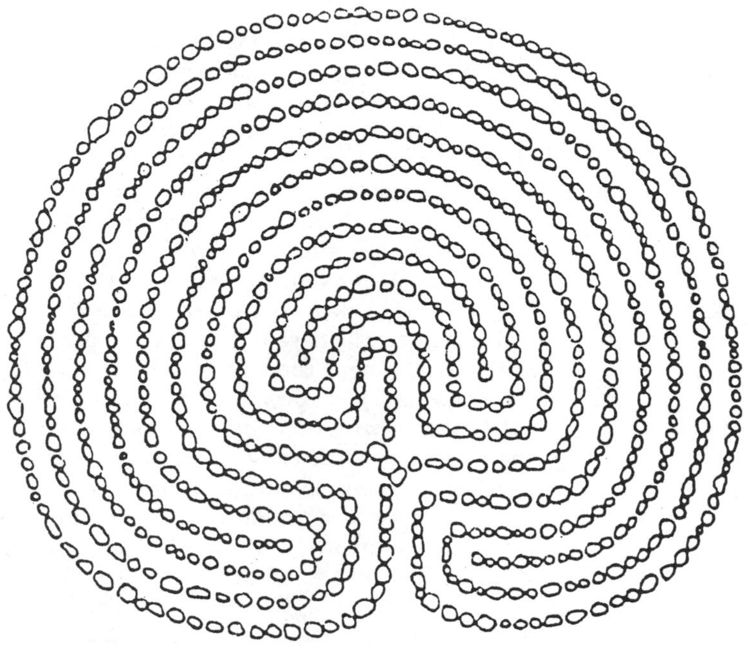

Exploring mathematics is rather like walking through a Troy-town labyrinth, with its path oscillating between the periphery and the center. Spreading out into numerous branches, mathematics is still a wholeness and its core has central themes and goals.

The name Troy-town, or Walls of Troy alludes to the defense walls around the ancient city of Troy (or cities as, in fact, nine cities of Troy have been excavated, one on top of the other). The origin of the Troy-town design is shrouded in mystery and might even antedate the fabled cities.

The oldest Troy-town designs known are carved into the stone wall of a 5000-year-old grave in Sicily. Similar designs are scattered around the world. Pregnant women in Java, Sumatra, and India have long used the pattern to search for peace in meditation; in the United States, Navajo Indians see the pattern as a model of the world's creation. The greatest number of Troy-town labyrinths outlined in stone have been found in Sweden, over 300, mainly along its eastern coastline and on Gotland, in the Baltic. The above design is from a Troy-town outside the medieval town of Visby on Gotland; the actual labyrinth is 19 meters wide.

One cannot get lost in a Troy-town labyrinth. At the endpoint one rests or, according to Nordic folklore, finds one's heart's desire.

Contents

Foreword: Mathematics in Our Culture

The unstated premise of this book – a premise that virtually all mathematicians would agree to – is that mathematics, like music, is worth doing for its own sake. The author is, by profession, a medical man, but he has a love of mathematics and wants others to share his enthusiasm.

This is not to deny the great usefulness of mathematics; this very usefulness, however, tends to conceal and disguise the cultural aspect of mathematics. The role of music suffers no such distortion, for it is clearly an art whose exercise enriches composer, performer, and audience; music does not need to be justified by its contribution to some other aspect of human existence. Nobody asks, after listening to a Beethoven symphony, 'What is the use of that?'. Moreover, mathematics does not gain in utility by having its inherent worth ignored – on the contrary, an appreciation of mathematics and an understanding of its inherent quality and dynamic are necessary in order to be able to apply it effectively.

The first serious error we often meet in considering the role of mathematics is the confusion of education with training. This error, of course, goes far beyond mathematics[1] – our bureaucrats and politicians now use the two terms quite synonymously – but it is particularly meretricious when applied to mathematics. For students, and their parents, believe that mathematics education should consist exclusively of the acquisition of a set of skills that will prove useful in their later careers; so the skills must be learned, that is, committed to memory, and no real understanding need occur. Of course, we cannot, in fact, predict what skills the student will need. What we can predict is that those skills will change and that the student will need to understand and not merely to remember. Adaptability to change is itself a hallmark of successful education, and it is change, not any specific technology, that most aptly characterizes life today and in the foreseeable future. A genuine education enables one to acquire, for oneself, the skills one happens, at a given stage of one's life, to need. A training, on its own, contributes almost nothing to education and produces distressingly ephemeral advantages.

The usefulness of mathematics leads to other, related abuses. Since mathematics is useful, its acquisition must be tested. Since, in the perverse view we are deprecating, it is a skill, it is tested as a skill. Since it is useful, it must be taught to all. Thus the testing problem becomes enormous, and grading by machine becomes commonplace. The result is that standard tests are applied that have almost nothing to do with the acquisition of mathematical understanding and put a premium on brute knowledge and memory, speed and slickness. They provide no opportunity for the student to explain his or her answer and treat all 'wrong' answers as equally wrong. They are, in short, inimical to mathematics itself.

---

[1] It is a wry commentary on the value-system in the United States that one speaks there of 'teacher training' and 'driver education'!

Further, the study of mathematics starts with the teaching of arithmetic, a horrible, wretched subject, far removed from real mathematics, but perceived to be useful. As a result vast numbers of intelligent people become 'mathematics avoiders' even though they have never met mathematics. Their desire to avoid the tedium of elementary arithmetic, with its boring, unappetizing algorithms and pointless drill-calculations, is perfectly natural and healthy.

To those intelligent people, it must seem absurd to liken mathematics to music as an art to be savored and enjoyed even in one's leisure time. Yet that is how it should appear and could appear if it were playing its proper role in our (otherwise) civilized society. Just as an appreciation of music is a hallmark of the educated person, so should be an appreciation of mathematics. The author of this book is a splendid example of an educated person bearing this hallmark.

An Educated Person

There is, we claim, a valid and valuable concept of an educated person. The ancient Greeks had this concept, and it included for them an appreciation of mathematics, especially geometry; on the other hand, the Romans, conspicuously, did not. As Philip Howard wrote, reporting on the 1989 meeting of the British Classical Association: 'The Romans were bad at science. They were practical men who followed intellectual pursuits only if they were useful and profitable, or, in the uncharming vogue phrase, "bankable skills". It is an attitude that is still with us.' Or perhaps one should say that it is again with us. For the broader concept of education was current in the 17th and 18th centuries in Britain and animated those who founded the Royal Society of London; other nations, too, in Europe and elsewhere in the world, had their Enlightenments, their Renaissances. However, the concept began to undergo a curious transition in Victorian England. Certainly, it continued to connote the desire and the ability to go on learning, by reading and other forms of study; and it implied a familiarity with, and appreciation for, poetry, literature, music, the arts, and architecture. However, when the transition was complete, it carried two rather unfortunate connotations as well. The term tended to be applied to members of the leisured class (and, naturally but sadly, predominantly to the masculine sex); and there was no implication of a knowledge or appreciation of science.

This last feature largely persists to this day in the English speaking countries. Exasperation with its manifestations led C. P. Snow to deliver his celebrated Rede Lecture, 'The Two Cultures', in which he deplored the prevalence, in positions of prominence and influence, of people having no knowledge or understanding of the Second Law of Thermodynamics. Of course, Snow was not the first to remark on this phenomenon, but his own popularity as a writer and reputation as a thinker and man of affairs undoubtedly broadened the discussion, if it did not always succeed in deepening it. It is important to recall that Snow's viewpoint that a person was only to be considered educated if he or she was versed in the arts *and* the sciences was by no means universally accepted at the time his lecture was delivered and published.

The problem is very different in certain contemporary societies where cultural philistinism is rampant. Now it is often necessary to argue that a technologically advanced society needs people with an understanding of history, an appreciation of language (theirs and other people's) and an awareness of the 'higher purposes' to which their increased affluence and computerized efficiency give them access. The tendency, to which we have already drawn attention, to confuse education with training has led, at least in the English-speaking world, to a marked down-grading of the study of the arts and the humanities, and to the emergence of the dangerous illusion that a modern industrial society should encourage applied science at the expense of pure science. Such an attitude, had it been widespread 30 years ago, would, for example, have seriously impeded the development of 'medium temperature' superconductors and laser technology. It is surely clear, moreover, that an educated person should have some understanding of both pure and applied aspects of science. If, for example, he or she is to appreciate the actual and potential roles of the computer in our and future societies, then the educated person must appreciate significant parts of science, technology, logic, and mathematics.

What Is Mathematics?

It is my special case that mathematics is common to the 'two cultures', and that therefore the educated person should appreciate it. However, it is not reasonable to expect lay persons to understand the details of sophisticated mathematical reasoning. Nevertheless, enough has surely been said to imply that our educated person must appreciate the role of mathematics in science and technology. Richard Feynman, echoing the thought of Galileo, has said: 'Nature talks to us in the language of mathematics', and it behoves educated people to understand just what this profound aphorism implies. Certainly such an understanding cannot be achieved without a far better insight into what mathematics actually is than is commonly found even among university-trained people today. Yet such insight would not be enough; for mathematics grows and develops in many ways unrelated to science, and thus plays a crucial role in the history of human thought. So I argue that the educated person must understand what mathematics is – but not in the sense of a dictionary definition. Such a person must have an appreciation of mathematical reasoning and of the role of mathematics in the evolution and development of human society. Such an appreciation requires one to understand something of what mathematicians *do* – this would provide a much better working description of what mathematics is, in practice, than any dictionary could provide[2].

Unfortunately very few people have this kind of appreciation of the true nature of mathematics. The most common fallacy, even among otherwise well-informed people, is, as we have said, to confuse mathematics with elementary arithmetic, and thus to suppose that progress in mathematics consists of performing ever more complicated calculations with speed, dexterity, and accuracy. Thus, for example, Dustin Hoffman received an Oscar in 1989 for his portrayal of the autistic brother in the film 'The Rain Man'. This person is an 'idiot savant', capable of performing rapid, totally unmotivated mental calculations such as computing 341×127, or taking $\sqrt{19}$ to 10 decimal places. However, he is described by various critics, in their reviews of the film, as a genius. It is surely unnecessary for me to belabor the point further that such an extraordinary ability, far from being evidence of genius, is usually an indication of stupidity – as in this case. There have been rare exceptions, such as the great German mathematician, astronomer, and physicist Karl Friedrich Gauss, the British civil engineer George Parker Bidder, and the statistician A. C. Aitken. However, it is interesting and significant to note that Gauss's powers of mental calculation declined as his genius grew, thus testifying to the antithesis between calculation and mathematical insight which we are claiming to exist.

In real life a characteristic example of the idiot savant was the Derbyshire agricultural laborer Jedediah Buxton, who was able to demonstrate that the Fermat number $2^{2^5} - 1$ is not prime by actually factorizing it when it was given to him in decimal notation. He performed this feat in his head while carrying out his everyday duties. Buxton was brought to London to be examined by a group of Fellows of the Royal Society. He was taken to the theater to see Garrick perform, to see how he would react to the experience. He reacted by compulsively counting the number of steps Garrick took during the performance! Thus indeed did Buxton symbolically demonstrate that arithmetical skill, of however high an order, is no part of our culture.

This justified conviction, on the part of many sensitive and 'educated' people, that arithmetic cannot be regarded as a part of the individual's cultural equipment, together with the erroneous belief that arithmetic is the essence of mathematics, has led to the widely-held view that mathematics itself is not to be regarded as a component of a liberal education. Thus many aesthetes are to be found positively glorying in their ignorance of, and ineptitude in, mathematics. Such people may proudly announce that they do not understand railway timetables, and are merely vexed by their difficulty in computing the tip in a

---

2 Bertrand Russell's famous dictum that 'mathematics is the subject in which you don't know what you're talking about, and don't care whether what you say is true' is merely a philosophical joke, though a good one!

restaurant. There are not to be found educated people who glory in their inability to use their language[3] or to read properly; anybody with such a difficulty would doubtless seek to conceal it.

Genuine mathematics, then, its methods and its concepts, by contrast with soulless calculation, constitutes one of the finest expressions of the human spirit. The great areas of mathematics – algebra, real analysis, complex analysis, number theory, combinatorics, probability theory, statistics, topology, geometry, and so on – have undoubtedly arisen from our experience of the world around us, in order to systematize that experience, to give it order and coherence, and thereby to enable us to predict and perhaps control future events. However, within each of these areas, and between these areas, progress is very often made with no reference to the real world, but in response to what might be called the mathematician's apprehension of the natural dynamic of mathematics itself.

Mathematics, while essential to science, as the Nobel prize-winning physicist Feynman has so vividly testified, has its own internal dynamic, powerful and subtle. Often, and today most especially, mathematics moves forward not under the stimulus of science but under the stimulus of its own recent advances. Applied mathematicians will often find a piece of mathematics, *developed for its own sake*, the precise tool they need for the expression and elucidation of their scientific problem. And applied mathematicians, once they have modeled the problem mathematically, proceed very much as the pure mathematician would.

Thus it emerges that there is no great difference between the procedures of pure and applied mathematics – there is really only one mathematics. Of course there is the difference that the source of the problem comes in one case from mathematics itself and in the other from the real world; but even here this difference is confined to the original source of the problem – the applied mathematician grappling with a differential equation is, at that point, behaving in a manner indistinguishable from that of a pure mathematician. Indeed, to strike a controversial note, it could be argued that the pure mathematician has opportunities for application that transcend those of the applied mathematician. One can apply mathematics to solve problems in physics – but it is difficult (though not, perhaps, absolutely impossible) to conceive of applying physics to solve problems in mathematics. However, within mathematics, it is perfectly clear, indeed commonplace, that one may, for example, apply algebra to solve a problem in geometry, or apply geometry to solve a problem in algebra.

The foregoing discussion is designed to show, in outline, what mathematics is. My own position, as a mathematician, is to be suitably humble about my own contributions to mathematics, but not modest at all about my claims for

---

[3] Regrettably, statistical evidence is accumulating to indicate that students, offered training rather than education, and fascinated by the potential of modern technology, are increasingly unable to use their language properly to convey their ideas.

mathematics itself. This was the position adopted by my teacher and friend, the great British topologist Henry Whitehead – though he had far less justification for his humility! Whitehead argued that there are relatively few pursuits in life that are inherently worth while – he instanced the making of music and the design of elegant and useful furniture – and that doing, or at least appreciating, mathematics is one of them. It is surely reasonable to equate Whitehead's concept of intrinsically valuable pursuits with our own concept of the desiderata of the educated person. There is, in fact, no doubt in my mind that mathematical appreciation is not only a component part of the education of civilized people, but a pillar of that education. I long for the day when, indeed, mathematics will be appreciated and enjoyed by educated laymen as an art and also respected as the mainstay of science. It has been so in the past, but it is not so now. Is it too optimistic to hope it might be so again?

The present text affords solid grounds for believing it may not be too optimistic. The author, imbued with the spirit that I have tried to convey, has sought to bring to his readers an understanding of the art and the science of mathematics. That he is not a professional mathematician makes his dedication all the more commendable – and remarkable. His text has been scrutinized by a number of leading mathematicians and judged to be of great merit in conveying both the content and the spirit of a true mathematical education. I hope it may enjoy the success it richly deserves and exert the influence it should on present and future generations of intelligent readers.

Peter Hilton

Distinguished Professor of Mathematics, Emeritus
State University of New York at Binghamton

Santa Clara, California, May 1996

Adapted from
"The Mathematical Component of a Good Education",
Miscellanea Mathematica, Springer-Verlag, Berlin, 1991.

Information for the Reader

Although all of the problems in this book have solutions to make it suitable for armchair reading, you may find it worthwhile to have pencil and paper handy. You choose whether to solve a given problem or prove a theorem, simply read and accept, or, perhaps, return to it at another time. There is often more than one way to find an answer or a proof – your approach may be as good as the one in the text.

To avoid tedious work, you should have a good scientific calculator.

• This mark denotes the beginning (left-hand side) of problems and examples, and sometimes, for clarity, the end (right-hand side).

The notations and symbols recommended by the International Standards Organization (ISO) are used throughout, with one exception: a decimal point is used instead of a decimal comma.

Instead of the American and British practice of separating digits with commas,

10,521 1,345,876 10,521.3672389124

in this text, following the recommendations of ISO, digits are marked off in groups of three by spaces, working in *both* directions from the decimal point:

10 521 1 345 876 10 521.367 238 912 4

Greek proper names are in their customary English versions within the text, and in transliterated Greek in the margin; *e.g.*, Euclid within the text, Evcleidis in the margin.

Names originating in Cyrillic, Arabic, and Hebrew alphabets may be difficult to locate in the index, as the English transliterations of such names vary between sources and the reader might be familiar only with a version not used in this text; such inconsistencies and difficulties are even more pronounced for Chinese and Japanese names.

You will not find descriptions of mathematicians' personalities or detailed biographies in this text. Generally only a mathematician's name, dates, nationality, and other fields of scientific activity are given along with a short account of his or her contribution to the topic under discussion. Dates are given to show the time period in history or chronology of events, and are often repeated for convenience and clarity. Titles of knighthood or nobility are generally omitted.

Readers of earlier printings have kindly responded to my invitation to comment on the text and so make it more dependable. I would like to give special thanks to Dennis E. Knopf, Herb Sachs, and Paul H. Stanford for their detailed and savvy comments.

JG

Chapter 1

NUMBERS AND LANGUAGE

Axel Ebbe (1868 - 1941):
"Breaking Away from the Darkness of Ignorance"
University of Lund, Sweden

... and then there was MATHEMATICS

1.0 The Origins of Reckoning

HOBBES Thomas
English philosopher
(1588 - 1679)

Without words there is no possibility of reckoning.

Thomas Hobbes, *Leviathan* (1651)

Scholars generally agree that our ability to count, and our vocabulary of counting, arose to meet practical needs and developed over many thousand years.

Numbers were originally expressed by reference to parts of the human body – nose, eyes, ears, arms, hands, feet, and particularly fingers and toes – in a specific order, making the names of these organs and members double as names of numbers.

These number names were eventually simplified in their spoken forms – long before the advent of writing – as the concept of abstract numeration came into being. Studies of "primitive peoples", who live in isolation from civilization, suggest several stages of development toward strict number names.

Hunting and gathering peoples – such as the aborigines of Australia, Tasmania, and Papua New Guinea – generally have had, or still have, specific names only for the numbers one and two and, sometimes, three, yet they can count to as many as six by combining numbers, for instance like this:

1	one	one
2	two	two
3	two-one	three
4	two-two	one-three
5	two-two-one	two-three
6	two-two-two	three-three

In these languages, speakers refer to everything beyond six as many, much, or plenty – more plenty or less plenty, as the case might be.

Analogies exist in contemporary English. When we count trees, we refer to each tree, counting one, two, three … seven, eight … until the trees cease to be important, *qua* trees; we then speak of a clump of trees, a coppice or copse, a grove, until we find ourselves in a wood, and finally in a forest.

There are many other ways of expressing an undefined number of, most commonly animate, objects: a gaggle of geese, a school of fish, a swarm of bees, a pride of lions. Similarly, many things are thought of, used, or counted in pairs, for instance a brace of fowls, a brace of pistols, a brace of oxen.

Dual, Trial, Quadrual

In many languages nouns and pronouns had, or still have, more than two forms of number, in some cases up to five: singular, dual (2), trial or trinal (3), quadrual (4), and plural.

Trial and quadrual forms are today found nearly exclusively in Austronesian languages (spoken in Australia, Polynesia, and Melanesia). A typical example of dual and trial forms of personal pronouns is offered by the language of one small island in the Melanesian archipelago:

	Singular		Dual		Trial		Plural	
1st person	ainjak	(I)	aijumrau	(we two)	aijumtai	(we three)	aijam	(we)
2nd person	aiek	(thou)	aijaurau	(ye two)	aijautaij	(ye three)	aijaua	(ye)
3rd person	aien	(he or she)	arau	(they two)	ahtaij	(they three)	ar	(they)

In Sanskrit, the classical language of ancient India, nouns had three forms: singular, dual, and plural. These are current in many of its descendant languages and in several unrelated languages:

	Singular (one man)	Dual (two men)	Plural (men, people)
Classical Greek	ántropos	antrópo	ántropoi
Arabic	radjul	radjulayn	radjjaal
Greenlandic (Inüpik)	inuq	innuq	inuit

While Arabic has dual forms for practically all nouns, Hebrew – also of the Semitic family of languages – uses the dual form chiefly for paired body parts (two ears, two feet) or objects consisting of two parts (tongs, scissors). Dual forms of nouns are also found in Old and Middle Irish, and in older forms of Russian.

Whereas Modern English and its precedent forms Old English (*c.* 550 - *c.* 1100) and Middle English (*c.* 1100 - *c.* 1500) lack dual forms in nouns, both Old English and Middle English had dual forms in the 1st and 2nd persons of personal pronouns, as in the table below for early Old English forms:

1st Person	Singular		Dual		Plural	
Nominative	ic	(I)	wit	(we two)	wē	(we)
Genitive	mīn	(mine)	uncer	(of us two)	ūre	(our)
Dative	mē	(to me)	un	(to us two)	ūs	(to us)
Accusative	mec	(me)	uncit	(us two)	usic	(us)

2nd Person	Singular		Dual		Plural	
Nominative	thū	(thou/you)	git	(ye/you two)	gē	(ye)
Genitive	thīn	(thine/your)	incer	(of you two)	ēower	(your)
Dative	thē	(thee/to you)	inc	(to you two)	ēow	(to you)
Accusative	thec	(thee/you)	incit	(you two)	ēowic	(you)

th indicates a single sound as in Modern English *then*, *either*.

After the 13th century, all traces of the dual form in English are gone.

1.1 Numbers and Numerals

von LINNÉ Carl (1707 - 1778)
Swedish botanist; originator of the scientific classification of plants and animals; held the first chair in Medicine at Uppsala for a year (1741), then held chair in Botany.

Although most often attributed to Linné, the maxim *Nomina si nescis* ... was originally used by the Spanish scholar **Isidore of Seville** (*c.* 560 - 636), foremost encyclopedist, composer of the encyclopedic dictionary *Etymologiae*.

Nomina si nescis, perit et cognitio rerum.

Carl von Linné, *Critica botanica* (1737)

"Who knoweth not the names, knoweth not the subject."

The earliest traces of reckoning, uncovered by archaeological excavations in the Middle East, date from about a hundred centuries before our time. Some of these finds suggest a form of counting where markers, counters, or tokens corresponded directly with the things or goods they represented. From these systems, symbols for abstract numbers and quantities eventually developed. These abstract symbols applied to any and every object, whereas symbols previously each represented a specific commodity.

Thus, the mathematical concept of pure quantity was born.

Old English for "number" is *rīm*, which also means "part"; Middle English got *no(u)mbre* via Old French from Latin *numerus*, "number", traced to Greek *νέμο*, "distribute", "share" – recognized in *νέμεσισ* (nemesis), "allotment", "portion" (as of Fate).

If mathematics is the language of science, the language of mathematics is numbers, its grammar being represented by the several kinds of numbers and by the ways in which we use them.

The terms **number** and **numeral** may often be used interchangeably, but in some instances finer distinctions or nuances are required. In this book, we shall observe the following definitions.

Numbers describe magnitude or position.

Cardinal Numbers

Cardinal numbers – zero, one, two, three ... fifty-six... – allow us to count the objects or ideas in a given collection.

Ordinal Numbers

Ordinal numbers – first, second, third ... fifty-sixth... – state the position of individual objects in a sequence.

Latin, *whole, entire*; literally *untouched*

Integers is the term used to denote zero and positive and negative whole numbers: ... $-3, -2, -1, 0, 1, 2, 3$ ....

Latin, *frangere*, "to break"

Fractions are numbers that represent a part, or several equal parts, of a whole: one-half, two-thirds, three-fifths, *etc*.

Numerals are symbols, or combinations of symbols, which describe numbers.

Digits are specific symbols used – alone or in combinations – to denote numbers.

Arabic Numerals

Arabic numerals are written with so-called **Arabic digits**, alone,

0, 1, 2, 3, 4, 5, 6, 7, 8, 9,

or in combinations,

10, 11, 12 ... 1995 ...,

which are modifications of the original Hindu-Arabic number signs.

Roman Numerals

Roman numerals are written with certain letters – I, V, X, L, C, D, M – of the Latin alphabet,

I, II, III, IV, V, VI, VII, VIII, IX, X, XI, XII ... MCMXCV ... .

Number Mystique

Many numbers have mystic or mythical associations: All things good and bad are said to be three; five is found in the pentagram of mystics and soothsayers, and in the pentacle of Solomon; the seven arms of the Hebrew candelabrum, and the Seven Wonders of the World (not always the same but always seven in number); the proverbial nine lives of a cat, and the cat-o'-nine-tails (though that is another kind of cat); the belief that the number thirteen forebodes bad luck; *etc.*

A more realistic, even mathematical, association attaches to the number forty, indicating "many" or "too many" with a touch of annoyance or boredom: We remember Ali Baba and his forty thieves; Moses was away from his people for forty days and forty nights when he was given the Ten Commandments on Mount Sinai; the Children of Israel were footslogging it for forty years through the wilds of the Sinai desert.

In many countries in the Near East a similar meaning attaches to the number sixty, and the "1001 Nights" signifies a kind of "finite infinity".

Arcane coincidence is attached to and between numbers, but:

"... With numbers you can do anything you like. Suppose I have the sacred number 9 and I want to get a number 1314, date of the execution of Jacques de Molay– a date dear to anyone who, like me, professes devotion to the Templar tradition of knighthood. What do I do? I multiply nine by one hundred and forty-six, the fateful day of destruction of Carthage. How did I arrive at this? I divided thirteen hundred and fourteen by two, three, et cetera, until I found a satisfying date. I could also have divided thirteen hundred and fourteen by 6.28, the double of 3.14, and I would have got two hundred and nine. That is the year Attalus I, king of Pergamon, ascended the throne. You see?"

Umberto Eco, *Foucault's Pendulum* (1989)

1.2 Number Names

BACON Roger
(English philosopher and scientist; introduced the concept of the "laws of nature"; *c.* 1220 - 1292)

... Bacon was right in saying that the conquest of learning is achieved through the knowledge of languages.

Umberto Eco, *The Name of the Rose* (1983)

In this section we shall see how numbers are expressed in various languages and how names for numbers developed in English and some other languages. We will also discuss a few items in the history of humankind's ability to count things, among them the days and the months of the year.

Greenlandic

It is hard to imagine a language surpassing Greenlandic in its practice of linking names of body parts with names of numbers. Greenlanders use a twenty-base (vigesimal) system of numeration, divided into recurring five-base (quinary) periods of reckoning referring to the fingers and toes of a person.

We begin by counting the fingers on one hand,

1	atuseq
2	mardluk
3	pingasut
4	sisamat
5	tatdlimat

and continue on the other hand,

6	arfineq	second hand
7	arfineq-mardluk	second-hand, two
8	arfineq-pingasut	second-hand, three
9	arfineq-sisamat	second-hand, four
10	arfineq-tatdlimat	second-hand, five

Ten is also called **qulit**.

We now go on to the toes of the first foot,

11	arkaneq	first foot
12	arkaneq-mardluk	first-foot, two
13	arkaneq-pingasut	first-foot, three
14	arkaneq-sisamat	first-foot, four
15	arkaneq-tatdlimat	first-foot, five

and then the toes of the second foot,

16	arfersaneq	second foot
17	arfersaneq-mardluk	second-foot, two
18	arfersaneq-pingasut	second-foot, three
19	arfersaneq-sisamat	second-foot, four
20	arfersaneq-tatdlimat	second-foot, five

Twenty is also known as **inuk nâvdlugo**, which means "man counted out". So you go on to the next person … and the next … and the next …

32	inûp áipagssâne	arqaneq-mardluk	2nd person, 12
48	inûp pingajugssâne	arfineq-pingasut	3rd person, 8
78	inûp sisamagssâne	arfersanek-pingasut	4th person, 18
94	inûp tatdlimagssâne	arqaneq-sisamat	5th person, 14

… and so on.

For 100 and 1000, Greenlanders have created words based on Danish names for these numbers,

100	=	untrite	Danish: hundrede
1000	=	tûsinte	tusinde
10^6	=	miliûne	million
10^{12}	=	piliûne	billion

Greenlanders also often turn to the Danish numbering system to avoid the longer and tongue-twisting Inüpik expressions for numerals higher than twenty, although this practice is considered not quite polite in good company.

Purists will use the proper Inüpik words, so when you are telling your friend Uvdluriaq about the ûvak (codfish) – sorry, ûvarssuaq (big codfish) – that you caught in

arfersaneq-sisamanik untritigdlit arfiniq-pingasunigdlo quligdit sisamatdlo

– that is, 1989 – be sure to say that it measured

inûp tatdlimagssâne arfersaneq-pingasut

– 98 centimeters, between the eyes, that is!

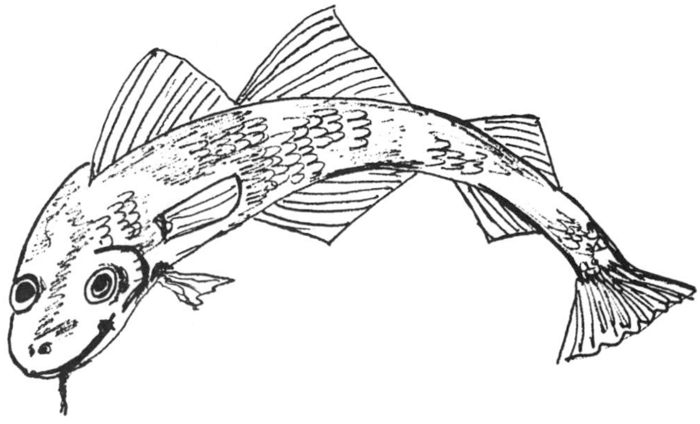

Maori Counting

New Zealand

New Zealand was first settled by a Polynesian people who settled mainly on the North Island. These inhabitants lived isolated for a long time and originally had no name for themselves; later they adopted the name Maori – meaning *normal* – to distinguish themselves from the European settlers.

The Maori used a decimal system of notation and generally could count correctly to 180, but were a bit vague about higher numbers – the word *rau* means 100, or "many"; *mano* is 1000, or "very many".

There are traces of vigesimal counting in the system:

- **hokotoru makete** means three-score and maybe a little more – that is, sixty-odd;
- **hokowhitu** is seven-score, that is, 140 – exactly or just about.

When counting *things*, cardinal numbers are preceded by the word **ka**; when counting *persons*, prefixed by **toko**. Ordinal numbers are preceded by **tua**.

	Cardinal numbers when used singly	**for things**	**for persons**	**Ordinal numbers**
one	tahi	ka tahi	(to)kotahi	tua tahi
two	rua	ka rua	tokorua	tua rua
three	toru	ka toru	tokotoru	tua toru
four	wha	ka wha	tokowha	tua wha
...	...	...	...	...

Many things are counted in pairs:

ko tahi pu	means	a pair, a brace, or just two
ko rua pu, tautahi	means	two brace and one, that is, five

Today, virtually all Maori speak English; only a minority of them also speak the Maori language.

Indo-European Languages

The majority of languages spoken today in Europe and the Americas belong to the Indo-European family of languages. The homeland was long believed to be on the Indian subcontinent, but research makes that unlikely. A modern hypothesis places the origin of Indo-European languages to southeastern Europe in the area between the Black and Caspian Seas – the Caucasus. The development from a common source could have begun some 10 000 years ago, but the indication of time is as controversial as the place of origin. Proto-Indo-European might, in turn, have been a sister language of an even more ancient family of languages.

Protolanguage: an assumed or recorded ancestral language
Merriam-Webster™

The oldest language of the Indic branch of the Indo-European family is **Sanskrit**, which was spoken until the 12th century and is still used as a literary language. Being the language of the sacred books of Hinduism and Buddhism, Sanskrit has remained virtually unchanged over the past 3000 years and is therefore of great importance to linguists studying the history and interrelations of Indo-European languages.

Number names are an important means in the research of the relationship between languages.

Most of the European languages that are descended from Proto-Indo-European belong to one of the following subgroups:

- Greek languages, in Southeastern Europe;
- Italic languages, in Southern Europe;
- Celtic languages, in Central Europe and in the British Isles;
- Slavic languages, in Eastern Europe;
- Germanic languages, in Northern and Western Europe.

Greek

Classical Greek and **Latin** occur in a wealth of scientific expressions, brought to us from the times when these languages were used in communications between learned people in the West. Today, the adoption of words from Classical Greek and Latin into scientific expressions is justified by the fact that the meaning of words in a "dead language" is less apt to change, permitting the formation of scientific terms that allow an exact and stable interpretation.

Greek has a longer documented history than any other Indo-European language – from Ancient Greek in the 14th century B.C. to Modern Greek today. Classical Greek dates from the 5th and 4th centuries B.C.

Greek number names meet us in a myriad of current English words:

monograph	pentagram	ennead
diphthong	hexameter	decathlon
triglyph	heptane	hendecagon
tetralogy	octopus	the Dodecanese

... hecatomb ... kilogram ...

In fact, *myriad* is also of Greek origin, possibly stemming from *mū́rmex* – Greek for "ant" – suggesting abundance; the word *mū́rmex* later acquired the meaning of "countless", and finally *mū́rioi* came to mean 10000.

Alphabets

The Phoenicians, who lived in what is today roughly Lebanon, were famous as tradesmen and for their skill as seafarers, but their most remarkable contribution was their **phonetic alphabet**, and with it the idea of sounding out words phonetically. This was developed between *c.* 1700 and *c.* 1500 B.C., and was a model and forerunner of the Hebrew and Greek alphabets.

The Greeks became acquainted with the Phoenician alphabet around the 10th century B.C. and adapted it to suit their own language.

The Greek Alphabet

The Greeks modified some of the Phoenician characters and added more letters to represent sounds in Greek not given in the earlier symbols. The Greek alphabet is still essentially a direct adoption of the ancient Phoenician alphabet, with the basic phonetic values of the original characters retained.

The Greeks also embraced the names of the Phoenician characters. For instance, the name of the letter *alpha* – originally shaped like an ox head – can be traced back to the Phoenician word for ox (*ălĕph*), chosen to represent the "a" sound; *beta* goes back to the word for house (*bēth*), representing the "b" sound.

Typeset: Upright		Typeset: Sloping			Handwritten	
Α	α	*Α*	*α*	alpha	*A*	*α*
Β	β	*Β*	*β*	beta	*B*	*β*
Γ	γ	*Γ*	*γ*	gamma	*Γ*	*γ*
Δ	δ	*Δ*	*δ*	delta	*Δ*	*δ*
Ε	ε, ∈	*Ε*	*ε, ∈*	epsilon	*E*	*ε* (∈)
Ζ	ζ	*Ζ*	*ζ*	zeta	*Z*	*ζ*
Η	η	*Η*	*η*	eta	*H*	*η*
Θ	ϑ, θ	*θ*	*ϑ, θ*	theta	*Θ*	*ϑ* (θ)
Ι	ι	*Ι*	*ι*	iota	*I*	*ι*
Κ	κ	*Κ*	*κ*	kappa	*K*	*ϰ* (κ)
Λ	λ	*Λ*	*λ*	lambda	*Λ*	*λ*
Μ	μ	*Μ*	*μ*	mu	*M*	*μ*
Ν	ν	*Ν*	*ν*	nu	*N*	*ν*
Ξ	ξ	*Ξ*	*ξ*	xi	*Ξ*	*ξ*
Ο	ο	*Ο*	*ο*	omicron	*O*	*o*
Π	π	*Π*	*π*	pi	*Π*	*π*
Ρ	ρ	*Ρ*	*ρ*	rho	*P*	*ρ*
Σ	σ, ς*	*Σ*	*σ, ς* *	sigma	*Σ*	*σ ς* *
Τ	τ	*Τ*	*τ*	tau	*T*	*τ*
Υ	υ	*Υ*	*υ*	upsilon	*Υ*	*υ*
Φ	φ, ϕ	*Φ*	*φ, ϕ*	phi	*Φ*	*φ* (ϕ)
Χ	χ	*Χ*	*χ*	chi	*X*	*χ*
Ψ	ψ	*Ψ*	*ψ*	psi	*Ψ*	*ψ*
Ω	ω	*Ω*	*ω*	omega	*Ω*	*ω*

* At the end of a word

Upright letters are used in science and technology as symbols for units of measurement (*e.g.*, Ω for ohm), and **sloping letters** (italics) as symbols for variable quantities (*e.g.*, λ for wavelength).

To distinguish between upright and sloping handwritten letters is hardly possible unless the writer indicates that, for instance, underlined letters signify sloping letters (italics).

Transliteration of Greek Character

Greek characters are transliterated into Latin in the following manner:

Greek			Latin	Greek			Latin
alpha	Α	α	A a	nu	Ν	ν	N n
beta	Β	β	B b	xi	Ξ	ξ	X x
gamma	Γ	γ	G g	omicron	Ο	ο	O o
delta	Δ	δ	D d	pi	Π	π	P p
epsilon	Ε	ε	E e	rho	Ρ	ρ	R r
zeta	Ζ	ζ	Z z	sigma	Σ	σ ς*	S s
eta	Η	η	Ē ē	tau	Τ	τ	T t
theta	Θ	θ	Th th	upsilon	Υ	υ	U u (or Y, y)
iota	Ι,	ι	I i	phi	Φ	φ	F f
kappa	Κ	κ	K k	chi	Χ	χ	Ch, ch (or H, h)
lambda	Λ	λ	L l	psi	Ψ	ψ	Ps ps
mu	Μ	μ	M m	omega	Ω	ω	Ō ō

* At the end of a word

In Classical Greek, a vowel at the beginning of a word has either a ʽ or a ʼ over it, indicating a rough and a smooth aspirate sound, respectively. In transliteration, only the rough aspirate is denoted, either by ʽ in front of the vowel or by beginning the word with an "h"; thus, ʽεπτα (seven) can be written either ʽeptá or, as in this book, heptá.

Greek Heritage

Notable descendants of the Greek alphabet were the Etruscan alphabet and the Cyrillic alphabet.

The **Etruscan alphabet** was modeled on the Greek alphabet in the 7th century B.C. by the Etruscans, who enjoyed the most highly developed civilization of all peoples in the Italian peninsula before the rise of Rome. The Etruscan alphabet, in turn, developed into the **Latin alphabet**, which is today employed as the script of most Western languages.

KYRILLOS (*c.* 827 - 869)

METHODOS (*c.* 825 - 884)

The original **Cyrillic alphabet** was invented in the 9th century by the Greek missionary St. Cyril. For his translation of Holy Writ into what became known as Old Church Slavonic, St. Cyril came to be referred to as one of the two "Apostles of the Slavs", the other being his elder brother St. Methodos.

The original Cyrillic alphabet had 43 characters; modern Slavic languages have 30 - 34 letters.

Names of Numbers in Classical Greek

	Cardinal Numbers	Ordinal Numbers	Numeral Adverbs	
1	heis, mía, hén	protos	hápax *	once, 1 time
2	dúo	deúteros	dís	twice, 2 times
3	treis, tría	trítos	trís	thrice, 3 times
4	téssares, téssara	tétartos	tetrákis	4 times
5	pénte	pémptos	pentákis	5 times
6	héx	héktos	hexákis	6 times
7	heptá	hébdomos	heptákis	7 times
8	októ	ógdoos	oktákis	8 times
9	ennéa	énatos	enákis	9 times
10	déka	dékatos	dekákis	10 times
11	héndeka	hendékatos	hendekákis	11 times
12	dódeka	dodékatos	dodekákis	12 times
13	treiskaídeka	triskaidékatos	triskaidekákis	13 times
14	tessareskaídeka	tessarakaidékatos	tessarakaidekákis	14 times
15	pentekaídeka	pentekaidékatos	pentakaidekákis	15 times
16	hekkaídeka	hekkaidékatos	hekkaidekákis	16 times
17	heptakaídeka	heptakaidékatos	heptakaidekákis	17 times
18	októkaídeka	oktokaidékatos	oktokaidekákis	18 times
19	enneakaídeka	enneakaidékatos	enneakaidekákis	19 times
20	eíkosi	eikostós	eikosákis	20 times
30	triā́konta	triākostós	triākontákis	30 times
40	tessarákonta	tessarakostós	tessarakontákis	40 times
50	pentékonta	pentekostós	pentekontákis	50 times
60	hexékonta	hexekostós	hexekontákis	60 times
70	hebdomékonta	hebdomekostós	hebdomekontákis	70 times
80	ogdoékonta	ogdoekostós	ogdoekontákis	80 times
90	enenékonta	enenekostós	enenekontákis	90 times
100	hekatón	hekatostós	hekatontákis	100 times
200	diākosioi, -ai, a	diākosiostós	diākosiákis	200 times
300	triākosioi, -ai, -a	triākosiostós	triākosákis	300 times
1000	chī́lioi, -ai, -a	chīliostós	chīliákis	1000 times
2000	dischī́lioi, -ai, -a	dischīliostós	dischīliákis	2000 times
10 000	mū́rioi, -ai, -a	mūriostós	mūriákis	10 000 times

* As in *hápax legómenon*, a word or form occurring only once

More than one form indicates inflection by gender – masculine, feminine, neuter.

The number noun **mónos**, English **monad**, meaning "single", "alone" is a common prefix signifying uniqueness or seclusion – monandry, monarch, monastery.

Greek mathematicians in classical times had no symbol for zero, nor any use for it, as they had not a full concept of its mathematical significance, but words denoting nothingness certainly existed:

oudeís, oudemía, oudén no one, nothing;

the three forms indicate inflection by gender – masculine, feminine, and neuter, respectively.

Many Greek words were taken over by the Romans and Latinized and have come down to us in these forms, which are the forms often used in current scientific terminology.

Names of numbers in Modern Greek and Classical Greek differ little from one another.

Latin

What is thy name? And he answered, saying, My name is Legion: for we are many.

Mark 5: 9

As with Greek, the imprints of Latin number names on the English language are legion:

unicorn	quintessence	nonagenarian
biceps	sextant	decade
tricycle	Septuagint	centennial
quartet	octave	millipede

Legion itself is also of Latin origin, from *legio* which means "army", suggesting untold numbers.

	Cardinal Numbers	**Ordinal Numbers**	**Numeral Adverbs**	**Distributives**
1	unus, una, unum	primus, *first*	semel, *once*	singuli, *one by one*
2	duo, duae, duo	secundus, alter, *second*	bis, *twice*	bini, *two by two*
3	tres, tres, tria	tertius	ter, *thrice*	terni *or* trini,
4	quattuor	quartus	quater	quaterni
5	quinque	quintus	quinquies	quini
6	sex	sextus	sexies	seni
7	septem	septimus	septies	septeni
8	octo	octavus	octies	octoni
9	novem	nonus14	novies	noveni
10	decem	decimus	decies	deni
11	undecim	undecimus	undecies	undeni
12	duodecim	duodecimus	duodecies	duodeni
13	tredecim	tertius decimus	ter decies	terni deni
14	quattuordecim	quartus decimus	quater decies	quaterni deni
15	quindecim	quintus decimus	quinquies decies	quini deni
16	sedecim	sextus decimus	sexies decies	seni deni
17	septendecim	septimus decimus	septies decies	septeni deni
18	duodeviginti	duodevicesimus	duodevicies	duodeviceni
19	undeviginti	undevicesimus	undevicies	undeviceni
20	viginti	vicesimus	vicies	viceni
21	viginti unus	vicesimus primus	semel et vicies	viceni singuli
30	triginta	tricesimus	tricies	triceni
40	quadraginta	quadragesimus	quadragies	quadrageni
50	quinquaginta	quinquagesimus	quinquagies	quinquageni
60	sexaginta	sexagesimus	sexagies	sexageni
70	septuaginta	septuagesimus	septuagies	septuageni
80	octoginta	octogesimus	octogies	octogeni
90	nonaginta	nonagesimus	nonagies	nonageni
100	centum	centesimus	centies	centeni
101	centum unus	centesimus primus	centies semel	centeni singuli
200	ducenti	ducentesimus	ducenties	duceni
300	trecenti	trecentesimus	trecenties	treceni
400	quadringenti	quadringentesimus	quadringenties	quadringeni
500	quingenti	quingentesimus	quingenties	quingeni
600	sescenti	sescentesimus	sescenties	sesceni
700	septingenti	septingentesimus	septingenties	septingeni
800	octingenti	octingentesimus	octingenties	octingeni
900	nongenti	nongentesimus	noningenties	nongeni
1000	mille	millesimus	milies	singula milie
2000	duo milia	bis millesimus	bis milies	bina milia
10 000	decem milia	decimus millesimus	decies milies	deni milia
10^5	centum milia	centies millesimus	centies milies	centena milia
10^6	decies centena milia	decies centies millesimus	decies centies milies	decies centena milia

More than one form indicates inflection by gender – masculine, feminine, neuter. Cardinal hundreds 200 - 900 are inflected by gender: -i, -ae, -a. All ordinal numbers are inflected by gender and number: -us, -a, -um; -i, -ae, -a.

Observe that

18 :	duodeviginti	means	two-from-twenty,
19 :	undeviginti	means	one-from-twenty;

analogously,

28 :	duodetriginta	means	two-from-thirty,
29 :	undetriginta	means	one-from-thirty, *etc.*

Etruscan Language

These forms were inherited from the **Etruscan language**, which was spoken in Etruria, the ancient country of western central Italy, now in Tuscany and part of Umbria:

17 :	ci-em zathrum	three-from-twenty
18 :	esl-em zathrum	two-from-twenty
19 :	thun-em zathrum	one-from-twenty
27 :	ci-em-ce-alch	three-from-thirty
28 :	esl-em-ce-alch	two-from-thirty
29 :	thun-em-ce-alch	one-from-thirty

millio

In Medieval Latin, the word **millio** appeared as an augmentative form of *mille*, meaning “great thousand”, that is to say a million (= 10^6).

Like the Greeks, Romans had neither a symbol for zero nor a clear concept of its mathematical significance, but words denoting nothingness existed:

nullus, –a, –um	no one
nulli (plural)	
nil, nihil	nothing

A comparison of the Greek and Latin names for lower numbers reveals several similarities pointing to early Greek influence. The schoolmaster scans:

Greek:		
heis, dúo, treis –	Unus, duo, tres – tu adire scholam debes.	*Unus, duo, tres –* you ought to go to school.
téssares, pénte, héx –	Quattuor, quinque, sex – ibi non es rex.	*Quattuor, quinque, sex –* there you are no ruler.
heptá, októ, ennéa –	Septem, octo, novem – te faciunt officiosum civem.	*Septem, octo, novem –* you be made a civil servant.
déka –	Decem – jam comprendis legem.	*Decem –* now you know the terms.

To express **fractions**, Latin has words for subunits of twelve:

			meaning	based on
	1/12	uncia	one-twelfth of unity	*unus*, one
1/6	2/12	sextans	one-sixth	*sex*, six
1/4	3/12	quadrans	one-quarter	*quattuor*, four
1/3	4/12	triens	one-third	*tres*, three
	5/12	quincunx	five-twelfths	*quinque*, five + *uncia*
1/2	6/12	semis	one-half	*semi-* (half-) + *as*
	7/12	septunx	seven-twelfths	*septem*, seven + *uncia*
2/3	8/12	bes	two-thirds	*bis-assi-s*; two (portions) short
3/4	9/12	dodrans	three-fourths	*de* + *quadrans*; one quarter short
5/6	10/12	dextans	five-sixths	*de* + *sextans*; one sixth short
	11/12	deunx	eleven-twelfths	*de* + *uncia*; one twelfth short
1	12/12	as	one, the whole, unity	

There are also subdivisions of the *uncia*:

semuncia	one-half	*semis*, one half
duella	one-third	diminutive of *duo*, two (sixths); also called *binae sextulae*
sicilicus	one-quarter	from *secula*, crescent, sector
sextula	one-sixth	diminutive of *sextus*
scripulum	1/24	*scrupulus*: a small chip of a stone

Roman measurements for weight and mass were the *uncia* and the *libra*:

1 *uncia* = 27 - 27.5 g

12 *unciae* = 1 *libra*

Roman copper *as* (Janus face)

The *as* – the whole, unity – was Rome's basic unit of money, originally worth one *libra* of copper, later depreciated to one-half *uncia*.

HORACE
(65 - 8 B.C.)

Mastery of the monetary system was of great importance in Roman education, as recounted by Horace (65 - 8 B.C.) in his *Ars poetica* (lines 325 - 330):

Romani pueri longis rationibus assem discunt in partis centum diducere.	Through lengthy reckonings, boys in Rome learn to divide the *as* into hundredths.
"dicat filius Albini: si de quincunce remota est uncia, quid superat? poteras dixisse."	"Tell me, son of Albinus: if from a *quincunx*, an *uncia* is taken, what is left? Please, say."
"triens."	"A *triens*."
"eu! rem poteris servare tuam. redit uncia, quid fit?"	"Good! You'll be able to manage your assets. An *uncia* added; what does it make?"
"semis."	"A *semis*."

Latin *uncia* gave both "inch" (Old English: *ynce*; Middle English: *inche*) and "ounce" (Middle English: *unce*).

The English abbreviation lb, the £ sign, and the word "pound" all have Latin forebears: £ and lb = *libra*; *pondo* means "in weight"; its Latin abbreviation P is for (*librae*) *pondo*.

p. 39
p. 168

The Roman **notations of numeration** will be described in Chapter 2, Roman use of the **abacus** in Chapter 4.

Mathematics in Ancient Rome

Tolle mathematicum ! est augur surdus, haruspex cecus et hariolus demens: praesentia scire Fas hominum est, soli Domino praescire futura.

From Hans Walther (Göttingen, 1963): *Proverbia Sententiaeque Latinitatis*

"Take the mathematician away: He is a stupid augur, blind prophet, a crazy soothsayer. Man may know the present; only God can foresee the future."

In contrast to the high regard in which mathematics and mathematicians were held in ancient Greece, Romans considered mathematics an unworthy subject for a noble Roman to pursue. Being more concerned with soldiering and politicking, only simple reckoning found favor with them.

Of old, the Romans used a **ten-month calendar** of 304 days, followed by a winter whose days nobody seems to have kept track of. The year began with March. The days of winter were later accounted for in the new months of January (at the end of the year!) and February (at the beginning of the year!), allegedly introduced by King Numa in 717 B.C. – eventually bringing the length of the calendar year up to 355, or 356, days, with every other year a leap year.

According to modern study, the calendar of Old Rome might not have been as bad as first thought; the description might have been spurious propaganda to enforce the introduction of a new calendar.

PRISCUS Tarquinius
(616 - 579 B.C.)

About 600 B.C., the Etruscan **Tarquinius Priscus** introduced the **Roman Republican Calendar**, a lunar calendar of 355 days, that is, 10 and 1/4 days short of the solar year. To compensate for this deficit, the Romans wedged in an extra month, Intercalarius, after 23 February, conveniently eliminating the remaining 5 days of February; in this manner, a calendar year averaging 366 and 1/4 days in a four-year period was achieved. This is mathematics after the Roman fashion.

The decision when and how to introduce the Intercalarius was the responsibility of a Roman Board of Magistrates, whose deliberations were theoretically secret, but who were not averse to receiving bribes, if ample enough – also a kind of Roman mathematics – causing the calendar to proceed by leaps and rebounds.

CAESAR Gaius Julius
(*c.* 100 - 44 B.C.)

SOSIGENES
(1st century B.C.)

In 46 B.C., **Julius Caesar** realized that drastic measures were required and called upon **Sosigenes** of Alexandria, a reputable Greek astronomer and mathematician, to bring order out of chaos. He immediately decided that the only possible remedy was to abandon the lunar calendar altogether, and he corrected the accumulated errors by inserting no fewer than three extra months in the year 46 B.C. Then, he introduced a 365-day year with an extra day placed between the 23rd and 24th of February every four years – the **punctum temporis**. This calendar came to be known as the **Julian calendar**.

The new calendar proved too much for the Romans and got out of hand; the last year of a four-year period was mistakenly taken as the first year of the following period. This went undetected for 36 years until, in 8 B.C., the Emperor Augustus ordered the canceling of three leap days, and in A.D. 4 the Julian calendar finally came abreast of time.

The Gregorian Calendar

GREGORIUS XIII
(1502 - 1585)

The Julian calendar contained a seemingly insignificant error of 11 minutes and 14 seconds annually, but as the years passed, this shortfall continued steadily to build up into days. When Ugo Buoncompagni, a former teacher of law at the University of Bologna, became Pope Gregory XIII in 1572, he inherited the problem of a calendar that was 10 days short; the vernal equinox fell on 11 March instead of its proper date of 21 March.

On the 24th of February 1582, acting on the advice of two renowned astronomers, Gregory XIII issued a papal edict directing that – to allow the calendar to catch up with the Lord's Time – the accumulated error of 10 days be removed by letting the day following the 4th of October 1582 be the 15th of October. The Gregorian calendar has an extra day (29 February) every fourth year, except – and that is what distinguishes it from the Julian calendar – in centennial years unless the year is exactly divisible by 400; so, *e.g.*, the year 1600 was a leap year, but not 1700, 1800, and 1900.

The Gregorian calendar, which became known as the New Style, was recognized immediately by nearly all countries of Catholic faith. Protestant countries were more resistant, with Protestant German states adopting the calendar in 1700, Britain in 1752, and Sweden in 1753. Countries of the Greek Orthodox persuasion in Eastern Europe adopted it in 1912 - 17, Greece in 1923, and atheist U.S.S.R. in 1918.

Monthly Number-Quandary

p. 17

cf. Latin number names, *p.* 14

The misleading names of our ninth, tenth, eleventh, and twelfth months are derived from the *ten-month calendar* – which started with March (*Martius*) – where *September*, *October*, *November*, and *December* were the seventh, eighth, ninth, and tenth months. There were also *Quinctilis* and *Sextilis*, later to became July (*Julius*; after Julius Caesar) and August (*Augustus*; after the first Roman emperor).

Modifications of Latin

The oldest evidence of Latin is an inscription in Greek characters dating from the 6th century B.C. Spoken originally by small groups of people living along the banks of the Tiber, Latin became the standard language of most of the Roman Empire, spreading throughout Southern and Western Europe and into large parts of the coastal regions of North Africa.

In A.D. 395 the Roman Empire split in two. The Western Empire fell in 476, the Eastern in 1453. Latin as a spoken language became extinct in the period of the 7th to 10th centuries, although it remained the official language of the Church. It also served as a language of general understanding among scientists, and it held a place as the language of diplomatic service well into the 18th century, when it was ousted by French.

Written Latin distanced itself from its models in form and substance until the 15th century, but the reawakening of an interest in science during the Renaissance saw a determined return to the Latin of the Golden Age (81 B.C. - A.D. 14).

GAUSS Carl Friedrich (1777 - 1855)

European scholars continued to use Latin extensively as a written language well into the first half of the 19th century. **Carl Friedrich Gauss**, the eminent German mathematician, published all his major works in Latin, and at Oxford all university business was transacted in Latin until 1854.

Fraktur

Fraktur is a style of letters formerly used in German manuscripts and printing. Capital letters of boldface Fraktur, *e.g.*,

$\mathfrak{A}$ (A), $\mathfrak{C}$ (C), $\mathfrak{D}$ (D), $\mathfrak{F}$ (F), $\mathfrak{P}$ (P), $\mathfrak{S}$ (S),

are sometimes used in mathematics for *sets of numbers*.

Romance Languages

The Romance languages – the major ones being Italian, French, Spanish, Portuguese, and Romanian – are descendants of Latin.

As the Roman Empire grew, soldiers, administrators, tradesmen, and settlers were sent into the conquered provinces, introducing not only Greek culture and Roman law, but also colloquial Latin, which continued to develop on its own long after Roman rule came to an end, thus giving rise to the languages we now refer to as Romance languages. As Latin is preserved in writing, Latin and the Romance languages constitute an unusually solid basis for the study of the relationships between the mother language and its descendents.

Latin, *Romanus*, "Roman"

The words "romance" and "romantic" did not originally connote love or (short-lived) enthusiasm. *Romance* was the Latin of everyday life, as distinguished from book Latin.

The common heritage of the Romance languages is evident when the names for numbers are compared. Spanish and Portuguese are linguistically very close; of the two, we choose, arbitrarily, Spanish.

Italian:
c before *i* and *e* pronounced like English *ch*

French:
Celtic influence on French number names will be discussed on *p.* 24

* Archaic and provincial forms:
septante = 70
huitante = 80
nonante = 90

Spanish:
† *y* means *and*

c and *z* before *i* and *e* in Spain pronounced as in English *th*ing, in Central and South America as in sing

c before *a*, *o*, *u* pronounced like *k* in English *k*ind; *ch* like English *ch*

Romanian:
c before *i* and *e* pronounced like English *ch*;
z is like an English voiced *s*
ş as in English *sh*

** spre: *above, higher*

	Italian	French	Spanish	Romanian
1	uno, una	un, une	uno, una	un, una
2	due	deux	dos	doi, două
3	tre	trois	tres	trei
4	quattro	quatre	cuatro	patru
5	cinque	cinq	cinco	cinci
6	sei	six	seis	şase
7	sette	sept	siete	şapte
8	otto	huit	ocho	opt
9	nove	neuf	nueve	nouă
10	dieci	dix	diez	zece
11	undici	onze	once	unsprezece**
12	dodici	douze	doce	doisprezece
13	tredici	treize	trece	treisprezece
14	quattordici	quatorze	catorce	patrusprezece
15	quindici	quinze	quince	cincisprezece
16	sedici	seize	diez y seis †	şasesprezece
17	diciassette	dix-sept	diez y siete	şaptesprezece
18	diciotto	dix-huit	diez y ocho	optsprezece
19	diciannove	dix-neuf	diez y nueve	nouăsprezece
20	venti	vingt	veinte	douăzecĭ
30	trenta	trente	treinta	treizecĭ
40	quaranta	quarante	cuarenta	patruzecĭ
50	cinquanta	cinquante	cincuenta	cincizecĭ
60	sessanta	soixante	sesenta	şasezecĭ
70	settanta	soixante-dix*	setenta	şaptezecĭ
80	ottanta	quatre-vingts*	ochenta	optzecĭ
90	novanta	quatre-vingt-dix*	noventa	nouăzecĭ
100	cento	cent	cien	o sută
1000	mille	mille	mil	o mie
2000	duemila	deux mille	dos mil	două mii

Rhaeto-Romance Languages

(1st century)

The ancient country of Rhaetia – comprising Vorarlberg and Tirol in Austria, parts of Northern Italy, the Eastern cantons of Switzerland, and parts of Bavaria and Baden-Württemberg in Germany – was invaded and made a province under Roman rule in the year 15 B.C., starting a Latinization of the native language. The Latinization was followed by German influence from the 6th century onward.

There are two main branches of the Rhaeto-Romance languages, Romansh and Ladin, with five dialects between them. The kinship with other Romance languages is demonstrated by comparison with the Italian names for the numbers 1 - 10 and 100.

	Italian	Romansh	Ladin
1	uno	in, egn	ûn
2	due	dus	duos
3	tre	tres, treis	trais
4	quattro	quatar, quater, catter	quatter
5	cinque	tschun, tschentg, tschintg	tschinch
6	sei	sis, seis	ses
7	sette	siat, seat, set	set
8	otto	otg	och, ot
9	nove	nov, nof	nouv
10	dieci	diesch, diasch	desch
100	cento	tschien, tsciant, tschient	tschient

Celtic Languages

Four score and seven years ago our fathers brought forth on this continent a new nation ...

Abraham Lincoln: *Gettysburg Address* (1863)

LINCOLN Abraham
(1809 - 1865)
16th U.S. President

Four score and seven is not an everyday expression in the English language. President Lincoln no doubt chose it wittingly because of its old-time flavor in an effort to create among his listeners an atmosphere of solemnity befitting the occasion.

The olden practice of counting in terms of twenties was in general use in most English-speaking countries as late as a few generations ago, and is still the rule in many existing languages.

Over a period from approximately 500 B.C. to A.D. 500, Celtic languages were spoken in a large area comprised by western, central, and southern Europe. Celtic conquerors and immigrants from Gaul (*Gallia*; in essence modern Belgium and France) spread to areas corresponding to modern northeastern central Spain and the northern part of Italy. Celts also penetrated into Asia Minor and created great destruction among Hellenistic states. At the beginning of the 3rd century B.C., Celts became settled in a region to be known as Galatia – centered on modern Ankara, Turkey; its name alluded to the Galatae, or Galatians, the appellation used for Celts by writers of that time. By the 2nd century A.D. the Galatians had become absorbed into the Hellenistic civilization. The Celtic languages in Europe were gradually repressed by Germanic and Roman languages and are presently spoken only in limited territories of the British Isles and in Brittany (Bretagne) where, in A.D. 5th or 6th century, a Celtic language appeared as a result of colonization from Britain.

Distribution of Celtic languages in near-modern times and, in a dissociated fashion, today.

Vigesimal counting is a feature of both the remaining groups of Celtic languages: the Goidelic, or Gaelic, group consisting of Irish, Scots Gaelic, and the recently extinct Manx; and the Britannic group comprising Welsh, long extinct Cornish, and Breton, still spoken in parts of Brittany.

For English speakers there is nothing difficult – except spelling and pronunciation – about the first ten numerals in the Celtic languages,

	Irish	**Scots**	**Welsh**
1	aon	aon	un
2	dó	dà	dau, dwy
3	trí	trì	tri, tair
4	ceathair	ceithir	pedwar, pedair
5	cúig	cóig	pump
6	sé	sia	chwech
7	seacht	seachd	saith
8	ocht	ochd	wyth
9	naoi	naoi	naw
10	deich	deich	deg

where double forms in Welsh denote masculine versus feminine gender.

Let's practice with some Welsh children's rhymes:

Mae gen i	I have
Un gaseg yn pori,	*Un* mare grazing,
Dwy yn y plwy'	*Dwy* in the parish,
Tair yn y ffair,	*Tair* in the fair,
Pedair heb ddim pedol,	*Pedair* without a shoe,
Pump yn eu crimp	*Pump* pinching them,
Chwech yn y frech,	*Chwech* vaccinated,
Saith yn y gwaith	*Saith* in the works,
Wyth yn y ffrwyth	*Wyth* in the effect,
Naw yn y galw,	*Naw* visiting,
A deg yn pori gwair Ton-teg.	And *deg* grazing the hay of Ton-teg.

* * *

Un, dau, eto tri	*Un*, *dau*, also *tri*
Mi wala' ar y tŷ.	I see on the house.
Pedwar, pump, eto chwech,	*Pedwar*, *pump*, also *chwech*,
Dyna blentyn yn rhoi sgrech.	Here is a screaming child.
Saith, wyth, naw, deg,	*Saith*, *wyth*, *naw*, *deg*,
Eto'n dilyn un ar ddeg.	Also *un ar ddeg* following.

Nor is there anything unexpected in the numbers 11 to 20 in Irish and Scots Gaelic, whereas Welsh has a more tortuous way of expressing these numbers, particularly from 16 to 19:

	Irish	**Scots**	**Welsh**	**meaning**
11	aondéag	aon deug	un ar ddeg	one and ten
12	dódhéag	dà dheug	deuddeg	two-ten
13	trídhéag	trì dheug	tri ar ddeg	three and ten
14	ceathairdéag	ceithir dheug	pedwar ar ddeg	four and ten
15	cúigdéag	cóig dheug	pymtheg	five-ten
16	sédéag	sia deug	un ar bymtheg	one and fifteen
17	seachtdéag	seachd deug	dau ar bymtheg	two and fifteen
18	ochtdéag	ochd deug	deunaw	twice nine
19	naoidéag	naoi deug	pedwar ar bymtheg	four and fifteen
20	fiche	fichead	ugain	twenty

	Irish	**Scots**	**Welsh**
20	fiche	fichead	ugain
30	deich fichead	deich air fichead	deg ar hugain
40	daichead	dà fhichead	deugain
50	deich ar daichead	dà fhichead ´s a deich	deg a deugain
60	trí fichead	trì fichead	trigain
70	deich is trí fichid	trì fichead ´s a deich	trigain a deg *
80	cheithre fichid	ceithir fichead	pedwar ugain
90	deich is cheithre fichid	ceithir fichead ´s a deich	pedwar ugain a deg **
100	céad	ceud	cant

* Welsh 70 is also: deg a thrigain

** Welsh 90 is also: deg a phedwar ugain

In addition to the older system, modern Welsh has a decimal system:

11	undeg un	
12	undeg day	*etc.*
20	daudeg	
30	trideg	*etc.*

Welsh zero is *dim*, "nothing".

Celtic Remnants in French and Danish

French

French still uses remnants of a vigesimal system of counting inherited from its Celtic predecessor, the language of the **Gaul**.

70	soixante-dix	sixty-ten
71	soixante-onze	sixty-eleven
72	soixante-douze	sixty-twelve
79	soixante-dix-neuf	sixty-ten-nine
80	quatre-vingts	four-twenties
90	quatre-vingt-dix	four-twenties-ten
99	quatre-vingt-dix-neuf	four-twenties-and-nineteen

Vigesimal counting in Old French is evidenced by another relic in the name of a Paris hospital built in the 13th century to accommodate 300 blind people, which is still called the

Hôpital des Quinze-Vingts,

that is, the Hospital of the Fifteen Score.

Danish

Danish, a North Germanic language, has a rather roundabout way of expressing numbers from fifty onward, by counting "halves", also a legacy from an otherwise forgotten vigesimal system.

50	halvtreds	or, in full:	halvtredsindstyve
60	tres		tresindstyve
70	halvfjerds		halvfjerdsindstyve
80	firs		firsindstyve
90	halvfems		halvfemsindstyve

where

treds	=	third	halv	=	half
fjerds	=	fourth	sinds	=	times
fems	=	fifth	tyve	=	twenty

All this becomes easier to understand if we realize that

$\frac{1}{2}$	is the first half	
$1\frac{1}{2}$	is the second half	Danish: halvanden
$2\frac{1}{2}$	is the third half	halvtreds
$3\frac{1}{2}$	is the fourth half	halvfjerds
$4\frac{1}{2}$	is the fifth half	halvfems

So, clearly, halvfjerds is $3\,\frac{1}{2} \times 20 = 70$.

Germanic Languages

Protolanguage: an assumed or recorded ancestral language
Merriam-Webster™

The original Germanic language – Proto-Germanic – separated into three subfamilies:

- East Germanic, via extinct Gothic, the parent language of Burgundic, Vandalic, Ostrogothic, and Visigothic;
- North Germanic, which turned into Old Norse from which developed Icelandic, Faeroese, Danish, two Norwegian languages, and Swedish;
- West Germanic, parent language of Old Saxon, Old High German, Old Frisian, and Old English, all well attested.

Icelandic:

þ indicates a voiceless lisping sound, as in *think* and *three*.

Swedish:

sj indicates a sound similar to the first sound in *shoe*.

å indicates a sound similar to *ou* in *four*. *y* is similar to German *ü*.

Dutch:

-ij is like English *ay*, as in *pay*.

-ig is like *ch* in Scottish *loch*.

German:

u is a sound like *u* in *bull*.

ü has no equivalent English vowel sound;

v is pronounced as f.

z is pronounced *ts* as in *its*.

s is a voiced *s* sound at the beginning of a word, as in *zeal*.

chs is pronounced *ks* as in *mix*.

cht is a sound similar to Scottish *ch* in *loch* plus the *t* sound.

-ig is a sound similar to the last sound in *catch*.

Number Names of Some Modern Germanic Languages

	Icelandic	**Swedish**	**Dutch**	**German**
0	núl	noll	nul	null
1	einn	ett	een	eins
2	tveir	två	twee	zwei
3	þrír	tre	drie	drei
4	fjórir	fyra	vier	vier
5	fimm	fem	vijf	fünf
6	sex	sex	zes	sechs
7	sjö	sju	zeven	sieben
8	átta	åtta	acht	acht
9	níu	nio	negen	neun
10	tíu	tio	tien	zehn
11	ellefu	elva	elf	elf
12	tólf	tolv	twaalf	zwölf
13	þrettan	tretton	dertien	dreizehn
14	fjórtan	fjorton	veertien	vierzehn
15	fimmtán	femton	vijftien	fünfzehn
16	sextán	sexton	zestien	sechzehn
17	sautján	sjutton	zeventien	siebzehn
18	átján	arton	achtien	achtzehn
19	nítján	nitton	negentien	neunzehn
20	tuttugu	tjugo	twintig	zwanzig
21	tuttugu og einn	tjugoett	eenëntwintig	einundzwanzig
22	tuttugu og tveir	tjugotvå	tweeëntwintig	zweiundzwanzig
23	tuttugu og þrír	tjugotre	drieëntwintig	dreiundzwanzig
...				
30	þrjátíu	trettio	dertig	dreissig
40	fjörtíu	fyrtio	veertig	vierzig
50	fimmtíu	femtio	vijftig	fünfzig
60	sextíu	sextio	zestig	sechzig
70	sjötíu	sjuttio	zeventig	siebzig

1.3 Etymology of English Number Names

Old English came to England in the 5th and 6th centuries with the Anglo-Saxons – a German people who settled in Britain when Roman rule crumbled. From the late 8th century to well into the 11th century Britain was ravaged and pillaged by Norse Vikings whose language, Old Norse, was quite similar to Old English.

An important event in the history of the English language was William the Conqueror's victory at the Battle of Hastings in 1066, which paved the way for the massive influx of French words into English that was to continue for more than 500 years. Curiously enough, this linguistic offensive made only scant inroads into well-established English names of numbers.

Zero

Zero derives from Hindu **sunya** – meaning "void", "emptiness" – via Arabic **sifr**, Latin **cephirum**, and Italian **zevero**; *sifr* is also the origin of **cipher**.

One ... Ten

Nearly all English names of numbers are of Germanic origin, based on the words **one**, **two**, **three**, **four**, **five**, **six**, **seven**, **eight**, **nine**, and **ten**, reflecting humankind's early use of fingers to illustrate numbers, and a forerunner of both number names and number symbols.

The word **ten** may actually have come from an old Indo-European word meaning "two hands", transformed into Gothic **tai-hun**, and into Old and Middle English **tien**.

	Sanskrit *	**Old Norse** *	**Old English** * ♣	**Middle English** * ♣	**Modern English**
1	eka	einn	ane, an	oon, on	one
2	dva	tveir	twa	twa, tu	two
3	tri	þrir	þri	þri	three
4	catur	fjörir	feower	four(e)	four
5	panca	fimm	fif	fif	five
6	sas	sex	siex	sex	six
7	sapta	sjau	seofan	seofan	seven
8	asta	atta	eachta	eighte	eight
9	nava	níu	nigon	nigon	nine
10	daca	tíu	tien	tien	ten

þ indicates a voiceless lisping sound, as in *think* and *three*.
Diacritic symbols have been left out in transliteration.

* Other transliterations are possible.

♣ Several other forms existed.

One and the Indefinite Article

The Old English (*c.* 550 - *c.* 1100) word **ane** served both as a counting number and as the indefinite article. Toward the end of the Old English period and the beginning of Middle English (*c.* 1100 - *c.* 1500), **ane** developed two pronunciations, the first being used for the number 1 and the other for the indefinite article **an**, **a**.

The existence of different words for the number one and the indefinite article seems to be a unique feature of the English language.

Revenge on William the Conqueror:
A baker's dozen

Eleven, Twelve (Dozen)

The names of numbers beyond *ten* are derived from combinations of number words.

Eleven is a derivative of Old English **endleofan**, where the terminal -fan is believed to carry the meaning of "to leave" or "left", that is, **one left** after counting ten (fingers).

Twelve has developed in a similar manner from Old English **twelfe** meaning **two left over** after counting ten; it may be more clearly seen from the Gothic **twailif**.

Dozen is one of the few French loans; *douze* means twelve, *douzaine* a dozen.

Thirteen ... Twenty (Score)

Thirteen comes of three-and-ten, and the remaining teens follow the same pattern.

Twenty was **twentig** or **twegentig** in Old and Middle English, where twen and twegen were forms meaning two, and the suffix -tig recalls German -zig.

cf. Swedish *skåra*, "a notch", "a cut"

Score originally meant a cut or scratch made on a tally-stick of wood, a tablet of clay, or the like, in order to keep a record. Twenty was marked by a longer scratch; hence the development of **a score** to mean *twenty*.

Thirty ... Ninety

All multiples of ten are formed in the same manner as twenty.

Hundred

Hundred derives from Old English **hund**, meaning *a hundred*, allied to the Gothic **taihun-taihund**, meaning *ten-tens*; the suffix **-red** is a later addition, a word for reckoning or ratio – a **hundred-number**.

In the past, hundred also had the meaning of 120, which was the original meaning of Old Norse **hund-rath**. English differentiated between the hundred for 100 and the **long hundred** for 120.

Thousand

Thousand is akin to Old Norse **thūshund**, giving the meaning of *great-hundred*. The prefix denoting **great** is of the same origin as in **thumb**, literally the *stronger finger*.

Million

A thousand times a thousand is called a **million**, from the Medieval Latin **millio**, which means **great-thousand**.

Beyond the Million

A thousand million is called a **milliard** in French, the word being used also in other Romance languages and in Germanic languages, though rarely in English.

Higher powers of ten, in steps of a million, are given names compounded from a Greek number prefix plus million, thus,

$$(10^6)^2 = \text{bi-million} = \text{billion} = 10^{12}$$
$$(10^6)^3 = \text{tri-million} = \text{trillion} = 10^{18}$$
$$(10^6)^4 = \text{quadri-million} = \text{quadrillion} = 10^{24}$$

etc.

The powers of ten will then be:

10^6	million	10^{54}	nonillion	10^{102}	septendecillion
10^{12}	billion	10^{60}	decillion	10^{108}	octodecillion
10^{18}	trillion	10^{66}	undecillion	10^{114}	novemdecillion
10^{24}	quadrillion	10^{72}	duodecillion	10^{120}	vigindecillion
10^{30}	quintillion	10^{78}	tredecillion	…	
10^{36}	sextillion	10^{84}	quattuordecillion	10^{600}	centillion
10^{42}	septillion	10^{90}	quindecillion		
10^{48}	octillion	10^{96}	sexdecillion		

The United States uses, unfortunately, another system – based on an old French system – of naming numbers greater than a million:

	U.S.	Non-U.S.	
$1000 \cdot 1000^1$ =	million =	million =	10^6
$1000 \cdot 1000^2$ =	billion =	thousand million =	10^9
$1000 \cdot 1000^3$ =	trillion =	billion =	10^{12}
$1000 \cdot 1000^4$ =	quadrillion =	thousand billion =	10^{15}
	etc.		

In mathematics and science, this ambiguity is apt to cause serious misunderstandings, and anybody of a scientific turn of mind is well advised to be wary of the "zillions" and keep to the purely metric numeric symbols of powers of ten.

– Move from the United States and your one billion deficit will be 1/1000 of a billion.
– Let's go for it!

– Move to the United States and your thousand million profit will be a billion.
– Let's go for it!

1.4 Numbers *vs.* Infinity

And the angel of the Lord said unto her, I will multiply thy seed exceedingly, that it shall not be numbered for multitude.

Genesis 16: 10

ARCHIMEDES
(287? - 212 B.C.)

Archimedes is the first person known to have discussed extremely large numbers. In his essay *The Sand Reckoner*, he built a system on the *myriad*, the name for 10 000. He called the numbers up to a *myriad myriads* (100 000 000) numbers of the *first order* of the *first period*.

With 100 000 000 as the unit of the *second order* of the first period, Archimedes could name numbers as large as $100\,000\,000^2$, and by continuing through the *myriad-myriadth order* of the first period, he could extend the naming to $100\,000\,000^{100\,000\,000}$.

Moving on to the *second period* (whose first order has $100\,000\,000^{100\,000\,000}$ for the unit), the *third period*, the *fourth period*, and so on through the *myriad-myriadth order* of the *myriad-myriadth period*, Archimedes had a system for naming all numbers up to $10^{80\,000\,000\,000\,000\,000}$, that is, a 1 followed by 80 000 billion (billion = a million million) zeros.

KASNER Edward
(1876 - 1955)

The name **googol** was coined in the late 1930s by the nine-year-old nephew of the American mathematician **Edward Kasner** when he was asked to come up with a name for a very large number; at the same time **googolplex** was suggested for a still larger number.

Kasner gave definitions for these two number names:

$$1 \text{ googol} = 10^{100},$$

that is, a one followed by 100 zeros, and

$$1 \text{ googolplex} = 10^{\text{googol}} = 10^{10^{100}},$$

a one followed by a googol of zeros.

Infinity

In *The Sand Reckoner*, **Archimedes** demonstrated that the number of sand grains the size of a poppy seed that could fill up a volume equal to that of the then known Universe was fewer than 10^{63}, that is, 1 followed by 63 zeros.

GAUSS Carl Friedrich
(1777 - 1855)

Carl Friedrich Gauss called

$$9^{9^{9^9}}$$

eine messbare Unendlichkeit

that is, "a measurable infinity". The number has something like $10^{369\,693\,100}$ digits, which makes it suitable for those who would like to put a tag on every electron in the Universe.

In ordinary conversation, *infinite* means something that is very great in comparison with everyday things. In mathematics, however, infinity is not a number but a concept of increase beyond bounds.

A collection like Archimedes' grains of sand, which can have no more than a certain number of elements, is finite; here number means an ordinary counting number: 1, 2, 3 ... through the googol and googolplex, through the last number of Archimedes' myriad-myriadth order of the myriad-myriadth period ... .

A collection such as that of *all* counting numbers, which continues indefinitely, is *infinite*, and a process that may be continued indefinitely is also infinite.

Although the concept of infinity is difficult to grasp, we may simply define it as *not finite*, where *finite* means something that – at least in theory – is completely determinable by counting or measurement.

Infinity has calculation rules of its own:

$$\infty + \infty = \infty; \qquad \infty \cdot \infty = \infty$$

p. 257

Further aspects on infinity are discussed in section 7.5, Transfinite Numbers.

WALLIS John ∞
(1616 - 1703)
BERNOULLI Jakob
(1654 - 1705)
BERNOULLI Nikolaus
(1662 - 1716)

The mathematical symbol of infinity was introduced in 1655 by **John Wallis** in his *Arithmetica infinitorum*, but did not appear anew in print until the *Ars conjectandi* by **Jakob Bernoulli**, published posthumously in 1713 by his nephew **Nikolaus Bernoulli** (Nikolaus I).

Preacher:

– *Eternity is infinitely long, indeed much longer than that. Just think of it, dear friends, that when millions and millions of years have passed and you look at your watches, breakfast will still be another hundred thousand years off.*

Albert Engström (1869 - 1940; Swedish artist and cartoonist)

Chapter 2

SYSTEMS OF NUMERATION

2.0 Forms of Notation

I cut every day a notch with my knife, and every seventh notch was as long again as the rest, and every first day of the month as long again as that long one; and thus I kept my calendar, or weekly, monthly, and yearly reckoning of time.

Daniel Defoe (1659–1731): *The Life and Strange Surprising Adventures of Robinson Crusoe of York, Mariner* (1719)

We distinguish between three essentially different methods of arranging numeral characters to form numbers: additive, multiplicative, and positional notation.

Most ancient number systems used characters whose meaning and value were always the same.

Additive Notation

In **additive notation**, composite numbers are formed by juxtaposing several single characters of each kind in a row, generally in order of descending value, and repeating each character as many times as required in order to form the desired composite number.

Subtractive Notation

To avoid number presentations of inordinate length, the additive method of notation was sometimes supplemented by **subtractive notation**, where a character of lower value preceding one of higher value signifies that it is to be deducted from the latter.

Of additive systems, we shall present Egyptian, Sumerian, Greek, and Roman numeration.

Multiplicative Notation

In **multiplicative notation**, two kinds of number symbols are employed, one kind being constant and its effective value being increased by multiplication by a symbol of the other kind, also of constant meaning and value.

The desired number is obtained by writing the multiplied numerals in a row, and adding them up.

Of multiplicative numeration systems, we shall describe the Chinese system, which appears in several forms, each with its unique sphere of application.

Positional Notation
Place-Value Notation

Positional notation, or **place-value notation**, is the method used exclusively by all modern number systems; it is a refinement of the multiplicative system, having abolished the value multiplier characters.

A character in a positional numeration system retains its meaning, but its actual value when part of a written number is determined by its position, or place, in the sequence of characters forming the composite number.

This will be illustrated by the Hindu-Arabic numeration system, by the Babylonian and Mayan systems, and by the computer-oriented binary system and its associated octal and hexadecimal forms.

2.1 Additive Notation

Egyptian Numeration

Egyptian hieroglyphics of some 5500 years ago are a numerical grouping system whose symbols were used to signify numbers which, when added together, could express any desired number.

The number 1 was represented by a vertical line, or a picture of a staff; 10 by a heel-bone sign; 100 by a coiled rope; 1000 by a lotus blossom; 10 000 by a bent finger; 100 000 by a tadpole; 1 000 000 by a kneeling genie with raised arms:

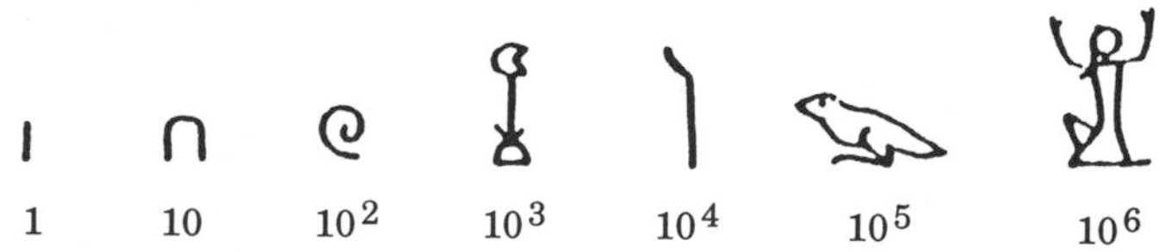

Each symbol could be repeated up to nine times, signifying addition. In a system of this kind the order of the symbols is of no consequence, but the Egyptians usually wrote the symbols in order of descending value, either from left to right or from right to left.

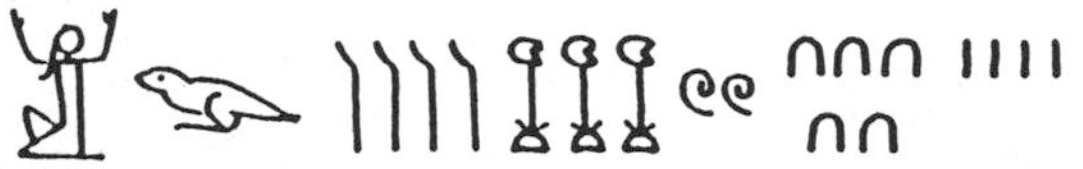
= 1 143 254

Latin, *decim*, "ten"
Greek, *deka*, "ten"
Latin, *denarius*, "containing ten"

A number system that uses ten as a base is called a **decimal** (decadic, denary) **system**; thus, the notation used by the ancient Egyptians is a strictly additive decimal notation.

The Egyptians generally used only so-called **unit fractions** having the number 1 as numerator, which they wrote by placing the symbol for an open mouth above the denominator:

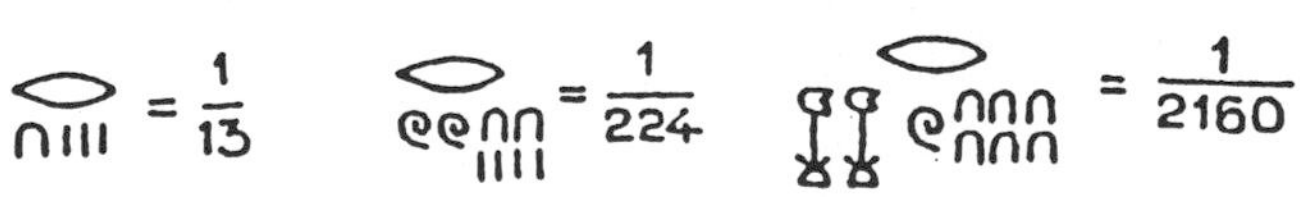

Special symbols were used for certain fractions:

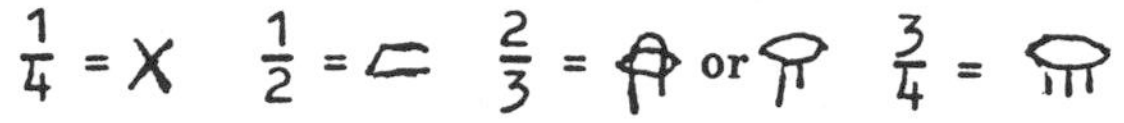

Sumerian Numeration

The Sumerians, who lived from about 3200 B.C. on the lower reaches of the river Euphrates, early developed a high level of culture and one of the oldest written languages known to us.

They had two systems of numeration: a sexagesimal system for astronomical observations, and a decimal system for everyday use. Both systems were additive: every symbol was repeated the number of times required by the desired number.

The characters were made with rods of two diameters which were pressed, at an angle or end-on, into tablets of clay which were then sun-dried or kiln-baked.

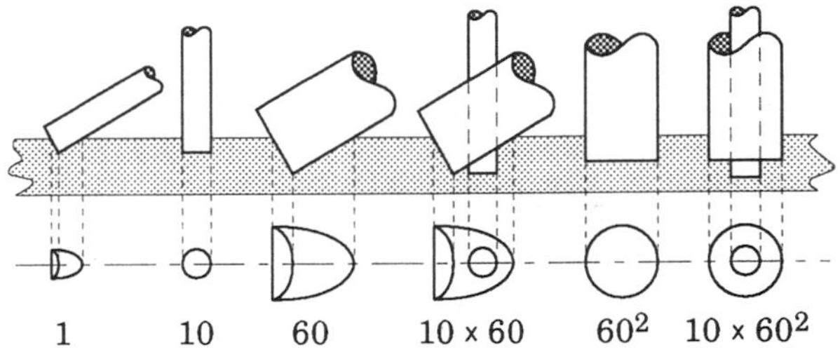

A subtraction symbol was sometimes used in order to reduce the number of symbols required, and to save space:

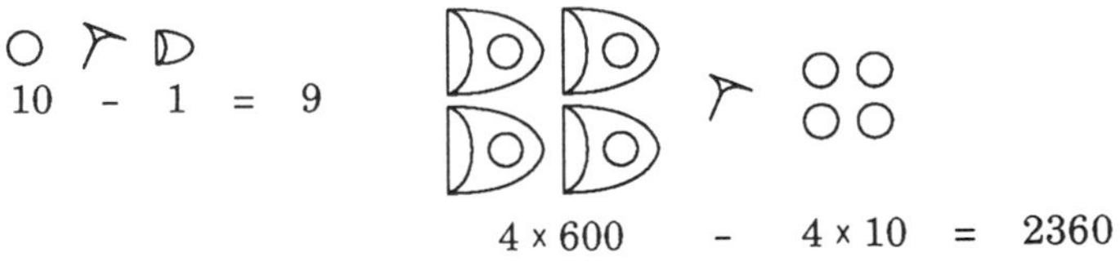

Cuneiform Symbols
Latin, *cuneus*, "a wedge"

From about 2750 B.C., these **archaic symbols** were gradually abandoned in favor of characters made with a sharp-edged tool which left wedge-shaped impressions in the clay; these **cuneiform symbols** were later traced with a fine stylus.

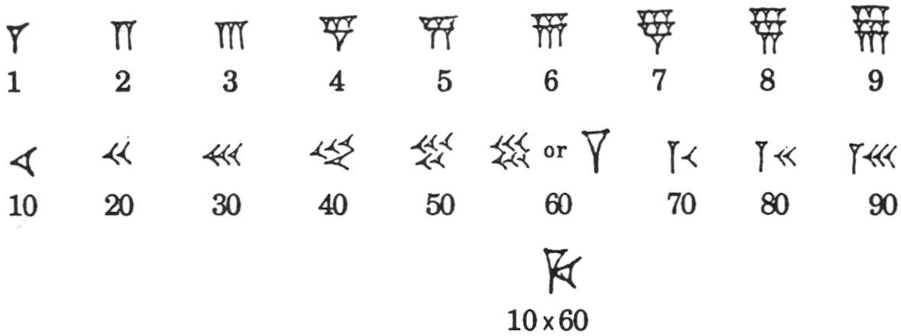

The symbol denoting $60^2 = 3600$ consisted of four wedges arranged as the sides of a square "60 squared". For $10 \times 60^2 = 36\,000$, a "corner" wedge denoting a factor of 10 is placed inside the symbol; for $60^3 = 216\,000$, a large vertical wedge, denoting a factor of 60, is placed inside.

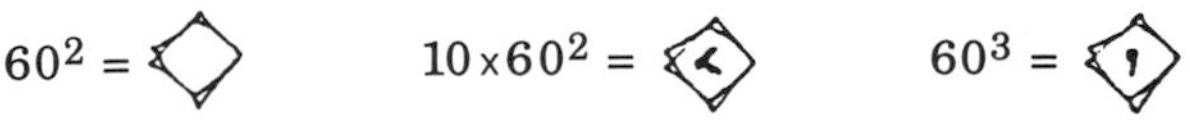

Akkadian Numeration

Around 2350 B.C. the Sumerians were invaded by the neighboring Akkadians, who assimilated the major part of the Sumerian culture, including their cuneiform characters which they adapted to the needs of their own language.

Like most Semitic peoples, the Akkadians were used to counting with hundreds and thousands, for which they found no suitable characters in the Sumerian numeration system, and had to devise new ones based on Sumerian cuneiform characters. Thus,

Sumerian  = 60 + 40 became Akkadian  = 100;
for 1000, they wrote  = 10 × 100.

Greek Numeration

The Phoenicians had special characters for numerals which the Greeks did not adopt when they copied the alphabet. They used, instead, various forms of alphabetic notation for numbers.

In an early form of numeration some letters were assigned a numerical value, but this was found to be impractical for mathematical operations and was replaced by two major systems of numeration in ancient Greece.

Attic Numeration

Athens and the surrounding province of Attica used a numeration system in which

| stood for unity,

Γ, an old form of Π (pi), for 5 (Πεντε), and

H (eta) for 100 (ʼεκατον),

while other numbers were denoted by the initial of the name of the number, thus

Δ (delta) for 10 (Δεκα),

X (chi) for 1000 (Χιλιοι), and

M (mu) for 10 000 (Μυριοι).

The symbols were combined in the following way:

Γ^{Δ}	Γ^{H}	Γ^{X}	Γ^{M}
50	500	5000	50 000

HERODIAN
(2nd century)

The Attic system is sometimes called **Herodianic**, named after **Herodian**, by profession a grammarian, who gave a description of the system toward the end of the 2nd century, long after the system had, in fact, gone out of use. Examples of its use can be found in Attic inscriptions from 454 to about 95 B.C.

Ionic Numeration

The Ionic system of numeration was a more suitable system for mathematical operations. There is evidence that the system was firmly established in the 8th century B.C.

The system's name refers to its origin in Ionia, the ancient Greek colony in Asia Minor, established around 1000 B.C., today part of Turkey.

For the numeration system, three otherwise obsolete symbols – *digamma*, *koppa*, and *san* – were revived and added to the Classical Greek alphabet. The twenty-seven characters thus obtained were divided into three groups, denoting units, tens, and hundreds:

Α, α	alpha	1	Ι, ι	iota	10	Ρ, ρ	rho	100
Β, β	beta	2	Κ, κ	kappa	20	Σ, σ	sigma	200
Γ, γ	gamma	3	Λ, λ	lambda	30	Τ, τ	tau	300
Δ, δ	delta	4	Μ, μ	mu	40	Υ, υ	upsilon	400
Ε, ε	epsilon	5	Ν, ν	nu	50	Φ, φ	phi	500
Ϛ, ϛ	digamma	6	Ξ, ξ	xi	60	Χ, χ	chi	600
Ζ, ζ	zeta	7	Ο, ο	omicron	70	Ψ, ψ	psi	700
Η, η	eta	8	Π, π	pi	80	Ω, ω	omega	800
Θ, θ	theta	9	Ϟ, ϙ	koppa	90	Ϡ, ϡ	san	900

It was common practice to use a horizontal bar above letters denoting numbers, to distinguish them from the surrounding text.

The numerals were written according to the additive principle, starting from the left with the higher denominations, *e.g.*,

$\overline{\text{ΡΚΔ}}$ $\overline{\text{ΦΝΕ}}$ $\overline{\text{ΨΟΖ}}$ $\overline{\text{ΩΗΓ}}$

and

$\overline{\rho\kappa\delta}$ $\overline{\varphi\nu\varepsilon}$ $\overline{\psi o\zeta}$ $\overline{\omega\eta\gamma}$

124 555 777 883

A mark at the upper left of the unit letters denoted thousands:

'Α	'Β	'Γ	...	'Θ
1000	2000	3000		9000

Ten-thousands and higher numbers were denoted by the letter M (mu) – for Μύριοι = 10 000 – with a letter or group of letters signifying the number of ten-thousands wanted placed above it:

$$\overset{\alpha}{\mathrm{M}} \quad \overset{\beta}{\mathrm{M}} \quad \overset{\gamma}{\mathrm{M}} \quad \ldots \quad \overset{\theta}{\mathrm{M}}$$

α M	β M	γ M	…	θ M
10 000	20 000	30 000		90 000

ϙθ M	ϡϙθ M	͵θϡϙθ M
990 000	9 990 000	99 990 000

These are merely examples of methods used by the ancient Greeks to express large numbers. Geographic variations occurred along with considerable ingenuity and individual preferences among writers.

Although the ancient Greek way of expressing large numbers by certain combinations suggests the possibility of a positional system of numeration, both the Attic and the Ionic notations are strictly additive decimal systems of numeration.

ARCHIMEDES
(*c.* 287 - 212 B.C.)

Fractions were written in various ways by the ancient Greeks, largely according to personal preferences. The famous mathematician and physicist-inventor **Archimedes** used an ordinary cardinal number for the numerator, followed by an accented number representing the denominator, *e.g.*,

$$\overline{\iota\varepsilon}\ '\ \overline{\xi\delta} = \frac{15}{64}$$

Ancient Hebrew and Arabic Numeration

The use of alphabetic characters in the Hebrew system of numeration evolved from the Ionic system. The oldest finds of the Hebrew system date from no farther back than the 1st century B.C., while Ionic numeration goes back to the 8th century B.C.

After the Hebrew language had adopted the Ionic model of numeration, the system was also adopted into the Arabic.

Roman Numeration

Roman numerals were modeled on the ancient Greek system, using letter symbols for powers of ten and for intermediate "fives". The primary symbols still in use today,

I	V	X	L	C	D	M
1	5	10	50	100	500	1000

were taken over, in part, from the Etruscans:

/	Λ	X	↑	C
1	5	10	50	100

The symbols D and M were not part of the original system of symbols. The number 1000 was initially written either CIↃ or, in a simplified form, (I); the symbol denoting 500 was obtained by splitting CIↃ down the middle and using either of the halves.

Starting with CIↃ, every additional CↃ or ( ) bracketing a number increased its value 10-fold:

CIↃ	CCIↃↃ	CCCIↃↃↃ	CCCCIↃↃↃↃ
(I)	((I))	(((I)))	((((I))))
1000	10 000	100 000	1000 000

Printers and stonemasons simplified these symbols further to D = 500 and M = 1000.

A German printer's guide published in 1835 has these rather lax rules for writing large numbers:

Handbuch
der
Buchdruckerkunst.
W. Gasper.
Carlsruhe und Baden,
1835.

M CIↃ	... 1,000
IICIↃ	... 2,000
IIICIↃ	... 3,000
XCIↃ CCIↃↃ ↃMC IMI	.. 10,000
CCCIↃↃↃ CM	100,000
CCM	200,000
DCCCCM	900,000
CCCCIↃↃↃↃ	1,000,000

There were also other ways of increasing the value of a number symbol. Bracketing a symbol by two vertical lines increased its value 100-fold; a bar placed above a number meant a 1000-fold increase of its value; a combination of both, a complete "doorframe", above a number increased its value 100 000 times:

|X|	$\overline{X}$	$\overline{|X|}$
1000	10 000	1 000 000

Only capital letters were used by the ancient Romans to represent numbers; lowercase letters came into use for this purpose in medieval times.

Additive Notation

JUSTINIANUS
Petrus Sabatius Flavius
(482 - 565)

To write other numbers, the Romans used the **additive principle** and simply ran the requisite number of symbols of each kind together in order of descending value, as in this reproduction from 1888 (Paul Krueger, Berlin) of part of the table of contents of the law book of Emperor Justinian I, the *Corpus Iuris Civilis*, compiled in the middle of the 6th century:

LIBER PRIMUS

XXXVII	**De officio praefecti Augustalis**
XXXVIII	**De officio vicarii**
XXXVIIII	**De officio praetorum**
XXXX	**De officio rectoris provinciae**
XXXXI	**Ut nulli patriae suae administratio sine speciali permissu principis permittatur**
XXXXII	**De quadrimenstruis tam civilibus quam militaribus brevibus**
XXXXIII	**De officio praefecti vigilum**
XXXXIIII	**De officio praefecti annonae**
XXXXV	**De officio civilium iudicum**
XXXXVI	**De officio iudicum militarium**
XXXXVII	**Ne comitibus rei militaris vel tribunis lavacra praestentur**
XXXXVIII	**De officio diversorum iudicum**
XXXXVIIII	**Ut omnes tam civiles quam militares iudices post administrationem depositam per quinquaginta dies in civitatibus vel certis locis permaneant**
L	**De officio eius qui vicem alicuius iudicis obtinet**
LI	**De adsessoribus et domesticis et cancellariis iudicum**

Subtractive Notation

To avoid fourfold repetition of a symbol, the **subtractive principle** of writing was introduced. Placing a symbol of a small value immediately in front of one of higher value implies that it is to be deducted from the higher number so that the combined symbol signifies the difference between the higher and the lower numeral; thus

	IV	IX	XL	XC	CD	CM
	4	9	40	90	400	900
instead of						
	IIII	VIIII	XXXX	LXXXX	CCCC	DCCCC

The subtractive principle was subject to two rules:

- The symbols V, L, and D must not be used as the number to be subtracted;
- Only one symbol I, X, C may be placed before a higher number symbol.

Thus XXC was not permitted for 80, but these rules were not strictly obeyed, particularly for higher numbers, as we have seen in Herr Hasper's handbook; and it is not uncommon to see in older texts, for instance, IIX denoting 8.

SCHEDEL Hartmann
(1440 - 1514)

The subtractive principle was introduced in the 13th century but took a long time to gain acceptance, as witnessed by an excerpt from *Buch der Cronicken* (Nuremberg, 1493) by **Hartmann Schedel** – physician and hagiographer – which also shows an alternative method of writing the calendar year:

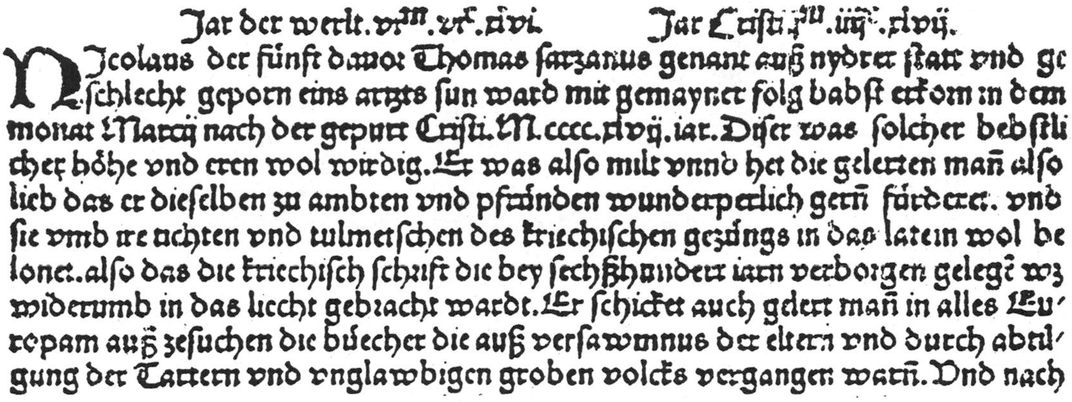
Jar der werlt. vi^{m}. vi^{c}. xlvi.   Jar Cristi. j^{m}. iiij^{c}. xlvij.

Nicolaus der fünft davor Thomas sarzanus genant auß nyderer statt vnd geschlecht geporn eins artzts sun ward mit gemayner folg babst erkorn in dem monat Marcij nach der geburt Cristi. M.cccc.xlvij. iar. Diser was solcher bebstlicher höhe vnd eren wol wirdig. Er was also milt vnnd het die gelerten man also lieb das er dieselben zu ambten vnd pfründen wunderperlich gern fürderet. vnd sie vmb ire achten vnd tulmetschen des kriechischen gezüngs in das latein wol belonet. also das die kriechisch schrifft die bey sechßhundert iarn verborgen gelegē was widerumb in das liecht gebracht wardt. Er schicket auch gelert man in alles Europam auß zesuchen die büecher die auß versawmnus der eltern vnd durch abtilgung der Tattern vnd vnglawbigen groben volcks vergangen warn. Vnd nach

Year of the World, vi^{m} · vi^{c} · xlvi *(6646).*

Year of Christ, j^{m} · iiij^{c} · xlvij *(1447).*

Nicolaus der fünft

Nicholas the Fifth, formerly known as Thomas of Sarzana, of low station and lineage, born a physician's son, was unanimously elected Pope in the month of March, in the year 1447 of Our Lord. He was well worthy of Papal Eminence and Glory.

In his clemency, he held learned men so dear that he was wondrously proud to exalt them to high offices and prebends, and he rewarded them well in recognition of their rendering of the Greek tongue into Latin, so that Greek texts that had lain incomprehensible for six hundred years were again brought into light.

He also despatched learned men into all Europe to search for books that had been lost through neglect on the part of forefathers and through the abuse by the Infidel and the coarse Heathen.

The "Year of the World" tells us that, by Papal Decree, the world was just 5199 years old at the time of the Star of Bethlehem.

We note the practice of substituting *j* for the final *i* in groups of more than one *i*, in order to minimize the risk of misreading: the single *j* of the thousands in *Jar Cristi* is used for the same reason: to distinguish it from the final *i* in *Cristi*.

In old ledgers of the Scottish Exchequer from the early 1300s one may find the sum of, say, £8697 14s. 8d.

m	c	xx				
viij	vj	iiij	xvj	łij	xiiijs̄	viijd̄
8^{000} +	6^{00} +	4×20 +	15 +	2£ +	14s +	8d

which also reflects Celtic counting by twenties,

och	mile	sia	ceud	ceithir	fichead	seachd deug	
eight	thousand	six	hundred	four	twenties	seven-ten	

Fractions

Latin fraction names: *p.* 16

Although the Roman numeral system was generally decimal, the method of representing fractions was based on the number twelve. Various systems of notation with groups of dots or lines were developed; in medieval times *S* was added to denote *semis* (one-half):

•	••	•••	∷	⁙	S
$\frac{1}{12}$	$\frac{2}{12}=\frac{1}{6}$	$\frac{3}{12}=\frac{1}{4}$	$\frac{4}{12}=\frac{1}{3}$	$\frac{5}{12}$	$\frac{6}{12}=\frac{1}{2}$
S•	S••	S•••	S∷	S⁙	I
$\frac{7}{12}$	$\frac{8}{12}=\frac{2}{3}$	$\frac{9}{12}=\frac{3}{4}$	$\frac{10}{12}=\frac{5}{6}$	$\frac{11}{12}$	$\frac{12}{12}=1$

Intractable Surrender

Roman numerals were used in bookkeeping well into the 18th century and beyond because of the ease of adding and subtracting written numbers by just keeping in mind that

five I make V	five X make L	five C make D
two V make X	two L make C	two D make M

Addition

M CCLXXX II	1282
DC VI	+ 606
MDCCCLXXXVIII	1888

Subtraction

MCCLXXXII	1282
D C VI	− 606
DC LXXVI	676

Roman characters were, however, cumbersome to write and clearly unsuitable for forms of calculation other than addition and subtraction. Perforce, they had to give way to the Hindu-Arabic numerals that we use today.

This was a long drawn-out process, however.

The city of Florence, in 1299, issued an ordinance concerning the use of the "new numerals", pointing out with what ease they could be falsified in the entries in the ledgers of the city and in tradesmen's books: a 0 might be made into a 6 or a 9, and a 1 into a 4, 7, or 9.

In 1348, a statute of the University of Padua ruled that the price lists of the University's *stationarii* (booksellers) must be clearly written *"non per cifras sed per literas claras"*, that is to say, with Christian characters, by which were understood Roman numerals.

The use of Roman numerals and the **abacus** is described in Chapter 4.

p. 168

Roman Revival

Roman numerals enjoyed a kind of revival in the 1920s. The Roman numeral X for 10 resembled the ends of a sawhorse, and a ten-dollar bill was called a sawbuck, or saw for short, with a one-dollar bill simply called a buck. A twenty-dollar bill then naturally became a double saw, and a five-dollar bill a half-saw.

A hundred-dollar bill was called a C-bill, a C-note, or just a C; with a departure from Roman practice, a thousand-dollar bill was known as a grand, or a G.

The idioms buck, C-bill, C-note, G, and grand are still perfectly good linguistic currency throughout the United States.

2.2 Multiplicative Notation

A multiplicative numeration system has two kinds of numeral symbols. By a multiplicative process, the characters of one kind of symbol increase the value of the other kind. The number is then expressed by the multiplied numerals, being added together.

For instance, using 1 through 9 to represent these values, and the letters A, B, C, and D to represent 10, 100, 1000, and 10 000, respectively, we can write

4D 5C 6B 1A 2 = 45 612

Chinese Numeration

The oldest discoveries of a Chinese written numerical system are from the Yin dynasty (*c.* 1523 - *c.* 1027 B.C.), but the system may be considerably older.

Japan, Korea, and several other countries in the Orient have adopted Chinese numeration.

We can distinguish at least three different styles of Chinese numerical symbols: traditional national numerals, official numerals, and mercantile numerals.

Traditional National Numerals

The traditional Chinese system of numeration is a decimal system employing thirteen basic characters, expressing the numbers 1 through 9, 10, 100, 1000, and 10 000:

一	二	三	四	五	六	七	八	九
1	2	3	4	5	6	7	8	9

十	百	千	萬 or 万
10	100	1000	10 000

These symbols, the **traditional national numerals**, are the symbols most commonly used today. They originated under the Han dynasty (206 B.C. - A.D. 220) and have kept their structure for more than 1700 years.

Traditionally, Chinese numerical signs were arranged vertically from top to bottom, but horizontal alignment from left to right is common practice today.

八萬九千五百六十七

$8 \times 10000 + 9 \times 1000 + 5 \times 100 + 6 \times 10 + 7 = 89567$

Official Numerals

The official numerals are beautifully ornamented forms used on banknotes, bonds, deeds, contracts, or other valuable documents as a safeguard against forgery:

壹	貳	參	肆	伍	陸	柒	捌	玖
1	2	3	4	5	6	7	8	9

拾	佰	仟	萬
10	100	1000	10 000

Mercantile Numerals

Mercantile numerals, traditionally used by shopkeepers, are falling into disuse, but can still be found both in Mainland China and in Taiwan.

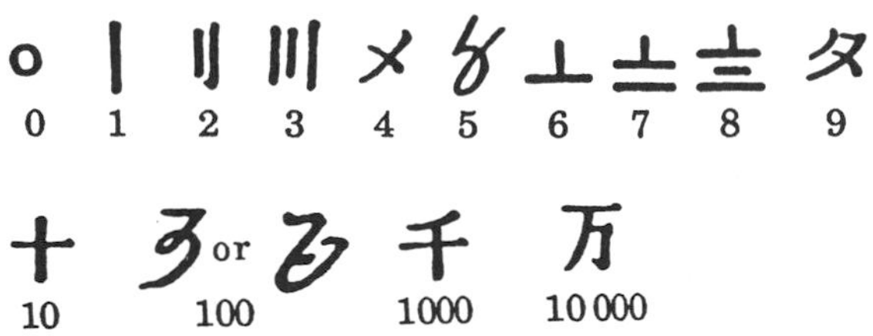

To denote zero (Chinese, *ling*), the zero sign of the mercantile numeral system is frequently used today with the traditional national system.

Besides the styles of Chinese numerals described above, there are cursive forms used for handwriting, as well as purely calligraphic styles.

p. 51 Chinese **rod numerals** will be discussed in their appropriate place, under Section 2.4, Decimal Position System.

2.3 Positional Notation

Blessed are the placemakers: for they shall be called the children of God.

Bible (1562 misprint); Matt. 5: 9

Names of Systems

In a positional numeration system (or **place-value system**), the numerical value of each symbol, or digit, depends on its position in the row of digits.

Every integer larger than 1 can serve as a **base** in a positional numeration system, and the system requires exactly as many digit symbols as indicated by the numerical value of the base. Positional numeration systems are named after the numerical value of the base.

Base	Name of System
2	binary (dyadic)
3	ternary
4	quaternary
5	quinary
6	senary
7	septenary
8	octal (octenary)
9	nonary
10	decimal (decadic, denary)
11	undenary
12	duodecimal (duodenary)
16	hexadecimal (hexadecadic)
20	vigesimal (vicenary)
60	sexagesimal

A Peep into Prehistory

Schoolchildren in the author's native Sweden used to read of a caveman and hunter named Ura-Kaipa. To the best of the author's recollection,

... Ura-Kaipa was the immensely proud owner of no less than twenty-seven stone axes. Of course, he didn't actually know in so many words that there were exactly twenty-seven of them, because

he could not count beyond three, but that did not trouble him at all.

He was a reasonably friendly old body with a neat and orderly mind, who liked to keep tabs on his goods and chattels, quick and dead. Assuredly, he could not write, because writing had not yet been invented, so he did the next best thing.

He arranged his axes in little groups of three, three to a row, and in three rows, like this:

so he could see at a glance at an audit if they were all present and correct, or if somebody had perhaps nicked one.

Not being a man of great learning, as we have said, he was blissfully unaware of the fact that he must have been the inventor of the ternary system of counting. If he had known Hindu-Arabic numerals as we know them, he would have totaled his hoard as

1000 axes

As we realize, twenty-seven was the maximum number of stone axes or anything else that anyone could possess – because what would you do if you had one more? You couldn't count it in the established manner, it just wouldn't fit in anywhere – in fact, it would constitute the closest approximation to infinity, and you would simply have to throw it away, wouldn't you?

2.4 Decimal Position System

Hindu-Arabic Numeration System

We use the base ten, or decimal, system daily, so it is the natural place to start if we want to understand positional systems generally.

Each of the digits 0, 1, 2, 3, 4, 5, 6, 7, 8, 9 represents a number on its own. The decimal positional system gives an unambiguous meaning to strings of digits as a sum of powers of ten, the rightmost digit being the number of ones (where we let $1 = 10^0$). The next digit to the left is the number of tens, so one ten and zero ones is written 10.

One of the wonders of this system is that while there are two ways to read 2300 (two thousand three hundred or twenty-three hundred), both name the same number. The canonical way to interpret 4321 is

$$\begin{aligned}&4 \text{ thousands} + 3 \text{ hundreds} + 2 \text{ tens} + 1 \text{ one}\\ &\qquad = 4 \times 10^3 + 3 \times 10^2 + 2 \times 10^1 + 1 \times 10^0,\end{aligned}$$

but we get the same result from

$$43 \times 10^2 + 2 \times 10^1 + 1 \times 10^0 \text{ or } 432 \times 10^1 + 1 \times 10^0.$$

The convention forced upon us by the desire to make the algebra of exponents consistent is that $a^0 = 1$ whenever $a \neq 0$.

The decimal system or any other positional numeration system may be extended beyond the integers to represent fractions. A point of reference is then selected to establish the positional values of the several digits, and characterized, for example, by a decimal point:

$$310.103 = 3 \times 10^2 + 1 \times 10^1 + 0 \times 10^0 + 1 \times 10^{-1} + 0 \times 10^{-2} + 3 \times 10^{-3}$$

Origin and Development

The oldest preserved samples of Hindu-Arabic, or Arabic, digits are inscriptions on a stone column found in India and believed to be from around the year 250 B.C. However, this and other early specimens display no zero and give no indication of the present-day use of Arabic digits. Scholars are uncertain about the earliest date for the use of these early digits and zero in a positional numeration system and whether the Hindu positional system developed entirely within India or under Phoenician and Persian influence.

al-KHOWARIZMI
Muhammad ibn Musa
(*c.* 780 - *c.* 850)

The oldest known writing using a fully developed numeration system appears in a book from A.D. 825 by the Persian mathematician **al-Khowarizmi**. The original is lost, but a

ROBERT OF CHESTER (*c.* 1100)

Latin translation from around 1120 exists, *Liber algorismi de numero indorum* ("The Book of al-Khowarizmi on Hindu Number") by the Englishman **Robert of Chester**. In this translation, the numerals were erroneously assumed to be of Arabic origin, a mistake carried through into present-day terminology. The European writing of the digits was influenced by Arabic manuscripts, but their European design also followed paths independent of Arabic models.

Algorithm

The importance of the 12th-century Latin translation of al-Khowarizmi's book on numerals is witnessed by the term **algorithm** (or **algorism**), a Latin corruption of al-Khowarizmi, which came to mean the art of computing with Hindu-Arabic numerals. Today *algorithm* refers to any systematic mathematical procedure that leads to a solution usually in a finite number of steps.

East Arabic Numerals

Based on the Hindu model, the Arabs developed two versions of numerals, East Arabic and West Arabic, of which the East Arabic is now by far the dominant. The numerals below represent modern **East Arabic numerals**:

٠	١	٢	٣	٤	٥	٦	٧	٨	٩
0	1	2	3	4	5	6	7	8	9

Although the Arabic language is written from right to left, Arabic numerals are arranged as in English, *e.g.*,

٥٠	٦٥	١٧٤٨	١٩٩٦
50	65	1748	1996

West Arabic Numerals

In Arab countries where the East Arabic numerals are prevalent, visitors are often surprised that the European version of "Arabic numerals" bears so little resemblance to the indigenous numerals. This is because European Arabic numerals were modeled primarily on the now extinct **West Arabic numerals**, *e.g.*,

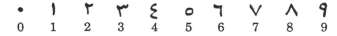

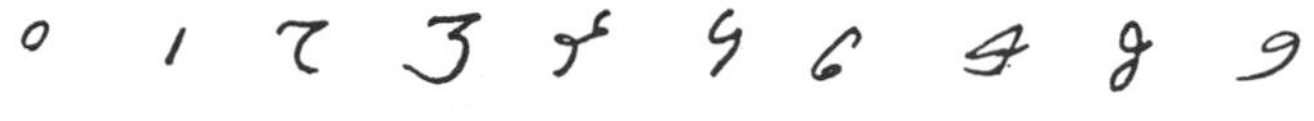

which varied considerably, however, geographically and with time.

Gobar

Their name, **Gobar**, is Arabic for *sand board* or *dust*, referring to the practice of performing calculations directly on the ground or on a board covered with sand, which could be wiped clean when a calculation was completed. The name is now attached to the early Hindu-Arabic numerals introduced into Europe. The oldest surviving sample of Gobar numerals in Europe is a Spanish manuscript from 976:

1 2 3 4 5 6 7 8 9

FIBONACCI Leonardo (*c.* 1170 - *c.* 1250)

Leonardo Fibonacci (Leonardo of Pisa) was an Italian merchant traveling in the Orient, where in addition to conducting business he also learned Oriental mathematics. Fibonacci wrote the *Liber abaci* (1202) – in all likelihood inspired by al-Khowarizmi's writings – and *Practica geometriae* (1220).

Liber abaci, dealing with arithmetic and algebra, played an important role in spreading the knowledge and use of the decimal positional system in Europe.

Significant Digits

The digits that define a numerical value are known as **significant digits**.

The significant digits of a given number

begin with a first non-zero integer digit or – if the number is less than unity – with the first, zero or non-zero, decimal digit;

and end with the final, zero or non-zero, decimal digit; the final zero or zeros of an integer (whole) number may or may not be significant.

Thus,

3200	has	at least	2	3.20	has	3
320	has	at least	2	0.320	has	3
32	has	at least	2	0.0320	has	4
32.0	has		3	0.032	has	3

significant digits.

p. 162

To make clear the number of significant digits of an integer ending in one or more zeros, it can be written in *scientific notation*. For instance, to clarify that 2 304 000 has only four significant digits, we write 2.304×10^6; to clarify that all digits are significant, we write $2.304\,000 \times 10^6$.

A number with four significant digits is said to have a **four-place accuracy**.

Chinese Rod Numerals

The rod-numeral system represents a positional decimal system created in China at least 2000 years ago, which uses bamboo rods to represent numbers. The system was adopted by Japan and Korea and used for arithmetic and algebraic calculations until about 150 years ago.

The numerals 1 … 9 can be reproduced by **two different** series of rod configurations:

First series

Second series

Writing the number 21 as  or  would risk confusion with 3; written as  it would be read as 201; with two series of symbols available, we can write

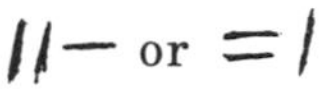

which are unambiguous.

For many centuries the system lacked a sign for zero. Simply leaving a space between symbols did not ensure unequivocal discrimination between numerals such as 36, 306, 3006, 3060, and 3600.

To overcome this difficulty, a traditional Chinese number symbol was sometimes used as a separator; at other times, a reckoning board was used, an empty square clearing up the matter, *e.g.*:

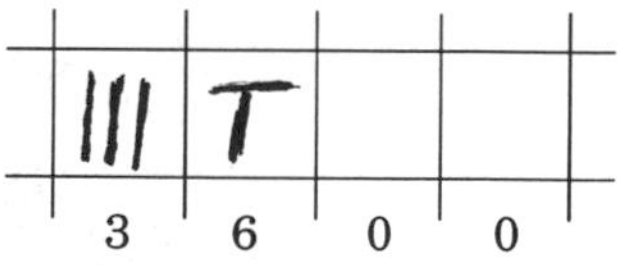

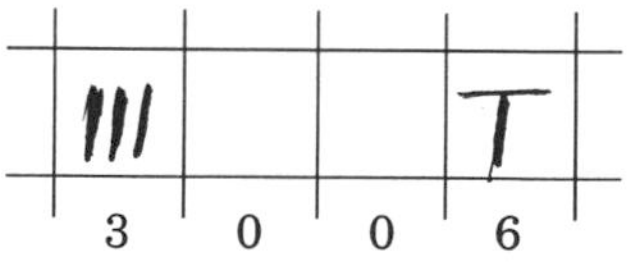

Through Indian influence in the 13th century, a symbol for zero – a circle – was incorporated into the rod-numeral system, making all numbers, and particularly fractions, easier to express (a leading zero can be assumed to designate a decimal marker):

0.15

0.0079

The Decimal System of Measurement

To get the full benefit of a positional numeration system, the system of units of measurement must agree with the numeration system.

Metric System
International System of Units, SI

The term **decimal system** is not limited to systems of numeration that employ base ten, but is also used to characterize *systems of units and measurement* whose multiples and submultiples of units are powers of ten. The **metric system** is based on multiples of 10 and the **International System of Units** (**SI**), which is convenient for science and technology, is organized around multiples of $1000 = 10^3$.

With the major exception of the United States, SI is today used almost exclusively in the industrialized world; it is also the standard of most developing countries. Aside from the basic sciences and some major industries, the U.S. has been resistant to adopting the metric system and its successor, the SI.

For all the technical and scientific advancement associated with the U.S., it is bizarre that it so staunchly abides by a conglomeration of basically incoherent measurement systems.

This is even more surprising since, in 1790, Thomas Jefferson, then secretary of state, suggested that a decimal system of measurement would be a way of achieving national uniformity of weights and measures. In 1875, the U.S. was one of the original signatories of the Metric Convention, and since 1893 the definitions of the U.S. foot and pound rely on the metric system; since 1959, a U.S. foot is defined as 30.48 centimeters.

"An unprecedented blizzard has buried the Eastern seaboard under one hundred and fifty-eight inches of snow. That's, let's see ... twelve into fifteen goes once, two from five is three, bring down the eight, twelve into thirty-eight goes three times with two left over ... "

Will he manage before the next thaw?

Scientific Notation

The handling of very large and very small numbers is always fraught with difficulties.

- The radius of the Earth is 6.37 million meters, and the height above Earth of a geostationary communications satellite is some 36 million meters; these distances are not too shocking to the reader.
- An *astronomical unit* – the mean distance from the Earth to the Sun – is about 150 thousand million meters, or more exactly,

  149 569 800 000 meters.
- The diameter of the Milky Way, the galaxy in which our Earth is an insignificant speck of dust, is on the order of 100 trillion (U.S.: 100 quintillion) meters – a 1 followed by 20 zeros.
- At the other end of the scale, the diameter of a poliomyelitis virus is some 28 thousandths of a millionth part of a meter; that of an atomic nucleus may be about 75 thousand-billionths (U.S.: 75 quintillionths) of a meter:

  0.000 000 028 or $0.0_7 28$

  0.000 000 000 000 075 or $0.0_{13} 75$

Whatever way we choose to write these numbers, they are unwieldy and may easily lead to miscalculations.

To avoid these difficulties, we observe that every positive real number a may be written in **scientific notation**,

$$a = p \cdot 10^n$$

where n is an integer, positive or negative; and $1 \leq p < 10$, that is, p is larger than or equal to 1, but less than 10.

To simplify calculation, specific prefixes to denote certain multiples and submultiples of units of physics have been introduced by the

General Conference on Weights and Measures

(CGPM, Conférence Générale des Poids et Mesures), which was founded by the Metric Convention of 1875. The majority of these prefixes date from the 1960 conference; in October 1991 yotta, zetta, zepto, and yocto were added.

The prefixes form part of the *International System of Units*, or SI (*Système International d'Unités*), which is, in fact, a modernized metric system.

By SI recommendations, prefix symbols should be in Roman (upright) type and attached to the SI unit symbol, *e.g.*, millimeter (mm), microampere (μA), decibel (dB).

The multiple should be chosen so that the numerical value preceding it will be between 0.1 and 1000. In tables, and in contexts where direct comparison between values is made, it may be advantageous not to change prefixes.

SI Multiples

Multiplier	SI Prefix Name	SI Prefix Symbol	Etymology		
10^{24}	yotta	Y	Greek:	oktákis, *eight times*	(8 x 3 = 24)
10^{21}	zetta	Z	Greek:	heptákis, *seven times*	(7 x 3 = 21)
10^{18}	exa	E	Greek:	hexákis, *six times*	(6 x 3 = 18)
10^{15}	peta	P	Greek:	pentákis, *five times*	(5 x 3 = 15)
10^{12}	tera	T	Greek:	téras, *monster*	
10^{9}	giga	G	Greek:	gígās, *giant*	
10^{6}	mega	M	Greek:	mégas, *big*	
10^{3}	kilo	k	Greek	chílioi, *thousand*	
10^{2}	* hecto	* h	Greek:	hekatón, *hundred*	
10^{1}	* deca	* da	Greek:	déka, *ten*	
10^{-1}	* deci	* d	Latin:	decim, *ten*	
10^{-2}	* centi	* c	Latin:	centum, *hundred*	
10^{-3}	milli	m	Latin:	mille, *thousand*	
10^{-6}	micro	μ (mu)	Greek:	mīkrós, *small*	
10^{-9}	nano	n	Greek:	nānos, *dwarf*	
10^{-12}	pico	p	Italian:	piccolo, *small*	
10^{-15}	femto	f	Danish:	femten, *fifteen*	
10^{-18}	atto	a	Danish:	atten, *eighteen*	
10^{-21}	zepto	z	Greek:	heptákis, *seven times*	(– 21 = – 7 x 3)
10^{-24}	yocto	y	Greek:	oktákis, *eight times*	(– 24 = – 8 x 3)

* Not advocated in scientific and technical writing; acceptable for domestic and lay use.

ÅNGSTRÖM, Anders Jonas (1814 - 1874; Swedish physicist)

With the introduction of SI prefixes some older measurement units have become obsolete, notably the ångström (Å) unit of length which is equivalent to 0.1 nanometer or 100 picometers.

Bits and Bytes

Alphanumeric or **alphameric characters** consist of alphabetic and numerical symbols, punctuation marks, and other symbols used in computer work.

In computer terminology **bit** stands for **bi**nary digi**t**. The bit is the smallest storage unit in a computer.

A series of consecutive bits representing one alphanumeric character in the storage unit ("memory") of the computer forms a **byte** (origin possibly from alteration of a bite, a small piece). The number of bits that form a byte consists of eight bits (eight on-off combinations, that is to say, a combination of eight binary numbers), and each byte can thus represent one of 2^8 (= 256) combinations.

A **kilobyte** is 2^{10} (= 1024) bytes and is denoted **K**. Kilo normally represents a 1000 multiple of a unit; consequently, the term kilobyte is not in line with **SI** or any other system of units. A **megabyte** (**MB**) is 2^{20} (= 1 048 576) bytes, a **gigabyte** (**GB**) 2^{30} bytes, *etc.*

Drawing by Levin; © 1989
The New Yorker Magazine, Inc.

2.5 Sexagesimal Numeration

The Babylonian, or Mesopotamian, sexagesimal system of numeration, which succeeded the Sumerian-Akkadian system, is the oldest known example of numeration using place value. We do not know exactly when this idea was first conceived, but archaeological finds suggest it originated around the beginning of the 2nd millennium B.C., with the creation of the Babylonian Empire.

Several Orders of Unity

Our decimal positional system requires ten individual signs to represent the numeral digits 0 through 9; a sexagesimal positional system, using 60 as a base, would need 59 different characters for the numerals within each span of sixty.

The Babylonian scientific numeration system is not a strictly positional system. It uses the additive principle of the Sumerians for all numbers below 60, and makes do with just two characters: a "corner" sign  signifying "ten" (10), and a "wedge" sign  for all forms of unity: 1, 60, 60^2, 60^3, *etc.*

Thus, a number below 60 is written, *e.g.*,

47 =  (40 + 7)

and a number greater than 60,

78 =  (60 + 10 + 8)

The Babylonian number

or, transcribed into Hindu-Arabic characters,

1; 54; 42; 40,

means

1×60^3	=		=	216 000
54×60^2	=	54×3600	=	194 400
42×60^1	=		=	2 520
40×60^0	=		=	40
				412 960

The Appearance of Zero

The Babylonians originally had no zero so they could not mark the place of a missing power of sixty in a number, which caused the same uncertainty as we would experience today without a zero to help distinguish between numbers like 61, 601, and 6001, for instance. The only means open to a Babylonian scribe was to leave an empty space where a zero would be.

An empty space has the disadvantage, however, of not being easily recognized as a single, double, or perhaps multiple empty space; even less is it identifiable at the beginning or end of a number. Babylonian mathematicians and scientists had to either invent some distinct, unmistakable place marker or invent zero. They used a double "corner" sign or a slanting double "wedge" as a place marker:

which could be used equally well in an initial, intermediate, or terminal position.

On tablets from the late 4th century B.C. – and believed to be copies of documents written some two centuries earlier – we encounter another character,

a corner sign with an elongated lower shank or tail, probably a scribe's shorthand version of a double corner sign, with one corner left out.

2; 0; 27

which in our way of thinking is

$$2 \times 60^2 + 0 \times 60^1 + 27 = 7227 .$$

This "zero" sign, from about 500 B.C., in many respects functions as our zero, but the Babylonians probably had no concept of zero as a number and instead used this symbol to mark a void.

Sexagesimal Fractions

The use of a double corner or a double wedge in an initial position provided the Babylonian astronomer and mathematician with a simple means of writing sexagesimal fractions (that is, fractions whose denominators are powers of sixty), important to Babylonian science.

 $= \frac{0}{60^0} + \frac{1}{60^1}$

0; 1

 $= \frac{0}{60^0} + \frac{0}{60^1} + \frac{30}{60^2}$

0; 0; 30

Such sexagesimal fractions survive to our day: an hour has 60 minutes, each of 60 seconds; a full circle has 6 x 60 = 360 degrees, each divisible in 60 minutes of arc, each of which is 60 seconds of arc.

2.6 Vigesimal Numeration

The vigesimal numeration system was the usual mode of counting in many ancient cultures because humans have ten fingers and ten toes on which to count. As we know from the Celtic languages and their remnants in French and Danish, vigesimal numeration has a long endurance, as attested by the lingering use of the word **score**.

Decimal-system numbers 2310 and 128 would be, in vigesimal notation,

$$2310 = 5 \times 20^2 + 15 \times 20^1 + 10 \times 20^0 = \text{five; fifteen; ten}$$
$$128 = 6 \times 20^1 + 8 \times 20^0 = \text{six; eight}$$

Mayan Numeration

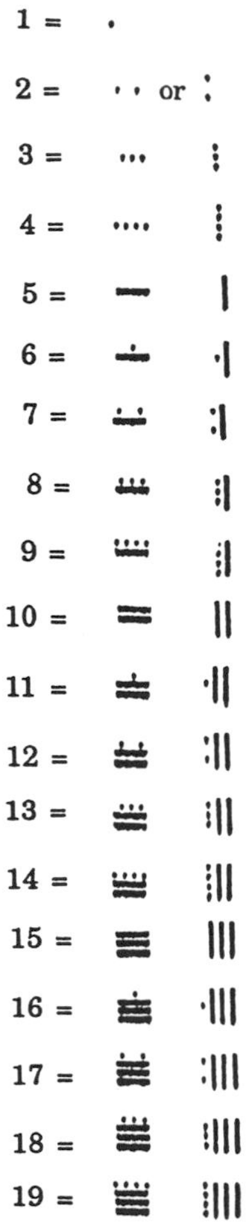

The **Maya Indians** lived in Central America – in the Yucatán Peninsula and neighboring Guatemala and Honduras – from perhaps 200 B.C. until A.D. 1540 when they disappeared from the pages of history.

During the formative period of their civilization – the first three centuries of our time – the early Maya seem to have devised a purely additive vigesimal system of counting for the everyday affairs of the general people. Indications of this are as many as they are vague, but as such trivial pursuits were seldom committed to writing, there are no actual, tangible proofs available of this, "the little man's arithmetic".

The **classical period** of Mayan civilization – from about A.D. 290 to 925 – is characterized by the attainment of a high level of culture, with truly remarkable achievements in arithmetic, astronomy, and art. Symbols relating to Mayan numeration, calendar, and deities have been understood for over a hundred years, but not until in the 1980s could epigraphers and linguists put together the vast body of scientific work for a comprehensive deciphering of Mayan writing.

For scientific purposes, the Maya had a written positional, near-vigesimal system of numeration possessing a definite zero. Besides highly ornate hieroglyphic numeral characters, depicting their gods, the Maya had a simpler system of three basic symbols:

- dots of value 1
- bars with value 5
- a stylized seashell, generally painted red, for zero

Numbers from 1 through 19 were made up of these bars and dots in the most "economical" manner, with the bars either horizontal or vertical: if the bars were horizontal, the dots were placed above them; if they were vertical, the dots were placed on their left.

The system was not strictly vigesimal, the second span having a base 18 instead of 20; a Mayan number is written in column form and reads from top to bottom:

$1 \times 20 \times 18 \times 20^2$	=	144 000
$0 \times 20 \times 18 \times 20$	=	0
$13 \times 20 \times 18 \times 20^0$	=	4 680
7×20	=	140
15×20^0	=	15
		148 835

The seashell zero could be used in intermediate and terminal positions of a number, but not in an initial position. The only use for a zero in our initial position would be for the analogue of decimal fraction notation, but the Maya knew only integer numbers and had no concept of fractions.

The Mayan Calendar

The Maya had two kinds of calendar: a religious calendar of 260 days consisting of 20 cycles of 13 days each; and a secular calendar of 365 days, in fact, a solar calendar of 18 *uinals* ("months") of 20 days each, plus an extra 5 days added at the end of the year.

The actual length of a solar year is the average time elapsing between successive passages of the Earth through the same point in its orbit around the Sun. Mayan astronomers were fully aware of the error in the length of the solar year, inherent in the official 365 days.

According to *our* latest measurements,

– the average solar year is	365.242 198 days
– the Mayas calculated it to be	365.242 000 days
– as against the Gregorian calendar's	365.242 500 days

which makes the error in the Mayan measurement just under two-thirds of that of the Gregorian calendar.

Mayan astronomers also observed the movements of the Sun and the Moon, of Venus, and possibly also of Mars, Jupiter, and Mercury; and they predicted solar eclipses with an astonishing precision.

To appreciate fully the excellence of these feats of Mayan astronomers, we must remember that they did not know how to make glass and thus had no lenses or telescopes or other optical equipment which we regard as indispensable in astronomical work.

Nor did they know the sandglass or the klepshydra (water clock), or any other means of measuring times shorter than the day, which they looked upon as the shortest unit of time – hours, minutes, and seconds being unknown quantities.

Yet they *did* measure the length of the true solar day – that is, the average time elapsing between two successive passages of the Sun through the observer's meridian, just by observing the shadow of a gnomon – the "hand" of a sundial.

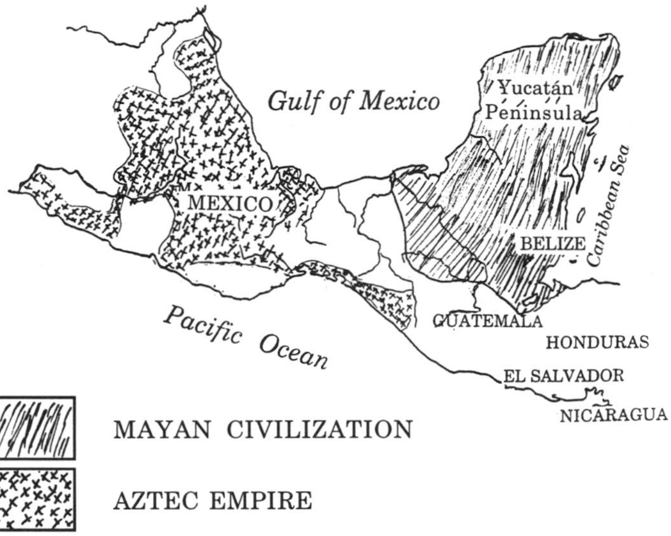

Aztec Numeration

From A.D. 925 the Maya went downhill, perhaps because of the clearing of the forests by burning and of overcultivating the soil, or perhaps because of social unrest and wars, and their culture declined. What remained was taken over by some of their neighbors, among them the Aztecs, known for their ritual human sacrifices. In the early 1500s, the Aztec Empire was conquered and many of its people massacred by the Spanish conquistadores.

We know their system of numeration through contemporary Spanish translations of Aztec books, most well known of which is the *Codex Mendoza*, named after Don Antonio de Mendoza, the first viceroy of New Spain.

The Aztec system of numeration was strictly vigesimal, with special characters for 1, a dot; for 20, a drawing of a hatchet; for $20^2 = 400$, a bird's feather; for $20^3 = 8000$, something looking like a string purse, *etc.*

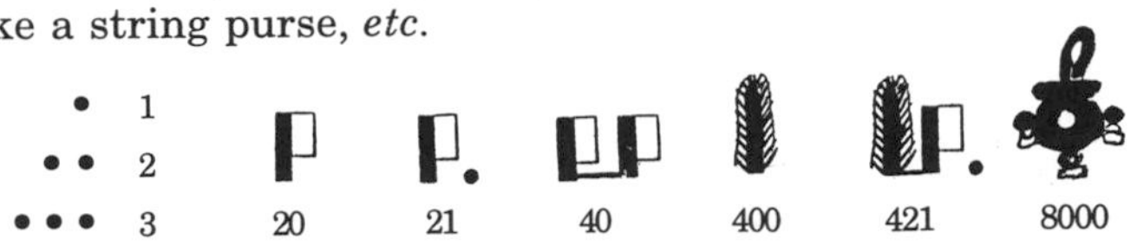

2.7 Duodecimal Numeration

Latin, *duodecim*, "twelve"
Old French, *dozaine*, "twelve"

The number twelve has set its mark on many aspects of our environment: there are twelve months in a year, and a clock dial shows twelve hours; twelve items make a dozen, and twelve dozen make a gross; a foot is twelve inches.

When the Babylonians were faced with the need for a smaller counting unit in their sexagesimal system, they settled for ten instead of twelve as a subsidiary base. They continued to use sexagesimal fractions, however, so the advantage afforded by the fact that twelve is divisible by 2, 3, 4, and 6 did not carry weight with them.

Dividing the day into twice twelve hours is not, as long thought, of Babylonian origin; Otto Neugebauer has shown that it was an Egyptian invention.

There are examples, in the late 19th century, of countries that redivided their old-time foot of twelve inches into ten "new" inches, before finally "going metric".

A really forceful – and final – argument against duodecimal numeration is the fact that we have a well-functioning decimal system.

An interesting combination of vigesimal and duodecimal counting was offered by the old British monetary system, a pound sterling being divided into twenty shillings of twelve pence each – until it was scrapped in 1971.

– It is, of course, well known among poultry farmers that Scandinavian hens are vigesimal and lay scores of eggs, whereas English-speaking hens are duodecimal and lay theirs by the dozen.

2.8 Binary, Octal, Hexadecimal

Binary Numeration

The binary positional numeration system uses 2 as base, and thus requires only two distinct symbols, say 0 and 1. In binary numeration, ten will be written 1010, because

$$10 = \underline{1} \times 2^3 + \underline{0} \times 2^2 + \underline{1} \times 2^1 + \underline{0} \times 2^0$$

where 2^0 equals 1, by definition.

The use of a binary system would be cumbersome for everyday manual calculation because of the inordinate length of most numbers, but it is ideal for electronic computers whose mechanical and electronic relays, such as transistors, know only two states of operation – **on/off**, **closed/open**, **yes/no**, **true/false**.

With 1 and 0 as operational characters, 1 will stand for on (closed circuit, true) and 0 for off (open, false).

von LEIBNIZ Gottfried Wilhelm (1646 - 1716)

Binary numbers, known to Chinese mathematicians since the 5th century or earlier, were investigated and set into a serious numerical system by the eminent German mathematician **Gottfried Wilhelm von Leibniz**, to whom is attributed this very unmathematic thesis of abstract theology:

> God, represented by the numeral 1, created the Universe out of nothing, represented by 0.

Binary Counting Rules

Addition and **subtraction** of binary numbers are governed by the following rules:

$0+0=0$	$0+1=1$	$1+0=1$	$1+1=10$	(0, carry 1)
$0-0=0$	$1-0=1$	$1-1=0$	$10-1=1$	(borrow 1)

The binary **multiplication** table looks like this:

$$0 \times 0 = 0 \qquad 0 \times 1 = 0 \qquad 1 \times 0 = 0 \qquad 1 \times 1 = 1$$

and the **division** table is even simpler:

$$0/1 = 0 \qquad 1/1 = 1; \qquad 1/0 \text{ and } 0/0 \text{ are undefined}$$

- Convert decimal number 23 to binary notation.

The relevant powers of 2 are

2^4	2^3	2^2	2^1	2^0
16	8	4	2	1

$$23_{dec} = 1 \times 2^4 + 0 \times 2^3 + 1 \times 2^2 + 1 \times 2^1 + 1 \times 2^0 = 10111_{bin}$$

- Convert binary 110111 to decimal form.

  We have

$$\begin{aligned} & 1 \times 2^5 + 1 \times 2^4 + 0 \times 2^3 + 1 \times 2^2 + 1 \times 2^1 + 1 \times 2^0 = \\ = \; & 32 + 16 + 0 + 4 + 2 + 1 = 55; \end{aligned}$$

  hence, binary 110111 = decimal 55

- Calculate and express the sum, difference, product, and quotient of binary numbers 1011 and 101 in binary and decimal form.

$1011_{bin} = 8 + 0 + 2 + 1 = 11_{dec}$ $\qquad$ $101_{bin} = 4 + 0 + 1 = 5_{bin}$

Sum

$$\begin{array}{r} 1011 \\ +\;\; 101 \\ \hline 10000_{bin} \end{array} \qquad \begin{array}{r} 11 \\ +\; 5 \\ \hline 16_{dec} \end{array}$$

Difference

$$\begin{array}{r} 1011 \\ -\;\; 101 \\ \hline 110_{bin} \end{array} \qquad \begin{array}{r} 11 \\ -\; 5 \\ \hline 6_{dec} \end{array}$$

Product

$$\begin{array}{r} 1011 \\ \times\;\; 101 \\ \hline 1011 \\ 0000 \\ +1011 \\ \hline 110111_{bin} \end{array} \qquad \begin{array}{r} 11 \\ \times\; 5 \\ \hline 55_{dec} \end{array}$$

Quotient

$$\begin{array}{r} 10.00110\ldots_{bin} \\ 101 \overline{)\, 1011.00000} \\ -101 \\ \hline 01\,000 \\ -\;\;\;\; 101 \\ \hline 110 \\ -101 \\ \hline 0010 \end{array} \qquad \begin{array}{r} 2.2_{dec} \\ 5 \overline{)\, 11.0} \\ -10 \\ \hline 1\;0 \\ -1\;0 \\ \hline 0 \end{array}$$

Binary Man. Drawing by Dedini; © 1987 The New Yorker Magazine, Inc.

Octal and Hexadecimal Numeration

In computer practice, octal and hexadecimal numbers are often used to represent large binary numbers which may be difficult to handle because of their length. The usefulness of these forms of notation arises from the ease of conversion between the number systems, as both eight (2^3) and sixteen (2^4) are powers of 2, the base of the binary system.

The **octal numeration system** requires only eight digits, which can all be supplied by Arabic digits (0, 1, 2 ... 7).

The **hexadecimal numeration system** needs sixteen symbols. As the decimal system provides only ten basic number symbols, the letters A, B, C, D, E, and F are used as additional symbols to make up sixteen symbols.

Decimal	Binary	Octal	Hexadecimal
0	0	0	0
1	1	1	1
2	10	2	2
3	11	3	3
4	100	4	4
5	101	5	5
6	110	6	6
7	111	7	7
8	1000	10	8
9	1001	11	9
10	1010	12	A
11	1011	13	B
12	1100	14	C
13	1101	15	D
14	1110	16	E
15	1111	17	F
16	10000	20	10
17	10001	21	11
⋮	⋮	⋮	⋮
23	10111	27	17
24	11000	30	18
25	11001	31	19
⋮	⋮	⋮	⋮
31	11111	37	1F
32	100000	40	20
⋮	⋮	⋮	⋮

Note that octal and hexadecimal numbers of, say, 13 and 20 do not read "thirteen" and "twenty", but "one-three" and "two-oh", respectively.

System Conversion

To convert a binary number to octal notation, divide the number into groups of three digits (triplets), beginning with the unity digit and working left – or both ways if the number contains a decimal part. Incomplete groups are filled in with zeros, and the several triplets are converted to octal notation.

To convert a binary number to hexadecimal notation, divide it into groups of four digits (quadruplets) in the same manner, and convert these groups, one by one, to hexadecimal notation.

- Convert binary 10 111 110 to octal, decimal, and hexadecimal notation.

Binary row	010	111	110	
	2	7	6	$= 276_{oct}$
	2×8^2 +	7×8^1 +	6×8^0	
	128 +	56 +	6	$= 190_{dec}$
Binary row		1011	1110	
		B	E	$= BE_{hexad}$

Hence,

$$10\,111\,110_{bin} = 276_{oct} = BE_{hexad} = 190_{dec}$$

- Explain how **(a):** $7 + 7 = 13$; **(b):** $7 + 8 = 12$; **(c):** $7 + 9 = 11$.

a: $7 + 7 = 1 \times 11^1 + 3 \times 11^0 = 13$; 11 is the base

b: $7 + 8 = 1 \times 13^1 + 2 \times 13^0 = 12$; 13 is the base

c: $7 + 9 = 1 \times 15^1 + 1 \times 15^0 = 11$; 15 is the base

- Explain how $2 \times 3^3 = 2000$.

$2 \times 3 \times 3 \times 3 = 2 \times 3^3 + 0 \times 3^2 + 0 \times 3^1 + 0 \times 3^0 = 2000$; 3 is the base

2.9 Special Forms of Notation

The Morse Code

This code uses a system of short and long signals – dots and dashes – to send messages by sound, as flashes of light, or on the telegraph. Combinations of dots and dashes make up letters, punctuation marks, and numerals.

MORSE Samuel F. B.
(1791 - 1872)

The original system was developed in 1838 by the American artist and inventor **Samuel Morse**; a modification – the International Morse Code – was devised in 1851.

1	•––––	6	–••••
2	••–––	7	––•••
3	•••––	8	–––••
4	••••–	9	––––•
5	•••••	0	–––––
period	•–•–•–	comma	––••––

••––– ––––– •–––– ––••• = 2017

––––– •–•–•– –•••• ••••• = 0.65

Braille

BRAILLE Louis
(1809 - 1852)

Braille is the written language of blind people, named after its inventor, the French educator **Louis Braille**, himself blind. Its characters are raised dots – read by touch – arranged in simple combinations in a two-by-three pattern.

The first ten letters, A through J, will double as numeral digits 1 ... 0 when preceded by a special numeral identifier sign.

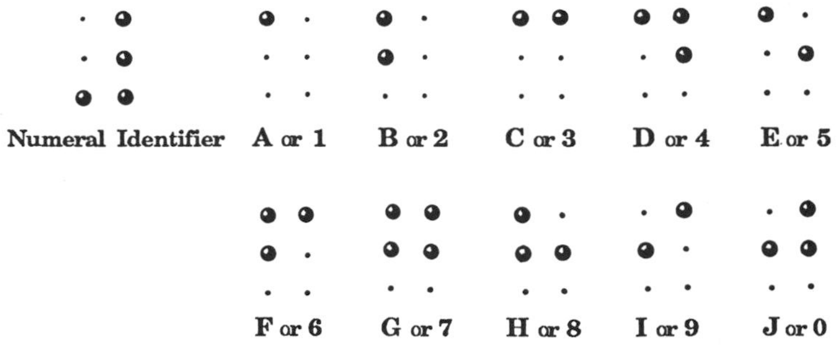

Manual Sign Language

d'ÉPÉE Charles Michel
(1712 - 1789)

Most national sign languages for the hearing impaired trace their ancestry to the French Sign Language, developed in the mid-18th century by a priest, Father **Charles Michel d'Épée**.

A true manual sign language develops independently from the language spoken around the deaf and, thus, has its own word stock and grammar. So, deaf Parisians had a manual sign language long before Father Michel learned it and, with the aid of some of his pupils, devised signs for spelling out words.

GALLAUDET Thomas Hopkins
(1787 - 1851)

Thomas Gallaudet, founder of the first American school for the hearing impaired, brought the French Sign Language to the United States in 1816. Several sign languages were current at the time; Gallaudet combined these with the French system to form the American Sign Language. However desirable, there is no international sign language for the hearing impaired.

To express numbers, Roman numerals C for hundred and M for thousand are brought into action.

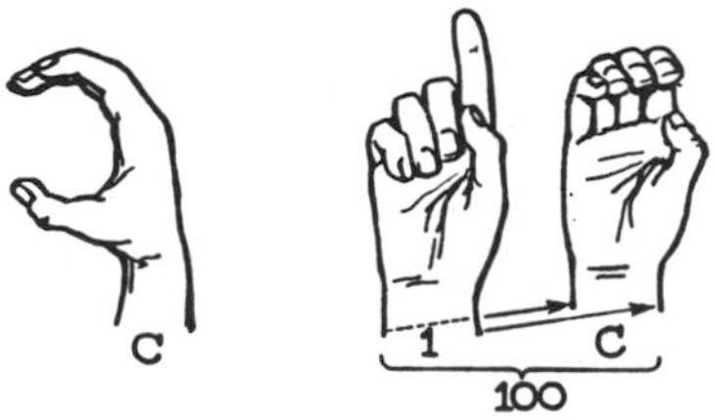

To make "hundreds", the person signs the *digit* 1, 2, 3 ..., followed by the sign for the *letter* C; to make "thousands", the person makes the proper *unit sign* followed by the *letter* M – the tips of three fingers of the right hand touching the left palm.

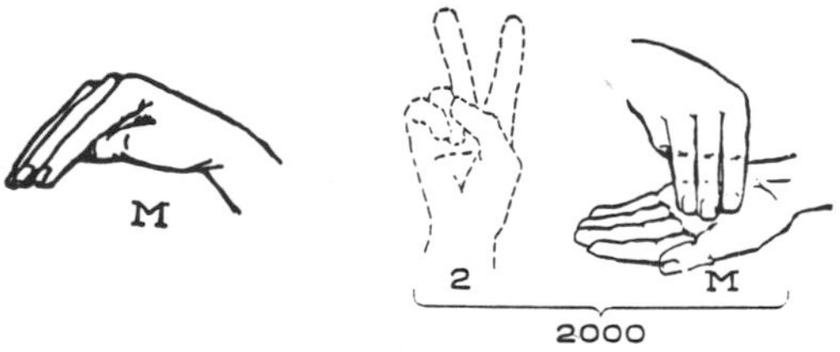

Semaphores

CHAPPE Claude
(1763 - 1805)

The semaphore visual telegraph was invented by the French engineer **Claude Chappe**, who, with the backing of the Legislative Assembly of the French Revolution, built between Paris and Lille (near the French-Austrian war front) a series of towers on high places, spaced 10 to 16 kilometers apart. Every tower was equipped with telescopes and a set of arm-like structures – a semaphore – which could be turned around on a pole; each arm had seven clearly discernible angular positions, making a total of 49 combinations that were assigned letters of the alphabet or other messages.

Greek, *sema*, "sign"; *pherein*, "to bear"

The word *semaphore* was a construction of Chappe's.

Semaphore telegraph towers were built also in other parts of France, and Chappe's invention was soon copied elsewhere in Europe.

Semaphore signaling with mechanically or electrically controlled semaphore arms, often with rows of lights, is still used to signal railroad trains.

Though currently at the point of extinction, semaphore signaling between ships has been of great importance. A person holds a small flag in each hand; with the arms extended, the person moves the flags to different angles. Succeeding a "numbers follow" signal, numbers from 1 to 9 are the same as the first nine letters of the alphabet; zero is the same as the tenth letter:

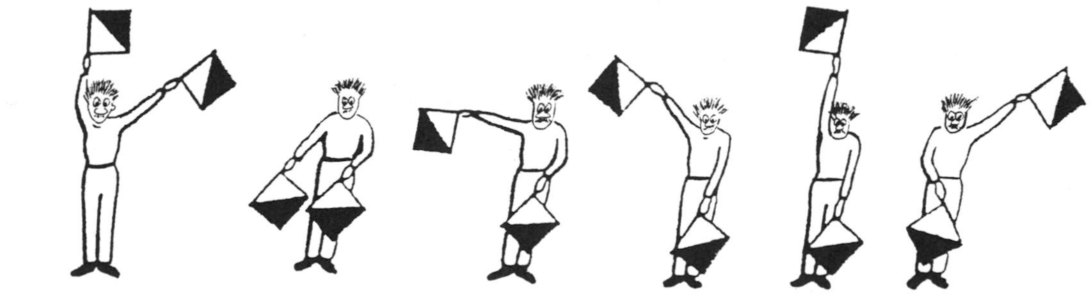

numbers follow A/1 B/2 C/3 D/4 E/5

F/6 G/7 H/8 I/9 J/zero

Chapter 3

TYPES OF NUMBERS

*** Indicates cross-reference

3.0 An Expanding Universe of Numbers

"Do you know what the foundation of mathematics is?" I ask. "The foundation of mathematics is numbers. If anyone asked me what makes me truly happy, I would say: numbers. Snow and ice and numbers. And do you know why?"

Peter Høeg, *Smilla's Sense of Snow* (1993)

Die Zahlen sind freie Schöpfungen des menschlichen Geistes, sie dienen als ein Mittel, um die Verschiedenheit der Dinge leichter und schärfer aufzufassen.

"Numbers are free creations of the human mind that serve as a medium for the easier and clearer understanding of the diversity of thought."

Richard Dedekind, *Was sind und was sollen die Zahlen?* (1887)

DEDEKIND
Julius Wilhelm Richard
(1831 - 1916)
German mathematician

Starting with the **natural numbers**, or **counting numbers**,

$$1, 2, 3 \ldots$$

we have numbers where addition and multiplication will always result in a natural number, but these numbers alone are not sufficient for subtraction. Supplementing with **zero** and **negative whole numbers**, thus obtaining the **integers**,

$$\ldots, -3, -2, -1, 0, 1, 2, 3, \ldots,$$

we have enough numbers for addition, multiplication, and subtraction of any integers but not for division, which necessitates that we include numbers that are *ratios of integers*, *e.g.*, $\frac{1}{2}, \frac{2}{3}, \frac{81}{82}$. We now have integers and ratios of integers, which all are known as **rational numbers**.

Rational numbers produce other rational numbers when added, multiplied, subtracted, or divided. Yet rational numbers are not always enough – no ratios of integers can exactly represent the solutions of, for instance, $x^2 - 2 = 0$, that is, $x = \pm\sqrt{2}$, nor can numbers such as π and e be represented by ratios of integers. Such numbers – not representable by ratios of integers – are known as **irrational numbers**.

Rational numbers and irrational numbers are together named **real numbers**.

We can visualize the real numbers on the number line, a straight line extending from negative infinity to positive infinity and graduated in **unit distances** on both sides of the **origin,** which symbolizes zero. Between the integers we have, on the number line, all the remaining real numbers.

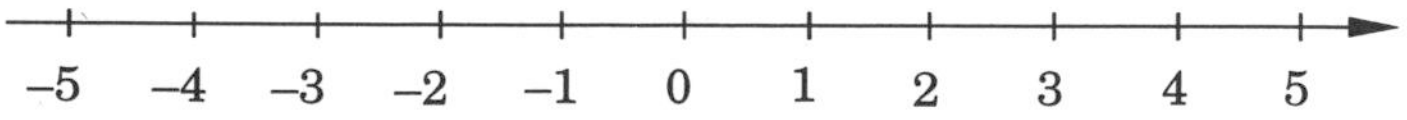

Absolute Value

The **absolute value** $|a|$ of a real number a is, by definition, either positive or zero – it cannot be negative.

Real numbers cannot describe the solutions of an equation such as $x^2 + 1 = 0$, that is, $x = \pm\sqrt{-1}$, for the product of any two real numbers that have the same sign is always positive or zero. Since real numbers cannot do the job, we take recourse to **complex numbers**,

$$a + b\,\mathrm{i}\,,$$

where a and b are real numbers and i is the imaginary unit, defined

$$\mathrm{i}^2 = -1\,.$$

If $a = 0$, we have **pure imaginary numbers**, *e.g.*,

$$\sqrt{-1} = \mathrm{i}\,; \quad \sqrt{-81} = 9\,\mathrm{i},$$

and if $b = 0$, we have real numbers; thus, we may make the whole "number universe" fit into the category of complex numbers.

Ring
Field

A number system in which addition, subtraction, and multiplication are always defined and the associative and distributive laws are valid is known as a **ring**; if division (except by 0) can also be carried out, one speaks of a **field**.

We shall deal first with rational numbers and their properties, before moving on to irrational numbers and, finally, imaginary and complex numbers.

* * *

KRONECKER Leopold
(1823 - 1891)

– The German mathematician **Leopold Kronecker** claimed that mathematical argumentation should be based only on integers and finite procedures; from Kronecker originates the dictum:

God made the integers; all the rest is the work of Man.

We take issue with this much-treasured quote and argue that the Ruler of Heaven and Earth might only have created the counting numbers – and in the reverse order:

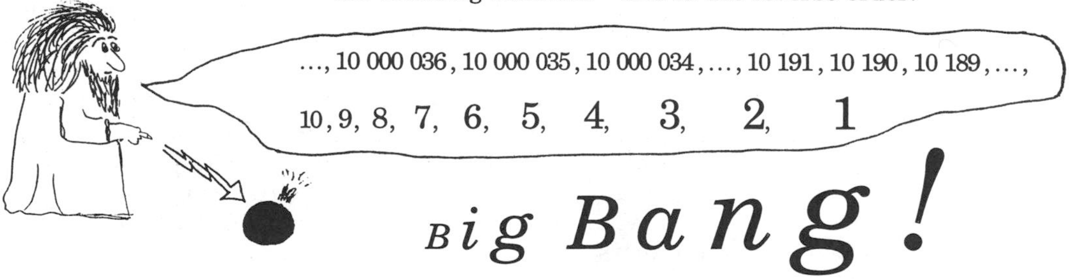

3.1 Rational Numbers

Rational numbers derive their name from the fact that they can all be written in the form of a ratio a/b where the numerator a may be any whole number and the **denominator** b any positive whole number except zero.

The notation a/b reads "a over b" or "a divided by b".

If b is equal to unity, $b = 1$, the ratio a/b is an **integer** $= a$; with arbitrary values a and b, we have a **fraction**.

Integers

All whole numbers, including **negative numbers**, **zero**, and **positive numbers**, are called *integers*.

Natural Numbers

Positive integers are also known as **natural numbers**.

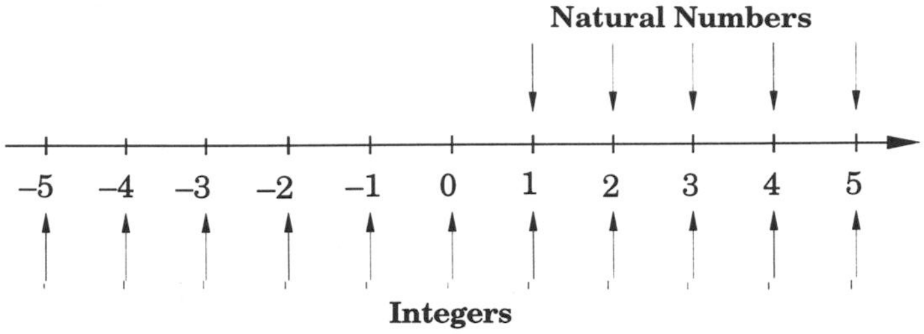

Zero was a long time in establishing itself as a number in its own right; it was hard for mathematicians to accept that "nothing" could be regarded as "something".

In one respect, zero is not equal in status to other numbers: division by zero is not permitted.

Identity for Addition
Identity for Multiplication

Zero is the **identity for addition** (and subtraction) of numbers, since $x + 0 = 0 + x = x$ for all x. *One* is the **identity for multiplication** of numbers, since $x \cdot 1 = 1 \cdot x = x$ for all x.

Negative numbers were also long denied legitimacy in mathematics. We have no evidence of negative numbers being recognized in Babylonian, Pharaonic, ancient Greek, or any other ancient civilization. On the contrary, the Greeks considered geometry the only acceptable form of mathematics, and since distance cannot be negative, they had no use for negative numbers.

BRAHMAGUPTA
(598 - after 665)

In the 7th century, negative numbers were used in bookkeeping in India; positive quantities denoted assets, negative ones debts. The Hindu astronomer **Brahmagupta**, in a chapter dealing with mathematics in his work on astronomy from about A.D. 630, shows a clear understanding of negative numbers.

CARDANO Girolamo (1501 - 1576)

The earliest documented evidence of the use of negative numbers in European mathematics is the *Ars magna*, published in 1545 by the Italian mathematician **Girolamo Cardano**, who, true to Renaissance custom, engaged in many other sciences as well; as a physician, he gave an early clinical description of typhoid fever.

HIERONYMI
CARDANI PRÆ-
STANTISSIMI MATHE-
mathici Philosophi, ac Medici
ARTIS MAGNAE, SIVE
DE REGVLIS ALGEBRAICIS LIB. VNus
qui et totius operis de Arithmetica, quod OPVS PER-
FECTVM inscripsit, est in ordine Decimus.

In the early 17th century, mathematicians began explicitly to use "negative numbers" but met with heavy opposition. Descartes called negative roots "false roots", and Pascal was convinced that numbers "less than zero" could not exist.

von LEIBNIZ Gottfried Wilhelm (1646 - 1716)

Leibniz admitted that negative numbers could lead to absurd conclusions and misconceptions, but defended them as useful aids in calculation.

The general acceptance and algebraic use of negative numbers came during the 18th century, although there were still mathematicians who did not feel at home with them and quite often tried to avoid using them.

– In some situations we still have not fully adopted the minus sign and the idea of negative numbers: The modern corporate report gives losses as (138 000), rather than a *gain* of – 138 000.

To facilitate the reading of numbers with many digits, these are commonly divided into groups of three, beginning on the right and working toward the left. The separation is often done by means of a small space rather than by a comma, a dot, or any other mark that might cause misunderstanding:

123 456 789

1 234 567

Four-digit numbers need not be divided: 5367

In India, traditionally, long numbers are divided so that the terminal three digits form one group and the remaining digits form *groups of two* or, for the initial group, only one. Thus, the number of the approximate 900 million inhabitants of India would be written 90, 00, 00, 000.

Old-Timers' Counting

As a sample of reading large numbers in your great-great-great- ... and a few more greats ... -great-grandparents' time, let us take an excerpt from the *Bamberger Rechenbuch* of 1483:

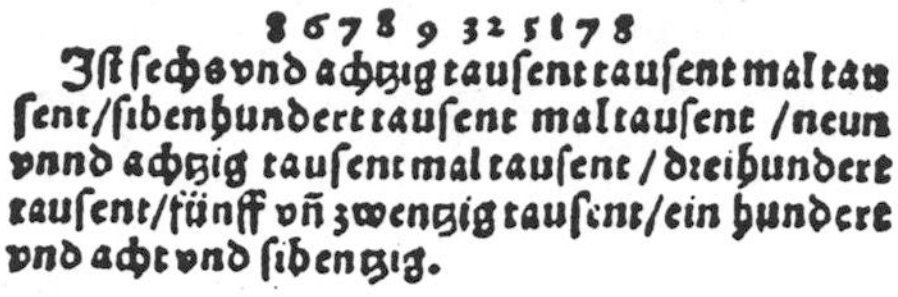
8 6 7 8 9 3 2 5 1 7 8
Jst sechsvnd achtzig tausent tausent mal tausent/sibenhundert tausent maltausent /neun vnnd achtzig tausent mal tausent / dreihundert tausent/fünff vñ zwentzig tausent/ein hundert vnd acht vnd sibentzig.

which tells us how to read out the number

$$86\,789\,325\,178\,;$$

like this:

six-and-eighty thousand thousand times thousand,
seven hundred thousand times thousand,
nine-and-eighty thousand times thousand,
three hundred thousand,
five-and-twenty thousand,
one hundred and eight-and-seventy.

Taken in installments, it isn't so very difficult, is it now?

Fractions

Common Fraction

Common fractions have the form $\frac{a}{b}$ or a/b, where the **numerator** a may be any integer and the **denominator** b any integer > 0; the notation reads "a over b" or "a divided by b".

Proper Fraction
Improper Fraction

If the absolute value of the numerator of a common fraction a/b is smaller than the denominator, $|a| < b$, we speak of a **proper fraction**; if it is not, $|a| \geq b$, we get an **improper fraction**, which can always be split into an integer and a proper fraction,

$$\frac{8}{3} = 2\frac{2}{3}\,; \quad \frac{5}{3} = 1\frac{2}{3}\,.$$

All common fractions have exactly defined positions on the number line:

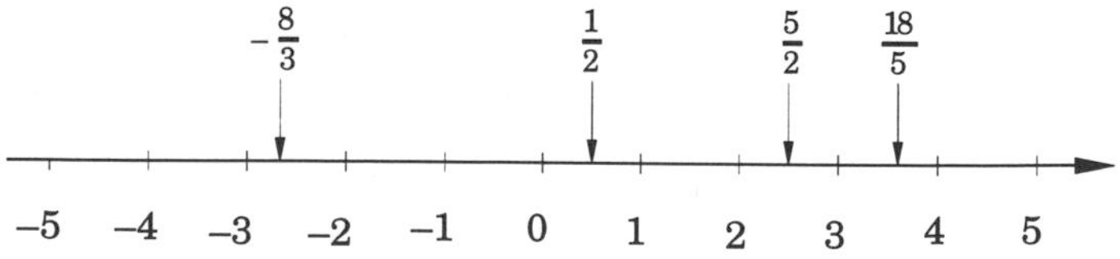

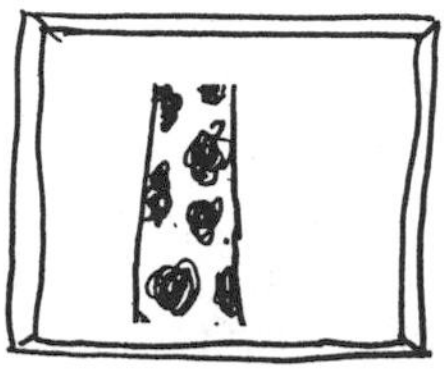

There is always another rational number between any two rational numbers p and q; their arithmetic mean, $\frac{1}{2}(p + q)$, is placed halfway between p and q on the number line. This interpolation can be repeated any number of times, so the number of rational numbers is infinite.

The sum, difference, product, and quotient of two fractions are always another fraction, or an integer.

Decimal Fractions

Decimal fractions are numbers that consist of an integer part – which may be zero – and a decimal part less than unity that follows the decimal marker, which may be a point or a comma.

An unfortunate practice has developed in English-speaking countries of curtailing the expression "decimal fraction" and calling the entire number by the name of "decimal", a word that should properly be reserved for any of the digits following the decimal marker – a practice respected by all other countries.

In this text, we shall call the complete number a decimal fraction, the digits following the integral part to be known as *decimals*.

Examples of decimal fractions are 0.90, 3.02, and 4.005, which read out "zero point ninety" or "oh point nine oh", "three point oh two", and "four point double-oh five", respectively; *point* is short for **decimal point**.

The SI – the International System of Units – recommends the use of a **decimal comma** on the line as **decimal marker** – 0,90 ; 3,02 ; 4,005 – and this practice is followed by most countries: in general, those countries that did use a comma already before the SI recommendation.

There is nothing to suggest that the "decimal-point" nations – among them the United States and English-speaking countries in general – will soon change from the decimal point to the comma. Even great institutions and large companies of world renown which do otherwise swear by SI still use the decimal point.

In the British press the decimal marker is sometimes a dot raised halfway above the line – a disturbing practice as it may be confused with a multiplication sign.

In this text, we use the decimal point, on the line.

To avoid missing the decimal point, decimal fractions whose absolute value is less than unity should be written showing the units zero in front of the decimal point; thus: 0.93 .

In long decimal fractions, digits are usually divided into groups of three, working both ways from the decimal marker:

$$12\,346\,582.006\,852\,3\ldots$$

In scientific tables, however, the digits are often in groups of five.

Conversion of Fractions

Every common fraction can be converted to a decimal fraction by carrying out the division implied by the fraction bar, with a decimal marker separating the integer part of the number from the decimal part.

The decimal part can be looked upon as a series of common fractions with denominators ten, a hundred, a thousand, *etc.*; for instance:

$$\frac{5}{3} = 1.666\ldots = 1 + \frac{6}{10} + \frac{6}{100} + \frac{6}{1000} \cdots$$

When converting common fractions to decimal form, we distinguish two cases.

Common fractions such as 2/5, 3/4, and 7/8 give decimal fractions 0.4, 0.75, and 0.875, respectively, that is, **finite** or **terminating decimal fractions**, whose sequences of decimals have a definite break-off point after which all the places are zeros.

Other fractions will produce endless sequences of decimals:

$$\begin{aligned} 5/6 &= 0.833\,333\,333\ldots \\ 1/96 &= 0.050\,505\,050\ldots \\ 1/27 &= 0.037\,037\,037\ldots \\ 1/74 &= 0.013\,513\,513\ldots \end{aligned}$$

We observe that the decimals repeat periodically, that is, we have **periodic nonterminating** (or **infinite**) **decimal fractions**. The periods are, in the above cases, 3; 05; 037; and 135; they may be much longer when the denominator of the common fraction is a prime number or a large number.

The fraction 1/29 gives a decimal number with a 28-digit cycle; 1/211 has a period of 45 digits.

The periodicity may be marked by placing a bar above the repeating cycle; thus:

$$\begin{aligned} 5/6 &= 0.8\overline{3} \\ 1/27 &= 0.\overline{037} \\ 1/74 &= 0.0\overline{135} \\ 1/29 &= 0.\overline{034\,482\,758\,620\,689\,655\,172\,413\,793\,1} \end{aligned}$$

To convert the periodic decimal expression $0.0\overline{135}$ to a common fraction, we write:

$$\begin{aligned} 1000x &= 13.5135\,135\,135\ldots \\ -\quad x &= -\ 0.0135\,135\,135\ldots \\ \hline 999x &= 13.5 \end{aligned}$$

$$x = \frac{135}{9990} = \frac{5 \cdot 9 \cdot 3}{9 \cdot 3 \cdot 74 \cdot 5} = \frac{1}{74}$$

3.2 Prime Numbers

Most integers are **composite numbers**, that is, they can be written as products of two or more integers, each larger than 1, *e.g.*,

$$120 = 2 \times 2 \times 2 \times 3 \times 5$$
$$462 = 2 \times 3 \times 7 \times 11$$
$$16\,873 = 47 \times 359$$

FIBONACCI Leonardo
(*c.* 1170 - *c.* 1250)

An integer larger than 1 that is divisible only by 1 and itself is called a **prime number**, or a **prime**. **Leonardo Fibonacci** of Pisa, in 1202, used the name *incomposite numbers*, as opposed to composite numbers.

Sieve of Eratosthenes

ERATOSTHENES
(*c.* 276 - 194 B.C.)
Greek mathematician, astronomer, and geographer

The simplest means of sifting out prime numbers from natural numbers is the **sieve of Eratosthenes**. In a sequence of natural numbers, beginning with 2, start by striking out *every second* number after 2 – this eliminates all multiples of 2. Next delete *every third* number after 3, then *every fourth* number after 4, *etc.* Several numbers will be canceled more than once.

	2	3	~~4~~	5	~~6~~	7	~~8~~	~~9~~	~~10~~
11	~~12~~	13	~~14~~	~~15~~	~~16~~	17	~~18~~	19	~~20~~
~~21~~	~~22~~	23	~~24~~	~~25~~	~~26~~	~~27~~	~~28~~	29	~~30~~
31	~~32~~	~~33~~	~~34~~	~~35~~	~~36~~	37	~~38~~	~~39~~	~~40~~
41	~~42~~	43	~~44~~	~~45~~	~~46~~	47	~~48~~	~~49~~	~~50~~
~~51~~	~~52~~	53	~~54~~	~~55~~	~~56~~	~~57~~	~~58~~	59	~~60~~
61	~~62~~	~~63~~	~~64~~	~~65~~	~~66~~	67	~~68~~	~~69~~	~~70~~
71	~~72~~	73	~~74~~	~~75~~	~~76~~	~~77~~	~~78~~	79	~~80~~
~~81~~	~~82~~	83	~~84~~	~~85~~	~~86~~	~~87~~	~~88~~	89	~~90~~
~~91~~	~~92~~	~~93~~	~~94~~	~~95~~	~~96~~	97	~~98~~	~~99~~	~~100~~

The numbers left standing are the prime numbers; those struck out are composite.

Thus, the sequence of primes begins

2, 3, 5, 7, 11, 13, 17, 19, 23, 29, 31, 37 ...

and goes on indefinitely,

...1 000 000 009 649, 1 000 000 009 651 ....

Primeness is a property of the number itself; *e.g.*, fifty-nine is prime whether written 59, binary 111011, or in any other system of numeration.

If a prime divides the product of two integers, it divides at least one of the factors.

EVCLEIDIS Proof: *p.* 159
(3rd century B.C.)

The first person known to have proved that the number of primes is infinite was **Euclid**, who called them "numbers measured by no number but a unit alone", where "measured" means divisible by a given number.

Fundamental Theorem of Arithmetic

The **fundamental theorem of arithmetic** states:

> Every positive integer greater than 1 is a prime or can be expressed as a unique product of primes and powers of primes.

For instance, with the primes in ascending order, 360 has the unique product of powers of primes

$$2^3 \cdot 3^2 \cdot 5^1 .$$

Fermat Numbers

Prime numbers have always intrigued mathematicians, who want to know how and if they might be predicted. Many supposedly prime-generating formulas have been suggested, but they have all failed in testing, and mathematicians are beginning to doubt that there will ever be a practical, simple formula guaranteed to produce exclusively prime numbers.

de FERMAT Pierre
(1601 - 1665)

Fermat numbers, named after **Pierre de Fermat**, a French lawyer and gentleman scholar, are described by the formula

$$F_p = 2^{2^p} + 1 .$$

Knowing that p values from 0 to 4, inclusive, did produce primes,

$$F_0 = 3$$
$$F_1 = 5$$
$$F_2 = 17$$
$$F_3 = 257$$
$$F_4 = 65\,537,$$

Fermat in 1640, somewhat rashly, prophesied that all numbers of this form would be primes, for all positive integers p – but he was proved wrong.

EULER Leonhard
(1707 - 1783)

In 1732, **Leonhard Euler** proved that the fifth Fermat number, F_5, can be factorized,

$$F_5 = 4\,294\,967\,297 = 641 \times 6\,700\,417,$$

and, consequently, is not a prime but a composite number.

In 1880 it was shown that F_6 has a prime factor 274 177; in 1905, F_7 was also proved to be composite, and in 1971 it was factorized into two prime factors of 17 and 22 digits, respectively.

F_8, with 78 digits, factorizes into two prime numbers, one with 16 digits and the other with 62 digits.

In 1992, F_9 was also factorized, and it is known that F_{10} to F_{21}, inclusive, and some other Fermat numbers are composite.

The belief is growing that all Fermat numbers greater than F_4 are composite numbers.

All numbers of the form

$$F_m = 2^m + 1,$$

where m is not a power of 2, are composite numbers.

Mersenne Numbers and Mersenne Primes

MERSENNE Marin
(1558 - 1648)
French mathematician and Franciscan friar

Fermat had communicated his mistaken belief that all Fermat numbers would be primes to **Marin Mersenne** who, in his work *Cogita physico-mathematica* (1644), studied numbers of the form

$$M_p = 2^p - 1,$$

where p is a prime number.

Mersenne was of the opinion that numbers of this kind – now known as **Mersenne numbers** – would be primes for the following prime values of p,

$$p = 2,\ 3,\ 5,\ 7,\ 13,\ 17,\ 19,\ 31,\ 67,\ 127,\ \text{and } 257;$$

and composite numbers for all other p smaller than 257.

LUCAS
François-Édouard-Anatole
(1842 - 1891)
French number theorist

It is not known on what grounds he made this assumption, but it was very nearly correct. It could not be properly substantiated till the advent of the electronic computer in 1947 when, against Mersenne's belief, it was shown that M_{67} and M_{257} are composite numbers, and that M_{61}, M_{89}, and M_{107} joined the ranks of **Mersenne primes**. The largest number proved to be prime, without the help of the electronic computer, is M_{127} (Lucas, 1876).

If p is not a prime number, Mersenne numbers are always composite numbers.

The rapid increase of the function 2^p causes Mersenne numbers to grow at a fascinating rate, and to lend themselves admirably to the generation of very large prime numbers.

In 1994 the known and proven Mersenne primes numbered 32; the most recently found are:

Prime	Number of Digits	Discovered by	Year
$M_{132\,049}$	39 751	David Slowinski	1983
$M_{216\,091}$	65 050	David Slowinski	1985
$391\,581 \cdot 2^{216\,193} - 1$	65 087	Noll *et al.*	1989
$M_{756\,839}$	227 832	David Slowinski	1992
$M_{859\,433}$	258 716	David Slowinski and Paul Gage	1994
$M_{1\,257\,787}$	378 632	David Slowinski and Paul Gage	1996
$M_{2\,976\,221}$	895 932	Gordon Spence	1997

The prime $391\,581 \cdot 2^{216\,193} - 1$ was found by means of an algorithm not entirely based on Mersenne numbers, and *is not* a Mersenne number.

The 895 932 digits of the latest found prime would fill up 450 pages of a paperback book in standard type.

Prime Testing Method of Fermat

Fermat has given us a method of finding out if a specific number p is prime or not: If p is a prime number and a is an arbitrary number smaller than p, then

$$a^{p-1}-1$$

will be divisible by p.

Thus, if it is not, p cannot be a prime.

If $2^{p-1}-1$ is divisible by p, that does not mean, however, that p must be a prime, as there may exist composite numbers with the same property, *e.g.*, $341 = 11 \cdot 31$; it can be proved that $2^{340} - 1$ is divisible by 341.

Pseudo-Primes

Such composite numbers are called **pseudo-primes**.

Thus, if $2^{p-1}-1$ is divisible by p, we can only say that p is either a prime or a pseudo-prime, and more likely the former. Of the first 1000 natural numbers, only three – 341, 561, and 645 – are pseudo-primes; of the first million, only 245.

The Prime Number Theorem

The number of primes is infinite, but they occur less frequently as one goes farther out in the number sequence. Since the time of Euclid, mathematicians have tried to formulate a law, the **prime number theorem**, relating the number – $\pi(n)$ – of primes lower than, or equal to, a given integer n, to n. Here π has nothing to do with the number π, but is used to indicate prime ("πrime").

LEGENDRE Adrien-Marie (1752 - 1833)

Adrien-Marie Legendre in 1778 published his work *Essai sur la théorie des nombres*, where he suggested a modified form of the first approximation, $\pi(n) \approx n/\ln n$,

$$\pi(n) \approx \frac{n}{\ln n - 1.08366},$$

Natural Logarithms: *p.* 155

where $\ln n$ is the natural logarithm of n, and 1.08366 is an empirically found correction term.

GAUSS Carl Friedrich (1777 - 1855)

Integrals: *pp.* 721 *et seq.*

Carl Friedrich Gauss's discovery – the **integral logarithm** – was published in 1863, eight years after his death,

$$\pi(n) \approx \mathrm{Li}(n) = \int_2^n \frac{dx}{\ln n}.$$

HADAMARD Jacques Salomon (1865 - 1963)

de la VALLÉE-POUSSIN C. J. (1866 - 1962)

These conjectures had to wait for final proof until 1896 when the Frenchman **Jacques Hadamard** and the Belgian **Jean de la Vallée-Poussin** independently proved that the Legendre formula is the better approximation of $\pi(n)$ for n values up to a million, the integral logarithm taking over for very large n.

n	$\pi(n)$	$\dfrac{n}{\ln n - 1.08366}$	$\mathrm{Li}(n)$
10^3	168	172	178
10^4	1 229	1 231	1 246
10^5	9 592	9 588	9 630
10^6	78 498	78 534	78 628
10^7	664 579	665 138	664 918
10^8	5 761 455	5 768 004	5 762 209

Goldbach Conjectures

Prime numbers exhibit many unexpected and unexplained properties:

GOLDBACH Christian
(1690 - 1764)
Born in Königsberg, Prussia; professor of mathematics and historian of the Imperial Academy at St. Petersburg

In a letter to Leonhard Euler in 1742, **Christian Goldbach** conjectured that every even integer greater than 2 can be written as the sum of two primes:

$4 = 2+2 \quad 8 = 5+3 \quad 12 = 7+5 \quad 16 = 11+5$

$6 = 3+3 \quad 10 = 7+3 \quad 14 = 11+3 \quad 18 = 11+7$ *etc.*

No exception to this conjecture is known; nor is a valid proof.

Another conjecture of Goldbach's maintains that every sufficiently large integer number – 6 or higher – can be written as a sum of three primes:

$6 = 2+2+2 \quad 8 = 2+3+3 \quad 10 = 2+3+5$

$7 = 2+2+3 \quad 9 = 3+3+3 \quad 11 = 3+3+5$ *etc.*

Also for this conjecture, no exception is known, nor is there a valid proof.

VINOGRADOV Ivan M.
(1891 - 1983)

It is known that every sufficiently large odd number can be expressed as the sum of three primes (**Vinogradov's theorem**, 1937).

Twin Primes

Many primes appear in pairs as **twin primes** that differ by 2:

3, 5 – 11, 13 – 17, 19 – 29, 31 – 41, 43 ...

Experimental evidence supports the infinitude of such pairs, but a strict proof is lacking.

Applications

Prime numbers and methods of factoring large composite numbers play a leading part in today's methods of concealing data from illegitimate access.

In speed-reduction gears, it is often advantageous to use gear wheels whose numbers of teeth are prime numbers – or any other numbers whose only common factor is 1. The same gear teeth will then mesh only at long intervals, limiting uncontrolled localized wear of gear flanks by high spots and other imperfections, and reducing gear noise.

3.3 Perfect and Amicable Numbers

Unlike prime numbers, composite numbers can be broken down into factors. Some composite numbers have attracted the attention of reputable mathematicians for serious investigations into number theory and have inevitably caught the fancy of various mystics, soothsayers, faith healers, and other such people to further their dubious pursuits.

Among these groups of composite numbers are **perfect numbers** and **amicable numbers**, or **friendly numbers**.

Perfect Numbers

Perfect Numbers

An integer number that is equal to the sum of all its possible divisors – except the number itself – is called a **perfect number**. If the sum is less than the number itself, the number is said to be **defective** or **deficient**; if greater, the number is **abundant**.

The lowest perfect numbers are 6, 28, 496, and 8128.

6 is divisible by 1, 2, 3:	28 is divisible by 1, 2, 4, 7, and 14:	496 is divisible by 1, 2, 4, 8, 16, 31, 62, 124, 248:
$1+2+3 = 6$	$1+2+4+7+14 = 28$	$1+2+4+8+16+31+62+124+248 = 496$

Euclid proved in his *Elements* (Book IX), that if $(2^p - 1)$ is what we now designate as a Mersenne prime, then

$$N = 2^{p-1}(2^p - 1)$$

is a perfect number.

NICOMACHOS (c. A.D. 100)
EULER Leonhard (1707 - 1783)

Nicomachos of Gerasa suggested that the converse of the above theorem was true, that is, that the formula gives *all* the perfect numbers without exception. **Euler** proved in 1750, a partial converse of that theorem: If N is an *even* number perfect number, then it has the form established by Euclid, $N = 2^{p-1}(2^p - 1)$. The full converse has not been proved as yet.

All perfect numbers known today – about 30 – are even. There is reason to believe that no odd perfect numbers exist; at all events, it has been proven that there is none smaller than 10^{100} (= 1 googol). The hunt for an odd perfect number or the proof that there is none goes on as of 1997.

Every perfect number except 6 can be written as a sum of the cubes of an unbroken sequence of consecutive odd integers:

$$28 = 1^3 + 3^3$$
$$496 = 1^3 + 3^3 + 5^3 + 7^3$$
$$8128 = 1^3 + 3^3 + 5^3 + 7^3 + 9^3 + 11^3 + 13^3 + 15^3; \textit{ etc.}$$

The single-digit cross sum of every perfect number, except 6, is 1:

$$28 \Rightarrow 2+8 = 10 \Rightarrow 1+0 = 1$$
$$496 \Rightarrow 4+9+6 = 19 \Rightarrow 1+9 \Rightarrow 10 \Rightarrow 1+0 = 1$$
$$8128 \Rightarrow 8+1+2+8 = 19 \Rightarrow 1+9 \Rightarrow 10 \Rightarrow 1+0 = 1 ; \textit{etc.}$$

The sum of the inverses of the factors of a perfect number, leaving out the unity factor but including the number itself, is also 1:

$$6 \Rightarrow \frac{1}{2} + \frac{1}{3} + \frac{1}{6} = 1$$

$$28 \Rightarrow \frac{1}{2} + \frac{1}{4} + \frac{1}{7} + \frac{1}{14} + \frac{1}{28} = 1$$

$$496 \Rightarrow \frac{1}{2} + \frac{1}{4} + \frac{1}{8} + \frac{1}{16} + \frac{1}{31} + \frac{1}{62} + \frac{1}{124} + \frac{1}{248} + \frac{1}{496} = 1$$

etc.

The table of possible perfect numbers begins:

p	2^{p-1}	$2^p - 1$			N	n
1	1	1			-	
2	2	3	=	prime	6	1
3	4	7	=	prime	28	2
4	8	15	=	3 x 5	-	
5	16	31	=	prime	496	3
6	32	63	=	3 x 3 x 7	-	
7	64	127	=	prime	8 128	4
8	128	255	=	3 x 5 x 17	-	
9	256	511	=	7 x 73	-	
10	512	1 023	=	3 x 11 x 31	-	
11	1 024	2 047	=	23 x 89	-	
12	2 048	4 095	=	3 x 3 x 3 x 5 x 91	-	
13	4 096	8 191	=	prime	33 550 336	5
14	8 192	16 383	=	3 x 43 x 127	-	
15	16 384	32 767	=	7 x 31 x 151	-	
16	32 768	65 535	=	3 x 5 x 17 x 257	-	

Amicable Numbers

Two integer numbers are said to be **amicable**, or friendly, if each is the sum of all the possible divisors of the other; the smallest pair of friendly numbers is 220 and 284:

- 220 is divisible by 1, 2, 4, 5, 10, 11, 20, 22, 44, 55, and 110, which add up to 284;
- 284 is divisible by 1, 2, 4, 71, and 142, whose sum is 220.

The pair 220, 284 was long the only known, and generally thought to be unique, until

in 1636, **Pierre de Fermat** found the pair 17 296 and 18 416; and

in 1638, **René Descartes** discovered the pair 9 363 584, 9 437 056.

In 1750, **Leonhard Euler** added a further 60 pairs of amicable numbers to the total; and in 1866, the pair 1184, 1210, until then overlooked, was found by the 16-year-old **Nicolò Paganini**, not to be confused with his violin virtuoso namesake (1782 - 1840).

The number of pairs of amicable numbers known today exceeds 1000.

3.4 Irrational Numbers

Irrational numbers are numbers that are not rational in the sense that they cannot be expressed as common fractions, that is, as a ratio of two integer numbers.

Writing an irrational number in decimal form will produce an endless sequence of decimal digits – a **nonperiodic nonterminating decimal number**.

Irrational numbers are either **algebraic numbers** or **transcendental**.

Algebraic Numbers

Algebraic numbers are roots of algebraic equations with integer coefficients; such equations have either rational roots

$$4a^2 - 9 = 0; \quad a = \pm 3/2$$

or irrational solutions; the roots of the equation

$$a^2 - 2 = 0$$

are the "square roots of 2", $a = \pm\sqrt{2}$.

In the sixth century B.C., the Pythagoreans – an ancient Greek philosophical school and religious brotherhood – encountered the number $\sqrt{2}$ (in our notation) as the length of the diagonal of a square whose sides are one unit in length. They lived by the dictum that *all is number*, believing that all things could be explained as relationships between numbers, which to them meant positive integers and common fractions.

This "number" did not fit into the ideas of the Pythagoreans, and they labeled it an anomaly of the square – only to find that the world was full of such anomalies, for instance, the square root of three,

$$\sqrt{3} = 1.732\,050\,807\ldots,$$

and the cube root of two,

$$\sqrt[3]{2} = 1.259\,921\,05\ldots.$$

Now $\sqrt{2}$ lies between 1.41 and 1.42, because $1.41^2 = 1.9881$ and $1.42^2 = 2.0164$; further calculation places $\sqrt{2}$ between 1.414 and 1.415 and closer to 1.414, because $1.414^2 = 1.999\,396$ and $1.415^2 = 2.002\,225$. Continued calculation with an electronic calculator will give values with an increasing accuracy,

$$\sqrt{2} = 1.414\,213\,562\ldots.$$

Transcendental Numbers

LIOUVILLE Joseph
(1809 - 1882)
French mathematician

Transcendental numbers are not the roots of any algebraic equation. The existence of transcendental numbers was proven in 1844 by **Joseph Liouville**.

Transcendental numbers do, in fact, make up the vast majority of all numbers.

Two transcendental numbers command special interest:

- e, the base of natural logarithms, and
- π, the ratio of the circumference of a circle to its diameter.

e is the limit value of the expression $\left(1+\frac{1}{n}\right)$ raised to the n-th power, when n increases indefinitely,

p. 774

$$e = \lim_{n \to \infty} \left(1+\frac{1}{n}\right)^n = 2.7\ 1828\ 1828\ 45\ 90\ 45\ldots.$$

The decimals have been grouped so as to enhance the peculiarity of the recurring groups 1828 and the two 45's and the 90 of the first fifteen decimals; these groups of digits may serve as a mnemonic.

EULER Leonhard
(1707 - 1783)

The symbol e was first used by the Swiss mathematician **Leonhard Euler** in accounts of his results, in letters written in 1727 or 1728 from St. Petersburg, and again in 1731. In print, e appeared in 1736 in his *Mechanica*, possibly inspired by the word *exponential*; today, it is generally regarded as a homage to Euler.

HERMITE Charles
(1822 - 1901)
French mathematician

Euler showed, in 1737, that e is irrational; in 1873 **Charles Hermite** added the proof that e is transcendental.

π is the symbol used today to represent the ratio of the circumference of a circle to its diameter.

JONES William
(1675 - 1749)

This designation was not introduced until 1706, when used by **William Jones** in his *Synopsis palmariorum matheseos*, probably after the initial letter of the Greek περιφέρεια, "periphery".

Until then, instead of π, one had to content oneself with the quaint Latin phrase

> *"quantitas, in quam cum multiplicetur diameter, provenient circumferentia"*,

meaning "the quantity which, when the diameter is multiplied by it, gives the circumference".

EULER Leonhard
(1707 - 1783)

It is due to the great prestige of the Swiss mathematician **Leonhard Euler** that we use π with today's meaning. In his early writings, Euler had frequently used p to denote the circumference-to-diameter ratio, but changed to π in his textbook *Mechanica*, published in 1736.

MECHANICA
SIVE
MOTVS
SCIENTIA
ANALYTICE
EXPOSITA
AVCTORE
LEONHARDO EVLERO
ACADEMIAE IMPER. SCIENTIARVM MEMBRO ET
MATHESEOS SVBLIMIORIS PROFESSORE.
TOMVS I.
INSTAR SVPPLEMENTI AD COMMENTAR.
ACAD. SCIENT. IMPER.
PETROPOLI
EX TYPOGRAPHIA ACADEMIAE SCIENTIARVM.
A. 1736.

PROPOSITIO 20.

Theorêma.

154. *Congruente puncti directione motus cum potentiae directione, erit incrementum celeritatis, vt potentia ducta in tempusculum et diuiſa per materiam ſeu quantitatem puncti.*

Demonſtratio.

Sint duo puncta ſeu corpuscula inaequalia A et B mota in rectis *AM*, BN. Sollicitentur ea a potentiis p et π reſpectiue, dum percurrunt ſpatiola *Mm*, N*n*, et ſint tempora, quibus ea percurruntur dt, $d\tau$. Manifeſtum eſt punctum B a potentia π eodem modo affici ac punctum A a potentia $\frac{A\pi}{B}$ (136.). Quare ſubſtituto loco B puncto ipſi A aequali, pro potentia π ſubſtitui debet potentia $\frac{A\pi}{B}$, hocque modo obtinemus caſum propoſitionis praecedentis, quo puncta ponuntur aequalia. Hanc ob rem incrementum celeritatis per M*m* eſt ad incrementum celeritatis per N*n* vt pdt ad $\frac{A\pi}{B}d\tau$, ſeu vt $\frac{pdt}{A}$ ad $\frac{\pi d\tau}{B}$ (150.). Ex quo conſtat propoſitum, quod celeritatis incrementum, ſit vt factum ex potentia et tempusculo diuiſum per puncti materiam ſeu quantitatem. Q. E. D.

In Euler's time, nothing definite was as yet known about the nature of π, although mathematicians in general agreed that e and π were probably not roots of algebraic equations.

von LINDEMANN
Carl Louis Ferdinand
(1852 - 1939)

The German mathematician **Ferdinand von Lindemann**, in 1882, succeeded in proving that π is transcendental. The area of a circle is radius$^2 \cdot \pi$, that of a square, side^2; consequently, the side of a square whose area is equal to that of a circle with radius 1 is $\sqrt{\pi}$. A construction with straightedge and compass alone can give only lengths that are algebraic numbers, so Lindemann's proof that π is transcendental was conclusive evidence that the age-old problem of *squaring the circle* is unsolvable.

The transcendentality of e^π was proved in 1929, that of $2^{\sqrt{2}}$ in 1930. Yet, even if we know today that the number of transcendental numbers is infinite, there are still many irrational numbers, *e.g.*, π^π, that defy our curiosity whether they are algebraic or transcendental.

3.5 Imaginary and Complex Numbers

The Imaginary Unit

CARDANO Girolamo (1501 - 1576)

BOMBELLI Raffaele (1526 - 1572)

Imaginary and complex numbers were long looked upon with suspicion, and their possible existence was politely ignored. **Cardano** called negative numbers "fictitious" and elevated imaginary numbers to "sophistic". The Italian mathematician **Raffaele Bombelli** introduced imaginary and complex numbers in his treatise *L'Algebra* (1572) in connection with the solving of cubic equations.

Negative numbers have no square roots that can be expressed by real numbers. To overcome this problem, the **imaginary unit** has been introduced, defined as the square root of minus one,

$$\mathrm{i} = \sqrt{-1}\,; \qquad \mathrm{i}^2 = -1\,.$$

Imaginary numbers have proved a very useful tool in a great many scientific and practical applications. To the scientist, imaginary numbers are indispensable and just as "real" as real numbers themselves.

The imaginary unit is generally represented by the symbol i – for *imaginary* – except in electrical engineering where j is used to avoid confusion with the symbol i, which denotes the instantaneous value of an electric current.

Complex Numbers

Complex numbers take the form

$$z = a + b\,\mathrm{i}\,,$$

where $a = \mathrm{Re}(z)$ is the **real part** and $b\,\mathrm{i} = \mathrm{Im}(z)$ the **imaginary part** of z.

Besides z, the letter w is often used to denote complex numbers.

Imaginary numbers, like real numbers, can be presented on a straight number line. This is not possible, however, with complex numbers, which require two independent coordinates for their representation. Thus, by bringing in geometry, the imaginary unit and complex numbers become just as "real" as real numbers themselves.

The Complex Number Plane

WALLIS John (1616 - 1703)

The idea of representing complex numbers with points in a coordinate plane originated with the English mathematician **John Wallis** in his *De algebra tractatus* (1685), but it was put forward in a rather vague manner and had no influence on contemporary mathematics.

WESSEL Caspar (1745 - 1818)

The first practical representation of complex numbers dates from 1797 when the Norwegian cartographer **Caspar Wessel** read a paper, "Om Directionens analytiske Betegning ..." ("On the Analytic Representation of Direction"), before the Danish Academy of Sciences, published the following year in the *Mémoires* of the Academy. Wessel's work went essentially unnoticed, however, until a French translation appeared a century later, in 1897.

ARGAND Jean Robert (1768 - 1822)

In 1806, the Geneva-born Parisian bookkeeper **Jean Robert Argand** published, anonymously, in a small privately printed edition, a method of representing imaginary numbers geometrically. This work might have suffered the same fate as Wessel's paper, except for J. F. Français, a professor of military engineering, who found a copy of it and invited the author to come forward and acknowledge his work.

GAUSS Carl Friedrich (1777 - 1855)

As often happens when "the time is ripe", Wessel and Argand had published their accounts of essentially the same idea independently and simultaneously; neither Wessel nor Argand explicitly mentioned a complex number plane, a name we owe to the German mathematician **Carl Friedrich Gauss**.

The usual form of representing a complex number $z = x + y\,i$ graphically is by presenting its real part x along the horizontal **real axis** and its imaginary part y on the vertical **imaginary axis** of a Cartesian coordinate system; the complex number will then appear as a point (x, y) in the coordinate plane.

The graph illustrates the positions of two complex numbers,

$$P_1 = 4 + 3\,i$$
$$P_3 = -2 - 5\,i$$

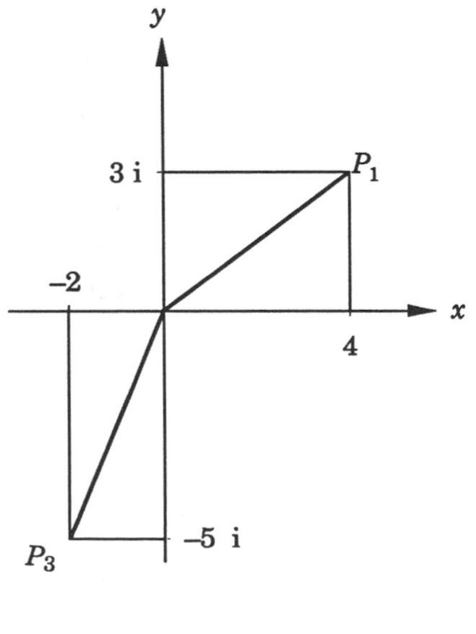

ESSAI

SUR UNE MANIÈRE

DE REPRÉSENTER

LES QUANTITÉS IMAGINAIRES

DANS LES CONSTRUCTIONS GÉOMÉTRIQUES.

A PARIS

M. D CCC. VI.

Title page of Argand's work on geometric representation of imaginary numbers. From an 1874 facsimile (Gauthier-Villars, Paris).

3.6 The Quest for π

The circle and everything pertaining to it have held captive humankind's interest, sometimes amounting to fascination, for several thousand years – far longer than any other single feature in mathematics.

People of all kinds – professional mathematicians, amateurs, and dilettantes – have felt the desire "to square the circle", that is, to construct a square with the same area as a given circle. To do this, one must know the ratio of a circle's circumference to its diameter. Since the time of Leonhard Euler, we denote this ratio by the Greek letter π.

The Ancients

Nearly all peoples of the ancient world used the number 3 for the ratio of a circle's circumference to its diameter as an approximation sufficient for everyday needs. In cases where a higher accuracy was needed, *e.g.*, when matters of law might be involved, our ancestors developed what they considered "more exact" values.

From a Babylonian tablet of sunbaked clay found in 1936 at Susa it appears that, besides 3, the Babylonians used the value

$$\pi = 3\frac{1}{8} = 3.125\,.$$

In the Rhind Papyrus of old Egypt we find a solved problem that states that the area of a circle of nine length units in diameter is the same as the area of a square whose side is eight units of length, which gives

$$\frac{\pi \cdot 9^2}{4} = 8^2; \qquad \pi = \frac{256}{81} = 3.160\,49\ldots.$$

The early Greeks also began with a $\pi = 3$ for everyday use; for matters more serious they evolved other values which were not much better, *e.g.*,

$$\pi = \sqrt{10} = 3.1622\ldots.$$

These values of π were all empirical in the sense that they were based exclusively on practical experience, not on theoretical considerations.

EUDOXOS
(405 - 355 B.C.)

The most important contribution of the early Greeks to our knowledge of π is the "method of exhaustion" attributed variously to Antiphon, to Euclid, and, more likely, to **Eudoxus**.

The perimeter of a regular polygon with n sides – an n-gon – inscribed in a circle is, of course, shorter than the circumference of the circle; the perimeter of a circumscribed regular n-gon is longer.

If n is made sufficiently large, the perimeters of the two polygons can be made to approach the periphery of the circle arbitrarily close – one from within, and one from without – and enclose it within ever narrower limits until the areas between each of the polygons and the circle periphery are "exhausted".

HIPPARCHOS
(*c.* 147 - after 127 B.C.)

In the 2nd century B.C., **Hipparchus** computed an extensive table of chords and proposed the value

$$\pi = \frac{377}{120} = 3.141\,66,$$

which is correct to four decimal places, a remarkably good approximation for his time.

Archimedes

ARCHIMEDES
(*c.* 287 – 212 B.C.)

Archimedes of Syracuse, regarded as the greatest scientist-mathematician of antiquity, applied his method for calculation of arc length to determine π.

Beginning with regular hexagons – inscribed in, and circumscribed to a circle – and doubling the number of sides four times until he had a pair of regular 96-gons, he calculated the lengths of the perimeters of the successive polygons,

$$P_6 \;\; p_6 \;\; P_{12} \;\; p_{12} \;\; P_{24} \;\; p_{24} \;\; P_{48} \;\; p_{48} \;\; P_{96} \;\; p_{96},$$

where P_n and p_n denote the perimeters of the circumscribed and inscribed n-gons, respectively, by the recursion formulas,

$$P_{2n} = \frac{2\,p_n\,P_n}{p_n + P_n}\,; \quad p_{2n} = \sqrt{p_n\,P_{2n}}\,,$$

that is, by taking alternately the harmonic and the geometric means.

For the 96-gons, Archimedes calculated the approximate limits for π,

$$3\,\frac{10}{71} < \pi < 3\,\frac{10}{70} = 3\,\frac{1}{7}\,,$$

or, in decimal notation,

$$3.1408\ldots < \pi < 3.1428\ldots.$$

These values appear in the extant part of Archimedes' book *On the Measurement of the Circle* as

> "The ratio of the circumference of any circle to its diameter is less than 3 1/7 but greater than 3 10/71 ."

After Archimedes, no essentially new ideas for the calculation of π were suggested until the development of calculus toward the end of the 17th century.

A Widened Outlook

While no scientific progress took place in Europe during the Dark Ages – the thousand years starting with the fall of the West Roman Empire in 476 – mathematics and other sciences progressed, albeit very slowly, in countries outside Europe where the Church had no influence.

In India, as in most countries of the day, astronomy was the foremost science, with mathematics considered only as an adjunct to it.

Aryabatha, in 499, published $\pi = 3.141\,6\ldots,$

Brahmagupta, born in 598, used $\pi = \sqrt{10} = 3.162\,2\ldots,$

Bhaskara, born in 1114, held that $\pi = 3.141\,56\ldots,$

all of them without telling why and wherefor.

In China, **Liu Hui** in A.D. 263 published the limits

$$3.141\,024\ldots < \pi < 3.142\,764\ldots$$

obtained for a pair of 96-gons; for a 3072-gon, he found

$$\pi = 3.141\,59\ldots.$$

An interesting π value was suggested by the astronomer **Tsu Chúng-chih** (430 - 501),

$$\pi = \frac{355}{113} = 3.\,141\,592\,9\ldots,$$

which is correct to six decimal places, and was not to be bettered in Europe until the 16th century, more than a thousand years later.

In Europe, around the year 1000, a new dawn began slowly to rise in the universities where the thirst for knowledge could no longer be denied.

When in 1085 Toledo was recaptured from the Moors, the authorities did not order the destruction of its library. Instead, scholars from all Europe were invited to come to Toledo and assist in translating the invaluable manuscripts of scientific books into Latin from the original Greek or from Arabic or Hebrew translations.

FIBONACCI
Leonardo of Pisa
(*c.* 1180 - 1250)

The first great mathematician of the Western world, marking the end of the Middle Ages, was **Fibonacci**, or **Leonardo of Pisa**, who traveled extensively in Arab countries and developed an admiration for their knowledge, which he brought back to Italy.

Like Archimedes, Fibonacci studied the 96-gon, from which he calculated the value – published in 1220 in *Practica geometriae* –

$$\pi = \frac{864}{275} = 3.141\,818\ldots$$

which is correct to three decimal places.

al-KASHI (*or* al-KASHANI)
Jamshid Masud
(? - 1429)

In Persia, **Jamshid Masud al-Kashi** in 1424 published *Risala al-muhitiyya* ("Treatise on the Circumference") with the result of his calculations for an inscribed $3 \cdot 2^{28}$-gon,

$$\pi = 3.141\,592\,653\,589\,793\,25\ldots$$

which is correct to sixteen decimal places, thereby surpassing all earlier determinations of π.

With this value of π, al-Kashi calculated the error to be expected in the circumference of a circle with a diameter 600 000 times that of the Earth – provided that this were known with the same accuracy – and found it to be less than "the thickness of a horse's hair" – an old Persian measure, $\approx$ 0.7 mm.

al-Kashi evidently had a good understanding of the meaninglessness of long chains of decimals.

European Polygonitis

The race for more π decimals did not start in Europe until 1593 when – evidently unaware of al-Kashi's result – **Valentinus Otho** determined π to six, **François Viète** to nine, and **Andriaan van Roomen** to sixteen decimal places.

van CEULEN Ludolph (1540 - 1610)

The hunt still went on using polygons with an ever-increasing number of sides. Best known are the results of **Ludolph van Ceulen** – a fencing master teaching arithmetic, surveying, and fortification at the engineering school at Leyden in Holland – who in 1596 gave 20 decimals of π in *Van den Circkel*, followed by 32 decimals in *Arithmetische en Geometische fondamenten*, published posthumously in 1615, and 35 decimals published in 1621 by his pupil Willebrord Snel,

$$\pi = 3.141\,592\,653\,589\,793\,238\,462\,643\,383\,279\,502\,88\ldots,$$

the last three digits of which were engraved as an epitaph on van Ceulen's tombstone.

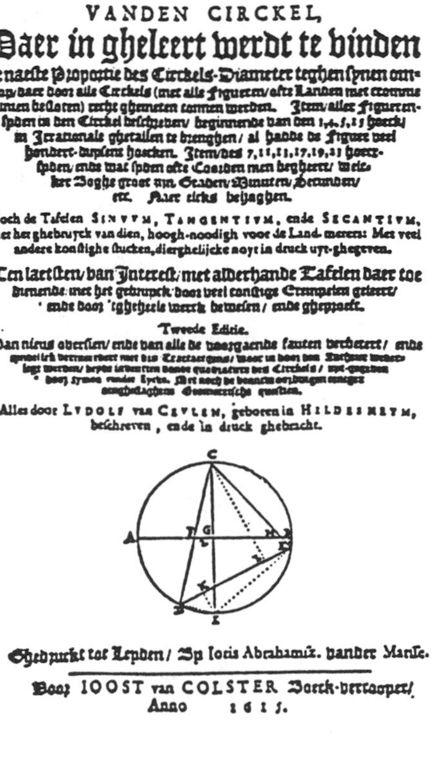
VANDEN CIRCKEL,
Daer in ghelcert werdt te vinden
de naeste Proportie des Circkels-Diameter teghen synen omloop, daer door alle Circkels (met alle Figueren, ofte Landen met cromme Linien besloten) recht ghemeten connen werden. Item, aller Figueren-syden in den Circkel beschreven, beginnende van den 3, 4, 5, 15 hoeck, in Irrationale ghetallen te brenghen, al hadde de Figuere veel hondert-duysent hoecken. Item, des 7, 11, 13, 17, 19, 23 hoeck-syden, ende wat syden ofte Coorden men begheert, welc-ker Boghe groot syn, Graden, Minuten, Secunden, etc. naer elcks behaghen.
Noch de Tafelen SINVVM, TANGENTIVM, ende SECANTIVM, met het ghebruyck van dien, hoogh-noodigh voor de Land-meters: Met veel andere konstighe stucken, dierghelijcke noyt in druck uyt-ghegeven.
Ten laetsten, van Interest, met alderhande Tafelen daer toe dienende: met het gebruyck door veel constige Exempelen geleert, ende door 't gheheele werck bewesen, ende gheproeft.
Tweede Editie.
Van nieus oversien, ende van alle de voorgaende fauten verbetert, ende
Alles door LVDOLF van CEVLEN, geboren in HILDESHEYM, beschreven, ende in druck ghebracht.
Ghedruckt tot Leyden, By Ioris Abrahamsz. vander Marsce.
Voor IOOST van COLSTER Boeck-vercooper,
Anno 1615.

Ludolphine Constant

van Ceulen's accomplishment so impressed his contemporaries that π was often called the **Ludolphine constant**.

The Archimedean approach to more and more decimals of π continued both within and outside Europe, but added nothing of theoretical importance or practical value.

Exit Polygonitis, Enter Analysis

VIÈTE François (1540 - 1603)

The first to devise an infinite product for π was the French mathematician **François Viète** with the formula

$$\frac{\pi}{2} = \frac{1}{\sqrt{\frac{1}{2}} \cdot \sqrt{\frac{1}{2}+\frac{1}{2}\cdot\sqrt{\frac{1}{2}}} \cdot \sqrt{\frac{1}{2}+\frac{1}{2}\cdot\sqrt{\frac{1}{2}+\frac{1}{2}\cdot\sqrt{\frac{1}{2}}}} \cdot \ldots}$$

published in 1593 in *Variorum de rebus mathematicis responsorum liber VIII*. The formula was obtained from a series of Archimedean polygons, starting with a square.

EULER Leonhard (1707 - 1783)

Calculation is easier with the trigonometric version of the formula given later by **Leonhard Euler,**

$$\frac{\pi}{2} = \frac{\sin(\pi/2)}{\cos(\pi/4)\cdot\cos(\pi/8)\cdot\cos(\pi/16)\ \ldots}.$$

The development of differential and integral calculus in the second half of the 17th century replaced the geometric methods of calculation by analytic methods using infinite products, continued fractions, and infinite series expansions of inverse trigonometric functions, permitting the calculation of π to any degree of accuracy. The formula

$$\frac{\pi}{2} = \frac{2\cdot2\cdot4\cdot4\cdot6\cdot6\cdot8\cdot8\cdot\ldots}{1\cdot3\cdot3\cdot5\cdot5\cdot7\cdot7\cdot9\cdot\ldots},$$

WALLIS John (1616 - 1703)

built on integration of trigonometric functions, was presented in 1655 by the English mathematician **John Wallis** in *Arithmetica infinitorum*. Here, for the first time, we have an expression for π as a product of rational numbers.

Plain Digit Hunting

Interest soon focused on the series expansion of the arctan function,

$$\arctan x = x - \frac{x^3}{3} + \frac{x^5}{5} - \frac{x^7}{7} + \ldots; \ (-1 < x < 1),$$

GREGORIE James (1638 - 1675)

von LEIBNIZ Gottfried Wilhelm (1646 - 1722)

found in 1670 by the Scottish mathematician-astronomer **James Gregorie** and, independently, by **Gottfried Wilhelm von Leibniz** in 1673. With $x = 1$, the series simplifies to

$$\frac{\pi}{4} = 1 - \frac{1}{3} + \frac{1}{5} - \frac{1}{7} + \ldots.$$

Since this series converges only very slowly, mathematicians looked for more rapidly converging expansions, generally in the form of sums or differences of arctan functions, starting another craze for ever-lengthening trains of decimals:

In	1705	Abraham Sharp computed	72	decimal places,
	1706	John Machin	100	
	1844	Zacharias Dase	200	
	1847	Thomas Clausen	248	
	1853	William Shanks	527	
	...			
	1945	D. F. Ferguson	710	
	1947	J. W. Wrench, Jr L. R. Smith	808	

spending endless hours in filling reams of writing paper with innumerable calculations of no practical value.

Computeritis

And there, perforce, human digit hunting came to an end because of the appearance of an immensely more powerful competitor – the **electronic computer** – with which no human brain could hope to compete successfully in speed and in memory capacity.

The first computer calculation of π in 1947 produced 2037 decimals in 70 hours of machine time; in 1955 the result improved to 10 000 decimals in 100 minutes.

100 000 decimals was reached in 1961; 250 000 in 1966; 500 000 in 1967; 1 000 000 in 1974; and in February 1992, 2180 million decimals.

In 1995 the fastest methods for computer calculation of π were realized with the help of *elliptic functions*.

With today's computer technology, there is no limit to the number of decimals of π that can be obtained except that set by time and money.

Besides the satisfaction of setting a record, and a faint hope for a chance appearance of some sequence of digits repeating itself – albeit irregularly – and providing a means for testing speed and accuracy of computers, the hunt for more decimals of the value of π makes no practical sense.

Scriptural π

In 1 Kings, 7.23, we find the following verse describing a large vessel in the courtyard of King Solomon's temple:

And he made a molten sea, ten cubits from the one brim to the other; it was round all about, and his height was five cubits: and a line of thirty cubits did compass it round about.

From this we infer a value of $\pi = 30/10 = 3$.

In his textbook on geometry from around A.D. 150, Rabbi Nehemiah taught:

> To measure the area of a circle, multiply the diameter into itself and throw away from it one seventh and the half of one seventh; the rest is the area.

This is, in today's notation,

$$A = d^2 - \frac{1}{7}d^2 - \frac{1}{2} \cdot \frac{1}{7}d^2 = 3\frac{1}{7}r^2,$$

where d denotes the diameter, r the radius.

Nehemiah tells us, if we may trust the sources, that the *scriptural circumference* is measured on the inside of the wall of the vessel whereas the "people of the world" measure their *secular circumference* on the outside.

The difference between **scriptural** π (= 3) and **secular** π (= 3 1/7) is – according to the learned rabbi – explained by the thickness of the wall which accounts for that seventh.

This explanation leaves us at a loss, however, to know whether the volume of the vessel – 125 π – may have been a sacred 375 or a worldly 398 cubic cubits.

πragmatic πractitioner's πreciπitous πlunge

The established fact that π is transcendental has not discouraged self-styled thinkers from believing otherwise.

Thus, it so happened, in 1896, that a physician of Indiana in his Solitude had given much thought to the age-old problem of **quadrature of the circle**. One of his thoughts was that this task might be simplified if one could only find a more propitious value for π, such as

$$\pi = 3.2 \,.$$

The good doctor wanted to make his great invention available to all schoolchildren, real estate agents, bureaucrats, and their likes, in Indiana and in the Great Big World outside – for a consideration, of course.

With visions in his mind of great tsunamis of dollars rolling in, he let it be known – being a good patriot – that he intended not to levy any fees on the use of his π in his home state, reserving for himself only a small cut of monies accruing from abroad.

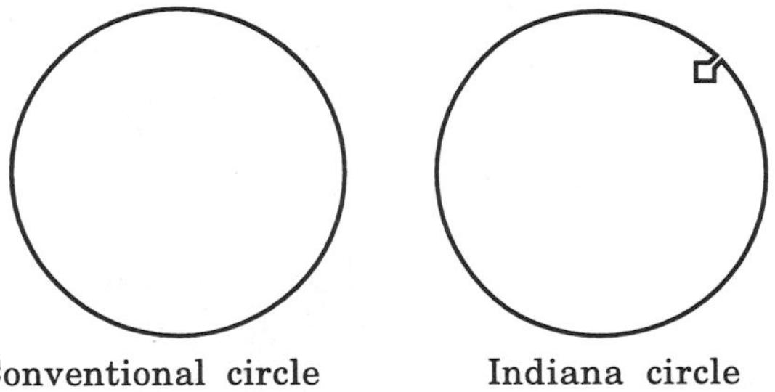

Conventional circle $\pi = \pi$ | Indiana circle $\pi = 3.2$

Moving in the very best circles in Indiana, Doc π managed to square, if not actually a geometric circle, at least some exalted personage to launch a state of Indiana bill laying down that π be henceforth equal to 3.2, exactly.

The bill was carried unanimously by a 67 to 0 vote in the House of Representatives, and thus was well on its way to becoming law – to wit, a law of Indiana, not a law of nature.

By a fortunate circumstance, a Purdue University professor of mathematics happened to be visiting the statehouse to lobby for an appropriation for his alma mater when the Senate was about to debate the proposed mathematic legislation.

During a lull in the debate he succeeded in convincing some of the senators of the horrendous absurdity of the bill – they might just as well pass a law that the Earth is flat (at least in Indiana) – and the debate of the bill was deferred "until a later date".

It still is.

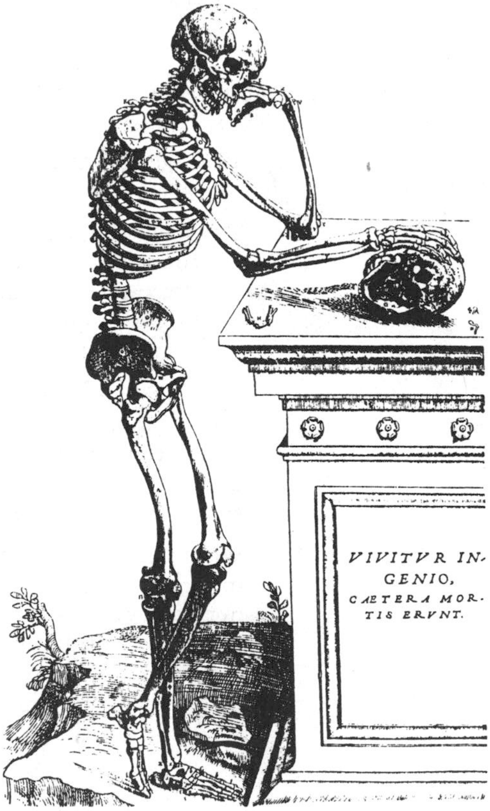

Vivitur ingenio, cætera mortis erunt.

"Genius lives on, all else is mortal."

From Andreas Vesalius, *De humani corporis fabrica* (1543).

VESALIUS Andreas
(1514 - 1564) Flemish anatomist

Chapter 4

CORNERSTONES OF MATHEMATICS

*** Indicates cross-reference

4.0 Beginnings

Egyptian Papyri

Rhind Papyrus

Our knowledge of ancient Egyptian mathematics has come to us mainly from the **Rhind Papyrus** – now in the British Museum – named after the Scottish archaeologist and antiquary Alexander Henry Rhind, who purchased it in 1858 at Luxor in Egypt.

The scroll is about half a meter wide by nearly five and a half meters long with many important fragments missing. By a lucky chance, however, several fragments were found in 1922 in the archives of the New York Historical Society, which had acquired them along with a noted medical papyrus.

The papyrus contains mathematical tables for the calculation of areas and the conversion of fractions, elementary sequences, linear equations, geometric problems, and extensive information about measurements.

The "title page" contains a preface which dates the copy of the scroll to "the fourth month of the inundation season of year 33, when ´A-user-Rê´ was King of Egypt". This means that the scroll dates from about 1650 B.C., when **Ahmes the Scribe** copied it from an older document dating back to the Twelfth Dynasty, 1849 - 1801 B.C.. In honor of Ahmes the Scribe, the scroll is sometimes referred to as the **Ahmes Papyrus**.

Detail from the Rhind Papyrus. © British Museum.

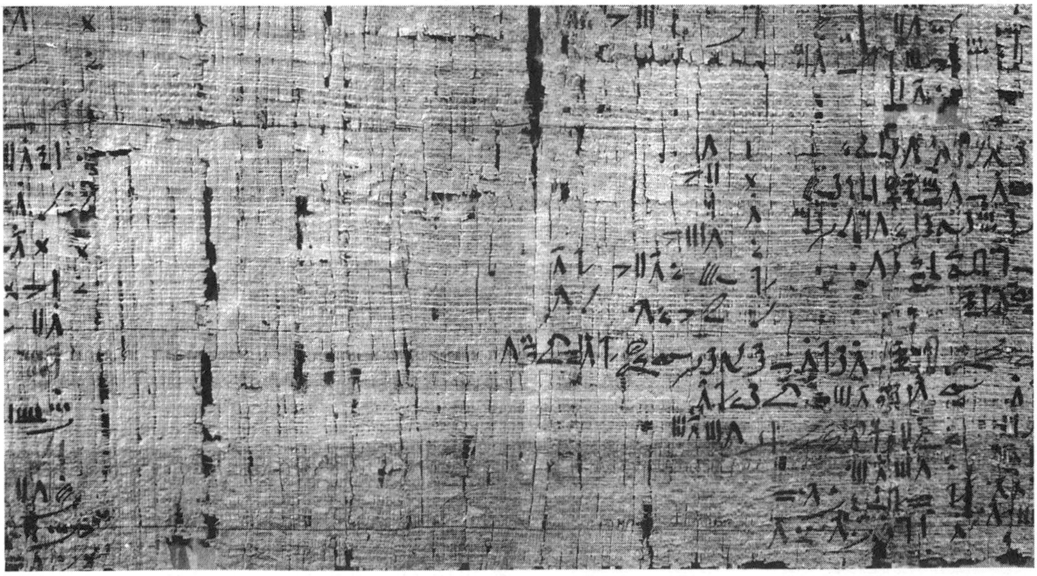

x

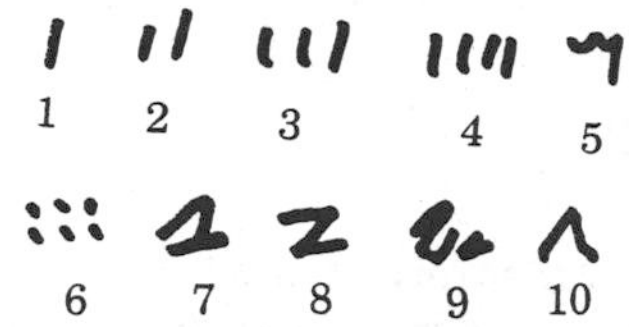

The Rhind Papyrus is in *hieratic script*, a form of writing developed from the *pictographic hieroglyphics* found on monuments and other Egyptian artifacts. The papyrus contains the earliest known symbols of mathematical operations:

Plus Symbol

- **plus** is denoted by a *pair of legs* walking *toward* the number to be added, with the general meaning of "to go up";

Minus Symbol

- **minus** is a *pair of legs* walking *away from* the number to be subtracted, meaning "to go out" in everyday language;

Equality Symbol

- the result of an operation is identified by the sign , meaning "together" – the earliest known **equality sign**, but placed at the very end of the problem;
- an unknown quantity is always indicated by the character , meaning "heap".

T. Eric Peet, *The Rhind Mathematical Papyrus*, 1923

Translated from the hieratic script into pictographic hieroglyphics, the section of the papyrus marked by an **x** at the right-hand edge becomes:

cf. p. 34

The problem reads, from right to left:

"Two-thirds of an unknown added [to the unknown], three-fourths of the unknown taken away [from the unknown], 10 remains",

which means, in modern notation,

$$x + \frac{2}{3}x - \frac{3}{4}x = 10\,.$$

Moscow Papyrus

Another notable Egyptian papyrus containing mathematical problems is the **Moscow Papyrus** – named after its current location. It dates from around 1850 B.C., that is, about the same time as the document copied by Ahmes.

The mathematical symbols used in the Moscow Papyrus are the same as in the Rhind Papyrus.

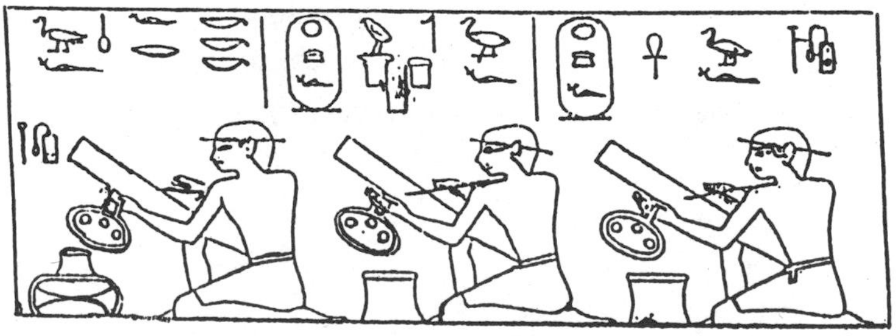

Scribes at work. From a relief on an Egyptian grave.

Arithmetic, Logistic, Algebra

Greek, *arithmetiké,* from *arithmós*, "number", and *techné*, "science"

Greek, *logós*, "reckoning", "account", "reason"

In ancient Greece, **arithmetic** meant theoretical work involving numbers, while the practical everyday calculations of the merchant were called **logistic** – a distinction that persisted well into the early 16th century. From then onward the term *arithmetic* has generally been used for both practical and theoretical operations involving numbers. Logistics, in the plural, has become a predominantly military term.

The advent of Islam, and the Arab conquests during the 8th century, led to Arab acquisition of Greek and Hindu scientific writings, many of which would have perished but for their translation into Arabic.

al-KHOWARIZMI (9th century)

Most important of the mathematical writings of this era were the works of the Persian **al-Khowarizmi**, who around 825, in Baghdad, wrote the

Hisab al-jabr w'al-musqabalah,

meaning literally,

"Science of the Reunion and the Opposition",

or more freely rendered,

"Science of the Transposition and Cancellation",

a monograph which became most influential in the introduction of algebra in Europe by the Moors in Spain.

The title refers to the two principal operations used by al-Khowarizmi in solving equations:

al-jabr = the transposition of terms from one side of an equation to the other; and

al-musqabalah = the cancellation of equal terms appearing on opposite sides of the equation.

The relation

$$bx + q = ax^2 + bx - 3q$$

transforms by *al-jabr* to

$$bx + q + 3q = ax^2 + bx$$

and by *al-musqabalah* to

$$4q = ax^2 .$$

Arabic, *jabara*, "to set", "consolidate"

The Arabic *al-jabr* was latinized to *algebra*.

There once was a fellow in Mosjøen,
Who felt math was lost in the ocean.
He wrote symbols galore
– A thousand pages or more –
So folks would acquire the notion.

4.1 Symbols Galore

We can distinguish three periods in the development of mathematics:

- the **rhetorical phase** when all words relating to mathematical operations and descriptions were written out in full;
- the **syncopation phase** when abbreviations of these words were beginning to be used; and
- the **symbolic phase** when abbreviations were fading out and giving place to symbols.

Mathematics symbols are of three kinds:

- **Symbols for numbers, quantities, variables, or objects**. Examples are symbols used for trigonometric functions, powers, roots and logarithms and their indices and exponents, and symbols used for variables.
- **Symbols of operation**, which describe things to be performed. These symbols include symbols indicating addition, subtraction, multiplication, division, and their attendant grouping symbols, summation and product symbols and factorial notation; also derivation and integration signs and certain other function symbols.
- **Symbols of relation**, which describe things established: symbols of equality and inequality, of ratio and proportion.

We tend today to regard the meaning of established mathematical symbols as non-negotiable and inviolable, but this has not always been so.

In times past,

- an × might have stood for addition;
- a "plaine lyne" — could have meant division;
- a double line ═ might have meant subtraction;
- multiplication and division could both be denoted by a dot, on the line or raised halfway above it, not forgetting that the same dot might also have been a decimal point;
- and the signs ∞ and ∝ both meant equality, when today ∝ signifies proportionality and ∞ is the designation used for infinity.

It must have led to many interesting debates when every worker in the mathematical vineyard could make up his own arsenal of symbols:

- "plus" might have been p or P or p̄ or e or —+— or ┼ or ⟂ or ✠ , or even +;
- "minus" might have been m or m̄ or de̅ or — or ∸ or ∸ or ÷ or ∸ or |—| or ∺ ;
- "equal to" might have been ∞ or ∝ or ∞ or [or || or ═ or ⌶ , or even =.

Symbols of Addition and Subtraction

For addition, the ancient Greeks simply juxtaposed the terms to be added together, without a connecting sign – as we are still doing today: $4\frac{1}{2} = 4 + \frac{1}{2}$. For subtraction, Diophantos of Alexandria seems to have used the symbol ⋔, believed to be an inverted and truncated letter ψ (psi).

Early European – Italian – symbols for *plus* (or *più*) were P, p, and p̄, the last of which was the more common; symbols for *minus* (or *meno*) were M, m, and m̄, the bar as usual indicating omission of one or more letters.

PACIOLI Luca (1445 - 1517)

Luca Pacioli wrote p̄ and sometimes e for plus, and m̄ or de̅ (Latin *demptus*, "taken away") for minus; this usage is found in many works in the 15th and 16th centuries.

The word *plus* was also used in full, but not before the end of the 15th century, considerably later than its fellow *minus*. which appeared in 1202, in Fibonacci's *Liber abaci*. German and Dutch writers continued to be rhetorical and wrote *signum additorum* and *signum subtractorum*.

The word *et*, Latin for "and", was found in many manuscripts, generally in a contracted form closely resembling the symbol +, leaving little room for doubt that + is a ligature for *et*.

The origin of the minus sign is less clear, but it may well be that it is the remainder of an m̄ whose bar has "moved down" and become a symbol of its own.

REGIOMONTANUS Johannes (1436 - 1476)

As far as we know, the symbols + and – first appeared in 1456 in an unpublished manuscript by the German mathematician and astronomer **Regiomontanus** (Johannes Müller, or Molitoris). Their first appearance in print is in the mercantile handbook *Behennde unnd hüpsche Rechnug auff allen Kauffmannschaften* ("Neat and Handy Calculations for All Tradesmen") by **Johannes Widmann**, printed in Leipzig in 1489.

Behennde vnnd
hübsche Rechnūg auff allen
Kauffmanschafften

4 + 5 Wiltu daß
4 — 17 wyssen ad
3 + 36 der gleichn
4 — 19 Szo sum=
3 + 44 mir die ct
3 + 22 vnd lb vn
Czentner 3 — 11 lb was — ist
3 + 50 dz ist mi⁹
4 — 16 dz secz besū
3 + 44 der vn wer
3 + 29 dē 4 5 3 9
3 — 12 lb (So du
3 + 9 die ct zcu lb
gemacht hast vnnd das + das ist mer
dar zu addirest) vnd 7 5 min⁹ Nu solt
du fur holcz abschlahn albeg fur eyn la=
gel 24 lb vn dz ist 13 mol 24 vn macht
312 lb dar zu addir dz — dz ist 75 lb
vnnd werden 387 Die subtrahir vonn
4539 Vnnd pleybn 4152 lb Nu
sprich 100 lb das ist 1 ct p. 4 fl $\frac{1}{8}$ wie

The above text concerns thirteen chests of goods assigned a certain price per hundredweight (cwt). The weight of each chest is given in whole cwt plus or minus a number of pounds.

The chest at the top weighs 4 cwt plus 5 lbs, the next chest 4 cwt less 17 lbs, *etc*.

Thus, the symbols + and – have no operational function, but merely indicate *excess* or *deficit*.

THE GROVND OF
ARTES:
Teaching the worke and practise of
Arithmetike, both in whole numbres
and Fractions, after a more easyer
and exacter sorte, then anye lyke
hath hytherto beene
set forth: with di-
vers new ad-
ditions.
Made by M. ROBERTE
RECORDE
Doctor of Physike.

1558 edition

RECORDE Robert (*c.* 1510 - 1558)

In England, **Robert Recorde**, influential author of mathematical textbooks, in 1543 published a textbook in arithmetic, *Grovnd of Artes*, in which he states that

> *thys figure, +, whiche betokeneth to muche, as this lyne, –, plaine without a crosse lyne, betokeneth to lyttle.*

As symbols of operation, most English writers reserved the notations + and – for use in algebra. They did not acquire today's operational functions until around 1630.

$\pm$ $\mp$ These pairs of symbols have several uses.

The notation $a \pm b$ implies that two cases exist: $a + b$ and $a - b$, with different numerical values except when b is 0.

The notation $\pm$ is also used to indicate reliability of the digits in a number and **limits of inaccuracy** of measurements.

A further use of the **plus-minus** and **minus-plus** notations is to indicate correspondence between mathematical expressions; instead of writing the two similar statements

$$\left.\begin{aligned} \cos(\theta + \phi) &= \cos\theta\cos\phi - \sin\theta\sin\phi \\ \cos(\theta - \phi) &= \cos\theta\cos\phi + \sin\theta\sin\phi \end{aligned}\right\}$$

we may combine them to one,

$$\cos(\theta \pm \phi) = \cos\theta\cos\phi \mp \sin\theta\sin\phi$$

where upper signs are taken together, and lower signs together.

Δ The increment Δx, "delta x", suggests a small amount added to, or subtracted from, a given value of a variable x ("change in x"). The increment is usually understood to be a minute quantity in comparison with the given value.

In mathematics, an upright typeface is used for the increment notation. The notation $\mathit{\Delta}$ (sloping delta) is generally used in physics to imply *quantity excess*, or *difference in quantity* (*e.g.*, mass excess, $\mathit{\Delta} m$, and temperature difference, $\mathit{\Delta} T$).

Symbols of Multiplication and Division

Symbols of multiplication and division were rather late in developing compared to those of addition and subtraction.

OUGHTRED William (1575 - 1660)

x The symbol x for **multiplication** was introduced by the English mathematician **William Oughtred** in 1631 in his *Clavis mathematicae* ("Key to Mathematics"). As a symbol, it was not exactly new having been employed in so-called "crosse multiplication" when dividing fractions.

HARRIOT Thomas (c. 1560 - 1621)

VLACQ Adriaan (1600 - 1666/67)

It was not readily accepted by arithmeticians, however, and did not come into general use in textbooks in elementary arithmetic until the latter half of the 19th century. Nor was it well received by algebraists because of its resemblance to the variable x, and so the dot – suggested by the English mathematician and astronomer **Thomas Harriot** in his *Artis analyticae praxis*, published posthumously in 1631 – came to be employed. **Adriaan Vlacq**, the Dutch publisher and computer of tables of logarithms, had suggested a dot in his *Aritmetica logarithmica* in 1628, though not as an active symbol of operation.

von LEIBNIZ Gottfried Wilhelm (1646 - 1716)

The first mathematician of undisputed prominence to use the dot in a general fashion for algebraic multiplication was **Leibniz** in 1686; later algebraists used the dot in cases where the absence of a symbol would not have sufficed.

The earliest method of writing a **division** or a quotient of two numbers was by placing the dividend above the divisor, thus

$$\begin{matrix}312\\256\end{matrix}$$

without any special division symbol between them, as in the *Bamberger Rechenbuch* of 1483. A fraction bar was later introduced between the numbers,

$$\frac{312}{256},$$

and finally the numbers themselves disappeared, leaving the symbol $\div$, where the two dots remained to represent the numbers.

RAHN Johann Heinrich (1622 - 1676)

$\div$ The symbol $\div$ appeared in print for the first time in *Teutsche Algebra* (Zürich, 1659) by **Johann Heinrich Rahn**; its first mention in English is in the 1668 edition of the translation of Rahn's algebra.

The symbol $\div$ had long been used in continental Europe and in Scandinavia to indicate subtraction, a use now fast receding. In English-speaking countries it always denotes division, and is generally found on the division key of electronic calculators.

Summation and Product Symbols

Σ Greek capital letter sigma, Σ, denotes summation.

If we let a_i (read "*a* subscript *i*" or "*a* sub *i*") be an algebraic expression, the notation

$$\sum_{i=m}^{n} a_i \ ,$$

which reads "the sum of all a_i as i goes from m to n", is a kind of shorthand notation for the sum

$$a_m + a_{m+1} + a_{m+2} + \ldots + a_{n-1} + a_n .$$

$\prod$ Greek capital letter pi, $\prod$, denotes a product,

$$\prod_{i=m}^{n} a_i \ ,$$

"the product of all a_i as i goes from m to n",

$$a_m \cdot a_{m+1} \cdot a_{m+2} \cdot \ldots \cdot a_{n-1} \cdot a_n .$$

EULER Leonhard (1707 - 1783)

DESCARTES René (1596 - 1650)

The summation notation was introduced by **Leonhard Euler**, the product notation by **René Descartes**.

The letter i is called the **index of the summation**, or **index of the product**; the letters m and n define the limits of the index, m being the **lower limit** and n the **upper limit** of i.

The use of the letter i is in no way crucial; j, k, and l are also commonly used as well as the Greek letters ι, ν, and μ. The index should not be confused with the imaginary unit, which is preferably written with an upright i.

$$\sum_{i=2}^{5}(3\,i+4) = (3\times2+4)+(3\times3+4)+(3\times4+4)+(3\times5+4)$$

$$= 10+13+16+19 = 58$$

$$\prod_{i=2}^{5}(3\,i+4) = (3\times2+4)(3\times3+4)(3\times4+4)(3\times5+4)$$

$$= 10\times13\times16\times19 = 39\,520$$

When there is no risk of ambiguity or misunderstanding, the notations may be simplified; thus

$$\sum_{i=0}^{n} a_i \Rightarrow \sum_{0}^{n} a_i \quad \text{and} \quad \prod_{i=0}^{n} a_i \Rightarrow \prod_{0}^{n} a_i$$

Factorial

$n\,!$ The factorial notation $n\,!$ represents the product of all positive integers from 1 to n, inclusive,

$$n\,! = 1\cdot2\cdot3\cdot4\cdot\ldots\cdot(n-1)\cdot n\,,$$

and thus may be considered a special case of the $\prod$ product notation.

We have, by definition,

$$(n\,!)\,(n+1) = (n+1)\,!$$

known as the **recursion formula**; inserting $n = 0$, we find that we should define factorial zero as follows:

$$0! = 1\,.$$

$6! =$	$6\times5\times4\times3\times2\times1$	$= 720$
$5! =$	$5\times4\times3\times2\times1$	$= 120$
$4! =$	$4\times3\times2\times1$	$= 24$
$3! =$	$3\times2\times1$	$= 6$
$2! =$	2×1	$= 2$
$1! =$	1	$= 1$
$0! =$	1	$= 1$

As we go down the lines, we find the next one by dividing, in turn, by 6, 5, 4, 3, 2, and 1.

KRAMP Christian (1760 - 1826)

The factorial symbol was introduced in 1808 by **Christian Kramp** in order to avoid printing difficulties caused by the earlier symbols $\underline{|n}$ and $\underline{n|}$.

Symbols of Equality

In manuscripts dating from before the advent of printing, and the first hundred years afterwards – until about 1600 – the equality of two mathematical expressions was denoted by the word *æqualis* or one of its inflected forms – written out in full.

From then on, until around 1680, various abbreviations of these words were used, or symbols chosen to signify equality, and generally distortions of the initial letter *æ* of the word *æqualis*. Descartes, as late as 1637, used ∞ to denote equality.

RECORDE Robert
(c. 1510 - 1558)

Robert Recorde, in his textbook of algebra, *The Whetstone of Witte* from 1557, introduced two long parallel lines as a symbol for equality.

The whetstone of witte, whiche is the seconde parte of Arithmetike: containyng thextraction of Rootes: The Cossike practise, with the rule of Equation: and the woorkes of Surde Nombers.

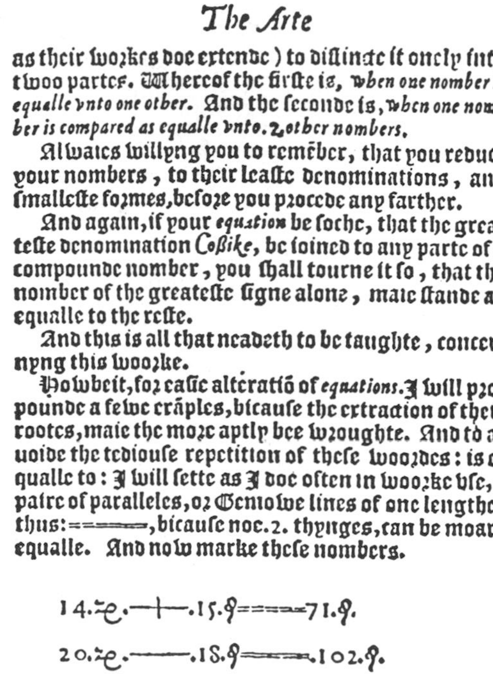
The Arte

as their woorkes doe extende) to distincte it onely into twoo partes. Whereof the firste is, *when one nomber is equalle unto one other*. And the seconde is, *when one nomber is compared as equalle unto .2. other nombers*.

Alwaies willyng you to remeber, that you reduce your nombers, to their leaste denominations, and smalleste formes, before you procede any farther.

And again, if your *equation* be soche, that the greateste denomination *Cosike*, be ioined to any parte of a compounde nomber, you shall tourne it so, that the nomber of the greateste signe alone, maie stande as equalle to the reste.

And this is all that neadeth to be taughte, concernyng this woorke.

Howbeit, for easie alteratio of *equations*. I will propounde a fewe exaples, bicause the extraction of their rootes, maie the more aptly bee wroughte. And to auoide the tediouse repetition of these woordes: is equalle to: I will sette as I doe often in woorke use, a paire of paralleles, or Gemowe lines of one lengthe, thus: ═══════, bicause noe .2. thynges, can be moare equalle. And now marke these nombers.

1. 14.ze.—+—.15.φ.═══════71.φ.
2. 20.ze.———.18.φ.═══════102.φ.

The symbols following the number symbols indicate variables and constants, and correspond to

$14x + 15 = 71; \; x = 4$

$20x - 18 = 102; \; x = 6$

Recorde provides an excellent argument in favor of his choice of symbol:

> *... And to auoide the tediouse repetition of these woordes: is equalle to : I will sette as I doe often in woorke use, a pair of parallels, or Gemowe lines of one lengthe, thus: ═══════ bicause noe .2. thynges, can be moare equalle...*

Gemowe is a distortion of Old French *gemeus*, meaning twin.

BUTEO Johannes
(*c.* 1492 - *c.* 1568)

XYLANDER Giulielmo
(1532 - 1576)

Recorde's proposal for an equality symbol was not without competition, however. In 1559, **Johannes Buteo**, otherwise Jean Borrel, in his *Opera geometrica* suggested a bracket [; and in 1575 **Xylander**, or Wilhelm Holzmann, in *Arithmetica* suggested two parallel vertical lines ||.

= This modern **equality notation** means that two mathematical expressions are equal.

It it also correct to employ this notation in the case of decimal fractions ending in three dots, indicating that further digits would follow, *e.g.*,

$$\pi = 3.141\,592\,653\ldots \quad \text{or} \quad 3.14\ldots.$$

≒ If the absence of further decimals has not been indicated by the three dots, the notation for approximate equality should be employed, *e.g.*,

$$\pi \fallingdotseq 3.14\,.$$

≡ An **identity** is a statement of equality which is true for all values of the variable, *e.g.*,

$$(x+y)^2 \equiv x^2 + 2\,x\,y + y^2\,.$$

This identity notation may be replaced by the equality symbol =, and the relationship is still valid.

$\stackrel{\text{def}}{=}$ This rather unusual symbol establishes the fact that the equality is *by definition*, *e.g.*,

$$a^0 \stackrel{\text{def}}{=} 1\,; \quad 0! \stackrel{\text{def}}{=} 1\,.$$

$0! \stackrel{\text{def}}{=} 1$ has a different meaning than $0\,! = 1$; the first must be taken as true, the second *might* be false.

According to the fundamental laws of arithmetic and algebra, the same number or quantity may be added to, or subtracted from, both members of an equation without affecting the equality. Nor will multiplication or division of both sides of an equation with the same number influence the correctness of the equation.

Symbols of Inequality

In mathematics, symbols are sometimes required to describe the relative magnitude of two or more quantities, even as a "degree of inequality".

≠ This symbol signifies **inequality** and is thus absolute in its meaning of "not equal to". It is often employed to emphasize a condition necessary for a certain statement to apply,

"$a\,/\,b$ is defined only if $b \neq 0$".

$>$ $<$

HARRIOT Thomas (c. 1560 - 1621)

OUGHTRED William (1575 - 1660)

These symbols mean **greater than** and **less than**, and were introduced by **Thomas Harriot** in his *Artis analyticae praxis*, published posthumously in 1631.

Harriot's symbols were not used very frequently until the early 1700s; many mathematicians and printers preferred the symbols ⊏ and ⊐ suggested by **William Oughtred** in 1631 in his *Clavis mathematicae*. The symbols reappeared slightly modified as ⊏ for *non majus* and ⊐ for *non minus* in the 1647 edition of *Clavis*, with the meanings **not greater than** and **not less than**, respectively; these definitions remained in following editions.

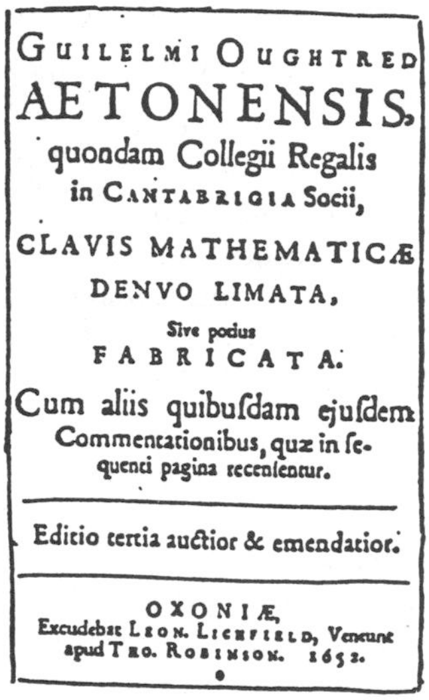

GUILELMI OUGHTRED AETONENSIS, quondam Collegii Regalis in CANTABRIGIA Socii, CLAVIS MATHEMATICÆ DENUO LIMATA, Sive potius FABRICATA. Cum aliis quibusdam ejusdem Commentationibus, quæ in sequenti pagina recensentur. Editio tertia auctior & emendatior. OXONIÆ, Excudebat LEON. LICHFIELD, Veneunt apud THO. ROBINSON. 1652.

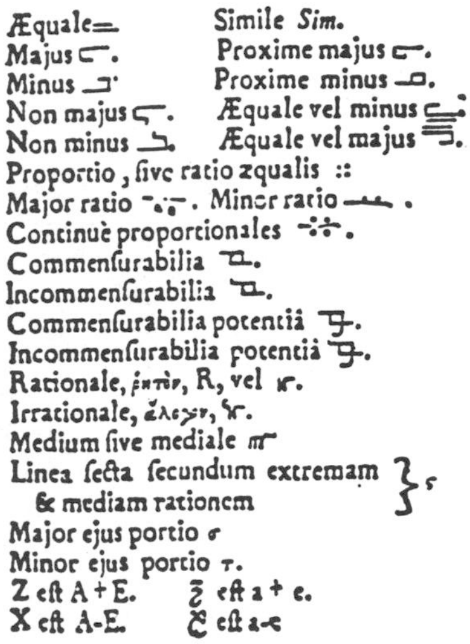

Æquale. Simile *Sim.* Majus. Proxime majus. Minus. Proxime minus. Non majus. Æquale vel minus. Non minus. Æquale vel majus. Proportio, five ratio æqualis :: Major ratio. Minor ratio. Continuè proportionales. Commensurabilia. Incommensurabilia. Commensurabilia potentiâ. Incommensurabilia potentiâ. Rationale, ῥητόν, R, vel. Irrationale, ἄλογον. Medium five mediale. Linea secta secundum extremam & mediam rationem. Major ejus portio. Minor ejus portio. Z est A+E. X est A-E.

Third edition, 1652

$\geq$ $\leq$

BOUGUER Pierre (1698 - 1758)

These symbols mean **greater than or equal to** and **less than or equal to**; they are fairly late additions to the family of symbols and were written $\geqq$ and $\leqq$ when introduced by the French scientist **Pierre Bouguer** in 1734.

$\ngtr$ $\nless$

These rather unusual symbols mean **not greater than** and **not less than**, respectively. We note that $\ngtr$ and $\leq$ have the same meaning.

$\gg$ $\ll$

These notations, **much greater than** and **much less than**, may be used to compare inequalities, *e.g.*, $a \gg b$; $c \ll b$.

Like equations, an inequality is unaffected if the same number is added to or subtracted from both sides or if each side is multiplied or divided by the same *positive* number. However, if each side of an inequality is multiplied or divided by a *negative* number, then the sense of inequality is reversed. For example, if $(x + 2)^2 > 3$, we have

$$4\,[(x+2)^2 - a] > 4\,(3-a)\,; \qquad -4\,[(x+2)^2 - a] < -4\,(3-a)\,.$$

Unconditional Inequality

Conditional Inequality

An **unconditional inequality**, *e.g.*, $5x^2 \geq 0$, is true for all values of x, whereas a **conditional inequality**, *e.g.*, $5x \geq 0$, is not.

Symbols of Ratio and Proportion

Nichomachos, in ancient Greece, looked upon **ratio** as a feature of arithmetic, Eudoxus as one of geometry; medieval writers regarded ratio and proportion as quite distinct from both arithmetic and geometry. Today, ratio and proportion are considered primarily part of algebra.

The actual meaning of the word *ratio* – from the past participle *ratus* of a Latin verb meaning "to estimate" – is reckoning, calculation, relation.

Ratio

The **ratio** of two numbers, or quantities,

$$\frac{a}{b}, \quad a/b, \quad \text{or} \quad a:b,$$

which reads "the ratio of a to b", is a measure of how many times a contains b. Thus, the terms **ratio**, **quotient**, and **fraction** are synonymous with the result of a division, a difference being that a ratio is generally not expressed in the form of a decimal fraction.

Antecedent
Consequent

The first term of a ratio is called the **antecedent**; the second term is the **consequent**. The antecedent equals the ratio times the consequent.

$\widehat{=}$ This notation is used exclusively to express a change of **scale**. In cartography, a scale of 1 : 100 000 means that 1 cm on the map corresponds to an actual distance of 100 000 cm, that is, 1 km. This may be written

$$1\text{ cm} \mathrel{\widehat{=}} 1\text{ km},$$

which reads "1 cm corresponds to 1 km". Another example is $212°\text{F} \mathrel{\widehat{=}} 100°\text{C}$.

Proportion

When two ratios a / b and c / d are equal, the four terms a, b, c, and d are said to form a **proportion**, or to **be in proportion**. Originally written

$$a - b - c - d,$$

OUGHTRED William (1575 - 1660)

this was changed in 1631 by **William Oughtred** in his work *Clavis mathematicae* to

$$a \,.\, b :: c \,.\, d$$

and in 1657 in *Canones sinuum*, a trigonometrical text, to

$$a : b :: c : d,$$

a form of notation which was used well into the 20th century. As there is no sense in having a different notation to signify equality in a proportion, today we write

$$a : b = c : d, \quad \text{or} \quad \frac{a}{b} = \frac{c}{d},$$

which can be read "a is to b as c is to d".

Proportionals
Extremes
Means

The terms in a proportion are called **proportionals**; a and d are known as the **extremes**, the intermediate ones b and c are the **means**.

We find by cross multiplication that $a\,d = b\,c$; thus:

- The product of the extremes equals the product of the means.

Regula de Tri

This rule has been known under many different names: **Regula de Tri**, *Regula de Tribus*, Rule of Three, Golden Rule, Merchants' Rule, Merchants' Key, *etc*. It was known by the Hindus in the 6th century A.D. and seems to have been of mercantile concern till the end of the 15th century, surrounded by mystique.

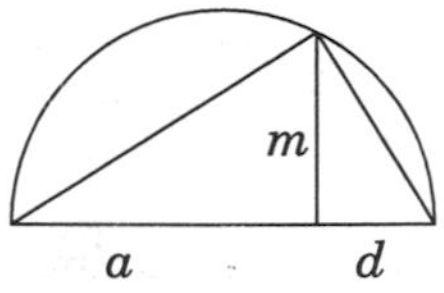

If the two means are equal,

$$\frac{a}{m} = \frac{m}{d},$$

Mean Proportional
Geometric Mean

m is the **mean proportional**, or **geometric mean**, of a and d,

$$m = \sqrt{a\,d}\ ; \text{ thus:}$$

- The mean proportional is the square root of the product of the extremes.

Two variables x and y may be so related that the ratio of one to the other is always the same, x/y = constant, which may be written

$\propto$ $\sim$

$$x \propto y, \quad \text{or} \quad x \sim y,$$

"x is proportional to y".

In **geometry**, $\sim$ has the meaning of similarity, that is, "not differing in shape but only in size".

Intervals

If a variable x satisfies both inequalities $a \le x$ and $x \le b$, where a and b are constants, then

$$a \le x \le b$$

and x is said to be located in the **closed interval** a and b; this interval $[a, b]$ includes a and b, if $a \le b$ and is empty otherwise.

If, on the other hand,

$$a < x < b,$$

x is located in the **open interval** a and b; this interval $]a, b[$ extends up to both a and b but includes neither.

If x satisfies either

$$a \le x < b \qquad \text{or} \qquad a < x \le b,$$

it is located in a **half-open interval** $[a, b[$ or $]a, b]$.

On the number line, open ends of intervals are marked by open circlets, closed ends by filled-in circlets.

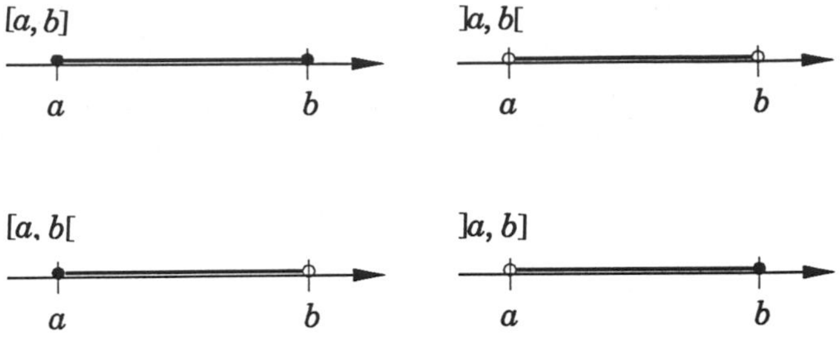

The use of parentheses instead of reversed brackets to indicate open ends of intervals is also common; thus,

$$(a, b] =]a, b]\ ; \quad [a, b) = [a, b[\ ; \quad (a, b) =]a, b[\ .$$

Closed ends split *a* and *b*:

If **both ends open**, *a* and *b* can split a pot of tea:

4.2 Fundamental Operations

Most textbooks recognize four fundamental operations of **arithmetic**: addition, subtraction, multiplication, and division, traditionally known as the **four simple rules of arithmetic**. Of these, subtraction can be considered the inverse of addition; multiplication by an integer is a repeated addition of like terms, and division the inverse of multiplication and related to repeated subtraction.

To these four operations we add three, making seven fundamental operations in all: raising to integer powers, which is a repeated multiplication of a given number of like factors; and its two inverse operations – root extraction, and the calculation of logarithms.

RIESE (RISEN) Adam
(1492 - 1559)

Adam Risen
Rechenbuch/ auff Linien
vnd Ziphren/ in allerley Hand
thierung/ Geschäfften vnnd Kauffman-
schafft. Mit neuwen künstlichen Regeln vnd
Exempeln gemehret/ Innhalt für-
gestellten Registers.
Visier vnd Wechselruthen künstlich
vnd gerecht zumachen/ auß dem Quadrat/
Durch die Arithmetic vnnd Geometri/ von
Erhart Helm/ Mathematico zu Franck-
furt/ beschrieben.
Alles von neuwem jetzunde widerumb erse-
hen vnd Corrigirt.

Franck. Bey. Chr. Egen. Erben. 1574.

Title page of the 1574 edition of Adam Riese's *Rechenbuch*, first published around 1520. Riese's books were of great importance in spreading the knowledge of arithmetic throughout Europe; "according to Adam Riese" became a household word guaranteeing accuracy of mathematical operations.

Constants

Unknowns

Algebra – the latinized form of Arabic *al-jabr* – is a **generalized arithmetic** which operates with letter symbols as well as numbers: letters from the beginning of the alphabet – a, b, c ... – are generally used to denote fixed or arbitrary numerical quantities, generally known as **constants**; letters at the end of the alphabet – x, y, z – denote **unknown quantities** to be determined by solving equations.

Coefficients

In equations – that is, statements of mathematical equality – constants defining multiples of unknowns, or powers of unknowns, are called **coefficients**:

$$\underset{\text{coefficient}}{\underset{\uparrow}{a}}x^2 + \underset{\text{coefficient}}{\underset{\uparrow}{b}}x + \underset{\text{constant}}{\underset{\uparrow}{c}} = 0$$

Variables

In equations and systems of equations containing several unknown quantities, these are often called **variables**.

LA PRIMA PARTE DEL
GENERAL TRATTATO DI NV-
MERI, ET MISVRE DI NICOLO TARTAGLIA,
NELLAQVALE IN DIECISETTE
LIBRI SI DICHIARA TVTTI GLI ATTI OPERATIVI,
PRATICHE, ET REGOLE NECESSARIE NON SOLA-
mente in tutta l'arte negotiaria, & mercantile, ma anchor in ogni altra
arte, scientia, ouer disciplina, doue interuenghi il calculo.

MALIGNITA'
NOIAR NON PVO
A' FORTEZZA

CON LI SVOI PRIVILEGII.

In Vinegia per Curtio Troiano dei Naui.
M D LVI.

TARTAGLIA Niccolò
(c. 1500 - 1557)

The Italian mathematician **Niccolò Tartaglia**'s *General trattato di numeri et misure* ("General Handling of Numbers and Measures") was published 1556 - 1560. Volumes 1 and 2 are considered two of the most important 16th-century texts on arithmetic; Volumes 3 - 5 deal with geometry; Volume 6 deals with algebra.

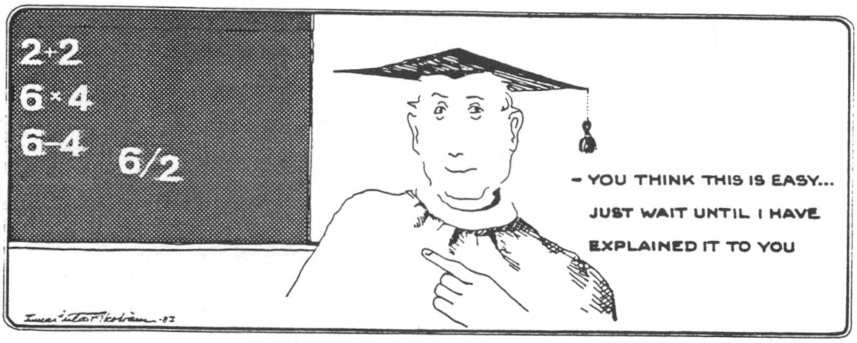

Addition

Addition was sometimes called "aggregation" in the 13th century; Fibonacci used the names "composition" and "collection" besides "addition". The *Margarita phylosophica* (1496) by **Gregor Reisch** illustrates addition

thus,		which is today:
4679	*numerus cui debet fieri additio*	addend
3232	*numerus addendus*	+ addend
7911	*numerus adductus* (or *collectus*)	= sum

Gregor Reisch: *Margarita phylosophica* (1496; 1503 reprint)

The technique of an addition has not changed materially since the introduction of Hindu-Arabic numerals.

The term "carry a digit" dates from the time of the abacus, when a counter was moved ("carried") from the lower part of the abacus to the space above; it appeared in English in the 17th century (Hodder, 1664); Robert Recorde writes "keepe in mynde" to describe a mentally carried digit.

The addition of positive numbers, 2 + 6 = 8, reading "two plus six equals (or makes) eight", and the addition of a negative number, 2 + (- 6) = - 4, "two plus negative six equals negative four", are illustrated on the number line thus,

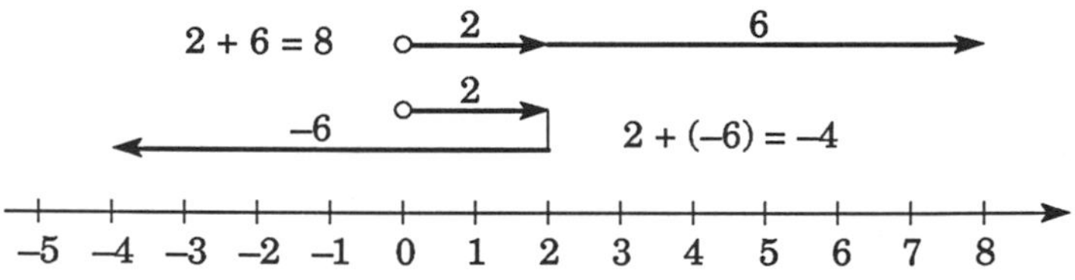

Subtraction

Subtraction has been called by many names: deduction, extraction, detraction, subduction; to subtract has been deduct, extract, detract, subduce, subtray, rebate. The *Margarita phylosophica* has

9 001 386	*numerus a quo fieri subtractio*
7 532 436	*numerus subtrahendus*
1 468 950	*numerus relictus*

GEMMA FRISIUS Reiner (1505 - 1555)

Gemma Frisius wrote in 1540

numerus ex quo subductus
numerus subducendus
numerus residius

which later became	and today
numerus minuendus (or *superior*)	minuend
numerus subtrahendus (or *inferior*)	– subtrahend
differentia	= difference

Instead of *difference*, the terms "rest" and "remainder" may be used.

Subtracting a positive number, the difference will be less than the minuend, as shown here on the number line:

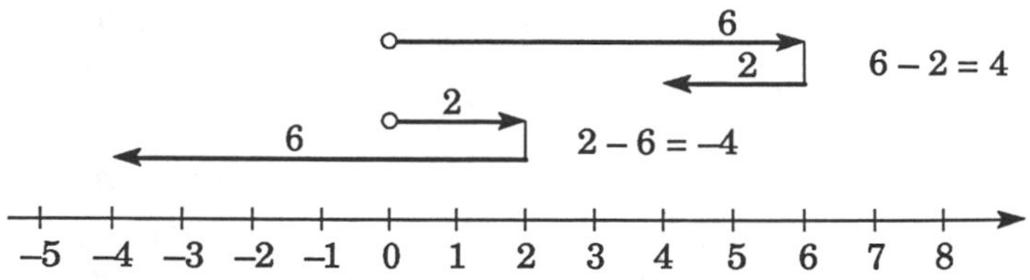

The subtraction of a negative number will, instead, cause the resultant difference to be greater than the original minuend:

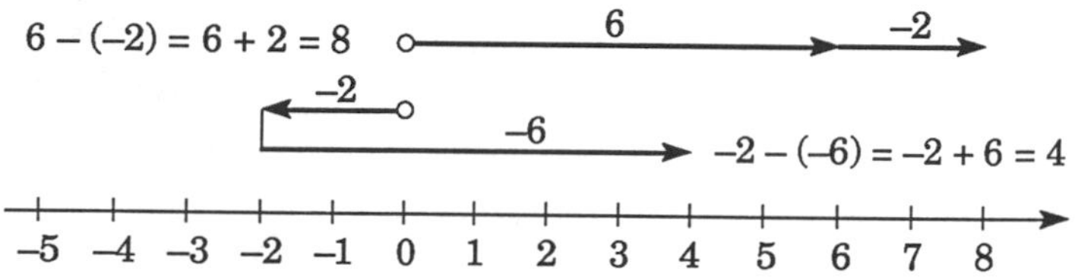

On the Mysteries of Subtraction

1.

2.

– *If you've got ten birds sitting on a telephone wire, and one takes flight – how many will there be left?*

– *None. All the others will also leave.*

This may not be mathematics, but is perfect logic.

Multiplication

Multiplication was thus defined by Robert Recorde (*c.* 1542):

(*ij* = ii = 2)

> *Multiplication is such an operaciō that by ij sumes producyth the thyrde, whiche thyrde sūme so manye times shall cōtaine the fyrst, as there are vnites in the second.*

The general form of an operation of multiplication

was, in Latin,	and is, in our terminology,
numerus multiplicandus	multiplicand
numerus multiplicans	x multiplier
numerus productus	product

where the word *product* means "something produced", and has been employed also in other operations, for instance, in addition with the meaning of "sum". The terms *multiplicand* and *multiplier* are today obsolescent, and are generally called just "factors".

Most people will recall from their schooldays a nuisance called the **multiplication table**. To refresh the reader's memory, let us reproduce "the foundation of all multiplying" from the *Bamberger Rechenbuch* of 1483:

Der grund alles Multiplicirt.

1	2	3	4	5	6	7	8	9
2	4	6	8	10	12	14	16	18
3	6	9	12	15	18	21	24	27
4	8	12	16	20	24	28	32	36
5	10	15	20	25	30	35	40	45
6	12	18	24	30	36	42	48	54
7	14	12	28	35	42	49	56	63
8	16	24	32	40	48	56	64	72
9	18	27	36	45	54	63	72	81

from which we learn, among other things, that 7 x 3 = 12 (*sic*!). Printer's errors have a long history.

Egyptian multiplication

	1	93	
+	2	186	+
	4	372	
+	8	744	+
+	16	1488	+
+	32	2976	+
	58	5394	

The earliest form of multiplication known is the **Egyptian method of duplation**, which reduces multiplication to a form of continued addition. The method – illustrated here by the operation 58 x 93 – was frequently copied by other peoples, and is commonly found in textbooks from the Renaissance.

"Russian peasant"
multiplication

	58	93	°	93	58 +
°	29	186 +		46	116
	14	372	°	23	232 +
°	7	744 +	°	11	464 +
°	3	1488 +	°	5	928 +
°	1	2976 +		2	1856
			°	1	3712 +
		5394			5394

The **"Russian peasant" method** is another form of multiplication by successive duplation and mediation. The operation 58 x 93 proceeds by successive doubling of 93 and halving 58, or doubling 58 and halving 93, each time ignoring fractions, and finally adding those numbers in the right-hand column which stand opposite *odd* numbers in the left-hand column:

$$186 + 744 + 1488 + 2976 = 5394$$
$$58 + 232 + 464 + 928 + 3712 = 5394$$

To remove the air of mystery, let us see what actually happens.

$93 = 2^0 \times 93$	$2^0 \times 58 = 58$
$= 2^0 + 2^1 \times 46$	
$= 2^0 + 2^2 \times 23$	$2^2 \times 58 = 232$
$= 2^0 + 2^2 + 2^3 \times 11$	$2^3 \times 58 = 464$
$= 2^0 + 2^2 + 2^3 + 2^4 \times 5$	$2^4 \times 58 = 928$
$= 2^0 + 2^2 + 2^3 + 2^4 + 2^5 \times 2$	
$= 2^0 + 2^2 + 2^3 + 2^4 + 2^6 \times 1$	$2^6 \times 58 = 3712$
	5394

Multiplication in its simplest form,

"four times six" $4 \times 6 = 6+6+6+6 = 24$

"six times four" $6 \times 4 = 4+4+4+4+4+4 = 24$,

is equivalent to a repeated addition of a specific number of like terms; on the number line, it looks like this:

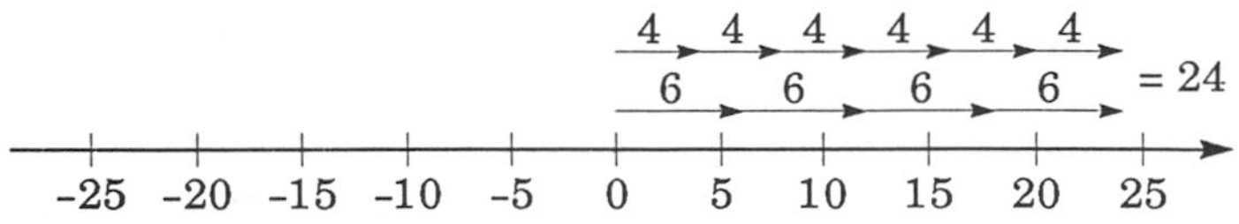

If either of the factors involved is negative,

$$4 \times (-6) = (-6) + (-6) + (-6) + (-6) = -24$$

$$6 \times (-4) = (-4) + (-4) + (-4) + (-4) + (-4) + (-4) = -24,$$

the representation on the number line will be

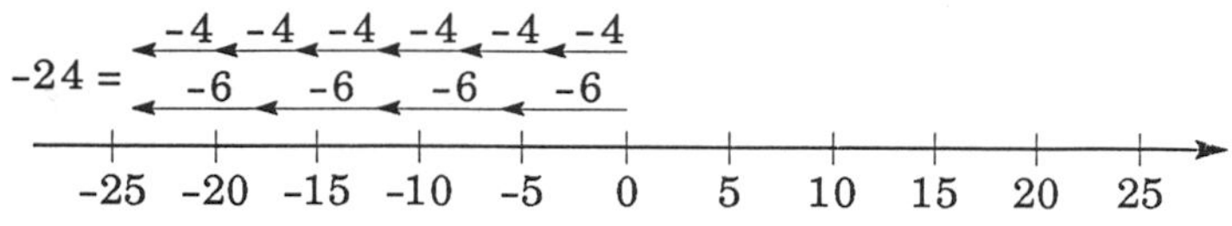

There is a choice of signs for multiplicative notation: a cross (x), or a dot (·) placed halfway above the line. When a dot is used as the decimal marker, the International System (SI) recommends that it not be used to double as a multiplication sign; the accepted notation will then be a cross,

$$4 \times 2.5 = 10.$$

In algebra, when multiplying symbols – a, b, c – or factorial expressions, no special multiplication sign is necessary:

$$a \times b = ab$$

$$(a + b) \times (a + b) = (a + b)(a + b).$$

Division

Division finds how many times a given number, the dividend, contains another given number, the divisor:

$$\frac{\text{dividend}}{\text{divisor}} = \text{quotient} + \frac{\text{remainder}}{\text{divisor}}$$

Mathematically, this is done by deducting, repeatedly, the divisor, or multiples of it, from the dividend. If the remainder is 0, the division is even.

Early mathematicians wrote

$$\frac{\textit{numerus dividendus}}{\textit{numerus divisor}}$$

but had no specific name for what is today called the quotient, except descriptions such as *numerus ex divisione proueniens* or *numerus querendus*; the remainder, if any, was *numerus residuus,* or just *residuus*.

Considered a difficult operation, we meet in 1600 the following characterization:

> *"Diuision is esteemed one of the busiest operations of Arithmetick, and such as requireth a mynde not wandering, or setled uppon other matters."*

On the number line, a division will look like this,

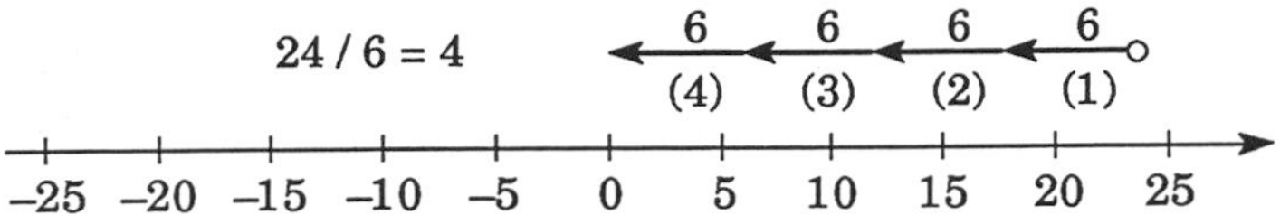

which shows that “24 divided by 6 equals 4”.

The International System (SI) recommends two ways of writing a division,

$$a/b \quad \text{or} \quad \frac{a}{b},$$

which read “a divided by b”. Other methods are $a : b$ and $a \div b$; the latter form is usually found on the division key of electronic calculators.

Rules of Divisibility

An integer N is evenly divisible

- by 2 if it is even – this depends only on the last digit;
- by 3 if the sum of its digits is divisible by 3;
- by 4 if the number formed by the two last digits is divisible by 4;
- by 5 if it ends in 5 or 0;
- by 6 if it is divisible by 2 and 3;
- by 7 if the number formed after
  - cancellation of the units digit and
  - subtraction of twice the value of the units digit

  is divisible by 7;
- by 8 if the number formed by the last three digits is divisible by 8;
- by 9 if the sum of its digits is divisible by 9;
- by 10 if it ends in 0;
- by 11 if the difference between the cross sums of alternate digits is divisible by 11 (presupposes a sufficiently high N);
- by 12 if it is divisible by 3 and 4 .

The number $N = 4893$ can be shown to be divisible by 7:

$$4893 \Rightarrow 489 - 2 \cdot 3 = 483 = 69 \cdot 7 .$$

Continuing the criterion,

$$48 - 2 \cdot 3 = 42 \Rightarrow 4 - 2 \cdot 2 = 0, \text{ which is divisible by 7 .}$$

The number $N = 6\,793\,522\,179\,658\,192$ can be shown to be divisible by 11:

$$6 + 9 + 5 + 2 + 7 + 6 + 8 + 9 = 52$$

$$7 + 3 + 2 + 1 + 9 + 5 + 1 + 2 = 30$$

$$\Delta = 22 = 2 \cdot 11$$

Sheepish Division

A newcomer visited a sheep station in Australia. Asked to estimate the number of sheep in the milling flock, she responded "10 461", almost immediately.

Stunned by the prompt and exact answer, the rancher asked how the visitor did it.

"That's easy," was the reply, "I just counted the hooves and divided by four."

Operational Precedence

Operations of multiplication and division always take precedence over addition and subtraction:

$$4 + 5 \times 3 - 6/2 - 7 = 4 + 15 - 3 - 7 = 9$$

If an addition and/or subtraction is to be made first, it must be enclosed by parentheses:

$$(4 + 5) \times 3 - 6/(5 - 2) = 9 \times 3 - 6/3 = 25$$

We also note the left-to-right rule of precedence for multiplication and division:

$$24/2 \times 6 = \frac{24}{2} \cdot 6 = 72; \quad 24/(2 \times 6) = \frac{24}{2 \times 6} = 2,$$

but as this principle of precedence is not universally respected, we recommend, for the first case:

$$(24/2) \times 6 = \frac{24}{2} \cdot 6 = 72$$

Where doubts might arise, the order of precedence is identified through the use of suitable **grouping notations**, also known as **aggregation symbols**:

( )	open parenthesis ... close parentheses
[ ]	open bracket ... close bracket
{ }	open brace ... close brace

These notations define the order in which operations are to be carried out within a mathematical expression. From the core of an expression, enclosed terms are handled successively:

$$x = 4 \left\{ 25 - [\, 3 \times 5 - 2\,(6 - 2)\,]\, 3 \right\}$$

```
                          └─4─┘
            └─15─┘   └──8───┘
            └─────7─────┘
                 └──────21──────┘
     └──────────4──────────────┘
```

$$x = 4 \times 4 = 16$$

Modular Arithmetic

GAUSS Carl Friedrich (1777 - 1855)

Congruent, Modulo

Calculation with remainders, or residues, is convenient for handling large numbers. This arithmetic, formalized by **Carl Friedrich Gauss**, is called **modular arithmetic** or **arithmetic of residue classes**.

We say "a is congruent to b, modulo c", and write

$$a \equiv b, \bmod c$$

when $b - a$ is a multiple of c.

For instance,

$$10 \equiv 1, \bmod 9\,; \quad 100 \equiv 10, \bmod 9\,; \quad 1000 \equiv 1, \bmod 9\,; \quad \dots,$$
$$10 \equiv 2, \bmod 8\,; \quad 100 \equiv 4, \bmod 8\,; \quad 1000 \equiv 0, \bmod 8\,; \quad \dots,$$

because $10 - 1$; $100 - 10$; $1000 - 1$ and $10 - 2$; $100 - 4$; $1000 - 0$ are all multiples of, respectively, 9 and 8.

The following calculation rules hold:

If

$$a_1 \equiv b_1, \bmod c\,; \quad a_2 \equiv b_2, \bmod c\,,$$

then

$$a_1 + a_2 \equiv b_1 + b_2, \bmod c$$
$$a_1 - a_2 \equiv b_1 - b_2, \bmod c$$
$$a_1 \times a_2 \equiv b_1 \times b_2, \bmod c\,.$$

- Find the least integer remainder of

$$(12\,345 \cdot 123\,456 \cdot 1\,234\,567) / 11.$$

Instead of multiplying out the product, replace each factor by its remainder when divided by 11.

We have, modulo 11,

$$12\,345 \equiv 3\,; \quad 123\,456 \equiv 3\,; \quad 1\,234\,567 \equiv 4\,,$$

and find

$$12\,345 \cdot 1\,23\,456 \cdot 1\,234\,567 \equiv 3 \cdot 3 \cdot 4 \equiv 36 \equiv 3, \bmod 11\,.$$

The sought remainder is 3.

Casting Out Nines

This is a method of checking the results of additions, subtractions, multiplications, and sometimes divisions.

Form single-digit sums of all terms, sums, differences, factors, and products. Perform the same mathematical operations as with the original terms, factors, *etc.*, and form new single-digit sums if required. If the underlined check digits do not agree, there is an error.

Addition:

3 756	$3+7+5+6 = 21$	$2+1 = 3$	$3+7+2 = 12$
17 395	$1+7+3+9+5 = 25$	$2+5 = 7$	$1+2 = \underline{3}$
+ 6 428	$6+4+2+8 = 20$	$2+0 = 2$	
27 579	$2+7+5+7+9 = 30$	$3+0 = 3$	$= \underline{3}$

Subtraction:

37 482	3 + 7 + 4 + 8 + 2 = 24	2 + 4 = 6	6 − 7 = − 1
− 25 963	2 + 5 + 9 + 6 + 3 = 25	2 + 5 = 7	− 1 + 9 = 8
11 519	1 + 1 + 5 + 1 + 9 = 17	1 + 7 = 8	= 8

Multiplication:

32756	3 + 2 + 7 + 5 + 6 = 23	2 + 3 = 5	
x 8975	8 + 9 + 7 + 5 = 29	2 + 9 = 11	
163780		1 + 1 = 2	
229292			
294804		5 x 2 = 10	1 + 0 = 1
262048			
293985100	2 + 9 + 2 + 9 + + 8 + 5 + 1 = 37	3 + 7 = 10	1 + 0 = 1

Forming cross digit sums is equivalent to casting out multiples of nine; a 1 in tens place, meaning 10, becomes a 1 in the digit sum and loses 10 − 1 = 9 in value; or, 10 ≡ 1, mod 9.

To speed up the cross digit addition, cancel all the nines and all cross digit sums equal to 9: 196 325 ≡ 1 ~~9~~ ~~63~~ 25 ≡ 8, mod 9.

When digit sums tally, the calculation *may be* correct but is not necessarily so; the check may mask an error of nine; if two digits have been accidentally reversed, *e.g.*, 15 119 instead of 11 519 in the subtraction remainder, the digit sums will tally – but the answer is wrong.

Fractions

From E.E. White, *A Complete Arithmetic*
Van Antwerp, Bragg & Co., Cincinnati (1870)

Common Fractions: p. 74

Complex Fractions

A fraction is a quotient of two quantities. **Common fractions** – described in Chapter 3 – are fractions whose numerator and denominator are both integers; **complex fractions** have a fraction for the numerator or denominator or both.

To **add or subtract** fractions formed from integers, begin by separating integers and proper fractions. If all fractions have the same denominator, add the numerators according to their signs and place the sum/difference over the common denominator:

$$\frac{13}{8} - \frac{1}{8} - \frac{5}{8} + \frac{7}{8} = \frac{13 - 1 - 5 + 7}{8} = \frac{14}{8} = \frac{7}{4} = 1\frac{3}{4}$$

If the fractions have different denominators,

$$1\frac{7}{8} + \frac{5}{12} - 1\frac{2}{9} - \frac{8}{15} = \frac{15}{8} + \frac{5}{12} - \frac{11}{9} - \frac{8}{15}$$

LCD

we must find a common denominator. The **least common denominator** (LCD) is the product of the several prime numbers occurring in the denominators, each taken with its greatest multiplicity:

8	=	2 x 2 x 2				
12	=	2 x 2 x	3			
9	=		3 x 3			
15	=		3 x	5		
LCD	=	2 x 2 x 2 x 3 x 3 x 5	=	360		

Extend the fractions by multiplying numerator and denominator of each fraction with factors of the LCD that are "missing" in the denominator of that fraction:

$$\frac{15 \times 45}{8 \times 45} + \frac{5 \times 30}{12 \times 30} - \frac{11 \times 40}{9 \times 40} - \frac{8 \times 24}{15 \times 24} = \frac{675 + 150 - 440 - 192}{360} = \frac{193}{360}$$

To **multiply** fractions, convert all mixed numbers to improper fractions, and place all fractions on a common bar; to reduce the labor, cancel like factors in numerator and denominator, and convert the result to a mixed number, if possible:

$$1\frac{7}{8} \cdot 2\frac{2}{5} \times \frac{3}{4} = \frac{15 \cdot 12 \cdot 3}{8 \cdot 5 \cdot 4} = \frac{3 \cdot \cancel{5} \cdot 3 \cdot \cancel{4} \cdot 3}{8 \cdot \cancel{5} \cdot \cancel{4}} = \frac{27}{8} = 3\frac{3}{8}$$

"Invert and multiply!"

Division is equivalent to interchanging the numerator and denominator of the devisor and multiplying by the dividend. Convert mixed numbers to improper fractions, place on a common bar, reduce the fraction by canceling out like factors as before, and convert the result to a mixed number:

$$6\frac{7}{8} / 3\frac{3}{4} = \frac{55}{8} / \frac{15}{4} = \frac{55 \cdot 4}{8 \cdot 15} = \frac{11 \cdot \cancel{5} \cdot \cancel{4}}{2 \cdot \cancel{4} \cdot 3 \cdot \cancel{5}} = \frac{11}{6} = 1\frac{5}{6}$$

p. 76 **Conversion of fractions** is described in Chapter 3.

Unit Fractions

Unit fractions are common fractions with unity for numerator and a positive integer as denominator,

$$N = \frac{1}{n};$$

e. g., $$\frac{2}{5} = \frac{1}{3} + \frac{1}{15} \quad \text{and} \quad \frac{2}{7} = \frac{1}{4} + \frac{1}{28}.$$

The Babylonians compiled tables of unit fractions in their sexagesimal representation in order to perform division by use of multiplication tables.

Unit fractions were used exclusively in Egypt; all other common fractions were written as sums of unit fractions.

Ahmes the Scribe in Egypt taught that

$$\frac{2}{13} = \frac{1}{8} + \frac{1}{52} + \frac{1}{104} \quad \text{and} \quad \frac{9}{10} = \frac{2}{3} + \frac{1}{5} + \frac{1}{30}$$

(the Egyptians had a special sign for 2/3);

and Heron of Alexandria wrote

$$\frac{25}{13} = 1 + \frac{1}{2} + \frac{1}{3} + \frac{1}{13} + \frac{1}{78}.$$

Unit fractions held their position in Europe for more than a thousand years after Heron; Fibonacci gave

$$\frac{98}{100} = \frac{1}{2} + \frac{1}{4} + \frac{1}{5} + \frac{1}{50} + \frac{1}{100}$$

$$\frac{99}{100} = \frac{1}{2} + \frac{1}{4} + \frac{1}{5} + \frac{1}{25}$$

in his book *Liber abaci*, from 1202, which contains tables for converting common fractions into unit fractions.

Euclidean Algorithm

Every number greater than 1 is either a prime number or a composite number, that is, a product of two smaller positive integers. To find the **greatest common divisor** (GCD) of two numbers, we use the **Euclidean algorithm**, a special sequence of divisions.

GCD

To find the GCD of two numbers, say, 102 and 30, begin by dividing 102 by 30,

$$102 = 3 \cdot 30 + 12;$$

continue by dividing 30 by the remainder 12,

$$30 = 2 \cdot 12 + 6,$$

and $$12 = 2 \cdot 6 + 0.$$

The last divisor (6) – the one that leaves zero remainder – is then the sought GCD of 102 and 30.

LAMÉ Gabriel
(1795 - 1870)

A theorem by the French mathematician and engineer **Gabriel Lamé** states that the number of steps in the Euclidean algorithm is never greater than *five* times the number of digits in the lesser of the two numbers.

One use of the Euclidean algorithm is to reduce a common fraction, another is to convert common fractions into continued fractions.

- Reduce $\frac{11\,661}{36\,777}$.

$$36\,777 = 3 \cdot 11\,661 + 1794\,;$$
$$11\,661 = 6 \cdot 1794 + 897\,;$$
$$1\,794 = 2 \cdot 897 + 0\,; \text{ the GCD is 897.}$$

$$\frac{11\,661}{36\,777} = \frac{11\,661/897}{36\,777/897} = \frac{13}{41}$$

Continued Fractions

All real numbers can be written as periodic or nonperiodic, terminating or nonterminating, decimal fractions. Another way of writing a rational number is in the form of a **continued fraction**,

$$N = a_0 + \cfrac{1}{a_1 + \cfrac{1}{a_2 + \cfrac{1}{a_3 + \cfrac{1}{a_4 + \ldots}}}},$$

or, in a kind of accepted mathematical "shorthand",

$$N = [a_0,\ a_1,\ a_2,\ a_3 \ldots],$$

where a_0, a_1, a_2, a_3, ... signify positive integers; consequently, every "part fraction", say,

$$\cfrac{1}{a_2 + \cfrac{1}{a_3 + \cfrac{1}{a_4 + \ldots}}},$$

will have a value between zero and unity.

The Euclidean algorithm for finding the greatest common denominator offers a simple method for converting common fractions into continued fractions.

Applied to 221/41, the Euclidean algorithm yields

$$221 = 5 \cdot 41 + 16, \qquad 41 = 2 \cdot 16 + 9,$$
$$16 = 1 \cdot 9 + 7, \qquad 9 = 1 \cdot 7 + 2,$$
$$7 = 3 \cdot 2 + 1, \qquad 2 = 2 \cdot 1 + 0,$$

and we derive the expressions:

$$\frac{221}{41} = 5 + \frac{16}{41} = 5 + \cfrac{1}{\frac{41}{16}} = 5 + \cfrac{1}{2 + \frac{9}{16}} = 5 + \cfrac{1}{2 + \cfrac{1}{\frac{16}{9}}}$$

$$= 5 + \cfrac{1}{2 + \cfrac{1}{1 + \frac{7}{9}}} = 5 + \cfrac{1}{2 + \cfrac{1}{1 + \cfrac{1}{\frac{9}{7}}}} = 5 + \cfrac{1}{2 + \cfrac{1}{1 + \cfrac{1}{1 + \frac{2}{7}}}}$$

$$= 5 + \cfrac{1}{2 + \cfrac{1}{1 + \cfrac{1}{1 + \cfrac{1}{\frac{7}{2}}}}} = 5 + \cfrac{1}{2 + \cfrac{1}{1 + \cfrac{1}{1 + \cfrac{1}{3 + \frac{1}{2}}}}}$$

$$= [5, 2, 1, 1, 3, 2]$$

Polynomials

Algebraic Expression

In the **algebraic expression**

$$ax^2 + bx + c$$

Coefficient

Variable; Constant Term

letters a and b are **coefficients**, that is, the numerical part of the term; x is a **variable**; and c is a **constant term**.

Degree of Term

The **degree** of a term is the sum of the exponents of the several variables of the term; the degree of a constant term is zero; that of x is one; of x^2 or xy, two; of $5xy^3z^2$, six.

Similar Terms

Terms of the same degree that differ only in their coefficients, *e.g.* , $3x^4y^3$ and $5x^3y^4$, both of degree seven, are called **similar terms**.

Monomial

Binomial Trinomial

Polynomial

An algebraic expression with only one term is a **monomial**; one with two terms a **binomial**; one with three, a **trinomial**, *etc.* A **polynomial** may include any number of terms; the monomial, binomial, trinomial, *etc.*, are just special cases of a polynomial.

A polynomial in one variable has the general form

$$a_n x^n + a_{n-1} x^{n-1} + \ldots a_1 x + a_0 ,$$

Degree of Polynomial

where the *highest exponent* is the **degree of the polynomial**.

We distinguish between **first-degree** or **linear polynomials**, **second-degree** or **quadratic polynomials**, **third-degree** or **cubic polynomials**, *etc.*; $2x^3 + y^2z + z$ is a third-degree polynomial in x, y, and z.

Variables are usually placed in alphabetical order and by descending degrees:

$$\underbrace{5x^5 + 2x^2y^3}_{\text{5th deg.}} + \underbrace{3x^2yz + z^4}_{\text{4th deg.}} \underbrace{- x^2 + y^2}_{\text{2nd deg.}} \underbrace{+ x - 2z}_{\text{1st deg.}}$$

Addition and Subtraction

After grouping like terms together, addition and subtraction are carried out within each group of the polynomials.

$$\begin{array}{r} 4x^3 + 8x^2y + 2xy^2 - 2y^3 - x^2 \qquad\quad - 7z \\ + 2x^2y \qquad\qquad\qquad\qquad\qquad - z^2 - \;\; z + 3 \\ \hline 4x^3 + 10x^2y + 2xy^2 - 2y^3 - x^2 - z^2 - 8z + 3 \end{array}$$

and

$$\begin{array}{r} 4x^3 + 8x^2y + 2xy^2 - 2y^3 - x^2 \qquad\quad - 7z \\ -(\quad +2x^2y \qquad\qquad\qquad\qquad\qquad - z^2 - \;\; z + 3) \\ \hline 4x^3 + 6x^2y + 2xy^2 - 2y^3 - x^2 + z^2 - 6z - 3 \end{array}$$

Multiplication

The product of two polynomials is the sum of all individual products of every term in one polynomial multiplied by every term in the other.

To find the product of more than two polynomials, begin by multiplying two of them, then the result and the next polynomial, *etc.*:

$$(3x^2 + 2x + 3)(2x - 4)(x + 2)$$
$$= (3x^2 + 2x + 3)(2x^2 - 8),$$

$$\begin{array}{r} 3x^2 + 2x + 3 \\ \times \quad 2x^2 \qquad\quad - 8 \\ \hline -24x^2 - 16x - 24 \\ 6x^4 + 4x^3 + 6x^2 \qquad\qquad\qquad \\ \hline 6x^4 + 4x^3 - 18x^2 - 16x - 24 \end{array}$$

Division

When dividing one integer by another, the division is carried on until the remainder is zero, or less than the divisor. Analogously, the division of one polynomial by another is continued until the remainder has as low a degree as possible.

To perform the division, start by organizing the terms of both polynomials in **descending order** of powers, and insert a coefficient zero for all missing powers.

To divide

$$\frac{a^3 + 2a^2b + 2ab^2 + b^3}{a+b} \quad \text{and} \quad \frac{x^4 - 18}{x^2 - x - 2}$$

we have

$$\begin{array}{l}
\qquad\qquad a^2 + ab + b^2 \\
(a+b)\ \overline{)\,a^3 + 2a^2b + 2ab^2 + b^3} \\
\qquad\qquad \underline{a^3 + a^2b} \\
\qquad\qquad\qquad a^2b + 2ab^2 \\
\qquad\qquad\qquad \underline{a^2b + ab^2} \\
\qquad\qquad\qquad\qquad ab^2 + b^3 \\
\qquad\qquad\qquad\qquad \underline{ab^2 + b^3} \\
\qquad\qquad\qquad\qquad\qquad 0
\end{array}
\qquad
\begin{array}{l}
\qquad\qquad\qquad x^2 + x + 3 \\
(x^2 - x - 2)\ \overline{)\,x^4 + 0x^3 + 0x^2 + 0x - 18} \\
\qquad\qquad\qquad \underline{x^4 - x^3 - 2x^2} \\
\qquad\qquad\qquad\quad x^3 + 2x^2 \\
\qquad\qquad\qquad\quad \underline{x^3 - x^2 - 2x} \\
\qquad\qquad\qquad\qquad 3x^2 + 2x - 18 \\
\qquad\qquad\qquad\qquad \underline{3x^2 - 3x - 6} \\
\text{Remainder:} \qquad\qquad\quad 5x - 12
\end{array}$$

- Divide the first term of the dividend by the first term of the divisor; enter the result as first term of the quotient.
- Multiply the divisor by the term found; enter the product under the dividend, like terms under each other; subtract.
- Move down as many terms as required in order to continue the operation.
- Go on until the remainder is of as low a degree as possible, or zero.

The method of division of polynomials with real and complex coefficients is the same.

Partial Fractions

Algebraic expressions containing a polynomial in a single variable in the denominator, or in the denominator and numerator, may be split into **partial fractions** for easier handling.

- If the numerator is of equal or higher degree than the denominator, divide the denominator into the numerator until the degree of the remainder is less than that of the denominator.
- Resolve the denominator into prime factors; prime factors higher than degree one are permitted only if they cannot be factored further.
- Expand the original fraction into partial fractions:

$$(ax+b)^n \qquad \frac{A_1}{ax+b} + \frac{A_2}{(ax+b)^2} + \dots + \frac{A_n}{(ax+b)^n}$$

$$(ax^2+bx+c)^n \qquad \frac{A_1x+B_1}{ax^2+bx+c} + \dots + \frac{A_nx+B_n}{(ax^2+bx+c)^n}$$

- Determine the coefficients of the numerators.

Example:

$$\frac{x^2+x+1}{2x^4+x^3+2x^2+x} = \frac{x^2+x+1}{x(2x+1)(x^2+1)} = \frac{A}{x} + \frac{B}{2x+1} + \frac{Cx+D}{x^2+1}$$

To determine A, B, C, and D, multiply both members of the equation by $x(2x+1)(x^2+1)$:

$$x^2+x+1 = A(2x+1)(x^2+1) + Bx(x^2+1) + (Cx+D)x(2x+1)$$

$$0x^3+1x^2+1x+1 = (2A+B+2C)x^3 + (A+C+2D)x^2 + (2A+B+D)x + A$$

Equate coefficients of like powers of x :

$$\left.\begin{aligned} 2A+B+2C &= 0 \\ A+C+2D &= 1 \\ 2A+B+D &= 1 \\ A &= 1 \end{aligned}\right\} \qquad \begin{aligned} A &= 1 \\ B &= -6/5 \\ C &= -2/5 \\ D &= 1/5 \end{aligned}$$

and

$$\frac{x^2+x+1}{2x^4+x^3+2x^2+x} = \frac{1}{x} - \frac{6}{5(2x+1)} + \frac{1}{5(x^2+1)} - \frac{2x}{5(x^2+1)}.$$

p. 745 This technique is used in finding integrals and in solving differential equations.

Complex Numbers

p. 87 The **sum** of complex numbers is obtained by adding their real parts and their imaginary parts, separately; for instance,

$$(5+3\mathrm{i}) + (3-2\mathrm{i}) = (5+3) + (3-2)\mathrm{i} = 8+\mathrm{i};$$

their **difference** is, analogously,

$$(5+3\mathrm{i}) - (3-2\mathrm{i}) = (5-3) + (3+2)\mathrm{i} = 2+5\mathrm{i}.$$

The **product** of two complex numbers is found by multiplying each term of the one by every term of the other, remembering that $\mathrm{i}^2 = -1$,

$$(5+3\mathrm{i}) \times (3-2\mathrm{i}) = (15-6\mathrm{i}^2) + (9-10)\mathrm{i} = 21 - \mathrm{i}.$$

Conjugate Complex Numbers
Notations $\bar{z}$ and z^* are equivalent.

A special case is offered by **conjugate complex numbers** $z = a + b\mathrm{i}$ and $z^* = a - b\mathrm{i}$, where a and b are assumed to be real numbers; their product is always real,

$$zz^* = (a+b\mathrm{i})(a-b\mathrm{i}) = a^2 - b^2\mathrm{i}^2 = a^2+b^2;$$

for instance,

$$(3+2\mathrm{i})(3-2\mathrm{i}) = 9+4 = 13.$$

The **quotient** of two complex numbers is obtained by multiplying numerator and denominator by the conjugate of the latter; thus,

$$\frac{5+3\mathrm{i}}{3-2\mathrm{i}} = \frac{(5+3\mathrm{i})(3+2\mathrm{i})}{(3-2\mathrm{i})(3+2\mathrm{i})} = \frac{9+19\mathrm{i}}{13} = \frac{9}{13} + \frac{19}{13}\mathrm{i}.$$

4.3 Laws of Arithmetic and Algebra

Definitions and rules that will be discussed here are valid for ordinary mathematical operations but do not all apply to special algebras – *e.g.*, Boolean algebra and algebra of transfinite numbers (Chapter 7), vector algebra (Chapter 16), and matrix algebra (Chapter 18).

- For every real number a there is a real number $-a$, called the additive inverse or the opposite or negative of a, although it is not necessarily itself a negative real number; for instance, if $a = -3$, then $-a = 3$.
- For every real number a, a real number $1/a$ exists, commonly called the "inverse of a" or the "reciprocal of a", instead of the longer "multiplicative inverse of a". Since division by 0 is not possible, $1/a$ exists only if $a \neq 0$.

The following laws are not likely to surprise anyone; the surprise is if the laws do not apply. The incentive to formulate these laws came from discoveries that seemingly self-evident rules did not apply to certain operations in new branches of mathematics.

Basic Laws of Identity and Equality	
Reflexive law	$a = a$
Law of symmetry	If $a = b$, then $b = a$
Transitive law	If $a = b$ and $b = c$, then $a = c$
Substitution law	If $a = b$, then b can replace a in any equation
Definition of $\lvert a \rvert$	$\lvert a \rvert = a$ if $a \geq 0$ $\lvert a \rvert = -a$ if $a < 0$
Identity laws of	
Addition and	$a + 0 = 0 + a = a$
Multiplication	$a \cdot 1 = 1 \cdot a = a$
Law of additive inverse	$a + (-a) = (-a) + a = 0$
Law of multiplicative inverse	$a \cdot \frac{1}{a} = \frac{1}{a} \cdot a = 1$, if $a \neq 0$
Commutative laws of	
Addition and	$a + b = b + a$
Multiplication	$a\,b = b\,a$

In a **commutative system**, the order of adding or multiplying terms is inconsequential; in a **non-commutative system** the order is crucial ($a + b \neq b + a$; $a\,b \neq b\,a$), as in multiplication of vectors and matrices.

ABEL Niels Henrik (1802 - 1829)

In group theory, a group following the *commutative laws* is an **abelian group**, named after **Niels Henrik Abel**; abelian groups are of central importance in branches of modern mathematics, notably algebraic topology.

Distributive law of Addition and Multiplication $a(b+c) = ab + ac$

In commutative operations, a product can be rendered as a sum and vice versa.

For non-commutative operations, we distinguish between

left distributivity, $a(b+c) = ab + ac$,

and right distributivity, $(b+c)a = ba + ca$.

Associative laws of Addition and Multiplication

$a + (b+c) = (a+b) + c$

$a(bc) = (ab)c$

In systems where the associative laws apply, any method of grouping may be used, that is, at any stage of addition (or multiplication) one may add (or multiply) two adjacent terms.

SERVOIS François-Joseph (1767 - 1847)

HAMILTON William Rowan (1805 - 1865)

The commutative and distributive laws were formulated by **François-Joseph Servois**, the associative laws by **William Rowan Hamilton**.

From the above, we may derive several other laws and definitions for real numbers:

Multiplication law of zero	1. $a \times 0 = 0 \times a = 0$
The zero product law	2. If $ab = 0$, then either $a = 0$ or $b = 0$
Laws of negation	1. $-(-a) = a$
	2. $(-a)b = a(-b) = -(ab)$
	3. $(-a)(-b) = ab$
Cancellation laws of	
Addition	If $a + x = a + y$, then $x = y$
Subtraction	If $a - x = a - y$, then $x = y$
Multiplication	If $a \neq 0$ and $ax = ay$, then $x = y$
Addition and subtraction laws of absolute values	$\lvert a \pm b \rvert \leq \lvert a \rvert + \lvert b \rvert$
Multiplication law of absolute values	$\lvert ab \rvert = \lvert a \rvert \, \lvert b \rvert$
Division law of absolute values	If $b \neq 0$, then $\left\lvert \frac{a}{b} \right\rvert = \frac{\lvert a \rvert}{\lvert b \rvert}$

Division Fallacy

To attempt to prove that $1 = 2$, let $x = 1$ and $x = y$,

multiply both sides by y: $xy = y^2$;

subtract x^2, $xy - x^2 = y^2 - x^2$;

factor, $x(y-x) = (x+y)(y-x)$;

and, finally, divide by $(y-x)$,

$$x = x + y.$$

In every mathematical game lies a trap; here, $(y-x)$ is equal to zero and cannot be used as a divisor.

4.4 Powers and Roots

History

Powers

A product of p equal factors $a \cdot a \cdot a \cdot \ldots a \cdot a = a^p$, where a is the **base** and p the **exponent**, is called the p-th power of a and reads "a raised to the p-th power".

The concept of powers was known to the old Babylonians and Egyptians; in the Rhind Papyrus, Ahmes the Scribe used a word meaning mass or quantity to denote the unknown quantity that we call x.

In the days of rhetorical mathematics, powers of the unknown had to have names before they could be described by symbols. The ancient Greeks, to whom mathematics meant geometry, called the square of the unknown a *tetragon number*, that is, a four-corner number.

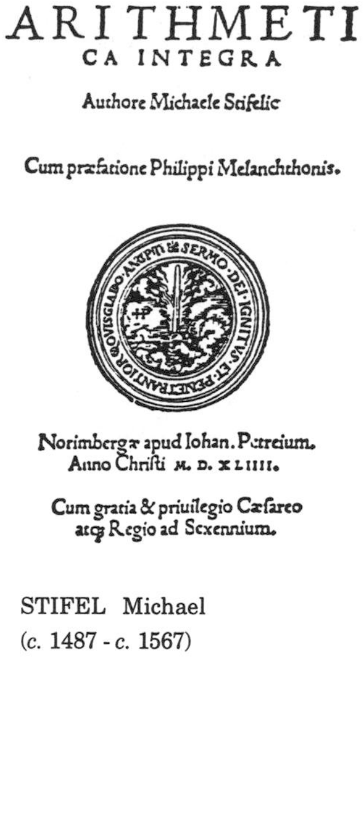
ARITHMETI
CA INTEGRA
Authore Michaele Stifelio
Cum præfatione Philippi Melanchthonis.
Norimbergæ apud Iohan. Petreium.
Anno Christi M. D. XLIIII.
Cum gratia & priuilegio Cæsareo atq; Regio ad Sexennium.

STIFEL Michael
(*c.* 1487 - *c.* 1567)

Diophantos of Alexandria used the word *power* (Greek: *dynamis*) for the square of the unknown; the third power was a *cube*; the fourth, a *power-power*; the fifth, a *power-cube*; and the sixth, a *cube-cube*.

To Arab writers, the unknown was *shai*, which means *thing* or *anything*; its square was *mal*, meaning *wealth*. The early Latin writers took over these terms as *res* meaning *thing*, and *census* for the evaluation of wealth, or tax. Algebra became *Ars rei* or *Ars rei et census*.

Italian writers translated *res* as *cosa*, which became *die Coss* in German and, in English, *cossike arte*.

The first algebraic power symbols corresponding to our x, x^2, x^3, x^4, *etc.*, to appear in print were found in the *Arithmetica integra* by the German mathematician **Michael Stifel**, to be followed by symbols invented by other writers:

1544	Stifel	A	AA	AAA	AAAA	AAAAA
1572	Bombelli	$\underset{\smile}{1}$	$\underset{\smile}{2}$	$\underset{\smile}{3}$	$\underset{\smile}{4}$	$\underset{\smile}{5}$
1585	Stevin	①	②	③	④	⑤
1591	Viète (Vieta)	A	Aq	Acu	Aqq	Aqcu
1631	Harriot	a	aa	aaa	aaaa	aaaaa
1634	Hérigone	a	a2	a3	a4	a5
1637	Descartes	a	a^2	a^3	a^4	a^5

The symbols used by Descartes agree with those used today, except that he recognized only positive integers as exponents; the introduction and acceptance of negative and arbitrary real exponents is the work of John Wallis and Isaac Newton, and imaginary and complex exponents are mainly Euler's work.

Roots

The p-th root of a number a, $\sqrt[p]{a}$, is a quantity that, raised to the p-th power, gives a; that is, it satisfies the relation

$$(\sqrt[p]{a})^p = a\,.$$

Like powers, the roots of numbers have a long history; the Egyptians knew how to calculate the square roots of numbers, including simple fractions, as early as about 4000 years ago.

Arab writers conceived of a square number as grown out of a root, while Greek and Latin scholars thought of roots as the sides of a geometrical square. Thus, when mathematics was still rhetorical, works translated from Arabic generally used the word *radix* for "root", while those derived from Greek and Latin had *latus* for "side".

The symbol most commonly employed by late medieval writers to denote a root was ℞ – a contraction of the word *radix* embodying its first and last letters – which, with numerous variations, held its place in manuscripts and printed books for more than a century:

		sq. root	cu. root	4th root
1484	Chuquet	℞ $℞^2$	$℞^3$	$℞^4$
1494	Pacioli	℞ · 2^a	℞ · 3^a	℞ · 4^a
1521	Ghaligai	℞ □	℞ ⧈	
1539	Cardano	℞	℞ . cu	
1572	Bombelli	℞ . q.	℞ . c.	

The modern symbol $\sqrt{\ }$ first appeared in print in 1525 in Christoff Rudolff's *Die Coss* ("Algebra"):

		sq. root	cu. root	4th root
1525	Rudolff	$\sqrt{\ }$	C $\sqrt{\ }\sqrt{\ }$	$\sqrt{\ }\sqrt{\ }$
1553	Stifel	Z/	ʓ/	ʓʓ/
1628	Vlacq	$\sqrt{\ }$	$\sqrt{C}$	$\sqrt{\ }\sqrt{\ }$
1659	Rahn	$\sqrt{\ }$	$\sqrt{③}$	$\sqrt{\ }\sqrt{\ }$; $\sqrt[4]{\ }$

Calculation Rules for Powers

The notation a^p means the product of p factors a,

$$a \cdot a \cdot a \ldots a \cdot a = a^p ,$$

Base, Exponent, Power

where a is the **base** and p the **exponent** of the **power** a^p.

The exponent p was originally assumed to be always a positive integer, but has since been generalized to include also negative integers, zero, arbitrary real numbers, and imaginary and complex numbers.

A product of, say, six factors of 2 may be written

$$2 \cdot 2 \cdot 2 \cdot 2 \cdot 2 \cdot 2 = 2^6 = 64 .$$

The first power of a number is the number itself, $2^1 = 2$; the second power is its square, $2^2 = 4$; the third power is the cube, $2^3 = 8$; *etc.*

The multiplication of powers of the same base,

$$\underbrace{a \cdot a \cdot a \cdot \ldots a \cdot a}_{(m \text{ factors})} \times \underbrace{a \cdot a \cdot a \cdot \ldots a \cdot a}_{(n \text{ factors})} = a^m \cdot a^n = a^{m+n}$$

gives the rule:

- To multiply powers of the same base, add their exponents.

Division of powers of the same base,

$$\begin{matrix}(m \text{ factors}) \\ (n \text{ factors})\end{matrix} \quad \frac{a \cdot a \cdot a \cdot \ldots a \cdot a}{a \cdot a \cdot a \cdot \ldots a} = \frac{a^m}{a^n} = a^{m-n}$$

provided that $a \neq 0$; that is:

- To divide powers of the same base, subtract their exponents.

This rule has two corollaries,

$$\frac{a^m}{a^m} = a^{m-m} = a^0 = 1 \quad (a \neq 0)$$

and

$$\frac{a^0}{a^n} = \frac{1}{a^n} = a^{-n} \quad (a \neq 0).$$

The n-th power of the m-th power of a is

$$\underbrace{\underbrace{(a \cdot a \cdot a \cdot \ldots a)}_{(m \text{ factors})} \times \underbrace{(a \cdot a \cdot a \cdot \ldots a)}_{(m \text{ factors})} \times \ldots \times \underbrace{(a \cdot a \cdot a \cdot \ldots a)}_{(m \text{ factors})}}_{n \text{ factor parentheses}}$$

$$= (a^m)^n = a^{mn}$$

and, analogously,

$$(a^m)^{1/n} = a^{m/n} .$$

$$a^m \cdot b^m = (a\,b)^m ; \qquad a^m / b^m = (a/b)^m \quad (b \neq 0).$$

Slightly more complicated power expressions are simplified thus,

$$\frac{(b-a)^{n+1}}{(a-b)^{2n+4}} = \frac{(a-b)^{n+1}}{(a-b)^{2n+4}}\,(-1)^{n+1}$$

$$= (-1)^{n+1}\,(a-b)^{-(n+3)}\,.$$

Powers of **imaginary and complex numbers** are defined in the same manner as powers of real numbers,

$$z^n = \underbrace{z\,z\,z\,\dots\,z\,z}_{n \text{ factors}}$$

and obey the same rules.

Positive integer powers of the imaginary unit i repeat in cycles of four,

$i^1 = i$	$i^5 = i$	$i^9 = i$	$i^{13} = i$
$i^2 = -1$	$i^6 = -1$	$i^{10} = -1$	$i^{14} = -1$
$i^3 = -i$	$i^7 = -i$	$i^{11} = -i$	$i^{15} = -i$
$i^4 = 1$	$i^8 = 1$	$i^{12} = 1$	$i^{16} = 1$

ad infinitum.

Since $a^0 = 1$ if $a \neq 0$, we have

$$i^0 = 1.$$

Negative integer powers of i are

$$i^{-1} = \frac{1}{i} = -i\,; \quad i^{-2} = \frac{1}{i^2} = -1\,; \quad i^{-3} = \frac{1}{i^3} = +i\,; \quad etc.$$

pp. 791 *et seq.*; *pp.* 793 *et seq.*

Powers and roots of **complex numbers**, and their products and quotients, will be described in Chapter 24.

p. 123

Arithmetic of residue classes was discussed in Section 4.2. For powers, we have: If

$$a \equiv b, \bmod m\,; \quad n \text{ is a positive integer,}$$

then $a^n \equiv b^n$, mod m.

- Find the least integer remainder of 3^{263} divided by 6567.

The 126 digits of the expansion of 3^{263} are too many to handle for most pocket calculators.

Use the following algorithm, in modulo 6567:

$$\begin{aligned}
3^2 &= 9 \\
3^4 &= (3^2)^2 = 81 \\
3^8 &= (3^4)^2 = 6561 \equiv -6 \\
3^{16} &= (3^8)^2 \equiv (-6)^2 = 36 \\
3^{32} &= (3^{16})^2 \equiv 36^2 = 1296 \\
3^{64} &= (3^{32})^2 \equiv 1296^2 = 1\,679\,616 = 255 \cdot 6567 + 5031 \equiv 5031 \\
3^{128} &= (3^{64})^2 \equiv 5031^2 = 3854 \cdot 6567 + 1743 \equiv 1743 \\
3^{256} &= (3^{128})^2 \equiv 1743^2 = 462 \cdot 6567 + 4095 \equiv 4095,
\end{aligned}$$

which is as close as we can get to 3^{263} by repeated squaring.

Continuing in modulo 6567, we find

$$3^{263} = 3^{256} \cdot 3^7 = 3^{256} \cdot 3^4 \cdot 3^2 \cdot 3 \equiv 4095 \cdot 81 \cdot 9 \cdot 3$$
$$= 8\,955\,765 = 1363 \cdot 6567 + 4944 \equiv 4944 .$$

The sought remainder is 4944 .

Calculation Rules for Roots

There are n distinct n-th roots of a number $N \neq 0$. Most of these will be complex numbers. They satisfy $r^n = N$ and might be written

$$\sqrt[n]{N} = r_1 ;\; r_2 ;\; r_3 \ldots r_n ,$$

Radical, Radicand, Root Index

where $\sqrt[n]{N}$ is the **radical**, N is the **radicand**, and n is the **root index**. More generally, we may allow the root to be real, imaginary, or complex.

When the alternative symbol √ – without the bar – is used on a composite radicand, this must be enclosed in parentheses,

$$\sqrt{a+b} = \surd\,(a+b) .$$

Second roots, $\sqrt{p}$, are called **square roots**, and are written without the root index; third roots, $\sqrt[3]{p}$, are **cube roots**.

Principal Roots, Secondary Roots

We distinguish between **principal roots** and **secondary roots**:

- The *principal root* of a *positive number* is the *positive root*; other roots, of which there are $n-1$, are secondary roots:

  $\sqrt{4} = +2$; -2 is a secondary root;

  $\sqrt[4]{1} = +1$; -1, $+\mathrm{i}$, and $-\mathrm{i}$ are secondary roots.

- The principal n-th *odd-index root* of a *negative number* is, by definition, the *negative real root*.

  Example: $\sqrt[3]{-8}$ has the principal root -2; the two other roots, $(1+\mathrm{i}\sqrt{3})$ and $(1-\mathrm{i}\sqrt{3})$, are secondary roots.

  Check: $(1 \pm \mathrm{i}\sqrt{3})^3 = -8$.

- Negative numbers have no even-index principal roots; the square roots of -4 are $+2\mathrm{i}$ and $-2\mathrm{i}$; neither is a real number and, hence, no principal root.

Using the power notation of radicals,

$$\sqrt[n]{N} = N^{1/n} \quad (n \neq 0) ,$$

we have the following calculation formulas for principal roots of $a, b > 0$:

- $\sqrt[n]{a} \cdot \sqrt[n]{b} = \sqrt[n]{a\,b} = (a\,b)^{1/n}$
- $\sqrt[m]{\sqrt[n]{a}} = \sqrt[m\,n]{a} = a^{1/m\,n}$
- $\sqrt[n]{a} \,/\, \sqrt[n]{b} = \sqrt[n]{a/b} = \left(\frac{a}{b}\right)^{1/n}$
- $\sqrt[m]{a^n} = \sqrt[m/n]{a} = a^{n/m}$

Surds

Latin, *surdus*, "deaf", "silent", "indistinct"

A radical expressing an irrational number is called a **surd**, qualified as **quadratic** (*e.g.*, $\sqrt{2}$), **cubic** ($\sqrt[3]{2}$), **quartic** ($\sqrt[4]{2}$), **quintic** ($\sqrt[5]{2}$), *etc.*, after the index of the radical.

(The term *surd* is sometimes used as a synonym for *irrational number*.)

A **pure surd**, or entire surd, contains no rational number, that is, all its factors or terms are surds, *e.g.*, $\sqrt{2}$ or $\sqrt{2} + \sqrt{3}$. A **mixed surd** contains at least one rational term, $2 + \sqrt{3}$, or factor, $3\sqrt{2}$.

In a **binomial surd**, at least one of the numbers must be a surd, $2 + \sqrt{2}$ or $\sqrt{2} + \sqrt{3}$; in a **trinomial surd**, at least two terms must be surds that cannot be reduced to a single surd, *e.g.*, $2 + \sqrt{2} + \sqrt{3}$ or $\sqrt{2} + \sqrt{3} + \sqrt{5}$.

Conjugate Binomial Surds

Conjugate binomial surds differ in the sign of one of the irrational terms; the product of two conjugate binomial surds is *rational*; *e.g.*,

$$(2 + \sqrt{2})(2 - \sqrt{2}) = 2 \; ; \qquad (\sqrt{2} + \sqrt{3})(\sqrt{2} - \sqrt{3}) = -1 \; ;$$

$$\frac{c}{\sqrt{a} - \sqrt{b}} = \frac{c\left(\sqrt{a} + \sqrt{b}\right)}{\left(\sqrt{a} - \sqrt{b}\right)\left(\sqrt{a} + \sqrt{b}\right)} = \frac{c\left(\sqrt{a} + \sqrt{b}\right)}{a - b} \; .$$

Simplified Form

A radical is said to be in **simplified form** if its radicand is not in fractional form and does not include powers of the same index as the root index:

$$\sqrt{\frac{5}{3}} = \sqrt{\frac{5 \times 3}{3 \times 3}} = \frac{1}{3}\sqrt{15} \; ;$$

$$\sqrt{72} = \sqrt{2 \times 2 \times 2 \times 3 \times 3} = 2 \times 3\sqrt{2} = 6\sqrt{2}$$

Rationalizing the Denominator

The process of eliminating surds from the denominator is called to **rationalize the denominator** or, simply, to rationalize.

$$\frac{\sqrt{3}}{2\sqrt{5} - \sqrt{3}} = \frac{\sqrt{3}\left(2\sqrt{5} + \sqrt{3}\right)}{\left(2\sqrt{5} - \sqrt{3}\right)\left(2\sqrt{5} + \sqrt{3}\right)} = \frac{\sqrt{3}\left(2\sqrt{5} + \sqrt{3}\right)}{\left(2\sqrt{5}\right)^2 - \left(\sqrt{3}\right)^2}$$

$$= \frac{2\sqrt{15} + 3}{20 - 3} = \frac{1}{17}\left(3 + 2\sqrt{15}\right) .$$

Simplification and rationalization were essential skills when roots had to be found in tables and computations done by paper and pencil alone. They still help us understand how to study the nature of specific numbers or expressions.

The Binomial Theorem

Integral powers of the binomial $(a + b)$ are

$$\begin{aligned}(a+b)^0 &= 1\\ (a+b)^1 &= 1a + 1b\\ (a+b)^2 &= 1a^2 + 2ab + 1b^2\\ (a+b)^3 &= 1a^3 + 3a^2b + 3ab^2 + 1b^3\\ (a+b)^4 &= 1a^4 + 4a^3b + 6a^2b^2 + 4ab^3 + 1b^4.\end{aligned}$$

Higher powers are simply expressed by using the **binomial theorem**, a way of expressing the n-th power of the binomial $(a + b)$ in symbolic form:

$$(a+b)^n = \binom{n}{0} a^n + \binom{n}{1} a^{n-1}b + \binom{n}{2} a^{n-2}b^2 + \dots + \binom{n}{n-1} ab^{n-1} + \binom{n}{n} b^n$$

or, abbreviated,

$$(a+b)^n = \sum_{i=0}^{n} \binom{n}{i} a^{n-i} b^i ,$$

where the coefficient of $a^{n-i}b^i$ is the number of ways we can chose b's from i of the factors; the other $(n - i)$ factors contribute a's. The formula is

! : p. 106

$\binom{n}{i}$: p. 196

$$\binom{n}{i} = \frac{n!}{(n-i)!\,i!} .$$

Pascal's Triangle

If we place the coefficients of the terms in the binomial expansions in a triangular array, with $(a + b)^0$ at the top, the coefficients of $(a + b)^1$ in the second row, those of $(a + b)^2$ in the third row, *etc.*, we obtain **Pascal's triangle**.

n = 0								1								
1							1		1							
2						1		2		1						
3					1		3		3		1					
4				1		4		6		4		1				
5			1		5		10		10		5		1			
6		1		6		15		20		15		6		1		
7	1		7		21		35		35		21		7		1	
⋮	⋮		⋮		⋮		⋮		⋮		⋮		⋮		⋮	

Every coefficient is the sum of the two coefficients straddling it in the row immediately above,

$$\frac{n!}{(n-i)!\;i!} + \frac{n!}{[n-(i+1)]!\;(i+1)!}$$

$$= \frac{n!\;[(i+1)+(n-i)]}{(n-i)!\;(i+1)!} = \frac{(n+1)!}{(n-i)!\;(i+1)!} ,$$

or, in symbolic notation,

$$\binom{n}{i} + \binom{n}{i+1} = \binom{n+1}{i+1}.$$

PASCAL Blaise (1623 - 1662)

Pascal's triangle is named after the French mathematician-philosopher-physicist **Blaise Pascal**, who described it in his posthumously published *Traité du triangle arithmétique* (1665). Although Pascal never claimed recognition for his discovery, his name is inseparably linked with it.

CHU Shih-chieh

The triangle is, in fact, much older; it appeared as early as 1303 in *Precious Mirror of the Four Elements* by the Chinese mathematician **Chu Shih-chieh**.

p. 51 The chart uses Chinese rod numerals.

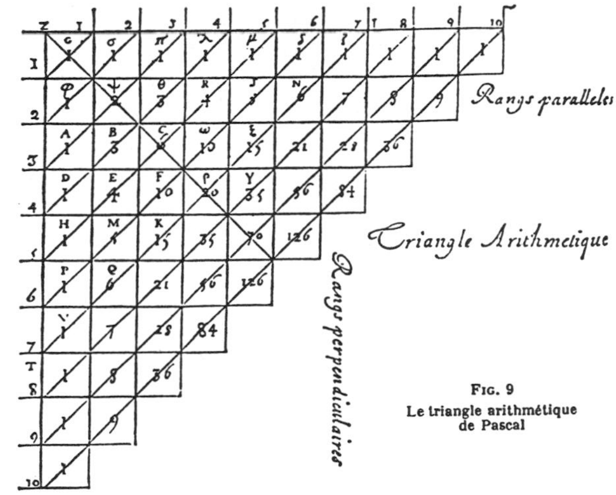

From Pascal's *Traité du triangle arithmétique* (1665).

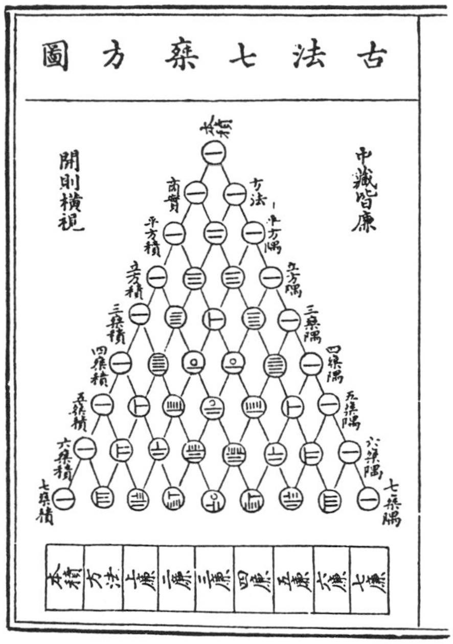

"Chu Shih-chieh's Triangle" (1303). From Joseph Needham, *Science and Civilization in China III: Mathematics and the Sciences of Heavens and the Earth* (Cambridge University Press, 1968).

Root Extraction

Square Roots

Before the time of the electronic calculator, square roots of real numbers were generally determined with the help of tables of squares or square roots. If these helpmates are not available, we can still find the square root of a real number by direct calculation.

Dividing and Averaging

The following simple method to find square roots of a real number goes back to the time of the ancient Babylonians.

If

$$\sqrt{N} \approx n,$$

a better approximation is

$$\sqrt{N} \approx \frac{1}{2}\left(\frac{N}{n} + n\right)$$

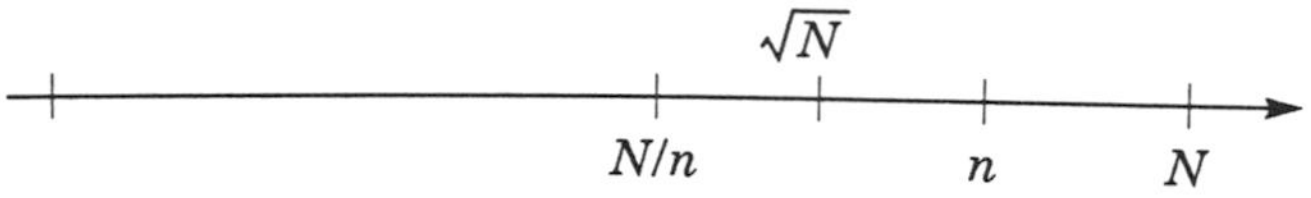

and increasing accuracy is obtained as we continue:

$$\sqrt{N} \approx \frac{1}{2}\left[\frac{N}{\frac{1}{2}\left(\frac{N}{n} + n\right)} + \frac{1}{2}\left(\frac{N}{n} + n\right)\right],$$

$$\sqrt{N} \approx \frac{1}{2}\left\{\frac{N}{\frac{1}{2}\left[\frac{N}{\frac{1}{2}\left(\frac{N}{n} + n\right)} + \frac{1}{2}\left(\frac{N}{n} + n\right)\right]} + \frac{1}{2}\left[\frac{N}{\frac{1}{2}\left(\frac{N}{n} + n\right)} + \frac{1}{2}\left(\frac{N}{n} + n\right)\right]\right\},$$

etc.

- Sought: $\sqrt{1000}$

A rough approximation (first guess) is

$$\sqrt{1000} \approx 30\,.$$

Continuing the approximations, we find:

$$\sqrt{1000} \approx \frac{1}{2}\left(\frac{1000}{30} + 30\right) = 31.666\,666\,66\ldots$$

$$\sqrt{1000} \approx \frac{1}{2}\left(\frac{1000}{31.666\,666\,66} + 31.666\,666\,66\right) = 31.622\,807\,01\ldots$$

$$\sqrt{1000} \approx \frac{1}{2}\left(\frac{1000}{31.622\,807\,01} + 31.622\,807\,01\right) = 31.622\,776\,60\ldots$$

$$\sqrt{1000} \approx \frac{1}{2}\left(\frac{1000}{31.622\,776\,60} + 31.622\,776\,60\right) = 31.622\,776\,60\ldots$$

Thus, from the initial guess, we achieve a result correct to eight decimal places in just three steps.

Continued Fractions

p. 127

To extend the method of **continued fractions** to irrational numbers, we choose $\sqrt{6}$, which lies between 2.4 and 2.5, since $2.4^2 = 5.76$ and $2.5^2 = 6.25$.

We can write

$$\sqrt{6} = b_0 + \cfrac{1}{b_1 + \cfrac{1}{b_2 + \cfrac{1}{b_3 + \dots}}} = [b_0\; b_1\; b_2\; b_3 \dots]\,. \qquad (1)$$

Since the fraction must be less than unity, we let $b_0 = 2$, and obtain

$$\frac{1}{[b_1\; b_2\; b_3 \dots]} = \sqrt{6} - 2 \qquad (2)$$

or

$$[b_1\; b_2\; b_3 \dots] = b_1 + \frac{1}{[b_2\; b_3\; b_4 \dots]} = \frac{1}{\sqrt{6} - 2}\,. \qquad (3)$$

Letting $b_1 = 2$, we have

$$\frac{1}{[b_2\; b_3\; b_4 \dots]} = \frac{1}{\sqrt{6} - 2} - 2 = \frac{\sqrt{6} - 2}{2} \qquad (4)$$

and

$$[b_2\; b_3\; b_4 \dots] = b_2 + \frac{1}{[b_3\; b_4\; b_5 \dots]} = \frac{2}{\sqrt{6} - 2} = \sqrt{6} + 2\,. \qquad (5)$$

With $b_2 = 4$, we obtain

$$\frac{1}{[b_3\; b_4\; b_5 \dots]} = \sqrt{6} + 2 - 4 = \sqrt{6} - 2\,, \qquad (6)$$

that is, we have the same relation in Eq. (6) that we had in Eq. (2); hence, the pattern repeats:

$$b_3 = 2\,; \quad b_4 = 4,\ \textit{etc.}$$

We can now write

$$\sqrt{6} = 2 + \cfrac{1}{2 + \cfrac{1}{4 + \cfrac{1}{2 + \cfrac{1}{4 + \cfrac{1}{2 + \dots}}}}}$$

or with the notation introduced for periodic decimal fractions,

$$\sqrt{6} = [2,\ 2,\ 4,\ 2,\ 4 \dots] = [2,\ \overline{2,\ 4}]\ .$$

Many algebraic irrational numbers produce relatively simple periodic continued fractions. A number's continued fraction is periodic precisely when the number is a quadratic surd:

$$\sqrt{2} = 1 + \cfrac{1}{2 + \cfrac{1}{2 + \cfrac{1}{2 + \cfrac{1}{2 + \dots}}}} = [1, 2, 2, 2, 2 \dots] = [1, \overline{2}]$$

$$\sqrt{3} = 1 + \cfrac{1}{1 + \cfrac{1}{2 + \cfrac{1}{1 + \cfrac{1}{2 + \cfrac{1}{1 + \dots}}}}} = [1, \overline{1, 2}]$$

$$\sqrt{11} = 3 + \cfrac{1}{3 + \cfrac{1}{6 + \cfrac{1}{3 + \cfrac{1}{6 + \dots}}}} = [3, \overline{3, 6}]$$

$$\sqrt{17} = [4, \overline{8}]$$

p. 776

Continued fractions converge more rapidly than **power series expansions** and are often useful means for finding approximative numerical values of irrational numbers.

Method for Real Numbers and Polynomials

We are required to find the square root of 9467.29 .

Step 1. Begin by dividing the radicand into groups of two digits, working both ways from the decimal point. Use a sign ' to separate the groups.

9		$\sqrt{94'67.'29} = 97.3$
+ 9	9 x 9 = 81	81
187		13 67
+ 7	187 x 7 = 1309	13 09
1943		58 29
3	1943 x 3 = 5829	58 29
		0

Step 2. Find the largest number whose square is less than the leading two-digit group (94) of the radicand; place this digit (9) in the root and in two positions on the left, in front of the extraction galley, and add these two digits (9 + 9 = 18).

Step 3. Calculate the square (9 x 9 = 81) of the first root digit, and subtract it from the first group (94) of digits in the radicand, leaving a remainder (13).

Step 4. Move the next two-digit group (67) down from the radicand to form a new minuend (1367). Assess the next digit of the root, place it (7) in two positions on the left as before.

Step 5. Carry out the addition (187 + 7 = 194) and the multiplication (187 x 7), and introduce the product (1309) in the galley (under 1367); subtract to obtain the remainder (58).

Step 6. Move the last two-digit group (29) down from the radicand to obtain a last minuend (5829). Determine the next digit of the the root (3), add it on the left, and perform the multiplication (1943 x 3) and place the product (5829) in the galley; subtract to find the remainder, zero.

To prove the correctness of this extraction, we let

$$a = 90$$
$$b = 7$$
$$c = 0.3$$

and find

a		$\sqrt{a^2 + b^2 + c^2 + 2ab + 2ac + 2bc} = a + b + c$
$+\ a$	$a \times a =$	a^2
$2a + b$		$0 \quad b^2 \quad 2ab$
b	$(2a + b)b =$	$b^2 \quad 2ab$
$2(a + b) + c$		$0 \quad c^2 \quad 0 \quad 2ac + 2bc$
c	$[2(a + b) + c]c =$	$c^2 \quad 2ac + 2bc$
		$0 \quad 0 \quad 0$

- Calculate $\sqrt{2}$ to five correct decimal places – a most tedious job that might make us appreciate the tenacity of the constructors of the square root tables.

Any fool can touch a button, but…

1		$\sqrt{2.'00'00'00'00'00}$ = 1.414213... = 1.41421
+1	1 x 1 =	1
24		1 00
+ 4	24 x 4 =	96
281		4 00
+ 1	281 x 1 =	2 81
2824		1 19 00
+ 4	2824 x 4 =	1 12 96
28282		6 04 00
+ 2	28282 x 2 =	5 65 64
282841		38 36 00
+ 1	282841 x 1 =	28 28 41
2828423		10 06 59 00
3	282842 x 3 =	8 48 52 69
		1 58 06 31

•

Using this method for a **polynomial**, we begin by arranging all terms by descending degrees.

x^2	$\sqrt{x^4 - 2x^3 + 3x^2 - 2x + 1} = x^2 - x + 1$
x^2	x^4
$2x^2 \quad - x$	$-2x^3 + 3x^2$
$-x$	$-2x^3 + x^2$
$2x^2 - 2x + 1$	$2x^2 - 2x + 1$
1	$2x^2 - 2x + 1$
	0

- Find the square root of the first term of the radicand; enter it in the quotient and twice in the control column to the left of the radical.
- Square the term and enter the result under the first term of the radicand; subtract.
- Add the terms in front of the radical to get the sum $2x^2$.
- Move the next two terms of the radicand down; divide by $2x^2$ to find the next term, $(-x)$, of the quotient.
- Add $(-x)$ as shown, multiply as before, and place under the dividend; subtract.
- Move the remaining two terms of the radicand down, find the last term of the quotient, +1, and complete the multiplication; subtract.

Cube Roots

Method of Trenchant

TRENCHANT Jean

The following method of extracting cube roots was used by the French mathematician and physician **Jean Trenchant** in his treatise *Arithmetique* (1557), but is probably older.

		$\sqrt[3]{100\,441\,020} = 464.8\ldots$
	(4)	64
		36 441
4 800	(6)	33 336
		3 105 020
634 800	(4)	2 561 344
		543 676 000
64 588 800	(8)	517 601 792
		26 074 208 000

etc.

We have found the good physician Trenchant's method an excellent **cure for insomnia**:

- Begin by dividing the number into groups of three digits, working from the last to the first digit of a whole number or, if a decimal fraction, in both directions from the decimal point. Here: 100 441 020

- Find the root of the greatest possible cube contained in the first group of three digits,

$$4^3 = 64 < 100 .$$

- Subtract the cube, 64, from 100 to obtain the remainder 36.
- Move down the next group of three digits, 441, to give a new dividend, 36 441.
- To find the second digit of the cube root, keeping in mind that shifting a digit one step toward the left means a tenfold increase of its value, we write $a = 40$ and assume the second digit to be b.
- The dividend 36 441 must then contain the sum $3a^2b + 3ab^2 + b^3$; disregarding the smaller terms, we compute

$$3a^2 = 3 \cdot 40^2 = 4800$$

and place it on the left of the extraction.

- We find that this divisor is contained 7 times in the dividend 36 441.

- Compute the terms

$$\begin{aligned} 3a^2b &= 3 \cdot 40^2 \cdot 7 &= 4800 \cdot 7 &= 33\,600 \\ 3ab^2 &= 3 \cdot 40 \cdot 7^2 &= 120 \cdot 49 &= 5\,880 \\ b^3 &= 7^3 &= &= \underline{343} \end{aligned}$$

and find the sum 39 823,

which, however, exceeds 36 441. The method then requires that one tries with the preceding, smaller, integer ($b = 6$); thus: $3 \cdot 40^2 \cdot 6 + 3 \cdot 40 \cdot 6^2 + 6^3 = 33\,336$, which we now enter below 36 441 and perform the subtraction to find the reminder 3105.

- Move down the next group of three digits, 020, to give the dividend, 3 105 020.
- If we now let $460 = c$, we have $3 \cdot c^2 = 3 \cdot 460^2 = 634\,800$, which we enter as the next divisor on the left.
- Division of 3 105 020 by 634 800 gives the next digit of the cube root, $d = 4$.
- We now compute

$$\begin{aligned} 3c^2d &= 3 \cdot 460^2 \cdot 4 &= 2\,539\,200 \\ 3cd^2 &= 3 \cdot 460 \cdot 4^2 &= 22\,080 \\ d^3 &= 4^3 &= \underline{64} \end{aligned}$$

and the sum 2 561 344

which we place under 3 105 020 and subtract to find the new remainder 543 676.

- Next, add as many groups of three zeros as you desire further decimals in the cube root, and move down the first of these groups to give the dividend 543 676 000.
- Continue the calculation and extraction in the manner described above.

- We find the next divisor $3 \cdot e^2 = 3 \cdot 4640^2 = 64\,588\,800$, which is contained $f = 8$ times in the dividend.
- Now compute

$$\begin{aligned} 3e^2f &= 3 \cdot 4640^2 \cdot 8 = 516\,710\,400 \\ 3ef^2 &= 3 \cdot 4640 \cdot 8^2 = 890\,880 \\ f^3 &= 8^3 = \underline{512} \end{aligned}$$

and the sum 517 601 792

etc.

Method of Heron

HERON
(*c.* A.D. 75)

The following method of calculating cube roots was used by Heron of Alexandria (1st century) but is much older, probably dating from the time of the ancient Babylonians.

- Sought: $n = \sqrt[3]{N}$

Enclose n between the consecutive integers a and b,

$$a < n < b\,; \quad a + 1 = b$$

where $n - a = p\,; \quad b - n = q\,; \quad p + q = 1\,.$

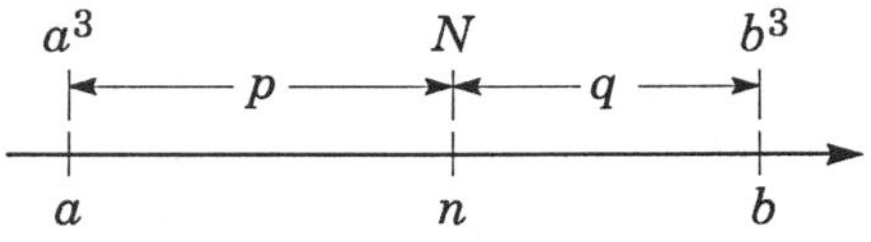

We then have

$$\left.\begin{aligned} a^3 &= (n-p)^3 = n^3 - 3n^2p + 3np^2 - p^3 \\ b^3 &= (n+q)^3 = n^3 + 3n^2q + 3nq^2 + q^3 \end{aligned}\right\}$$

Since

$$p < 1\,;\ p^3 \ll 1 \text{ and } q < 1\,;\ q^3 \ll 1,$$

we discard the terms p^3 and q^3, and obtain

$$\left.\begin{aligned} P &= N - a^3 \approx 3np(n-p) = 3anp \\ Q &= b^3 - N \approx 3nq(n+p) = 3bnq \end{aligned}\right\}$$

and

$$\frac{p}{q} \approx \frac{b \cdot P}{a \cdot Q}\,; \quad \frac{p}{p+q} = p \approx \frac{b \cdot P}{b \cdot P + a \cdot Q}\,.$$

We can now write

$$n = a + p \approx a + \frac{b \cdot P}{b \cdot P + a \cdot Q}\,.$$

Example: $N = 90;\ a = 4,\ a^3 = 64;\ b = 5,\ b^3 = 125$

$P = 90 - 64 = 26$

$Q = 125 - 90 = 35$

$$n \approx 4 + \frac{5 \cdot 26}{5 \cdot 26 + 4 \cdot 35} = 4 + \frac{130}{270} = 4.4\overline{81}$$

which is correct to three decimal places.

- Sought: $\sqrt[3]{400}$

We have

$$\left.\begin{aligned} a &= 7 \\ b &= 8 \end{aligned}\right\} \quad \left.\begin{aligned} a^3 &= 343 \\ b^3 &= 512 \end{aligned}\right\} \quad \left.\begin{aligned} P &= 400-343 = 57 \\ Q &= 512-400 = 112 \end{aligned}\right\} \quad \left.\begin{aligned} b \cdot P &= 456 \\ a \cdot Q &= 784 \end{aligned}\right\}$$

$$b \cdot P + a \cdot Q = 1240$$

$$n \approx 7 + \frac{456}{1240} = 7.3677\ldots$$

(The correct value of the cube root is

$$\sqrt[3]{400} = 7.368\,062\ldots.)$$

Other Methods

There are several methods of approximating the cube root of a number $N = a^3 + b$:

- Hindu mathematicians used the approximation,

  $$\sqrt[3]{a^3+b} \approx a + \frac{b}{3\,a^2}$$

  as early as the 5th century;

- Tartaglia (*c.* 1500 - 1557) used a closer approximation,

  $$\sqrt[3]{a^3+b} \approx a + \frac{b}{3\,a^2 + 3\,a},$$

  which can also be found in some 15th-century manuscripts;

- Fibonacci (*c.* 1170 - *c.* 1250) used

  $$\sqrt[3]{a^3+b} \approx a + \frac{b}{3\,a^2 + 3\,a + 1}.$$

Through the use of *calculus*, we have even more powerful methods for finding cubic and other roots.

The above methods all work as well as one might want. We make a our best current approximation, and b the remainder; using **Newton's binomial series**, we have

cf. pp. 775 *et seq.*

$$\sqrt[3]{a^3\left(1+\frac{b}{a^3}\right)} = a\left(1+\frac{b}{a^3}\right)^{1/3} = a\left(1+\frac{1}{3}\cdot\frac{b}{a^3}+\ldots\right),$$

of which the ancient Hindu mathematicians had an approximation of the first term.

4.5 Logarithms

History

The rebirth of science after the state of rigor mortis imposed by the Church in the Middle Ages was particularly evident in the awakened interest in astronomy and in the attendant development of trigonometry, generated also by other forms of world exploration – land surveying, cartography, and navigation.

Scientists everywhere began to spend enormous amounts of time in calculating tables of trigonometric functions, and it became important to find methods of replacing the often laborious operations of multiplication and division with addition and subtraction, – *e.g.*, by employing formulas such as

$$2 \cdot \sin\alpha \sin\beta = \cos(\alpha - \beta) - \cos(\alpha + \beta) .$$

NAPIER John (1550 - 1617)

John Napier, laird of Merchiston, near Edinburgh, Scotland, voiced his opinion thus:

> There is nothing more troublesome in mathematics than the multiplications, divisions, square and cubic root extractions of great numbers which involve a tedious expenditure of time, as well as being subject to "slippery errors".

Napier presumably had in mind a passage in Michael Stifel's *Arithmetica integra* (1544) where he compares a sequence of consecutive integer numbers with a sequence of corresponding powers of 2 having these integers as exponents:

0	1	2	3	4	5	6	7	8	9	10
1	2	4	8	16	32	64	128	256	512	1024

From this it is evident that sums and differences of the power indices correspond to products and quotients of the powers themselves – but equally that the number 2 as a base would be impracticable for purposes of computation because of the large gaps between successive terms, which would make interpolation altogether too inaccurate. One would have to choose a number much closer to 1.

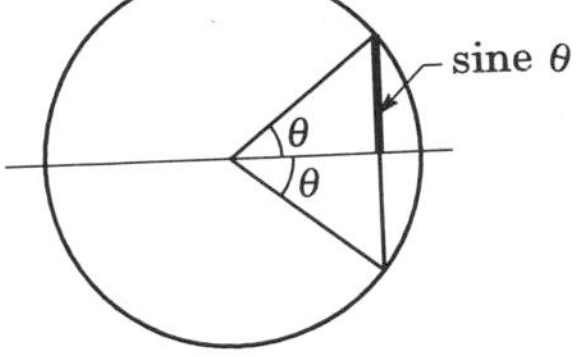

Napier's stated aim was to facilitate the calculation of natural sines and other trigonometric functions. In Napier's time, the sine of an arc, or an angle, was not thought of as a ratio, but rather as half the length of the chord subtending the double angle at the center of a circle of a given radius – actually, of course, the ratio of this semichord to the radius of the circle.

To avoid introducing fractions, Napier chose 10^7 as a value for the radius of the circle. As base for his powers, he decided on $(1 - 10^{-7}) = 0.999\,999\,9$.

By multiplying all powers by 10^7, we have the relation

$$N = 10^7 (1 - 10^{-7})^L ,$$

where L is the "Naperian logarithm" of the number N.

373 A
DESCRIPTION OF THE ADMIRABLE TABLE OE LOGARITHMES:

WITH

A DECLARATION OF THE MOST PLENTIFVL, EASY, and ſpeedy vſe thereof in both kindes of Trigonometrie, as alſo in all *Mathematicall calculations.*

INVENTED AND PVBLISHED IN LATIN BY THAT Honorable L. IOHN NEPAIR, Baron of *Marchiſton*, and tranſlated into Engliſh by the late learned and famous Mathematician *Edward Wright.*

With an Addition of an Inſtrumentall Table to finde the part proportionall, inuented by the Tranſlator, and deſcribed in the end of the Booke by HENRY BRIGS *Geometry reader at Greſham-houſe in London.*

All per... chalenged approued by the Author, & publiſh... ...urs. Thi... ...death of ... Tranſlator.

...ſcended from

...le by educ...V,

Printed by ...NICHOLAS OKES.

1616

THE FIRST BOOKE.

CHAP. I.
Of the Definitions.

A LINE *is ſaid to increaſe equally, when the poynt deſcribing the ſame, goeth forward equall ſpaces, in equall times, or moments.* 1. Definition.

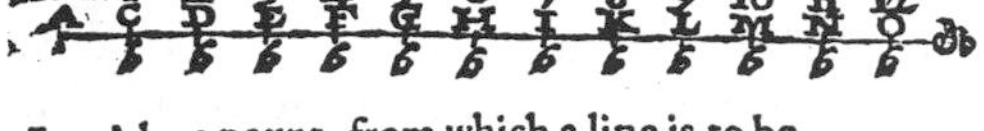

Let A be a poynt, from which a line is to be drawne by the motion of another poynt, which let be B.

Now in the firſt moment, let B moue from

B

2 *The firſt Booke.* CHAP. I

A to C. In the ſecond moment from C to D. In the third moment from D to E, & ſo forth infinitely, deſcribing the line A C D E F, &c. The ſpaces A C, C D, D E, E F, &c. And all the reſt being equall, and deſcribed in equall moments (or times.) This line by the former definition ſhall be ſaid to increaſe equally.

A Corollary or conſequent. *Therefore by this increaſing, quantities equally differing, muſt needes be produced, in times equally differing.*

As in the Figure before, B went forward from A to C in one moment, and from A to E in three moments. So in ſixe moments from A to H: and in 8 moments from A to K. And the differences of thoſe moments, one and three, and of theſe 6 and 8 are equall, that is to ſay two.

So alſo of thoſe quantities A C, and A E, and of theſe, A H, and A K, the differences C E, and H K are equall, and therefore differing equally, as before.

2. Definition. *A Line is ſaid to decreaſe proportionally into a ſhorter, when the poynt deſcribing the ſame in equall times, cutteth off parts continually of the ſame proportion to the lines from which they are cut off.*

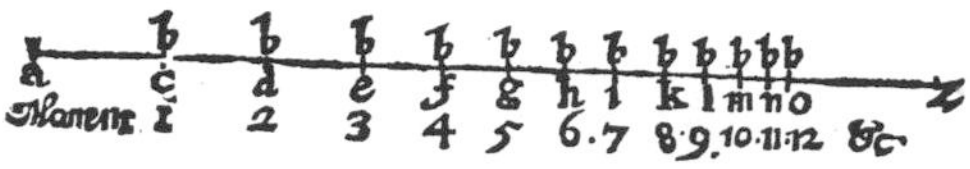

For examples ſake. Let the line of the whole ſine a Z be to bee diminiſhed proportionally: let the poynt diminiſhing the ſame by his motion

CHAP. I. *The firſt Booke.* 3

motion be *b*: and let the proportion of each part to the line from wch it is cut off, be as Q R to Q S. Therefore in what proportion Q S is cut in R, in the ſame proportion (by the 10 of the 6 of *Euclid*) Let *a* Z be cut in *c*, and ſo let *b*. running from *a* to *c* in the firſt moment, cut off *a c* from *a* Z, the line or ſine *c* Z remaining.

And from this *c* Z let *b* proceeding in the ſecond moment, cut off the like ſegment, or part, as Q R to Q S: and let that bee *c d*, leauing the ſine. *d* Z. From which therefore in the third moment, let *b* in like manner, cut off the ſegment *d e*, the ſine *e* Z being left behinde. From which likewiſe in the fourth moment, by the motion of *b*, let the ſegment *e f* be cut off, leauing the ſine *f* Z. From this *f* Z in the fifth moment, let *b* in the ſame proportion cut off the ſegment *f g*, leauing the ſine *g* Z, and ſo forth infinitly. I ſay therfore out of the former definition, that here the line of the whole ſine *a* Z, doth proportionally decreaſe into the ſigne *g* Z, or into any other laſt ſine, in which *b* ſtayeth, and ſo in others.

Hence it followeth that by this decreaſe in equall moments (or times) there muſt needes alſo bee left proportionall lines of the ſame proportion. A Corolary.

Logarithm is the name chosen by Napier from the Greek words *logos*, meaning “ratio”, and *arithmos*, “number”, and may be translated as “ratio number”.

The relation shows that the logarithm of 10^7 is zero, and the logarithm of $10^7(1-10^{-7}) = 0.999\,999\,9$ is -1, which in many ways is highly inconvenient for general calculation.

By dividing Naperian logarithms and their corresponding numbers (antilogarithms) by 10^7, we nearly obtain a system of logarithms to the base 1/e, because

$$\left(1-\frac{1}{10^7}\right)^{10^7} \approx \lim_{n\to\infty}\left(1-\frac{1}{n}\right)^n = \frac{1}{e}.$$

Thus, the often-repeated statement that Naperian logarithms are “natural” logarithms – that is, logarithms to the base e – is not true. It must be remembered, however, that Napier had no thought of a base for his system of logarithms but arrived at it from a comparison of uniform motion and retarded motion.

Napier published his results in 1614 under the title *Mirifici logarithmorum canonis descriptio* (“Description of the Wonderful Law of Logarithms”); an English translation appeared in 1616 (see previous page). They were the outcome of about 20 years of calculations, for he had communicated his intentions in 1594 in a letter to Tycho Brahe, who told Johannes Kepler.

BRIGGS Henry (1561 - 1630)

In the summer of 1615, and again in 1616, Napier received visits from **Henry Briggs**, then professor of geometry at Gresham College in London. On a suggestion by Napier, it was agreed that a change to a decimal base would be an improvement, and a first installment of logarithms for the numbers 1 to 1000, inclusive, to base 10 was published in 1617 in London with the title *Logarithmorum chilias prima*.

VLACQ Adriaan (c. 1600 - after 1655)

DECKER Ezechiel de

Briggs's *Arithmetica logarithmica* (1624) contained Briggsian, or “common”, logarithms for the numbers 1 to 20 000 and 90 000 to 100 000, inclusive, the gap of 70 000 logarithms being bridged in 1627 by the two Dutchmen **Adriaan Vlacq** and **Ezechiel de Decker** with *Het tweede deel van de Nieuwe telkonst*, issued in English the following year as *Arithmetica logarithmica*.

KEPLER Johannes (1571 - 1630)

Johannes Kepler also prepared extensive logarithm tables, used for accurate calculations of the revolutions of the planets around the Sun. These tables appear in Kepler's *Tabulae Rudolphinae*, printed in Germany in 1627; they are named in honor of Kepler's patron, Emperor Rudolph II of Austria and King of Bohemia.

Napier's description of the development of his ideas of his logarithms, left in manuscript form, was published in 1619 as *Mirifici logarithmorum canonis constructio*.

As often happens when the time is "ripe" for a new discovery or invention, Napier was not alone in developing improved methods of converting operations of multiplication into addition.

BÜRGI Joost (Jobst)
(1552 - 1632)

The Swiss watchmaker **Joost Bürgi**, maker of astronomical instruments and an indefatigable computer and assistant to Johannes Kepler, Imperial astronomer in Prague, also conceived a system of logarithms to facilitate the multiplication of large numbers.

Where Napier decided on powers of $(1 - 10^7)$, Bürgi made the better choice, $(1 + 10^{-4})$; this made his power indices increase as the power numbers increased, while Napier's decreased.

There was another difference between the work of the two men: where Napier multiplied his powers by 10^7, Bürgi chose 10^8; he also multiplied his logarithms by 10 in his tables.

$$N = 10^8(1 + 10^{-4})^L .$$

Bürgi called $10L$ the "red number" corresponding to the "black number" N. If all black numbers are divided by 10^8 and all red numbers by 10^4, we obtain what is nearly a system of logarithms to the base e, because

$$\left(1 + \frac{1}{10^4}\right)^{10^4} \quad \text{comes near} \quad \lim_{n \to \infty} \left(1 + \frac{1}{n}\right)^n = e ;$$

it agrees to four significant figures.

Bürgi published his system in 1620 as *Arithmetische und geometrische Progress-Tabulen*, effectively a table of antilogarithms; it may be that he had begun his work in 1588, or even 1584. Napier is reported to have discussed his own results with Brahe in 1594.

The matter of who was "first" must remain unsettled, but the official priority belongs to Napier because of his publishing date, 1614, six years before Bürgi.

Definitions

We have

$$x = a^{\log_a x},$$

Logarithm, Exponent
Base
Antilogarithm

that is, a **logarithm** is an **exponent** which defines to what power the **base** a must be raised in order to give x, called the **antilogarithm**.

Only positive real numbers $a > 1$ are acceptable as bases for workable systems of logarithms. With bases $a < 1$, the higher the antilogarithm, the smaller will be the logarithm – as Napier came to realize. The number 1 is unthinkable as a base, because logarithms will be undefined or, for 1, indeterminate.

With a base $a > 1$:

$$\log x > 0, \quad \text{if } x > 1$$
$$\log x < 0, \quad \text{if } 0 < x < 1$$
$$\log 1 = 0$$

The following practicable systems of logarithms are used:

$\lg x$	The **common** logarithm of x (to base 10) (also $\log x$, especially in older texts and on electronic calculators)
$\ln x$	The **natural** logarithm of x (to base e)
$\mathrm{lb}\, x$	The **binary** logarithm of x (to base 2)
$\log_a x$	The logarithm of x to the base a (also ${}^a\!\log x$, especially in older texts)

We note the following equivalent notations:

$$\log_a{}^2 x = (\log_a x)^2$$
$$\log_a \log_a x = \log_a (\log_a x)$$

Calculation Rules

Logarithms being exponents, calculation rules for logarithms to all bases are the same as those of exponents:

$a^{m+n} = a^m \cdot a^n$	$\log xy = \log x + \log y\,;$	$x > 0,\ y > 0$
$a^{m-n} = a^m / a^n$	$\log x/y = \log x - \log y\,;$	$x > 0,\ y > 0$
$(a^m)^n = a^{mn}$	$\log x^p = p \log x\,;$	$x > 0$
$\sqrt[n]{a^m} = a^{m/n}$	$\log \sqrt[p]{x} = \dfrac{1}{p} \log x\,;$	$x > 0,\ p \neq 0$

Common Logarithms

Logarithms to the decimal base 10 – symbol lg or log – are known as **common logarithms** or Briggs's logarithms.

A common logarithm may have the form

$$\lg 3300 = 3.5185\ldots$$
$$\lg 0.033 = 0.5185\ldots - 2 = -1.4815\ldots.$$

Mantissa
Characteristic

where the decimal fraction is the **mantissa** and the integer part is the **characteristic** of the logarithm. Both were terms suggested by Henry Briggs in his *Arithmetica logarithmica* (1624).

Mantissa is a late Latin word of Etruscan origin, meaning "addition" or "makeweight" – that is, something added to make up the weight; it later came to acquire also the meaning of "appendix".

The two logarithms quoted above demonstrate between them the unique feature of the mantissa: it is linked exclusively to the sequence, and order, of the digits of the antilogarithm, irrespective of the position of the decimal point, if any.

Similarly, the characteristic has nothing whatever to do with the sequence of digits, but depends exclusively on where the decimal point lies.

The explanation is simple. We write:

$$\begin{aligned} 3300 &= 1000 \times 3.3 \\ \lg 3300 &= \lg 1000 + \lg 3.3 \\ &= 3 + 0.5185\ldots \\ &= 3.5185\ldots \end{aligned} \qquad \begin{aligned} 0.033 &= 3.3 \times 0.01 \\ \lg 0.033 &= \lg 3.3 + \lg 0.01 \\ &= 0.5185\ldots - 2 \\ (&= -1.4815\ldots) \end{aligned}$$

For logarithms of N...	the characteristic is
$N > 1$	one unit less than the number of integer units of N
$N < 1$	negative, one unit greater than the number of leading decimal zeros of N

The method of writing negative logarithms with a separate, negative characteristic facilitates calculation; electronic calculators may use either form of representation.

Natural Logarithms

Natural logarithms – symbol ln – are logarithms to the base e, defined as

limits: *p.* 355 *et seq.*

$$e = \lim_{n \to \infty} \left(1 + \frac{1}{n}\right)^n.$$

In honor of Napier, they are also (wrongly) called Napierian or Napier's logarithms.

The exponential function e^x and logarithms to the base e appear extensively in mathematics and physics. Since e^x is its own derivative, the laws of growth and decay, absorption in optics, acoustics, and radioactivity, vibration and oscillation phenomena in dynamics and electricity, *etc.*, are most easily expressed and worked with in terms of e^x and ln. Otherwise, there are no advantages to the specific use of natural logarithms; as an instance, logarithms to the base 2 are customary in reference to radioactivity.

Logarithms of Negative and Imaginary Numbers

In Chapter 24, we shall define e^z when z is a complex number, and we shall then see that

p. 792

$$e^{i\pi} = -1,$$

so we may take

$$\ln(-1) = i\pi .$$

As the imaginary unit i equals $\sqrt{-1}$, and $\ln\sqrt{-1} = \ln(-1)^{1/2} = (1/2)\ln(-1)$, it follows that we may write

$$\ln i = \frac{i\pi}{2} .$$

In fact, all numbers – real, imaginary, and complex – except 0 have logarithms.

Changing the Base

It is sometimes desirable to change the base of a logarithm.

We have, by definition,

$$a^{\log_a m} \equiv m .$$

The logarithm to the base b of both sides is

$$\log_a m \cdot \log_b a = \log_b m ,$$

from which

$$\log_a m = \frac{\log_b m}{\log_b a} .$$

Example:

To find the natural logarithm of x, if its common logarithm is 2.77815 ..., we have

$$\log_a m = \ln x ;$$
$$\log_b m = \lg x = 2.778\,15 \ldots ;$$
$$a = e ;$$
$$b = 10 ;$$
$$\log_b a = \lg e ,$$

and

$$\ln x \approx \frac{2.778\,15}{\lg e} .$$

Using an electronic calculator, we find $\lg e \approx 0.434\,29$, and

$$\frac{2.778\,15}{0.434\,29} \approx 6.397\,0 .$$

4.6 Mathematical Proof

Greek, *axíoma*, "worth", "quality"
Latin, *postulatum*, "a thing demanded"

Statements that are accepted without discussion or proof are known as **axioms** or **postulates**.

Greek, *hypó*, "under"; *thesis*, "a thing laid down"
Greek, *theórema*, "a subject for contemplation"
Greek, *lémma*, "a thing taken"

In mathematics and other fields of logical reasoning, axioms are used as a basis for the formulation of statements called **premises** or **hypotheses**, which may result in propositions called **theorems**. A **lemma** is an *ancillary theorem* whose result is not the target for the proof.

The Peano Axioms

PEANO Giuseppe
(1858 - 1932)

The natural numbers were axiomatically defined in 1899 by the Italian mathematician and logician **Giuseppe Peano** in his *Aritmetices principia, nova methodo exposita*. Later, Peano modified his axioms to include also zero:

1. Zero is a number.
2. Every natural number or zero, a, has an immediate successor $a + 1$.
3. Zero is not the successor of a natural number.
4. No two numbers have the same immediate successor.
5. The **axiom of induction**: Any property that belongs to zero, and also to the immediate successor of any natural number to which it belongs, belongs to all natural numbers.

Peano's first statement, after the axioms, is the most basic and important numerical sentence of mathematics:

$$1 + 1 = 2.$$

1 + 1 = 2. One of Nicaragua's series of 10 stamps honoring the world's most celebrated mathematical formulas.

Proof by Deduction

A well-argued discussion of hypotheses based on established postulates will lead to a **conclusion** whose **validity** depends exclusively on the validity of the premises and on the correctness of the reasoning, which then constitutes a **proof by deduction** of the thesis put forward.

With deductive reasoning, a conclusion is reached by working from established facts to particulars; that is, what applies for, say, all polynomials in algebra will apply for any and every particular polynomial, and in geometry, what applies for all triangles will equally be true for any and every particular triangle.

Proof by Induction

de MORGAN Augustus (1806 - 1871)

With the inductive method of proof, known from earliest times but first described in 1838 by the British logician and mathematician **Augustus de Morgan**, we base our reasoning on particular cases in order to arrive at a general conclusion.

For an inductive proof to be valid, that is, to be universally true, it must be *complete*, which is to say that what is known to apply for a particular case must be shown to apply for any and every particular case of the given kind.

Proof by mathematical induction, or proof from n to $(n + 1)$, is not a method to discover new formulas or new truths but rather one of reasoning, from an assumption, to results that may confidently be expected to hold true.

To prove a theorem concerning a natural number n by mathematical induction, we must

- establish that the theorem is true for some starting value N;
- assume that it is valid for a certain value $n = p \geq N$ (the induction hypothesis); and
- prove that it then also holds for the next higher value of n.

The theorem is then true for all $n \geq N$.

- Prove by induction that, for all natural numbers n,

$$\sum_{i=1}^{n} i = 1 + 2 + 3 + \ldots + n = \frac{n\,(n+1)}{2}.$$

It is true for $n = 1$, so we start with $N = 1$:

$$1 = \frac{1\,(1+1)}{2}$$

Let us now assume that the statement is true for an arbitrary number $n = p > N = 1$:

$$\sum_{i=1}^{p} i = 1 + 2 + 3 + \ldots + p = \frac{p\,(p+1)}{2}$$

Under the induction hypothesis, the theorem must then also be true for $n = p + 1 = q$:

$$\sum_{i=1}^{p+1} i = 1 + 2 + 3 + \ldots + p + (p+1) = \frac{p\,(p+1)}{2} + (p+1)$$

$$= \frac{p\,(p+1)}{2} + \frac{2\,(p+1)}{2} = \frac{(p+1)\,(p+2)}{2} = \frac{q\,(q+1)}{2}$$

We know that the formula is true for $n = 1$; choosing $p = 1$, it will be true also for $n = 1 + 1 = 2$, $n = 2 + 1$, *etc.*; that is, the formula is true for all integers $n \geq 1 = N$. QED

QED

QED is the abbreviation of *Quod erat demonstrandum*, Latin for "which was to be demonstrated". Euclid, in the 3rd century B.C., used the Greek equivalent: ὅπερ ἔδει δεῖξαι

Proof by *Reductio ad Absurdum*

Indirect Proof

Indirect proof, or proof by *reductio ad absurdum*, establishes the truth of a statement by showing that the *contradiction* of it is false and that, therefore, the statement must be true.

EVCLEIDIS (*c.* 300 B.C.)
PYTHAGORAS (*c.* 500 B.C.)

Let us illustrate by quoting two theorems dating back to the days of Euclid and Pythagoras.

Hypothesis: The number of primes is finite.

Assume that p is the highest prime number in existence. Let P be the finite product of *all* prime numbers,

$$P = 2 \times 3 \times 5 \times 7 \times 11 \ldots \times p\,.$$

Now consider

$$Q = P + 1,$$

which evidently cannot be composite as it would then be divisible by *one or more* of the prime numbers making up the product P. This is impossible, because every such division leaves the remainder 1.

Q must then be a prime number greater than p or divisible by a prime number greater than p. This contradicts the original assumption, which must, therefore, be false.

Hence, the hypothesis is false; the number of primes is infinite.

Hypothesis: The square root of 2 is rational.

If the hypothesis were true, then $\sqrt{2} = p/q$, where p and q are integers, and $2 = p^2/q^2$ so

$$2\,q^2 = p^2.$$

p. 78

By the *fundamental theorem of arithmetic* the integers on the left and right hand sides have unique and identical prime factorizations. But, on the left side $2q^2$ must have an odd number of factors of 2 since q^2 must have an even number; on the right side p^2 must have an even number of factors of 2.

This is a contradiction.

Hence the hypothesis, which is that $\sqrt{2}$ is rational, must be false. So, $\sqrt{2}$ is rational. •

If uncomfortable with indirect proofs, one is in good company with many mathematicians who take such proofs as a challenge to find the positive version of them.

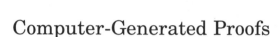

But couldn't medicine, for instance, also benefit from indirect proof? If an investigation was designed to prove that a hypothesis for cause or cure is *false*, one might be spared "truths" that cause worrying, unwarranted hope, and waste of time and money – the miracle of oat bran being just one, curing everything from hemorrhoids and clogged arteries to buck teeth.

Computer-Generated Proofs

Modern proofs involve not only increasingly specialized and complex mathematical methods but oftentimes also high-speed computers, usually generating reams and reams of data. The necessary hands-on verification of such proofs has become cumbersome and time-consuming – sometimes close to unmanageable – causing a dilemma of modern mathematics.

Theorems that long, in some cases hundreds of years, have defied mathematicians have now been proved with the help of

p. 203

the computer – notably the *four-color map theorem*.

The Incompleteness Theorem

There are many propositions in mathematics to which no exceptions have been found and yet, no proofs have been obtained to demonstrate whether they are true or not. As an instance, all perfect numbers known today are even numbers; we have no reason to believe that odd perfect numbers exist, but no proof has been found to demonstrate their nonexistence.

GÖDEL Kurt (1906-1976)

TARSKI Alfred (1902 - 1983)

Before 1931 it was believed that the axioms of arithmetic were consistent and adequate to prove or disprove any mathematical conjecture. However, in 1931 **Kurt Gödel**, Czech-Austrian-American mathematician and logician, published his famous incompleteness, or undecidability, theorem stating that any consistent formal system adequate to describe arithmetic must contain statements which can neither be proved nor disproved within this system. At the same time, **Alfred Tarski**, Polish-American mathematician and logician, studied the notion of truth in formal systems.

Tarski's results together with those of Gödel's show that there is no systematic way to list the true statements in arithmetic.

4.7 Reliability of Digits and Calculations

Approximate Results and Errors

Teacher: *John, what is four plus five?*

John: *... ahm... twelve.*

Teacher: *Wrong. You must stay after class and study!*

John: *Well, ma'am ... haven't you heard of tolerance for some margin of error?*

An **approximate** value is not exact but might be accurate enough for some specific consideration.

Accuracy

The **accuracy** is stated in the magnitude of the absolute or relative error of the approximated value.

Absolute Error

The **absolute error** is the difference between the approximate value and the exact value.

Relative Error

The **relative error** is the absolute value of the quotient of the absolute error divided by the exact value; it is often expressed in percent (parts per hundred, %).

- The exact values $x_1 = 2/3$ and $x_2 = 2/15$ are approximated to $a_1 = 0.67$ and $a_2 = 0.13$, respectively. Determine the absolute, relative, and percent relative errors of a_1 and a_2.

$$e_1 = a_1 - x_1 = \frac{67}{100} - \frac{2}{3} = \frac{3 \times 67 - 100 \times 2}{3 \times 100} = \frac{1}{300}$$

$$|e_1/x_1| = \frac{1/300}{2/3} = \frac{1 \times 3}{2 \times 300} = 0.005 = 0.5\,\%$$

$$e_2 = a_2 - x_2 = \frac{13}{100} - \frac{2}{15} = \frac{15 \times 13 - 100 \times 2}{15 \times 100} = -\frac{1}{300}$$

$$|e_2/x_2| = \left|\frac{-(1/300)}{2/15}\right| = \frac{1 \times 15}{2 \times 300} = \frac{1}{40} = 0.025 = 2.5\,\%$$

Exact value	Approximate value	**Absolute error**	**Relative error**	**Relative error, %**
$x_1 = 2/3$	$a_1 = 0.67$	1/300	0.005	0.5 %
$x_2 = 2/15$	$a_2 = 0.13$	– 1/300	0.025	2.5 %

Although the absolute errors are equal, a_1 is a five times more accurate approximation for x_1 than a_2 is for x_2.

Reliable Digits

When a number is given in decimal notation, the *absolute error* should not exceed a half unit of the last digit retained. The approximate value 1.7 should not vary beyond ± 0.05; the approximate value 1.70 should not vary beyond ± 0.005 .

If the **absolute error does not exceed a half unit** in the last digit, this digit is usually referred to as **reliable**. For instance, if 1299.6 (exact) is rounded to the closest 10 interval, 1300 is obtained with absolute error

$$1300 - 1299.6 = 0.4,$$

indicating that all the digits of 1300 are reliable.

If, on the other hand, 1287 (exact) is rounded to the nearest boundary of the 100 interval in which the number is situated, we again obtain 1300, but this time the absolute error is

$$1300 - 1287 = 13,$$

showing that only the first two digits are reliable; the purpose of the last two digits of 1300 is to define the magnitude of the number.

If results are expressed in **scientific notation,** $p \times 10^n$, it is generally understood that p represents digits that are reliable.

Thus, by expressing 1300 as 1.3×10^3 it is understood that only 1 and 3 are reliable digits in 1300, which can be stated to have an **accuracy of two significant digits**.

Similarly, 1.30×10^3 indicates that 1300 has an accuracy of three significant digits; 1.30×10^{-3} means 0.00130, that is, a number with three significant digits.

Calculation with Approximate Values

When calculating with **exact values**, all significant digits can be reported in the answer. For instance, when adding, subtracting, or multiplying 5000 (exact) and 0.003 (exact), it is correct to state the result as 5000.003, 4999.997, or 15, respectively.

When calculating with **approximate values** or with blended exact and approximate values, it is important to make sure that the result does not include errors amplified by the calculations.

Generally, a result should be rendered with no more significant digits than there are significant digits in the value that could carry the greatest absolute error.

The three numbers 57.1, 3.304, and 34.16 are approximate values with reliable digits, that is, with a maximum absolute error corresponding to one-half of the last digit reported. 57.1 is evidently the term that could have the greatest absolute error, ± 0.05 – compared with ± 0.0005 for 3.304 and ± 0.005 for 34.16. The **sum** should therefore be rounded to *one* decimal:

$$\begin{array}{r} 57.1 \\ 3.304 \\ +\ 34.16 \\ \hline 94.564 \end{array} \Rightarrow 94.6$$

Of the approximate values 1.005 and 0.37, the greater absolute error could be carried by the term 0.37; their **difference** should be given with *two* decimals:

$$\begin{array}{r} 1.005 \\ -\ 0.37 \\ \hline 0.635 \end{array} \Rightarrow 0.64$$

Of the approximate values 3.65 and 0.1020, the former could present the greater absolute error; their **product** and **quotient** should be rounded to *three* significant digits:

$$3.65 \times 0.1020 = 0.3723 \rightarrow 0.372$$

$$3.65 / 0.1020 = 35.784\ldots \rightarrow 35.8$$

The Case for Approximation

$\sqrt{2}$ is an exact notation, but if $\sqrt{2}$ or any other **irrational number** is represented by **nonperiodic infinite decimal fractions** it cannot be presented exactly as a finite decimal expression. The greater the number of decimal digits given, the closer the approximation will approach the true value, but it will never be exact.

Periodic, or repeating, infinite decimal fractions also have an interminable succession of decimals, and also represent approximations of the true value; unlike nonperiodic fractions they can, however, be written as common fractions, and represent exact values.

To convert the infinite decimal fraction 1.166 666 ... to a common fraction x, we write

$$\begin{array}{rcl} 10x &=& 11.6\overline{6} \\ x &=& 1.1\overline{6} \\ \hline 9x &=& 10.50 \end{array} \qquad x = \frac{105}{90} = \frac{7}{6}$$

For convenience, **finite decimal fractions** can be rendered as approximations by limiting the sequence of decimals.

Measurements of physical quantities are generally subject to accidental or systematic errors; these translate themselves into inaccuracies of the results obtained, whose later digits must be regarded with reserve, and may have to be removed by further approximation.

We distinguish between two forms of approximation: **truncation** and **rounding**.

Truncation

Three dots following a row of decimals indicate that further digits have been suppressed. The number π, for instance, can be written with nine correct decimals,

$$\pi = 3.141\,592\,653\ldots,$$

a degree of accuracy beyond the needs of practical calculation. If only six – or four, or two – decimals are required, we obtain by **truncation**,

$$\pi = 3.141\,592\ldots \qquad \pi = 3.141\,5\ldots \qquad \pi = 3.14\ldots,$$

or, replacing the equality sign = by the "approximately equal" sign $\approx$:

$$\pi \approx 3.141\,592 \qquad \pi \approx 3.141\,5 \qquad \pi \approx 3.14$$

Valid Digits

In a truncated expression, all digits are **valid digits**, that is, they agree with those in the unabridged sequence, up to the point of the cut. Thus, a truncated value will always be lower than the exact value.

Rounding

Rounding of a number means replacing it by another number having fewer significant decimal digits or, for integer numbers, fewer value-carrying (non-zero) digits.

Thus, 123.7032 may be rounded successively to

$$123.703 \qquad 123.70 \qquad 123.7 \qquad 124 \qquad 120 \qquad 100,$$

that is, to the nearest thousandth, hundredth, tenth, unit, ten, and hundred, respectively.

Rounding may be carried out in two ways: by **rounding down**, which is equivalent to truncation, and by **rounding up** the last digit to be retained by one unit according to the **rounding rules** of the *International Standard Organization* (ISO), ISO standard 31/0.

The decision whether to round down or to round up depends on the value of the leading digit of the sequence to be rounded off; this is best illustrated by a practical example, using the real number line.

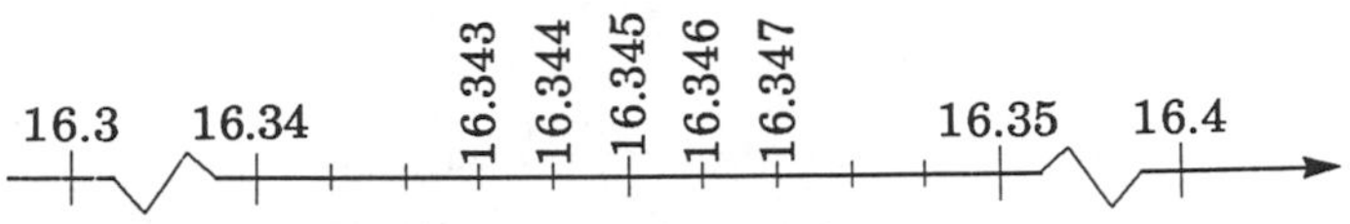

Assume that we are to round the numbers

16.343, 16.344, 16.345, 16.346, 16.347

to the nearest hundredth: 16.34 or 16.35. We find immediately that

- 16.343 and 16.344 shall be rounded down to 16.34;
- 16.346 and 16.347 shall be rounded up to 16.35; and
- 16.345 – halfway between 16.34 and 16.35 – can be rounded either way.

Rounding Errors

If we decide to round all ...5 numbers consistently down or consistently up, we introduce a systematic **rounding error**, negative or positive.

Scientific Rounding

To minimize the risk of systematic errors, ISO advocates the use of **scientific rounding**, that is, all ...5 numbers are to be rounded to the nearest **even** rounding boundary; thus,

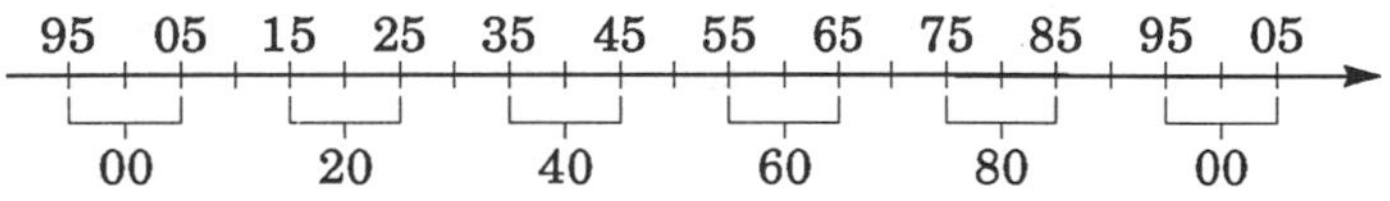

Shopkeepers´ Rounding

In commerce, so-called **shopkeepers' rounding** is the general rule; all "5" pennies, cents, centavos, *etc.*, are consistently **rounded up**, which brings more coppers and nickels into the till.

Error may be introduced if a number that has already been approximated is again rounded. In the example above, 16.347 was – correctly – rounded up to 16.35; if rounded again, 16.35 becomes 16.4, which is evidently wrong, as 16.347 is nearer 16.3 than 16.4.

Thus, it is essential that all rounding processes be executed **in one step**.

A way of guarding against mistakes of this kind is to place a bar under or above a digit 5 that has resulted from a rounding-up operation, *e.g.*,

$$26.146 \approx 26.1\underline{5} = 26.1\bar{5} \approx 26.1\,;$$

a 5 that remains unchanged after a rounding-down operation may be marked by a dot above it, *e.g.*,

$$24.153 \approx 24.1\dot{5} \approx 24.2\,.$$

Calculation with Boundary Values

Maximum accuracy in calculations with approximate values is achieved when **lower and upper boundaries** are observed.

Assume the initial values

$$x = a \pm \Delta a\ ;\quad y = b \pm \Delta b\ .$$

Sum:

The lower boundary value is obtained by using the lower values $(a - \Delta a)$ and $(b - \Delta b)$ of the addends; the upper boundary value by using the upper values $(a + \Delta a)$ and $(b + \Delta b)$:

$$[(a - \Delta a) + (b - \Delta b)] \leq (x + y) \leq [(a + \Delta a) + (b + \Delta b)]\ .$$

Difference:

The lower boundary value is obtained by using the lower value of the minuend $(a - \Delta a)$ and the upper value $(b + \Delta b)$ of the subtrahend; the upper boundary value by using the upper value $(a + \Delta a)$ of the minuend and the lower value $(b - \Delta b)$ of the subtrahend:

$$[(a - \Delta a) - (b + \Delta b)] \leq (x - y) \leq [(a + \Delta a) - (b - \Delta b)]$$

Product:

The lower boundary value is obtained by using the lower values $(a - \Delta a)$ and $(b - \Delta b)$ of the factors; the upper boundary value by using the upper values $(a + \Delta a)$ and $(b + \Delta b)$ of the factors, where the intervals contain only positive values:

$$(a - \Delta a)(b - \Delta b) \leq xy \leq (a + \Delta a)(b + \Delta b)$$

Quotient:

Again, assuming all the numbers are positive, the lower boundary value is obtained by using the lower value $(a - \Delta a)$ of the dividend and the upper value $(b + \Delta b)$ of the divisor; the upper boundary value, by using the upper value $(a + \Delta a)$ of the dividend and the lower value $(b - \Delta b)$ of the divisor:

$$(a - \Delta a)/(b + \Delta b) \leq x/y \leq (a + \Delta a)/(b - \Delta b)$$

Lower boundaries may only be rounded down; upper boundaries only rounded up.

- The approximate values 4.376 and 2.356 are given, whose digits are reliable, that is,

$$\left.\begin{aligned} x &= 4.3760 \pm 0.0005 \\ y &= 2.3560 \pm 0.0005 \end{aligned}\right\} \qquad \left.\begin{aligned} 4.3755 \leq x \leq 4.3765 \\ 2.3555 \leq y \leq 2.3565 \end{aligned}\right\}$$

Find their sum, difference, product, and quotient with their proper boundaries.

Sum:

$$(4.3755 + 2.3555) \le (x + y) \le (4.3765 + 2.3565)$$
$$6.7310 \le (x + y) \le 6.7330$$
$$6.731 \le (x + y) \le 6.733$$

Difference:

$$(4.3755 - 2.3565) \le (x - y) \le (4.3765 - 2.3555)$$
$$2.0190 \le (x - y) \le 2.0210$$
$$2.019 \le (x - y) \le 2.021$$

Product:

$$(4.3755 \times 2.3555) \le x\,y \le (4.3765 \times 2.3565)$$
$$10.30649\ldots \le x\,y \le 10.313\,22\ldots$$
$$10.306 \le x\,y \le 10.313$$

Quotient:

$$(4.3755 / 2.3565) \le x / y \le (4.3765 / 2.3555)$$
$$1.856779\ldots \le x / y \le 1.857\,991\ldots$$
$$1.8568 \le x / y \le 1.8580$$

The boundary value for sheep
Counts only when grass isn't deep.
With head, more or less,
It can just be a guess,
Rounding up, rounding down as they leap.

4.8 Simple Calculating Devices

The Abacus

[An abacus! Wow! It can count!]

Counting on the fingers and writing in the sand at one's feet were early forms of manual counting. The first device actually invented to assist in arithmetic calculations was probably the *counting board*, or *abacus*, which appeared independently in various forms and in several parts of the ancient world.

The early counting boards were tablets or trays of sunbaked clay, or boards or blocks of wood, which were spread with a thin layer of fine sand or dust in which the symbols were traced. After a calculation had been completed and was no longer needed, the designs could be easily effaced leaving the board clean – *tabula rasa* – in readiness for the next operation.

Notable improvements of the counting board were the use of more lasting materials, such as marble with the Greeks and bronze with the Romans, and the provision of parallel grooves in which markers in the form of pebbles or small balls could be placed to serve as number tokens, symbolizing ones, tens, hundreds, *etc.*

The Romans introduced a further improvement by providing additional grooves between the existing ones, intended for the "fives" of the Roman number system, making it ideally suited to the Roman way of counting, which helped to maintain the Roman numbers in operation for several hundred years – well through the Middle Ages.

The origin of the word *abacus* can be traced to the Arabic *abq* – meaning "dust" or "fine sand" – which became *abax* for "sand tray" in Greek, and *abacus* in Latin. The Latin word for "pebbles" is *calculi* (plural of *calculus*), a diminutive form of *calx*, which means "stones".

To add 2769 (MMDCCLXVIIII) and 1987 (MDCCCCLXXXVII) on the counting board, proceed in the following manner.

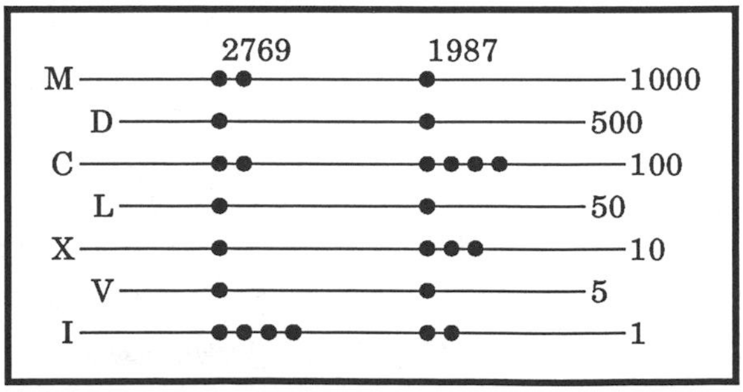

Begin by assembling all counters of the two numbers on the left; this adds up to six I (1); two V (5); four X (10); two L (50); six C (100); two D (500); and three M (1000).

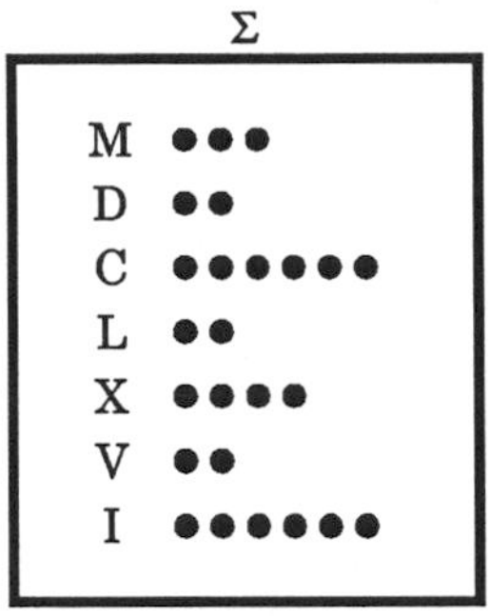

Counter arrangement, intermediate

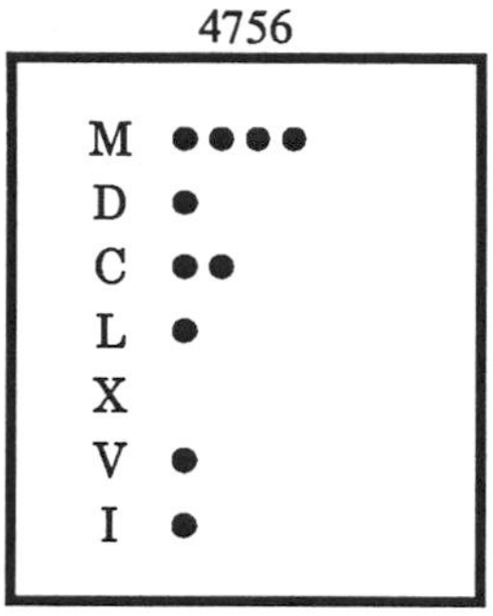

Counter arrangement, final

Abace, fac!
[Get on with it!]

- Convert five I (1) to one V (5); carry it to the V groove, making three V.
- One I (1) counter is left in the units groove.
- Convert two V to one X (10), carry it to the X line, making five X counters.
- One V (five) remains in the "fives" groove.
- Carry the five X as one L (50) to the L groove, making three L counters.
- No X (10) counter remains on the tens line.
- Convert two L (50) to one C (100), carry it to the C groove, making seven C counters.

Difficile erit facile!
[Difficult gets easy!]

- One L (50) remains in the "fifties" groove.
- Convert five C (100) to one D (500), move it to the D line, making three D counters.
- Two C (100) counters are left in the hundreds groove.

Facilior quam facilis!
[Easier than easy!]

- Convert two D (500) to one M (1000) and carry it to the M groove.
- One D (500) counter remains on the "five-hundreds" line.
- There are now four M (1000) on the thousands line.

Rapide factum!
[Swiftly done!]

The sum is MMMM D CC L V I = 4756 .

Suan-pan
Soroban

The form of abacus that we know today developed first in the Far East – the *suan-pan* in China as early as the 11th century, and the *soroban* in the 14th century in Japan. They consist essentially of a wooden frame in which are mounted a number of thin bamboo or metal rods, each with nine or ten colored beads, corresponding to the counters of the counting board. As many beads as there are units, tens, hundreds, *etc.*, in the numbers to be entered are collected in groups at the ends of the rods.

The method of adding and subtracting with the abacus is the same as for the counting board. Instructions how to use it for multiplication and division, and how to extract square and cube roots, can be found in old Chinese and Japanese works, and in some early Western books, notably the 1610 edition of Robert Recorde's *Ground of Artes*.

The abacus used today in Far Eastern countries has a frame divided into two sections, one with four "Earth" beads on each rod and the other with a single "Heaven" bead, equivalent to five beads. The Russian abacus, the *s'choty*, has ten beads on each rod. The beads shown below may represent the whole number 1 096 503 or a decimal fraction.

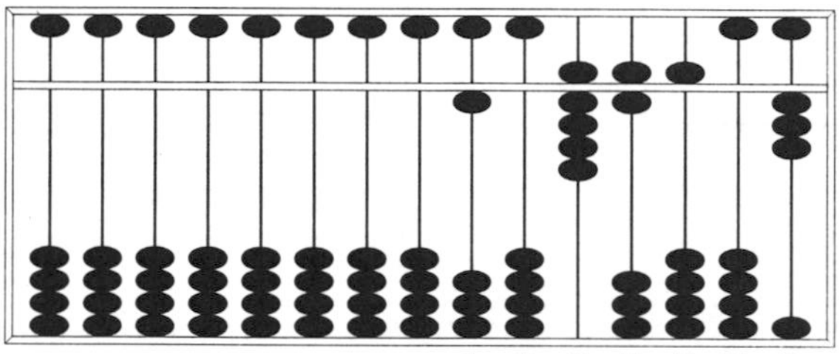

Modern soroban

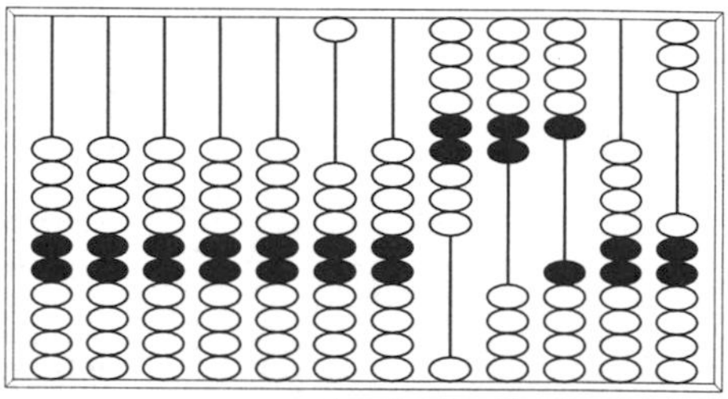

Russian s'choty

The use of the abacus for multiplication and division is described here for the Japanese soroban, which operates with two kinds of counters in the form of balls or beads: Heaven counters move down from above, and Earth counters come up from below, to meet and be read at a "horizon" line.

そろばんを
弾きましょう！

Soroban o hajikimasho!

[Let's work the soroban!]

Multiplication

がんばりま
しょう！

Gambarimasho!

[Give it a good try!]

Task: Multiply 23 x 613

Begin by placing the multiplier 23 at the left end of the soroban, the multiplicand 613 to the right of it; and reserve space for a five-digit product at the right end of the counting board.

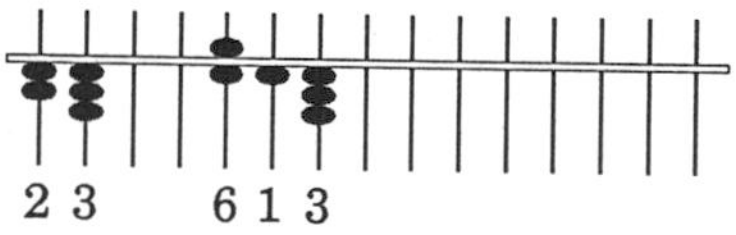

23 x 613

1. Multiply the units digit of the factor 23 by the units digit of the factor 613: $3 \times 3 = 9$

   Place counters to the value of nine units – one Heaven counter plus four Earth counters – on the units line of the product.

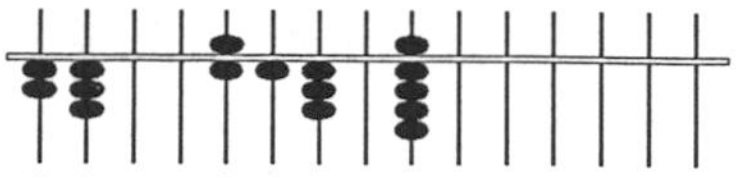

23 x 613
• •

2. Multiply the tens digit of the factor 23 with the units digit of 613: 2(0) x 3 = 6(0)

Add one Heaven and one Earth counter on the tens line of the product; then, having completed operations with the units position 3 of the multiplicand 613, remove that digit.

よくできました。

Yoku dekimashita.

[Good!]

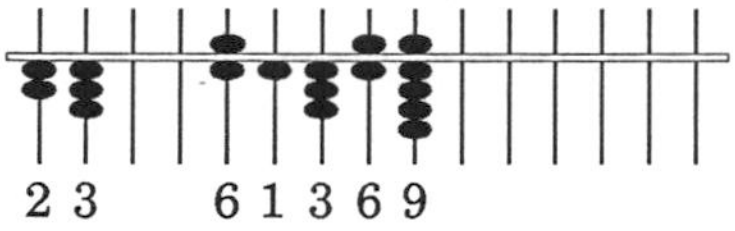

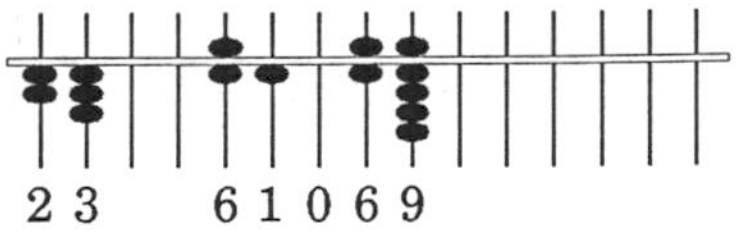

23 x 613 23 x 613
• • • •

3. Multiply the units and tens digits of 23 by the tens digit of 613: 3 x 1(0) = 3(0); 2(0) x 1(0) = 2(00)

Add three Earth counters on the product tens line, and two on the hundreds line; then, having completed operations with the tens position 1 of the multiplicand 613, remove that digit.

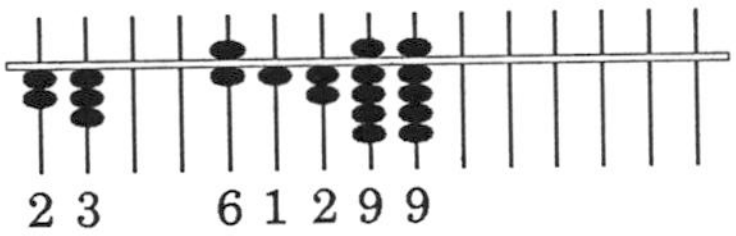

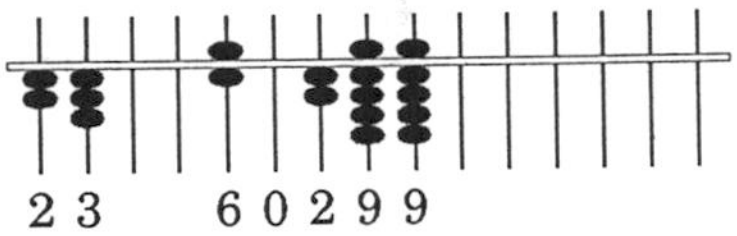

23 x 613
• •

4. Multiply the units digit of 23 by the hundreds of 613: 3 x 6(00) = 18(00); 18 = 20 – 2

Adding 18 on the hundreds line is equivalent to adding two counters on the thousands line and removing the two counters on the hundreds line.

注意！

Choi!

[Pay attention here!]

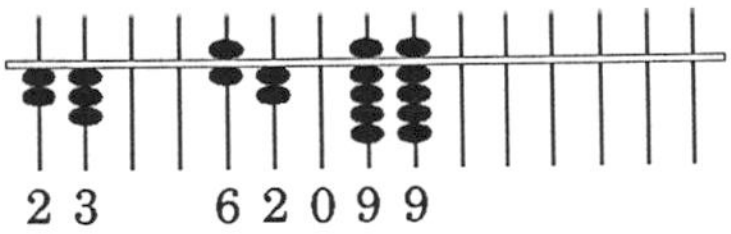

23 x 613
• •

5. Multiply the tens of 23 by the hundreds of 613: 2(0) x 6(00) = 12(000)

Operations completed with the hundreds position 6 of the multiplicand 613, remove that digit. Add two counters on the thousands line, and one on the ten-thousands line.

簡単ですね。

Kantan desu ne?

[Isn't this easy?]

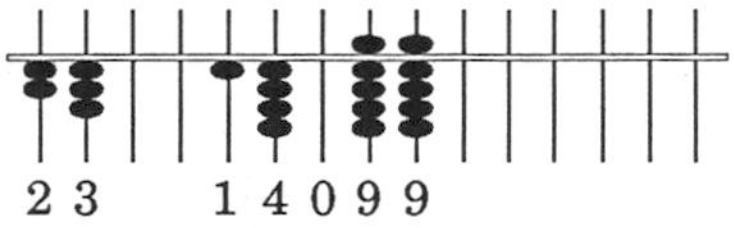

6. The sought product is 14 099.

よくできました！

Yoku dekimashita!

[Yes, you did it!]

割り算をやってみましょう！

Warizan o yatte mimasho!

[Let's try division!]

Division

Task: 377 / 26

Place the divisor 26 to the left of the dividend 377. Beginning with the units position of the 377, count off toward the left the number of columns which are contained in 26; the second column (▲) to the left of those columns will be the units position of the quotient.

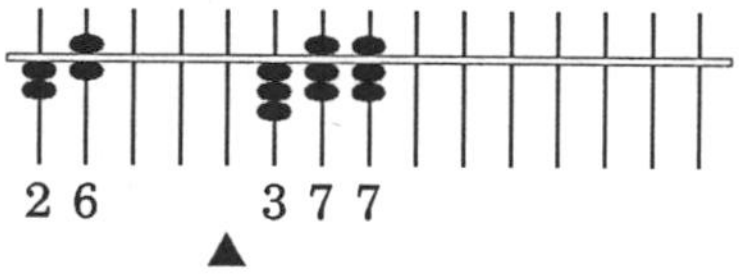

We shall use a method that follows ordinary longhand division.

1. 26 goes into 37 once. Raise one Earth bead in the column to the left of the units position of the quotient.

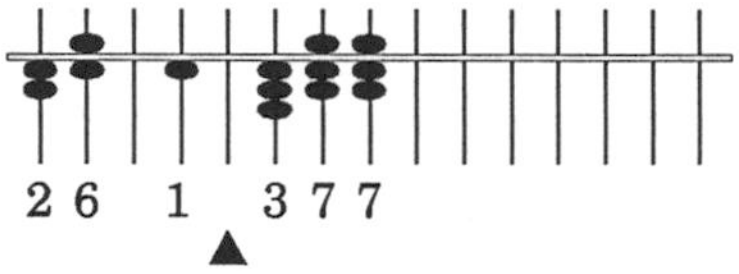

2. Multiplying the temporary quotient, 1, by the first digit of the divisor 26 gives 1 x 2 = 2; return two Earth beads from the hundreds position in 377.

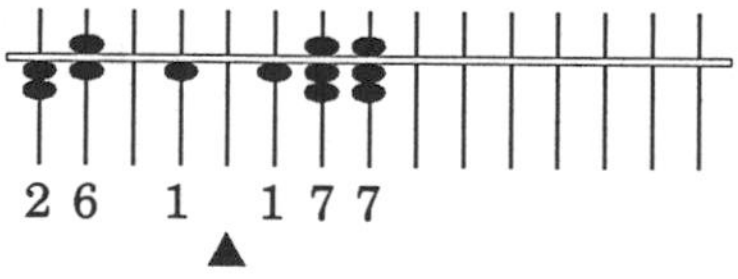

3. Multiplying the 1 of the temporary quotient by the second digit of 26 gives 1 x 6 = 6; return one Heaven and one Earth bead from the tens position in the dividend 377.

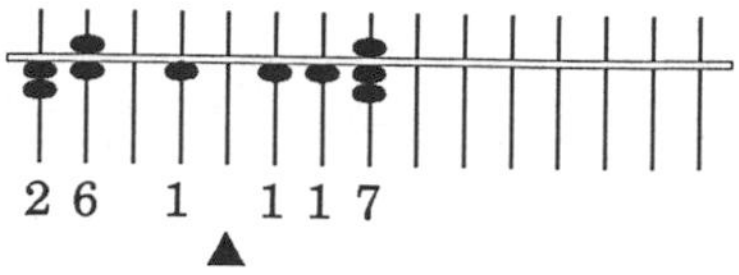

上手ですね。

Jozu desu ne.

[You are very skillful.]

4. As 26 does not go into 11, include one more digit to the right; 26 goes four times into 117. Raise four Earth beads in the column of the units position in the temporary quotient.

わかりましたか。

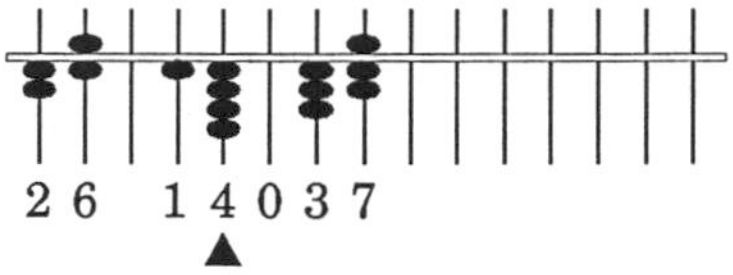

Wakarimashita ka?

[Did you understand it?]

5. Multiplying the 4 of the temporary quotient by the second digit of 26 gives 4 x 6 = 24; return two Earth beads from the tens position and four from the units position in 377.

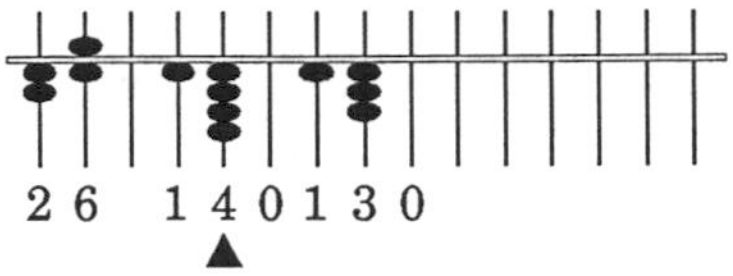

6. As 26 does not go into 13, include a 0 immediately to the right of the divisor 26; 26 goes five times into 130. Lower one Heaven bead in the column immediately to the right of the units position in the quotient.

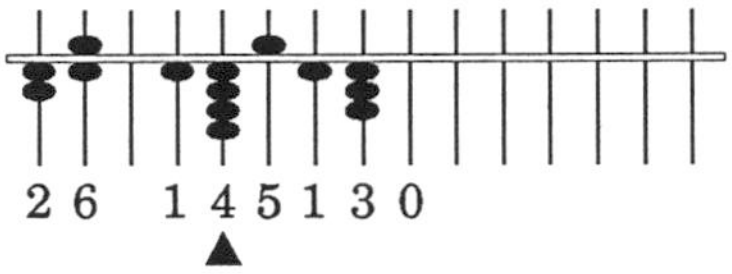

終り

Owari!

[Finished!]

7. Multiplying the 5 of the quotient by the second digit of 26 gives 5 x 6 = 30; return three Earth beads from the tens position in 377. Multiplying the 5 by the first digit of 26 gives 5 x 2 = 10; return one Earth bead from the hundreds position in 377.

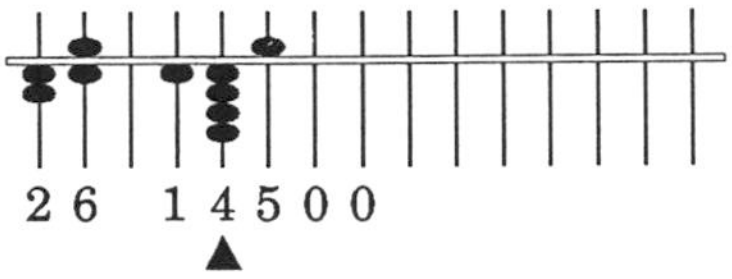

答えは同じでしたか。

Kotae wa onaji deshita ka?

[Did you get the same answer?]

8. The sought quotient is 14.5 .

A disadvantage common to all abaci is that a calculation can never be checked except by doing it all over again, or by keeping a record of it on paper. In old times, paper was handmade from rags and consequently very expensive – hardly within reach for most people. Also, although the Chinese had known for about a thousand years how to make paper, it was not introduced into Europe until the 12th century and did not become common in the Western world until the 19th century.

Among those who master the abacus, its use is still often preferred for basic operations – even over electronic calculators. The abacus is very useful for teaching children the fundamental operations of addition and subtraction, and the meaning of place value in position arithmetic.

Napier's Bones

NAPIER John
(1550 - 1617)
Greek: *rhabdós*, "staff", "rod"

Many mechanical devices have been invented for carrying out specifically multiplication, among them Napier's bones, also called Napier's rods, described by **John Napier**, Scottish mathematician and inventor of logarithms, in his *Rabdologiae* in 1617.

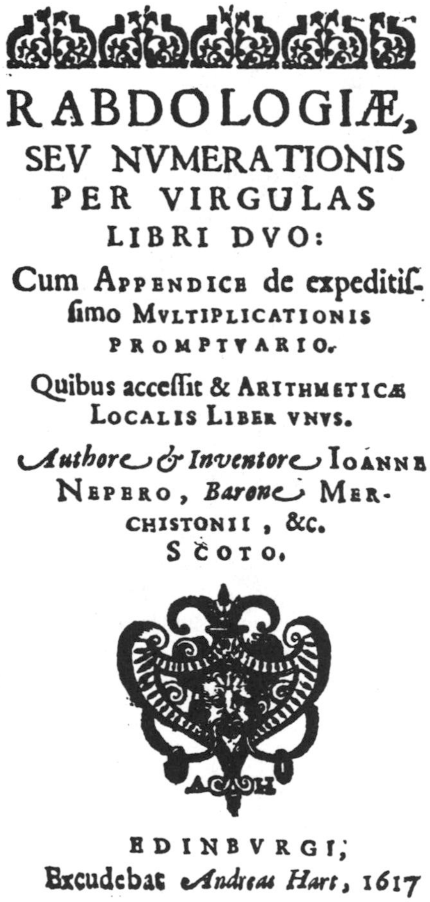

RABDOLOGIÆ,
SEV NVMERATIONIS
PER VIRGULAS
LIBRI DVO:
Cum Appendice de expeditiſſimo Mvltiplicationis
Promptvario.
Quibus acceſſit & Arithmeticæ
Localis Liber vnvs.
Authore & Inventore Ioanne
Nepero, *Barone* Merchistonii, &c.
Scoto.

EDINBVRGI,
Excudebat *Andreas Hart*, 1617

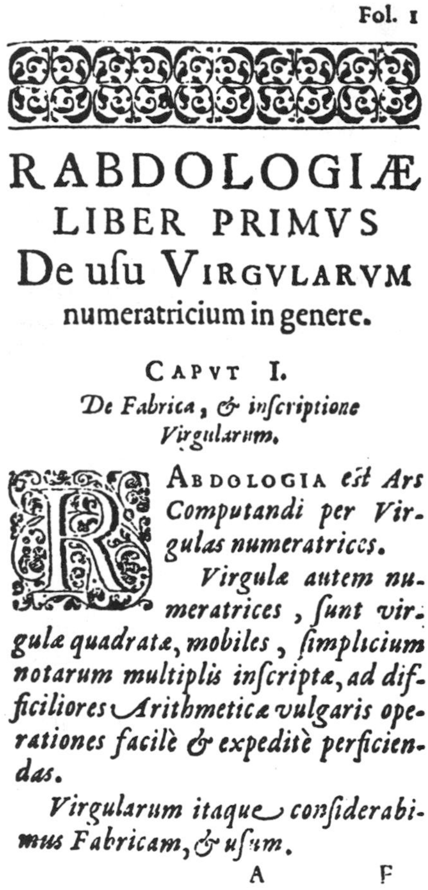

Fol. 1

RABDOLOGIÆ
LIBER PRIMVS
De uſu Virgvlarvm
numeratricium in genere.

Capvt I.
De Fabrica, & inſcriptione Virgularum.

Abdologia *eſt Ars Computandi per Virgulas numeratrices.*
Virgulæ autem numeratrices, ſunt virgulæ quadratæ, mobiles, ſimplicium notarum multiplis inſcriptæ, ad difficiliores Arithmeticæ vulgaris operationes facilè & expeditè perficiendas.
Virgularum itaque conſiderabimus Fabricam, & uſum.
A F

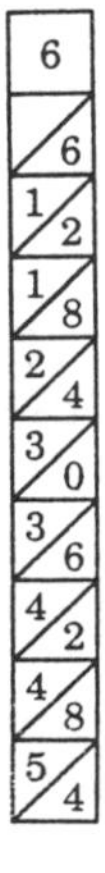

Napier's rods constitute in essence a mechanical multiplication table.

Multiplications were performed with the aid of rectangular strips of bone, wood, metal, or some other suitable material. A column of the multiples of the digit heading the rod was inscribed or engraved on each rod, with a diagonal line separating the tens – to be "carried" – and the units, as shown in the drawing.

To illustrate the use of the rods, we choose the multiplication of 1615 by 365 – the example described by Napier in *Rabdologiae*. Place rods 1 - 6 - 1 - 5 beside each other as shown; the results of the multiplication of 1615 by 3, 6, and 5 are found to be 4845, 9690, and 8075, respectively – representative of the partial products 484 500, 96 900, and 8075, which are then added in the usual manner.

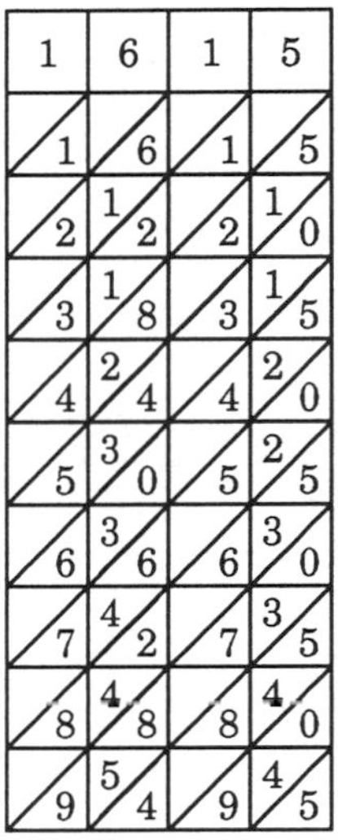

$3 \cdot 1615 = 4845$

$5 \cdot 1615 = 8075$

$6 \cdot 1615 = 9690$

```
   1615
 x  365
 ------
   8075
  9690
 4845
 ------
 589475
```

The Slide Rule

History

Mathematical computations can be performed with the help of graduated scales which measure numbers as distances. In their simplest forms, linear scales are used for direct addition and subtraction of numbers. Linear scales have equal distances between consecutive integer graduations.

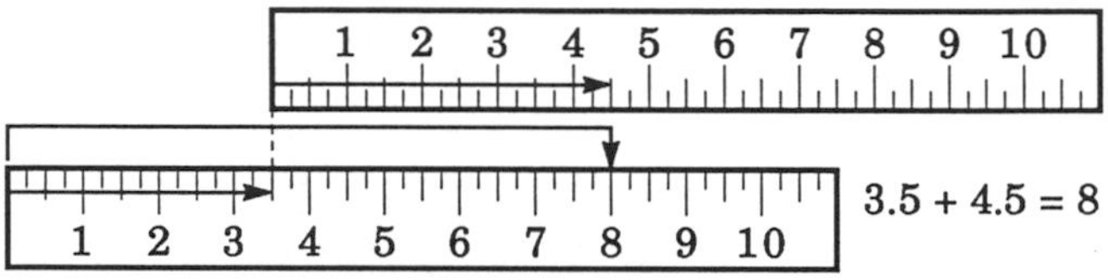

3.5 + 4.5 = 8

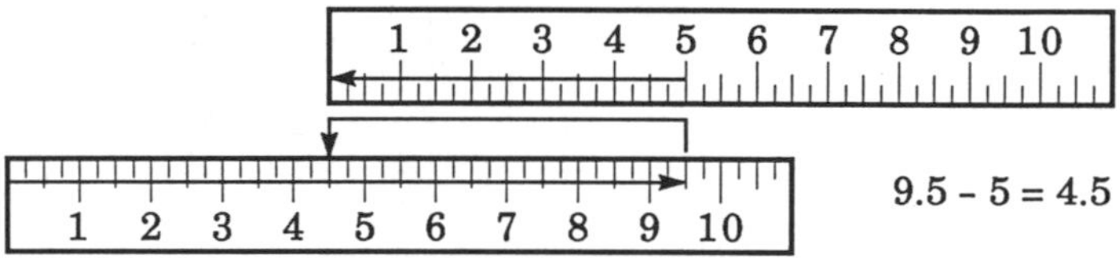

9.5 - 5 = 4.5

Soon after Napier's publication of *Mirifici* in 1614, logarithmic scales were beginning to be used for mechanical computation. The drawing below shows a logarithmic scale to base 10 matched with a *linear* scale, which reads the logarithms of the numbers on the *logarithmic* scale.

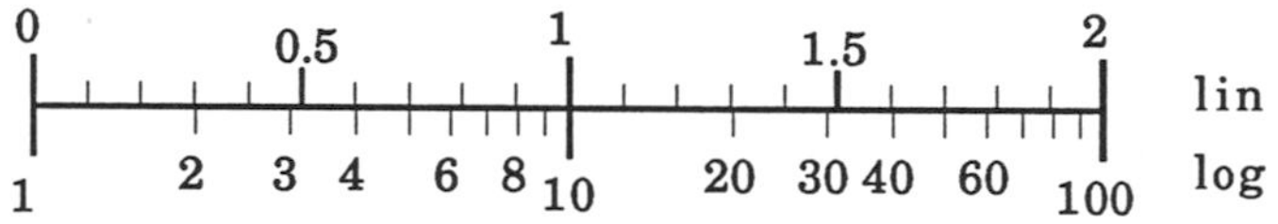

GUNTER Edmund (1581 - 1626)

In 1620, **Edmund Gunter**, English mathematician and designer/maker of navigation instruments, constructed a "line of numbers" on which distances proportional to the logarithms of numbers were marked off. Distances were added and subtracted on the scale with the help of compasses, so that multiplication and division of numbers could be performed.

OUGHTRED William (1574 - 1660)

The English mathematician **William Oughtred** invented, in 1621, a rectilinear slide rule with two movable logarithmic scales, and a circular slide rule, a device having two circular logarithmic scales moving against each other to permit the addition and subtraction of logarithms.

DELAMAIN Richard (? - *c.* 1645)

In 1630, Oughtred's former pupil **Richard Delamain** – a joiner by trade but a capable mathematician, tutor to King Charles I – published a 32-page pamphlet, *Grammologia, or the Mathematicall Ring*, in which he presented the circular slide rule as his own invention. He had sent his pamphlet to the king in 1629 for approval before publishing it.

William Oughtred then published his invention, in 1632, as *The Circle of Proportion and the Horizontal Instrument*, which started a lifelong quarrel between them. Delamain countered, in the same year, with *The Making, Description and Use of a Small portable Instrument for the Pocket ... Called a Horizontal Quadrant.* The war of words and pamphlets went on, several people joining in, till the death of Delamain in the civil war about 1645.

Oughtred's priority to the rectilinear slide rule has never been in doubt, however, and his two logarithmic scales are the undisputed precursors of the C and D scales of modern slide rules.

ROGET Peter Mark (1779 - 1869)

The development of the slide rule gave it a few more logarithmic scales, including a reciprocal scale, a cubic scale, and trigonometric scales; in 1815, **Peter Mark Roget** – known from *Roget's Thesaurus* – added so-called loglog scales for calculations of and with exponential functions.

Experiments were made with better material – seasoned wood from pear trees was found to be particularly suitable – and the development of improved engraving machines raised the accuracy of slide rules.

Many new scales were introduced during the decades immediately following World War II, arranged on both the front and the back of stock and slide in an ingenious manner to maintain constant connection with other scales, especially the basic C and D scales.

The advent of fast, low-cost, handheld electronic calculators has largely forced the engineer's beloved slide rule to beat a retreat into the science museums. But the satisfaction of owning an electronic calculator and the fun of experimenting with its buttons hardly measure up to the pride felt possessing and mastering an advanced slide rule!

Berke Breathed; Washington Post Writer's Group; © 1987

Construction

A slide rule consists of a stock, a slide (or slider), and a transparent cursor (or runner) with a hairline to facilitate the reading of the settings, or a system of hairlines, as the case may be, for cooperation between the scales.

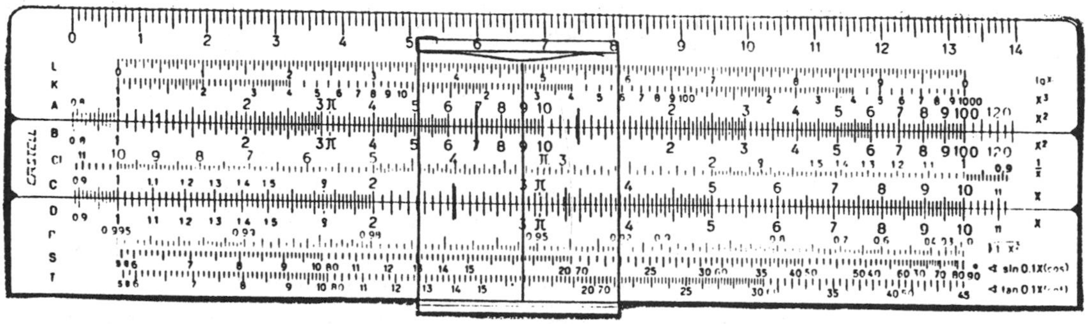

The stock and the slide carry several scales for a variety of purposes:

C D are the basic scales – C on the slide and D on the stock. They are graduated 0 - 10 in logarithmic measure: if the length of the scales is 250 mm – a normal value – the distance from point 1, the origin of the scale, is,

to point 2, 250 x lg 2 = 75.26 mm
to point 3, 250 x lg 3 = 119.28 mm
to point 5, 250 x lg 5 = 174.74 mm.

A B These scales – A on the stock and B on the slide – are also logarithmic, and graduated 1 - 100; this means that every number on scale A, or B, is the square of a corresponding number on scale D, or C, permitting direct squaring of any number and extraction of square roots.

These computations may, of course, be incorporated into longer sequences of operations by developing some ingenuity on the part of the user.

CI This standard scale, on the slide, is the reverse of the C scale, running from right to left; the letter I stands for **inverse**.

DI AI BI More sophisticated slide rules may also have one or more reciprocal scales DI, AI, and BI.

K This scale, a fixed scale on the stock, is calibrated logarithmically 1 - 1000, making it represent the cubes of numbers on the D scale.

K' Some slide rules may also have a moving cubic scale K' on the slide.

S The S scale – S for *sinus* – has a double function. Read from left to right, numbers in black represent the sines of angles from 5° 8 to 90°; read right to left, numbers in red represent cosines of angles from 0° to 84° 2 . Sines and cosines are read off the D scale.

P The P scale – where P stands for *Pythagorean* – gives the cosine of an angle when the sine is known, and the sine from a known cosine. It is used in conjunction with the S scale to find better values of sine and cosine.

T The T scale – T for *tangens* – is used to find tangents of angles; black numbers refer to angles from 5° 8 to 45°, red numbers to angles 45° - 86°. Tangents (and cotangents) are read on the D scale and the CI scale.

ST Some slide rules have an ST scale, also known as an arc scale. For very small angles, the sine and tangent are very nearly equal, and equal to the arc of the angle, in radian measure.

Cursor hairline on ST 2° 25 reads

sin = tan = arc = 0.0395.

lg *x* Most slide rules also have a fixed *linear* scale on the stock, which corresponds with the D scale and gives the mantissas of D numbers.

LL Most slide rules have at least one, generally three LL – for loglog – or exponential scales, used for computing powers with fractional exponents, solving exponential equations, and calculating the values of hyperbolic functions.

In their heyday, slide rules were designed for the most diversified special uses – for mechanical strength calculations of truss bridges, steel wire ropes, and reinforced concrete structures; for electric engineers designing generators and motors, and building power transmission lines; for architects and surveyors; for economists and businesspeople; and even simplified versions ... for schoolteachers and physicians.

Advanced Deluxe Slide Rules

There were also double-sided *deluxe* slide rules for those who demanded the utmost in mathematical versatility. Besides all the usual scales mentioned above, they had several of the special scales described below.

CF DF One side of the stock and slide of the slide rule was provided with F – for *folded* – scales; if C and D were graduated in x, scales CF and DF showed πx, and went from π through 10 to 10π.

Besides providing scales for the handy calculation of all relations involving π, the folded scales gave the additional advantage that you never risked getting "out of bounds" with multiplication.

CI DI on these slide rules were reciprocal scales graduated in $1/x$ or $10/x$.

CIF is a reciprocal scale of $1/\pi x$ or $10/\pi x$.

W 1
W 1'
W 2
W 2' For increased accuracy, these scales have twice the length of the corresponding C and D scales by being divided into parts; W1 and W1' are graduated in $\sqrt{x}$, W2 and W2' in $\sqrt{10x}$.

LL The most advanced slide rules featured up to eight exponential scales e^x and e^{-x}, covering powers from 0.000 01 to 0.999 1 on four scales and from 1.000 9 to 100 000 on another four scales.

Calculations

For maximum accuracy, multiplication and division made with a slide rule should be carried out an scales C and D, whenever possible. Operations 82 x 62 and 0.082 x 620 are performed in the same way, and provide answers that are just an array of digits; the place of a decimal, if any, must be determined separately.

Let's dust off a slide rule and illustrate with a few problems.

Find product 2.1 x 3.45 .

- Place 1 of scale C over 2.1 on scale D.
- Move cursor hairline to C 3.45 .
- Read answer, approx. 7.25 on D scale.

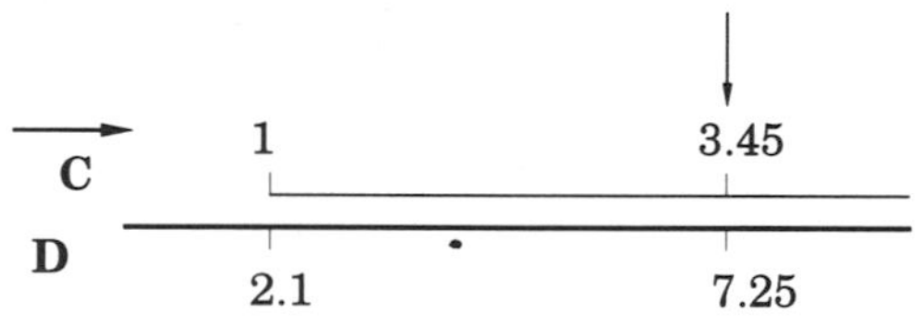

The exact answer is 7.245 .

Find product 3.2 x 6.3 .

If C 1 is placed at D 3.2, the factor C 6.3 will extend beyond the end of the D scale. To avoid this,

- Place cursor hairline on D 3.2 .
- Place CI 6.3 under hairline to *divide* by 1 / 6.3 .
- Read answer, approx. 202, on D below C 1.

A rough estimate gives 3 x 7 = 21 to place the decimal point: 20.2. The exact answer is 20.16 .

Find quotient 19.2 / 89 .

- Place C 1 on D 1.92 .
- Move hairline to CI 8.9, to *multiply* by 1 / 89 .
- Read answer, approx. 216, on D under hairline.

A rough estimate is 20 / 90 $\approx$ 0.2. The answer is 0.2157... .

Find quotient 7.1 / 3.4 .

- Place cursor hairline on D 7.1 .
- Place C 3.4 under hairline.
- Read answer, approx. 2.09, on D at C 1 .

The result is based on lg (7.1 / 3.4) = lg 7.1 – lg 3.45.

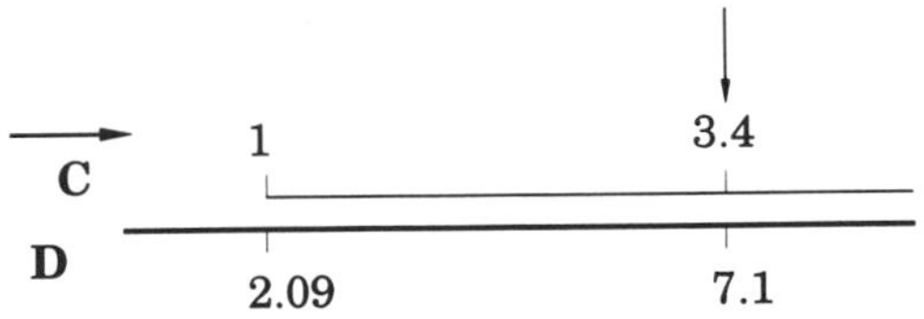

(The answer is 2.088... .)

Find product 312 x 27 x 45 x 175 x 26 .

- Place hairline on D 312 .
- Place CI 27 under hairline.
- Move hairline to C 45 .
- Place CI 175 under hairline.
- Move hairline to C 26 .
- Read answer, approx. 172 on D under hairline.

The answer is approximately $1.72 \cdot 10^9$.

Find $\dfrac{15 \times 36 \times 37}{23 \times 29}$.

- Place hairline on D 15 .
- Place C 23 under hairline.
- Move hairline to C 36 .
- Place C 29 under hairline.
- Move hairline to C 37 .
- Read answer 29.9 on D under hairline.

Find $\dfrac{24 \times 87 \times 37}{34 \times 11}$.

Rearrange factors to $\dfrac{37}{34} \cdot \dfrac{24}{11} \cdot 87$.

- Place hairline on D 37 .
- Place C 34 under hairline.
- Move hairline to C 24.
- Place C 11 under hairline.
- Move hairline to C 1.
- Place CI 87 under hairline.
- Read answer, approx 206.6, on D at C 1 .

Find $19 \cdot \sqrt{35}$.

- Place hairline on A 35 .
- Place CI 19 under hairline.
- Read answer, approx. 112.4, on D scale.

Find $28 \cdot \sqrt[3]{240}$.

- Place hairline on K 240 .
- Place CI 28 under hairline.
- Read answer, approx. 174.0, on D scale.

The Quipu

Quipu

The **quipu** is generally associated with the Incas, who ruled Peru before the Spanish conquest. Similar devices were used by several other Indian tribes and are also described in Chinese and Persian documents from the 6th and 5th centuries B.C.

The Inca quipu consists of a number of color-coded cords. Knots of various kinds are tied on these cords to represent a variety of information. The Incas had no written language; instead, the administration of their mighty empire depended heavily on the use of the quipu.

An important use of the quipu was to *record* numbers, as in trade, keeping accounts, and for calendars, but the knotted strings might also have served as mnemonics for the recording of important historical events, astronomical data, mythology, and legends.

There is evidence to support the theory that the knots and the cords of the old Inca quipu were in a decimal system of numeration: A knot tied farthest away from the cord designated units, a nearer knot specified the next position in the system of numeration, *etc.*; absence of a knot symbolized zero; larger knots were multiples of smaller ones.

Still, today, devices related to the Inca quipu are used by shepherds in the Andes for keeping an account of their herds.

Chapter 5

COMBINATORICS

5.0 Historical Notes

Combinatorics is the name of a branch of mathematics that is concerned with the selection of objects – called **elements** – from a given set of elements.

Combinatorics traces its history back to ancient times when it was often closely associated with number mysticism, as in the Chinese book *I Ching* from 2200 B.C.

In the Western world, interest in combinatorial mathematics was awakened in the 17th and 18th centuries and largely stimulated by questions being raised about odds in gambling and other games of chance, which led to the creation and development of **probability theory**.

GALILEI Galileo (1564 - 1642)

de FERMAT Pierre (1601 - 1665)

PASCAL Blaise (1623 - 1662)

von LEIBNIZ Gottfried Wilhelm (1646 - 1716)

BERNOULLI Nicolaus (1695 - 1726)

BERNOULLI Daniel (1700 - 1782)

EULER Leonhard (1707 - 1783)

KEPLER Johannes (1571 - 1630)

The Italian astronomer, mathematician, and physicist **Galileo Galilei** studied the relative probability of various sums of points occurring when rolling dice. Other important contributors in the field of probability theory were the Frenchmen **Pierre de Fermat** and **Blaise Pascal**, the German **von Leibniz**, and the Swiss brothers **Nicolaus** and **Daniel Bernoulli**.

In 1736, the Swiss mathematician **Leonhard Euler** solved the celebrated problem of the Königsberg bridges – the question whether it would be possible to make a tour of the city and return to the starting point by crossing all of its seven bridges just once; this feat marks the beginning of **graph theory**.

Kepler's Conjecture

Combinatorial geometry includes problems of covering, packing, and symmetry. A celebrated conjecture in packing theory, posed in 1611 by **Johannes Kepler**, is that the most compact way to stack spheres is into four-sided pyramids. This method has been used for the display of fruit long before Kepler's time, but the mathematical proof is as yet elusive.

Combinatorial methods are widely used in most branches of mathematics and in many areas of applied mathematics; they are of special interest in **probability** and **statistics**, and in **computer science**.

As with most branches of mathematics, the exact scope of combinatorics cannot be defined; only in textbooks does it exist as an independent subject.

Saint Peter's Game

In a well-renowned pub in Scotland a party had assembled consisting of fifteen whiskey drinkers and fifteen temperance apostles – emissaries of the Guild of Whiskey Distillers and the League of Teetotalitarians who were arguing volubly about the excellence or perniciousness, as the case might be, of the pub's drinkables – to which, understandably, the publican took exception.

So he ordered everybody on the premises to form a line, counted off nine several times, booting every "niner" out before resuming his counting ... until there were finally only fifteen emissaries left and – lo and behold! – they were all whiskey drinkers. – Uisghe Beata!

Now, how did the publican manage this little piece of fiddling?

Our story is a variation of the "not so chancy" *Ludus Sancti Petri*, or Saint Peter's Game, known from medieval times – typically a ship was struck by a heavy storm and, to lighten and possibly save the vessel and some of her passengers, half the passengers were thrown overboard. In Europe the fifteen favored persons were most often Christians, the other fifteen Jews. St. Peter, who usually presided, arranged the passengers in such a way that the Christians were saved.

In a Jewish version, Saint Peter was replaced by the Spanish Jew Ibrahim ben Meir Ezra – a foremost mathematician, astronomer, and translator of science from Arabic to Hebrew – who saves his fifteen pupils, the other fifteen passengers being drowned. An Arab version favors the Muslims and throws the infidels overboard; an East Indian story lets fifteen good men be saved and fifteen thieves be drowned. A Japanese land-based version describes a stepmother who wishes to deprive her stepchildren of their inheritance.

Beginning with four favored (f) and five unfavored (u), the complete arrangement is

4 f; 5 u; 2 f; 1 u; 3 f; 1 u; 1 f; 2 u; 2 f; 3 u; 1 f; 2 u; 2 f; 1 u.

Speakers of English could memorize this arrangement – if able to spell – from the order of vowels in a mnemonic rhyme, where $a = 1$, $e = 2$, $i = 3$, $o = 4$, and $u = 5$,

From numbers' aid and art,
Never will fame depart.

5.1 Multiplication Principle

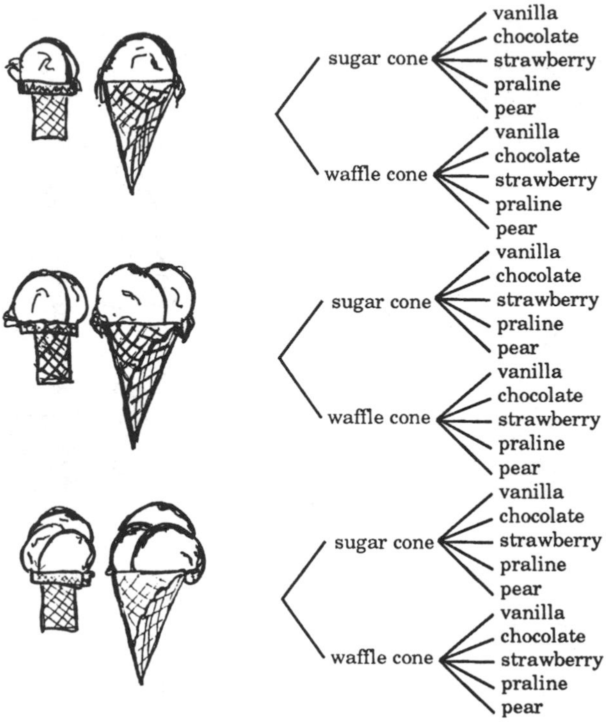

The following selections of ice cream cones are available:

A: 1, 2, or 3 scoops, all the same flavor

B: Sugar cone or waffle cone

C: Vanilla, chocolate, strawberry, praline, or pear flavor

Tree Diagram (Product Tree, Factorial Tree)

The **tree diagram** (**product tree**, or **factorial tree**) illustrates the number of possibilities.

As the illustration shows, if no flavors are mixed the total number of varieties available is

$$3 \cdot 2 \cdot 5 = 30 .$$

This is an example of the **multiplication principle**:

> If k different selections can be made in succession, and the first selection can be made in n_1 ways, the second selection in n_2 ways, the third selection in n_3 ways, *etc.*, then the total number of selections is
>
> $$n_1 \cdot n_2 \cdot n_3 \cdot \ldots \cdot n_k .$$

***

Handshakes; Tennis Matches

Sixty-four tennis players meet at a welcome party before the start of a tournament. They greet each other with handshakes; how many? And how many matches will be played to find a winner?

Answer: 2016 handshakes; 63 matches

Ordered Sample

If the order of selection is noted, as in the example below, the collection is called an **ordered sample**.

- ## *Dinner Menu*

Appetizers
- Potato Skins
- Escargot
- French Onion Soup
- Shrimp Cocktail
- Oysters on the Half Shell
- Marinated Mushrooms

Entrees
- Broiled Orange Roughy
- Chicken Kiev
- New York Strip Steak
- Breaded Jumbo Shrimp
- Filet Mignon
- Crab Louie
- Fillet of Sole
- Fettuccine Alfredo

Desserts
- Black Forest Torte
- Chocolate Chambard Cake
- Cherry Cheesecake
- Deep Dish Apple Pie
- Cheese and Fruit Plate
- Vanilla Ice Cream
- Peppermint Sherbet

When the waiter asks if you are ready to order, you answer that you want a three-course meal but want to study all the possible selections before you order.

How many three-course meals can you compose from the menu?

6 appetizers $\cdot$ 8 entrees $\cdot$ 7 desserts

= 336 three-course dinner selections.

(The waiter won't like you very much.)

- In days past, a physician made house calls in four villages **A**, **B**, **C**, and **D**.

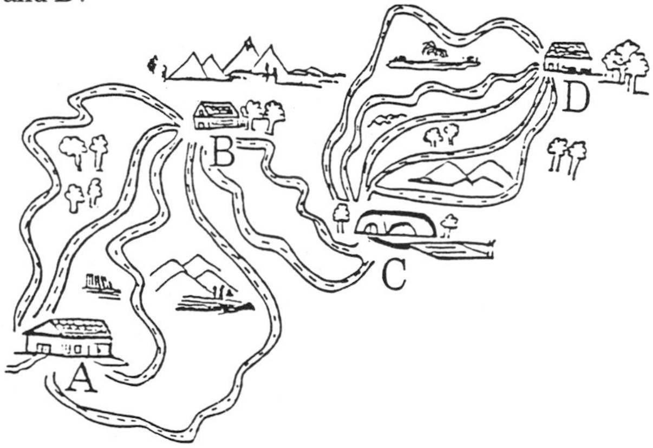

Find in how many ways the physician can travel

a: between **A** and **D**, and

b: from **A** to **D** and back to **A**

a:

$4 \cdot 2 \cdot 5 = 40$ different ways.

b:

$4 \cdot 2 \cdot 5 \cdot 5 \cdot 2 \cdot 4 = 1600$ different ways.

5.2 Permutations

Left to right. top row: Arthur Tylor Winston, Tylor Winston Arthur, Winston Arthur Tylor. *Bottom row:* Tylor Arthur Winston, Winston Tylor Arthur, Arthur Winston Tylor.

Latin, *per*, "thoroughly"
mutare, "to change"

Permutations are *ordered arrangements* of a finite number of elements, either of all of the available n elements or of a part p of them. Two permutations that contain exactly the same elements but not in the same order are regarded as different.

The notation

$${}_nP_p$$

reads, "The number of permutations of n elements, selected p at a time" or "The number of ordered p-tuples that can be formed from n elements".

Equivalent notations are ${}^nP_p$, P_p^n, and $P(n, p)$.

Size
Tuple

The number of elements (p) used at a time is the **size** of the **tuples** (permutation).

Braces are used for unordered sets, *parentheses* for ordered tuples, thus: {**A**, **B**, **C**} = {**B**, **C**, **A**}; (**A**, **B**, **C**) ≠ (**B**, **C**, **A**).

There are six possible permutations of the three letters of the triple {**A**, **B**, **C**}:

(**A**, **B**, **C**); (**A**, **C**, **B**); (**B**, **A**, **C**); (**B**, **C**, **A**); (**C**, **A**, **B**); (**C**, **B**, **A**) ; ${}_3P_3 = 6$,

that is, the number of permutations of all the elements of the collection of 3 elements equals 6.

And here are the six ordered pairs taken from {**A**, **B**, **C**}:

(**A**, **B**) (**A**, **C**) (**B**, **A**) (**B**, **C**) (**C**, **A**) (**C**, **B**) ; ${}_3P_2 = 6$.

Mapping
Map
Functions: Chapter 10

We think of the permutation of (**A**, **B**, **C**) to (**B**, **A**, **C**) as an **action**, **function**, or **mapping** that takes a triple and transposes its first two elements. This allows us to compose several permutations by following one action by another and gives a dynamic element to the study.

The **map** that sends **A** to **B** and **B** to **A**, and leaves **C** fixed, may be displayed:

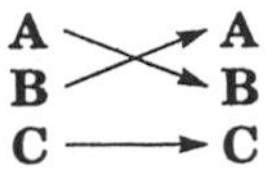

All Elements Distinct

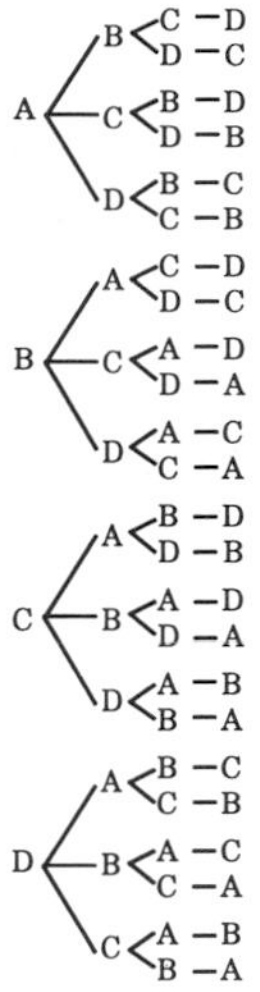

The first element of a permutation of n elements can be chosen in n different ways, the second element in $(n - 1)$ ways, the third in $(n - 2)$ ways, *etc.*, until all n positions have been filled. Thus, the number of permutations that are possible with n elements is the product of all natural numbers from 1 to n, inclusive, that is, ***n* factorial**,

$$_nP_n = n(n-1)(n-2)\dots 3\cdot 2\cdot 1 = n!$$

In order to find all possible permutations of a given set of elements, proceed systematically: for numerical elements in their natural order, by magnitude; for letter arrangements, in lexicographical order.

The tree diagram on the left shows that {**A**, **B**, **C**, **D**} has 24 permutations if all four letters are selected in every instance.

The diagram illustrates that

$$_4P_4 = 4\times 3\times 2\times 1 = 4! = 24.$$

- The word NUMBERS has seven letters, all different. When all letters are selected in every instance, we have

$$_7P_7 = 7!\,(= 1\times 2\times 3\times 4\times 5\times 6\times 7) = 5040 \text{ permutations.}$$

- If 12 gates are available for 12 racehorses, the number of ways the gates can be assigned is

$$12! = 479\,001\,600.$$

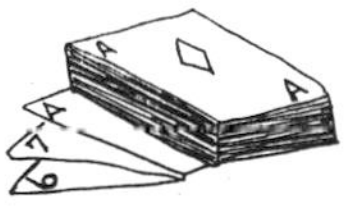

- The 52 cards of a conventional deck of playing cards can be arranged in

$$52\cdot 51\cdot 50\dots 3\cdot 2\cdot 1 = 52!$$

$$= 80\,658\,175\,170\,943\,878\,571\,660\,636\,856\,403\,766\,975\,289\,505\,440\,883\,277\,824\,000\,000\,000\,000$$

different ways.

Permuting the cards at the improbable rate of one permutation a second round the clock, year in and year out, would require $2.5\cdot 10^{60}$ years to run through the entire lot. This is more than 10^{50} times the age of the Universe. •

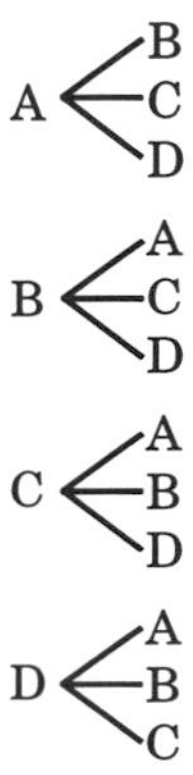

Sometimes a **limited number** of elements are to be selected from the original collection. The tree diagram on the left demonstrates the 12 permutations of the collection {**A, B, C, D**} when only two letters are selected:

$$_4P_2 = 4\times 3 = 12$$

The **general formula** for the number of permutations of a collection when **all elements are different** is

$$_nP_p = n(n-1)\dots[n-(p-1)] = \frac{n!}{(n-p)!}$$

where n and p are positive integers with $n \geq p$.

For the above (**A**, **B**, **C**, **D**), when only two letters were selected, we have

$$_4P_2 = 4 \cdot 3 = \frac{4\,!}{(4-2)\,!} = \frac{24}{2} = 12,$$

which shows that the formula works.

Why? Let n elements be available for p positions.

At the second selection, $n - 1$ elements remain to be placed in $p - 1$ positions.

For the second element, $n - 1$ selections can be made, for the third $n - 2$ selections, *etc.*; for the last element the number of selections is $n - (p - 1)$.

We have

$$_nP_p = n\,(n-1)\,(n-2)\,\ldots\,[n - (p-1)]\,,$$

which may be rearranged to

$$_nP_p = n\,(n-1)\,(n-2)\,\ldots\,(n-p+1)\,.$$

Rewriting,

$$_nP_p = \frac{n(n-1)\,(n-2)\,\ldots\,(n-p+1)\,[(n-p)!]}{(n-p)!}$$

gives

$$_nP_p = \frac{n!}{(n-p)!}\,,$$

which is the **general formula** for the number of permutations of a collection when **all elements are different**.

- When in NUMBERS only two letters are selected at a time, the number of permutations is

$$_7P_2 = \frac{7\,!}{(7-2)\,!} = 42\,.$$

- In how many ways can the starting order be posted in an eight-member relay cross-country skiing team if

  a: all 8 members will take part in a race;

  b: only 4 members – chosen from the 8 members – will take part?

a:

$8\,! = 40\,320$ different ways.

b:

$\frac{8\,!}{(8-4)\,!} = 1680$ different ways.

- A group of Australian businessmen, traveling by air, plans a visit to eight states in the continental U.S.

In how many ways can the order vary if

a: the states may be visited in any order;

b: the group decides to visit New York State first;

c: the group decides to start in New York State and finish in California?

a:

8 ! = 40 320 permutations.

Thus, the order of visiting the eight states may vary in 40 320 different ways.

b:

Since the first state has already been chosen, we are left with the number of permutations of seven states.

7 ! = 5 040 different ways.

c:

Since the first as well as the last state to be visited have already been chosen, we are left with the number of permutations of six states, all six taken at a time:

6 ! = 720 different ways.

• In how many ways can seven people be placed

a: at a bar counter;

b: at a round table?

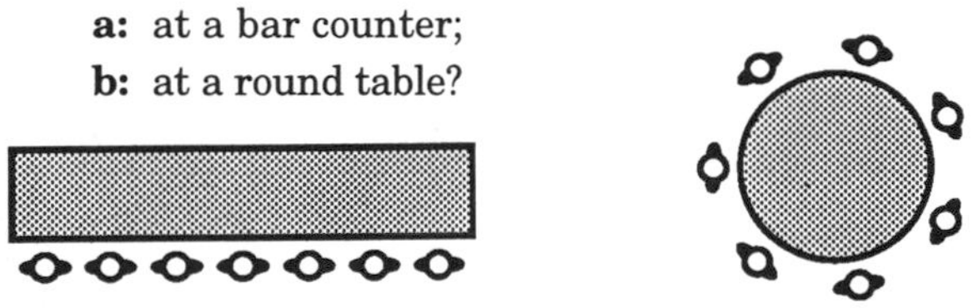

a:

7 ! = 5 040 different ways.

b:

Rings

Making a *ring* instead of a row is equivalent to starting out with one element whose position has already been determined.

6 ! = 720 different ways.

Inversions

If two elements in a permutation of distinct elements are in reverse order relative to their normal or natural order, they constitute an **inversion**.

Even
Odd

A permutation is said to be **even** if it contains an even number of inversions; it is **odd** if the number of inversions is odd.

The number of transpositions that are required to return a sequence of elements to their natural order is even or odd according to the number of inversions in the arrangement.

In the permutation $(c\ e\ d\ b\ a)$,

element c	precedes	b and a	$\Rightarrow$	2 inversions
e	precedes	d, b, and a	$\Rightarrow$	3 inversions
d	precedes	b and a	$\Rightarrow$	2 inversions
b	precedes	a	$\Rightarrow$	1 inversion,

making *eight* inversions in all; hence, $(c\ e\ d\ b\ a)$ is an even permutation.

A transposition interchanges two *elements* in a permutation.

A transposition always alters the number of inversions by an odd number: if c and a in $(c\ e\ d\ b\ a)$ are interchanged, we have a new permutation $(a\ e\ d\ b\ c)$, where

element e	precedes	d, b, and c	$\Rightarrow$	3 inversions
d	precedes	b and c	$\Rightarrow$	2 inversions,

making *five* inversions, *three less* than before.

Transposing two *adjacent elements* alters the number of inversions by one unit. Changing reverse-order elements $d\ b$ to natural-order $b\ d$ permutes $(a\ e\ d\ b\ c)$ to $(a\ e\ b\ d\ c)$, where

element e	precedes	b, d, and c	$\Rightarrow$	3 inversions
d	precedes	c	$\Rightarrow$	1 inversion,

making *four* inversions, *one less* than the former five inversions.

If, instead, we reverse the normal-order $a\ e$ to $e\ a$, in the new permutation $(e\ a\ d\ b\ c)$,

element e	precedes	a, d, b, and c	$\Rightarrow$	4 inversions
d	precedes	b and c	$\Rightarrow$	2 inversions,

making a total of *six* inversions, one more than the former five inversions.

Cyclic Permutations

The shifting of an entire ordered sequence of elements one or more steps forward or backward – the first element taking the position of the last, or vice versa – without changing the order of the elements in the sequence is a **cyclic permutation**:

$$(a\,b\,c\,d\,e) \Rightarrow (b\,c\,d\,e\,a) \Rightarrow (c\,d\,e\,a\,b) \Rightarrow (d\,e\,a\,b\,c) \Rightarrow (e\,a\,b\,c\,d) \Rightarrow (a\,b\,c\,d\,e)$$

Degree

The number of elements in the collection being permuted is the **degree** of the permutation.

Transposition

A cyclic permutation of degree two – $(a\ b) \Rightarrow (b\ a)$ – is a **transposition**.

The Fifteen Puzzle

This popular puzzle was invented in the latter part of the 19th century. Initially in random order and with the empty space in the lower right-hand corner, the fifteen counters are to be rearranged in numerical order.

There are more than $20 \cdot 10^{12}$ possible permutations but only half of them admit a solution: the puzzle can be solved only if the *total* number of inversions in a given display is *even*. This is because each move is a transposition of the blank (we may think of it as 16) with some other square.

2	1	3	4
8	5	6	7
9	10	11	12
15	13	14	

Solvable

2 precedes 1 ⇒ 1 inversion
8 precedes 5, 6, 7 ⇒ 3 inversions
15 precedes 13, 14 ⇒ 2 inversions
Total: 6 inversions

1	2	3	4
5	6	7	8
9	10	11	12
13	14	15	

Solved

No inversions

1	2	4	3
8	5	6	7
9	10	11	15
12	13	14	

Unsolvable

4 precedes 3 ⇒ 1 inversion
8 precedes 5, 6, 7 ⇒ 3 inversions
15 precedes 12, 13, 14 ⇒ 3 inversions
Total: 7 inversions

Rubik's Cube

Of course, a recent craze – enlightenment or frustration – is **Rubik's cube**, consisting of 26 colored, rotatable cubelets (no cubelet at the center) whose faces are of different colors, yet every cubelet is similar, giving more than 10^{19} possible arrangements (permutations). The challenge is to arrange the cubelets to form a cube with every face displaying only one of the six colors.

Identical Elements

The number of permutations is reduced when a collection contains **identical elements**. The number of permutations of n objects, q of one type, r of another, s of another, *etc.* to a *total of p objects*, is given by

$$\frac{n\,!}{q\,!\;r\,!\;s\,!\;\ldots}\,.$$

- Find the number of distinct permutations of the letters in

  a: MISSISSIPPI **b:** MATHEMATICS **c:** COMBINATORICS

  when all the original letters are used once each.

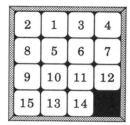

a: In MISSISSIPPI the letter I occurs 4 times, S 4 times, P twice, M once.

$$\frac{11\,!}{(11-11)\,!\;4\,!\;4\,!\;2\,!\;1\,!} = \frac{11\,!}{0\,!\;4\,!\;4\,!\;2\,!\;1\,!}$$

$$= 34\,650 \text{ permutations.}$$

b: For MATHEMATICS,

$$\frac{11\,!}{(11-11)\,!\;2\,!\;2\,!\;2\,!\;1\,!\;1\,!\;1\,!\;1\,!\;1\,!}$$

$$= \frac{11\,!}{0\,!\;2\,!\;2\,!\;2\,!\;1\,!} = 4\,989\,600 \text{ permutations.}$$

c: For COMBINATORICS,

$$\frac{13\,!}{0\,!\;2\,!\;2\,!\;2\,!\;1\,!} = 778\,377\,600 \text{ permutations.}$$

- On analysis, the active part of an insulin molecule is shown to consist of 30 amino-acid links, bound together in a chain-like manner with two free ends. The number of different amino-acid links is:

3 Glycine	1 Serine	1 Glutamine	3 Phenylalanine
1 Alanine	2 Threonine	1 Lysine	2 Tyrosine
3 Valine	1 Aspartic acid	1 Arginine	2 Histidine
4 Leucine	2 Glutamic acid	2 Cysteine	1 Proline

Find the number of permutations for linking the amino acids together.

$$\frac{30\,!}{(30-30)\,!\;3\,!\;1\,!\;3\,!\;4\,!\;1\,!\;2\,!\;1\,!\;2\,!\;1\,!\;1\,!\;1\,!\;2\,!\;3\,!\;2\,!\;2\,!\;1\,!}$$

$$\approx 1.598\,987\,629 \times 10^{27} \text{ permutations.}$$

- In the Welsh island Anglesey, a few miles west of Bangor, there is a village with the quaint name

Llanfairpwllgwyngyllgogerychwyrndrobwll-llantysiliogogogoch,

which means, in English,

> The Church of St. Mary in the hollow of white hazel near the rapid whirlpool, and the Church of St. Tysilio of the red cave.

The fifty-eight letters of the name include eleven l's, seven g's, six o's, five y's, four each of n, r, and w, three each of a and i, two each of c and h, and one each of b, d, e, f, p, s, t –

$$\frac{58\,!}{(1\,!)^7\,(2\,!)^2\,(3\,!)^2\,(4\,!)^3\,5\,!\,6\,!\,7\,!\,11\,!} \approx 6.793\,215 \cdot 10^{55}$$

– clearly a name that will appeal to people with acute permutationitis.

Latin Squares

Standard Form

A Latin square of the n-th order is a permutation of n symbols arranged in n rows and n columns; it is in **standard form** if the first row and first column consist of the natural (original) order:

A	*B*	*C*	*D*
B	*A*	*D*	*C*
C	*D*	*A*	*B*
D	*C*	*B*	*A*

Diagonal Latin Square

If the diagonals also have a permutation of the n symbols, it is called a **diagonal Latin square**:

A	*B*	*C*	*D*
C	*D*	*A*	*B*
D	*C*	*B*	*A*
B	*A*	*D*	*C*

Mutually Orthogonal

Euler Square

If all pairs of symbols are different when two Latin squares of the same order are superimposed, the Latin squares are **mutually orthogonal** and the square formed is known as an **Euler square**:

A	*B*	*C*	*D*		*a*	*b*	*c*	*d*		*Aa*	*Bb*	*Cc*	*Dd*
B	*A*	*D*	*C*	+	*c*	*d*	*a*	*b*	=	*Bc*	*Ad*	*Da*	*Cb*
C	*D*	*A*	*B*		*d*	*c*	*b*	*a*		*Cd*	*Dc*	*Ab*	*Ba*
D	*C*	*B*	*A*		*b*	*a*	*d*	*c*		*Db*	*Ca*	*Bd*	*Ac*

BOSE Raj Chandra (b. 1901)

SHRIRKHANDE S. S.

PARKER Ernest Tilden (b. 1926)

In 1959 **Bose**, **Shrirkhande**, and **Parker**, working in the U.S., demonstrated that Euler squares exist for any order, except two and six.

Latin squares are of importance in the design of experiments that will be subjected to statistical analysis. If four species of peas A, B, C, D are to be tested with four separate fertilizers a, b, c, d, the plots can be laid out in the combined way shown, so each cultivar and each fertilizer appear in each row and column.

***

Home Cooked Combinatorics

A few minutes before their train was due to leave, six people ordered six eggs, over easy, one minute frying time each side. The cook at the Home Cooked Combinatorics Café, a great fan of Martin Gardner, had one frying pan, with room only for four eggs. She needed only three minutes to carry out the order; how?

Beginning first minute: She fried eggs 1, 2, 3, and 4 on one side;

End first, beginning second minute: She turned eggs 1 and 2 over, put 3 and 4 aside, and placed 5 and 6 into the pan;

End second, beginning third minute: Eggs 1 and 2 done, removed; 3 and 4 now returned to the pan, sunny side down; 5 and 6 were turned over;

End third minute: Eggs 3, 4, 5, and 6 done.

5.3 Combinations

Combinations are concerned only with the **selection** of objects from a collection and, unlike permutations, **disregard order**.

The current mathematical terminology is unfortunate and confusing, since in everyday language the word *combination* often conveys a sense of order. For instance, the *combination* of a combination lock is, in mathematical terms, not a combination but simply a specified order of digits.

We know that the six possible *permutations* of two letters from the sequence **A**, **B**, **C** are

$$(\mathbf{A}, \mathbf{B}) \quad (\mathbf{A}, \mathbf{C}) \quad (\mathbf{B}, \mathbf{A}) \quad (\mathbf{B}, \mathbf{C}) \quad (\mathbf{C}, \mathbf{A}) \quad (\mathbf{C}, \mathbf{B})\,.$$

The above (**A**, **B**) and (**B**, **A**) are made up from the same *combination*, the unordered set {**A**, **B**}. Likewise, (**A**, **C**), (**C**, **A**) both arise from {**A**, **C**}, and (**B**, **C**), (**C**, **B**) from {**B**, **C**} Consequently, the three possible *combinations* of two letters from the sequence **A**, **B**, **C** are:

$$(\mathbf{A}, \mathbf{B}) \quad (\mathbf{A}, \mathbf{C}) \quad (\mathbf{B}, \mathbf{C})\,.$$

The notations

$$_nC_p\ ; \quad \binom{n}{p}$$

both denote the number of combinations, and read

"The number of combinations of n elements selected p at a time" or, in short, "n choose p"; or, in the case of the latter notation, simply "n over p".

Equivalent notations are ^nC_p, C_p^n , and $C(n, p)$.

Binomial Coefficient *p.* 140

Because $\binom{n}{p}$ is the number of ways we can choose p x's from the factors $(1 + x)^n$, it is the **binomial coefficient** for x^p in this expansion.

Size

The number (p) of elements used at a time is the **size** of the combination.

We found that the collection {**A**, **B**, **C**, **D**} has 24 *permutations* when all 4 elements are selected at a time. Since the order is disregarded in combinations, the collection {**A**, **B**, **C**, **D**} has only one *combination* when all 4 elements are selected at a time. If, on the other hand, only 3 elements are selected at a time, there are 4 *combinations*:

$$\{\mathbf{A}, \mathbf{B}, \mathbf{C}\} \quad \{\mathbf{A}, \mathbf{B}, \mathbf{D}\} \quad \{\mathbf{A}, \mathbf{C}, \mathbf{D}\} \quad \{\mathbf{B}, \mathbf{C}, \mathbf{D}\}$$

Since each combination of a collection has the potential of forming permutations, there is a *simple relation between combinations and permutations.*

In the collection {**A**, **B**, **C**, **D**}, the 4 *combinations*

{**A**, **B**, **C**} {**A**, **B**, **D**} {**A**, **C**, **D**} {**B**, **C**, **D**}

would each form 3! = 6 ordered triples, giving a total of 4 x 6 = 24 ordered triples formed from {**A**, **B**, **C**, **D**}. We obtain the relation

$$_nP_p = ({_nC_p})\, p! \quad \text{or} \quad {_nC_p} = \frac{_nP_p}{p\,!}.$$

Each of the $_nC_p$ combinations may be ordered in $p\,!$ ways to make p-tuples, and each of these p-tuples can be completed in $(n - p)\,!$ ways to make a permutation of n. So,

$$n\,! = {_nC_p}\,p!\,(n - p)\,!$$

and we now have the general formula for the number of combinations of a collection of **different elements** :

$$_nC_p = \frac{n\,!}{(n-p)\,!\;p\,!}\;;\quad \binom{n}{p} = \frac{n\,!}{(n-p)\,!\;p\,!},$$

where n and p are positive integers with $n \geq p$.

We have

$$\binom{7}{3} = \frac{7\,!}{(7-3)\,!\;3\,!} = 35\,;\quad \binom{9}{0} = \frac{9\,!}{(9-0)\,!\;0\,!} = 1\,.$$

- What is the total number of matches in a round-robin tennis tournament with thirteen contestants?

Every one of the 13 contestants meets 12 other players and plays a total of 12 matches. Each meeting is mutual between 2 players, so the total number is

$$\frac{13 \cdot 12}{2} = 78 \text{ matches}$$

or, using the formula,

$$\binom{13}{2} = \frac{13\,!}{(13-2)\,!\;2\,!} = \frac{13 \cdot 12}{2} = 78\,.$$

- Find the number of combinations in the word NUMBERS, selecting at a time

a: 2 letters; **b:** 6 letters.

a:

$$_7C_2 = \frac{7\,!}{(7-2)\,!\;2\,!} = 21 \text{ combinations.}$$

b:

$$_7C_6 = \frac{7\,!}{(7-6)\,!\;6\,!} = 7 \text{ combinations.}$$

- A committee-crazed organization has 18 members; every committee is composed of 4 members, and no same 4 members may form more than one committee.

Find the maximum number of committees.

$$\frac{18\,!}{(18-4)\,!\;4\,!} = 3060 \text{ committees.}$$

- 12 men and 14 women apply for a total of 8 assignments in a research project that will take place in a secluded underground site, placing extreme demands on congeniality and cooperation between all participants.

  Find the number of possible combinations of the work crew if
  a: there is no stipulated men-women quota;
  b: as many men as women should be employed.

  a:
  All applicants may be considered indistinct.

  $$\frac{(12+14)!}{[(12+14)-8]!\ 8!} = 1\,562\,275 \text{ combinations.}$$

  b:
  Select the women and men separately, then apply the multiplication principle.

  $$\left(\frac{14!}{(14-4)!\ 4!}\right)\left(\frac{12!}{(12-4)!\ 4!}\right) = 495\,495 \text{ combinations.}$$

- 18 mice were placed in two experimental groups and one control group, with all groups equally large. In how many ways can the animals be placed into the three groups?

  There are 18 animals to select from for the first group, 12 for the second, and 6 for the third. To find the total number of combinations, the multiplication principle is applied.

  $$\left(\frac{18!}{(18-6)!\ 6!}\right)\left(\frac{12!}{(12-6)!\ 6!}\right)\left(\frac{6!}{(6-6)!\ 6!}\right)$$

  $$= 17\,153\,136 \text{ combinations.}$$

Systems of Blocks of Elements

The digits of 123456789 can be arranged in *blocks* of triples so that *each pair* of digits appears only once:

{1, 2, 3}	{1, 4, 7}	{1, 5, 9}	{1, 6, 8}
{4, 5, 6}	{2, 5, 8}	{2, 6, 7}	{2, 4, 9}
{7, 8, 9}	{3, 6, 9}	{3, 4, 8}	{3, 5, 7}

KIRKMAN Thomas Penyngton (1806 - 1895)

Amateur mathematician and English clergyman **Thomas Penyngton Kirkman** showed in 1847 that the requirement for n elements to be arranged in a system of blocks of triples where each pair of elements appears once – no more or less – is that the number n divided by 6 gives a remainder of 3 or 1. In our example, we have 9 elements: 9 divided by 6 gives the remainder 3.

Since the 1950s much research has been done in this area of combinatorics, resulting in criteria to form other systems of blocks of elements. Applications exist in data processing and telecommunications.

5.4 Samples with Replacement

In sampling with replacement, we imagine a bag containing n objects. A sample of size p is made by picking an item from the bag, recording it, and tossing it back into the bag, and so on until an ordered p-tuple of values has been recorded.

Every sample may be selected as many times as there are objects in the collection. A sample is replaced by a like sample *before* the next sample is selected.

The number of samples formed with n objects, p at a time, **with replacement**, is

$$n^p.$$

- Find the number of three-letter formations of the word CUP, employing

  a: samples with replacement;
  b: permutations;
  c: combinations.

a: $3^3 = 27$ samples with replacement; these are

(**C**, **C**, **C**)	(**C**, **C**, **U**)	(**C**, **U**, **U**)	(**C**, **C**, **P**)	(**C**, **P**, **P**)
(**U**, **U**, **U**)	(**U**, **U**, **C**)	(**U**, **C**, **C**)	(**U**, **U**, **P**)	(**U**, **P**, **P**)
(**P**, **P**, **P**)	(**P**, **P**, **C**)	(**P**, **C**, **C**)	(**P**, **P**, **U**)	(**P**, **U**, **U**)
(**C**, **U**, **P**)	(**P**, **U**, **C**)	(**U**, **P**, **C**)	(**C**, **P**, **U**)	(**U**, **C**, **P**)
(**P**, **C**, **U**)	(**P**, **C**, **P**)	(**P**, **U**, **P**)	(**U**, **P**, **U**)	(**U**, **C**, **U**)
(**C**, **U**, **C**)	(**P**, **C**, **U**)			

b: $3! = 6$ permutations.

c: 1 combination.

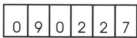

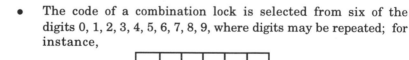

- The code of a combination lock is selected from six of the digits 0, 1, 2, 3, 4, 5, 6, 7, 8, 9, where digits may be repeated; for instance,

0	9	0	2	2	7

The number of possible lock combinations is

$$10^6 = 1\,000\,000.$$

- Using the digits

  a: 1, 3, 5, 7, 9 **b:** 0, 2, 4, 6, 8

find the maximum number of positive integer numbers composed of one, two, three, four, or five digits that can be formed when the same digit may be repeated 5 times in each number.

a: $5 + 5 \times 5 + 5 \times 5^2 + 5 \times 5^3 + 5 \times 5^4 = 3905$ integer numbers.

b: A zero leading an integer number has no place value, so

$$4 + 4 \times 5 + 4 \times 5^2 + 4 \times 5^3 + 4 \times 5^4 = 3124 \text{ integer numbers.}$$

- A licensing agency uses a system of any two letters from the English alphabet, followed by three digits (1 through 9), followed by any one letter from the English alphabet; for instance,

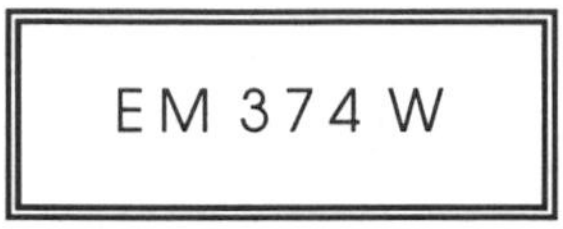

The greatest possible number of licenses that can be issued using this system is

$$(26^2)(9^3)(26^1) = 12\,812\,904 .$$

- The five lamps of an optical paging system can be switched to show a steady light, a flashing light, or no light. When all lights are off, nobody is being paged.

How many people can be individually paged by the system?

There are 3 choices (**steady**, **flashing**, and **off**) for the five lamps; thus,

$$3^5 \text{ possibilities.}$$

When no light is on, nobody is being paged; thus,

$$3^5 - 1 = 242 \text{ individuals can be separately paged.}$$

- **Braille** is the language of the blind, who read, with their fingertips, characters composed of raised dots in the paper, arranged in a two-by-three pattern.

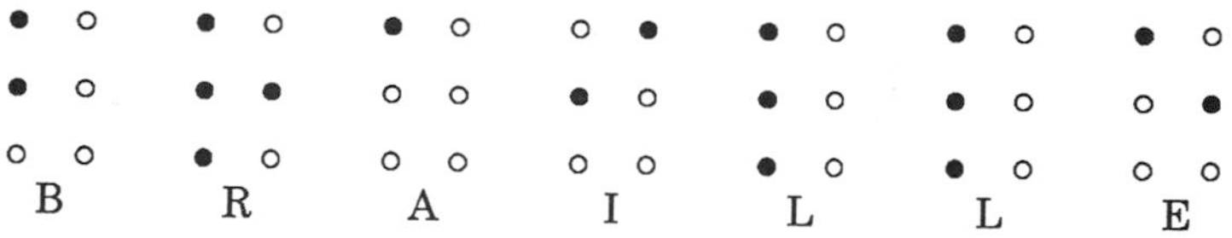

This system is an example of permutations, with repetition, of two elements – *dot* and *no-dot*,

$$2^6 = 64 .$$

Deducting one unit for the "six-no-dot" configuration, we are left with 63 characters, which are enough to represent the letters of conventional alphabets plus ten digits and the necessary punctuation marks and diacritic signs.

5.5 Graph Theory

The Königsberg Bridges

The former German city of Königsberg, now Russian Kaliningrad, on the banks of the Pregel and on two islands in the river, is famous for its bridges and for two of its sons: **Immanuel Kant** and **David Hilbert**.

KANT Immanuel (1724 - 1804)

HILBERT David (1862 - 1943)

Kant, the great philosopher, based his entire thinking on non-negotiable mathematical decrees; although sometimes placed among mathematicians, Kant did not advance mathematics. Hilbert was a mathematical universalist, one of the leading mathematicians of the 20th century.

Of old, Königsberg had seven bridges connecting the several parts of the town. This started a debate whether it would be possible to make a complete tour of the town and return to the starting point by crossing all of the bridges just once.

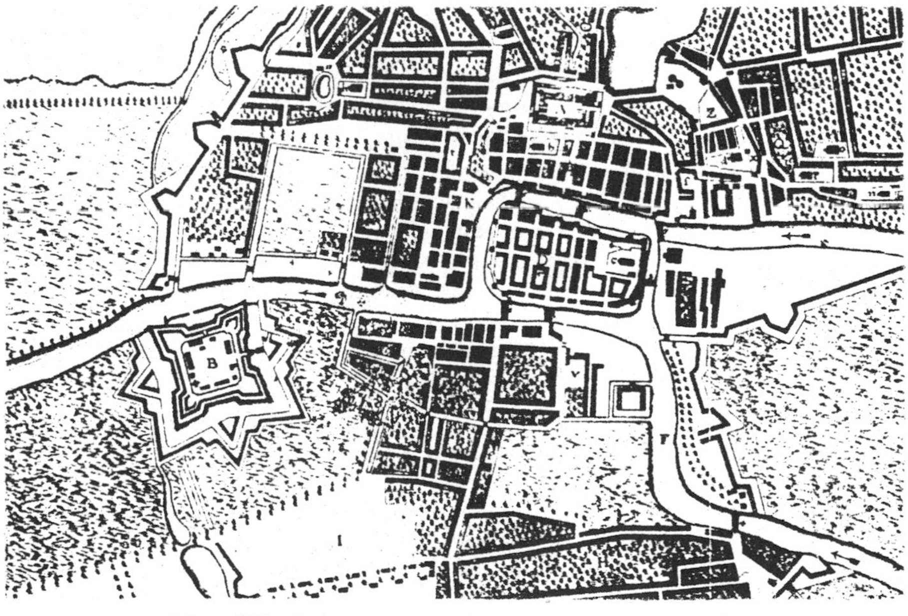

Map of Königsberg, circa 1740

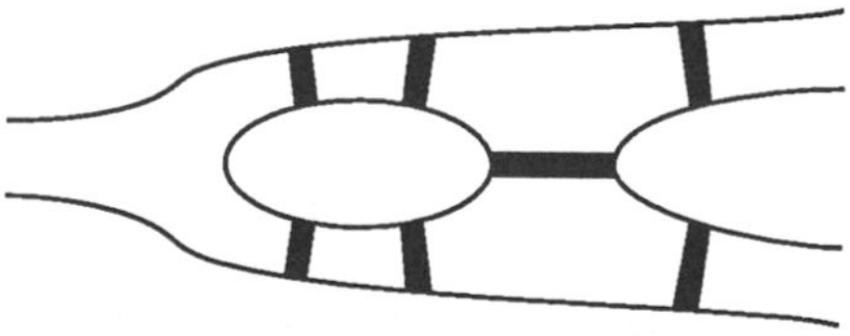

Location of the bridges

EULER Leonhard
(1707 - 1783)

The problem was settled in 1736 by the Swiss mathematician **Leonhard Euler**, who demonstrated the impossibility of the task – at least one bridge would have to be crossed twice.

Graph
Network
Vertex
Arc
Valence

Euler reduced the city plan to what is today known as a **graph** or **network**, in which land areas are represented by dots – called **vertices** (singular: a vertex) – and bridges are shown as lines between the vertices, called **arcs**. The **valence** of a vertex is the number of arcs that originate or end at the vertex.

Euler formulated the following general law for the solvability of the problem:

> A path traversing the network by crossing every segment just once is possible *only if* the network is connected and has at most two vertices with odd valence.

Networks of Königsberg and Kaliningrad

7 bridges

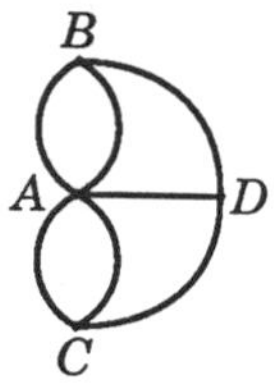

8 bridges

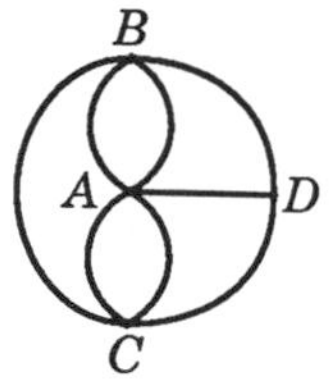

The Königsberg network had four vertices – $A\ B\ C\ D$ – all of odd valence: A with five; B, C, D with three segments each. Consequently, the suggested tour is impossible.

An eighth bridge was built between B and C in 1875, directly connecting the banks of the river. Of the four vertices, A still has five segment connections, and D has three; but B and C have four; that is, only two vertices have an odd number of segments, and the suggested tour of the city is possible – starting at one of the vertices with an odd number of connections and ending at the other. If we build a ninth bridge between A and D, we can also meet the requirement that the tour should begin and end in the same place.

- Which of the figures below can be drawn without lifting the pen and without repeating lines? Give an Euler argument.

A.  **B.** 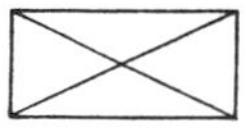 **C.** 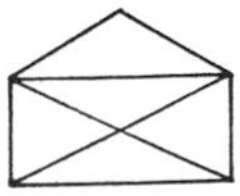

A: Possible; every vertex has an even valence.
B: Not possible; 4 (> 2) vertices with odd valence.
C: Possible; only 2 vertices with odd valence.

How Many Possible Connections?

CAYLEY Arthur
(1821 - 1895)

Digressing from the Euler paths, we ask how many ways exist to connect a specified number of dots, representing, *e.g.*, cities that need road and telephone connections. **Arthur Cayley** showed that the number of connections of n dots is n^{n-2}. To connect five dots, there are $5^{5-2} = 125$ possibilities; to connect eight, 262 144 possibilities.

Euler Paths *vs.* Hamilton Paths

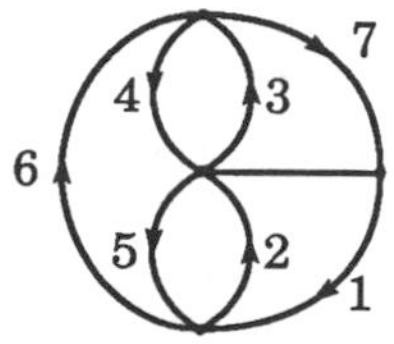

Euler path

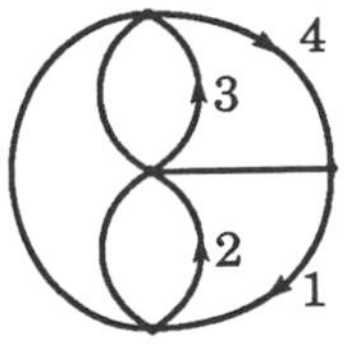

Hamilton path

HAMILTON William Rowan
(1805 - 1865)
Irish mathematician

An **Euler path** traverses *every segment* of a network *once*, with no restrictions as to the number of times each vertex may be passed through; a **Hamilton path** goes through *every vertex once only*, with no obligation to traverse all segments but ending at the starting point. They are important in traffic planning, electric circuit design, and many other applications of networks when the task is to find the *shortest* path for the least expensive installation.

There is no simple necessary and sufficient condition for the existence of a Hamilton path.

The Four-Color Map Theorem

If a map can be drawn as a continuous closed curve (or broken line) on a sphere or a plane – returning to the starting point – it requires only *two colors*; if it does not return to the starting point, it requires *three colors*.

Chromatic Number

The question arises: What is the minimum number of colors needed for *any* map on a sphere or a plane? The least number of colors that is sufficient so that regions with common boundary-line segments on a surface are distinguished by different colors is known as the **chromatic number** of that surface.

GUTHRIE Francis
(1831 - 1899)

In October 1852, **Francis Guthrie** – then a young mathematics student, later to become professor of mathematics at the University of Cape Town – found that he could color a map using just four colors so that no two neighboring countries would have the same color.

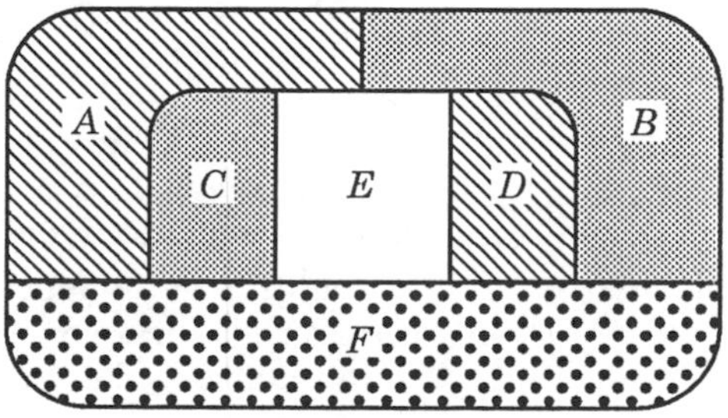

Guthrie conjectured that four colors would do, but he could not present a satisfactory mathematical proof of this thesis.

de MORGAN Augustus
(1806 - 1871)

The problem was brought to the attention of the eminent English mathematician **Augustus de Morgan,** who proved that, irrespective of individual size and shape, five countries cannot occupy such positions on a map that every one of them will have a border common with all of the other four.

This suggests but does not prove that four colors might suffice for the coloring of any map; the difficulty of the problem lies in the fact that the proof must apply for all imaginable configurations of any number of countries, of any size and shape.

CAYLEY Arthur
(1821 -1895)

Mentioning Guthrie as originator of the problem, de Morgan submitted it to several well-known contemporary mathematicians, but it failed to attract general interest until, in June 1878, **Arthur Cayley** brought it before the London Mathematical Society. Many amateur and professional mathematicians have since tried their luck at the problem.

KEMPE Alfred Bray

HEAWOOD P. J.
(1861 - 1955)

pp. 448 - 50

YOUNGS J.W.T.

RINGEL Gerhard
(b. 1919)

Klein bottle: *p.* 380

In 1879 the British lawyer **Alfred Bray Kempe** claimed to have proved the four-color map conjecture. Kempe substituted nodes for countries and connected nodes of bordering countries by arcs to reduce the problem to a set of "unavoidable cases". For more than a decade Kempe's proof stood but, in 1890, **P. J. Heawood** pointed out that Kempe had missed a particular case. Heawood used Kempe's approach and *Euler's polyhedron formula* (Chapter 12) to prove that *five* colors are always enough for maps on the plane or sphere; moreover, he gave a general formula for the chromatic number, computed to be less than or equal to a number p. In 1968 **J. W. T. Youngs** and **Gerhard Ringel** proved that p equals the chromatic number, except for a Klein bottle.

APPEL Kenneth
(b. 1932)

HAKEN Wolfgang
(b. 1928)

The four-color problem was solved by an approach similar to that of Kempe's in 1976 by **Kenneth Appel** and **Wolfgang Haken** of the University of Illinois. Their proof was made with the assistance of a computer, which allowed analysis of a large collection of cases. After correcting the initial argument, there was now – almost a hundred years after Cayley's invigorating presentation of the problem – an infallible mathematical proof of the four-color map conjecture.

Mathematicians still hope for an easily accessible proof – obtained without the need for computer assistance – but no such proof has yet been found, if in fact it is possible. The present proof is available only to the most determined.

Möbius strip: *p.* 380

FRANKLIN Philip
(1898 - 1965)

Thus, the plane and sphere have the chromatic number 4; it has further been proved that the torus, the cylinder, and the Möbius strip all have the chromatic number 7. These numbers can all be computed by Heawood's formula, given in 1890. In 1934, **Philip Franklin** proved that the chromatic number for a Klein bottle is 6. A condition for all chromatic numbers is that a colony must not require the same color as its parent country.

Twelve-Color Map Problem

In addition to the four-color map problem – now exalted to the position of a theorem – there is a **twelve-color map problem**: If on a plane or sphere each country has at the most one colony, requiring the same color as its parent country, at most twelve different colors are needed to distinguish the political regions on a map (chromatic number: 12).

5.6 Magic Squares and Their Kin

肆	玖	貳
參	伍	柒
捌	壹	陸

4	9	2
3	5	7
8	1	6

A square with 9, 16, 25 ... n^2 boxes, called *cells*, filled with integer numbers – all different – is called a **magic square** if the sums of the numbers in the horizontal rows, vertical columns, and main diagonals are all equal.

The magic square is of Chinese origin, being first mentioned in a manuscript from the time of Emperor Yu around 2200 B.C. This square had 3 x 3 = 9 cells, each with Chinese characters equivalent to 1 through 9, and giving the sum 15 in all directions.

It was inevitable that a square with these "magical", and therefore mysterious, qualities should appeal to astrologers and cranks of all descriptions.

Thus, a square of one cell containing a digit 1 – exhibiting the magic of producing the sum 1 in all directions – was considered to represent the eternal perfection of God (= Number One), as explained to the lay mind by one *Cornelius Agrippa* (1486 - 1535), an astrologer by profession.

The unfortunate fact that a magic square with 2 x 2 cells cannot be constructed was considered proof of the imperfection of the four elements: air, earth, fire, and water; some self-styled Great Thinkers attributed the failure to Original Sin.

If the integers in a magic square are the consecutive numbers from 1 to n^2, the square is said to be of the n-th order, and the magic number is equal to

$$\frac{n\,(n^2+1)}{2}\,.$$

The simplest magic square possible is one of the 3rd order, with 3 x 3 = 9 cells containing the first nine integers, 1 ... 9, and with the magic sum 15 along eight lines: 3 horizontal, 3 vertical, and 2 diagonal.

Only one arrangement of digits, and its mirror image, is possible for a 3rd-order square:

4	9	2
3	5	7
8	1	6

and mirror image

2	9	4
7	5	3
6	1	8

They may be rotated into four positions each, permitting eight 3rd-order magic squares in all.

DÜRER Albrecht
(1471 - 1528)

The first magic square to appear in the Western world was, in all probability, the one depicted in the upper right-hand corner of a copperplate engraving called "Melencolia § I" by the German artist-mathematician **Albrecht Dürer**, who also managed to enter the year of engraving – 1514 – in the two middle cells of the bottom row.

16	3	2	13
5	10	11	8
9	6	7	12
4	15	14	1

The Dürer magic square is of the 4th order, that is, it has 4 x 4 = 16 cells containing the first sixteen natural numbers so arranged that adding any one of four horizontal rows, four vertical columns, and two main diagonals will produce the magic sum,

$$\frac{4\,(16+1)}{2} = 34,$$

as in all 4th-order magic squares.

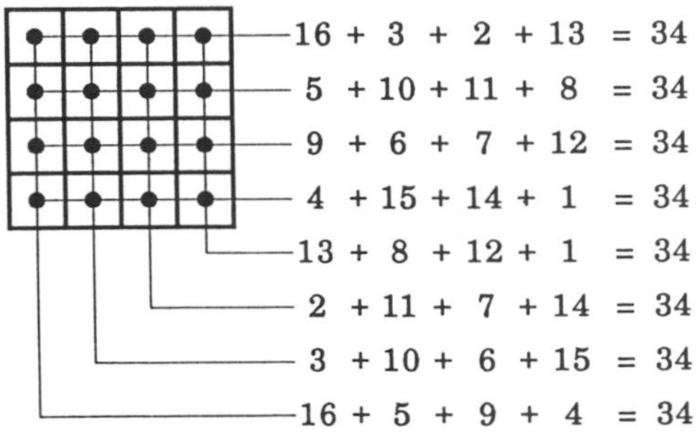
16 + 3 + 2 + 13 = 34
5 + 10 + 11 + 8 = 34
9 + 6 + 7 + 12 = 34
4 + 15 + 14 + 1 = 34
13 + 8 + 12 + 1 = 34
2 + 11 + 7 + 14 = 34
3 + 10 + 6 + 15 = 34
16 + 5 + 9 + 4 = 34

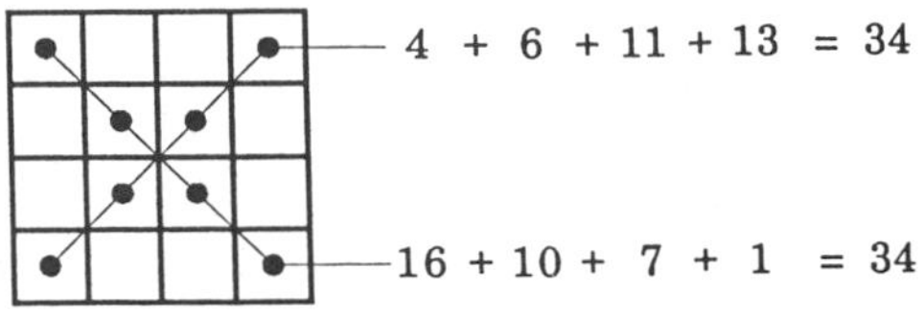
4 + 6 + 11 + 13 = 34
16 + 10 + 7 + 1 = 34

Besides these sums, the Dürer square yields the same magic sum when adding other configurations of cells as shown in the drawings below, where, for clarity, filled circles denote the numbers to be added.

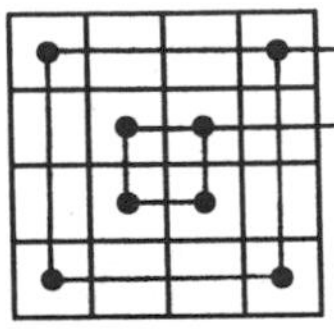

16 + 13 + 1 + 4 = 34
10 + 11 + 7 + 6 = 34

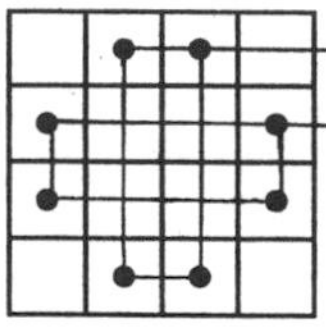

3 + 2 + 14 + 15 = 34
5 + 8 + 12 + 9 = 34

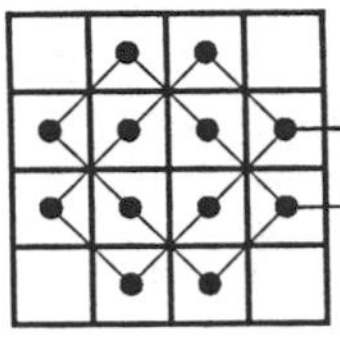

2 + 8 + 15 + 9 = 34
5 + 3 + 12 + 14 = 34

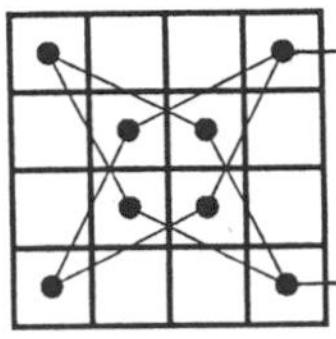

13 + 10 + 4 + 7 = 34
16 + 11 + 6 + 1 = 34

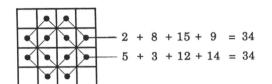

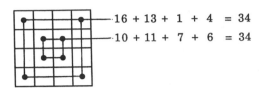

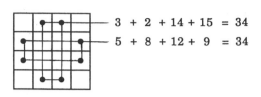

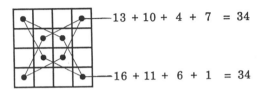

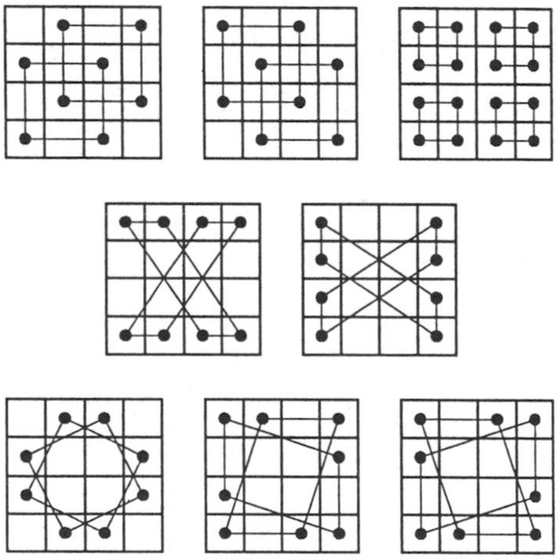

12	9	2
1	6	36
18	4	3

In addition to the magic squares already described, there are also magic multiplication squares and magic division squares. In the former, the products of all numbers in every horizontal row, vertical column, and main diagonal are the same; in the square illustrated here, the product is 216, *e.g.*,

$$12 \cdot 1 \cdot 18 = 216$$
$$18 \cdot 4 \cdot 3 = 216, \textit{ etc.}$$

3	9	18
1	6	36
2	4	12

If we reverse the order of the numbers in the main diagonals, we obtain a magic "division square", in which the first figure of every row, column, and main diagonal divided by the quotient of the second and third figures will always give the same result, and so will the last figure divided by the quotient of the second and first figures, *e.g.*,

$$\frac{3}{\frac{9}{18}} = 6 \qquad \frac{18}{\frac{9}{3}} = \frac{18}{3} = 6, \textit{ etc.}$$

CHÊNG TA-WEI

In 1593, the Chinese mathematician **Chêng Ta-wei** published a book from which we have culled the following *6th-order* magic square:

27	29	2	4	13	36
9	11	20	22	31	18
32	25	7	3	21	23
14	16	34	30	12	5
28	6	15	17	26	19
1	24	33	35	8	10

Every row, column, and main diagonal adds up to 111.

The fascination of magic squares does not stop with 6th-order squares.

FRANKLIN Benjamin (1706 - 1790)

52	61	4	13	20	29	36	45
14	3	62	51	46	35	30	19
53	60	5	12	21	28	37	44
11	6	59	54	43	38	27	22
55	58	7	10	23	26	39	42
9	8	57	56	41	40	25	24
50	63	2	15	18	31	34	47
16	1	64	49	48	33	32	17

Franklin's Magic Square

The American statesman and scientist-inventor **Benjamin Franklin** constructed a magic square of the 8th order, whose every horizontal row and vertical column add up to the magic number 260 or, if you stop halfway, to 130 for its quarter-squares.

But the magic does not end there: Every four-cell mini-square, say,

$$\begin{matrix} 3 & 62 \\ 60 & 5 \end{matrix} \quad \text{or} \quad \begin{matrix} 6 & 59 \\ 58 & 7 \end{matrix},$$

sums to 130; every four cells symmetrically placed in relation to a mini-square, *e.g.*,

$$\begin{matrix} 52 & 13 \\ 11 & 54 \end{matrix}; \quad \begin{matrix} 61 & 4 \\ 6 & 59 \end{matrix}; \quad \begin{matrix} 14 & 51 \\ 55 & 10 \end{matrix}; \quad \text{and} \quad \begin{matrix} 53 & 12 \\ 9 & 56 \end{matrix}; \quad \begin{matrix} 60 & 5 \\ 8 & 57 \end{matrix}; \quad \begin{matrix} 11 & 54 \\ 55 & 10 \end{matrix},$$

add further to the magic by presenting the sum 130.

But there is more to come: Adding the numbers on the rising and descending diagonal dotted lines will also give the expected sum, 260 – all in all, a magnificent combinatorical feat, worthy of the highest praise.

From the ordinary 8th-order magic square it is but a short step to the refinement of letting the numbers follow the movements of a chess knight across a chessboard – "up two and over one" or "up one and over two".

EULER Leonhard
(1707 - 1783)

The famous Swiss mathematician **Leonhard Euler** published a paper in 1759, where he showed how to complete *partial knight's tours* and to construct *symmetric knight's tours*. Euler's chessboard magic squares inspired many amateur and professional mathematicians to design chess-knight magic squares.

1	48	31	50	33	16	63	18
30	51	46	3	62	19	14	35
47	2	49	32	15	34	17	64
52	29	4	45	20	61	36	13
5	44	25	56	9	40	21	60
28	53	8	41	24	57	12	37
43	6	55	26	39	10	59	22
54	27	42	7	58	23	38	11

Beverley's Magic Square

First to design a *magic knight's tour* was **William Beverley** (1814 - 1889), an artist and designer of theatrical effects; his publication appeared in the *Philosophical Magazine* in 1848. Like Franklin's square, the Beverley magic square shown here adds up to 260 – 130 in its quarter-squares – for all horizontal rows and vertical columns; also like Franklin's square, all four-cell mini-squares sum to 130, but the agreement ends there. The Beverley magic square can be considered four 4th-order magic squares put together. The more limited versatility of the Beverley magic square compared with that of Franklin is a consequence of the restriction imposed by the formal pattern of movement of the chess knight, whereas Franklin was free to arrange his numbers in the square largely at his own discretion.

5	14	53	62	3	12	51	60
54	63	4	13	52	61	2	11
15	6	55	24	41	10	59	50
64	25	16	7	58	49	40	1
17	56	33	42	23	32	9	48
34	43	26	57	8	39	22	31
27	18	45	36	29	20	47	38
44	35	28	19	46	37	30	21

Feisthamel's Magic Square

The second to construct magic knight's tours was **Carl Wenzelides** (1770 - 1852); his first one was published in the German chess magazine *Schachszeitung* in 1849.

The magic square of **Feisthamel** fulfills the requirement of adding up to 260 in all horizontal rows, vertical columns, and, unlike Beverley's magic square, also in the two main diagonal lines.

The chessboard square is characterized by a kind of axial symmetry with reference to the vertical midline; similarly placed cells on opposite sides always add up to 65.

The path of the knight across the chessboard is not continuous but consists of two halves, 1 ... 32 and 33 ... 64, cells 32 and 33 being both on the 5th horizontal line from the top with no possibility for the knight to bridge the gap except by riding piggyback on a rook.

The (1) IN and (64) OUT cells are at either end of the 4th line from the top of the board.

46	55	44	19	58	9	22	7
43	18	47	56	21	6	59	10
54	45	20	41	12	57	8	23
17	42	53	48	5	24	11	60
52	3	32	13	40	61	34	25
31	16	49	4	33	28	37	62
2	51	14	29	64	39	26	35
15	30	1	50	27	36	63	38

Jaenisch's Magic Square I

One of **Jaenisch**'s magic squares (I in the margin) has an uninterrupted numerical path 1 ... 64 for the knight with the finishing cell 64 inside the board but so placed that the knight can reach the starting point 1 from it for a new round, an elegant feature. All horizontal and vertical lines of this square add up to 260, but the diagonals do not, the bottom left to top right diagonal giving 256 and the top left to bottom right giving 264, which makes the square only *semi-magic*.

50	11	24	63	14	37	26	35
23	62	51	12	25	34	15	38
10	49	64	21	40	13	36	27
61	22	9	52	33	28	39	16
48	7	60	1	20	41	54	29
59	4	45	8	53	32	17	42
6	47	2	57	44	19	30	55
3	58	5	46	31	56	43	18

Jaenisch's Magic Square II

Another Jaenisch magic square shown here (II in the margin) is also semi-magic; the horizontal and vertical sums are 260, but the diagonals add up to 192 and 328, respectively. The path of the knight is continuous 1 ... 64, with a regular jump back to 1. The square exhibits a form of radial symmetry, in that numbers contained in cells symmetrically placed relative to the mid-point of the board always have a difference of 32 units.

Composite Magic Squares

Sometimes you can construct large magic squares by fitting together magic squares of lesser orders.

As an illustration, consider the simple 3rd-order magic square

(1)

2	9	4
7	5	3
6	1	8

magic sum = 15

and build it up into similar squares by adding, repeatedly, 9 to every cell, obtaining successively magic 3rd-order squares (2) ... (9) with the magic sums 42, 69, 96, 123, 150, 177, 204, and 231, respectively.

11	18	13
16	14	12
15	10	17

(2)

20	27	22
25	23	21
24	19	26

(3)

29	36	31
34	32	30
33	28	35

(4)

38	45	40
43	41	39
42	37	44

(5)

47	54	49
52	50	48
51	46	53

(6)

56	63	58
61	59	57
60	55	62

(7)

65	72	67
70	68	66
69	64	71

(8)

74	81	76
79	77	75
78	73	80

(9)

(2)	(9)	(4)
(7)	(5)	(3)
(6)	(1)	(8)

Combine these in the same order as the natural numbers in magic square (1) to form a 9th-order magic square with the magic sum

$$\frac{9\,(81+1)}{2} = 369\,.$$

11	18	13	74	81	76	29	36	31
16	14	12	79	77	75	34	32	30
15	10	17	78	73	80	33	28	35
56	63	58	38	45	40	20	27	22
61	59	57	43	41	39	25	23	21
60	55	62	42	37	44	24	19	26
47	54	49	2	9	4	65	72	67
52	50	48	7	5	3	70	68	66
51	46	53	6	1	8	69	64	71

The drawing below depicts another 9th-order composite magic square which has some interesting features besides meeting the standard requirement that all horizontal rows, vertical columns, and main diagonals add up to the magic sum 369:

2	11	12	13	77	78	79	81	16
6	18	27	26	61	62	65	28	76
7	59	30	35	51	53	36	23	75
8	58	32	38	45	40	50	24	74
73	57	49	43	41	39	33	25	9
72	22	48	42	37	44	34	60	10
68	19	46	47	31	29	52	63	14
67	54	55	56	21	20	17	64	15
66	71	70	69	5	4	3	1	80

By "peeling off" the outer frame of numbers, we obtain a true 7th-order magic square with the horizontal, vertical, and diagonal magic sum 287:

18	27	26	61	62	65	28
59	30	35	51	53	36	23
58	32	38	45	40	50	24
57	49	43	41	39	33	25
22	48	42	37	44	34	60
19	46	47	31	29	52	63
54	55	56	21	20	17	64

Continuing the process of removing the border of numbers, we get, in turn, a true 5th-order magic square adding up to 205 in all directions, and a 3rd-order square with the magic sum 123:

30	35	51	53	36
32	38	45	40	50
49	43	41	39	33
48	42	37	44	34
46	47	31	29	52

38	45	40
43	41	39
42	37	44

15-by-15 Magic Square

FAULHABER Johann
(1580 - 1635)

The magic square shown below, with 15-by-15 number cells, by an unknown author, is from the *Arithmetischer Cubic-cossischer Lustgarten* ("Arithmetic-Algebraic Pleasure Garden"), by the German *Rechenmeister* **Johann Faulhaber** (Tübingen, 1604). It provides an instructive example of the general method of constructing *odd*-order magic squares; there is no corresponding general rule, however, for the construction of *even*-order magic squares.

8	121	24	137	40	153	56	169	72	185	88	201	104	217	120
135	23	136	39	152	55	168	71	184	87	200	103	216	119	7
22	150	38	151	54	167	70	183	86	199	102	215	118	6	134
149	37	165	53	166	69	182	85	198	101	214	117	5	133	21
36	164	52	180	68	181	84	197	100	213	116	4	132	20	148
163	51	179	67	195	83	196	99	212	115	3	131	19	147	35
50	178	66	194	82	210	98	211	114	2	130	18	146	34	162
177	65	193	81	209	97	225	113	1	129	17	145	33	161	49
64	192	80	208	96	224	112	15	128	16	144	32	160	48	176
191	79	207	95	223	111	14	127	30	143	31	159	47	175	63
78	206	94	222	110	13	126	29	142	45	158	46	174	62	190
205	93	221	109	12	125	28	141	44	157	60	173	61	189	77
92	220	108	11	124	27	140	43	156	59	172	75	188	76	204
219	107	10	123	26	139	42	155	58	171	74	187	90	203	91
106	9	122	25	138	41	154	57	170	73	186	89	202	105	218

We begin by entering the number **1** in the cell immediately to the right of the center cell (8th row, 9th column), and then proceed diagonally upward toward the right - in a "north-easterly" diretion – entering successively the consecutive numbers, in this case 2 through 7, in the cells of the diagonal.

A diagonal that reaches the right-hand border column of the square is continued at the left-hand side of the square as if nothing untoward had happened; analogously, when a diagonal reaches the top row of the square , it is similarly continued in the bottom row. These rules are illustrated by the transitions **7** - **8** and **8** - **9**, respectively.

When a diagonal encounters a cell that is already occupied, and thus obstructs the direct continuation of the diagonal, a re-start is made in the cell that is located two steps to the right in the same horizontal row. This is shown by the shift **15** - **16**, where the number **1** stands in the way of the direct progress of the diagonal.

A special case is illustrated by the cell **120** at the upper right-hand corner of the square, where the progress of the diagonal is actually blocked by the cell **106** at the lower left-hand corner; **120** being the last cell in the top row, the two-step displacement of the diagonal will have to be made at the left end of the top row, and **121** is entered in the second cell from the left. The diagonal is then continued from cell **122** in the bottom row.

The procedures described above are repeated as often as required until the last number – **225** – falls into the cell immediately to the left of the center cell **113**.

The sum total of all numbers in a 15th-order magic square is

$$\frac{n^2\,(n^2+1)}{2} = \frac{225 \cdot 226}{2} = 25\,425,$$

and the sum of every row, column, and main diagonal is

$$\frac{26\,425}{15} = 1695\,.$$

The number magic may be extended also to other geometric figures, for instance, to pentagrams, hexagrams, circles, spheres, and cubes.

Magic Pentagrams and Hexagrams

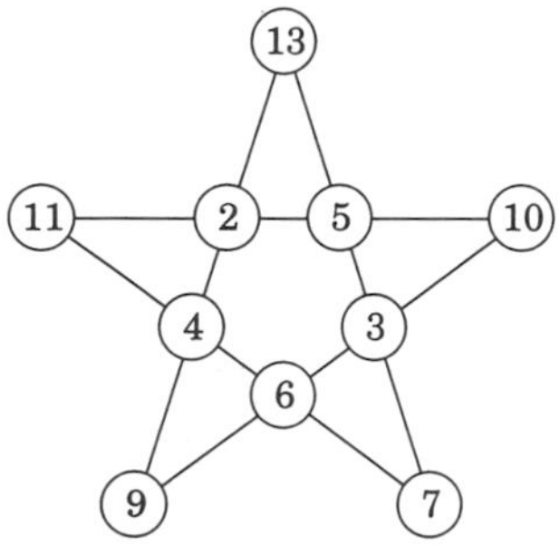

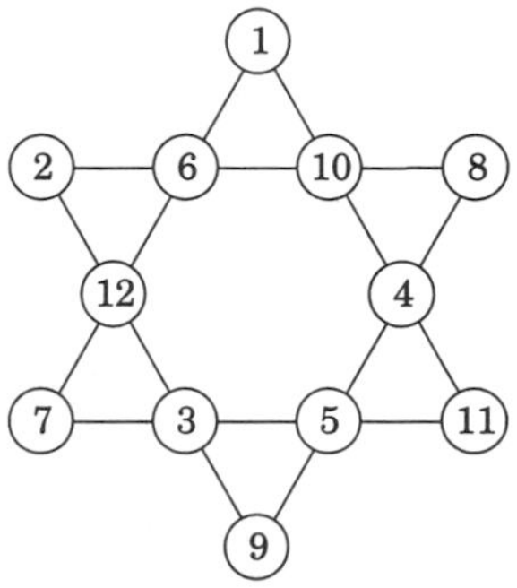

In the **magic pentagram** shown here, the numbers are not consecutive; the sum of the figures in every line is 28.

Of the 479 001 600 ways of arranging the consecutive numbers from 1 to 12 at the nodes of a **magic hexagram**, as shown above, there are 192 possible magic hexagrams (John R. Hendricks, "The Magic Hexagram", *Journal of Recreational Mathematics*, 25:1, 1993); the sum of the figures in every line is 26.

Magic Circles and Spheres

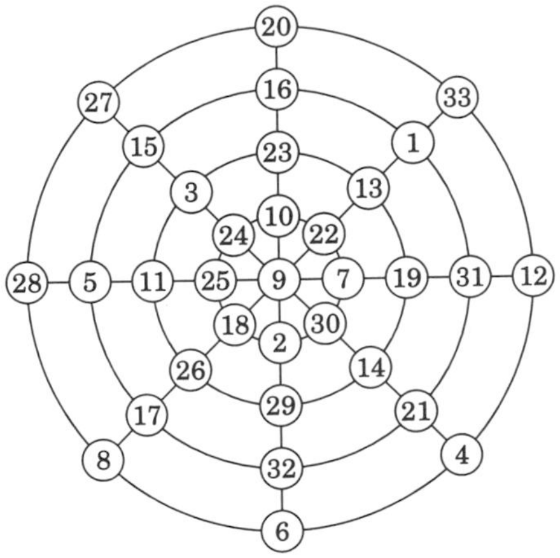

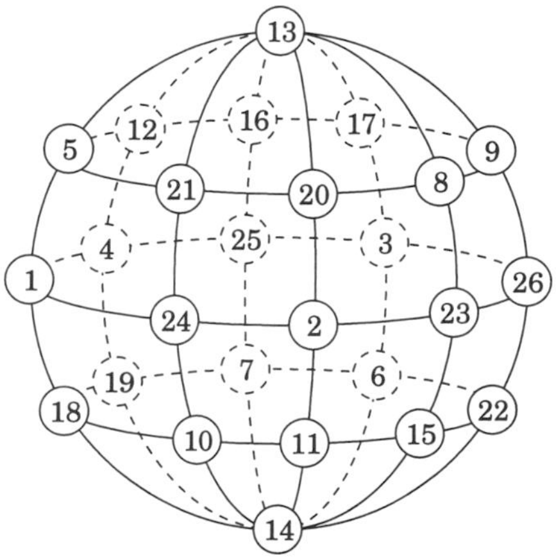

YANG HUI

The **magic circle** shown here was designed in 1275 by the Chinese **Yang Hui**. The sum of any circle is 138, that of any diameter also 138 if we disregard the central 9; with it, we find the sum 147 for all diameters.

The **magic sphere**, as shown above, has five great circles (equator and four double meridians) and two small (latitude) circles, presenting 26 crossing points in all. At these, the numbers 1 - 26, inclusive, are entered, so that the eight numbers on every circle add up to 108, diametrically opposite numbers having the sum of 27.

Magic Cubes

PAO CHHI-SHOU

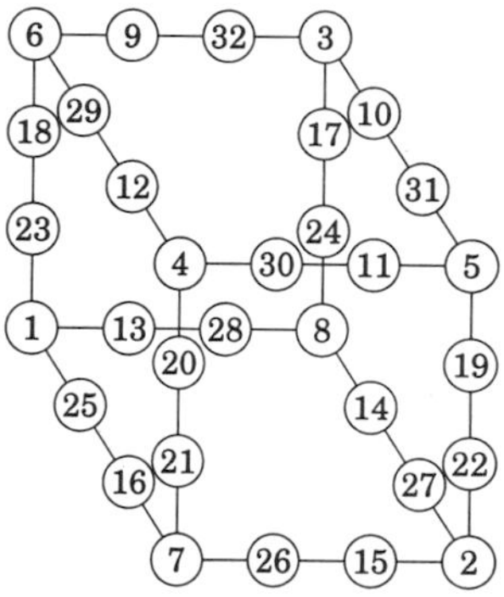

The cube, as shown here to the left, was published at the end of the 19th century by the Chinese **Pao Chhi-shou** in his *Pi Nai Shan Fang Chi* ("Pi Nai Mountain Hut Records"). It has 32 numbers, while established magic cubes of the n-th order consist of n^3 numbers. All edges of our cube have the same sum (41) *between* the vertices, which makes it, at best, semi-magic.

In a true magic cube, the sum shall be the same along all edges, vertices *included*.

Assuming that the placing of the first eight numbers at the vertices is correct, we shall change the order of some of the other numbers to make Pao Chhi-shou's cube into a true magic cube. We have

$$\sum_{i=1}^{32} = 528$$

$$\sum_{i=1}^{8} = 36$$

$$492 = 12 \times 41$$

This net sum shall now be split up into twelve part sums determined by the respective sums of the vertex pairs:

$6+3=9$	(41)	$6+4=10$	(40)
$7+2=9$	(41)	$8+2=10$	(40)
$1+7=8$	(42)	$1+8=9$	(41)
$3+5=8$	(42)	$4+5=9$	(41)
$1+6=7$	(43)	$4+7=11$	(39)
$2+5=7$	(43)	$3+8=11$	(39)

好了

Hao le *Good!*

Thus, the task is to compose 12 two-component sums,

39, 39 – 40, 40 – 41, 41, 41, 41 – 42, 42 – 43, 43,

from the sequence of numbers

9, 10, 11, 12 … 31, 32.

找到了

Zhao dao la *I found it!*

The sum along every edge – vertices included – is 50.

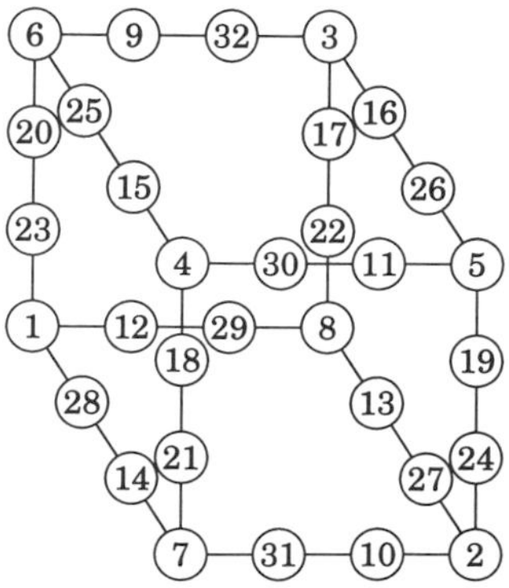

完了

Won le *Finished!*

Chapter 6

SYMBOLIC LOGIC

Knowledge of symbols and concepts introduced in this chapter is not needed to understand subsequent chapters. Some concepts that originated in the field of symbolic logic are described in the next chapter, "Set Theory".

6.0 Historical Notes

ARISTOTELES
(384 - 322 B.C.)

Greek, *lógos*, "speech", "reasoning"

Aristotle, Greek philosopher and scientist – pupil of Plato, who was a pupil and friend of Socrates – held that any logical argument could be reduced to two premises and a conclusion, and laid down three basic laws, or principles, of logical reasoning, often referred to as **classical logic** or **Aristotelian logic**:

1. **The principle of identity.**
   A thing is itself: A is A.
2. **The principle of the excluded middle.**
   A proposition is either true or false: either A or *not* A.
3. **The principle of contradiction.**
   No proposition can be both true and false. A cannot be both A and not A.

Aristotelian logic dominated scientific reasoning in the Western world for 2000 years; principles of most modern schools of logic are extensions of Aristotelian logic.

von LEIBNIZ Gottfried Wilhelm
(1646 - 1716)

The partnership between mathematics and logic was initiated by the German mathematician and philosopher **G. W. von Leibniz**, who tried to find a *lingua universalis* – a language where errors in thinking would be equivalent to arithmetical errors – and Leibniz thereby laid the foundation for **mathematical**, or symbolic, **logic**. A decisive stimulus to its further development was the need at the end of the 19th century to bring to order an abundance of collections of axioms that were being used in various systems of mathematics in an intuitive manner rather than by logical deduction.

BOOLE George
(1815 - 1864)

The English mathematician and logician **George Boole** demonstrated in his 1847 *Mathematical Analysis of Logic* that a system of algebra can be used to express logical relations, and expanded his system in *An Investigation into the Laws of Thought, on Which are Founded the Mathematical Theories of Logic and Probabilities*, published in 1854. Originally devised as a system for logical reasoning, Boole's concepts, today referred to as **Boolean algebra**, have wide applications in electronic network design.

FREGE Gottlob
(1848 - 1925)
Predicate Logic

The German mathematician and philosopher **Gottlob Frege**, founder of modern symbolic logic, constructed an elaborate logico-mathematical system, now known as **predicate logic**, which he described in 1879 in *Begriffsschrift: Eine der aritmetischen nachgebildete Formelschprache des reinen Denkens* ("Concept-Script: A Formal Language of Pure Thought on the Pattern of Arithmetic"). Frege also wrote *Grundsetze der Arithmetik* ("Basic Laws of Arithmetic"), published, in two volumes, in 1893 and 1903.

DISSERTATIO
De
ARTE COMBI-
NATORIA,
In qua
Ex Arithmeticæ fundamentis Complicationum ac Transpositionum
Doctrina novis præceptis exstruitur, & usus ambarum per uni-
versum scientiarum orbem ostenditur; nova etiam
Artis Meditandi,
Seu
Logicæ Inventionis semina
sparguntur.
Præfixa est Synopsis totius Tractatus, & additamenti loco
Demonstratio
EXISTENTIÆ DEI,
ad Mathematicam certitudi-
nem exacta.
AUTORE
GOTTFREDO GUILIELMO
LEIBNüZIO Lipsensi,
Phil. Magist. & J. U. Baccal.

LIPSIÆ,
Apud Joh. Simon. Fickium et Joh.
Polycarp. Seuboldum
in Platea Nicolaea,
Literis Spörelianis.
A. M. DC. LXVI.

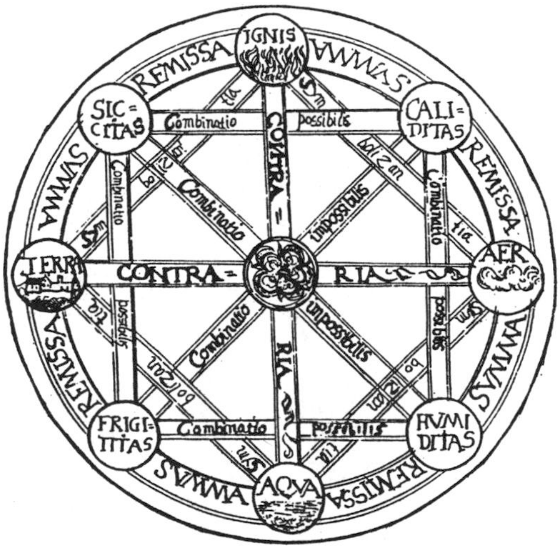

Leibniz's interests in his childhood and youth did not include higher mathematics, but rather philosophy, logic, and law, subjects that also became the basis for his career as a diplomat. In 1666, Leibniz earned a doctorate in law; the same year, his interest in logic had led him to publish his first mathematical undertaking, *Dissertatio de arte combinatoria*, the aim of which was to reduce all truths of reason to a simple system of arithmetic. Understanding that permutations and combinations were part of such a system led Leibniz to advance also these fields of mathematics, and that work directed him toward ideas for the development of infinitesimal calculus.

RUSSELL Bertrand (1872 - 1970)
WHITEHEAD Alfred North (1861 - 1947)

The English philosopher, mathematician, and social reformer **Bertrand Russell** published *Principles of Mathematics* in 1903 and, in collaboration with the English mathematician and philosopher **Alfred North Whitehead**, the three-volume *Principia Mathematica* in 1910, 1912, and 1913; carrying on the work of Frege, they tried to derive mathematics from self-evident logical principles. Russell and Whitehead did not completely reach their goal, but their work has been important for the development of logic and mathematics.

cf. p. 160

The vision – first projected by Leibniz – to reduce *all* truths of reason to incontestable arithmetic was shattered in the 1930s with **Gödel's incompleteness theorem**, stating that any consistent formal system adequate to describe arithmetic must contain statements which can neither be proved nor disproved within this system.

Today, symbolic, or mathematical, logic is used not only in genuinely logical or mathematical domains but also in the natural sciences, and in disciplines such as linguistics, law, and computer technology.

TURING Alan Mathison (1912 - 1954)
von NEUMANN John (Johann) (1903 - 1957) *p.* 301

A prominent figure of contemporary mathematical logic and computer technology was the English mathematician **Alan Turing**, who, with **John von Neumann**, opened the door to the computer era.

Venn diagrams and **Boolean algebra** originated as tools of logic; they will be described in the following chapter, "Set Theory".

Georg Reisch, *Margarita phylosophica* (first edition, 1496; here reproduced from a 1583 reprint).

The huntswoman stands on the steady rock of Aristotelian logic. Behind her is the Greek philosopher **Parmenides** (5th century B.C.), who, as maintained by legend, sitting on a cliff of the Caucasus, invented logic. Before her is a copse of *insolubilia*, and the two dogs *Véritas* (Truth) and *Fálcitas* (Falsity) chasing the hare *Probléma*. In the background is the forest of doctrines (*silva opiniónem*).

From Anders Piltz, *Die Gelehrte Welt des Mittelalters*, 1982.

6.1 Pitfalls

... Further, since I had been with my master I had become aware, and was to become even more aware in the days that followed, that logic could be especially useful when you entered it but then left it.

Umberto Eco, *The Name of the Rose* (1983)

A hippie walks down the street; friendly person remarks:
– *Young man, you've lost a shoe.*
– *Oh no, I've found one.*

***

This statement is not true.

Liar Paradox
Epiminides' Paradox

This is called the **liar paradox** or **Epiminides' paradox** after Epiminides of Crete, who lived in the 6th century B.C. and, allegedly, was the first to record paradoxes of this kind.

If the quotation is true, then it is false; if it is false, then it is true.

***

The Epistle of Paul to Titus, 1.12:

One of themselves, even *a prophet of their own, said, The Cretians* are *alway liars, evil beasts, slow bellies.*

If a *Cretian* says, "The Cretians are alway liars", then the statement must include the Cretian who was quoted; though grammatically correct, this makes no sense within the bounds of the epistle.

***

The statement "I am lying" is true only if it is false, and false if it is true.

***

The following story is one of many versions on the same theme:

The prisoner was told that by making a statement, he could choose the method of his execution; if the statement was true, he would be shot, and if false, he would be hanged. The prisoner made the statement, "I shall be hanged".

***

– *Have you lived here all your life?*
– *No, not yet.*

***

p. 235

Russell's paradox will be described in the following chapter.

6.2 Propositions

Proposition, Statement

A **proposition**, or **statement**, describes and communicates one or more facts; it must have a **truth value**.

Two-Value Principle

The **two-value principle** encompasses only two kinds of truth values: **truth** and **falsity**; it is assumed that no proposition is both true and false (*principle of the excluded middle*).

Propositions are usually denoted by lowercase letters, such as p, q, r, and s. In

$$p: 2+5 = 7; \quad q: 2+4 = 5; \quad r: x+5 = 8$$

- proposition p is **true**;
- q is **false**;
- r is not a proposition as the equation $x + 5 = 8$ alone is neither true nor false; it does not become a proposition until we substitute a number for x.

Negation

$\neg$ $\sim$

The symbol denoting a **negation** is $\neg$ or, alternatively, $\sim$.

$\neg p$ reads, "Not p", "Non p", or "Negation of p".

If a proposition is false, then its negation is true.
If a proposition is true, then its negation is false.

The negation of $p(x)$: $x \leq 0$, where x denotes real numbers, is

$$\neg p(x): x > 0 .$$

If $x = 5$, $p(x)$: $x \leq 0$ is **false**; $\neg p(x)$: $x > 0$ is **true**;

if $x = -2$, $p(x)$: $x \leq 0$ is **true**; $\neg p(x)$: $x > 0$ is **false**.

Truth Values
Boolean Constants
Truth Table

Statements may be expressed by the **truth values, or Boolean constants**, 1 denoting truth, 0 falsity. Truth values are often arranged as a **truth table**,

p	$\neg p$
1	0
0	1

Universal Quantifier

$\forall$

The **universal quantifier** $\forall$ indicates "for every" or "for all".

The expression $\forall\, x\, A, p(x)$ reads:

"For every x belonging to A, the proposition $p(x)$ is valid", *or* "For all x belonging to A, the proposition $p(x)$ is valid."

$\in$ The expression gains in clarity if we employ the symbol $\in$ of set theory, meaning "belonging to" or "is an element of",

$$x \in A, p\,(x)\,.$$

With **R** denoting the collection of all real numbers, the proposition

$$\forall x \in \mathbf{R}, x^2 + 1 > 0 \text{ is true;}$$

$$\forall x \in \mathbf{R}, x + 1 > 0 \text{ is } \textbf{false}.$$

Existential Quantifier

$\exists$ The **existential quantifier** $\exists$ indicates "there exists".

The expression $\exists\, x\, A, p\,(x)$, or $\exists\, x \in A, p\,(x)$, reads:

"There exists an x, belonging to A, for which the proposition $p\,(x)$ is valid", *or*

"There exists at least one x, belonging to A, for which the proposition $p\,(x)$ is valid".

With **R** denoting the collection of all real numbers,

$$\exists\, x \in \mathbf{R}, 2\,x - 5 > 7 \text{ is } \textbf{true};$$

$$\exists\, x \in \mathbf{R}, x^2 < 0, \text{ is } \textbf{false}.$$

Propositions can be combined in several ways to form a **compound proposition**, or **composite proposition**.

Conjunctions

$\wedge$ The **conjunction** symbol $\wedge$ indicates "and".

The expression $p \wedge q$ reads, "p and q".

A conjunction is **true only when both components are true**:

p	q	$p \wedge q$
1	1	**1**
1	0	**0**
0	1	**0**
0	0	**0**

Given the propositions

p: 4 is an even number
q: 7 is an even number
r: 3 is an odd number
s: 5 is an odd number,

q is obviously false; p, r, and s, true. Consequently, the conjunctions

$(p \wedge r)$; $(p \wedge s)$; $(r \wedge s)$ are **true**;

$(p \wedge q)$; $(r \wedge q)$; $(s \wedge q)$, are **false**.

Disjunctions

$\vee$

The **disjunction** symbol $\vee$ indicates "or".

The expression $p \vee q$ reads, "p or q".

A disjunction is **true when at least one of its components is true**:

p	q	$p \vee q$
1	1	**1**
1	0	**1**
0	1	**1**
0	0	**0**

Given the propositions

p: 4 is an even number
q: 7 is an even number
r: 3 is an odd number
s: 5 is an odd number
t: 4 is an odd number

q and t are false; p, r, and s are true. As a result, the disjunctions

$$q \vee t$$
$$t \vee q$$
$$q \vee q$$
$$t \vee t$$
$$q \vee (q \vee t)$$
$$t \vee (q \vee t)$$

have no true components and, consequently, are false; all the other disjunctions have at least one true component, giving them a true value.

Implications

$\Rightarrow$

The **implication** symbol $\Rightarrow$ indicates "implies".

The expression $p \Rightarrow q$ reads, "p implies q" or "if p then q".

Antecedent, Hypothesis
Conclusion

In $p \Rightarrow q$, we call p the **antecedent**, or **hypothesis**, and q the **conclusion**:

antecedent $\Rightarrow$ conclusion

$p \Rightarrow q$ is **false *only* when the antecedent is true and the conclusion is false**:

p	q	$p \Rightarrow q$
1	1	**1**
1	0	**0**
0	1	**1**
0	0	**1**

Principle of Detachment

The principle, or axiom, of detachment states that

if p is true and $p \Rightarrow q$ is true, then q is true:

p	$p \Rightarrow q$	q
1	1	**1**

Converse

The **converse** of $p \Rightarrow q$ is $q \Rightarrow p$.

p	q	$p \Rightarrow q$	$q \Rightarrow p$
1	1	1	1
1	0	0	1
0	1	1	0
0	0	1	1

Inverse

The **inverse** of $p \Rightarrow q$ is $\neg p \Rightarrow \neg q$.

p	$\neg p$	q	$\neg q$	$p \Rightarrow q$	$\neg p \Rightarrow \neg q$
1	0	1	0	**1**	**1**
1	0	0	1	**0**	**1**
0	1	1	0	**1**	**0**
0	1	0	1	**1**	**1**

Logical Equivalence

Two propositions are **logically equivalent** or, simply, equivalent when they have true or false in the same rows of their truth tables. The *converse* and the *inverse* of a given implication have identical truth values and, consequently, are *logically equivalent*.

Contrapositive

The **contrapositive** of $p \Rightarrow q$ is $\neg q \Rightarrow \neg p$.

$p \Rightarrow q$ and its contrapositive have identical truth values:

p	$\neg p$	q	$\neg q$	$p \Rightarrow q$	$\neg q \Rightarrow \neg p$
1	0	1	0	**1**	**1**
1	0	0	1	**0**	**0**
0	1	1	0	**1**	**1**
0	1	0	1	**1**	**1**

Thus, an *implication* and its *contrapositive* are *logically equivalent*.

The implication $x > 10 \Rightarrow x > 5$, where x represents a real number, is **true** for all real numbers x. Its converse,

$$x > 5 \Rightarrow x > 10,$$

is **false**; for instance, $x = 8$ makes the antecedent, $8 > 5$, true but the conclusion, $8 > 10$, false. The inverse,

$$x \le 10 \Rightarrow x \le 5,$$

is also **false**, which could be predicted as the converse and the inverse are logically equivalent.

The contrapositive of $x > 10 \Rightarrow x > 5$,

$$x \le 5 \Rightarrow x \le 10,$$

is **true**, which could also be predicted, as an implication and its contrapositive are logically equivalent.

Law of Substitution

The law of substitution states that in a deductive process, one proposition may at any time be substituted for an equivalent proposition. For instance, at any point in an argument, we may replace the contrapositive $\neg q \Rightarrow \neg p$ with $p \Rightarrow q$, or the converse $q \Rightarrow p$ with its inverse $\neg p \Rightarrow \neg q$.

Equivalence

$\Leftrightarrow$ The **equivalence** symbol $\Leftrightarrow$ indicates "is equivalent to".

The expression $p \Leftrightarrow q$ reads, "p is equivalent to q".

$p \Leftrightarrow q$ is **true** when the propositions **have identical truth values** (*both* propositions are true, or *both* false).

$p \Leftrightarrow q$ is **false** when the propositions **do not have identical truth values**.

An equivalence has the same truth value whether read from left to right or right to left; the equivalence $p \Leftrightarrow q$ is true **if and only if** $p \Rightarrow q$ and $q \Rightarrow p$:

p	q	$p \Leftrightarrow q$	$q \Leftrightarrow p$
1	1	1	1
1	0	0	0
0	1	0	0
0	0	1	1

The equivalence

$$p\text{: } x \text{ is 2 or a multiple of 2} \Leftrightarrow q\text{: } x \text{ is an even number},$$

where x denotes a real-number integer, is **true** for any natural number, because whatever number we choose to represent x, both components of the equivalence are either true or false (identical truth values).

– *Should I walk to work or carry my lunch?*

walk to work $\Leftrightarrow$ money to buy lunch
carry lunch $\Leftrightarrow$ money for transportation

6.3 Tautologies

"Let's talk film or let's not talk film–I'm easy"

Greek, *tauto*, "the same thing"; *lógos*, "a discourse", "a reasoning"

A tautology is a compound proposition which is true regardless of the truth values of its components.

p. 216

The disjunction $p \vee (\neg p)$, an application of the *principle of the excluded middle*, is a tautology and asserts that **a proposition is either true or false –** the two-value principle; there are no values between true (1) and false (0).

p	$\neg p$	$p \vee (\neg p)$
1	0	1
0	1	1

Given the proposition

(p): The book is in the library,

the disjunction

$$p \vee (\neg p),$$

meaning, "The book either **is or is not** in the library",

is a **true** proposition.

Given (q): $a + 2 = 7$, the disjunction

$$q \vee (\neg q),$$

meaning, "$a + 2$ either **is or is not** equal to 7",

is a **true** proposition.

A proposition $p \wedge (\neg p)$ is **always false**, whereas its negation, the tautology $\neg[p \wedge (\neg p)]$, is **always true**:

p	$\neg p$	$p \wedge (\neg p)$
1	0	**0**
0	1	**0**

p	$\neg p$	$p \wedge (\neg p)$	$\neg[p \wedge (\neg p)]$
1	0	0	**1**
0	1	0	**1**

The *principle of contradiction*, stating that a proposition and its negation cannot be both true and false, is expressed in symbolic form as the tautology

$$\neg[p \wedge (\neg p)] .$$

Given the proposition

(p): The book is in the library,

the conjunction

$$p \wedge (\neg p)$$

– meaning, "The book both **is and is not** in the library" – is a **false** proposition.

On the other hand, the proposition

$$\neg[p \wedge (\neg p)]$$

– "It is **false** that the book both **is and is not** in the library" – is **true**.

Given (q): $a + 2 = 7$, the proposition

$$q \wedge (\neg q)$$

is **false**, but

$$\neg[q \wedge (\neg q)]$$

– "It is **false** that $a + 2$ both **is and is not** equal to 7" – is a **true** proposition.

* * *

Wishful thinking has no use for logic – symbolic or not:

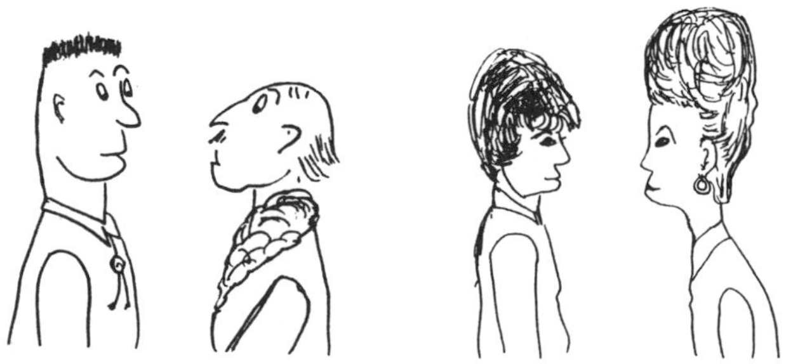

– *If you could choose between your spouse and a film star, which film star would you choose?*

6.4 Syllogisms and Proofs

DODGSON Charles Lutwidge
English mathematician, logician, photographer; as Lewis Carroll, author of children's books; 1832 - 1898)

"Contrariwise," continued Tweedledee, "if it was so, it might be; and if it were so, it would be; but as it isn't, it ain't. That's logic."

Lewis Carroll, *Through the Looking-Glass* (1872)

Greek, *syllogismós*, "a reasoning"

A **syllogism** is a valid deductive argument consisting of **two premises**, usually called the major premise and the minor premise, and **one conclusion**.

Major Premise:	Only even numbers are divisible by 2
Minor Premise:	624 is divisible by 2
Conclusion:	624 is an even number

Principle of Syllogisms

If the two implications $p \Rightarrow q$ and $q \Rightarrow r$ are true, then the implication $p \Rightarrow r$ must also be true. This may be stated

$$[(p \Rightarrow q) \wedge (q \Rightarrow r)] \Rightarrow (p \Rightarrow r),$$

which reads, "If p implies q, and q implies r, then p implies r". Such an inference of conclusions from two propositions complies with the **principle**, or **law**, **of syllogisms**, also called the **chain rule of inference**.

Chain Rule of Inference

p	q	r	$p \Rightarrow q$	$q \Rightarrow r$	$p \Rightarrow r$
1	1	1	1	1	**1**
1	1	0	1	0	**0**
1	0	1	0	1	**1**
1	0	0	0	1	**0**
0	1	1	1	1	**1**
0	1	0	1	0	**1**
0	0	1	1	1	**1**
0	0	0	1	1	**1**

For a **direct proof**, arrange a chain of implications from the hypothesis to the theorem.

- Given the true proposition p and the true implications $p \Rightarrow q$, $q \Rightarrow r$, and $r \Rightarrow t$, prove the validity of the proposition that t is true.

By the *principle of syllogisms*, we have

$$[(p \Rightarrow q) \wedge (q \Rightarrow r)] \Rightarrow (p \Rightarrow r).$$

As p is true and $p \Rightarrow r$ is a true implication, we have, by the *principle of detachment*, that r is true; and as $r \Rightarrow t$ is a true implication, we have, again by the *principle of detachment*, that t is true. •

cf. p. 159

For an **indirect proof**, or proof by *reductio ad absurdum*, assume that the *negation* of the sought conclusion is true, proceed with a chain of implications, and arrive at a contradiction.

- Given the true proposition q and the true implications $p \Rightarrow \neg q$ and $\neg p \Rightarrow r$, prove the validity of the proposition that r is true.

Assume that $\neg r$ is true.

True by hypotheses: $q;\ p \Rightarrow \neg q;\ \neg p \Rightarrow r$

True by assumption: $\neg r$

As $\neg p \Rightarrow r$ is true, the contrapositive, $\neg r \Rightarrow \neg\neg p$, must also be true. Using the axiom

$\neg\neg p$ is true if and only if p is true,

we may state that $\neg r \Rightarrow p$.

As $\neg r \Rightarrow p$ and $p \Rightarrow \neg q$ are true, we have, by the *principle of syllogisms*, that the proposition $\neg r \Rightarrow \neg q$ is true, and thus

$$(\neg r \Rightarrow p) \Rightarrow (p \Rightarrow \neg q) \Rightarrow (\neg r \Rightarrow \neg q)$$

is true.

The negation $\neg r$ is true by assumption and the implication $\neg r \Rightarrow \neg q$ is true by proof.

If $\neg r$ were true, then, by the *principle of detachment*, the negation $\neg q$ would also be true. However, q is true by hypothesis, and by the *two-value principle* $\neg q$ cannot also be true; therefore, it is a *false assumption* that $\neg r$ is true.

Hence, the proposition r is true. •

Technically, there is no distinction between a direct proof and an indirect proof. The indirect proof uses the rule

from $p \Rightarrow q$ and $\neg q$, infer $\neg p$.

***

Forest of Sideways Logic

After the witch was killed, Hansel and Gretel wandered around in the forest trying to find their way home. At a fork in the path was stationed a forest ogre's guard; one path lead out of the forest, the other to the ogre's supper table, and not as a guest!

By order of the ogre, the guard was to be asked only one question by each traveler or group of travelers. The dead witch's grateful cat had warned Hansel and Gretel that the guard lied every other time. Hansel and Gretel walked, one by one, Hansel first, then Gretel.

– What question did Gretel ask the guard to be sure to infer the correct direction?

– *What direction did you tell Hansel?*

The guard's statement to Gretel is true only if the statement to Hansel was false, and false only if the statement to Hansel was true. Consequently, if the reply to Gretel is "right", then

"*right*" $\Rightarrow$ the road to the *left* is the correct one

and, if "left", then

"*left*" $\Rightarrow$ the road to the *right* is the correct one.

6.5 Logic Circuits

Graphic designs of circuits can be used not only for electric systems but also for any system whose components constitute a flow pattern, such as a transportation system, a computer system, a political system, or an economic system.

Series Switch-Circuit System

A system that quits functioning if only one component is out of order is comparable to a **series switch-circuit system**. The circuit below works, because all three switches are closed:

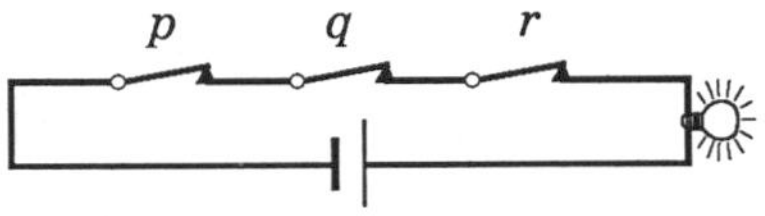

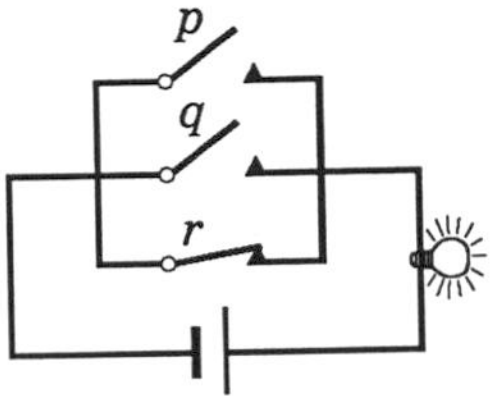

Parallel Switch-Circuit System

A **parallel switch-circuit system**, as shown above to the right, functions as long as at least one component works.

Complex Switch-Circuit System

A system composed of both series and parallel networks is called a **complex switch-circuit system**.

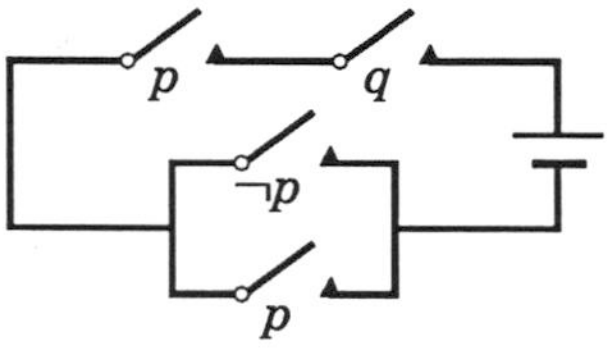

Two switches p and q in a series circuit may be expressed as $p \wedge q$; in a parallel circuit, $p \vee q$:

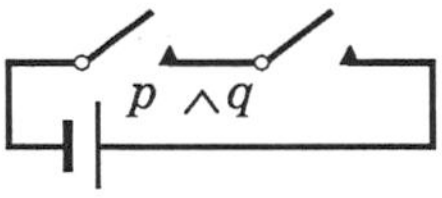

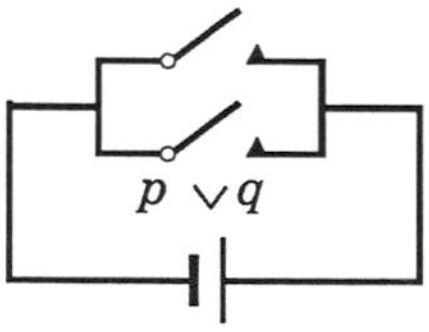

A switch that is always open when another switch, p, is closed, and vice versa, may be described as $\neg p$.

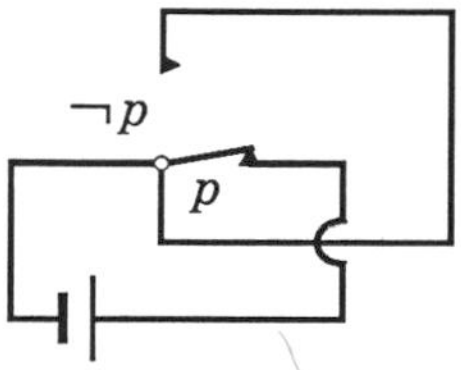

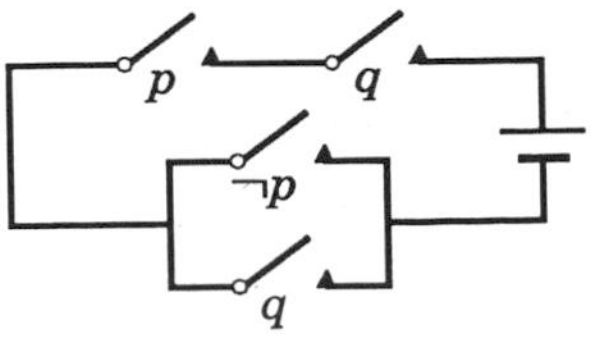

A switch may appear in more than one location. The same switch can only be in one of two states: **open** or **closed**. The switch p is both in series circuit with q and in parallel circuit with $\neg p$; at the same time, p must – in both locations – be either open or closed. A closed switch has the truth value 1; an open switch, 0:

p	$\neg p$	q	$p \wedge q$	$\neg p \vee q$	$(p \wedge q) \wedge (\neg p \vee q)$
1	0	1	1	1	**1**
0	1	0	0	1	**0**
1	0	0	0	0	**0**
0	1	1	0	1	**0**

Equivalent Switch Circuits

If two switch networks have the same truth value, then they are referred to as equivalent switch circuits:

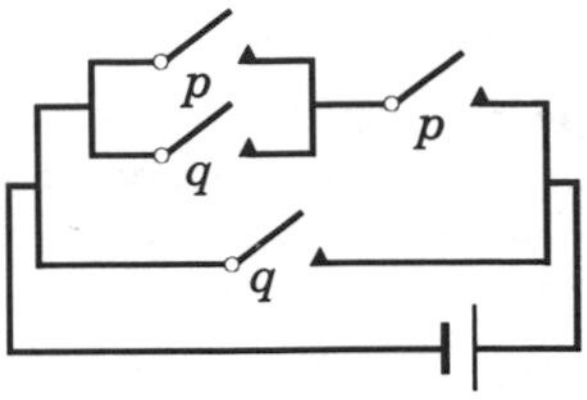

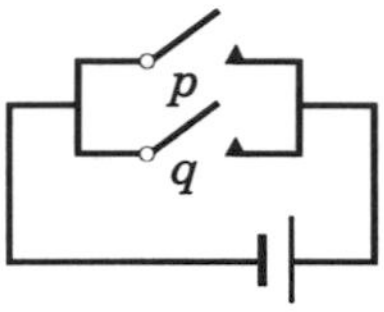

The above two systems have the same truth values and, consequently, are equivalent:

p	q	$p \vee q$	$(p \vee q) \wedge p$	$[(p \vee q) \wedge p] \vee q$
1	1	**1**	1	**1**
1	0	**1**	1	**1**
0	1	**1**	0	**1**
0	0	**0**	0	**0**

- **a:** $(p \wedge q) \vee (r \wedge t)$ **b:** $(p \vee q) \wedge (r \vee t)$

  c: $(p \vee q) \wedge [p \vee (q \wedge r)] \wedge r$

  Draw the corresponding switch circuits.

a:

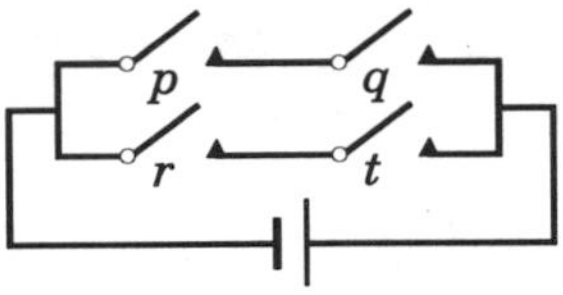

b:

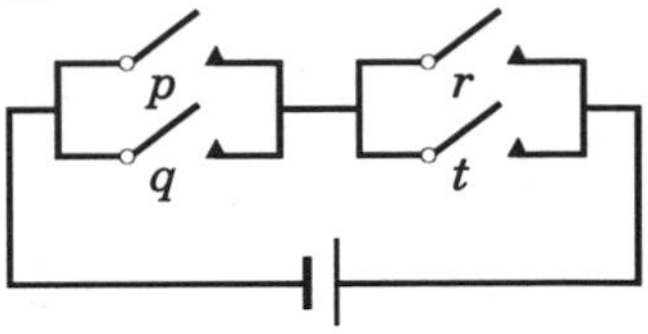

c:

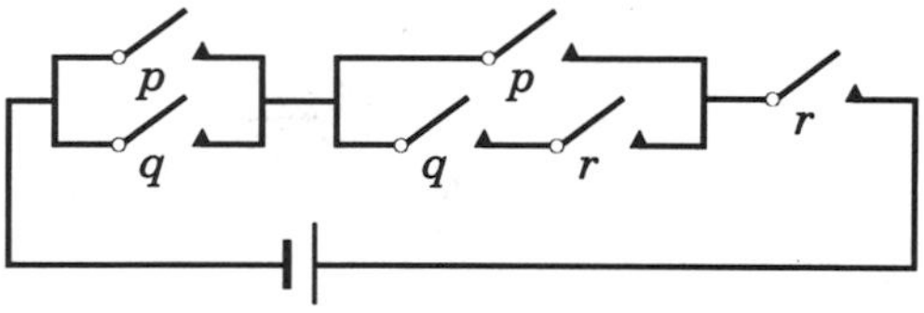

- 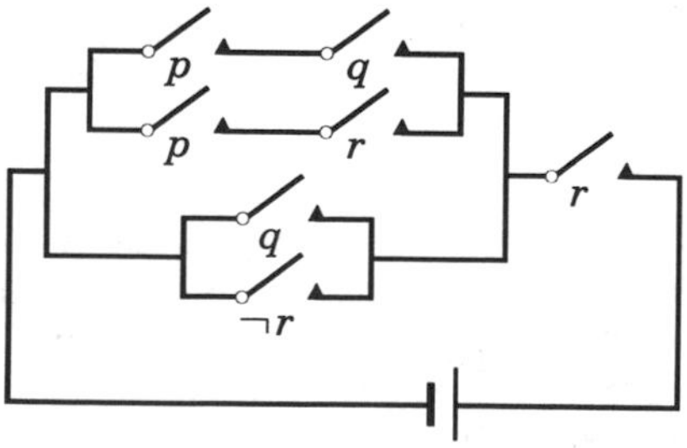

By the use of symbols, describe the circuit.

$\{[(p \wedge q) \vee (p \wedge r)] \vee [q \vee (\neg r)]\} \wedge r$.

Chapter 7

SET THEORY

Knowledge of symbols and concepts introduced in this chapter is not needed to understand subsequent chapters.

7.0 Introduction

Elements, Members

Set theory deals with the properties of well-defined collections, or **sets**, of entities – the **elements** or **members** of the set – conceived as a whole. The elements may be of a mathematical nature,

{0, 1, 2, 3, 4} ; {2 x; x represents any natural number} ,

or non-mathematical,

{Robin, Kelly, Kim} ; {all humans with webbed toes} .

CANTOR Georg (1845 - 1918)

The mathematical **theory of sets** grew out of the German mathematician **Georg Cantor**'s study of infinite sets of real numbers. The *language of sets* has become an important tool for all branches of mathematics – as a basis for precise definition of higher concepts, and for mathematical reasoning – but is of very little relevance to the practice of mathematics in everyday life.

In the 1950s educators/reformers introduced the language of sets as the basis for mathematical studies in schools. Many children started studying sets before they could count the number of elements in the sets. The language of sets and the surrounding "New Math" created havoc in schools in the 1950s, '60s, and '70s. It was a frustrating time in education. Strange symbols were introduced for seemingly simple things; teachers had to be retrained and most parents had no idea what their children were doing in mathematics.

The concepts of set theory are simple, but they require a precision and maturity of language that is beyond the power of many students. An idea that was meant to simplify in fact complicated matters. A dull but useful drill was replaced by a dull and useless drill. Sadly, in Sweden and other countries, the New Math created a generation with sometimes very limited residual arithmetic skill.

And in high school it got even harder. There we had to learn to count to 15 ... by heart.

7.1 Sets and Their Contents

Notation	Example	Reads
$\{\ ,\dots,\ \}$		"set"
	$\{x_1, x_2 \dots x_n\}$	"Set with elements $x_1, x_2 \dots x_n$"
$\in$		"belongs to" *or* "is an element of"
	$x \in A$	"x belongs to A" *or* "x is an element of the set A"
$\notin$		"does not belong to" *or* "is not an element of"
	$x \notin A$	"x does not belong to A" *or* "x is not an element of the set A"
$\ni$		"contains"
	$A \ni x$	"The set A contains x as an element"
$\not\ni$		"does not contain"
	$A \not\ni x$	"The set A does not contain x as an element"

PEANO Giuseppe
(1858 - 1932)

Like many other notations in symbolic logic and set theory, $\in$ was introduced, in 1889, by the Italian mathematician **Giuseppe Peano**. $\in$ is a stylized form of the Greek *epsilon*, ε, first letter of the Greek word ἐστι, which means "is".

The notations $\in$ and $\notin$, on one hand, and $\ni$ and $\not\ni$, on the other hand, represent the same qualities seen from opposite points of view;

if $B = \{a, c, k\}$, then

$a \in B$, $c \in B$, and $k \in B$, ***or*** $B \ni a$, $B \ni c$, and $B \ni k$

and

$b \notin B$ ***or*** $B \not\ni b$.

One set may be an element of another set; *e.g.*, $C = \{1, 2, \{3, 4\}, 5\}$ has an element that is a set, namely, $\{3, 4\}$, which we may write $\{3, 4\} \in C$.

In the sets

$$A = \{1, 2, 3\};\quad B = \{1, \{2, 3\}\}\ ;\quad C = \{\{1,2\}, 3\},$$

1 is an element of A and B but not of C,

$$1 \in A\ ;\quad 1 \in B\ ;\quad 1 \notin C\ .$$

Sets of Numbers

Notation	Reads	Definition
Ø	"The empty set"	**Ø** = { }
N	"The set of natural numbers and zero"	**N** = {0, 1, 2, 3 ...}
N*	"The set of natural numbers"	**N*** = {1, 2, 3 ...}
Z	"The set of integers"	**Z** = {... −2, −1, 0, 1, 2 ...}
$\mathbf{Z}_+$	"The set of positive integers"	$\mathbf{Z}_+$ (= **N***) = {1, 2, 3 ...}
$\mathbf{Z}_-$	"The set of negative integers"	$\mathbf{Z}_-$ = {... −3, −2, −1}
Z*	"The set of all integers except zero	**Z*** = {... −2, −1, 1, 2 ...}
Q	"The set of rational numbers"	
$\mathbf{Q}_+$	"The set of positive rational numbers"	
$\mathbf{Q}_-$	"The set of negative rational numbers"	
Q*	"The set of all rational numbers except zero"	
R	"The set of real numbers"	
$\mathbf{R}_+$	"The set of positive real numbers"	
$\mathbf{R}_-$	"The set of negative real numbers"	
R*	"The set of all real numbers except zero"	
C	"The set of complex numbers"	
C*	"The set of all complex numbers except zero"	

The exclusion of zero from the set is generally denoted by an asterisk, as in the above **N***, **Z***, **Q***, **R***, and **C***.

Sets of positive or negative numbers are generally denoted by a subscript of a positive or negative sign, as in $\mathbf{Z}_+$, $\mathbf{Z}_-$, $\mathbf{Q}_+$, $\mathbf{Q}_-$, $\mathbf{R}_+$, and $\mathbf{R}_-$.

The sets **N**, **Z**, **Q**, **R**, and **C** may be denoted by an outlined typeface, for instance, $\mathbb{Z}$, $\mathbb{Q}$, $\mathbb{R}$.

The empty set, **Ø**, contains no elements.

Finite Sets

A **finite set** is empty or contains a finite number of elements.

Infinite Sets

An **infinite set** contains an infinite number of elements. The number sets **N**, **Z**, **Q**, **R**, and **C** are all infinite.

Examples:

Set	Number of elements
A = {1, 2, 3}	3
B = {0, 2, 3}	3
C = {0, 1, {2, 3}}	3
D = {0}	1
E = **Ø**	0
F = {**Ø**}	1
G = **Z**	∞
H = {**Z**}	1

Set Builder

Notation	Example	Reads
$\{\mid\}$ (set builder) Also used: $\{:\}, \{;\},$ *or* $\{.\}$	$\{x \in A \mid p(x)\}$	"The set of those elements x of A for which the proposition $p(x)$ is true"

Subset

Universal Set, Universe

If each element of a set A is contained within a set B, then A is a **subset** of B. Every set is a subset of the **universal set**, or **universe**, which contains all elements capable of being accepted to the problem. The universal set of the problem is usually denoted U.

If the universal set is $\mathbf{Z}$, then $\{x \in \mathbf{Z} \mid x \leq 3\}$ is the set of all integers less than or equal to 3, that is, $\{\ldots, -2, -1, 0, 1, 2, 3\}$.

Russell's Paradox

FREGE Gottlob (1848 - 1925)

RUSSELL Bertrand (1872 - 1970)

In 1902, in a letter to Gottlob Frege, Bertrand Russell drew attention to the danger of unrestricted use of abstraction when forming sets.

Suppose that there are two kinds of sets,

- **Normal sets** – sets that do not contain themselves as elements; and
- **Non-normal sets** – sets that contain themselves as elements.

"The set of all dogs" is a normal set, for obviously the set itself is not a dog.

"The set of all sets" and "the set of all things that are not dogs" are non-normal sets, for the sets are elements of themselves.

Now consider the set S whose elements x are sets that are not elements of themselves,

$$S = \{x \mid x \notin x\}.$$

Is S an element of S?

If $S \notin S$, then S meets the requirement $(x \notin x)$ to be a member of S and, paradoxically, $S \in S$.

If $S \in S$, then S fails to meet the requirement $(x \notin x)$ to be a member of S and, paradoxically, $S \notin S$.

The contradiction,

$$\text{if } S \notin S \text{ then } S \in S; \quad \text{if } S \in S \text{ then } S \notin S,$$

is known as **Russell's paradox**.

We may see the above as a proof of the theorem that

$$S = \{x \mid x \notin x\} \textit{ does not exist.}$$

To enact his discovery, Russell proposed, in 1918, the **barber paradox**.

A barber has a sign:

> *I shave all those men in town, and only those men, who do not shave themselves.*

– *Shave me, shave me not, shave me, shave me not shave me, shave me not, shave me ... take the sign down.*

A non-member of the set of all men in town who do not shave themselves – a *woman* – can, of course, advertise as above with impunity.

If a man, and he shaved *himself*, the barber would belong to the set of men who shave themselves, but the sign implies that he shaves *only* those who do not shave themselves, so he cannot shave himself – and neither can anybody else, as he, the barber, shaves *all* those men who do not shave themselves. If he decided to wear a full beard, the contradiction prevails, for the sign says that *he* shaves *all* those men who do not shave themselves.

Russell's paradox was not the first paradox to be discovered in set theory, but its simplicity and directness had an immense impact on the development of the ideas and foundations of mathematics at the beginning of the 20th century; before that time the concept "class of all classes" or "set of all sets" had been used in an unhampered manner.

Subsets

Notation	**Example**	**Reads**
$\subseteq$ (subset)	$B \subseteq A$	"B is included in A" *or* "B is a subset of A"
$\nsubseteq$ (no subset)	$C \nsubseteq A$	"C is not included in A" *or* "C is not a subset of A"
$\subset$ (proper subset)	$B \subset A$	"B is a proper subset of A" *or* "B is properly included in A"
$\supseteq$ (include as a subset)	$A \supseteq B$	"A includes B (as a subset)"
$\nsupseteq$ (not include as a subset)	$A \nsupseteq C$	"A does not include C (as a subset)"
$\supset$ (include properly)	$A \supset B$	"A includes B properly"
$\mathcal{P}$, P, or P (power set)	$\mathcal{P}(A)$	"The power set of A"

The notations $\subseteq$, $\nsubseteq$, $\subset$ and $\supseteq$, $\nsupseteq$, $\supset$ represent the same qualities seen from opposite points of view.

If $A = \{1, 2, 3, 4, 5\}$, $B = \{3, 5\}$, $C = \{2, 6\}$, and $D = \{4, 5, 1, 3, 2\}$, then "*B* is a subset of *A*" and "*D* is a subset of *A*", written

$$B \subseteq A; \quad D \subseteq A,$$

or "*A* includes *B*" and "*A* includes *D*",

$$A \supseteq B; \quad A \supseteq D,$$

but "*C* is not a subset of *A*", or "*A* does not include *C*",

$$C \nsubseteq A; \quad A \nsupseteq C;$$

and, as the order of the elements in a set carries no weight,

$$D = A.$$

We note that

- the empty set is a subset of every set; every set is a subset of itself.

The correct use of the notations $\in$, $\notin$ ("element of", "not an element of", respectively) and $\subseteq$, $\nsubseteq$ is here illustrated for the set {12}, which has one element, 12, and two subsets, $\emptyset$ and the set itself:

$$12 \in \{12\}; \quad \{12\} \subseteq \{12\}; \quad 12 \nsubseteq \{12\}$$
$$\emptyset \subseteq \{12\}; \quad \emptyset \notin \{12\}; \quad \{12\} \notin \{12\}$$

$\subsetneq, \subset;\ \supsetneq, \supset$

We note that the notations $\subset$, $\subseteq$ and $\supset$, $\supseteq$ used here, correspond to $\subsetneq$, $\subset$ and $\supsetneq$, $\supset$, often encountered in other texts.

Proper and Improper Subsets

A **proper subset** includes elements of the set to which it is matched, but is not equal to that set; a proper subset may also be an empty set.

$B = \{2, 5\}$ and $C = \emptyset$ are proper subsets of $A = \{1, 2, 4, 5\}$.

What sometimes is called an **improper subset** is supposed to include all the elements of the set to which it is matched; thus

$C = \{2, 1, 5, 4\}$ would then be an improper subset of $A = \{1, 2, 4, 5\}$.

The notations $\subseteq$ and $\supseteq$ are used in connection with subsets of a given set, whether proper or improper:

$$B \subseteq A; \quad A \supseteq B; \quad C \subseteq A; \quad A \supseteq C.$$

Although not the practice in all texts, the symbols $\subset$ and $\supset$ are usually preferred in connection with proper subsets,

$$B \subset A; \quad A \supset B.$$

As the order of the elements in a set carries no weight, the very concept of improper subsets seems unnecessary.

Power Sets

A **power set** is the set of all subsets of a given set, containing the empty set and the original set.

If $A = \{a, b, c\}$, then

$$\mathcal{P}(A) = \{\emptyset, \{a\}, \{b\}, \{c\}, \{a, b\}, \{a, c\}, \{b, c\}, \{a, b, c\}\}.$$

"Power" in *power sets* means "exponent"; if the number of elements contained in a set A is n, then $\mathcal{P}(A)$ contains 2^n elements.

Union

Notation	Example	Reads
$\cup$ (union or "cup")	$A \cup B$	"The union of A and B" *or* "A union B"
$\bigcup$ (union of collection of sets)	$\bigcup_{i=1}^{n} A_i$	"The union of a collection of sets $A_1, \dots, A_n$"

p. 222 The union notation is analogous to the disjunction notation $\vee$ of symbolic logic.

A **union** is a set consisting of all elements that appear at least once in the original sets.

If $B = \{2, 3, 5\}$ and $C = \{1, 2, 3, 6\}$,

$$B \cup C = \{1, 2, 3, 5, 6\}.$$

$\bigcup_{i=1}^{n} A_i$ denotes a new set formed by all elements that appear at least once in any of the sets $A_1, \dots, A_n$.

The union of a set with itself does not form a new set; if $B = \{2, 3, 5\}$,

$$B \cup B = B = \{2, 3, 5\}.$$

- Given: The universal set $\mathbf{R}$, and the subsets

$$A = \{x \in \mathbf{R} \mid -2 < x \leq 4\}; \; B = \{x \in \mathbf{R} \mid x > 0\}$$

Find $A \cup B$.

A is the set of all real numbers x which are greater than -2 but smaller than or equal to 4:

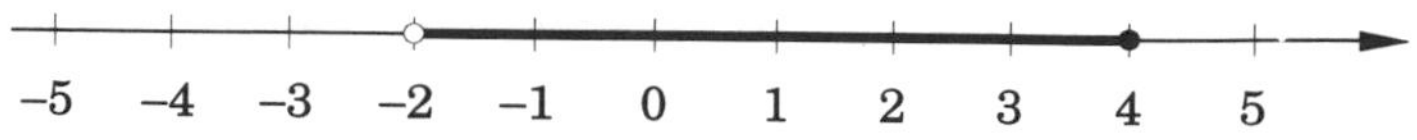

B is the set of all real numbers x which are greater than 0:

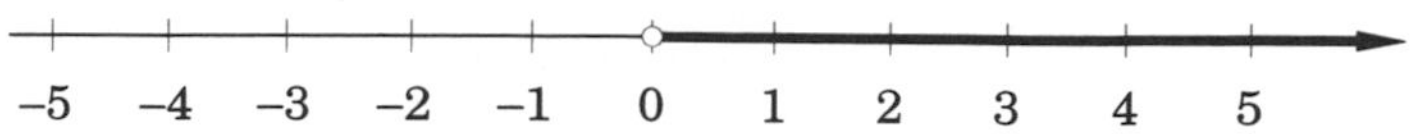

$A \cup B = \{x \in \mathbf{R} \mid x > -2\}$

Intersection

Notation	Example	Reads
$\cap$ (intersection, or "cap")	$A \cap B$	"The intersection of A and B" *or* "A intersection B"
$\bigcap$ (intersection of a collection of sets)	$\bigcap_{i=1}^{n} A_i$	"The intersection of a collection of sets $A_1, \ldots, A_n$"

p. 221

The intersection notation is analogous to the conjunction notation $\wedge$ of symbolic logic.

An **intersection** is a set consisting of the elements that are common to the original sets.

If $B = \{2, 3, 5\}$ and $C = \{1, 2, 3, 6\}$,

$$B \cap C = \{2, 3\} .$$

$\bigcap_{i=1}^{n} A_i$ denotes a new set formed by all the elements that are common to all the sets $A_1, \ldots, A_n$.

The intersection of a set with itself does not form a new set; if $B = \{2, 3, 5\}$,

$$B \cap B = B = \{2, 3, 5\} .$$

- Given: The universal set $\mathbf{R}$, and the subsets

$$A = \{x \in \mathbf{R} \mid -2 < x \le 4\} \text{ and } B = \{x \in \mathbf{R} \mid x > 0\} .$$

Find $A \cap B$.

$A = \{x \in \mathbf{R} \mid -2 < x \le 4\}$:

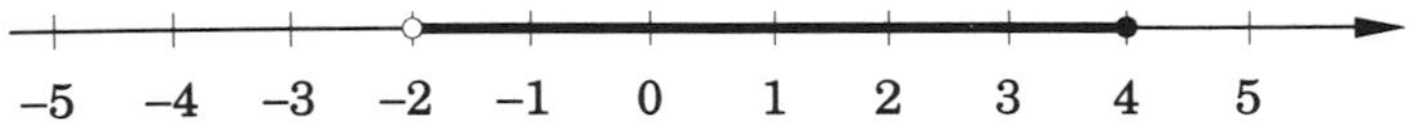

$B = \{x \in \mathbf{R} \mid x > 0\}$:

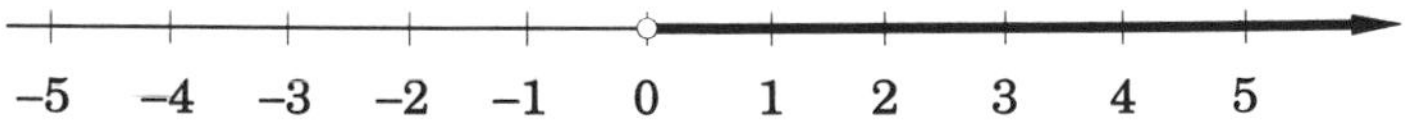

$A \cap B = \{x \in \mathbf{R} \mid 0 < x \le 4\}$

Difference

Notation	Example	Reads
$\setminus$ (difference) Also used: $-$	$A \setminus B$ $A - B$	"The difference of A and B" *or* "A minus B"

The **difference** $A \setminus B$ of sets A and B is the set of all elements that belong to A but not to B. If $A = \{1, 2, 4, 5\}$, $B = \{2, 5\}$, $C = \{2, 3\}$:

$$A \setminus B = \{1, 4\} ; \quad A \setminus C = \{1, 4, 5\}$$

Complement

Notation	**Example**	**Reads**
$\complement$ (complement) Also used: $'$, $\tilde{}$, $\bar{}$	$\complement_U A$ A', $\tilde{A}$, $\bar{A}$	"The complement of subset A of U"

p. 220 The complement notation is analogous to the negation notation $\neg$ of symbolic logic.

Often, the notation of the universal set is not written in the complement notation; thus, instead of $\complement_U A$, it is usually correct to write the complement of a subset A as $\complement A$.

The complement is defined as the difference between the universal set and the subset,

$$\complement_U A = U \setminus A .$$

If $U = \{1, 2, 4, 5, 8, 9\}$ and $A = \{2, 5, 8\}$,

$$\complement_U A = \{1, 4, 9\} .$$

While the definitions of "union" and "intersection" do not require specification of the universe, the definition of "complement" does.

- Match $B \cup \complement B$; $A \cup \complement A$; $B \cap \complement B$; $A \cap \complement A$; $\complement(\complement B)$; $\complement(\complement A)$

  with $\varnothing$, U, A, B, $\complement A$, $\complement B$,

  where A and B are subsets of the universal set U.

  $B \cup \complement B = U$; $A \cup \complement A = U$; $B \cap \complement B = \varnothing$; $A \cap \complement A = \varnothing$;
  $\complement(\complement B) = B$; $\complement(\complement A) = A$.

- Given: The universal set $\mathbf{R}$, and the subset

  $$A = \{x \in \mathbf{R} \mid -2 < x \leq 4\} .$$

  Find $\complement_{\mathbf{R}} A$.

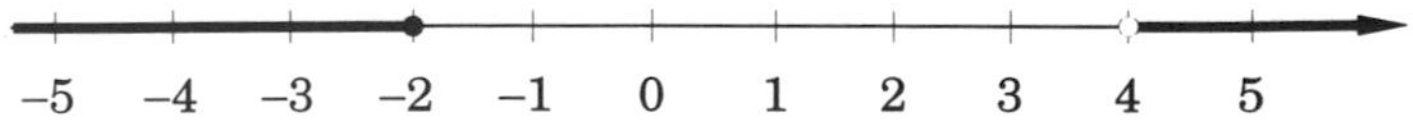

The complement of A contains -2, all real numbers smaller than -2, and all real numbers greater than 4 (but does not include 4),

$$\complement_{\mathrm{R}} A = \{x \mid x \leq -2 \text{ or } x > 4\} .$$

Ordered Components and Product Sets

In an **ordered pair** (a, b), or an **ordered *n*-tuplet** $(a_1, a_2, \ldots, a_n)$, the order is of significance:

$$(a, b) = (b, a) \textbf{ only if } a = b$$

Components

The elements within an ordered pair or an ordered n-tuplet are called **components**. In (a, b), a is the *first component* and b is the *second component*.

Product Sets, Cartesian Products

If sets are combined into sets of ordered pairs the result is called **product sets** or **Cartesian products**. To indicate the forming of a product set, the symbol $\times$ is used.

The product sets of $A = \{1, 2, 3\}$ and $B = \{a, b\}$ are

$$A \times B = \{(1, a), (1, b), (2, a), (2, b), (3, a), (3, b)\}$$
$$B \times A = \{(a, 1), (a, 2), (a, 3), (b, 1), (b, 2), (b, 3)\}$$
$$A \times A = \{(1, 1), (1, 2), (1, 3), (2, 1), (2, 2), (2, 3), (3, 1), (3, 2), (3, 3)\}$$
$$B \times B = \{(a, a), (a, b), (b, a), (b, b)\}$$

7.2 Venn Diagrams

A **Venn diagram** is a rectangle – the universal set – that includes circles depicting the subsets.

von LEIBNIZ Gottfried Wilhelm (1646 - 1716)

EULER Leonhard (1707 - 1783)

VENN John (1834 - 1923)

The first to systematically use diagrams to represent statements of logic reasoning was the German mathematician and philosopher **G. W. von Leibniz**. The diagrams used today are sometimes referred to as **Euler diagrams** after the Swiss mathematician **Leonhard Euler**, who devised them, but are usually called **Venn diagrams** after the English logician **John Venn**, who, in 1880, greatly improved the diagrams and popularized their use.

A Venn diagram can be employed for any number of subsets, but more than three defeat the purpose of gaining increased clarity.

Two Subsets

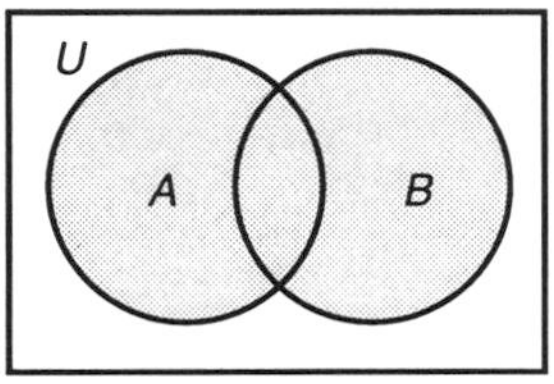

Union. A and B are subsets of the universal set U. The shaded area represents $A \cup B$.

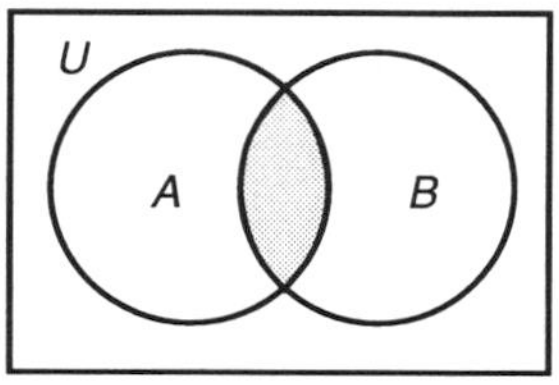

Intersection. A and B are subsets of the universal set U. The shaded area represents $A \cap B$.

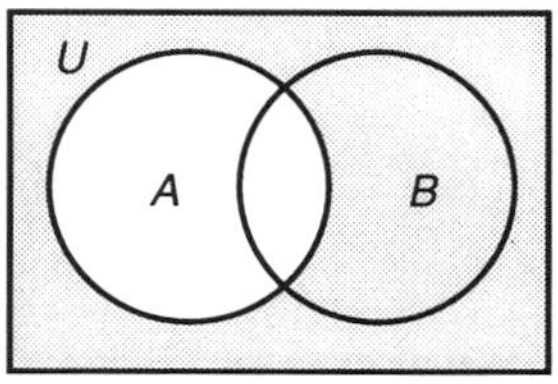

Complement. A and B are subsets of the universal set U. The shaded area represents $\complement A$.

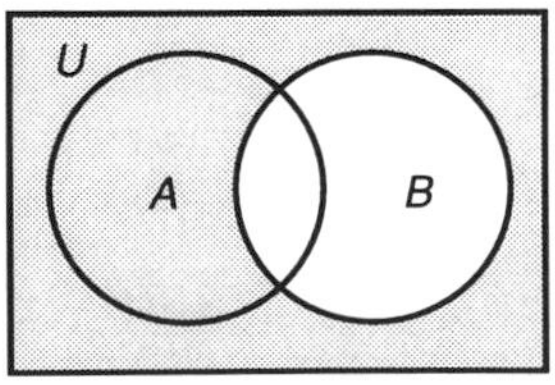

Complement. A and B are subsets of the universal set U. The shaded area represents $\complement B$.

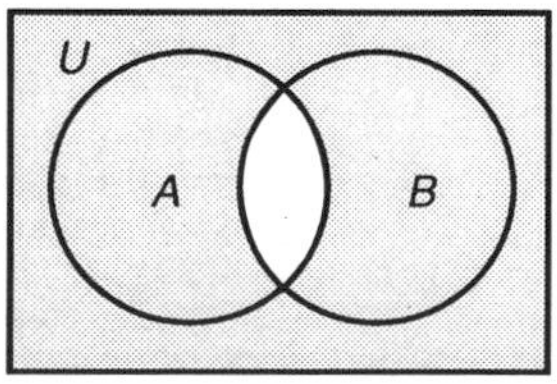

Union of Complements. A and B are subsets of the universal set U. The shaded area represents $\complement A \cup \complement B$.

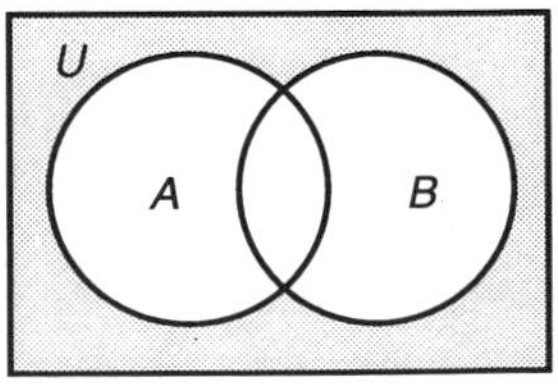

Intersection of Complements. A and B are subsets of the universal set U. The shaded area represents $\complement A \cap \complement B$.

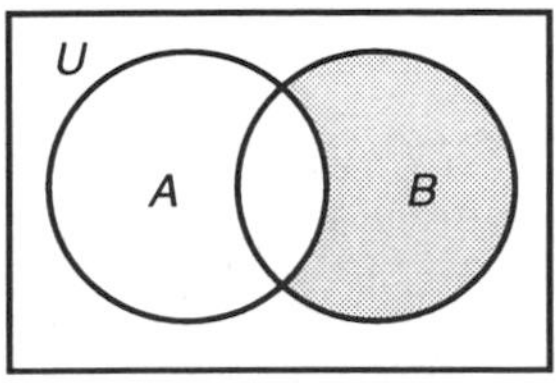

Intersection of complement and subset. A and B are subsets of the universal set U. The shaded area represents $(\complement A) \cap B$.

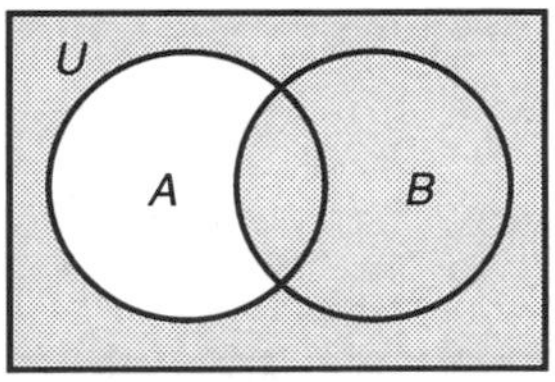

Union of complement and subset. A and B are subsets of the universal set U. The shaded area represents $(\complement A) \cup B$.

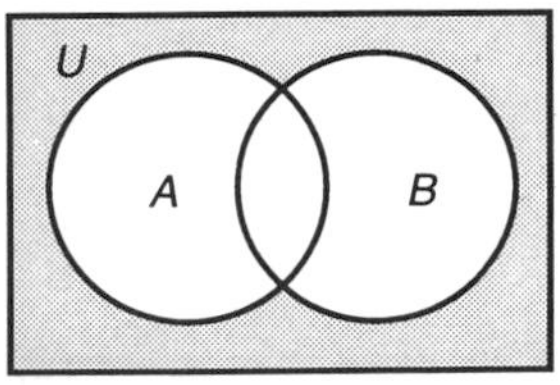

Complement of union of subsets. A and B are subsets of the universal set U. The shaded area represents $\complement(A \cup B$.

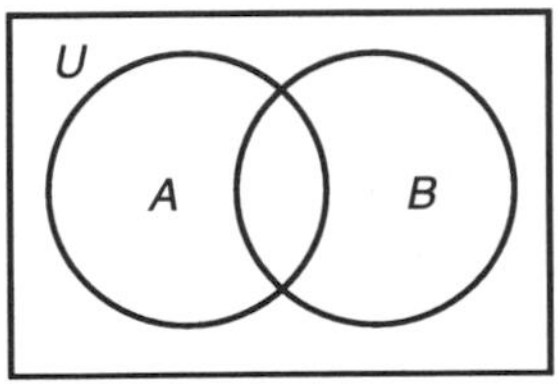

Intersection of subset and its complement. A and B are subsets of the universal set U. $B \cap \complement B = \varnothing$.

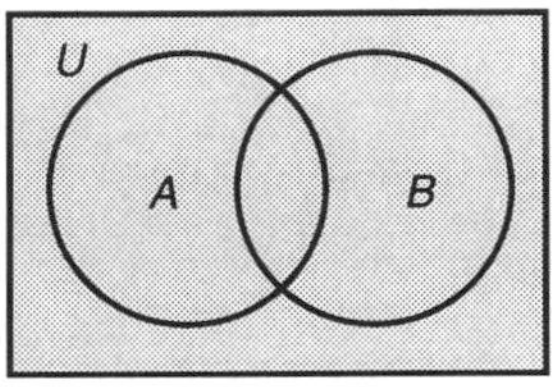

Union of subset and its complement. A and B are subsets of the universal set U. $B \cup \complement B$.

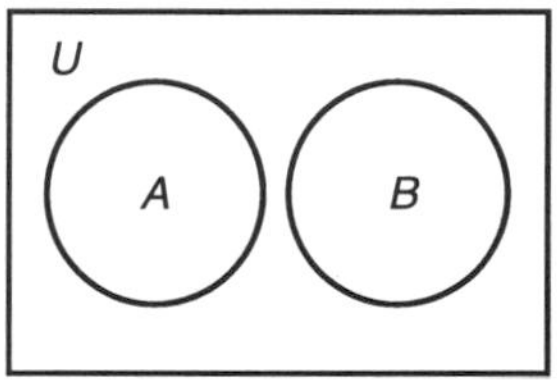

Disjoint sets. A and B are subsets of the universal set U. A and B have no elements in common. Sets that have no elements in common are called **disjoint sets**.

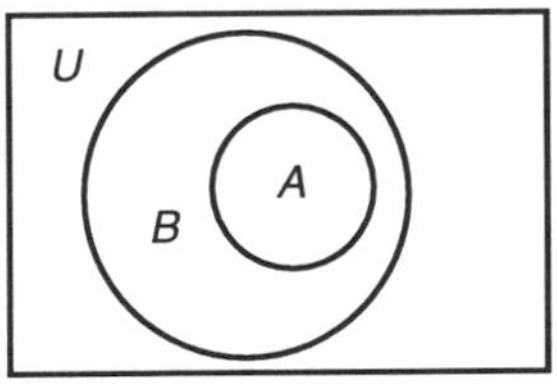

Proper subset. A and B are subsets of the universal set U. A is a proper subset of B, which may be expressed $A \subset B$.
Also, *e.g.*,
$A \cup B = B$ and $A \cap B = A$.

•

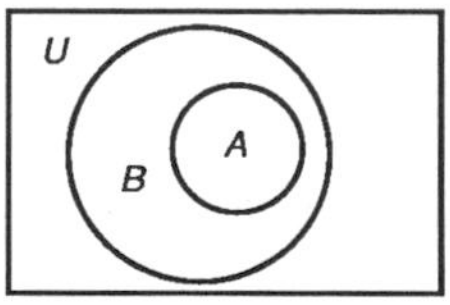

Match each of

$$A \cup B, \quad A \cap B, \quad A \cap \complement B, \quad A \setminus B,$$
$$\complement A \cup B, \quad \complement A \cap \complement B, \quad \complement A \cup \complement B$$

with one of

$$\varnothing, A, B, U, \complement B, \complement A.$$

$A \cup B = B$; $A \cap B = A$; $A \cap \complement B = \varnothing$; $A \setminus B = \varnothing$;
$\complement A \cup B = U$; $\complement A \cap \complement B = \complement B$; $\complement A \cup \complement B = \complement A$.

- Construct Venn diagrams showing

  a: $A \cap B = \emptyset$ **b:** $A \cup B = A$ **c:** $A \setminus B = A$,

  where A and B are subsets of the universal set U.

a:

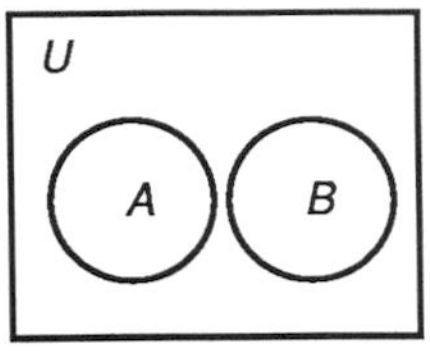

b:

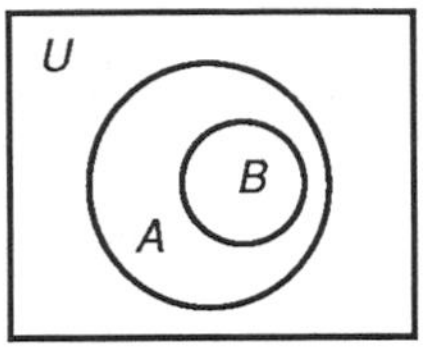

c:

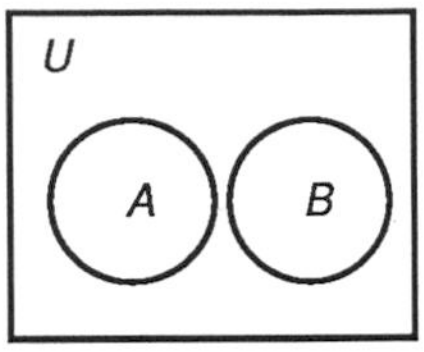

- Given: $U = \{1, 2, 3, 4, 5, 6, 7, 8, 9, 10\}$
  $A = \{1, 2, 4\}$
  $B = \{4, 5, 6, 7\}$

  Display the sets in a Venn diagram and determine

  a: $A \cup B$ **b:** $A \cap B$ **c:** $A \setminus B$ **d:** $B \setminus A$
  e: $\complement A$ **f:** $\complement B$ **g:** $\complement (A \cup B)$ **h:** $\complement (A \cap B)$

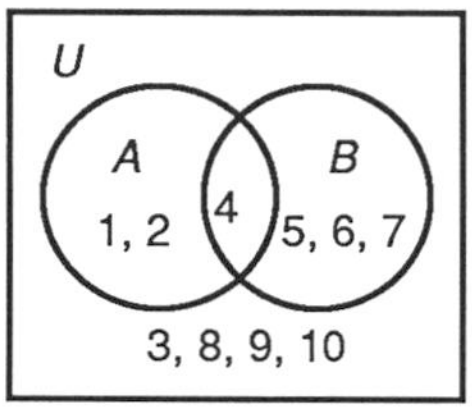

a: $A \cup B = \{1, 2, 4, 5, 6, 7\}$

b: $A \cap B = \{4\}$

c: $A \setminus B = \{1, 2\}$;

d: $B \setminus A = \{5, 6, 7\}$

e: $\complement A = \{3, 5, 6, 7, 8, 9, 10\}$

f: $\complement B = \{1, 2, 3, 8, 9, 10\}$

g: $\complement (A \cup B) = \{3, 8, 9, 10\}$

h: $\complement (A \cap B) = \{1, 2, 3, 5, 6, 7, 8, 9, 10\}$

- Of 160 individuals with a skin disorder, 100 had been exposed to chemical **A** (individuals A), 50 to chemical **B** (individuals B), and 30 to chemicals **A** and **B.**

  Use symbols and Venn diagrams to describe the number of individuals exposed to

  a: chemicals **A** and **B**
  b: chemical **A** but not chemical **B**
  c: chemical **B** but not chemical **A**
  d: chemical **A** or chemical **B**
  e: neither chemical **A** nor chemical **B**

Solution:

Of individuals A,

$100 - 30 = 70$

had been exposed only to chemical **A**.

Of individuals B,

$50 - 30 = 20$

had been exposed only to chemical **B**.

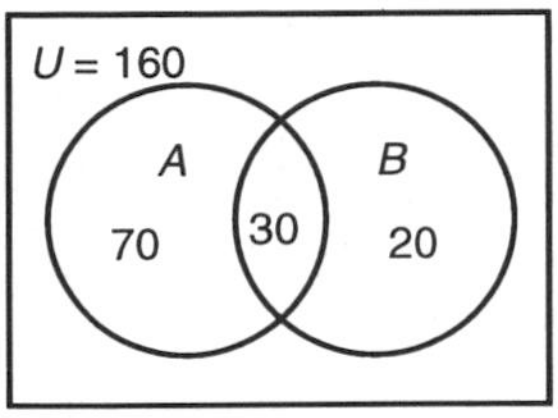

a:

The number of individuals exposed to both chemical **A** and chemical **B** can be expressed as

$|A \cap B| = 30$.

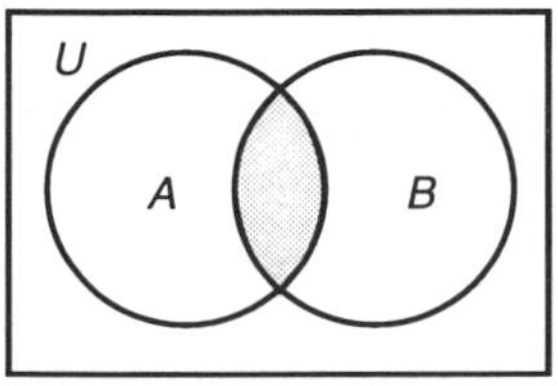

b:

The number of individuals exposed to chemical **A** but not to chemical **B** can be expressed as

$|A \cap \complement B| = 70$.

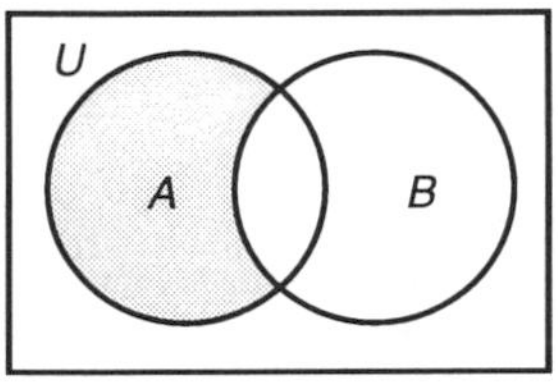

c:

The number of individuals exposed to chemical **B** but not to chemical **A** can be expressed as

$|\complement A \cap B| = 20$.

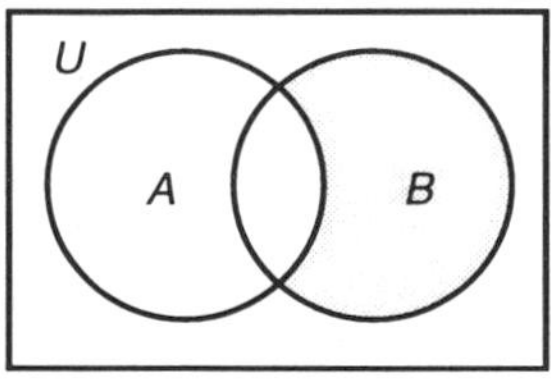

d:

The number of individuals exposed to chemical **A** or chemical **B** can be expressed as

$|A \cup B| = 120$.

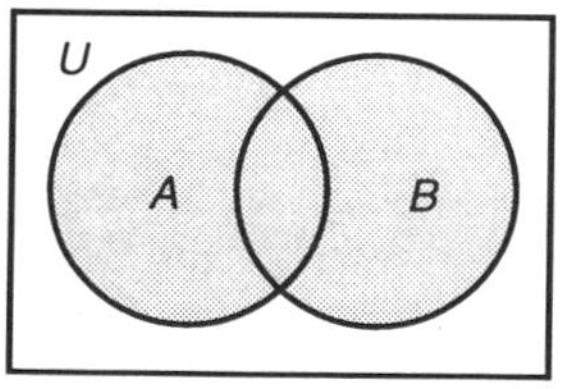

e:

The number of individuals exposed to neither chemical **A** nor chemical **B** can be expressed as *either*

$|\complement A \cap \complement B| = 40$

or $|\complement(A \cup B)| = 40$.

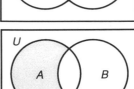

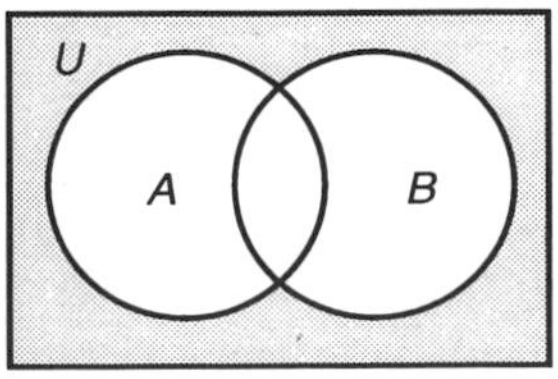

Three Subsets

When **three circular areas** are used to represent **three subsets** of the universal set and the areas overlap, eight regions may be formed:

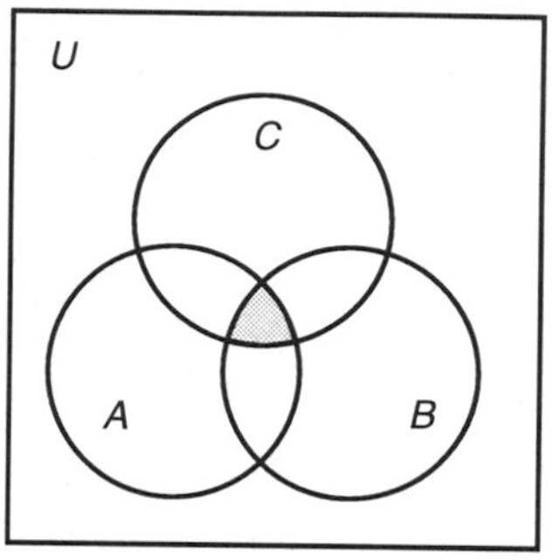

A, B, and C are subsets of the universal set U. The shaded area represents $A \cap B \cap C$.

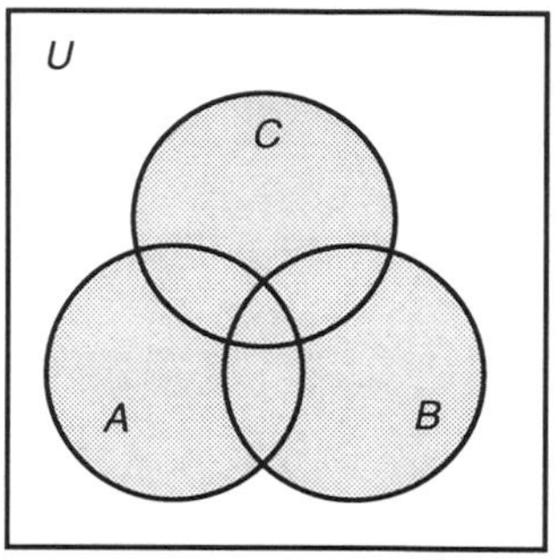

A, B, and C are subsets of the universal set U. The shaded area represents $A \cup B \cup C$.

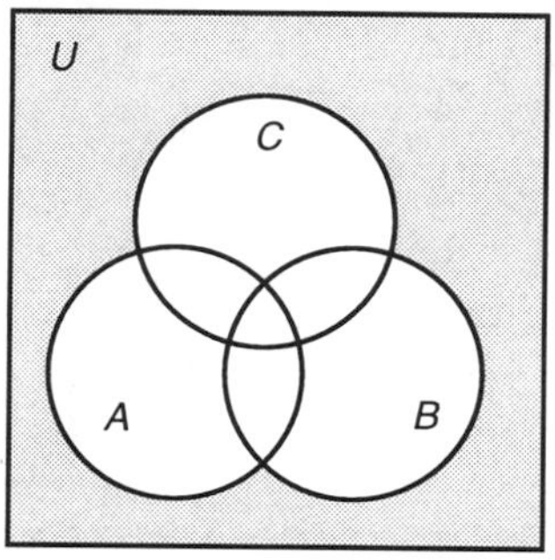

A, B, and C are subsets of the universal set U. The shaded area represents

$\complement(A \cup B \cup C$.

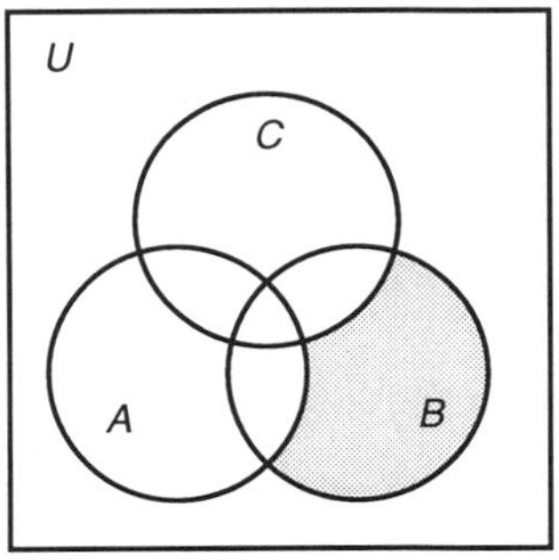

A, B, and C are subsets of the universal set U. The shaded area represents

$\complement(A \cup C) \cap B$.

- 800 individuals were examined for antigens called α_1, α_2, and α_3.

Number of individuals positive for

α_1	500
α_2	350
α_3	400
α_1 and α_2	250
α_2 and α_3	150
α_1 and α_3	200
α_1, α_2, and α_3	50

Find **a**: The number of individuals negative for all three antigens, α_1, α_2, and α_3.

b: The number of individuals positive for antigen α_2, but negative for both α_1 and α_3.

Solution:

Construct Venn diagrams with three circles and let

circle A hold elements representing individuals positive for antigen α_1 (500);

circle B, individuals positive for antigen α_2 (350);

circle C, individuals positive for antigen α_3 (400):

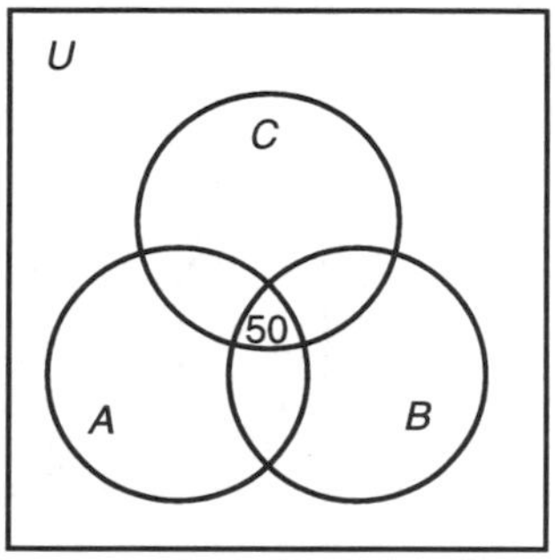

1. 50 individuals are positive for all three antigens.

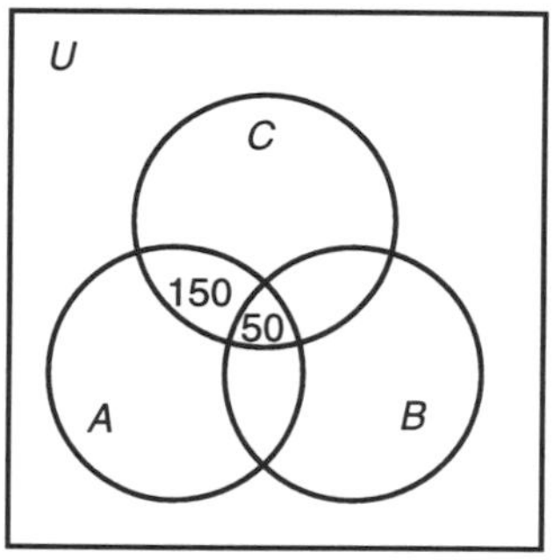

2. 200 individuals are positive for α_1 and α_3; of these individuals, 50 are already displayed in the figure to the left, and 150 remain.

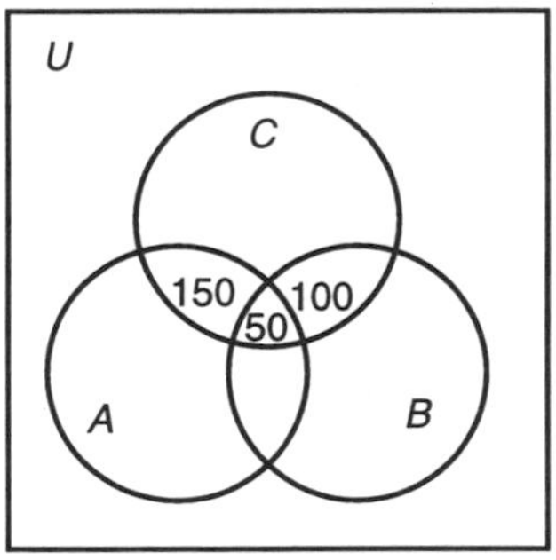

3. 150 individuals are positive for α_2 and α_3; of these individuals, 50 are already displayed in the above, and 100 remain.

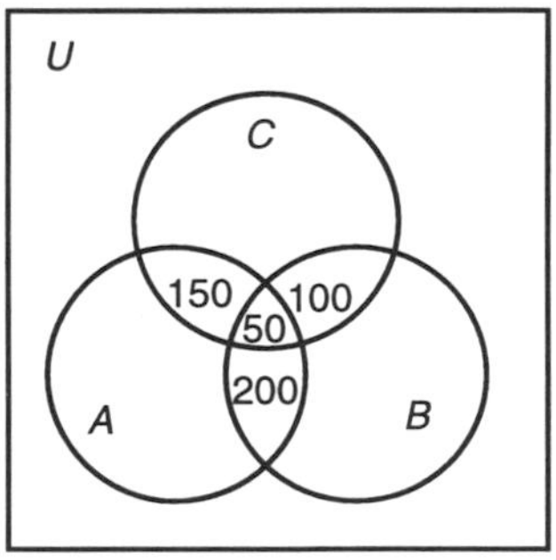

4. 250 individuals are positive for α_1 and α_2; of these, 50 are already displayed in previous figures, and 200 remain.

5. The remaining elements of region A are

$$500-(50+150+200) = 100.$$

The remaining elements of region B are

$$350-(50+100+200)=0.$$

The remaining elements of region C are

$$400-(50+100+150)=100.$$

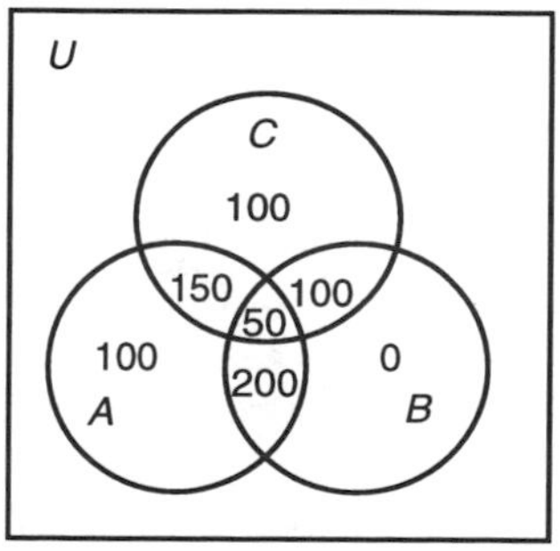

The number of individuals negative for all three antigens α_1, α_2, and α_3 is $\complement(A \cup B \cup C)$.

Since the universal set is 800, and $|A \cup B \cup C| = 700$, we have

$$|\complement(A \cup B \cup C)| = 800 - 700 = 100,$$

which is the answer to **a**.

We can now complete the diagram:

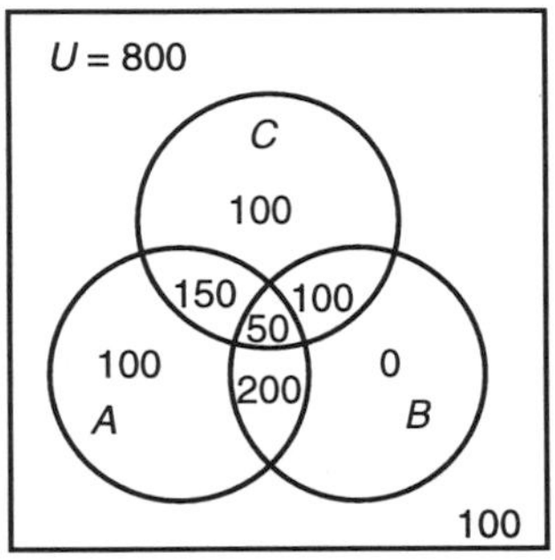

The number of individuals positive for antigen α_2 but negative for α_1 and α_3 is $B \cap \complement(A \cup C)$.

The above diagram shows that the set of individuals positive for antigen α_2 but negative for α_1 and α_3 is empty; thus,

$$B \cap \complement(A \cup C) = \emptyset,$$

which is the answer to **b**.

a: 100 of the 800 examined individuals have no antigen α_1, α_2, or α_3.

b: No individuals positive for antigen α_2 are negative for α_1 and α_3 (that is, all individuals with antigen α_2 also have either α_1 or α_3, or α_1 and α_3).

* * *

A shoemaker died, leaving behind 17 pairs of men's shoes and boots. In his last will and testament he disposed of them as follows: 1/2 of the collection to his youngest, hardworking, son; 1/3 to the middle son; and 1/9 to "Lazybones", the eldest son.

Understanding that it was not possible to divide the 17 pairs into halves, thirds, and ninths, the exasperated executor of the will tossed in his own footwear, making it a total of eighteen pairs, which he divided as follows: 1/2, or 9 pairs, to the youngest son; 1/3, or 6 pairs, to the middle son; and 1/9, or 2 pairs, to the eldest son.

Upon realizing that adding up 9, 6, and 2 made 17, the executor elatedly retrieved his own pair of shoes and closed the estate.

7.3 Algebra of Sets

Fundamental Laws

There are similarities but also significant differences between conventional algebra and the algebra of sets.

Commutative Laws

1. $A \cup B = B \cup A$

2. $A \cap B = B \cap A$

These identities follow directly from the definition of union and intersection:

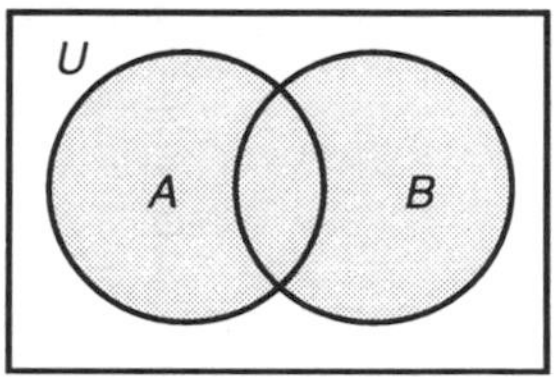

$A \cup B = B \cup A$

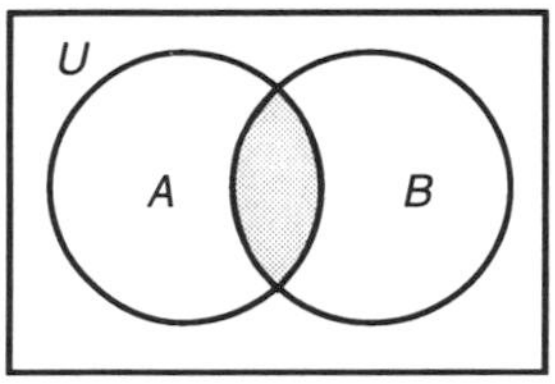

$A \cap B = B \cap A$

Idempotent Laws

1. $A \cup A = A$

2 $A \cap A = A$

Associative Laws

1. $A \cup (B \cup C) = (A \cup B) \cup C$

2. $A \cap (B \cap C) = (A \cap B) \cap C$

The associative laws of set theory follow directly from the definitions of union and intersection:

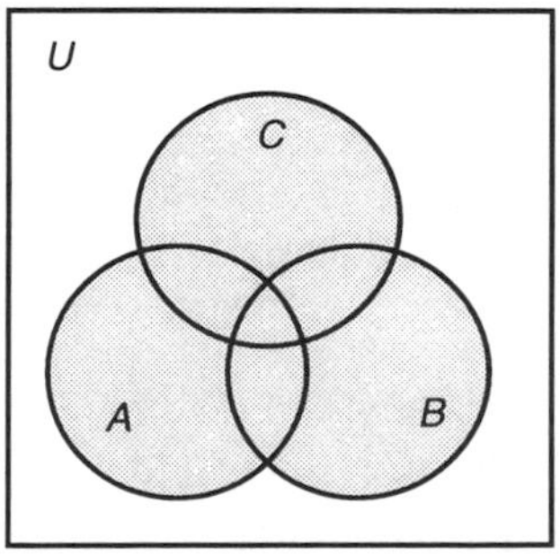

$A \cup (B \cup C) = (A \cup B) \cup C$

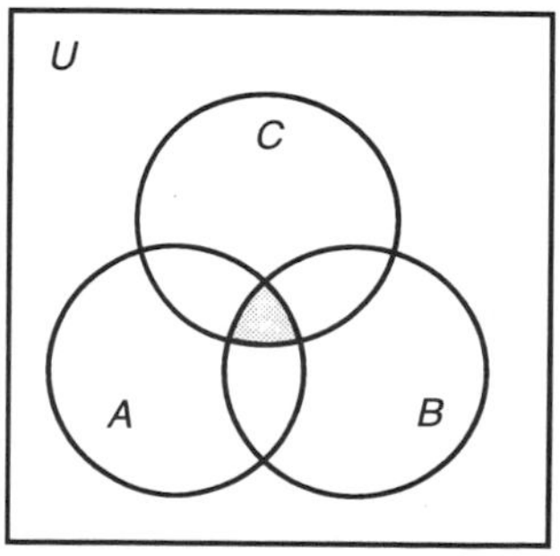

$A \cap (B \cap C) = (A \cap B) \cap C$

Distributive Laws

1. $A \cup (B \cap C) = (A \cup B) \cap (A \cup C)$

2. $A \cap (B \cup C) = (A \cap B) \cup (A \cap C)$

These laws can be proved by successively building on to the Venn diagrams of the expressions on either side of the equality sign.

For instance, to prove $A \cup (B \cap C) = (A \cup B) \cap (A \cup C)$, we have:

Left Side **Right Side**

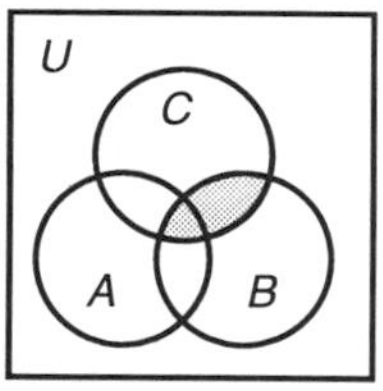

$B \cap C$

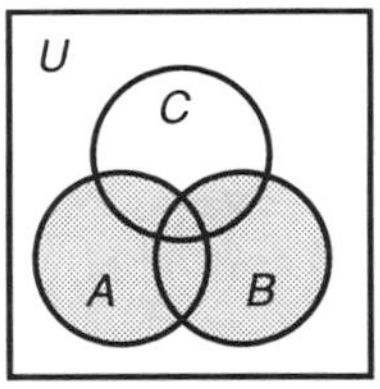

$A \cup B$

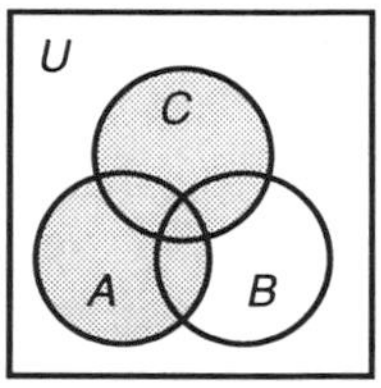

$A \cup C$

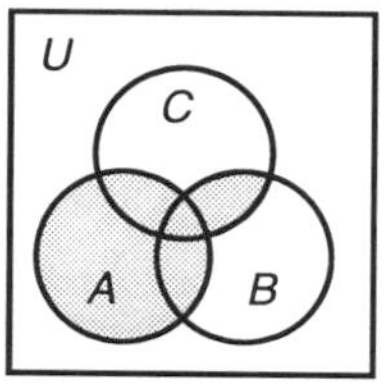

$A \cup (B \cap C)$

=

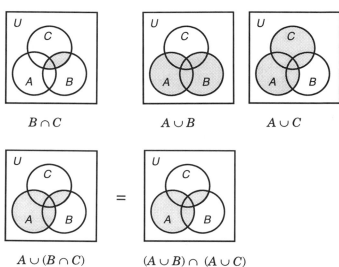

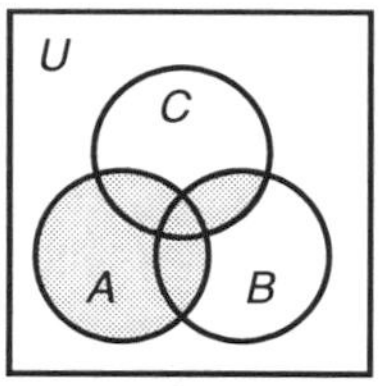

$(A \cup B) \cap (A \cup C)$

Identities Involving Empty Sets

1. $A \cup \emptyset = A$

2. $A \cap \emptyset = \emptyset$

de Morgan's Laws

1. $\complement(A \cup B) = (\complement A) \cap \complement B$

2. $\complement(A \cap B) = (\complement A) \cup \complement B$

de MORGAN Augustus (1806 - 1871)

These laws, named after the British logician and mathematician **Augustus de Morgan**, read, respectively:

- The complement of the union of two sets is the intersection of the complements.
- The complement of the intersection of two sets is the union of the complements.

de Morgan's laws can be proved by successively building on to Venn diagrams of the expressions on either side of the equality sign.

For instance, to prove $\complement(A \cup B) = \complement A \cap \complement B$, we have:

Left Side **Right Side**

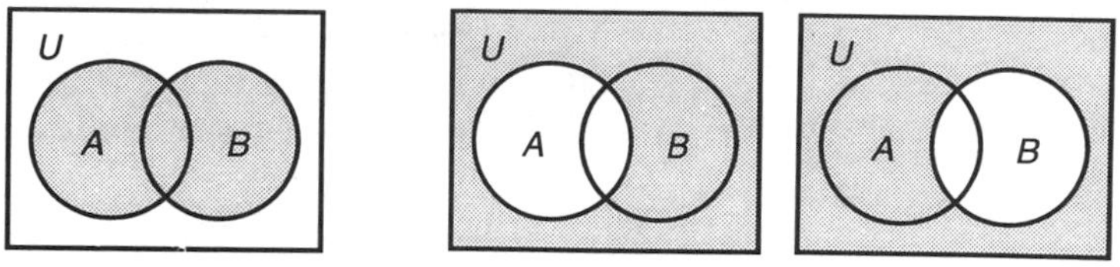

$A \cup B$ $\complement A$ $\complement B$

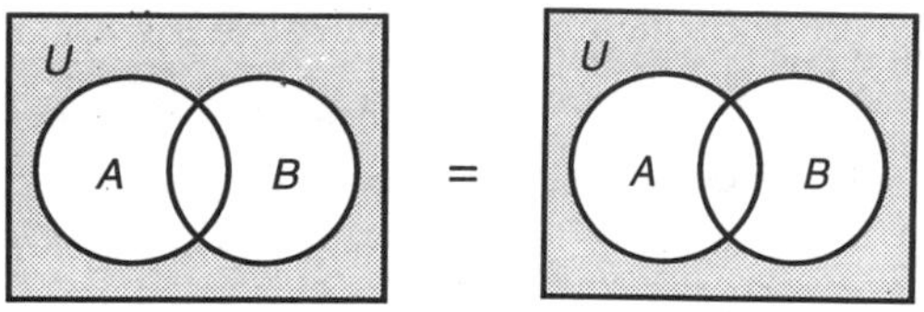

$\complement(A \cup B)$ $\complement A \cap \complement B$

To demonstrate the unwieldiness of Venn diagrams for depicting more than three subsets, the instructor brought his pet elephant.

7.4 Boolean Algebra

Boolean algebra is concerned with ideas or objects that have *only two possible stable states* – *e.g.*, **on/off**, **closed/open**, **yes/no**, **true/false**.

BOOLE George (1815 - 1864) p. 217

This algebra was devised by the English mathematician and logician **George Boole** in 1847, but it was not until the middle of the 20th century it was applied to the analysis of relay and switching circuits.

Notations and Definitions

Notations previously used in this chapter may be used in Boolean algebra, but it is customary to use:

' or $^{-}$	for negation or complement (∁)	a'; $\bar{a}$
⊕ or +	for union (∪)	$a \oplus b$; $a + b$
⊗, ·, or ×	for intersection (∩)	$a \otimes b$; $a \cdot b$; $a \times b$

We favor the notations ', ⊕, and ⊗; criticisms of this use could be raised, however, as ⊕ and ⊗ are established notations for *direct sum* and *tensor product*, respectively.

a' is the **complement** of a.

The results of operations ⊕ and ⊗ are referred to as **sum** and **product**, respectively.

As in conventional arithmetic and algebra, parentheses indicate precedence of operations; otherwise, complement (') has precedence over ⊗, and ⊗ has precedence over ⊕. To avoid confusion, we will be generous with the use of parentheses.

A Boolean set is commonly referred to as B.

0 and 1 denote two distinct identity elements of B. 0 is the **zero** element; 1 is the **unit** element. The zero element of a Boolean algebra corresponds to **Ø** of general set theory; the unit element corresponds to the universal set.

Fundamental Laws

Commutative Laws

cf. p. 249

$a \oplus b = b \oplus a$ $\qquad$ $a \otimes b = b \otimes a$

Identity Laws

$a \oplus 0 = a$ $\qquad$ $a \otimes 1 = a$

The set $a \oplus 0$ is the set of elements belonging to either the set a ***or*** to the empty set 0. This set is identical with the set 0. This set is also identical with the set a, so $a \oplus 0 = a$.

$a \otimes 1 = a$, because set a is contained in the universal set; consequently, the elements common to both sets are precisely the elements of set a.

Distributive Laws

cf. p. 250

$a \oplus (b \otimes c) = (a \oplus b) \otimes (a \oplus c)$ $\quad$ $a \otimes (b \oplus c) = (a \otimes b) \oplus (a \otimes c)$

Complement Laws

A statement a is either true ***or*** false; therefore,

$$a \oplus a' = 1 .$$

A statement a cannot be both true ***and*** false; therefore,

$$a \otimes a' = 0 .$$

Supplementary Laws

Supplementary Complement Laws

1. $0' = 1$ $\qquad$ $1' = 0$

2. $(a')' = a$ (involution law)

3. If $a \oplus x = 1$ and $a \otimes x = 0$, then $x = a'$

Boundedness Laws

$a \oplus 1 = 1$ $\qquad$ $a \otimes 0 = 0$

Using the terms of general set theory, equation $a \oplus 1 = 1$ states that the union of elements contained in set a ***or*** the universal set (1) cannot exceed what is already contained in the universal set.

Equation $a \otimes 0 = 0$ implies that there are no elements common to set a and an empty set; that is, a set formed by elements belonging to set a ***and*** an empty set is an empty set.

Duality Principle

By interchanging the operations $\oplus$ and $\otimes$ and interchanging the identity elements 0 and 1 of a valid expression, we obtain another valid expression, called the **dual** of the original expression.

The dual of $(1 \oplus b) \otimes (a \oplus 0) = a$ is

$$(0 \otimes b) \oplus (a \otimes 1) = a .$$

Absorption Laws

$a \oplus (a \otimes b) = a$ $\qquad$ $a \otimes (a \oplus b) = a$

Associative Laws

cf. p. 249

$(a \oplus b) \oplus c = a \oplus (b \oplus c)$; $\quad$ $(a \otimes b) \otimes c = a \otimes (b \otimes c)$

Idempotent Laws

A product $x \otimes x$, sometimes written x^2, is the set which has the properties of set x; therefore,

$$x \otimes x \;(= x^2) = x .$$

Similarly,

$$x \otimes x \otimes x\ (= x^3) = x,$$
$$x \otimes x \otimes x \otimes x\ (= x^4) = x, \text{ etc.};$$

and

$$x \oplus x = x,$$
$$x \oplus x \oplus x = x,$$
$$x \oplus x \oplus x \oplus x = x, \text{ etc.}$$

de Morgan's Laws

cf. p. 250

$(a \oplus b)' = a' \otimes b'$; $(a \otimes b)' = a' \oplus b'$

Simplifying Boolean Expressions

It often requires considerable patience, ingenuity, or luck to simplify a Boolean expression.

Examples:

- $x' \oplus (y \otimes x)$ can be simplified to $x' \oplus y$ by the

  distributive law,

  $$x' \oplus (y \otimes x) \Rightarrow (x' \oplus y) \otimes (x' \oplus x)\ ;$$

  complement law,

  $$(x' \oplus y) \otimes (x' \oplus x) \Rightarrow (x' \oplus y) \otimes 1\ ;\ \text{and}$$

  identity law,

  $$(x' \oplus y) \otimes 1 \Rightarrow x' \oplus y\ .$$

- $x \otimes (x \oplus y)$ can be simplified to x by the

  identity law,

  $$x \otimes (x \oplus y) \Rightarrow (x \oplus 0) \otimes (x \oplus y)\ ;$$

  distributive law,

  $$(x \oplus 0) \otimes (x \oplus y) \Rightarrow x \oplus (0 \otimes y)\ ;$$

  boundedness law,

  $$x \oplus (0 \otimes y) \Rightarrow x \oplus 0\ ;\ \text{and}$$

  identity law,

  $$x \oplus 0 \Rightarrow x\ .$$

- $(x \oplus y)' \oplus (x \otimes y')$ can be simplified to y' by

  de Morgan's law $(a \oplus b)' = a' \otimes b'$,

  $$(x \oplus y)' \oplus (x \otimes y') \Rightarrow (x' \otimes y') \oplus (x \otimes y')\ ;$$
  $$(x' \otimes y') \oplus (x \otimes y') \Rightarrow (y' \otimes x') \oplus (y' \otimes x)\ ;$$

  the distributive law,

  $$(y' \otimes x') \oplus (y' \otimes x) \Rightarrow y' \otimes (x' \oplus x)\ ;$$

  the complement law $x' \oplus x = 1$, and thus,

  $$y' \otimes (x' \oplus x) \Rightarrow y' \otimes 1\ ;$$

  and, finally, the identity law,

  $$y' \otimes 1 \Rightarrow y'\ .$$

Busting the Brainbuster

Problems like the one discussed here are frequently encountered in “brainbusters”. Boolean algebra may offer a reliable approach, but simple logic or common sense more often provides the optimal solution.

- A congeniality contest in the forest has four prizes, 1st, 2nd, 3rd, and 4th. Three judges (I, II, III) announce the prizewinners (A, B, C, D) by each giving one true and one false:

	1st	2nd	3rd	4th
I	A	B		
II		A	D	
III		C		D

A = 

B = 

C = 

D = 

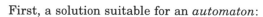

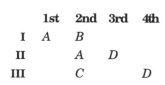

Find the order of the prizewinners.

First, a solution suitable for an *automaton*:

While, in a Boolean algebra, a disjunction (a $\oplus$ operation) is true if at least one of its components is true, a conjunction (a $\otimes$ operation) is true only if both components are true. Therefore, for the announcements of, for instance, judge **I**, we have

$$A_1 \oplus B_2 = 1; \quad A_1 \otimes B_2 = 0,$$

where the indices indicate the announced order of the prizewinners.

If the conjunction of announcement A_1 and the complement of B_2 is true, then the conjunction of announcement B_2 and the complement of A_1 is false, and vice versa; therefore,

$$A_1 \otimes (B_2)' \oplus B_2 \otimes (A_1)' = 1\,.$$

Similarly, we have for the announcements of judge **II**,

$$A_2 \otimes (D_3)' \oplus D_3 \otimes (A_2)' = 1\,;$$

and for judge **III**,

$$C_2 \otimes (D_4)' \oplus D_4 \otimes (C_2)' = 1\,.$$

The conjunction of any number of true propositions is a true proposition; therefore,

$$[A_1 \otimes (B_2)' \oplus B_2 \otimes (A_1)'] \otimes [A_2 \otimes (D_3)' \oplus D_3 \otimes (A_2)'] \otimes [C_2 \otimes (D_4)' \oplus D_4 \otimes (C_2)'\,] = 1\,.$$

The process of multiplying out the left-hand side of the equation

$$[A_1 \otimes (B_2)' \oplus B_2 \otimes (A_1)'] \otimes [A_2 \otimes (D_3)' \oplus D_3 \otimes (A_2)'] \otimes [C_2 \otimes (D_4)' \oplus D_4 \otimes (C_2)'] = 1$$

is more easily visualized if instead of $\otimes$ we use juxtaposition; thus,

$$\begin{aligned}
&[A_1 (B_2)' \oplus B_2 (A_1)'] [A_2 (D_3)' \oplus D_3 (A_2)'] [C_2 (D_4)' \oplus D_4 (C_2)'] \\
&= A_1 (B_2)' A_2 (D_3)' C_2 (D_4)' \oplus A_1 (B_2)' D_3 (A_2)' D_4 (C_2)' \\
&\quad \oplus A_1 (B_2)' D_3 (A_2)' C_2 (D_4)' \oplus A_1 (B_2)' A_2 (D_3)' D_4 (C_2)' \\
&\quad \oplus B_2 (A_1)' A_2 (D_3)' C_2 (D_4)' \oplus B_2 (A_1)' D_3 (A_2)' D_4 (C_2)' \\
&\quad \oplus B_2 (A_1)' D_3 (A_2)' C_2 (D_4)' \oplus B_2 (A_1)' A_2 (D_3)' D_4 (C_2)' = 1 .
\end{aligned}$$

The *first* term is false (0), for A cannot be both 1st and 2nd; the *second* is false, for D cannot be both 3rd and 4th; the *fourth* is false, for A cannot be both 1st and 2nd; the *fifth* is false, for A and B cannot both be 2nd; the *sixth* is false, for D cannot be both 3rd and 4th; the *seventh* is false, for B and C cannot both be 2nd; the *eighth* is false, for B and A cannot both be second.

We are left with the **third** term,

$$A_1 (B_2)' D_3 (A_2)' C_2 (D_4)' = 1 ,$$

which reads: A is 1st, B is not 2nd, D is 3rd, A is not 2nd, C is 2nd, D is not 4th.

The order of the prizewinners is:

Hedgehog (A) - Rabbit (C) - Squirrel (D) - Fox (B)

Machines do not take shortcuts; human beings do!

⇒, "implies" (*p.* 222)

$B_2 \Rightarrow D_3$ and D_4, so B_2 is false and A_1 is correct; then,

$A_1 \Rightarrow D_3$, so B_4 and C_2 .

– *Child's play, my dear Robert.*

7.5 Transfinite Numbers

Since ancient times the subject of infinity has occupied the minds of men and women of philosophy, theology, and mathematics. Today the concept of infinity becomes especially intriguing when we compare the infinitude of all natural numbers, all even numbers, all odd numbers, all rational numbers, all irrational numbers, all real numbers, *etc.* Then the question arises:

Are there different sizes, or degrees, of infinity?

GALILEI Galileo (1564 - 1642)

Galileo Galilei, the Italian mathematician, astronomer, and physicist, argued in his *Dialogue Concerning the Two Chief World Systems* (1632):

There are as many squares as there are natural numbers because they are just as numerous as their roots.

In view of the fact that there are natural numbers that are not squares – *e.g.*, 2, 3, 5, 6, 7, 8, 10, 11, 12, 13, 14, 15, 17 – Galilei's statement seems to be a paradox.

CANTOR Georg (1845 - 1918)

The question of infinity and the possibility of different degrees of infinity was given new life when the German mathematician **Georg Cantor**, the inventor of set theory, in 1874 presented a method of investigating the concept of infinity in *Über eine Eigenschaft des Inbegriffes aller reellen Zahlen* ("On the Characteristic Property of All Real Numbers").

Cantor published several other mathematically revolutionary papers on set theory and infinity, culminating in 1895 - 97 in his best-known work, *Beiträge zur Begründung der transfiniten Mengenlehre*, published in *Matematische Annalen*; a French translation appeared in 1899; an English translation, by Philip E. B. Jourdain, in 1915: *Contributions to the Founding of the Theory of Transfinite Numbers.*

Although mainly examining infinite sets, Cantor developed most of his ideas from concepts of finite sets. Two finite sets contain the same number of elements when there is an exact correspondence between the two sets. For example, there is a **one-to-one correspondence** between six people and six chairs around a dining table, between the fingers of the hands, or between the fingers of one hand and the toes of one foot.

Cantor showed that the concept of one-to-one correspondence can be as readily applied to infinite sets as to finite sets.

Denumerably Infinite Sets

Sets that are equivalent to the set of all natural numbers are called **denumerably infinite sets**. Thus, the members of a denumerably infinite set can be put in one-to-one correspondence with the infinitude of natural numbers {1, 2, 3, ...}, as shown in the following examples.

Our "finite intellects" tell us that the number of natural numbers ought to be larger than the number of even numbers. However, to prove the proposition that the set of all natural numbers and the set of all positive even numbers are denumerably infinite sets, we need only make the following one-to-one correspondence diagram:

1	2	3	4	5	6	7	...	n	...
⇓	⇓	⇓	⇓	⇓	⇓	⇓		⇓	
2	4	6	8	10	12	14	...	$2n$	...

In the same way, the sets of all odd numbers can be shown to be denumerably infinite:

1	2	3	4	5	6	7	...	n	...
⇓	⇓	⇓	⇓	⇓	⇓	⇓		⇓	
1	3	5	7	9	11	13	...	$2n-1$	...

By making a one-to-one correspondence diagram, we can prove Galilei's statement that there are as many squares as there are natural numbers:

1	2	3	4	5	6	7	8	9	...
⇓	⇓	⇓	⇓	⇓	⇓	⇓	⇓	⇓	
1^2	2^2	3^2	4^2	5^2	6^2	7^2	8^2	9^2	...

In the same way we can show that the sets of all cubes, natural numbers to the fourth power, fifth power, *etc.*, are denumerably infinite.

In the above examples the set of the *infinitude* of natural numbers is put in a one-to-one correspondence with a subset of itself: **the whole is equal to part of itself**. This is a glaring contrast to the axiom of *finite* magnitudes, where the whole is equal to the sum of its parts and, therefore, is greater than any of them.

DEDEKIND Richard (1831 - 1916)

We may, with Cantor, define an **infinite set** as one that can be put in **one-to-one correspondence with a proper subset of itself**. This had been proposed in 1872 by the German mathematician **Richard Dedekind** in *Stetigkeit und irrationale Zahlen* ("Continuity and Irrational Numbers"), but Cantor realized that not all infinite sets are the same.

$\aleph_0$

Cantor denoted the manyness of elements of a denumerably infinite set with the cardinal number $\aleph_0$, which reads "aleph-null", "aleph-zero", or "aleph-naught"; $\aleph$ is the first letter of the Hebrew alphabet. Larger infinite sets – those that are not in one-to-one correspondence with the infinitude of natural numbers – are denoted $\aleph_1$, $\aleph_2$, $\aleph_3$, *etc.*

Latin, *trans*, "across", "beyond"; *finire*, "to end"

ℵ$_0$, ℵ$_1$, ℵ$_2$, ℵ$_3$, *etc.*, are the cardinal numbers of infinite sets, where each successive set has a higher degree of infinity; such cardinal numbers are called **transfinite numbers**.

Cantor proved that the set of all integers, the set of all natural numbers, the set of all rational numbers, and the set of all algebraic numbers are denumerably infinite.

As one can interpose an infinity of rational numbers between any two given rational numbers, one would be tempted to believe that the set of all rational numbers is non-denumerable. However, by arranging the set of all rational numbers as below, we can prove that the set of all rational numbers is denumerable and, thus, has the cardinal number ℵ$_0$.

0

1/1 = 1	−1/1= −1	2/1 = 2	−2/1 = −2	3/1 = 3	−3/1 = −3	4/1 = 4	...
1/2	−1/2	2/2 = 1	−2/2 = −1	3/2	−3/2	4/2	...
1/3	−1/3	2/3	−2/3	3/3 = 1	−3/3 = −1	4/3	...
1/4	−1/4	2/4 = 1/2	−2/4 = −1/2	3/4	−3/4	4/4	...
1/5	−1/5	2/5	−2/5	3/5	−3/5	4/5	...
⋮	⋮	⋮	⋮	⋮	⋮	⋮	

In the first column all *numerators* are 1, in the second −1, third 2, fourth −2, *etc.*; in the first row, after 0, all *denominators* are 1, in the second 2, third 3, *etc.*

Delete all fractions that have a factor common to the numerator and he denominator; every rational number will now appear only once:

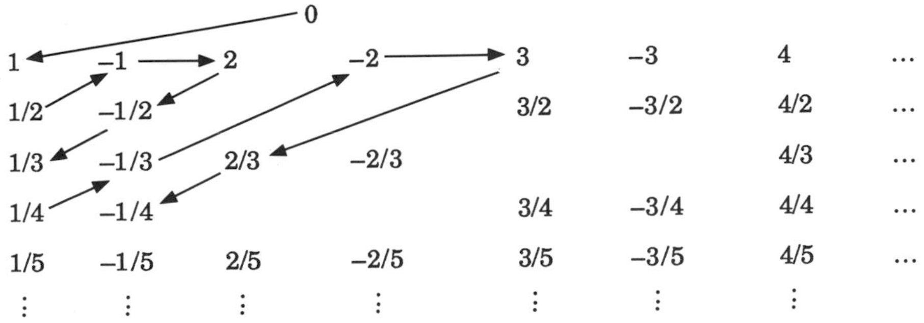

The arrows indicate the order in which we set up a one-to-one correspondence between the natural numbers and the rational numbers so that no rational number will be omitted:

1	2	3	4	5	6	7	8	9	10	11	12	13	14	...
⇓	⇓	⇓	⇓	⇓	⇓	⇓	⇓	⇓	⇓	⇓	⇓	⇓	⇓	
0	1	1/2	−1	2	−1/2	1/3	1/4	−1/3	−2	3	2/3	−1/4	1/5	...

Thus, with Cantor, we have proved that the set of all rational numbers is denumerably infinite.

Non-Denumerably Infinite Sets

Cantor also showed that there are transfinite numbers that are larger than $\aleph_0$. Specifically, he proved that the set of all real numbers is **non-denumerably infinite**; that is, there is no one-to-one correspondence between the set of all real numbers and the set of all natural numbers.

The proof of the non-denumerability of the set of all real numbers is necessarily indirect. For the proof, we need only an interval of real numbers; we choose the interval that is strictly between 0 and 1.

For the proof, write all numbers to be considered in the form of infinite decimal expansion, *e.g.*,

$$1 = 0.999\,9\ldots\,; \quad 0.367 = 0.366\,999\,9\ldots\,; \quad 0.001 = 0.000\,999\,9\ldots$$

and assume that the infinite decimal expansion of all real numbers between 0 and 1 can be put in one-to-one correspondence with the set of all natural numbers, $\mathbf{N}^* = \{1, 2, 3 \ldots\}$:

N*		**decimal expansion of all real numbers between 0 and 1**											
1	⇒	0.	a_1	a_2	a_3	a_4	a_5	a_6	a_7	a_8	…	a_n	…
2	⇒	0.	b_1	b_2	b_3	b_4	b_5	b_6	b_7	b_8	…	b_n	…
3	⇒	0.	c_1	c_2	c_3	c_4	c_5	c_6	c_7	c_8	…	c_n	…
4	⇒	0.	d_1	d_2	d_3	d_4	d_5	d_6	d_7	d_8	…	d_n	…
5	⇒	0.	e_1	e_2	e_3	e_4	e_5	e_6	e_7	e_8	…	e_n	…
6	⇒	0.	f_1	f_2	f_3	f_4	f_5	f_6	f_7	f_8	…	f_n	…
7	⇒	0.	g_1	g_2	g_3	g_4	g_5	g_6	g_7	g_8	…	g_n	…
⋮		⋮	⋮	⋮	⋮	⋮	⋮	⋮	⋮	⋮	…	⋮	…
n	⇒	0.	ξ_1	ξ_2	ξ_3	ξ_4	ξ_5	ξ_6	ξ_7	ξ_8	…	ξ_n	…
⋮		⋮	⋮	⋮	⋮	⋮	⋮	⋮	⋮	⋮	…	⋮	…

Consider a number

$$0.\,z_{\mathrm{I}} z_{\mathrm{II}} z_{\mathrm{III}} z_{\mathrm{IV}} \ldots$$

such that

- z_{I} is 8 if a_1 is 0, 1, 2, 3, 4, 5, 6, or 7 ; z_{I} is 1 if a_1 is 8 or 9
- z_{II} is 8 if b_2 is 0, 1, 2, 3, 4, 5, 6, or 7 ; z_{II} is 1 if b_2 is 8 or 9
- z_{III} is 8 if c_3 is 0, 1, 2, 3, 4, 5, 6, or 7 ; z_{III} is 1 if c_3 is 8 or 9

  etc.

The number $0.\,z_{\mathrm{I}} z_{\mathrm{II}} z_{\mathrm{III}} z_{\mathrm{IV}} \ldots$ cannot be equal to any decimal expansion of any row in the diagram. Thus, there is no equality between $0.\,z_{\mathrm{I}} z_{\mathrm{II}} z_{\mathrm{III}} z_{\mathrm{IV}} \ldots$ and $0.\,a_1 a_2 a_3 a_4 \ldots$, for by design z_{I} differs from a_1, or between $0.\,z_{\mathrm{I}} z_{\mathrm{II}} z_{\mathrm{III}} z_{\mathrm{IV}} \ldots$ and $0.\,b_1 b_2 b_3 b_4 \ldots$, for z_{II} differs from b_2, *etc.*

Thus, it is a false assumption that the infinite decimal expansion of all real numbers between 0 and 1 can be put in one-to-one correspondence with the set of all natural numbers; the cardinal number of the set of all real numbers lying between 0 and 1 is greater than $\aleph_0$.

Cantor also proved that there is an infinitude of other infinite sets whose size is larger than the set of all natural numbers.

Although the above investigation shows that the cardinality of the set that contains all real numbers between 0 and 1 is greater than $\aleph_0$, it is not possible to tell whether this greater number is $\aleph_1$, $\aleph_2$, $\aleph_3$, or any other transfinite number greater than $\aleph_0$.

Algebra of Transfinite Numbers

Being an algebra of sets, the algebra of transfinite numbers offers no real surprises at this stage:

$$\aleph_0 + \aleph_0 = \aleph_0$$
$$\aleph_0 + \aleph_0 + \ldots + \aleph_0 = \aleph_0$$
$$\aleph_0 \times \aleph_0 = \aleph_0$$
$$\aleph_0 \times \aleph_0 \times \ldots \times \aleph_0 = \aleph_0$$

Transfinite numbers and *transfinite induction* have been applied in the form of "non-standard analysis" for proofs of results associated with the Boltzmann equation in kinetic gas theory.

... sed libera pueros ludi a Cantu Cantoris.

"... and deliver our small schoolchildren from set theory."

The Continuum Hypothesis

c

The transfinite number expressing the cardinality of the set that contains all real numbers between 0 and 1 is denoted **c**. The letter "c" stands for *continuum* and refers to Cantor's **continuum hypothesis**.

According to the continuum hypothesis, any infinite subset of real numbers *either* is denumerable, that is, in one-to-one correspondence with the set of natural numbers, *or* can be put in a one-to-one correspondence with the set that contains all real numbers between 0 and 1.

GÖDEL Kurt
(1906 - 1976); *p.* 160

COHEN Paul Joseph
(b. 1933)

Cantor's efforts to prove the continuum hypothesis were fruitless. In 1940, **Kurt Gödel** showed that the continuum hypothesis *cannot be disproved*, and in 1963 **Paul J. Cohen**, an American mathematician with an interest in logic, showed that the continuum hypothesis is, in fact, *undecidable*: besides true statements and false statements, set theory also has undecidable statements.

* * *

MITTAG-LEFFLER Gösta
(1846 - 1927)

During a time when editors were generally unsympathetic to articles about transfinite numbers, Swedish mathematician **Gösta Mittag-Leffler** – founder and editor of *Acta Mathematica* – took great interest in Cantor's discoveries and, thus, much of Cantor's work was published in *Acta Mathematica*.

Inscribed on the mantelpiece in Gösta Mittag-Leffler's home – now a research institute for mathematics – in Djursholm, Sweden, we read:

Talet är tänkandets början och slut.
Med tanken föddes talet.
Utöfver talet når tanken icke.
ML 1903
"Number is the beginning
and end of thought.
Thought gave birth to number
but reaches not beyond.
ML 1903"

Chapter 8

INTRODUCTION TO SEQUENCES AND SERIES

8.1 Terminology

Sequence, Progression

A **sequence** or **progression** of numbers is a set of numbers arranged in an orderly fashion such that the preceding and following numbers are completely specified; in the sequence

$$\ldots, 7, 10, 13, 16, 19, \ldots$$

the difference between numbers is 3, which indicates that the number preceding 7 is 4 and the number following 19 is 22.

Term, Element
General Term

The numbers of a sequence are called **terms** or **elements**. The **general term** defines the rule of the sequence; if n is the ordinal number of a term in the sequence, 1 is the first term, and the general term a_n is $2n - 1$, the sequence is

$$1, 3, 5, 7, \ldots, 2n-1, \ldots\ ;$$

if the general term a_n is n^2, the sequence is

$$1, 4, 9, \ldots,\ n^2, \ldots .$$

Series

A **series** is the sum of the terms of a sequence,

$$S_n = a_1 + a_2 + a_3 + a_4 + \ldots + a_n .$$

Unlike a **series of positive terms**, such as

$$\sum_{n=0}^{5} \left(\frac{1}{2}\right)^n = 1 + \frac{1}{2} + \frac{1}{4} + \frac{1}{8} + \frac{1}{16} + \frac{1}{32},$$

Alternating Series

an **alternating series** has alternately positive and negative terms,

$$\sum_{n=0}^{5} \left(\frac{-1}{2}\right)^n = 1 - \frac{1}{2} + \frac{1}{4} - \frac{1}{8} + \frac{1}{16} - \frac{1}{32} .$$

Arithmetic Sequence
Common Difference
Arithmetic Series

In the **arithmetic sequence** 1, 3, 5, 7, 9, 11 ... the terms have a **common difference** of 2 units; the corresponding series is an **arithmetic series**.

Harmonic Sequence
Harmonic Series

A sequence whose terms are the reciprocals of an arithmetic sequence is a **harmonic sequence**; the corresponding series is a **harmonic series**. The terms of the harmonic sequence 1, 1/2, 1/3, 1/4, ... are the reciprocals of the terms of the arithmetic sequence 1, 2, 3, 4, ...

Geometric Sequence
Common Ratio
Geometric Series

In the **geometric sequence** 1, 2, 4, 8, 16, 32, ... the terms have a **common ratio** of 2 units; the corresponding series is a **geometric series**.

Finite Sequences and Series

Finite sequences and series have defined first *and* last terms, *e.g.*,

$$1 + 3 + 5 + 7 + 9 + 11 + 13 + 15\,;$$

Infinite Sequences and Series

infinite sequences and series continue indefinitely,

$$1 + 3 + 5 + 7 + 9 + 11 + 13 + 15 + \ldots.$$

Infinite series whose terms increase in magnitude have no attainable sum.

For an infinite series to have a sum its terms must get ever smaller, but this is not the only requirement; a series whose terms decrease in magnitude toward zero may have a sum or may have no attainable sum.

Convergent Series
Divergent Series

If an infinite series has a finite **sum**, it is referred to as a **convergent series**; we say that the series converges. If it has **no sum**, it is known as a **divergent series**; it diverges.

We note that, in mathematical terminology, the expressions "convergent" and "divergent", when applied to a series, refer exclusively to the existence of a sum or no sum.

8.2 Finite Sequences and Series

Arithmetic Sequences and Series

A finite arithmetic sequence may be written

$$a_1,\ a_1+d,\ a_1+2d,\ a_1+3d,\ \ldots,\ a_1+(n-1)d,$$

where a_1 denotes the first term, n is the number of terms, and d is the common difference.

In an arithmetic sequence or series, the n-th term is

$$a_n = a_1 + (n-1)\,d\,.$$

- Find the 17th term and the last term of the arithmetic series

$$52+56+60+\ldots+a_{32}.$$

The common difference is

$$56-52 = 4,$$

so

$$a_{17} = 52+(17-1)\,4 = 116$$

and

$$a_{32} = 52+(32-1)\,4 = 176\,.$$

Harmonic Sequences and Series

To find a specific term of a harmonic sequence or series, we compare it to a corresponding arithmetic sequence or series whose terms are the reciprocals of those of the harmonic sequence.

- Find the 36th term of the harmonic series

$$1+\frac{1}{4}+\frac{1}{7}+\frac{1}{10}+\frac{1}{13}+\ldots.$$

The reciprocals of the given terms form an arithmetic series, 1 + 4 + 7 + 10 + 13 + …, whose common difference is 3. The 36th term of the arithmetic series is

$$1+(36-1)\,3 = 106\,;$$

the corresponding term in the harmonic series is $\frac{1}{106}$. •

Harmonic Tones

In *music*, vibrating strings of the same material and with equal diameter, equal torsion, and equal tension and whose lengths are proportional to terms in a harmonic sequence generate **harmonic tones**.

The Sum of a Finite Arithmetic Series

The **sum S_n of a finite arithmetic series** consisting of n terms is given by

$$S_n = \frac{n}{2}(a_1 + a_n);$$

that is, the sum of an arithmetic series equals half the product of the number of terms and the sum of the first and last terms.

Why? Write the sum of the arithmetic series twice, the second time in *reverse* order; then add the terms in pairs and solve for S_n :

$$S_n = a_1 + (a_1 + d) + (a_1 + 2d) + \ldots + [a_1 + (n-1)d]$$
$$S_n = [a_1 + (n-1)d] + [a_1 + (n-2)d] + [a_1 + (n-3)d] + \ldots + a_1$$

$$2S_n = [2a_1 + (n-1)d] + [2a_1 + (n-1)d] + [2a_1 + (n-1)d] + \ldots + [2a_1 + (n-1)d]$$
$$2S_n = n\,[2a_1 + (n-1)d] = n\,[a_1 + a_1 + (n-1)d]$$
$$S_n = \frac{n}{2}[a_1 + a_1 + (n-1)d]$$

Since the last term of the series is a_n, and $a_n = a_1 + (n-1)d$, we have

$$S_n = \frac{n}{2}(a_1 + a_n).$$

To illustrate with *numerical values*, find

$$S_{50} = \sum_{i=1}^{50} 2i,$$

the sum of all positive, even integers from 2 to 100.

Again, write the sum of the series twice – the second time in reverse order – add the sums, and divide by 2 :

$$S_{50} = 2 + 4 + 6 + 8 + \ldots + 94 + 96 + 98 + 100$$
$$S_{50} = 100 + 98 + 96 + 94 + \ldots + 8 + 6 + 4 + 2$$

$$2S_{50} = 102 + 102 + 102 + 102 + \ldots + 102 + 102 + 102 + 102$$

$$S_{50} = \frac{50}{2}(2 + 100) = 2550.$$

- Find the sum of the series

$$1 + \frac{5}{3} + \frac{7}{3} + \ldots + 201.$$

The common difference is $\frac{2}{3}$; if the number of terms is x, then

$$201 = 1 + (x-1)\frac{2}{3}; \qquad x = 301.$$

And the sum is

$$\frac{301}{2}(1 + 201) = 30\,401.$$

Geometric Sequences and Series

In general terms, a **finite geometric sequence** is written

$$a_1, \; a_1 r, \; a_1 r^2, \; a_1 r^3, \; \ldots, \; a_1 r^{n-1},$$

where a_1 denotes the first term, n is the number of terms, and r is the common ratio; the geometric sequence

$$\frac{1}{3}, \; 1, \; 3, \; 9, \; 27$$

has the common ratio 3.

In a geometric sequence or series, the n-th term is

$$a_n = a_1 r^{n-1},$$

where r is the common ratio.

- The series

$$4 + 8 + 16 + 32 + \ldots + a_{64}$$

has the common ratio

$$\frac{8}{4} = 2 .$$

The 17th term is

$$a_{17} = 4 \times 2^{(17-1)} = 262\,144 .$$

Geometric Mean

The geometric mean m of two numbers a and b is the square root of their product,

$$\frac{a}{m} = \frac{m}{b} \; ; \quad m = \sqrt{a\,b} \; .$$

The geometric mean of 4 and 4096 is

$$\sqrt{4 \cdot 4096} = \sqrt{16\,384} = 128 .$$

The Sum of a Finite Geometric Series

The **sum S_n of a finite geometric series** consisting of n terms is given by

$$S_n = a_1 \frac{1 - r^n}{1 - r} = \frac{a_1 - r a_n}{1 - r}; \; r \neq 1,$$

where a_1 denotes the first term, n the number of terms, and r the common ratio.

To determine the sum of a finite geometric series, we must know the number of terms, the first term, and the common ratio.

Why? If the sum of the finite geometric series is

$$S_n = a_1 + a_1 r + a_1 r^2 + a_1 r^3 + \dots + a_1 r^{n-1},$$

then

$$rS_n = a_1 r + a_1 r^2 + a_1 r^3 + \dots + a_1 r^{n-1} + a_1 r^n .$$

Subtracting, all terms on the right-hand side disappear except two,

$$S_n (1-r) = a_1 (1-r^n) ;$$

and

$$S_n = a_1 \frac{1-r^n}{1-r}; \quad r \neq 1 .$$

- Determine $\sum_{i=0}^{19} 4\,(2^i)$.

We are asked to complete the sum of a geometric series where the first term is 4, the number of terms is 20, and the common ratio is 2,

$$4 \cdot 2^0 + 4 \cdot 2^1 + 4 \cdot 2^2 + \dots + 4 \cdot 2^{19} .$$

$$\sum_{i=0}^{19} 4\,(2^i) = 4\frac{1-2^{20}}{1-2} = 4\,(2^{20}-1) = 4\,194\,300 .$$

- Find the sum of the geometric series

$$2 + 6 + 18 + \dots + 39366 .$$

The common ratio is 3.

If the number of terms is x, then

Exponential Equations: p. 314

$$39\,366 = 2 \cdot 3^{(x-1)};\ 19\,683 = 3^{(x-1)} .$$

$$(x-1)\lg 3 = \lg 19\,683 = 9 \cdot \lg 3$$

$$x = 9 + 1 = 10 .$$

The sum of the series is

$$2\frac{1-3^{10}}{1-3} = 3^{10} - 1 = 59\,048 .$$

- Assuming that one generation corresponds to 25 years, find the number of ancestors – parents, grandparents, great-grandparents, *etc.* – a person might have over a 600 year period.

600 / 25 = 24 generations

2 parents + 4 grandparents + 8 great-grandparents + …

+ (2) 2^{24-1} great- …great-grandparents.

$$\sum_{i=1}^{24} 2^i = 2 + 4 + 8 + \dots + 2^{24} = 2\frac{1-2^{24}}{1-2} = 2\,(2^{24}-1)$$

$$= 33\,554\,430 \text{ ancestors.}$$

- Legend has it that the game of chess was invented for the amusement of a Persian shah – or an Indian maharajah, or a Chinese emperor – who became so enthusiastic that he wanted to reward the inventor, who desired only one grain of wheat on the first square of the chessboard, two grains on the second square, four on the third, and so on, doubling the number of grains for each successive square on the 64-square chessboard.

  The shah – or maharajah, or emperor, or whatever he was – smiled to himself at this modest wish; but the smile froze when next morning his Keeper of the Granary told him – with the help of his electronic calculator, no doubt – that he would have to shell out

$$1+2+4+8+16+32+\ldots+2^{63} = 1\frac{1-2^{64}}{1-2} = 2^{64}-1 \text{ grains,}$$

that is,

18 446 744 073 709 551 615 grains of wheat.

With about 100 grains to a cubic centimeter, the total volume of wheat would be nearly two hundred thousand million cubic meters, or two hundred cubic kilometers, to be loaded on two thousand million railway wagons, which would make up a train reaching a thousand times around the Earth.

8.3 Infinite Series

As more terms are added to a convergent series, its terms become arbitrarily small. We say that the terms of a convergent series **tend to 0**. The sum of such a convergent series is referred to as the **sum to infinity**, and is denoted

$$\lim_{n\to\infty} S_n,$$

Further discussion: *pp.* 355 *et seq.*

where "lim" stands for "limit" (Latin: *limes*) and indicates "boundless approach" or "nearness".

For a series to converge its terms must tend to 0, but this is not the only requirement for convergence; even if the terms tend to 0, the series does not necessarily converge.

The n-th Term Test

The following theorem, often referred to as the n-th term test, provides a simple test to determine the limit of the terms of a sequence or series.

If a series $\sum_{n=1}^{\infty} a_n$ is convergent, then

$$\lim_{n\to\infty} a_n = 0.$$

Why? We have

$$\lim_{n \to \infty} S_n = \lim_{n \to \infty} (a_1 + a_2 + \ldots + a_{n-1} + a_n)$$

$$= \lim_{n \to \infty} (S_{n-1} + a_n) = \lim_{n \to \infty} S_{n-1} + \lim_{n = \infty} a_n .$$

Solving for a_n, we find

$$\lim_{n \to \infty} a_n = \lim_{n \to \infty} (S_n - S_{n-1}) = 0 .$$

Convergence and Sums of Selected Series

We shall demonstrate in the following examples that there are other requirements for convergence of an infinite series than that the terms must tend to zero.

1. Consider the series

$$\sum_{n=1}^{\infty} \left(\frac{1}{n} - \frac{1}{n+1}\right) = \lim_{n \to \infty} \left[\left(\frac{1}{1} - \frac{1}{2}\right) + \left(\frac{1}{2} - \frac{1}{3}\right) + \left(\frac{1}{3} - \frac{1}{4}\right) + \left(\frac{1}{4} - \frac{1}{5}\right) + \ldots + \left(\frac{1}{n} - \frac{1}{n+1}\right)\right],$$

where

$$\lim_{n \to \infty} \left(\frac{1}{n} - \frac{1}{n+1}\right) = 0;$$

the terms tend to zero, a prerequisite for convergence.

Partial Sums

The series has the **partial sums** $S_1, S_2, S_3, S_4, \ldots, S_n$:

$$S_1 = 1 - \frac{1}{2} = \qquad = \frac{1}{2}$$

$$S_2 = 1 - \frac{1}{2} + \frac{1}{2} - \frac{1}{3} = 1 - \frac{1}{3} = \qquad = \frac{2}{3}$$

$$S_3 = 1 - \frac{1}{2} + \frac{1}{2} - \frac{1}{3} + \frac{1}{3} - \frac{1}{4} = 1 - \frac{1}{4} = \qquad = \frac{3}{4}$$

$$S_4 = 1 - \frac{1}{2} + \frac{1}{2} - \frac{1}{3} + \frac{1}{3} - \frac{1}{4} + \frac{1}{4} - \frac{1}{5} = 1 - \frac{1}{5} = \frac{4}{5}$$

$$\ldots$$

$$\lim_{n \to \infty} S_n = \lim_{n \to \infty} \left[1 - \frac{1}{n+1}\right] = 1 - 0 = 1 .$$

Consequently, the series is **convergent** and its sum is 1.

2. Consider the series

$$\sum_{n=1}^{\infty} \frac{1}{\sqrt{n}} = 1 + \frac{1}{\sqrt{2}} + \frac{1}{\sqrt{3}} + \frac{1}{\sqrt{4}} + \ldots + \frac{1}{\sqrt{n}} + \ldots .$$

As

$$\lim_{n \to \infty} \frac{1}{\sqrt{n}} = 0,$$

the terms tend to zero.

The partial sum S_n has n terms, the smallest being $1/\sqrt{n}$, so

$$S_n \geq n\left(\frac{1}{\sqrt{n}}\right) = \sqrt{n} .$$

When n tends to infinity, so does $\sqrt{n}$. The series is thus **divergent**; it has no sum.

3. Consider the **harmonic series**

$$\sum_{n=1}^{\infty} \frac{1}{n} = 1 + \frac{1}{2} + \frac{1}{3} + \frac{1}{4} + \frac{1}{5} + \frac{1}{6} + \ldots + \frac{1}{n} + \ldots .$$

As

$$\lim_{n \to \infty} \frac{1}{n} = 0,$$

the terms tend to zero.

Starting with the third term, we add blocks of terms, containing 2, 4, 8, 16, 32, 64, ... terms, *etc.*

The sum S_n of each block will be greater than the product of the number of terms times the smallest term of the block:

1st term	$1 = 1$
2nd term	$1/2 = 1/2$
3rd and 4th terms	$1/3 + 1/4 > 2\,(1/4) = 1/2$
5th to 8th terms	$1/5 + 1/6 + 1/7 + 1/8 > 4\,(1/8) = 1/2$
9th to 16th terms	$1/9 + 1/10 + \ldots + 1/16 > 8\,(1/16) = 1/2$
17th to 32nd terms	$1/17 + 1/18 + \ldots + 1/32 > 16\,(1/32) = 1/2$

We can continue indefinitely to find blocks of terms whose sum exceeds 1/2; S_n will thus tend to infinity as n does and, consequently, the series is **divergent**.

4. Consider the series

$$\sum_{n=1}^{\infty} (-1)^n = -1 + 1 - 1 + 1 - 1 + 1 - 1 + \ldots + (-1)^n + \ldots .$$

As the limit value

$$\lim_{n \to \infty} (-1)^n$$

does not exist, the terms cannot tend to zero; the series is **divergent**.

It might be tempting to assume that the sum is either (– 1) or 0, depending on the last term. However, **there is no last term** – the terms continue indefinitely.

5. It is obvious by the very definition of an arithmetic series that the magnitude of its terms becomes ever greater. Consequently, infinite arithmetic series expand beyond bounds and never converge.

Summarizing, we have the following **facts concerning infinite series**:

- A series whose terms do not tend to zero is divergent.
- A series whose terms tend to zero may or may not be convergent.

Euler's Constant

Although the harmonic series

$$\sum_{n=1}^{\infty} \frac{1}{n} = 1 + \frac{1}{2} + \frac{1}{3} + \frac{1}{4} + \frac{1}{5} + \frac{1}{6} + \ldots + \frac{1}{n} + \ldots$$

diverges, there is a formula for an approximate value of the sum of a finite number of its terms:

$$\sum_{n=1}^{m} \frac{1}{n} \approx \ln m + \gamma,$$

where $\ln m$ is the natural logarithm of the number of terms, and γ is *Euler's constant*; the accuracy of the formula increases with increasing values of m.

Chapter 22: "Power Series", *pp.* 765 *et seq.*

Euler discovered this formula in 1731 by starting with the logarithmic function

$$\ln\left(1 + \frac{1}{x}\right) = \frac{1}{x} - \frac{1}{2x^2} + \frac{1}{3x^3} - \ldots, \text{ or } \frac{1}{x} = \ln\left(\frac{x+1}{x}\right) + \frac{1}{2x^2} - \frac{1}{3x^3} + \ldots; \; x \geq 1,$$

and letting $x = 1, 2, 3, \ldots n$:

$$\frac{1}{1} = \ln 2 + \frac{1}{2} - \frac{1}{3} + \ldots; \quad \frac{1}{2} = \ln\frac{3}{2} + \frac{1}{2 \cdot 4} - \frac{1}{3 \cdot 8} + \ldots; \quad \frac{1}{n} = \ln\frac{n+1}{n} + \frac{1}{2n^2} - \frac{1}{3n^3} + \ldots$$

Adding terms on both sides, we get

$$1 + \frac{1}{2} + \ldots + \frac{1}{n} = \ln(n+1) + \left[\frac{1}{2}\left(1 + \frac{1}{4} + \ldots + \frac{1}{n^2}\right) - \frac{1}{3}\left(1 + \frac{1}{8} + \ldots + \frac{1}{n^3}\right) + \ldots\right]$$

or, substituting C_n for the sum of the infinite number of terms enclosed by brackets above,

$$1 + \frac{1}{2} + \frac{1}{3} + \ldots + \frac{1}{n} = \ln(n+1) + C_n.$$

Subtracting $\ln n$ from both sides gives

$$1 + \frac{1}{2} + \frac{1}{3} + \ldots + \frac{1}{n} - \ln n = \ln\left(1 + \frac{1}{n}\right) + C_n,$$

$$\lim_{n \to \infty} \ln\left(1 + \frac{1}{n}\right) = 0$$

and as n tends to ∞, we obtain

$$\gamma = \lim_{n \to \infty}\left(1 + \frac{1}{2} + \frac{1}{3} + \ldots + \frac{1}{n} - \ln n\right),$$

where γ is **Euler's constant**.

Thus far, no one has been able to determine whether Euler's constant is rational or irrational; to do so would be akin to grabbing the mathematicians' brass ring.

Euler's constant has been calculated to more than 20 000 decimals; to 26 places of decimals, we have

$$\gamma = 0.577\,215\,664\,901\,532\,860\,606\,512\,09\ldots.$$

- Find an approximate value of $\displaystyle\sum_{n=1}^{1000} \frac{1}{n}$.

$$\sum_{n=1}^{1000} \frac{1}{n} \approx \ln 1000 + 0.5772 \approx 7.48$$

Infinite Geometric Series

Convergence

An infinite geometric series

$$a_1 + a_1 r + a_1 r^2 + a_1 r^3 + \ldots$$

is **convergent** *only* if the common ratio r lies strictly between -1 and $+1$.

More specifically, denoting the common ratio r:

- If $|r| < 1$, the infinite series converges.
- If $|r| > 1$, the value of $|r|^n$ increases without limit; the infinite series diverges.
- If $|r| = 1$, the terms of the series either are constant or differ by successive sign changes; again, the infinite series diverges.

The Sum of Infinite Geometric Series

To find the sum of an infinite geometric series, we must know the first term of the series and the common ratio.

The sum S_n of the *finite* geometric series is, as we have seen,

$$S_n = a_1 \frac{1 - r^n}{1 - r}; \; r \neq 1,$$

where a_1 is the first term and r is the common ratio.

When n approaches infinity, the series is convergent only if $|r| < 1$, in which case r^n will tend to zero; thus,

$$\lim_{n \to \infty} S_n = \lim_{n \to \infty} \frac{a_1 (1 - r^n)}{1 - r} = \frac{a_1 (1 - 0)}{1 - r};$$

that is, the **sum of an infinite, convergent geometric series** is

$$\lim_{n \to \infty} S_n = \frac{a_1}{1 - r}.$$

- Find the sum

$$\frac{1}{4}+\frac{1}{8}+\frac{1}{16}+\frac{1}{32}+\frac{1}{64}+\dots .$$

The first term of this geometric series is $\frac{1}{4}$ and the common ratio is $\frac{1}{2}$; the series is convergent and, consequently,

$$\sum_{n=0}^{\infty}\frac{1}{4}\left(\frac{1}{2}\right)^n = \lim_{n\to\infty} S_n = \frac{1/4}{1-1/2} = \frac{1}{2} .$$

- Find the product

$$\prod_{n=0}^{\infty} 10^{2^{-n}}$$

The product may be written

$$10^{1+1/2+1/4+1/8+1/16+\dots},$$

where the exponents form an infinite geometric series whose first term is 1 and whose common ratio is 1/2; the sum of the exponents is

$$\frac{1}{1-1/2}=2$$

and the sought product is

$$10^2 = 100 .$$

- Express the periodic infinite decimal fraction

$$0.167\,171\,71\dots$$

as a fraction.

$$0.167\,171\,71\dots = \frac{16}{100}+\frac{71}{10\,000}+\frac{71}{1\,000\,000}+\frac{71}{100\,000\,000}+\dots .$$

The terms succeeding 16/100 form an infinite geometric series with the first term 71/10 000 and the common ratio 1/100; thus,

$$0.167\,171\,71\dots = \frac{16}{100}+\frac{71/10\,000}{1-1/100} = \frac{1584+71}{9900} = \frac{331}{1980} .$$

Achilles and the Tortoise

ZENO
(*c.* 490 - *c.* 435 B.C.)

The Greek philosopher and mathematician **Zeno** of Elea propounded that in a footrace between Achilles and a tortoise, Achilles would lose the race – no matter how fast he runs – if the tortoise had been given a head start – no matter how short.

Zeno argued that, although Achilles runs faster and would get closer and closer, he would never quite catch up to the tortoise since, while Achilles covered the distance from his starting point, a, to that of the tortoise, b, the tortoise moved to c, and while Achilles dashed to c, the tortoise scuttled off to d, and so on in intervals that became shorter and shorter but never ceased to be produced.

It is likely that Zeno fabricated this and other anecdotes to demonstrate mathematical shortcomings of his time, but we are not aware of any resolution of Zeno's to this or any other of his paradoxes.

What *is* the fallacy of the paradox of Achilles and the tortoise?

Using concepts established some 2500 years after Zeno, here is the explanation why Achilles can finally catch up to and pass the tortoise:

> Although the number of time intervals is infinite, the total amount of time is not necessarily infinite.

- If the tortoise is given a head start of 3 meters and advances at the speed of 3 m/s (cheating – pulled by a tractor) and Achilles ambles along at 6 m/s, Achilles will catch up to the tortoise at the end of

$$\frac{1}{2} + \frac{1}{4} + \frac{1}{8} + \frac{1}{16} + \dots = \frac{1/2}{1 - 1/2} = 1 \text{ second.}$$

Alternatively, you could solve $6t = 3t + 3$ for the time t.

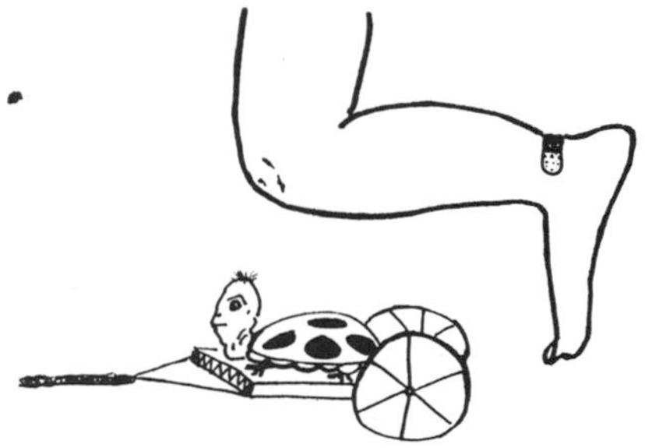

Convergence Tests

While geometric infinite series are easily assessed as to convergence and sum, infinite series in general need further consideration.

Series of Positive Terms

We will discuss two convergence tests for series with positive terms only – the term comparison test, and the ratio test.

Comparison Test

- A series of positive terms is **convergent** if the value of each term is equal to or less than the value of the corresponding terms of another series of positive terms which is known to be convergent.

- o A series of positive terms is **divergent** if the value of each term is equal to or greater than the value of the corresponding terms of another series of positive terms which is known to be divergent.

Compare the series

$$(1) \qquad 1 + \frac{1}{2 \cdot 3} + \frac{1}{3 \cdot 3^2} + \ldots + \frac{1}{n\, 3^{n-1}} + \ldots$$

with the geometric series

$$(2) \qquad 1 + \frac{1}{3} + \frac{1}{3^2} + \ldots + \frac{1}{3^{n-1}} + \ldots$$

whose common ratio is 1/3 and therefore converges.

Corresponding terms of (1) are less than the values of the terms of (2):

$$1 = 1; \quad \frac{1}{2 \cdot 3} < \frac{1}{3}; \quad \frac{1}{3 \cdot 3^2} < \frac{1}{3^2}; \quad \frac{1}{n\, 3^{n-1}} < \frac{1}{3^{n-1}}$$

Consequently, (1) is convergent.

Ratio Test

d'Alembert's Ratio Test

It is often difficult to find a series suitable for convergence comparison. A special case of the comparison test, called **d'Alembert's ratio test**, provides an easier solution when it is applicable.

d'ALEMBERT Jean Le Rond (1717 - 1783)

CAUCHY Augustin Louis (1789 - 1857)

Although named after the French mathematician and physicist **d'Alembert**, the test was not formally stated and proved until after d'Alembert's time, by the French mathematician **Cauchy**.

The series $\sum_{n=1}^{\infty} a_n = a_1 + a_2 + \ldots + a_n$ (assume positive terms)

(A) is **convergent** if $\lim_{n \to \infty} \frac{a_{n+1}}{a_n} < 1$

(B) is **divergent** if $\lim_{n \to \infty} \frac{a_{n+1}}{a_n} > 1$.

Why? Consider the two cases $r < 1$ and $r > 1$.

A. $r < 1$

Assuming that $\lim_{n \to \infty} \frac{a_{n+1}}{a_n} = r$ and that $r < 1$, let P be any real number such that $r < P < 1$; there must then exist a positive integer S such that

$$\frac{a_{n+1}}{a_n} < P \text{, or } a_{n+1} < P a_n \text{, for any integer } n \geq S.$$

Thus,

$$a_{S+1} < P a_S,$$
$$a_{S+2} < P a_{S+1} < P^2 a_S,$$
$$a_{S+3} < P a_{S+2} < P^3 a_S,$$

or, in general,

$$a_{S+k} < P^k a_S$$

for any integer $k > 0$.

The geometric series

$$(1) \quad P a_S + P^2 a_S + P^3 a_S + \ldots + P^k a_S + \ldots$$

is convergent since the common ratio $P < 1$.

Each term of the series

$$(2) \quad a_{S+1} + a_{S+2} + a_{S+3} + \ldots + a_{S+k} + \ldots$$

is smaller than the corresponding term of (1); by the comparison test, (2) is convergent, and so, since adding a finite number of terms does not affect convergence, the original series $a_1 + a_2 + \ldots + a_n$ is convergent as well.

B. $r > 1$

If $\lim\limits_{n \to \infty} \dfrac{a_{n+1}}{a_n} = r$ and $r > 1$, there must exist a positive integer S such that

$$\frac{a_{n+1}}{a_n} > 1, \text{ or } a_{n+1} > a_n,$$

for any integer $n \geq S$, and we have

$$a_n \neq 0.$$

Consequently, if $\lim\limits_{n \to \infty} \dfrac{a_{n+1}}{a_n} > 1$, then $\sum\limits_{n=1}^{\infty} a_n$ diverges.

The ratio test is **inconclusive** if $\lim\limits_{n \to \infty} \dfrac{a_{n+1}}{a_n} = 1$ or if the limit does not exist; in such cases, we must resort to other tests for convergence.

- Does the series $\displaystyle\sum_{n=1}^{\infty} \frac{n^2 3^{n+1}}{4^n}$ converge?

$$\lim_{n \to \infty} \frac{a_{n+1}}{a_n} = \lim_{n \to \infty} \frac{3 \cdot \left(\frac{3}{4}\right)^{n+1} \cdot (n+1)^2}{3 \cdot \left(\frac{3}{4}\right)^n \cdot n^2}$$

$$= \frac{3}{4} \lim_{n \to \infty} \left(1 + \frac{2}{n} + \frac{1}{n^2}\right) = \frac{3}{4}(1 + 0 + 0) = \frac{3}{4}.$$

As $\lim\limits_{n \to \infty} \dfrac{a_{n+1}}{a_n} = \dfrac{3}{4} < 1$, the series converges.

- Does the series $\sum_{n=1}^{\infty} n^{-1}\left(\frac{6}{5}\right)^n$ converge?

$$\lim_{n\to\infty} \frac{a_{n+1}}{a_n} = \lim_{n\to\infty} \frac{6}{5}\cdot\frac{n}{n+1}$$

$$= \frac{6}{5}\lim_{n\to\infty}\left(\frac{n+1}{n+1} - \frac{1}{n+1}\right)$$

$$= \frac{6}{5}(1-0) = \frac{6}{5}\,.$$

As $\lim_{n\to\infty} \frac{a_{n+1}}{a_n} = \frac{6}{5} > 1$, the series diverges.

- Does the series $\sum_{n=1}^{\infty} \frac{1}{\sqrt{n}}$ converge?

$$\lim_{n\to\infty} \frac{a_{n+1}}{a_n} = \lim_{n\to\infty} \frac{\frac{1}{\sqrt{n+1}}}{\frac{1}{\sqrt{n}}} = \lim_{n\to\infty} \sqrt{\frac{n}{n+1}}$$

$$= \lim_{n\to\infty} \sqrt{\frac{n+1}{n+1} - \frac{1}{n+1}}$$

$$= \sqrt{1-0} = 1\,.$$

Consequently, the ratio test is inconclusive. We have already seen, however, that the series diverges. p. 272

Convergence of Alternating Series

While a series with continually decreasing positive terms, approaching zero, converges only when special conditions are met, an alternating series with terms whose absolute values are continually decreasing towards zero will always converge. The latter part of this statement is restated in the following theorem.

The alternating series

$$\sum_{n=1}^{\infty} (-1)^{n-1} a_n = a_1 - a_2 + a_3 - a_4 + a_5 - a_6 + \dots \quad (a_n > 0)$$

converges if

$$\left.\begin{array}{l} a_{n+1} \le a_n \text{ for all } n \\ \lim\limits_{n \to \infty} a_n = 0 \end{array}\right\} .$$

Why? We content ourselves with an intuitive explanation.

Consider the partial sums

$$\begin{aligned} S_1 &= a_1, \\ S_2 &= (a_1 - a_2) < S_1, \\ S_2 < S_3 &= (a_1 - a_2 + a_3) < S_1, \\ S_4 &= (a_1 - a_2 + a_3 - a_4) < S_3, \\ S_4 < S_5 &= (a_1 - a_2 + a_3 - a_4 + a_5) < S_3, \\ S_6 &= (a_1 - a_2 + a_3 - a_4 + a_5 - a_6) < S_5, \textit{etc.} \end{aligned}$$

Placing the values of $S_1, S_2, S_3, \dots$ on the real number line, we see that for each addition or subtraction of a term the values of the partial sums, in an oscillating manner, are closing in on a definite value S_∞ .

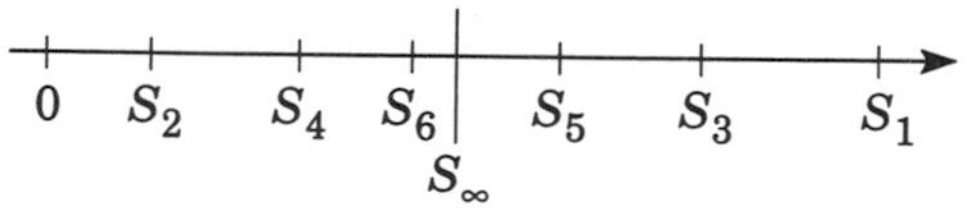

Hence, the alternating infinite series converges.

Convergence, Absolute or Conditional

A series is said to be **absolutely convergent** if it converges when the terms are replaced by their absolute values. If it is convergent but diverges when its terms are replaced by their absolute values, it is said to be **conditionally convergent**.

Absolute convergence leads to convergence, for if $\sum_1^\infty |a_n|$

p. 276 converges then – according to the comparison criterion –

$\sum_1^\infty \frac{1}{2}(|a_n| \pm a_n)$ will converge, as will the difference $\sum_1^\infty a_n$.

The harmonic alternating series

$$(1)\qquad \sum_{n=1}^{\infty} \frac{(-1)^{n+1}}{n} = 1 - \frac{1}{2} + \frac{1}{3} - \frac{1}{4} + \ldots$$

converges.

Replacing the terms by their absolute values, we have the divergent harmonic series

$$\sum_{n=1}^{\infty} \left| \frac{(-1)^{n+1}}{n} \right| = 1 + \frac{1}{2} + \frac{1}{3} + \frac{1}{4} + \ldots .$$

Consequently, (1) is conditionally convergent.

On the other hand, the alternating series

$$(2)\qquad \sum_{n=0}^{\infty} \frac{1}{4}\left(\frac{-1}{2}\right)^{n} = \frac{1}{4} - \frac{1}{8} + \frac{1}{16} - \frac{1}{32} + \frac{1}{64} - \ldots$$

is absolutely convergent, for the series

$$\sum_{n=0}^{\infty} \left| \frac{1}{4}\left(\frac{-1}{2}\right)^{n} \right| = \frac{1}{4} + \frac{1}{8} + \frac{1}{16} + \frac{1}{32} + \frac{1}{64} + \ldots$$

is convergent.

Methods for testing series of positive terms for convergence are also applicable for testing alternating series for absolute convergence.

Approximating Sums and Remainders

The **remainder of an infinite convergent series** is the sum of terms remaining after adding any chosen number of the first terms of the series. If we add the first four terms of

$$\sum_{n=1}^{\infty} a_n = a_1 + a_2 + a_3 + a_4 + a_5 + a_6 + \ldots ,$$

we have an **approximate value of the sum,**

$$\sum_{n=1}^{\infty} a_n \approx a_1 + a_2 + a_3 + a_4 ,$$

whose **error** is the **remainder**

$$\sum_{n=5}^{\infty} a_n = a_5 + a_6 + a_7 + a_8 + a_9 + a_{10} + \ldots .$$

Series of Positive Terms

Since the exact sum of a convergent **geometric series** can always be determined, the remainder of such a series is also readily calculable. The series

$$\sum_{n=0}^{\infty} \frac{1}{4}\left(\frac{1}{2}\right)^n = \frac{1}{4} + \frac{1}{8} + \frac{1}{16} + \frac{1}{32} + \frac{1}{64} + \ldots$$

has the sum

$$\frac{1/4}{1 - 1/2} = \frac{1}{2}\,.$$

The first 10 terms give the sum

$$\sum_{n=0}^{\infty} \frac{1}{4}\left(\frac{1}{2}\right)^n \approx \left(\frac{1}{4}\right)\frac{1-(1/2)^{10}}{1-1/2} = \frac{1023}{2048}\,.$$

The **error** is then the **remainder**

$$\frac{1}{2} - \frac{1023}{2048} = \frac{1}{2048}\,.$$

With the exception of geometric series, it is generally cumbersome to find the sum of a convergent infinite series of positive terms, especially if it converges slowly. ("Slow" and "rapid" convergence are relative notions, referring to the number of terms needed to attain a desired accuracy.)

However, with the assistance of computers, we can easily obtain approximate values of virtually any convergent series; the accuracy is determined by the number of terms added.

To determine the sum

$$\sum_{n=1}^{\infty} \frac{n^2\, 3^{n+1}}{4^n} = \frac{1\,(3^2)}{4} + \frac{2^2\,(3^3)}{4^2} + \frac{3^2\,(3^4)}{4^3} + \frac{4^2\,(3^5)}{4^4} + \ldots$$

p. 278 tested for convergence in a previous example, we program a computer to add 1000 terms, displayed with 10 decimals:

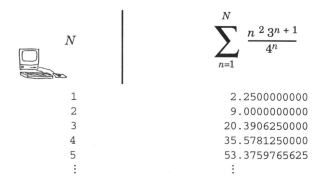

N	$\sum_{n=1}^{N} \frac{n^2\, 3^{n+1}}{4^n}$
1	2.2500000000
2	9.0000000000
3	20.3906250000
4	35.5781250000
5	53.3759765625
⋮	⋮

10	146.5811004639
⋮	⋮
20	235.2179464192
⋮	⋮
100	251.9999999687
⋮	⋮
119	251.9999999998
122	251.9999999999
123	251.9999999999
124	252.0000000000
⋮	⋮
999	252.0000000000
1000	252.0000000000

One must be cautious with this kind of numerical calculation, for divergence may be so slow that it does not show for what, at first, may seem a reasonable number of terms.

Alternating Series

Unlike convergent infinite series of positive terms, the magnitude of the remainder of a convergent alternating series may always be estimated at any point in the succession of terms.

The absolute value of the remainder of the sum of an alternating series,

$$\sum_{n=1}^{\infty}(-1)^{n+1}a_n \ ,$$

is equal to or less than the value of the first deleted term.

Consequently, if the sum of the series is S, the sum of the first j terms is S_j, and the sum of the remaining terms is R_{j+1}, then

$$|S - S_j| = |R_{j+1}| \le a_{j+1} \ .$$

Why? We have

$$R_{j+1} = S - S_j = \sum_{n=1}^{\infty}(-1)^{n+1}a_n - \sum_{n=1}^{j}(-1)^{n+1}a_n$$

$$= (-1)^j a_{j+1} + (-1)^{j+1} a_{j+2} + (-1)^{j+2} a_{j+3} + \ldots$$

$$= (-1)^j (a_{j+1} - a_{j+2} + a_{j+3} - \ldots)$$

or

$$|R_{j+1}| = a_{j+1} - a_{j+2} + a_{j+3} - a_{j+4} + a_{j+5} - \ldots$$

$$= a_{j+1} - (a_{j+2} - a_{j+3}) - (a_{j+4} - a_{j+5}) - \ldots$$

and, since every expression in parentheses yields a positive term,

$$|R_{j+1}| \le a_{j+1} \ .$$

Consider, for instance, the convergent alternating series

$$\sum_{n=1}^{\infty} (-1)^{n+1}\left(\frac{1}{n}\right) = 1 - \frac{1}{2} + \frac{1}{3} - \frac{1}{4} + \dots .$$

The absolute value of the 10th term is $|-1/10|$; consequently, adding the first 9 terms gives an "accuracy" of only about $\pm 1 \cdot 10^{-1}$, and adding the first 19 terms, about $\pm 0.5 \cdot 10^{-1}$.

To ensure an accuracy of $\pm 1 \cdot 10^{-5}$ requires that the first 100 000 terms be added.

(By methods of integration, it can be shown that the sum is, in fact, $\ln 2$.)

8.4 The Tower of Hanoi

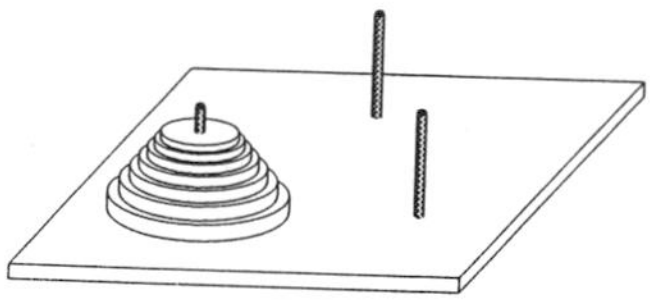

The game of *the tower of Hanoi* has been played since the latter part of the 19th century but in all likelihood is much older. The task is to transfer and rebuild the tower around one of the initially empty pegs in such a way that

- only one ring be moved at a time, and
- no ring must rest on a smaller ring.

For 2 and 3 rings a minimum of, respectively, 3 and 7 moves are required:

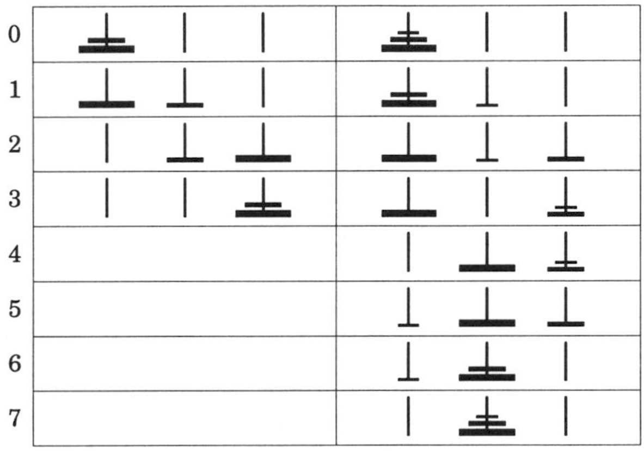

The increase from 2 to 3 rings required that the two rings on the top be piled up twice, the second time on the moved bottom ring. We discern the pattern

$$\begin{aligned} 1 \text{ ring} &\Rightarrow 1 \text{ move} \\ 2 \text{ rings} &\Rightarrow 2 \cdot 1 + 1 = 3 \text{ moves} \\ 3 \text{ rings} &\Rightarrow 2 \cdot 3 + 1 = 7 \text{ moves.} \end{aligned}$$

With similar reasoning, a 4-ring tower means that 3 rings must be built up twice, a 5-ring tower requires that 4 rings be built up twice, *etc.*, and every time the second edifice is erected on the moved bottom ring, giving

$$\begin{aligned} 4 \text{ rings} &\Rightarrow 2 \cdot 7 + 1 = 15 \text{ moves} \\ 5 \text{ rings} &\Rightarrow 2 \cdot 15 + 1 = 31 \text{ moves} \\ 6 \text{ rings} &\Rightarrow 2 \cdot 31 + 1 = 63 \text{ moves, } etc. \end{aligned}$$

We have

$$1,\ 3,\ 7,\ 15,\ 31,\ 63, \ldots, A_{n-1},\ A_n\,,$$

where $A_n = 2A_{n-1} + 1$, that is, $A_n + 1 = 2(2A_{n-1} + 1)$.

To find the n-th term, rewrite the sequence thus:

$$2^1 - 1,\ 2^2 - 1,\ 2^3 - 1,\ 2^4 - 1, 2^5 - 1,\ 2^6 - 1,\ \ldots,\ 2^n - 1$$

To rebuild a tower of 10 rings requires

$$2^{10} - 1 = 1023 \text{ moves.}$$

8.5 The Fibonacci and Related Sequences

FIBONACCI Leonardo
(1170 - *c.* 1250)

One of the most celebrated problems concerning sequences appears in the *Liber abaci*, published in 1202 by the Italian merchant and mathematician Leonardo di Pisa, or **Fibonacci** (Figlio dei Bonacci, "Son of the Bonaccis"); the problem can be related as follows:

Mathematics of Rabbit Breeding

A pair of newly born rabbits, male and female, were placed in a hutch. In two months these rabbits began their breeding cycle and produced one pair of rabbits, one male and one female. The original rabbits and their offspring continued to breed in this manner, that is, the first pair of offspring appearing at the parental age of two months and then a new pair every month thereafter – always one male and one female. All rabbits survived the first year.

What then is the total number of pairs of rabbits at the beginning of each month during the first year?

The original pair had their first pair of offspring at the beginning of the third month, and another pair at the beginning of the fourth month. At the beginning of the fourth month, 2 pairs were fertile, resulting in 2 pairs of offspring at the beginning of the fifth month, and so on:

Month:	Pairs of Rabbits:		
Beginning of	**Productive**	**Nonproductive**	**Total**
1st	0	1	1
2nd	1	0	1
3rd	1	1	2
4th	2	1	3
5th	3	2	5
6th	5	3	8
7th	8	5	13
8th	13	8	21
9th	21	13	34
10th	34	21	55
11th	55	34	89
12th	89	55	144

A look at the resulting sequence,

$$1, 1, 2, 3, 5, 8, 13, 21, 34, 55, 89, 144, 233, 377, \ldots,$$

reveals that, from the third term onward, every successive term is the sum of the two immediately preceding terms; that is, **Fibonacci numbers** satisfy the recursion formula

Fibonacci Numbers

$$\left.\begin{aligned} F_n + F_{n+1} &= F_{n+2} \\ F_1 = 1;\ F_2 &= 1 \end{aligned}\right\}$$

We have no evidence that Fibonacci further explored the sequence; nor did his name become attached to it until the 19th century – to the best of our knowledge, in a paper by the French mathematician **Edouard Lucas** in a publication devoted to recreational mathematics. (Lucas is otherwise best known for the fact that, in 1876, he discovered that the 39-digit Mersenne number $M_{127} = 2^{127} - 1$ is a prime.)

LUCAS
François Edouard Anatole
(1841 - 1891)

In 1753, **Robert Simson** of the University of Glasgow showed that the ratio of one Fibonacci number to the one preceding it,

SIMSON Robert
(1687 - 1768)

$$\frac{1}{1}, \frac{2}{1}, \frac{3}{2}, \frac{5}{3}, \frac{8}{5}, \frac{13}{8}, \ldots.$$

draws progressively nearer, alternately from above and from below, to the **golden number** Φ,

Golden Number

$$\Phi = \tfrac{1}{2}(\sqrt{5}+1) = 1 + \cfrac{1}{1+\cfrac{1}{1+\cfrac{1}{1+\cfrac{1}{1+\ldots}}}} = 1.618\,03\ldots.$$

Ascending integer powers of the golden number yield the following sequence:

$$\Phi = \tfrac{1}{2}(\sqrt{5}+1), \quad \Phi^4 = \tfrac{1}{2}(3\sqrt{5}+7), \quad \Phi^7 = \tfrac{1}{2}(13\sqrt{5}+29),$$

$$\Phi^2 = \tfrac{1}{2}(\sqrt{5}+3), \quad \Phi^5 = \tfrac{1}{2}(5\sqrt{5}+11), \quad \Phi^8 = \tfrac{1}{2}(21\sqrt{5}+47),$$

$$\Phi^3 = \tfrac{1}{2}(2\sqrt{5}+4), \quad \Phi^6 = \tfrac{1}{2}(8\sqrt{5}+18), \quad \Phi^9 = \tfrac{1}{2}(34\sqrt{5}+76)\ldots,$$

where the coefficients of the irrational $\sqrt{5}$ terms form the Fibonacci sequence, and those of the rational terms form the **Lucas sequence**

Lucas Sequence

$$1,\ 3,\ 4,\ 7,\ 11,\ 18,\ 29,\ 47,\ 76,\ 123,\ 199, \ldots,$$

whose terms satisfy the recursion formula

$$\left.\begin{aligned} L_n + L_{n+1} &= L_{n+2} \\ L_1 = 1;\ L_2 &= 3 \end{aligned}\right\}.$$

That is, like the Fibonacci sequence, every term of the Lucas sequence is the sum of the two immediately preceding ones.

Binet Formula

BINET Jacques-Phillippe-Marie (1786 - 1855)

The general terms of the Fibonacci and Lucas sequences are given by the **Binet formulas** for F_n and L_n :

$$F_n = \frac{1}{\sqrt{5}}\left[\left(\frac{1+\sqrt{5}}{2}\right)^n - \left(\frac{1-\sqrt{5}}{2}\right)^n\right]$$

$$L_n = \left(\frac{1+\sqrt{5}}{2}\right)^n + \left(\frac{1-\sqrt{5}}{2}\right)^n$$

Pell Number Sequence

PELL John (1610 - 1685)

The **Pell number sequence**

$$P_n: \ 1, 2, 5, 12, 29, 70, 169, \ldots,$$

named after the English mathematician **John Pell**, and

$$Q_n: \ 1, 3, 7, 17, 41, 99, 239, \ldots$$

satisfy the recursion formulas

$$\left.\begin{array}{l} P_n + 2 \cdot P_{n+1} = P_{n+2} \\ P_1 = 1;\ P_2 = 2 \end{array}\right\} \qquad \left.\begin{array}{l} Q_n + 2 \cdot Q_{n+1} = Q_{n+2} \\ Q_1 = 1;\ Q_2 = 3 \end{array}\right\}$$

and the Binet formulas

$$P_n = \frac{1}{2\sqrt{2}}\ [(1+\sqrt{2})^n - (1-\sqrt{2})^n]$$

$$Q_n = \frac{1}{2}\ [(1+\sqrt{2})^n + (1-\sqrt{2})^n]$$

Fibonacci numbers and related number sequences appear as natural phenomena, such as the shape of snail shells and the heads of sunflowers, and in phyllotaxy. They have also proved relevant and useful in branches of mathematics, for instance, the theory of equations, and in the study of genetics, in electronics, and in data handling of statistics.

The Fibonacci Association was founded in 1962 in California to further interest in Fibonacci numbers and related topics and has published, since 1963, *The Fibonacci Quarterly*.

THE FIBONACCI ASSOCIATION

8.6 Figurate Numbers

Arrangements of dots to represent numbers as geometrical figures, found as far back as Stone Age rock carvings, were of special importance to the Pythagoreans (*c.* 6th century B.C.), who imparted numbers with specific characteristics and personalities and believed that everything could be explained by numbers. Mystic or divine attributes of numbers were prevalent also among the Babylonians, ancient Maya, and most other ancient cultures.

The Pythagoreans demonstrated many of the arithmetic features of figurate numbers.

Beyond serving as number games, figurate numbers lead to interesting and useful progressions and series of numbers, and they give us ways to visualize and geometrize relations between various sorts of numbers.

Triangular Numbers

Triangular numbers are the natural numbers which can be drawn as dots and arranged in triangular shape: 1, 3, 6, 10, 15, 21, 28, 36, 45, 55, 66, *etc.*

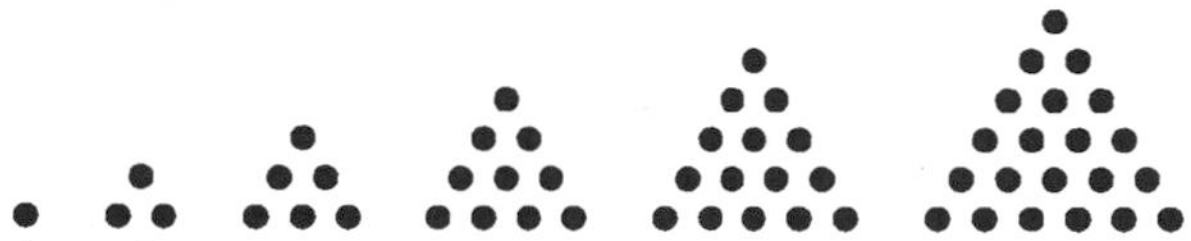

1 3 6 10 15 21

Τετρακτύσ

Of all numbers, 10 was held in greatest reverence by the Pythagoreans; the sum 1 + 2 + 3 + 4 = 10 was named *tetraktys*, "the holy fourfoldness", representing the four elements: fire, water, air, and earth.

1							
2	3						
4	5	6					
7	8	9	10				
11	12	13	14	15			
16	17	18	19	20	21		
22	23	24	25	26	27	28	
29	30	31	32	33	34	35	36

By writing the numbers in rows of increasing length, as shown in the figure to the left, we see the progression of triangular numbers.

The n-th triangular number T_n is given by the sum of an *arithmetic progression of natural numbers*,

$$T_n = 1 + 2 + 3 + \ldots + n,$$

which may be written

$$2\,T_n = n\,(n+1)$$

$$T_n = \frac{n\,(n+1)}{2};$$

the figure shows why:

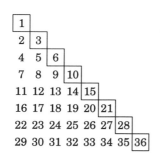

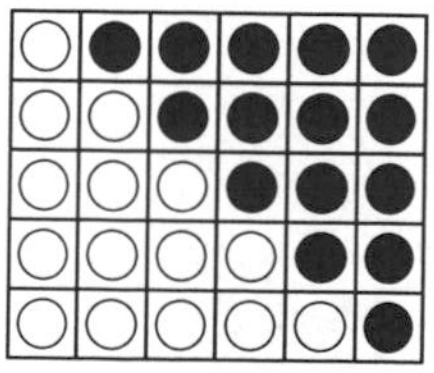

$$2\,T_5 = 5 \cdot 6$$

The sum of two consecutive triangular numbers is the sum of consecutive odd integers starting at 1,

$$T_n + T_{n+1} = 1 + 3 + 5 + \ldots + (2n + 1);$$

and here is why:

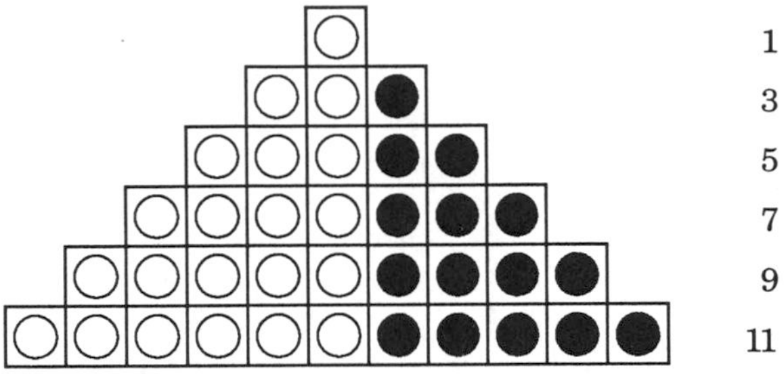

It has been proved that every integer is either a triangular number or the sum of two or three triangular numbers.

Square Numbers

Square numbers, 1, 4, 9, 16, 25, 36, 49, 64, 81, 100, 121, 144, *etc.*, are figurate in this way:

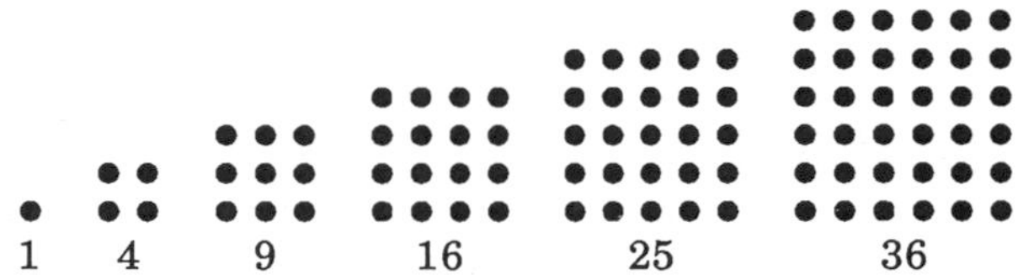

The sum of two consecutive triangular numbers is always a square number:

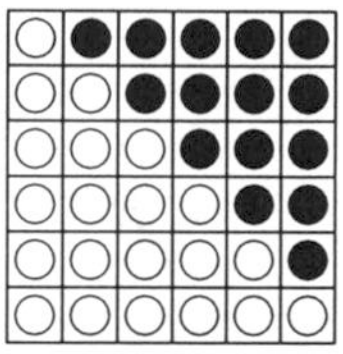

$$T_n + T_{n-1} = n^2$$

Since the sum of two consecutive triangular numbers is equal to the sum of consecutive odd integers starting at 1, we also have

$$n^2 = T_n + T_{n-1} = 1 + 3 + 5 + \ldots + (2n - 1).$$

The main diagonal of the multiplication table is composed of square numbers:

1	2	3	4	5	6	7	8	9	10	11	12
2	4	6	8	10	12	14	16	18	20	22	24
3	6	9	12	15	18	21	24	27	30	33	36
4	8	12	16	20	24	28	32	36	40	44	48
5	10	15	20	25	30	35	40	45	50	55	60
6	12	18	24	30	36	42	48	54	60	66	72
7	14	21	28	35	42	49	56	63	70	77	84
8	16	24	32	40	48	56	64	72	80	88	96
9	18	27	36	45	54	63	72	81	90	99	108
10	20	30	40	50	60	70	80	90	100	110	120
11	22	33	44	55	66	77	88	99	110	121	132
12	24	36	48	60	72	84	96	108	120	132	144

LAGRANGE Joseph Louis (1736 - 1813)

Lagrange proved in 1770 that every natural number is the sum of no more than four squares.

Gnomons

Gnomons are the geometric representations of odd numbers as dots on equally long legs of a right angle. The name refers to the angle's likeness to the Babylonian sundial, the gnomon.

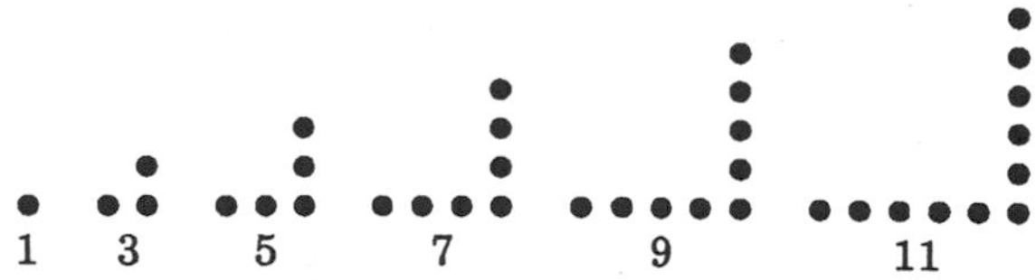

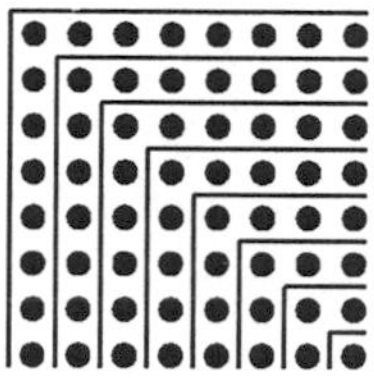

By adding gnomons, the Pythagoreans built larger squares from which they deduced many interesting connections between numbers.

A square thus formed demonstrates the relationships

$$1+3=2^2,$$
$$1+3+5=3^2,$$
$$1+3+5+7=4^2,$$
$$1+3+5+7+9=5^2,$$
$$1+3+5+7+9+11=6^2,$$
$$1+3+5+7+9+11+13=7^2,$$
$$\dots$$

and generally, $1+3+5+7+\ \dots\ +2n-1=n^2$.

Oblong Numbers

Oblong numbers, 2, 6, 12, 20, 30, 42, 56, 72, 90, 110, 132, *etc.*, have a number of dots that can be placed in a rectangular pattern:

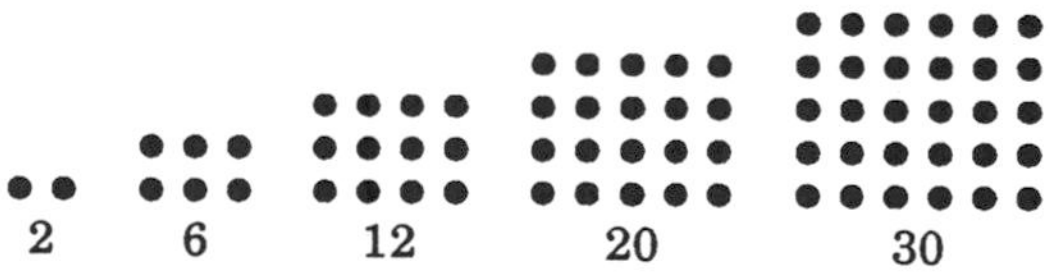

These numbers lack particular interest in mathematics, but justify attention because "square - oblong" is one of the ten "cosmic opposites" in the Pythagorean *doctrine of opposites.*

The doubling of a triangular number yields an oblong number,

$$O_n = 2\,T_n = n\,(n+1)\,.$$

Pentagonal Numbers

Pentagonal numbers, 1, 5, 12, 22, 35, 51, 70, 92, 117, 145, *etc.*, are figurate in this way:

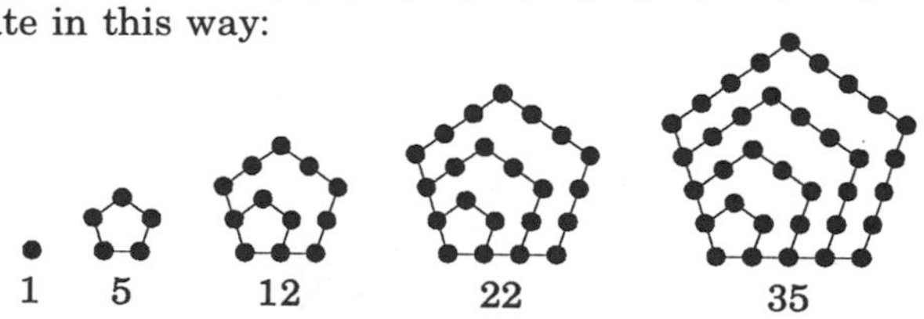

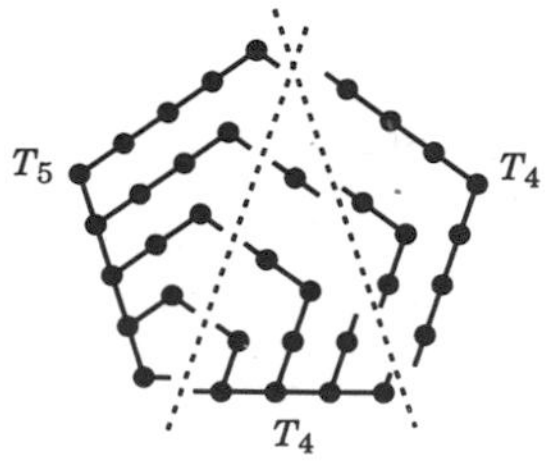

$P_5 = T_5 + 2\,T_4$

Pentagonal numbers can be generated by triangular ones, as seen here to the left, yielding the general formula

$$P_n = 1+4+7+\dots+(3n-2) = T_n + 2\,T_{n-1}$$
$$= \frac{1}{2}n\,(n+1) + n\,(n-1) = \frac{1}{2}n\,(3\,n-1)\,.$$

We now have the relations:

natural numbers n,	n	$= T_n + (-1)\, T_{n-1}$
triangular numbers T_n,	T_n	$= T_n + \ 0 \cdot T_{n-1}$
square numbers n^2,	n^2	$= T_n + \ 1 \cdot T_{n-1}$
pentagonal numbers P_n,	P_n	$= T_n + \ 2 \cdot T_{n-1}$

and may continue:

hexagonal numbers Hex_n,	Hex_n	$= T_n + \ 3 \cdot T_{n-1}$
heptagonal numbers Hep_n,	Hep_n	$= T_n + \ 4 \cdot T_{n-1}$
octagonal numbers Oct_n,	Oct_n	$= T_n + \ 5 \cdot T_{n-1}$

and the polygonal number with p sides, n in each side, is

$$T_n + (p-3) \cdot T_{n-1} .$$

Three Dimensions

We may extend the scope of figurate numbers to three-dimensional space: tetrahedral numbers, pyramidal numbers, cubic numbers, *etc.*

Cubic Numbers

To build **cubic numbers**, 1, 8, 27, 64, 125, 216, *etc.*, we successively stack $n \cdot n$ squares n high.

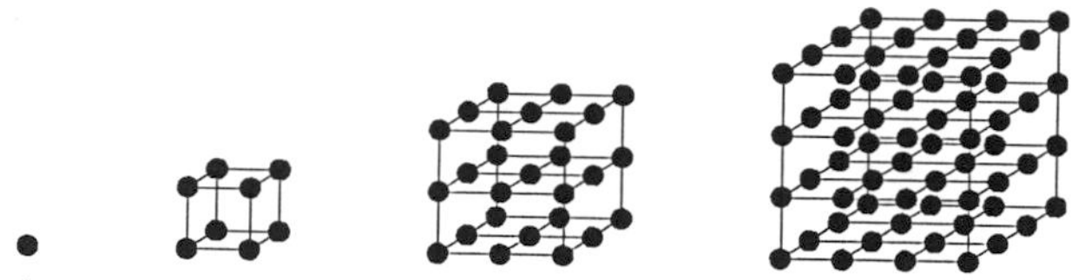

1 8 27 64

In the succession of positive odd numbers,

$$1,\ 3,\ 5,\ 7,\ 9,\ 11,\ 13,\ 15,\ 17, 19, \ldots,$$

we have

$$\begin{aligned} 1 &= 1^3, \\ 3+5 &= 2^3, \\ 7+9+11 &= 3^3, \\ 13+15+17+19 &= 4^3, \\ &etc., \end{aligned}$$

the general formula being

$$[n\,(n-1)+1] + [n\,(n-1)+3] + \ldots + [n\,(n-1)+2n-1] = n^3 .$$

Tetrahedral Numbers

To build **tetrahedral numbers**, 1, 4, 10, 20, 35, 56, *etc.*, we successively stack triangular numbers, 1, 3, 6, 10, 15, ..., one at a time.

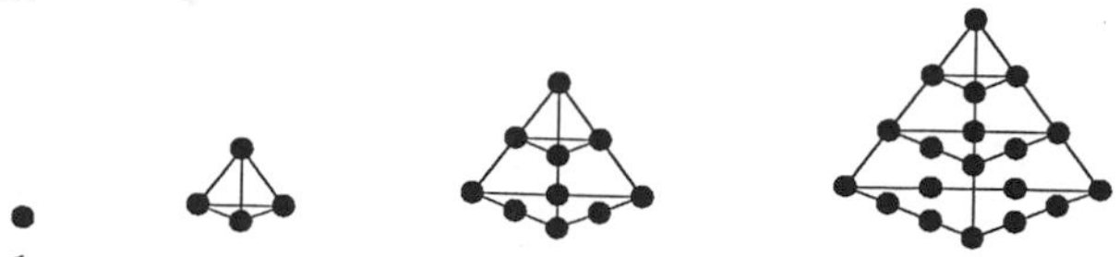

1 4 10 20

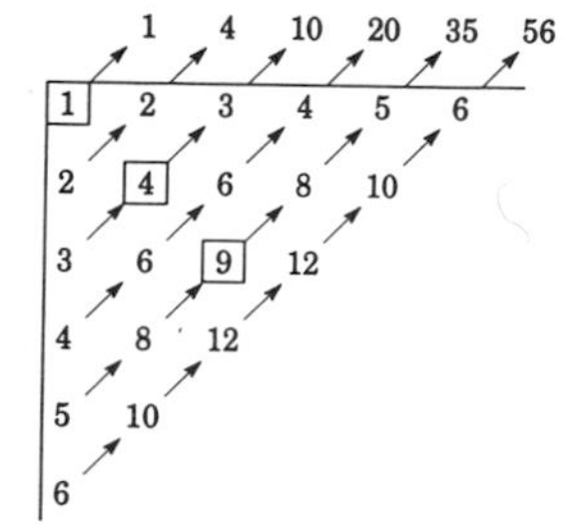

The general formula for a tetrahedral number of n layers is

$$Th_n = n \cdot 1 + (n-1) \cdot 2 + (n-2) \cdot 3 + (n-3) \cdot 4 + \ldots + 2(n-1) + 1 \cdot n$$
$$= \frac{1}{6}\, n\,(n+1)\,(n+2)\,.$$

By adding diagonals of the multiplication table, as shown here to the left, tetrahedral numbers can easily be found.

Square Pyramidal Numbers

Similarly, to build **square pyramidal numbers**, 1, 5, 14, 30, 55, 91, *etc.*, we successively stack the square numbers, 1, 4, 9, 16, 25, 36, *etc.*, one at a time.

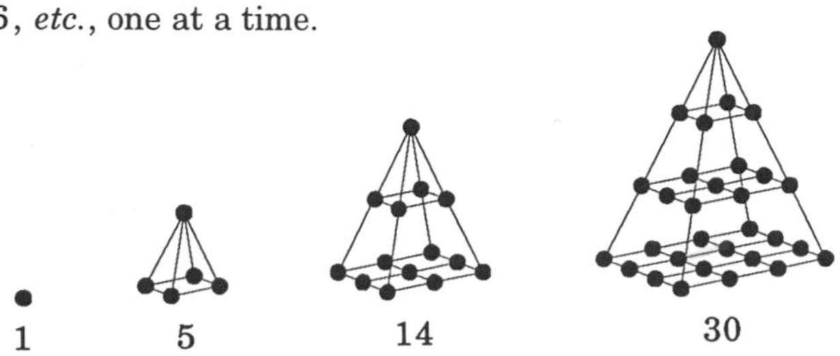

To obtain the n-layer square pyramid, $Pyram_n$, add the first n squares,

$$Pyram_n = 1^2 + 2^2 + 3^2 + \ldots + n^2,$$

whose sum can be found by a formula derived from the general formula for a tetrahedral number: The sum of two consecutive triangular numbers is a square, so the sum of two consecutive tetrahedral numbers is a square pyramidal number,

$$Pyram_n = \frac{1}{6}\, n\,(n+1)\,(n+2) + \frac{1}{6}\,(n-1)\,n\,(n+1)$$
$$= \frac{1}{6}\, n\,(n+1)\,(2n+1)\,.$$

Four Dimensions

Figurate numbers may be formed in four or more dimensions, but these shapes are extremely hard, if not impossible, to visualize.

Sums of Cubes

Adding on cubes, we have

$$1^3 + 2^3 + 3^3 + \ldots + n^3\,.$$

Writing

$$1^3 = 1(1)$$
$$2^3 = 2(1+2+1)$$
$$3^3 = 3(1+2+3+2+1)$$
$$4^3 = 4(1+2+3+4+3+2+1)$$
$$5^3 = 5(1+2+3+4+5+4+3+2+1)$$
$$\ldots$$
$$n^3 = n\,[1+2+3+4+5+\ldots+n+(n-1)+\ldots+4+3+2+1],$$

the sum total of the right-hand column can be expressed as

$$(1+2+3+\ldots+n)(1+2+3+\ldots+n)\,.$$

Hence, the sum of n cubes

$$1^3 + 2^3 + 3^3 + \ldots + n^3$$

is equal to the square of the n-th triangular number,

$$\left[\frac{1}{2} n\,(n+1)\right]^2 = \frac{1}{4}\,n^2\,(n+1)^2\,.$$

"Supertetrahedral Numbers"

By piling up tetrahedral numbers, 1, 4, 10, 20, 35, 56, *etc.*, we make four-dimensional numbers:

$$\begin{aligned} 1 &= 1, \\ 1+4 &= 5, \\ 1+4+10 &= 15, \\ 1+4+10+20 &= 35, \\ 1+4+10+20+35 &= 70, \end{aligned}$$

etc.

Adding on Dimensions

Extending from one dimension (1-D) to five (5-D), we have

1-D the n-th counting number:

$$1+1+1+\ldots+1 = n\,;$$

2-D triangular numbers:

$$1+2+3+\ldots+n = \frac{1}{2}\,n\,(n+1)\,;$$

3-D tetrahedral numbers:

$$1+3+6+\ldots+\frac{1}{2}\,n\,(n+1) = \frac{1}{6}\,n\,(n+1)\,(n+2)\,;$$

and we can go on to

4-D $$1+4+10+\ldots+\frac{1}{6}\,n\,(n+1)\,(n+2) = \frac{1}{24}\,n\,(n+1)\,(n+2)\,(n+3)$$

and on,

5-D $$1+5+15+\ldots+\frac{1}{24}\,n\,(n+1)\,(n+2)\,(n+3) = \frac{1}{120}\,n\,(n+1)\,(n+2)\,(n+3)\,(n+4)$$

and on ... and there will always be another dimension.

Difficulty imagining dimensions of higher order places us in safe and illustrious company:

> *It is often helpful to think of the four coordinates of an event as specifying its position in a four-dimensional space called space-time. It is impossible to imagine a four-dimensional space. I personally find it hard enough to visualize three-dimensional space!*
>
> Stephen W. Hawking, *A Brief History of Time* (1988)

Chapter 9

THEORY OF EQUATIONS

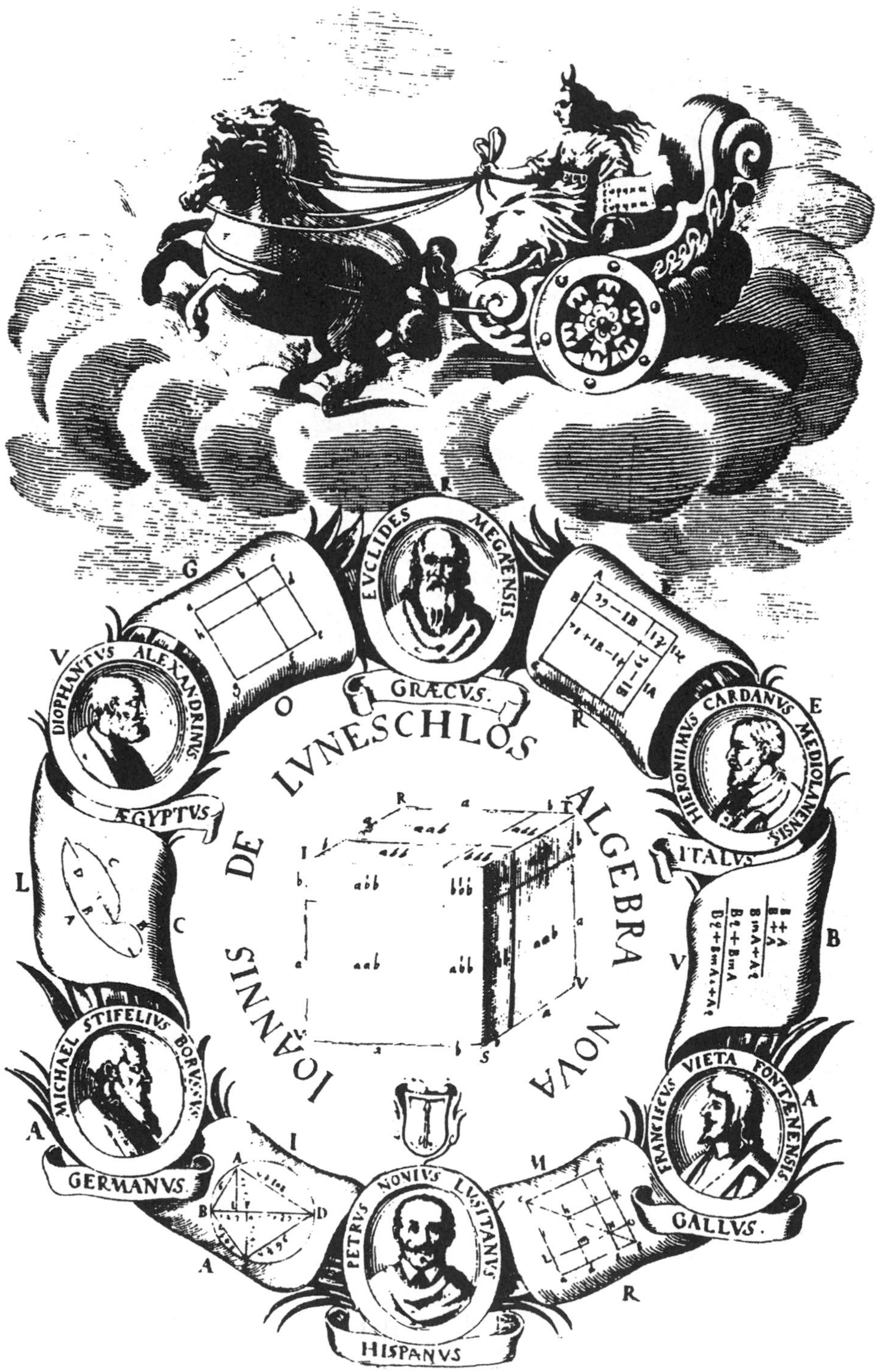

Frontispiece of *Thesaurus Mathematum Reservatus per Algebram Novam* (Passau, 1646) by **Johann de Luneschlos**, physician of great and varied learning who in 1649 became professor of mathematics and physics at Heidelberg.

9.01 History

The Rhind Papyrus – dating from around 1650 B.C. but probably based on a document 200 years older – contains a problem reading:

A quantity and its 1/7 part become 19. What is the quantity?

Regula Falsi

The problem is solved in the Egyptian manner of *regula falsi*; that is, one assumes a – probably wrong – solution and makes the following calculation:

Assume that the answer is	7,
a seventh part is then	1,
making a total of	8.

This is obviously not the right answer, and the assumed answer 7 must be multiplied as many times as 8 must be multiplied to give 19; that is, $8 \cdot \frac{19}{8} = 19$ and $7 \cdot \frac{19}{8} = 16\frac{5}{8}$.

Today we would write

$$x + \frac{x}{7} = 19; \quad x = 7 \cdot \frac{19}{8} = 16\frac{5}{8}\ .$$

The Ahmes papyrus and other ancient Egyptian scrolls are mainly concerned with problems leading to first-degree equations, but also describe some second-degree equations relating to land surveying. Babylonian clay tablets from the time of the Hammurabi dynasty (about 1800 - 1600 B.C.) deal with quadratic equations and their solutions by the method of "completing the square", and also describe numerical methods of solving all quadratic equations and some simpler forms of cubic equations.

Much of the knowledge built up by the old civilizations of Egypt and Babylonia was passed down to the Greeks, who, in turn, gave mathematics scientific form.

PYTHAGORAS
(*c.* 582 - *c.* 507 B.C.)
EVCLEIDIS
(*c.* 330 - *c.* 275 B.C).

Between about 540 and 250 B.C., the ancient Greeks, represented by **Pythagoras**, his followers the **Pythagoreans**, and **Euclid**, gave strict geometric proofs to algebraic problems, using lines and areas for numbers and products:

$$(a+b)(a+b) = a^2 + ab + ab + b^2 = a^2 + 2ab + b^2$$
$$(a-b)(a-b) = a^2 - ab - ab + b^2 = a^2 - 2ab + b^2$$
$$(a+b)(a-b) = a^2 + ab - ab - b^2 = a^2 - b^2,$$

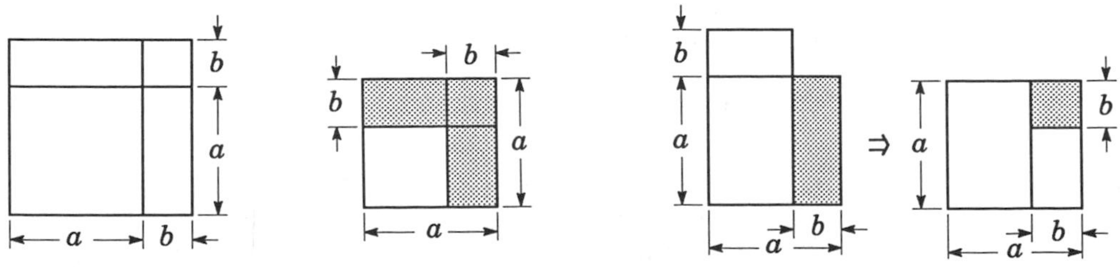

$a^2 + 2ab + b^2$ $\qquad$ $a^2 - 2ab + b^2$ $\qquad$ $a(a-b) + b(a-b) =$ $\qquad$ $a^2 - b^2$

To the Greeks, $x^2 + 2x = 8$ meant a combination of a square whose side was x length units and two rectangles of side lengths x and 1 length units; to develop this figure into a full square, they added a one-by-one small square.

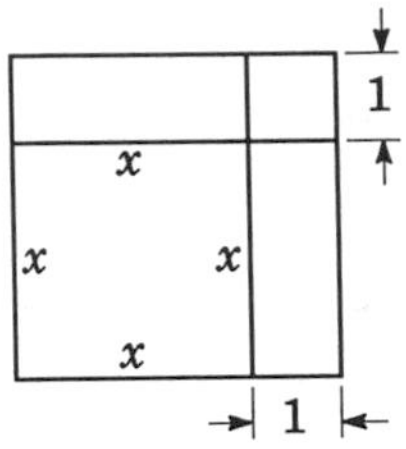

This form of reasoning corresponds to our method of *completing the square*,

$$\begin{aligned} x^2 + 2x &= 8 \\ x^2 + 2x + 1^2 &= 8 + 1 \\ (x+1)^2 &= 9 \\ x + 1 &= \pm 3;\ x_1 = 2,\ x_2 = -4 \end{aligned}$$

but with the significant difference that the Greeks would have regarded -4 as an impossible solution, as no geometric distance can have a negative length.

The Greek system of numeric notation, using letters of their alphabet sometimes built several stories high, and littered with accents, apostrophes, and other diacritic signs, may explain why they turned to geometric interpretations instead of arithmetic calculations.

The Greeks had considerable difficulty in solving cubic equations since their practice of treating algebraic problems as problems of geometry led to complicated three-dimensional constructions.

DIOPHANTOS
(c. A.D. 250)

Greek algebra did not separate from geometric methods until around A.D. 250. This algebra, as presented by **Diophantus** of Alexandria, resembles Babylonian algebra. Unlike Babylonian mathematicians, however, who usually gave only approximate solutions, Diophantus gave exact solutions of his equations. He introduced symbols for frequently recurring quantities and operations, and to denote powers of variables. This **syncopated style** contrasts with the older **rhetorical style** – generally used by other Greek mathematicians and by the Arabs – which employs no mathematical symbols.

Greek, *synkope,* "cutting short"
Greek, *rhetor,* "speaker", "orator"

BRAHMAGUPTA
(c. 630)

In ancient India, algebra was influenced by the Babylonian and Greek schools of mathematics. The Hindu **Brahmagupta** – around A.D. 630 – had a clear understanding of negative numbers and of the concept of zero; he gave solutions to second-degree equations, and outlined general methods for solving equations containing several variables.

al-KHOWARIZMI
(9th century)

Around 825, in Baghdad, **al-Khowarizmi** composed the *Hisab al-jabr w'al-musqabalah* ("Science of the Transposition and Cancellation"), whose title is the source of the term *algebra*; this monograph greatly influenced the introduction of algebraic equations in Europe.

BHASHKARA
(1114 - *c.* 1185)

FIBONACCI Leonardo
Leonardo of Pisa
(*c.* 1170 - *c.* 1250)

In India, around 1150, **Bhashkara**, using letters to represent unknown quantities, wrote the *Lilavati* ("The Beautiful") and the *Vija-ganita* ("Seed Counting"), which suggested that positive numbers have two square roots and negative numbers no roots – or, with present-day terminology, no *real* roots.

Arabic, Persian, and Hindu achievements in algebra were funneled to Europe – especially to Italy – in the 13th and 14th centuries. An important link in this process was *Liber abaci* ("Book of the Abacus") of 1202 by the Italian mathematician and merchant **Leonardo Fibonacci**, also known as **Leonardo of Pisa**.

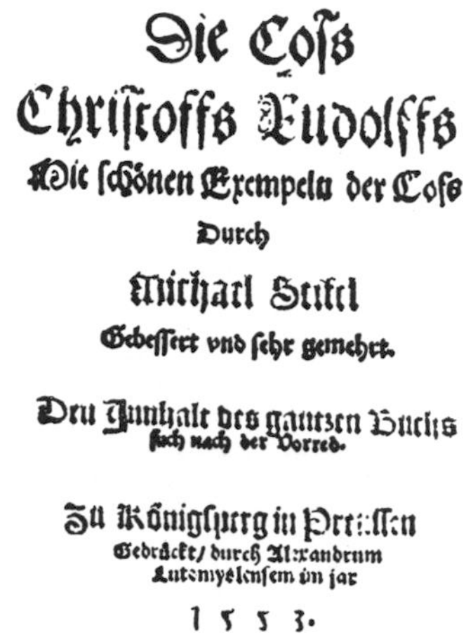
Die Coss
Christoffs Rudolffs
Die schönen Exempeln der Coss
Durch
Michael Stifel
Gebessert vnd sehr gemehrt.
Den Innhalt des gantzen Buchs
such nach der Vorred.
Zu Königsperg in Preussen
Gedrückt / durch Alexandrum
Lutomyslensem im jar
1553.

Influenced by al-Khowarizmi and later Islamic writers who called the unknown *shai*, Arabic for "thing", Latin texts used *res* and those in Italian used *cosa* ("thing"). In Italy, algebra became known as *l'arte della cosa*, in England as *cossike arte* or *the rule of coss*, and in Germany, *die Coss*.

In 1545, Cardano published his work

Artis magnae sive de regulis algebraicis,

generally known as *Ars magna* – which included solutions of the cubic and the quartic equations as well as other mathematical discoveries. The title *Artis magnae* ... is not to be viewed as reference to the grandeur of his work but simply to algebra itself – in that era often referred to as *ars magna* as opposed to the lesser art of arithmetic.

HIERONYMI
CARDANI PRÆ-
STANTISSIMI MATHE-
mathici Philosophi, ac Medici
ARTIS MAGNAE, SIVE
DE REGVLIS ALGEBRAICIS LIB. VNus.
qui et totius operis de Arithmetica, quod OPVS PER-
FECTVM inscripsit, est in ordine Decimus.

No single publication has promoted interest in algebra like Cardano's *Ars magna*, which, however, provides very boring reading to a present-day peruser by consistently devoting pages of verbose rhetoric to a solution which would be much better served by a few symbolic equations. With the untiring industry of an organ grinder, Cardano monotonously reiterates the same solution for a dozen or more near-identical problems, where just one would do.

CARDANO Girolamo
(1501 - 1576)

VIÈTE François
(1540 - 1603)

DESCARTES René
(1596 - 1650)

François Viète, "the father of modern algebraic notation" described the connection between positive roots – the only roots taken note of in his time – and the coefficients of the powers of the unknown quantities. His *In artem analyticem isagoge* (1591; "Introduction to the Art of Analysis") greatly resembles modern texts in elementary algebra.

We owe to **Descartes**, however, the present usage of denoting unknowns by the last letters (x, y, z) of the alphabet, and known quantities by the first (a, b, c). Descartes was one of the main inventors of *analytic geometry*, the method for analyzing geometric problems algebraically, which has contributed greatly to the development of both geometry and algebra.

GAUSS Carl Friedrich
(1777 - 1855)

The **fundamental theorem of algebra** – which states that every polynomial of degree $n \geq 1$, with real or complex coefficients, has n real or complex roots – was proved by the German mathematician **Carl Friedrich Gauss** in 1797. The proof was published in 1799.

DEMONSTRATIO NOVA

THEOREMATIS

OMNEM FVNCTIONEM ALGEBRAICAM
RATIONALEM INTEGRAM

VNIVS VARIABILIS

IN FACTORES REALES PRIMI VEL SECVNDI GRADVS

RESOLVI POSSE

AVCTORE

CAROLO FRIDERICO GAVSS.

HELMSTADII
APVD C. G. FLECKEISEN. 1799.

Gauss returned to the theorem many times, and published his fourth and last proof in 1849, in which he extended the coefficients of the unknown quantities to include complex numbers.

EULER Leonhard
(1707 - 1783)

The fact that the general solution of the quartic equation depended on that of a resolvent cubic equation inspired **Leonhard Euler**, in about 1750, to attempt the solution of the general quintic equation by reducing it to the solution of a related quartic equation. The attempt failed, which led to a general belief that the task might not be possible.

BRING Edvard Samuel
(1736 - 1798)

LAGRANGE Joseph Louis
(1736 - 1813)

Edvard Samuel Bring, a Swedish university professor of history, in 1786 managed to reduce the general quintic equation to the trinomial $x^5 + p\,x + q = 0$ without the quartic and cubic terms, but did not carry the attempt for a solution further. **Lagrange**, the French mathematician and physicist, had, at about the same time, already shown that the solution of the general quintic equation depends on the solution of a sixtic equation, *"Lagrange's resolvent sixtic"*.

RUFFINI Paola
(1765 - 1822)

The first serious attempt to prove that a quintic equation could not be solved by a finite number of algebraic operations was made in 1803 - 1813 by the Italian mathematician and physician **Paola Ruffini**, in communications to a mathematical journal (*Della insolubilità delle equazioni algebraiche generali di grado superiore al quarto*).

ABEL Niels Henrik
(1802 - 1829)

The actual **proof** that the roots of the general equation of fifth degree or higher cannot be expressed in terms of radicals was given by the Norwegian mathematician **Niels Henrik Abel**. At its first appearance, in 1824, in an unpretentious pamphlet published at Abel's own expense, the proof still had some inaccuracy. The correct proof was published in 1826 in the very first volume of *Journal für die reine und angewandte Mathematik* ("Journal for Pure and Applied Mathematics"), often referred to as *Crelle's Journal* after its founder, the

CRELLE August Leopold (1780 - 1855)

German engineer and mathematician **August Leopold Crelle**, whose recognition of mathematical prowess contributed greatly to the advancement of mathematics. A comparable work on algebraic equations was left by the Frenchman **Évariste Galois** in a scientific testament which, when deciphered, was found to give the criteria that an algebraic equation must satisfy in order to be solvable by radicals, a branch of mathematics now known as **Galois** – or **group** – **theory**.

GALOIS Évariste (1811 - 1832)

Galois Theory, Group Theory

Even if general polynomial equations of fifth degree and beyond cannot be solved *by the extraction of roots*, they do have solutions – according to the fundamental theorem of algebra.

HILBERT David (1862 - 1943)

A monumental personage in all main branches of modern mathematics is the German mathematician **David Hilbert**, who in 1895 was appointed to a professorship of mathematics at Göttingen, which, since Gauss's days, was a leading university in mathematical research and produced many mathematicians of great inspirational leadership.

NOETHER Emmy (Amalie) (1882 - 1935)

One of the many eminent figures that Hilbert had attracted to Göttingen was the German mathematician **Emmy Noether**, who, after the Nazi takeover in Germany, settled in the United States in 1933. Noether made important contributions to our understanding of properties of algebraic operations; among other things, she developed the theory of non-commutative algebras, that is, algebras where the order in which terms are multiplied affects the answer. She is recognized as one of the most creative abstract algebraists of the 20th century. The Noether legacy extends beyond her own published works; her influence is evident in the accomplishments of many contemporary mathematicians for whom she was a great inspiration and an unflagging teacher and collaborator.

GREGORIE James (1638 - 1675)

NEWTON Isaac (1642 - 1727)

High-speed electronic computers today allow us to find *approximate* values of roots of polynomial equations of any degree by **numerical analysis**. Instead of exact analytical methods, numerical analysis uses "number-crunching" techniques – sometimes held in contempt by mathematical purists, who do not even recognize numerical analysis as mathematics, although some of its techniques go back to methods of interpolation conceived in the 17th century by **James Gregorie**, Scottish mathematician and astronomer, and by **Isaac Newton**. The term "numerical analysis" has been in use only since 1947, when the Institute of Numerical Analysis was founded in California.

WIENER Norbert (1894 - 1964)

TURING Alan Mathison (1912 - 1954)

von NEUMANN John (Johann) (1903 - 1957)

Prominent figures in the design and use of electronic computers were the American mathematician **Norbert Wiener**, the English mathematician **Alan Turing** (celebrated for breaking the German code Enigma during World War II), and the Hungarian-born **John von Neumann**, considered one of the most all-round and creative mathematicians of the 20th century.

WOLFRAM Stephen (b. 1959)

Due to work by **Stephen Wolfram** and others, easily managed computer software is now available that can provide answers in surds, π, and the base of the natural logarithm, e, making those mathematicians who stay away from computer assistance an endangered species.

9.02 Groundwork

Types of Equations

In **algebraic equations**, variables appear as terms with coefficients that are subject only to the fundamental mathematical operations of addition, subtraction, multiplication, and division.

Algebraic equations may also have terms with non-algebraic coefficients; an equation with a term $2\pi x^2$, for instance, is algebraic in x. A trigonometric equation of the type

$$4 \sin^2 x - \sin x - 2 = 0,$$

while not algebraic in x, may quite well be regarded as an algebraic equation in $\sin x$, and be solved by algebraic methods.

Root equations, which feature terms in which the variable appears in the form of radicands, also belong under the title of algebraic equations.

Transcendental equations – that is, equations that are not algebraic – are concerned with relationships between non-algebraic numbers and quantities, but often have algebraic coefficients.

There are three kinds of transcendental equations: exponential, logarithmic, and trigonometric. To solve such equations, graphic and numerical approximation methods are often required.

In **exponential equations** the variables generally appear in the exponents of power terms, but may also be found in bases and coefficients. Exponential equations are generally solvable by algebraic methods when the variable is part of an exponent. Since logarithms are a form of exponent, there is no clear distinction between exponential and **logarithmic equations**, or between the methods of solving them.

Trigonometric equations containing trigonometric terms with algebraic coefficients can often be solved as algebraic equations.

Trigonometric equations have developed into a convenient means of solving complicated algebraic equations, *e.g.*, the *casus irreducibilis* of the general cubic equation.

Terminology

Latin, *æquare*, "to make equal"

An **equation** is a statement which shows, in mathematical symbols, that two mathematical expressions are equal:

$$1 + x = 3 - x^2$$

Roots, Solutions
Satisfy an Equation

The numerical values of variables which make the equation true are known as the **roots** or **solutions** of the equation; they are said to **satisfy the equation**: $x + 3 = 5$ is satisfied by $x = 2$.

Equations are basically of two types: identities and conditional equations.

Identities

An **identity** is a statement of equality that holds true for all values of the variables in a universal set; to underline this, the identity notation ≡ may be employed:

$$\left(\sqrt{x}\right)^2 \equiv x\,; \quad y^2 - 1 \equiv (y-1)(y+1)$$

Conditional Equations

A **conditional equation**, on the other hand, is true only for certain values of the variables. Thus, the equations

$$x^2 - 1 = 0\,; \quad y + 3 = 5$$

are satisfied only by $x = 1$ or $x = -1$, and by $y = 2$.

Factoring

Suddenly Christopher Robin began to tell Pooh about some of the things: People called Kings and Queens and something called Factors, and a place called Europe, and an island in the middle of the sea where no ships came, and how to make a Suction Pump (if you want to), and when Knights were Knighted, and what comes from Brazil.

A. A. Milne: *The House at Pooh Corner* (1928). Illustrated by Ernest H.Shepard.

Factor

A **factor** is one of two or more quantities that when multiplied together yield a given product. **Factoring** is the process of breaking down polynomial expressions into factors, an essential part of the solving of polynomial equations.

Common Factor
Common Divisor

If a polynomial Q is a factor in every polynomial $P_1 \ldots P_k$, then Q is considered a **common factor**, or a **common divisor**.

Prime Polynomial
Prime Factor

A polynomial with integer coefficients that cannot be further factored to lesser polynomials with integer factor coefficients is a **prime polynomial** or a **prime factor**; for instance, $x + 1$ and $x^2 + x + 1$ are prime polynomials.

The Remainder Theorem

- If a polynomial in an unknown quantity x is divided by a first-degree expression in the same variable, $x - c$, where c may be any real or complex number, the remainder to be expected will be equal to the sum obtained when the numerical value of c is inserted for x in the polynomial.

Dividing the polynomial $(x^3 - 3x^2 + 6x - 3)$ by $(x - 2)$ will give a quotient $(x^2 - x + 4)$ and the remainder 5:

$$2^3 - 3\cdot 2^2 + 6\cdot 2 - 3 = 5\,.$$

$$\begin{array}{r|l} & x^2 - x + 4 \\ \hline x - 2 & x^3 - 3x^2 + 6x - 3 \\ & \underline{x^3 - 2x^2} \\ & \quad -x^2 + 6x \\ & \quad \underline{-x^2 + 2x} \\ & \qquad\quad 4x - 3 \\ & \qquad\quad \underline{4x - 8} \\ & \qquad\qquad\quad 5 \end{array}$$

The Factor Theorem

- If the division of a polynomial by $(x - c)$ results in a remainder of zero, $(x - c)$ is a factor of the polynomial.

BÉZOUT Étienne (1730 - 1783)

The remainder theorem and the factor theorem were suggested by the French mathematician **Étienne Bézout**.

Factoring Guidelines

The factoring of a polynomial is complete when it has been turned into a product of prime factors.

There are no universal rules for how to achieve a complete factoring of a polynomial, but the following suggestions may be useful.

1. If all terms have a **factor in common**, pull it out of the polynomial:

$$x^3 y - 4x^2y^2 + 4xy^3 = xy(x^2 - 4xy + 4y^2)$$

$$5x^6y^2 - 5y^2 = 5y^2(x^6 - 1) = 5y^2(x^3 + 1)(x - 1)(x^2 + x + 1)$$

2. Try the following **formulas** for breaking up the polynomial:

$$x^2 \pm 2xy + y^2 = (x \pm y)^2$$
$$x^2 - y^2 = (x - y)(x + y)$$
$$x^3 \pm y^3 = (x \pm y)(x^2 \mp xy + y^2)$$
$$x^3 \pm 3x^2y + 3xy^2 \pm y^3 = (x \pm y)^3$$
$$x^2 + y^2 + z^2 + 2xy \pm 2yz \pm 2zx = (x + y \pm z)^2$$

3. Use the **factor theorem** to find first-degree factors; if the coefficients of a polynomial are all integers, consider integer factors of the constant term including 1.

In the polynomial $(x^2 - 2x - 15)$, the values to be tried are

$$\pm 1,\ \pm 3,\ \pm 5, \pm 15\,.$$

$+1$	$\Rightarrow$	1^2	$-2\cdot 1$	-15	$=$	-16	$\neq$	0	
-1		$(-1)^2$	$+2\cdot 1$	-15	$=$	-12	$\neq$	0	
$+3$		3^2	$-2\cdot 3$	-15	$=$	-12	$\neq$	0	
-3		$(-3)^2$	$+2\cdot 3$	-15	$=$	0			★
$+5$		5^2	$-2\cdot 5$	-15	$=$	0			★
-5		$(-5)^2$	$+2\cdot 5$	-15	$=$	20	$\neq$	0	
$+15$		15^2	$-2\cdot 15$	-15	$=$	180	$\neq$	0	
-15		$(-15)^2$	$+2\cdot 15$	-15	$=$	240	$\neq$	0	

Hence, $x^2 - 2x - 15 = (x - 5)(x + 3)$.

Analyzing the polynomial

$$x^4 - 5x^3 + 5x^2 + 5x - 6$$

in the same manner we find that the roots are –1, +1, +2, and +3; and the polynomial is factored as

$$(x + 1)(x - 1)(x - 2)(x - 3)\,.$$

4. **Group terms** for "factoring by parts":

$$x^2 - y^2 + 2y - 1 = x^2 - (y^2 - 2y + 1) = x^2 - (y-1)^2 = (x + y - 1)(x - y + 1)$$

$$3xy - 2x - 12y + 8 = (3xy - 12y) - (2x - 8) = (3y - 2)(x - 4)$$

5. **Add and subtract** equal terms to provide groups that are more easily factored:

$$4x^2 - 12xy + 8y^2 = (4x^2 - 12xy + 9y^2) - y^2 = (2x - 3y)^2 - y^2 = [(2x - 3y) + y]\,[(2x - 3y) - y] = 4(x - y)(x - 2y)$$

$$x^4 - 12x^2 + 144 = (x^4 + 24x^2 + 144) - 36x^2 = (x^2 + 12)^2 - (6x)^2 = (x^2 + 6x + 12)(x^2 - 6x + 12) = (x + 3 - i\sqrt{3})(x + 3 + i\sqrt{3}) \cdot (x - 3 - i\sqrt{3})(x - 3 + i\sqrt{3})$$

Factoring Frustration

More intricate problems of factoring are only rarely encountered in practice. Yet many curricula call for seemingly endless drills of factoring. Instead of nurturing an interest in or at least some curiosity about mathematics, such "education" – bordering on harassment – could block the path to more exciting mathematical challenges.

... and then, as Pooh seemed disappointed, he added quickly, "but it's grander than Factors."

A. A. Milne, *The House at Pooh Corner* (1928)

Root/Coefficient Relationships

General Form

An algebraic polynomial equation is said to be in **general form** if terms of all degrees below the highest degree are present; to achieve this, it may be necessary to complete the equation with additional terms with zero coefficients.

Standard Form

To bring the equation into **standard form**, terms in the right member are transposed to the left side of the equality sign, where all terms are arranged in order of descending degrees:

$$a_n x^n + a_{n-1} x^{n-1} + \dots a_1 x + a_0 = 0$$

Monic Form

The equation is in **monic form** when all terms are divided by the coefficient of the highest-degree term:

$$x^n + p_{n-1} x^{n-1} + \dots p_1 x + q = 0$$

Fundamental Theorem of Algebra

The **fundamental theorem of algebra** states that every polynomial of degree $n \geq 1$, with real or complex coefficients, has n real or complex roots. Hence, every polynomial equation of degree n can be written in the form

$$(x - r_1)(x - r_2)(x - r_3) \dots (x - r_n) = 0,$$

where the factors are individually equated to zero in order to find the roots $r_1, r_2, r_3, \dots, r_n$.

The degree of a polynomial equation is defined as that of the highest-degree term of the polynomial. A first-degree equation is called a **linear equation**, since its representation in a coordinate system is a straight line. A second-degree equation is a **quadratic equation**, a third-degree equation a **cubic equation**, a fourth-degree equation a **quartic equation**, a fifth-degree equation a **quintic equation**, *etc.*

A second-degree equation

$$x^2 + px + q = 0$$

with the roots r_1 and r_2 can be written

$$(x - r_1)(x - r_2) = 0$$

or, developed,

$$x^2 - (r_1 + r_2)x + r_1 r_2 = 0,$$

which gives

$$\left.\begin{aligned} r_1 + r_2 &= -p \\ r_1 r_2 &= q \end{aligned}\right\} .$$

Similarly, a third-degree equation

$$x^3 + p\,x^2 + q\,x + r = 0$$

with the roots r_1, r_2, and r_3 is

$$(x - r_1)(x - r_2)(x - r_3) = 0\,,$$

or

$$x^3 - (r_1 + r_2 + r_3)\,x^2 + (r_1 r_2 + r_2 r_3 + r_3 r_1)\,x - r_1 r_2 r_3 = 0.$$

from which we have

$$\left.\begin{aligned} r_1 + r_2 + r_3 &= -p \\ r_1 r_2 + r_2 r_3 + r_3 r_1 &= q \\ r_1 r_2 r_3 &= -r \end{aligned}\right\} .$$

The same kind of relationship exists between the roots and coefficients of all polynomial equations of higher degrees in monic form, with the coefficient 1 for the highest-degree term.

- The coefficient of the term with the second-highest degree, $n - 1$, is the **negative** of the **sum of the roots**:

  $$\sum r_i = r_1 + r_2 + r_3 + \ldots + r_n$$

- The coefficient of the third-highest term, of degree $n - 2$, is the **sum of all double products** of the roots:

  $$\sum r_i r_j = r_1 r_2 + r_1 r_3 + \ldots + r_1 r_{n-1} + r_1 r_n + r_2 r_3 + \ldots + r_{n-1} r_n$$

- The coefficient of the fourth-highest term, $n - 3$, is the **negative** of the **sum of all triple products** of the roots:

  $$\sum r_i\, r_j\, r_k = r_1 r_2 r_3 + \ldots + r_{n-2} r_{n-1} r_n$$

- *Etc.*
- The coefficient of the constant term of the polynomial is the **product of all roots**,

  $$\prod r_i = r_1 r_2 r_3 \ldots r_n\,,$$

  with a plus sign if the degree of the polynomial is even and a minus sign if the degree is odd.

9.1 Linear Equations

The simplest forms of equations are linear equations, in which the unknown variable appears only in first-degree terms.

A linear equation in two variables generally has no unique solution unless there is some other condition that the variables must satisfy.

Polynomial Equations

We distinguish two types of polynomial equations: **linear equations with integer coefficients** and **fractional equations**.

Linear Equations with Rational Coefficients

The equation

$$3x - (2x + 1) - 5 = 4x + (4 - 2x) - 3x$$

is solved in the following manner.

– Remove the parentheses:

$$3x - 2x - 1 - 5 = 4x + 4 - 2x - 3x$$

– Transpose terms:

$$3x - 2x - 4x + 2x + 3x = 4 + 5 + 1$$

– Combine:

$$2x = 10; \quad x = 5$$

Fractional Equations

$$2y + \frac{3y+3}{5} - 3 = 11 - \frac{5y+4}{8}$$

– Multiply both members of the equation by the least common denominator, $5 \cdot 8 = 40$:

$$80y + 24y + 24 - 120 = 440 - 25y - 20$$

– Transpose and combine terms:

$$80y + 24y + 25y = 440 - 20 - 24 + 120$$

$$129y = 516; \quad y = 4$$

9.2 Equations with Absolute Values

Although first-degree equations in one variable generally have only one solution, those containing an *absolute-value expression* have more than one.

Examples:

$$|x+4| = 3 \qquad \left.\begin{array}{r} x+4 = 3;\quad x_1 = -1 \\ -(x+4) = 3;\quad x_2 = -7 \end{array}\right\}.$$

$$|x+2| = |2x-3| \qquad \left.\begin{array}{r} x+2 = 2x-3;\quad x_1 = 5 \\ x+2 = 3-2x;\quad x_2 = \frac{1}{3} \end{array}\right\}.$$

There once was a salesman named Ness
With high expenses, but sales much less.
He was known for his slumbers,
But used absolute numbers,
So then he seemed one of the best.

9.3 Quadratic Equations

A polynomial equation of the second degree is known as a quadratic equation,

$$a x^2 + b x + c = 0,$$

where $a x^2$ is the **quadratic term**, $b x$ the **linear term**, and c the **constant term**.

We distinguish between **pure quadratic equations**, without the linear term, and **complete quadratic equations**, with quadratic, linear, and constant terms.

A pure quadratic equation,

$$3 x^2 - 32 = x^2,$$

is easily solved; it reduces immediately to

$$2 x^2 = 32 ; \quad x = \pm 4 .$$

We shall describe three methods of solving a complete quadratic equation: by factoring, by "completing the square", and by employing the "quadratic formula".

In all three cases, we consider the quadratic equation reduced to its **monic form** with the x^2 coefficient = 1,

$$x^2 + p x + q = 0 ,$$

where p and q are real numbers.

Factoring

The equation

$$x^2 - x - 2 = 0$$

may be conveniently solved by factoring into

$$x^2 - x - 2 = (x + 1)(x - 2) = 0,$$

where the binomial expressions are individually equated to zero:

$$x + 1 = 0 ; \quad x = -1,$$

$$x - 2 = 0 ; \quad x = 2$$

Lost Roots

In factoring polynomials, there is a risk that **a root may be lost** if both members of the equation are divided by a factor that contains the variable.

The equation

$$x^2 + x - 2 = 4x - 4$$

may be handled in two ways:

Right	Wrong
$x^2 - 3x + 2 = 0$	$(x-1)(x+2) = 4(x-1)$
$(x-2)(x-1) = 0$	$x + 2 = 4$
$x_1 = 2;\quad x_2 = 1$	$x = 2$

The method shown on the right loses us the root $x = 1$, since division by $(x - 1)$ is equivalent to dividing by zero, which is not permissible.

A check of the solution does *not* discover a lost root.

Completing the Square

The idea of this method of solution is to make the binomial expression

$$x^2 + px$$

a *perfect square* by adding a quadratic supplement $\left(\frac{p}{2}\right)^2$ to *complete the square*:

$$x^2 + px + \left(\frac{p}{2}\right)^2 = \left(x + \frac{p}{2}\right)^2 .$$

To solve the equation

$$x^2 - 3x = 4,$$

add $\left(\frac{3}{2}\right)^2$ to both members of the equation,

$$x^2 - 3x + \frac{9}{4} = 4 + \frac{9}{4} = \frac{25}{4};$$

$$\left(x - \frac{3}{2}\right)^2 = \left(\frac{5}{2}\right)^2; \quad x = \frac{3}{2} \pm \frac{5}{2};$$

$$x_1 = 4; \quad x_2 = -1 .$$

General Solution of Quadratic Equations

To solve the equation

$$ax^2 + bx + c = 0,$$

bring it into monic form

$$x^2 + px + q = 0,$$

or

$$x^2 + px = -q .$$

Complete the square by adding $(p/2)^2$ to both members of the equation, as before,

$$\left(x + \frac{p}{2}\right)^2 = \frac{p^2}{4} - q .$$

Quadratic Formula

By root extraction and rearranging the terms, we obtain as solution the **quadratic formula**,

$$x = -\frac{p}{2} \pm \frac{\sqrt{p^2 - 4q}}{2},$$

or, for the general equation before reduction,

$$x = -\frac{b}{2a} \pm \frac{\sqrt{b^2 - 4ac}}{2a}.$$

The radicand in these expressions,

$$D = p^2 - 4q \quad \text{or} \quad D = b^2 - 4ac,$$

Discriminant

known as the **discriminant**, decides the nature of the roots of the equation.

If

$D > 0$ the equation has two real, distinct roots;
$D = 0$ the equation has two real, duplicate roots;
$D < 0$ the equation has two conjugate complex roots.

To solve the **fractional quadratic equation**

$$1 - \frac{x-1}{x+2} = \frac{x-2}{2x-3},$$

we multiply all terms by the least common denominator $(x+2)(2x-3)$,

$$(x+2)(2x-3) - (x-1)(2x-3) = (x-2)(x+2),$$

which simplifies to

$$3(2x-3) = x^2 - 4$$

and

$$x^2 - 6x + 5 = 0$$

$$x = 3 \pm \sqrt{9-5} = 3 \pm 2.$$

Check: $x_1 = 5;$ $\qquad 1 - \frac{4}{7} - \frac{3}{7} = 0$

$x_2 = 1;$ $\qquad 1 - \frac{0}{3} - \frac{-1}{-1} = 1 - 1 = 0$

Both values are roots of the equation.

Biquadratic Equation

The name **biquadratic equation** is sometimes used as a misnomer for the general quartic equation, but should correctly be reserved for the special case

$$x^4 + px^2 + q = 0,$$

which is actually a quadratic equation in x^2, and is solved as such.

$$x^4 - 25x^2 + 144 = 0$$

$$x^2 = \frac{25}{2} \pm \frac{1}{2}\sqrt{625 - 576} = \begin{cases} 16 \\ 9 \end{cases}$$

$$x_{1,2} = \pm 4; \quad x_{3,4} = \pm 3$$

9.4 Inequalities

Conditional Inequality

A **conditional inequality** is an inequality that is true only for certain values of the variables. Thus,

$$(x + 3) > 5$$

is a conditional inequality, true only if $x > 2$.

Unconditional (Absolute) Inequality

An **unconditional inequality**, or **absolute inequality**,

$$(x^2 + 1) > 0,$$

is true for all values of the variable.

Inequalities are solved in much the same way as equations.

- $5x - 6 > 2x + 3; \quad 3x > 9; \quad x > 3$

  The inequality is satisfied by all real numbers > 3 .

- $\frac{3x - 6}{2} > \frac{5x + 1}{3}; \quad 9x - 18 > 10x + 2; \quad -x > 20; \quad x < -20$

  Changing the order of inequality members reverses the sign.

- $x^2 - 3x - 10 > 0$

  Inequalities with nonlinear members are solved by factoring:

  $$(x + 2)(x - 5) > 0$$

  The numerical values of the factors must have the *same* sign; that is, they must be both positive or both negative.

  Two cases must be examined.

  $$\left.\begin{matrix} x + 2 > 0 \\ x - 5 > 0 \end{matrix}\right\} \quad \left.\begin{matrix} x > -2 \\ x > 5 \end{matrix}\right\} \quad x_1 > 5 .$$

  $$\left.\begin{matrix} x + 2 < 0 \\ x - 5 < 0 \end{matrix}\right\} \quad \left.\begin{matrix} x < -2 \\ x < 5 \end{matrix}\right\} \quad x_2 < -2 .$$

- $x^2 - 3x - 10 < 0; \quad (x + 2)(x - 5) < 0$

  In this case, the numerical values of the two factors must have *opposite* signs.

  $$\left.\begin{matrix} x + 2 > 0 \\ x - 5 < 0 \end{matrix}\right\} \quad \left.\begin{matrix} x > -2 \\ x < 5 \end{matrix}\right\} \quad -2 < x < 5 .$$

  $$\left.\begin{matrix} x + 2 < 0 \\ x - 5 > 0 \end{matrix}\right\} \quad \left.\begin{matrix} x < -2 \\ x > 5 \end{matrix}\right\} \quad \text{Impossible and rejected.}$$

9.5 Root, Exponential, and Logarithmic Equations

Root Equations

Many equations require, for their solution, the squaring of members of the equation, or parts of members; this procedure may introduce **extraneous roots**, also called **foreign roots**, belonging to an equation closely resembling the given equation.

It is therefore often necessary, for these equations, to check if the solutions obtained are, in fact, true roots of the given equation; if not, they shall be rejected.

- $x - 9 + \sqrt{x-3} = 0\,.$

Rearrange the terms and square both sides:

$$9 - x = \sqrt{x-3}\,; \quad 81 - 18x + x^2 = x - 3\,; \quad x^2 - 19x + 84 = 0$$

$$x = \frac{1}{2}\left(19 \pm \sqrt{361-336}\right) = \frac{1}{2}(19 \pm 5) = \begin{cases} 12 \\ 7 \end{cases}$$

Check: $x = 12;\ 12 - 9 + 3 = 6$ — A false root; rejected.

$x = 7;\ 7 - 9 + 2 = 0$ — $x = 7$ is a true root of the equation.

- $\sqrt{5x^2 + 10x - 6} = 2x + 3\,.$

Square both sides,

$$5x^2 + 10x - 6 = 4x^2 + 12x + 9\,; \quad x^2 - 2x - 15 = 0,$$

which factors to

$$(x-5)(x+3) = 0\,; \quad x_1 = 5\,; \quad x_2 = -3\,.$$

Check: $x = 5;\ \left.\begin{array}{r} \sqrt{125 + 50 - 6} = \sqrt{169} = 13 \\ 2 \cdot 5 + 3 = 13 \end{array}\right\}$ $x = 5$ is a true root.

$x = -3;\ \left.\begin{array}{r} \sqrt{45 - 30 - 6} = \sqrt{9} = 3 \\ 2(-3) + 3 = -3 \end{array}\right\}$ $x = -3$ does not satisfy the equation; rejected.

Exponential Equations

A number of exponential equations can be solved by algebraic methods when the variable is part of an exponent; we can distinguish three types of exponential equations:

- $a^x = a^n$ Since the bases are equal, the exponents must be equal, and $x = n$ (if $a \neq 1$).
- $a^x = b^x$ When different bases, both $\neq 0$, raised to the same power are equal, the exponent must be $= 0$.
- $a^x = b^n$ Both bases are assumed to be $\neq 0$, and $n \neq 0$; the equation is solved by taking the logarithm of both sides, to the same base:

$$x \cdot \lg a = n \cdot \lg b\,; \quad x = n \cdot \frac{\lg b}{\lg a}$$

- $\sqrt[3]{a^{x+3}} = \sqrt{a^{2x-2}}\,; \quad a^{\frac{x+3}{3}} = a^{\frac{2x-2}{2}}\,; \quad \frac{x+3}{3} = x-1\,; \quad x = 3\,.$

- $325^x = 200^3\,; \quad x \cdot \lg 325 = 3 \cdot \lg 200\,; \quad x = 3 \cdot \frac{\lg 200}{\lg 325}\,.$

- $3^{2x+1} \cdot 4^{2x-1} \cdot 5^{x+2} = 6^{2x+1}\,.$

$$3 \cdot 3^{2x} \cdot \frac{1}{4} \cdot 2^{4x} \cdot 25 \cdot 5^x = 6 \cdot 2^{2x} \cdot 3^{2x}\,,$$

which simplifies to

$$2^{2x} \cdot 5^x = \frac{24}{75}\,.$$

Take the logarithm of both sides,

$$x = \frac{\lg 0.32}{\lg 20} = -0.380\,352\,5\ldots.$$

- $3^{2+x} + 3^{2-x} = 82\,.$

$$9 \cdot 3^x + \frac{9}{3^x} = 82\,; \quad 9 \cdot 3^{2x} - 82 \cdot 3^x + 9 = 0,$$

which can be factored to

$$(9 \cdot 3^x - 1)(3^x - 9) = 0\,.$$

$$\left.\begin{aligned} 3^x &= \frac{1}{9}\,;\ x = -2 \\ 3^x &= 9\,;\ x = 2 \end{aligned}\right\}\,.$$

Logarithmic Equations

Since logarithms are a form of exponent, there is no clear distinction between exponential and logarithmic equations, or between the methods of solving them.

- $x^{\lg x - 1} = 100\,.$

Take the logarithm of both sides,

$$(\lg x - 1)\lg x = 2$$

$$\lg^2 x - \lg x - 2 = 0\;; \quad (\lg x + 1)(\lg x - 2) = 0\;;$$

$$\lg x = -1,\; x_1 = 0.1 \quad\Big|\quad \lg x = 2,\; x_2 = 100\,.$$

- $\lg(2x^2 - 1) - 2\cdot\lg(2 + x) = 0\,.$

Rearrange terms and equate antilogarithms,

$$2x^2 - 1 = (2 + x)^2$$

rewritten,

$$2x^2 - 1 = 4 + 4x + x^2, \text{ or}$$

$$x^2 - 4x - 5 = 0,$$

$$(x + 1)(x - 5) = 0\;; \quad x_1 = -1\;; \quad x_2 = 5\,.$$

Check: $x_1 = -1$; $\lg 1 - 2\cdot\lg 1 = 0 - 0 = 0$,

$x_2 = 5$; $\lg 49 - 2\cdot\lg 7 = 2\cdot\lg 7 - 2\cdot\lg 7 = 0\,.$

Both roots belong to the original equation.

- $\ln(3x - 8) + 2 = \sqrt{2\cdot\ln(3x - 8) + 7}\,.$

Substitute $\ln(3x - 8) = u$,

$$u + 2 = \sqrt{2u + 7}\,.$$

Square both members,

$$u^2 + 4u + 4 = 2u + 7$$

$$u^2 + 2u - 3 = 0$$

$$(u - 1)(u + 3) = 0$$

$$\ln(3x - 8) = 1 \qquad\Big|\qquad \ln(3x - 8) = -3$$

$$3x - 8 = e \qquad\Big|\qquad 3x - 8 = e^{-3}$$

$$x_1 = \frac{1}{3}(e + 8) \approx 3.573 \qquad\Big|\qquad \left(x_2 = \frac{1}{3}(e^{-3} + 8) \approx 2.683\right)$$

By squaring the members of the equation, one runs the risk of acquiring an extraneous root. Checking the roots one finds that only x_1 satisfies the original equation; x_2 is an extraneous root.

9.6 Cubic Equations

Historical Notes

The earliest occurrence of cubic (third-degree) equations may have been in antiquity in connection with the famous problems of duplicating the cube and trisecting the arbitrary angle, which both would lead to cubic equations if stated analytically. The task was to solve these problems using only an unmarked ruler and a pair of compasses, which we now know is impossible. However, several mathematicians of antiquity solved them geometrically using other curves as tools.

HERON
(1st century)

An important advance was made by the mathematician-inventor **Heron** of Alexandria by reviving and developing old Babylonian and Egyptian practices of extracting roots by successive approximation.

KHAYYAM Omar
(1048? - 1122)

ARCHIMEDES
(*c.* 287 - 212 B.C.)

The cubic equation was a cherished topic of study among mathematicians of the Muslim world. Best known is the work of the Persian poet, mathematician, and astronomer **Omar Khayyam**. Building on the Greek tradition, he obtained solutions to cubic equations as intersections between conic sections; such studies had, however, been made by **Archimedes** about 1000 years before Khayyam. In mathematics, the importance of Omar Khayyam and other early Muslim mathematicians lies mainly in their perpetuation of ancient Greek and Hindu knowledge.

Outstanding among mathematical discoveries during the Renaissance were the general solutions of the cubic and quartic equations by radical expressions – an emancipation from a 2000-year-old framework of knowledge set up by the Greeks, Indians, Arabs, and Persians.

del FERRO Scipione
(1465 - 1526)

About 1515 **Scipione del Ferro**, a teacher of mathematics at the University of Bologna, solved a cubic equation lacking the quadratic term,

$$x^3 + q\,x \;=\; r\,.$$

After the manner of his time, he kept his discovery secret, except for telling it in confidence to one of his pupils, Antonio Fiore.

TARTAGLIA Niccolò
Italian, *tartaglia*, "stammerer"
(*c.* 1500 - 1557)

CARDANO Girolamo
(1501 - 1576)

Some time around 1535 **Niccolò Tartaglia**, a mathematics teacher in Venice, solved a cubic equation in another form, without the linear term,

$$x^3 + q\,x^2 \;=\; r\,.$$

He, too, guarded his solution as a dead secret until he divulged it, in the form of a cipher in verse and against a solemn promise of silence, to **Girolamo Cardano**, of Milan and Pavia, a mathematician of note and a famous physician.

In 1545 Cardano published his great work

Artis magnae sive de regulis algebraicis

– generally known as *Ars magna* – in which, breaking his promise, he included Tartaglia's solution of the cubic equation.

Solutions of Cubic Equations

The general form of a cubic equation,

$$a\,x^3 + b\,x^2 + c\,x + d = 0,$$

can be reduced to the *normal form*,

$$x^3 + p\,x^2 + q\,x + r = 0,$$

where x^3 is the cubic, $p\,x^2$ the quadratic, $q\,x$ the linear, and r the constant term.

Every cubic equation with real coefficients has at least one real root r_1; the other two roots may also be real or a conjugate complex pair. The cubic expression can be written as a product of a binomial $(x - r_1)$ and a second-degree polynomial which gives the roots r_2 and r_3 when equated to zero.

A **pure cubic equation**, $x^3 + t = 0$, has the three roots

$$r_1 = \sqrt[3]{-t}$$

$$r_2 = w_2\sqrt[3]{-t}$$

$$r_3 = w_3\sqrt[3]{-t},$$

where

$$w_2 = \frac{1}{2}\left(-1 + i\sqrt{3}\right)$$

$$w_3 = \frac{1}{2}\left(-1 - i\sqrt{3}\right)$$

(*cf. pp.* 793 - 94) are the complex cubic roots of unity.

A symmetric cubic equation,

$$a\,x^3 + b\,x^2 + b\,x + a = 0,$$

can be reduced to monic form,

$$x^3 + p\,x^2 + p\,x + 1 = 0,$$

and factored to

$$(x + 1)\,(x^2 - x + 1 + p\,x) = 0\,.$$

Consequently, the equation has a root $r_1 = -1$. The other two roots are obtained by equating the polynomial $(x^2 - x + 1 + p\,x)$ to zero; to solve that equation, we write

$$x^2 + (p - 1)\,x + 1 = 0\,.$$

Then, using the quadratic formula, we find

$$r_2, r_3 = -\frac{p-1}{2} \pm \frac{\sqrt{p^2 - 2\,p - 3}}{2}\,.$$

Example:

$$y^3 + 6\,y^2 + 6\,y + 1 = 0$$

$$(y + 1)\,(y^2 + 5\,y + 1) = 0$$

$$y_1 = -1$$

$$y_2, y_3 = -\frac{5}{2} \pm \frac{\sqrt{21}}{2}\,.$$

General Solution of Cubic Equations

The cubic equation in *general form*,

$$a x^3 + b x^2 + c x + d = 0,$$

can be transformed by substituting

$$x = y - \frac{b}{3a}$$

into the *reduced form*,

$$y^3 + p y + q = 0 .$$

We assume that this equation has a root

$$y = u + v ,$$

which, inserted in the cubic equation, gives

$$(u^3 + v^3) + (3 u v + p)(u + v) + q = 0 .$$

This equation is satisfied if

$$3 u v = -p ; \qquad \left. \begin{aligned} u^3 \cdot v^3 &= \left(-\frac{p}{3}\right)^3 \\ u^3 + v^3 &= -q \end{aligned} \right\} .$$

Thus, u^3 and v^3 are the roots of a quadratic equation,

$$t^2 + q t - \left(\frac{p}{3}\right)^3 = 0 ; \qquad \left. \begin{matrix} u^3 \\ v^3 \end{matrix} \right\} = -\frac{q}{2} \pm \sqrt{\left(-\frac{q}{2}\right)^2 + \left(\frac{p}{3}\right)^3} .$$

u and v being interchangeable, we select

$$\left. \begin{aligned} u_1 &= \sqrt[3]{-\frac{q}{2} + \sqrt{\left(\frac{q}{2}\right)^2 + \left(\frac{p}{3}\right)^3}} \\ v_1 &= \sqrt[3]{-\frac{q}{2} - \sqrt{\left(\frac{q}{2}\right)^2 + \left(\frac{p}{3}\right)^3}} \end{aligned} \right\} \left. \begin{aligned} u_2 &= u_1 \omega_2 \\ v_2 &= v_1 \omega_3 \end{aligned} \right\} \left. \begin{aligned} u_3 &= u_1 \omega_3 \\ v_3 &= v_1 \omega_2 \end{aligned} \right\} ,$$

where ω_2 and ω_3 are the complex cube roots of unity,

$$\omega_2, \ \omega_3 = \frac{1}{2} \left(-1 \pm i \sqrt{3}\right) .$$

The three solutions of the original equation are then

$$\left. \begin{aligned} x_1 &= u_1 + v_1 - \frac{b}{3a} \\ x_2 &= u_2 + v_3 - \frac{b}{3a} \\ x_3 &= u_3 + v_2 - \frac{b}{3a} \end{aligned} \right\} .$$

The radicand

$$D = \left(\frac{q}{2}\right)^2 + \left(\frac{p}{3}\right)^3$$

Discriminant

is the **discriminant** that determines the nature of the roots of the reduced equation.

If $D > 0$	we have	one real root, and two conjugate complex roots;
$D = 0$	we have	three real roots, of which at least two are equal;
$D < 0$	we have	three distinct real roots.

To solve a cubic equation whose discriminant is negative, one must extract cube roots of complex numbers. For a long time all attempts to solve, algebraically, such an equation – the *casus irreducibilis* – came to naught despite the fact that this is the very case when the equation has only real roots. This deadlock was resolved by trigonometric means around 1600 by **François Viète**.

VIÈTE François (1540 - 1603)

A particularly elegant solution of the cubic equation was presented by Viète in *De aequationum recognitione et emendatione*, published posthumously in 1615.

In the cubic equation without the square term

$$x^3 + 3ax = 2b$$

he substituted

$$x = \frac{a}{y} - y\,;$$

the equation then becomes

$$y^6 + 2by^3 - a^3 = 0\,,$$

that is, a quadratic equation in y^3; solve for y^3, extract the cube root y, and compute the x values.

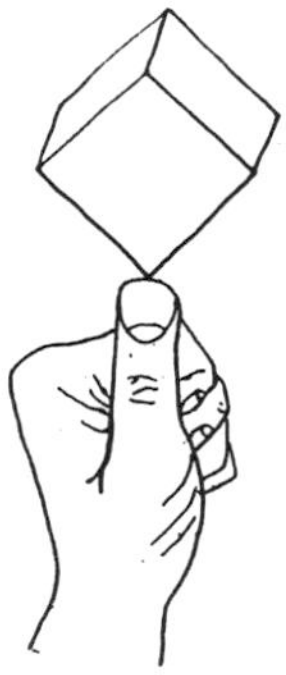

9.7 Quartic Equations

FERRARI Ludovico (or Luigi) (1522 - 1565)

The general solution – with rational operations and radical expressions – of the quartic equation proposed by **Ludovico Ferrari** is, as we shall see, rather complicated and cumbersome because it relies, in several steps, on the solution of equations of lower degrees.

By a stroke of good luck, Nature has ordained that problems presenting themselves in physics seldom lead to general quartic equations; the quartic equations that do occur – for instance, the irradiation of light and heat from a black body – are of a kind more easily coped with, numerically or graphically, if need be.

al-KASHI Masud (? - 1429)

Actually, the enthusiasm raised by the 16th-century solutions of the cubic and quartic equations was of no practical significance, since the Persian **al-Kashi** – a century before del Ferro, Tartaglia, Cardano, and Ferrari – could have solved any cubic equation to any desired degree of accuracy by successive approximation.

The significance of the del Ferro/Tartaglia/Cardano/Ferrari formulas was primarily their contributions to an accelerated development of algebra.

The Symmetric Quartic Equation

- The equation

$$2x^4 - 9x^3 + 14x^2 - 9x + 2 = 0$$

is solved by grouping terms with like coefficients together and dividing all terms by the square of the variable,

$$2\left(x^2 + \frac{1}{x^2}\right) - 9\left(x + \frac{1}{x}\right) + 14 = 0,$$

which we rewrite

$$\left(x^2 + 2 + \frac{1}{x^2}\right) - \frac{9}{2}\left(x + \frac{1}{x}\right) + 5 = 0,$$

or

$$\left(x + \frac{1}{x}\right)^2 - \frac{9}{2}\left(x + \frac{1}{x}\right) + 5 = 0,$$

that is, a quadratic equation in the new variable $\left(x + \frac{1}{x}\right)$, with the solutions $\left(x + \frac{1}{x}\right) = \frac{9}{4} \pm \sqrt{\frac{81}{16} - 5}$:

$x + \frac{1}{x} = \frac{5}{2}$	$x + \frac{1}{x} = 2$
$x^2 - \frac{5}{2}x + 1 = 0$	$x^2 - 2x + 1 = 0$
$x_1 = 2, \; x_2 = \frac{1}{2}$	$x_3, x_4 = 1$

- The near-symmetric equation

$$x^4 - 8x^3 + 14x^2 + 8x - 8 = 0$$

can be rearranged as

$$(x^4 - 8x^3 + 16x^2) - 2(x^2 - 4x) - 8 = 0,$$

that is,

$$(x^2 - 4x)^2 - 2(x^2 - 4x) - 8 = 0,$$

a quadratic equation in the binomial $y = (x^2 - 4x)$ with the solution

$$y^2 - 2y - 8 = 0; \quad y = 1 \pm \sqrt{9} = 1 \pm 3 = \left.\begin{matrix} 4 \\ -2 \end{matrix}\right\} .$$

$x^2 - 4x - 4 = 0$	$x^2 - 4x + 2 = 0$
$x_1, x_2 = 2 \pm 2\sqrt{2}$	$x_3, x_4 = 2 \pm \sqrt{2}$

Solution by Factoring

Certain quartic equations may be reduced to equations of a lower degree, generally quadratic equations, by factoring.

If one or more roots of the equation are known, or easily identified, we may reduce the quartic equation by dividing out the known root.

- $x^4 - 6x^3 + 5x^2 + 12x - 12 = 0$.

We note that the equation is satisfied by $x_1 = 1$ and $x_1 = 2$, and **divide** the fourth-degree polynomial by the product $(x-1)(x-2) = (x^2 - 3x + 2)$, which gives the equation

$$x^2 - 3x - 6 = 0;$$

that is, we have reduced the original polynomial to a product of two quadratic polynomials,

$$(x^2 - 3x + 2)(x^2 - 3x - 6),$$

which we equate to zero separately:

$x^2 - 3x + 2 = 0$	$x^2 - 3x - 6 = 0$
$x_1 = 1, \quad x_2 = 2$	$x_3, x_4 = \frac{1}{2}\left(3 \pm \sqrt{33}\right)$

Solution by Rearranging

- $x^4 + 3x^3 + 12x - 16 = 0$.

The equation can be rearranged as

$$(x^4 - 16) + (3x^3 + 12x) = 0$$

or

$$(x^2 + 4)(x^2 - 4) + 3x(x^2 + 4) = 0$$

and

$$(x^2 + 3x - 4)(x^2 + 4) = 0 .$$

$x^2 + 3x - 4 = 0$	$x^2 + 4 = 0$
$x_1 = 1, \quad x_2 = -4$	$x_3, x_4 = \pm 2\mathrm{i}$

- $x^4 - 9x^2 + 12x - 4 = 0$.

The equation can be rearranged as

$$x^4 - (9x^2 - 12x + 4)$$
$$= x^4 - (3x-2)^2 = (x^2+3x-2)(x^2-3x+2) = 0$$

$x^2 - 3x + 2 = 0$	$x^2 + 3x - 2 = 0$
$x_1 = 1,\ x_2 = 2$	$x_3, x_4 = -\frac{3}{2} \pm \frac{\sqrt{17}}{2}$

General Solution of Quartic Equations

The Kindling Proposal

In 1540 a mathematics dilettante, Zuanne de Tonino da Coi – also known as Ioannes Colla – proposed this problem,

> "Divide 10 into three parts such that they shall be in continued proportion and that the product of the first two shall be 6",

which he gave to Cardano, who, himself unable to solve it, handed it to his pupil Ludovico Ferrari, who could – and did.

The problem can be stated,

$$a+b+c = 10; \quad a/b = b/c; \quad ab = 6,$$

which gives

$$a = 6/b; \ c = b^2/a = b^3/6,$$

and thus,

$$\frac{6}{b} + b + \frac{b^3}{6} = 10$$

$$b^4 + 6b^2 - 60b + 36 = 0,$$

a quartic equation, without a third-degree term, which Ferrari rewrote,

$$(b^2+6)^2 = 6b^2 + 60b,$$

and, again,

$$(b^2+6+z)^2 = (2z+6)\cdot b^2 + 60b + (12z + z^2).$$

For the right-hand member of the equation to be a perfect square, the discriminant of the quadratic polynomial must equal zero,

$$4\cdot(2z+6)\cdot(z^2+12z) = 60^2;$$

this gives the *resolvent cubic* equation

$$z^3 + 15z^2 + 36z = 450,$$

which appears in Cardano's *Ars magna* thus:

> *habebimus 1 cubum p: 15 quadratis*
> *p: 36 positionibus aequalia 450.*

By substituting

$$z = y - 5$$

we have

$$y^3 - 39y - 380 = 0 .$$

A comparison with

p. 318

$$y^3 + py + q = 0$$

gives:

$$p = -39 \qquad p/3 = -13 \qquad (p/3)^3 = -2\,197$$

$$q = -380 \qquad q/2 = -190 \qquad (q/2)^2 = 36\,100$$

$$\overline{33\,903}$$

$$y = \sqrt[3]{190 + \sqrt{33\,903}} + \sqrt[3]{190 - \sqrt{33\,903}} \approx 9.009\,8 \approx 9 ;$$

$$z \approx 9 - 5 = 4 .$$

Inserting $z \approx 4$ in

$$(b^2 + 6 + z)^2 = (2z + 6) \cdot b^2 + 60\,b + (12\,z + z^2)$$

we obtain

$$(b^2 + 10)^2 \approx 14\,b^2 + 60\,b + 64,$$

whose right-hand member must be a perfect square,

$$m^2 b^2 + 2\,mn \cdot b + n^2 = (m\,b + n)^2$$

$$m \approx \sqrt{14} ; \quad n \approx \sqrt{64} = 8 .$$

We now have

$$b^2 + 10 \approx b\sqrt{14} + 8 ; \quad b^2 - b\sqrt{14} + 2 \approx 0$$

$$b \approx \frac{\sqrt{14}}{2} \underset{(-)}{+} \frac{\sqrt{14 - 8}}{2} = \frac{1}{2}\left(\sqrt{14} + \sqrt{6}\right) \approx 3.095\,57;$$

$$a \approx 6/b = \frac{12}{\sqrt{14} + \sqrt{6}} \approx 1.938\,25;$$

$$a/b = b/c \quad \Rightarrow \quad c = b^3/a,$$

$$c = \frac{\left(\sqrt{14} + \sqrt{6}\right)^3}{48} \approx 4.943\,93 .$$

Checking:

$$a + b + c = 1.938\,25 + 3.095\,57 + 4.943\,93 = 9.98 \approx 10$$

$$a/b = \frac{1.938\,25}{3.095\,57} \approx 0.626\,1 ; \quad b/c = \frac{3.095\,57}{4.943\,93} \approx 0.626\,1$$

$$a \cdot b = 1.938\,25 \cdot 3.095\,57 = 5.999\,88 \ldots \approx 6$$

General Formula for Solution of Quartic Equations

The **general quartic equation**

$$a\,x^4 + b\,x^3 + c\,x^2 + d\,x + e = 0$$

can be reduced by substituting $x = y - b/4\,a$ and dividing by a to the form

$$y^4 + p\,y^2 + q\,y + r = 0,$$

a quartic without a third-degree term, which is solved as above.

Rewriting

$$(y^2 + p)^2 = p\,y^2 - q\,y + (p^2 - r)$$

and

$$(y^2 + p + z)^2 = (p + 2\,z)\,y^2 - q\,y + (p^2 - r + 2\,p\,z + z^2),$$

where the discriminant of the quadratic expression is equated to zero,

$$4 \cdot (p + 2\,z) \cdot (p^2 - r + 2\,p\,z + z^2) = q^2,$$

gives the *resolvent cubic* equation

$$z^3 - \frac{5}{2}p\,z^2 + (2\,p^2 - r)\,z + \frac{1}{2}\left(p^3 - p\,r - \frac{q^2}{4}\right) = 0\,.$$

pp. 322 - 23 We may check this equation using Colla's proposal, and have

$$b^4 + 6\,b^2 - 60\,b + 36 = 0;\ \text{thus, } p = 6;\ \text{q} = -60;\ r = 36\,.$$

$$z^3 - \frac{5}{2}\,6z^2 + (2 \cdot 6^2 - 36)\,z + \frac{1}{2}\left(6^3 - 6 \cdot 36 - \frac{(-60)^2}{4}\right) = 0,$$

or

$$z^3 + 15\,z^2 + 36\,z = 450,$$

which agrees with the problem on *p.* 322.

If the roots of the resolvent cubic equation are		the roots of the quartic equation will be
all real and > 0	⇒	all real;
all real, one > 0, two < 0; all real, two > 0, one < 0; or all real, three < 0	⇒	two pairs of conjugate complex roots;
one real, two conjugate complex roots	⇒	two real roots, one pair of conjugate complex roots.

For Cardano, mathematician-physician,
Equations were the greater ambition.
But with his crystal ball hazy,
And his patron quite crazy,
He gave typhoid its modern description.

9.8 Systems of Equations

Solving an equation in more than one variable usually calls for additional conditions being imposed on the variables. For a first-degree equation, *e.g.*,

$$3x-4y = 3,$$

the number of pairs of x and y that satisfy the equation is unlimited.

For a unique solution to exist, yet another equation with the same variables is required, *e.g.*,

$$x+2y = 11,$$

which gives us the common solution

$$x = 5; \quad y = 3.$$

In a system of **simultaneous equations**, all of the equations must be true at the same time. A complete solution with identification of all roots of the several equations is possible only if the system has as many equations as there are variables.

We distinguish three cases:

- The system has a unique solution, as above.
- The system is not uniquely solvable, but has infinitely many solutions:

$$\left.\begin{aligned} 6x+8y &= 20 \\ 3(x-1)+4y &= 7 \end{aligned}\right\} \quad \Rightarrow \quad \left.\begin{aligned} 3x+4y &= 10 \\ 3x+4y &= 10 \end{aligned}\right\}$$

- The equations of the system may be contradictory; the system then has no solution:

$$\left.\begin{aligned} 6x+8y &= 24 \\ 3(x-1)+4y &= 7 \end{aligned}\right\} \quad \Rightarrow \quad \left.\begin{aligned} 3x+4y &= 12 \\ 3x+4y &= 10 \end{aligned}\right\}$$

Systems of Linear Equations

To solve a system of two linear equations in two variables,

$$\left.\begin{aligned} 3x+2y &= 19 \\ 2x-3y &= 4 \end{aligned}\right\},$$

we have a choice of three methods: adding, substituting, and equating.

Addition

Multiply each equation by a suitable quantity so as to make the coefficients of the two variables agree between them; adding or subtracting the resultant two equations eliminates one variable.

$$\begin{array}{llll} \cdot 3 \Rightarrow & 9x + 6y = 57 & \text{or} & \cdot 2 \Rightarrow \quad 6x + 4y = 38 \\ \cdot 2 \Rightarrow & \underline{4x - 6y = 8} & \text{or} & \cdot(-3) \Rightarrow \quad \underline{-6x + 9y = -12} \\ & 13x \quad = 65 & & \qquad 13y = 26 \\ & \left.\begin{array}{l} x = 5 \\ y = 2 \end{array}\right\} & & \qquad \left.\begin{array}{l} x = 5 \\ y = 2 \end{array}\right\} \end{array}$$

Substitution

Solve one equation for one of the variables and insert this value in the other equation.

$$y = \frac{2x-4}{3}; \qquad 3x + \frac{2(2x-4)}{3} = 19;$$

$$9x + 4x - 8 = 57; \qquad 13x = 65; \ etc.$$

Equating

Solve both equations for one of the variables and equate the solutions:

$$y = \frac{19-3x}{2} = \frac{2x-4}{3}; \quad 57 - 9x = 4x - 8; \quad 13x = 65; \quad etc.$$

Practical problems often lead to systems of linear equations in many unknowns. To solve such problems a method called **Gaussian elimination** is convenient; it will be described in Chapter 18, where we employ **matrices** and **determinants** to facilitate the solution of unwieldy systems of linear equations.

p. 661

Systems of Linear and Quadratic Equations

From the multitude of systems of equations that include nonlinear equations, we select two frequently occurring types.

One Quadratic, One Linear Equation

$$\left.\begin{array}{ll}(1) & x^2+y^2=26 \\ (2) & x-y=4\end{array}\right\} \qquad y=x-4.$$

Inserting (2) in (1),

$$x^2+(x-4)^2=26$$
$$2x^2-8x+16=26$$
$$x^2-4x-5=0$$
$$(x-5)(x+1)=0.$$
$$\left.\begin{array}{l}x_1=5\\ y_1=1\end{array}\right\}; \qquad \left.\begin{array}{l}x_2=-1\\ y_2=-5\end{array}\right\}.$$

Two Linear Equations, One Quadratic

$$\left.\begin{array}{ll}(1) & x-2y-z=14 \\ (2) & 2x-y-z=18 \\ (3) & x^2+y^2+z^2=84\end{array}\right\}$$

Using the linear equations, express x and y in terms of z:

$$\left.\begin{array}{ll}(1) & x-2y-z=14 \\ (2) & 4x-2y-2z=36\end{array}\right\} \qquad x=\frac{z+22}{3}$$

$$\left.\begin{array}{ll}(1) & 2x-4y-2z=28 \\ (2) & 2x-y-z=18\end{array}\right\} \qquad y=-\frac{z+10}{3}$$

Insert x and y in (3):

$$\left(\frac{z+22}{3}\right)^2+\left(-\frac{z+10}{3}\right)^2+z^2=84$$

$$z^2+44z+484+z^2+20z+100+9z^2=756$$

$$z^2+\frac{64}{11}z-\frac{172}{11}=0$$

$$z=-\frac{32}{11}\pm\frac{\sqrt{2916}}{11}=\frac{-32\pm 54}{11}$$

$$\left.\begin{array}{l}z_1=2\\ x_1=8\\ y_1=-4\end{array}\right\} \qquad \left.\begin{array}{l}z_2=-\frac{86}{11}\\ x_2=\frac{52}{11}\\ y_2=-\frac{8}{11}\end{array}\right\}$$

Systems of Two Second-Degree Equations

$$\left.\begin{array}{ll}(1) & x^2+y^2+x+y = 8\\ (2) & xy+x+y = 5\end{array}\right\}$$

The following approach is one of several possible.

Multiply (2) by 2 and add to (1):

$$(x+y)^2+3\,(x+y)-18 = 0$$

$$(x+y) = \frac{1}{2}\left(-3\pm\sqrt{9+72}\right) = \frac{1}{2}\,(-3\pm 9) = \begin{cases}3\\-6\end{cases}$$

$$\begin{array}{r|r} x+y = 3 & x+y = -6\\ xy = 2 & xy = 11\end{array}$$

which lead to

$$x^2-3x+2 = 0 \quad\Big|\quad x^2+6x+11 = 0$$

and the solutions

$$\left.\begin{array}{l}x_1 = 2\\ y_1 = 1\end{array}\right\};\qquad \left.\begin{array}{l}x_2 = 1\\ y_2 = 2\end{array}\right\};\qquad \left.\begin{array}{l}x_3, x_4 = -3\pm i\sqrt{2}\\ y_3, y_4 = -3\mp i\sqrt{2}\end{array}\right\}.$$

Systems with One Homogeneous Equation

All terms in a homogeneous equation have the same degree, which makes it possible to determine the ratio of the variables.

Consider a system of equations in two variables; one of the equations is homogeneous:

$$\left.\begin{array}{lll}(1) & 6x^2-xy-y^2 & = 0\\ (2) & x^2-2y & = 27\end{array}\right\}$$

Substitute

$$y = cx$$

in (1),

$$x^2\,(6-c-c^2) = 0$$

$$c^2+c-6 = 0;$$

$$c_1 = 2; \qquad c_2 = -3.$$

Insert in (2),

$$y = 2x \qquad\qquad y = -3x$$

$$x^2-4x-27 = 0 \qquad\qquad x^2+6x-27 = 0$$

$$\left.\begin{array}{l}x_1, x_2 = 2\pm\sqrt{31}\\ y_1, y_2 = 4\pm 2\sqrt{31}\end{array}\right\};\qquad \left.\begin{array}{l}x_3 = 3\\ y_3 = -9\end{array}\right\};\qquad \left.\begin{array}{l}x_4 = -9\\ y_4 = 27\end{array}\right\}.$$

Systems with Symmetric Equations

Variables in symmetric equations are interchangeable. The simultaneous equations

$$\left.\begin{aligned} (1)\quad x^2+y^2 &= 13 \\ (2)\quad xy &= 6 \end{aligned}\right\}$$

are symmetric and homogeneous.

As with most systems of equations, the solution can be arrived at by different approaches; below, we account for two.

Insert $y = \dfrac{6}{x}$ from (2) into (1):

$$x^2 + \left(\frac{6}{x}\right)^2 = 13$$

$$x^4 - 13x^2 + 36 = 0$$

$$(x^2 - 9)(x^2 - 4) = 0$$

$x = \pm 3$	$x = \pm 2$
$y = \pm \dfrac{6}{3}$	$y = \pm \dfrac{6}{2}$

Multiply (2) by 2 and add to (1):

$$x^2 + y^2 + 2xy = 25$$

$$(x+y)^2 = 25$$

$x + y = 5$	$x + y = -5$
$xy = 6$	$xy = 6$
$x(5 - x) = 6$	$x(-5 - x) = 6$
$x^2 - 5x + 6 = 0$	$x^2 + 5x + 6 = 0$
$(x-3)(x-2) = 0$	$(x+3)(x+2) = 0$

$$\left.\begin{aligned} x_1, x_2 &= \pm 3 \\ y_1, y_2 &= \pm 2 \end{aligned}\right\} ; \qquad \left.\begin{aligned} x_3, x_4 &= \pm 2 \\ y_3, y_4 &= \pm 3 \end{aligned}\right\} .$$

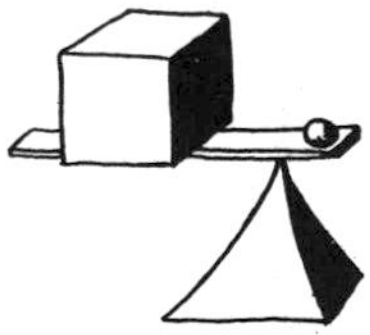

9.9 Diophantine Equations

DIOPHANTOS
(c. A.D. 250)

Diophantus of Alexandria carried out extensive studies of problems relating to indeterminate equations, which he published in *Arithmetica*; this work may have consisted, originally, of thirteen books, six of which survive through Arabic translations.

Diophantus accepted any solution in rational numbers, but the name **Diophantine equations** today refers exclusively to equations with **integer solutions**.

The prototype equation

$$ax + by + c = 0$$

Euclidean Algorithm, GCD: p. 126

has integer solutions only if c is a multiple of the GCD (a, b). It is solved for one of the variables, preferably the one that has the lowest coefficient or whose coefficient is a factor of the constant term; one can then determine integer values of the other variable.

- Find the positive integer values of x and y that satisfy the equation

$$12x + 7y = 220\,.$$

The GCD of 12 and 7 is 1. We can solve the equation either for x or for y:

$x = \dfrac{220 - 7y}{12} = 18 - \dfrac{7y - 4}{12}$,	$y = \dfrac{220 - 12x}{7} = 31 - \dfrac{12x - 3}{7}$,
where $\dfrac{7y-4}{12}$ must be an integer ≤ 17, that is,	where $\dfrac{12x-3}{7}$ must be an integer ≤ 30, that is,
$0 < y \le 29$ with $\Delta y = 12$.	$0 \le x \le 18$ with $\Delta x = 7$.
The lowest possible y is 4, as $4 \cdot 7 - 4 = 24 = 2 \cdot 12$.	The lowest possible x is 2, as $2 \cdot 12 - 3 = 21 = 3 \cdot 7$.

$$\left.\begin{matrix} x_1 = 2 \\ y_1 = 28 \end{matrix}\right\}; \quad \left.\begin{matrix} x_2 = 9 \\ y_2 = 16 \end{matrix}\right\}; \quad \left.\begin{matrix} x_3 = 16 \\ y_3 = 4 \end{matrix}\right\}.$$

- Find all natural numbers x and y that satisfy

$$2x + 5y = 32\,.$$

Solving for x gives

$$x = 16 - \frac{5y}{2}\,;$$

x is > 0, if $y < \dfrac{32}{5} = 6\dfrac{2}{5}$, and even.

$$\left.\begin{matrix} y_1 = 6 \\ x_1 = 1 \end{matrix}\right\}; \quad \left.\begin{matrix} y_2 = 4 \\ x_2 = 6 \end{matrix}\right\}; \quad \left.\begin{matrix} y_3 = 2 \\ x_3 = 11 \end{matrix}\right\}.$$

- The membership of an archaeological society is limited to 100; members pay an annual fee of \$125, reduced to \$75 for senior members of 60 years of age and older. One year, the total fees paid by members below 60 exceeded those paid by senior members by \$10 575.

  How many members were under 60 years of age, and how many were older?

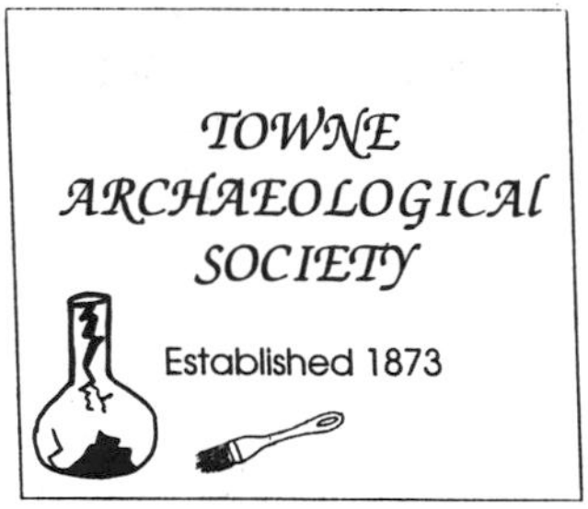

If members under 60 number x, senior members y, we have

$$125x - 75y = 10\,575$$

or, simplified,

$$5x - 3y = 423,$$

from which

$$y = \frac{5x - 423}{3}.$$

For $y > 0$, we must have

$$x > \frac{423}{5} = 84\frac{3}{5}$$

and x divisible by 3:

$$\left.\begin{array}{l} x_1 = 87 \\ x_2 = 90 \end{array}\right\} ; \quad \left.\begin{array}{l} y_1 = 4 \\ y_2 = 9 \end{array}\right\} ; \quad \begin{array}{l} 87 + 4 = 91 < 100 \\ 90 + 9 = 99 < 100 . \end{array}$$

The society had either

91 members, 87 of whom were below 60 years of age and 4 aged 60 or more, or

99 members, 90 below 60 and 9 senior members of 60 or more.

- In the equation

$$xy - 5x + 4y = 0,$$

the variables x and y are natural numbers or zero; solving for y gives

$$y = \frac{5x}{x+4} = 5 - \frac{20}{x+4}.$$

y is a positive integer, or zero, only if

$$\frac{20}{x+4} \leq 5,$$

where x is a positive integer or zero. We find the following solutions,

$$\left.\begin{array}{l} x_0 = 0 \\ y_0 = 0 \end{array}\right\} ; \quad \left.\begin{array}{l} x_1 = 1 \\ y_1 = 1 \end{array}\right\} ; \quad \left.\begin{array}{l} x_2 = 6 \\ y_2 = 3 \end{array}\right\} ; \quad \left.\begin{array}{l} x_3 = 16 \\ y_3 = 4 \end{array}\right\} ,$$

of which the solution $x_0\,y_0$ is trivial.

- Eight students were given a list of synonymous words that were to be sorted into pairs. Within the allotted time, the best student had 21 correct pairs; the next best three students had two-thirds of the possible pairs each; and the remaining four students each had half the possible number of pairs plus four.

What was the total number of possible pairs?

With N as the sought number, we have

$$\left.\begin{array}{l} \dfrac{21}{N} > \dfrac{2}{3} \\[2ex] \dfrac{2}{3}N > \dfrac{1}{2}N + 4 \end{array}\right\} \text{ giving } 24 < N < \frac{63}{2}.$$

N is an integer divisible by 2 and 3; thus, the total number of possible pairs of synonyms was 30.

The Pythagorean Number Theorem

Pythagorean Numbers
Pythagorean Triples

Pythagorean numbers, or **triples**, consist of positive integers a, b, c, satisfying the relation

$$a^2 + b^2 = c^2;$$

e.g., 3, 4, 5 form the Pythagorean triple of $3^2 + 4^2 = 5^2$.

Basic Pythagorean triples have no common factor; thus, every Pythagorean triple is basic or a multiple of a basic triple.

Analysis of Babylonian tablets dating before 1600 B.C. by Otto Neugebauer and his co-workers shows that not only did the scribes of that era know the relation $a^2 + b^2 = c^2$ but they also knew how to find all triples of integers satisfying it.

To obtain Pythagorean triples one may choose any odd positive number and divide its square into two integers that are as equal as possible in size; *e.g.*, $9^2 = 81 = 40 + 41$, and the triples are 9, 40, and 41. One may also choose any even positive number h; the Pythagorean triples are then $2h$, $h^2 - 1$, and $h^2 + 1$.

The **Pythagorean number theorem** gives all the *basic triples* of positive integers a, b, c, having no common factors, and satisfying

$$a^2 + b^2 = c^2,$$

such that

$$a = 2mn; \quad b = m^2 - n^2; \quad c = m^2 + n^2,$$

where one but not both of m and n must be even, and m and n are positive integers with $m > n > 0$.

Why? If m and n both were even, then a, b, c would all be even, which is ruled out by the premise of no common factors; and if both were odd, $a^2 + b^2$ would have a remainder of 2 on division by 4, while c^2 must have remainder 0, which is impossible.

Suppose b is odd; move it to the right side, and factor to get

$$a^2 = c^2 - b^2 = (c - b)(c + b),$$

where $(c - b)$ and $(c + b)$ must have in common only a single factor of 2, otherwise c and b would have common factors.

But then $(c - b) / 2$ and $(c + b) / 2$ must be perfect squares; the only possibility is that

$$c - b = 2n^2; \quad c + b = 2m^2$$

for some integers m and n without common factors and satisfying $m > n > 0$. So all basic Pythagorean triples are given by

$$a = 2mn; \quad b = m^2 - n^2; \quad c = m^2 + n^2,$$

with m and n as above. Substitution shows such triples satisfy

$$a^2 + b^2 = c^2.$$

This result is given in Euclid's *Elements* (c. 300 B.C.).

Fermat's Last Theorem

de FERMAT Pierre (1601 - 1665)

Fermat's last theorem, also called **Fermat's great theorem**, has been one of the most famous mathematical conjectures.

In 1637, in the margin of a copy of Diophantus's *Arithmetica*, Fermat stated – in our notation – that

if $n > 2$, the equation

$$x^n + y^n = z^n$$

cannot be solved in positive integers x, y, z.

Fermat's last theorem is an *impossibility theorem*.

Fermat claimed, "and I have assuredly found a marvellous proof of this, but the margin is too narrow to contain it".

We know that Fermat had a **proof for the case $n = 4$**, but he did not mention a general method of proof when he wrote down the proof for $n = 4$.

When the theorem once has been proved true for $n = 4$, we only need to prove the case where n is an odd prime, because in $x^n + y^n = = z^n$, n is either (**I**) a power of 2 or (**II**) divisible by an odd prime p.

Why? **I**:

If $n = 4k$, where k is a natural number, we have $x^{4k} + y^{4k} = z^{4k}$, which is the same as

$$(1) \qquad (x^k)^4 + (y^k)^4 = (z^k)^4.$$

Substitute $a = x^k$, $b = y^k$, $c = z^k$. If there existed a triple of natural numbers (x, y, z) satisfying (**1**), then $a^4 + b^4 = c^4$ would also exist, which is impossible since the (impossibility) theorem is true for $n = 4$.

II:

If $n = pk$, where p is an odd prime and k a natural number, then $x^n + y^n = z^n$ may be written

$$(2) \qquad (x^k)^p + (y^k)^p = (z^k)^p.$$

Substitute $a = x^k$, $b = y^k$, $c = z^k$. If there existed a triple of natural numbers (x, y, z) satisfying (**2**), then $a^p + b^p = c^p$ would also exist.

EULER Leonhard
(1707 - 1783)

LEGENDRE Adrien Marie
(1752 - 1833)

LAMÉ Gabriel
(1795 - 1870)

Euler proved Fermat's last theorem for $n = 4$ and attempted, around 1753, a proof for the validity of the theorem for exponents of 3, but some details had been left out; later, **Gauss** gave a correct proof. In 1825 **Legendre** proved the theorem for exponents of 5 and, in 1839, the French mathematician and engineer **Gabriel Lamé** for exponents of 7.

With computer technology Fermat's theorem has been proved for exponents less than 4 000 000 by **Joe Buhler** and **Richard Crandall**; such investigations cannot tell us whether the next exponent might falsify the theorem.

GERMAIN Sophie
(1776 - 1831)

The French mathematician **Sophie Germain** found some results on more general cases; she showed that if n is an odd prime < 100, the equation $x^n + y^n = z^n$ is insolvable in integers not divisible by n. In 1941, **D. H. Lehmer** and **E. Lehmer** extended Germain's theorem to all primes less than 253 747 889.

WILES Andrew
(b. 1953)

TAYLOR Richard Lawrence
(b. 1962)

Until recently all claims to a general proof of Fermat's last theorem proved false, but in June 1993 the Englishman **Andrew Wiles** of Princeton University gave the essential steps toward such a proof. The proposition that Wiles set out to prove was, however, not Fermat's but one of much greater scope, to which the proof of Fermat's last theorem came only as an incidental consequence. Wiles applied methods of algebra and geometry; he made no use of computers. A gap in the argument of the alleged proof was soon revealed, but, in October 1994, Wiles, with the English mathematician **Richard Taylor**, filled the gap and gave a proof of certain crucial properties of the algebras used in Wiles's general proof. The completed proof, more than 130 pages, was published in May 1995 in the *Annals of Mathematics*. Although some caution was still recommended in 1996, most experts are confident that the Wiles-Taylor proof will withstand further scrutiny.

If Wiles and Taylor have succeeded, a great advance in number theory has been accomplished, but it will not end the search for the other proof – the one that Fermat had "assuredly found" about 360 years ago. Experts seem to believe, however, that a relatively brief proof of Fermat's last theorem simply isn't attainable.

– *Only marvellous, not brief. Mon Dieu! I told you long ago it wouldn't fit in the margin!*

Messieurs Wiles et Taylor, mes compliments!

Chapter 10

INTRODUCTION TO FUNCTIONS

10.0 Historical Notes

The term "function" – Latin *functio* – first appeared in 1692 in a mathematical article in the *Acta Eruditorem* to denote various tasks that a straight line may accomplish with respect to a curve, such as forming a chord, tangent, or normal. The article was signed O.V.E. but is attributed to **Gottfried von Leibniz**, the German mathematician. In an article from 1694, also in *Acta Eruditorem*, Leibniz gave the term "function" a more specific meaning by letting it denote the slope of a curve, a definition that has very little in common with the present-day mathematical definition of a function.

von LEIBNIZ Gottfried
(1646 - 1716)

The Swiss mathematician **Leonhard Euler** in 1749 defined a function as a variable quantity that is dependent upon another quantity, thereby approaching today's definition.

EULER Leonhard
(1707 - 1783)

Euler's definition was challenged when the French physicist and mathematician **Joseph Fourier** in 1822 presented his work on heat flow (*Théorie analytique de la chaleur*). For his investigations, Fourier introduced series with sines and cosines as terms, which led to the concept that a given representation of a function may be valid for only a certain range of values.

FOURIER Joseph
(1768 - 1830)

Based on Fourier's investigations, the German mathematician **Lejeune Dirichlet** in 1837 proposed that, from the mathematical point of view, a function is a correspondence that *assigns a unique value of the dependent variable to every permitted value of an independent variable*. There will be reason to return to this definition many times in this text.

DIRICHLET
Peter Gustav Lejeune
(1805 - 1859)

10.1 Groundwork

When we buy coffee and pay by mass, we can say that the price is a *function* of mass. If the cost of coffee is \$9.00 per kilogram (kg), half a kg is \$4.50, 2 kg is \$18.00, *etc.* We have the formula

$$f(x) = 9x,$$

where x designates the purchased amount (kg) and $f(x)$ is the price (in \$). Here, $f(x)$ is defined only for $x \geq 0$.

Independent Variable, Argument
Dependent Variable

Expressed in a more mathematical way, we say that a **function** is an association between two or more variables, in which to every value of each of the **independent variables**, or **arguments**, corresponds *exactly one value* of the **dependent variable** in a specified set called the domain of the function.

Map, Mapping, Operator, Transformation

Map, **mapping**, **operator**, and **transformation** are other names for a function.

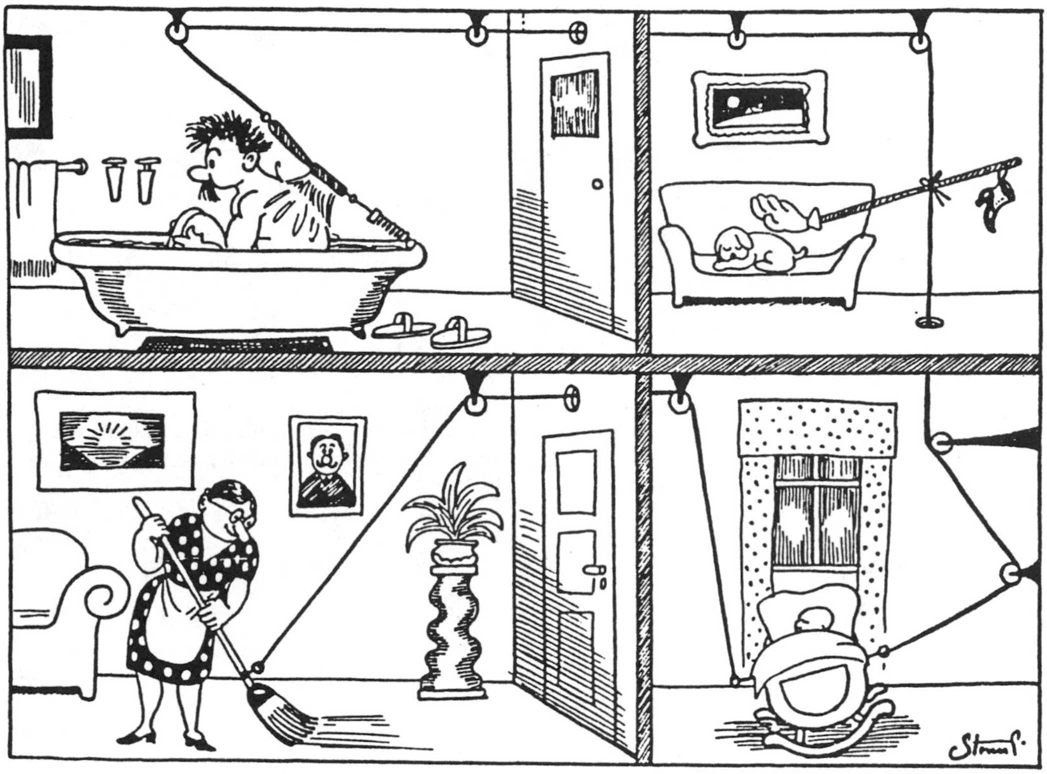

STORM PETERSEN Robert
(*alias* Storm P.; 1882 - 1949)

A function of one variable x may be written

$$f(x),$$

which reads "f of x" or, in full, "the value of the function f at x"; a function of several independent variables is written, analogously,

$$f(x, y, z \ldots).$$

Letters of the Latin or Greek alphabet may be used to designate functions; the letters f, F, g, G, h, H, u, v, and φ (phi), ψ (psi) are the most common.

It is standard practice to write the **dependent variable** on the left-hand side of the equality sign of an equation; thus, in

$$y = x + 1 \quad \text{or} \quad f(x) = x + 1,$$

y or $f(x)$ is the dependent variable, x the **independent variable**.

On the other hand, in the expression

$$x = y + 1 \quad \text{or} \quad f(y) = y + 1,$$

x or $f(y)$ is the dependent variable, y the independent.

Orthogonal Coordinate System: *p.* 472

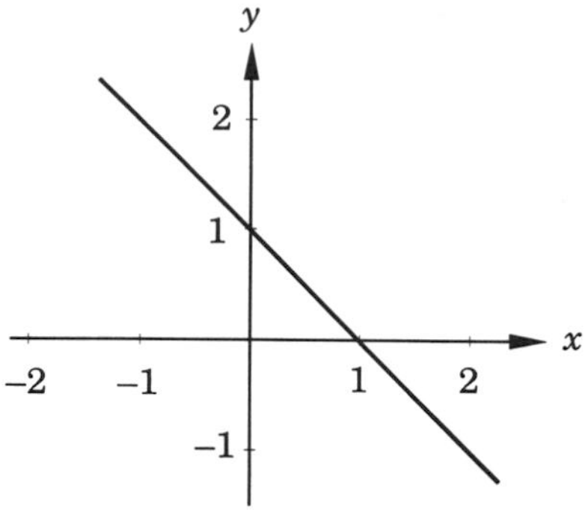

The equation $x + y = 1$ denotes a straight line, as shown by the graph; the line may be represented by either of the two functions

$$\left.\begin{aligned} f(x) &= 1 - x \\ f(y) &= 1 - y \end{aligned}\right\} .$$

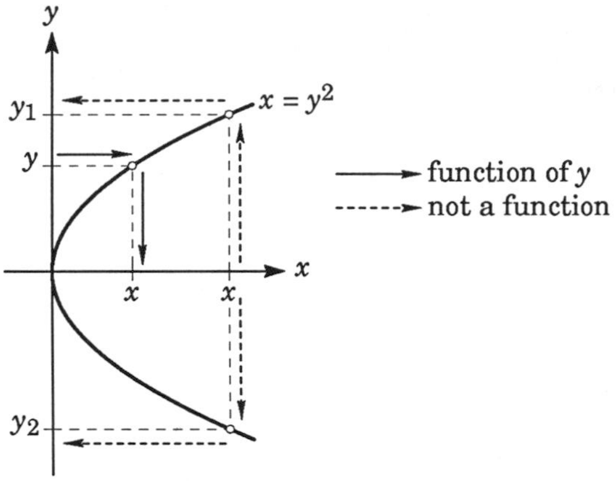

The above curve represents the equation

$$x = y^2.$$

There is *only one* value of x for a value assigned to y, which means that x is a *function* of y.

If, instead, we consider y to be dependent on x, we solve the equation for y, obtaining

$$y = \pm\sqrt{x} ,$$

which shows that there are *two* values of y for every value of x, indicating that y is *not* a function of x.

Restricting the equation to either

$$y = \sqrt{x} \quad \text{or} \quad y = -\sqrt{x} ,$$

representing the upper and lower half of the parabola, respectively, there is only one value of x for every y,

$$f_+(x) = \sqrt{x} ; \quad f_-(x) = -\sqrt{x} .$$

Explicit and Implicit Forms

The equations – and functions –

$$y = f(x) = 2x + 5 \quad \text{and} \quad z = g(x) = 3x - x^2,$$

where the dependent variable is given in terms of the independent variable, are said to be in **explicit form**.

In contrast, an **implicit form** is characterized by the occurrence of dependent as well as independent variables on one side of the equality sign:

$$2x - y = 6 \,; \quad x^2 + y^2 = 9 \,.$$

The above equations may be rendered into explicit form,

$$y = 2x - 6 \,; \quad y^2 = 9 - x^2,$$

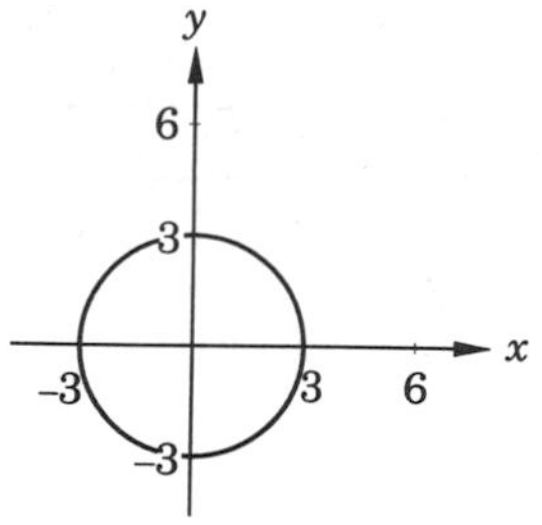

but only $y = 2x - 6$ is a function.

As $x^2 + y^2 = 9$ represents a circle with radius 3 about the origin of the coordinate system, we can visualize that for every value of x or y between -3 and 3, there are two values y and x, respectively, that satisfy the equation. On the other hand, the *explicit forms*

$$y = f(x) = \sqrt{9 - x^2} \quad \text{and} \quad y = g(x) = -\sqrt{9 - x^2},$$

representing the upper and lower semicircles, are *functions*.

Domains and Ranges of Functions of One Independent Variable

Domain
Range

The collection of all values that the *independent variable* can assume is the **domain** of the function; all values taken by the *dependent variable* represent the **range** of the function. A real-value **function of one real variable** (that is, one independent variable) can be represented in a two-dimensional orthogonal coordinate system, the domain referring to the x-axis, the range to the y-axis.

- Find the domain and range of $f(x) = 1 - x^2$, where x is a real number.

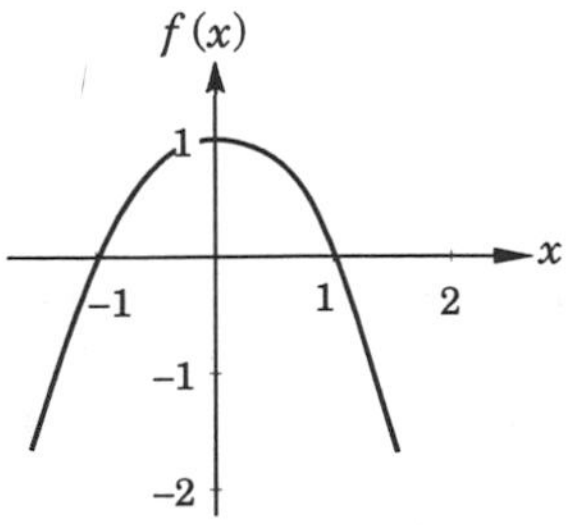

Since the only restriction imposed on x is that it must be a real number, the *domain* of the function must be the collection of all real numbers,

$$]-\infty, \infty[\,.$$

To find the *range* of the function, consider the expression

$$1 - x^2 \,;$$

the numerical value of x^2 will always be zero or positive. As x goes toward ∞ or $-\infty$, $f(x)$ approaches $-\infty$. The function will reach its greatest value when x is 0. Hence, the range of the function contains the number 1 and all real numbers less than 1,

$$-\infty < f(x) \leq 1 \quad \text{or} \quad]-\infty, 1] \,.$$

- Find the largest possible domains of

$$f(x) = \frac{x}{x-3}\,; \quad g(x) = \frac{x-3}{x}\,,$$

where x represents a real number.

Since division by 0 is not permissible, the domain of f must be all real numbers *except* 3; and the domain of g must be all real numbers *except* 0.

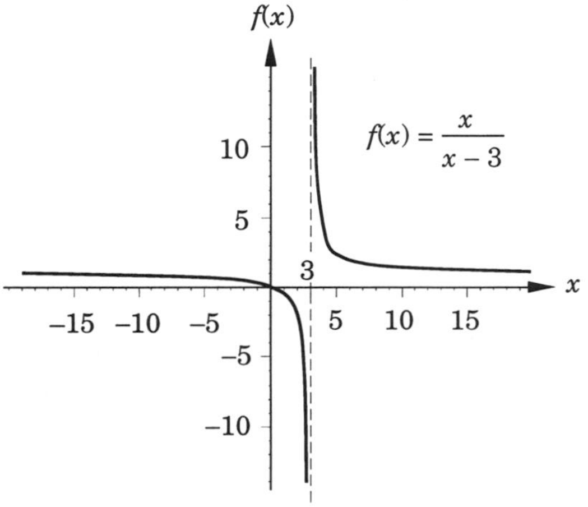

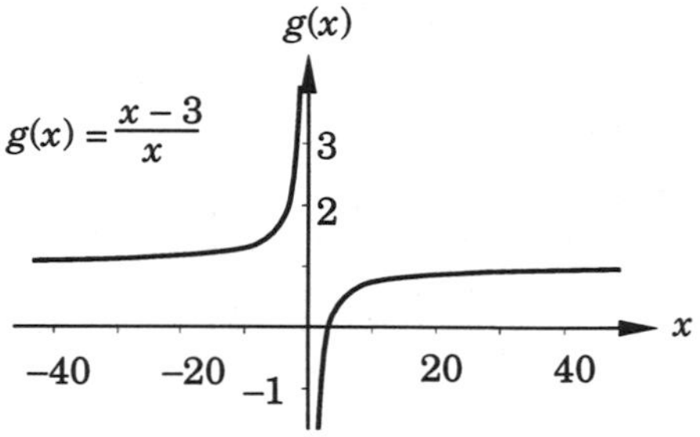

Even and Odd Functions

A function is said to be *even* if $f(-x) = f(x)$ for every value of x in the domain of definition. The graph of an even function is *symmetrical about the ordinate axis.*

A function is *odd* if $f(-x) = -f(x)$ for every value of x in the domain of definition; the graph is *symmetrical with respect to the origin.*

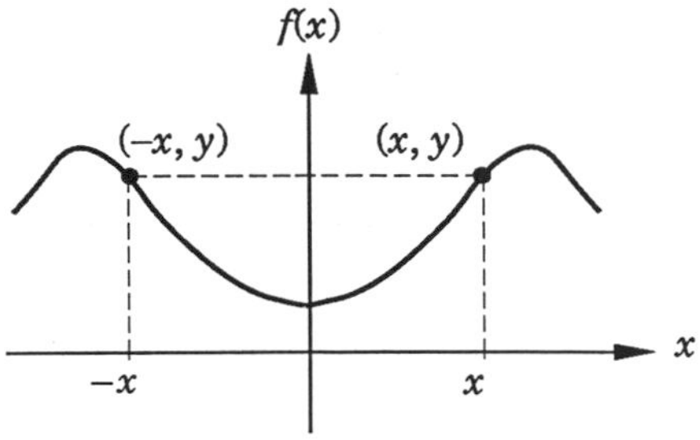

Even Function

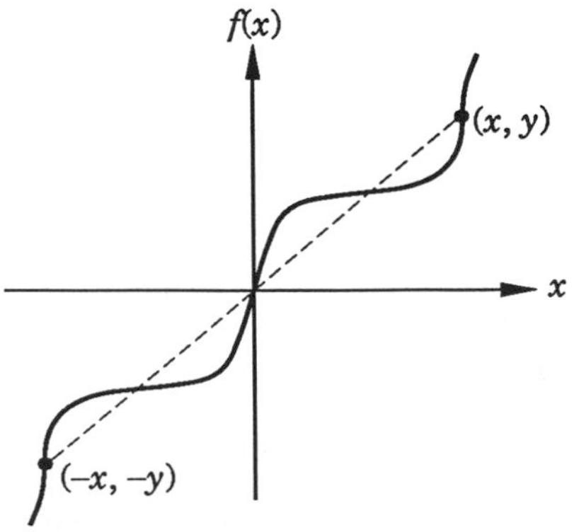

Odd Function

Monotonicity, Extrema, Inflection Points

A real-value function is said to be **increasing** over an interval if greater values of the independent variable within the interval produce greater values of the dependent value, and to be **decreasing** if the dependent variable decreases with increasing values of the independent variable. Such intervals – over which a function *either* increases *or* decreases – are **intervals of monotonicity**.

Below, the interval $]x_1, x_2[$ is an interval of increasing monotonicity, while the interval $]x_2, x_3[$ is an interval of decreasing monotonicity of the function:

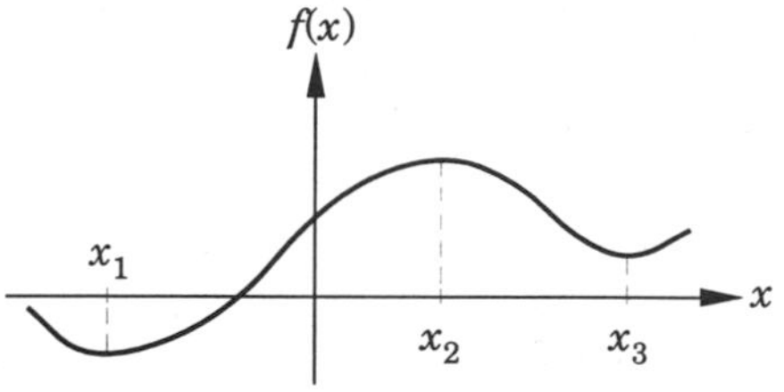

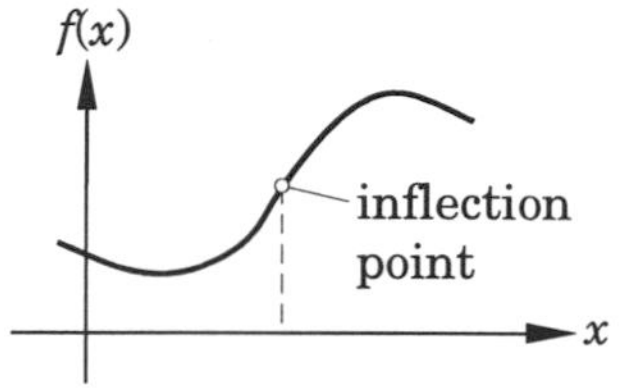

Inflection Point

A point within an interval of monotonicity where the curve changes from concavity to convexity, or vice versa, is an **inflection point**.

Greek, *asymptotos,* "not falling together"; *a,* "not"; *sym*, "together"; *ptotos,* "falling"

As the independent variable proceeds toward positive or negative infinity, the corresponding points on the graph of a function may approach successively nearer to a straight line, the **asymptote**.

(Absolute) Maximum Value

(Absolute) Minimum Value

Relative (Local) Maximum Value

Relative (Local) Minimum Value

The value of a function may attain its greatest (**maximum** or **absolute maximum**) and lowest (**minimum** or **absolute minimum**) values on an interval either at the endpoints or at points between the endpoints of the interval. The greatest and lowest values occurring *between* the endpoints are also **relative** (or **local**) **maxima** and **relative** (or **local**) **minima**, respectively, meaning that they are points where the value of the function is greater or lower than at all nearby points.

At a **relative maximum**, an interval of increasing monotonicity changes into an interval of decreasing monotonicity; at a **relative minimum** an interval of decreasing monotonicity continues into an interval of increasing monotonicity.

A relative maximum or a relative minimum can also be the absolute maximum or the absolute minimum of a function.

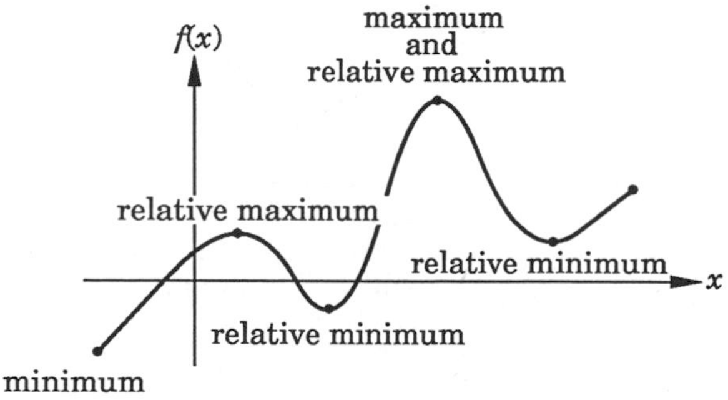

Extreme Values, Extrema

A collective term for maxima and minima, whether absolute or relative, is **extreme values** or **extrema** (singular: *extremum*).

It is common practice to refer to an absolute maximum or a relative minimum as simply *maximum* or *minimum* when it is evident from the context what kind of extremum one is dealing with.

Domains and Ranges of Functions of More than One Independent Variable

The domain of a function of two real variables may be represented graphically in an *orthogonal coordinate plane*.

A domain defined by $]-\infty < x < \infty[$ and $]-\infty < y < \infty[$ corresponds to the whole (x, y)-plane, whereas a domain defined by $]-\infty < x < \infty[$ and $]-\infty < y \le 0]$ corresponds only to the lower half-plane and the x-axis:

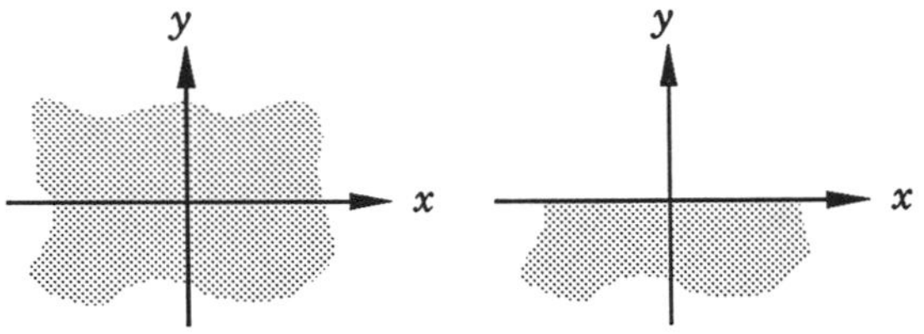

With the condition $\frac{x^2}{16} + \frac{y^2}{9} < 1$, the domain of definition is the interior of an **ellipse centered at the origin**, the major axis of the ellipse 8 units along the x-axis, the minor axis 6 units along the y-axis:

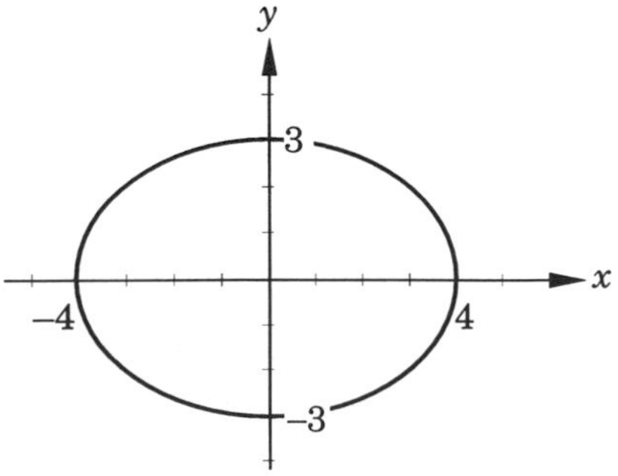

To represent graphically both the **range** and the **domain** of a function of two independent variables, we must use a *three-dimensional orthogonal coordinate system*. The domain is still represented in the (x, y)-plane, but may be conceived as an $(x, y, 0)$-plane.

Assume a function $z = f(x, y)$. The *domain* of f may be represented in the (x, y)-plane; the *range* of f is the collection of values of z coordinates of all points on the graph. The collection of all points (x, y, z) that satisfy the equation $z = f(x, y)$ is the graph of f in 3-space.

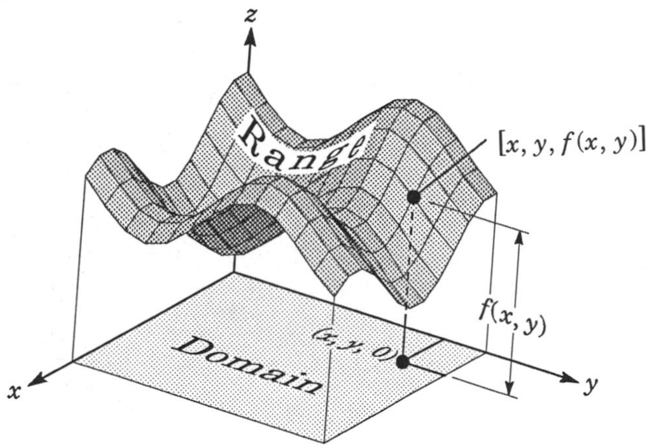

Calculations with Functions

Functions with a common domain can be combined algebraically:

$$f(x) + g(x) = (f + g)(x) \qquad f(x) - g(x) = (f - g)(x)$$

$$f(x)\,g(x) = (fg)(x) \qquad \frac{f(x)}{g(x)} = \left(\frac{f}{g}\right)(x), \text{ if } g(x) \neq 0$$

If $f(x) = x^2$ and $g(x) = x - 1$,

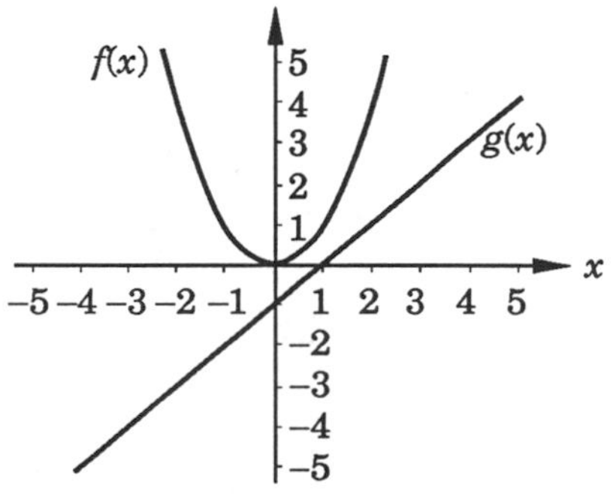

we have

the sum:

$f(x) + g(x) = x^2 + (x - 1) = x^2 + x - 1$

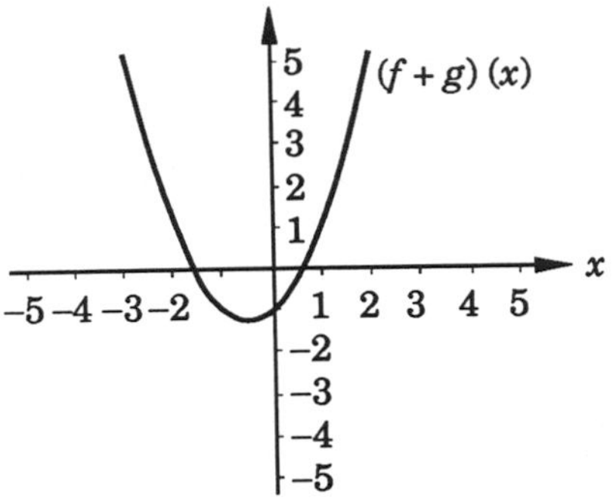

the product:

$f(x)\,g(x) = x^2(x - 1) = x^3 - x^2$

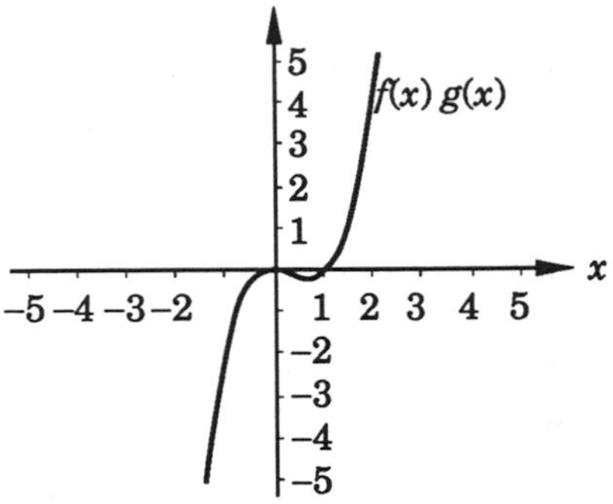

the difference:

$f(x) - g(x) = x^2 - (x - 1) = x^2 - x + 1$

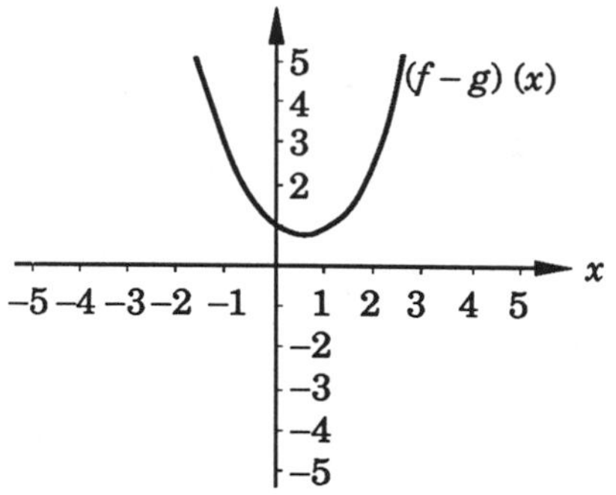

the quotient:

$\frac{f(x)}{g(x)} = \frac{x^2}{x - 1}$

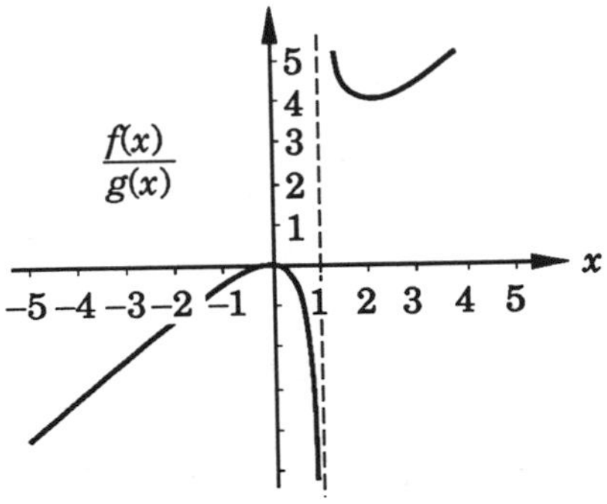

the difference:

$g(x) - f(x) = (x - 1) - x^2 = -x^2 + x - 1$

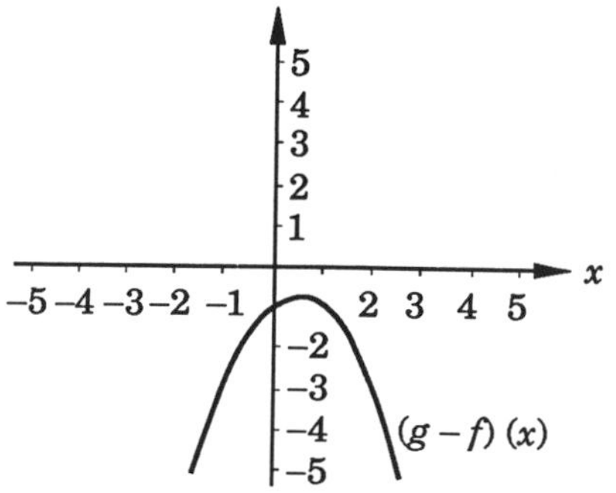

the quotient:

$\frac{g(x)}{f(x)} = \frac{x - 1}{x^2}$

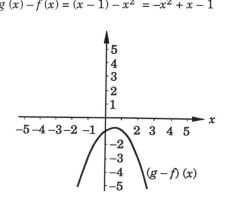

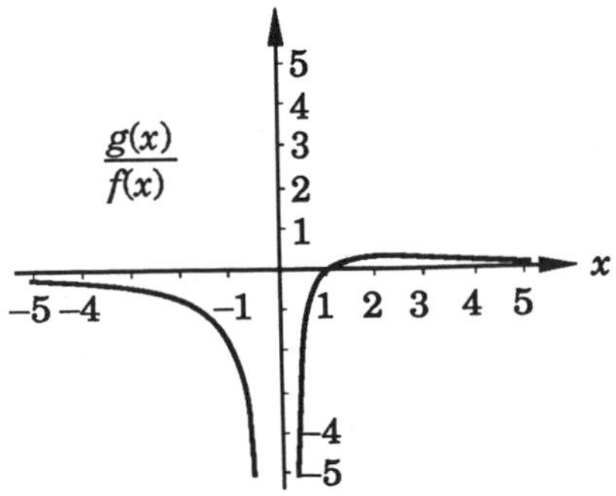

Composite Functions

COURS D'ANALYSE

DE

L'ÉCOLE ROYALE POLYTECHNIQUE;

PAR M. AUGUSTIN-LOUIS CAUCHY,

Ingénieur des Ponts et Chaussées, Professeur d'Analyse à l'École polytechnique, Membre de l'Académie des sciences, Chevalier de la Légion d'honneur.

I.re PARTIE. *ANALYSE ALGÉBRIQUE.*

DE L'IMPRIMERIE ROYALE.

Chez Debure frères, Libraires du Roi et de la Bibliothèque du Roi, rue Serpente, n.° 7.

1821.

§. 3. *Des Fonctions composées.*

Les fonctions qui se déduisent d'une variable à l'aide de plusieurs opérations prennent le nom de *fonctions composées*; et l'on distingue parmi ces dernières les *fonctions de fonctions* qui résultent de plusieurs opérations successives, la première opération étant effectuée sur la variable, et chacune des autres sur le résultat de l'opération précédente. En vertu de ces définitions,

$$x^2,\ \sqrt[3]{x},\ \frac{lx}{y},\ \&c....$$

sont des fonctions composées de la variable x; et

$$l(\sin. x),\ l(\cos. x),\ \&c....$$

des fonctions de fonctions, dont chacune résulte de deux opérations successives.

CAUCHY Augustin-Louis (1789 - 1857)

The French mathematician **Augustin-Louis Cauchy** was a pioneer who introduced rigor and clarity into modern mathematics during the 1820s in his research, lectures, and textbooks.

If

$$y = f(z) = z^2 \quad \text{and} \quad z = g(x) = 2x - 1,$$

where y, z, and x are variables such that y depends on z and z depends on x, then y will depend on x, and we may write

$$y = (2x - 1)^2 \qquad \text{or} \qquad (f \circ g)(x) = (2x - 1)^2 .$$

The symbol $f \circ g$ reads "The composite, or composition, of the two functions g and f".

A composite function is a **function of a function**.

In a function such as $(f \circ g)(x)$, f is the **external function** and g the **core function**.

In the example below, we show that, generally, $(f \circ g)(x)$ does not equal $(g \circ f)(x)$.

- For $f(x) = x^2 + 3x - 2$; $g(x) = x - 1$, find

$$(f \circ g)(x); \quad (g \circ f)(x); \quad (g \circ g)(x).$$

$$(f \circ g)(x) = f[g(x)] = f(x-1) = (x-1)^2 + 3(x-1) - 2$$
$$= x^2 + x - 4.$$

$$(g \circ f)(x) = g[f(x)] = g(x^2 + 3x - 2) = (x^2 + 3x - 2) - 1$$
$$= x^2 + 3x - 3.$$

$$(g \circ g)(x) = g(x-1) = (x-1) - 1 = x - 2.$$

Inverses

If $y = f(x)$ is equivalent to $x = g(y)$, the function g is said to be the inverse of f, and f the inverse of g. Denoting these inverses by f^{-1}, g^{-1}, we have

$$f^{-1}[f(x)] = x; \quad g^{-1}[g(y)] = y.$$

One-to-One

To have an inverse, the function must be **one-to-one**, that is, a functional relationship must also exist when the dependent and independent values are interchanged. The domain of f^{-1} is the range of f, and the range of f^{-1} is the domain of f.

- Find the inverse (if it exists) of $g(x) = x^2 - 2$.

$$y = g(x) = x^2 - 2; \; x^2 = y + 2; \; x = \pm\sqrt{y+2},$$

which shows that $g(x)$ is not a one-to-one function and, consequently, has no inverse function.

- Find the inverse (if it exists) of

$$f(x) = \sqrt{x}; \; x \geq 0.$$

Solve for x and check if $y = f(x) = \sqrt{x}$ is a one-to-one function:

$$x = y^2.$$

Interchanging x and y in $y = f(x) = \sqrt{x}$ and solving for y, we have

$$x = \sqrt{y}; \; y = x^2,$$

and f^{-1} is the function whose domain is $[0, \infty[$, such that

$$f^{-1}(x) = x^2.$$

Plotting the graphs for $f(x)$ and $f^{-1}(x)$ for values $x \geq 0$, we obtain

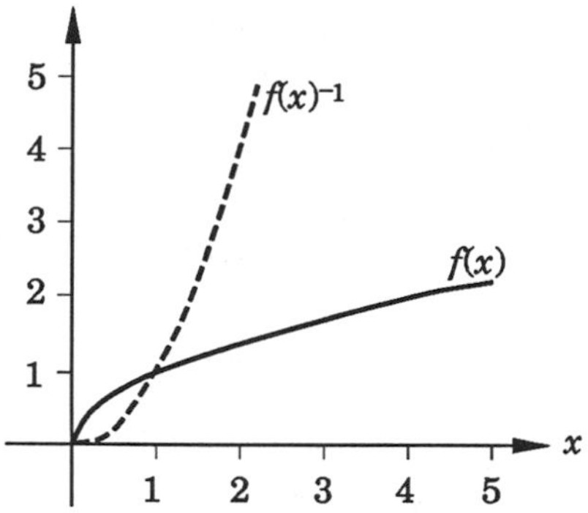

Similarly:

$f(x) = \dfrac{x+1}{5} \Leftrightarrow f^{-1}(x) = 5x - 1;$

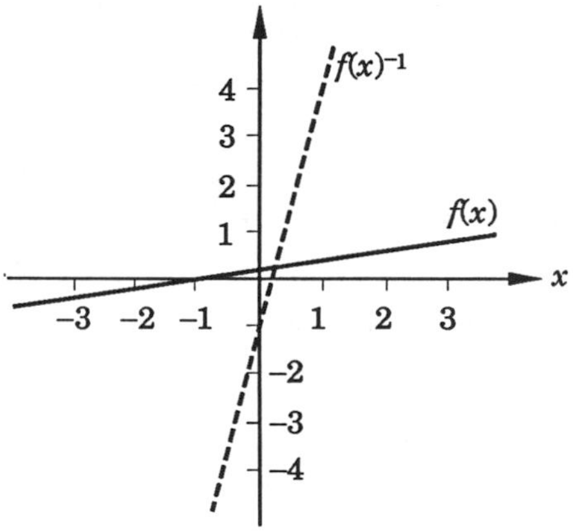

$f(x) = x^3 \Leftrightarrow f^{-1}(x) = \sqrt[3]{x}$

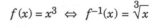

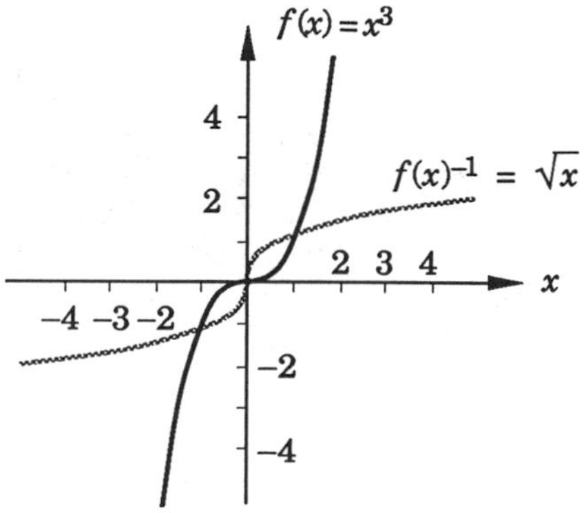

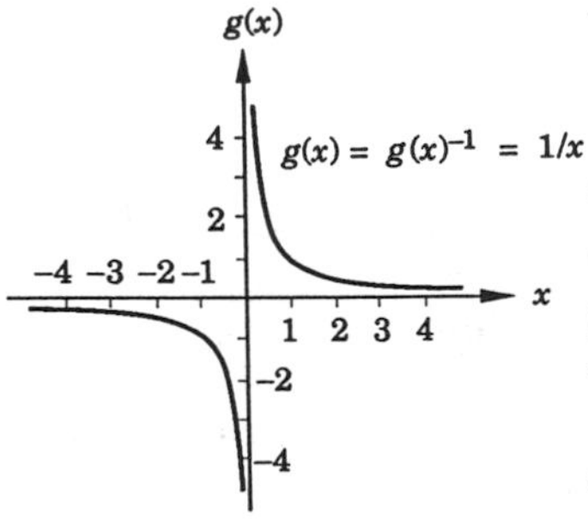

A function and its inverse may be equal, *e.g.*,

$$g(x) = \frac{1}{x} = g^{-1}(x);\ x \neq 0.$$

To verify this, first solve for x and check if $y = g(x) = 1/x$ is a one-to-one function: $x = 1/y$

Interchanging x and y in $y = g(x) = 1/x$ and solving for y, we have

$$x = 1/y;\ y = 1/x,$$

and thus,

$$g^{-1}(x) = \frac{1}{x}.$$

Iteration and Chaos

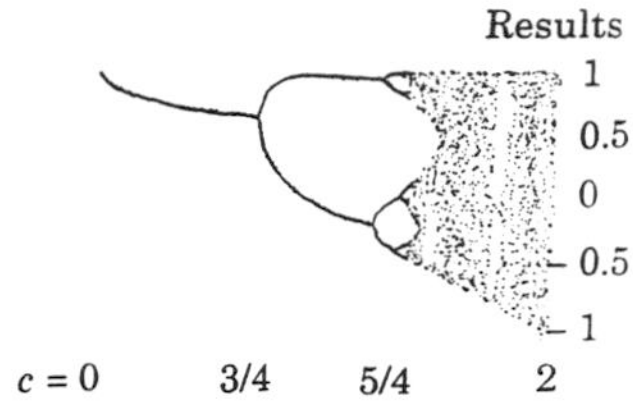

If a nonlinear function is iterated so that the value from every iteration is included sequentially into the next one – $f(x)$, $f[f(x)]$, $f\{f[f(x)]\}$... – amazing results may be expected.

Let's look at the iteration of $f(x) = 1 - cx^2$ with an initial value $(-1 < x < 1)$ and c a constant, here called a *parameter*. Whatever the initial value of x (within the given interval), only the value of the parameter – sufficiently small – will determine the general behavior of the iterated function. Iterating this function (with the use of a computer) for $c = 0$ to 2, the results shown in the diagram to the left are obtained.

Stable Point
Bifurcation Point
Chaos

For $c = 0$ to 3/4, every result tends to a **stable point**; at $c = 3/4$ we have a **bifurcation point**, that is, a point in the *parametric space* where the behavior mutates so that every result oscillates between two values in the *dynamical system* being studied; at $c = 5/4$ the bifurcation doubles and this repeats at higher values of c, until finally a so-called **chaos** is reached, where results jump around in an *apparently* haphazard pattern that is dependent on the initial value of x. This is called "sensitive dependence of behavior dependent on the initial conditions".

Dear Mom + Dad,

Last month I reached a fork in the road – a bifurcation point. My money oscillated – enough, not enough, enough, not .... The bifurcation points have now doubled. To avoid CHAOS, please SEND MORE MONEY. Since my behavior depends sensitively on the initial amount of money, please double the amount.

With all my love, Pat

The mathematical model of chaos has attracted researchers in the study of certain predicted changes in population growths (bacteria, viruses, animals, plants) that failed to occur; disastrously misleading weather forecasts; sudden major financial crises, *etc.* Insufficient data are usually the cause of miscalculated results of a succession of events in a dynamical system, but a "sensitive dependence of behavior dependent on the initial conditions" may also be a reason for a chaotic behavior that had not been predicted.

p. 633

Iteration of functions with complex numbers will be discussed in Chapter 17, "Fractals".

10.2 Elementary Functions

Elementary functions are real-value algebraic functions or transcendental functions (trigonometric, hyperbolic, exponential, logarithmic) and all those functions that can be obtained from these functions through addition, subtraction, multiplication, division, or by the process of forming composite functions or taking the inverse of the functions.

Algebraic Functions

Algebraic functions are definable in terms of a finite number of polynomials and roots.

A **polynomial function** f is defined as

$$f(x) = a_0 x^n + a_1 x^{n-1} + \ldots + a_{n-1} x + a_n ,$$

where n is a non-negative integer and $a_0, a_1, \ldots a_n$ are real numbers.

A **rational function** f is an algebraic function defined as

$$f(x) = \frac{P(x)}{Q(x)},$$

where P and Q are polynomials and the domain contains all x for which $Q(x) \neq 0$. For instance,

$$r(x) = \frac{3x^4 - 2x + 4}{x^3 + 5x}$$

is a quotient of two polynomials and, therefore, a rational function.

Transcendental Functions

Transcendental functions cannot be expressed in terms of a finite number of polynomials. They include exponential, logarithmic, trigonometric, inverse trigonometric, hyperbolic, and inverse hyperbolic functions.

Linear Functions

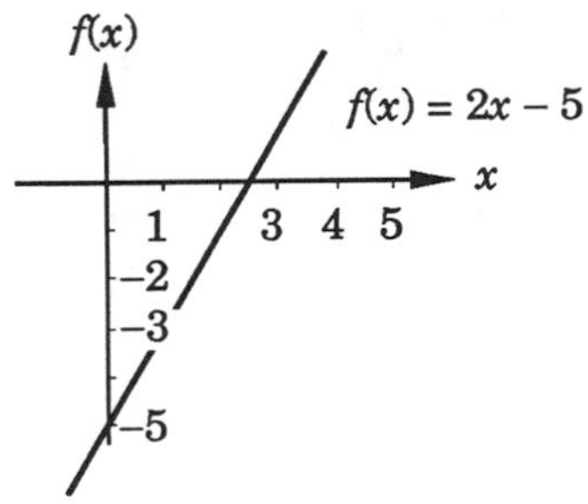

Since the graph of the function $y = f(x) = ax + b$ is a straight line, the function is called a **linear function**.

A straight line is determined by any two of its points, for instance, by its intercepts with the coordinate axes.

The line $f(x) = y = 2x - 5$ intersects the y-axis at

$$y = 2 \cdot 0 - 5 = -5$$

and the x-axis at

$$0 = 2x - 5; \; x = 2.5.$$

As shown at left, a line is drawn through the points (0, – 5) and (2.5, 0).

Direct Proportionality

Two variables y and x are directly proportional if their functional relationship is given by

$$y = f(x) = a\,x,$$

Proportionality Constant

where a is the **proportionality constant**. This is a special case of a linear function $f(x) = a\,x + b$; $b = 0$.

The graph is a straight line through the origin of the coordinate system; the proportionality constant a represents the *slope of the line*.

The graph below represents the function $f(x) = ax$ for $a = -1/2$, $a = 1$, and $a = 2$:

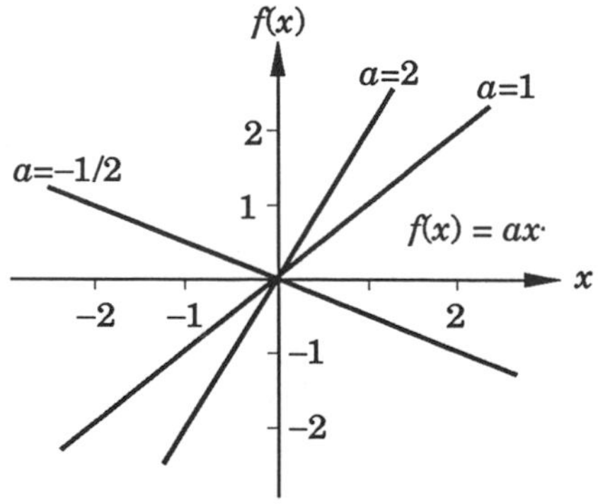

Power Functions

A function $y = f(x) = x^n$ is a **power function of degree n**.

If n is an **even natural number**, then $f(x) = x^n$ is an even function and the graph of the function is symmetrical about the y-axis; curves for different values of n all contain the origin and the points (–1, 1) and (1, 1).

If n is an **odd natural number**, then $f(x) = x^n$ is an odd function with a graph that is symmetrical about the origin; curves for different values of n all contain the origin and the points (–1, –1) and (1, 1):

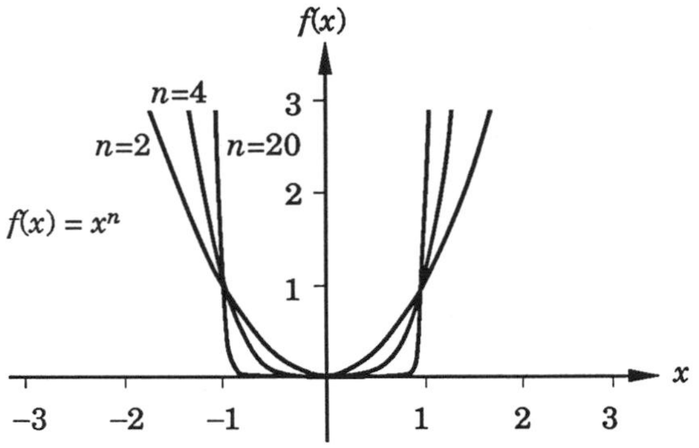

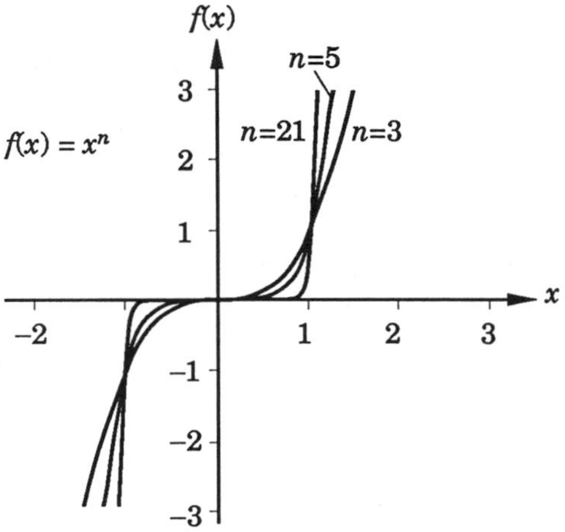

Inverse Proportionality

For $n = -1$, that is, $f(x) = a\,x^{-1}$ $(a \neq 0)$, the curve is a hyperbola, expressing an **inverse proportionality**, where a is the proportionality constant; the asymptotes are the ordinate and the abscissa:

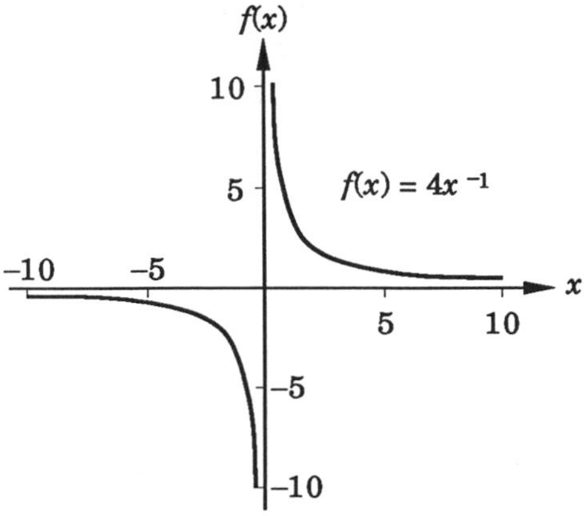

Quadratic Functions

A function

$$f(x) = a\,x^2 + b\,x + c,$$

where a, b, and c are constants and $a \neq 0$, is called a quadratic function.

If $b = c = 0$, then we have the simplest case of a quadratic function, $f(x) = ax^2$, a power function of the second degree whose graph is a parabola with its vertex at the origin of the orthogonal coordinate system. With a positive coefficient the parabola opens upward; with a negative coefficient the parabola opens downward:

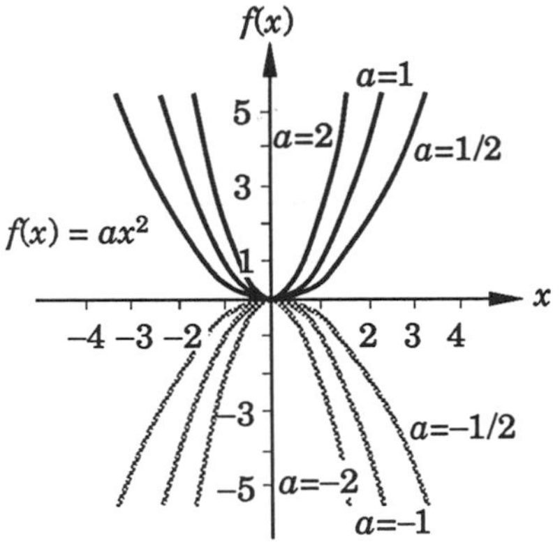

The vertex of the graph of a quadratic function represents the extreme of the function; a positive coefficient a gives an absolute minimum, a negative a an absolute maximum.

For the same value a, the graph of

$$f(x) = ax^2 + bx + c$$

is congruent with that of $g(x) = ax^2$, but with its vertex at

$$\left(-\frac{b}{2\,a}\,,\ c - \frac{b^2}{4\,a}\right):$$

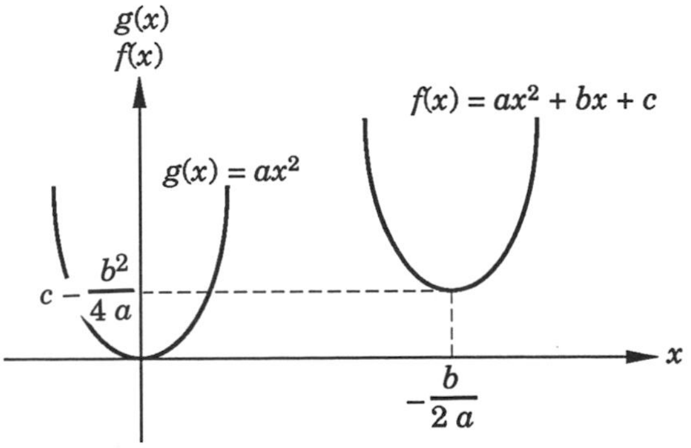

The vertex of the graph of

$$f(x) = \frac{1}{2}x^2 + 4x + 6$$

is located at

$$x = -\frac{b}{2a} = -4; \quad y = c - \frac{b^2}{4a} = -2.$$

Since a is positive (1/2), the graph opens upward.

$f(x) = \frac{1}{2}x^2 + 4x + 6$ is congruent with $g(x') = \frac{1}{2}(x')^2$ in a coordinate system whose origin coincides with the position (–4, –2) of the original system:

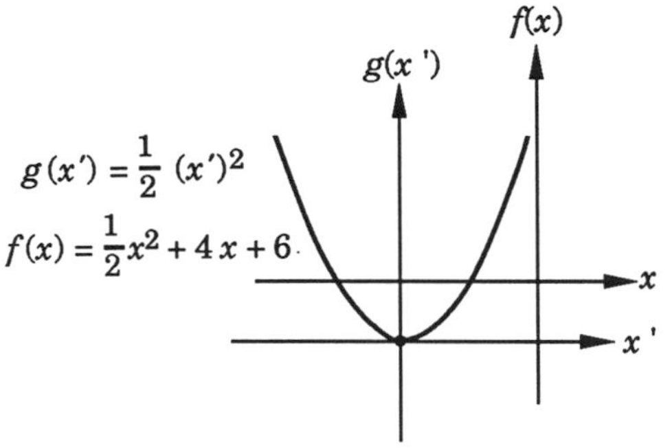

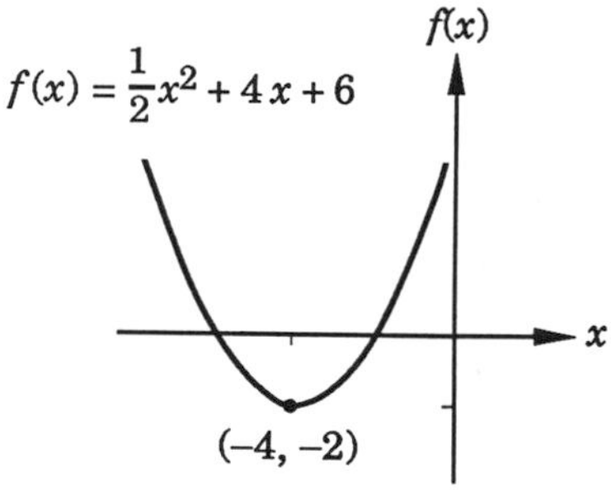

Instead of using the graph of $g(x') = \frac{1}{2}(x')^2$, we can plot the graph from values obtained by computing the function for various values of x.

The vertex of the graph of

$$f(x) = x^2 - 4x + 3; \quad -1 \le x \le 5$$

is located at (2, –1).

Using the information $-1 \le x \le 5$, we may determine the following points:

x	$f(x) = x^2 - 4x + 3$
-1	8
0	3
1	0
2	-1
3	0
4	3
5	8

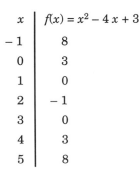

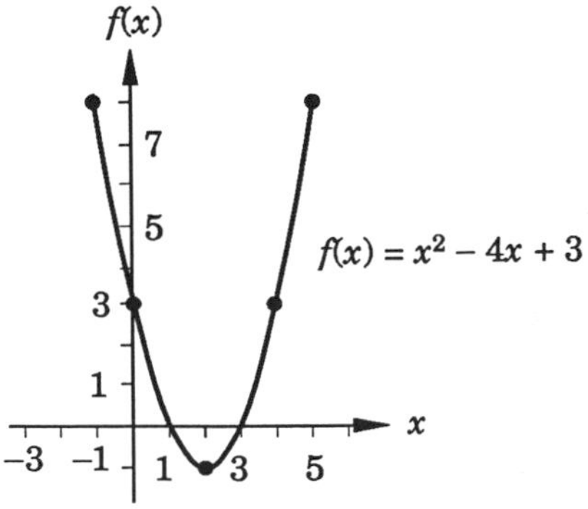

Exponential Functions

A function $f(x) = a^x$, where $a > 0$, is an **exponential function to the base a**.

All curves of exponential functions pass through the point (0, 1) of the coordinate plane. For $a > 1$, the curves rise toward the right; for $a < 1$, they fall. The functions $f(x) = a^x$ and $g(x) = a^{-x}$ are symmetrical to each other about the y-axis:

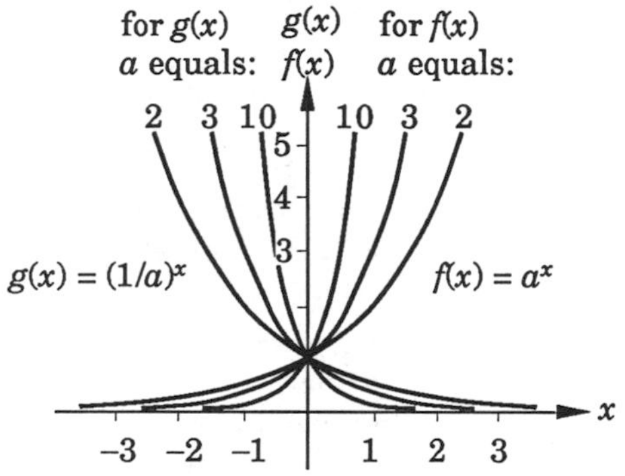

The exponential function to the base e of x is denoted

$$e^x \text{ or } \exp x .$$

Generally, when reference is made to "the exponential function", it is understood that the function is to the base e.

The curves of all exponential functions contain the point (0, 1); exponential functions are conveniently compared by determining the slope of the tangent at that point.

The base of natural logarithms, e, is the base of the exponential function whose slope at point (0, 1) is exactly 1.

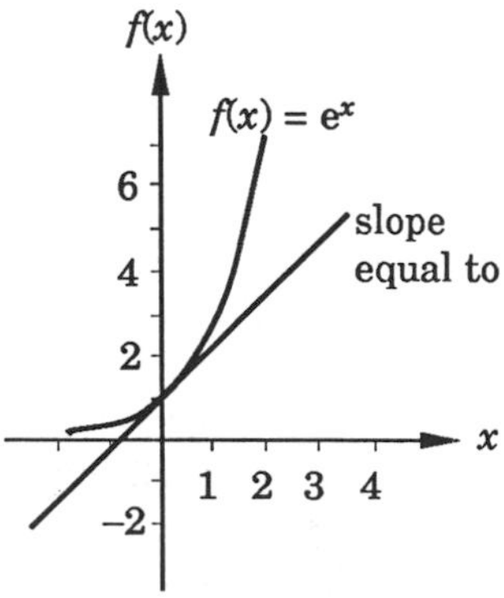

Logarithmic Functions

Functional Equation

Logarithmic functions may be defined by their *desired property*: to convert multiplication into addition. This is expressed by a **functional equation** as follows:

$$f(x \cdot y) = f(x) + f(y),$$

pp. 154 - 56

where f is the function we seek to define and where $x, y > 0$ range freely. From this functional equation and the assumption that f is continuous, it follows that $f(x^2) = 2f(x)$ and, more generally, that $f(x^p) = p\,f(x)$.

If we know that $f(a) = 1$, then $f(a^p) = p\,f(a) = p$, so that f must be the *logarithm* to the base a, which has the defining property

$$y = \log_a x \quad \text{if and only if} \quad x = a^y .$$

The logarithm to the base a of x is the exponent p to which a must be raised to yield the quantity x; thus, these are equivalent:

$$x = a^p \quad \text{or} \quad \log_a x = p .$$

Functional equations of logarithms, together with simple properties of $\log_{10}$, were used by Briggs and others to compute the early tables of logarithms. For example, from $2^{10} = 1028 \approx 10^3$, we see that $10 \cdot \log_{10} 2 \approx 3$, so that $\log_{10} 2 \approx 3$, with the logarithm being slightly larger than the approximation.

The graphs of the logarithmic and exponential functions $\log_e x = \ln x$ and $e^x = \exp(x)$ are shown here for the special base e, which has the unique property of giving each curve a slope of 1 where it crosses the axis:

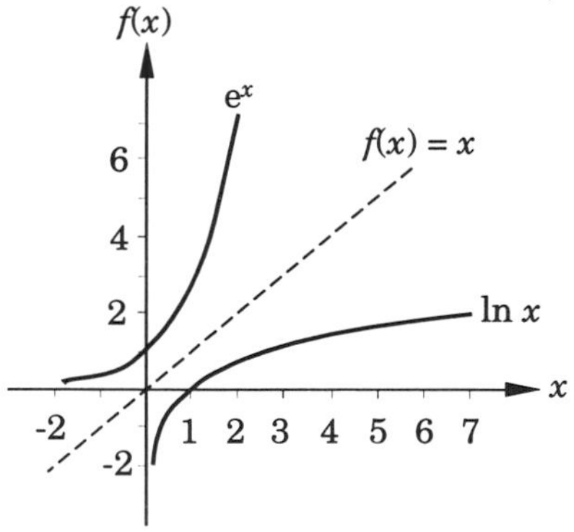

While $y = \log_a x$ always contains the point (1, 0), the slope is 1 here only for the curve $y = \log_e x$. This property makes $\log_e x$ and e^x useful and is one of the reasons why we call $\ln = \log_e x$ the natural logarithm.

The *domain* of e^x corresponds to the range of the natural logarithm of x ($\ln x$), that is, *all real numbers*, and the *range* of e^x corresponds to the domain of $\ln x$, that is, *all positive real numbers*. If $f(x) = e^x$, then $\ln x$ is the inverse function of $f(x)$.

The domain of e^x is $]-\infty, \infty[$; its range, $]0, \infty[$.

The domain of $\ln x$ is $]0, \infty[$; its range, $]-\infty, \infty[$.

10.3 Continuity and Limits

Definitions

Below, f is a **continuous** function; g is **discontinuous**:

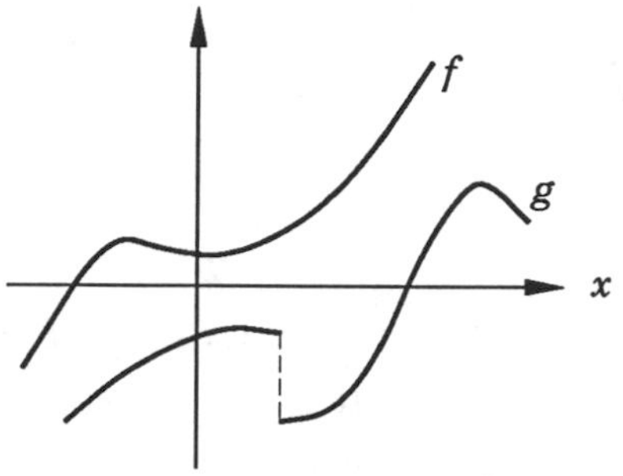

A function f is continuous at a point x_0 if

$$\lim_{x \to x_0} f(x) = f(x_0).$$

Thus:

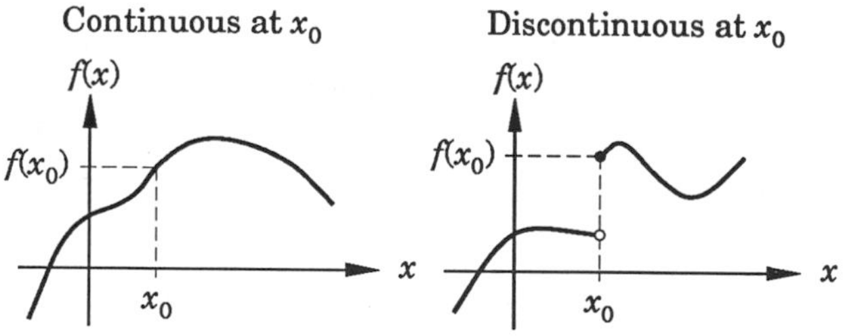

$f(x) = \dfrac{1}{x-4}$ is not defined at $x = 4$, for division by 0 is not permissible. Similarly, $f(x) = \dfrac{x^2-9}{x-3}$ is not defined at $x = 3$:

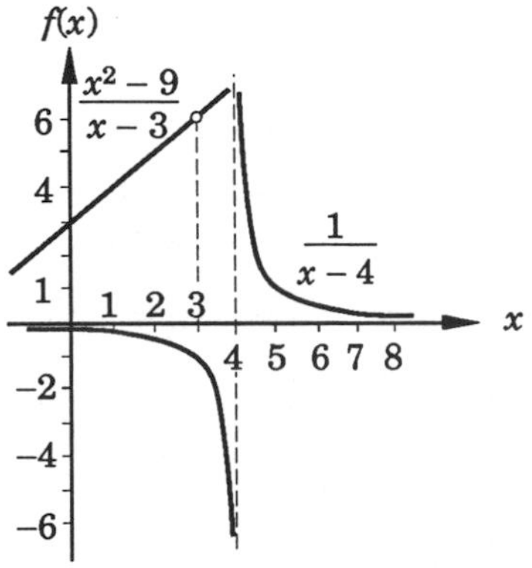

The expression

$$\lim_{x \to a} f(x) = b$$

reads:

"The limit value of the function $f(x)$, as x approaches (tends toward) a, is b",

which means, more precisely, that

for every number $\varepsilon > 0$ there exists a number $\delta > 0$ such that $|f(x) - b| < \varepsilon$ when $0 < |x - a| < \delta$.

We shall not expound on this definition. For the purpose of this text, an informal or intuitive concept of a limit value is sufficient.

For

$$\lim_{x \to a} f(x)$$

to exist, $f(x)$ must be defined for all x near a.

It is entirely possible, however, that $\lim_{x \to a} f(x)$ exists but $f(a)$ is not defined. For instance, if

$$f(x) = \frac{x^2 - 4}{x - 2},$$

then $f(2)$ is undefined, since division by 0 is not permitted.

But

$$\lim_{x \to 2} f(x) = \lim_{x \to 2} \frac{(x + 2)(x - 2)}{x - 2} = \lim_{x \to 2} (x + 2) = 4.$$

In expressions such as

$$\lim_{x \to 0} \frac{a_m \cdot x^m \ldots + a_1 \cdot x + a_0}{b_n \cdot x^n \ldots + b_1 \cdot x + b_0}$$

all powers of x disappear when $x = 0$, and the limit value is $\frac{a_0}{b_0}$.

If $a_0 = 0$ but $b_0 \neq 0$, we have

$$\lim_{x \to 0} \frac{a_m \cdot x^m \ldots + a_1 \cdot x + a_0}{b_n \cdot x^n \ldots + b_1 \cdot x + b_0} = \frac{0}{b_0} = 0,$$

and if $b_0 = 0$ and $a_0 \neq 0$,

$$\lim_{x \to 0} \frac{a_m \cdot x^m \ldots + a_1 \cdot x + a_0}{b_n \cdot x^n \ldots + b_1 \cdot x + b_0} = \lim_{x \to 0} \frac{a_1 \cdot x + a_0}{b_1 \cdot x}$$

does not exist, since $\frac{a_1 \cdot x + a_0}{b_1 \cdot x}$ takes on values which are increasingly large in size; these values will have one sign for $x > 0$ and the opposite sign if $x < 0$.

The lowest power of x will always decide the limit value, as $x \to 0$

$$|a_1 x| > |a_2 x^2| > \ldots > |a_m x^m|,$$

and similarly,

$$|b_1 x| > |b_2 x^2| > \ldots > |b_n x^n|.$$

To determine $\lim\limits_{x \to 0} \dfrac{x^3 - 8\,x}{5\,x^2 + 4\,x}$,

we have, as $x \to 0$, $x^3 << 8\,x$ and $5\,x^2 << 4\,x$, and thus

$$\lim_{x \to 0} \frac{x^3 - 8\,x}{5\,x^2 + 4\,x} = \lim_{x \to 0} \frac{-8\,x}{4\,x} = -2\,.$$

Similarly,

$$\lim_{x \to 0} \frac{x^3 + 3\,x}{9\,x^2 + 4\,x} = \lim_{x \to 0} \frac{3\,x}{4\,x} = \frac{3}{4}\,.$$

In expressions of the kind

$$\lim_{x \to \infty} \frac{a_m \cdot x^m \, ... + a_1 \cdot x + a_0}{b_n \cdot x^n \, ... + b_1 \cdot x + b_0}\,,$$

where x tends toward infinity, we can distinguish three cases:

$$m > n\,; \quad m = n\,; \quad m < n$$

$m > n$ $\qquad \lim\limits_{x \to \infty} \dfrac{7\,x^4 - 2\,x}{6\,x^3 - 5\,x} = \lim\limits_{x \to \infty} \dfrac{7\,x^4}{6\,x^3} = \lim\limits_{x \to \infty} \dfrac{7\,x}{6} = \infty\,;$

$m = n$ $\qquad \lim\limits_{x \to \infty} \dfrac{7\,x^3 - 2\,x}{6\,x^3 - 5\,x} = \lim\limits_{x \to \infty} \dfrac{7\,x^3}{6\,x^3} = \dfrac{7}{6}\,;$

$m < n$ $\qquad \lim\limits_{x \to \infty} \dfrac{7\,x^2 - 2\,x}{6\,x^3 - 5\,x} = \lim\limits_{x \to \infty} \dfrac{7\,x^2}{6\,x^3} = \lim\limits_{x \to \infty} \dfrac{7}{6\,x} = 0\,.$

As for $x \to \infty$,

$$7\,x^4 >> 2\,x\,; \quad 7\,x^3 >> 2\,x\,; \quad 7\,x^2 >> 2\,x\,; \quad 6\,x^3 >> 5\,x\,.$$

- Evaluate

$$\lim_{x \to 0} \frac{\sqrt{1 + 8\,x} - 1}{x}\,.$$

Since division by 0 is not permissible, it is necessary to rearrange the given expression so that x is not a factor of the denominator.

$$\lim_{x \to 0} \frac{\sqrt{1 + 8\,x} - 1}{x} = \lim_{x \to 0} \frac{\left(\sqrt{1 + 8\,x} - 1\right)\left(\sqrt{1 + 8\,x} + 1\right)}{x\left(\sqrt{1 + 8\,x} + 1\right)}$$

$$= \lim_{x \to 0} \frac{8\,x}{x\left(\sqrt{1 + 8\,x} + 1\right)}$$

$$= \lim_{x \to 0} \frac{8}{\sqrt{1 + 8\,x} + 1} = 4\,.$$

One-Sided Limits

On the real number line, the independent variable can approach a fixed value either from the right or from the left, which may be denoted $x \to a_+$ and $x \to a_-$, respectively.

If x in $f(x) = \dfrac{1}{x}$ tends to 0 through *positive* values,

$$f(x) = \lim_{x \to 0_+} \frac{1}{x} = \infty,$$

which reads: "The limit value of $f(x)$ as x tends to 0 from the right is infinity."

If x tends to 0 through *negative* values,

$$f(x) = \lim_{x \to 0_-} \frac{1}{x} = -\infty,$$

which reads: "The limit value of $f(x)$ as x tends to 0 from the left is minus infinity."

Right-Hand Continuity

A function f has **right-hand continuity** at a point $x = a$ if

$$\lim_{x \to a_+} f(x) = f(a)$$

Left-Hand Continuity

and **left-hand continuity** if

$$\lim_{x \to a_-} f(x) = f(a)\,.$$

To examine a function for continuity at a specified point, it is sometimes necessary to determine the one-sided values at that point. $f(x)$ is continuous at a point $x = a$ if

$$\lim_{x \to a_+} f(x) = \lim_{x \to a_-} f(x) = f(a)\,.$$

To draw the graph and assess continuity of

$$f(x) = e^{1/x}\,; \quad x \neq 0$$

several values of x are tested with the help of an electronic calculator:

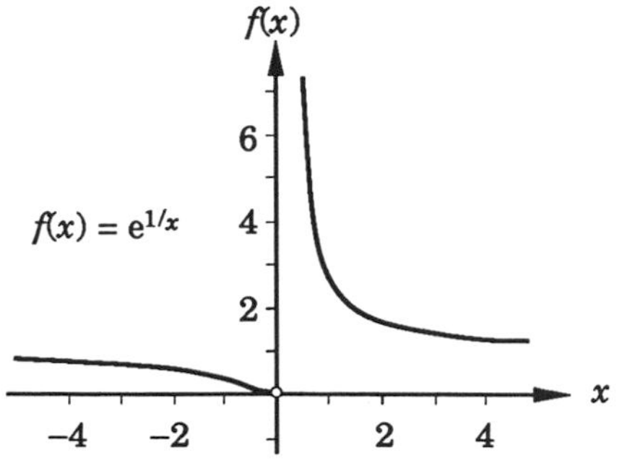

x	$e^{1/x}$	x	$e^{1/x}$
4.5	1.24 ...	−4.5	0.80...
4	1.28 ...	−4	0.77...
3	1.39...	−3	0.71...
2.5	1.49...	−2.5	0.67...
2	1.64...	−2	0.60...
1.7	1.80...	−1.7	0.55...
1.5	1.94...	−1.5	0.51...
1.3	2.15...	−1.3	0.46...
1.1	2.48...	−1.1	0.40...
1	2.71...	−1	0.36...
0.9	3.03...	−0.9	0.32...
0.7	5.17...	−0.7	0.23...
0.6	5.29...	−0.6	0.18...
0.4	12.1...	−0.4	0.08...
0.3	28.0...	−0.3	0.03...
0.2	$1.4... \times 10^{2}$	−0.2	$0.67... \times 10^{-2}$
0.1	$2.2... \times 10^{4}$	−0.1	$0.45... \times 10^{-4}$
0.01	$2.6... \times 10^{43}$	−0.01	$3.7... \times 10^{-44}$
0.001	$1.9... \times 10^{434}$	−0.001	$5.0... \times 10^{-435}$

The function fails to be defined at a point where $x = 0$.

Testing the **one-sided limit values** as x tends to 0 through *positive* values, we find that $f(x)$ tends to positive infinity:

$$x \to 0_+, \; e^{1/x} \to \infty .$$

On the other hand, as x tends to 0 through *negative* values, $f(x)$ tends to 0:

$$x \to 0_-, \; e^{1/x} \to 0 .$$

- Assess the continuity of

$$f(x) = \begin{cases} -x^2 + x + 1; & x < 2 \\ -x + 1; & x \geq 2 \end{cases}$$

at $x = 2$.

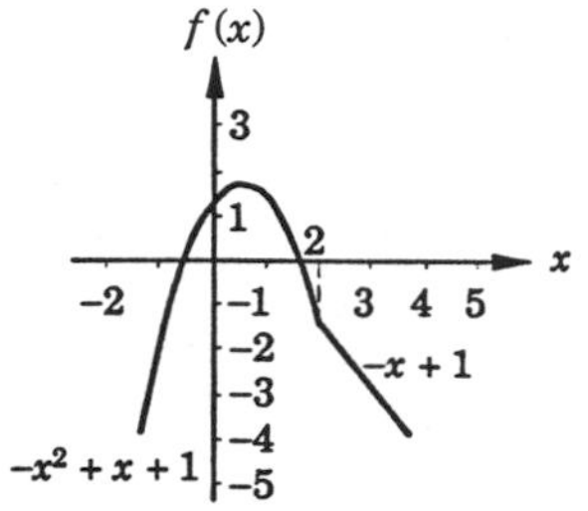

$$\lim_{x \to 2_-} f(x) = \lim_{x \to 2_-} (-x^2 + x + 1) = -4 + 2 + 1 = -1 .$$

$$\lim_{x \to 2_+} f(x) = \lim_{x \to 2_+} (-x + 1) = -2 + 1 = -1 .$$

Moreover, $f(2) = -2 + 1 = -1$.

Thus,

f is continuous at $x = 2$.

- Assess the continuity of

$$g(x) = \begin{cases} 2 \; ; & -2 \leq x \leq -1 \\ x^2 \; ; & -1 < x < 1 \end{cases}$$

at $x = -1$.

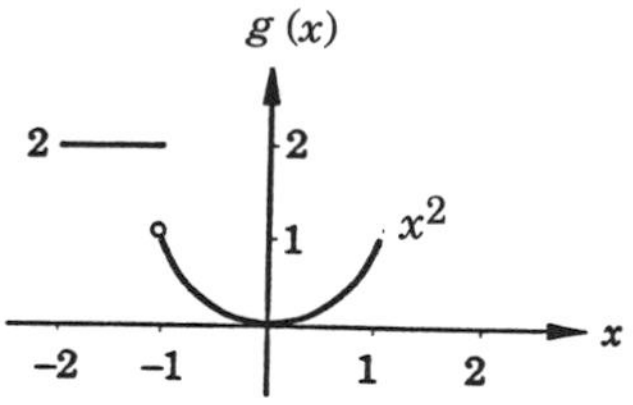

$$\lim_{x \to -1_-} g(x) = \lim_{x \to -1_-} 2 = 2 .$$

$$\lim_{x \to -1_+} g(x) = \lim_{x \to -1_+} x^2 = (-1)^2 = 1 .$$

Thus,

g is discontinuous at $x = -1$.

- Find a coefficient a for which

$$h(x) = \begin{cases} 5 - x \; ; & x < 1 \\ ax - x^2; & x \leq 1 \end{cases}$$

is continuous at $x = 1$.

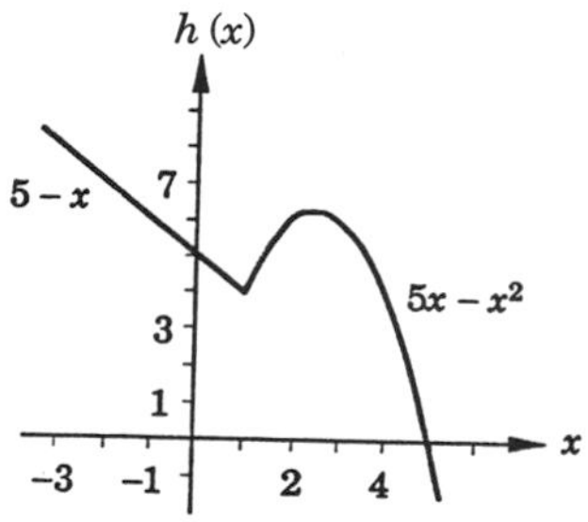

$$\lim_{x \to 1_-} h(x) = \lim_{x \to 1_-} (5 - x) = 5 - 1 = 4 .$$

$$\lim_{x \to 1_+} h(x) = \lim_{x \to 1_+} (ax - x^2) = a - 1 .$$

Continuity at $x = 1$ exists if $\lim_{x \to 1_-} h(x) = \lim_{x \to 1_+} h(x) = h(1)$, that is, if

$$a - 1 = 4; \; a = 5 .$$

Thus,

h is continuous for $a = 5$.

Calculations with Limits

The following rules of calculation with limits apply, provided that the limits exist for f and g:

$$\lim_{x \to a} [f(x) \pm g(x)] = \lim_{x \to a} f(x) \pm \lim_{x \to a} g(x)$$

$$\lim_{x \to a} [f(x) \cdot g(x)] = \lim_{x \to a} f(x) \cdot \lim_{x \to a} g(x)$$

$$\lim_{x \to a} [q \cdot f(x)] = q \left[\lim_{x \to a} f(x)\right]$$

$$\lim_{x \to a} \left[\frac{f(x)}{g(x)}\right] = \left[\frac{\lim_{x \to a} f(x)}{\lim_{x \to a} g(x)}\right], \text{ if } \lim_{x \to a} g(x) \neq 0$$

$$\lim_{x \to a} [f(x)^n] = \left[\lim_{x \to a} f(x)\right]^n$$

$$\lim_{x \to a} [f(x)^{1/n}] = \left[\lim_{x \to a} f(x)\right]^{1/n}$$

$$\lim_{x \to a} q^{f(x)} = q^{\left[\lim_{x \to a} f(x)\right]}$$

$$\lim_{x \to a} f[g(x)] = f\left[\lim_{x \to a} g(x)\right], \text{ if } f \text{ is continuous at } b = \lim_{x \to a} g(x)$$

$$\lim_{x \to a} [\log_q f(x)] = \log_q \left[\lim_{x \to a} f(x)\right]$$

Indeterminate Forms

A **quotient**

$$\lim_{x \to a} \frac{f(x)}{g(x)},$$

where $f(x)$ and $g(x)$ both approach 0 or $\pm\infty$ as x tends to a, is called an **indeterminate form**,

$$\frac{0}{0} \text{ or } \frac{\infty}{\infty}.$$

Indeterminate indicates that such limits may or may not exist.

Indeterminate forms also occur as products, differences, and powers.

A **product** $\lim_{x \to a} f(x) \cdot g(x)$, where

$$\lim_{x \to a} f(x) = 0 \; ; \quad \lim_{x \to a} g(x) = \infty,$$

gives rise to the indeterminate form $0 \cdot \infty$, and may be rewritten

$$\lim_{x \to a} \frac{f(x)}{1/g(x)} \quad \text{or} \quad \lim_{x \to a} \frac{g(x)}{1/f(x)},$$

that is, as an indeterminate form $\frac{0}{0}$ or $\frac{\infty}{\infty}$.

A **difference** $\lim_{x \to a} [f(x) - g(x)]$, that is, an indeterminate form $(\infty - \infty)$, may be rendered as

$$\lim_{x \to a} \frac{1/g(x) - 1/f(x)}{\frac{1}{f(x)} \cdot \frac{1}{g(x)}},$$

that is, as a form $\frac{0}{0}$.

A **power expression** $\lim_{x \to a} f(x)^{g(x)}$ may be transformed into

$$\lim_{x \to a} e^{g(x) \cdot \ln f(x)},$$

where the exponent has the form $0 \cdot \infty$.

The products

$$\left.\begin{aligned} 0 \cdot \ln 0 &= 0 \cdot (-\infty) \\ 0 \cdot \ln \infty &= 0 \cdot \infty \\ \infty \cdot \ln 1 &= \infty \cdot 0 \end{aligned}\right\}$$

and, consequently, the power expressions 0^0, ∞^0, and 1^∞ are also indeterminate.

Summing up, the following expressions are indeterminate:

$$\frac{0}{0},\ \frac{\infty}{\infty},\ 0 \cdot \infty,\ \infty - \infty,\ 0^0,\ \infty^0,\ 1^\infty.$$

p. 782 Many limits of an indeterminate form can be evaluated by a formula, **L'Hospital's rule**, which uses methods of differential calculus.

Determinate Forms

The forms

$$\infty \cdot \infty = \infty;\ \infty + \infty = \infty;\ -\infty - \infty = -\infty$$

are evidently determinate, in the sense that, for instance, if

$$\lim_{x \to a} f(x) = \lim_{x \to a} g(x) = 0,$$

then

$$\lim_{x \to a} f(x) \cdot g(x) = 0.$$

Other determinate forms are

$$0^\infty;\ \sqrt[n]{\infty}.$$

Undefined Forms

Expressions of the form

$$\frac{a}{0}, \text{ where } a \text{ is a non-zero real number, or } \infty$$

are undefined, because if y is very small, then $\frac{a}{y}$ will be very large in size but positive *or* negative according to the sign of y and a.

Also,

$$0^{-\infty} \text{ is undefined,}$$

because

$$\lim_{x \to 0_+} x^{-1/x^2} = +\infty,$$

but

$\lim_{x \to 0_-} x^{-g(x)} = -\infty$, if $g(x)$ equals, for instance, the greatest odd integer $\leq \frac{1}{|x|}$.

- Determine the limit (if it exists) of

$$\lim_{x \to 0} \frac{x^2 + 3x + 2}{9x^2 + 4x}.$$

If

$$f(x) = \frac{x^2 + 3x + 2}{9x^2 + 4x}$$

then

$$\lim_{x \to 0_+} f(x) = \infty \; ; \quad \lim_{x \to 0_-} f(x) = -\infty,$$

where neither statement says that the limit exists and has the value ∞ (or $-\infty$), but rather each statement says that the limit *fails to exist* in a particular way.

Chapter 11

OVERTURE TO THE GEOMETRIES

De Maleficis et Mathematicis et Ceteris Similibus

Artem geometriae discere atque exerceri publice intersit, ars autem mathematica dammabilis interdicta est. ... Haruspex ... qui huic ritui adsolent ministrare ... concremando illo haruspice ...

Corpus Juris Civilis, Codex Justinianus,
Book IX, XVIII, 2,3 (AD *circa* 650)

Or, in other words:

Concerning Mathematicians and Soothsayers and Kindred Evildoers

The study and teaching of the science of geometry are in the public interest, but whosoever practices the damnable art of mathematical divination, shall be put to the stake.

Read on at your own peril!

11.0 History

Greek, *ge*, "Earth"; *metría*, "measurement"

The oldest suggestions of an ordered system of measurements go back to the ancient Babylonians, who developed methods of land surveying embodying calculations of the area of simple geometric figures bounded by straight lines and arcs of circles. This is reflected in the name "geometry", whose literal meaning is *Earth measuring*.

The assumption of Babylonian astronomers that the year had 360 days is very likely the origin of our system of measuring angles in degrees; the fact that the angles of equilateral triangles are 60° may explain, in part, the sexagesimal method of counting. The Babylonians laid the first foundations of the science of geometry by their purposeful study of the properties of circles.

Unlike the Egyptians, whose interest in geometry lay exclusively in the practical considerations of land measurement, the Greeks devoted their energies to a systematic study of geometrical figures and their properties to establish a new science.

ΑΓΕΩΜΕΤΡΗΤΟΣ ΜΙΔΕΙΣ ΕΙΣΙΤΩ

PLATON
(427 - 348 or 347 B.C.; original name ARISTOCLES; at school given the nickname Platon in reference to his broad shoulders; Greek, *platon*, "broad")

Greatest among ancient Greek scientists were Plato and his pupil Aristotle. In an olive grove, or park, near Athens, **Plato** conducted a school of philosophy, *Achademya*, where he is said to have erected an ornamental gateway with a notice proclaiming "No Admittance" for those who knew no geometry.

ARISTOTELES
(384 - 322 B.C.)

Aristotle belonged to those in the know; eventually the most famous polymath of his time, his teachings came to be regarded as absolute and inviolable truths, which made them lasting obstacles to scientific progress for more than 1500 years, all through the Middle Ages.

EVDOXOS
(*c.* 408 - 355 B.C.)

Though no original works of **Eudoxus** – a pupil of Plato's – have survived, we know of them through references made by Euclid and Archimedes. Eudoxus is remembered for the theory of proportions, described in Euclid's *Elements*, and is usually credited with the invention of the method of exhaustion to find approximate areas and volumes of curvilinear forms and shapes – a forerunner of integral calculus.

Together with Eudoxus, the greatest mathematicians of antiquity, towering over all their contemporaries, were *Euclid* and *Archimedes* – the former in Alexandria and the latter at Syracuse in Sicily, both cities to be considered culturally as Grecian.

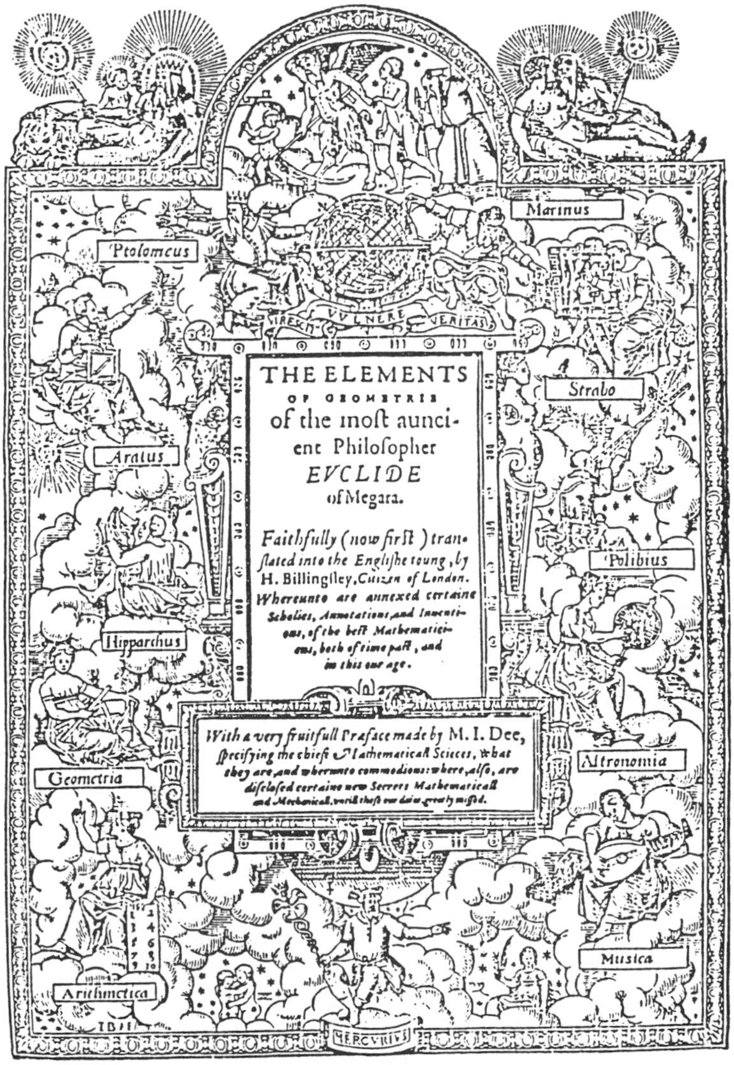

Title page of the first complete publication in English of Euclid's *Elements* (1570). Euclid (of Alexandria) is incorrectly referred to as "Euclide of Megara", also a geometer, and a contemporary of the rightful author of *Elements*.

EVCLEIDIS
(*c.* 330 - *c.* 275 B.C.)
Elementary Geometry

Euclid is best known for his 13-book treatise *Stoicheia* (*Στοιχεῖα*, "Elements"), a collection of all the geometric knowledge of his day, often referred to as **elementary geometry** after the name of the book.

Euclid's problems are all solved by logical reasoning from a central core of postulates, or axioms. This is the **classical axiomatic method of Euclid**. All his problems, solutions, and proofs rely on straight lines and circles, and his constructions are carried out using only an unmarked ruler and a pair of compasses. Euclid concerned himself with problems of **plane geometry**, that is, with geometric figures constructed on a plane surface, and with polyhedra, that is, solids bounded by plane polygonal regions.

ARCHIMEDES
(287 - 212 B.C.)

Solid geometry was the province of the mathematician and scientist-inventor **Archimedes**, whose contributions to geometrical science overshadow those of any other person. He studied, in depth, the properties of the sphere, cylinder, and cone, and the relations between them, which brought him very close to the foundations of calculus; he systematized his findings after the manner of Euclid, and his books are still useful.

APOLLONIOS
(c. 260 - after 200 B.C.)

By the 3rd century B.C., Greek mathematician **Apollonius** of Perga studied and named the **conic sections**. Like Archimedes, Apollonius used longitude, latitude, and altitude to define the position of a point.

HIPPARCHOS
(? - after 127 B.C.)

Hipparchus of Nicea and Rhodes is usually credited with inventing **trigonometry**, using trigonometric methods in his calculations of the distances to astronomic objects.

Originally, functions of chords of isosceles triangles were tabulated for use in spherical trigonometry, in response to the need for accurate information in astronomy. Today, trigonometric functions have been generalized to apply to any type of triangle, and are widely used to simplify calculations in several other areas of mathematics.

PTOLEMAEUS Claudius
(Roman citizen and, therefore, his name in Latinized form; *c.* A.D. 100 - *c.* 170)

The geographer, mathematician, and astronomer **Ptolemy** of Alexandria is mainly remembered for his work on projective geometry, which came to be the foundation of **descriptive geometry**, and for his 13-book treatise *Almagest* dealing with the application of geometry to astronomy, which became the standard textbook for over a thousand years, until Copernicus published *De revolutionibus*.

PAPPOS of Alexandria
(*c.* 300 - *c.* 350)

PROCLUS
(412 - 485)

Pappus of Alexandria calculated the surfaces and volumes of solids of revolution, particularly those generated by the rotation of conic sections. After Pappus, Alexandria produced no geometers of note, but a native Greek, **Proclus**, wrote a history of geometry summing up the work of Alexandrian geometers. After Proclus, the history of the University of Alexandria offers nothing of interest to chroniclers of geometry. Oppression by Christians and other warring religious factions caused the university and the science library to decline toward the end of the 3rd century.

HYPATIA
(370 - 415)

Hypatia, head of the Neoplatonist school of philosophy at Alexandria and one of the most learned and eloquent teachers of antiquity, wrote commentaries on Apollonius's *Conics* and on several other works in mathematics, all of them now vanished. She was barbarously murdered by a mob of so-called Christians who equated science and learning with paganism.

The murder of Hypatia was followed by an exodus from Alexandria of many Greek scientists and mathematicians who settled in Persia and in Arab countries – mainly in the city of Baghdad – where they became teachers. Many scientific books from the library were saved from destruction, among them Euclid's *Elements* and Ptolemy's *Almagest*, which were translated into Arabic and eventually into Latin and became standard works in mathematics, geometry, trigonometry, and astronomy.

The Arabs carefully tended and cultivated the knowledge that they acquired from the Greek scientists of Alexandria. Some of the world's most famous and learned scholars of the six centuries following the fall of Alexandria, in A.D. 641, were Arabs.

Arab scientists were very successful in applying their acquired knowledge of geometry and trigonometry to astronomy. They also made an important contribution to the history of geometry – and to science in general – by the preservation of the scientific knowledge of antiquity through the Dark Ages of European science. They translated learned works of science, and it is through this work that, in the 12th century, the science of geometry finally found its way into Europe via Moorish Spain.

ALBERTI Leone (1404 - 1472)

DESARGUES Girard (1591 - 1661)

PONCELET Jean Victor (1788 - 1867)

The Italian architect **Leone Alberti** was the first who, when discussing perspective in art, formulated ideas that pointed the way toward a future science of **projective geometry**. **Girard Desargues**, in his turn, was the first professional mathematician who, exploring the logical foundations of Euclidean geometry, initiated a formal study which extended Euclid's methods to projective geometry. **Jean Victor Poncelet** in 1822 published a treatise which revitalized interest in the subject of projective geometry as a science that treats form and position distinct from size.

MONGE Gaspard (1746 - 1818)

LAMBERT Johann Heinrich (1728 - 1777)

Descriptive geometry, in itself a form of projective geometry, may be said to have been created in 1794 when **Gaspard Monge** published his *Géométrie descriptive*. It takes up many features already known from works by Desargues and **Lambert**, but the credit for having developed projective geometry into a science in its own right belongs to Monge.

DESCARTES René (1596 - 1650)

René Descartes published, in 1637, *La géométrie*, showing how geometric figures could be analyzed algebraically. **Analytic geometry** – the geometry where positions are represented in a coordinate system and their properties analyzed algebraically – evolved under the influence of Descartes's work, but *La géométrie* itself does not use "Cartesian" coordinate axes or any other coordinate system; it is as much concerned with the rendition of algebra into terms of geometry as vice versa. As understood by Descartes, every *algebraic* step in an argument had to correspond to a *geometrical* construction. A remarkable feature of the text is that Descartes departs from the Greek practice of regarding a^2 and a^3 as area and volume; he represented them as *lines*.

La géométrie is the earliest mathematical text that a modern student of mathematics could read without stumbling over an abundance of obsolete notations.

DISCOURS
DE LA METHODE
Pour bien conduire sa raison, & chercher la verité dans les sciences.

PLUS

LA DIOPTRIQVE.

LES METEORES.

ET

LA GEOMETRIE.

Qui sont des essais de cete METHODE.

A LEYDE

De l'Imprimerie de IAN MAIRE.

CIↃ IↃ C XXXVII.

Auec Priuilege.

LA
GEOMETRIE.
LIVRE PREMIER.

Des problesmes qu'on peut construire sans y employer que des cercles & des lignes droites.

TOus les Problesmes de Geometrie se peuuent facilement reduire a tels termes, qu'il n'est besoin par aprés que de connoistre la longeur de quelques lignes droites, pour les construire.

Et comme toute l'Arithmetique n'est composée, que de quatre ou cinq operations, qui sont l'Addition, la Soustraction, la Multiplication, la Diuision, & l'Extraction des racines, qu'on peut prendre pour vne espece de Diuision : Ainsi n'a-t'on autre chose a faire en Geometrie touchant les lignes qu'on cherche, pour les preparer a estre connuës, que leur en adiouster d'autres, ou en oster, Oubien en ayant vne, que ie nommeray l'vnité pour la rapporter d'autant mieux aux nombres, & qui peut ordinairement estre prise a discretion, puis en ayant encore deux autres, en trouuer vne quatriesme, qui soit à l'vne de ces deux, comme l'autre est a l'vnité, ce qui est le mesme que la Multiplication; oubien en trouuer vne quatriesme, qui soit a l'vne de ces deux, comme l'vnité

est a l'autre, ce qui est le mesme que la Diuision; ou enfin trouuer vne, ou deux, ou plusieurs moyennes proportionnelles entre l'vnité, & quelque autre ligne; ce qui est le mesme que tirer la racine quarrée, ou cubique, &c. Et ie ne craindray pas d'introduire ces termes d'Arithmetique en la Geometrie, affin de me rendre plus intelligibile.

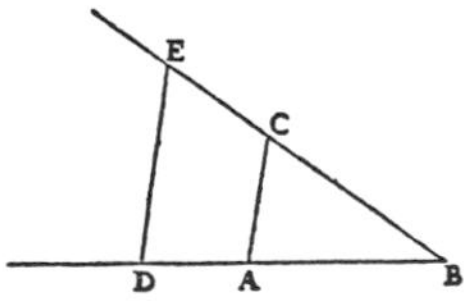

Soit par exemple AB l'vnité, & qu'il faille multiplier BD par BC, ie n'ay qu'a ioindre les poins A & C, puis tirer DE parallele a CA, & BE est le produit de cete Multiplication.

Oubien s'il faut diuiser BE par BD, ayant ioint les poins E & D, ie tire AC parallele a DE, & BC est le produit de cete diuision.

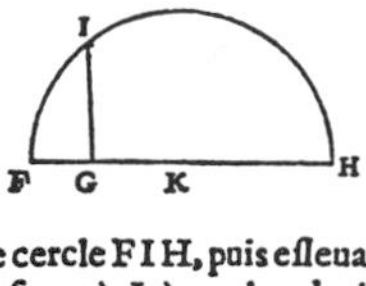

Ou s'il faut tirer la racine quarrée de GH, ie luy adiouste en ligne droite FG, qui est l'vnité, & diuisant FH en deux parties esgales au point K, du centre K ie tire le cercle FIH, puis esleuant du point G vne ligne droite iusques à I, à angles droits sur FH, c'est GI la racine cherchée. Ie ne dis rien icy de la racine cubique, ny des autres, à cause que i'en parleray plus commodement cy aprés.

Mais souuent on n'a pas besoin de tracer ainsi ces ligne

gnes sur le papier, & il suffist de les designer par quelques lettres, chascune par vne seule. Comme pour adiouster la ligne BD a GH, ie nomme l'vne *a* & l'autre *b*, & escris $a+b$; Et $a--b$, pour soustraire *b* d'*a*; Et ab, pour les multiplier l'vne par l'autre; Et $\frac{a}{b}$, pour diuiser *a* par *b*; Et aa, ou a^2, pour multiplier *a* par soy mesme; Et a^3, pour le multiplier encore vne fois par *a*, & ainsi a l'infini; Et $\sqrt{a^2+b^2}$, pour tirer la racine quarrée d'a^2+b^2; Et $\sqrt{C.\, a^3--b^3+abb}$, pour tirer la racine cubique d'a^3--b^3+abb, & ainsi des autres.

Où il est a remarquer que par a^2 ou b^3 ou semblables, ie ne conçoy ordinairement que des lignes toutes simples, encore que pour me seruir des noms vsités en l'Algebre, ie les nomme des quarrés ou des cubes, &c.

Il est aussy a remarquer que toutes les parties d'vne mesme ligne, se doiuent ordinairement exprimer par autant de dimensions l'vne que l'autre, lorsque l'vnité n'est point determinée en la question, comme icy a^3 en contient autant qu'abb ou b^3 dont se compose la ligne que i'ay nommée $\sqrt{C.\, a^3--b^3+abb}$: mais que ce n'est pas de mesme lorsque l'vnité est determinée, a cause qu'elle peut estre sousentendue par tout ou il y a trop ou trop peu de dimensions: comme s'il faut tirer la racine cubique de $aabb--b$, il faut penser que la quantité $aabb$ est diuisée vne fois par l'vnité, & que l'autre quantité *b* est multipliée deux fois par la mesme.

After the discovery that geometric problems could be transformed into algebraic problems, mathematicians, around 1700, began to apply calculus to the study of geometrical curves and surfaces. This eventually led to a new branch of geometry known as **differential geometry**, now of paramount importance in science.

SACCHERI Girolamo (1667 - 1733)

First to question the validity of the Euclidean parallel axiom was the Italian mathematician **Girolamo Saccheri**; his ideas were forgotten, however, until they were rediscovered in 1889 by his compatriot Beltrami. Another early attempt to devise a new geometry was made by Lambert in his *Theorie der Parallellinien*, published posthumously in 1786.

LOBACHEVSKI Nicolai (1792 - 1856)
BÓLYAI János (1802 - 1860)
GAUSS Carl Friedrich (1777 - 1855)
RIEMANN Bernhard (1826 - 1866)

Other mathematicians persisted in developing non-Euclidean geometries. **Nicolai Lobachevski** in 1829 and **János Bólyai** in 1832 independently described *hyperbolic geometry*; here the ubiquitous **Gauss** also cast his shadow. In 1854, **Bernhard Riemann** presented *elliptic geometry* – so named in 1871 by Felix Klein – and general Riemannian geometry, which, incidentally, is the mathematical foundation of Albert Einstein's general theory of relativity.

Greek, *tópos*, "place"

While the branches of geometry mentioned above depend on measurement of length and angle and therefore are referred to as metric, **topology** is a **non-metric geometry**.

MÖBIUS Augustus Ferdinand (1790 - 1868)
LISTING J. B. (1808 - 1882)

An offshoot of 17th- and 18th-century findings by **Descartes** and **Euler**, correlating the number of vertices, edges, and faces of polyhedra, topology is mainly an invention of the late 19th century, initiated when one-sided surfaces were discovered, independently, by the astronomers **Augustus Möbius** and **J. B. Listing**. In 1848 Listing published, at Göttingen, *Vorstudien zur Topologie* ("Introductory Studies in Topology"), thereby introducing the term "topology" into mathematics.

BETTI Enrico (1823 - 1892)
JORDAN Camille (1838 - 1922)
POINCARÉ Jules Henri (1854 - 1912)
HILBERT David (1862 - 1943)

Other pioneers in topology were **Bernhard Riemann**, **Enrico Betti**, **Camille Jordan**, and **Jules Henri Poincaré** – "the last mathematical universalist", who in *Analysis situs* (1895) launched the foundations for the development of topology.

A dominant figure in modern geometry was the German mathematician **David Hilbert**, professor at the University of Göttingen from 1895 until his death. In *Grundlagen der Geometrie* (1899; "Foundations of Geometry"), Hilbert gave a logical examination of Euclidean geometry, and in 1900, at an international congress of mathematics in Paris, he submitted 23 important mathematical problems to be targeted during the 20th century; many of *Hilbert's problem*s have been solved. Solved or unsolved, these problems have been and remain an enormous stimulation for the general development of mathematics.

Greek, *homos*, "same"; *morphé*, "shape", "form"

Topological Invariant

Objects are **topologically equivalent**, or **homeomorphic**, if one object can be transformed into another by **topological transformation**, that is, by bending, stretching, or twisting, but not by overlapping, tearing, or cutting. A **topological invariant** is a property held in common by all homeomorphic objects.

Topology today is not limited to classical geometrical configurations but is used in connection with invariant properties of equations and functions. It has generalized mathematical application to the study of *continuity*.

KLEIN Felix
(1849 - 1925)
Erlangen Program

LIE Marius Sophus
(1842 - 1899)
Galois theory: *p.*: 301

The German mathematician **Felix Klein** presented in 1872, as an inaugural lecture for a chair of mathematics at the University of Erlangen, his celebrated **Erlangen program**, a systematization – from the standpoint of invariants – of all the then known fields of geometry; this systematization was based on his and the Norwegian mathematician **Sophus Lie**'s expansion of the *group theory* to include geometry. After Erlangen, Klein attained the chair of mathematics at the University of Göttingen, made famous by Gauss. Besides his pioneering role in many fields of mathematical research, Klein was a leader of reforms of education; he introduced the *heuristic method* in mathematical education, that is, the use of a process of reasoning that, without pretensions to rigor, often leads to a correct result.

Greek, *heuriskein*, "to discover"
cf. Eureka!, "I have found it!"

BROUWER
Luitzen Egbertus Jan
(1881 - 1966)

David Hilbert's work on invariants between 1890 and 1893 and the Dutch mathematician **L.E.J. Brouwer**'s 1911 work on **topological invariants** are important landmarks of modern topology, whose foundations had been laid by Poincaré.

VEBLEN Oswald
(1880 - 1960)
WHITEHEAD
John Henry Constantine
(1904 - 1960)

Greatly influential in modern topological research and teaching were the American **Oswald Veblen** and the Englishman **Henry Whitehead**, whose many joint publications include the *Foundations of Differential Geometry* (1932), a classic in modern geometric research.

The 20th century has seen an increasing role of geometry in science, from relativity theory and the differential geometry of fundamental particles and fields to catastrophe theory in biology and psychology and the geometry of complex dynamical systems, chaos, and fractals. These applications have led to new and surprising developments in pure mathematics.

11.1 Geometric Abstraction

Geometry exists as an innate phenomenon in our consciousness.

Peter Høeg, *Smilla's Sense of Snow* (1993)

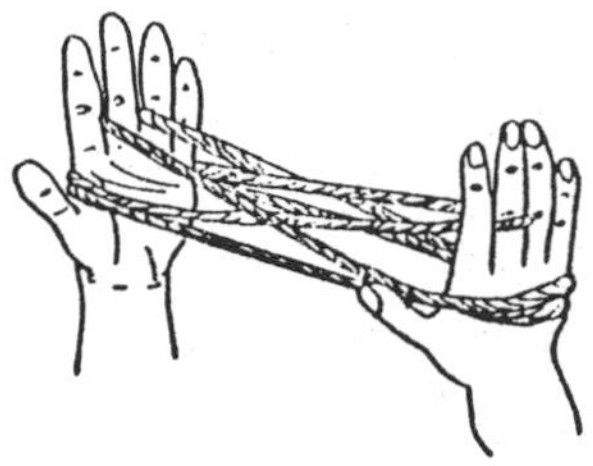

A purely spatial abstraction of the visible world in three dimensions – length, breadth, and height – generally presupposes that color, surface texture, and other essential qualities be ignored.

Many of us have taken a circle of string, wrapped it around our hands, and asked a friend, "Can you take this?" The game, of course, is cat's cradle.

Around the world, children and adults can be seen playing cat's cradle and making other string models. In the industrialized world it is mostly a game, but in those isolated societies where cat's cradle has retained its traditional significance, this kind of string weaving is sometimes used to teach about animals, leaves, and other objects, or to illustrate the narrative of a legend or a recent hunt.

Australian Aboriginal open-air classroom. The pattern of a turtle is formed with bark twine. Photo: Belinda Wright © National Geographic Society

String model of seal (Greenland).
Photo: Ivars Silis © National Geographic Society

It is noteworthy that such abstract models of nature are used and understood by people all over our planet; identical figures of string weaving may be presented by people as far apart as the Batwa pygmies of Africa, the Navajos, Apaches, Cherokees, and Eskimos of North America, and the tribes of Borneo, the Philippines, New Guinea, New Zealand, Australia, and Hawaii.

Pick up a piece of string, make a triangle, a square, a rectangle, a parallelogram, a trapezoid ... or a temporary work of art!

11.2 Perspective and Projection

Focused Perspective
Latin, *perspicere*, "to look through"

Developing a system of **focused perspective**, painters of the Renaissance departed from the traditional lack of depth and created a *visual geometry* as opposed to the prevailing *tactile geometry*. In a focused perspective, objects are depicted as they appear to the eye – the objects shrink with distance and *parallel lines seem to converge* to vanishing points.

Perspectiva.

Ein schön nützlich büchlin vnd vnderweisung der kunst des Messens / mit dem Zirckel / Richtscheidt oder Linial. Zu nutz allen kunstliebhabern / fürnemlich den Malern / Bildhawern / Goldschmiden / Seidensticker / Steinmetzen / Schreinern / auch allen andern / so sich der kunst des Messens (Perspectiua zu latein genant) zu gebrauchen lust haben. Darumb man auch solche kunst leichter / dann auß etlichen hievor getruckten büchern / begreiffen vnd lernen mag / mit viel schönen darzu dienenden figuren.

ROLLER Hieronymus
(1539 - ?)

Title page of **Hieronymus Roller**'s *Perspectiva*, a textbook for the craftsman; published in Frankfurt in 1546.

Projection

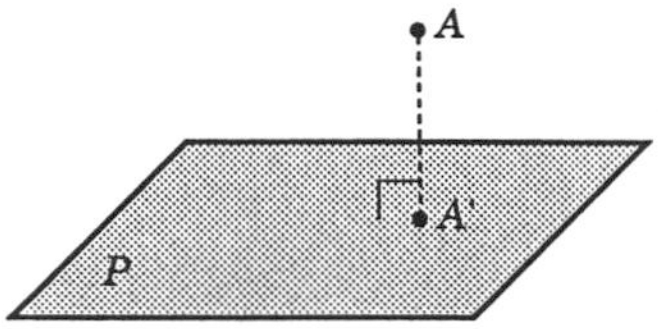

The fundamental principle of focused perspective is the projection of the object on the plane.

The perpendicular $\overline{AA'}$ is at right angles to every straight line in plane P that passes through the point of intersection between the plane and the perpendicular. The point of intersection, A' – called the **foot** of the perpendicular – is the **projection** of point A onto plane P. $\overline{A'B'}$ is the projection of the distance AB.

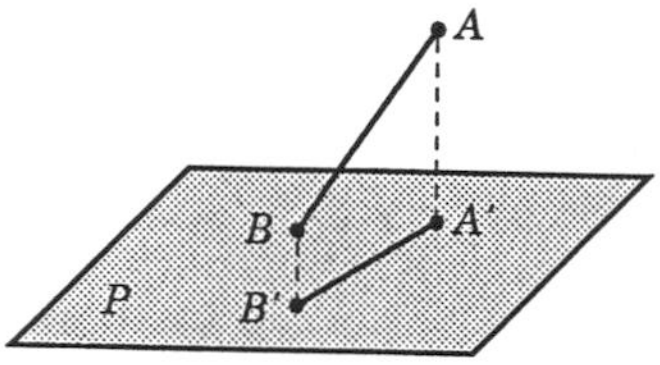

$\overline{A'BC}$ is the projection of all points on the **perimeter** of ABC, illustrating *perpendicular projection*, just one limited application of projection. The Albrecht Dürer (1471 - 1528) woodcut below suggests another and more intuitive projection.

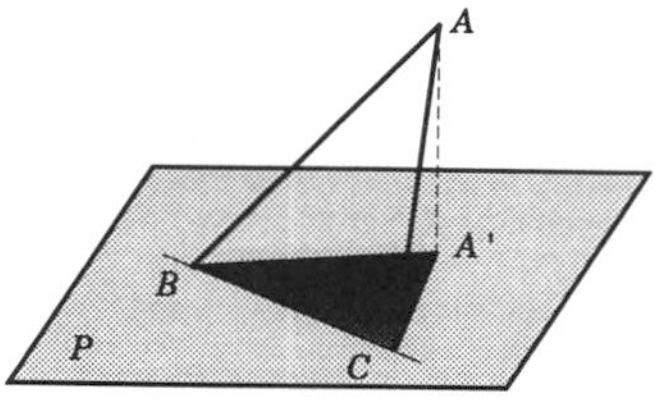

STÖR Lorenz
(? - *c.* 1621)
German woodcarver and painter; author and illustrator of *Geometria et Perspectiva* (1556).

Phantasmagoric Geometries

REUTERSVÄRD Oscar
(b. 1915)

The Swedish artist and art historian **Oscar Reutersvärd**'s 1934 drawing of an "impossible triangle" (Opus 1), and the many works that followed, opened an entire world of undecidable figures and shapes never before imagined. Reutersvärd uses a "Japanese perspective", where all parallel lines remain parallel and do not meet at points of visual convergence.

PENROSE Lionel Sharples
(b. 1898)

PENROSE Roger
(b. 1931)

A revival of interest in this art form and a wider interest in Reutersvärd's work came in 1958 after the article "Impossible Objects: A Special Type of Visual Illusion", published in *The British Journal of Psychology*, by the English geneticist **L. S. Penrose** and his son **Roger**, a mathematician, later known for his work, with Stephen Hawking, on black holes in space, and for the discovery of aperiodic tilings; the article was illustrated with a "tribar".

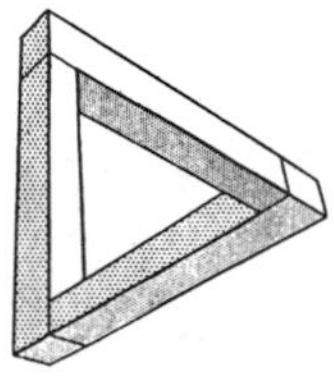

"Tribar"
(L.S. and R. Penrose, 1958)

Reutersvärd's explorations of a world of the impossible, or undecidable, have gained the attention not only of artists but scientists as well, primarily for the study of visual perception in psychology.

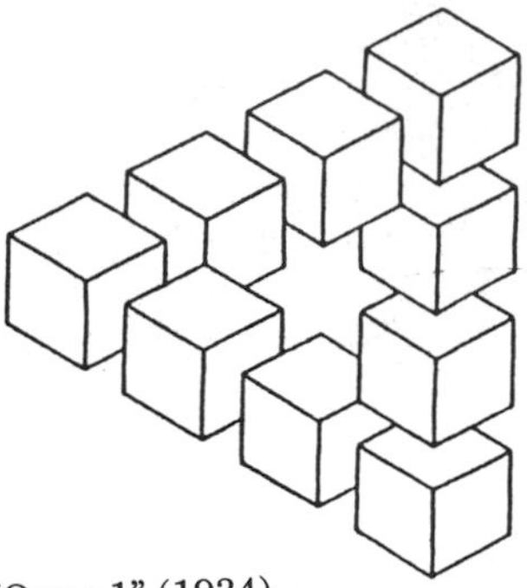

"Opus 1" (1934)

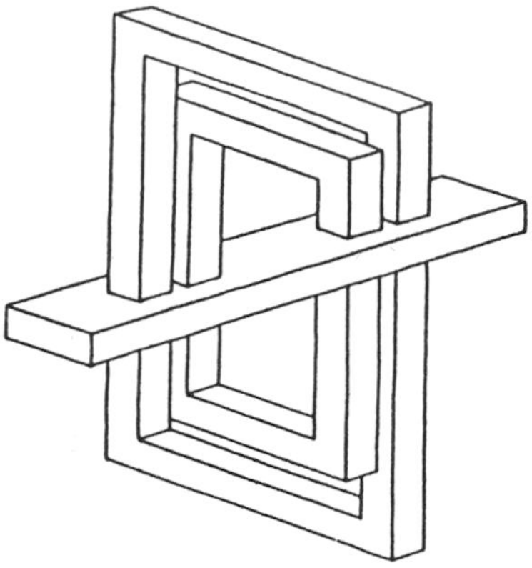

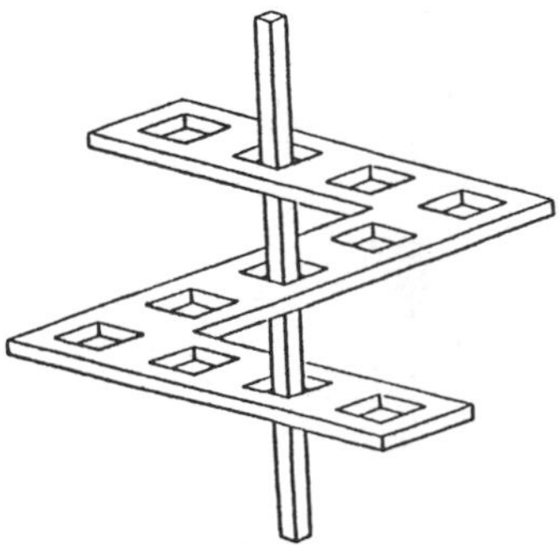

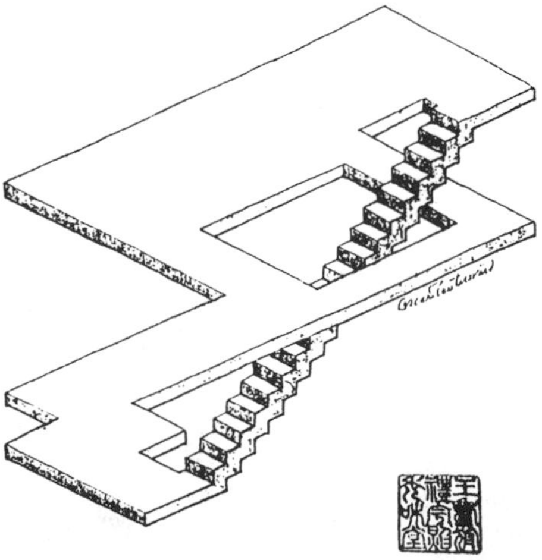

Drawings by Oscar Reutersvärd

ESCHER Maurits Cornelis (1898 - 1970)

No presentation seems needed here of the paradoxical visual and perspective effects of the images of Dutch graphic artist **M. C. Escher**, whose art is widely reproduced.

Op Art

Works of optical art, or **op art**, which emerged in the 1950s, are creations of illusory movement, devised by simple repetitive configurations with gradual, small changes of form and pattern or by "chromatic tension" from the juxtaposition of complementary colors.

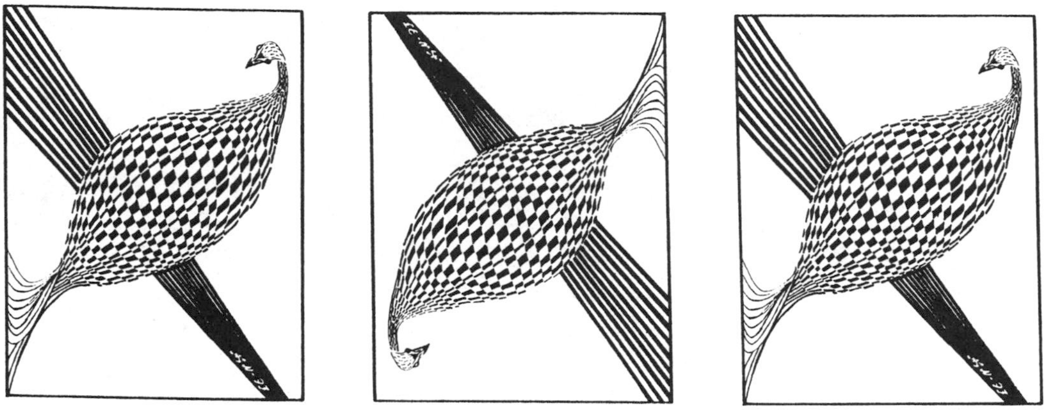

How to Catch Elephants – an Elephantasy

If you want to catch elephants, this is how you do it according to the Copenhagen newspaper Berlingske Tidende.

All you need is a blackboard, a piece of chalk, a mariner's telescope, a pair of tweezers, and an empty jampot.

You begin by writing 1 + 1 = 3 *on the blackboard and set it up in a place rich in elephants, and hide yourself in a nearby tree.*

Very soon an inquisitive little baby elephant will draw near to muse upon this remarkable statement, and by and by several of her elders will join her.

When there are enough of them, you just turn your telescope the wrong way around, so that the assembled elephants become quite small. Then you pick the small elephants up with your tweezers, one by one, and put them in the jampot.

11.3 Form and Shape

Plane and Space

In geometry, a straight line has *length* only, and is of **one dimension**; a geometrical shape that has *area*, but no volume, is of **two dimensions** – length and breadth (width); a *volume* has **three dimensions**. A three-dimensional region is called **space** or **3-space**.

A *point* – which may be regarded as the intersection of two lines – has the **dimension zero**.

Planar

The adjective **planar** refers to a two-dimensional quality *in the plane*. Surfaces are always two-dimensional but only rarely planar ("flat").

Plane Geometry

Plane geometry is concerned with the properties of plane figures – geometrical shapes of two dimensions: angles, triangles, squares, higher polygons, conic sections, *etc*.

Solid Geometry

Solid geometry deals with shapes of three dimensions – figures in space – such as pyramids, prisms, *etc*., and angles between planes. Plane sections of solids belong to plane geometry but may be studied profitably in space.

Euclidean geometry is based on the assumptions of Euclid, who, about 300 B.C., collected the mathematical knowledge of that time in the 13-volume *Elements*. Problems are solved by logical reasoning from an initial core of postulates (axioms), a method commonly referred to as the *classical axiomatic method of Euclid*. Alluding to the *Elements*, Euclidean geometry is often referred to as **elementary geometry**.

Any theory of geometry that denies one or more of the Euclidean axioms is a **non-Euclidean geometry**. Most commonly a non-Euclidean geometry is a theory where the *Euclidean parallel axiom* is rejected. *p.* 381

Congruence, Similarity, Correspondence

When superimposed, **congruent figures** can be made to coincide exactly; thus, congruent figures have **equal size and shape**, differing only in position.

$\cong$ $\equiv$ The notation $\cong$, or $\equiv$, reads "congruent to".

Similar figures have the **same shape**, but differ in size; the ratio of the lengths of any two corresponding lines in similar figures is the same, and all corresponding angles are equal.

$\sim$ The notation $\sim$ reads "similar to".

$\widehat{=}$ The notation $\widehat{=}$ reads "corresponds to". If 1 mm on a map corresponds to a distance of 1 km, this can be written

$$1 \text{ mm} \widehat{=} 1 \text{ km}.$$

11.4 Survey of Geometries

Euclidean geometry is based on definitions and axioms described in Euclid's *Elements*. This geometry is mainly concerned with points, lines, circles, polygons, polyhedra, and the conic sections. The concepts of *congruence* and *similarity* are fundamental in Euclidean geometry.

A geometry not based on the assumptions of Euclid is a **non-Euclidean geometry**; in particular, a non-Euclidean geometry does not depend on the *parallel postulate* of Euclid.

In learning the basics of mathematics, somewhat contrived divisions of disciplines apply, while no such boundaries exist for the mature user of mathematics.

p. 373

Projective geometry is the study of those properties of plane figures that are unchanged when a given set of points is projected onto a second plane.

p. 457

Trigonometry is the specialized geometry of the triangle, originally defined in terms of the angles and sides of a right-angled triangle but, as generalized, applicable to any type of triangle.

Plane trigonometry is concerned with plane triangles, *spherical trigonometry* with triangles on the surface of a sphere.

Trigonometric concepts are also widely used to simplify calculations in fields of mathematics that are not primarily geometrical.

p. 547

Analytic geometry investigates geometric problems by means of coordinate systems, thereby transforming them into algebraic problems.

Plane analytic geometry is devoted primarily to the analysis of equations in two variables, *solid analytic geometry* to equations in three variables. The methods are applicable to any number of variables and to any dimensions.

p. 599

Vector analysis deals specifically with the study of quantities that have both magnitude and direction. Geometrically, a vector is drawn as an arrow in a specific direction, its length representing the magnitude of the vector. This field includes the study of flows and uses generalized methods of the calculus. A generalization of vector analysis is **tensor analysis**, a study of components, motions, and the like in n-dimensional space.

Differential geometry applies differential and integral calculus to curves, surfaces, and other geometrical entities.

p. 625

The mathematics of **fractals** is concerned with shapes having generalized self-similarity, where patterns of the parts are miniatures of the whole.

p. 378

Topology deals with those properties – often associated with calculable invariant qualities – which are not altered by continuous deformations.

11.5 Topology

Topological Equivalence

Greek, *homos*, "same"; *morphé*, "shape", "form"

Objects are **topologically equivalent**, or **homeomorphic**, if one can be changed into another by **topological transformation**, that is, by bending, stretching, twisting, or the like.

Mapping: *p.* 188

Objects X and Y are homeomorphic if and only if there is a continuous one-to-one **mapping** from X onto Y whose inverse mapping is also continuous. Informally we may think of bending, stretching, or twisting; what is not allowed is folding that brings once distant points into direct contact/overlap or cutting unless followed by a regluing that reestablishes the preexisting relationships of continuity.

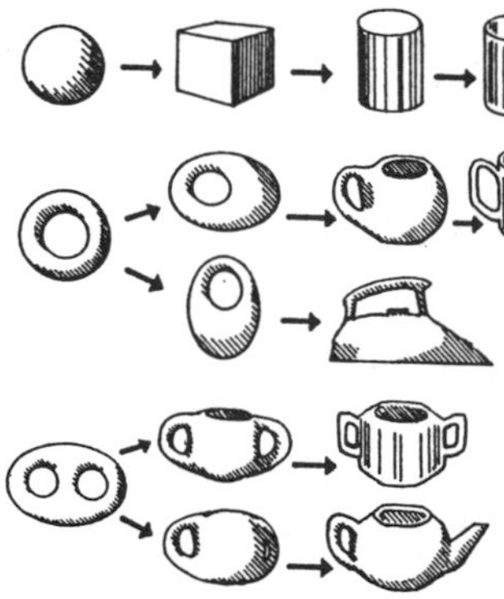

A sphere is topologically equivalent to a cube or a handleless mug but not to an anchor ring, whereas an anchor ring is topologically equivalent to a cup or an iron; an object with two holes is topologically equivalent to a sugar bowl or a lidless teapot, but not to a sphere or an anchor ring.

Despite the distortion of the mirror, there is homeomorphism between the dog and its reflection.

Knots

Mathematically, **knots** are curves formed in space by first interlacing a piece of string and then joining the ends together. Knots may be classified according to the number of crossings (overcrossing and undercrossing points). The lowest possible number of crossings is 3. Knots of 3 and 4 crossings have only one type each, knots of 5 crossings have two types, knots of 6 crossings have three types; thereafter the number of knot types increases rapidly.

Two knots are topologically equivalent if, by continuous movement (**deformation**) and not breaking the string, one knot can be deformed into the other. In addition, just as a triangle is equivalent (congruent) to its mirror image, a knot and its mirror reflection are considered topologically equivalent, although no continuous movement (deformation) can change one of the two knots into the other.

The mathematical study of knots is an area of active research. Practical applications are found in endeavors such as the analysis of electrical circuits, molecular structure analysis, and the planning of street and highway networks.

Genus of a Surface; the Jordan Curve

Jordan Curve

JORDAN Camille (1838 - 1922)

VEBLEN Oswald (1880 - 1960)

A continuous curve that does not intersect itself and has no endpoints – *e.g.*, a circle or a polygon – is a **simple closed curve**, or **Jordan curve**; it divides the coordinate planes into two regions, a theorem stated in 1893 by the French mathematician **Camille Jordan**. The theorem is intuitively self-evident, but a mathematical proof is another matter, and was not given until 1905 by the American mathematician **Oswald Veblen**.

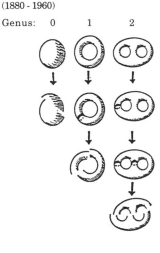

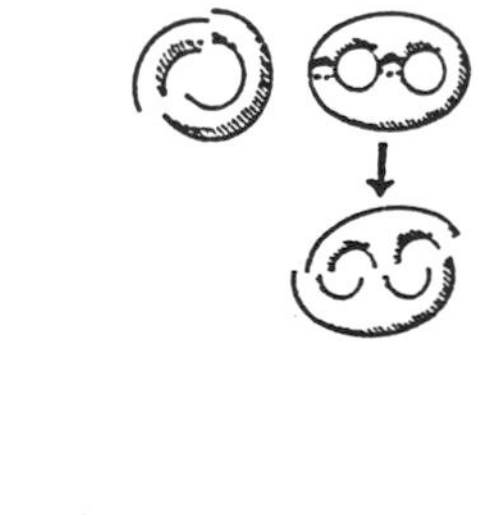

The **genus** of a surface is the largest number of Jordan curves that can be drawn on the surface without cutting it into two unconnected parts. For instance, the genus of a sphere is 0, for any simple closed curve separates it into two parts. Similarly, an ellipsoid and a convex polyhedron are of genus 0. An anchor ring is of genus 1, a two-holed object of genus 2 – the former always becoming disconnected by more than one cut, the latter by more than two cuts.

A convenient means of determining whether a given point is located inside or outside a convoluted Jordan curve is to draw a straight line from that point to a point that is undoubtedly located outside. If the line crosses the curve transversally an odd number of times, the point is located inside the curve; if it crosses an even number of times, it is outside. The crossing number is used in computer graphics programs to decide which pixels to shade when filling the inside of a region.

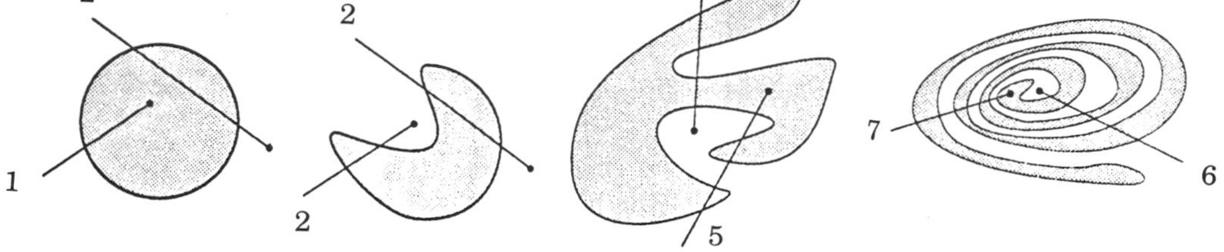

One-Sided Surfaces

The property of being two-sided or one-sided is a topological invariant; thus, a one-sided surface cannot be changed into a two-sided one by topological transformation, and vice versa. A one sided surface may be of any genus ≥ 1.

The Möbius Strip

MÖBIUS Augustus (1790 - 1868)

An interesting topological discovery, simple but surprising, was made about 1865 by the German mathematician **Augustus Möbius**, who showed that it is possible for a sheet to have only one surface. This is demonstrated by the **Möbius strip**: Cut a long rectangular strip of paper, give the paper a half twist, and paste the ends together. The Möbius strip has only one surface and one edge.

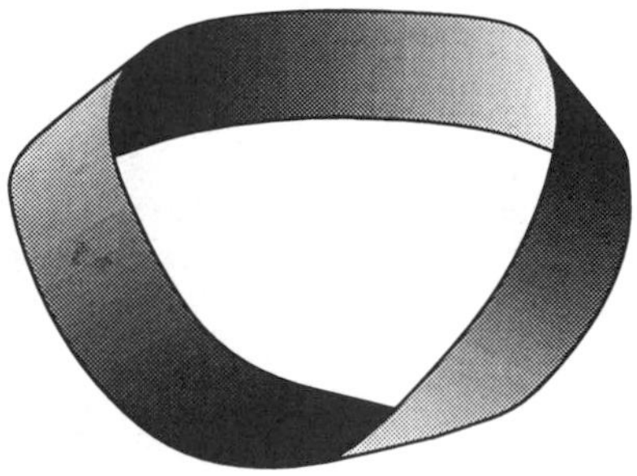

The experiment can continue: By cutting the Möbius strip lengthwise in the middle, *an edge is added* and we obtain a *single* twisted double-faced ring; thus, the Möbius strip is of genus 1. A third lengthwise cut results in two interlocking double-faced rings.

Fan belts and other belts for mechanical drives are sometimes made with a half twist – as Möbius strips – to provide more uniform wear.

The Klein Bottle

KLEIN Felix (1849 - 1925)

Named after the German mathematician **Felix Klein**, the **Klein bottle** has one surface with no edges; thus, the "bottle" has neither an outside nor an inside.

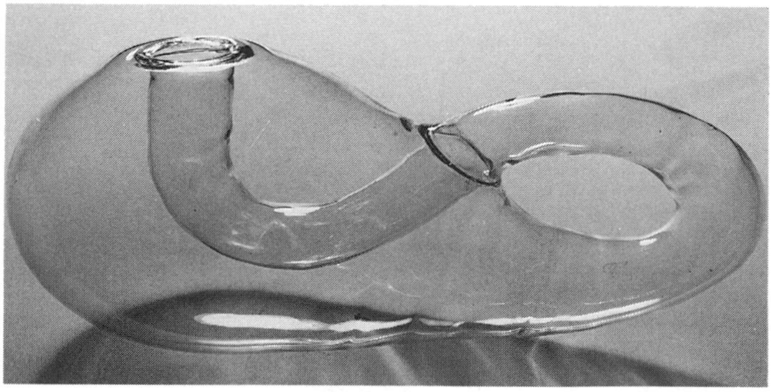

Topologically, cutting the bottle into mirror-symmetrical halves forms two Möbius strips; vice versa, the edge of one Möbius strip attached to the edge of its mirror-image twin forms a Klein bottle. The Klein bottle is a one-sided suface of genus 2.

11.6 Euclidean and Non-Euclidean Geometries

Euclidean Postulates and Common Notions

EVCLEIDIS
(*c.* 330 - *c.* 275 B.C.)

Euclid's presentation of plane geometry is based on a number of theorems that can all be derived from **five postulates** (axioms) and **five common notions** (the phrasing below does not always strictly follow Euclid's own).

The **postulates** are:

1. Exactly one straight line can be drawn between any two points.

Euclidean Postulate on Indefinite Continuation of a Straight Line

2. A straight line can be continued indefinitely.
3. With any point as center, a circle with any radius may be described.
4. All right angles are equal.

Euclidean Parallel Postulate

5. Through a given point outside a given straight line, there passes only one line parallel to the given line; that is, such a line *does not intersect* the given line.

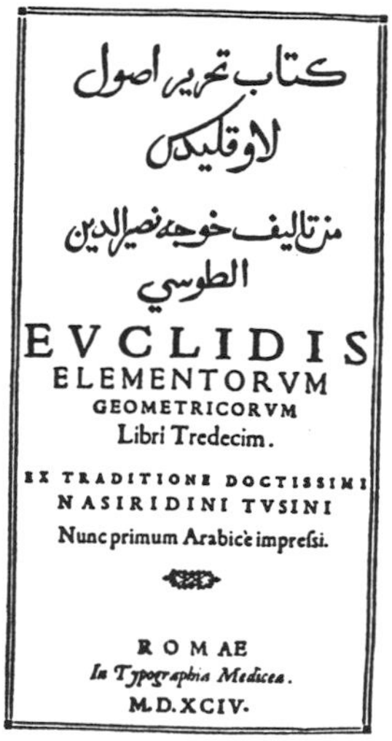
كتاب تحرير اصول
لاوقليدس
من تاليف خوجه نصير الدين
الطوسي
EVCLIDIS
ELEMENTORVM
GEOMETRICORVM
Libri Tredecim.
EX TRADITIONE DOCTISSIMI
NASIRIDINI TVSINI
Nunc primum Arabicè impreſsi.
ROMAE
In Typographia Medicea.
M.D.XCIV.

The **common notions** are:

1. Things equal to the same thing are equal.
2. If equals are added to equals, the wholes are equal.
3. If equals are subtracted from equals, the remainders are equal.
4. Things which coincide with one another are equal.
5. The whole is greater than a part.

From his postulates and notions, Euclid deduced 465 theorems.

Non-Euclidean Geometries

For 2000 years, Euclid's system was held as the only possible foundation of geometry, thought to be supported by logical reasoning and empirical proof.

In the early 19th century, mathematicians demonstrated that geometries that were just as valid and consistent as the age-old Euclidean geometry could be produced by replacing two of Euclid's postulates – the one about infinite continuation of a straight line, and the one about parallel lines – by other postulates that were employed with the remaining Euclidean postulates and the common notions.

The new concepts of geometry were later named *hyperbolic geometry* and *elliptic geometry*. These names, founded on analogies with **conic sections**, were suggested by the German mathematician **Felix Klein** in his notable program for classifying geometries (1872); for Euclidean geometry, Klein coined the name *parabolic geometry*, a term which is, however, rarely used.

KLEIN Felix
(1849 - 1925)

Hyperbolic Geometry

Hyperbolic geometry substitutes the Euclidean parallel postulate with the following: Through a given point outside a given straight line pass *more than one line* not intersecting the given line.

As one might expect, many theorems of hyperbolic geometry contradict the theorems of Euclidean geometry. While in Euclidean geometry the sum of angles of a plane triangle is 180°, in hyperbolic geometry it is less than 180° and varies with the size of the triangle: the smaller the area of the triangle, the closer the angle sum is to 180°. Two triangles are similar in hyperbolic geometry only if they are congruent.

LOBACHEVSKI Nikolai Ivanovitch
(1793 - 1856)

BÓLYAI János
(1802 - 1860)

The first publication on hyperbolic geometry, by the Russian mathematician **Lobachevski**, appeared in 1829. The Hungarian mathematician **Bólyai** independently published similar results in an appendix to a treatise written by his father, with 1829 as the printing year, but apparently not published until 1832. Lobachevski followed his first publication with further writings on hyperbolic geometry, but the works of Lobachevski and Bólyai were almost completely neglected by their contemporaries.

GAUSS Carl Friedrich
(1777 - 1855)

Independently of Lobachevski and Bólyai, **Gauss**, the prominent German mathematician, had already formulated a hyperbolic geometry, but had decided that the results should not be published in his lifetime. When they were published, about thirty years after the writings of Lobachevski and Bólyai, Gauss's fame made mathematicians take notice of the new geometry and opened the door for appreciation of the works of Lobachevski and Bólyai.

POINCARÉ Jules Henri
(1854 - 1912)

The French mathematician-physicist **Jules Henri Poincaré**, active in the entire field of mathematics, used the inside of a circle as a *model for hyperbolic geometry*. In Poincaré's model, lines are either diameters or circular arcs whose ends are perpendicular to the circumference of the circle; two such lines which do not meet correspond to parallel lines.

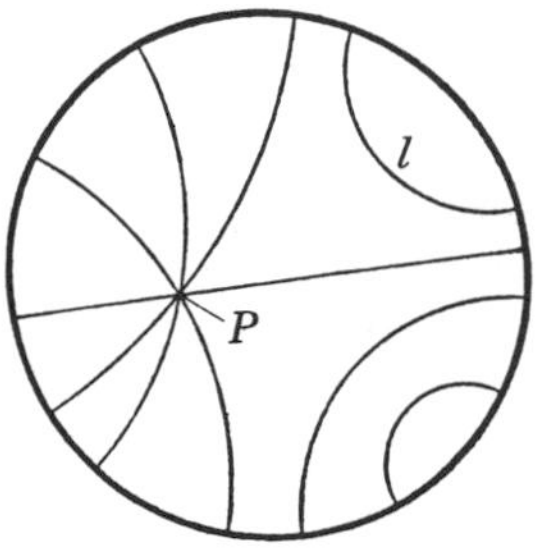

If, in the circle to the left, P is assumed to be a point in the hyperbolic plane and l a line that does not contain P, then an infinite number of lines parallel to l can be drawn through P.

Elliptic Geometry

Elliptic geometry rejects the Euclidean parallel postulate on the assumption that there are *no parallel lines* and, if extended far enough, any two straight lines in a plane will meet.

RIEMANN Bernhard (1826 - 1866)

In 1854, in a paper entitled *Über die Hypothesen, welche der Geometrie zu Grunde liegen* ("On the Hypotheses which Form the Foundation of Geometry"), the idea of elliptic geometry was first conceived by the German mathematician **Riemann**; as a consequence, elliptic geometry is commonly referred to as **Riemann geometry**, though this might also be taken to be the more general geometry of Riemannian manifolds.

EINSTEIN Albert (1879 - 1955)

An important concept in elliptic geometry is that boundlessness does not necessarily imply infinitely long lines: If we follow a straight line we do not come to an end, nor would we if we followed the equator of the Earth. Thus, experience does not convince us of the infinitude of a straight line – it only tells us that a straight line is boundless. This concept, that space could be unbounded without being infinite, was adopted by **Albert Einstein** in his general theory of relativity (1916).

In Riemann geometry, *all perpendiculars to a straight line meet at a point*, and (as in hyperbolic geometry) *triangles that are similar are also congruent.*

Whereas in hyperbolic geometry the sum of the angles of any triangle is less than 180°, in Riemann geometry *the sum of the angles of any triangle is greater than 180°.*

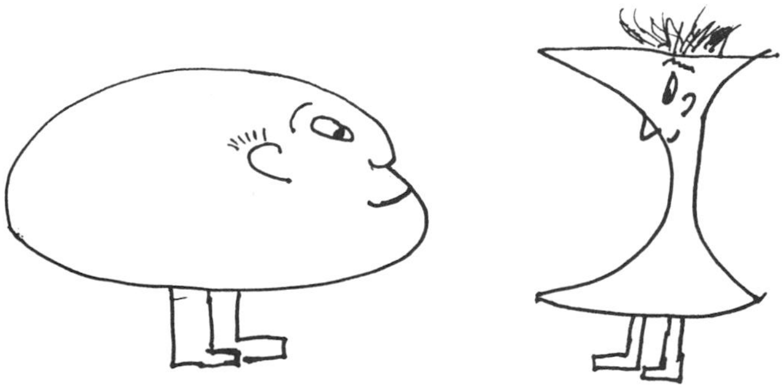

Elliptic Geometrist:

– *If Euclid were alive today, do you think he would be looked upon as a remarkable person?*

Hyperbolic Geometrist:

– *By all means. He would be over 2000 years old.*

The Infinitude of Geometries

There is no such thing as only one unassailable, mathematically true, geometry. From a mathematical viewpoint, any geometry – or any other branch of mathematics – that does not produce contradictions is acceptable. Another matter of concern is, however, to find the geometry that gives the most accurate representation of the physical world.

Despite its inherent flaws, Euclidean geometry is still the basis for most practical applications of geometry – it has taken human beings to the Moon and beyond.

Chapter 12

ELEMENTARY GEOMETRY

12.0 What Do We Mean by "Elementary Geometry"?

The main topic of this chapter is Euclidean geometry. Based on definitions and axioms described in Euclid's *Elements*, *p.* 381 **Euclidean geometry** is sometimes referred to as *elementary geometry*. This chapter is also elementary, or basic, in the sense that the mathematical function concept is not required, and a coordinate system is not used.

12.1 Geometric Elements and Figures

We will study concepts and names of geometric elements and figures, a sometimes dreary pursuit but giving the necessary preparation for more gratifying study, such as geometric construction (Section 12.3) and theorems (Section 12.4).

Points

A geometric **point** has no dimension – is void of quantity – and therefore cannot be drawn as "just a point". Thus, the concept of a geometric point is **axiomatic**.

x o A point is often represented by a small cross – signifying the intersection of two lines – or a small circlet.

Points associated with plane curves often have special names; see, *e.g.*, the definition of *tangent*, on page 388.

Cusp

A **cusp** is a double point on a curve where the curve has two coincident tangents.

A **cusp of the first kind** is a point where the curve has a branch on each side of the common tangent near the point of contact.

A **cusp of the second kind** is a point where the two parts of the curve lie on the same side of the common tangent near the point of contact.

A double cusp – or **point of osculation** – is a point where the two branches of the curve have a common tangent, each branch extending in both directions of the tangent.

Node, Crunode

A **node**, or **crunode**, is a point where two branches of a curve *cross* and have different tangents.

Salient point

At a **salient point** two branches of the curve *meet and stop*, and have different tangents.

Locus

A **locus** (plural: loci) is a geometric figure for which all of its points satisfy a given condition. For instance, the locus of all points in a plane at equal distance from a given point is a circle; the locus of all points in space with the same distance from a given point is a sphere.

Lines

A moving point describes a **line** that has only length, but no breadth. As with the point, the concept of a line is **axiomatic.**

We find ourselves with the slightly paradoxical truth that

two nonparallel lines define a point

and

two points define a line;

Dual Elements

Dual Operations

in this sense points and lines are said to be **dual elements**, and the intersection of lines to give points and the connection of points to give lines are **dual operations.**

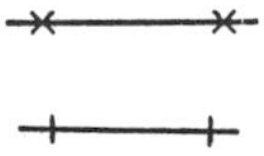

A **straight line** – usually called just a **line** – is the shortest distance between two points. It may extend without limit, or be a **determinate straight line** – also called a **line segment** – which contains both of the endpoints and all points between them; the length of the line segment is the **distance** between the endpoints.

A succession of straight line segments is a **broken line**.

Two or more lines in the same plane are **parallel lines** if they do not meet however far they extend.

Transversal

A line that intersects two or more (often parallel) lines is a **transversal**.

Perpendicular

Latin, *per*, "through" *pendere*, "to weigh"

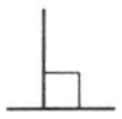

A line that meets another line at right angles is a **perpendicular**, indicated by a small square in the angle.

Crossing Line

Skew Line

Lines which lie in different planes and do not intersect each other are **crossing lines** or **skew lines**.

Ray (Half-Line)

Origin

A straight line issuing from an **initial point**, often denoted O for **origin**, is a **ray** or a **half-line**; a **closed ray** includes the origin, an **open ray** does not.

Curve, Arc

A **curve** is any continuous image of a line or segment; an **arc** is a portion of a curve, as in an *arc of a circle*. Curves are usually curved, but they can have straight sections.

Secant

Chord

A straight line that intersects a curve is called a **secant**; the part of the secant contained between the points of intersection is a **chord**.

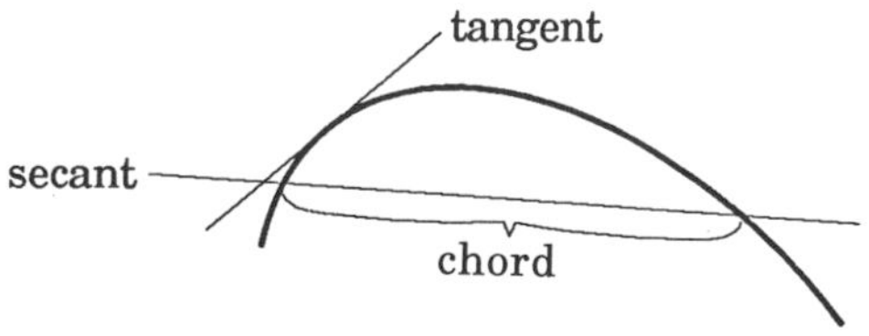

Tangent

A straight line that just touches a curve – a limit position of the secant – is a **tangent**; it has a double point of contact with the curve.

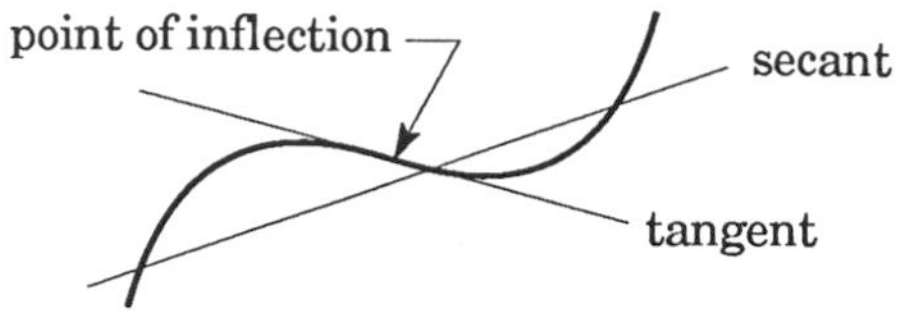

Inflection Tangent

A tangent to a curve in a point of inflection – where the curve changes its direction of curvature – is an **inflection tangent**; it has three points in common with the curve (these notions are best understood in the context of *calculus*, which will be discussed in later chapters).

Perpendicular
Normal

A straight line at right angles to the tangent in its point of contact with the curve is a **perpendicular** to the tangent and a **normal** to the curve.

Angles

∠

Latin, *vortex*, "whirl", "summit", "top of the head"

A **plane angle** (∠), or simply an **angle**, is formed by two **rays** – sides or legs of the angle – which extend from a common point, the **vertex** of the angle.

Angles have different names:

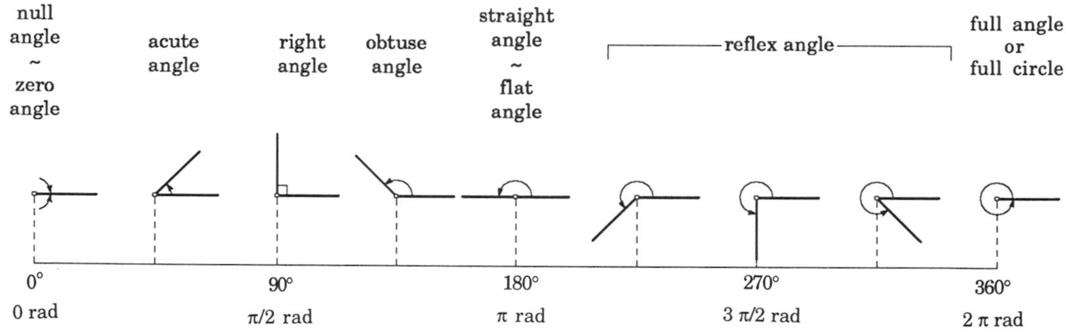

Two angles with a leg in common are **adjacent angles**.

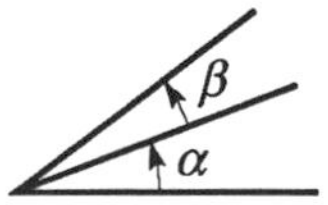

Complementary Angles
Supplementary Angles

Two angles whose sum is a right angle are **complementary angles**; two angles whose sum is a straight angle are **supplementary angles**:

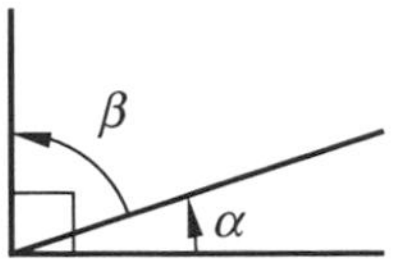

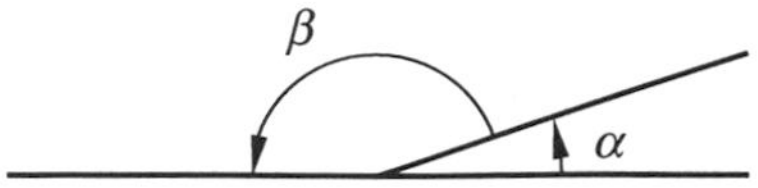

Bisector

A **bisector** is a straight line which divides an angle into two equal angles.

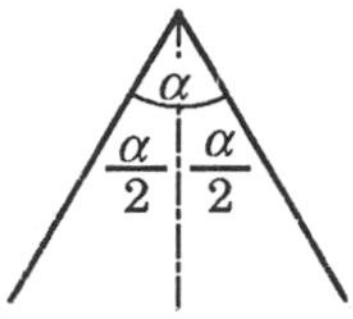

Vertical Angles

Two intersecting straight lines form two pairs of **vertical angles**; vertical angles are equal.

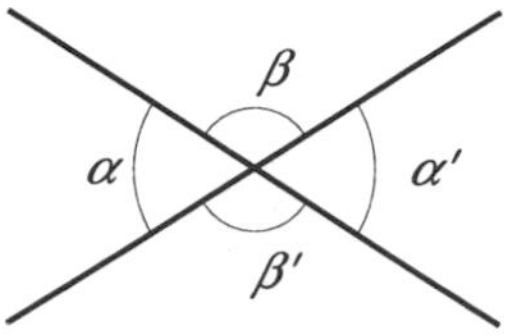

vertical angles $\begin{cases} \alpha = \alpha' \\ \beta = \beta' \end{cases}$

These angles are called *vertical* because each side of one is an extension through the *vertex* of a side of the other.

A transversal which intersects a pair of parallel lines produces many pairs of angles, some of which are equal, as shown below:

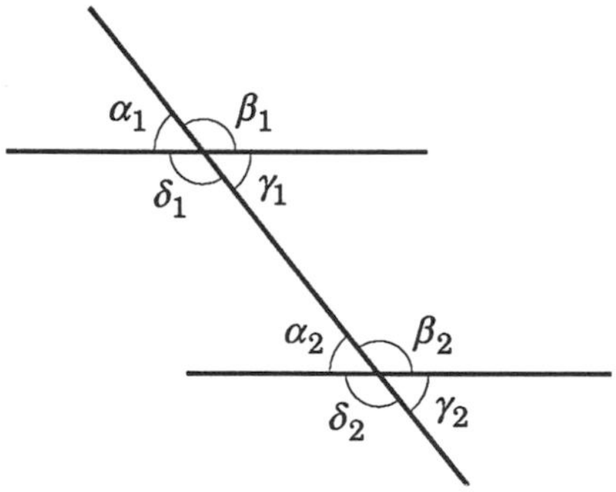

$\alpha_1 \quad \beta_1 \quad \gamma_2 \quad \delta_2$	are exterior angles
$\alpha_2 \quad \beta_2 \quad \gamma_1 \quad \delta_1$	are interior angles
$\gamma_1 = \alpha_2;\ \delta_1 = \beta_2$	are alternate interior angles
$\alpha_1 = \gamma_2;\ \beta_1 = \delta_2$	are alternate exterior angles
$\alpha_1 = \alpha_2;\ \beta_1 = \beta_2$ $\gamma_1 = \gamma_2;\ \delta_1 = \delta_2$	are corresponding angles
$\alpha_1 = \gamma_1;\ \beta_1 = \delta_1$ $\alpha_2 = \gamma_2;\ \beta_2 = \delta_2$	are opposite angles

* * *

Joe was out hoeing beans when he noticed smoke from the hay field. He grabbed two buckets from the shed and, via the edge of the irrigation ditch, took the shortest way to the burning hay. Living and working by the maxim "The shortest distance between two points is a straight line", how did Joe set his course?

[Answer on page 391]

Ellipses and Circles

Ellipses

An ellipse is a plane curve where the sum of the distances between two fixed points (the **foci**) and any point on the periphery is constant; below, F_1 and F_2 are fixed points and P_1 and P_2 are arbitrary points on the periphery:

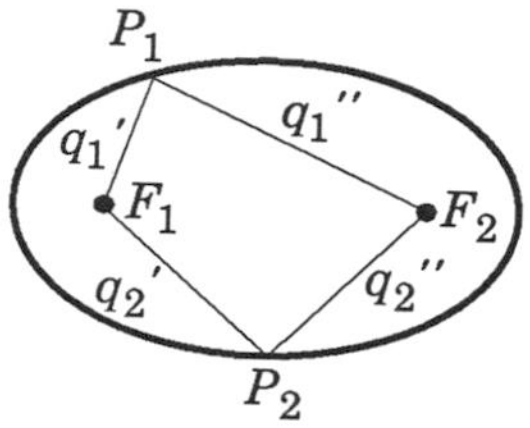

$$q_1' + q_1'' = q_2' + q_2''$$

Major Axis
Minor Axis

An ellipse has two axes of symmetry, a **major axis**, between the **vertices** of the ellipse, and a **minor axis**, which intersect at the center of the ellipse:

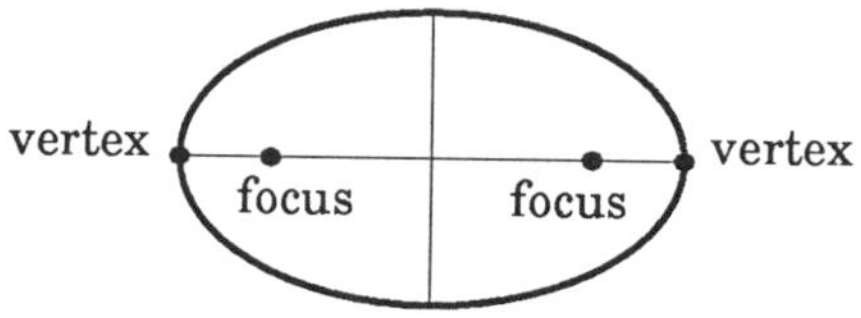

Perimeter

Periphery

The boundary of a figure in a plane is the **perimeter**. A curved perimeter, such as is found in an ellipse, is called the **periphery**.

Circumference

The **circumference** is the *length* of a perimeter or periphery.

Circles

A **circle** is a plane curve that is the locus of all points in the plane equidistant from a given point, the **center** of the circle.

Arc
Periphery

An **arc** of the circle is bounded by two distinct points on the **periphery**.

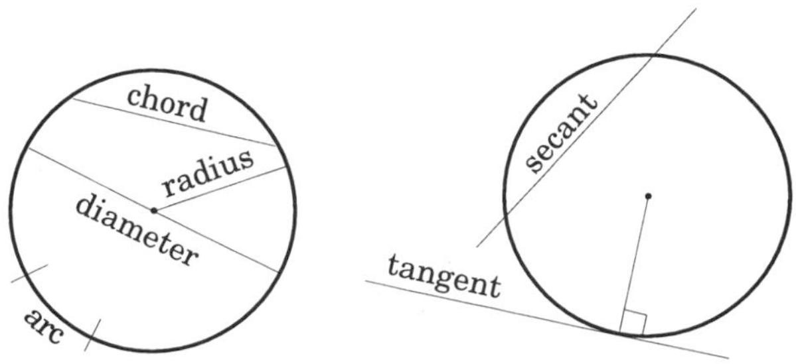

Secant
Chord
Diameter

A **secant** of a circle is a straight line intersecting the periphery in two points; a **chord** is the part of the secant within the circle. A chord passing through the center is the **diameter** – the longest chord of the circle.

Tangent
Radius

A **tangent** is a straight line which has a double point of contact with the periphery, and may be regarded as the limit position of a secant. The tangent is perpendicular to the **radius** of the circle at the point of contact.

Sector
Segment
Subtend: to be opposite to and mark off

A **sector** of the circle is bounded by two radii and the included arc; a **segment** is bounded by a chord and the arc subtending the chord:

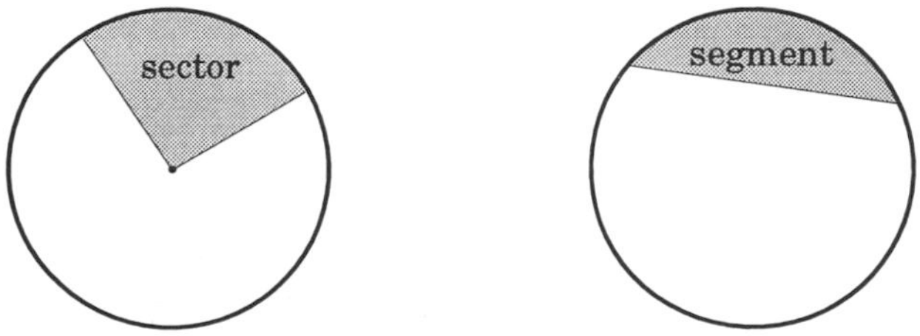

Angle Subtended by an Arc

A **central angle**, θ, **subtended by an arc**, a, as shown below, is an angle whose vertex is at the center and whose sides are radii. The **peripheral angle**, ψ, subtended by the same arc a, has half the measure of θ:

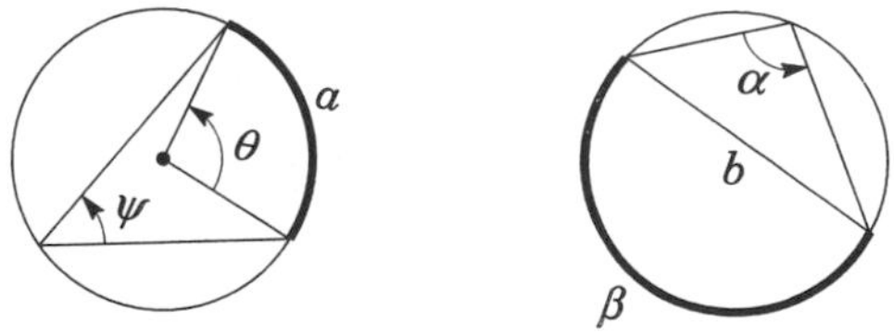

Angle Subtended by a Chord

The **angle** α **subtended by a chord** b has its vertex on the periphery and its two sides are chords that together with b form a triangle as shown below. Because α is the peripheral angle for the arc β, its measure is independent of where its vertex falls in the complement of β.

* * *

Answer to the haystack problem from page 389: To hit the point on the edge of the irrigation ditch that gives the shortest total distance to the haystack, Joe envisioned a mirror image of the haystack (the edge of the ditch closest to Joe being the "mirror") and a straight line between the shed and the mirror image.

Polygons

Greek, *póly*, "many"; *gonía*, "angle", from *góny*, "knee"

A **polygon** is a plane figure with three or more angles and as many sides. It is bounded by a broken line forming a simple closed curve, a circuit without self-intersections.

Diagonal

A **diagonal** is a straight line connecting two opposite vertices.

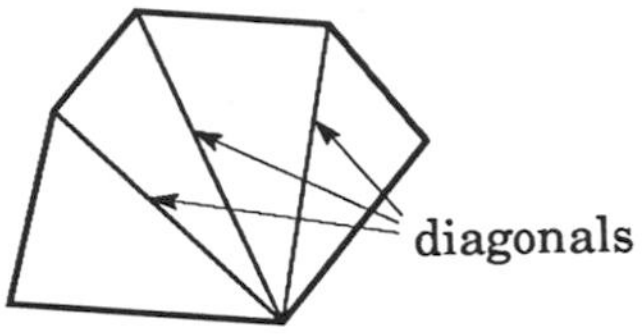

Polygons are named after the number of sides or vertices:

No. of sides	Name	No. of sides	Name
3	Triangle	9	Nonagon
4	Quadrilateral Tetragon	10	Decagon
5	Pentagon	11	Undecagon
6	Hexagon	12	Dodecagon
7	Heptagon		
8	Octagon	n	n-gon

Regular Polygon

In a **regular polygon**, all sides have the same length, and all **interior angles** are equal.

Convex
Concave

In a **convex** polygon each interior angle is less than a flat angle (180°); a polygon is **concave** if an interior angle exceeds 180°:

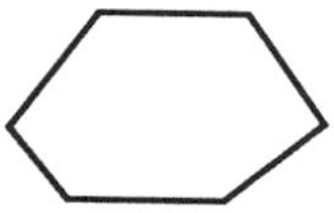

Convex Hexagon

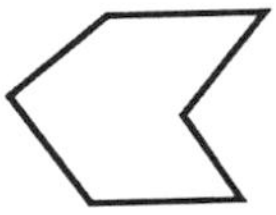

Concave Hexagon

Reentrant Angle
Salient Angle

The inward-pointing angle of the concave polygon is a **re-entrant angle**; the other angles are **salient angles**.

Triangles

Greek, *skalinós*, "uneven"
Greek, *iso-*, "equal", *skélos*, "leg"

A triangle is a polygon with three sides. Every triangle is contained in a plane. If no two sides are equal, it is a **scalene triangle**; if two sides are equal, it is an **isosceles triangle**; and if all sides are equal, it is an **equilateral triangle**.

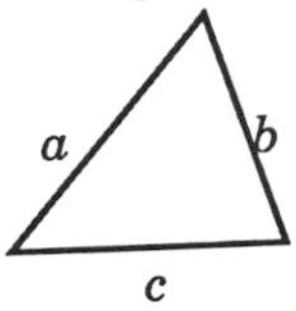

Scalene Triangle

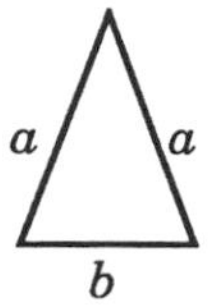

Isosceles Triangle

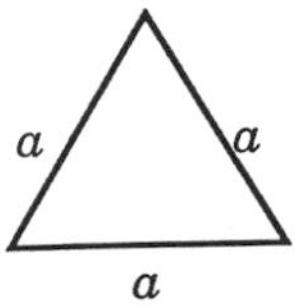

Equilateral Triangle

If each interior angle of a triangle is less than a right angle, we speak of an **acute triangle**; if one angle is greater than a right angle, we have an **obtuse triangle**. Acute and obtuse triangles are called **oblique triangles**. A triangle with a right angle is called a right-angled triangle or a **right triangle**.

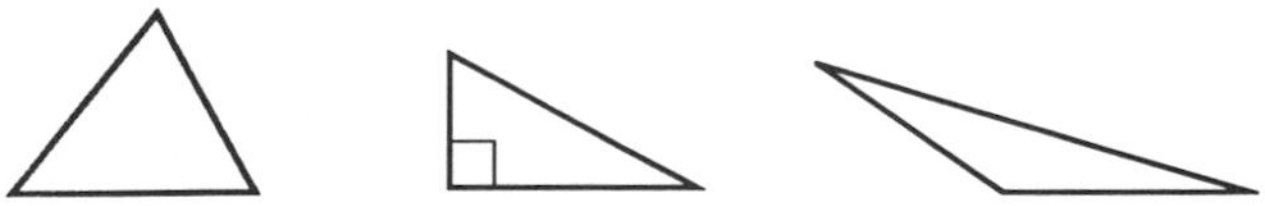

Acute Triangle Right Triangle Obtuse Triangle

Egyptian Triangle

A right triangle with the sides of 3, 4, and 5 units is known as an **Egyptian triangle**.

Leg Base
Hypotenuse
Greek, *hypoteínoysa*, "stretching under"

The two equal sides of an isosceles triangle are known as the **legs**, the third side being the **base**. In a right triangle, the side opposite the right angle is the **hypotenuse**, the other two sides being the legs.

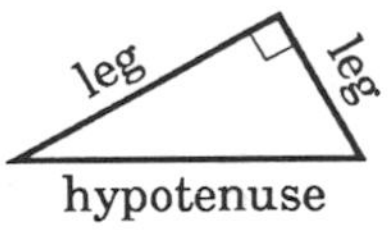

Median

A **median** of a triangle is a straight line from a vertex to the midpoint of the opposite side:

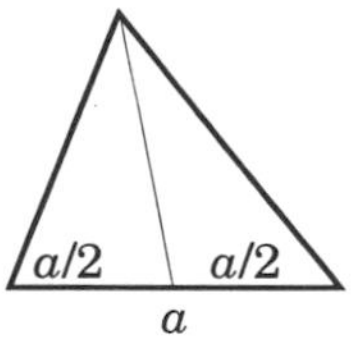

Height, Altitude
Normal, Perpendicular

The **height**, or **altitude**, of a triangle is a **normal**, or **perpendicular**, from a vertex to the opposite side or to its extension. In the triangle ABC below, h_a is the height from vertex A to the extension of side a; h_b is the height from vertex B to side b; and h_c is the height from vertex C to the extension of side c:

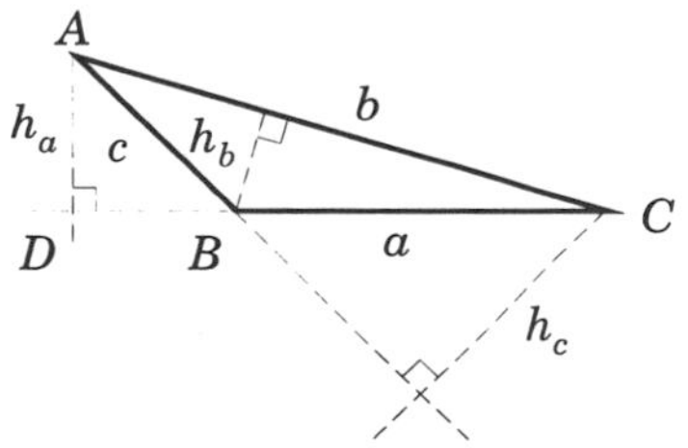

Quadrilaterals

Quadrangle
Tetragon

A **quadrilateral** is a polygon with four sides; other names, less common, are **quadrangle** and **tetragon**.

Trapezoid, Trapezium

A quadrilateral with no two sides parallel is known as a **trapezoid** in the U.K., a **trapezium** (plural: trapezia) in the U.S.; if two sides are parallel, it is a trapezium in the U.K. and a trapezoid in the U.S.

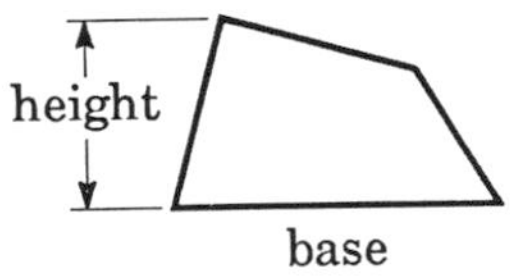

The **base** of a trapezium or trapezoid is generally its longest side, on which it is supposed to stand. If two sides are parallel, they are usually both called bases.

The **height**, or **altitude**, of a trapezium or trapezoid is the perpendicular from the highest point toward the base, or the distance between the two parallel bases.

Leg
Isosceles Trapezoid

The two nonparallel sides are called **legs**. If the legs are of equal length, we have an **isosceles** trapezium or trapezoid:

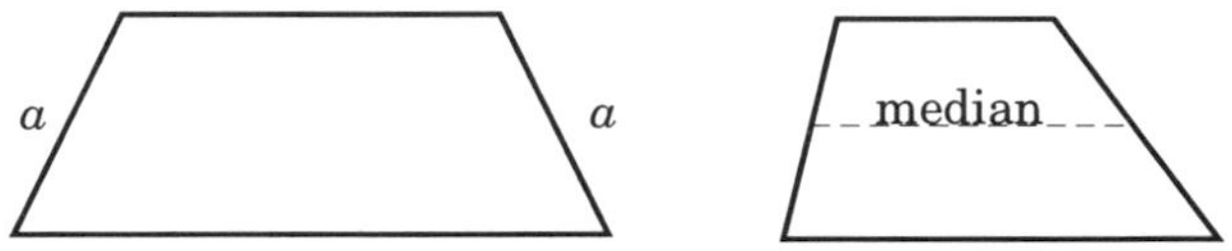

Median

A **median**, or midline, is a straight line segment joining the midpoints of the nonparallel sides.

Parallelograms

A general parallelogram is a quadrilateral in which both pairs of opposite sides are parallel.

Rhomboid

A **rhomboid** is a parallelogram whose adjacent sides are unequal:

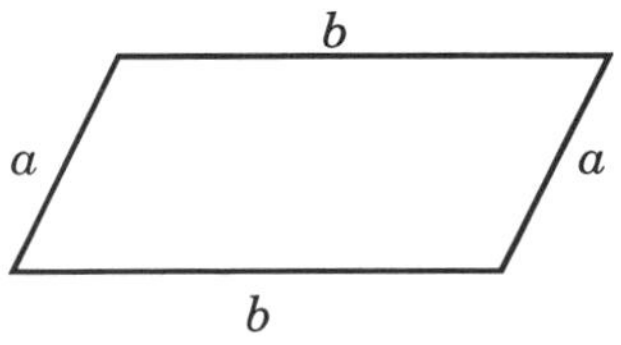

Rhombus

A rhomboid with all sides equal is a **rhombus** (plural: rhombi):

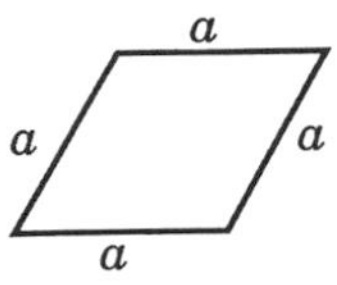

Rectangle
Square

A **rectangle** is a right-angled parallelogram; a **square** is a rectangle with all sides equal – a regular quadrilateral:

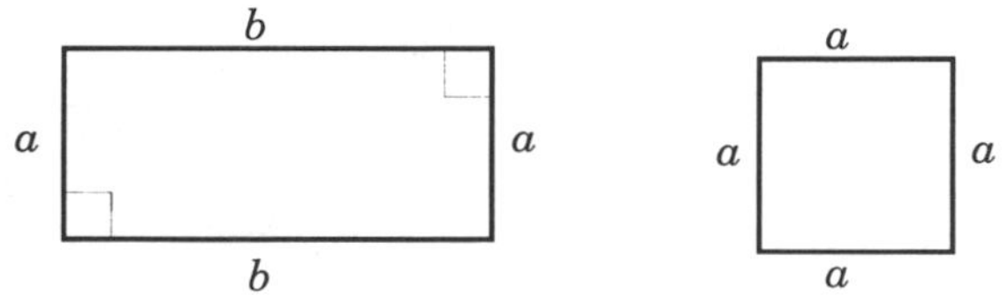

* * *

How would you construct a quadrilateral that, after joining the midpoints of its adjacent sides, includes a parallelogram?

Answer: Any quadrilateral will do. Try, and look for yourself!

Kites and Deltoids

Kites and **deltoids** are quadrilaterals whose adjacent sides are equal in pairs.

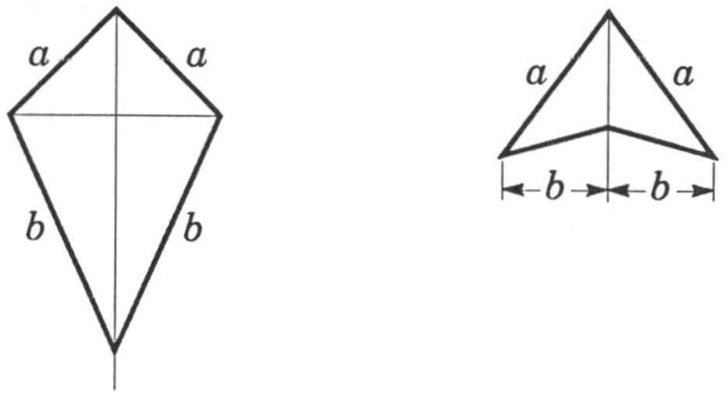

kite deltoid

Kites and deltoids have mirror symmetry, as do rhombi.

Tiling

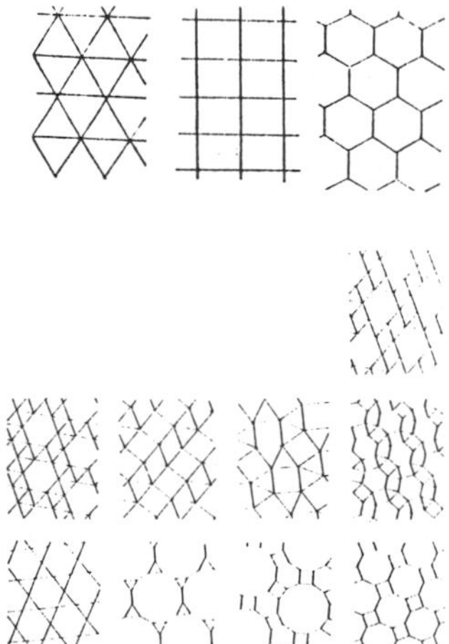

The mathematical study of tiling (tessellation) is concerned with how shapes can be placed to completely fill a plane or space.

The only regular polygons that – one kind at a time – can tile the plane are equilateral triangles, squares, and hexagons (*cf.* honeycombs and the crystal pattern of snowflakes, both originally investigated by Kepler).

By allowing two or more types of regular polygons to meet at each vertex but requiring the vertex configurations to be the same, you allow an additional nine tilings; two of these are mirror images of each other. Allowing more general shapes leads to a wealth of possibilities.

Tiling – in the plane and in space – forms a basis of active and challenging mathematical research with applications in graphics, networks, the study of chemical structure, and geology.

Three-Dimensional Figures

Polyhedra

STÖR Lorenz
(? - c. 1621) p. 373

Title page of Lorenz Stör's *Geometria et Perspectiva* (1567).

Greek, *pöly*, "many"; *hédra*, "base"

A solid whose faces are plane polygons is known as a **polyhedron** (plural: polyhedrons, polyhedra). Polyhedra are named according to the number of faces: a polyhedron with four faces is a tetrahedron; one with six faces is a hexahedron; and so on.

Convex Polyhedron
Concave Polyhedron

A **convex** polyhedron lies entirely on one side of a plane that contains any one of its faces; a **concave** polyhedron has at least one face so located that there are parts of the polyhedron on both sides of a plane containing that face.

Edge
Vertex
Face Diagonal
Space Diagonal

The faces of a polyhedron intersect each other along the **edges**, which meet at the **vertices**, or corners. A straight line joining two vertices of a face is a **face diagonal**; a line between two vertices that are not in the same face is a **space diagonal**.

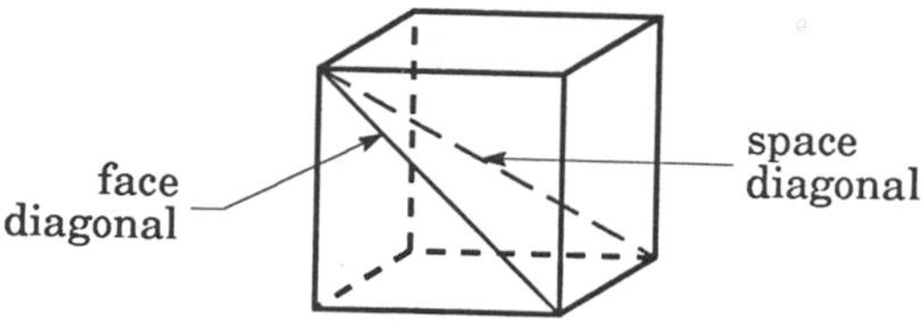

Regular Polyhedra

Regular Polyhedra

The faces of a **regular polyhedron** are identical regular polygons, each of whose vertices are surrounded by congruent arrangements of faces. Taken together, the faces intersect only along their edges and surround a topological ball.

Platonic Solids

PLATON
(427 - 348 or 347 B.C.)

There are five regular polyhedra, all of them convex, often referred to as **Platonic solids** in a tribute to Plato, who used them as metaphors in his cosmology: the tetrahedron, the hexahedron or cube, the octahedron, the dodecahedron, and the icosahedron.

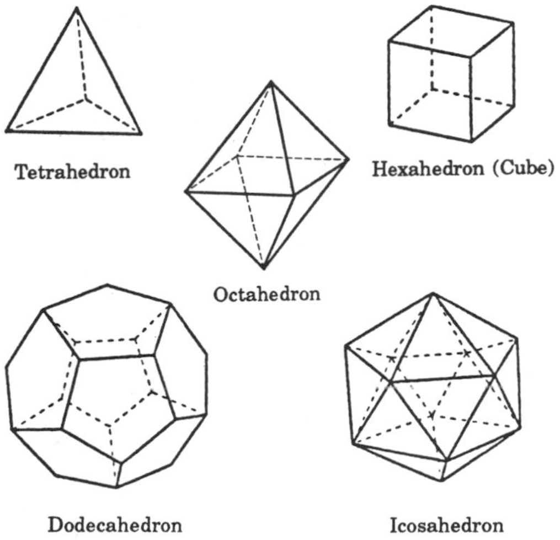

Tiling: *p.* 395

The cube is the only regular polyhedron to completely fill space.

The **Pythagorean brotherhood**, about 500 B.C., was originally aware of only four regular polyhedra – tetrahedron, cube, octahedron, and icosahedron – considered to represent the four basic elements fire, earth, air, and water. Measuring vessels from the Egyptian Old Kingdom (3110 - 2884 B.C.) provide evidence, however, that all five regular polyhedra had been known long before the Pythagorean brotherhood.

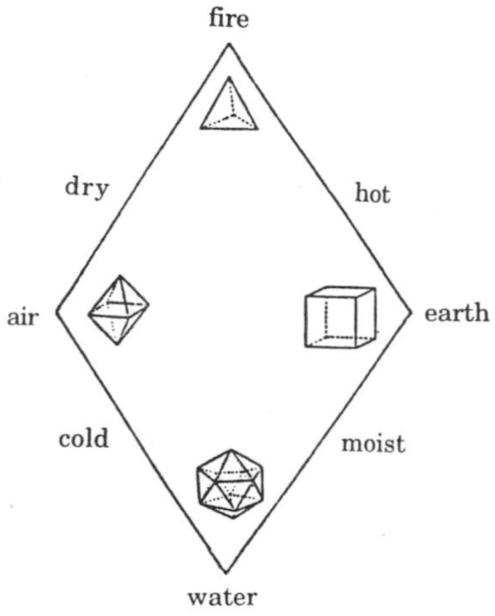

When the Pythagoreans realized that there was also a fifth regular polyhedron, the dodecahedron, they considered it to represent a fifth element (*quinta essentia*) – *aether* or universe.

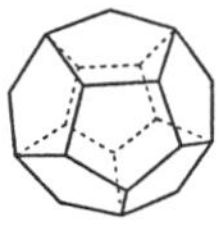

Quinta essentia
dodecahedron

Pythagorean and Platonic ideas continued to rule for centuries – the universe and its components had to be in beautiful and harmonious order, as in the geometric forms of circles and Platonic solids. A notable example is **Johannes Kepler**'s *Mysterium cosmographicum*, published in 1596, in which Kepler described our solar system as having the Sun in the center of six *circular* planetary orbits (only six of the nine planets were known at that time); the Platonic bodies, one within the other, could be fitted between the orbits of the planets.

KEPLER Johannes
(1571 - 1630)

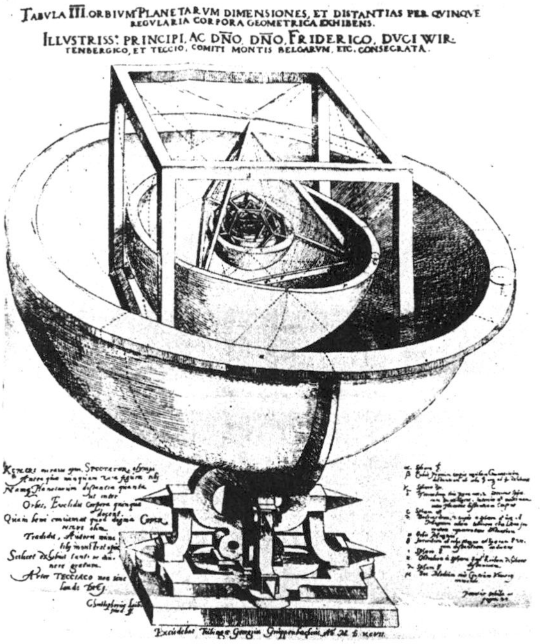

From Johannes Kepler's
Mysterium cosmographicum (Tübingen, 1596).

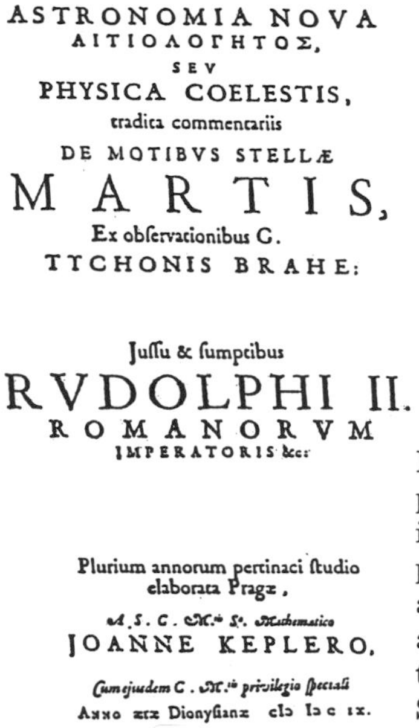

ASTRONOMIA NOVA
ΑΙΤΙΟΛΟΓΗΤΟΣ,
SEV
PHYSICA COELESTIS,
tradita commentariis
DE MOTIBVS STELLÆ
MARTIS,
Ex obſervationibus G.
TYCHONIS BRAHE:

Juſſu & ſumptibus
RVDOLPHI II.
ROMANORVM
IMPERATORIS &c:

Plurium annorum pertinaci ſtudio elaborata Pragæ,
A. S. C. M.tis S. Mathematico
JOANNE KEPLERO,
Cum ejuſdem C. M.tis privilegio ſpeciali
Anno æræ Dionyſianæ clɔ Iɔ c ix.

In his *Astronomia nova* of 1609, Kepler demonstrated that the planet Mars travels in an *elliptic* orbit with the Sun in one of its two foci. He stated that, like the Earth, the "heavenly" planets were material bodies and, thus, did not have to revolve along "perfectly" circular paths. Instead of the previously assumed equal lengths of *arcs* of a circle for equal intervals of time, Kepler found an alternative form of harmony: with the Sun in one focus of the elliptic orbit, the planet would sweep, in equal intervals of time, equal *areas* of the ellipse.

Semi-Regular Polyhedra

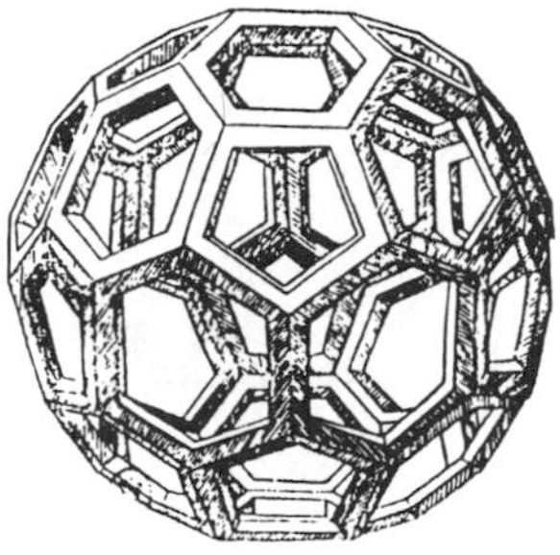

Leonardo da Vinci's illustration of a semi-regular polyhedron in Luga Pacioli's *De Divina Proportione* (1509).
See text above, right.

LEONARDO DA VINCI (Italien painter, sculptor, anatomist, architect, engineer; 1452 - 1519)

Like regular polyhedra, semi-regular polyhedra are bounded by regular polygons, but of more than one kind, generally two kinds, and they can all be inscribed in a sphere.

The polyhedron formed by 12 pentagons and 20 hexagons – a modern *soccer ball* – is illustrated to the left; it was described by Archimedes among a total of 13 semi-regular polyhedra, illustrated below in a reproduction from the *Harmonices mundi* (1619) by Johannes Kepler.

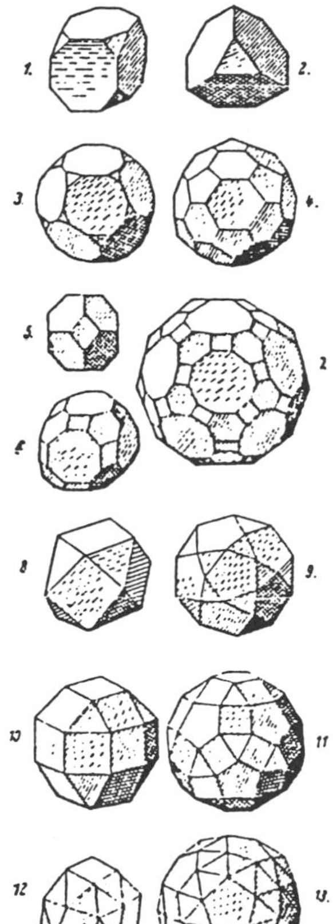

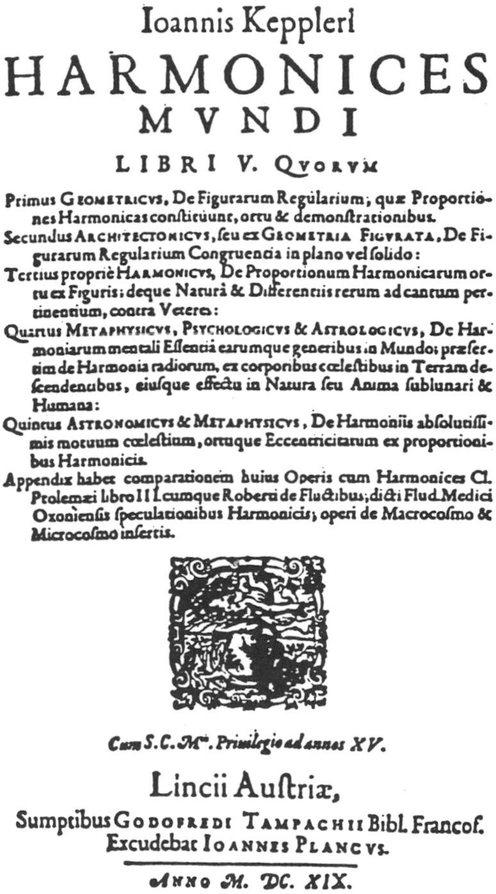

Ioannis Kepleri
HARMONICES
MVNDI
LIBRI V. QVORVM

Primus GEOMETRICVS, De Figurarum Regularium, quæ Proportiones Harmonicas constituunt, ortu & demonstrationibus.
Secundus ARCHITECTONICVS, seu ex GEOMETRIA FIGVRATA, De Figurarum Regularium Congruentia in plano vel solido:
Tertius propriè HARMONICVS, De Proportionum Harmonicarum ortu ex Figuris; deque Naturâ & Differentiis rerum ad cantum pertinentium, contra Veteres:
Quartus METAPHYSICVS, PSYCHOLOGICVS & ASTROLOGICVS, De Harmoniarum mentali Essentiâ earumque generibus in Mundo; præsertim de Harmonia radiorum, ex corporibus cœlestibus in Terram descendentibus, eiusque effectu in Natura seu Anima sublunari & Humana:
Quintus ASTRONOMICVS & METAPHYSICVS, De Harmoniis absolutissimis motuum cœlestium, ortuque Eccentricitatum ex proportionibus Harmonicis.
Appendix habet comparationem huius Operis cum Harmonices Cl. Ptolemæi libro III. cumque Roberti de Fluctibus, dicti Flud Medici Oxoniensis speculationibus Harmonicis, operi de Macrocosmo & Microcosmo insertis.

Cum S.C. M^{tis}. Privilegio ad annos XV.

Lincii Austriæ,
Sumptibus GODOFREDI TAMPACHII Bibl. Francof.
Excudebat IOANNES PLANCVS.

ANNO M. DC. XIX.

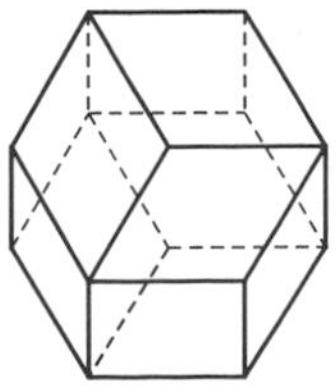

HAÜY René-Just pronounced "Ah oui") (1743 - 1822)

Another type of polyhedron bounded by equal polygons occurs in crystallography. An example is the **rhombic dodecahedron**, described in 1822 by the French crystallographer **René-Just Haüy**. The cube is the nucleus of the rhombic dodecahedron.

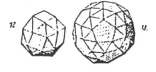

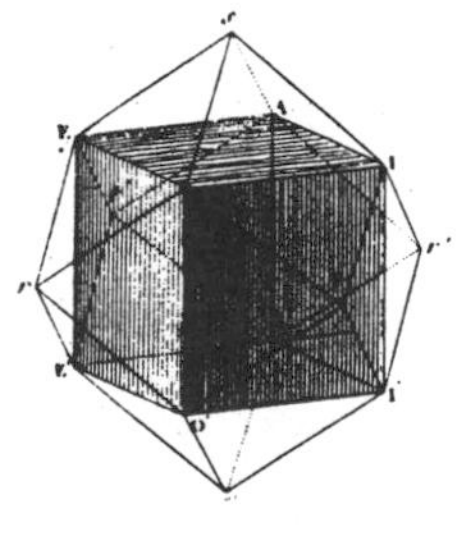

From Haüy's *Traité de christallographie* (1822).

Prisms

Base
Lateral Face

A prism is a polyhedron with two congruent and parallel faces (**bases**), whose remaining faces (**lateral faces**) are parallelograms:

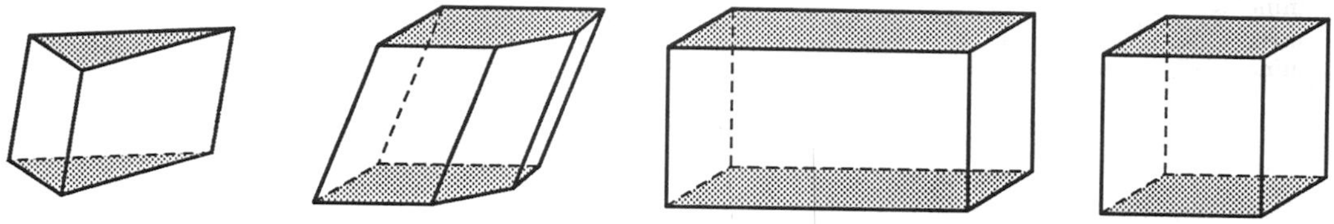

Parallelepiped
Rectangular Parallelepiped
Cube

Prisms are named after the shape of their base faces: a prism whose base is a triangle – for instance, a wedge – is a *triangular prism*; a prism with a pentagon as a base is a *pentagonal prism*; and so on. A **parallelepiped** is a prism with six faces, all parallelograms. A **rectangular parallelepiped** is bounded by six rectangles, a **cube** by six equal squares.

Right Prism
Oblique Prism
Regular Prism

The lateral faces of a **right prism** are perpendicular to the base; those of an **oblique prism** are not. A **regular prism** is a right prism whose bases are regular polygons.

Altitude, Height

The **altitude**, or **height**, of a prism is the perpendicular distance between the bases.

Truncated Prism

A **truncated prism** is the portion of a prism contained between the base and a plane that is not parallel to the base:

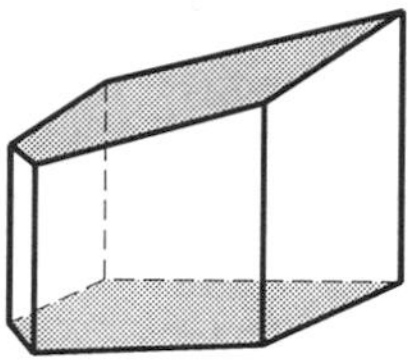

Pyramids

A pyramid is a polyhedron whose **base** is a polygon and whose **lateral faces** are triangles with a common vertex.

Altitude, Height

Frustum
Latin, *frustum*, "a piece"

A pyramid may be triangular, quadrangular, pentagonal, *etc.*, depending on the shape of its base. The **altitude**, or **height**, of a pyramid is the perpendicular distance from vertex to base. A **regular pyramid** has a regular polygon for its base; the height meets the base at its center. A **frustum** (plural: frusta) of a pyramid is the part between the base and any transecting plane parallel to the base.

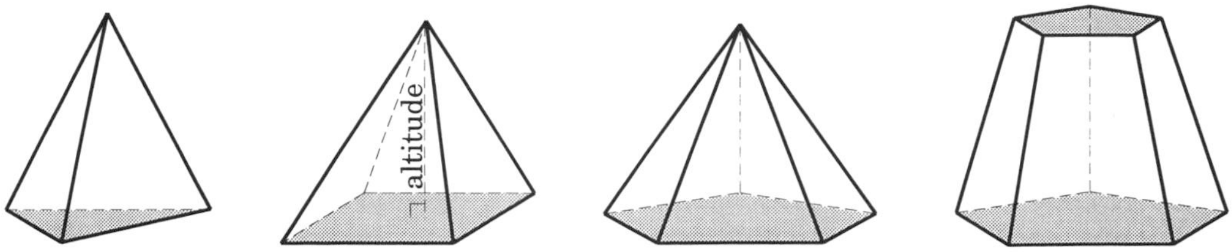

Cylinders

Base
Lateral Surface
Generator
Generatrix
(feminine form of *generator*)
Directrix
(feminine form of *director*)

A cylinder is a solid with three faces. The two **bases** are parallel plane surfaces having the shape of any closed plane curve. The third face, or **lateral surface**, is generated by a moving straight line, the **generator** or **generatrix**, which traces the perimeter of the base, the **directrix**, at the same time keeping parallel to its original direction in space.

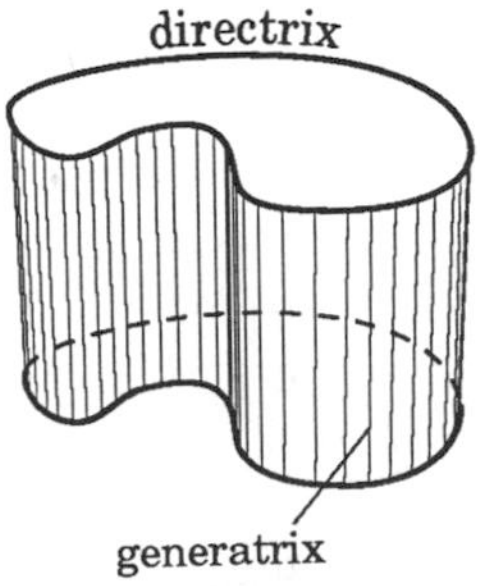

Right Cylinder
Oblique Cylinder

The lateral surface of a **right cylinder** is orthogonal to the bases; other cylinders are **oblique cylinders**. The **height**, or **altitude**, of the cylinder is the perpendicular distance between the bases.

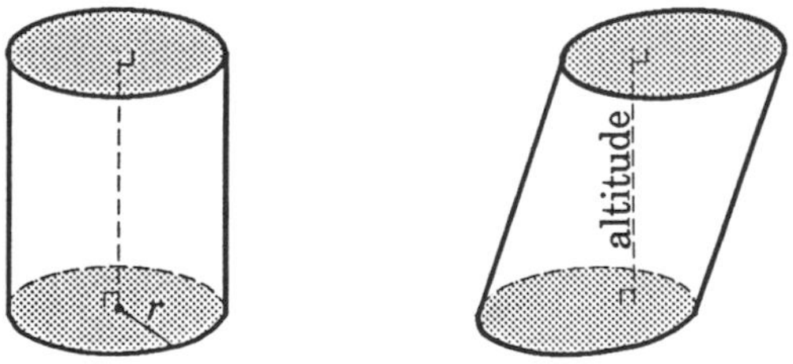

Right Circular Cylinder   Oblique Circular Cylinder

Cylinders are also classified according to the shape of their bases: **circular**, **elliptic**, *etc.* Cylinders may be **hollow**, having one or more holes parallel to the generatrix; a cylinder with one coaxial hole is a pipe, or a tube.

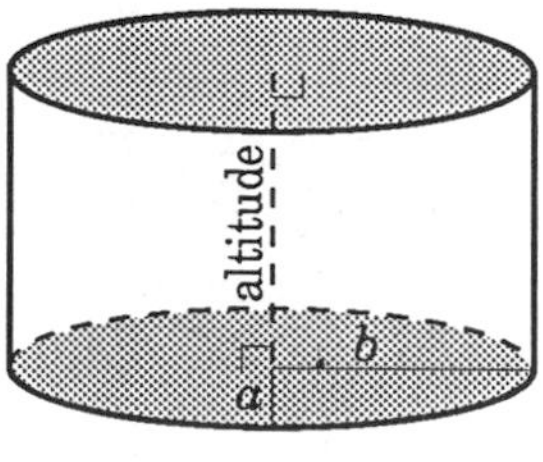

Right Elliptic Cylinder

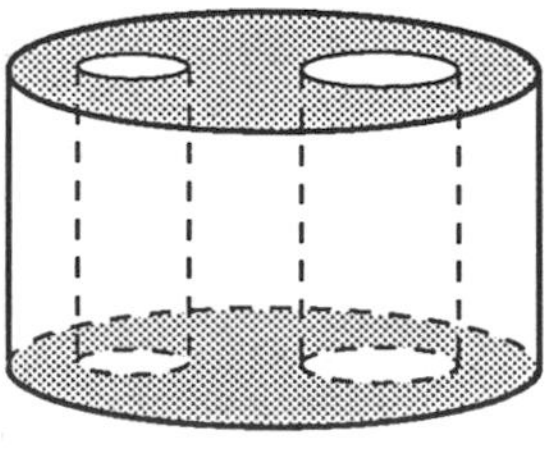

Cones

Generator, Generatrix
Vertex, Apex
Directrix
Nappe
French, *nappe*, "sheet", "surface"

A conical surface is generated by a straight line (the **generator** or **generatrix**) passing through a fixed point (**vertex**, or **apex**) which moves so as to trace a given curve (the **directrix**). The two parts of the conical surface are called **nappes** (singular: nappe) – one on either side of the vertex.

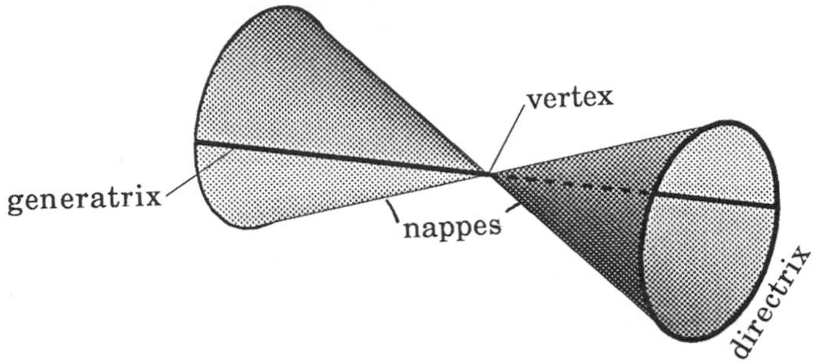

Half-Cone

The *nappe* is also known as a **half-cone**; when there is no risk of ambiguity, it is generally called just a **cone**.

Base
Lateral Surface

A cone is thus bounded by a plane surface, the **base**, which may be any closed curve, the directrix, and a nappe which forms the **lateral surface** of the cone.

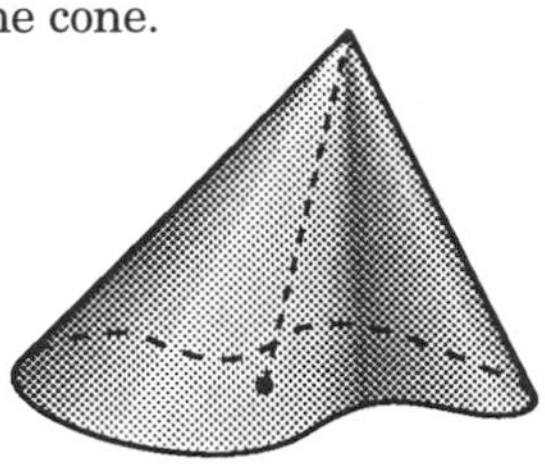

Axis
Height

A line from the vertex of the cone to the center of its base is the **axis** of the cone. The **height**, or **altitude**, is the perpendicular distance from vertex to base (see illustrations below).

Right Cone
Oblique Cone

In a **right cone** the base is perpendicular to the axis; in an **oblique cone** it is not. Like cylinders, cones are classified according to their bases – circular, elliptic, *etc.*

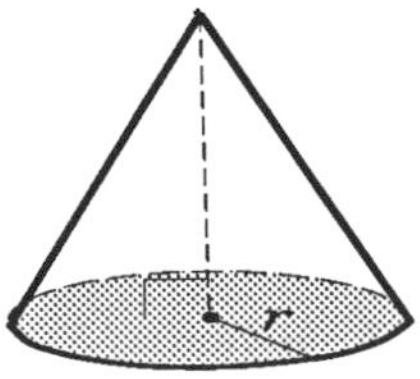

Right Circular Cone

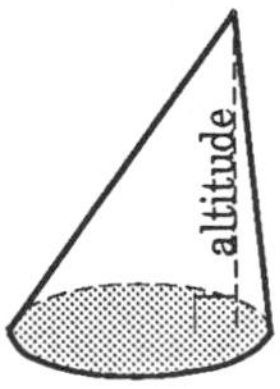

Oblique Circular Cone

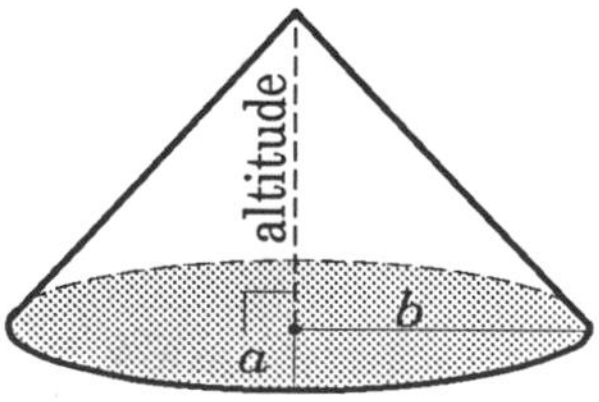

Right Elliptic Cone

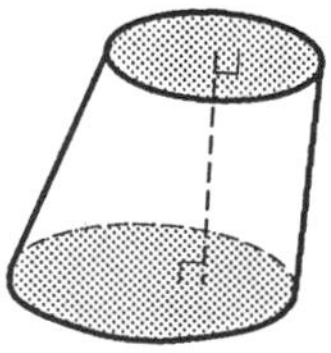

Frustum of a Cone

Frustum
Latin, *frustum*, "a piece"

The **frustum** (plural: frusta) of a cone is a part between the base and a plane parallel to the base.

Solids of Revolution

Axis of Rotation

pp. 401, 402

A plane figure revolving around a straight line in the plane – the **axis of rotation** – generates a solid of revolution.

A **right circular cylinder** and a **right circular cone**, already discussed, are examples of solids of revolution.

Sphere

Radius

A **sphere** is a solid generated by a circle that revolves about its diameter. It has only one surface; all points of the surface are at the same distance – the **radius** – from the **center**.

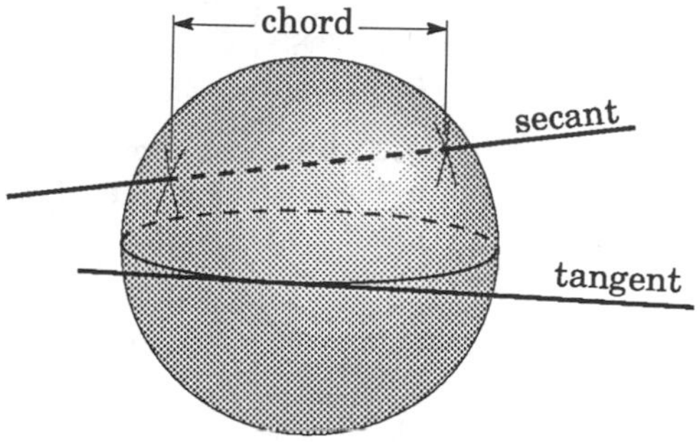

Secant
Chord
Diameter
Radius

A **secant** is a straight line that cuts the surface of the sphere in two points; the portion of the secant that lies between these points is a **chord**. The longest chord, passing through the center of the sphere, is the **diameter**. The distance from the center to the spherical surface is the **radius** of the sphere.

Great Circle
Small Circle

If a plane transects a sphere through the center, the intersection will be a **great circle**; an off-center intersection constitutes a **small circle**.

Tangent

A **tangent** to the sphere is the tangent of the periphery of a great circle.

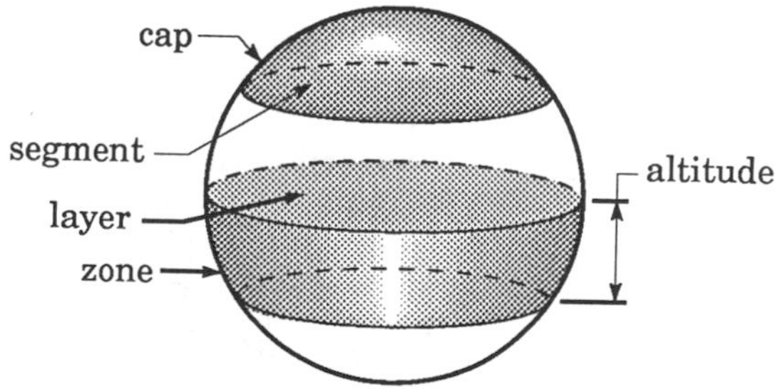

Spherical Segment
Spherical Cap
Diametral plane
Hemisphere

A plane that cuts a sphere divides it into two **spherical segments** and its surface into two **spherical caps**. If the plane is a **diametral plane**, cutting along a great circle, the segments are **hemispheres**.

Spherical Layer
Spherical Zone

Parallel planes divide the sphere into several segments; a segment between two parallel planes is a **spherical layer**. The spherical part of its surface is a **spherical zone**.

Altitude, Height,

The thickness of a spherical segment, or layer, measured at right angles to the transecting planes, is its **altitude**, or **height**.

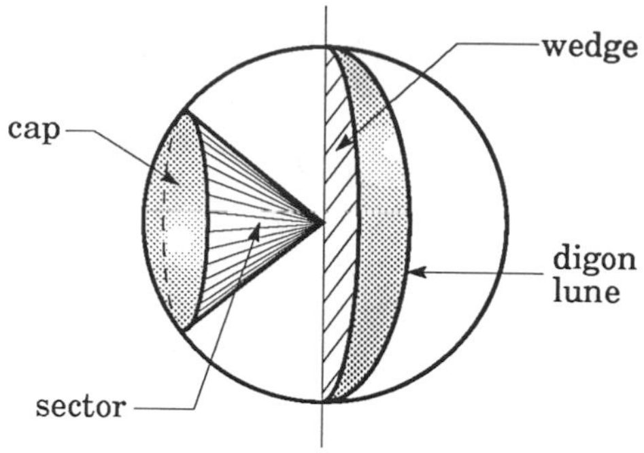

Spherical Wedge
Lune, Digon

Two diametral planes that form an angle divide the sphere into two **spherical wedges**, opposite wedges being congruent. The surfaces of wedges are **spherical lunes,** or **digons** (*cf.* German *Zweieck*, "two-corner").

A radius that moves along a circle on the surface of a sphere generates a **spherical sector**, consisting of a spherical segment and a right circular cone. If the contour is a great circle, the sectors become *hemispheres*.

Spherical Polygon
Semilune

A part of a spherical surface bounded by arcs of three or more great circles is a **spherical polygon**; a special form is the **spherical triangle** – and the **semilune** is a special form of spherical triangle.

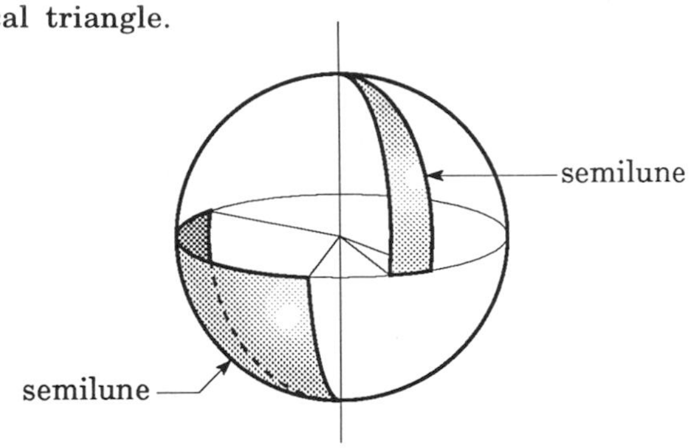

The sum of the angles of a spherical triangle is always greater than 180°, and may be as high as 540°; a spherical triangle may have two or even three right angles or obtuse angles. A spherical triangle with two right angles is called **birectangular**, one with three is **trirectangular**.

Three mutually perpendicular planes passing through the center of a sphere divide the surface of the sphere into eight trirectangular spherical triangles.

Spherical Excess

Spherical excess is the sum of the angles of a spherical triangle, less 180°.

Ellipsoid

Prolate Ellipsoid

Oblate Ellipsoid, Spheroid

An ellipse rotating about its major axis generates a **prolate ellipsoid of revolution**; if rotated around its minor axis, an **oblate ellipsoid of revolution** will develop. Another term for oblate ellipsoid of revolution is **spheroid**.

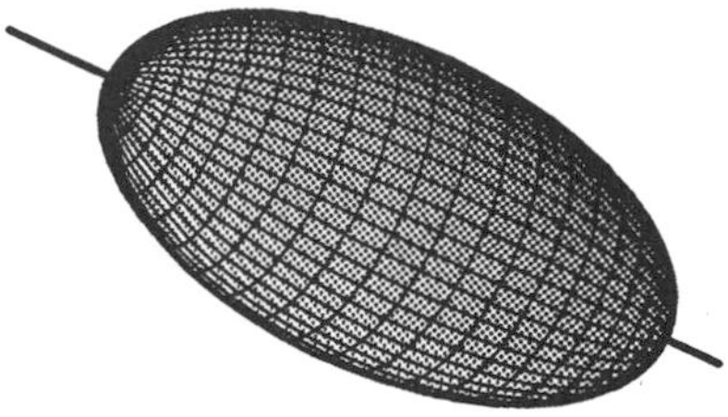

A "geoid" would emerge if the Earth were altered by removing all parts above mean sea level and filling in all gaps below mean sea level. As the Earth is "flattened at the poles", the geoid fits the description of a spheroid.

Paraboloid

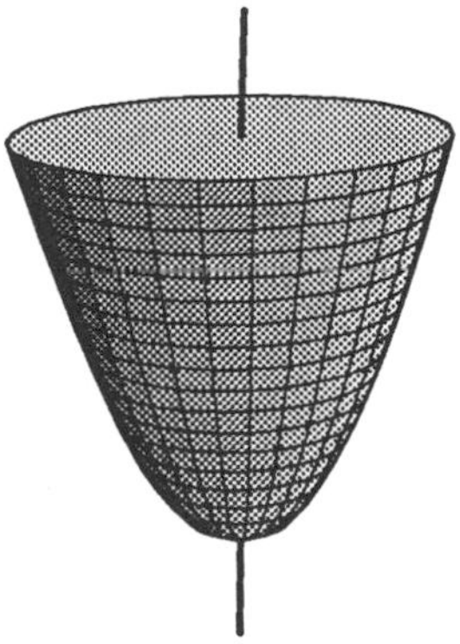

A **paraboloid of revolution** is formed by revolving a parabola about its axis.

Torus

A **torus** (**anchor ring** or doughnut) is generated by a circle rotating on an axis outside it but in the same plane.

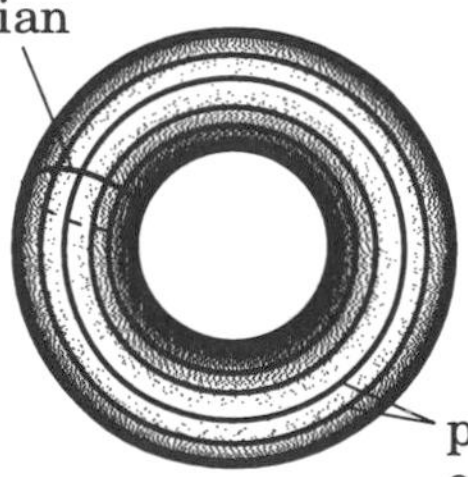

Meridian

Parallel

A plane containing the axis of rotation intersects the surface of the torus, or any other solid of revolution, along **meridians**; a plane at right angles to the axis intersects the surface along **parallel circles**, or **parallels** for short.

Conic Sections

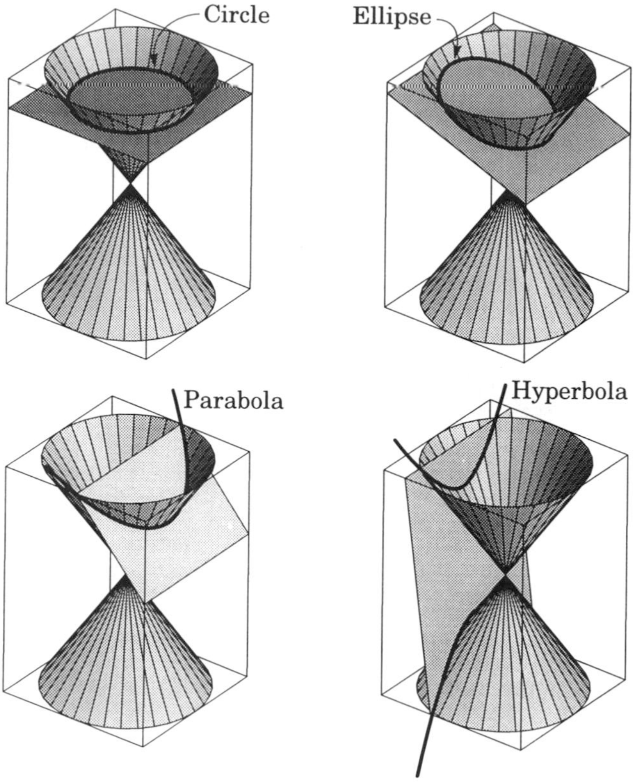

This family of curves may be described by slicing a cone as shown above, or as

Focus
Directrix
Eccentricity

> the locus of all points in the plane whose distances – r from a fixed point, the **focus**; and a from a given straight line, the **directrix** – have a constant ratio.

This ratio is called the **eccentricity** and is denoted e:

$$\frac{r}{a} = e \,.$$

We distinguish four cases, depending upon the numerical value of e:

$e = 0$	corresponds to	a circle
$0 < e < 1$	corresponds to	an ellipse
$e = 1$	corresponds to	a parabola
$e > 1$	corresponds to	a hyperbola

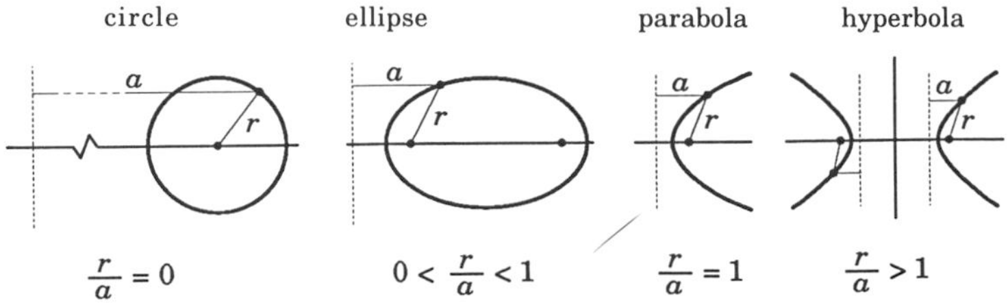

The relation $r/a = 0$ for the circle shows that the directrix is placed at infinite distance $a = \infty$ from the circle.

The ellipse and hyperbola have an extra focus and directrix; for reasons of symmetry, the definition will apply also for them.

Strictly speaking, the parabola also has two foci and two directrices, one pair of which has receded into infinity.

The eccentricity of the circle is zero, $e = 0$. If the eccentricity is allowed to increase from near zero – corresponding to an ellipse closely resembling a circle in shape – the ellipse will continually lengthen until the right-hand side of it vanishes into infinity, and the ellipse turns into a parabola, $e = 1$, with just one open branch.

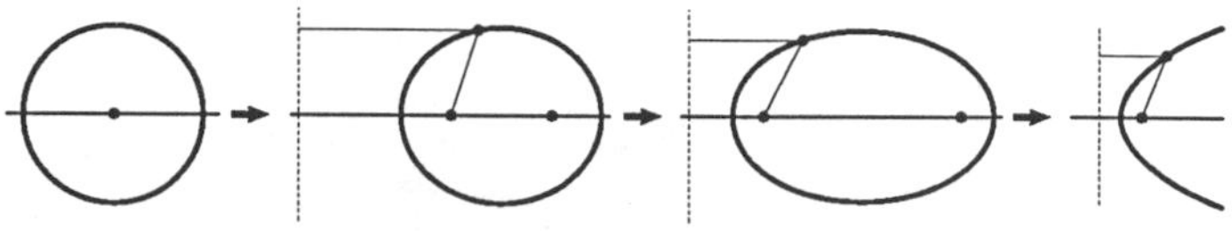

$e = 0$ $e = 1/2$ $e = 3/4$ $e = 1$

When the eccentricity increases still further, the "lost" right-hand end of the ellipse makes a reappearance "from the other side of infinity", as it were, and turns into the left-hand branch of a hyperbola.

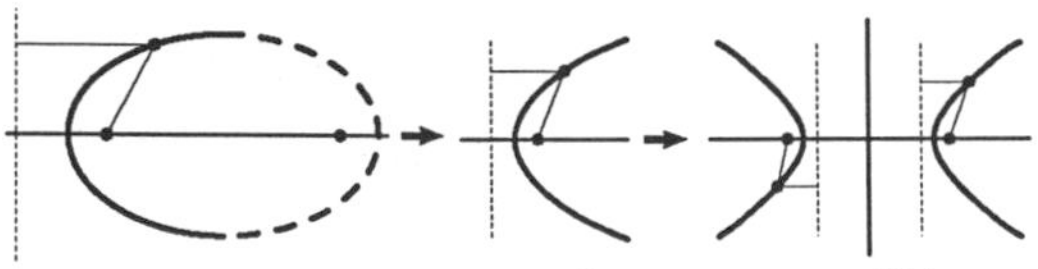

$e = 3/4$ $e = 1$ $e = 5/4$

We can formulate another set of definitions for the curves, spelled out individually:

The curve of	is the locus of
– the circle	– all points that have the same distance from a given point, the center of the circle;
– the ellipse	– all points whose distances to two given points, the **foci**, have a *constant sum*;
– the parabola	– all points that have the same distance to a given point, the focus, and a given line, the directrix;
– the hyperbola	– all points whose distances to two given points, the foci, have a *constant difference*.

The definitions given for the ellipse and the hyperbola apply also to the circle and the parabola, if we understand that the two foci of the circle coincide and that the second focus of the parabola is located at infinite distance:

$$\infty \pm r = \infty$$

We come full circle to again view those curves as conic sections – or **conics** – produced by a plane slicing a circular cone at different angles to the axis of that cone.

A plane cutting the cone at an angle		will produce
– orthogonal to the axis	$\Rightarrow$	a circle
– between right angles to the axis and parallel to the generatrix	$\Rightarrow$	an ellipse
– parallel to the generatrix	$\Rightarrow$	a parabola
– between a plane parallel to the generatrix and a plane parallel to the axis	$\Rightarrow$	a hyperbola

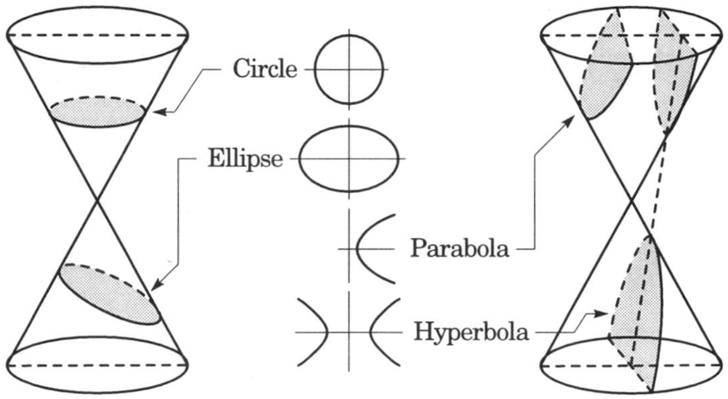

APOLLONIOS
(c. 255 - 170 B.C.)

The name *conic sections* derives from the 8-volume treatise *Conics* (*Κωνικά*) by **Apollonius** of Perga, renowned mathematician and astronomer of antiquity. Of the work, 7 volumes are still extant (Books I - IV in the original Greek, Books V - VII in Arabic translations); some ideas of book VIII may be gained from the encyclopedic work of Pappus (4th century). Apollonius is also the originator of the names *ellipse*, *parabola*, and *hyperbola*.

The curves have another geometric property that can be made use of for the reflection of light rays and beams of sound.

The curve of	reflects rays issuing from a/the focus
– the circle	– back to the center of the circle;
– the ellipse	– into the other focus;
– the parabola	– as a parallel outgoing beam;
– the hyperbola	– as if coming from the other focus.

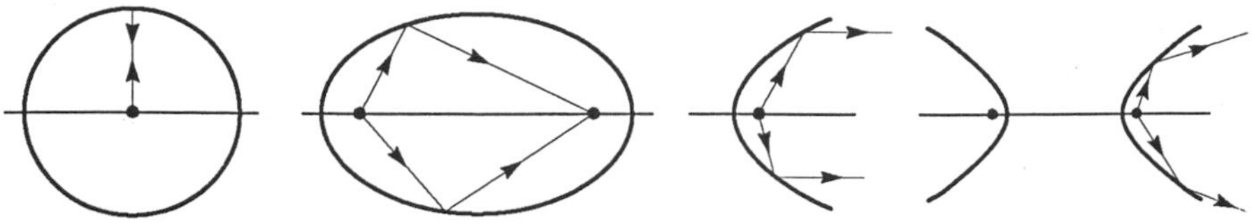

12.2 Units of Measurement

Length, Area, and Volume

Length

meter In the *International System of Units* (SI), the basic unit of **length** (l) is the **meter** (m), defined ("ISO 31-1:1992"):

"The meter is the length of the path travelled by light in vacuum during a time interval of 1/299 792 458 of a second."

1 foot = 12 inches = 0.305 m

1 yard = 3 feet = 0.914 m

1 mile (statute) = 5 280 feet = 1.609 km

1 mile (nautical) = 1.852 km (= 1 minute angle on a great circle)

Area

square meter The derived unit of **area** (A) is the **square meter** (m^2):

"1 square meter is the area of a square with sides of length 1 meter."

1 acre = 43 560 square feet = 4 047 m^2

1 square mile = 640 acres = 2.590 km^2

Volume

cubic meter The derived unit of **volume** (V) is the **cubic meter** (m^3):

"1 cubic meter is the volume of a cube with edges of length 1 meter."

1 000 liters (l) = 1 m^3

1 000 milliliters (ml) = 1 l

1 fluid ounce (British Imperial) = 28.413 ml

1 fluid ounce (U.S.) = 29.573 ml

1 pint (liquid and dry, British Imperial) = 568.245 ml

1 pint (liquid, U.S.) = 0.473 l

1 pint (dry, U.S.) = 0.551 l

1 quart = 2 pints

1 gallon (British Imperial, liquid or dry; U.S., liquid measurement only) = 4 quarts, corresponding to

British Imperial:	4.546 l
U.S.:	3.785 l

Angles

Plane Angles

There are several systems of units currently in use for measuring and expressing plane angles, all of them derived from the division of the full circle.

Radian The **radian** is the standard angular measure in the *International System of Units* (SI):

"1 rad is the angle between two radii of a circle which cut off on the circumference an arc equal in length to the radius."

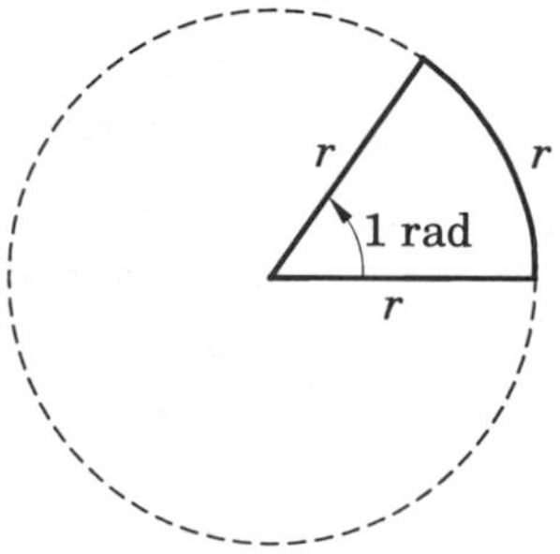

A full angle is 2π radians.

Degree The **degree** is a non-SI unit approved in 1969 by the *General Conference on Weights and Measures.*

An established unit with a history of over 4 000 years, dating from the ancient Babylonian astronomers, the degree is 1/360th of the full circle. It is divided into 60 minutes, each of 60 seconds:

"minutes" = small units
"seconds" = "second-order minutes" or "second minutes" or, briefly, "seconds"

$$1^\circ = 60' ; \quad 1' = 60''$$

For ease of calculation, the degree may also be divided decimally, the degree symbol either taking the place of the decimal marker (*e.g.*, 112°37, or being placed after the last decimal digit, 112.37°).

Gon The **gon**, or **centesimal degree**, is 1/400th of the full angle, a right angle being 100 gons.

The gon is the generally accepted "surveyor's angle ", and is also used in aircraft navigation.

A gon is divided decimally either into 100 **centesimal minutes**, each minute into 100 **centesimal seconds**,

$$1^g = 100^c ; \quad 1^c = 100^{cc},$$

or into **centigons** and **milligons**,

$$1 \text{ gon} = 100 \text{ cgon} = 1\,000 \text{ mgon}.$$

Other names for the gon are **grad** and **grade**.

Radian measure must be used in certain formulas in mathematics and physics to avoid errors or the need for additional scale factors. Otherwise, radians and degrees are at the user's discretion.

$$2\pi \text{ rad} = 360^\circ$$

$$1 \text{ rad} \approx 57.3^\circ$$

- To convert $3\pi/5$ rad to degrees, we have

$$\frac{3\pi}{5} \text{ rad} = \frac{3\pi}{5} \text{ rad} \cdot \frac{360^\circ}{2\pi \text{ rad}} = 108^\circ .$$

- To give 58.26° in degrees, minutes, and seconds, we have

$$0.26^\circ = 0.26 \cdot \frac{60''}{1^\circ} = 15.6'; \; 0.6' = 0.6 \cdot 60'' = 36'',$$

$$58.26^\circ = 58^\circ \, 15' \, 36'' .$$ •

Units are easily converted with an electronic calculator with modes for degrees and radians; results in radians are approximate unless a highly sophisticated calculator giving results in π is used.

In radian measure, the word *radian(s)* or its abbreviation, *rad*, is generally omitted, whereas the symbol °, representing *degree(s)*, is always written.

Solid Angles

The measure of a **solid angle** (Ω) is the area it cuts out on the unit sphere, that is, a sphere whose radius is unity:

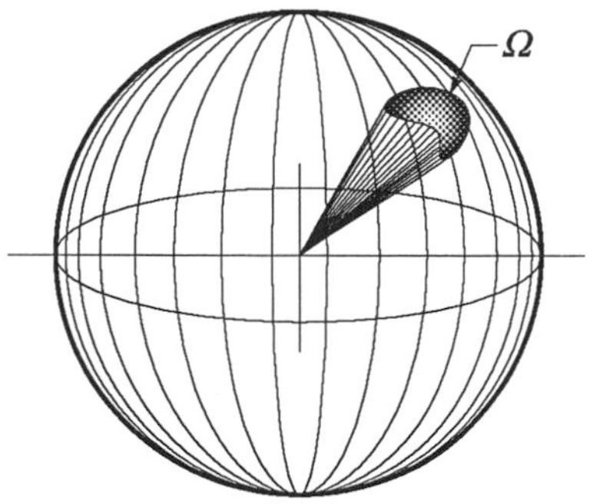

A solid angle need not have any particular shape; it is defined by the conical surface of the nappe enclosing it. Thus, solid angles of equal magnitude may have entirely different shapes.

Steradian (sr)

The SI *unit of measurement* of solid angles is the **steradian** (**sr**), which is a solid angle with its vertex in the center of a sphere and which cuts off from the spherical surface an area equal to the square of the radius of the sphere. The full sphere represents a spherical angle of 4π steradians.

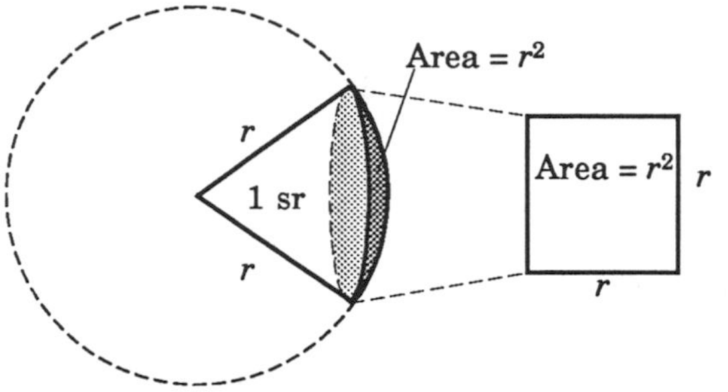

The main importance of the solid angle is to measure, in physics, the fraction of the total emission of radiation from a source in space, whether an electric lightbulb, a radioactive preparation, or the Sun.

- The radius of the Earth is approximately 6 370 km; consequently, the area of the surface it presents toward the Sun is $\pi \cdot 6\,370^2$ km^2. The radius of the Earth's orbit around the Sun – the average distance between the Sun and Earth, or the astronomical unit (AU) – is some 150 million km.

Thus, the solid angle occupied by the Earth is

$$\Omega = \frac{\pi \cdot 6\,370^2}{150^2 \cdot 10^{12}} \approx 5.6 \cdot 10^{-9} \text{ steradian,}$$

and the share of the total energy – light and heat – irradiated by the Sun that is received by the Earth is the ratio of Ω to the 4π steradians of a full sphere:

$$\frac{\Omega}{4\pi} = \frac{\pi \cdot 6\,370^2}{4 \cdot \pi \cdot 150^2 \cdot 10^{12}} = \frac{406}{900} \cdot 10^{-9} \approx 0.45 \cdot 10^{-9},$$

that is, less than one two-thousand-millionth of the total energy emitted by the Sun, but this suffices to sustain us.

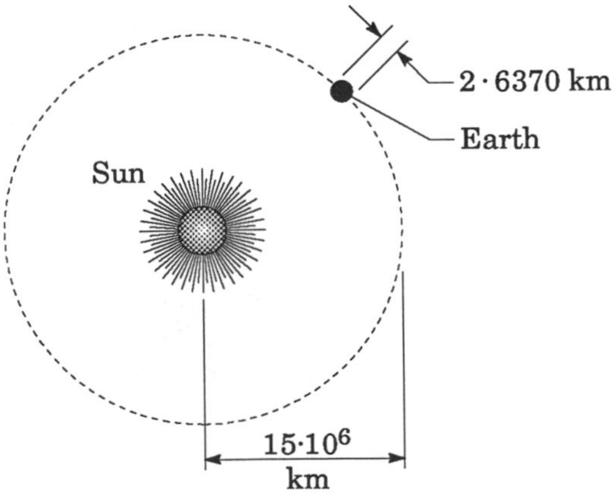

12.3 Euclidean Construction

Construction in Euclid's plane geometry is founded on the use of an unmarked straightedge and a pair of compasses.

Title page from Henric Coets, *Euclidis Elementorum* ... (1705), intended for self-instruction. **Henric Coets** was a mathematics teacher at the University of Leiden, The Netherlands.

COETS Henric
(end 17th century - 1730)

To construct a **line segment equal to a given segment** AB, draw a line l, longer than AB, using a *modern compass*; we thus deviate from the classical method which assumed the compass to be collapsible.

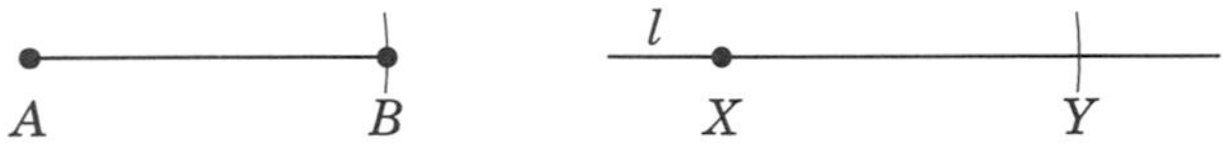

Choose a point X on l and set the compasses for radius AB. Use X as a center and describe an arc intersecting line l at Y.

The segment XY is equal to AB.

To construct the **midpoint of a segment** AB, use the compasses to describe about each of the points A and B circles of equal radius, greater than half the length of AB:

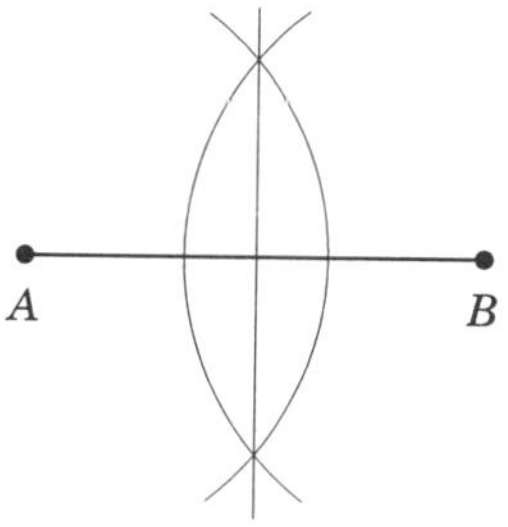

The line through the intersections of the arcs intersects AB at its midpoint and is perpendicular to AB.

The **center of a given circle** is located at the intersection of perpendiculars raised from the midpoint of arbitrary chords.

To construct an **angle equal to a given angle** ABC, draw a ray OY.

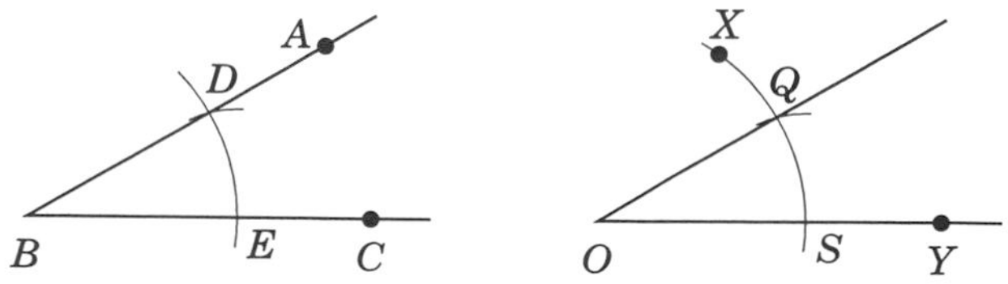

- With B as center, describe an arc of any convenient radius intersecting BA and BC at D and E, respectively.
- With the same radius, and using O as center, describe an arc XS, where S is at the point of intersection with OY.
- Using S as center and with the radius equal to the distance DE, describe an arc that intersects arc XS; call the point of intersection Q.
- Draw a ray from O through Q.

$\angle O$ is equal to $\angle B$.

To construct the **bisector of an arbitrary angle** ABC, use B as a center and any convenient radius to describe arcs intersecting BA and BC; call the points of intersection X and Y. Using X and Y as centers of arcs of any convenient radius, describe intersecting arcs; call the point of intersection Z.

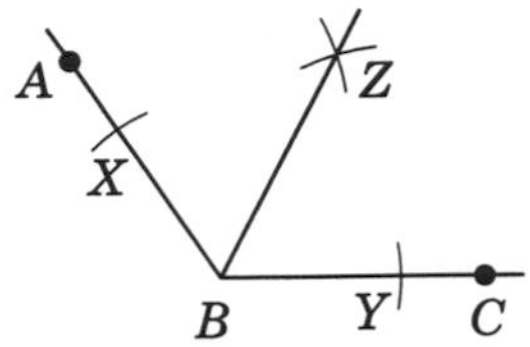

Ray BZ is the bisector.

To raise the **perpendicular to a line** l, at a point A on l, bisect the straight angle whose vertex is at A.

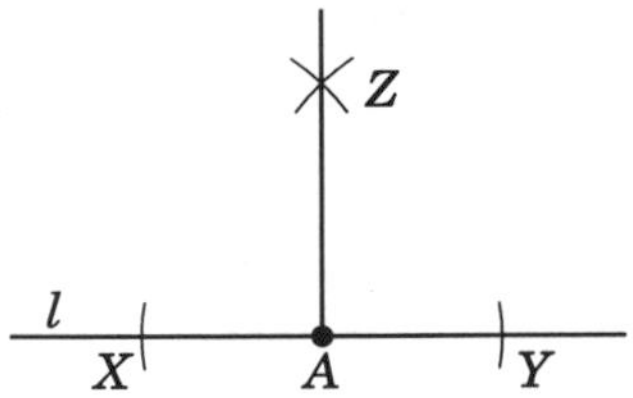

Bisector AZ is perpendicular to l at A.

To construct a **perpendicular from a point** B, **outside a line** l, use B as a center and any convenient radius to describe arcs that intersect l; call the points of intersection X and Y.

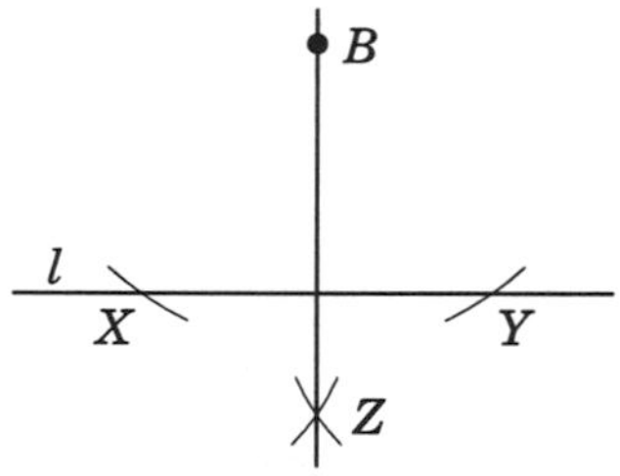

Use X and Y as centers and any convenient radius to describe intersecting arcs; call the point of intersection Z.

Line BZ is perpendicular to l.

To construct the **parallel through a given point** P **to a given line** l, draw through P any line t that intersects l and make angle γ_1 a *corresponding angle* to γ_2.

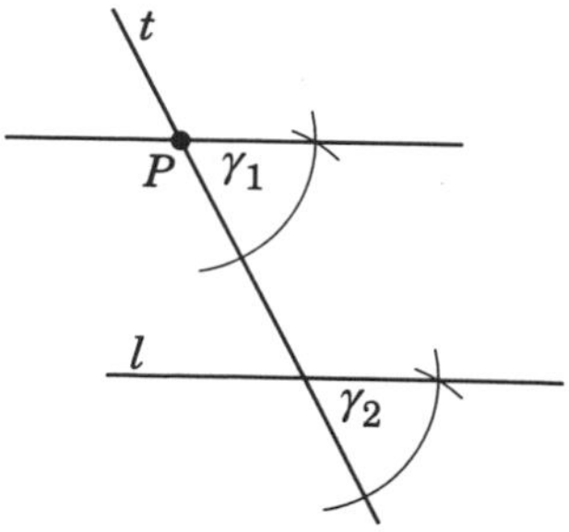

Golden ratio: *p.* 418

To construct an **isosceles triangle with the base angles double the size of the third angle**, divide a line segment AB at a point D in the **golden ratio**. Centered at A, describe a circle of radius AB. Set off chord $BC = AD$.

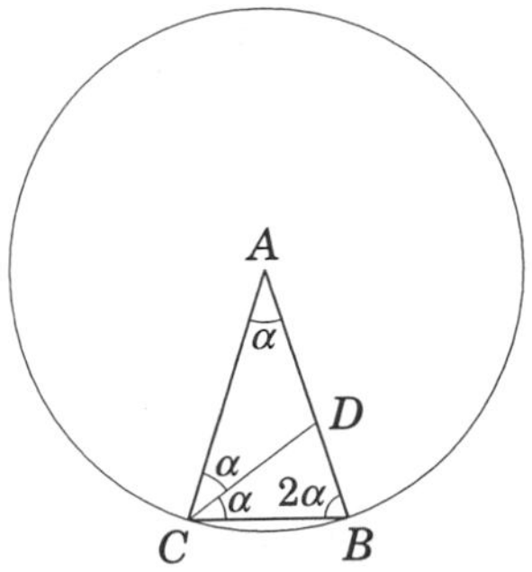

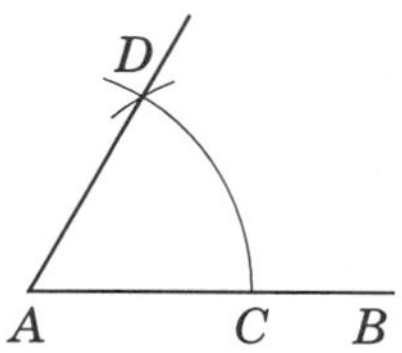

To construct a **60° angle**, start by describing an arc of arbitrary radius about endpoint A of the straight line AB; the arc intersects AB at C. With the radius unchanged, describe an arc about C. The arcs intersect at D.

Angle BAD is 60°, because triangle ADC is equilateral and equiangular.

Regular Polygons

Mathematicians in ancient Greece could construct polygons with 3, 4, 5, and 15 sides using only an unmarked straightedge and a pair of compasses, as described in Euclid's *Elements*. As all angles can be bisected, polygons with even multiples of the above numbers of sides are also constructible, a fact known to Euclid.

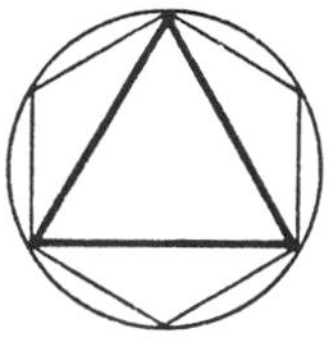

6-gon

The sides of a **regular hexagon** inscribed in a circle are equal to the radius of the circle; copying consecutive chords – with the length of the radius – around the circle will give the sides and vertices of the regular hexagon.

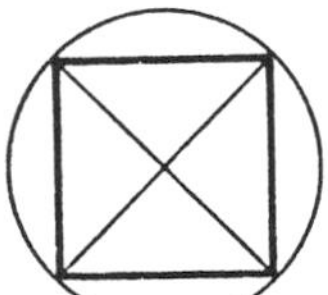

3-gon

Using the straightedge to connect alternate vertices of the hexagon produces an **equilateral triangle**.

4-gon

The corners of a **square** inscribed in a circle are formed by the intersections with the periphery of two mutually perpendicular diameters of the circle.

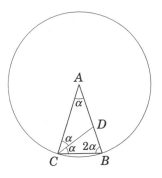

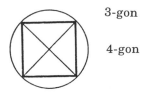

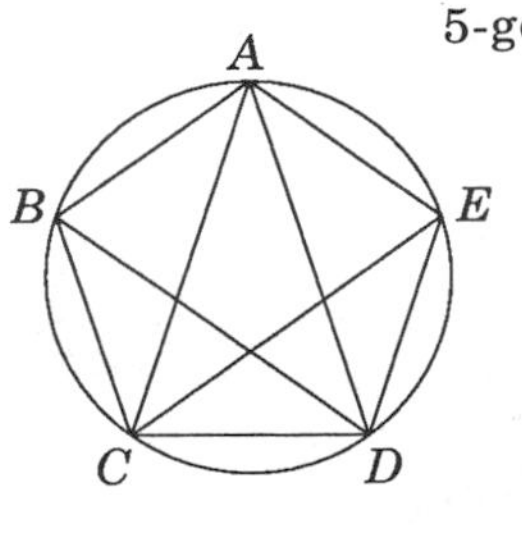

5-gon

To construct a **regular pentagon**, start with an isosceles triangle ACD whose base angles ACD and ADC are double the size of angle CAD (see the top of this page).

Bisect the base angles. Using CD as the radius, describe two circles centered at C and D, respectively, intersecting the bisectors of the base angles; call these points of intersection B and E, respectively. Construct the regular pentagon $ABCDE$ by joining points A and B, points B and C, points D and E, and points E and A.

The regular pentagon may be constructed more easily by connecting every other corner of a regular decagon.

10-gon

To construct a **regular decagon** draw an isosceles triangle whose base angles are double the size of the third angle as previously described. Copy the chord BC around the circle – with end and starting points of consecutive chords coinciding.

p. 415

15-gon

A **regular pentadecagon**, or 15-gon, may be constructed by inscribing in a circle an equilateral triangle ABC and a regular pentagon $AEFGH$.

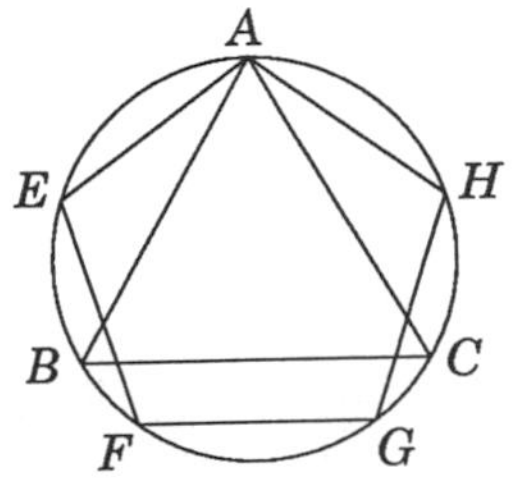

Since arc AF is 2/5 and arc AB is 1/3 of the circumference of the circle, arc BF is the difference $2/5 - 1/3 = 1/15$ and, consequently, chord BF is the length of one side of a regular pentadecagon inscribed in the circle. Copying this chord around the circle – with end and starting points of consecutive chords coinciding – completes the construction of the sought polygon.

BARROW Isaac
(1630 - 1677)

For more than 2000 years, it seemed that the pentadecagon and its even multiples were the acme of Euclidean construction of regular polygons. The English mathematician **Isaac Barrow** expressed himself with a prudent "as yet" at the end of the Fourth Book in his "conveniently portable" edition of Euclid, wisely keeping an eye open toward future achievements.

EUCLID's

ELEMENTS.

The Whole

FIFTEEN BOOKS

Compendiouſly Demonſtrated

By Mr ISAAC BARROW, Fellow of *Trinity College* in CAMBRIDGE.

And Tranſlated out of the Latin.

Καθαρμοὶ ψυχῆς λογικῆς εἰσιν αἱ μαθηματικαὶ ἐπιστῆμαι.
Hierocl.

The *Second Edition*, very carefully corrected.

LONDON.

Printed for *Chriſtopher Huſſey* and *C. P.* in *Little Britain.* MDCLXXXVI.

PROP. XVI.

In a circle given AEBC to inſcribe a quindecagone (or fifteen ſided figure) equilateral and equiangular.

[a] Inſcribe an equilateral pentagone AEFGH in the circle given, and [b] alſo an equilateral triangle ABC, then I ſay BF is the ſide of the quindecagone required.

For the arch AB [c] is $\frac{1}{3}$ or $\frac{5}{15}$ of that periphery, whereof AF is $\frac{2}{5}$ or $\frac{6}{15}$: therefore the remaining part BF is $\frac{1}{15}$ of the periphery. And therefore the quindecagone, whoſe ſide is BF, is equilateral; but it is equiangular alſo, [d] becauſe all the angles inſiſt on equal arches of a circle, whereof every one $\frac{1}{15}$ of the whole circumference. Therefore, &c.

Scholium.

A circle is geometrically divided into parts,
- 4,8,16,&c by 6,4 and 9,1.
- 3,6,12,&c. by 15,4 and 9,1.
- 5,10,20,&c. by 11,4 and 9,1.
- 15,30,60,&c. by 16,4 and 9,1.

Any other way of dividing the circumference into any parts given is as yet unknown; wherefore in the conſtruction of ordinate figures we are forced to have recourſe to mechanick artifices, concerning which you may conſult the Writers of Practical Geometry.

The End of the Fourth Book.

GAUSS Carl Friedrich
(1777 - 1855)

In 1796, **Carl Friedrich Gauss** in his first scientific publication demonstrated that regular n-gons can be constructed with an unmarked straightedge and compasses if

$$n = 2^l p_1 \cdot p_2 \cdot \ldots \cdot p_k ,$$

where l is any positive integer or zero, and the p_i's are distinct primes of the form

$$2^{2^s} + 1,$$

where s is an integer.

The key numbers are the primes of this special form. The first few (s = 0, 1, 2, 3, and 4) are:

3 5 17 257 65 537

On the other hand, $s = 5$, 6, and 7 give composite numbers and do not give us new constructible regular polygons.

17-gon

There are several methods to construct a **regular heptadecagon**, or 17-gon, besides the one suggested by Gauss. One of the simpler methods, due to H. W. Richmond, is shown here.

Describe a circle on the line segment AOB, with O as center:

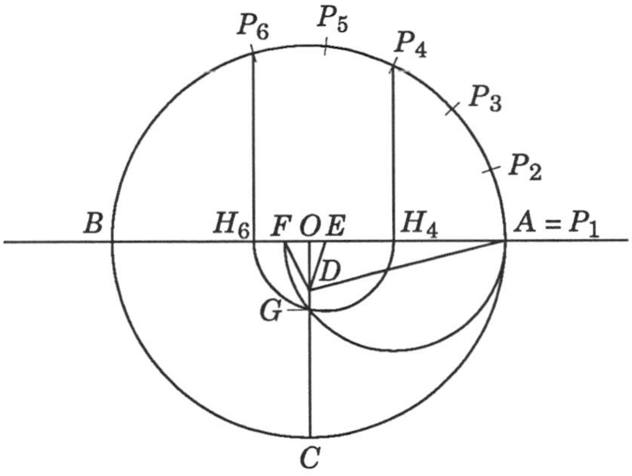

- o Draw the perpendicular radius OC.
- o Bisect the perpendicular OC, and repeat the bisection to produce $OD = \frac{1}{4} OC$.
- o Draw line AD.
- o Bisect angle ODA twice to make angle $ODE = \frac{1}{4} ODA$.
- o Draw a perpendicular to DE at D; bisect the generated right angle to obtain angle $EDF = 45°$.
- o Describe a circle with AF as diameter to intersect OC at G.
- o Describe a circle with E as its center and EG as its radius; this circle intersects diameter AOB at H_4 and H_6.
- o Raise perpendiculars to AOB at H_4 and H_6.
- o These perpendiculars intersect the periphery of the original circle at points P_4 and P_6.
- o P_4 and P_6 are the 4th and 6th vertices of a regular 17-gon, with $A = P_1$; bisect arc P_4P_6 to obtain P_5.
- o Copy chord P_4P_5 around the circle to complete the regular 17-gon.

DISQVISITIONES

ARITHMETICAE

AVCTORE

D. CAROLO FRIDERICO GAVSS

LIPSIAE

IN COMMISSIS APVD GERH. FLEISCHER, IVN.

1801.

Disquisitiones arithmeticae (1801) is one of the great standard works of number theory. The text is primarily concerned with prime numbers, which were shown to cut into geometry and other branches of mathematics in the most surprising way; it includes Gauss's proof of the constructibility of the 17-gon and other regular polygons.

The Golden Section

A line segment that is divided into two segments, a greater a and a smaller b such that the length of $a + b$ is to a as a is to b, is divided in the **golden section** or the **golden ratio**.

We have the relation

$$\frac{a+b}{a} = \frac{a}{b}\,; \quad \left(\frac{a}{b}\right)^2 - \frac{a}{b} - 1 = 0,$$

whose positive solution for $\frac{a}{b}$ is

Golden Number: *p.* 287

$$\frac{a}{b} = \frac{\sqrt{5}+1}{2} = \frac{2}{\sqrt{5}-1}\,.$$

Divine Section
(*sectio divina*)
PACIOLI Luca
(1445 - 1514)

The golden section is also known as the **divine section** after its Latin appellation *sectio divina*, which was first used by the Franciscan monk and mathematician **Luca Pacioli** in his work *De divina proportione*, published in 1509 in Venice.

The golden section was, for about 2500 years, a recognized aesthetic guide in art, even a decree absolute, governing the shape and disposition of drawings and paintings, and in architecture, where the façades of many public edifices were proportioned according to this ratio.

The golden section has nowadays lost most of its appeal to architects and artists.

A golden section may be constructed in several ways; we choose the following.

Given: on a straight line RS, a line segment AB of length a.

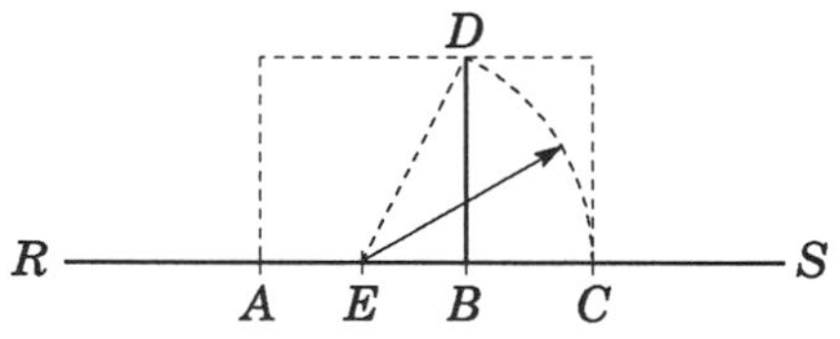

- Construct at B a perpendicular BD of length a.
- Bisect AB at E; with E as center and ED as radius, describe an arc that intersects RS at C. By the *Pythagorean theorem*, the radius here is $\sqrt{5}\frac{a}{2}$.

p. 435

- We now have

$$\frac{AB}{AC} = \frac{a}{\frac{a}{2}\,(1+\sqrt{5})} = \frac{\sqrt{5}-1}{2}\,.$$

- The line segment AC is divided in the golden ratio by the point B.

Golden Triangle

A **golden triangle** is an isosceles triangle whose side is to its base in the golden ratio; its angles are 72°, 72°, 36°. If a base angle is bisected, two isosceles triangles are generated, one of which is a new golden triangle. Using this method, new golden triangles may be generated *ad infinitum*. If arcs are drawn through the vertices as shown below, we obtain an equiangular, or logarithmic, spiral. This is one of an infinite family of logarithmic spirals having various pitches.

p. 580

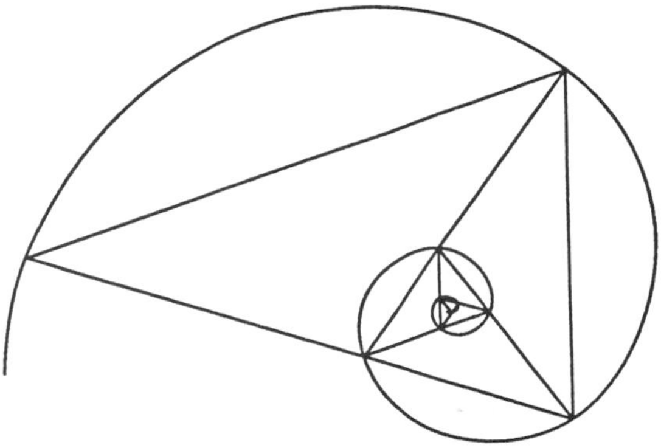

Golden Rectangle

A rectangle whose long side is in a golden ratio to the short side is a **golden rectangle**:

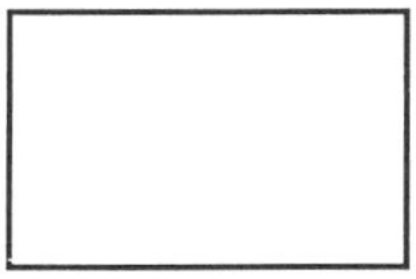

By removing from a golden rectangle a square one side of which coincides with the shorter side of the original rectangle, we generate another golden rectangle. By this method, new golden rectangles may be generated *ad infinitum*. If arcs are drawn between nonadjacent corners of the square as shown below, we obtain an approximation of a logarithmic spiral; the true logarithmic spiral does not touch the sides of the rectangles.

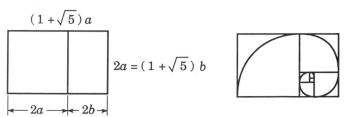

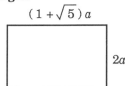

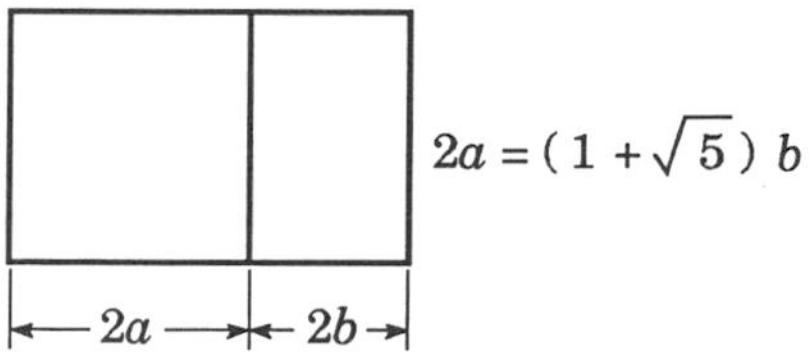

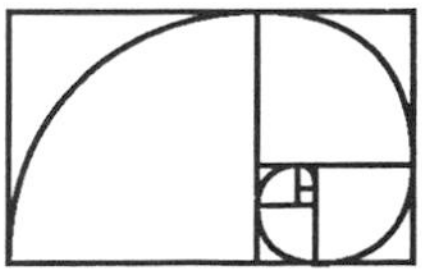

Popular books on "the wonders and joys of mathematics" often haul out the golden rectangle as unique in producing a logarithmic spiral. However, any rectangle can be split into parts, one of which is similar to the whole rectangle. The splitting can be continued *ad infinitum*, giving rise to a logarithmic spiral analogous to the one of the golden rectangle but of a different pitch.

Pentagram, Pentacle

The five-pointed star described by the diagonals of a regular pentagon is a **pentagram**, or **pentacle**, whose central part is another pentagon, the diagonals of which form yet another pentagon, *etc*., in an endless succession of pentagons and pentagrams. The triangles forming the points of the stars are golden triangles. Each of the ratios AC/CD, AB/BC, and AD/AE is equal to $\frac{1}{2}(1+\sqrt{5})$.

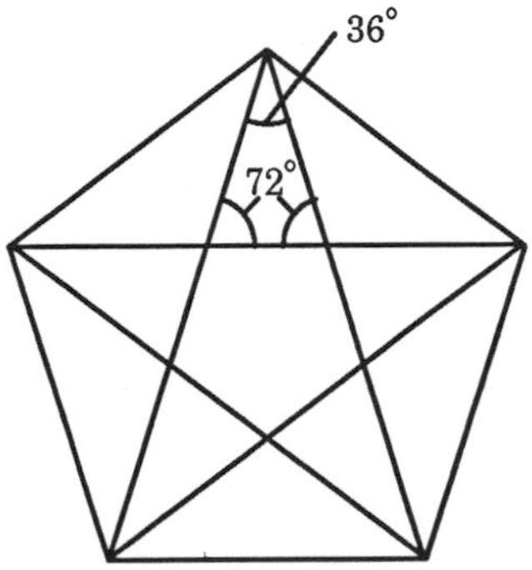

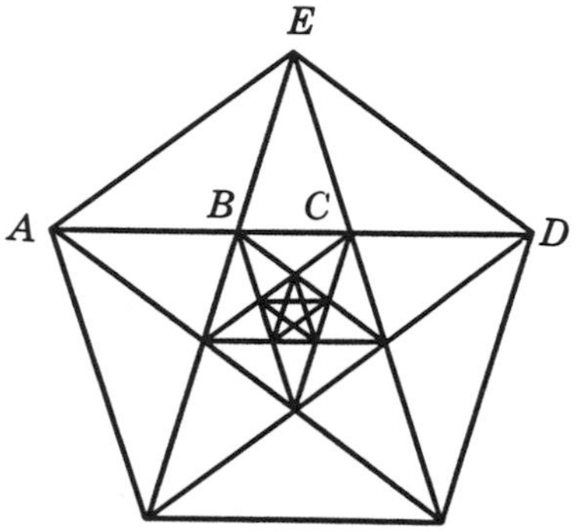

The **Pythagoreans** used the pentagram as a secret identification emblem; later it became a trademark of alchemists and, perhaps because of its repeating properties, a sign of the occult. According to legend, it was used by Doctor Faustus to exorcize Mephistopheles.

A progression of diminishing pentagons and pentagrams, linking vertices together as shown below, is **Pythagoras's lute**; the diagram is replete with lines in golden ratio:

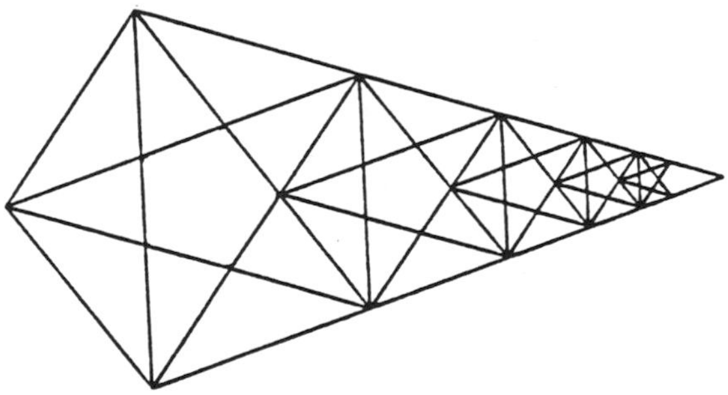

A hexagon is a polygon with six sides. The internal angles of a regular hexagon measure 120°.

By extending the sides of a regular hexagon to the points of intersection, we obtain a **hexagram**, that is, the six-pointed **Star of David**.

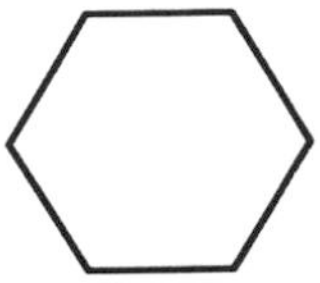

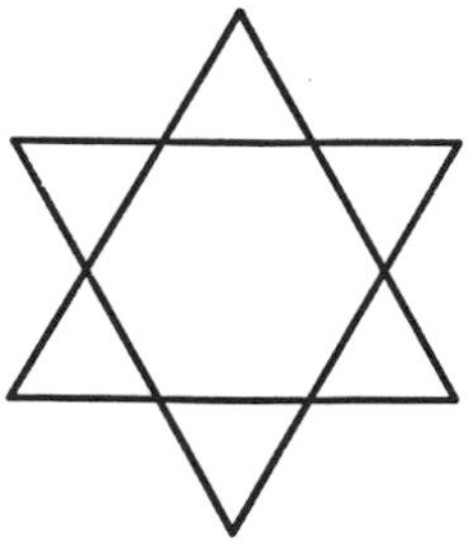

Three Impossible Problems

Three geometrical problems have attracted the interest of mathematicians and geometers since ancient times:

- **Duplication of the cube** – to construct a cube whose volume shall be twice that of a given cube.
- **Squaring the circle** – to construct a square whose area shall be equal to the area of a given circle.
- **Trisection of an arbitrary angle** – to construct an angle that is exactly one-third of a given angle.

These tasks are "impossible problems" since they cannot be solved by Euclidean methods, using exclusively an unmarked straightedge and a pair of compasses. This has not, however, restrained mathematicians from producing solutions, exact and approximate, *by other means*.

Duplication of the Cube

The **Delian problem**, or **duplication of the cube**, is the task of constructing a cube whose volume shall be twice that of a given cube. One of the many legends surrounding the problem tells how the people of Athens, in 430 B.C., appealed to the Oracle of Delos for advice to rid their city of the plague and were instructed to double the size of their cubic altar to Apollo, whose wrath they had supposedly incurred.

If we assume the edge of the altar (cube) to be a units in length, we are faced with the task of solving the equation $x^3 = 2a^3$, and – with $x = b$ – to construct a length $b = a\sqrt[3]{2}$.

If a rational solution of this problem could be found, where a and b are integers with *no factor in common*, we would have that $b^3 = 2a^3$ is an even integer; since the cube of an even integer is even, b must also be even, say, $b = 2p$, which gives $b^3 = 8p^3 = 2a^3$, and

$$a^3 = 4p^3,$$

that is, a^3 is an even integer and, consequently, a is also even, and a and b have a common factor, 2, which contradicts the original hypothesis. We can conclude that the polynomial $x^3 - 2 = 0$ is irreducible over the field of rational numbers; thus, $\sqrt[3]{2}$ is an irrational number.

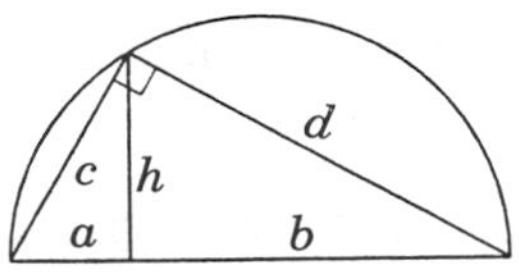

Theorem of Thales: *p.* 430
Pythagorean Theorem: *p.* 435

$$\left.\begin{aligned} b^2 + h^2 &= d^2 \\ a^2 + h^2 &= z^2 \\ d^2 + c^2 &= (a+b)^2 \end{aligned}\right\} h = \sqrt{ab}$$

The barrier to the construction of a length, by the exclusive use of an unmarked straightedge and a pair of compasses, is not the irrationality of a number as such. Given any constructible lengths l_1 and l_2 we can construct their sum and difference, and by proportionality their product. We can also construct $\sqrt{l}$, given l: construct a semicircle (as to the left) with $a = l$, $b = 1$; if $a = 2$ and $b = 1$, we can construct the irrational $\sqrt{2}$. This is, however, all that we can achieve to construct lengths.

$\sqrt[3]{2}$ cannot be constructed from rational numbers by square roots, and so, as proven in 1837 by P. L. Wanzel, the Delian problem is not solvable.

HIPPOCRATES of Chios
(2nd half of 5th century B.C.)

ARCHYTAS of Tarentum
(about 375 B.C.)

Many of the ancient Greeks concerned themselves with the solution of the Delian problem. **Hippocrates** of Chios and **Archytas** of Tarentum (now Taranto, Italy) were the first to demonstrate that the problem could be reduced to the insertion of two mean proportionals x and y between the edge length a of a cube and twice its length,

$$a/x = x/y = y/2a;$$

x is then the sought edge length of the doubled cube.

Squaring the Circle

Squaring the circle is the expression used to describe various procedures for the geometric construction of a square that is equal in area to a circle of given radius.

The problem of constructing a square whose area shall be exactly equal to that of a given circle,

$$a^2 = \pi \cdot r^2,$$

becomes a task of constructing a line segment whose length is proportional to the square root of π,

$$a = r \cdot \sqrt{\pi}.$$

Theorem of Thales: *p.* 430

Pythagorean Theorem: *p.* 435

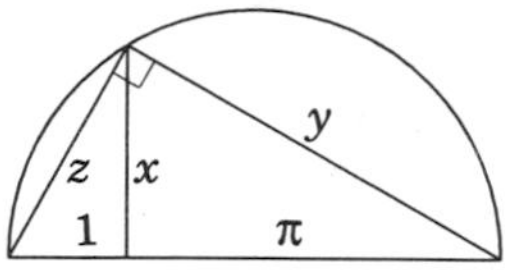

$$\left.\begin{aligned} \pi^2 + x^2 &= y^2 \\ 1 + x^2 &= z^2 \\ y^2 + z^2 &= (\pi + 1)^2 \end{aligned}\right\} x = \sqrt{\pi}$$

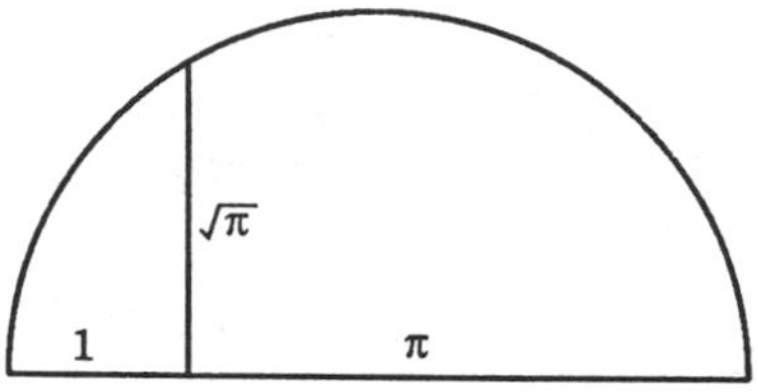

This would be possible if we could describe a semicircle on a diameter compounded of two line segments whose lengths are 1, and π units, respectively. A perpendicular raised at the point where the two sections of the diameter meet will have the length $\sqrt{\pi}$, but classical constructions can only produce algebraic numbers. Since π is a transcendental number, the segment π cannot be constructed, so this approach fails.

Squaring Lunes

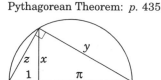

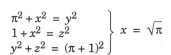

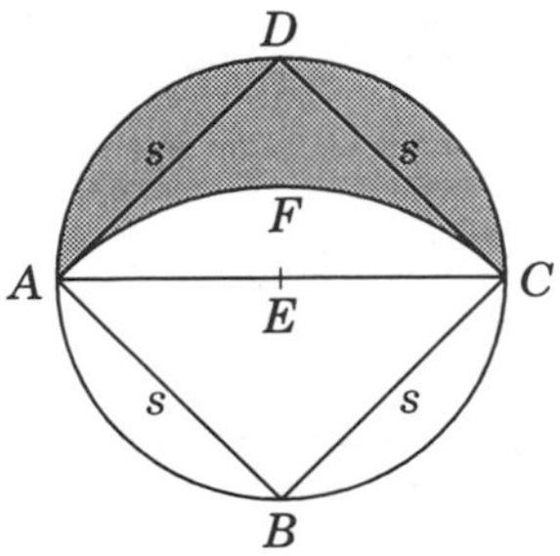

Hippocrates of Chios constructed, in the 5th century B.C., crescent-shaped figures – lunes – whose area, in contrast to that of the circle, may be squared; thus, there exists a square whose area is exactly equal to that of a lune. That particular lunes could be squared gave hope to those engaged in trying to square the circle. The lune to the left is one of several lunes that may be squared.

- Make a square *ABCD* with sides *s*.
- Form a lune *ADC* from intersecting lines of one circle centered at *B* with radius *s* and another circle centered at *E* (midpoint of *AC*) with radius half the length of *AC*.

$$\text{Area}_{\text{lune } ADC} = \text{Area}_{\text{semicircle } ADC} - \text{Area}_{\text{circle segment } AFC} \cdot$$

Pythagorean Theorem: *p.* 435

$\text{Radius}_{\text{semicircle } ADC} = \frac{1}{2} s\sqrt{2}$; and thus:

Area of a Circle: *p.* 442

$$\text{Area}_{\text{semicircle } ADC} = \frac{1}{4}\pi s^2$$

$$\text{Area}_{\text{circle segment } AFC} = \text{Area}_{\text{circle sector } ABC} - \text{Area}_{\text{triangle } ABC} \cdot$$

Circle sector ABC is one-fourth of the circle; thus,

$$\text{Area}_{\text{circle sector } ABC} = \frac{1}{4}\pi s^2; \quad \text{Area}_{\text{triangle } ABC} = \frac{s^2}{2}.$$

We find

$$\text{Area}_{\text{lune } ADC} = \frac{1}{4}\pi s^2 - \left(\frac{1}{4}\pi s^2 - \frac{s^2}{2}\right) = \frac{s^2}{2},$$

which corresponds to the area of a right triangle with legs s, an area equal to that of a square with sides $\frac{1}{2} s\sqrt{2}$, which *can be constructed.*

Trisecting an Angle

Certain specific angles can be easily trisected, but there is no general procedure that permits the construction, with Euclidean tools, of an angle that is exactly one-third of a given, arbitrary angle.

A 90° angle can be trisected, since a 30° angle is easily constructed by using only a straightedge and a pair of compasses. Since an obtuse, or larger, angle may always be split into one or more right angles plus an acute angle, the problem of trisecting an arbitrary angle can be reduced, without loss of validity of the thesis, to the task of trisecting an acute angle. It is then sufficient that we find just one angle that cannot be trisected according to Euclid for the thesis to be proved.

We assume that the angle $3\theta = 60°$ is to be trisected, and jump ahead to information given in Chapter 13 and use the trigonometric identity

p. 511

$$\cos 3\theta = 4\cos^3\theta - 3\cos\theta,$$

where we insert $\cos 3\theta = 60° = \frac{1}{2}$, and substitute

$$\left.\begin{aligned} \cos\theta = \frac{y}{2}; \quad 3\cos\theta &= \frac{3y}{2} \\ 4\cos^3\theta &= \frac{y^3}{2} \end{aligned}\right\}$$

and obtain

$$y^3 - 3y - 1 = 0,$$

an irreducible cubic equation, whose roots cannot be constructed using Euclidean methods.

Therefore, cos 20° cannot be constructed with straightedge and compasses and, hence, the angle 20° is not constructible.

The angle 60° cannot be trisected, and the thesis is proved.

Non-Classical Trisection

ARCHIMEDES
(287 - 212 B.C.)

An interesting method of trisecting an arbitrary angle by non-classical means is attributed to **Archimedes**.

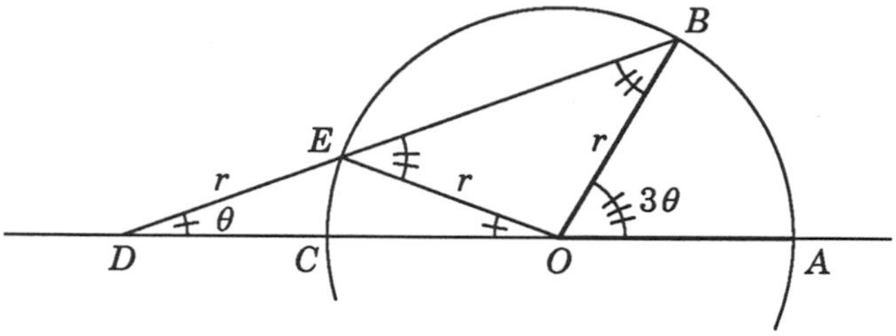

Let AOB be the angle to be trisected.

- Construct a circle with O as its center and with any convenient length as its radius.
- Next, construct a secant through B that intersects the extension of diameter AC produced at a point D and the circle at E, such that DE is equal to the radius of the circle.
  The point D must be found by trial and error; a straightedge is rotated around B until $ED = EO$.
- The angle ADB is then one-third of AOB.

For proof, we observe that base angles ODE and DOE of isosceles triangle DEO are equal and, similarly, base angles OEB and OBE of isosceles triangle EOB are equal. OEB is an external angle to triangle OED and, consequently, equal to the sum of angles ODE and DOE. Analogously, AOB is the sum of OBD and ODB.

Consequently, ADB is the sought one-third of AOB.

As a curiosity, we present the simple instrument below:

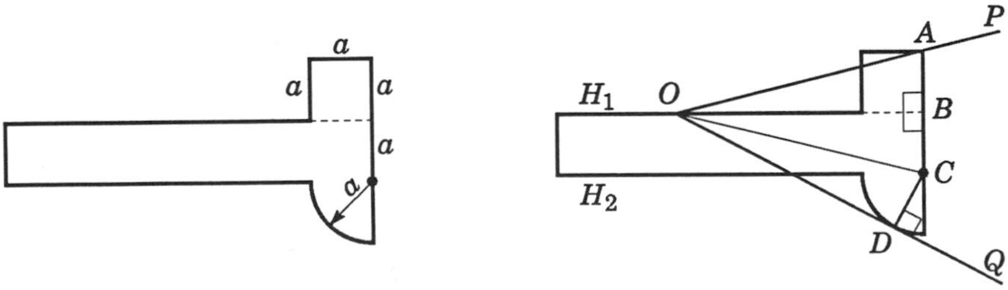

Let POQ be the angle to be trisected. Place the tool with its edge H_1 on the vertex O of the given angle, one leg of the angle passing through the corner point A and the other leg as a tangent to the quadrant at D.

Since the triangles OAB, OCB, and OCD are congruent, the angle POQ, or AOD, is trisected by the points B and C on the extensions of tool edges H_1 and H_2, respectively.

p. 579

Trisecting an angle is also accomplished by use of the **spiral of Archimedes**.

12.4 Theorems and Formulas

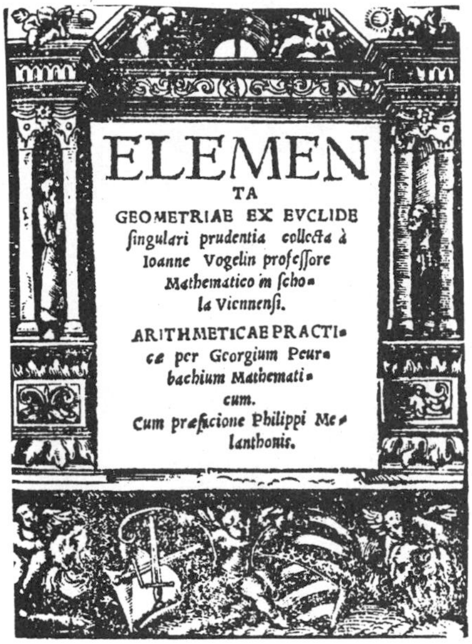
ELEMENTA
GEOMETRIAE EX EVCLIDE
singulari prudentia collecta à
Ioanne Vogelin professore
Mathematico in schola Viennensi.
ARITHMETICAE PRACTICAE per Georgium Peurbachium Mathematicum.
Cum praefatione Philippi Melanthonis.

Johann Vögelin's (? - 1549) *Elementa geometriae ex Evclide*, first published in 1528, was a popular textbook and served as a model for many later texts. This title page is from the 1536 edition (Wittenberg).

Εὐκλείδου στοιχείων
βιβλία ἕξ.
EVCLIDIS
ELEMENTORVM GEO-
METRICORVM LIBRI SEX
conuersi in latinum sermonem à
Ioach. Camerario.
QVIBVS ADIECTÆ SVNT
trium priorum librorum Demonstrationes, atque
editæ in gratiam & vtilitatem studioforum Ma-
thematices in Acad. Lipſ. à
Mauritio Steinmetz Gersb.
Medicinæ Licentiato.
LIPSIÆ
Anno M. D LXXVII.

Joachim Camerarius's (1500 - 1574) *Evclidis* ... was first published without proofs and mainly intended as a Greek primer. This title page is from the 1577 edition (Leipzig), where proofs in Latin had been added to many of the theorems.

This section is principally concerned with theorems from **Euclidean geometry**. Strict proofs are given only for a limited number of theorems; those given are not always of Euclid.

Notations

$\angle$	angle	P	perimeter
Δ	triangle	p	$\frac{1}{2}P$ = semiperimeter
$a, b, c, \ldots$	sides of a triangle, quadrilateral, *etc.*	r	radius of any circle
$\alpha, \beta, \gamma, \ldots$	peripheral angles	r	radius of inscribed circle
θ	central angle of a circle	R	radius of circumscribed circle
A	area		
B	area of the base of a solid	s	arc
C	circumference of a circle	S	surface area (boundary) of a solid
d	segment of a line; diagonal of a polygon; diameter of a circle	S_L	lateral surface area of a solid
h	height (altitude)	S_T	total surface area of a solid
h_c	height to side c		
m	median	t	bisector
m_c	median to side c	t_c	bisector to side c
		V	volume

12.41 Plane Geometry

Angles

complementary angles: *p.* 388

If two angles are **complements of equal angles**, then the two angles are equal.

supplementary angles: *p.* 388

If two angles are **supplements of equal angles**, then the two angles are equal.

Vertical Angles

Vertical angles are equal:

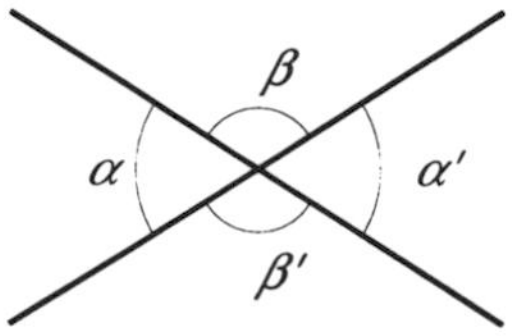

$\alpha = \alpha'$; $\beta = \beta'$

Why? $\alpha + \beta = 180°$ (definition of a straight line);

$\alpha' + \beta = 180°$.

α is supplementary to β; α' is supplementary to β.

Since supplements of the same angle are equal, $\alpha = \alpha'$.

Alternate Interior Angles
Alternate Exterior Angles
Corresponding Angles
Opposite Angles

The pairs of **alternate interior**, **alternate exterior**, **corresponding**, and **opposite angles**, formed by a transversal intersecting a pair of parallel lines, are equal:

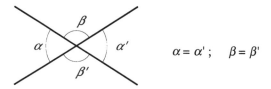

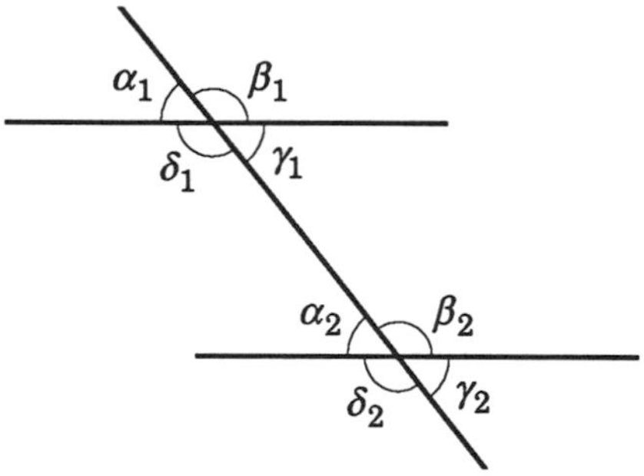

alternate interior angles $\gamma_1 = \alpha_2$; $\delta_1 = \beta_2$

alternate exterior angles $\alpha_1 = \gamma_2$; $\beta_1 = \delta_2$

corresponding angles $\begin{cases} \alpha_1 = \alpha_2; & \beta_1 = \beta_2 \\ \gamma_1 = \gamma_2; & \delta_1 = \delta_2 \end{cases}$

opposite angles $\begin{cases} \alpha_1 = \gamma_1; & \beta_1 = \delta_1 \\ \alpha_2 = \gamma_2; & \beta_2 = \delta_2 \end{cases}$

Congruent and Similar Triangles

Two triangles are **congruent** ($\cong$) if the following elements of one triangle are equal to the corresponding parts of the other triangle:

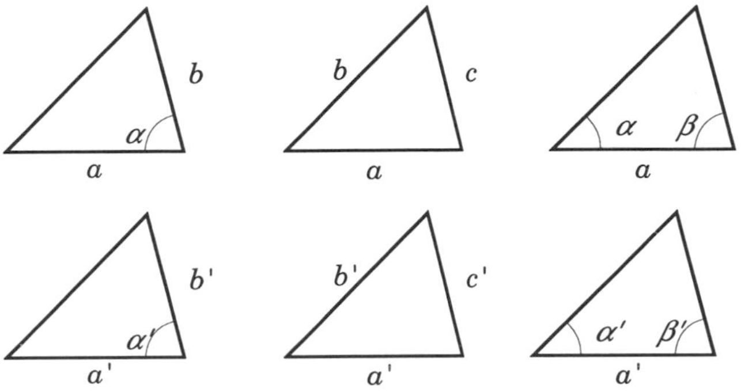

$a = a'$
$b = b'$
$\alpha = \alpha'$

Two sides and the included angle are equal.

$a = a'$
$b = b'$
$c = c'$

Three sides are equal.

$a = a'$
$\alpha = \alpha'$
$\beta = \beta'$

One side and adjoining angles are equal.

Two triangles are **similar** ($\sim$) if any of the following conditions are satisfied:

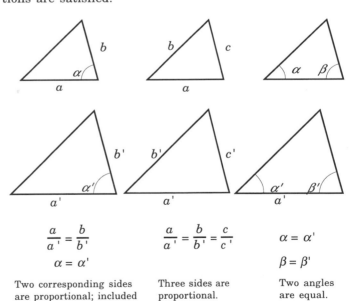

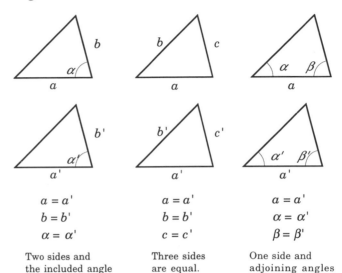

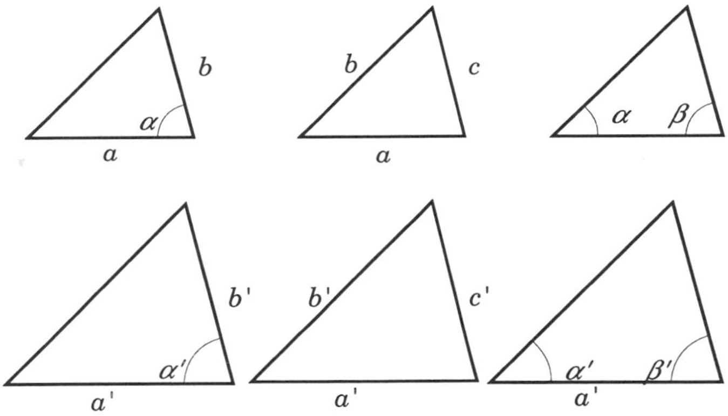

$$\frac{a}{a'} = \frac{b}{b'}$$
$$\alpha = \alpha'$$

Two corresponding sides are proportional; included angles are equal.

$$\frac{a}{a'} = \frac{b}{b'} = \frac{c}{c'}$$

Three sides are proportional.

$\alpha = \alpha'$
$\beta = \beta'$

Two angles are equal.

Base Angles of an Isosceles Triangle

The base angles of an isosceles triangle are equal.

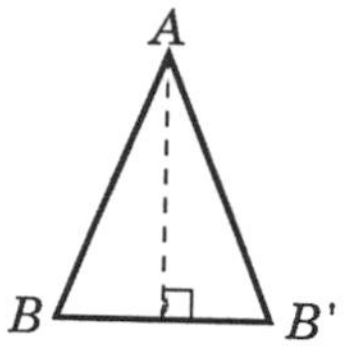

Why? $\Delta ABB' \cong \Delta AB'B$ by two sides and included angle being equal, so $\angle B = \angle B'$.

Sum of Angles of Triangle

The sum of the angles of a triangle is 180°.

Why? ABC is an arbitrary triangle; DE is parallel to BC and has a point coinciding with point A.

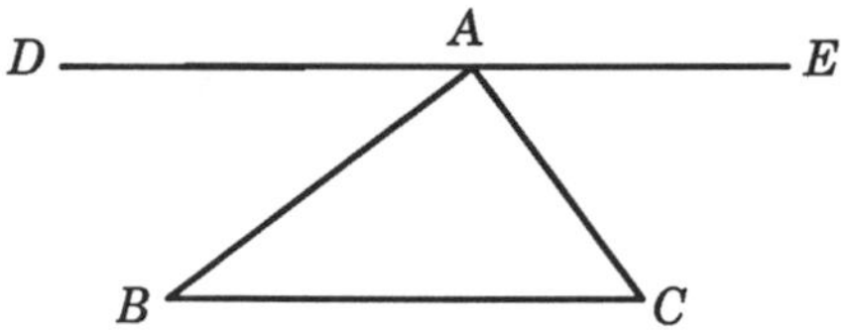

$\angle DAB + \angle BAC + \angle EAC = 180^\circ$, or, since alternate angles $\angle DAB = \angle ABC$ and $\angle EAC = \angle ACB$,

$$\angle ABC + \angle BAC + \angle ACB = 180^\circ .$$

External and Opposite Angles of a Triangle

An external angle of a triangle equals the sum of the two opposite interior angles.

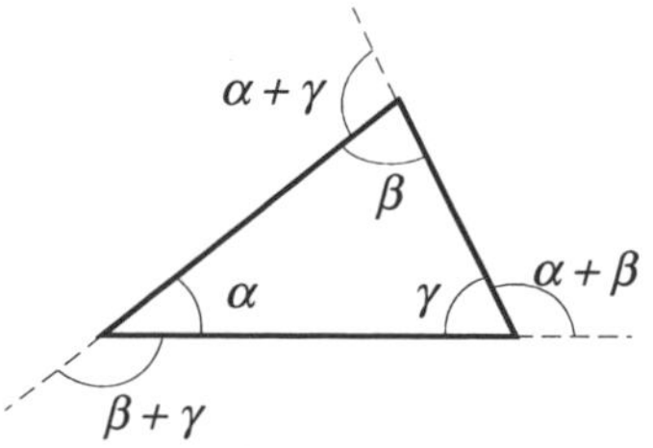

Why? The sum of the angles of a triangle is 180°, as is the angle of a straight line.

Angles Subtended by the Same Arc

If a **central angle** and a **peripheral angle** are subtended by the same arc, then the central angle is twice as large as the peripheral angle.

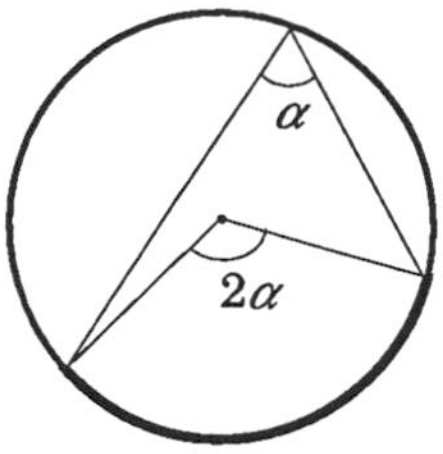

Why? We distinguish three cases:

1. The center of the circle, O, lies on one of the sides of the peripheral angle.
2. O lies between the sides of the peripheral angle.
3. O lies outside the sides of the peripheral angle.

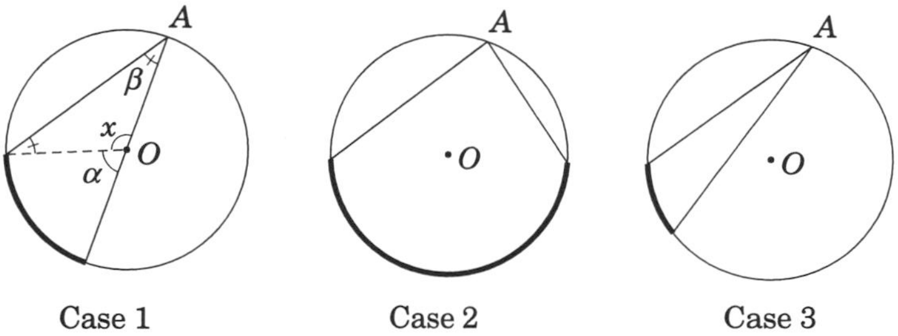

We start with Case 1 and the observation that

$$2\beta + x = 180^\circ = \alpha + x;$$

so the peripheral angle is half the central one.

By drawing diameters AD_2 and AD_3, Case 2 and Case 3 are reduced to Case 1.

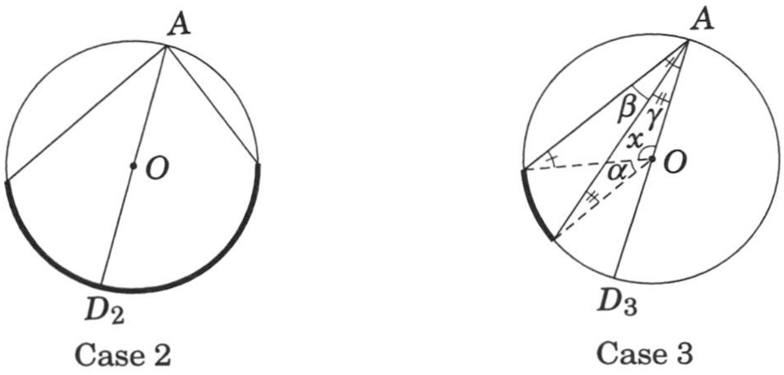

$$\left.\begin{aligned} \alpha + x + 2\gamma &= 180^\circ \\ x + 2(\beta + \gamma) &= 180^\circ \end{aligned}\right\} \quad \alpha = 2\beta$$

From the above we may also deduce that the peripheral angle depends only on the intercepted arc:

> Angles inscribed in a circle and subtended by the same arc are equal.

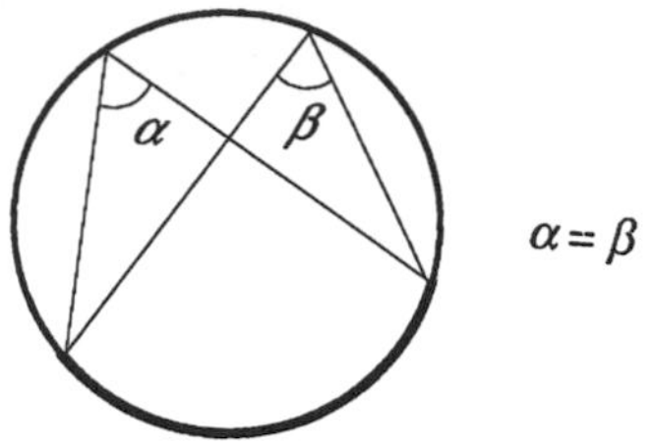

Theorem of Thales

THALES of Miletus (*c.* 624 - 547 B.C.)

Looking at the symmetric case, with the intercepted arc equal to a semicircle, we see that the peripheral angle is 90°, and we have arrived at the **theorem of Thales,** named for the mathematician **Thales** of Miletus:

> Angles inscribed in a semicircle are right angles.

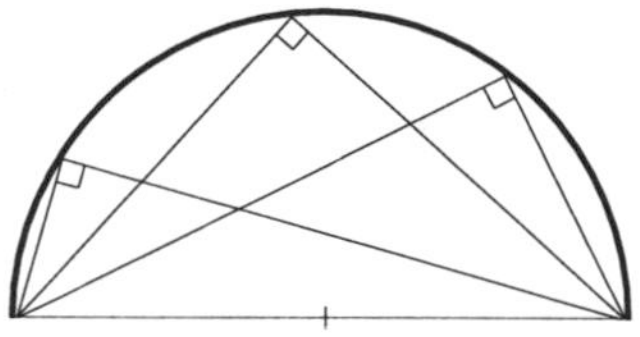

Quadrilateral in a Circle

Opposite angles of a quadrilateral inscribed in a circle are supplementary.

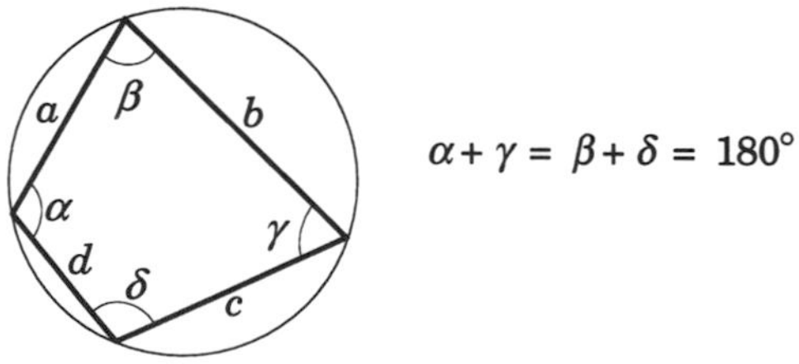

p. 429

Why? This is a corollary of the theorem stating that the central angle is twice as large as the peripheral angle if the two angles are subtended by the same arc.

Angle Subtended by a Chord of a Circle *vs.* Angle Formed between Chord and Tangent

The angle formed between a tangent to a circle and a chord equals the inscribed angle on the other side of the chord and subtended by the chord.

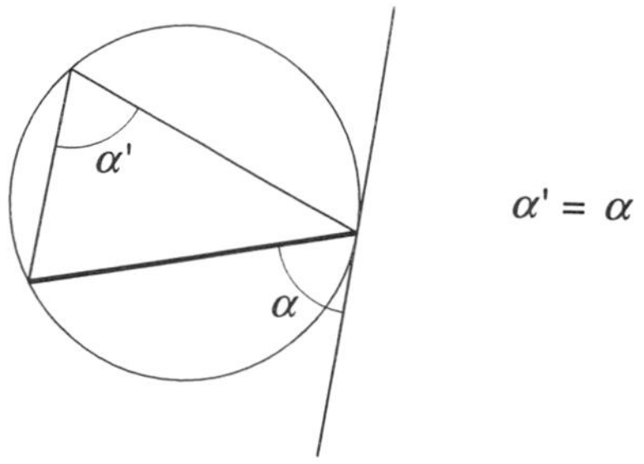

Why? We realize the above by using a limiting argument based on the previous theorem. As C moves toward D, $\bar{T}$ tends to the tangent while $\bar{\alpha}' = \bar{\alpha}$ at all points and $\bar{\alpha}'$ tends to α' while $\bar{\alpha}$ tends to α:

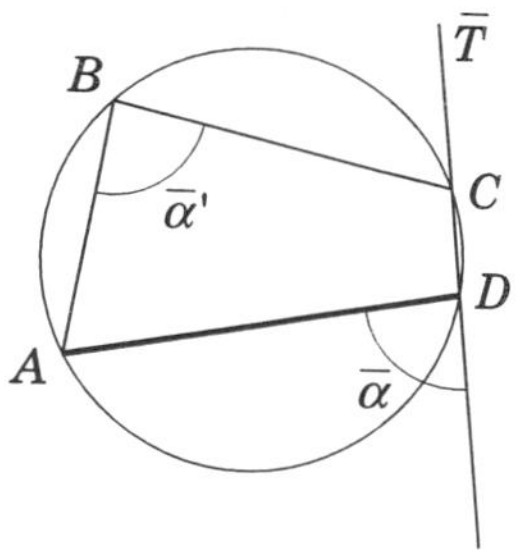

Ellipsoid Reflectors

Radii from the foci of an ellipse to a point on the periphery make equal angles with the tangent to the ellipse at that point.

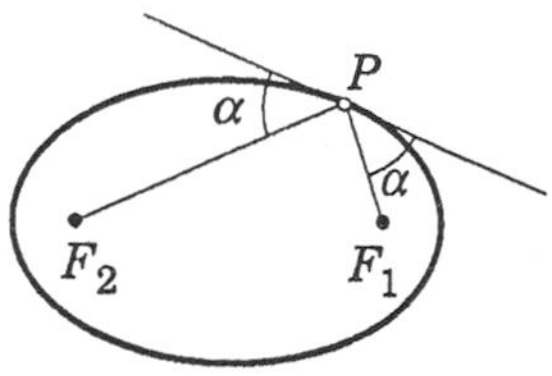

This fact is the basis of ellipsoid reflectors; when emitted at one focus of an ellipsoid, light rays and beams of sound are reflected in such a manner that they converge into the other focus.

Paraboloid Reflectors

The angle between a tangent to a parabola and a radius from the focus to the tangent point is equal to the angle between the tangent and a line parallel to the axis of the parabola that intersects the parabola at the tangent point.

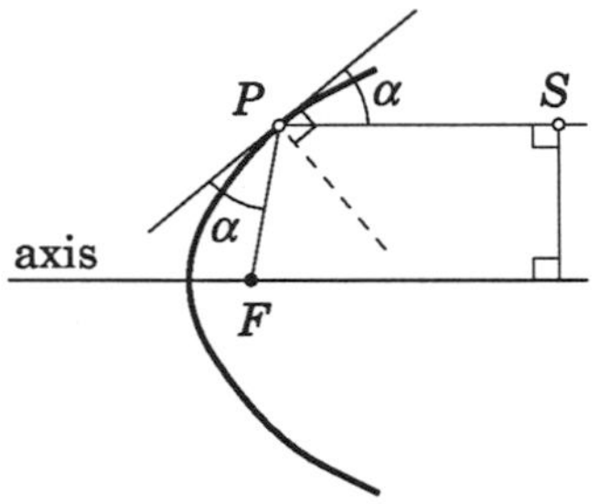

This fact forms the basis of paraboloid reflectors; light rays and beams of sound emitted from the focus of a paraboloid reflector are reflected as a parallel beam. Conversely, a parallel beam received by a paraboloid reflector is converged into its focus; this is the principle behind the reflecting telescope, where the image of the object appears at the focus of the paraboloid, and parabolic satellite disks.

Concurrent Points and Lines

Intersection of Bisectors of a Triangle

The bisectors of the angles of a triangle are concurrent, and intersect at the center of an inscribed circle:

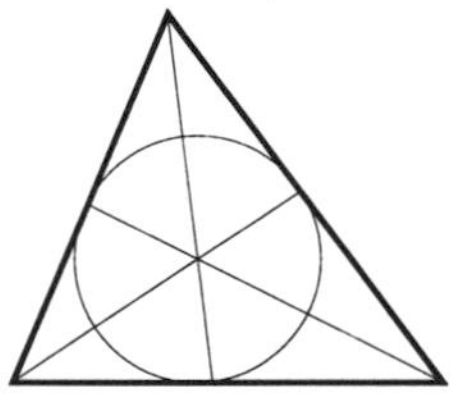

Circumcenter of a Triangle

Normals from the midpoints of the sides of a triangle are concurrent, and intersect at the circumcenter, the center of a circumscribed circle:

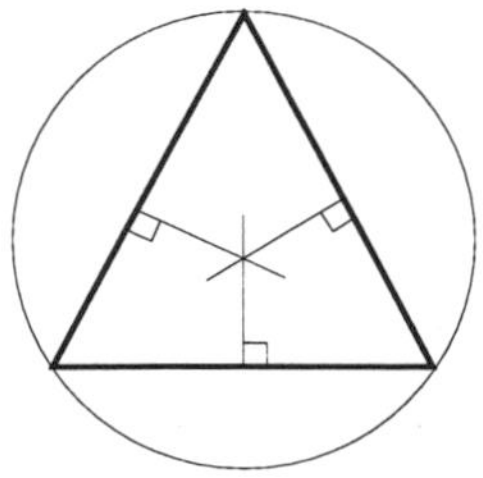

The 9-Point Circle

The midpoints of the three sides, the base points of the three heights, and the midpoints of the line segments between the corners of a triangle and the intersection of the heights are on a circle:

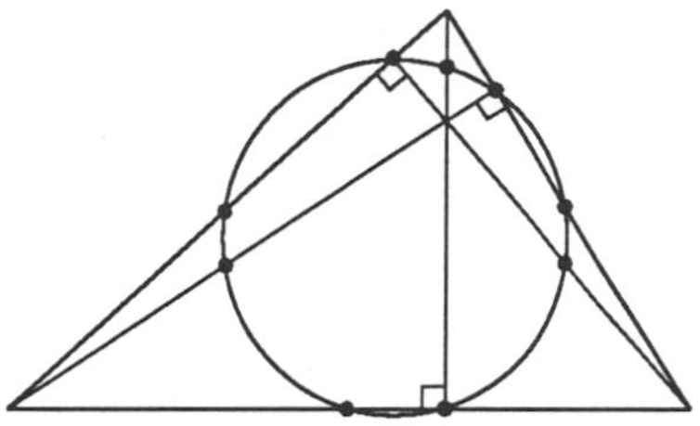

This fact is surprising and is, like the following theorem, one of the high points of triangle geometry.

Euler's Line

In every triangle, the intersection of the medians, the intersection of the heights, and the center of the circumscribed circle are on a straight line.

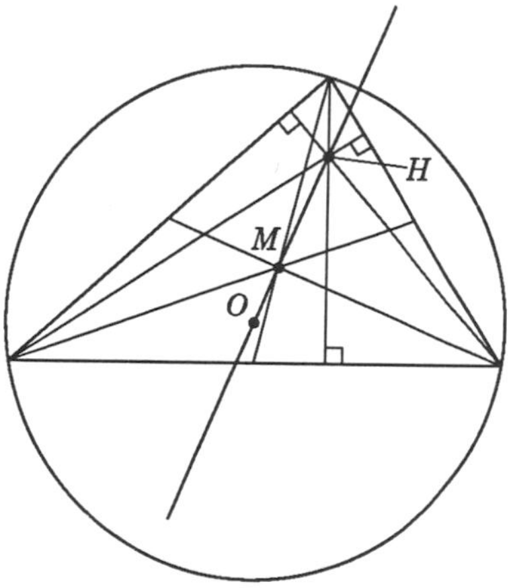

Intersection of Medians of a Triangle

The medians of a triangle are concurrent, intersecting at a point which is two-thirds of the distance from any vertex to the midpoint of the opposite side; this is the center of gravity of the triangle:

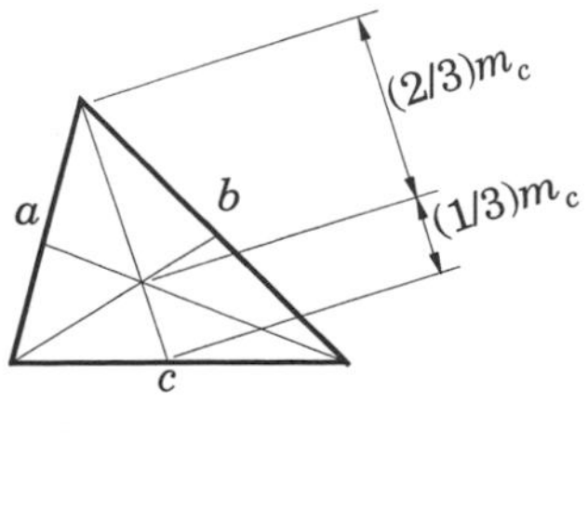

Dual Theorems of Projective Geometry

DESARGUES Girard (1591 - 1661)

The French mathematician and engineer **Girard Desargues** is the founder of the formal study of *projective geometry*. His *Brouillon projet* ... (1639) and other works of fundamental importance for the theory of projective geometry were, however, largely ignored until the 19th century when another French mathematician and engineer, **Poncelet**, revived the study of projective geometry.

PONCELET Jean Victor (1788 - 1867)

Dual Theorem

Dual Elements

Principle of Duality

A **dual theorem** (or simply **dual**) of projective geometry is a theorem that can be obtained from another by interchanging the **dual elements** – lines and points (vertices). The **principle of duality** of projective geometry is that if one of two dual theorems is true, the other theorem is also true.

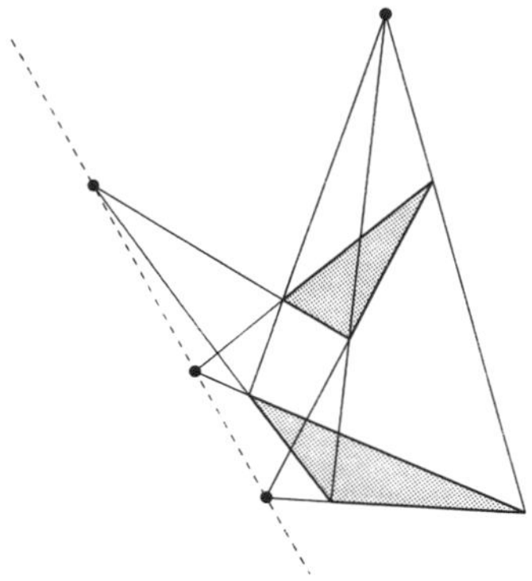

Desargues's Theorem

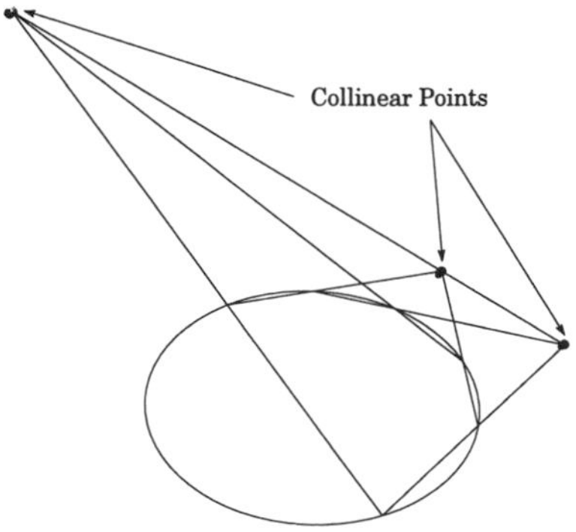

Pascal's Theorem

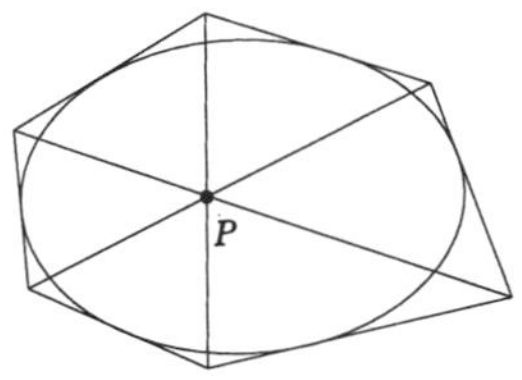

Brianchon's Theorem

The lines joining opposite vertices concur at P.

Desargues's Theorem: If the lines joining corresponding vertices of two triangles are **concurrent** (that is, they meet in one *point*), then the intersections of corresponding sides are **collinear** (that is, they lie on a *line*). The dual is also true.

Pascal's Theorem: If a hexagon is inscribed in a conic section, the three points of intersection of pairs of *opposite sides* are *collinear*.

PASCAL Blaise (1623 - 1662)

This theorem was first published in *Essay pour les coniques* ("Study of Conics") by the French mathematician-physicist-philosopher **Blaise Pascal** in 1639, who had been inspired by Desargues's recent work.

Brianchon's Theorem: If a hexagon is *circumscribed* about a conic section, the *lines* of intersection of pairs of *opposite vertices* are *concurrent*.

BRIANCHON Charles Julien (1783 - 1864)

This theorem is named after the French artillery officer and geometer **Brianchon**, who described it around 1806.

Brianchon's theorem is the dual of Pascal's theorem, and vice versa.

The Pythagorean Theorem

In a right triangle, the sum of the squares of the lengths of the legs is equal to the square of the length of the hypotenuse.

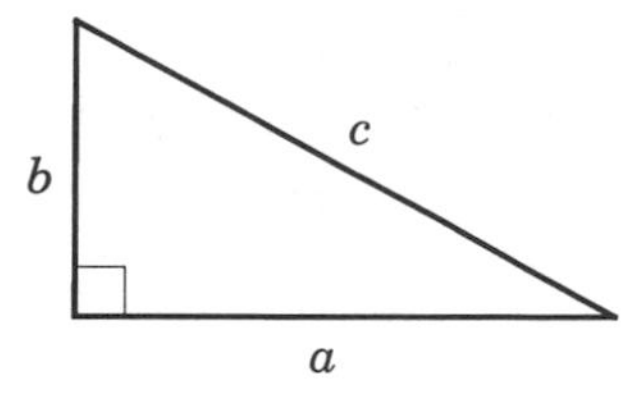

$$a^2 + b^2 = c^2$$

PYTHAGORAS of Samos (c. 580 - c. 500 B.C.)

The **Pythagorean theorem**, or **Pythagoras's theorem** – perhaps the most renowned of all mathematical theorems – was known by the Babylonians as far back as 2000 B.C., that is, about 1500 years before the time of Pythagoras and the Pythagoreans.

Classical Proof

Consider the following figure – as Euclid did about 300 B.C.:

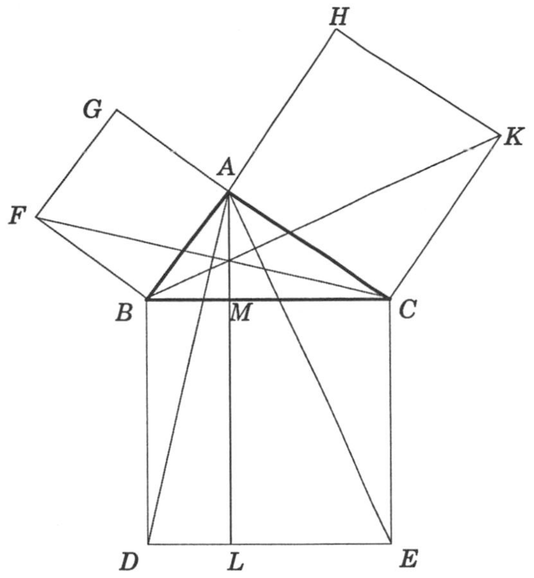

We make the following four assumptions:

- On a given straight line, it is possible to construct a square; in Euclid's *Elements*, the proof of this fact immediately preceded that of the Pythagorean theorem. The existence of squares requires the Euclidean parallel postulate.
- The area of a rectangle is the product of two adjacent sides.
- The area of a triangle is half the product of the height and the base.
- Two triangles are congruent when they have two sides and the included angle equal.

Let *ABC* be a triangle with *BAC* a right angle.

Let there be described on *BC* the square *BDEC*, on *BA* the square *ABFG*, and on *AC* the square *AHKC*.

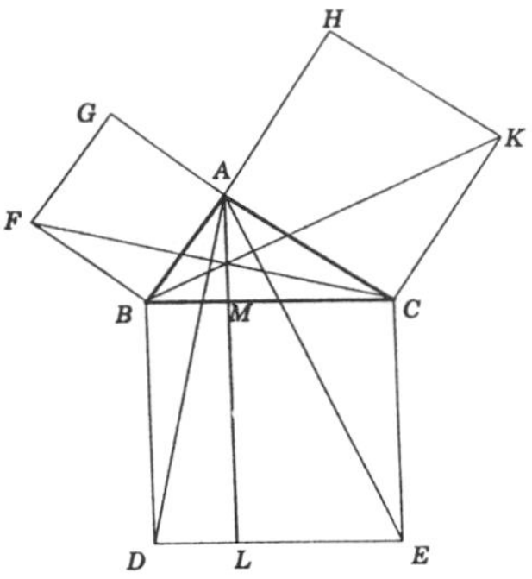

Through A, let AL be drawn parallel to BD and CE, and draw AD and AE.

The angles CBF and DBA are equal, for they are obtained when the angle ABC is added to either of the equal (right) angles DBC or FBA.

Since $BF = BA$ and $BC = BD$, and the angles CBF and DBA are equal, the triangles CBF and DBA are congruent.

The height of the triangle DBA – with respect to the base BD – is DL (which is equal to BM). The height of the triangle CBF – with respect to the base FB – is FG (which is equal to AB).

The area of the triangle DBA is half the area of the rectangle $BDLM$ (same base, BD, and same height, BM). Likewise, the area of the triangle CBF is half the area of the rectangle $ABFG$ (same base, BF, and same height, AB).

As shown above, $CBF \cong DBA$, and consequently the area of the square $ABFG$ equals the area of the rectangle $BDLM$.

By an argument corresponding to the above, we can also show that the area of the square $AHKC$ equals the area of the rectangle $CMLE$.

The area of square $BDEC$ is composed of the rectangles $BDLM$ and $CMLE$, whose areas we have shown to equal the squares $ABFG$ and $AHKC$, respectively. Therefore, the sum of the areas of the squares $ABFG$ and $AHKC$ equals the area of the square $BDEC$. QED

Bhaskara
(1114 - c. 1185)

There are many other verifications of the Pythagorean theorem; the following geometrical proof, given about 1150 by the Indian mathematician **Bhaskara**, provides a visual feeling of correctness.

Bhaskara's Proof

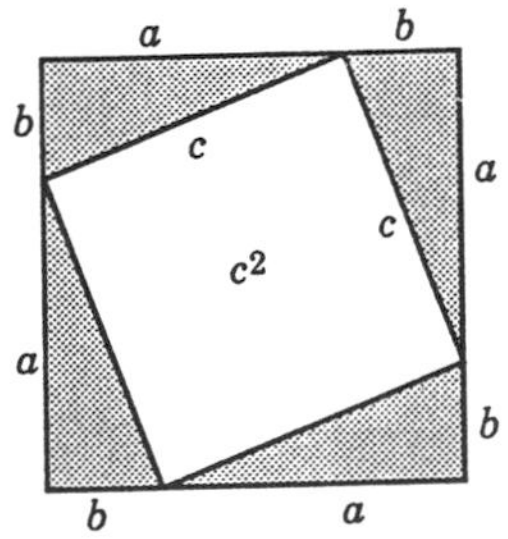

Assume that the area of a square is the square of its side, and that the area of a right-angled triangle is half the product of its smaller sides.

Inscribe a square in another square so that each corner of the inscribed square coincides with a point of one side of the circumscribed square.

The circumscribed square contains four right-angled triangles (shaded areas) whose hypotenuse, c, is a side of the inscribed square, and whose smaller sides are a and b.

The total area of the four triangles is

$$4\frac{a\,b}{2} = 2\,a\,b\,.$$

The area of the circumscribed square whose side is $(a + b)$ may be expressed as $(a + b)^2$, *or* as the sum of the area of the inscribed square and the total area of the four right-angled triangles; thus:

$$(a + b)^2 = 2\,a\,b + c^2$$
$$a^2 + 2\,a\,b + b^2 = 2\,a\,b + c^2,$$

and we have

$$a^2 + b^2 = c^2\,.$$

QED

The Quick and the Dead

One pleasant summer evening Fred invited a group of fellow pure mathematicians to a garden party. Chatting about surjections, cohomology rings, unknotting conjecture, immersion of projective spaces, and other relaxing topics, Andy suddenly remarked:

"You've got a pretty tall flagpole, Fred. I bet it's about 10 meters high."

"Actually, if I remember right, it's 14."

They all got quite excited about this nontrivial problem. The wagers came fast.

Fred borrowed a measuring rod from a neighbor.

As the mathematicians were about three-quarters of the way to the top of the flagpole, climbing up on each other's shoulders, Fred's young daughter Anna came home. When she found out what they were up to, she blurted:

"Haven't you heard of the Pythagorean theorem?"

The heap of mathematicians tumbled to the ground.

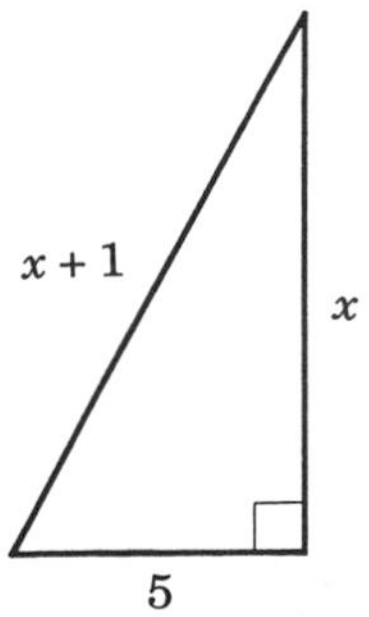

This is what Anna did:

She measured the double string of halyard that *exceeded* the length of the flagpole, and found the excess to be 1 meter. The double string just reached the level ground when tightened at a point 5 meters from the foot of the flagpole.

Letting the height of the flagpole be x meters:

$$x^2 + 5^2 = (x+1)^2$$

$$x = 12 \text{ meters}$$

•

The clue to proofs for theorems of geometry usually lies in *finding similar triangles* and, with right-angled triangles, the *use of the Pythagorean theorem*.

Diagonals and Sides of a Parallelogram

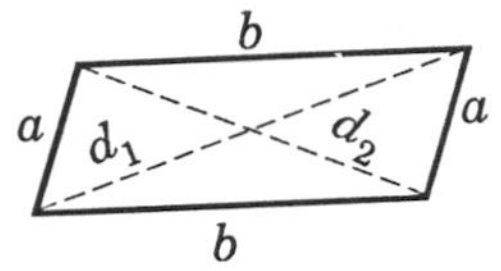

$$d_1^2 + d_2^2 = 2(a^2 + b^2)$$

The diagonals bisect each other.

Ptolemy's Theorem

The sum of the products of the two pairs of opposite sides of a convex quadrilateral inscribed in a circle is equal to the product of the lengths of the diagonals.

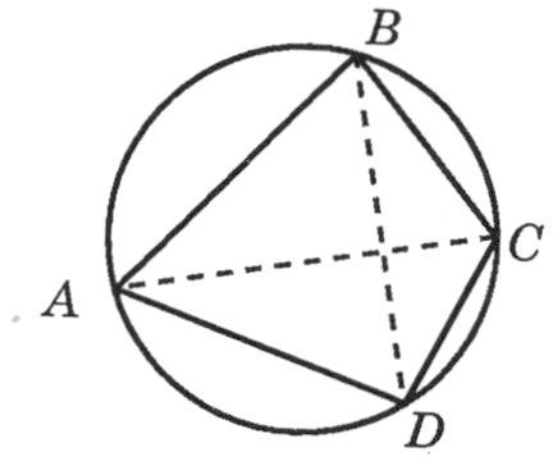

$$AB \times CD + AD \times BC = AC \times BD$$

PTOLEMAEUS Claudius (*c.* 100 - *c.* 168)

The theorem is named after the mathematician, astronomer, and geographer **Ptolemy** of Alexandria.

Quadrilateral About a Circle

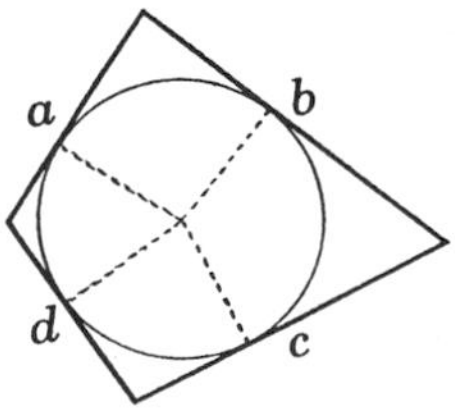

$$a + c = b + d$$

(For proof, note that the radii of the circle join the points of contact.)

Chord Theorem

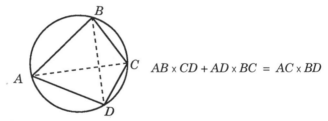

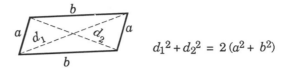

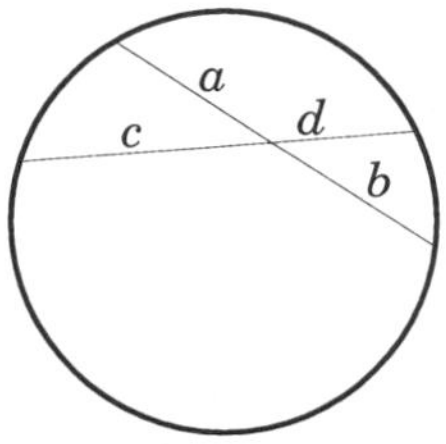

$$ab = cd$$

Secant Theorem

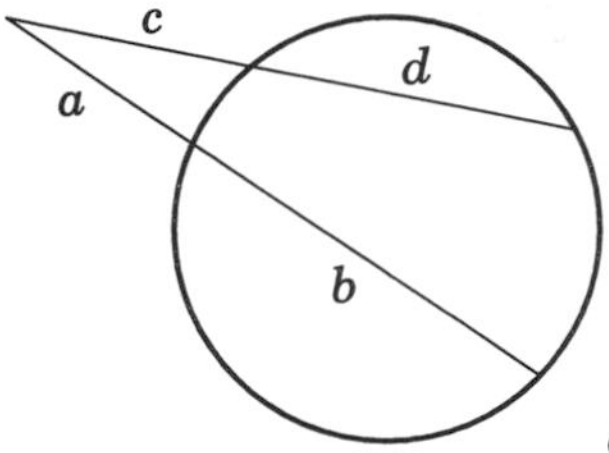

$$a\,(a+b) = c\,(c+d)$$

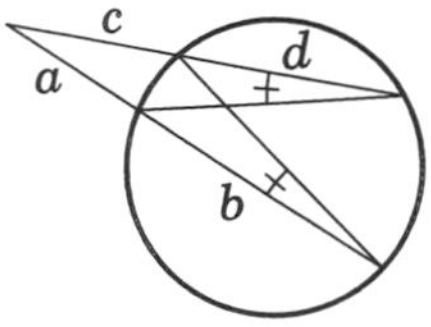

Why? With similar triangles, we have

$$\frac{c}{a+b} = \frac{a}{c+d}\,.$$

Secant-Tangent Theorem

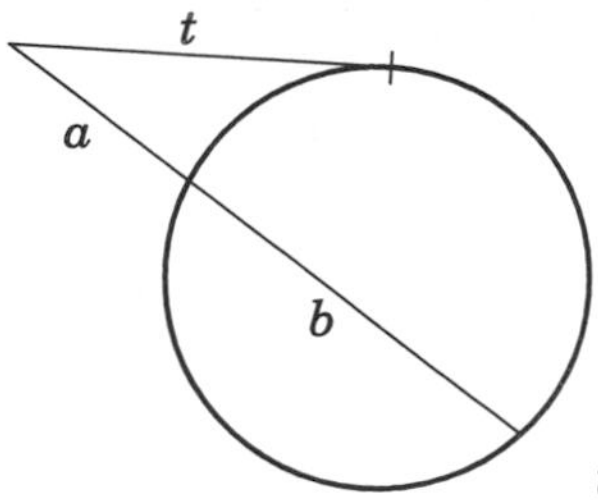

$$t^2 = a\,(a+b)$$

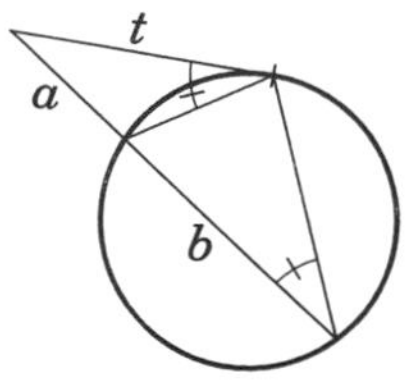

Why? Again, with similar triangles:

cf. *p.* 431

$$\frac{t}{a+b} = \frac{a}{t}\,.$$

Height, Median, and Bisector of a General Triangle

cf. Heron's formula: *p.* 444

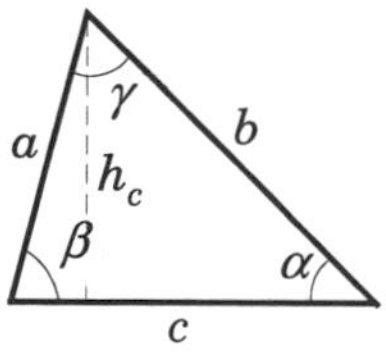

$$h_c = \frac{2\sqrt{p\,(p-a)\,(p-b)\,(p-c)}}{c}$$

where $p = \frac{1}{2}\,(a+b+c)$.

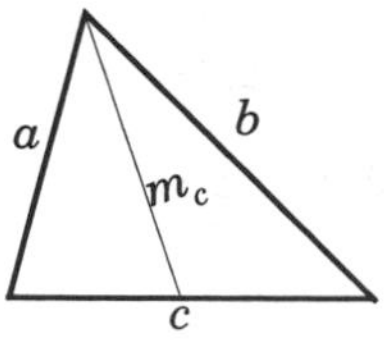

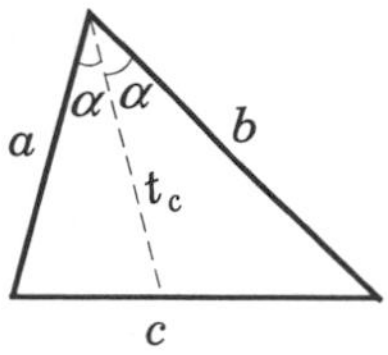

$$m_c = \frac{1}{2}\sqrt{2\,a^2 + 2\,b^2 - c^2} \qquad t_c = \sqrt{a\,b\left[1 - \left(\frac{c}{a+b}\right)^2\right]}$$

Radii of Circumscribed and Inscribed Circles

Cyclic Quadrilateral

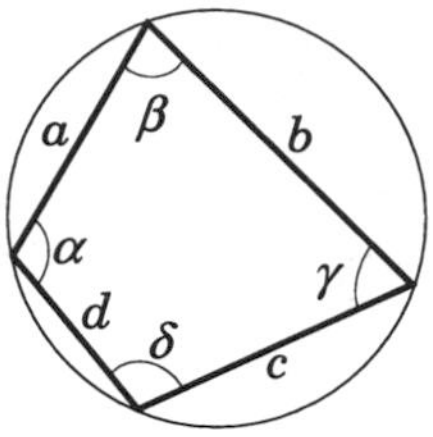

Square

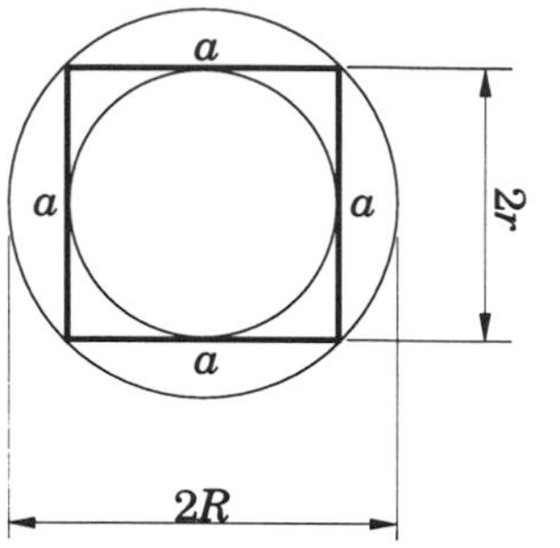

$$R = \frac{1}{4}\sqrt{\frac{(a\,c + d\,b)\,(a\,d + b\,c)\,(a\,b + c\,d)}{(p - a)\,(p - b)\,(p - c)\,(p - d)}}$$

where $p = \frac{1}{2}(a + b + c + d)$; *cf.* Heron's formula: *p.* 444.

$$R = \frac{a\sqrt{2}}{2};\quad r = \frac{1}{2}a$$

General Triangle

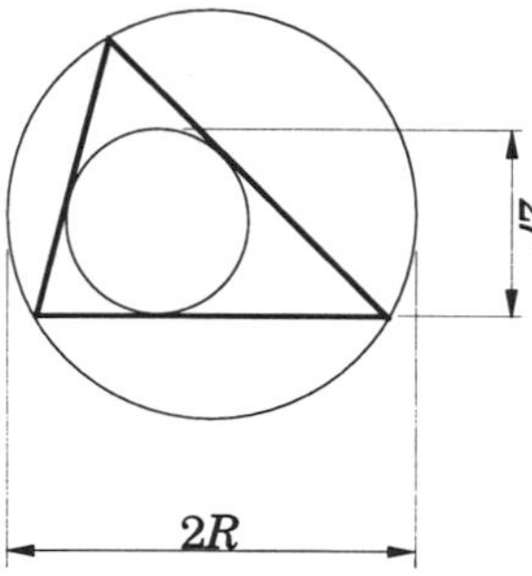

$$r = \frac{2 \cdot A_{\text{Triangle}}}{a + b + c}$$

$$R = \frac{a\,b\,c}{4 \cdot A_{\text{Triangle}}}$$

Equilateral Triangle

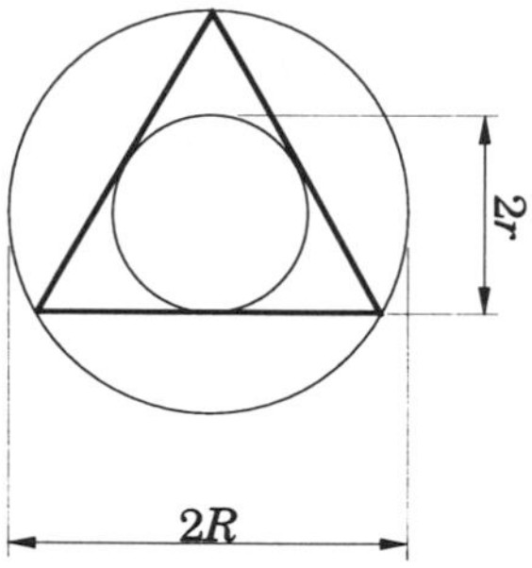

$$r = \frac{a\sqrt{3}}{6}$$

$$R = \frac{a\sqrt{3}}{3}$$

Right Triangle

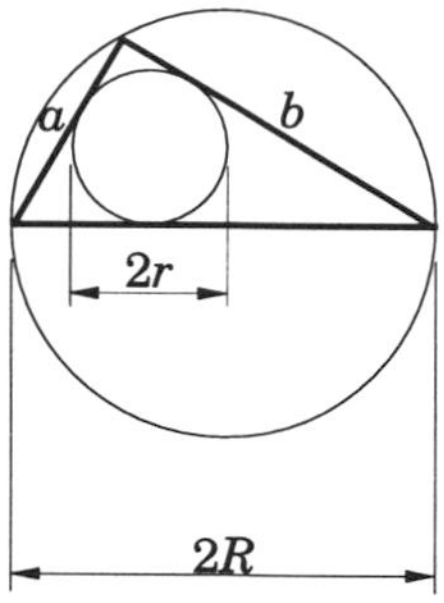

$$r = \frac{a\,b}{a + b + c}$$

$$R = \frac{\text{hypotenuse}}{2}$$

Curvilinear Figures

The principle of finding curve lengths will be discussed in the chapters dealing with calculus.

Circle

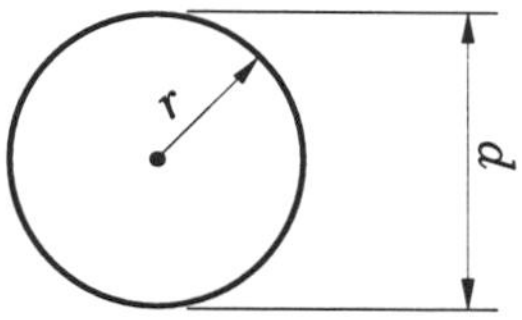

$$C = 2\pi r = \pi d$$

Arc of a Circle

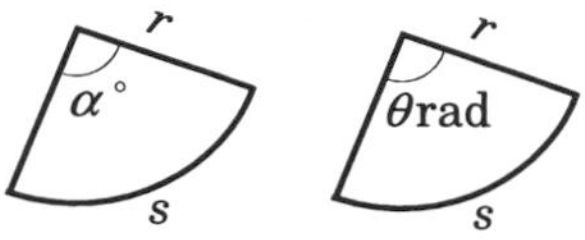

$$s = \frac{\alpha\,\pi\,r}{180} = \theta r$$

Ellipse

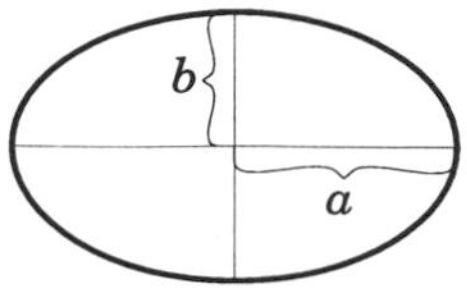

$$P \approx 2\pi\sqrt{\tfrac{1}{2}(a^2 + b^2)}$$

Areas

Areas of Rectilinear Figures

Rectangle The area of a rectangle is the product of two adjacent sides,

$$A = a\,b\,,$$

which is evident from the drawing below, where b rows, each of a unit squares, fill the rectangle:

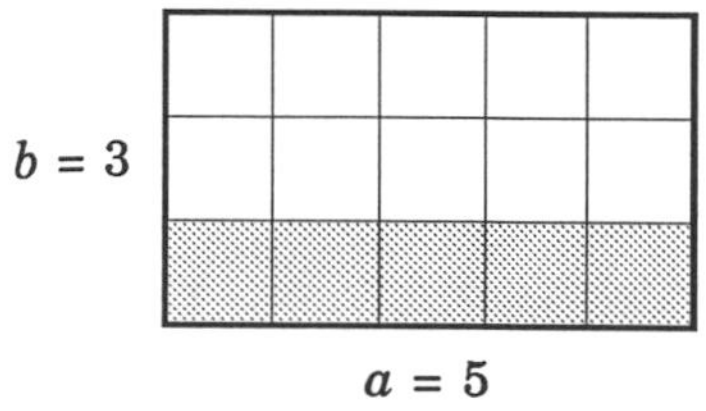

Square A square may be considered a special case of a rectangle with all sides equal. It has the area

$$A = a \cdot a = a^2\,.$$

Rhomboid or Parallelogram The area of a **rhomboid** (**parallelogram**), is the product of the length of one side and the corresponding height; for a rhomboid with sides a and b and corresponding heights h_a and h_b, we have

$$A = a \cdot h_a = b \cdot h_b,$$

as depicted for $A = a \cdot h_a$ in the illustration below, where a right-angled triangle has been transferred from one end of the rhomboid to the other, transforming the rhomboid into a rectangle:

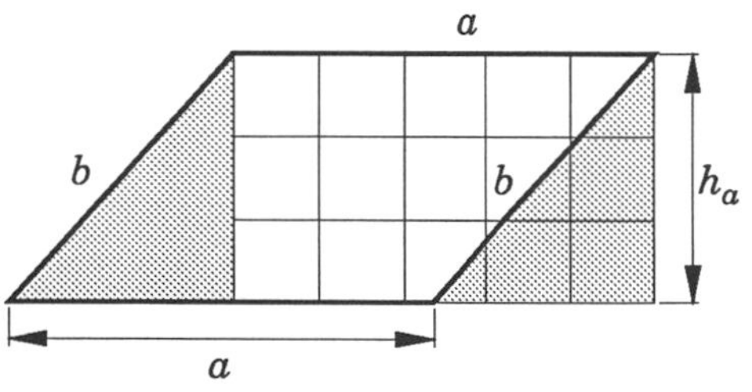

Rhombus A rhombus is a rhomboid with all sides equal; the area is obtained in the same way as for the rhomboid.

Triangle The area of a triangle with sides a, b, and c and the corresponding heights h_a , h_b, and h_c is half the products of base and height,

$$A = \frac{1}{2}\,a\,h_a = \frac{1}{2}\,b\,h_b = \frac{1}{2}\,c\,h_c\ ;$$

a triangle may always be considered one-half of a corresponding rhomboid.

Trapezium (U.K.) **Trapezoid** (U.S.) The area of a trapezium (trapezoid) is half the product of the sum of the parallel sides times its height; we have for a trapezium/trapezoid with parallel sides a and b, and height h,

$$A = \frac{1}{2}\,(a + b)\,h\,.$$

To demonstrate the correctness of this statement, rotate the quadrilateral 180° and place the upside-down replica beside the original to transform the trapezium/trapezoid into a rhomboid of twice the area:

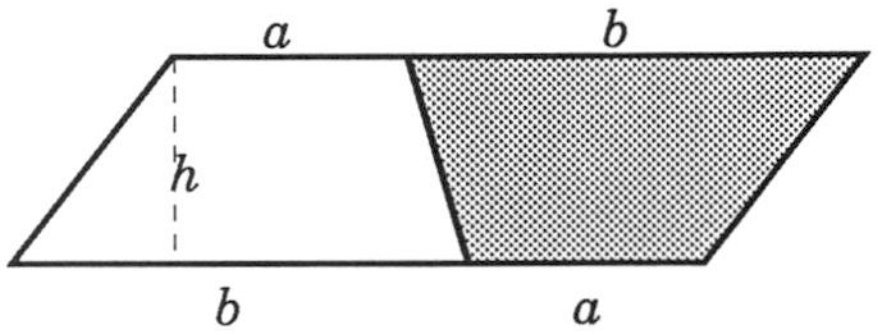

Areas of Curvilinear Figures

The principle of finding areas of curvilinear figures will be discussed in the chapters dealing with calculus.

Circle

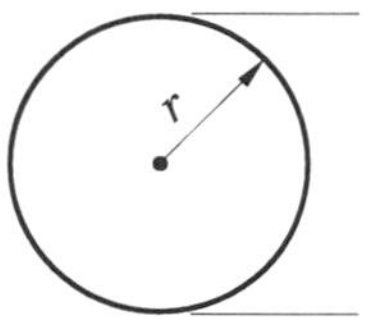

$A = \pi r^2$

Ellipse

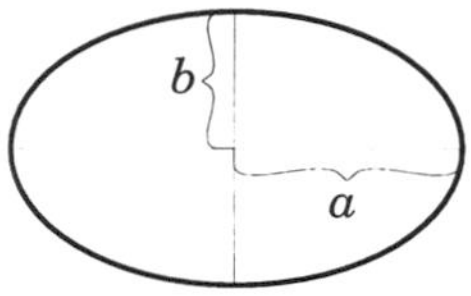

$A = \pi a b$

Sector of a Circle

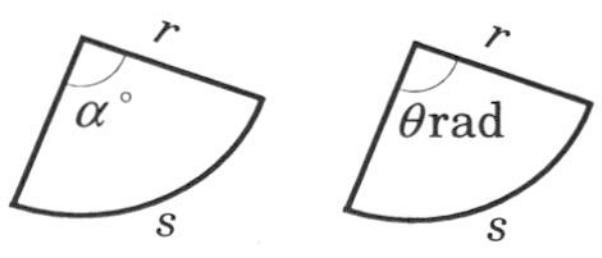

$$A = \frac{1}{2}\,s\,r = \frac{\alpha \pi r^2}{360^\circ} = \frac{\theta r^2}{2}$$

Segment of a Circle

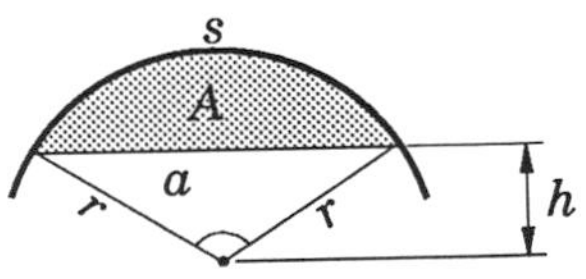

$$A = \frac{1}{2}\,(s\,r - a\,h)$$

Segment of a Parabola

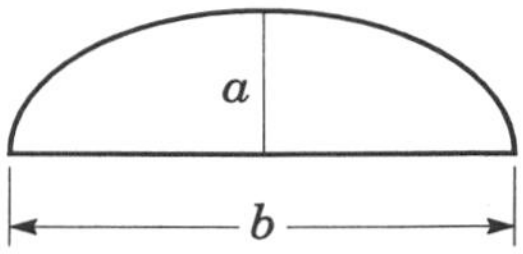

$$A = \frac{2}{3}\,a\,b$$

Problems

We return to rectilinear figures.

- An isosceles triangle has a perimeter of 12 units and a height of 4 units with respect to the base.

  Find the legs, base, and area of the triangle.

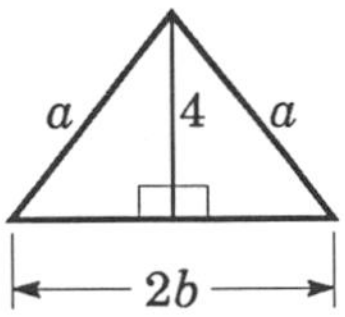

With the base of the triangle $2b$, we have, with Pythagoras,

$$\left.\begin{aligned} a &= \sqrt{16+b^2} \\ 2a+2b &= 12 \end{aligned}\right\} \quad \left.\begin{aligned} a &= \sqrt{16+b^2} \\ a &= 6-b \end{aligned}\right\}$$

$$\sqrt{16+b^2} = 6-b; \qquad 16+b^2 = 36-12b+b^2$$

$$12b = 20, \qquad b = \frac{5}{3}; \qquad \left.\begin{aligned} 2b &= 3\frac{1}{3} \\ a &= 4\frac{1}{3} \end{aligned}\right\}$$

$$A = bh = \frac{5}{3}\cdot 4 = \frac{20}{3} = 6\frac{2}{3} \text{ square units.}$$

- Find the length of a side of an equilateral triangle with an area equal to that of a square whose side is 5 cm.

  Let the side of the triangle be $2a$, and h its height.

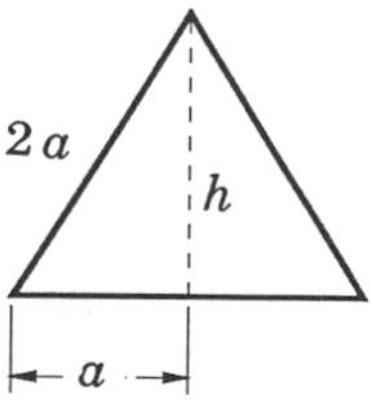

$$4a^2 = h^2 + a^2$$
$$h = a\sqrt{3}.$$

$$A = ah \quad \Rightarrow \quad a^2\sqrt{3} = 25$$

$$a = 5\cdot\frac{\sqrt[4]{27}}{3}$$

$$2a = 10\cdot\frac{\sqrt[4]{27}}{3} \text{ cm.}$$

- A square is inscribed in a right triangle with legs 7 and 14 cm; one side of the square is coincident with the hypotenuse and its two remaining corners are on the legs of the triangle.

  Find the area of the square.

  Let the side of the square be x.

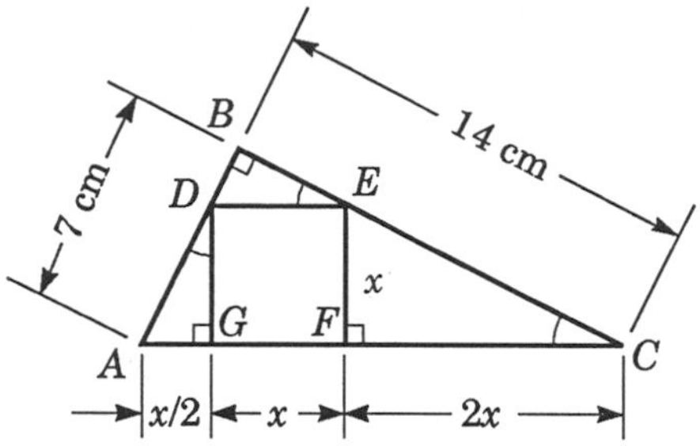

We have, with Pythagoras,

$$AC = \sqrt{14^2 + 7^2} = \sqrt{245} = 7\sqrt{5} \text{ cm}.$$

Since triangles ACB, ADG, and EFC are similar, we have

$$CF = 2x \; ; \; AG = \frac{x}{2},$$

and

$$\frac{x}{2} + x + 2x = 3\frac{1}{2}x = \sqrt{245} = 7\sqrt{5}$$

$$x = 2\sqrt{5}\,;$$

and the area of the square

$$x^2 = \left(2\sqrt{5}\right)^2 = 20 \text{ cm}^2.$$

Heron's Formula

HERON
(Probably 1st century AD)

Heron of Alexandria – one of the greatest engineers and inventors of antiquity – pioneered the science of applied mathematics and was a prolific writer and compiler of the knowledge of mathematics and engineering of his time.

His name is attached to formulas for the calculation of the areas of triangles, quadrilaterals, and certain regular polygons of higher orders, although – as is often the case – they may have been known by some of his predecessors.

Area of a Triangle

The **area of a triangle** can be calculated if the lengths of its three sides are known,

$$A = \sqrt{p\,(p-a)\,(p-b)\,(p-c)},$$

where a, b, and c are the sides of the triangle, $2p = a + b + c$ is the perimeter, and p is called the semiperimeter.

Proof The height h, perpendicular to the base c of the triangle, can be calculated by the Pythagorean theorem:

$$h^2 = a^2 - (c - c_1)^2 = b^2 - c_1^2$$

$$a^2 = b^2 + c^2 - 2\,c\,c_1$$

$$c_1 = \frac{b^2 + c^2 - a^2}{2\,c}$$

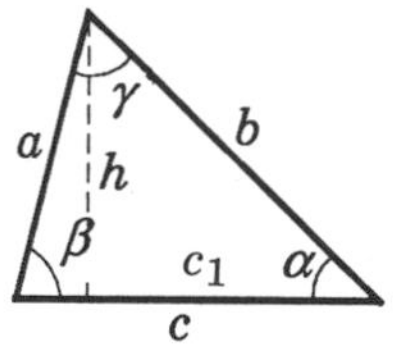

$$h^2 = b^2 - \left(\frac{b^2 + c^2 - a^2}{2\,c}\right)^2$$

$$4\,c^2\,h^2 = 4\,b^2\,c^2 - (b^2 + c^2 - a^2)^2$$

$$= [2\,b\,c + (b^2 + c^2 - a^2)] \cdot [2\,b\,c - (b^2 + c^2 - a^2)]$$

$$= [(b + c)^2 - a^2] \cdot [a^2 - (b - c)^2]$$

$$= (a + b + c)\,(-a + b + c)\,(a - b + c)\,(a + b - c)$$

where we introduce

$$a + b + c = 2\,p \qquad -a + b + c = 2\,(p - a)$$

$$a - b + c = 2\,(p - b)$$

$$a + b - c = 2\,(p - c)\,.$$

$$4\,c^2\,h^2 = 16\,p\,(p - a)\,(p - b)\,(p - c)$$

and find the area

$$A = \frac{c\,h}{2} = \sqrt{p\,(p - a)\,(p - b)\,(p - c)}\,.$$ QED

Heron has also given formulas for the areas of quadrilaterals whose sides and angles are known.

Arbitrary Quadrilateral

The area of an **arbitrary quadrilateral** is

$$A = \sqrt{(p - a)\,(p - b)\,(p - c)\,(p - d)\ - a\,b\,c\,d\,\cos^2 \alpha}\,,$$

where

a, b, c, d	are	the sides of the quadrilateral
$2p$	=	the perimeter, $a + b + c + d$
α	=	half the sum of two opposite angles.

Cyclic Quadrilateral

For a **cyclic quadrilateral** – a quadrilateral that can be inscribed in a circle – the formula simplifies to

$$A = \sqrt{(p - a)\,(p - b)\,(p - c)\,(p - d)}\,.$$

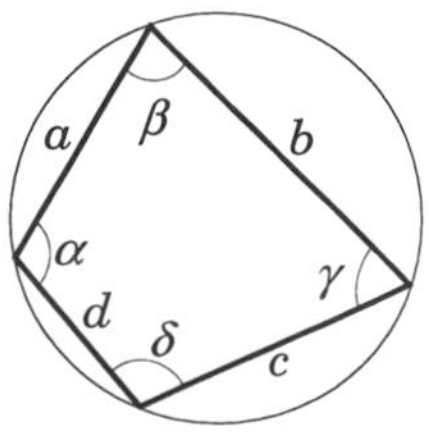

12.42 Solid Geometry

Volumes

Parallelepipeds The volume V of a rectangular parallelepiped with edges a, b, c is the product of its edge lengths,

$$V = a \cdot b \cdot c .$$

A parallelepiped with edges a, b, c can be subdivided into c layers with b rows of a unit cubes each:

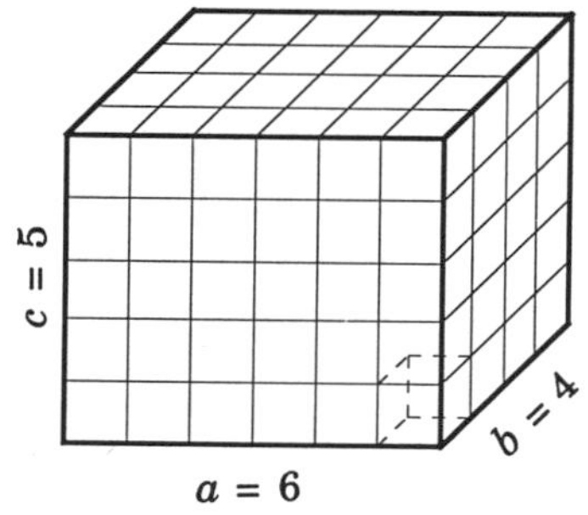

The bottom of the rectangular parallelepiped has an area ab, and the height is c; if we call the base area B and the height h, we have the volume

$$V = B \cdot h .$$

Cube Analogously, the volume of a cube with $a = b = c$ is

$$V = a \cdot a \cdot a = a^3 .$$

Oblique Parallel-Epipeds For the volume of an oblique parallelepiped, we can extend the method of reasoning used for oblique parallelograms, or use Cavalieri's theorem to find

$$V = B \cdot h .$$

Cavalieri's Theorem Solids of equal height have equal volumes if sections parallel to and equidistant from their bases have equal areas, that is,

$$V = B \cdot h .$$

CAVALIERI Bonaventura (1598 - 1647)

The theorem is named for the Italian physicist and mathematician **Bonaventura Cavalieri**.

Prisms, Cylinders The volume of a prism is equal to the product of the base area B and the perpendicular height h,

$$V = B \cdot h .$$

This also applies for all cylinders, which can be considered as prisms with an infinite number of lateral faces.

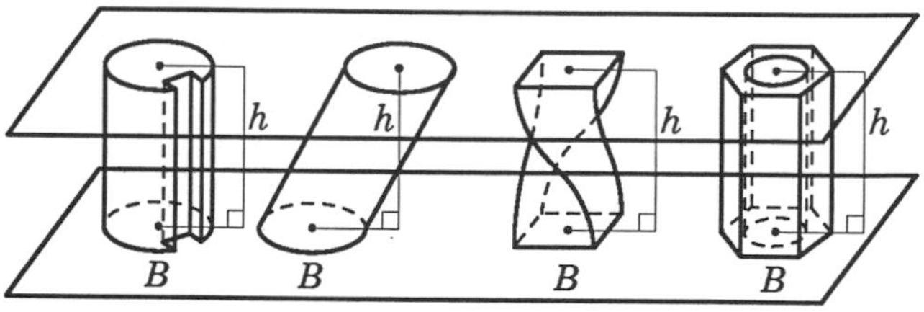

The solids shown in the above drawing all have the same base area at both ends, and the same sectional area at all levels between the ends. By Cavalieri's theorem, their volumes are equal.

Pyramids, Cones The volume of all pyramids – or cones, which may be looked upon as pyramids with an infinite number of lateral faces – is one-third of the product of the base area B and the corresponding height h,

$$V = \frac{1}{3} \cdot B \cdot h \, .$$

A triangular prism can be split into *three* pyramids of equal volume by two plane sections.

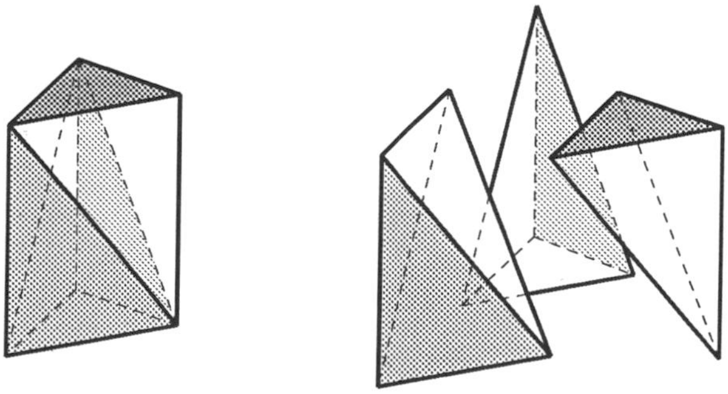

***

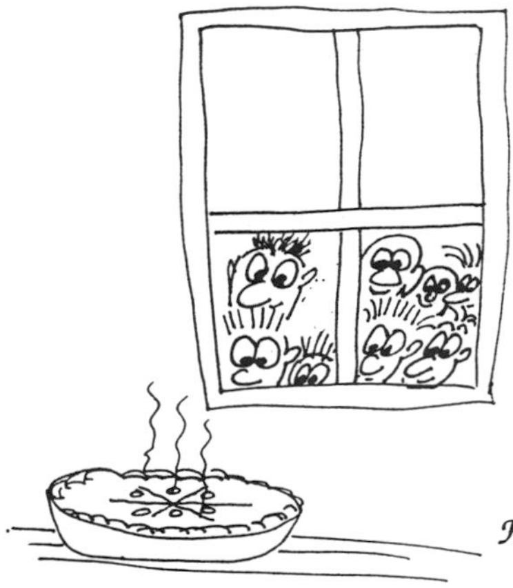

How do you cut a pie into eight equal pieces with only three cuts?

[Answer on page 456]

The Five Regular Polyhedra

To construct a vertex of a **regular polyhedron**, two conditions must be met:

- a minimum of three planes of regular polygons is required;
- the sum of the angles of the polygonal vertices must be less than a full angle of 360°.

Regular polygon	No. of vertex polygons	Polygon angle	Sum of angles	Regular polyhedron
Triangle	3	60°	180°	Tetrahedron
	4		240°	Octahedron
	5		300°	Icosahedron
	6		360°	None
Square	3	90°	270°	Cube (hexahedron)
	4		360°	None
Pentagon	3	108°	324°	Dodecahedron
	4		432°	None
Hexagon	3	120°	360°	None

It is not possible to construct a regular polyhedron with regular hexagons or polygons of higher order.

Thus, there are just five regular polyhedra.

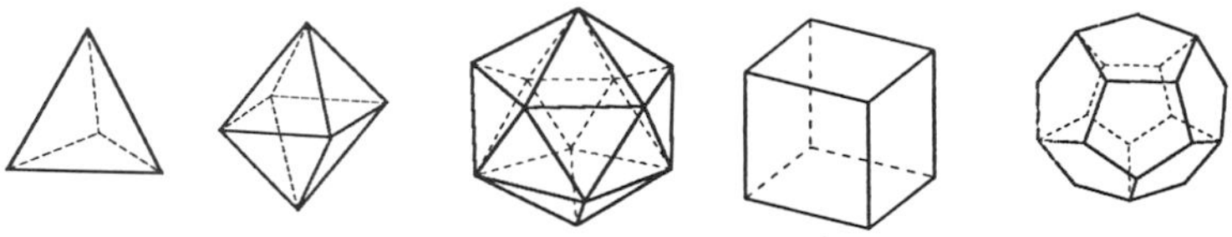

The Euler Characteristic

If a polyhedron has V vertices, F faces, and E edges and is topologically equivalent to the sphere, one will always find

$$V + F - E = 2,$$

DESCARTES René (1596 - 1659)

EULER Leonhard (1707 - 1783)

where the 2 is the **Euler characteristic** of the polyhedron. The formula – originally devised by **René Descartes** and later resurrected by **Leonhard Euler** – is often referred to as **Euler's polyhedron formula**. We will see how the Euler characteristic for other shapes can be calculated by breaking them into polyhedra.

Consider a polyhedron with the least possible number of faces – a tetrahedron:

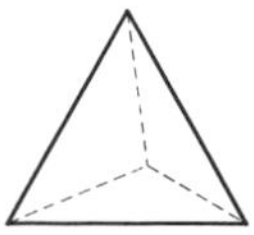

Topologically, the tetrahedron is equivalent to a sphere. On a sphere, denote the vertices of the tetrahedron with dots, the edges with arcs; the areas between the arcs correspond to the faces of the tetrahedron:

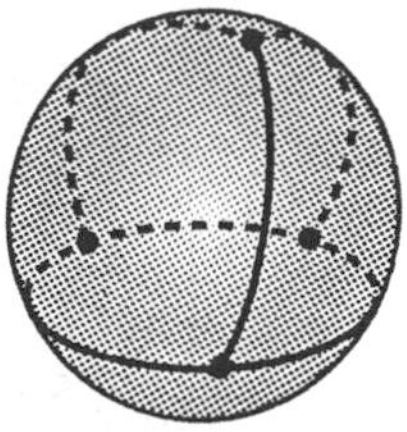

We may displace the vertices, edges, and faces so that they are all exposed from one direction:

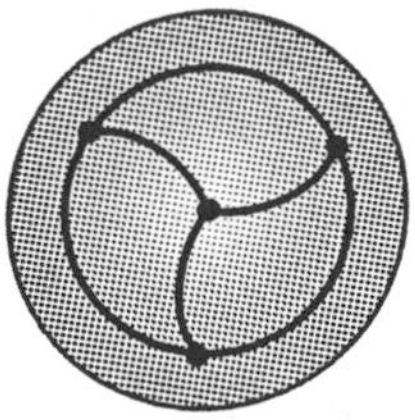

We have 4 dots (V), 4 fields (F), and 6 arcs (E), which gives

$$4_V + 4_F - 6_E = 2 .$$

An arc can be added in two ways inside a triangular field:

- from a dot to an opposite arc
  – giving one new dot, one new field, and two new arcs;
- between adjacent arcs
  – forming two new dots, one new field, and three new arcs.

Thus, whichever way an arc is added, the sum of the increase in dots and fields ($V + F$) equals the increase in arcs (E) and, consequently, $V + F - E$ remains equal to 2.

The formula $V + F - E = 2$ is true for any polyhedron, as long as its interior is a topological ball:

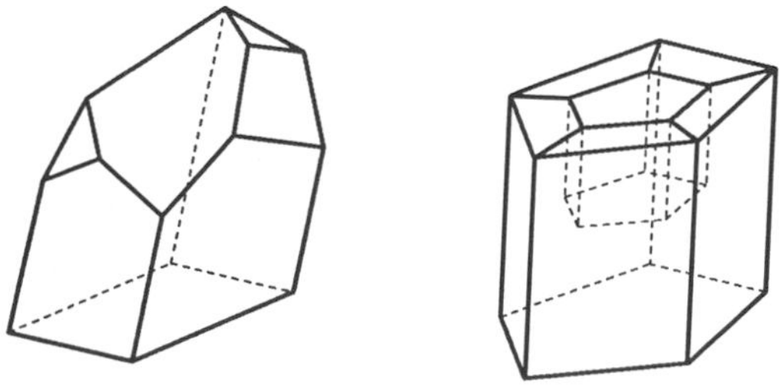

For the polyhedron at the right we must count the 6 faces in the well at the top as well as the 5 quadrilateral faces that ring that well, giving a total of 17 faces; thus,

$$20_V + 17_F - 35_E = 2 .$$

As we have shown, the Euler characteristic is 2 for a surface topologically equivalent to the sphere. If topologically equivalent to a torus the surface has the Euler characteristic 0.

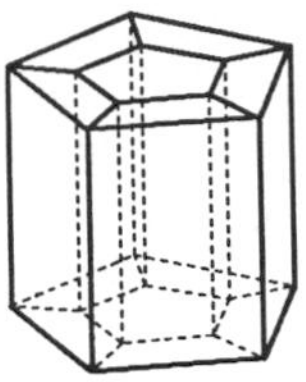

Topologically equivalent to a torus, the prism with a hole straight through its center has an Euler characteristic of 0,

$$20_V + 20_F - 40_E = 0\,.$$

To classify a polyhedron topologically, we may determine its Euler characteristic, and to do so we make the paradoxical use of counting the number of vertices, faces, and edges, none of which need to be preserved under topological equivalence.

For two-sided surfaces the Euler characteristic is

$$V + F - E = 2 - 2p,$$

Genus: *p.* 379

where p is the *genus* of the surface, while for one-sided surfaces

$$V + F - E = 2 - p,$$

with p the genus as before.

Selection of Formulas from Solid Geometry

Many of the following formulas will be further discussed in later chapters, concerned with calculus.

V = Volume
S = Lateral surface
R = Radius of circumscribed sphere
r = Radius of inscribed sphere

Regular Polyhedra

Tetrahedron

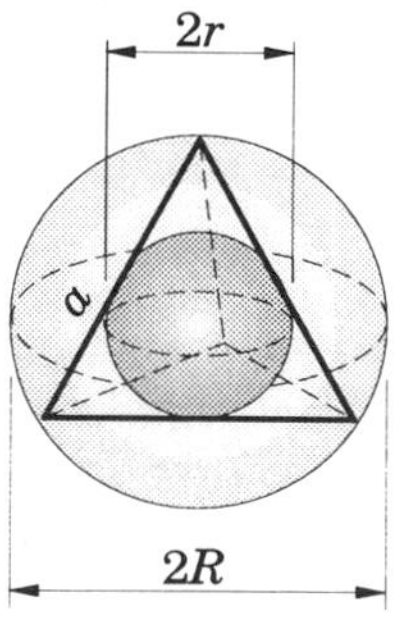

$$V = \frac{a^3\sqrt{2}}{12}$$

$$S = a^2\sqrt{3}$$

$$R = \frac{a\sqrt{6}}{4}$$

$$r = \frac{a\sqrt{6}}{12}$$

Cube

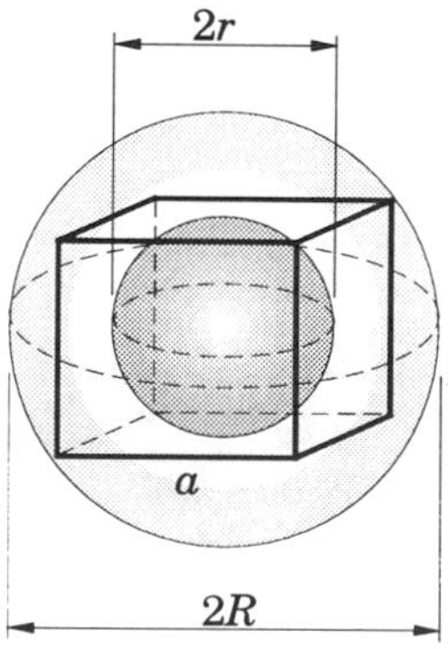

$$V = a^3$$

$$S = 6a^2$$

$$R = \frac{a\sqrt{3}}{2}$$

$$r = \frac{a}{2}$$

Octahedron

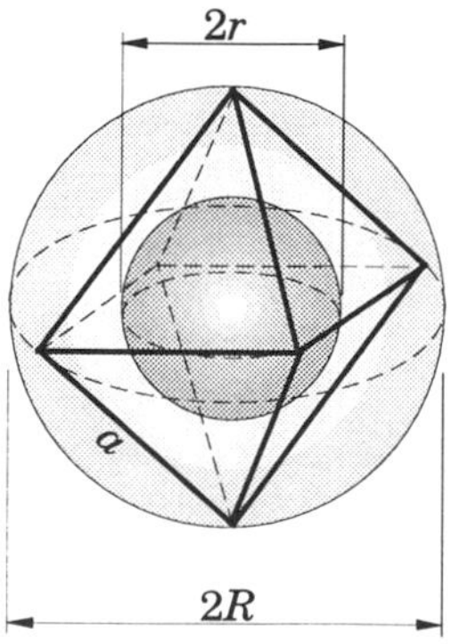

$$V = \frac{a^3\sqrt{2}}{3}$$

$$S = 2a^2\sqrt{3}$$

$$R = \frac{a\sqrt{2}}{2}$$

$$r = \frac{a\sqrt{6}}{6}$$

Dodecahedron

Icosahedron

Dodecahedron

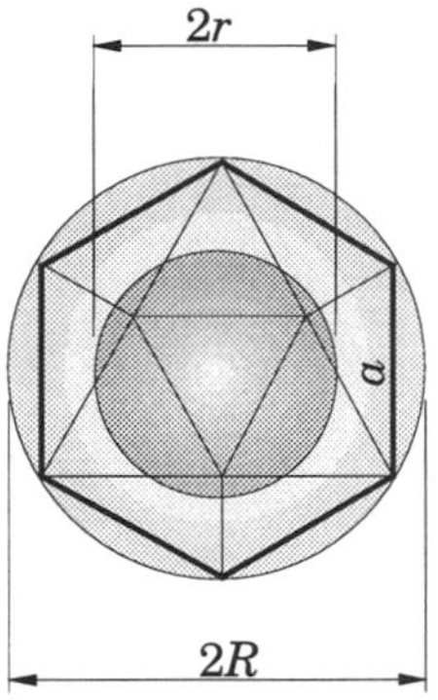

$$V = \frac{a^3\,(15 + 7\sqrt{5})}{4}$$

$$S = 3\,a^2\sqrt{5\,(5 + 2\sqrt{5})}$$

$$R = \frac{a\,(1 + \sqrt{5})\ \sqrt{3}}{4}$$

$$r = \frac{a}{4}\sqrt{\frac{50 + 22\sqrt{5}}{5}}$$

Icosahedron

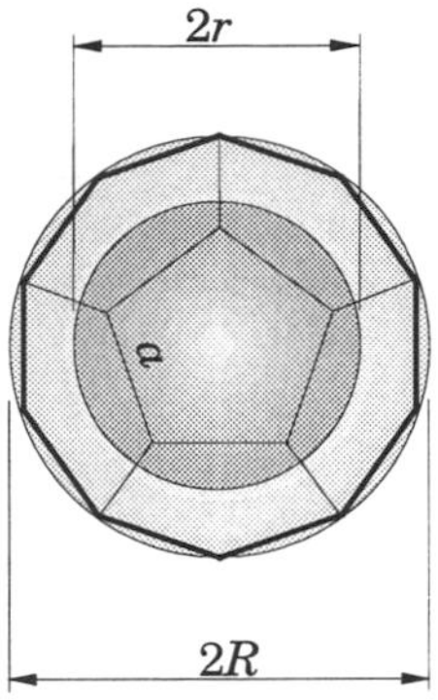

$$V = \frac{5\,a^3}{12}\,(3 + \sqrt{5})$$

$$S = 5\,a^2\sqrt{3}$$

$$R = \frac{a}{4}\sqrt{2\,(5 + \sqrt{5})}$$

$$r = \frac{a}{2}\sqrt{\frac{7 + 3\sqrt{5}}{6}}$$

Prisms

General Prism

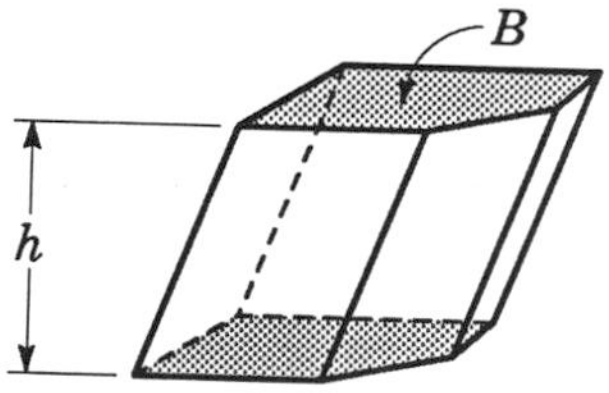

$$V = B\,h$$

Parallelepiped

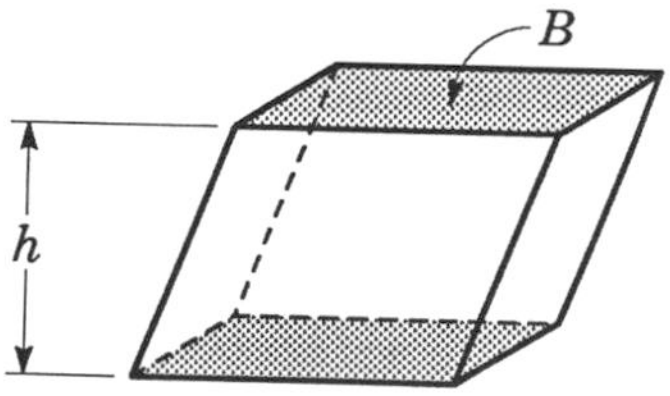

$$V = B\,h$$

Rectangular Parallelepiped

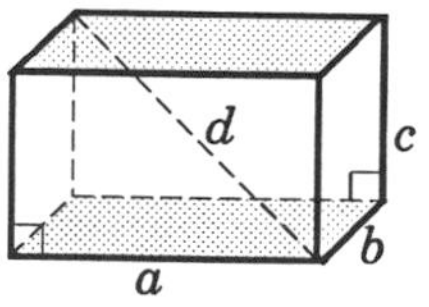

$$V = a\,b\,c$$

$$d = \sqrt{a^2 + b^2 + c^2}$$

Pyramids

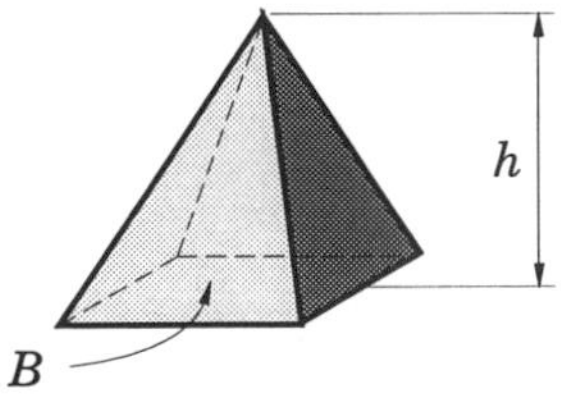

$$V = \tfrac{1}{3}\,B\,h$$

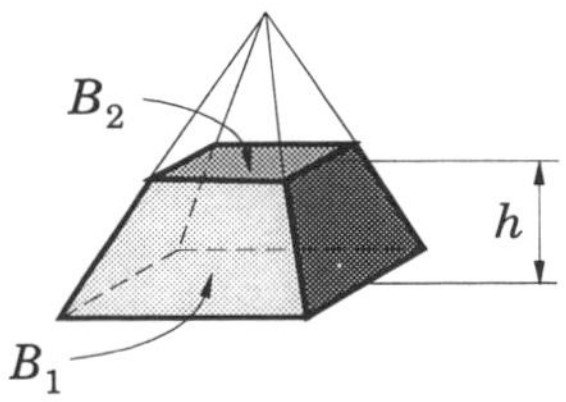

$$V = \tfrac{1}{3}\,h\,\left(B_1 + B_2 + \sqrt{B_1 B_2}\right)$$

Cylinders

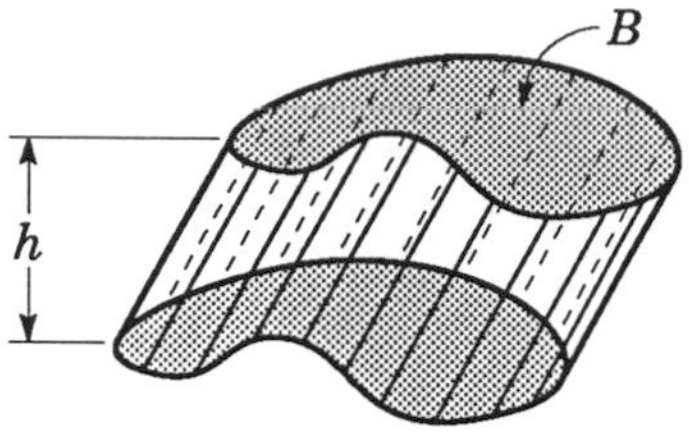

$V = Bh$

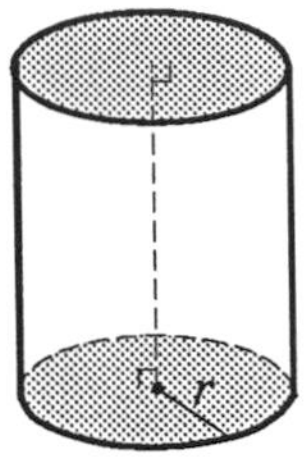

$V = \pi r^2 h$

$S = 2\pi r(r + h)$

Cones

General Cone

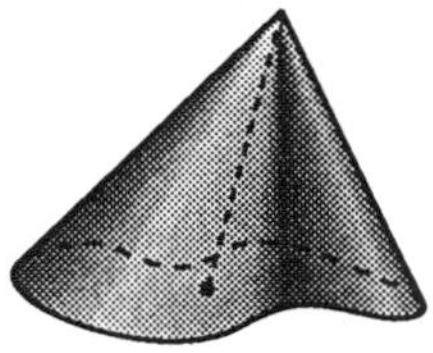

$V = \frac{1}{3} Bh$

Right Circular Cone

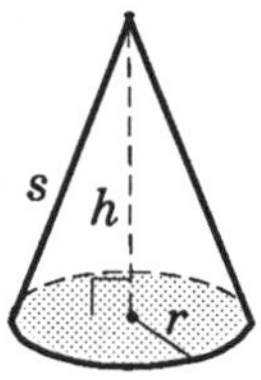

$V = \frac{1}{3} \pi r^2 h$

$S = \pi r(r + s)$

Lateral surface area $= \pi r s$

Frustum of a Right Circular Cone

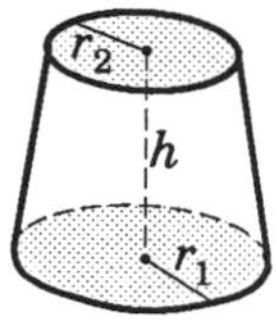

$V = \frac{1}{3} \pi h (r_1^2 + r_1 r_2 + r_2^2)$

$S = \pi [r_1^2 + s(r_1 + r_2) + r_2^2]$

Lateral surface area $= \pi s (r_1 + r_2)$

Spheres and Spherical Segments

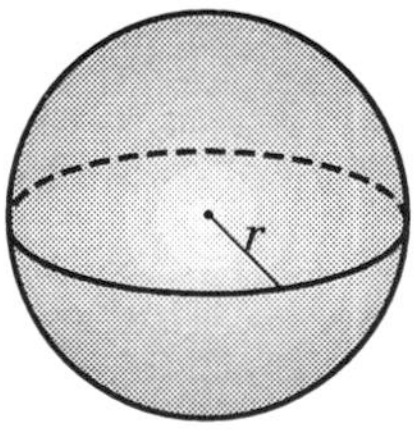

$V = \frac{4}{3} \pi r^3$

$S = 4\pi r^2$

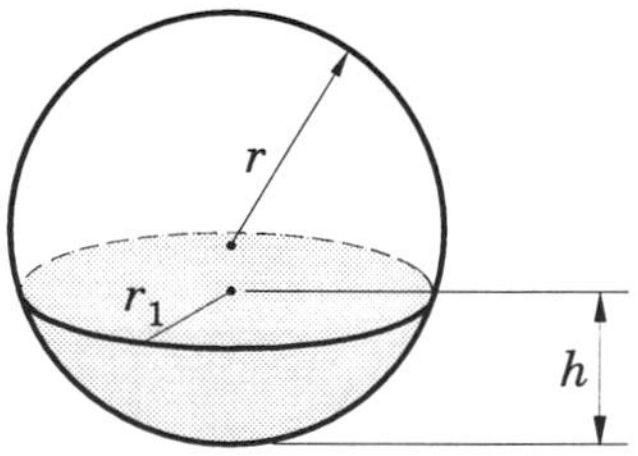

$V = \frac{1}{6} \pi h (3 r_1^2 + h^2)$

$S = 2\pi r h$

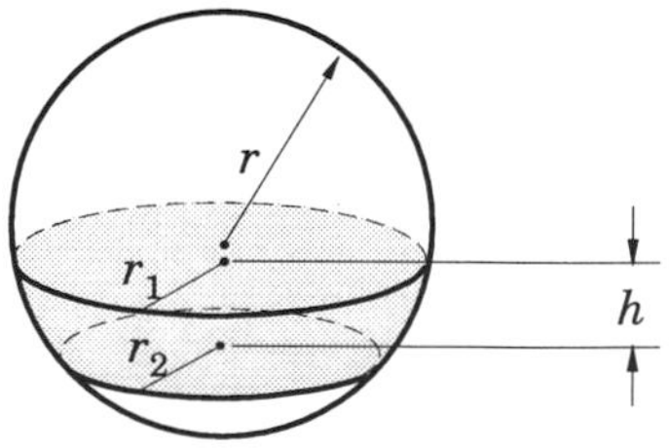

$V = \frac{1}{6} \pi h (3 r_1^2 + 3 r_2^2 + h^2)$

$S = 2\pi r h$

Spherical Triangle

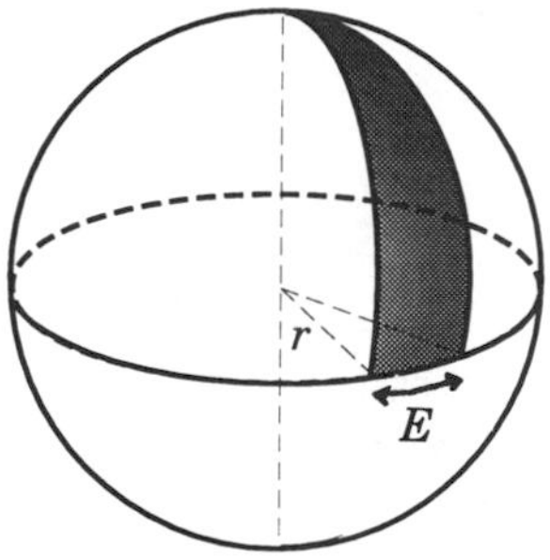

$$S = \pi r^2 \frac{E_{\text{degrees}}}{180} = r^2 E_{\text{radians}},$$

where E is the spherical excess; this is shown here for a semi-lune but is true for all spherical triangles.

Torus

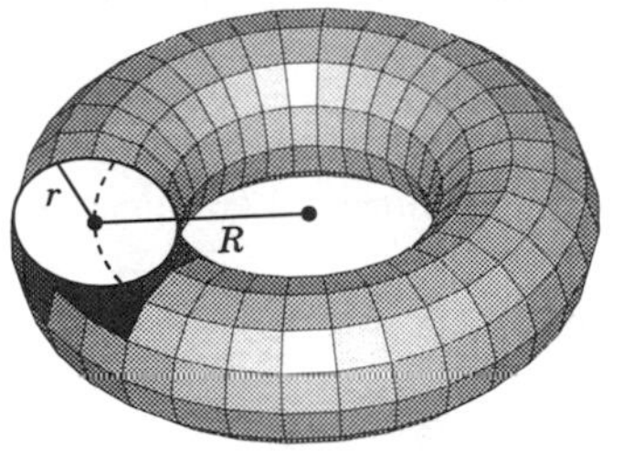

p. 884 $V = 2\pi^2 r^2 R$

p. 866 $S = 4\pi^2 r R$

Problems

- The corners of a cube are truncated so as to present a solid bounded by six regular octagons and eight equilateral triangles.

  Determine the area and volume of the truncated cube.

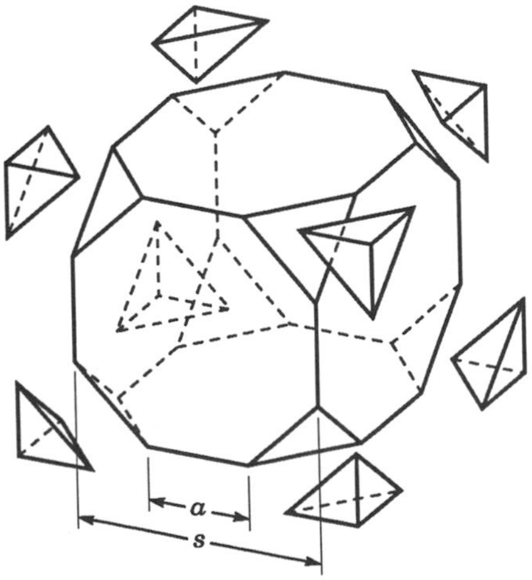

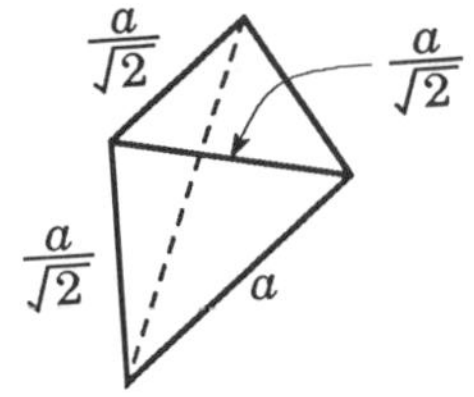

Let the side of the octagon be a and the cube edge s.

We then have

$$s = a + 2\frac{a}{\sqrt{2}} = a\left(\sqrt{2} + 1\right); \qquad a = s\left(\sqrt{2} - 1\right)$$

$$a^2 = s^2\left(3 - 2\sqrt{2}\right)$$

$$a^3 = s^3\left(5\sqrt{2} - 7\right)$$

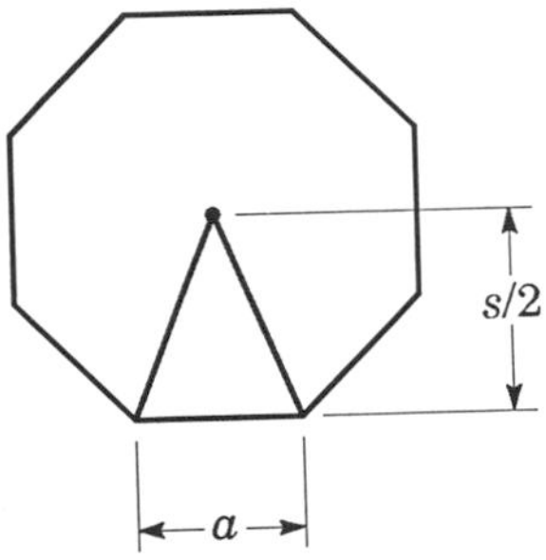

The area A of the truncated cube is the sum of six octagon areas A_8 and eight areas A_3 of the equilateral triangles.

An octagon consists of eight isosceles triangles with base a and height $s/2$; their aggregate area is

$$6A_8 = 6\cdot\left(8\cdot\frac{1}{2}\cdot a\cdot\frac{1}{2}\cdot s\right) = 12a^2\cdot\left(\sqrt{2}+1\right).$$

With the perimeter $2p$ of an equilateral triangle of side a,

$$2p = 3a; \qquad p = \frac{3a}{2}; \qquad (p-a) = \frac{a}{2},$$

p. 444

we have, with Heron, the aggregate area of eight triangles,

$$8\cdot A_3 = 8\cdot\sqrt{\frac{3a}{2}\left(\frac{a}{2}\right)^3} = 2a^2\sqrt{3},$$

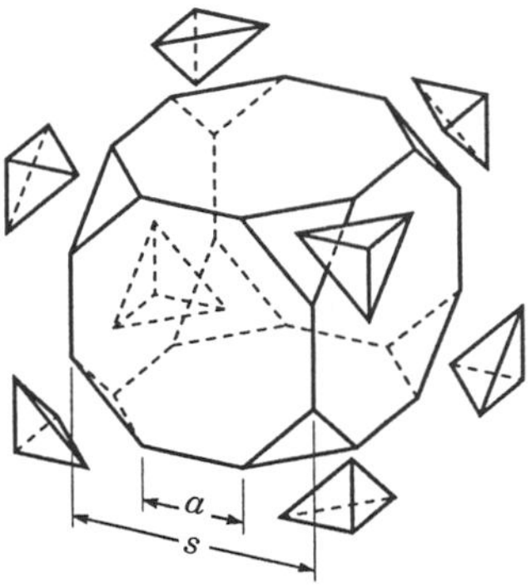

and the **total area**

$$A = 6A_8 + 8A_3 = a^2\left(12\sqrt{2} + 12 + 2\sqrt{3}\right),$$

or, in terms of s,

$$A = s^2\left(12\sqrt{2} + 12 + 2\sqrt{3}\right)\left(3 - 2\sqrt{2}\right)$$

$$\approx 5.565\, s^2.$$

The volume V of the truncated cube equals the volume V_0 of the intact cube less the aggregate volume $8V_3$ of eight right-angled corner pyramids,

$$V_0 = s^3$$

$$8V_3 = 8\cdot\frac{1}{3}\cdot\frac{1}{2}\cdot\left(\frac{a}{\sqrt{2}}\right)^3 = a^3\frac{\sqrt{2}}{3}$$

$$= s^3\frac{1}{3}\left(5\sqrt{2} - 7\right)\sqrt{2} = \frac{1}{3}s^3\left(10 - 7\sqrt{2}\right),$$

and the **total volume** is

$$V = V_0 - 8V_3 \approx 0.966\, s^3.$$

•

- Some chemical compounds which normally crystallize in a cubic structure will sometimes form “twin cubes”. These have a space diagonal in common, around which one cube has rotated through 60° relative to the other.

  Determine the total area and volume of the twin cube.

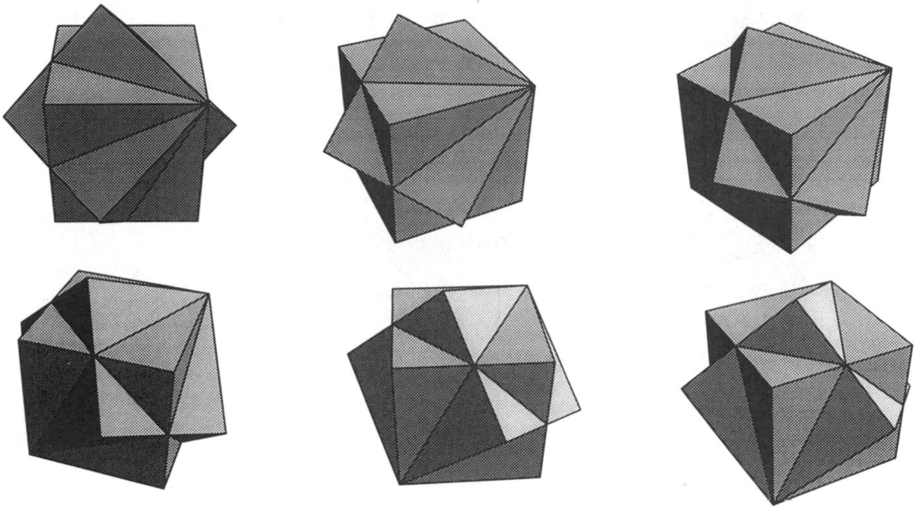

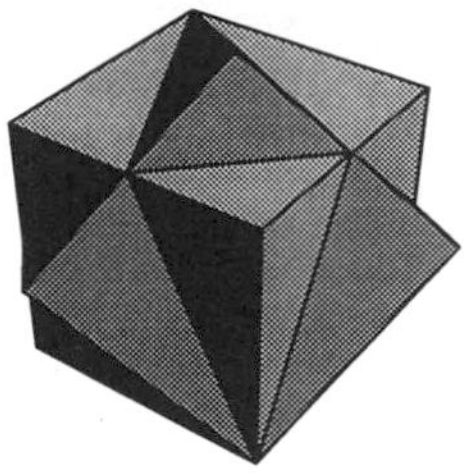

The twin cube may be regarded as a conventional cube, each of whose six faces carries a ridge formed by an edge and corner of the other cube. Each ridge adds three right triangles to the surface area of the twin cube,

two side flanks, with legs s and $\frac{s}{2}$;

one end face, with legs $\frac{s}{2}$ and $\frac{s}{2}$,

corresponding to an aggregate area

$$A_+ = 2\left(\frac{1}{2} \cdot s \cdot \frac{s}{2}\right) + \frac{1}{2} \cdot \frac{s}{2} \cdot \frac{s}{2} = \frac{s^2}{2} + \frac{s^2}{8} = \frac{5}{8}\, s^2 .$$

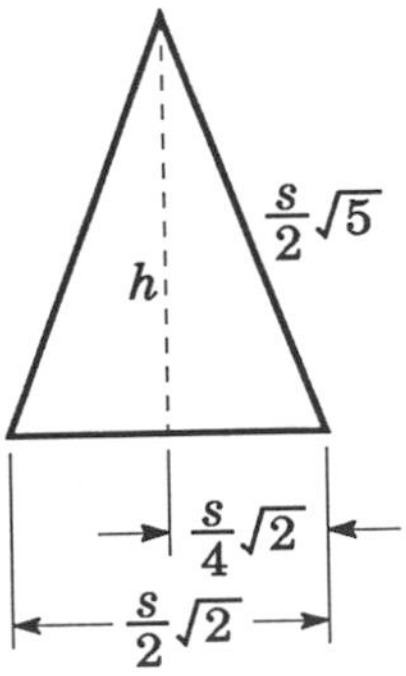

From this must be deducted the area A_- of the base surface on which the ridge is sitting, an isosceles triangle with

the base $\frac{s}{2}\sqrt{2}$

and the legs $\frac{s}{2}\sqrt{5}$, from $\sqrt{\left(\frac{s}{2}\right)^2 + s^2}$.

The area of this base surface can be calculated in two ways, $\frac{1}{2}$ base × height or by Heron's formula; we will demonstrate both.

The height of the triangle is

$$h = \frac{s}{4}\sqrt{\left(2\sqrt{5}\right)^2 - \left(\sqrt{2}\right)^2} = \frac{s}{4}\sqrt{18}$$

$$A_- = \frac{b \cdot h}{2} = \frac{1}{2} \cdot \frac{s}{2}\sqrt{2} \cdot \frac{s}{4}\sqrt{18} = \frac{6 \cdot s^2}{16} = \frac{3}{8} \cdot s^2 .$$

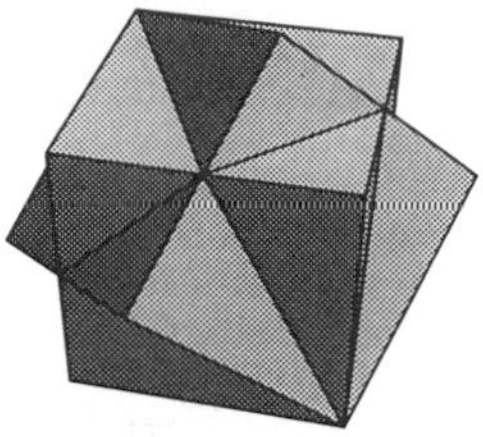

Heron's formula gives

$$2p = \frac{s}{2}\left(\sqrt{5} + \sqrt{5} + \sqrt{2}\right); \quad p = \frac{s}{4}\left(2\sqrt{5} + \sqrt{2}\right)$$

$$p-a = \frac{s}{4}\left(2\sqrt{5} + \sqrt{2} - 2\sqrt{2}\right) = \frac{s}{4}\left(2\sqrt{5} - \sqrt{2}\right)$$

$$p-b = p-c = \frac{s}{4}\left(2\sqrt{5} + \sqrt{2} - 2\sqrt{5}\right) = \frac{s}{4}\sqrt{2};$$

$$A_{-} = \frac{s^2}{16}\sqrt{\left(2\sqrt{5} + \sqrt{2}\right)\left(2\sqrt{5} - \sqrt{2}\right)\left(\sqrt{2}\right)^2}$$

$$= 6\frac{s^2}{16} = \frac{3}{8}s^2 .$$

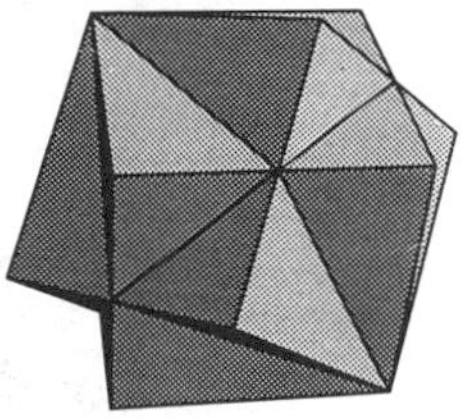

The **total area** of the twin cube is

$$6s^2 + 6\left(\frac{5}{8}s^2 - \frac{3}{8}s^2\right) = 7\frac{1}{2}s^2 .$$

The **total volume** of the twin cube is

$$s^3 + 6 \cdot \frac{1}{2} \cdot \frac{s}{2} \cdot \frac{s}{2} \cdot \frac{1}{3} \cdot s = 1\frac{1}{4}s^3 .$$

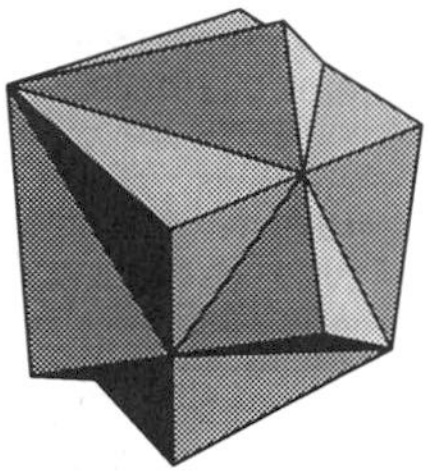
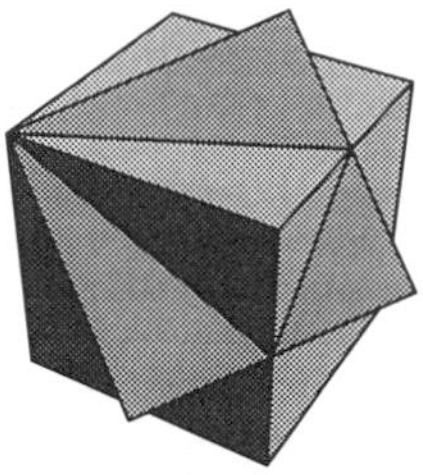
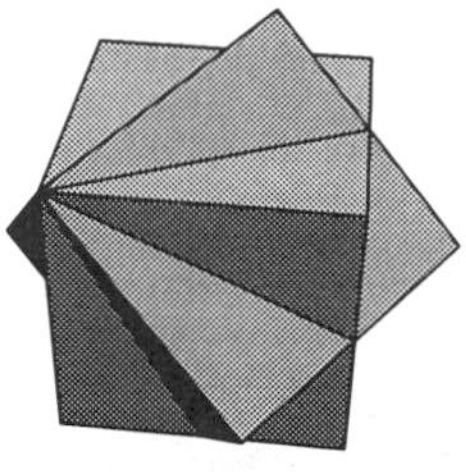
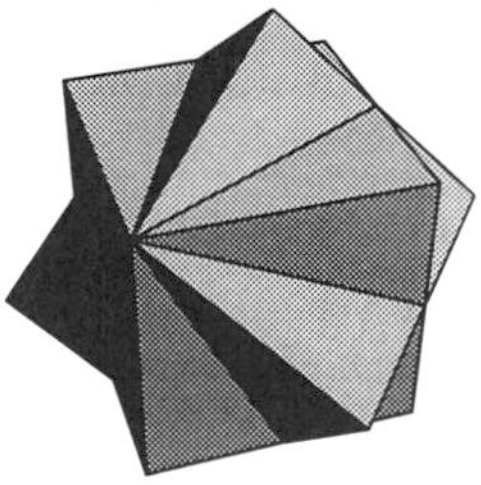
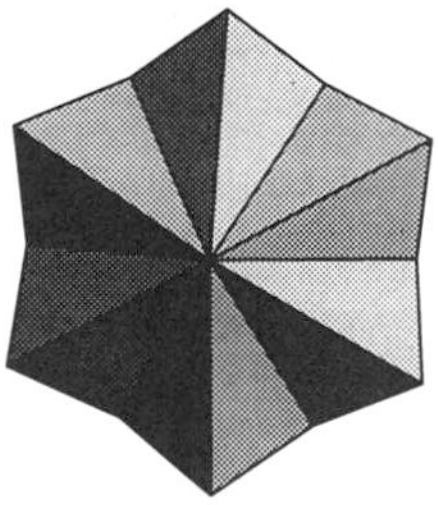
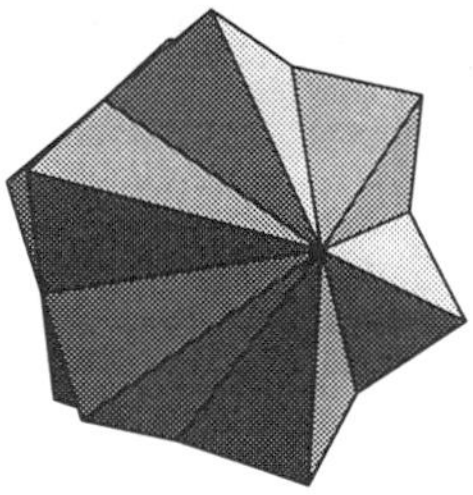

***

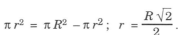

Answer to pie problem from page 447:

The pie has radius R; the inner circle has radius r.

$$\pi r^2 = \pi R^2 - \pi r^2 ; \quad r = \frac{R\sqrt{2}}{2} .$$

Chapter 13

TRIGONOMETRY

*** Indicates cross-reference

13.0 Scope and History

Greek, *trígonon*, "triangle"; *-metria*, "measurement"

Trigonometry developed from the study of right-angled triangles by applying their relations of sides and angles to the **study of similar triangles**.

Plane Trigonometry

Plane trigonometry relates exclusively to triangles in the two dimensions of the plane, but has applications in nearly every field of physics, such as radiation, the propagation of light and sound, alternating current, and all other periodic phenomena.

Spherical Trigonometry

Spherical trigonometry is concerned with triangles on the surfaces of spheres; its uses are more limited, the principal applications being to astronomy and long-range navigation.

The primary use of trigonometry is for operations of surveying, cartography, astronomy, and navigation. Modern mathematics has extended the uses of trigonometric functions far beyond a simple study of triangles to make trigonometry indispensable in many other areas.

PITISCUS Bartholomaeus (1561 - 1613)

Although made up of Greek phonemes, the term "trigonometry" is actually not a native Greek word. To the best of our knowledge, the term was invented by the German mathematician and astronomer **Bartholomaeus Pitiscus** and first appeared in his work *Trigonometria sive de solutione triangularum tractatus brevis et perspicius* ..., first published in 1595; it was revised in 1600 and published as *Trigonometria sive de dimensione triangulae*.

Trigonometric functions, which are today dimensionless numbers describing ratios of the sides of right-angled triangles, have a varied history.

The ancients had no clear conception of trigonometric functions as elements of a geometric science, but they knew how to use them to advantage in the furtherance of their chosen professions.

The old Egyptians looked upon trigonometric functions as features of similar triangles which were useful in land surveying and when building pyramids; the old Babylonian astronomers related trigonometric functions to arcs of circles and to the lengths of the chords subtending the arcs.

The ancient Greeks who took over from the Babylonians as astronomers developed trigonometry into an ordered science and constructed tables of chords expressed in terms of 60th "small parts" of the radius for every $\frac{1}{2}$ degree from 0° to 180° of the central angle.

Centuries later, trigonometric functions acquired a geometric interpretation when they came to be looked upon as the *lengths* of specific line segments *related to the central angle*.

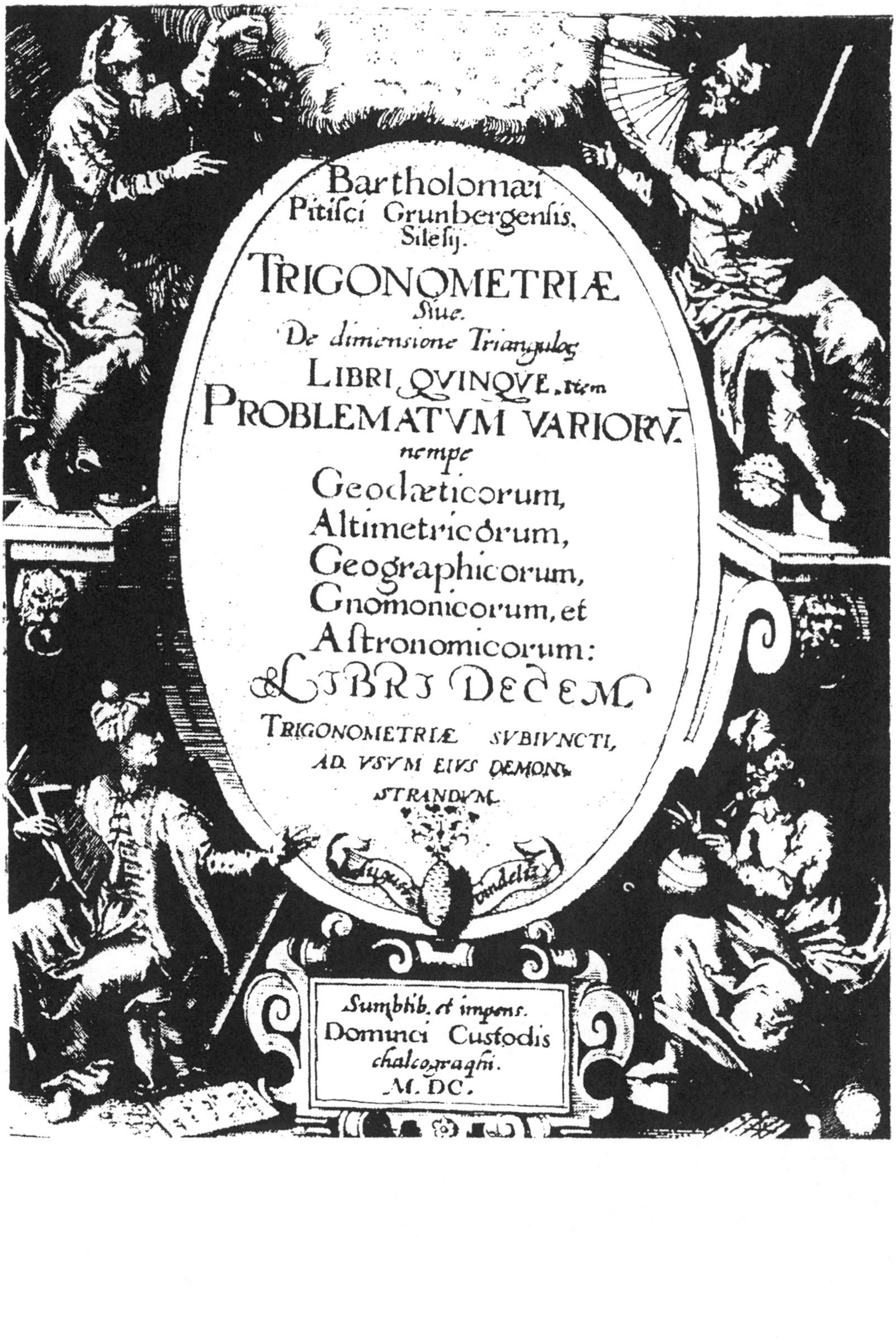
Bartholomæi
Pitisci Grunbergensis.
Silesij.
TRIGONOMETRIÆ
Siue.
De dimensione Triangulorum
LIBRI QVINQVE. Item
PROBLEMATVM VARIORV.
nempe
Geodæticorum,
Altimetricorum,
Geographicorum,
Gnomonicorum, et
Astronomicorum:
LIBRI DECEM
TRIGONOMETRIÆ SVBIVNCTI,
AD VSVM EIVS DEMON-
STRANDVM.
Sumtib. et impens.
Dominici Custodis
chalcographi.
M. DC.

In a circle whose radius is 1 unit of length – a unit circle – the radius OP is assumed to rotate in a counterclockwise direction. When moving from its initial position OA to the terminal position OP, it sweeps the angle $AOP = \theta$.

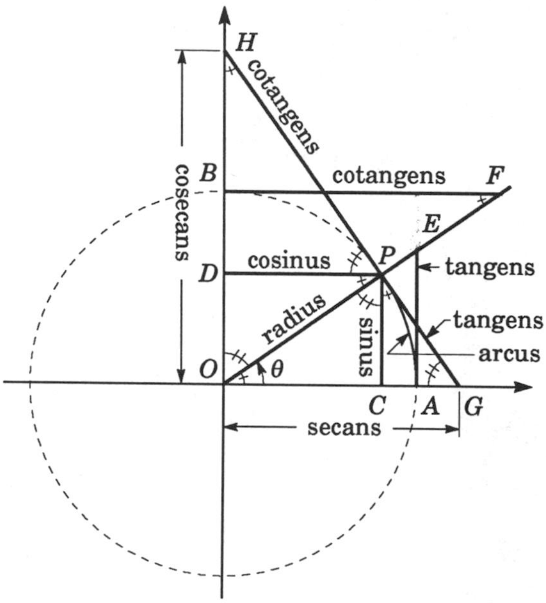

The tangents to the circle at points A and B meet the extension of the radius OP at E and F, respectively. The line segments PC, perpendicular to OA, and PD, at right angles to OB, help to form triangles OCP, OAE, and OBF, which are similar, angles being equal as marked.

The tangent GPH at P is divided into two segments: PG is equal to the tangent line segment AE, and PH is equal to the cotangent segment BF.

With $OA = OB = OP = 1$, and observing that the angle BOF is the complementary angle of $AOP = \theta$, we have the following equalities:

radius	$= OP = 1$	*tangens* θ	$= \dfrac{AE}{OA} = AE = PG$
arcus	$= AP$	*cotangens* θ	$= \dfrac{BF}{OB} = BF = PH$
sinus θ	$= \dfrac{PC}{OP} = PC = OD$	*secans* θ	$= \dfrac{OE}{OA} = OE = OG$
cosinus θ	$= \dfrac{OC}{OP} = OC = PD$	*cosecans* θ	$= \dfrac{OF}{OB} = OF = OH$

ARYABHATA Kusumapura (476 - c. 550)

The origin of the term "sinus" – our "sine" – is Indian. The Hindu mathematician and astronomer **Aryabhata** the Elder called it *ardha-jya* ("half-chord"), later abbreviated to *jya* ("chord"). Arab translators turned this phonetically into *jiba* – a meaningless word in Arabic – and, according to Arabic practice of omitting the vowels in writing, wrote it *jb*.

GERARD of Cremona
(*c.* 1114 - 1187)

Arabic-to-Latin translators, with no knowledge of the Sanskrit origin, assumed *jb* to be an abbreviation of *jaib*, which is good Arabic for "cove", "bay", "bulge", "bosom". **Gerard of Cremona**, who translated the *Almagest* in the late 12th century, substituted for *jaib* its Latin equivalent *sinus*.

The names of the other trigonometric functions have less complicated interpretations. In the figure, *tangens* of θ is represented by the length of the *tangent* segment (AE), *secans* of θ by the *secant* segment (OE). The *cotangens* and *cosecans* are, of course, the *tangens* and *secans* of the complementary angle.

FINCKE Thomas
(1561 - 1656)

The *tangens*, *cotangens*, *secans*, and *cosecans* functions have been known by various names; the present names – *tangens* from Latin *tangere*, "to touch", and *secans* from *secare*, "to cut" – were introduced in 1583 by the Dane **Thomas Fincke**.

GUNTER Edmund
(1581 - 1626)

The names *cosinus* and *cotangens*, suggested in 1620 by the English mathematician and astronomer **Edmund Gunter**, replace the older terms *sinus complementi* and *tangens complementi* – *sinus* and *tangens* of the complementary angles.

Masters of the Angle

There are problems in the Rhind Papyrus from about 1650 B.C. that describe the measurement of face slope angles of pyramids, involving what is today known as the cotangent function of the dihedral angle at the base of the pyramid.

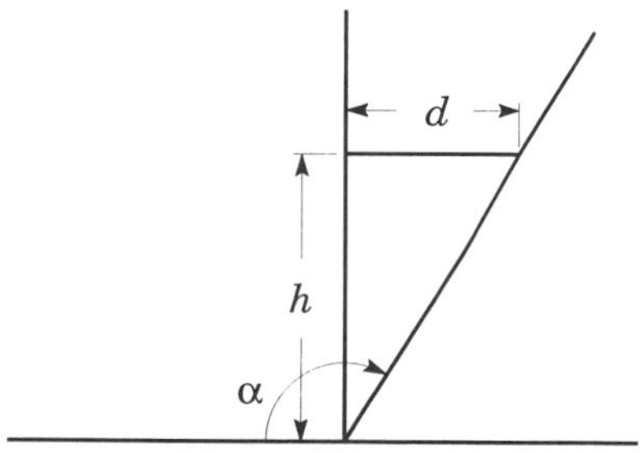

A Babylonian clay tablet from around 1900 - 1600 B.C. contains, in cuneiform script, a table of the secant functions of fifteen angles between 30° and 45°. The wealth of astronomical data collected by the Babylonians was passed on to the ancient Greeks and gave rise to spherical trigonometry, essential in astronomy and navigation.

THALES
(*c.* 625 - *c.* 547 B.C.)

The Greek philosopher **Thales** of Miletos is said to have made use of similar triangles to determine the height of the Cheops pyramid by comparing the length of its shadow with the length of the shadow cast by a rod of known length; he also determined the distances offshore of ships at sea.

ARISTARCHOS
(*c.* 310 - 250 B.C.)

Aristarchus of Samos, mathematician and astronomer, and the first man to propound a heliocentric theory of the Universe – eighteen centuries before Copernicus – made an attempt to compare the distances from Earth to the Sun and to the Moon.

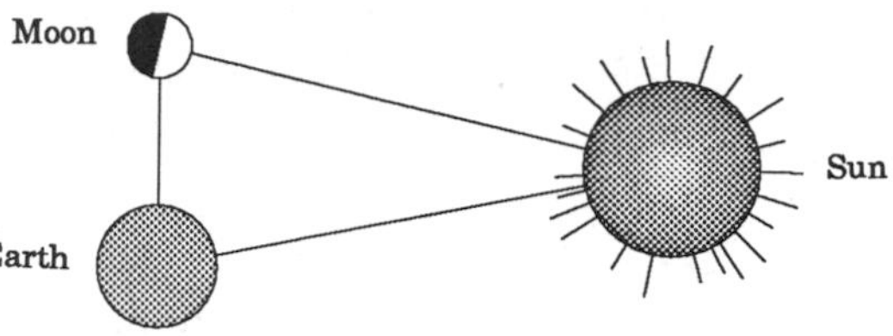

Although his reasoning was perfectly sound, the instrument he used to determine the angle of sight between the Sun and the half Moon failed him by faulty calibration – he found the distance to the Sun to be about 18 to 20 times that to the Moon, instead of the correct figure of approximately 390 times.

ERATOSTHENES
(*c.* 276 - 194 B.C.)

The first known attempt to calculate the circumference of the Earth was made by **Eratosthenes** of Alexandria. Having heard that, at the summer solstice, the Sun was at zenith at Syene – modern Aswan – he decided to determine the height of the Sun also at Alexandria. From his measurements he deduced, correctly, that the distance between Alexandria and Syene must equal 1/50th of the Earth's circumference, but all his other data were inaccurate or pure guesswork.

HIPPARCHOS
(? - after 127 B.C.)

Apart from delivering severe criticism of the disdain for fundamental data and the rather cavalier treatment of them evinced by Eratosthenes, **Hipparchus** of Nicea and Rhodes is justly famous for his manifold contributions to astronomy; his achievements include the determination of the length of the lunar month to within one second of today's accepted value and accurate calculations of the inclination of the ecliptic and of the changes of the equinoxes.

As a mathematician, Hipparchus introduced to Greek science the Babylonian method of dividing the circle into 360°.

It was common practice in ancient Greece, when solving triangles, to assume them to be inscribed in a circle, their sides being chords of the circle. Hipparchus is known to have prepared extensive tables of chords at $\frac{1}{2}$-degree intervals for all central angles from 0.5° to 180°, calculated to three positions by the Babylonian sexagesimal system.

The radius of the circle was divided into 60 equal segments, called "small parts" used as unit of length; the first figure of the chord length refers to the number of small parts needed to make up the length of the chord, the second figure is the number of 60ths of those parts, and the third figure is the number of 3600ths.

Like all of Hipparchus's numerous works, except one, this table is lost; knowledge of most of Hipparchus's work has come down to us through Ptolemy.

MENELAOS
(*c.* A.D. 100)

Menelaus of Alexandria, in the first century A.D., wrote a six-book treatise on chords – mentioned by Theon of Alexandria, but now lost – and also made important contributions to spherical trigonometry, providing later writers with a basis for further development of the subject. It is interesting to note that early achievements in trigonometry were predominantly directed toward spherical trigonometry, in response to the need for accurate information in astronomy.

A three-book work by Menelaus, called *Sphaerica*, is extant in an Arabic translation and provides excellent information about Greek development of trigonometry.

PTOLEMAEUS Claudius
(*c.* 100 - *c.* 168)

Arabic,. *al majisti*, "the greatest"; Arabicized superlative form of Greek *megas*, "great" and *magiste*, "greatest"

In the second century A.D., the great mathematician and astronomer **Ptolemy** of Alexandria presented an extensive treatise on all aspects of astronomy in his 13-book *Syntaxis mathematica* ("Mathematic Collection"), also known as "the greatest of the great", and commonly referred to as the *Almagest*, from a Latin distortion of the Arabic *al-majisti*, meaning "the greatest".

Gregor Reisch, *Margarita phylosophica* (1496; 1503 reprint). Escorted by Dame Astronomy, Ptolemy is using a quadrant, an instrument for measuring altitudes. Ptolemy is wrongly identified as one of the 15 Ptolemy kings of Egypt (323 - 30 B.C.) and, therefore, drawn wearing a crown. In the left lower corner is an *armillary sphere* (an ancient instrument composed of rings showing the positions of important circles of the *celestial sphere* (that is, the imaginary sphere against which the celestial bodies appear).

The first book of the *Almagest* contains tables of chords for all arcs 0° to 180° at 0.5-degree intervals to at least five places of decimals, believed to be the table prepared by Hipparchus and mentioned by Theon. The tables are equivalent to a table of sines for all central angles 0° to 90° at 15' intervals.

The *Almagest* also contains theorems corresponding to the present-day law of sines and compound-angle and half-angle identities.

That so much of early Greek work on astronomy has been lost could be a result of the completeness and elegance of presentation of Ptolemy's *Almagest*, making all earlier works appear superfluous.

In the 9th to 14th centuries, trigonometry was further developed by Arabic and Persian scholars. As with Babylonian and Greek trigonometry, the incentive came from interest in its application to astronomy.

ABU AL-WAFA
(940 - 997 or 998)

umbra versa: " shadow (ghost) behind *arcus*"

Building further on Ptolemy's *Almagest*, in the 10th century, the Persian **Abu al-Wafa** of Baghdad systematized theorems and proofs of trigonometry and prepared extensive trigonometric tables of sines and tangents for 15' intervals; he is believed to have introduced the concept of the tangent function – called *umbra versa* – and possibly also the secant and the cosecant.

NASIR AD-DIN
(1201 - 1274)

The first presentation of trigonometry as a science independent of astronomy is credited to the Persian **Nasir ad-Din** in the 13th century. In his writings we meet, also for the first time, plane trigonometry as a discipline in its own right, separate from spherical trigonometry.

ADELARD of Bath
(active *c.* 1116 - 1142)

JOHN of Seville

ROBERT of Chester

PLATO of Tivoli
(*c.* 1100)

GERARD of Cremona
(*c.* 1114 - 1187)

In the 12th century, many Greek, Hebrew, and Arabic texts were translated into Latin, often by Jewish translators working in Spain. Notable translators were **Adelard of Bath**, **John of Seville**, **Robert of Chester**, **Plato of Tivoli**, and **Gerard of Cremona** – the translator of the *Almagest* into Latin.

FIBONACCI Leonardo
(*c.* 1170 - *c.* 1250)

Leonardo Fibonacci, also known as **Leonardo Pisano**, became acquainted with trigonometry during his extensive travels in Arab countries; his newly acquired knowledge was presented in *Practica geometriae* in 1220.

REGIOMONTANUS
(1436 - 1476)

Around 1464 the German astronomer and mathematician **Regiomontanus**, also known as Iohannes Molitoris, compiled *De triangulis omnimodis*, a compendium of the trigonometry of that time. After being printed in 1533, it became an important medium for spreading knowledge of trigonometry all over Europe.

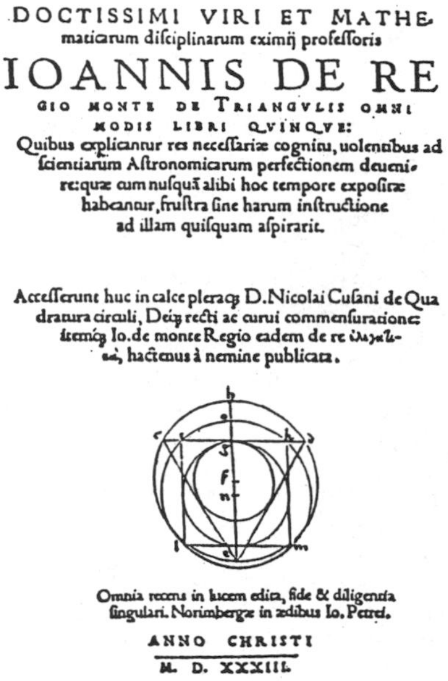
DOCTISSIMI VIRI ET MATHE-
maticarum disciplinarum eximij professoris
IOANNIS DE RE
GIO MONTE DE TRIANGVLIS OMNI
MODIS LIBRI QVINQVE:
Quibus explicantur res necessariae cognitu, uolentibus ad
scientiarum Astronomicarum perfectionem deueni-
re: quae cum nusquã alibi hoc tempore expositae
habeantur, frustra sine harum instructione
ad illam quisquam aspirarit.

Accesserunt huc in calce pleraq; D. Nicolai Cusani de Qua
dratura circuli, Deq; recti ac curui commensuratione:
itemq; Io. de monte Regio eadem de re [illegible]
[illegible], hactenus à nemine publicata.

Omnia recens in lucem edita, fide & diligentia
singulari. Norimbergae in aedibus Io. Petrei.
ANNO CHRISTI
M. D. XXXIII.

RHÆTICUS Georg Joachim
(1514 - 1574)

The German mathematician and astronomer **Georg Joachim Rhæticus** in 1542 published, separately, under the title *De lateribus et angulis triangulorum*, the section of Copernicus's *De revolutionibus* dealing with plane and spherical trigonometry, with an added table of half-chords subtended by arcs of a circle – that is, in effect, a table of sines, although both he and Copernicus avoided the term *sinus*.

Rhæticus's sine table differs from that of Copernicus in two respects: its radius has been increased from 100 000 to 10 000 000, and the interval of the central angle is 1' instead of 10'. The high values of the radii were used to avoid fractions.

Rhæticus's *Canon doctrinae triangulorum*, published at Leipzig in 1551, was the first publication to give all six of the fundamental trigonometric functions, and the first in which they were defined as ratios of the lengths of the sides of a right triangle and related directly to the angles – not, as before, as lengths of line segments or ratios of an arc.

Rhæticus managed to reduce by half the space occupied by his tables by equating functions of angles greater than 45° with the corresponding cofunctions of their complementary angles.

OTHO Valentinus
(c. 1550 - 1605)

After the publication of *Doctrinae*, Rhæticus embarked upon the project which was to occupy him and a number of computing collaborators for at least 12 years; on the death of Rhæticus in 1574, **Valentine Otho**, a mathematics student at Wittenberg, took over the enormous task of compiling the manuscript for the *Opus palatinum de triangulis*, to consist of a canon of sines for the range of 0° to 90° at 10" intervals, and for every 1" within the ranges of 0° to 1° and 89° to 90°, all to 15 decimal places, and a complete canon of sines, tangents, and secants to 10 or 15 decimal places.

The *Opus palatinum* was eventually published in 1596, but it was marred by systematic errors in the tangents and secants between 83° and 90°, later to be corrected by Bartholomaeus Pitiscus.

FINCKE Thomas
(1561 - 1656)

The Danish physician and mathematician **Thomas Fincke** published in 1583, at Basle, a 14-book treatise called *Geometriae rotundi...* with discourses on plane and spherical trigonometry, in which he introduced the terms *tangens* and *secans*, and several new formulas, *e.g.*, the law of tangents.

PITISCUS Bartholomaeus
(1561 - 1613)

The German mathematician **Bartholomaeus Pitiscus** published at Heidelberg, in 1595, a treatise named *Trigonometria: sive de solutione triangularum tractatus brevis et perspicuus*, noted for the first appearance in print of the word **trigonometry**.

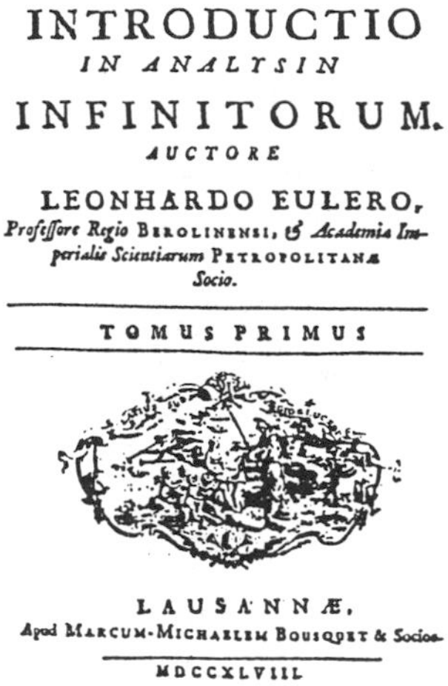
INTRODUCTIO
IN ANALYSIN
INFINITORUM.
AUCTORE
LEONHARDO EULERO,
Professore Regio Berolinensi, & Academiæ Imperialis Scientiarum Petropolitanæ Socio.
TOMUS PRIMUS
LAUSANNÆ,
Apud Marcum-Michaelem Bousquet & Socios.
MDCCXLVIII

A revised version of the work appeared at Augsburg, in 1600, as *Trigonometriae sive de dimensione triangulae*; it treats all aspects of plane and spherical trigonometry: definitions of functions, methods of solving triangles, trigonometric identities, and tables for the sine, tangent, and secant functions. Enlarged versions were published at Augsburg in 1609 and at Frankfurt/Main in 1614.

Pitiscus's works became valuable textbooks; English translations appeared in 1614, French in 1619. The English text – *A Canon of Triangles: or the Tables of Sines, Tangents and Secants, the Radius Assumed to be 100 000* – is based on the 1595 *Trigonometria* and on trigonometric tables from later publications by Pitiscus.

EULER Leonhard
(1707 - 1783)

The transformation to modern trigonometry is to a great extent attributed to work by the Swiss mathematician **Leonhard Euler**, whose name is also firmly linked with fundamental contributions to calculus and topology. For almost a hundred years, Euler's *Introductio in analysin infinitorum* (1748) was the dominating textbook of trigonometry and other fundamental areas of mathematics.

The history of trigonometry, and of geometry in general, is linked to technical advancement in instrument making. For practical land measure, the ancient surveyors' simple instruments – a knotted rope, a wooden rod, and a tool similar to an ordinary carpenter's square – usually gave satisfactory results. But we have also seen – in Aristarchus's calculations of interplanetary distances – how an inaccurate instrument can invalidate what is right in theory.

HULSIUS Levinus
(? - 1606)

Born in Flanders; lived and worked in The Netherlands, England, and Germany. Maker of fine instruments; publisher and printer; linguist and lexicographer; wrote extensively on the construction of geometrical instruments.

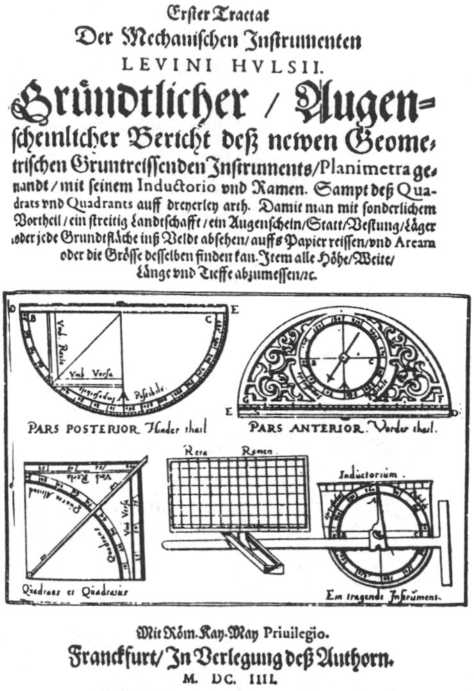

Erster Tractat
Der Mechanischen Instrumenten
LEVINI HVLSII.
Gründtlicher / Augenscheinlicher Bericht deß newen Geometrischen Gruntreissenden Instruments/Planimetra genandt / mit seinem Inductorio vnd Ramen. Sampt deß Quadrats vnd Quadrants auff dreyerley arth. Damit man mit sonderlichem Vortheil / ein streitig Landtschafft / ein Augenschein/Statt/Vestung/Läger oder jede Grundtfläche inß Veldt absehen/ auffs Papier reissen/vnd Aream oder die Grösse desselben finden kan. Item alle Höhe/Weite/ Länge vnd Tieffe abzumessen/rc.

Mit Röm. Kay. May Priuilegio.
Franckfurt/ In Verlegung deß Authorn.
M. DC. IIII.

The Swedish astronomer **Anders Celsius** (1701 - 1744), was one of the members of the 1736 expedition to Lapland to measure the length of the degree along the meridian. The results of this expedition, led by French mathematician and astronomer **Pierre-Louis Maupertuis** (1698 - 1759), verified Isaac Newton's hypothesis that the Earth is flattened at the poles and thus resembles a spheroid (*p.* 405) more than a sphere.

"Celsius the Astronomer" Linoleum cut by Knut Erik Lindberg (1921 - 1988).

In his left hand, Celsius holds a sextant pointed at the North Star; in his right hand is a thermometer.

Wangled Angles

ERATOSTHENES
(*c.* 276 - 194 B.C.)

Eratosthenes, head librarian of the science library at Alexandria – the largest in Antiquity – was also active as a mathematician and a geographer.

As a geographer, he is best known for his *Map of the World*, with a zero meridian and a "zero" parallel both passing through Alexandria.

STRABON
(64 or 63 B.C. - *c.* A.D. 25)

As a mathematician, he was known for his boasts that he had solved the (unsolvable) Delian problem of doubling the cube. The eminent Greek geographer **Strabo** was later to characterize him as "a mathematician among geographers and a geographer among mathematicians".

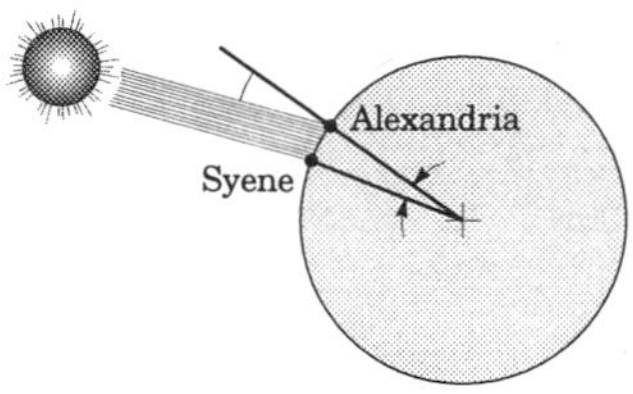

Around 230 B.C. Eratosthenes was told that at Syene in southern Egypt the Sun was at zenith at the time of the summer solstice; its rays were penetrating vertically down to the bottom of a deep well on the Elephantine Island in the Nile without illuminating the walls of the well.

He assumed that Syene was situated on the Tropic of Cancer (the most northerly latitude at which the Sun can be seen at zenith) and that it was located due south of Alexandria – it said so on the world map that he had himself prepared. Actually, Syene (now Aswan) is about 3 degrees east of Alexandria, corresponding to a difference of 12 minutes of the clock in the Sun's passing through the meridian – of no practical importance for the measurements, but a fact.

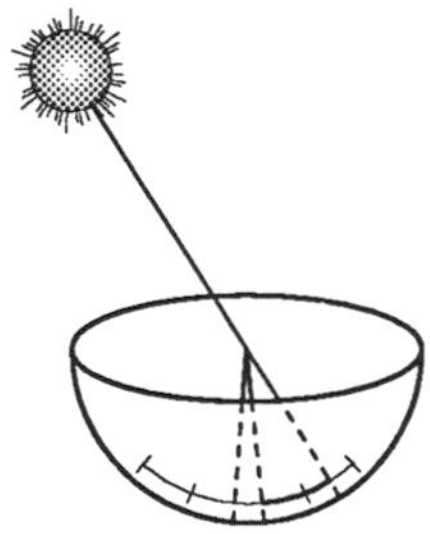

At Alexandria, he measured the angle of incidence of the Sun's rays with a **scaphe** – a hemispherical bowl with a needle-shaped gnomon placed at its center. The inside surface of the bowl is graduated to permit the length of the arc of shadow cast by the gnomon to be read.

From the observed length of this shadow Erathostenes could deduce that his baseline of measurement – the arc of the Earth's surface between Alexandria and Syene – must occupy 1/50th of the Earth's circumference along the meridian. Referring to his world map, it is open to some doubt what he meant by "circumference" and "meridian", as the map shows his world as a large flat mass of land floating in the Oceanus.

What the distance between Alexandria and Syene might be, he did not know; nor did anybody else at the time. Without recourse to trigonometric methods – which he certainly did not have – it was impossible at his time to measure such distances.

Erathostenes had been told, however, by wayfarers and camel drivers that one could go by caravan from Syene to Alexandria in 25 days – or even in 20 days by fast camel. Erathostenes therefore decided on a distance of 5000 stadia – divisible by both 20 and 25 – though there were knowledgeable people who held that the distance was 4530 stadia. It could be that both parties were right – or both wrong – as there were at least seven different lengths to choose from for a stadion; we have no means today of knowing which one he used for his calculations.

Having decided in favor of 5000 stadia, he calculated the Earth's circumference as 50 x 5000 = 250 000 stadia. This figure had the disadvantage, however, of not being divisible by 60, disagreeing with Egyptian practice. So – as attested by Hipparchus and Strabo – he added another 2000 stadia for good measure, to bring the supposed circumference up to 252 000 stadia, which *is* divisible by 60.

Eratosthenes' assumed 5000 stadia was pure guesswork, and so was his 250 000 stadia, not to speak of the 2000 extra stadia and his reason for adding them, a swindle on top of his surmise. The only thing that Eratosthenes actually *knew* was that the distance between Syene and Alexandria occupied 1/50th of the Earth's circumference.

Eratosthenes' method of calculating the circumference of the Earth is scientifically sound, but most of his data were inaccurate or pure guesswork – if not worse. His contemporaries recognized him for an eclectic and an opportunist, but many later-day chroniclers have allowed themselves to be fooled by the fact that his various guesses, mistakes, and errors largely managed to cancel each other out.

Knowing the correct figure of the Earth's circumference to be very near 40 000 km and the *actual* distance between Alexandria and Syene/Aswan to be just under 800 km, they make 5000 stadia correspond to 800 km, and so,

$$50 \times 5000 \text{ stadia} = 50 \times 800 = 40\,000 \text{ km}.$$

Some people call this kind of science "fidelity to ancient tradition".

Mystery Angles

Ale's Stones. A stone ship ...

Located on the southern coast of Sweden, *Ale's Stones* were long thought to be only a stone ship raised to honor a Viking chief named Ale (pronounced *a* as in "f*a*ther", *e* as in p*e*t).

If, in fact, just a stone ship, it is uniquely composed of antipodal *parabolas* – other stone ships generally have sides shaped in an arc of a circle. Though the stones are not in perfect parabolic order today, investigations confirm an original parabolic design; stones that had fallen were reerected slightly out of line with the original design.

[Roslund, Curt: See Bibliography]

A captivating question now is whether this majestic arrangement of stones might represent an astronomical instrument, like the 4700-year-old construction at Stonehenge, near Salisbury in Wiltshire, England. Among the supporting testimonies are the parabolic outline and the stern of the ship pointing toward the winter solstice, but crucial evidence of remains from an era matching that of Stonehenge still falls short; in fact, radiocarbon dating of charred wood found on the site sets the age of that wood at about A.D. 400 – considerably younger than the supposed age of Stonehenge.

Investigations and speculations will no doubt continue.

… or an astronomical instrument?

13.1 Fundamental Trigonometric Functions

The ratios of the lengths of any two sides in a right triangle will depend on the magnitude of the acute angles of the triangle; these ratios, of which there are six, are called **fundamental trigonometric functions** of the angles.

Referred to the angle α in the triangle below, these functions are:

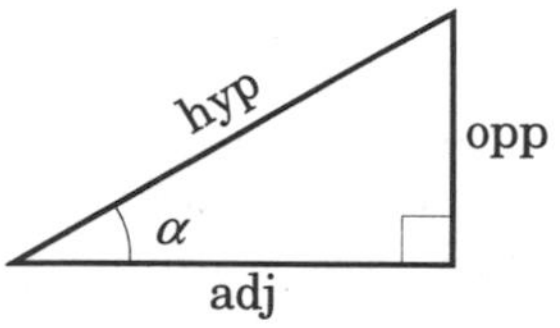

sine	$\sin\alpha = \dfrac{\text{opposite side}}{\text{hypotenuse}}$ read: "sine of α"	*sinus*
cosine	$\cos\alpha = \dfrac{\text{adjacent side}}{\text{hypotenuse}}$	*cosinus*
tangent	$\tan\alpha = \dfrac{\text{opposite side}}{\text{adjacent side}}$	*tangens*
cotangent	$\cot\alpha = \dfrac{\text{adjacent side}}{\text{opposite side}}$	*cotangens*
secant	$\sec\alpha = \dfrac{\text{hypotenuse}}{\text{adjacent side}}$	*secans*
cosecant	$\csc\alpha = \dfrac{\text{hypotenuse}}{\text{opposite side}}$	*cosecans*

These definitions lead to the following relations between the functions:

$$\tan = \frac{\sin}{\cos} \qquad \cot = \frac{\cos}{\sin} \qquad \sec = \frac{1}{\cos} \qquad \csc = \frac{1}{\sin}$$

The cotangent is the reciprocal of the tangent function, the secant of the cosine, and the cosecant of the sine:

$$\cot\alpha = \frac{1}{\tan\alpha} \qquad \sec\alpha = \frac{1}{\cos\alpha} \qquad \csc\alpha = \frac{1}{\sin\alpha}$$

Exponential Form

The fundamental trigonometric functions can be defined in terms of exponential functions and the imaginary unit; thus,

$$\sin x = \frac{e^{ix} - e^{-ix}}{2i} \qquad \cos x = \frac{e^{ix} + e^{-ix}}{2}$$

$$\tan x = \frac{-i\,(e^{ix} - e^{-ix})}{e^{ix} + e^{-ix}} \qquad \cot x = \frac{i\,(e^{ix} + e^{-ix})}{e^{ix} - e^{-ix}}$$

where e denotes the base of the natural logarithms and i denotes the imaginary unit.

Complementary Angles

Two angles whose sum is a right angle – 90° – are called **complementary angles**; *e.g.*, the acute angles of a right-angled triangle:

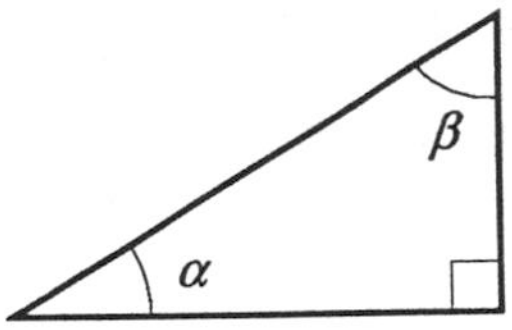

$$\left.\begin{aligned}\alpha &= 90^\circ - \beta\\ \beta &= 90^\circ - \alpha\end{aligned}\right\}\quad \alpha+\beta = 90^\circ$$

Consequently, the *sine* of one of the acute angles is always equal to the *cosine* of the other, complementary, angle; and correspondingly for all other functions and cofunctions:

$$\sin\alpha = \cos\beta \qquad \tan\alpha = \cot\beta \qquad \sec\alpha = \csc\beta$$
$$\cos\alpha = \sin\beta \qquad \cot\alpha = \tan\beta \qquad \csc\alpha = \sec\beta$$

The Orthogonal Coordinate System

Origin

x-axis, *y*-axis

The fundamental trigonometric functions, so far defined only for acute angles, may be extended to apply for all angles with the help of unit circles, of radius 1, centered at the **origin** of an **orthogonal coordinate system** with two real-number axes, a horizontal ***x*-axis** and a vertical ***y*-axis**, intersecting at right angles at the origin.

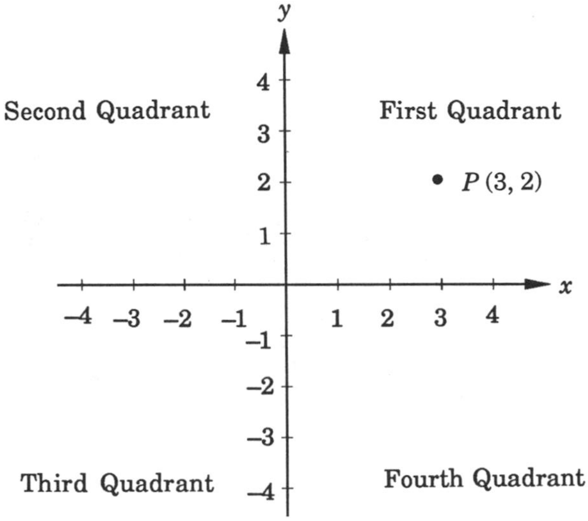

Abscissa

Ordinate

An arbitrary point on the coordinate plane is defined by its perpendicular distances from each of the two axes: the *x*-coordinate, or **abscissa**, is its horizontal distance from the *y*-axis; its *y*-coordinate, or **ordinate**, is the vertical distance from the *x*-axis. The point *P* in the graph above has the abscissa = 2, and the ordinate = 3.

Radius

Positive Angle

Negative Angle

A ray issuing from the origin of the coordinate system is known as a **radius**; the **angle** it forms with the positive *x*-axis counts **positive** if produced by *counterclockwise* (CCW) rotation and **negative** if caused by *clockwise* (CW) rotation of the radius:

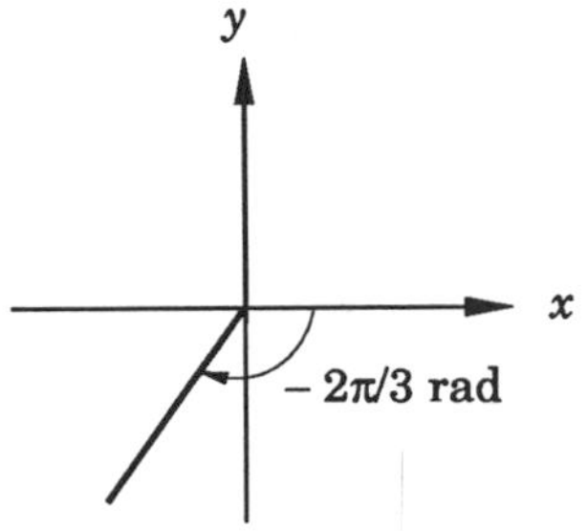

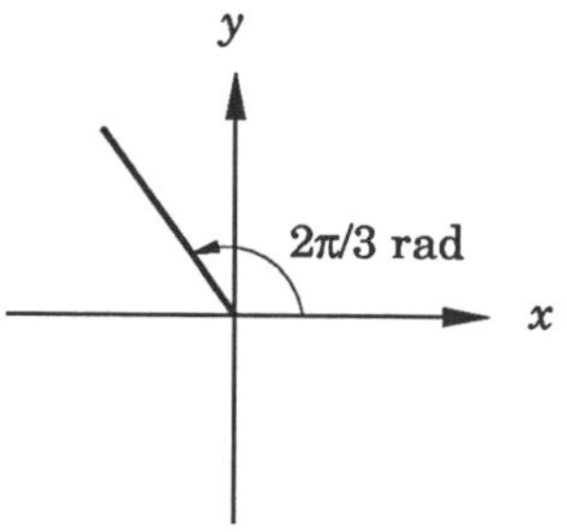

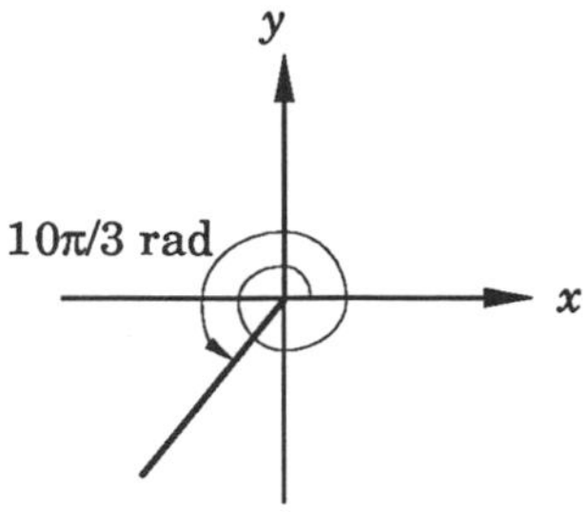

Observing that the radius of a unit circle is $r = 1$, we have by our previous definition that, for acute positive angles θ,

$$\sin\theta = \frac{y}{r} \qquad \tan\theta = \frac{y}{x}$$

$$\cos\theta = \frac{x}{r} \qquad \cot\theta = \frac{x}{y}$$

and we may declare these equations to be definitions for *all* possible values of θ:

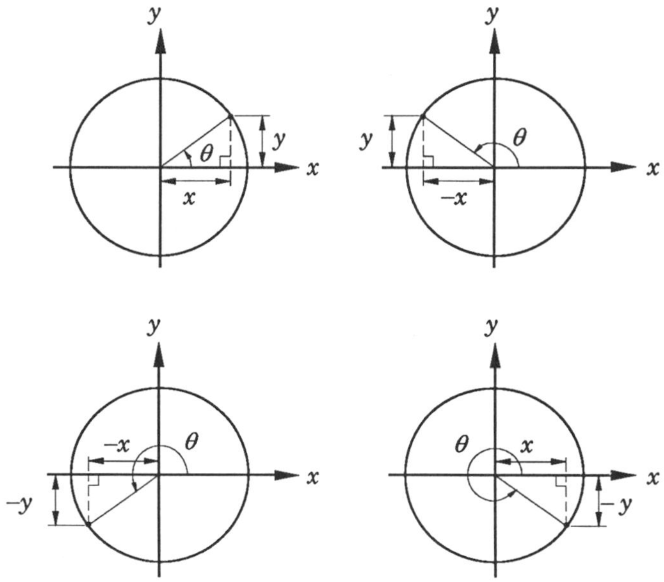

As an angle increases through all four quadrants, the definitions remaining the same, the sign of the trigonometric functions will change:

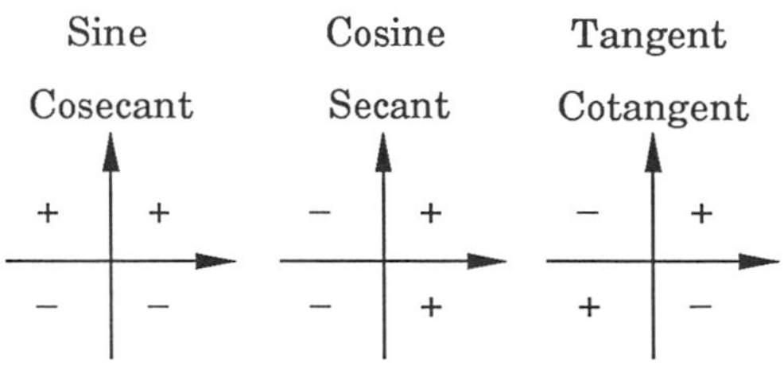

Values of Trigonometric Functions for Any Angle

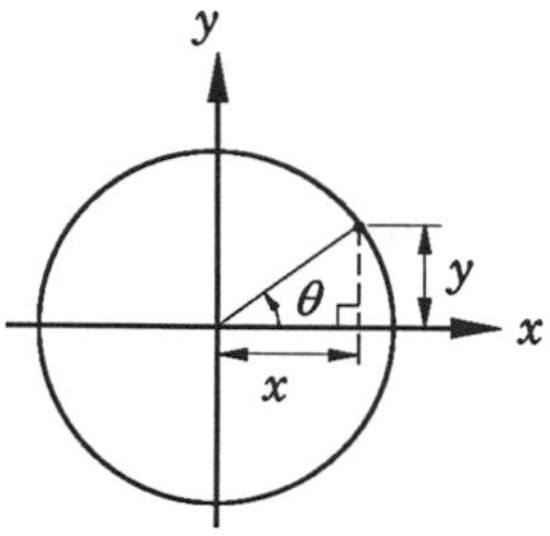

First Quadrant: Acute Angles

$$\sin\theta = \frac{y}{1} = y \qquad \csc\theta = \frac{1}{y}$$

$$\cos\theta = \frac{x}{1} = x \qquad \sec\theta = \frac{1}{x}$$

$$\tan\theta = \frac{y}{x} \qquad \cot\theta = \frac{x}{y}$$

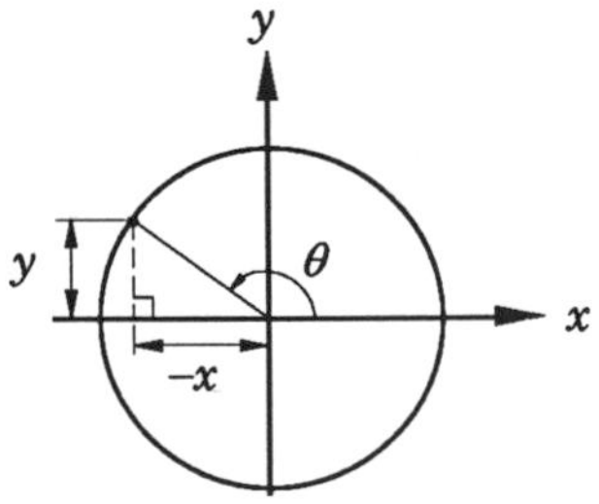

Second Quadrant: Obtuse Angles, Supplementary Angles

$$\sin\theta = \sin(180^\circ - \theta) \qquad \csc\theta = \csc(180^\circ - \theta)$$

$$\cos\theta = -\cos(180^\circ - \theta) \qquad \sec\theta = -\sec(180^\circ - \theta)$$

$$\tan\theta = -\tan(180^\circ - \theta) \qquad \cot\theta = -\cot(180^\circ - \theta)$$

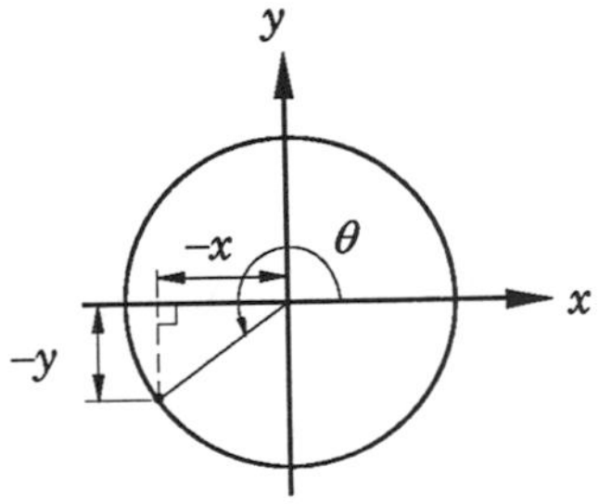

Third Quadrant: Reflex Angles

$$\sin\theta = -\sin(\theta - 180^\circ) \qquad \csc\theta = -\csc(\theta - 180^\circ)$$

$$\cos\theta = -\cos(\theta - 180^\circ) \qquad \sec\theta = -\sec(\theta - 180^\circ)$$

$$\tan\theta = \tan(\theta - 180^\circ) \qquad \cot\theta = \cot(\theta - 180^\circ)$$

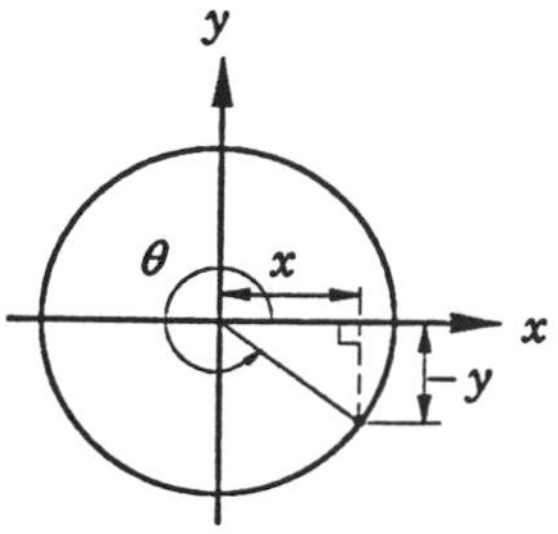

Fourth Quadrant: Reflex Angles

$$\sin\theta = -\sin(360^\circ - \theta) \qquad \csc\theta = -\csc(360^\circ - \theta)$$

$$\cos\theta = \cos(360^\circ - \theta) \qquad \sec\theta = \sec(360^\circ - \theta)$$

$$\tan\theta = -\tan(360^\circ - \theta) \qquad \cot\theta = -\cot(360^\circ - \theta)$$

Analogously, for any negative angle $(-\theta)$:

$$\sin(-\theta) = -\sin\theta \qquad \csc(-\theta) = -\csc\theta$$
$$\cos(-\theta) = \cos\theta \qquad \sec(-\theta) = \sec\theta$$
$$\tan(-\theta) = -\tan\theta \qquad \cot(-\theta) = -\cot\theta$$

Finding All the Fundamental Trigonometric Functions of an Angle

If the numerical value of one trigonometric function of an angle is known, and we know in which quadrant the angle is located, the remaining trigonometric functions can be calculated.

- With $\cos\theta = -\frac{24}{25}$, find the sine, tangent, and cotangent.

We have, with Pythagoras,

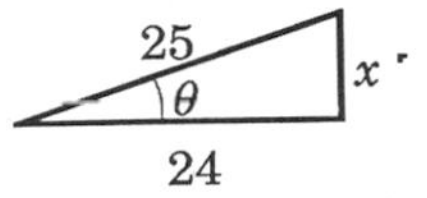

$$24^2 + x^2 = 25^2\ ; \quad x = \pm 7 .$$

There are two solutions, with θ in the 2nd or 3rd quadrant,

2nd quadrant: $\sin\theta = \frac{7}{25}$ $\quad \tan\theta = -\frac{7}{24}$ $\quad \cot\theta = -\frac{24}{7}$

3rd quadrant: $\sin\theta = -\frac{7}{25}$ $\quad \tan\theta = \frac{7}{24}$ $\quad \cot\theta = \frac{24}{7}$

Conversion Table

	sine	**cosine**	**tangent**	**cotangent**
$\sin x =$	–	$\pm\sqrt{1-\cos^2 x}$	$\pm\frac{\tan x}{\sqrt{1+\tan^2 x}}$	$\pm\frac{1}{\sqrt{1+\cot^2 x}}$
$\cos x =$	$\pm\sqrt{1-\sin^2 x}$	–	$\pm\frac{1}{\sqrt{1+\tan^2 x}}$	$\pm\frac{\cot x}{\sqrt{1+\cot^2 x}}$
$\tan x =$	$\pm\frac{\sin x}{\sqrt{1-\sin^2 x}}$	$\pm\frac{\sqrt{1-\cos^2 x}}{\cos x}$	–	$\frac{1}{\cot x}$
$\cot x =$	$\pm\frac{\sqrt{1-\sin^2 x}}{\sin x}$	$\pm\frac{\cos x}{\sqrt{1-\cos^2 x}}$	$\frac{1}{\tan x}$	–

Values of Some Trigonometric Functions

radian: *p.* 410

Degrees	Radians	sine	cosine	tangent	cotangent
0	0	0	1	0	undefined
30	$\frac{\pi}{6}$	$\frac{1}{2}$	$\frac{\sqrt{3}}{2}$	$\frac{\sqrt{3}}{3}$	$\sqrt{3}$
45	$\frac{\pi}{4}$	$\frac{\sqrt{2}}{2}$	$\frac{\sqrt{2}}{2}$	1	1
60	$\frac{\pi}{3}$	$\frac{\sqrt{3}}{2}$	$\frac{1}{2}$	$\sqrt{3}$	$\frac{\sqrt{3}}{3}$
90	$\frac{\pi}{2}$	1	0	undefined	0
120	$\frac{2\pi}{3}$	$\frac{\sqrt{3}}{2}$	$-\frac{1}{2}$	$-\sqrt{3}$	$-\frac{\sqrt{3}}{3}$
135	$\frac{3\pi}{4}$	$\frac{\sqrt{2}}{2}$	$-\frac{\sqrt{2}}{2}$	-1	-1
150	$\frac{5\pi}{6}$	$\frac{1}{2}$	$-\frac{\sqrt{3}}{2}$	$-\frac{\sqrt{3}}{3}$	$-\sqrt{3}$
180	π	0	-1	0	undefined
210	$\frac{7\pi}{6}$	$-\frac{1}{2}$	$-\frac{\sqrt{3}}{2}$	$\frac{\sqrt{3}}{3}$	$\sqrt{3}$
225	$\frac{5\pi}{4}$	$-\frac{\sqrt{2}}{2}$	$-\frac{\sqrt{2}}{2}$	1	1
240	$\frac{4\pi}{3}$	$-\frac{\sqrt{3}}{2}$	$-\frac{1}{2}$	$\sqrt{3}$	$\frac{\sqrt{3}}{3}$
270	$\frac{3\pi}{2}$	-1	0	undefined	0
300	$\frac{5\pi}{3}$	$-\frac{\sqrt{3}}{2}$	$\frac{1}{2}$	$-\sqrt{3}$	$-\frac{\sqrt{3}}{3}$
315	$\frac{7\pi}{4}$	$-\frac{\sqrt{2}}{2}$	$\frac{\sqrt{2}}{2}$	-1	-1
330	$\frac{11\pi}{6}$	$-\frac{1}{2}$	$\frac{\sqrt{3}}{2}$	$-\frac{\sqrt{3}}{3}$	$-\sqrt{3}$
360	2π	0	1	0	undefined

Calculators and Trigonometric Values

Tables of trigonometric functions have largely been replaced by electronic calculators; as these generally do not possess keys for the cotangent, secant, and cosecant functions, they have to be computed with the reciprocal functions:

$$\cot\theta = \frac{1}{\tan\theta}\ ; \quad \sec\theta = \frac{1}{\cos\theta}\ ; \quad \csc\theta = \frac{1}{\sin\theta}\ .$$

13.2 Inverse Trigonometric Functions

Arcus Function

An **inverse trigonometric function**, or **arcus function**, is the reverse of a fundamental trigonometric function.

For instance,

sine of 30° is 0.5 $\Rightarrow$ arc sine of 0.5 is 30° .

The inverse trigonometric functions are:

arc sine	denoted	arcsin
arc cosine		arccos
arc tangent		arctan
arc cotangent		arccot (*or* arcctg)
arc secant		arcsec
arc cosecant		arccsc (*or* arccosec)

We read $\arcsin x$ as "The arc sine of x" or "The inverse sine of x".

The practice of denoting inverse trigonometric functions by $\sin^{-1} x$, $\cos^{-1} x$, *etc.*, is discouraged because of possible confusion with the expressions $(\sin x)^{-1}$ and $(\cos x)^{-1}$, equivalent to $1/\sin x$ and $1/\cos x$.

Domains and Ranges: *p.* 339

p. 505

For a trigonometric function to have an inverse which is also a function, the domain of f, and, as a consequence, the range of the inverse, must be restricted. Domains and ranges of fundamental and inverse trigonometric functions will be discussed in Section 13.4.

Geometrically – with measurements in radians – the arc sine function is the *arc length* of the unit circle for a given value of a fundamental trigonometric function, as reflected in the Latin expression *arcus cuius sinus x est* ("the arc whose sine is x"), and thus accounting for the *arc* part of the names of inverse trigonometric functions.

- Find angles α and β of the triangle:

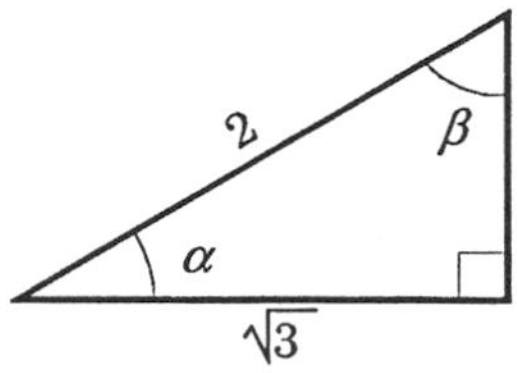

$$\cos\alpha = \frac{\sqrt{3}}{2} \qquad \alpha = \arccos\frac{\sqrt{3}}{2} = 30^\circ.$$

$$\sin\beta = \frac{\sqrt{3}}{2} \qquad \beta = \arcsin\frac{\sqrt{3}}{2} = 60^\circ.$$

13.3 Solving Triangles

Right-Angled Triangles

To solve a **right-angled triangle** we must know one side and one acute angle, or two sides.

- Find the diameter of a circle which circumscribes a right triangle with one leg equal to 5 cm and the opposite angle 48°.

By the theorem of Thales, the hypotenuse is the diameter D of the circle.

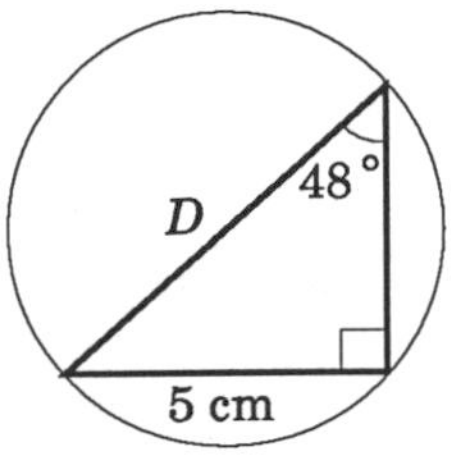

$$\sin 48^\circ = \frac{5}{D} \;;\; D = \frac{5}{\sin 48^\circ} \approx 6.73 \text{ cm}.$$ •

An **isosceles triangle** may be solved by using one of the two right triangles formed when the angle between the equal sides is bisected. Similarly, a **regular polygon** may be solved after forming an isosceles triangle by connecting the center of the polygon with the endpoints of one side:

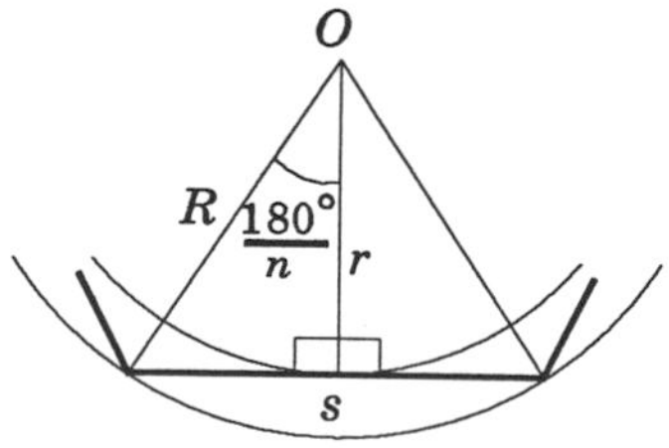

In the above n-gon, O is its center, R the radius of a circumscribed circle, r the radius of an inscribed circle, and s the side of the n-gon. The central angle of the isosceles triangle is $360°/n$. The central angle of each right triangle formed by bisecting the isosceles triangle is $\frac{360°/2}{n} = 180°/n$.

We have

$$\sin \frac{180°}{n} = \frac{s/2}{R} \;;\; \cos \frac{180°}{n} = \frac{r}{R} \;;\; \tan \frac{180°}{n} = \frac{s/2}{r} = \frac{s}{2r} \,.$$

ARCHIMEDES
(c. 287 - 212 B.C.)

Archimedes, eminent Greek mathematician, physicist, and inventor/engineer of the 3rd century B.C., used circumscribed and inscribed regular 96-gons to determine the numerical value of π, though not by trigonometric means, to between $3\frac{10}{71}$ and $3\frac{1}{7}$.

We shall use circumscribed and inscribed regular polygons to determine the numerical value of π between even closer boundaries than those achieved by Archimedes.

- Let the side of the circumscribed 180-gon be $2a$, that of the inscribed polygon $2b$, and the radius of the circle r.

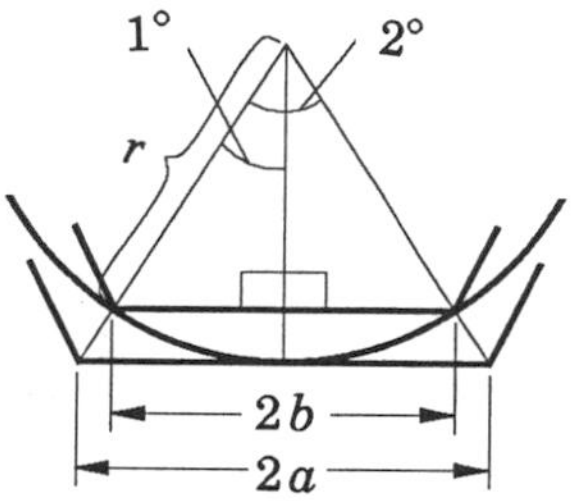

We have

$$180 \cdot 2b < 2\pi r < 180 \cdot 2a$$

where

$$\left.\begin{aligned} a &= r \cdot \tan 1^\circ \\ b &= r \cdot \sin 1^\circ \end{aligned}\right\}$$

$$180 \cdot 2r \cdot \sin 1^\circ < 2\pi r < 180 \cdot 2r \cdot \tan 1^\circ$$

$$180 \sin 1^\circ < \pi < 180 \tan 1^\circ$$

$$3.1414 < \pi < 3.1419\,.$$

Arbitrary Triangles

Law of Sines

To solve a triangle when we know

two angles and one opposite side

or

two sides and an opposite angle,

PTOLEMAEUS Claudius (*c.* 100 - *c.* 168)

we may use the law of sines, demonstrated about A.D. 150 by **Ptolemy** of Alexandria:

$$\frac{\sin A}{a} = \frac{\sin B}{b} = \frac{\sin C}{c}$$

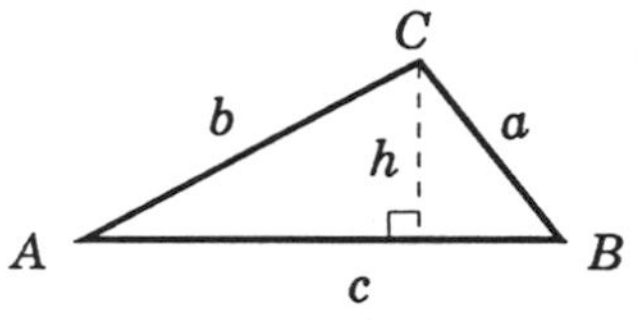

Why? The altitude h is

$$h = b \sin A \ , \quad \text{and} \quad h = a \sin B,$$

which gives

$$b \sin A = a \sin B$$

$$\frac{\sin A}{a} = \frac{\sin B}{b} \ .$$

In the same manner, it can be shown that

$$\frac{\sin A}{a} = \frac{\sin C}{c} \ .$$

- Find the length of the longest side of a triangle with angles 66.0° and 52.0°, if the shortest side is 35 cm.

The third angle is 62.0°, < 66.0° but > 52.0°. Thus, the longest side must be opposite the angle 66.0° and the shortest side opposite 52.0°.

Let the longest side be x cm.

By the law of sines,

$$\frac{x}{\sin 66.0°} = \frac{35}{\sin 52.0°}$$

$$x = \frac{35 \sin 66.0°}{\sin 52.0°} \approx 40.6 \text{ cm.}$$

The Ambiguous Case

Inspector Sally Miles found the body of the intruder, electrocuted when trying to inactivate the alarm system. In his hand was a parched but still readable note:

> *At the site of the old millstone, a triangle. Two sides: 15 and 10 yards; 30° angle opposite the 10-yard side: corner located at the center of the millstone. Third side: due east of the millstone. Money: 5 feet down in a dry well at the corner of the 10-yard side and the third side.*

"Again, trigonometry to the fore," exulted the inspector. She found the old millstone, read the compass, marked off a triangle with the given measurements, and dug a good 5-foot hole.

But the money was not there!

"Not to worry," said the inspector to her assistant, Jonathan Bell. "This must be the famed Ambiguous Case. The money can be in only one other place. One more hole will do it."

And, as so many times before, Inspector Sally Miles was right ....

***

If two sides and an opposite angle are known, one of the other two angles may be determined by the law of sines and the third from the sum of angles. Since $\sin\theta = \sin(\pi - \theta)$, a unique solution is obtained only for a right triangle. When dealing with other triangles, we find two triangles, a phenomenon known as the **ambiguous case**.

- Back to Inspector Miles: A triangle with two sides 15 and 10 units and an angle of 30° opposite the shorter side. Find the two remaining angles.

Let the angle opposite the side measuring 15 units be α.

Apply the law of sines:

$$\frac{\sin\alpha}{15} = \frac{\sin 30^\circ}{10}$$

$$\sin\alpha = \frac{15}{10}\cdot\sin 30^\circ \approx 0.75$$

$$\alpha \approx 48.6^\circ,\ \text{or } 131.4^\circ.$$

The angle opposite the unknown side is, alternatively,

$$180^\circ - (30^\circ + 48.6^\circ) = 101.4^\circ$$

or

$$180^\circ - (30^\circ + 131.4^\circ) = 18.6^\circ$$

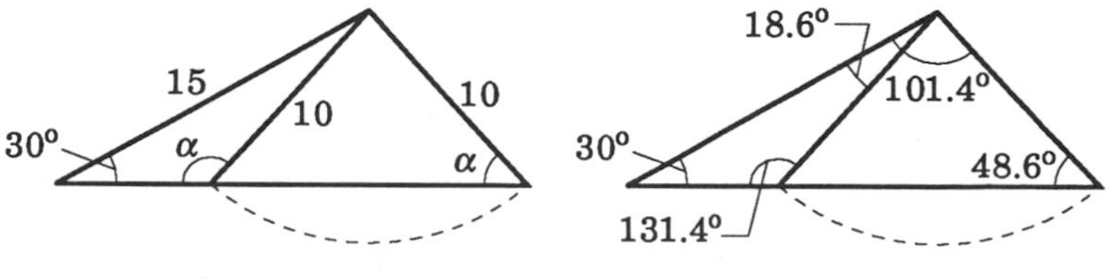

The angles of the triangle are either

$$\left.\begin{matrix} 30^\circ & 48.6^\circ & 101.4^\circ \\ 30^\circ & 131.4^\circ & 18.6^\circ \end{matrix}\right\}.$$

(second row preceded by "or")

Law of Sines for Spherical Triangles

For spherical triangles, that is, triangles whose sides are arcs of great circles, we have

$$\frac{\sin A}{\sin a} = \frac{\sin B}{\sin b} = \frac{\sin C}{\sin c},$$

where A, B, and C are angles formed by tangent lines and sides a, b, and c are measured by their subtended central angles.

To prove

$$\frac{\sin A}{\sin a} = \frac{\sin B}{\sin b},$$

let ABC be any spherical triangle on a sphere with the center O. From any point P on *the radius* OC, draw line PQ perpendicular to plane OAB.

Through PQ and perpendicular to OA and OB pass planes which meet radius OA and line OB in A' and B', respectively, forming angle $PA'Q = A$ and angle $PB'Q = B$.

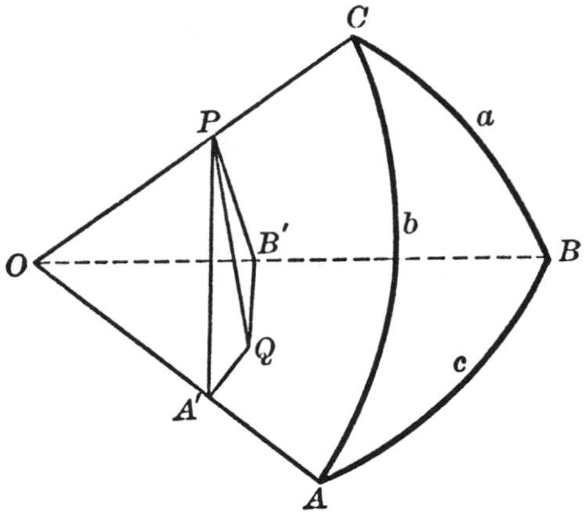

In the right triangle $A'PQ$, we find

$$\sin A' = \sin A = \frac{PQ}{PA'};$$

in the right triangle $B'PQ$,

$$\sin B' = \sin B = \frac{PQ}{PB'};$$

in the right triangle OPB',

$$\sin a = \frac{PB'}{OP};$$

in the right triangle OPA',

$$\sin b = \frac{PA'}{OP}.$$

Consequently,

$$\frac{\sin A}{\sin a} = \frac{PQ}{PA'} \cdot \frac{OP}{PB'} \quad \text{and} \quad \frac{\sin B}{\sin b} = \frac{PQ}{PB'} \cdot \frac{OP}{PA'}$$

and we have

$$\frac{\sin A}{\sin a} = \frac{\sin B}{\sin b}.$$

Analogously,

$$\frac{\sin A}{\sin a} = \frac{\sin C}{\sin c}\ .$$

For a spherical triangle, the **ambiguous case** occurs in the solution of an oblique triangle if given two sides and an angle opposite one of them *or* given two angles and the side opposite one of them.

Law of Cosines

To solve a triangle when given

two sides and the included angle

or

three sides,

VIÈTE François (1540 - 1603)

we use the law of cosines, first described by the French mathematician **François Viète**:

$$a^2 = b^2 + c^2 - 2\,b\,c\cos A$$
$$b^2 = a^2 + c^2 - 2\,a\,c\cos B$$
$$c^2 = a^2 + b^2 - 2\,a\,b\cos C$$

To prove that

$$a^2 = b^2 + c^2 - 2\,b\,c\cos A,$$

we apply the Pythagorean theorem to the two right triangles inside the triangle ABC:

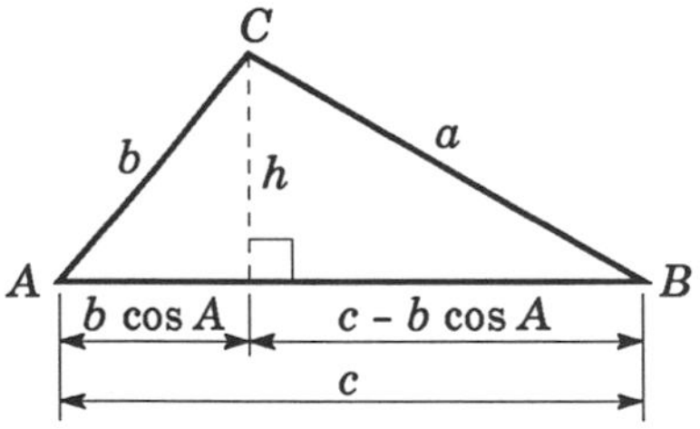

$$b^2 = h^2 + b^2\cos^2 A$$

and

$$a^2 = h^2 + (c - b\cos A)^2$$

Subtraction gives

$$b^2 - a^2 = h^2 + b^2\cos^2 A - [h^2 + (c - b\cos A)^2]$$
$$a^2 = b^2 + c^2 - 2bc\cos A\ .$$

The formulas

$$b^2 = a^2 + c^2 - 2ac\cos B$$
$$c^2 = a^2 + b^2 - 2ab\cos C$$

are proved in the same manner.

- Find the length of the third side of a triangle with two sides a and b, and an angle α opposite side a.

Let the third side be x, and let the unknown angles be β and γ; draw an arbitrary triangle:

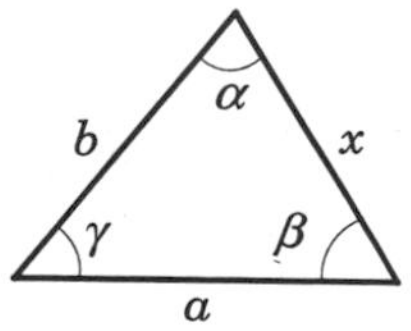

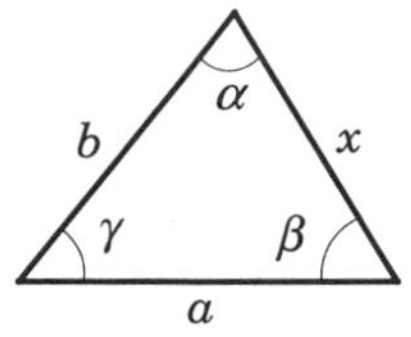

Apply the law of cosines:

$$x = \sqrt{a^2 + b^2 - 2ab\cos\gamma}$$

To find γ, first determine β by the law of sines:

$$\frac{\sin\beta}{b} = \frac{\sin\alpha}{a}$$

$$\sin\beta = \frac{b\sin\alpha}{a}$$

We have the possibility that β is an acute angle such that

$$\beta = \arcsin\left(\frac{b\sin\alpha}{a}\right)$$

or

$$\beta' = \left[\pi - \arcsin\left(\frac{b\sin\alpha}{a}\right)\right].$$

The angle γ is determined by the equation $\gamma = \pi - (\alpha + \beta)$; since β is known, we find

$$\gamma = \left[\pi - \alpha - \arcsin\left(\frac{b\sin\alpha}{a}\right)\right]$$

or

$$\gamma' = \left[\arcsin\left(\frac{b\sin\alpha}{a}\right) - \alpha\right].$$

By substituting for γ in $x = \sqrt{a^2 + b^2 - 2ab\cos\gamma}$, we obtain the third side,

$$\sqrt{a^2 + b^2 - 2\,ab\cos\left[\pi - \alpha - \arcsin\left(\frac{b\sin\alpha}{a}\right)\right]}$$

or

$$\sqrt{a^2 + b^2 - 2\,ab\cos\left[\arcsin\left(\frac{b\sin\alpha}{a}\right) - \alpha\right]}.$$

- Find the length of the diagonals of a rhomboid with sides 35.0 cm and 46.2 cm and the acute angle 50°.

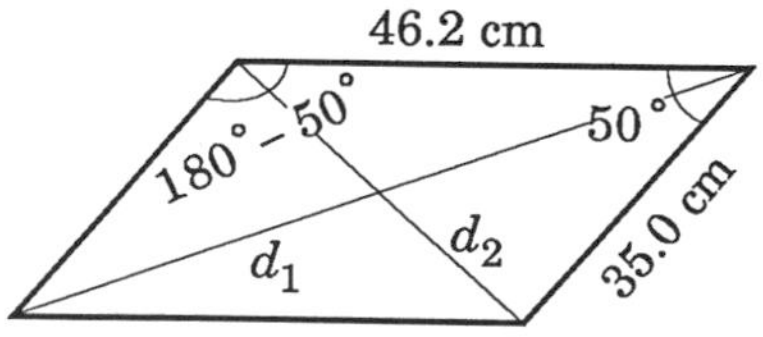

Applying the law of cosines, we obtain

$$d_1 = \sqrt{46.2^2 + 35.0^2 - 2\cdot 46.2\cdot 35.0\cos 130^\circ} \approx 73.7\text{ cm},$$

$$d_2 = \sqrt{46.2^2 + 35.0^2 - 2\cdot 46.2\cdot 35.0\cos 50^\circ} \approx 35.8\text{ cm}.$$

Law of Cosines for Spherical Triangles

Sides of a Spherical Triangle

For spherical triangles, we have

$$\cos c = \cos a \cos b + \sin a \sin b \cos C .$$

where sides a, b, and c, being arcs of great circles, are measured by their subtended central angles.

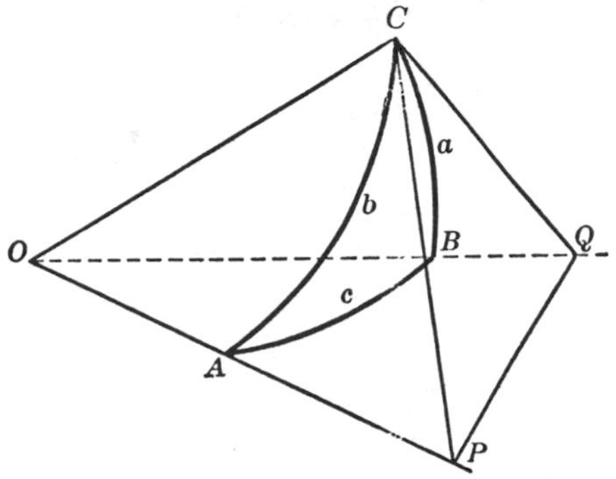

Why? Let ABC be any spherical triangle on a sphere with the center O. Draw the radii of the sphere to vertices ABC of the triangle. Draw tangents to sides a and b at C and let the tangents meet the extensions of OA and OB at P and Q, respectively.

Applying the rule of cosines for the plane triangles OPQ and CPQ and subtracting, we find

$$(PQ)^2 = (OQ)^2 + (OP)^2 - 2\,(OQ)\,(OP)\cos c$$

$$\underline{(PQ)^2 = (CQ)^2 + (CP)^2 - 2\,(CQ)\,(CP)\cos C}$$

$$0 = (OQ)^2 - (CQ)^2 + (OP)^2 - (CP)^2 + 2\,(CQ)\,(CP)\cos C - 2\,(OQ)\,(OP)\cos c,$$

and rearranging,

$$2\,(OQ)\,(OP)\cos c = [(OQ)^2 - (CQ)^2] + [(OP)^2 - (CP)^2] + 2\,(CQ)\,(CP)\cos C .$$

By the Pythagorean theorem,

$$(OQ)^2 - (CQ)^2 = (OC)^2 \quad \text{and} \quad (OP)^2 - (CP)^2 = (OC)^2,$$

and, consequently,

$$2\,(OQ)\,(OP)\cos c = 2(OC)^2 + 2\,(CQ)\,(CP)\cos C .$$

Dividing through by $2\,(OQ)\,(OP)$,

$$\cos c = \frac{OC}{OQ}\cdot\frac{OC}{OP} + \frac{CQ}{OQ}\cdot\frac{CP}{OP}\cos C$$

or

$$\cos c = \cos a \cos b + \sin a \sin b \cos C .$$

Cyclic permutation gives

$$\cos a = \cos b \cos c + \sin b \sin c \cos A$$

$$\cos b = \cos c \cos a + \sin c \sin a \cos B .$$

Angles of a Spherical Triangle

The law of cosines concerning angles of a spherical triangle is here quoted without proof:

$$\cos A = -\cos B \cos C + \sin B \sin C \cos a$$
$$\cos B = -\cos C \cos A + \sin C \sin A \cos b$$
$$\cos C = -\cos A \cos B + \sin A \sin B \cos c$$

Law of Tangents

If we know

two sides of a triangle and the included angle,

then, *to find the remaining angles,*

first use the law of cosines for the third side and

then – when all three sides are known – find the angles by the law of sines or the law of cosines.

FINCKE Thomas (1561 - 1656)

For a direct solution of the problem, we use the law of tangents, first described in 1583 in *Geometriæ rotundi* by the Danish mathematician and physician **Thomas Fincke**:

$$\frac{a-b}{a+b} = \frac{\tan \frac{1}{2}(A-B)}{\tan \frac{1}{2}(A+B)}$$

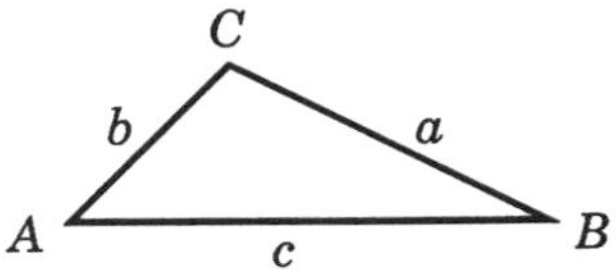

To prove the law of tangents, consider a triangle ABC where angle A is greater than angle B. The external angle at C is bisected by DE, which forms one side of a rectangle $BDEF$ about the triangle:

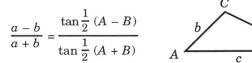

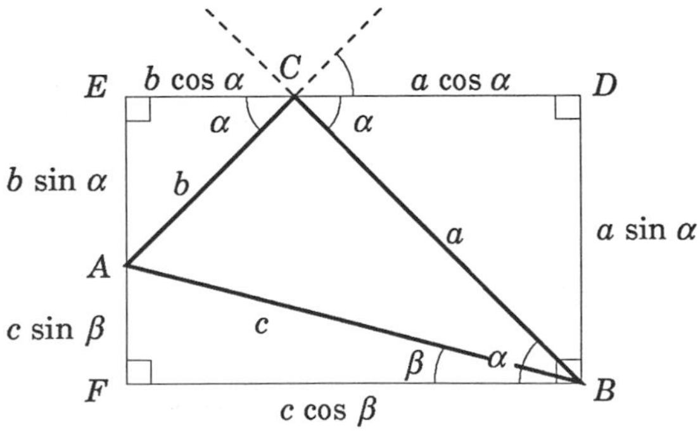

If half the external angle of C is α, then alternate angle CBF is equal to α.

Let angle ABF be β.

In the above figure, the three right-angled triangles yield

$$CD = a \cos \alpha \qquad BD = a \sin \alpha \qquad CE = b \cos \alpha$$
$$AE = b \sin \alpha \qquad AF = c \sin \beta \qquad BF = c \cos \beta\,.$$

Opposite sides of a rectangle are equal,

$$\left.\begin{aligned} a \sin \alpha &= b \sin \alpha + c \sin \beta \\ c \cos \beta &= a \cos \alpha + b \cos \alpha \end{aligned}\right\},$$

and, after rearrangement,

$$\left.\begin{aligned} (a-b) \sin \alpha &= c \sin \beta \\ (a+b) \cos \alpha &= c \cos \beta \end{aligned}\right\}.$$

Division gives

$$\frac{(a-b)\sin\alpha}{(a+b)\cos\alpha} = \frac{\sin\beta}{\cos\beta}$$

$$\frac{a-b}{a+b}\tan\alpha = \tan\beta$$

$$\frac{a-b}{a+b} = \frac{\tan\beta}{\tan\alpha}.$$

The external angle at C equals the sum of the two opposite interior angles of the triangle; therefore,

$$2\alpha = A+B\ ;\quad \alpha = \frac{1}{2}(A+B).$$

Since $\beta = \alpha - B$, and $\alpha = \frac{1}{2}(A+B)$, we have

$$\beta = \frac{1}{2}(A+B) - B$$

$$\beta = \frac{1}{2}(A-B),$$

and finally

$$\frac{a-b}{a+b} = \frac{\tan\frac{1}{2}(A-B)}{\tan\frac{1}{2}(A+B)}.$$

- Find the remaining angles of a triangle with two sides 9 and 6 and an included angle of 43°.

Use the notations of the triangle to the right and apply the law of tangents:

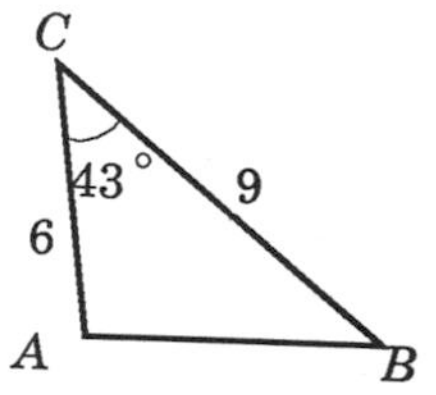

$$\frac{9-6}{9+6} = \frac{1}{5} = \frac{\tan\frac{1}{2}(A-B)}{\tan\frac{1}{2}(A+B)}$$

$$A+B = 180° - 43° = 137°\ ;\quad \frac{1}{2}(A+B) = 68.5°.$$

$$\frac{1}{5} = \frac{\tan\frac{1}{2}(A-B)}{\tan 68.5°}$$

$$\tan\frac{1}{2}(A-B) = \frac{\tan 68.5°}{5} = 0.5077\ldots\ ;\quad \frac{1}{2}(A-B) \approx 26.9°.$$

We solve the system of equations

$$\left.\begin{aligned}\frac{1}{2}(A+B) &= 68.5^\circ \\ \frac{1}{2}(A-B) &= 26.92^\circ\end{aligned}\right\},$$

$$A = 95.4^\circ$$
$$B = 41.6^\circ .$$

The sought angles are 95.4° and 41.6°.

- The angles of a triangle are 50°, 60°, and 70°; the sum of its two longer sides is 18 length units.

  Find the lengths of the sides.

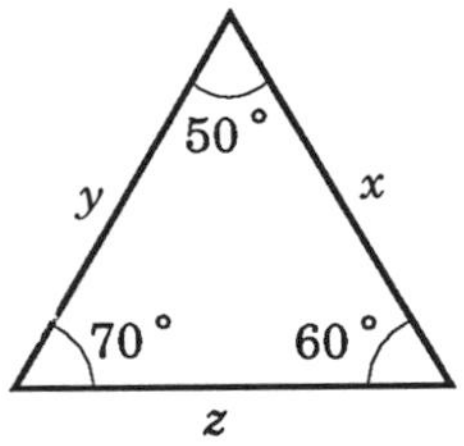

Let the longer sides be x and y, and let the third side be z.

By the law of tangents,

$$\frac{x-y}{18} = \frac{\tan\dfrac{70^\circ - 60^\circ}{2}}{\tan\dfrac{70^\circ + 60^\circ}{2}}$$

$$x - y = 18 \cdot \frac{\tan 5^\circ}{\tan 65^\circ} .$$

$$\left.\begin{aligned}x-y &= 18 \cdot \frac{\tan 5^\circ}{\tan 65^\circ} \\ x+y &= 18\end{aligned}\right\}$$

$$x = 9 + \frac{9 \tan 5^\circ}{\tan 65^\circ} \approx 9.37 ; \quad y \approx 18 - 9.37 = 8.63 .$$

By the law of sines,

$$\frac{z}{\sin 50^\circ} = \frac{9.37}{\sin 70^\circ}$$

$$z \approx 7.64 \text{ cm.}$$

The sides of the triangle are

9.37, 8.63, and 7.64 .

Area of a Triangle

For a triangle *ABC* of sides *a*, *b*, and *c*, we have

$$\text{Area} = \frac{a\,b\sin C}{2} = \frac{a\,c\sin B}{2} = \frac{b\,c\sin A}{2}\,.$$

This theorem is a corollary to the theorem stating that the area of a triangle is half the product of the base and the height. In the triangles *ABC* and *DEF* below, the height equals the product of the sine of an angle (*e.g.*, *A* and *E*, respectively) and the length of the hypotenuse of a triangle whose side opposite the employed angle is the height:

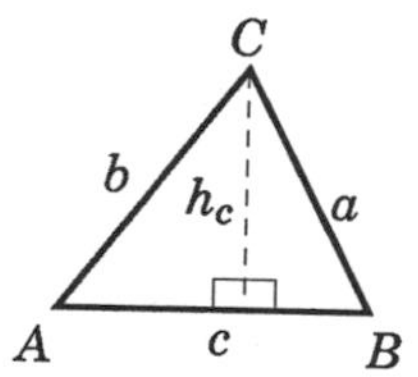

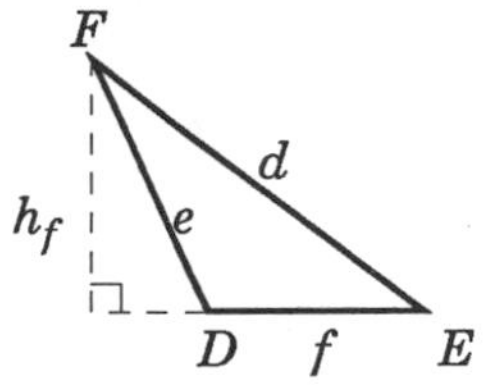

$$h_c = b\sin A \qquad\qquad h_f = d\sin E$$

$$\text{Area}_{ABC} = \frac{b\,c\sin A}{2} \qquad\qquad \text{Area}_{DEF} = \frac{d\,f\sin E}{2}$$

The area theorem applies to both acute and obtuse angles. Therefore, when the area and two sides are known, there are two different triangles that satisfy the given conditions.

- Find the angle *C* in a triangle *ABC* of area 24 cm^2, where *AC* is 16 cm and *BC* is 6 cm.

$$\frac{16 \cdot 6\sin C}{2} = 24; \quad \sin C = \frac{1}{2}; \quad C = 30° \text{ or } 150°\,.$$

Two triangles are possible:

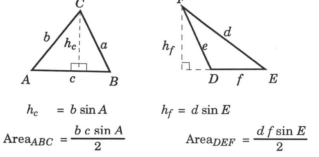

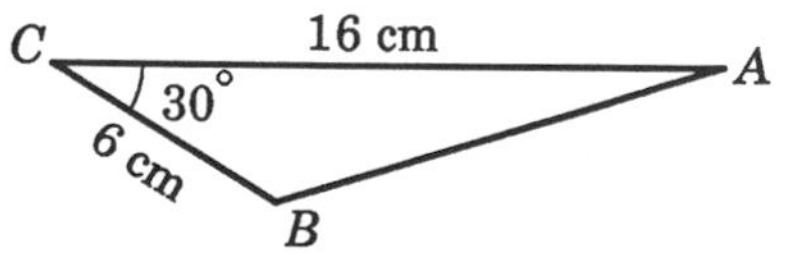

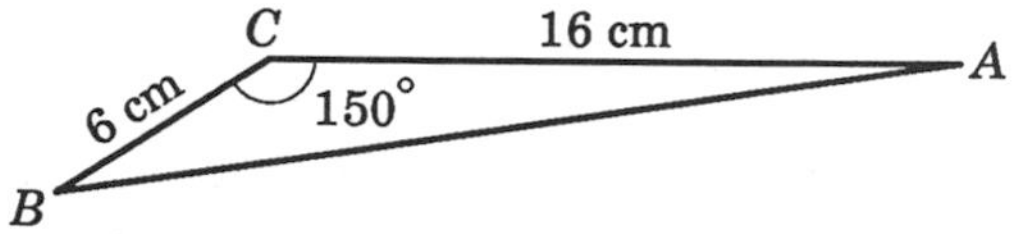

Surveying and Navigation

de JODE (or JUDAEIS) Cornelis
(*c.* 1568 - *c.* 1600)

Dutch geographer and copper-plate engraver

DE QVADRANTE GEOMETRICO
LIBELLVS.
IN QVO QVIDQVID
AD LINEARVM ET SVPERFICIERVM,
VTPOTE ALTITVDINVM ET LATITVDINVM, DI-
menſiones facit lucidiſsime demonſtratur.

Addita figura anea XXXVII. ad maiorem doctrina intelligentiam & lucem hactenus non ita expoſita.

Sumptibus & expenſis CORNELII DE IVDAEIS editus.

NORIBERGAE,
TYPIS CHRISTOPHORI LOCHNERI.
M. D. XCIIII.

Greek, *geo*, "Earth"; *daiesthai*, "to divide"

Geodesy is a branch of geology applying mathematics to determine the shape and size of the Earth and its varying magnetism and gravity.

Surveying is the science of making accurate measurements of the Earth's surfaces. **Land surveying** includes both **plane surveying**, which deals with areas sufficiently small to allow the surveyor to disregard of the Earth's curvature, and **geodetic surveying**, which takes into account the curvature of the Earth's surface. **Hydrographic surveying** deals with contours of the Earth under bodies of water.

Latin, *navis*, "ship; –*igare* (from *agere*), "to drive"

Although the literal meaning of **navigation** is sailing, the term refers to the art or science of setting a course at sea, on land, or in the atmosphere.

Locating Points and Measuring Distances

Ellipsoid, Spheroid: *p.* 405
Geoid
Greek, *geoeides*, "earthlike"

The Earth has an irregular, roughly ellipsoidal shape, called a spheroid and generally referred to as a **geoid**. For the problems in this text, however, we shall regard the Earth as a sphere.

The poles are on the **axis** of rotation of the Earth. The plane of the **equator** is imagined as passing through the center of the Earth, perpendicular to the Earth's axis. Thus, all points on the equator are equidistant from the poles.

Parallels
Meridians

Parallels (of latitude) are curves parallel to the equator; **meridians** (of longitude) are curves perpendicular to the equator. The parallels and meridians form circles of 360° in circumference around the Earth. Meridians are all great circles, whereas, among parallels, only the equator is a great circle.

Greenwich Meridian

By international agreement in 1864, the 0° meridian passes through the former London observatory at Greenwich, England, and is generally known as the **Greenwich meridian**.

Longitude

The **longitude** of a point is the angle between the plane of its meridian and the plane of the Greenwich meridian. Longitude is measured from 0° to 180° East and West of Greenwich.

Latitude

The **latitude** of a point is the smallest angle formed between the radius from the point to the Earth's center and the plane of the equator. Latitude is measured from 0° to 90° North and South of the equator. 0° is the equator itself; 90° North and 90° South are the poles.

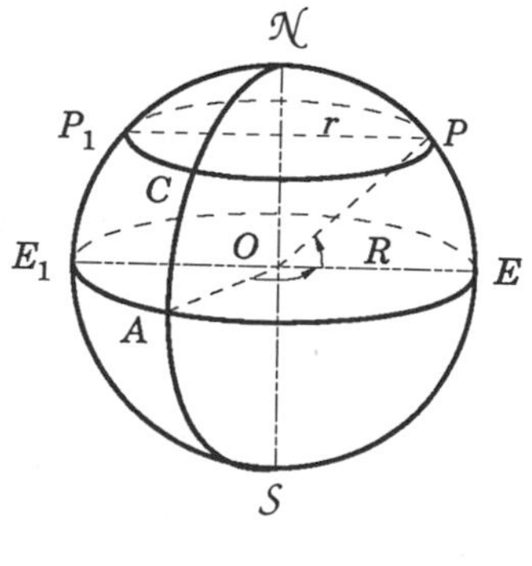

In the figure, NS is the axis of the Earth; point A is the intersection of the Greenwich meridian and the equator; $NPESE_1P_1$ is a meridian through a point P on the surface of the Earth; EE_1 is the diameter of the equator; PP_1 is the diameter of the parallel through P. Angle POE is the latitude of point P, and AOE is the longitude of point P.

Geodesic

A **geodesic** (or geodetic) **curve** describes the shortest distance between two points on a spherical surface, and is always an arc of a great circle.

Bearing

Bearings, or **courses**, are angles made with the meridians.

- Two points A and B are located on latitude 40° North, longitudes 10° West and 20° East, respectively.

  Find the surface distance between A and B, along the North 40° parallel, if the circumference of the Earth at the equator is approximately 40 000 km.

  Let the radius of the Earth be R and the radius of the parallel of latitude 40° be r:

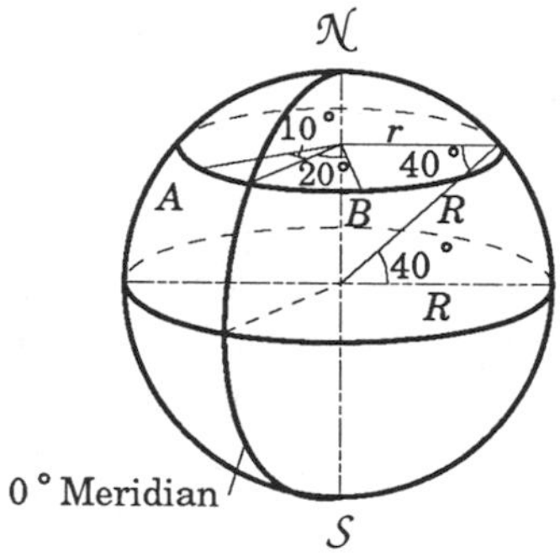

We have

$$\text{arc } AB = \frac{10 + 20}{360} \cdot 2\pi r\,,$$

where

$$r = R \cos 40^\circ\,.$$

Thus,

$$AB = \frac{30}{360} \cdot 2\pi \frac{40\,000}{2\pi} \cdot \cos 40^\circ \approx 2553.5 \approx 2550\,.$$

The sought distance is

$$2.55 \cdot 10^3 \text{ km}.$$

We have determined the surface distance along the 40° North parallel, which is not on the arc of a great circle and, therefore, does not mean the shortest surface distance between A and B.

- Calgary, Canada, is located on latitude 51.03° North, longitude 114.05° West, and Gatwick Airport, London, U.K. is on latitude 51.09° North, longitude 0.21° West, that is, nearly on the same latitude.

  The Earth's circumference is approximately 40 000 km.

  Find the distance between Calgary and Gatwick

  1. along the 51st parallel;
  2. along the geodesic line connecting the cities.

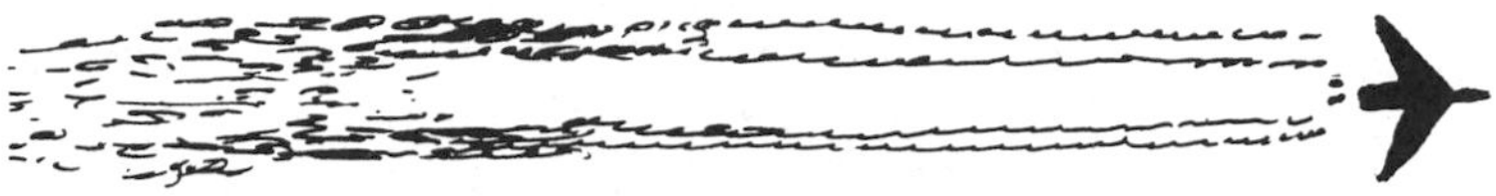

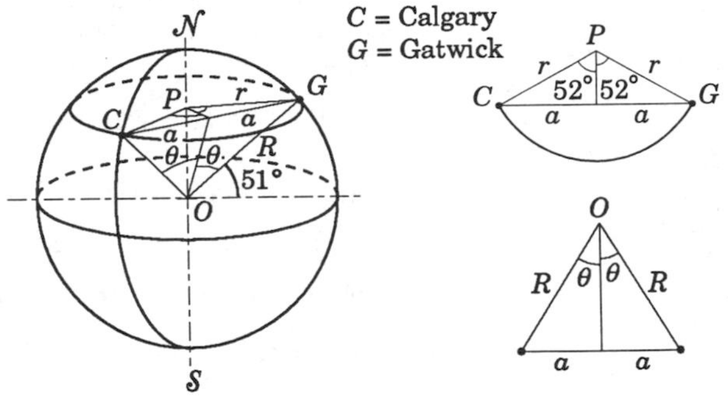

The **arc** *CG* of the 51st parallel is

$$\text{arc } CG = \frac{114.05 - 0.21}{360} (2\pi r)$$

where

$$r = R\cos 51^\circ ; \; R = \frac{40\,000}{2\pi} \; ; \; 2\pi r = 40\,000 \cdot \cos 51^\circ .$$

Thus,

$$\text{arc } CG \approx \frac{113.84}{360} \cdot 40\,000 \cos 51^\circ \approx 7960 \approx 8.0 \cdot 10^3 \text{ km.}$$

p. 485

To find the length of the distance **along the geodesic line**, apply the law of cosines concerning sides of a spherical triangle,

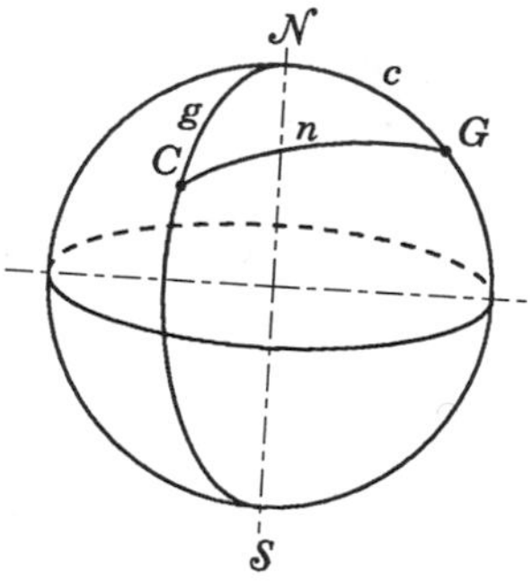

$$\cos n = \cos c \cos g + \sin c \sin g \cos N .$$

Central angle $c = 90^\circ - 51.09^\circ = 38.91^\circ$;

$$g = 90^\circ - 51.03^\circ = 38.97^\circ .$$

Included angle $N = 114.05^\circ - 0.21^\circ = 113.84^\circ$.

Thus,

$$\cos n = \cos 38.91^\circ \cos 38.97^\circ + \sin 38.91^\circ \sin 38.97^\circ \cos 113.84^\circ$$

$$n = 63.556^\circ$$

and the distance is

$$\frac{63.556 \cdot 40\,000}{360} \approx 7\,061 \approx 7.1 \cdot 10^3 \text{ km.}$$

The distance between Calgary and Gatwick is

1. along the 51st parallel is $8.0 \cdot 10^3$ km;
2. along the geodesic line is $7.1 \cdot 10^3$ km.

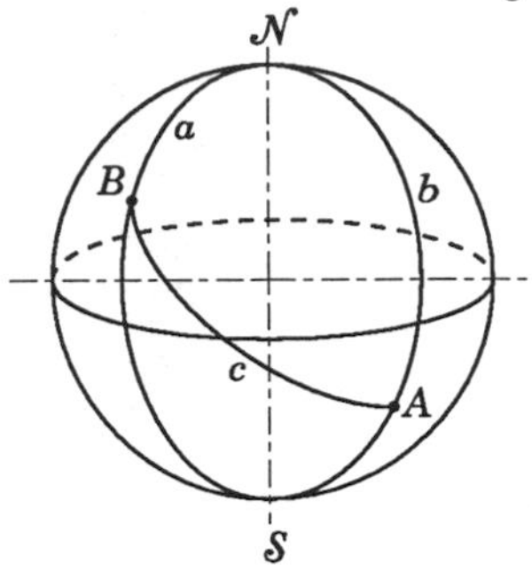

- Find the distance along the geodesic line between a point A located on latitude 34.84° South, longitude 59.09° West, and a point B on latitude 34.82° North, longitude 139.31° East.

  The Earth's circumference is approximately 40 000 km.

  Central angle $b = (90° + 34.84°) = 124.84°$;

  $a = 90° - 34.82° = 55.18°$.

  Included angle $c = 180° - 139.31° + (180° - 59.09°) = 161.60°$.

  $\cos c = \cos 124.84° \cos 55.18° + \sin 124.84° \sin 55.18° \cos 161.60°$

  $\approx -0.965\,55$.

  $c = 164.92°$, so the distance is

  $$\frac{164.92 \cdot 40\,000}{360} \approx 18\,320 \approx 1.8 \cdot 10^4 \text{ km}.$$

Working All the Angles

In order to determine width, height, and depth of a distant object – and its distance – we have a choice of measuring angles of sight, of elevation, and of depression.

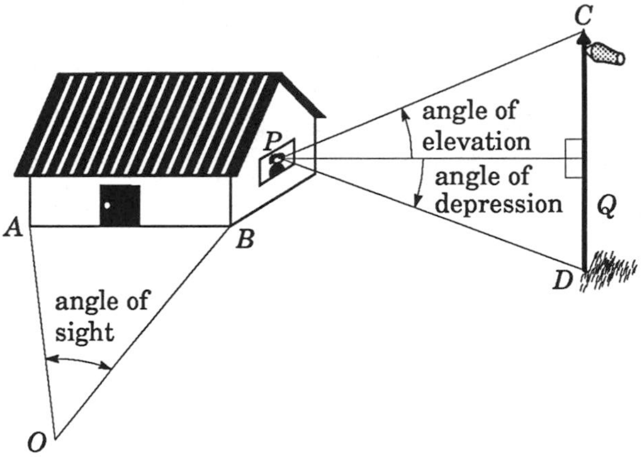

Lines of Sight
Sighting Angle

When observing an object AB from a point O, OA and OB are **lines of sight**, and the angle AOB is the **angle of sight**, or **sighting angle**.

Angle of Elevation
Angle of Depression

If we observe a vertical object CD from a point P, PC and PD are sighting lines; C is sighted at an **angle of elevation**, D at an **angle of depression**.

In the following examples we disregard the curvature of the Earth and the fact that light does not travel in a straight line when passing through air of varying density.

- A radio tower is observed from a point 20 m above ground level; the top of the tower is at an elevation of 30.53°, the base at an angle of depression of 8.04°.

  Find the height of the tower.

Let the distance between the observation point and the tower be x, the height $h + 20$ meters.

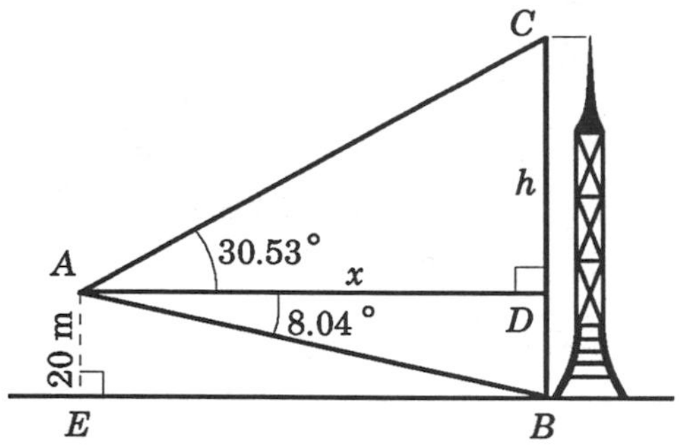

We have

elevation: $h = x \tan 30.53°$

depression: $20 = x \tan 8.04°$

$$\frac{h}{20} = \frac{\tan 30.53°}{\tan 8.04°}.$$

$$h + 20 = 20 \cdot \frac{\tan 30.53°}{\tan 8.04°} + 20 \approx 103.5 \text{ m}.$$

- A farmstead B is to be supplied with electricity by underground cable from a distribution box A. To determine the length of the cable, sightings are taken from two points on the opposite bank of a river: point O situated on the extension of line BA, and point P at a distance of 1530 m from O, as shown in the drawing.

  Sightings have given the angles

$$OPA = 32.3°$$
$$APB = 48.1°$$
$$POA = 82.8°.$$

  Find the minimum length of cable required.

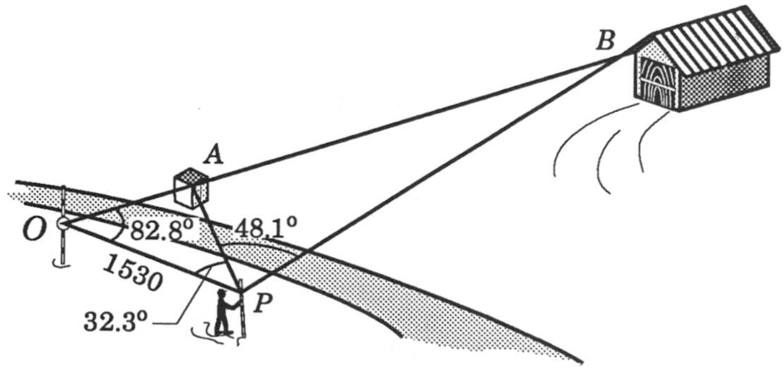

We have the following angles:

$$PAB = 82.8^\circ + 32.3^\circ = 115.1^\circ$$
$$OAP = 180^\circ - 115.1^\circ = 64.9^\circ$$
$$OPB = 32.3^\circ + 48.1^\circ = 80.4^\circ$$
$$AOP + OPB = 82.8^\circ + 80.4^\circ = 163.2^\circ$$
$$ABP = 180^\circ - 163.2^\circ = 16.8^\circ$$

Apply the law of sines to

triangle OAP: triangle BAP:

$$AP = 1530 \cdot \frac{\sin 82.8^\circ}{\sin 64.9^\circ} \qquad BA = AP \cdot \frac{\sin 48.1^\circ}{\sin 16.8^\circ}$$

$$BA = 1530 \cdot \frac{\sin 82.8^\circ \cdot \sin 48.1^\circ}{\sin 64.9^\circ \cdot \sin 16.8^\circ} \approx 4317 \text{ m.}$$

- A hot-air balloon is observed in the Serengeti due south, from a point A at 22.23° elevation and, simultaneously, from a point B 1500 m due south of A at 48.11° elevation.

  Find the height of the hot-air balloon.

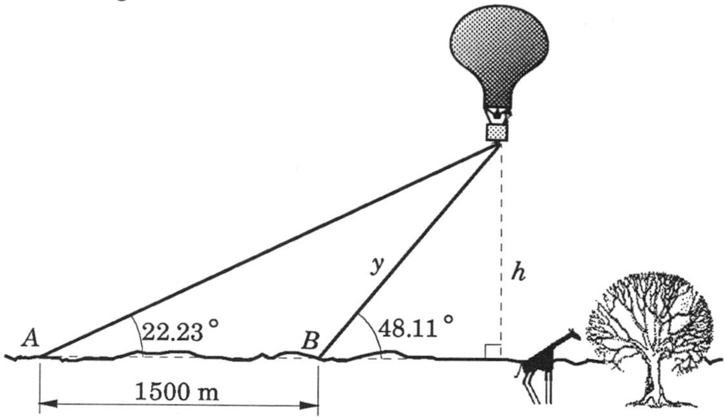

We have the "sighting angle" from the balloon,

$$48.11^\circ - 22.23^\circ = 25.88^\circ,$$

and

– by definition: – by the law of sines:

$$h = y \cdot \sin 48.11^\circ \qquad y = 1500 \cdot \frac{\sin 22.23^\circ}{\sin 25.88^\circ}$$

$$h = 1500 \cdot \frac{\sin 22.23^\circ \cdot \sin 48.11^\circ}{\sin 25.88^\circ} \approx 970 \text{ m.}$$

- In an aircraft flying at constant speed v km/h and at an altitude of 3200 m, the navigator observes an approaching coastline at an angle of depression of -15.2° at time zero seconds; 120 seconds later, an electronic circuit freezes the inclinometer reading at -52.6°.

  Find the speed of the aircraft.

Let the distance between the two observation points be x m, and the distance between the first observation point and the coastal strip y m:

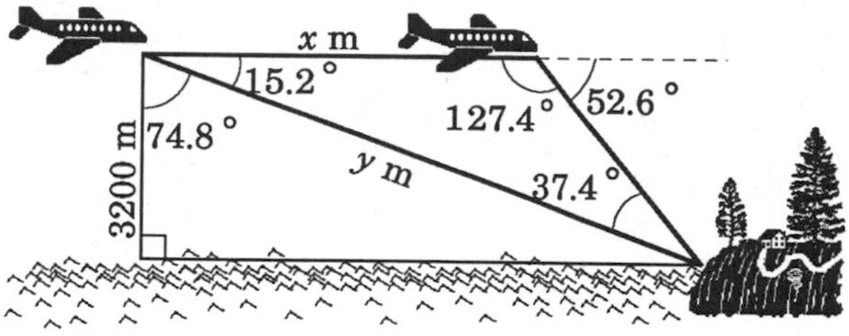

We have

$$x = y \cdot \frac{\sin 37.4^\circ}{\sin 127.4^\circ} = y \cdot \frac{\sin 37.4^\circ}{\sin 52.6^\circ}$$

and

$$y = \frac{3200}{\cos 74.8^\circ} .$$

$$x = 3200 \cdot \frac{\sin 37.4^\circ}{\sin 52.6^\circ \cdot \cos 74.8^\circ} .$$

$$v = 3200 \cdot \frac{\sin 37.4^\circ}{\sin 52.6^\circ \cdot \cos 74.8^\circ} \cdot \frac{1}{1000} \cdot \frac{3600}{120} \approx 280 \text{ km/h}.$$

- A 20-m-high mast is placed on top of a cliff whose height above sea level is unknown. An observer at sea sees the top of the mast at an elevation of 46° 42', the foot at 38° 23'.

  Find the height of the cliff.

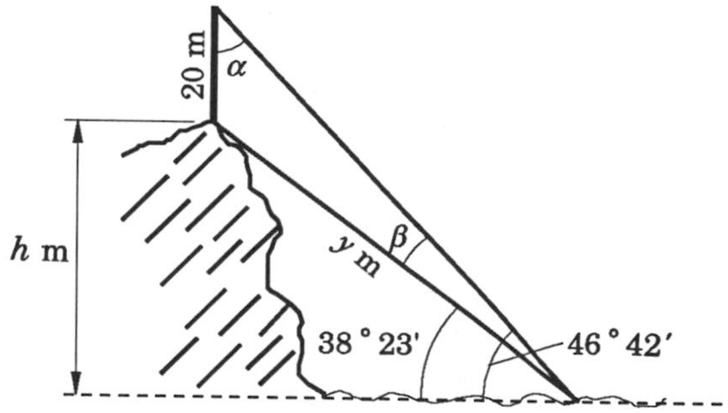

$\alpha = 90^\circ - 46^\circ 42' = 43^\circ 18'$; $\beta = 46^\circ 42' - 38^\circ 23' = 8^\circ 19'$.

We have

$$h = y \sin 38^\circ 23' ; \quad y = 20 \cdot \frac{\sin \alpha}{\sin \beta} = 20 \cdot \frac{\sin 43^\circ 18'}{\sin 8^\circ 19'} .$$

$$h = 20 \cdot \frac{\sin 43^\circ 18'}{\sin 8^\circ 19'} \cdot \sin 38^\circ 23'$$

$$= 20 \cdot \frac{\sin 43.30^\circ \cdot \sin 38.38^\circ}{\sin 8.32^\circ} \approx 59 \text{ m}.$$

Alternative solution:

Let z be the distance from the observer to the plumb line through the cliff-top pole.

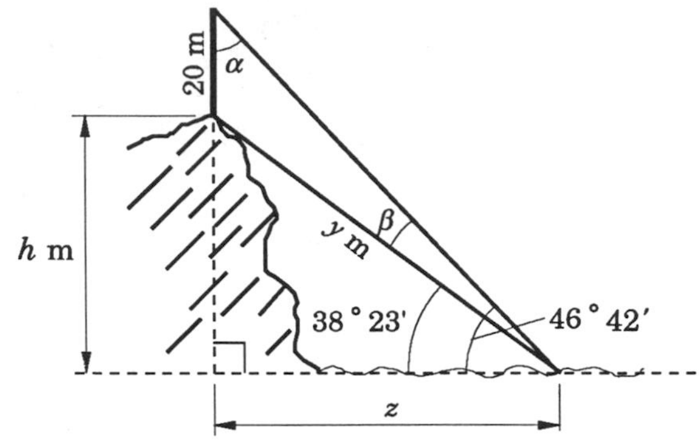

We have

$$\begin{aligned} h+20 &= z\tan 46^\circ\,42' \\ -h &= -z\tan 38^\circ\,23' \\ \hline 20 &= z(\tan 46^\circ\,42' - \tan 38^\circ\,23') \end{aligned}$$

$$z = \frac{20}{\tan 46^\circ\,42' - \tan 38^\circ\,23'};$$

$$h = \frac{20\tan 38^\circ\,23'}{\tan 46^\circ\,42' - \tan 38^\circ\,23'}$$

$$= \frac{20\tan 38.38^\circ}{\tan 46.70^\circ - \tan 38.38^\circ} \approx 59 \text{ m.}$$

- Determine the free space between a ski-lift chair and a terraced area near the top of the hill. At the foot of a 35° incline, the elevation to a lift chair directly over the terraced area measures 46°; 60 m up the incline, it is 53°.

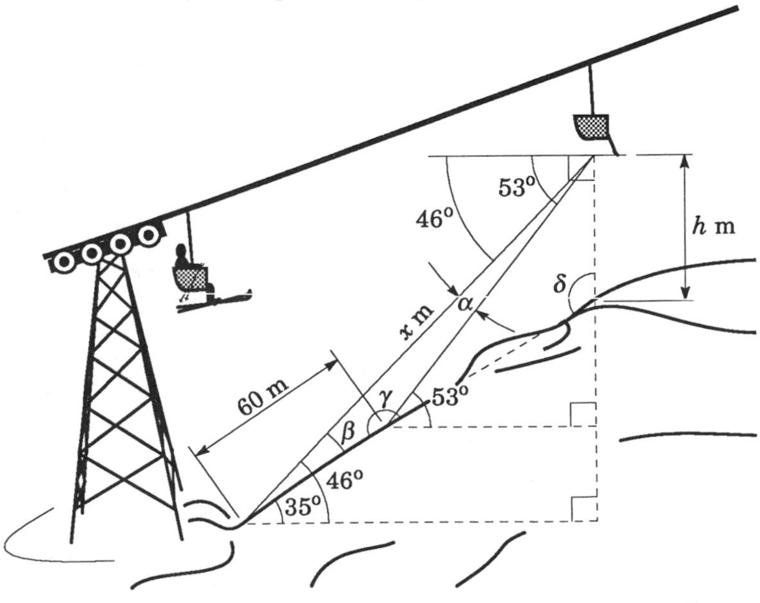

$\alpha = 53° - 46° = 7°$. $\gamma = 180° - (\alpha + \beta) = 162°$.

$\beta = 46° - 35° = 11°$. $\delta = 180° - (90° - 46°) - \beta = 125°$.

By the law of sines,

$$\frac{x}{\sin \gamma} = \frac{60}{\sin \alpha}; \quad x = \frac{60 \sin \gamma}{\sin \alpha}.$$

Also,

$$\frac{h}{\sin \beta} = \frac{x}{\sin \delta}.$$

$$h = \frac{x \sin \beta}{\sin \delta}$$

$$= 60 \cdot \frac{\sin 162° \cdot \sin 11°}{\sin 7° \cdot \sin 125°} = 60 \cdot \frac{\sin 18° \cdot \sin 11°}{\sin 7° \cdot \sin 55°} \approx 35.4 \text{ m}.$$

Bearings and Cardinal Points

The **bearing** of, or direction to, a point on land or at sea may be stated as the angle that the line of sight to the point forms with the **cardinal points** of the compass – North, East, South, and West.

There are two ways of stating a bearing.

The points P_1 P_2 P_3 P_4 in the chart below have the following bearings relative to the observation point O:

P_1	North 62° East	or	N 62 E
P_2	North 20° West		N 20 W
P_3	South 18° West		S 18 W
P_4	South 46° East		S 46 E

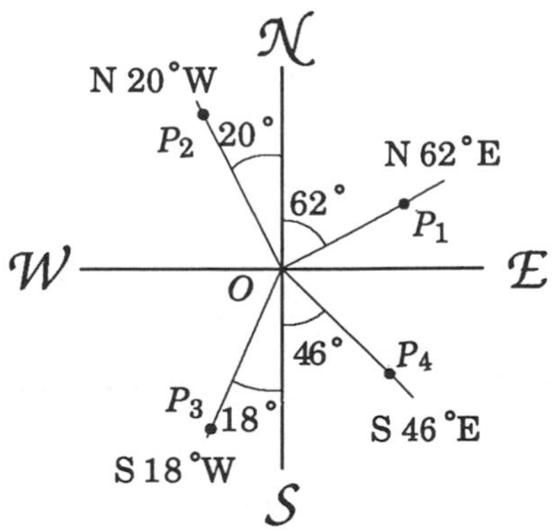

The other method of stating a bearing, or course, is by giving the clockwise (CW) angle from true North:

P_1	bearing 62
P_2	bearing 340
P_3	bearing 198
P_4	bearing 134

- A ship at A sights a lighthouse L on bearing N65E and, after sailing 8 nautical miles on course S45E to position B, the same lighthouse at N32E.

  Determine the distance between the ship when at B and the lighthouse. (1 n.mi. = 1.852 km.)

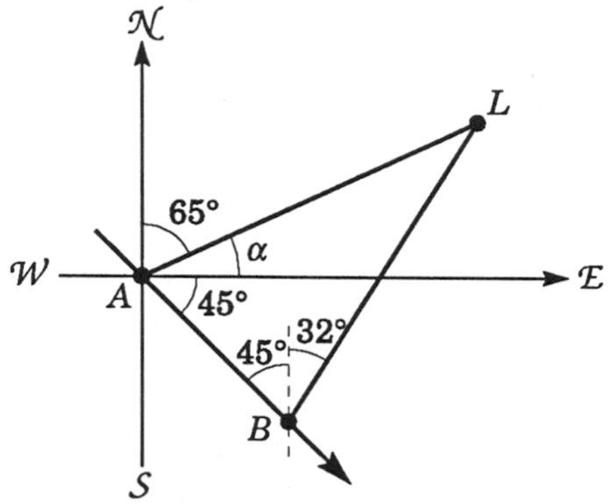

$$\alpha = 90° - 65° = 25°$$

Angles $BAL = 45° + 25° = 70°$

$ABL = 45° + 32° = 77°$

$ALB = 180° - (70° + 77°) = 33°$

According to the law of sines,

$$\frac{BL}{\sin 70°} = \frac{8}{\sin 33°}\ ; \quad BL \approx 13.8 \text{ n.mi.} \approx 25.6 \text{ km.}$$

- A ship heading due north at a logged speed of 14 knots (kt) sights at position A a lighthouse L at N20E and, at the same time, a known shipwreck W at N45E; 30 minutes later, the lighthouse is due east of the ship's new position B and the shipwreck is at N68E.

  Find the distance LW between the lighthouse and the shipwreck. (1 n.mi. = 1852 m; 1 kt = 1 n.mi./h.)

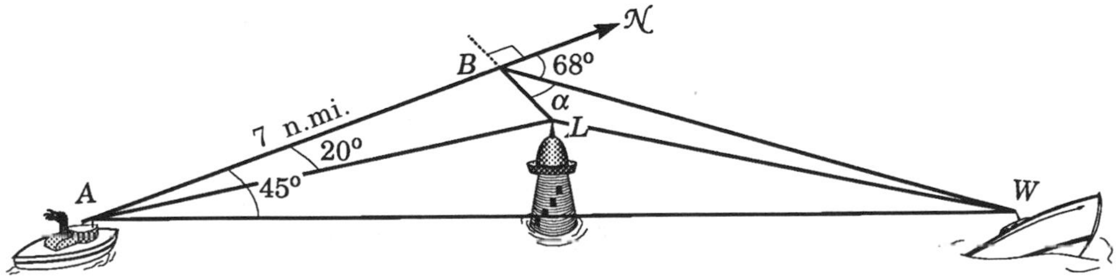

The distance traveled by the ship is 7 n.mi.

We have, by the law of sines,

$$\frac{BW}{\sin 45°} = \frac{7}{\sin AWB},$$

where the angle $AWB = 68° - 45° = 23°$; thus,

$$BW = 7 \cdot \frac{\sin 45°}{\sin 23°}.$$

Observing that

$$BL = 7 \tan 20^\circ$$

and

$$LBW = 90^\circ - 68^\circ = 22^\circ,$$

and applying the law of cosines to triangle BLW,

$$LW^2 = BL^2 + BW^2 - 2\, BL\, BW \cos 22^\circ$$

gives

$$LW = 7\sqrt{\tan^2 20^\circ + \frac{\sin^2 45^\circ}{\sin^2 23^\circ} - 2\,\frac{\tan 20^\circ \cdot \sin 45^\circ \cdot \cos 22^\circ}{\sin 23^\circ}}$$

$$\approx 10.3 \text{ n.mi.} \approx 19.1 \text{ km}.$$

- A ship S sights two navigation beacons, A at N 51.2 W and B at N 32.0 E, situated 32 km apart on a line bearing N 78 E. To steer clear of a reef off the spit of land at A, a course must be set to take the ship past A at a distance of not less than 6.4 km.

  Set the ship's course.

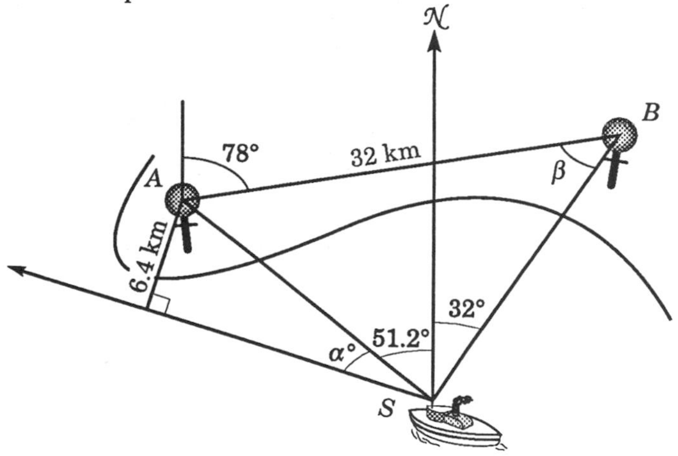

The sought course is N $(51.2^\circ + \alpha)$ W.

We have

$$\sin \alpha = \frac{6.4}{SA}.$$

Noting that

$$\beta = 180^\circ - 32.0^\circ - (180^\circ - 78^\circ) = 46.0^\circ,$$

we can determine SA by the law of sines applied to triangle ABS,

$$\frac{SA}{\sin 46.0^\circ} = \frac{32}{\sin(51.2^\circ + 32.0^\circ)} = \frac{32}{\sin 83.2^\circ}$$

$$SA = \frac{32 \cdot \sin 46.0^\circ}{\sin 83.2^\circ}.$$

We have

$$\sin \alpha = \frac{6.4}{SA} = \frac{6.4}{32} \cdot \frac{\sin 83.2^\circ}{\sin 46.0^\circ} \approx 0.276; \quad \alpha \approx 16.0^\circ,$$

and

$$51.2^\circ + \alpha = 67.2^\circ.$$

The ship's course should be south of N 67.2 W.

13.4 Graphs, Domains, Ranges

Sine Curve

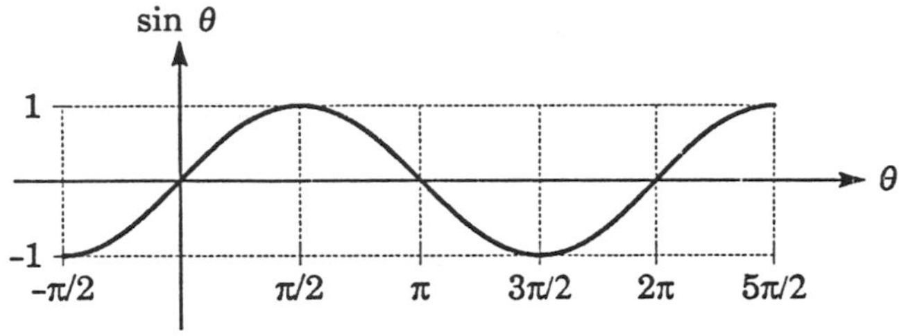

A point-by-point construction of the curves of the fundamental trigonometric functions can be made from the unit circle; below, that method is used to construct the sine curve:

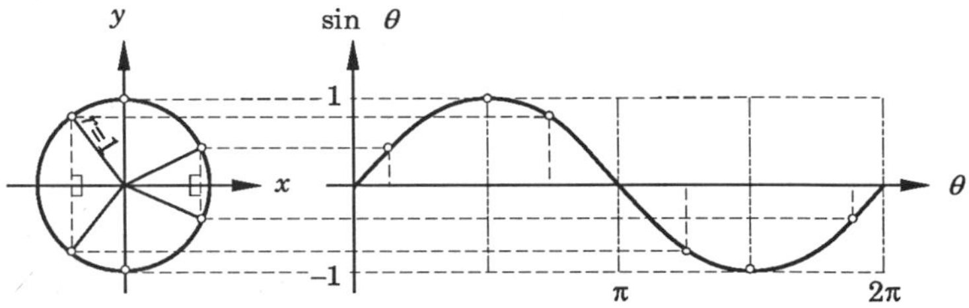

Since the circumference of a circle equals the radius multiplied by 2π, and the radius of the unit circle is 1, the circumference of the circle is 2π, and the length of the curve – representing one period – is 2π radians.

Cosecant Curve

$$\csc\theta = \frac{1}{\sin\theta}$$

The cosecant is the reciprocal of the sine of an angle, and we obtain the following graph:

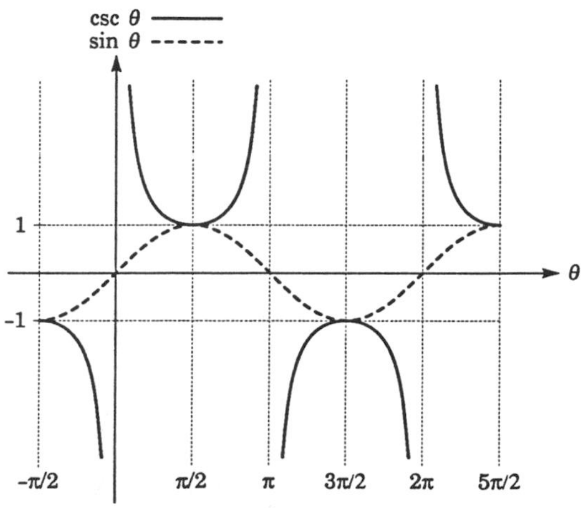

Greek, *asymptotos*, "not falling together"

An **asymptote** is that straight line to which a curve moves closer and closer.

As shown in the above graph, the asymptotes of the cosecant curve coincide with the points where sine of θ is 0. At these points we have 1/0, which is undefined.

As with the sine, the length of one period of the cosecant is 2π radians.

Cosine Curve

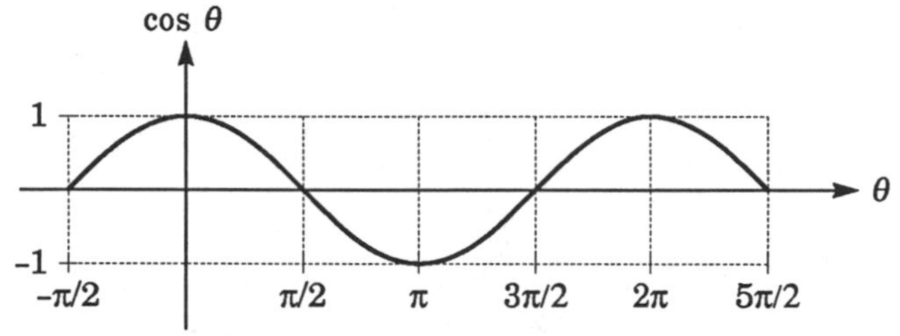

The length of one period of the cosine function is 2π radians.

Secant Curve

$$\sec\theta = \frac{1}{\cos\theta}$$

The asymptotes of the secant curve coincide with the points where cosine of θ is 0:

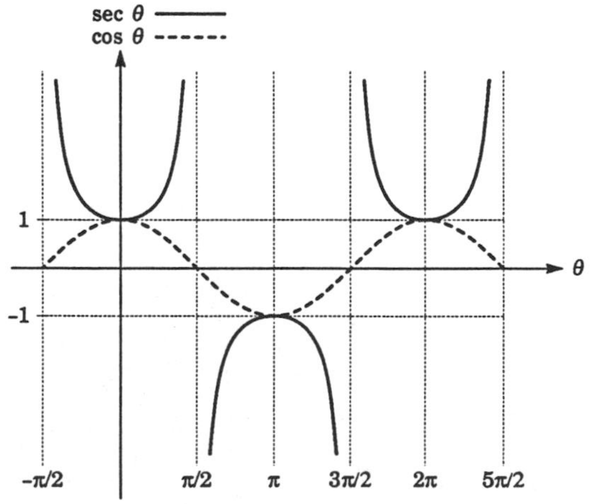

The length of one period of the secant function is 2π radians.

Tangent Curve

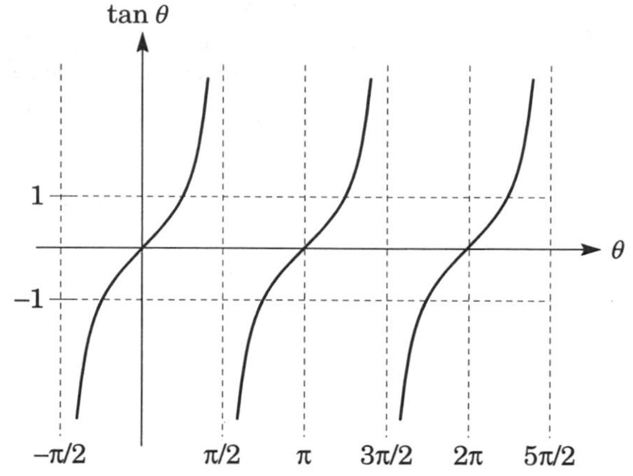

The tangent curve approaches the vertical line = π/2 asymptotically, and this is repeated in periods of π radians; we note that this is only half the length of one period of the sine, cosine, secant, and cosecant functions.

Cotangent Curve

$$\cot \theta = \frac{1}{\tan \theta}$$

As a consequence of the cotangent being the reciprocal of the tangent of an angle, the following graph is generated:

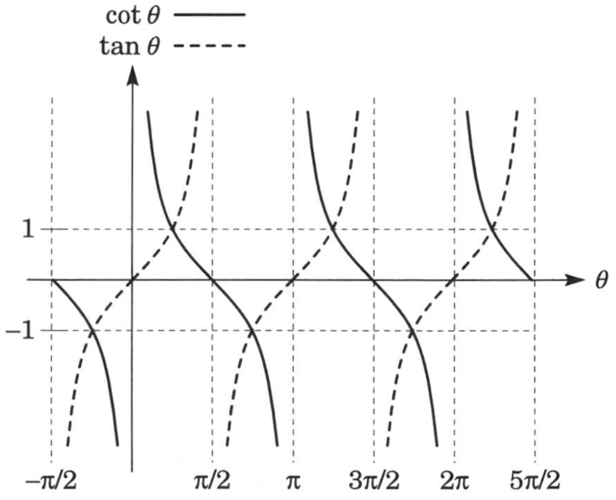

The length of one period of the cotangent function is π radians.

Domains and Ranges

Trigonometric functions are periodic, which means that every value of a trigonometric function corresponds to an infinite number of arguments. To cover the entire interval [−1 , 1] of sine and cosine, we can **restrict the angle** to the closed interval $\left[-\frac{\pi}{2}, \frac{\pi}{2}\right]$ for $\sin x$, and to the closed interval $[0, \pi]$ for $\cos x$. The corresponding inverse trigonometric functions are given values in these intervals:

arcsin x: $\left[-\frac{\pi}{2}, \frac{\pi}{2}\right]$

arccos x: $[0, \pi]$

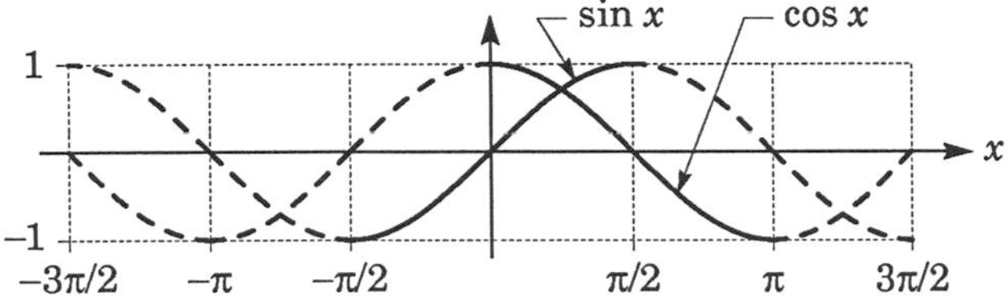

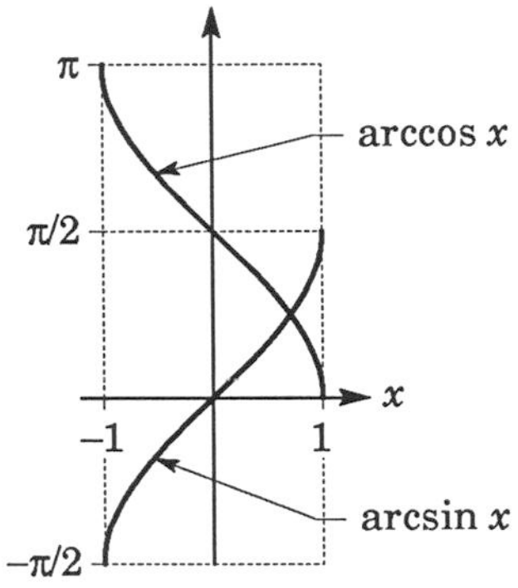

Much the same argumentation applies for the functions $\tan x$ and $\cot x$ and their corresponding inverse functions. Values of $\tan x$ are restricted to the given open interval $\left]-\frac{\pi}{2}, \frac{\pi}{2}\right[$, $\cot x$ to the open interval $]0, \pi[$; the corresponding inverse trigonometric functions are

arctan x: $\left]-\frac{\pi}{2}, \frac{\pi}{2}\right[$

arccot x: $]0, \pi[$.

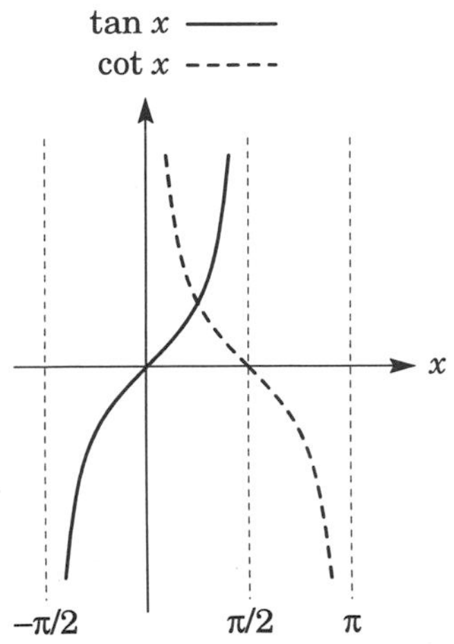

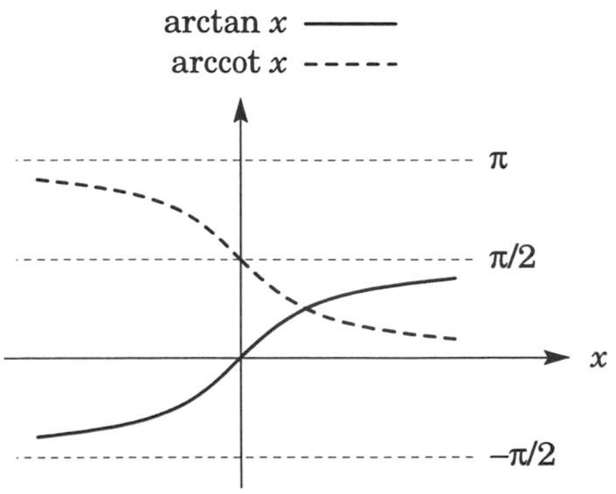

There is no agreement regarding definitions of restricted intervals for the functions sec x and csc x and their inverses; often used are

$$\text{arcsec } x: \qquad \left[-\pi, -\frac{\pi}{2}\right[, \left[0, \frac{\pi}{2}\right[$$

$$\text{arccsc } x: \qquad \left]-\pi, -\frac{\pi}{2}\right], \left]0, \frac{\pi}{2}\right].$$

The domain of the trigonometric function, restricted to the given interval, is the range of its inverse, and vice versa, for instance:

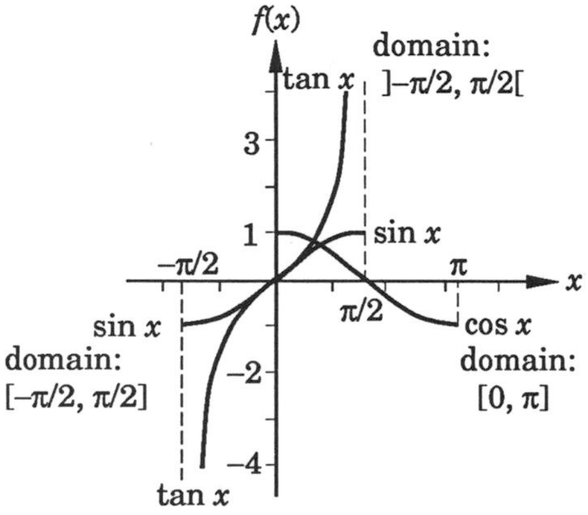

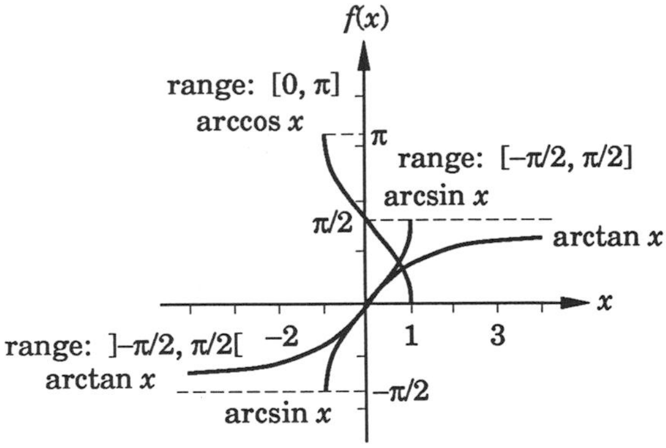

13.5 Trigonometric Identities

Trigonometric identities are equations that express relations among trigonometric functions which are true for all values of the variables involved. Such identities are of importance when solving trigonometric problems and are powerful media in calculus.

Fundamental Trigonometric Functions

Fundamental Identities

$$\sin\theta = \frac{1}{\csc\theta} \qquad \tan\theta = \frac{\sin\theta}{\cos\theta} \qquad \csc\theta = \frac{1}{\sin\theta}$$

$$\cos\theta = \frac{1}{\sec\theta} \qquad \cot\theta = \frac{\cos\theta}{\sin\theta} \qquad \sec\theta = \frac{1}{\cos\theta}$$

Euler's Formula

The identity

$$e^{ix} = \cos x + i\sin x\,,$$

where i is the imaginary unit and x a real number, is known as **Euler's formula**, named after the Swiss mathematician **Leonhard Euler**. In a later chapter, we shall show its validity by using power series.

EULER Leonhard
(1707 - 1783)

p. 792

Since $\cos\pi = -1$ and $\sin\pi = 0$, we obtain

$$e^{i\pi} = -1,$$

by many regarded as the most beautiful formula in all of mathematics.

Thus, the transcendental number e raised to the power of the product of the imaginary unit and π, also a transcendental number, yields exactly -1.

Similarly, since $\cos 2\pi = 1$, $\sin 2\pi = 0$, and $e^0 = 1$,

$$e^{2i\pi} = 1\,.$$

Cofunction Identities

The value of a fundamental trigonometric function of an acute angle is numerically equal to the value of the cofunction of the complementary angle.

$$\sin\theta = \cos\left(\frac{\pi}{2} - \theta\right) \qquad \tan\theta = \cot\left(\frac{\pi}{2} - \theta\right) \qquad \csc\theta = \sec\left(\frac{\pi}{2} - \theta\right)$$

$$\cos\theta = \sin\left(\frac{\pi}{2} - \theta\right) \qquad \cot\theta = \tan\left(\frac{\pi}{2} - \theta\right) \qquad \sec\theta = \csc\left(\frac{\pi}{2} - \theta\right)$$

Even-Odd Identities

The *cosine* is an **even function**: $\cos\theta = \cos(-\theta)$.

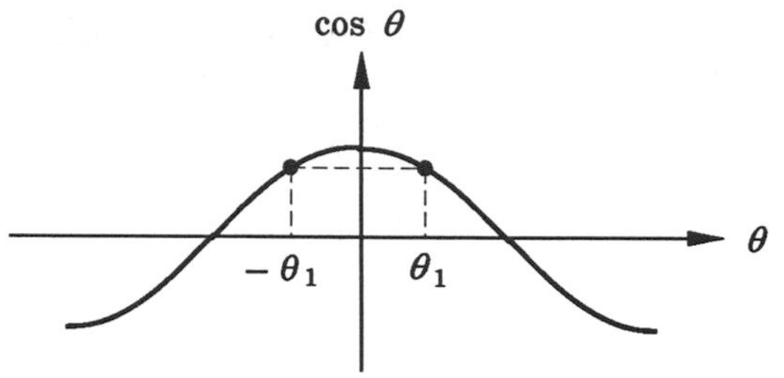

Being the inverse of the cosine, the *secant* is also an even function.

The *sine*, *cosecant*, *tangent*, and *cotangent*, on the other hand, are all **odd functions**: they have the same numerical values for equal positive and negative angles, although with opposite signs.

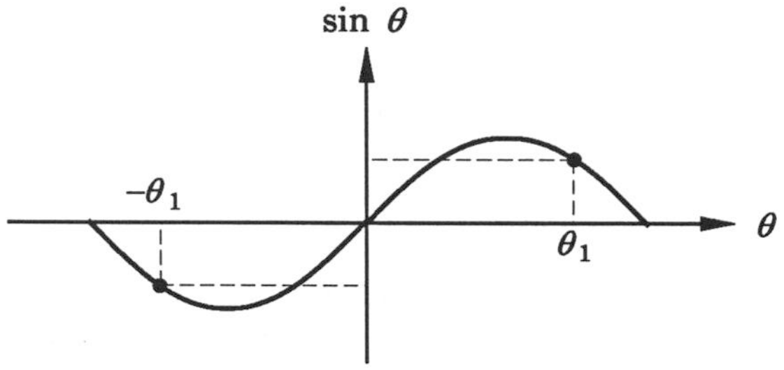

The **even-odd identities** are:

$\sin(-\theta) = -\sin\theta$ $\tan(-\theta) = -\tan\theta$ $\csc(-\theta) = -\csc\theta$

$\cos(-\theta) = \cos\theta$ $\cot(-\theta) = -\cot\theta$ $\sec(-\theta) = \sec\theta$

Pythagorean Identities

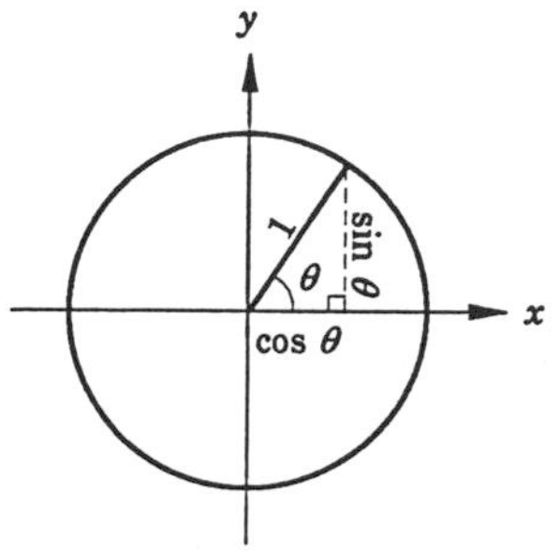

Applying the Pythagorean theorem to a triangle in the unit circle gives

$$\sin^2\theta + \cos^2\theta = 1.$$

Dividing both sides by $\cos^2\theta$,

$$\frac{\sin^2\theta}{\cos^2\theta} + \frac{\cos^2\theta}{\cos^2\theta} = \frac{1}{\cos^2\theta} \text{ and } 1 + \tan^2\theta = \sec^2\theta.$$

Dividing by $\sin^2\theta$,

$$\frac{\sin^2\theta}{\sin^2\theta} + \frac{\cos^2\theta}{\sin^2\theta} = \frac{1}{\sin^2\theta} \text{ and } 1 + \cot^2\theta = \csc^2\theta.$$

Compound-Angle Identities

To establish formulas for the sine, cosine, and tangent functions of the sum $(\theta + \phi)$ and difference $(\theta - \phi)$ of angles θ and ϕ in terms of functions of the constituent angles, we refer to the compound figure below.

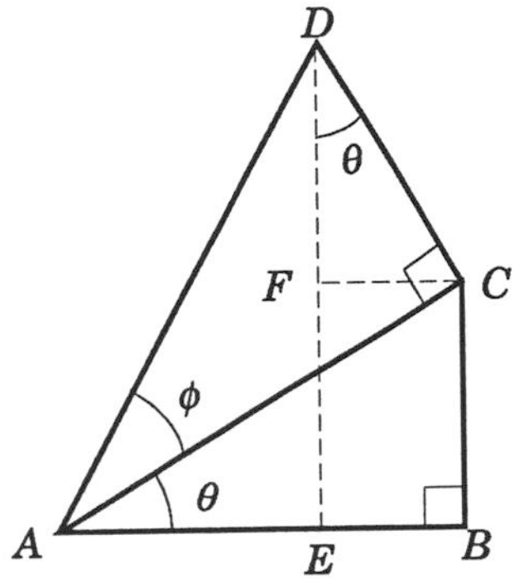

We have

$$\sin(\theta + \phi) = \frac{DE}{AD} = \frac{EF + FD}{AD} = \frac{BC + FD}{AD}$$

$$= \frac{BC}{AD}\frac{AC}{AC} + \frac{FD}{AD}\frac{CD}{CD}$$

$$= \frac{BC}{AC}\frac{AC}{AD} + \frac{FD}{CD}\frac{CD}{AD}$$

$$\sin(\theta + \phi) = \sin\theta\cos\phi + \cos\theta\sin\phi\,. \quad (1)$$

Replace ϕ in (1) by $(-\phi)$ to obtain

$$\sin(\theta - \phi) = \sin\theta\cos\phi - \cos\theta\sin\phi\,. \quad (2)$$

Replace θ in (2) by $\left(\frac{\pi}{2} - \theta\right)$ to obtain

$$\sin\left(\frac{\pi}{2} - \theta - \phi\right) = \sin\left(\frac{\pi}{2} - \theta\right)\cos\phi - \cos\left(\frac{\pi}{2} - \theta\right)\sin\phi$$

$$\cos(\theta + \phi) = \cos\theta\,\cos\phi - \sin\theta\,\sin\phi\,. \quad (3)$$

Replace ϕ in (3) by $(-\phi)$ to obtain

$$\cos(\theta - \phi) = \cos\theta\,\cos\phi + \sin\theta\,\sin\phi\,. \quad (4)$$

To find a formula for $\tan(\theta+\phi)$, divide (1) by (3):

$$\frac{\sin(\theta+\phi)}{\cos(\theta+\phi)} = \frac{\sin\theta\ \cos\phi + \cos\theta\ \sin\phi}{\cos\theta\ \cos\phi - \sin\theta\ \sin\phi}$$

$$= \frac{\dfrac{\sin\theta\ \cos\phi}{\cos\theta\ \cos\phi} + \dfrac{\cos\theta\ \sin\phi}{\cos\theta\ \cos\phi}}{\dfrac{\cos\theta\ \cos\phi}{\cos\theta\ \cos\phi} - \dfrac{\sin\theta\ \sin\phi}{\cos\theta\ \cos\phi}}$$

$$\tan(\theta+\phi) = \frac{\tan\theta + \tan\phi}{1 - \tan\theta\ \tan\phi}\ . \qquad (5)$$

Replace ϕ in (5) by $(-\phi)$ to obtain

$$\tan(\theta-\phi) = \frac{\tan\theta - \tan\phi}{1 + \tan\theta\ \tan\phi}\ . \qquad (6)$$

The cotangent is the inverse of the tangent function,

$$\cot(\theta+\phi) = \frac{1}{\tan(\theta+\phi)} = \frac{1-\tan\theta\ \tan\phi}{\tan\theta + \tan\phi}$$

$$= \frac{1 - \dfrac{1}{\cot\theta\ \cot\phi}}{\dfrac{1}{\cot\theta} + \dfrac{1}{\cot\phi}}$$

$$\cot(\theta+\phi) = \frac{\cot\theta\ \cot\phi - 1}{\cot\theta + \cot\phi}\ . \qquad (7)$$

Replace ϕ in (7) by $(-\phi)$ to obtain

$$\cot(\theta-\phi) = \frac{\cot\theta\ \cot\phi + 1}{\cot\theta - \cot\phi}\ . \qquad (8)$$

- Find the exact numerical value of $\sin 105°$.

$$\sin(60° + 45°) = \sin 60° \cos 45° + \cos 60° \sin 45°$$

$$= \frac{\sqrt{3}}{2}\left(\frac{\sqrt{2}}{2}\right) + \frac{1}{2}\left(\frac{\sqrt{2}}{2}\right) = \frac{\sqrt{6}+\sqrt{2}}{4}\ .$$

Double-Angle and Multiple-Angle Identities

Formulas for the trigonometric functions of the double angle and of multiple angles may be computed from the compound-angle formulas.

Functions of 2θ

Equation (1),

$$\sin(\theta+\phi) = \sin\theta\ \cos\phi + \cos\theta\ \sin\phi\ ,$$

gives, with $\phi = \theta$,

$$\sin 2\theta = 2\sin\theta\ \cos\theta\ . \qquad (9)$$

Equation (3),

$$\cos(\theta+\phi) = \cos\theta\cos\phi - \sin\theta\sin\phi,$$

gives, with $\phi = \theta$,

$$\cos 2\theta = \cos^2\theta - \sin^2\theta \quad (10a)$$

$$= 2\cos^2\theta - 1 \quad (10b)$$

$$= 1 - 2\sin^2\theta. \quad (10c)$$

From Equation (5),

$$\tan(\theta+\phi) = \frac{\tan\theta + \tan\phi}{1 - \tan\theta\tan\phi},$$

we obtain

$$\tan 2\theta = \frac{2\tan\theta}{1-\tan^2\theta}$$

$$\theta \neq (2n-1)\frac{\pi}{2}; \qquad 2\theta \neq (2n-1)\frac{\pi}{2}; \quad (11)$$

and from (7),

$$\cot(\theta+\phi) = \frac{\cot\theta\cot\phi - 1}{\cot\theta + \cot\phi},$$

$$\cot 2\theta = \frac{\cot^2\theta - 1}{2\cot\theta} \qquad \theta \neq n\cdot\pi;\ 2\theta \neq n\cdot\pi. \quad (12)$$

Functions of 3θ, 4θ, and 5θ

$$\sin 3\theta = \sin(2\theta+\theta)$$

$$= \sin 2\theta\cos\theta + \cos 2\theta\sin\theta$$

$$= (2\sin\theta\cos\theta)\cos\theta + (1-2\sin^2\theta)\sin\theta$$

$$= 2\sin\theta\cos^2\theta + \sin\theta - 2\sin^3\theta$$

$$= 3\sin\theta - 4\sin^3\theta.$$

$$\sin 4\theta = 4\sin\theta\cos\theta - 8\sin^3\theta\cos\theta.$$

$$\sin 5\theta = 16\sin^5\theta - 20\sin^3\theta + 5\sin\theta.$$

$$\cos 3\theta = 4\cos^3\theta - 3\cos\theta.$$

$$\cos 4\theta = 8\cos^4\theta - 8\cos^2\theta + 1.$$

$$\cos 5\theta = 16\cos^5\theta - 20\cos^3\theta + 5\cos\theta.$$

$$\tan 3\theta = \frac{3\tan\theta - \tan^3\theta}{1 - 3\tan^2\theta}.$$

$$\tan 4\theta = \frac{4\tan\theta - 4\tan^3\theta}{1 - 6\tan^2\theta + \tan^4\theta}.$$

$$\tan 5\theta = \frac{5\tan\theta - 10\tan^3\theta + \tan^5\theta}{1 - 10\tan^2\theta + 5\tan^4\theta}.$$

$$\cot 3\theta = \frac{\cot^3\theta - 3\cot\theta}{3\cot^2\theta - 1}.$$

$$\cot 4\theta = \frac{\cot^4\theta - 6\cot^2\theta + 1}{4\cot^3\theta - 4\cot\theta}.$$

$$\cot 5\theta = \frac{\cot^5\theta - 10\cot^3\theta + 5\cot\theta}{5\cot^4\theta - 10\cot^2\theta + 1}.$$

Functions of the Half-Angle

Equation (10c) may be rearranged as

$$\sin^2\theta = \frac{1-\cos 2\theta}{2}\,; \quad \sin^2\frac{\theta}{2} = \frac{1-\cos\theta}{2}$$

$$\sin\frac{\theta}{2} = \pm\sqrt{\frac{1-\cos\theta}{2}} \tag{13}$$

and Equation (10b) as

$$\cos^2\theta = \frac{1+\cos 2\theta}{2}\,; \quad \cos^2\frac{\theta}{2} = \frac{1+\cos\theta}{2}$$

$$\cos\frac{\theta}{2} = \pm\sqrt{\frac{1+\cos\theta}{2}}\,. \tag{14}$$

Dividing (13) by (14), we have

$$\tan\frac{\theta}{2} = \pm\sqrt{\frac{1-\cos\theta}{1+\cos\theta}}\,; \quad \cot\frac{\theta}{2} = \pm\sqrt{\frac{1+\cos\theta}{1-\cos\theta}}\,. \qquad (15)\ (16)$$

Conversion of Functions to the Half-Angle Tangent

To solve trigonometric equations, it is often useful, or necessary, to have all trigonometric quantities expressed in one function only, usually the tangent of the half-angle.

We have, from Equation (9),

$$\sin\theta = 2\sin\frac{\theta}{2}\cos\frac{\theta}{2}\,,$$

where we introduce a denominator

$$\sin^2\frac{\theta}{2} + \cos^2\frac{\theta}{2} = 1$$

to the right-hand side of the equation and divide both the numerator and the new denominator by $\cos^2\frac{\theta}{2}$ to obtain

$$\sin\theta = \frac{2\sin\frac{\theta}{2}\cos\frac{\theta}{2}}{\sin^2\frac{\theta}{2}+\cos^2\frac{\theta}{2}} = \frac{\dfrac{2\sin\frac{\theta}{2}\cos\frac{\theta}{2}}{\cos^2\frac{\theta}{2}}}{\dfrac{\sin^2\frac{\theta}{2}+\cos^2\frac{\theta}{2}}{\cos^2\frac{\theta}{2}}} = \frac{2\tan\frac{\theta}{2}}{1+\tan^2\frac{\theta}{2}}\,. \tag{17}$$

We further have, from Equation (10a)

$$\cos\theta = \cos^2\frac{\theta}{2} - \sin^2\frac{\theta}{2}\,,$$

which, by analogous treatment, is transformed into

$$\cos\theta = \frac{\cos^2\frac{\theta}{2} - \sin^2\frac{\theta}{2}}{\sin^2\frac{\theta}{2} + \cos^2\frac{\theta}{2}} = \frac{\dfrac{\cos^2\frac{\theta}{2} - \sin^2\frac{\theta}{2}}{\cos^2\frac{\theta}{2}}}{\dfrac{\sin^2\frac{\theta}{2} + \cos^2\frac{\theta}{2}}{\cos^2\frac{\theta}{2}}} = \frac{1 - \tan^2\frac{\theta}{2}}{1 + \tan^2\frac{\theta}{2}}. \quad (18)$$

The remaining formulas are:

$$\tan\theta = \frac{2\tan\frac{\theta}{2}}{1 - \tan^2\frac{\theta}{2}} \qquad \csc\theta = \frac{1 + \tan^2\frac{\theta}{2}}{2\tan\frac{\theta}{2}} \qquad (19)\ (21)$$

$$\cot\theta = \frac{1 - \tan^2\frac{\theta}{2}}{2\tan\frac{\theta}{2}} \qquad \sec\theta = \frac{1 + \tan^2\frac{\theta}{2}}{1 - \tan^2\frac{\theta}{2}} \qquad (20)\ (22)$$

Since $\tan\frac{\pi}{2}$ is undefined, these conversion formulas are not valid for angles $\theta = n \cdot \pi$.

Converting Sums of Functions to Products

Add and subtract the compound-angle formulas (1) and (2), in the form

$$\begin{aligned} (1):&\quad \sin(p+q) = \sin p\cos q + \cos p\sin q \\ (2):&\quad \sin(p-q) = \sin p\cos q - \cos p\sin q, \\ (1+2):&\quad \sin(p+q) + \sin(p-q) = 2\sin p\cos q \\ (1-2):&\quad \sin(p+q) - \sin(p-q) = 2\cos p\sin q \end{aligned}$$

where we substitute

$$p+q = \theta; \quad p-q = \phi; \quad p = \frac{\theta+\phi}{2}; \quad q = \frac{\theta-\phi}{2}$$

and obtain

$$\sin\theta + \sin\phi = 2\sin\left(\frac{\theta+\phi}{2}\right)\cos\left(\frac{\theta-\phi}{2}\right) \qquad (23)$$

$$\sin\theta - \sin\phi = 2\cos\left(\frac{\theta+\phi}{2}\right)\sin\left(\frac{\theta-\phi}{2}\right). \qquad (24)$$

In an analogous manner, Equations (3) and (4) will produce the equations

$$\cos\theta + \cos\phi = 2\cos\left(\frac{\theta+\phi}{2}\right)\cos\left(\frac{\theta-\phi}{2}\right) \qquad (25)$$

$$\cos\theta - \cos\phi = -2\sin\left(\frac{\theta+\phi}{2}\right)\sin\left(\frac{\theta-\phi}{2}\right). \qquad (26)$$

We quote without proof:

$$\tan\theta + \tan\phi = \frac{\sin(\theta+\phi)}{\cos\theta\cos\phi}; \quad \cot\theta + \cot\phi = \frac{\sin(\theta+\phi)}{\sin\theta\sin\phi} \qquad (27)\ (29)$$

$$\tan\theta - \tan\phi = \frac{\sin(\theta-\phi)}{\cos\theta\cos\phi}; \quad \cot\theta - \cot\phi = -\frac{\sin(\theta-\phi)}{\sin\theta\sin\phi} \qquad (28)\ (30)$$

Breaking up Trigonometric Products

Formulas for breaking up products $\sin\theta\cos\phi$ and $\cos\theta\sin\phi$ may be obtained by adding and subtracting the compound-angle formulas (1) and (2),

$$(1)\quad \sin\theta\cos\phi + \cos\theta\sin\phi = \sin(\theta+\phi)$$
$$(2)\quad \sin\theta\cos\phi - \cos\theta\sin\phi = \sin(\theta-\phi)\,.$$

Addition and division by 2 gives

$$(1+2)\qquad \sin\theta\cos\phi = \frac{1}{2}\,[\sin(\theta+\phi)+\sin(\theta-\phi)]\,. \qquad (31)$$

Subtraction and division by 2 gives

$$(1-2)\qquad \cos\theta\sin\phi = \frac{1}{2}\,[\sin(\theta+\phi)-\sin(\theta-\phi)]\,. \qquad (32)$$

To break up products $\sin\theta\sin\phi$ and $\cos\theta\cos\phi$, we add and subtract Equations (3) and (4),

$$(3)\quad \cos\theta\cos\phi - \sin\theta\sin\phi = \cos(\theta+\phi)$$
$$(4)\quad \cos\theta\cos\phi + \sin\theta\sin\phi = \cos(\theta-\phi)\,.$$

Subtraction and division by 2 gives

$$(4-3)\qquad \sin\theta\sin\phi = \frac{1}{2}\,[\cos(\theta-\phi)-\cos(\theta+\phi)]\,. \qquad (33)$$

Addition and division by 2 gives

$$(4+3)\qquad \cos\theta\cos\phi = \frac{1}{2}\,[\cos(\theta-\phi)+\cos(\theta+\phi)]\,. \qquad (34)$$

Inverse Trigonometric Functions

As a corollary of the trigonometric cofunction identities, we have the inverse trigonometric identities, here quoted without proof.

$$\sin(\arcsin x) = x\,; \qquad \arcsin(\sin x) = x \quad \text{if} \quad -\frac{\pi}{2} \le x \le \frac{\pi}{2}$$

$$\cos(\arccos x) = x\,; \qquad \arccos(\cos x) = x \quad \text{if} \quad 0 \le x \le \pi$$

pp. 505 - 6 *etc.* intervals: *pp.* 505 - 6

$$\arcsin(-x) = -\arcsin x$$
$$\arccos(-x) = \pi - \arccos x$$
$$\arctan(-x) = -\arctan x$$
$$\operatorname{arccot}(-x) = \pi - \operatorname{arccot} x$$

$$\arcsin x = \frac{\pi}{2} - \arccos x = \arccos\sqrt{1-x^2} = \arctan\frac{x}{\sqrt{1-x^2}} = \operatorname{arccot}\frac{\sqrt{1-x^2}}{x} \qquad x>0$$

$$\arccos x = \frac{\pi}{2} - \arcsin x = \arcsin\sqrt{1-x^2} = \arctan\frac{\sqrt{1-x^2}}{x} = \operatorname{arccot}\frac{x}{\sqrt{1-x^2}} \qquad x>0$$

$$\arctan x = \frac{\pi}{2} - \operatorname{arccot} x = \arcsin\frac{x}{\sqrt{1+x^2}} = \arccos\frac{1}{\sqrt{1+x^2}} = \operatorname{arccot}\frac{1}{x} \qquad x>0$$

$$\operatorname{arccot} x = \frac{\pi}{2} - \arctan x = \arcsin\frac{1}{\sqrt{1+x^2}} = \arccos\frac{x}{\sqrt{1+x^2}} = \arctan\frac{1}{x} \qquad x>0$$

$$\operatorname{arccot} x = \begin{cases} \arctan \dfrac{1}{x} & x > 0 \\ \arctan \dfrac{1}{x} + \pi & x < 0 \end{cases}$$

$$\arcsin x = -\mathrm{i} \cdot \ln\left(\mathrm{i}\,x + \sqrt{1 - x^2}\right)$$

$$\arccos x = -\mathrm{i} \cdot \ln\left(x + \sqrt{x^2 - 1}\right)$$

$$\arctan x = \frac{1}{2\,\mathrm{i}} \cdot \ln \frac{1 + \mathrm{i}\,x}{1 - \mathrm{i}\,x}$$

$$\operatorname{arccot} x = -\frac{1}{2\,\mathrm{i}} \cdot \ln \frac{\mathrm{i}\,x + 1}{\mathrm{i}\,x - 1} = \frac{1}{2\,\mathrm{i}} \cdot \ln \frac{\mathrm{i}\,x - 1}{\mathrm{i}\,x + 1} \qquad \text{if } x > 0$$

Sums and Differences

$$\arcsin x \pm \arcsin y \begin{cases} = \arcsin\left(x\sqrt{1 - y^2} \pm y\sqrt{1 - x^2}\right) & x^2 + y^2 \le 1\,;\ xy \le 0 \\ = \pi - \arcsin\left(x\sqrt{1 - y^2} \pm y\sqrt{1 - x^2}\right) & x > 0\,;\ y > 0\,;\ x^2 + y^2 > 1 \\ = -\pi - \arcsin\left(x\sqrt{1 - y^2} \pm y\sqrt{1 - x^2}\right) & x < 0\,;\ y < 0\,;\ x^2 + y^2 > 1 \end{cases}$$

$$\arccos x + \arccos y = \arccos\left[xy - \sqrt{(1 - x^2)(1 - y^2)}\right] \qquad (x + y) \ge 0$$

$$\arccos x - \arccos y \begin{cases} = \arccos\left[xy + \sqrt{(1 - x^2)(1 - y^2)}\right] & x < y \\ = -\arccos\left[xy + \sqrt{(1 - x^2)(1 - y^2)}\right] & x \ge y \end{cases}$$

$$\arctan x + \arctan y \begin{cases} = \arctan \dfrac{x + y}{1 - xy} & xy < 1 \\ = \pi + \arctan \dfrac{x + y}{1 - xy} & x > 0\,;\ xy > 1 \\ = -\pi + \arctan \dfrac{x + y}{1 - xy} & x < 0\,;\ xy > 1 \end{cases}$$

$$\arctan x - \arctan y \begin{cases} = \arctan \dfrac{x - y}{1 + xy} & xy > -1 \\ = \pi + \arctan \dfrac{x - y}{1 + xy} & x > 0\,;\ xy < -1 \\ = -\pi + \arctan \dfrac{x - y}{1 + xy} & x < 0\,;\ xy < -1 \end{cases}$$

$$\operatorname{arccot} x + \operatorname{arccot} y = \operatorname{arccot} \frac{xy - 1}{x + y}$$

$$\operatorname{arccot} x - \operatorname{arccot} y = \operatorname{arccot} \frac{xy + 1}{y - x}$$

13.6 Trigonometric Equations

In trigonometric equations, the unknown quantity appears in the form of trigonometric functions of one or more variables.

If only one kind of trigonometric function is present, the equation is said to be **basic** and may be solved in the same manner as algebraic equations. If the equation contains several functions of one angle or several angles, it must be reduced to basic form.

Because of the periodicity of trigonometric functions, trigonometric equations have an infinite number of solutions unless restricted by a **side condition**; all solutions must satisfy *both* the equation *and* the side condition.

Positive solutions less than 2π (360°) should always be determined first; all other solutions will follow automatically.

If $\sin x = a \qquad x = \left\{ \begin{matrix} \arcsin a \\ \pi - \arcsin a \end{matrix} \right\} + n \cdot 2\pi$

$\csc x = a \qquad x = \left\{ \begin{matrix} \operatorname{arccsc} a \\ \pi - \operatorname{arccsc} a \end{matrix} \right\} + n \cdot 2\pi$

$\cos x = a \qquad x = \pm \arccos a + n \cdot 2\pi$

$\sec x = a \qquad x = \pm \operatorname{arcsec} a + n \cdot 2\pi$

$\tan x = a \qquad x = \arctan a + n \cdot \pi$

$\cot x = a \qquad x = \operatorname{arccot} a + n \cdot \pi$

where n is an integer.

In a **pure** trigonometric equation, the unknowns appear only in the form of trigonometric functions; in a **mixed** trigonometric equation, they will be found also in algebraic form.

Mixed trigonometric equations can generally be solved only by graphical methods or by numerical analysis.

What is said here about trigonometric equations applies also for trigonometric inequalities.

Basic Trigonometric Equations

- Solve the equation

$$\sin x = \frac{1}{2},$$

in the general case and with the side condition $0 \le x < \pi$.

$$x = \left\{ \begin{matrix} \arcsin \frac{1}{2} \\ \pi - \arcsin \frac{1}{2} \end{matrix} \right\} + n \cdot 2\pi = \left\{ \begin{matrix} \frac{\pi}{6} \\ \frac{5\pi}{6} \end{matrix} \right\} + n \cdot 2\pi,$$

where n is an integer.

Of these, the solutions

$$x_1 = \frac{\pi}{6} = 30^\circ \; ; \; x_2 = \frac{5\pi}{6} = 150^\circ$$

satisfy the side condition.

- Solve the equation $\tan x = -\sqrt{2}$.

An electronic calculator gives

$$x \approx -0.955 + n \cdot \pi \approx -54.717^\circ + n \cdot 180^\circ$$

$$\approx 125.28^\circ + n \cdot 180^\circ, \text{ where } n \text{ is an integer.}$$

- Solve the equation $\frac{\cot^2 x}{4} = 1$.

$\cot x = \pm 2$.

Since the arc cotangent function is usually not available on calculators, we use its reciprocal,

$$x = \arctan\left(\pm\frac{1}{2}\right) + n \cdot \pi \approx \pm 0.464 + n \cdot \pi,$$

where n is an integer.

- Solve the equation $\cos 3x = \frac{\sqrt{3}}{2}; \quad 0 \le x < 2\pi$.

$$3x = \pm\frac{\pi}{6} + n \cdot 2\pi$$

$$x = \pm\frac{\pi}{18} + n \cdot \frac{2\pi}{3} = \frac{12n \pm 1}{18} \cdot \pi .$$

$$x = \frac{\pi}{18}; \; \frac{11\pi}{18}; \; \frac{13\pi}{18}; \; \frac{23\pi}{18}; \; \frac{25\pi}{18}; \; \frac{35\pi}{18}.$$

Reducing the Trigonometric Equation

The simplification of trigonometric equations involving several functions of several angles to basic form is an integral part of the task of solving trigonometric equations.

To effect this transformation, we have recourse to a number of trigonometric conversion formulas and Pythagorean identities which are helpful when handling trigonometric terms that include multiples of the unknown variable; the decision of which identity to choose generally depends on the form of the equation.

The introduction of a new variable is sometimes beneficial in order to make calculations simpler, although this method should be used with discretion. The purpose is often to bring the equation into a factorable form and, subsequently, to equate the individual factors to zero.

To solve a trigonometric equation that contains different trigonometric functions, one must arrange the equation so that it contains terms of *one and the same* trigonometric function (exclusively sine, exclusively cosine, exclusively tangent, *etc.*).

Converting all trigonometric functions present into tangent functions of the half-angle is often a profitable form of substitution, provided that it does not lead to equations higher than the second degree; it must be kept in mind, however, that this method cannot be employed for an angle π, as $\tan\frac{\pi}{2}$ is undefined.

There is no general rule, however, that will unconditionally provide the solution of a trigonometric equation; ingenuity and perseverance are often the prerequisites of success.

If algebraic methods fail, we might fall back on graphical solution or numerical analysis.

Employing Double-Angle Identities

- Solve the equation

$$\sin x \cos x = \frac{1}{4}; \quad 0 \le x < \frac{5\pi}{2}.$$

We have

p. 510, Eq. (9)

$$2 \sin x \cdot \cos x = \sin 2x = \frac{1}{2},$$

which gives

$$2x = \left\{ \begin{matrix} \frac{\pi}{6} \\ \frac{5\pi}{6} \end{matrix} \right\} + n \cdot 2\pi; \quad x = \left\{ \begin{matrix} \frac{\pi}{12} \\ \frac{5\pi}{12} \end{matrix} \right\} + n \cdot \pi.$$

In the interval $0 \le x < \frac{5\pi}{2} = \frac{30\pi}{12}$, we have the following solutions:

$$x_1 = \frac{\pi}{12} \qquad x_2 = \frac{5\pi}{12}$$

$$x_3 = \frac{13\pi}{12} \qquad x_4 = \frac{17\pi}{12}$$

$$x_5 = \frac{25\pi}{12} \qquad x_6 = \frac{29\pi}{12}$$

Converting to Half-Angle Tangent

Equations that contain different trigonometric functions of one and the same angle may be rendered into a form containing exclusively tangent functions of the half-angle. This method is advantageous except when it leads to equations higher than the second degree.

It is important to observe that the method is **not valid for an angle π rad**; and since $\tan\frac{\pi}{2}$ does not exist when half-angle tangent identities are employed, one must test the equation for the validity of a possible solution $= \pi$.

- Solve the equation

$$\cot x - \csc x - \frac{\sqrt{3}}{3} = 0; \; 0 < x < 2\pi .$$

p. 512 By substituting half-angle tangents, we have

$$\frac{1 - \tan^2\frac{x}{2}}{2\tan\frac{x}{2}} - \frac{1 + \tan^2\frac{x}{2}}{2\tan\frac{x}{2}} = \frac{\sqrt{3}}{3}$$

and, after reduction,

$$1 - \tan^2\frac{x}{2} - 1 - \tan^2\frac{x}{2} = 2\cdot\frac{\sqrt{3}}{3}\cdot\tan\frac{x}{2}$$

$$\tan^2\frac{x}{2} + \frac{\sqrt{3}}{3}\cdot\tan\frac{x}{2} = 0$$

$$\tan\frac{x}{2}\left(\tan\frac{x}{2} + \frac{\sqrt{3}}{3}\right) = 0 .$$

$\tan\frac{x}{2} = 0$	$\tan\frac{x}{2} = -\frac{\sqrt{3}}{3}$
$\frac{x}{2} = 0 + n\cdot\pi$	$\frac{x}{2} = \frac{5\pi}{6}$
$x = 0 + n\cdot 2\pi .$	$x = \frac{5\pi}{3} .$
There is no solution $< 2\pi$; rejected.	

$x = \pi$ cannot be a solution since $\cot \pi$ and $\csc \pi$ are undefined.

Testing for $x = \frac{5\pi}{3}$ gives

$$\cot\frac{5\pi}{3} - \csc\frac{5\pi}{3} - \frac{\sqrt{3}}{3} = -\frac{\sqrt{3}}{3} + \frac{1}{\frac{\sqrt{3}}{2}} - \frac{\sqrt{3}}{3} = 0 .$$

In the interval $0 < x < 2\pi$, the equation has only the solution

$$x = \frac{5\pi}{3} .$$

Factoring

Like algebraic equations, trigonometric equations can often be solved by factoring with the help of trigonometric identities. Which identity should be chosen depends entirely on the type of equation.

Employing Pythagorean Identities

A Pythagorean identity may be conveniently employed to convert terms into one and the same trigonometric function and render the equation suitable for factoring. Conversion into tangent functions of the half-angle would, in this case, produce a fourth-degree equation.

• Solve the equation

$$2\cos^2 x - \sin x = 1; \quad 0 \le x < 2\pi .$$

Write

$$2\cos^2 x - \sin x - 1 = 0$$

and obtain, successively,

$$2(1 - \sin^2 x) - \sin x - 1 = 0$$
$$2\sin^2 x + \sin x - 1 = 0$$
$$(2\sin x - 1)(\sin x + 1) = 0 .$$

Equate each factor with 0 :

$2\sin x - 1 = 0$	$\sin x + 1 = 0$
$\sin x = \frac{1}{2}$.	$\sin x = -1$.
$x = \frac{\pi}{6} ; \frac{5\pi}{6}$.	$x = \frac{3\pi}{2}$.

The solutions are

$$x_1 = \frac{\pi}{6} ; \; x_2 = \frac{5\pi}{6} ; \; x_3 = \frac{3\pi}{2} .$$

Converting Sums of Functions to Products

When every trigonometric term features a **different multiple of the variable**, it is sometimes possible to accomplish factoring by converting sums of trigonometric functions to products.

• Solve the equation

$$\sin 8x - \cos 5x = \sin 2x .$$

Rearrange,

$$(\sin 8x - \sin 2x) - \cos 5x = 0 ;$$

p. 513 and transform $(\sin 8x - \sin 2x)$ into a product,

$$2\cos\left(\frac{8x + 2x}{2}\right)\sin\left(\frac{8x - 2x}{2}\right) - \cos 5x = 0 ;$$
$$2\cos 5x \sin 3x - \cos 5x = 0 ;$$
$$\cos 5x\,(2\sin 3x - 1) = 0 .$$

Equate each factor with 0 :

$\cos 5x = 0$	$\sin 3x = \frac{1}{2}$
$5x = \pm\frac{\pi}{2} + n \cdot 2\pi$	$3x = \frac{\pi}{6} + n \cdot 2\pi ; \; \frac{5\pi}{6} + n \cdot 2\pi ;$
$x_1 = \pm\frac{\pi}{10} + \frac{n \cdot 2\pi}{5}$	$x_2 = \frac{\pi}{18} + \frac{n \cdot 2\pi}{3} ; \; x_3 = \frac{5\pi}{18} + \frac{n \cdot 2\pi}{3}$

The infinitude of solutions is

$$x_1 = \pm\frac{\pi}{10} + \frac{n \cdot 2\pi}{5} ; \; x_2 = \frac{\pi}{18} + \frac{n \cdot 2\pi}{3} ; \; x_3 = \frac{5\pi}{18} + \frac{n \cdot 2\pi}{3} ,$$

where n is an integer.

Conversion to Functions of the Same Angle

Trigonometric identities are employed to produce functions of equal angles.

- Solve the equation

$$\frac{4\cot 4x}{1-3\cot 2x} = 1; \quad 0 \le x < 2\pi$$

and give the answer to three reliable digits.

The equation contains only one trigonometric function, but of two multiples of the unknown.

We consider $2x$ as the unknown variable, and write

p. 511

$$4\cot 2(2x) = 1 - 3\cot 2x,$$

where we insert

$$\cot 2(2x) = \frac{\cot^2 2x - 1}{2\cot 2x}$$

and obtain

$$4 \cdot \frac{\cot^2 2x - 1}{2\cot 2x} = 1 - 3\cot 2x$$

$$4\cot^2 2x - 4 = 2\cot 2x - 6\cot^2 2x.$$

By rearranging and dividing by 2,

$$5\cot^2 2x - \cot 2x - 2 = 0,$$

we have a second-degree equation in $\cot 2x$ with the solution

$$\cot 2x = \frac{1 \pm \sqrt{41}}{10}$$

$$2x = \operatorname{arccot}\frac{1 \pm \sqrt{41}}{10} + n \cdot \pi.$$

Most calculators have no key for the arccot function, and we write instead

$$2x = \arctan\frac{10}{1 \pm \sqrt{41}} + n \cdot \pi.$$

In the interval $0 \le x < 2\pi$,

$x = \frac{1}{2}\left(\arctan\frac{10}{1+\sqrt{41}} + n \cdot \pi\right)$	$x = \frac{1}{2}\left(\arctan\frac{10}{1-\sqrt{41}} + n \cdot \pi\right)$
$x \approx 0.4668 + n \cdot \frac{\pi}{2}.$	$x \approx -0.5377 + n \cdot \frac{\pi}{2}.$
$x_1 \approx 0.4668 + 0 = 0.4668$	$x_5 \approx -0.5377 + 1.5708 = 1.0331$
$x_2 \approx 0.4668 + 1.5708 = 2.0376$	$x_6 \approx -0.5377 + 3.1416 = 2.6039$
$x_3 \approx 0.4668 + 3.1416 = 3.6084$	$x_7 \approx -0.5377 + 4.7124 = 4.1747$
$x_4 \approx 0.4668 + 4.7124 = 5.1792$	$x_8 \approx -0.5377 + 6.2832 = 5.7454$

The eight solutions of the equation in the interval $0 \le x < 2\pi$ are:

$x_1 = 0.467$	$x_5 = 1.03$
$x_2 = 2.04$	$x_6 = 2.60$
$x_3 = 3.61$	$x_7 = 4.17$
$x_4 = 5.18$	$x_8 = 5.75$

Introduction of a New Variable

- Solve the equation

$$2\,(1 - \sin x) = \cos x\ ; \quad 0 \le x \le 90^\circ .$$

Rewrite,

$$\sin x + 0.5 \cos x = 1 .$$

Substitute

$$0.5 = \tan u\ ; \quad u \approx 26.57^\circ .$$

$$\sin x + \frac{\sin u}{\cos u} \cos x = 1$$

$$\sin x \cos u + \cos x \sin u = \cos u$$

$$\sin (x + u) = \cos u \approx 0.8944.$$

$$x + u \approx 63.43^\circ ;\ 116.57^\circ$$

$$x_1 = 36.87^\circ\ ;\ x_2 = 90^\circ .$$

Alternative solution:

$$2\,(1 - \sin x) = \cos x$$
$$4 - 8 \sin x + 4 \sin^2 x = 1 - \sin^2 x$$
$$5 \sin^2 x - 8 \sin x + 3 = 0$$
$$\sin x = \frac{4}{5} \pm \frac{1}{5}$$
$$x_1 = 36.87^\circ\ ;\ x_2 = 90^\circ .$$

- Solve the equation

$$2 \cos x - \sin x = -2; \quad 0 \le x < 15 .$$

Rewrite,

$$\sin x - 2 \cos x = 2.$$

Substitute

$$\tan u = 2\ ; \quad u \approx 1.1071 .$$

$$\sin x - \cos x \frac{\sin u}{\cos u} = 2$$

$$\sin x \cos u - \cos x \sin u = 2 \cos u$$

$$\sin (x - u) = 2 \cos u = \frac{2}{\sqrt{5}}$$

$$x - u \approx \begin{Bmatrix} 1.1071 \\ \pi - 1.1071 \end{Bmatrix} + n \cdot 2\pi$$

$$x \approx \begin{Bmatrix} 2.2142 \\ \pi \end{Bmatrix} + n \cdot 2\pi,$$

where n is an integer.

In the interval $0 \le x < 15$, we have:

$$\begin{aligned}
x_1 &= 2.21\ldots + 0 \cdot 2\pi && \approx 2.21 \\
x_2 &= \pi \ldots + 0 \cdot 2\pi && = \pi \\
x_3 &= 2.21\ldots + 2\pi && \approx 8.50 \\
x_4 &= \pi \ldots + 2\pi = 3\pi && \approx 9.42 \\
x_5 &= 2.21\ldots + 4\pi && \approx 14.78
\end{aligned}$$

Alternative solution:

After substituting the half-angle identities, we have

$$2\left(\frac{1-\tan^2\frac{x}{2}}{1+\tan^2\frac{x}{2}}\right)-\left(\frac{2\tan\frac{x}{2}}{1+\tan^2\frac{x}{2}}\right)+2=0$$

and, after reduction,

$$\tan\frac{x}{2} = 2.$$

$\frac{x}{2} \approx 1.1071 + n \cdot \pi$

$x \approx 2.2142 + n \cdot 2\pi$, where n is an integer.

Testing the original equation for $x = \pi$ (where $\tan x$ is undefined), we have

$$2\cos\pi - \sin\pi = 2(-1) - 0 = -2,$$

which shows that

$$x = \pi + n \cdot 2\pi = (2n+1)\pi$$

is also a solution.

In the interval $0 \le x < 15$, we have:

$$\begin{aligned} x_1 &= 2.21\ldots + 0\pi && \approx 2.21 \\ x_2 &= (2 \cdot 0 + 1)\pi = \pi && \approx 3.14 \\ x_3 &= 2.21\ldots + 2\pi && \approx 8.50 \\ x_4 &= (2 \cdot 1 + 1)\pi = 3\pi && \approx 9.42 \\ x_5 &= 2.21\ldots + 4\pi && \approx 14.78 \end{aligned}$$

Extraneous Solutions

Extraneous solutions may occur when both sides of an equation are raised to a power or when both sides are multiplied by an expression that contains the variable. Solutions obtained after such manipulations must be verified by substitution into the original equation.

- Solve the equation

$$\sin x = \sqrt{1-\cos x}; \quad 0 \le x < 2\pi.$$

Square both sides of the equation,

$$\sin^2 x = 1 - \cos x$$
$$\sin^2 x - 1 + \cos x = 0;$$

Pythagorean Identities: *p.* 508

we obtain

$$-\cos^2 x + \cos x = 0$$
$$\cos x\,(1-\cos x) = 0.$$

Equate each factor with 0 :

$\cos x = 0$	$\cos x = 1$
$x = \pm\frac{\pi}{2} + n\pi.$	$x = 0 + n \cdot 2\pi.$

In the interval $0 \le x < 2\pi$, we have the potential solutions

$$x_1 = \frac{\pi}{2};\ x_2 = \frac{3\pi}{2} \qquad \Big| \qquad x_3 = 0$$

To verify the solutions, insert x_1, x_2, and x_3 in the original equation:

$$\sin\frac{\pi}{2} - \sqrt{1 - \cos\frac{\pi}{2}} = 1 - 1 = 0$$

$$\sin\frac{3\pi}{2} - \sqrt{1 - \cos\frac{3\pi}{2}} = -1 - 1 = -2 \neq 0$$

$$\sin 0 - \sqrt{1 - \cos 0} = 0$$

The results show that $x_2 = \frac{3\pi}{2}$ is an extraneous solution and that the equation has only the solutions

$$x = 0;\ x = \frac{\pi}{2}.$$

- Solve the equation

$$\sec x - \tan x - 1 = 0; \quad 0 \le x < 2\pi.$$

$$\tan x + 1 = \sec x$$

$$(\tan x + 1)^2 = \sec^2 x$$

Pythagorean Identities: *p.* 508

$$2\tan x = \sec^2 x - (\tan^2 x + 1) = \sec^2 x - \sec^2 x$$

$$\tan x = 0.$$

In the interval $0 \le x < 2\pi$, we have the potential solutions

$$x_1 = 0 \qquad \Big| \qquad x_2 = \pi.$$

Checking the results,

$\sec 0 - \tan 0 - 1$	$\sec \pi - \tan \pi - 1$
$= 1 - 0 - 1 = 0$	$= -1 - 0 - 1 = -2,$

shows that π is an extraneous solution and that the equation has only the solution $x = 0$.

Alternative solution:

$$\frac{1}{\cos x} - \frac{\sin x}{\cos x} = 1$$

$$1 - \sin x = \cos x$$

$$(1 - \sin x)^2 = 1 - \sin^2 x$$

$$1 - 2\sin x + \sin^2 x = 1 - \sin^2 x$$

$$\sin x\,(\sin x - 1) = 0$$

$\sin x = 0$	$\sin x = 1$
$x = 0.$	$x = \frac{\pi}{2}$; rejected since secant and tangent are undefined at $\frac{\pi}{2}$.

Impossible Solutions

- Solve the equation

$$2\cos^2 x + 5\cos x = 3 .$$

$$2\cos^2 x + 5\cos x - 3 = 0$$

$$(2\cos x - 1)(\cos x + 3) = 0$$

$$\cos x = \frac{1}{2}$$

$$x = \pm\frac{\pi}{3} + 2\pi\cdot n$$

where n is an integer.

$\cos x = -3$;

rejected,
as $|\cos x|$ must be ≤ 1 .

- Solve the equation

$$2\sec x - \cos x + 1 = 0 ; \quad 0 \le x < 2\pi .$$

$$\frac{2}{\cos x} - \cos x + 1 = 0 .$$

Multiply both sides of the equation by $\cos x$ and rearrange,

$$2 - \cos^2 x + \cos x = 0$$

$$\cos^2 x - \cos x - 2 = 0$$

$$(\cos x - 2)(\cos x + 1) = 0$$

$\cos x = 2$;
rejected.

$$\cos x = -1$$

$$x = \pi .$$

Checking,

$$2\sec\pi - \cos\pi + 1$$

$$= -2 + 1 + 1 = 0 .$$

The equation has only the solution

$$x = \pi .$$

Lost Solutions

When both sides of an equation are erroneously divided by the same factor, one or more solutions might get lost.

- Solve the equation

$$\tan x \cos^2 x = \sin^2 x ; \quad 0 \le x < 2\pi .$$

$$\frac{\sin x \cos^2 x}{\cos x} - \sin^2 x = 0$$

$$\sin x(\cos x - \sin x) = 0 .$$

$$\sin x = 0$$

$$x = 0 ; \pi .$$

$$\cos x - \sin x = 0$$

$$\tan x = 1$$

$$x = \frac{\pi}{4} ; \frac{5\pi}{4} .$$

Alternative solution:

$$\tan x = \frac{\sin^2 x}{\cos^2 x} = \tan^2 x$$

(which *must not* be reduced to $\tan x = 1$),

$$\tan x\,(1 - \tan x) = 0\,.$$

$\tan x = 0$	$\tan x = 1$
$x_1 = 0$	$x_3 = \frac{\pi}{4}$
$x_2 = \pi$	$x_3 = \frac{5\pi}{4}$

Systems of Trigonometric Equations

- Solve the system of equations

$$\left.\begin{aligned} \sin x + \sin y &= 1 \\ x + y &= \frac{5\pi}{3} \end{aligned}\right\}.$$

p. 513 Transform the left-hand side of the first equation into a product,

$$\sin x + \sin y = 2 \sin \frac{x+y}{2} \cos \frac{x-y}{2},$$

where

$$2 \sin \frac{x+y}{2} = 2 \sin \frac{5\pi}{6} = 1\,.$$

Consequently,

$$\cos \frac{x-y}{2} = 1\ ; \quad \frac{x-y}{2} = n \cdot 2\pi.$$

We have

$$\left.\begin{aligned} x - y &= n \cdot 4\pi \\ x + y &= \frac{5\pi}{3} \end{aligned}\right\},$$

which gives

$$\left.\begin{aligned} x &= \frac{5\pi}{6} + n \cdot 2\pi \\ y &= \frac{5\pi}{6} - n \cdot 2\pi \end{aligned}\right\},$$

where n is an integer.

- Solve the system of equations

$$\left.\begin{aligned} \sin x \sin y &= \frac{1}{2} \\ x + y &= \frac{2\pi}{3} \end{aligned}\right\}.$$

$\sin x \sin\left(\frac{2\pi}{3} - x\right) = \frac{1}{2}$, which may be written

$$\sin x\left(\sin\frac{2\pi}{3}\cos x - \sin x \cos\frac{2\pi}{3}\right) = \frac{1}{2}$$

$$\sin x\left(\frac{\sqrt{3}}{2}\cos x + \frac{1}{2}\sin x\right) = \frac{1}{2}$$

$$\sin x\left(\sqrt{3}\cos x + \sin x\right) = 1$$

Pythagorean Identities: *p.* 508

$$\sqrt{3}\sin x\cos x - (1 - \sin^2 x) = 0$$

$$\sqrt{3}\sin x\cos x - \cos^2 x = 0$$

and, finally,

$$\cos x\left(\sqrt{3}\sin x - \cos x\right) = 0.$$

Equating each factor with 0, we have

$$\begin{aligned} \cos x &= 0 \\ x_{1,2} &= \pm\frac{\pi}{2} + n\cdot 2\pi. \\ y_1 &= \frac{2\pi}{3} - x_1 \\ &= \frac{\pi}{6} - n\cdot 2\pi. \\ y_2 &= \frac{2\pi}{3} - x_2 \\ &= \frac{7\pi}{6} - n\cdot 2\pi. \end{aligned}$$

$$\begin{aligned} \sqrt{3}\sin x - \cos x &= 0 \\ \tan x &= \frac{\sqrt{3}}{3}. \\ x_3 &= \frac{\pi}{6} + n\cdot\pi. \\ y_3 &= \frac{2\pi}{3} - x_3 \\ &= \frac{\pi}{2} - n\cdot\pi. \end{aligned}$$

The solutions are

$$\left.\begin{aligned} x_1 &= \frac{\pi}{2} + n\cdot 2\pi \\ y_1 &= \frac{\pi}{6} - n\cdot 2\pi \end{aligned}\right\} \quad \left.\begin{aligned} x_2 &= -\frac{\pi}{2} + n\cdot 2\pi \\ y_2 &= \frac{7\pi}{6} - n\cdot 2\pi \end{aligned}\right\} \quad \left.\begin{aligned} x_3 &= \frac{\pi}{6} + n\cdot\pi \\ y_3 &= \frac{\pi}{2} - n\cdot\pi \end{aligned}\right\},$$

where n is an integer.

13.7 Limits

The Inclusion Theorem

The **inclusion theorem**, intuitively obvious and here quoted without proof, is often useful when evaluating complicated limits.

If $\lim_{x \to a} f(x) = \lim_{x \to a} g(x) = A$

and $f(x) \le h(x) \le g(x)$, for all x near a,

then $\lim_{x \to a} h(x) = A$.

- Given
$$\lim_{x \to \infty} \left(-\frac{1}{x}\right) = \lim_{x \to \infty} \frac{1}{x} = 0,$$
show that
$$\lim_{x \to \infty} \frac{\sin x}{x} = 0.$$

The value of a sine function varies between –1 and 1; thus, for $x > 0$,

$$-\frac{1}{x} \le \frac{\sin x}{x} \le \frac{1}{x}.$$

Since $\lim_{x \to \infty} \left(-\frac{1}{x}\right) = \lim_{x \to \infty} \frac{1}{x} = 0$, we have

$$\lim_{x \to \infty} \frac{\sin x}{x} = 0.$$

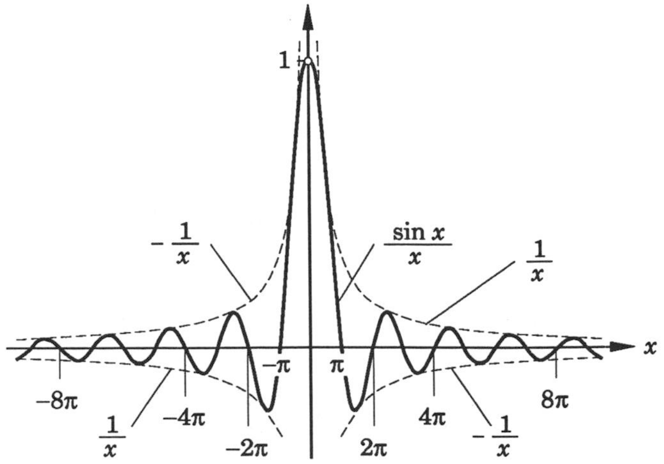

Two Crucial Limits

We shall discuss two limits that are of special importance for finding derivatives of trigonometric functions.

$$\lim_{x \to 0} \frac{\sin x}{x} = 1 \text{ ; } x \text{ in radians.}$$

Proof: In a unit circle, with

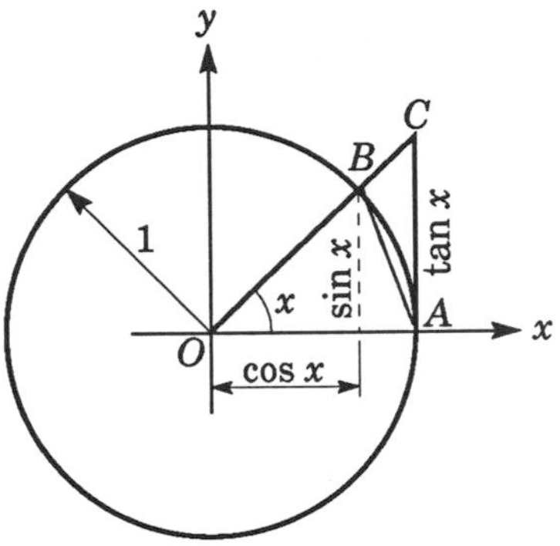

O at the center,

OAB a sector of a unit circle,

C the intersection of an extension of line segment OB and a tangent to the circle at A ,

$\angle AOC = x$ rad,

we have

$$OA = OB = 1 ,$$

$$AC = \tan x ,$$

and

$$\text{area of triangle } AOB = \frac{\sin x}{2} ,$$

$$\text{area of sector } AOB = \frac{x}{2\pi} \cdot \pi = \frac{x}{2} ,$$

$$\text{area of triangle } AOC = \frac{1}{2} (1) \tan x = \frac{\tan x}{2} .$$

Comparing these areas, we find

$$\frac{\sin x}{2} < \frac{x}{2} < \frac{\tan x}{2} \Rightarrow \sin x < x < \frac{\sin x}{\cos x}$$

$$\Rightarrow 1 < \frac{x}{\sin x} < \frac{1}{\cos x} \Rightarrow 1 > \frac{\sin x}{x} > \cos x .$$

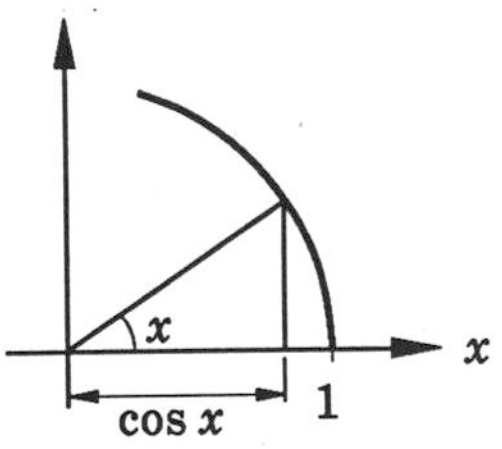

As x tends to 0, $\cos x$ tends to 1 and since sine is an odd function, with

$$\frac{\sin(-x)}{-x} = \frac{-\sin x}{-x} = \frac{\sin x}{x} ,$$

we have for positive and negative values of x that

$$\lim_{x \to 0} \frac{\sin x}{x} = 1 . \qquad \text{QED}$$

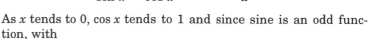

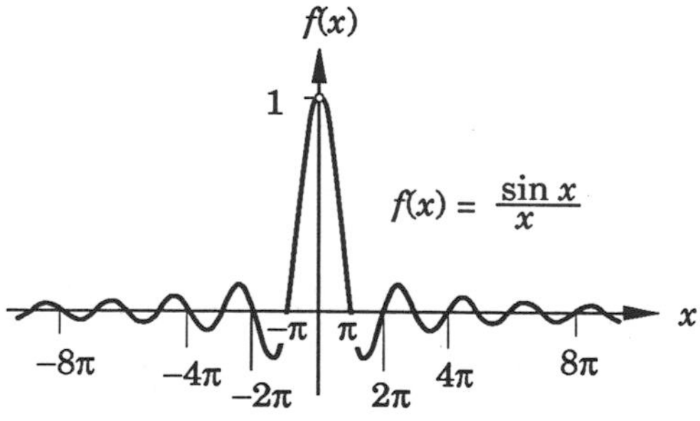

- $\lim_{x \to 0} \frac{1 - \cos x}{x} = 0$; x in radians.

Proof: Multiply the numerator and denominator by $(1 + \cos x)$:

Pythagorean Identities: *p.* 508

$$\lim_{x \to 0} \frac{1 - \cos x}{x} = \lim_{x \to 0} \frac{1 - \cos^2 x}{x\,(1 + \cos x)} = \lim_{x \to 0} \frac{\sin^2 x}{x\,(1 + \cos x)}$$

$$= \lim_{x \to 0} \left(\frac{\sin x}{x} \cdot \frac{\sin x}{1 + \cos x} \right).$$

As $\sin 0 = 0$, $\cos 0 = 1$, and $\lim_{x \to 0} \frac{\sin x}{x} = 1$, we find

$$\lim_{x \to 0} \frac{1 - \cos x}{x} = 1\left(\frac{0}{1 + 1} \right) = 0.$$

QED

Building New Limits

- To evaluate

$$\lim_{x \to 0} \frac{\tan x}{x},$$

use the values

$$\lim_{x \to 0} \frac{\sin x}{x} = 1; \quad \lim_{x \to 0} \cos x = 1$$

and write

$$\lim_{x \to 0} \frac{\tan x}{x} = \lim_{x \to 0} \frac{\frac{\sin x}{\cos x}}{x} = \lim_{x \to 0} \left[\frac{\sin x}{x} \cdot \frac{1}{\cos x} \right] = 1.$$

- To evaluate

$$\lim_{x \to 0} \frac{\sin 5\,x}{2\,x},$$

substitute $y = 5\,x$, use the value $\lim_{y \to 0} \frac{\sin y}{y} = 1$, and write

$$\lim_{x \to 0} \frac{\sin 5\,x}{2\,x} = \lim_{y \to 0} \frac{\sin y}{\frac{2\,y}{5}} = \lim_{y \to 0} \left(\frac{\sin y}{y} \cdot \frac{5}{2} \right) = \frac{5}{2}.$$

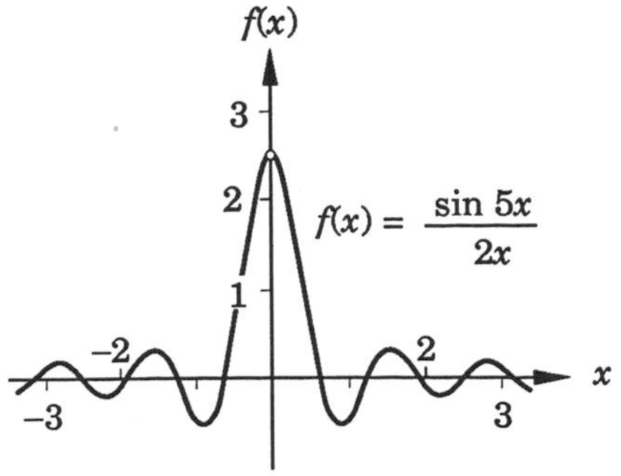

Oscillating Values

Some oscillating functions fail to have a limit value when x tends to 0. We shall discuss the behavior of the functions $f(x) = \sin(1/x)$ and $f(x) = \cos(1/x)$ as x tends to 0.

$$f(x) = \lim_{x \to 0} \sin \frac{1}{x} \text{ does not exist.}$$

Why? As

$$\sin \frac{\pi}{2} = 1;$$

$$\sin \frac{3\pi}{2} = -1;$$

$$\sin x = \sin(x + n \cdot 2\pi) \text{ for any integer } n,$$

we have

$\sin \frac{1}{x} = 1$ if $\frac{1}{x} = \frac{\pi}{2} + 2n\pi$ or $x = \frac{2}{\pi(1 + 4n)}$, where n is an integer.	$\sin \frac{1}{x} = -1$ if $\frac{1}{x} = \frac{3\pi}{2} + 2n\pi$ or $x = \frac{2}{\pi(3 + 4n)}$, where n is an integer.

Thus, if x tends to 0, then $f(x)$ oscillates between -1 and 1, and no limit value exists for $f(x) = \lim_{x \to 0} \sin \frac{1}{x}$.

By a similar line of reasoning, one can also show that $\lim_{x \to 0} \cos \frac{1}{x}$ oscillates between -1 and 1 and, thus, does not exist.

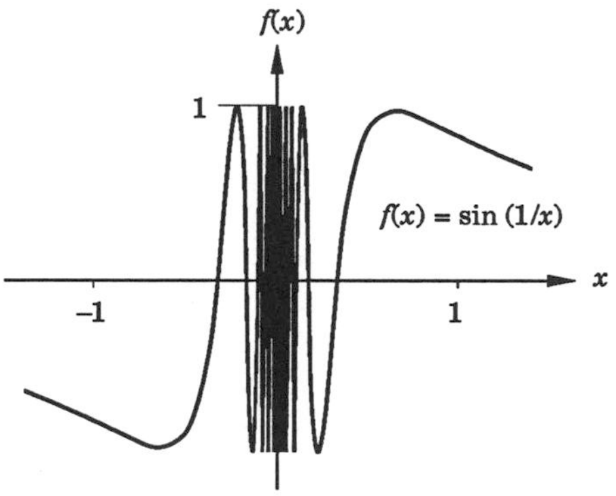

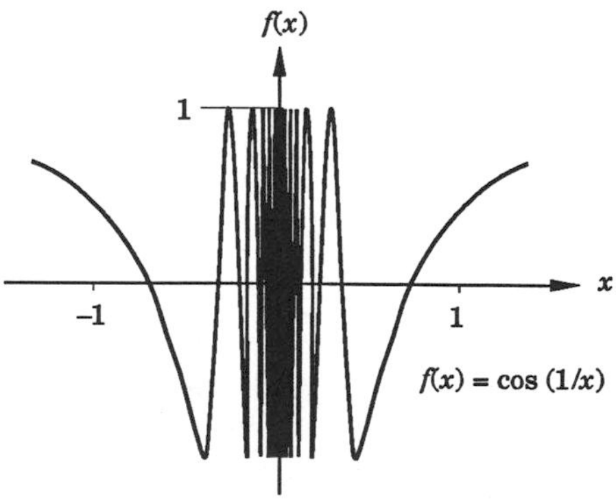

Chapter 14

HYPERBOLIC FUNCTIONS

There once was a knitter from Wales
Who wanted to increase her sales.
She knit fat and thin,
And she never gave in.
Sweaters hyperbolic sold in bales.

14.0 Introduction

The analogy between the equation $x^2 + y^2 = 1$, which represents a circle whose radius is 1 and whose center is at the origin of an orthogonal coordinate system, and that of a hyperbola, $x^2 - y^2 = 1$, started to attract interest from mathematicians in the latter part of the 17th century.

Since the equations of the circle and hyperbola differ only by a minus sign, and their areas may be expressed in trigonometric and logarithmic functions, respectively, the idea developed that the imaginary unit was involved in a relation between trigonometric functions and the logarithmic function, the circle being represented by trigonometric functions and the hyperbola by *hyperbolic functions* – that is, factors of the imaginary unit and trigonometric functions.

LAMBERT Johann Heinrich (1728 - 1777)

Many participated in the development of hyperbolic functions, but the most comprehensive early publications (1768, 1770) were by the German mathematician **J. H. Lambert**, who introduced the names and notations that we still use.

		Pronounce	Read
sinus hyperbolicus	sinh x *or* sh x	sinsh	"Hyperbolic sine of x"
cosinus hyperbolicus	cosh x *or* ch x	cosh	"Hyperbolic cosine of x"
tangens hyperbolicus	tanh x *or* th x	tansh	"Hyperbolic tangent of x"
cotangens hyperbolicus	coth x *or* cth x	cotansh	"Hyperbolic cotangent of x"
secans hyperbolicus	sech x	secansh	"Hyperbolic secant of x"
cosecans hyperbolicus	cosech x *or* csch x	cosecansh	"Hyperbolic cosecant of x"

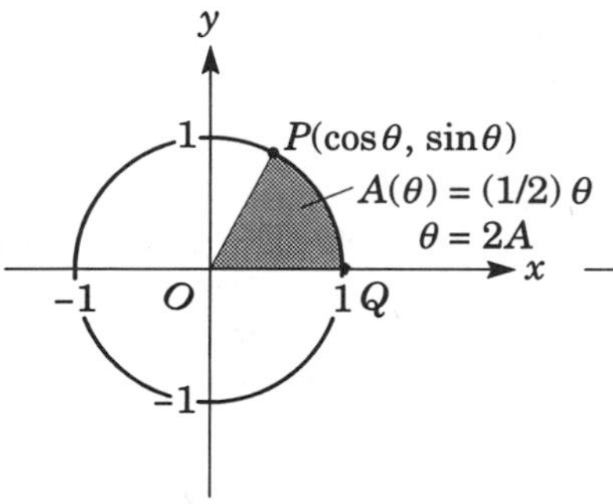

$x^2 + y^2 = 1$

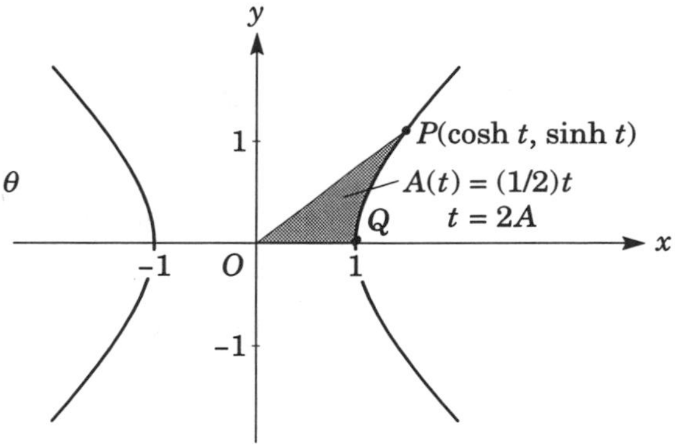

$x^2 - y^2 = 1$

The appellation *hyperbolic functions* is due to a comparison between the Pythagorean identities $\sin^2\theta + \cos^2\theta = 1$ for trigonometric, or *circular*, functions and $\cosh^2 t - \sinh^2 t = 1$ for *hyperbolic* functions:

- While $\cos\theta = x$ and $\sin\theta = y$ are the parametric equations of the circle $x^2 + y^2 = 1$, $\cosh t = x$ and $\sinh t = y$ are the parametric equations of one branch of the hyperbola $x^2 - y^2 = 1$.
- In the trigonometric case, θ is a radian measure of $\angle POQ$, which represents *twice the area* of the sector POQ. In the hyperbolic case, t is not a measure of an angle; yet, with the help of integral calculus, t can be shown to represent *twice the area* of the hyperbolic sector POQ.

Hyperbolic functions have applications in science and engineering, where they may express gradual absorption or decay, *e.g.,* of light, sound, electricity, or radioactivity. They are of importance for finding integrals.

14.1 Fundamental Hyperbolic Functions

Hyperbolic functions, or **fundamental hyperbolic functions**, are generated by the function e^x. Their names and notations are explained and justified by analogies between hyperbolic functions and trigonometric functions.

The hyperbolic sine and cosine are defined as

$$\sinh x = \frac{e^x - e^{-x}}{2}\ ; \qquad \cosh x = \frac{e^x + e^{-x}}{2}\ ,$$

where **e** is the base of natural logarithms and x represents a real number. Analogously, we have

$$\tanh x = \frac{\sinh x}{\cosh x} = \frac{e^x - e^{-x}}{e^x + e^{-x}} = \frac{e^{2x} - 1}{e^{2x} + 1}\ ;$$

$$\coth x = \frac{\cosh x}{\sinh x} = \frac{e^x + e^{-x}}{e^x - e^{-x}} = \frac{e^{2x} + 1}{e^{2x} - 1}\ ;$$

$$\operatorname{sech} x = \frac{1}{\cosh x} = \frac{2}{e^x + e^{-x}}\ ;$$

$$\operatorname{cosech} x = \frac{1}{\sinh x} = \frac{2}{e^x - e^{-x}}\ .$$

Hyperbolic *vs.* Trigonometric Functions

The connections between hyperbolic and trigonometric functions become evident if we compare

$$\left.\begin{aligned} \sin x &= \frac{e^{ix} - e^{-ix}}{2i} \\ \sinh ix &= \frac{e^{ix} - e^{-ix}}{2} \end{aligned}\right\}\ ,$$

which gives

$$\sinh ix = i \sin x\ ; \quad \sinh x = -i \sin ix$$

and, by analogy,

$$\begin{aligned} \cosh ix &= \cos x\ ; & \cosh x &= \cos ix \\ \tanh ix &= i \tan x\ ; & \tanh x &= -i \tan ix \\ \coth ix &= -i \cot x\ ; & \coth x &= i \cot ix\ . \end{aligned}$$

p. 507 Euler's formula, $e^{ix} = \cos x + i \sin x$, has its hyperbolic analogue,

$$e^x = \cosh x + \sinh x\ .$$

Graphs

Graphs of $\sinh x$ and $\cosh x$ may be constructed by combining the graphs of $\frac{1}{2} \cdot e^x$ and $\frac{1}{2} \cdot e^{-x}$; $\operatorname{sech} x$ as $1/\cosh x$; $\operatorname{cosech} x$ as $1/\sinh x$; $\tanh x$ as $\sinh x/\cosh x$; and $\coth x$ as $\cosh x/\sinh x$.

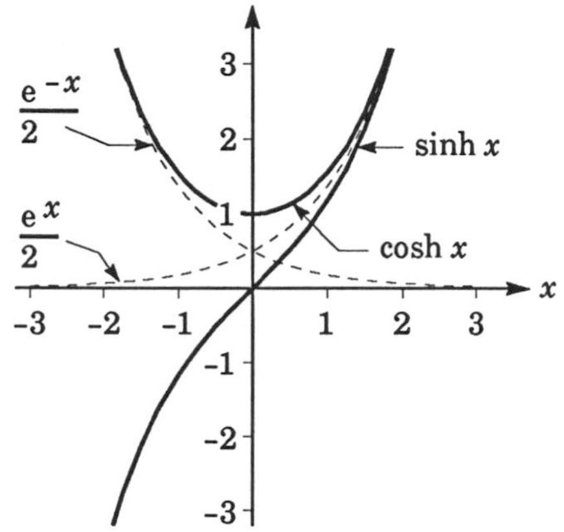

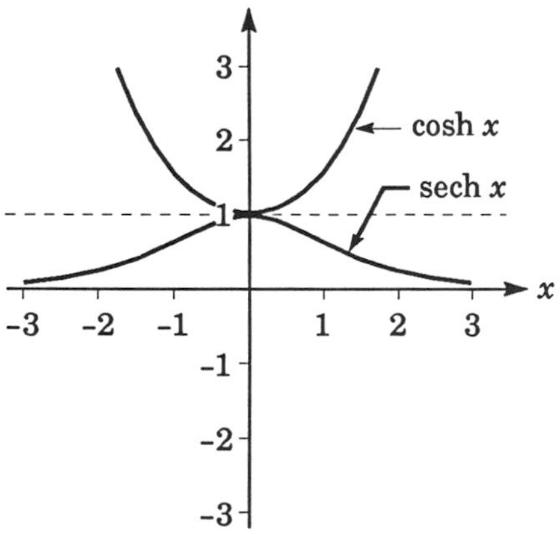

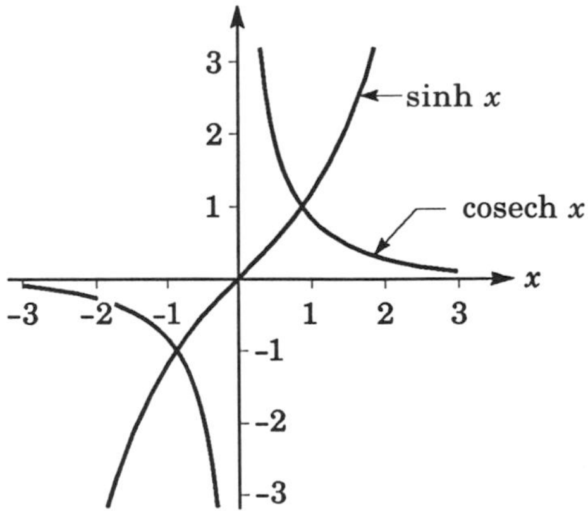

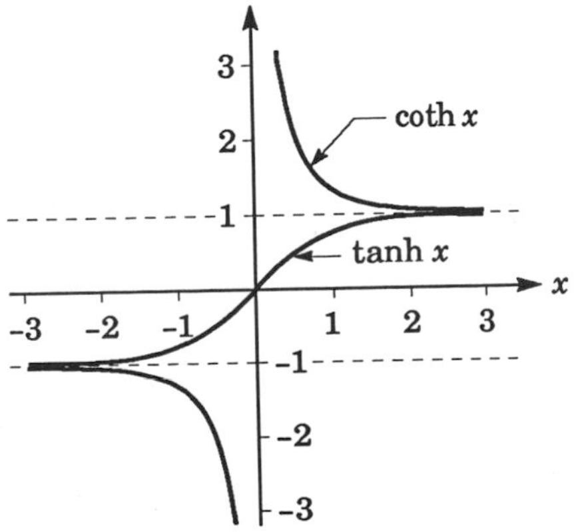

The Catenary

A uniform, non-elastic, and infinitely flexible cable or wire suspended from two fixed points under the influence of gravity forms a **catenary**.

Latin, *catena,* "chain"

The catenary bears great similarity to the parabola and, in fact, Galileo asserted that what later became known as the catenary was indeed a parabola. By applying laws of statics and conceiving the catenary as a succession of equal masses connected by weightless cords of equal length, **Huygens** could, in 1646, refute Galileo's assertion, and thus demonstrate that the catenary is not a parabola. In the 1690s **Jakob Bernoulli** determined the equation of the catenoid.

HUYGENS Christiaan (1629 - 1695)

BERNOULLI Jakob (1654 - 1705)

The catenary may be expressed in the form of a hyperbolic cosine function,

$$f(x) = a\cosh\frac{x}{a} = a\left(\frac{e^{x/a} + e^{-x/a}}{2}\right),$$

where a is a positive constant equal to the y-intercept:

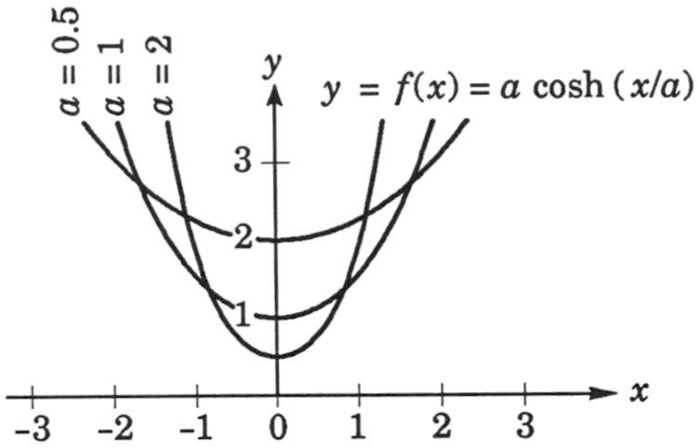

A **catenoid** is the surface generated when a catenary is rotated about its *directrix*. The catenoid is a **minimal surface**, meaning that it takes the shape of least area when bounded by a given closed space. Soap film between two empty circular rings takes the shape of a catenoid.

Catenoid
Minimal Surface
Directrix: *p.* 406

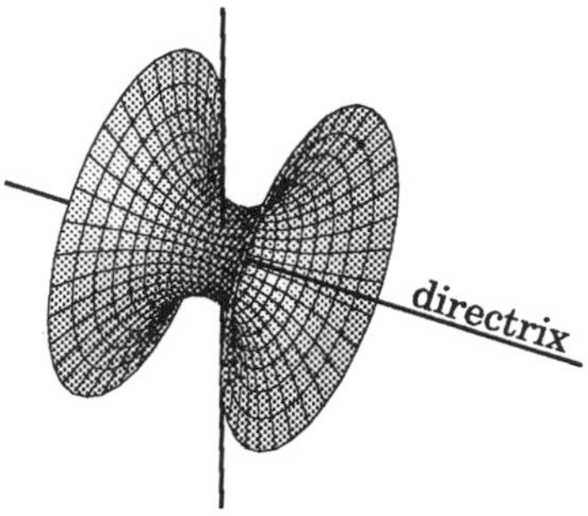

Catenoid

14.2 Inverse Hyperbolic Functions

Area Function

There are six inverse hyperbolic functions, or **area functions**, one corresponding to each of the six fundamental hyperbolic functions.

		Read
arsinh x	(*or* arsh x)	Inverse hyperbolic sine of x.
arcosh x	(*or* arch x)	Inverse hyperbolic cosine of x.
artanh x	(*or* arth x)	Inverse hyperbolic tangent of x.
arcoth x	(*or* arcth x)	Inverse hyperbolic cotangent of x.
arsech x		Inverse hyperbolic secant of x.
arcsch x	(*or* arcosech x)	Inverse hyperbolic cosecant of x.

Inverse hyperbolic functions are sometimes denoted $\sinh^{-1} x$, $\cosh^{-1} x$, *etc.*; this form of notation should be discouraged because of possible confusion with the expression $1/\sinh x$, *etc.*

Another form of notation, arcsinh x, arccosh x, *etc.*, is a practice to be condemned as these functions have nothing whatever to do with arc, but with area, as is demonstrated by their full Latin names,

arsinsh	*area sinus hyperbolicus*
arcosh	*area cosinus hyperbolicus*, *etc.*

p. 537

Graphs of hyperbolic functions show that sinh, tanh, coth, and cosech are odd functions which are one-to-one; cosh and sech are even functions and are thus *not* one-to-one. To compute inverse functions of cosh and sech, their domains must be restricted, generally by allowing only positive arguments, $x \geq 0$.

With this restriction, we obtain the following graphs of inverse hyperbolic functions.

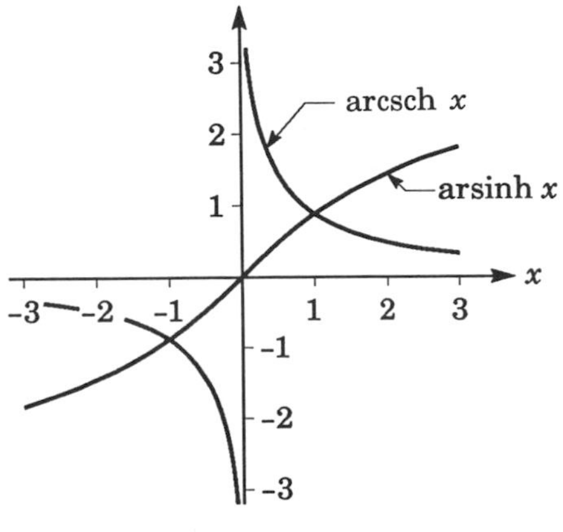

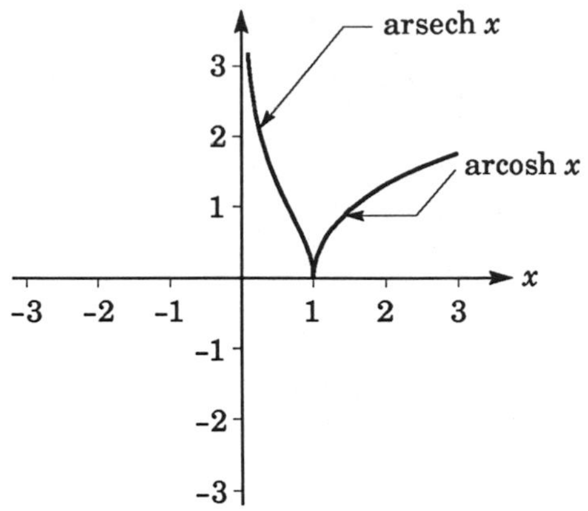
arsech x
arcosh x
x
-3 -2 -1 1 2 3
3 2 1 -1 -2 -3

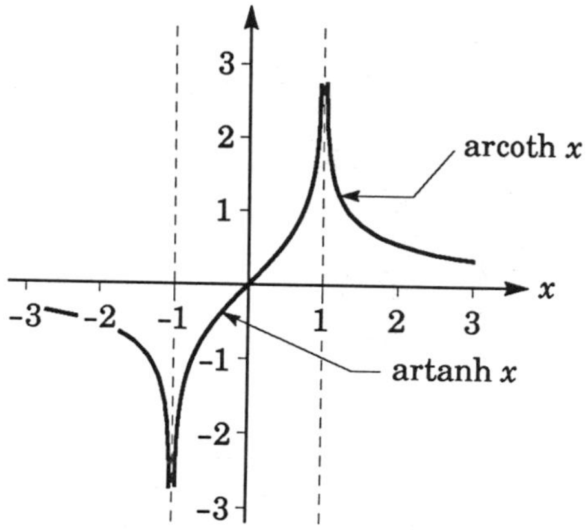
arcoth x
artanh x
x
-3 -2 -1 1 2 3
3 2 1 -1 -2 -3

14.3 Identities

As with trigonometric identities, hyperbolic identities are powerful media in calculus.

Fundamental Hyperbolic Functions

Fundamental hyperbolic identities may be transcribed from trigonometric identities, sinh and cosh replacing sin and cos, and plus or minus sign being changed before any product of two sines. The rule can be extended to the remaining fundamental hyperbolic functions, $\tanh x = \dfrac{\sinh x}{\cosh x}$, $\coth x = \dfrac{\cosh x}{\sinh x}$, $\operatorname{sech} x = \dfrac{1}{\cosh x}$, and $\operatorname{cosech} x = \dfrac{1}{\sinh x}$.

p. 507

If desired, the reader may check the following identities by transcribing them from the corresponding *trigonometric identities*, or just consult them when needed.

Negative Arguments

$\sinh(-x) = -\sinh x$; $\cosh(-x) = \cosh x$

$\tanh(-x) = -\tanh x$; $\coth(-x) = -\coth x$

$\operatorname{sech}(-x) = \operatorname{sech} x$; $\operatorname{cosech}(-x) = -\operatorname{cosech} x$

Pythagorean Identities

$$\cosh^2 x - \sinh^2 x = \left(\frac{e^x + e^{-x}}{2}\right)^2 - \left(\frac{e^x - e^{-x}}{2}\right)^2$$

$$= \frac{e^{2x} + 2 + e^{-2x}}{4} - \frac{e^{2x} - 2 + e^{-2x}}{4}$$

from which

$$\cosh^2 x - \sinh^2 x = 1,$$

and, similarly,

$$\operatorname{sech}^2 x + \tanh^2 x = 1,$$

$$\coth^2 x - \operatorname{cosech}^2 x = 1.$$

Sums and Differences

$$\sinh x \pm \sinh y = 2 \cdot \sinh \frac{x \pm y}{2} \cdot \cosh \frac{x \mp y}{2}$$

$$\cosh x + \cosh y = 2 \cdot \cosh \frac{x + y}{2} \cdot \cosh \frac{x - y}{2}$$

$$\cosh x - \cosh y = 2 \cdot \sinh \frac{x + y}{2} \cdot \sinh \frac{x - y}{2}$$

$$\tanh x \pm \tanh y = \frac{\sinh(x \pm y)}{\cosh x \cdot \cosh y}$$

$$\coth x \pm \coth y = \frac{\sinh(x \pm y)}{\sinh x \cdot \sinh y}$$

Products

$$\sinh x \cdot \sinh y = \frac{1}{2}\left[\cosh(x+y) - \cosh(x-y)\right]$$

$$\cosh x \cdot \cosh y = \frac{1}{2}\left[\cosh(x+y) + \cosh(x-y)\right]$$

$$\sinh x \cdot \cosh y = \frac{1}{2}\left[\sinh(x+y) + \sinh(x-y)\right]$$

$$\tanh x \cdot \tanh y = \frac{\tanh x + \tanh y}{\coth x + \coth y}$$

Double Arguments

$$\sinh 2x = 2\cosh x \cdot \sinh x; \qquad \tanh 2x = \frac{2\tanh x}{1 + \tanh^2 x}$$

$$\cosh 2x = \cosh^2 x + \sinh^2 x; \qquad \coth 2x = \frac{\coth^2 x + 1}{2\coth x}$$

Half Arguments

$$\sinh\frac{x}{2} \overset{\text{if } x \ge 0}{=} \sqrt{\frac{\cosh x - 1}{2}} = \frac{\sinh x}{\sqrt{2(\cosh x + 1)}}$$

$$\cosh\frac{x}{2} = \sqrt{\frac{\cosh x + 1}{2}} \overset{\text{if } x > 0}{=} \frac{\sinh x}{\sqrt{2(\cosh x - 1)}}$$

$$\tanh\frac{x}{2} = \frac{\sinh x}{\cosh x + 1} \overset{\text{if } x \ne 0}{=} \frac{\cosh x - 1}{\sinh x} \overset{\text{if } x \ge 0}{=} \sqrt{\frac{\cosh x - 1}{\cosh x + 1}}$$

$$\coth\frac{x}{2} \overset{\text{if } x \ne 0}{=} \frac{\sinh x}{\cosh x - 1} \overset{\text{if } x \ne 0}{=} \frac{\cosh x + 1}{\sinh x} \overset{\text{if } x > 0}{=} \sqrt{\frac{\cosh x + 1}{\cosh x - 1}}$$

Multiple Arguments

$$\sinh 3x = \sinh x\,(4\cosh^2 x - 1)$$
$$\sinh 4x = \sinh x \cdot \cosh x\,(8\cosh^2 x - 4)$$
$$\sinh 5x = \sinh x\,(1 - 12\cosh^2 x + 16\cosh^4 x)$$

$$\cosh 3x = \cosh x\,(4\cosh^2 x - 3)$$
$$\cosh 4x = 1 - 8 \cdot \cosh^2 x + 8\cosh^4 x$$
$$\cosh 5x = \cosh x\,(5 - 20\cosh^2 x + 16\cosh^4 x)$$

$$\sinh nx = \binom{n}{1} \cdot \cosh^{n-1} x \cdot \sinh x + \binom{n}{3} \cdot \cosh^{n-3} x \cdot \sinh^3 x + \binom{n}{5} \cdot \cosh^{n-5} x \cdot \sinh^5 x + \ldots$$

$$\cosh nx = \cosh^n x + \binom{n}{2} \cdot \cosh^{n-2} x \cdot \sinh^2 x + \binom{n}{4} \cdot \cosh^{n-4} x \cdot \sinh^4 x + \ldots$$

Compound Arguments

$$\sinh(x \pm y) = \sinh x \cdot \cosh y \pm \cosh x \cdot \sinh y$$
$$\cosh(x \pm y) = \cosh x \cdot \cosh y \pm \sinh x \cdot \sinh y$$
$$\tanh(x \pm y) = \frac{\tanh x \pm \tanh y}{1 \pm \tanh x \cdot \tanh y}$$
$$\coth(x \pm y) = \frac{1 \pm \coth x \cdot \coth y}{\coth x \pm \coth y}$$

Powers of Arguments

$$\sinh^2 x = \frac{1}{2}(\cosh 2x - 1)$$
$$\cosh^2 x = \frac{1}{2}(\cosh 2x + 1)$$
$$\sinh^3 x = \frac{1}{4}(-3\sinh x + \sinh 3x)$$
$$\cosh^3 x = \frac{1}{4}(3\cosh x + \cosh 3x)$$
$$\sinh^4 x = \frac{1}{8}(3 - 4\cosh 2x + \cosh 4x)$$
$$\cosh^4 x = \frac{1}{8}(3 + 4\cosh 2x + \cosh 4x)$$
$$\sinh^5 x = \frac{1}{16}(10\sinh x - 5 \cdot \sinh 3x + \sinh 5x)$$
$$\cosh^5 x = \frac{1}{16}(10\cosh x + 5 \cdot \cosh 3x + \cosh 5x)$$
$$\sinh^6 x = \frac{1}{32}(-10 + 15 \cdot \cosh 2x - 6\cosh 4x + \cosh 6x)$$
$$\cosh^6 x = \frac{1}{32}(10 + 15 \cdot \cosh 2x + 6\cosh 4x + \cosh 6x)$$

Complex Arguments

$$\sinh(x \pm \mathrm{i}\,y) = \sinh x \cdot \cos y \pm \mathrm{i} \cdot \cosh x \cdot \sin y$$
$$\cosh(x \pm \mathrm{i}\,y) = \cosh x \cdot \cos y \pm \mathrm{i} \cdot \sinh x \cdot \sin y$$
$$\tanh(x \pm \mathrm{i}\,y) = \frac{\sinh 2x \pm \mathrm{i} \cdot \sin 2y}{\cosh 2x + \cos 2y}$$
$$\coth(x \pm \mathrm{i}\,y) = \frac{\sinh 2x \mp \mathrm{i} \cdot \sin 2y}{\cosh 2x - \cos 2y}$$
$$\sin(x \pm \mathrm{i}\,y) = \sin x \cdot \cosh y \pm \mathrm{i} \cdot \cos x \cdot \sinh y$$
$$\cos(x \pm \mathrm{i}\,y) = \cos x \cdot \cosh y \mp \mathrm{i} \cdot \sin x \cdot \sinh y$$
$$\tan(x \pm \mathrm{i}\,y) = \frac{\sin 2x \pm \mathrm{i} \cdot \sinh 2y}{\cos 2x + \cosh 2y} = \frac{\sin 2x \pm \mathrm{i} \cdot \sinh 2y}{2(\cos^2 x + \sinh^2 y)}$$
$$\cot(x \pm \mathrm{i}\,y) = -\frac{\sin 2x \mp \mathrm{i} \cdot \sinh 2y}{\cos 2x - \cosh 2y} = \frac{\sin 2x \mp \mathrm{i} \cdot \sinh 2y}{2(\sin^2 x + \sinh^2 y)}$$

$$\sinh(x + \mathrm{i} \cdot 2n\pi) = \sinh x$$
$$\cosh(x + \mathrm{i} \cdot 2n\pi) = \cosh x$$
$$\tanh(x + \mathrm{i} \cdot n\pi) = \tanh x$$
$$\coth(x + \mathrm{i} \cdot n\pi) = \coth x$$

Conversion Table

	sinh	cosh	tanh	coth
$\sinh x =$	–	$\pm\sqrt{\cosh^2 x - 1}$	$\dfrac{\tanh x}{\sqrt{1-\tanh^2 x}}$	$\pm\dfrac{1}{\sqrt{\coth^2 x - 1}}$
$\cosh x =$	$\sqrt{1+\sinh^2 x}$	–	$\dfrac{1}{\sqrt{1-\tanh^2 x}}$	$\dfrac{\lvert\coth x\rvert}{\sqrt{\coth^2 x - 1}}$
$\tanh x =$	$\dfrac{\sinh x}{\sqrt{1+\sinh^2 x}}$	$\dfrac{\sqrt{\cosh^2 x - 1}}{\cosh x}$	–	$\dfrac{1}{\coth x}$
$\coth x =$	$\dfrac{\sqrt{1+\sinh^2 x}}{\sinh x}$	$\dfrac{\cosh x}{\sqrt{\cosh^2 x - 1}}$	$\dfrac{1}{\tanh x}$	–

Inverse Hyperbolic Functions

$$\sinh(\operatorname{arsinh} x) = \operatorname{arsinh}(\sinh x) = x$$

$$\cosh(\operatorname{arcosh} x) = x; \qquad \operatorname{arcosh}(\cosh x) = x, \text{ if } x \geq 0$$

Sums and Differences

$$\operatorname{arsinh} x \pm \operatorname{arsinh} y = \operatorname{arsinh}\left(x \cdot \sqrt{1+y^2} \pm y \cdot \sqrt{1+x^2}\right)$$

$$\operatorname{arcosh} x + \operatorname{arcosh} y = \operatorname{arcosh}\left(x\,y + \sqrt{(x^2-1)(y^2-1)}\right)$$

$$\operatorname{arcosh} x - \operatorname{arcosh} y = \operatorname{arcosh}\left(x\,y - \sqrt{(x^2-1)(y^2-1)}\right) \cdot \operatorname{sgn}(x-y),$$

where sgn $(x - y)$ refers to the sign of the difference (that is, 1 or -1)

$$\operatorname{artanh} x \pm \operatorname{artanh} y = \operatorname{artanh} \frac{x \pm y}{1 \pm x\,y}$$

$$\operatorname{arcoth} x \pm \operatorname{arcoth} y = \operatorname{arcoth} \frac{1 \pm x\,y}{x \pm y}$$

Conversion Table

	arsinh	arcosh $(x \geq 0)$	artanh	arcoth
$\operatorname{arsinh} x =$	–	$\operatorname{arcosh} \sqrt{x^2+1}$	$\operatorname{artanh} \dfrac{x}{\sqrt{x^2+1}}$	$\operatorname{arcoth} \dfrac{\sqrt{x^2+1}}{x}$
$\operatorname{arcosh} x =$	$\operatorname{arsinh} \sqrt{x^2-1}$	–	$\operatorname{artanh} \dfrac{\sqrt{x^2-1}}{x}$	$\operatorname{arcoth} \dfrac{x}{\sqrt{x^2-1}}$
$\operatorname{artanh} x =$	$\operatorname{arsinh} \dfrac{x}{\sqrt{1-x^2}}$	$\operatorname{arcosh} \dfrac{1}{\sqrt{1-x^2}}$	–	$\operatorname{arcoth} \dfrac{1}{x}$
$\operatorname{arcoth} x =$	$\operatorname{arsinh} \pm \dfrac{1}{\sqrt{x^2-1}}$	$\operatorname{arcosh} \dfrac{x}{\sqrt{x^2-1}}$	$\operatorname{artanh} \dfrac{1}{x}$	–

Inverse Hyperbolic Functions *vs.* Natural Logarithms

Since hyperbolic functions are combinations of exponential functions e^x, we may expect simple relations to exist also between inverse hyperbolic functions and natural logarithms.

If

$$y = \operatorname{arsinh} x,$$

where

$$x = \sinh y = \frac{1}{2}\left(e^y - e^{-y}\right),$$

we have

$$e^y - e^{-y} = 2x$$

$$(e^y)^2 - 2x \cdot e^y - 1 = 0$$

$$e^y = x \underset{(-)}{+} \sqrt{x^2+1},$$

where the minus sign is discarded, since $e^y > 0$. The natural logarithm of both members of the square gives

$$[y =]\ \operatorname{arsinh} x = \ln\left(x + \sqrt{x^2+1}\right).$$

Similarly, we find:

$$\operatorname{arcosh} x = \ln\left(x + \sqrt{x^2-1}\right); \qquad x \geq 1$$

$$\operatorname{artanh} x = \frac{1}{2} \cdot \ln \frac{1+x}{1-x}; \qquad |x| < 1$$

$$\operatorname{arcoth} x = \frac{1}{2} \ln \frac{x+1}{x-1}; \qquad |x| > 1$$

$$\operatorname{arsech} x = \ln \frac{1+\sqrt{1-x^2}}{x}; \qquad 1 \geq x > 0$$

$$\operatorname{arcsch} x = \begin{cases} \ln \dfrac{1+\sqrt{1+x^2}}{x}; & x > 0 \\ \ln \dfrac{1+\sqrt{1+x^2}}{|x|} \cdot \operatorname{sgn} x; & x \neq 0, \end{cases}$$

where sgn x refers to the sign of x (that is, 1 or – 1)

Chapter 15

ANALYTIC GEOMETRY

15.0 Scope and History

Coordinate Geometry

In **analytic**, or **coordinate**, **geometry**, geometric problems are made accessible to algebraic reasoning by introducing the connection between *points* and *numbers* – the fundamental elements of geometry and algebra, respectively. Once a coordinate system is defined, every point of a plane curve is uniquely represented by an ordered pair of real numbers, every point of a three-dimensional surface by an ordered triplet of real numbers.

APOLLONIOS (255 - 170 B.C.)

ARCHIMEDES (287 - 212 B.C.)

By the 3rd century B.C., Greek mathematicians **Apollonius** of Perga and **Archimedes** had already used longitude, latitude, and altitude to define the position of a point, but it was not until the 17th century that the French mathematicians **René Descartes** and **Pierre de Fermat** directed this idea into systematic use.

DESCARTES René *CARTESIUS* *p.* 367 (1596 - 1650)

Descartes – Latinized ***Cartesius*** – presented the fundamental concepts of coordinate geometry in *La géométrie* (1637), which, with Newton's *Principia*, is one of the most influential scientific texts of the 17th century. By representing a point by a pair of real numbers, and straight lines and curves by equations, Descartes provided a link between geometry and algebra. Yet, rather than devising a system for algebrizing geometry, Descartes's purpose was to advance the methods of geometric construction.

de FERMAT Pierre (1601 - 1665)

Fermat had outlined the principles of analytic geometry even before the publication of *La géométrie*, but his text was circulated for many years among mathematicians in manuscript form until it was finally printed, in 1679, in *Varia opera mathematica* as "Ad locos planos et solidos isagoge" ("Introduction to Plane and Solid Loci"). In addition to being one of the originators of analytic geometry, Fermat – a lawyer by profession – was the founder of modern number theory and, with Blaise Pascal, of probability theory; Fermat, "the king of amateurs", also devised and applied the principal idea of differential calculus.

NEWTON Isaac (1643 - 1727)

von LEIBNIZ Gottfried Wilhelm (1646 - 1716)

Descartes and Fermat did not consider negative values for distances and, consequently, their works contain no orthogonal *Cartesian* coordinate systems in the modern sense. The concept of negative distances is due to **Newton** and **Leibniz**.

Newton is also considered the originator of polar coordinates, where a point is defined by a pair of real numbers, one representing its distance from a fixed point, the origin, the other an angle between a fixed ray and a radius from the origin to the given point.

The development of analytic geometry and calculus in the 17th century signaled the beginning of modern mathematics.

15.1 Rectilinear Figures

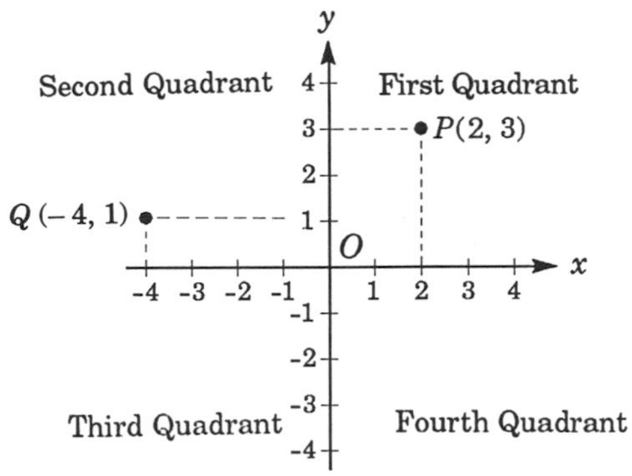

x-axis
y-axis
Origin

A point in the *plane* can be located in an **orthogonal** (**rectangular**, or **Cartesian**) **coordinate system** formed by two perpendicular real number lines, the ***x*-axis** and the ***y*-axis**. The point of intersection of the axes is called the **origin**, denoted O. Each point in the plane is determined by an ordered pair of real numbers (x, y). The axes divide the plane into four quadrants.

Abscissa

The x-coordinate, or **abscissa**, of a point is its horizontal distance from the y-axis measured parallel to the x-axis, and is positive if the point lies to the right of the y-axis and negative if it lies to the left of the y-axis. The abscissa of P is 2, of Q is -4.

Ordinate

The y-coordinate, or **ordinate**, of a point is similarly defined to be its vertical distance from the x-axis measured parallel to the y-axis. The ordinate of P is 3, of Q is 1.

A point in *space* can be located by determining its distance to three perpendicular planes, intersecting at the origin. Any such planes intersect in a line and, consequently, the coordinate system in space has a total of three coordinate axes, usually denoted x, y, and z:

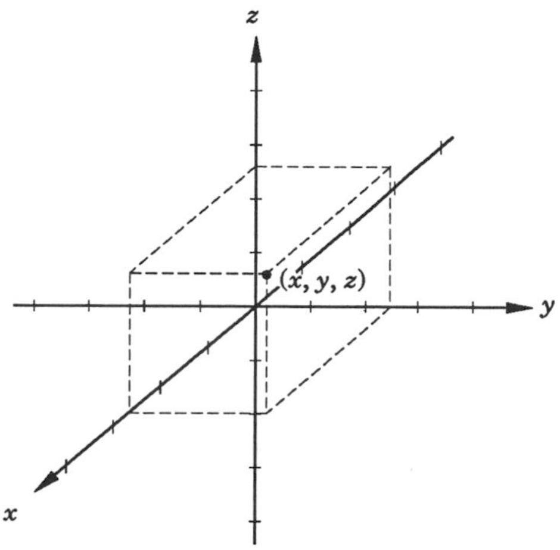

Distance Formulas

In the Plane

In a two-dimensional orthogonal coordinate system, the distance between point (x_1, y_1) and point (x_2, y_2) is

$$d = \sqrt{(x_2 - x_1)^2 + (y_2 - y_1)^2}\ .$$

Why?

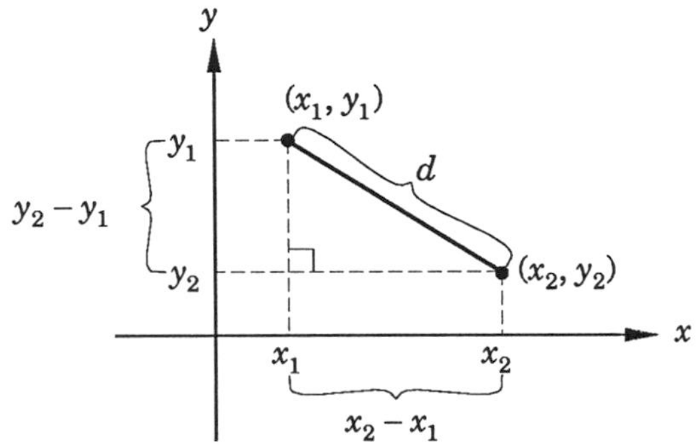

By the Pythagorean theorem, we find

$$d^2 = (x_2 - x_1)^2 + (y_2 - y_1)^2$$
$$d = \sqrt{(x_2 - x_1)^2 + (y_2 - y_1)^2}\ .$$

In Space

In a three-dimensional orthogonal coordinate system, the distance between point (x_1, y_1, z_1) and point (x_2, y_2, z_2) is

$$d = \sqrt{(x_2 - x_1)^2 + (y_2 - y_1)^2 + (z_2 - z_1)^2}\ .$$

Why?

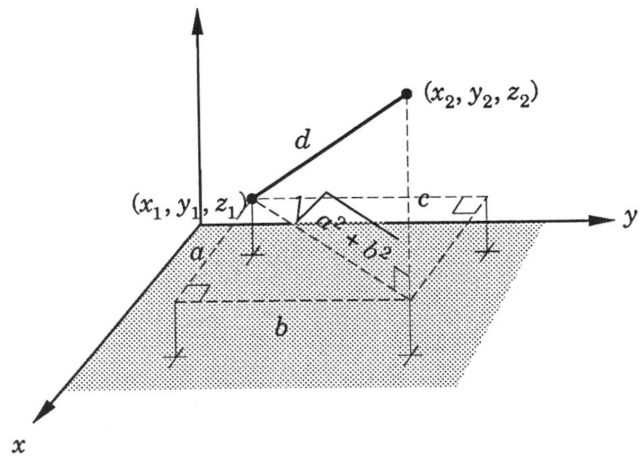

These triangles have one side in common, the length of which, with Pythagoras, is

$$\sqrt{a^2 + b^2}\ .$$

The sought distance d is the hypotenuse of a right triangle whose other sides are $\sqrt{a^2+b^2}$ and c; using the Pythagorean theorem, we obtain

$$d^2 = \left(\sqrt{a^2+b^2}\right)^2 + c^2$$

$$d = \sqrt{a^2+b^2+c^2}\,.$$

Since $a = x_2 - x_1$; $b = y_2 - y_1$; $c = z_2 - z_1$, we obtain

$$d = \sqrt{(x_2-x_1)^2 + (y_2-y_1)^2 + (z_2-z_1)^2}\,.$$

Midpoint of a Line Segment

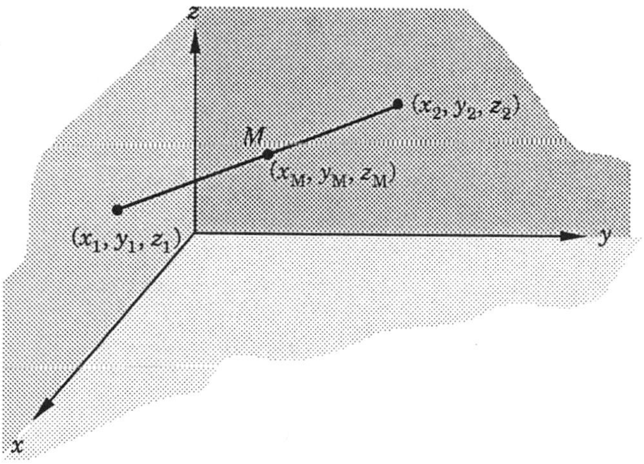

Whether in the plane or in space, each coordinate of the midpoint of a line segment is the average of the corresponding coordinates of the endpoints of the segment; thus,

$$x_M = \frac{x_2+x_1}{2} \qquad y_M = \frac{y_2+y_1}{2} \qquad z_M = \frac{z_2+z_1}{2}\,.$$

Area of a Triangle

The formula for the **area** (A) **of a triangle** with vertices at points (x_1, y_1), (x_2, y_2), (x_3, y_3) is

$$A = \left| \frac{1}{2} \left[x_1 (y_2 - y_3) - x_2 (y_1 - y_3) + x_3 (y_1 - y_2) \right] \right| .$$

Why? $$A_{\text{triangle}} = \frac{1}{2} (\text{base} \times \text{height})$$

$$A_{\text{trapezoid}} = \frac{1}{2} (\text{sum of parallel sides} \times \text{height})$$

Let the vertices of a triangle be $P_1(x_1, y_1)$, $P_2(x_2, y_2)$, $P_3(x_3, y_3)$, and assume P_2 to be the lowermost vertex.

Let Q_1 and Q_2 be projections on a line parallel to the x-axis and passing through P_2:

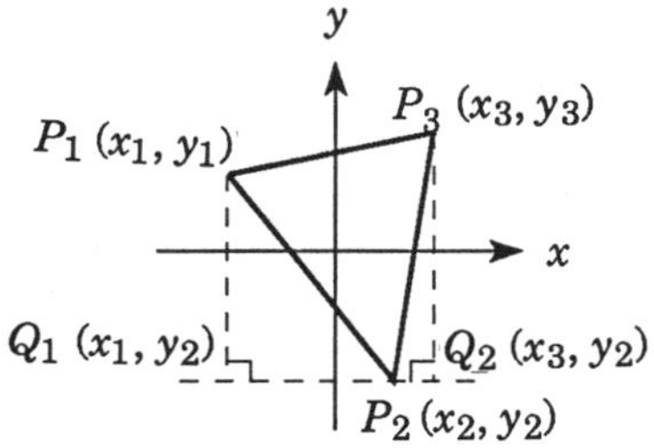

The area of triangle $P_1P_2P_3$ equals the area of trapezoid $P_1Q_1Q_2P_3$ less triangles $P_1Q_1P_2$ and $P_2Q_2P_3$:

$$A_{\text{triangle}P_1P_2P_3} =$$

$$\frac{1}{2} (x_3 - x_1) \left[(y_1 - y_2) + (y_3 - y_2) \right] - \frac{1}{2} (x_2 - x_1)(y_1 - y_2) - \frac{1}{2} (x_3 - x_2)(y_3 - y_2)$$

$$= \frac{1}{2} \left[x_1 (y_2 - y_3) + x_2 (y_3 - y_1) + x_3 (y_1 - y_2) \right] .$$

Instead of being in a point of segment Q_1Q_2, vertex P_2 may be to the left of Q_1 or to the right of Q_2:

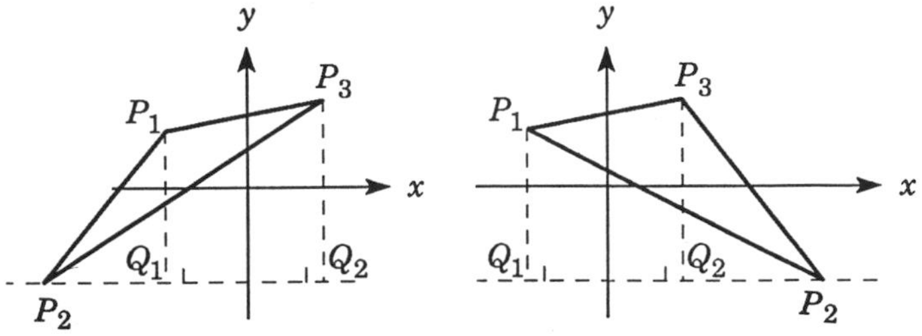

In the left-hand diagram, the area of triangle $P_1P_2P_3$ equals the area of trapezoid $P_1Q_1Q_2P_3$ plus the area of triangle $P_1Q_1P_2$ less the area of triangle $P_2Q_2P_3$.

In the right-hand graph, triangle $P_1P_2P_3$ equals the area of trapezoid $P_1Q_1Q_2P_3$ plus the area of triangle $P_3Q_2P_2$ less the area of triangle $P_1Q_1P_2$.

These two cases also lead to

$$A_{\text{triangle}P_1P_2P_3} = \frac{1}{2}\left[x_1\left(y_2 - y_3\right) + x_2\left(y_3 - y_1\right) + x_3\left(y_1 - y_2\right)\right] .$$

If we use the absolute value of the result,

$$A = \left| \frac{1}{2}\left[x_1\left(y_2 - y_3\right) - x_2\left(y_1 - y_3\right) + x_3\left(y_1 - y_2\right)\right] \right| ,$$

the vertices of the triangle may be read in any order.

The formula expressing the area of a triangle in terms of the coordinates of its vertices may be applied to find the **area of any polygon**.

Equations and Graphs of Straight Lines

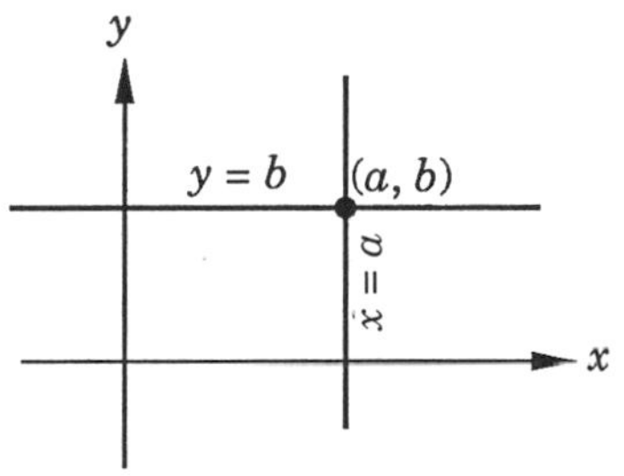

Horizontal and vertical lines may be identified by their constant y-coordinates and x-coordinates, respectively.

A straight vertical line through point (a, b) is identified by the **equality**

$$x = a;$$

a straight horizontal line through (a, b) by the **equality**

$$y = b .$$

Point-Slope and Two-Point Equations

Slope, Inclination

"Rise over Run"

The **slope**, or **inclination**, of a line is defined as the ratio of the vertical distance to the horizontal distance from any one point to any other point on the line. This ratio is sometimes conveniently referred to as "rise over run".

A straight line that has the slope m and contains the point (x_1, y_1) has the **point-slope equation**

$$y - y_1 = m\left(x - x_1\right) .$$

Consider two points (x_1, y_1) and (x_2, y_2) on a line. The slope m is calculated as the ratio of $\Delta y = y_2 - y_1$ units in the vertical direction and $\Delta x = x_2 - x_1$ units in the horizontal direction, where Δ denotes an **increment**:

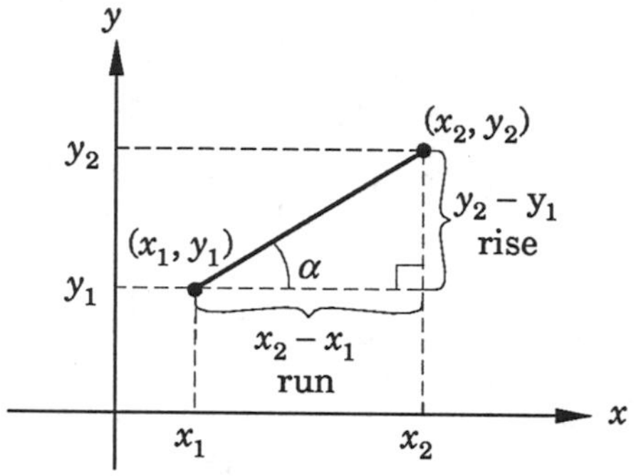

$$m = \tan\alpha = \frac{\text{rise}}{\text{run}} = \frac{\Delta y}{\Delta x} = \frac{y_2 - y_1}{x_2 - x_1};\ x_1 \neq x_2$$

If we know two points (x_1, y_1) and (x_2, y_2) of a straight line, the line has the **two-point equation**

$$y - y_1 = \frac{y_2 - y_1}{x_2 - x_1}(x - x_1).$$

Slope-Intercept Equation

If a straight line is not parallel to the y-axis, it must intersect the y-axis at some point $(0, b)$, called the **y-intercept** of the line:

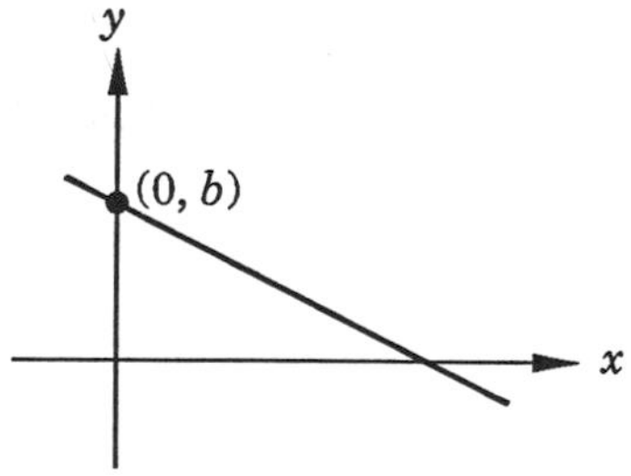

As point $(0, b)$ is on the y-axis, the point-slope equation, $y - y_1 = m(x - x_1)$, may be simplified to the **slope-intercept equation**

$$(y - b) = m(x - 0),$$

or

$$y = mx + b.$$

- The equation of a line with the slope $m = -\frac{1}{5}$ and a point $(-6, 9)$ is

$$y - 9 = -\frac{1}{5}[x - (-6)]$$
$$y = -\frac{1}{5}x + \frac{39}{5}.$$

General Form of the Equation for a Line

If the slope-intercept form of a line

$$y = mx + b$$

is rearranged in a form where

$$m \text{ is } -\frac{A}{B} \quad \text{and } b \text{ is } -\frac{C}{B},$$

we obtain the **general form**, or **standard form**, of the equation for a line,

$$Ax + By + C = 0.$$

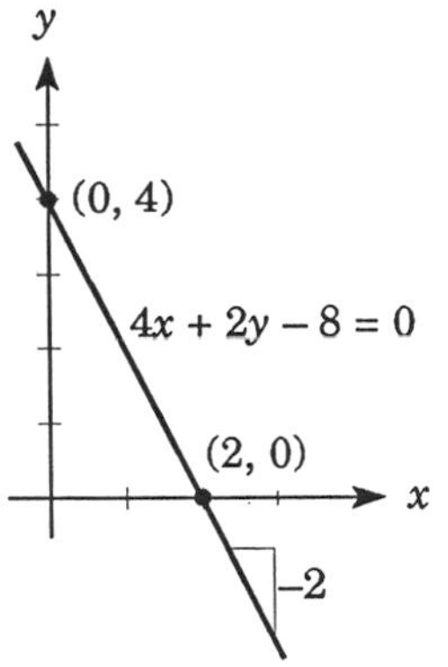

- Find the slope and the y- and x-intercepts of a line

$$4x + 2y - 8 = 0.$$

To find the slope m and the y-intercept b, we have

$$m = -\frac{4}{2} = -2; \quad b = -\frac{-8}{2} = 4.$$

The x-intercept is determined by setting $y = 0$,

$$4x = 8; \quad x = 2.$$

Parallel or Perpendicular Lines

Parallel lines have the same slope.

Lines

L_1: $3y = 5x + 4$ and L_2: $3y = 5x + 9$ are parallel

but

L_1: $2y = 3x + 4$ and L_2: $y = 3x + 4$ are not.

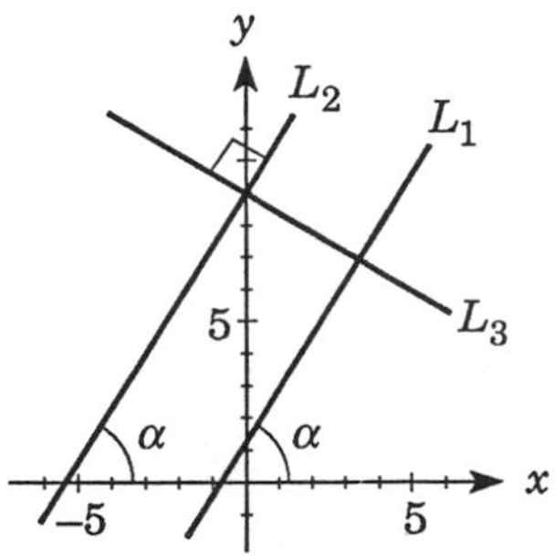

Perpendicular lines have slopes that are negative reciprocals of one another.

Lines

L_1: $3y = 5x + 4$ and L_3: $5y = -3x + 9$ are perpendicular

but

L_1: $3y = 5x + 4$ and L_2: $3y = 5x + 9$ are not.

Intersecting Lines and Their Angles

The angle between two intersecting straight lines is defined as the smallest positive angle through which the first line must rotate to coincide with the second line.

The angle between two lines may be determined from the values of the slopes of the lines.
Two types exist. In the figures below, we have

to the left: $\phi_{L1} < \phi_{L2}$
to the right: $\phi_{L2} < \phi_{L1}$

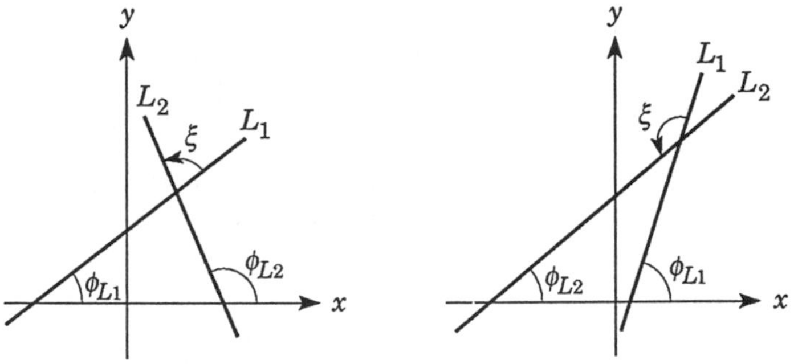

Using the notations of the figures, the **angle ξ between two lines**, whose slopes are m_1 and m_2, is

$$\xi = \arctan \frac{m_2 - m_1}{1 + m_2 m_1} \qquad \text{if} \quad \frac{m_2 - m_1}{1 + m_2 m_1} > 0$$

and

$$\xi = 180^\circ + \arctan \frac{m_2 - m_1}{1 + m_2 m_1} \qquad \text{if} \quad \frac{m_2 - m_1}{1 + m_2 m_1} < 0 .$$

Why? For $\phi_{L1} < \phi_{L2}$, we have

$$\phi_{L2} = \phi_{L1} + \xi$$
$$\xi = \phi_{L2} - \phi_{L1} ;$$

for $\phi_{L2} < \phi_{L1}$,

$$\phi_{L1} = \phi_{L2} + (180^\circ - \xi)$$
$$\xi = 180^\circ + (\phi_{L2} - \phi_{L1}) .$$

p. 504 Since the tangent function has period 180° (π rad), we have in either case that

$$\tan \xi = \tan(\phi_{L2} - \phi_{L1}) = \frac{\tan \phi_{L2} - \tan \phi_{L1}}{1 + \tan \phi_{L2} \tan \phi_{L1}} ,$$

provided that neither ϕ_{L1} nor ϕ_{L2} is 90° (tan 90° is not defined).

Let the slopes of L_1 and L_2 be m_1 and m_2, respectively.

As

$$m_1 = \tan \phi_{L1} \quad \text{and} \quad m_2 = \tan \phi_{L2},$$

we have

$$\tan \xi = \frac{m_2 - m_1}{1 + m_2\, m_1}\,.$$

If the quantity on the right is positive, ξ is simply its arc tangent function. If the quantity on the right is negative, then its arc tangent function is negative, and the positive angle ξ is obtained by adding 180°.

- Determine the angles between the intersecting lines

  a: L_1: $y = 2x+3$ and L_2: $2y = -3x+8$

  b: L_1: $y = 2$ and L_2: $3y = -x\sqrt{3}$

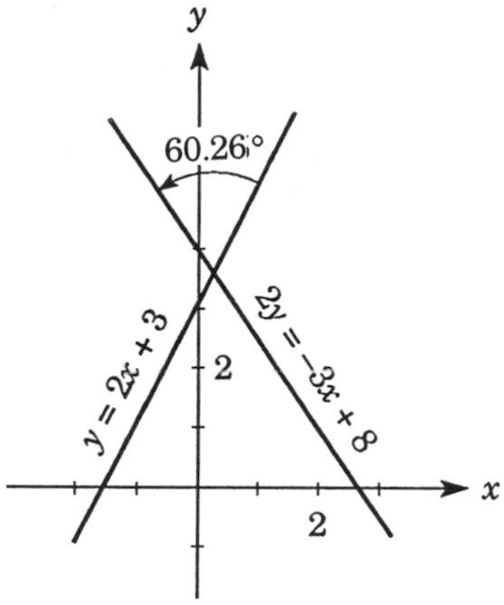

a:

Let the slope of lines $y = 2x + 3$ and $2y = -3x + 8$ be m_1 and m_2, respectively; then,

$$m_1 = 2 \quad \text{and} \quad m_2 = -\frac{3}{2}\,.$$

$$\xi = \arctan \frac{-\frac{3}{2} - 2}{1 + \left(-\frac{3}{2}\right)2} = \arctan \frac{7}{4} \approx 60.26°\,.$$

b:

$$m_1 = 0\,;\; m_2 = -\frac{\sqrt{3}}{3}$$

$$\xi = 180° + \arctan \frac{-\frac{\sqrt{3}}{3} - 0}{1 + \left(-\frac{\sqrt{3}}{3}\right)0}$$

p. 476

$$= 180° + \arctan\left(-\frac{\sqrt{3}}{3}\right) = 150°\,.$$

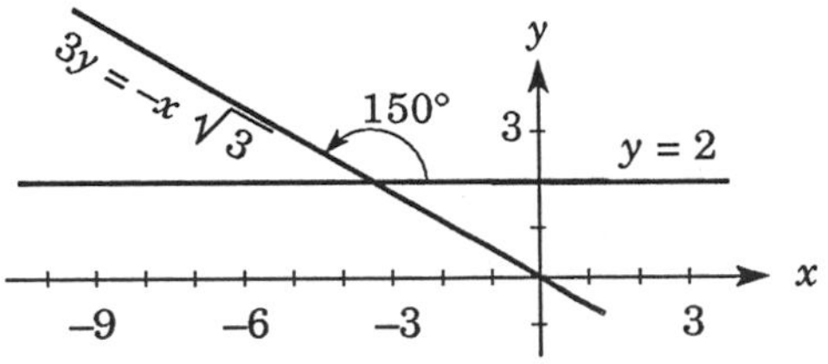

If one of L_1, L_2 is vertical, the formula does not apply, but the situation is easy to handle.

- Determine the angle between the intersecting lines

$$L_1: \ y = x\sqrt{3} \quad \text{and} \quad L_2: \ x = 2$$

L_2 is parallel to the y-axis; the sought angle is equal to the angle between L_1 and the y-axis.

Let the slope of line $y = x\sqrt{3}$ be m_1,

$$m_1 = \frac{\text{run}}{\text{rise}} = \frac{\sqrt{3}}{1},$$

and let the sought angle be ϕ,

$$\tan\phi = \frac{1}{\sqrt{3}}$$

$$\phi = 30°.$$

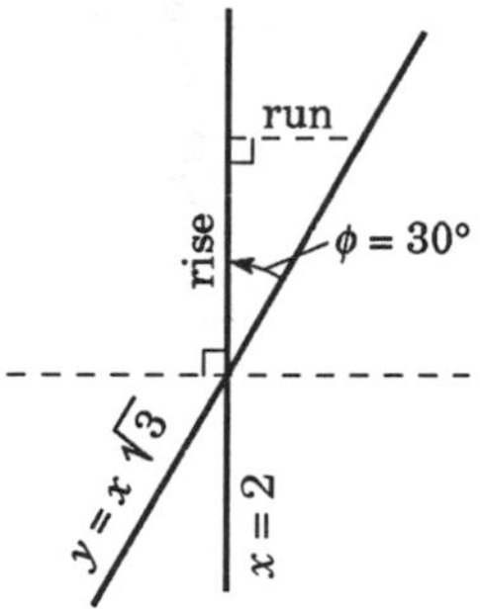

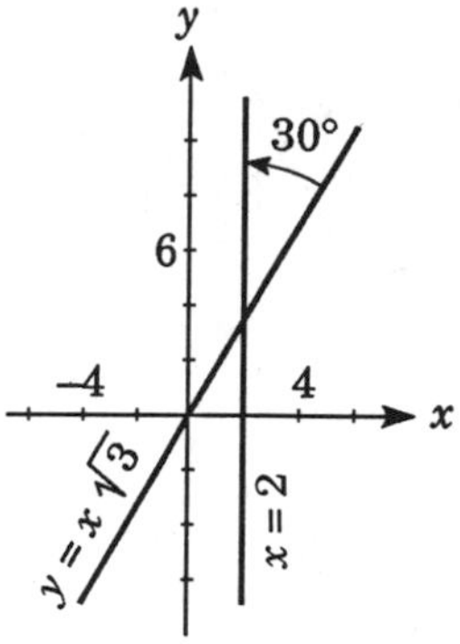

15.2 Conic Sections

The Circle

Latin, *circulus*, diminutive form of *circus*, "ring", akin to Greek *kirkos*, "ring"

A circle is the locus (set) of points (x, y), in the plane, that are at a given distance, called the **radius** (r), from a fixed point, the **center** (u, v):

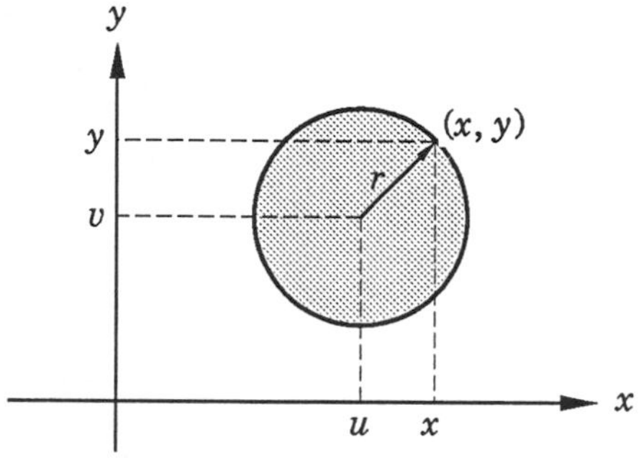

Equations of the Circle

General Equation

The equation of a circle centered at any point in the coordinate system – the **general equation** – is obtained by expressing the radius r in terms of the distance formula; thus,

$$r = \sqrt{(x-u)^2 + (y-v)^2}\ ,$$

leading to

$$(x-u)^2 + (y-v)^2 = r^2\ .$$

Central Equation

The equation of a circle centered at the origin of the coordinate system – the **central equation** – is

$$x^2 + y^2 = r^2\ .$$

- A circle centered at $(3, -5)$ and with the radius 3 units of length has the equation

$$(x-3)^2 + (y+5)^2 = 9\ .$$

The Ellipse

Greek, *ellipsis*, "a defect", indicating that the angle between the ellipse and the base of the cone is less than the angle between the parabola and the base of the cone

An ellipse is the locus of points in the plane for which the *sum* of their distances from two fixed points (foci) is constant.

In the graph below F_1 and F_2 denote the foci of the ellipse; P represents any point on the ellipse.

The sum of the distances $|F_1 P|$ and $|F_2 P|$ is constant; this constant value equals the length of the major axis, $2\,a$.

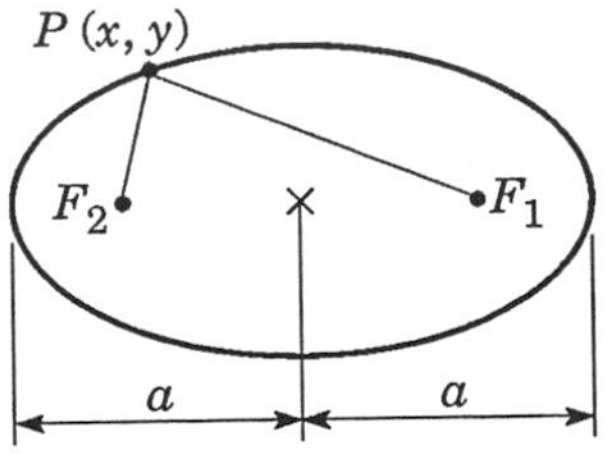

Equations of the Ellipse

Central Equations

Central equations concern an ellipse whose axes meet at right angles at the origin of the orthogonal coordinate system.

Two cases occur: the one for the ellipse whose major axis coincides with the x-axis, the other for the ellipse whose major axis coincides with the y-axis:

$$\frac{x^2}{a^2} + \frac{y^2}{b^2} = 1$$ center at the origin, major axis coinciding with the x-axis;

$$\frac{x^2}{b^2} + \frac{y^2}{a^2} = 1$$ center at the origin, major axis coinciding with the y-axis,

where $2a$ and $2b$ are the axes of the ellipse.

General Equations

If, instead, the center of the ellipse is at a point (u, v), we have the **general equations**:

$$\frac{(x-u)^2}{a^2} + \frac{(y-v)^2}{b^2} = 1$$ major axis parallel to or coinciding with the x-axis;

$$\frac{(x-u)^2}{b^2} + \frac{(y-v)^2}{a^2} = 1$$ major axis parallel to or coinciding with the y-axis.

To deduce the **central equations**, let $P(x, y)$ represent any point on the periphery of an ellipse **centered at the origin** and major axis $2\,a$ coinciding with the x-axis of an orthogonal coordinate system; let the minor axis of the ellipse equal $2b$, and the distance between the center and a focus (F_1 or F_2) equal c.

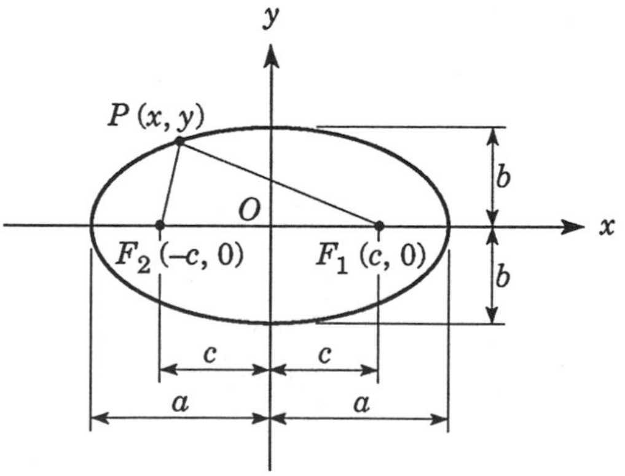

The distance formula gives

$$|F_1 P| = \sqrt{(x-c)^2 + (y-0)^2}$$

and

$$|F_2 P| = \sqrt{[x-(-c)]^2 + (y-0)^2}\,.$$

By definition of the ellipse, the sum of the distances $|F_1 P|$ and $|F_2 P|$ is equal to the length $2a$ of the major axis; thus,

$$\sqrt{(x-c)^2+(y-0)^2} + \sqrt{[x-(-c)]^2+(y-0)^2} = 2a,$$

which may be rewritten

$$\sqrt{(x-c)^2+y^2} = 2a - \sqrt{(x+c)^2+y^2}\,.$$

Square both sides,

$$x^2-2cx+c^2+y^2 = 4a^2-4a\sqrt{(x+c)^2+y^2} + x^2+2cx+c^2+y^2,$$

and simplify,

$$a\sqrt{(x+c)^2+y^2} = a^2+cx.$$

Again, square both sides and simplify,

$$a^2x^2+2a^2cx+a^2c^2+a^2y^2 = a^4+2a^2cx+c^2x^2;$$

$$(a^2-c^2)x^2+a^2y^2 = a^2(a^2-c^2)\,. \tag{1}$$

Since the sum of $|F_1 P|$ and $|F_2 P|$ equals the length of the major axis, $2a$, the distance between either focus and an end point of the minor axis is a.

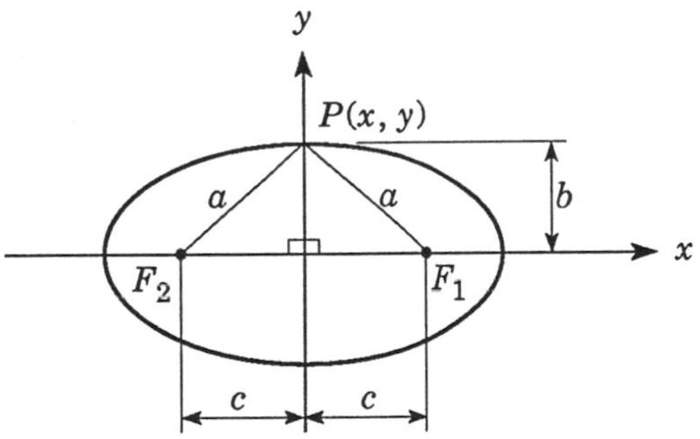

By the Pythagorean theorem:

$$b^2+c^2 = a^2$$
$$a^2-c^2 = b^2$$

Substituting b^2 for a^2-c^2 in (1) gives

$$b^2x^2+a^2y^2 = a^2b^2$$

and, after dividing both sides by a^2b^2, we find the central equation

$$\frac{x^2}{a^2}+\frac{y^2}{b^2} = 1\,.$$

For the **general equation** of an ellipse whose major axis coincides with or is parallel to the x-axis of the coordinate system, consider the graph of the ellipse below, whose center is at a point (u, v):

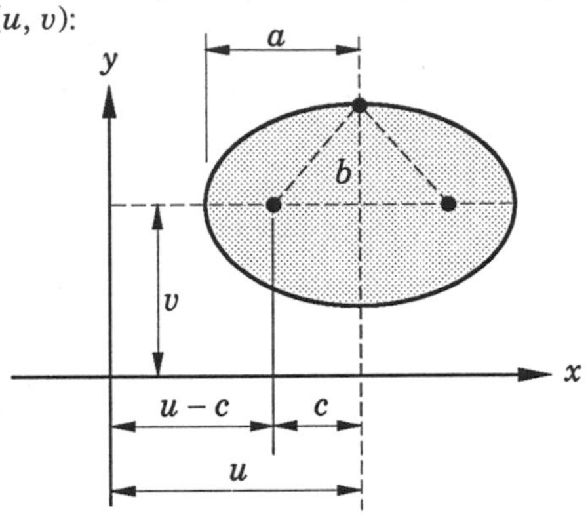

By definition of the ellipse, we have

$$\sqrt{\left[x-(u-c)\right]^2+(y-v)^2}+\sqrt{\left[x-(u+c)\right]^2+(y-v)^2}=2a$$

which, analogously to the central equation, may be simplified and rewritten

$$\frac{(x-u)^2}{a^2}+\frac{(y-v)^2}{b^2}=1.$$

To derive the central and general equations of an ellipse whose major axis coincides with or is parallel to the y-axis, one follows a similar pattern.

The Parabola

A parabola is the locus of points (x, y), in the plane, whose distances to a given straight line (the **directrix**) and from a given point (**focus**) outside the line are equal, as illustrated by the distances d_1 and d_2 below:

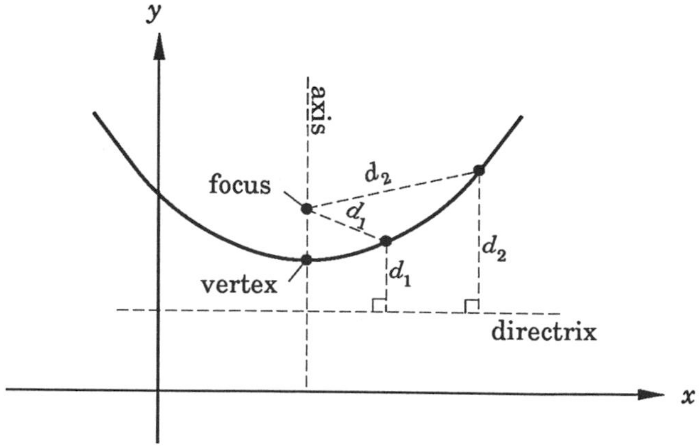

The parabola is symmetrical with respect to an **axis** passing through the focus and perpendicular to the **directrix**. The vertex of the parabola is defined to be located at the midpoint of the segment of the axis extending from the focus to the directrix.

Equations of the Parabola

Central Equations

Central equations concern a parabola whose axis coincides with one of the axes of the coordinate system and whose vertex is at the origin. Two cases occur: the one for the parabola whose axis coincides with the y-axis, the other for the parabola whose axis coincides with the x-axis:

$x^2 = 4py$ axis coinciding with the y-axis;
$y^2 = 4px$ axis coinciding with the x-axis,

where p is the distance between the vertex and the focus.

General Equations

If, instead, the vertex of the parabola is at a point (u, v), we have the **general equations**:

$(x-u)^2 = 4p(y-v)$ vertex at (u, v), axis parallel to or coinciding with the y-axis;
$(y-v)^2 = 4p(x-u)$ vertex at (u, v), axis parallel to or coinciding with the x-axis.

To deduce the **central equations**, let $P(x, y)$ be a point on the parabola with its **vertex at the origin**, axis coinciding with the y-axis, and focus at a distance p from the vertex ($p > 0$):

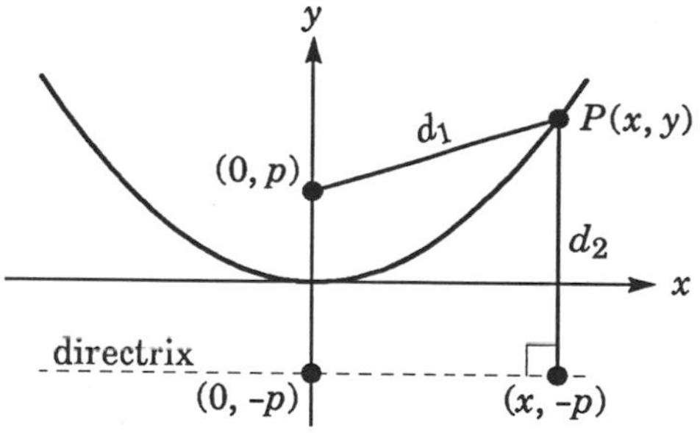

The directrix is $y = -p$.

$d_1 = \sqrt{(x-0)^2 + (y-p)^2}$; $d_2 = y - (-p)$

By the definition of the parabola, $d_1 = d_2$; thus,

(1) $\sqrt{(x-0)^2 + (y-p)^2} = y - (-p)$

$$x^2 = 4py.$$

For the **general equation** of a parabola whose axis is parallel to or coincides with the y-axis, consider the graph of the parabola below, whose vertex is at (u, v):

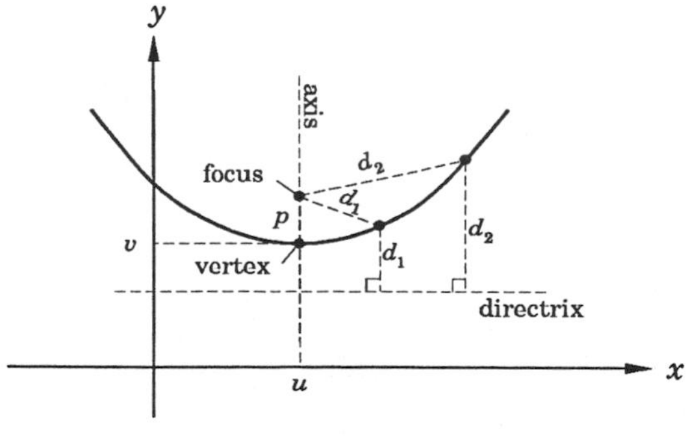

By definition of the parabola, we have

$$\sqrt{(x-u)^2 + [y-(v+p)]^2} = y-(v-p)$$
$$(x-u)^2 = 4p(y-v).$$

For the central and general equations of a parabola whose axis coincides with or is parallel to the x-axis, one follows a similar pattern.

The Hyperbola

A hyperbola is the locus of points, in the plane, for which the *difference* of their distances from two fixed points (foci) is constant.

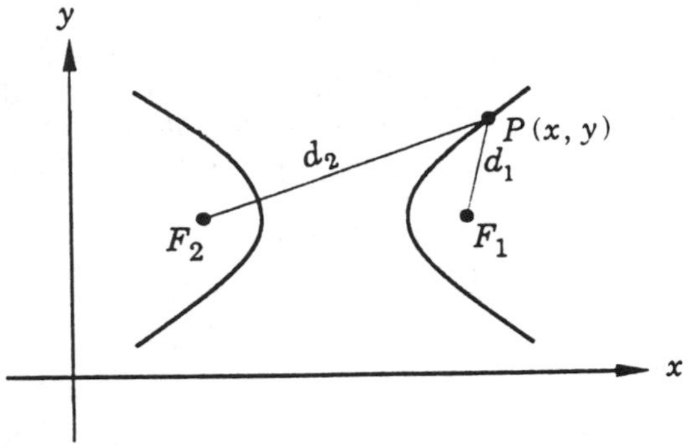

In the above graph, where F_1 and F_2 denote the foci of the hyperbola, the difference of the distances d_1 and d_2 from each focus to any point on the curves of the hyperbola is constant.

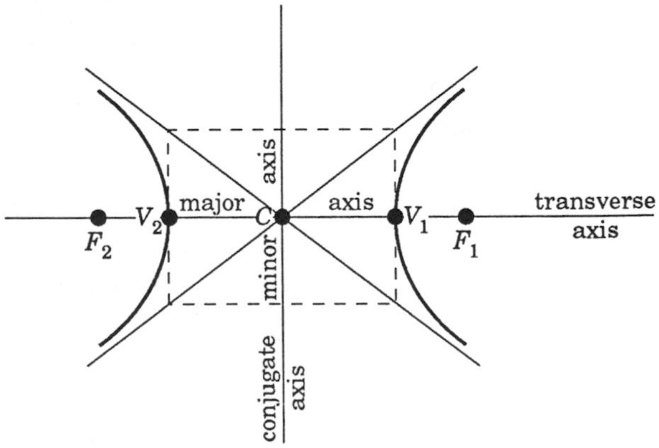

Referring to the above graph of the hyperbola, we give the following definitions:

Vertices
Transverse Axis

- The line of indefinite length through the two foci and intersecting the **vertices** (V_1, V_2) is the **transverse axis**.

Major Axis

- The segment of the transverse axis between the vertices is the **major axis**.

Center

- The midpoint of the major axis is the **center** (C).

Conjugate Axis

- The line of indefinite length through the center and perpendicular to the transverse axis is the **conjugate axis**.

Asymptote

Greek, *asymptotos,* "not falling together"

- An **asymptote** is a straight line which, indefinitely produced, does not meet the hyperbola but comes arbitrarily closer to it. The asymptotes of the hyperbola intersect at its center.

Minor Axis

- The segment of the conjugate axis that has the same length as a line parallel to the conjugate axis, connecting the asymptotes and one vertex, is the **minor axis**.

Principal Axis

The terminology may vary from text to text. Thus, another term for transverse axis is **principal axis**. The terms **transverse axis** and **conjugate axis** are in some texts used for the line *segments* that in the above are referred to as *major axis* and *minor axis*, respectively.

Equations of the Hyperbola

Central Equations

Central equations concern a hyperbola whose axes intersect at the origin of the coordinate system. Two cases occur: the one with the major axis coinciding with the x-axis, the other with the major axis coinciding with the y-axis.

If

major axis = $2a$,

minor axis = $2b$,

$b = \sqrt{c^2 - a^2}$, where c is the distance between the center and a focus of the hyperbola,

then the central equations of the hyperbola are

$$\frac{x^2}{a^2} - \frac{y^2}{b^2} = 1$$ center at the origin, foci on the x-axis;

$$\frac{y^2}{a^2} - \frac{x^2}{b^2} = 1$$ center at the origin, foci on the y-axis.

General Equations

If, instead, the center of the hyperbola is at a point (u, v), we have the **general equations**:

$$\frac{(x-u)^2}{a^2} - \frac{(y-v)^2}{b^2} = 1$$ foci on a line parallel to or coinciding with the x-axis;

$$\frac{(y-u)^2}{a^2} - \frac{(x-v)^2}{b^2} = 1$$ foci on a line parallel to or coinciding with the y-axis.

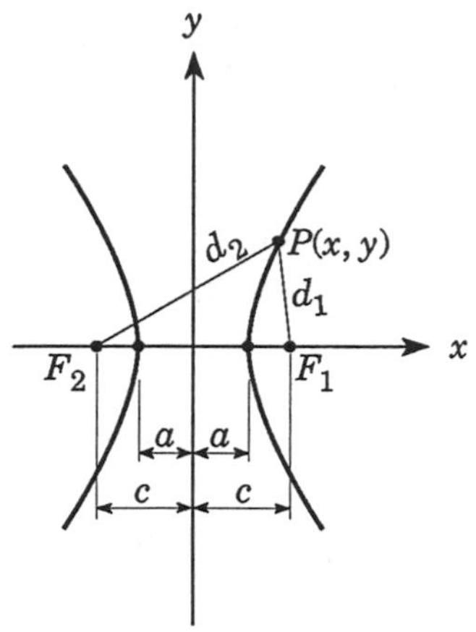

To deduce the central equations, let $P(x, y)$ represent any point on a hyperbola **centered at the origin** and with foci on the x-axis.

Using the distance formula and letting the distance between center and focus (F_1 or F_2) be c, then:

$$d_1 = \sqrt{(x-c)^2 + (y-0)^2} = \sqrt{(x-c)^2 + y^2}$$

$$d_2 = \sqrt{[x-(-c)]^2 + (y-0)^2} = \sqrt{(x+c)^2 + y^2}$$

From the definition of the hyperbola, we have that the difference between $P(x, y)$ and the foci equals the length of the major axis, $2a$; thus, $d_2 - d_1 = 2a$, or

$$\sqrt{(x+c)^2 + y^2} - \sqrt{(x-c)^2 + y^2} = 2a\,,$$

rewritten

$$\sqrt{(x+c)^2 + y^2} = 2a - \sqrt{(x-c)^2 + y^2}\,.$$

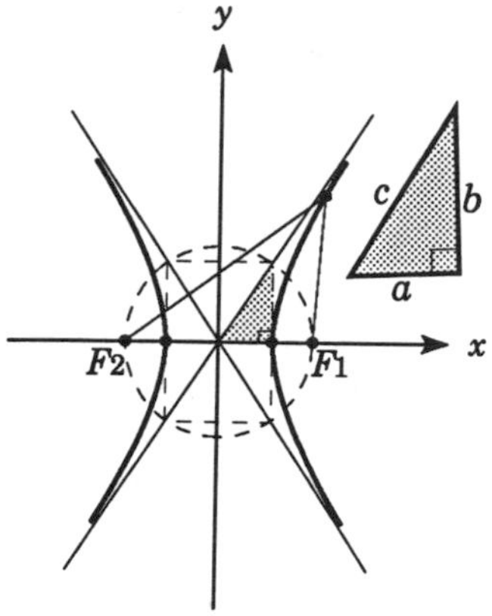

Square both sides,

$$x^2 + 2cx + c^2 + y^2 = 4a^2 - 4a\sqrt{(x-c)^2 + y^2} + x^2 - 2cx + c^2 + y^2,$$

and simplify,

$$a\sqrt{(x-c)^2 + y^2} = cx - a^2\,.$$

Again, square both sides and simplify,

$$a^2x^2 - 2a^2cx + a^2c^2 + a^2y^2 = a^4 - 2a^2cx + c^2x^2;$$

(1) $\qquad (c^2 - a^2)\,x^2 - a^2y^2 = a^2(c^2 - a^2)\,.$

Referring to the figures, we obtain $a^2 + b^2 = c^2$, and can define

$$b = \sqrt{c^2 - a^2}\,;$$

after substituting b^2 for $c^2 - a^2$ in (1), we obtain

$$b^2x^2 - a^2y^2 = a^2b^2$$

and, after dividing both sides by a^2b^2,

$$\frac{x^2}{a^2} - \frac{y^2}{b^2} = 1\,.$$

For the **general equation** of a hyperbola with foci on a line coinciding with or parallel to the x-axis, consider the graph of the hyperbola below, centered at a point (u, v):

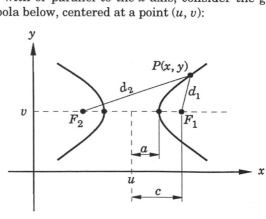

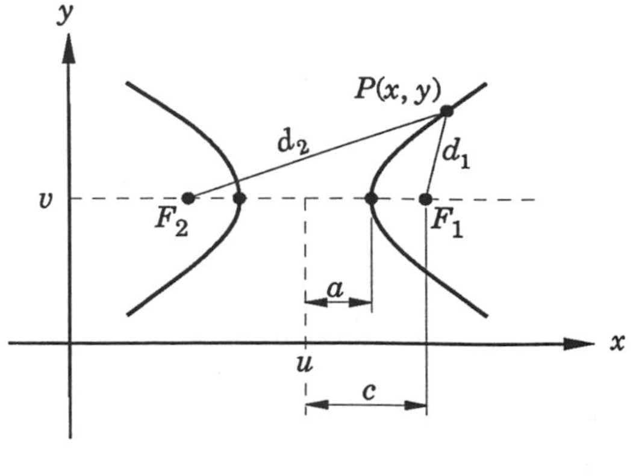

From the definition of the hyperbola, we have

$$\sqrt{\left[x - (u - c)\right]^2 + (y - v)^2} + \sqrt{\left[x - (u + c)\right]^2 + (y - v)^2} = 2a,$$

which, after some algebra, may be rewritten

$$\frac{(x - u)^2}{a^2} - \frac{(y - v)^2}{b^2} = 1.$$

- A hyperbola is centered at the origin and has its transverse axis coinciding with the x-axis of an orthogonal coordinate system; the distance between the center and a vertex is 2; the distance between the center and a focus is 3.

Determine the equations of the hyperbola and its asymptotes.

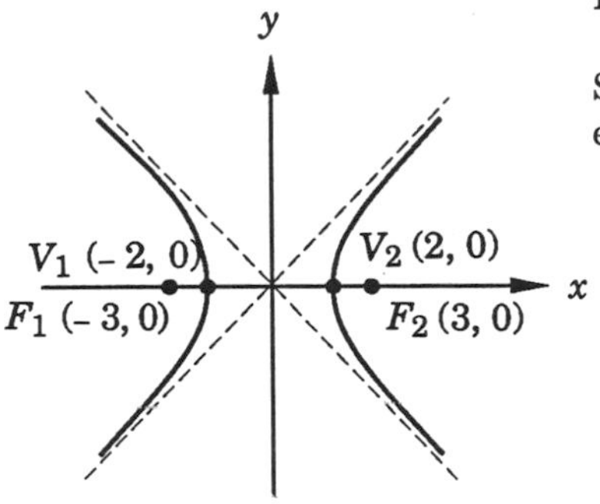

Since the hyperbola is centered at the origin, we use the equation

$$\frac{x^2}{a^2} - \frac{y^2}{b^2} = 1,$$

where

major axis = $2a$
minor axis = $2b$
$b = \sqrt{c^2 - a^2}$
c is the distance between the center and a focus of the hyperbola.

$b = \sqrt{c^2 - a^2}$ yields $b^2 = 3^2 - 2^2$; $b = \sqrt{5}$.

The equation of the given hyperbola is

$$\frac{x^2}{4} - \frac{y^2}{5} = 1.$$

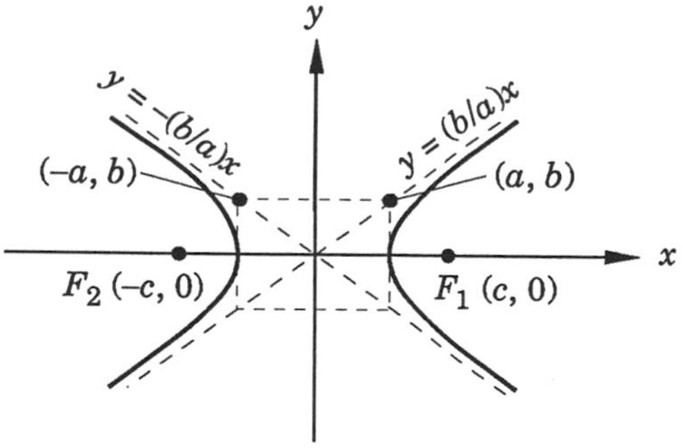

Written in *slope-intercept form*, the equations of the asymptotes of a hyperbola, centered at the origin and with its major axis coinciding with the x-axis, are $y = \frac{b}{a}x$; $y = -\frac{b}{a}x$, and for the given hyperbola

$$y = \frac{\sqrt{5}}{2}x \; ; \; y = -\frac{\sqrt{5}}{2}x.$$

The Vertex Equation

The general definition of a conic section is:

Directrix

Eccentricity

> The locus of all points in the plane whose distances to a given point, the **focus**, and to a given straight line, the **directrix**, in the same plane, are in a specific, constant ratio, known as the **eccentricity**.

As might be expected, the equations of the conic sections can all be brought into a common form.

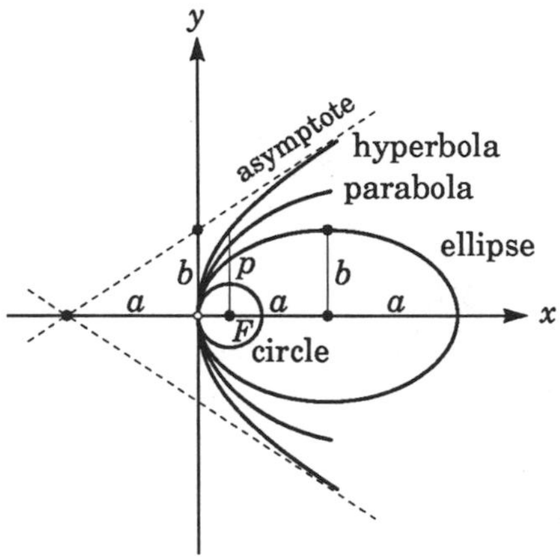

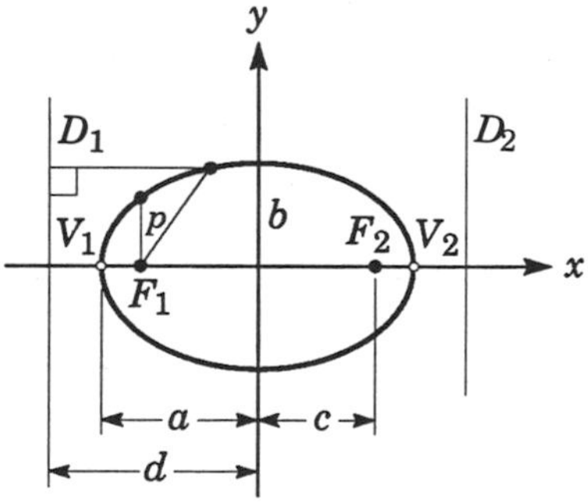

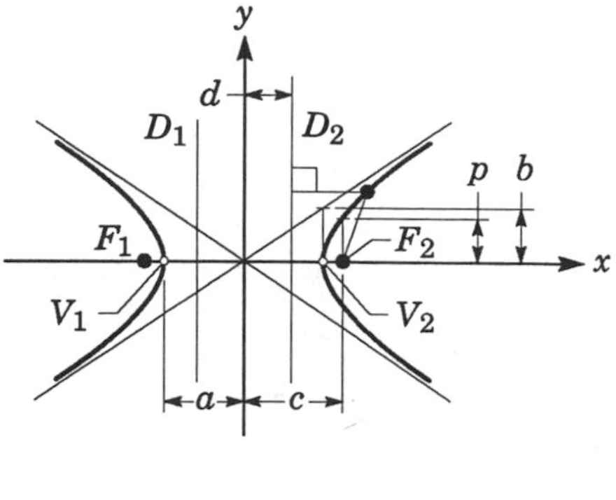

The central equation is

for the **ellipse**	for the **hyperbola**
$\frac{x^2}{a^2} + \frac{y^2}{b^2} = 1$	$\frac{x^2}{a^2} - \frac{y^2}{b^2} = 1$

where

$2a$	= the major axis	$2a$	= the transverse axis
$2b$	= the minor axis	$2b$	= the conjugate axis
$2c$	= the interfocal distance	$2c$	= the interfocal distance
a^2	$= b^2 + c^2$	a^2	$= c^2 - b^2$
d	= the distance between the center and the directrix	d	= the distance between the center and the directrix
$D_1 D_2$	= the directrices	$D_1 D_2$	= the directrices
$F_1 F_2$	= the foci	$F_1 F_2$	= the foci
$V_1 V_2$	= the vertices	$V_1 V_2$	= the vertices

The general definition applies for all points on the curves; specifically for the vertices, we have the distances:

	for the **ellipse**	for the **hyperbola**
vertex V_1 to focus F_1,	$a - c$	$c - a$
vertex V_1 to directrix D_1,	$d - a$	$a - d$
vertex V_2 to focus F_1,	$a + c$	$c + a$
vertex V_2 to directrix D_1,	$d + a$	$a + d$

We then have the eccentricity,

$$e = \frac{a-c}{d-a} = \frac{a+c}{d+a} \qquad\qquad e = \frac{c-a}{a-d} = \frac{c+a}{d+a}$$

or

$$ad + a^2 - cd - ac = ad + cd - a^2 - ac$$

$$a^2 = cd;$$

and $d = \frac{a^2}{c}$

or

$$ac - a^2 + cd - ad = ac + a^2 - cd - ad$$

$$a^2 = cd;$$

and $d = \frac{a^2}{c}$

which insert in

$$e = \frac{a+c}{d+a} = \frac{a+c}{\frac{a^2}{c} + a} = \frac{c\,(a+c)}{a\,(a+c)} \qquad\qquad e = \frac{c+a}{a+d} = \frac{c+a}{a + \frac{a^2}{c}} = \frac{c\,(a+c)}{a\,(a+c)}$$

and $e = c/a$. and $e = c/a$.

We introduce in the central equations the point P,

$$P_{\text{ellipse}} = (-c\,,\ p)\,;\quad P_{\text{hyperbola}} = (c\,,\ p),$$

Latex Rectus, Focal Chord

where $2p$ is the **latex rectus** (plural: *latera recta*), or **focal chord**,

$$\frac{(-c)^2}{a^2} + \frac{p^2}{b^2} = 1\,; \qquad \frac{c^2}{a^2} - \frac{p^2}{b^2} = 1\,;$$

or

$$p^2 = \frac{b^2\,(a^2 - c^2)}{a^2} = \left(\frac{b^2}{a}\right)^2; \quad p^2 = \frac{b^2\,(c^2 - a^2)}{a^2} = \left(\frac{b^2}{a}\right)^2;$$

and

$$p = \frac{b^2}{a}. \qquad p = \frac{b^2}{a}.$$

Shift the curve a distance a in the x direction so that one vertex will be at the origin,

the **ellipse** toward the right,

$$\frac{(x-a)^2}{a^2} + \frac{y^2}{b^2} = 1\,;$$

$$\frac{x^2}{a^2} - \frac{2\,a\,x}{a^2} + \frac{a^2}{a^2} + \frac{y^2}{b^2} = 1\,;$$

$$y^2 = \frac{2\,b^2}{a}\,x - \frac{b^2}{a^2}\,x^2\,;$$

$$y^2 = 2\,p\,x - (1 - e^2)\,x^2\,;$$

the **hyperbola** toward the left,

$$\frac{(x+a)^2}{a^2} - \frac{y^2}{b^2} = 1\,;$$

$$\frac{x^2}{a^2} + \frac{2\,a\,x}{a^2} + \frac{a^2}{a^2} + \frac{y^2}{b^2} = 1\,;$$

$$y^2 = \frac{2\,b^2}{a}\,x - \frac{b^2}{a^2}\,x^2\,;$$

$$y^2 = 2\,p\,x - (1 - e^2)\,x^2\,;$$

that is, the *same equation*, in vertex form, for the ellipse and the hyperbola; it also applies for the parabola, with $e = 1$,

$$y^2 = 2\,p\,x,$$

and for the circle, with $e = 0$ and $p = r$,

$$y^2 = 2\,r\,x - x^2\,;$$

or $$x^2 - 2\,r\,x - y^2 = 0$$

and $$(x - r)^2 + y^2 = r^2\,.$$

Consequently, the **vertex equation**

$$y^2 = 2\,p\,x - (1 - e^2)\,x^2$$

describes

		The Greek word	means	
for $e = 0$:	a circle	*kirkos*	a ring	$-\ 1 \cdot x^2$
$0 < e < 1$:	an ellipse	*elleipein*	to deduct	$-\ (1 - e^2)\,x^2$
$e = 1$:	a parabola	*paraballein*	to equate	$+\ 0 \cdot x^2$
$e > 1$:	a hyperbola	*hyperballein*	to exceed	$+\ (e^2 - 1)\,x^2$

Another illustration of the meaning of the names *elliptic*, *parabolic*, and *hyperbolic* is offered by the Euclidean and non-Euclidean geometries in the interpretation of the Euclidean parallel axiom; through a given point outside a given straight line a number of other straight lines can be drawn, parallel to the first line:

– in elliptic geometry (Riemann)	less than one, that is, none;
– in Euclidean geometry ("parabolic")	just one;
– in hyperbolic geometry (Gauss, Bólyai, Lobachevski)	more than one

Beyond the Conics

With $n = 2$,

$$\left|\frac{x}{a}\right|^n + \left|\frac{y}{b}\right|^n = 1$$

is the formula for the ellipse, but with other values of n, we have

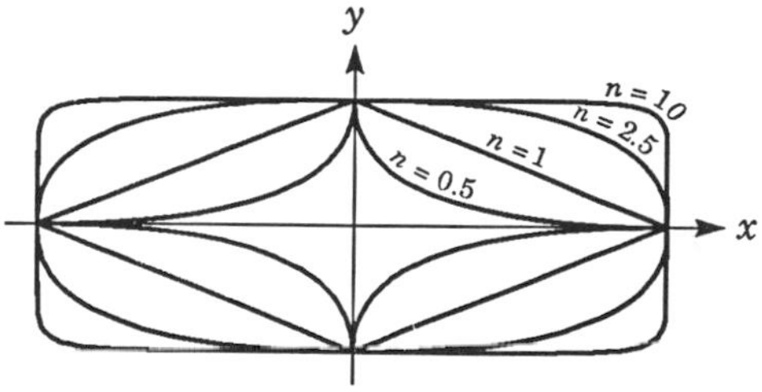

HEIN Piet
(1905 - 1996)

The graph of the formula with $n = 2.5$, once considered an ideal design for a tabletop, was named "super-ellipse" by **Piet Hein**, Danish architect, cartoonist/satirist, and author of *Grooks*.

With $n = 2$,

$$\left|\frac{x}{a}\right|^n - \left|\frac{y}{b}\right|^n = 1$$

is the formula for the hyperbola. Examining for other values of n, we have

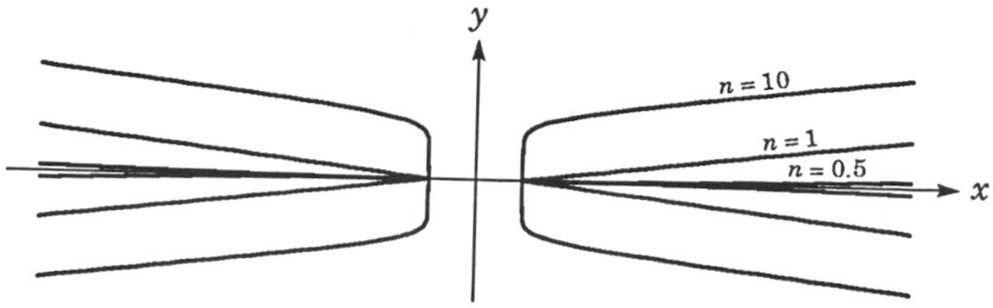

15.3 Shifting Orthogonal Coordinates

By imposing a new orthogonal system on the original system, an equation can sometimes be made more accessible.

We distinguish two forms of transposition:

- **Translation:** A parallel displacement of the original system along one or more of its axes.
- **Rotation:** The origin of the new system coincides with the origin of the original system.

Translation

If after translation the coordinates x, y, and z of a point P become x', y', and z', then the **translation equations**

$$x = x' + a \qquad y = y' + b \qquad z = z' + c$$

imply that a, b, and c are the coordinates of the origin of the x', y', z'-system with reference to the x, y, z-system:

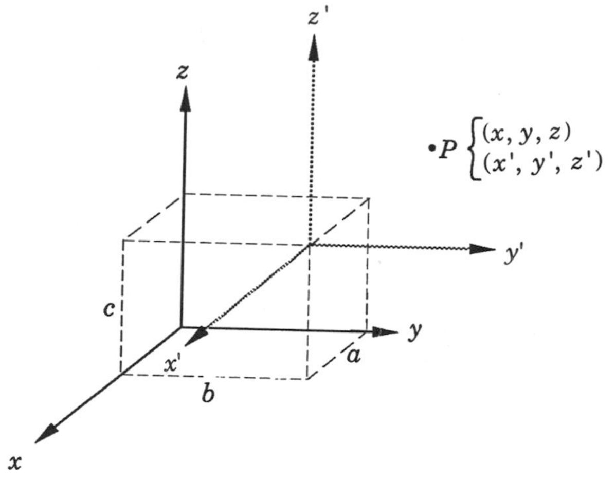

- A circle has the equation $(x - 3)^2 + (y - 2)^2 = 1$. After translation, the origin of the new coordinate system (x', y') is located at (3, 2) with reference to the original (x, y) coordinate system. Determine the equation of the circle expressed in the coordinates (x', y').

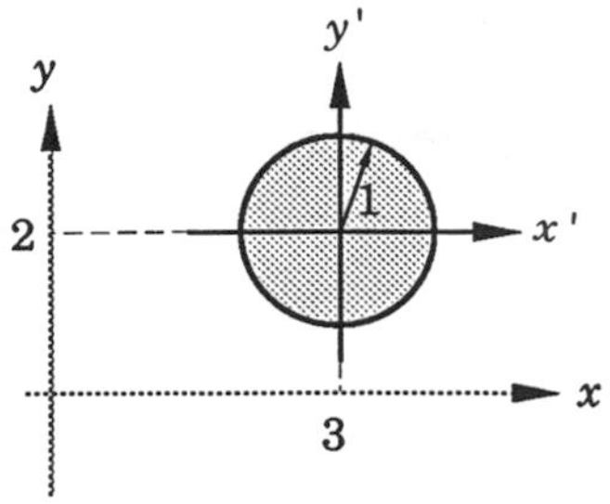

The original equation represents a circle of radius 1 and center at (3, 2) in the original system.

After translation, we have

$$x = x' + 3\,; \quad y = y' + 2\,.$$

The new equation is

$$[(x' + 3) - 3]^2 + [(y' + 2) - 2]^2 = 1$$

or

$$(x')^2 + (y')^2 = 1\,.$$

Rotation

Expressing New Coordinates in Original Coordinates

With the origin fixed, let the x, y-axes rotate counterclockwise through an angle θ into a system of x', y'-axes:

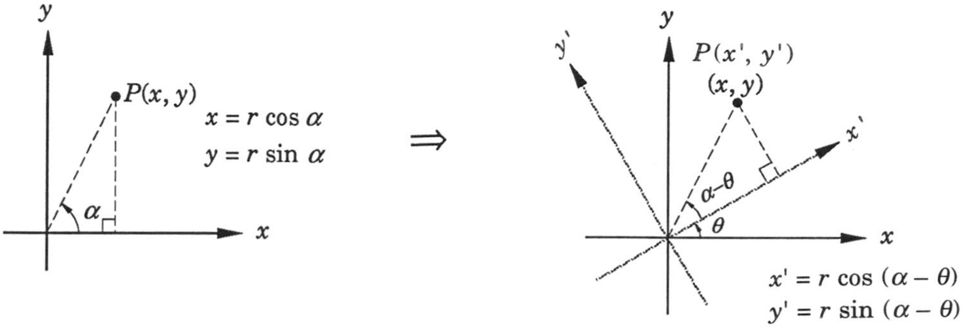

Expressed with reference to the original coordinates (x, y), the new coordinates (x', y') are

$$x' = x\cos\theta + y\sin\theta;\quad y' = -x\sin\theta + y\cos\theta.$$

Proof: $\cos(\alpha-\theta) = \dfrac{x'}{r}$

$$x' = r\cos(\alpha-\theta).$$

p. 509 Using the compound-angle identity, we have

$$x' = r\cos\alpha\cos\theta + r\sin\alpha\sin\theta;$$

and, as $x = r\cos\alpha$ and $y = r\sin\alpha$,

$$x' = x\cos\theta + y\sin\theta.$$

Similarly,

$$\sin(\alpha-\theta) = \frac{y'}{r}$$

$$y' = r\sin(\alpha-\theta);$$

$$= r\sin\alpha\cos\theta - r\cos\alpha\sin\theta;$$

and, as $x = r\cos\alpha$ and $y = r\sin\alpha$,

$$y' = -x\sin\theta + y\cos\theta.$$

Expressing Original Coordinates in New Coordinates

After coordinate rotation with the origin fixed, the original coordinates (x, y) expressed with reference to the new coordinates (x', y') are

$$x = x'\cos\theta - y'\sin\theta \quad\text{and}\quad y = x'\sin\theta + y'\cos\theta.$$

To derive the formulas that express the original coordinates in the new coordinates, we solve for x and y in the known formulas:

$$\left.\begin{aligned} (1) \qquad x' &= x\cos\theta + y\sin\theta \\ (2) \qquad y' &= -x\sin\theta + y\cos\theta \end{aligned}\right\}$$

Multiply (1) by $\cos\theta$, and (2) by $\sin\theta$; thus,

$$(1) \Rightarrow (3) \qquad x'\cos\theta = x\cos^2\theta + y\sin\theta\cos\theta$$

$$(2) \Rightarrow (4) \qquad y'\sin\theta = -x\sin^2\theta + y\sin\theta\cos\theta$$

Subtract (4) from (3); thus,

$$\begin{aligned} x'\cos\theta - y'\sin\theta &= x\cos^2\theta + y\sin\theta\cos\theta - (-x\sin^2\theta + y\sin\theta\cos\theta) \\ &= x\cos^2\theta + y\sin\theta\cos\theta + x\sin^2\theta - y\sin\theta\cos\theta \\ &= x(\cos^2\theta + \sin^2\theta), \end{aligned}$$

p. 508 simplified to

$$x = x'\cos\theta - y'\sin\theta\,.$$

Similarly, multiply (1) by $\sin\theta$, and (2) by $\cos\theta$; thus,

$$(1) \Rightarrow (5) \qquad x'\sin\theta = x\cos\theta\sin\theta + y\sin^2\theta$$

$$(2) \Rightarrow (6) \qquad y'\cos\theta = -x\sin\theta\cos\theta + y\cos^2\theta$$

Add (5) and (6):

$$\begin{aligned} x'\sin\theta + y'\cos\theta &= x\cos\theta\sin\theta + y\sin^2\theta - x\sin\theta\cos\theta + y\cos^2\theta \\ &= y(\cos^2\theta + \sin^2\theta)\,; \\ y &= x'\sin\theta + y'\cos\theta\,. \end{aligned}$$

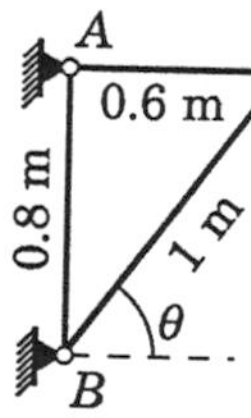

- A truss is a structural load-carrying system made from geometrical arrangement of straight structural elements which are connected at their ends only. The figure shows a 3-member truss.

When a specified load is applied to connection C (node C), C is displaced vertically to C'. With the displacement CC' the truss assumes a new configuration which is shown exaggerated by dashed lines:

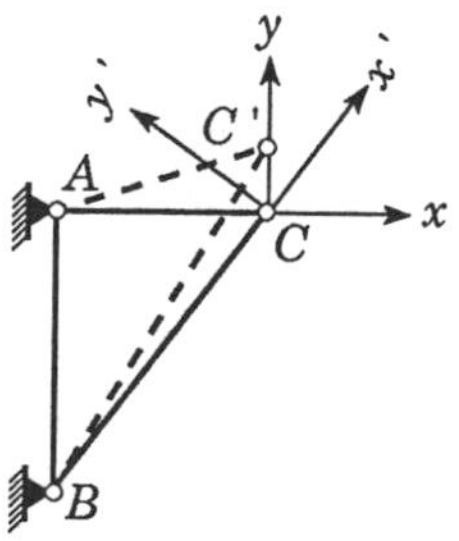

From precision theodolite measurement we know that the displacement CC' is 4.53 mm.

Determine the components of CC' in x', y'-coordinates where x' is taken through the axis of member BC.

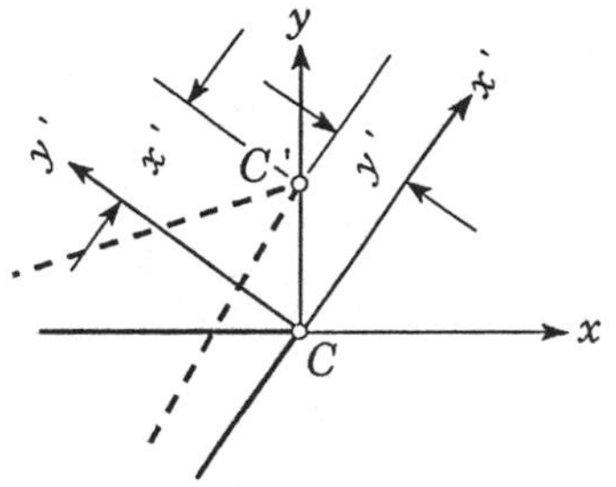

The basic relationships are

$$x' = x \cos\theta + y \sin\theta$$

$$y' = -x \sin\theta + y \cos\theta .$$

From the measurement we have $x = 0$, $y = 4.53$ mm and from the geometry $\sin\theta = 0.8$ and $\cos\theta = 0.6$ (complementary angles); thus,

$$x' = 0 + 4.53\,(0.8) \approx 3.62 \text{ mm}$$

$$y' = 0 + 4.53\,(0.6) \approx 2.72 \text{ mm}.$$

Note: The 3.62-mm elongation of member BC is produced by the stress occurring in BC as a consequence of the external load applied at C, which in its turn causes the displacement CC'. Since the deformations are small in relation to the overall dimensions of the truss, it is customary to carry out a first-order analysis, that is, to use the original geometry even for the deformed configuration of the truss. Thus, θ remains unchanged.

15.4 Polar Coordinate Systems

In previous discussions the position of points has been specified by means of orthogonal coordinates – ordered real numbers in a plane (x, y) or in space (x, y, z). In some situations, however, it is more convenient to locate a point by means of **polar coordinates**, that is, a system of coordinates where the position of a point is determined by

Radius Vector
Pole

- the length of the ray segment (the **radius vector**) from a fixed origin (the **pole**), that is, the distance between the point and the pole, and

Polar Angle, Vector
Polar Axis

- the angle (the **polar angle**) the ray (the **vector**) makes with a fixed line (the **polar axis**):

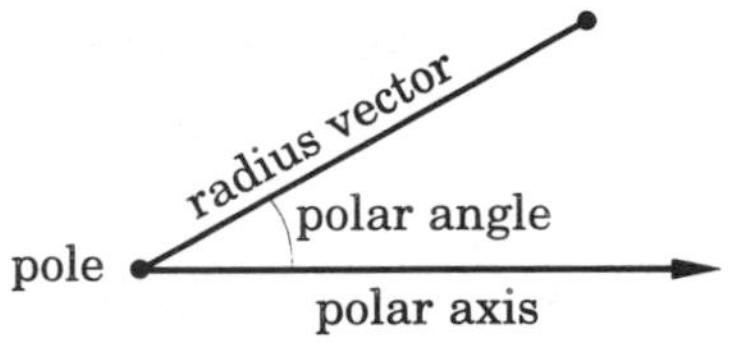

The pole is usually denoted O; the length of the radius vector, r or ρ; the polar angle, θ or ϕ:

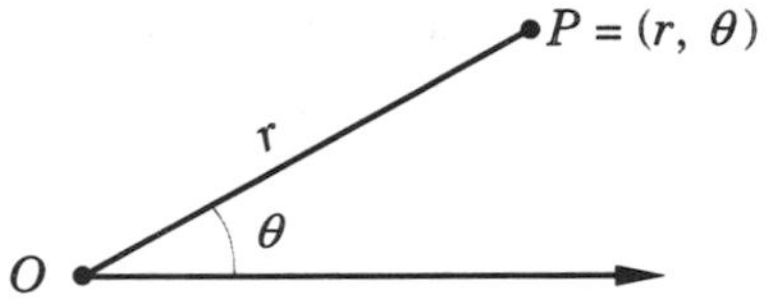

Thus, in a polar coordinate system the position of a point is determined by **distance** (r) and **direction** (θ).

Vectorial Angle, Argument, Amplitude, Azimuth

Arabic, *as-samūt*, "points of the horizon"

The polar angle is sometimes called the **vectorial angle**, the **argument**, the **amplitude**, or the **azimuth** of the point.

Polar Coordinates and the Plane

In polar coordinates, each point P in the plane is specified by two coordinates (r, θ), where r is the distance OP, and – as in trigonometry – positive θ is the angle measured counterclockwise.

Contrary to the coordinates of a point in an orthogonal coordinate system, polar coordinates are not unique; $(4, 50°)$, $(4, 410°)$, and $(4, -310°)$ all represent the same point.

Polar Grid

A **polar grid** is formed by concentric circles and by rays with endpoints at the center of the circles and at P. The first (innermost) ring is of radius 1, the second of radius 2, and so on.

In the graph below, angles are expressed in radians:

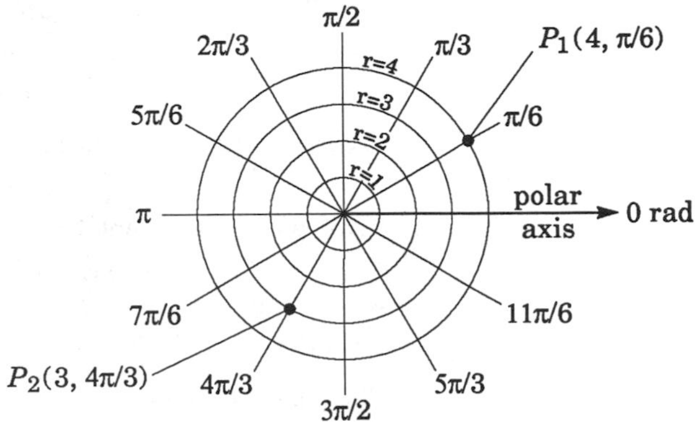

Shifting between Orthogonal and Polar Coordinates in the Plane

To convert between orthogonal and polar coordinates, let the pole coincide with the origin of the orthogonal system and the polar axis coincide with the x-axis:

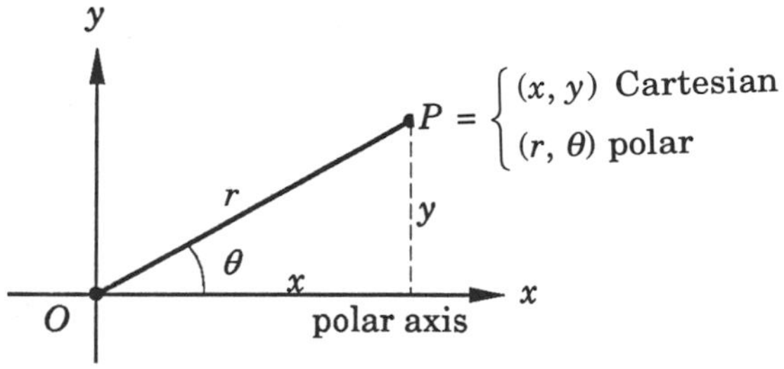

The graph enables us to see that the **conversions between orthogonal and polar coordinates** are

$$x = r \cos \theta$$

$$y = r \sin \theta$$

$$r = \sqrt{x^2 + y^2} \quad \text{(the Pythagorean theorem)}$$

$$\theta = \arctan \frac{y}{x}, \text{ if } P \text{ is in the first or fourth quadrant.}$$

Polar Equations and Two-Dimensional Graphs

An equation involving the variables of the radius vector (r) and the polar angle (θ) is referred to as a **polar equation**.

Straight Lines

We use the conversion formulas

$$x = r\cos\theta; \quad y = r\sin\theta$$

to convert the general form of an equation of a **straight line**,

$$ax + by + c = 0,$$

to its polar equivalent,

$$r(a\cos\theta + b\sin\theta) + c = 0.$$

Circles

The polar equation

$$r = a,$$

where a is a constant > 0, represents a **circle** with its center at the pole and radius a:

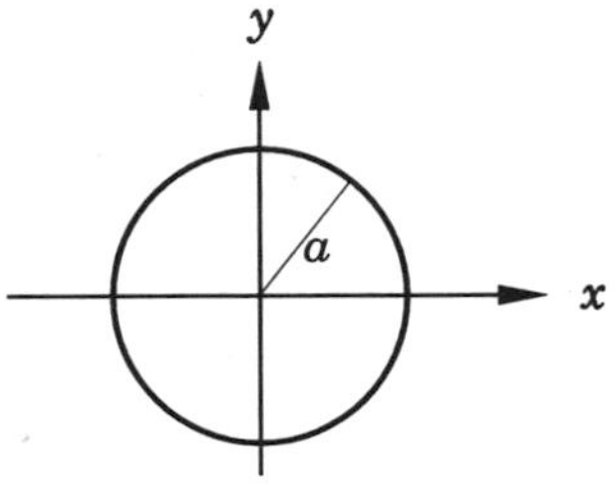

- Find the orthogonal equation that corresponds to the polar equation

$$r = 2\cos\theta.$$

Multiply both sides of the equation by r:

$$r^2 = 2r\cos\theta$$

Using the conversion formulas

$$x = r\cos\theta; \quad r = \sqrt{x^2 + y^2} \text{ or } r^2 = x^2 + y^2$$

we obtain

$$r^2 = 2x$$
$$x^2 + y^2 = 2x$$
$$x^2 - 2x + y^2 = 0.$$

Rewriting the equation,

$$x^2 - 2x + 1 + y^2 = 1,$$

and completing the square,

$$(x-1)^2 + (y-0)^2 = 1,$$

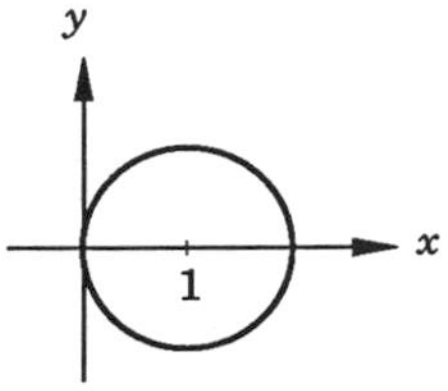

we see that the equation represents a circle centered at (1, 0) and with radius 1.

Notable Spirals in the Plane

A spiral is a curve that winds around a fixed center and gradually recedes from the center.

Contrary to popular assumption, a spider does not build its web in concentric circles but in the form of a spiral. An approximate logarithmic spiral is observed in some spiderwebs.

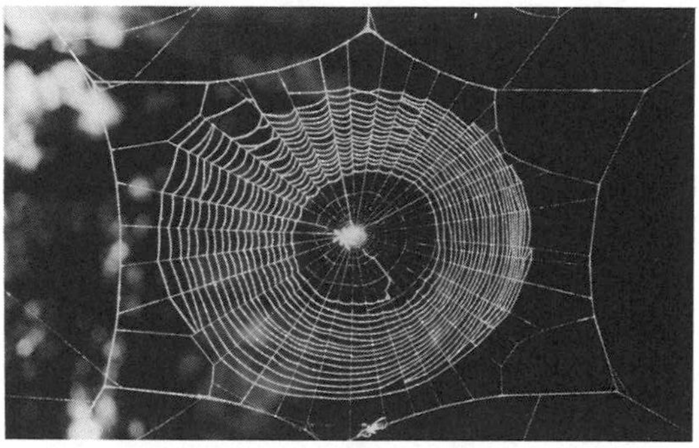

Courtesy Jonathan Coddington, Smithsonian Institution

Spiral of Archimedes

ARCHIMEDES
(c. 287 - 212 B.C.)

In the polar equation

$$r = a\,\theta,$$

where a is a proportionality constant, radius vector r increases with the polar angle θ, forming a curve discovered by and named after **Archimedes**:

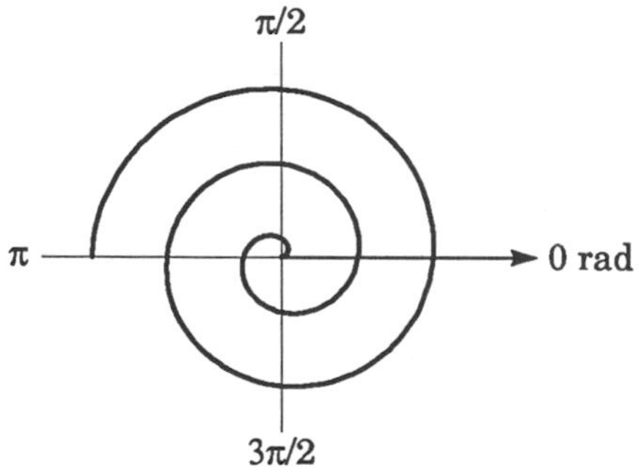

$r = a\,\theta;\ \theta > 0$

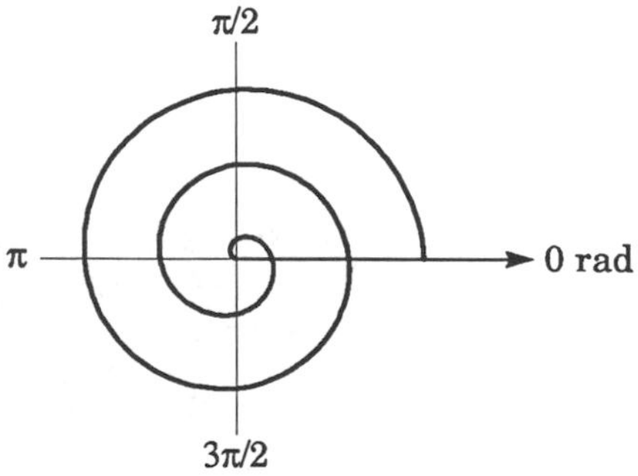

$r = a\,\theta;\ \theta < 0$

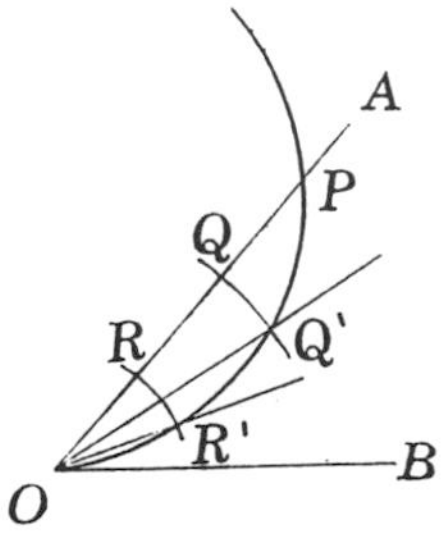

Since r is proportional to θ, the spiral can be used to divide an arbitrary angle into any number of equal parts. To trisect the angle AOB, which includes a portion OP of a spiral of Archimedes,

- trisect OP at R and Q,
- describe circles with O as center and OR and OQ as radii; the circles intersect the spiral at R' and Q',
- then OR' and OQ' trisect the angle.

Hyperbolic Spiral

In the polar equation

$$r\,\theta = a\,,$$

where a is a constant of proportionality, the radius vector r varies inversely with the polar angle θ, so that the curve forms a **hyperbolic spiral**. The spiral is asymptotic to a straight line that is parallel to the polar axis and located at the distance a above the axis:

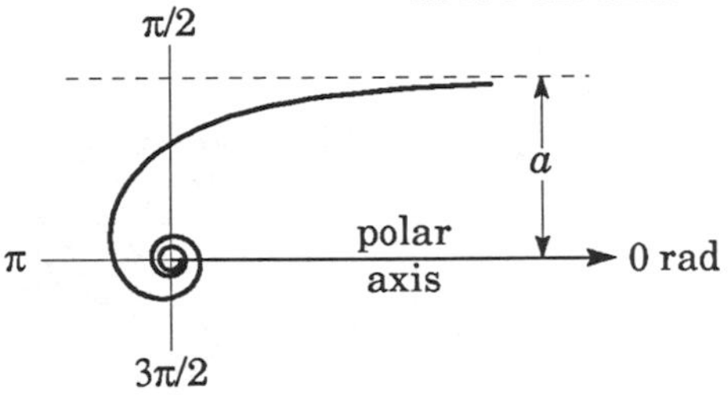

$r\,\theta = a\,;\ \theta > 0$

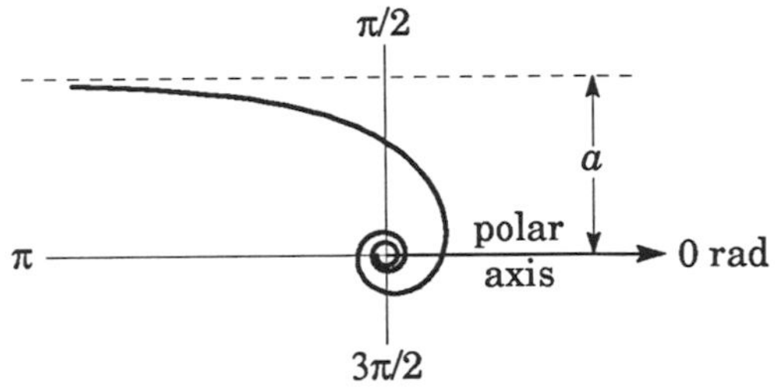

$r\,\theta = a\,;\ \theta < 0$

VARIGNON Pierre
(1654 - 1722)

The hyperbolic spiral was first conceived in 1704 by the French mathematician **Pierre Varignon**.

Logarithmic Spiral

A polar equation written either in logarithmic form,

$$\log_b r = a\,\theta,$$

where a is a constant of proportionality > 0, or in exponential form,

$$r = b^{a\theta},$$

represents a curve that is referred to as a **logarithmic spiral** or an **equiangular spiral**:

Neg. No. 2A12055 Photo: A Singer
Courtesy Department of Library Services,
American Museum of Natural History

The **chambered nautilus** does not grow in all directions but only adds on to its open end, thereby maintaining its logarithmic spiral.

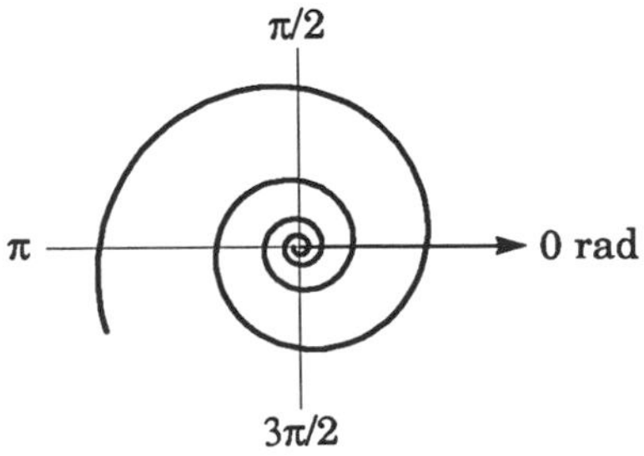

$\log_b r = a\,\theta \qquad r = b^{a\theta}$

A notable property of the logarithmic spiral is that it intersects its radii everywhere at the same angle – thus the name *equiangular spiral*.

DESCARTES René (1596 - 1650)

BERNOULLI Jakob (1654 - 1705)

The logarithmic spiral was first discussed by **Descartes** in 1638; in 1698 its properties were further studied by the Swiss mathematician **Jakob Bernoulli**, who was particularly intrigued by its *self-similarity*: if any portion is magnified or reduced, it is identical to any other portion of the curve.

The Lituus

Latin, *lituus*, "a crooked staff"

The lituus (plural: litui) is asymptotic to the polar axis, and winds around and gets increasingly closer to the pole but never reaches it.

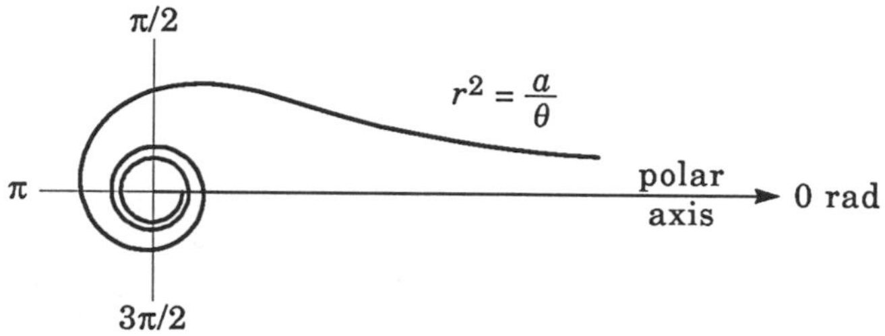

COTES Roger (1682 - 1716)

The lituus originated with the English mathematician **Roger Cotes**; it was published posthumously in a collection of Cotes's mathematical papers, *Harmonia mensurarum* (1722).

Parabolic Spiral (Fermat's Spiral)

de FERMAT Pierre (1601 - 1665)

In the parabolic spiral the square of the radius vector is proportional to the vector angle. First discussed by the French mathematician **Pierre de Fermat** in 1636, this spiral is also referred to as Fermat's spiral.

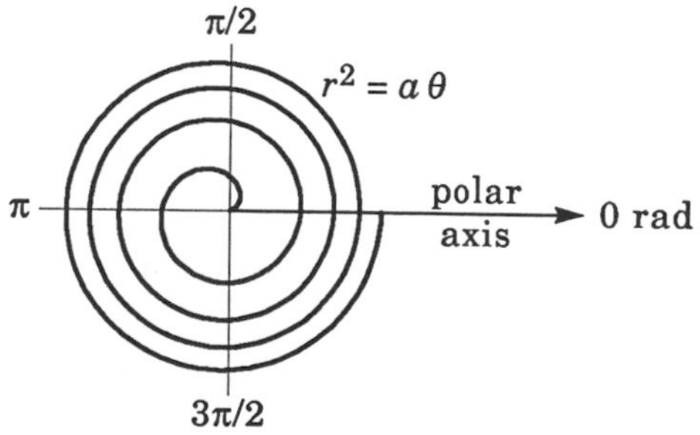

The Lemniscate

Latin, *lemniscatus*, "with hanging ribbons"; *lemniscus*, "ribbon"

The lemniscate, or lemniscate of Bernoulli, was conceived by **Jakob Bernoulli** in an article on tides (1694).

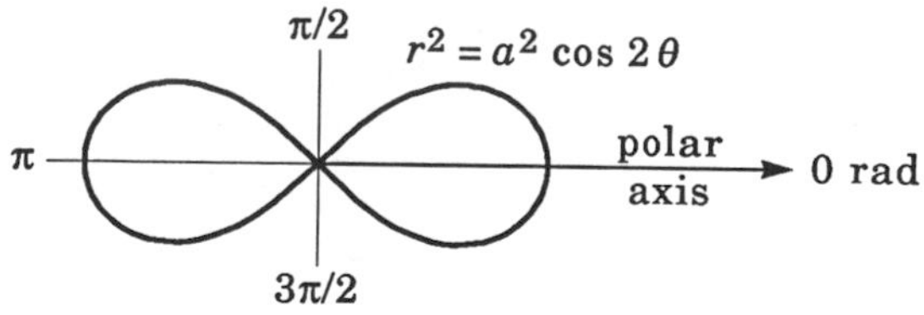

Polar form: $r^2 = a^2 \cos 2\,\theta$ Cartesian form: $(x^2 + y^2)^2 = a^2\,(x^2 - y^2)$

Polar Coordinates and Space

Cylindrical Coordinates

A system of **cylindrical coordinates** combines the use of polar coordinates in the plane with the z-coordinate of three-dimensional orthogonal coordinates.

The location of a point $P(r, \theta, z)$ is its projection (r, θ) onto the x, y-plane and its distance z from the x, y-plane:

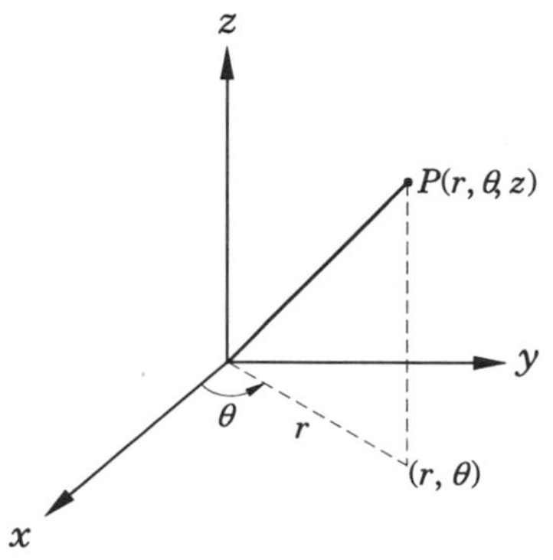

Cylindrical Coordinates and Their Graphs

Axis of Symmetry

Cylindrical coordinates are useful primarily to reproduce figures that have an **axis of symmetry**, which may be conveniently placed at the z-axis of the graph.

- If θ and z vary with r constant, a **right circular cylinder** develops.
- If r and z vary with θ constant, a **half-plane** with one side at the z-axis develops.
- If z is constant and r and θ vary, a **half-plane** with one side at the y-axis develops.

In the graphs below, c denotes a constant:

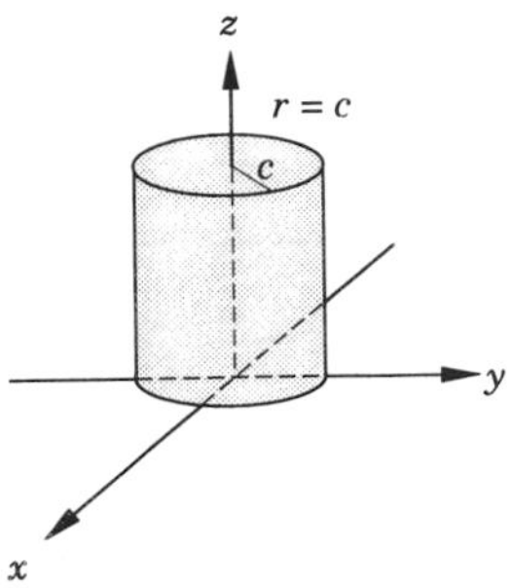

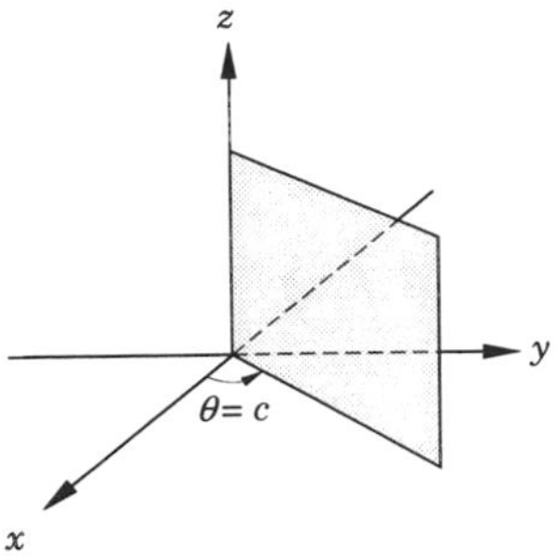

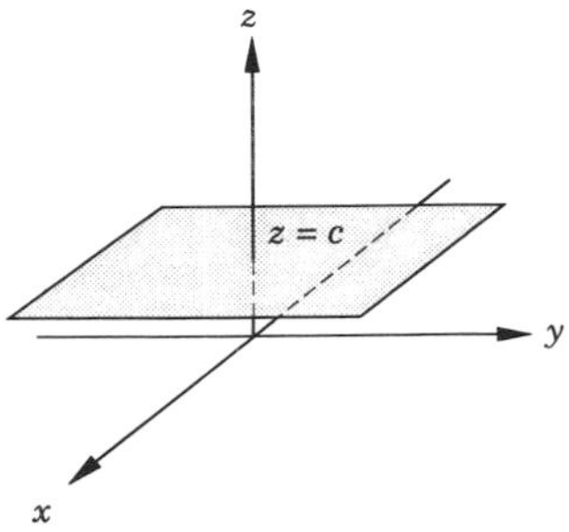

Conversions between Orthogonal and Cylindrical Coordinates

The equations connecting orthogonal and polar coordinates in the plane also apply in space. The z-coordinates of the orthogonal coordinate system and the cylindrical coordinate system coincide. We obtain the equations:

$$x = r\cos\theta$$

$$y = r\sin\theta$$

$$z = z$$

$$r = \sqrt{x^2 + y^2}$$

$$\theta = \arctan\frac{y}{x}, \text{ if } (x, y) \text{ is in the first or fourth quadrant.}$$

Spherical Coordinates

In a system of **spherical coordinates** a point $P(\rho, \theta, \phi)$ is located by the distance of P from a fixed origin or pole O (distance ρ) and two angles, denoted θ and ϕ. Angle θ is the same as in polar coordinates in the plane; ϕ is the angle between the positive z-axis and the line segment OP:

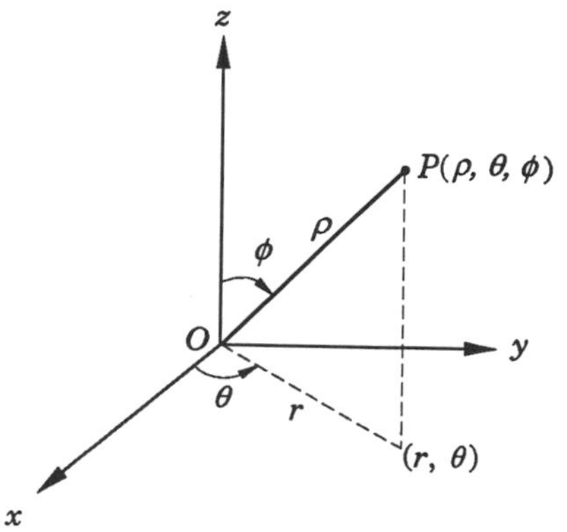

Spherical Coordinates and Their Graphs

Center of Symmetry

Spherical coordinates are useful primarily to reproduce figures that have a **center of symmetry**, which may be conveniently placed at the pole.

- If θ and ϕ vary with ρ constant, a **sphere** develops.
- If ρ and ϕ vary with θ constant, a **half-plane** with one side at the z-axis develops.
- If ρ and θ vary with ϕ constant, a **nappe of a right circular cone** develops.

In the graphs below, c denotes a constant:

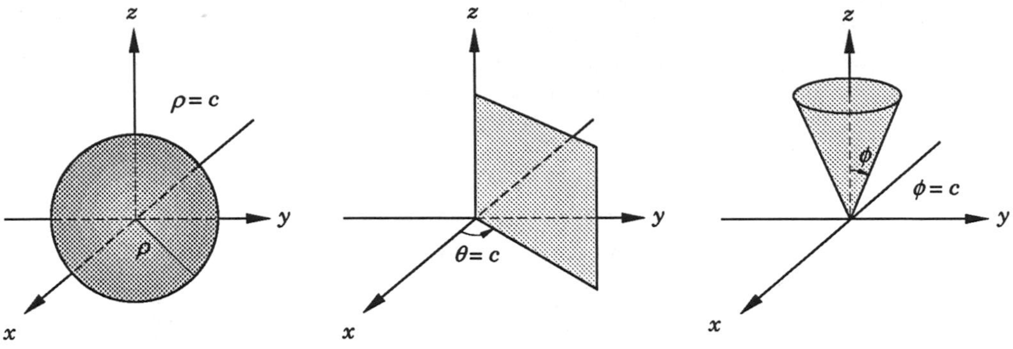

Conversions between Spherical and Orthogonal Coordinates

Formulas Transforming Spherical Coordinates into Orthogonal Coordinates

To transform spherical coordinates into orthogonal coordinates, we employ the formulas

$$x = \rho \sin \phi \cos \theta$$

$$y = \rho \sin \phi \sin \theta$$

$$z = \rho \cos \phi$$

Why?

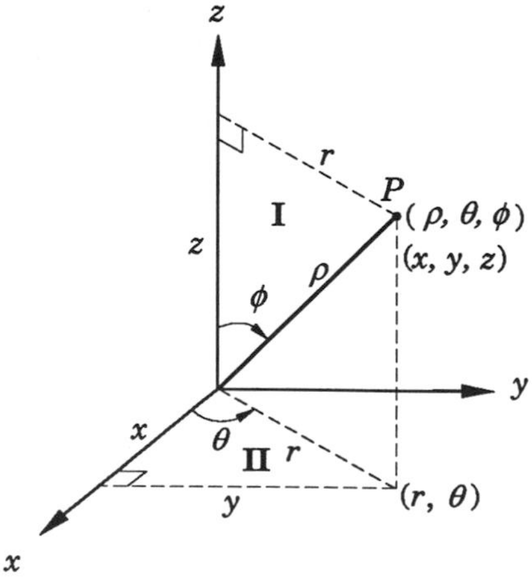

Solving the triangle marked **I**, we find

$$\sin \phi = \frac{r}{\rho}$$

(1) $$r = \rho \sin \phi$$

and that

$$\cos \phi = \frac{z}{\rho}$$

$$z = \rho \cos \phi .$$

Using Equation (1) when solving the triangle marked **II**, we also find

$$\cos\theta = \frac{x}{\rho \sin\phi}$$

$$x = \rho \sin\phi \cos\theta$$

and, finally,

$$\sin\theta = \frac{y}{\rho \sin\phi}$$

$$y = \rho \sin\phi \sin\theta .$$

Formulas Transforming Orthogonal Coordinates into Spherical Coordinates

To transform orthogonal coordinates into spherical coordinates, we employ the formulas

$\rho = \sqrt{x^2 + y^2 + z^2}$

$\theta = \arctan \dfrac{y}{x}$, if (x, y) is in the first or fourth quadrant

$\phi = \arccos \dfrac{z}{\sqrt{x^2 + y^2 + z^2}}$

Why?

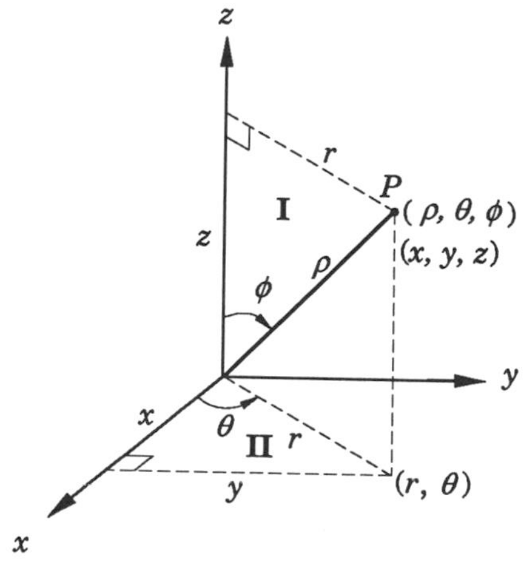

Using the formula for distance in space and referring to the above graph, we find

$$\rho = \sqrt{x^2 + y^2 + z^2} ,$$

and from the triangle marked **II**,

$$\tan\theta = \frac{y}{x} ;$$

from the triangle marked **I**,

$$\cos\phi = \frac{z}{\sqrt{x^2 + y^2 + z^2}} ; \quad \phi = \arccos \frac{z}{\sqrt{x^2 + y^2 + z^2}} .$$

15.5 Parametric Equations

Instead of representing a two-dimensional graph by *one* equation in *two* variables, it may be represented by *two* equations, each of which gives the coordinates of x and y in terms of a *third variable*, called a **parameter** and usually denoted t.

Equations with parameters are called **parametric equations**; they often facilitate the handling of complex graphs.

Consider the equation

$$y^2 = 4x.$$

If we let $y = 2t$, then we have the parametric equations

$$x = t^2 \text{ and } y = 2t,$$

representing the parabola.

Similarly, a three-dimensional graph may be represented by *three* parametric equations, each of which gives the coordinates of x, y, and z in terms of the parameter.

There is no general method of transforming an equation representing an orthogonal or polar graph into parametric form; such transformation often takes considerable persistence and ingenuity.

The Circle

The parametric representation of a circle with radius r and centered at the origin is

$$\left.\begin{aligned} x &= r\cos t \\ y &= r\sin t \end{aligned}\right\}$$

Why? Consider the equation $x^2 + y^2 = r^2$, that is, a circle with radius r and its center at the origin of an orthogonal coordinate system:

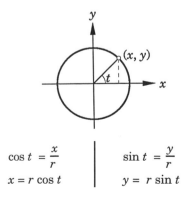

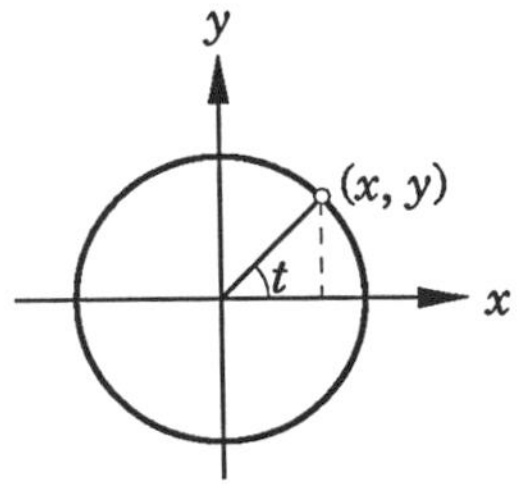

$\cos t = \dfrac{x}{r}$	$\sin t = \dfrac{y}{r}$
$x = r\cos t$	$y = r\sin t$

The procedure for finding the parametric representation of a circle centered outside the origin of an orthogonal coordinate system follows the same course of reasoning as for a circle centered at the origin.

- Express x and y of

$$(x-4)^2 + (y-3)^2 = 4$$

as parametric equations in t, where t is the central angle of the circle.

The equation represents a circle whose radius is $\sqrt{4} = 2$ and whose center in the orthogonal coordinate system is at (4, 3):

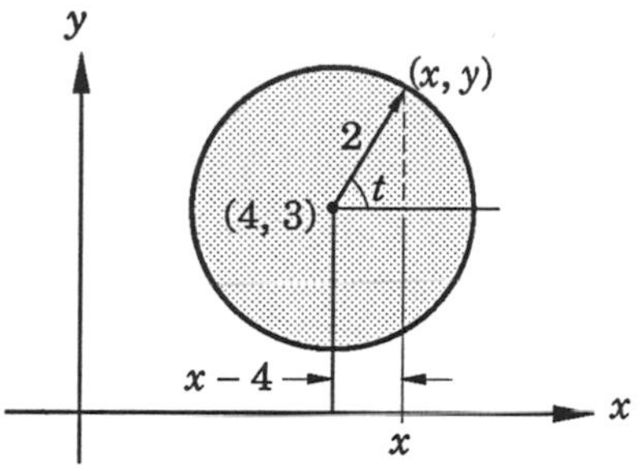

$\cos t = \dfrac{x-4}{2}$	$\sin t = \dfrac{y-3}{2}$
$x = 4 + 2\cos t$	$y = 3 + 2\sin t$

The sought parametric equations are

$$\left.\begin{aligned} x &= 4 + 2\cos t \\ y &= 3 + 2\sin t \end{aligned}\right\}$$

The Ellipse

The parametric representation of an ellipse centered at the origin is

$$\left.\begin{aligned} x &= a\cos t \\ y &= b\sin t \end{aligned}\right\}$$

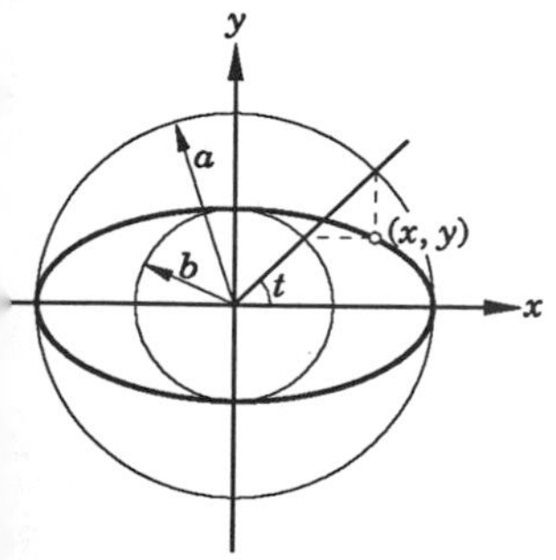

where $2a$ and $2b$ are axes of the ellipse, such that if $a > b$, the foci of the ellipse are on the x-axis, and if $b > a$, the foci are on the y-axis; if $a = b$, then the curve is a circle.

To find a parametric representation of an ellipse, use auxiliary concentric circles and let an ellipse whose major axis is $2a$ and minor axis is $2b$ be placed with its center at the origin of an orthogonal coordinate system, and let t be the angle that an arbitrary ray from the origin makes with the x-axis; then

$$\left.\begin{aligned} x &= a\cos t \\ y &= b\sin t \end{aligned}\right\}$$

The Hyperbola

A hyperbola centered at the origin and with the transverse axis $2a$ (coinciding with the x-axis of an orthogonal coordinate system) and conjugate axis $2b$ has the parametric equation

$$\left.\begin{aligned} x &= a \sec t \\ y &= b \tan t \end{aligned}\right\}$$

To deduce this system of equations, consider two concentric circles centered at the origin and with diameters $2a$ and $2b$, equal to the transverse axis and the conjugate axis of the hyperbola, respectively:

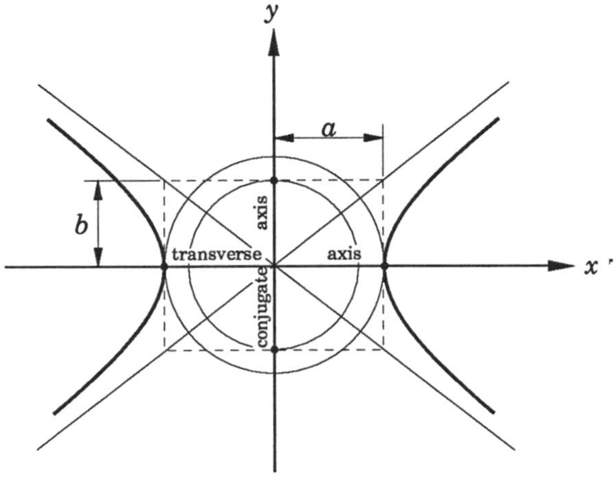

- The x-axis intersects the inner circle at a point K.
- Let the x-axis and an arbitrary ray from the origin form an angle t.
- A tangent to the inner circle at K intersects the arbitrary ray at L; the ray intersects the outer circle at M.
- The tangent to the outer circle at M intersects the x-axis at N.
- Lines parallel to the x-axis and y-axis and passing through L and N, respectively, intersect at a point $P(x, y)$ on the hyperbola.

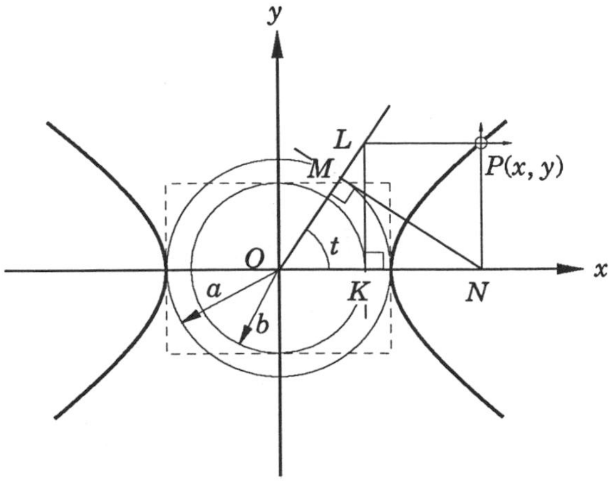

From the right triangle *OMN*, we find

$$\sec t = \frac{x}{a}$$

$$x = a \sec t\,.$$

Similarly, from the right triangle *OLK*, we find

$$\tan t = \frac{y}{b}$$

$$y = b \tan t\,.$$

The Parabola

p. 586

In the introduction to this section, we found that a parabola whose orthogonal equation is $y^2 = 4x$ may be represented by the parametric equations

$$\left.\begin{aligned} x &= t^2 \\ y &= 2t \end{aligned}\right\}$$

which represent a parabola whose y-axis is a tangent at the conjoining points of the origin of the orthogonal coordinate system and the vertex of the parabola:

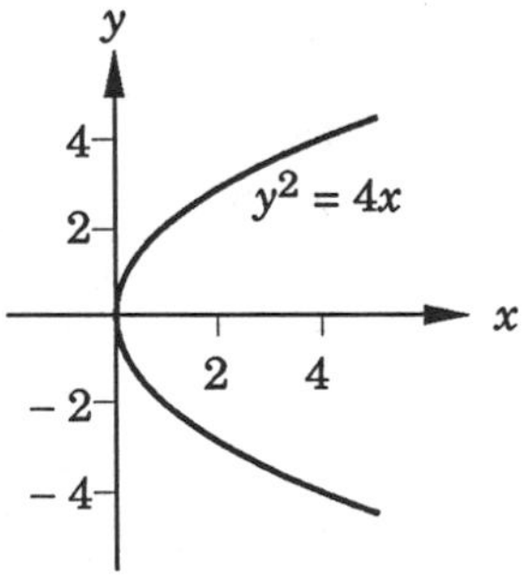

Path of a Projectile

Verification: *pp.* 901 - 902

Different forms of parametric equations may be derived to suit different types of problems. An example is the parametric equations of the parabola, used to determine the **path of a projectile**, here quoted without proof; thus,

$$\left.\begin{aligned} x &= v_o t \cos\theta \\ y &= v_o t \sin\theta - \frac{1}{2} g t^2 \end{aligned}\right\}$$

where v_o is the initial velocity of the projectile, θ the initial angle between the projectile and the horizontal plane, g the acceleration of gravity, and t the time elapsed.

Cycloids

A cycloid is the curve generated by a point on the circumference of a circle which rolls on a straight line in its plane. At a completed revolution of the circle and the beginning of the following cycle a double point – a **cusp** – is formed:

Cusp
Latin, *cuspis*, "a point"

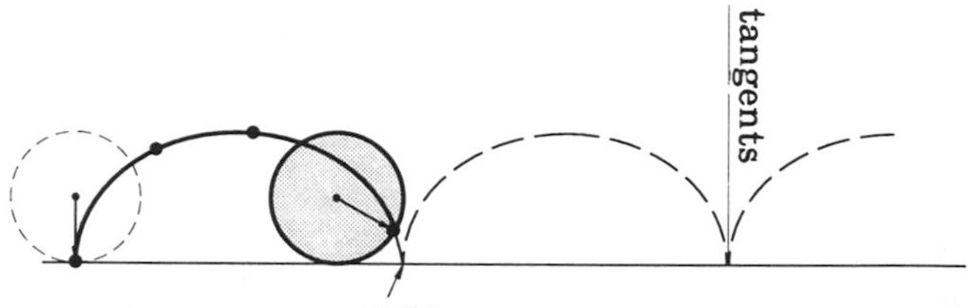

Properties of the Cycloid

The **length of the cycloid baseline** – the distance between successive cusps – is equal to the circumference of the generating circle.

The **length of a cycloid** between successive cusps is four times the diameter of the generating circle. Alternatively, the length of one complete arc of a cycloid may also be determined by multiplying the corresponding distance along the baseline by 4 and dividing the product by π. For instance, if the baseline distance between two cusps is l, then the length of the arc is $4l/\pi$.

WREN Christopher
(1632 - 1723)

In 1658 the English architect and mathematician **Christopher Wren** proved that the length of one complete arc of a cycloid equals the perimeter of a square circumscribed about the generating circle. This agrees with the results above.

GALILEI Galileo
(In English literature referred to by first name, "Galileo"; 1564 - 1642)

ROBERVAL Gilles Personne de
(1602 - 1675)

By the use of models and comparing the weights of the circle and the cycloid it generates, **Galileo** demonstrated in 1599 that one complete cycloidal arch is about three times the generating circle in area; in 1634 the French mathematician **Roberval** proved that the area of one complete cycloidal arch is indeed three times that of the generating circle.

The Path of Most Rapid Descent

BERNOULLI Johann
(1667 - 1748)

In 1696 the Swiss mathematician **Johann Bernoulli** challenged fellow mathematicians to find the curve along which a particle will slide – under the influence of gravity – in the shortest time between two points at different altitudes but not on the same vertical line, a problem referred to as the **brachistochrone problem**.

Greek, *brachys*, "short"; *chronos*, "time"

Johann Bernoulli had found the solution; among those who picked up the gauntlet and solved the brachistochrone problem were Newton, Leibniz, L'Hospital, and Johann's brother Jakob Bernoulli.

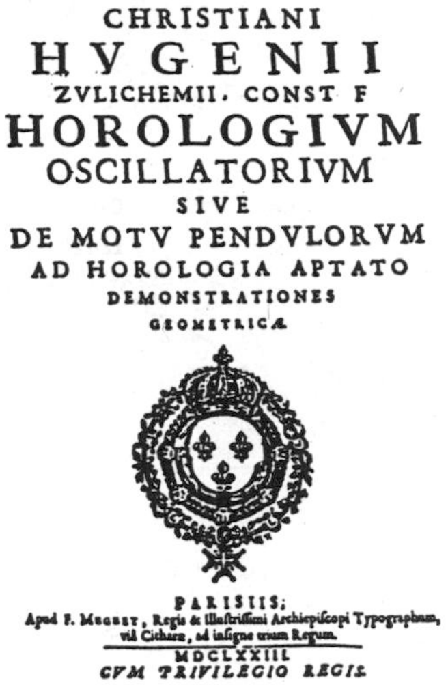
CHRISTIANI
HVGENII
ZVLICHEMII. CONST. F.
HOROLOGIVM
OSCILLATORIVM
SIVE
DE MOTV PENDVLORVM
AD HOROLOGIA APTATO
DEMONSTRATIONES
GEOMETRICÆ.

PARISIIS;
Apud F. Muguet, Regis & Illustrissimi Archiepiscopi Typographum, viâ Citharæ, ad insigne trium Regum.
MDCLXXIII.
CVM PRIVILEGIO REGIS.

HUYGENS Christiaan
(1629 - 1695)

Greek, *isos*, "equal"; *chronos*, "time"

One might feel inclined to guess that the shortest distance would also be the quickest, but the answer is that the quickest path is along a cycloid.

Path along a straight line

Path along a cycloid

In *Horologium Oscillatorium* (1673) – a work of fundamental importance for the development of dynamics – the Dutch mathematician, astronomer, and physicist **Christiaan Huygens** described the **isochronous property** of the cycloid, that is, the property that wherever a particle is placed on the concave side of a cycloid, the time for sliding down to the lowest point of the cycloid is – disregarding friction – always the same:

Pendulum Property

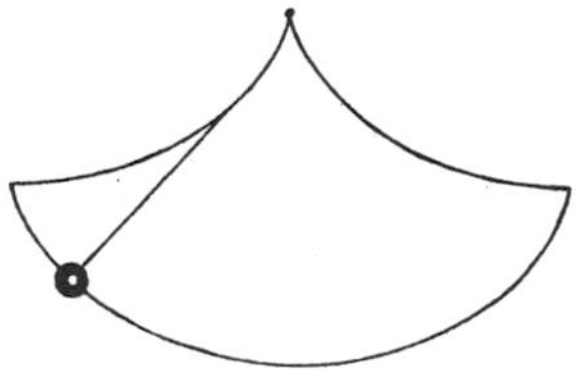

Another famed quality of the cycloid is its **pendulum property**: If a pendulum is hung at the cusp of an inverted cycloid and made to wrap around the arcs of the cycloid as it swings, the end of the pendulum describes another cycloid and the period of oscillation of the pendulum is independent of the amplitude; as shown by Huygens, the period of the pendulum becomes dependent only on its length and its shape.

Parametric Equations of the Cycloid

A cycloid whose generating circle has the radius a has the parametric equations

$$\left.\begin{aligned} x &= at - a\sin t \\ y &= a - a\cos t \end{aligned}\right\}$$

To deduce the equations, let the circle roll on the x-axis of an orthogonal coordinate system:

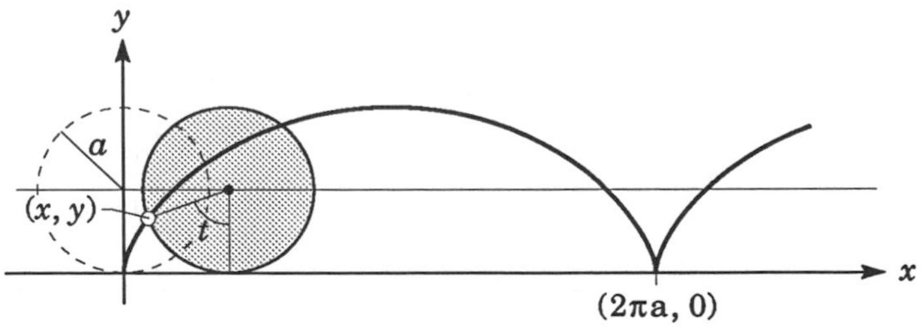

Use the **angle of rotation** (t) as a parameter to express the curve of the cycloid:

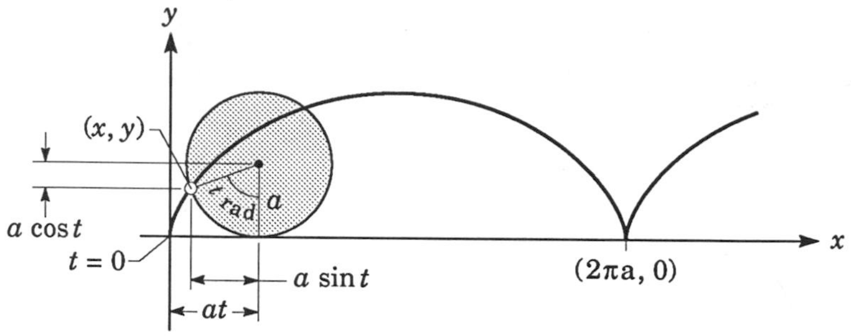

By examining the above graph, we find

$$\left.\begin{aligned} x &= at - a\sin t \\ y &= a - a\cos t \end{aligned}\right\}$$

Epicycloids

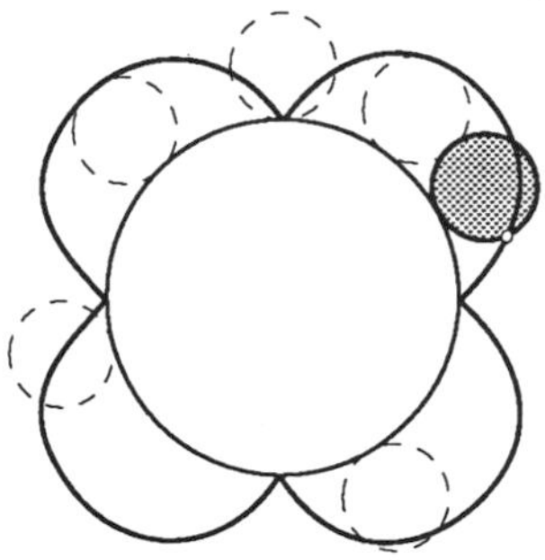

An epicycloid is a curve generated by a point on the circumference of a circle (the **epicycle**) which rolls on the *outside* of a fixed circle (the **deferent**). More complicated curves can be generated by having a 3rd circle roll on the 2nd, a 4th on the 3rd, *etc*. Epicycloids were introduced by the ancient Greeks and, until the 16th century, formed the basis for ideas about the paths of the Moon and planets. They still have significance in engineering and present interesting mathematical properties.

Parametric Equations of the Epicycloid

An epicycloid whose generating fixed and rolling circles have the radii a and b, respectively, has the parametric equations, here quoted without proof,

$$\left.\begin{aligned} x &= (a+b)\cos t - b\cos\left(\frac{a+b}{b}\,t\right) \\ y &= (a+b)\sin t - b\sin\left(\frac{a+b}{b}\,t\right) \end{aligned}\right\}$$

where a is the radius of the fixed circle and b the radius of the circle rolling on the fixed circle. The parameter t is the angle formed by the x-axis and the ray from the center of the fixed circle to the point of contact with the rolling circle:

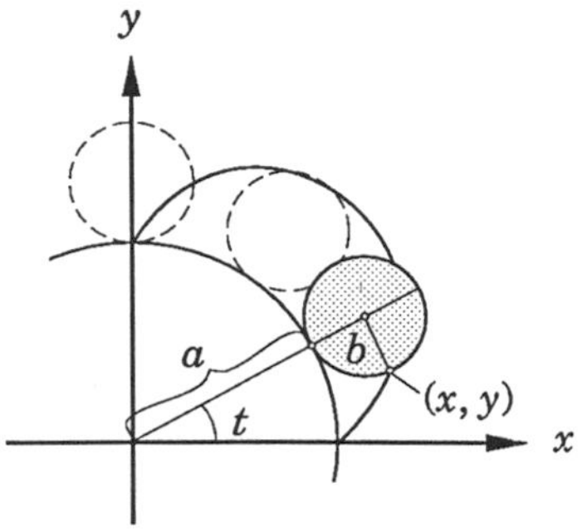

The epicycloid has a cusp at every point where it meets the fixed circle.

The epicycloid describes one arc when the radius of the fixed circle and the radius of the rolling circle are equal ($a = b$), two arcs when $a = 2b$, and n arcs when $a = nb$.

Hypocycloids

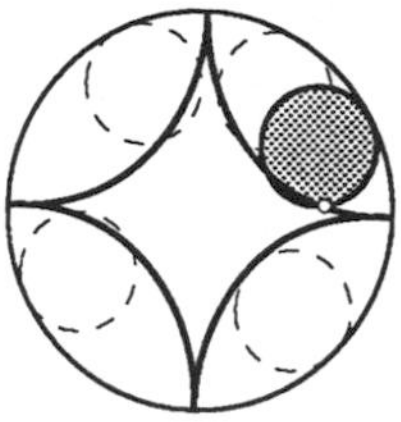

A hypocycloid is a curve generated by a point on the circumference of a circle which rolls on the *inside* of a fixed circle.

Parametric Equations of the Hypocycloid

A hypocycloid whose generating fixed and rolling circles have the radii a and b, respectively, has the parametric equations, here quoted without proof,

$$\left.\begin{aligned} x &= (a-b)\cos t + b\cos\left(\frac{a-b}{b}\,t\right) \\ y &= (a-b)\sin t - b\sin\left(\frac{a-b}{b}\,t\right) \end{aligned}\right\}$$

The parameter t is the angle formed by the x-axis and the ray from the center of the fixed circle to the point of contact with the rolling circle:

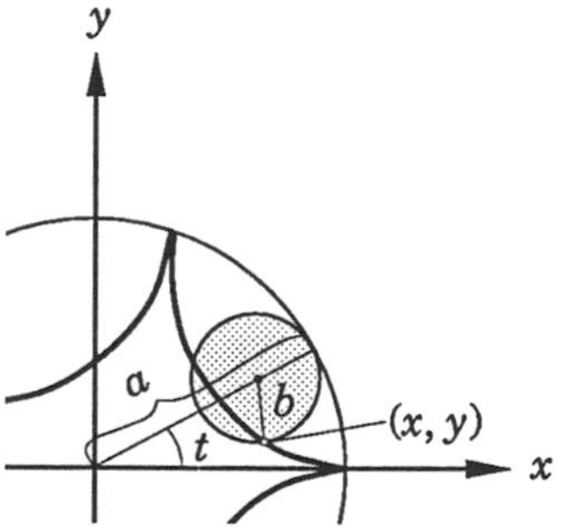

Like the epicycloid, the hypocycloid has a cusp at every point where it meets the fixed circle; it describes two arcs when the radius of the fixed circle is twice the length of the radius of the rolling circle ($a = 2b$), and n arcs when $a = nb$.

* * *

Let the bottom coin roll half way around the fixed upper one.

– Which way will George Washington's head point?

Answer: Upward (as at the starting point).

The Astroid

A hypocycloid whose rolling circle has a diameter one-fourth of that of the fixed circle is called an **astroid**.

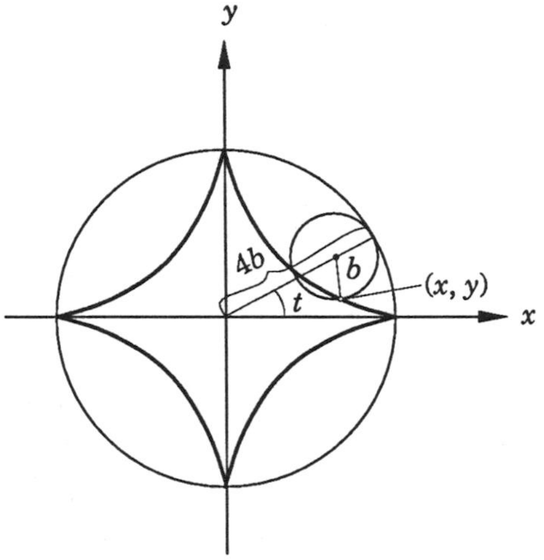

By applying the parametric equations of the hypocycloid to the astroid and then simplifying the equations, we come to the result that the parametric equations of the astroid are

$$\left.\begin{aligned} x &= a\cos^3 t \\ y &= a\sin^3 t \end{aligned}\right\}$$

where a is the radius of the fixed circle.

- Using the general parametric equations of the hypocycloid, derive the formula for the astroid; a and b are the radii of the fixed and rolling circles, respectively.

$$\left.\begin{aligned} x &= (4b-b)\cos t + b\cos\left(\frac{4b-b}{b}\,t\right) = b\,(3\cos t + \cos 3t) \\ y &= (4b-b)\sin t - b\sin\left(\frac{4b-b}{b}\,t\right) = b\,(3\sin t - \sin 3t) \end{aligned}\right\}.$$

As

p. 511 $\cos 3t = 4\cos^3 t - 3\cos t;\quad \sin 3t = 3\sin t - 4\sin^3 t\,,$

we find

$$\left.\begin{aligned} x &= b\,(3\cos t + \cos 3t) = 4b\cos^3 t \\ y &= b\,(3\sin t - \sin 3t) = 4b\sin^3 t \end{aligned}\right\},$$

and since $a = 4b$,

$$\left.\begin{aligned} x &= a\cos^3 t \\ y &= a\sin^3 t \end{aligned}\right\}.$$

Notable Three-Dimensional Spirals

The Cylindrical Helix

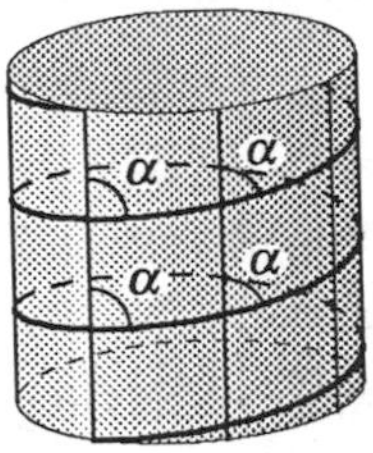

Greek, *helix*, "a spiral", "a twist"

The **cylindrical helix** – often simply called a **helix** – is a three-dimensional curve formed as if lying on a right circular cylinder, where it cuts the generators of the surface at a constant angle α.

APOLLONIOS
(*c*. 255 - 170 B.C.) *p*. 408

The helix is mentioned in a text by Geminus dating from the 1st century B.C., but a passage in the text suggests that the helix was already known to the Greek mathematician **Apollonius**, author of *Conics*.

Familiar structures of helix-like form are the threads of bolts and lightbulbs, climbing plants, the human umbilical chord, and the the famed double helix of DNA.

The **parametric representation** of the cylindrical helix is

$$\left.\begin{aligned} x &= a \sin t \\ y &= a \cos t \\ z &= bt \end{aligned}\right\}$$

p. 586

where a is the radius of the cylinder, b is a constant, and t is the parameter, obtained as for the circle.

As t increases from 0 to 2π, the height z steadily increases from 0 to $2\pi b$, completing one full turn of the helix.

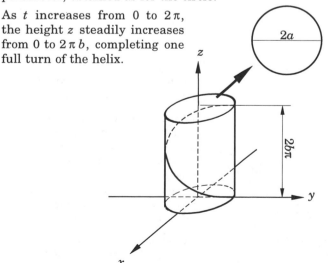

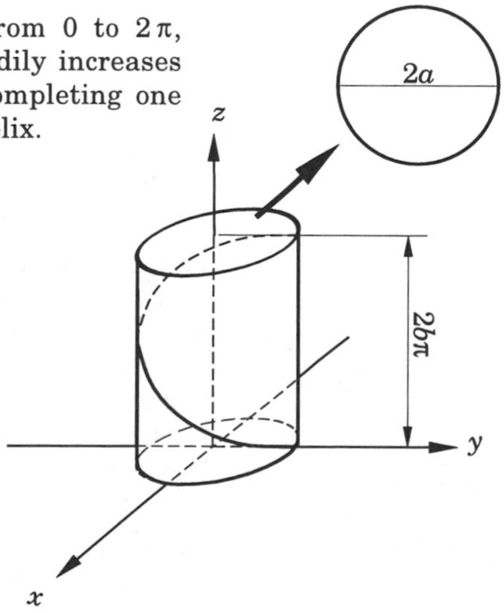

The Conical Helix

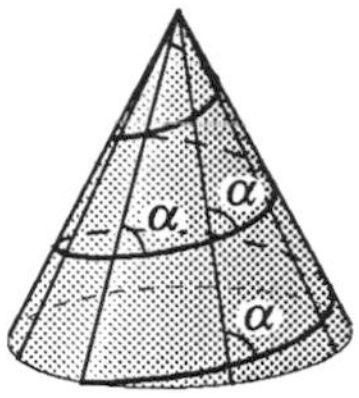

The **conical helix** is a three-dimensional curve formed as if lying on a right circular cone, where it cuts the generators of the surface at a constant angle α.

The **parametric representation** of the conical helix is here quoted without proof:

$$\left.\begin{aligned} x &= b\,e^t \sin t \\ y &= b\,e^t \cos t \\ z &= e^t \end{aligned}\right\}$$

where b is a constant and t is the parameter.

The Spherical Helix

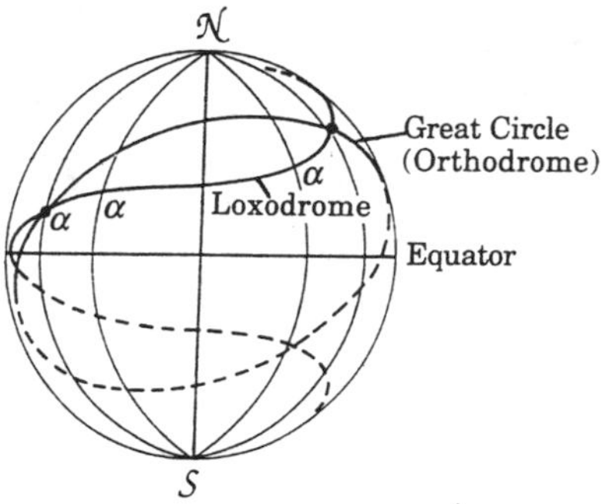

Greek, *loxos*, "oblique"; *dromos*, "course"

The spherical helix – better known as a **loxodromic spiral**, or just **loxodrome**, or a **rhumb line** – cuts the meridians of a sphere at a constant angle α, not equal to a right angle. The loxodrome coils around the poles of the Earth, but never reaches the poles.

The shortest distance of traveling between two points on a sphere is along the arc of a great circle (the *orthodrome*) and to stay the course, one must every instant adjust to the direction shown by the compass. Along the *loxodrome*, the path to the destination is longer, but it is always directed to the same point of the compass.

NUÑEZ SALACIENCE Pedro (In English often referred to as Peter Nunes; 1502 - 1578)

The loxodrome was invented by the Portuguese mathematician and geographer **Pedro Nuñez** after observations reported to him in 1533 by Admiral Martim Alfonso de Sousa.

MERCATOR Geradus (1512 - 1594)

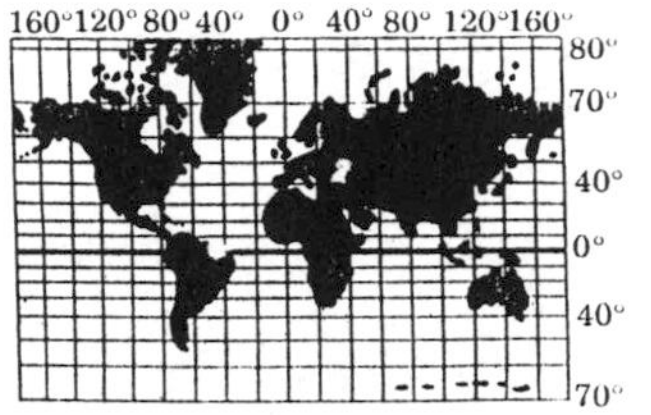

Mercator projection

In 1569 the Flemish cartographer **Geradus Mercator** designed a world map with a cylindrical projection such that meridians and parallels were straight lines, intersecting at right angles, and whose distance from each other increased with their distance from the equator. The **Mercator projection** distorts the image (especially at high latitudes) but has the great advantage of showing loxodromes as straight lines; such maps are still important tools for navigation at sea and in the air.

The loxodrome can be written in terms of the longitude and latitude of a point on the curve and its angle with the meridians; there is no advantage in using parametric equations.

Eliminating the Parameter

A real bear ... eliminating the parameter.

To eliminate a parameter from a parametric equation and thereby obtain an orthogonal or polar equation is often at least as complex as forming a parametric equation.

- Find the corresponding orthogonal equation of the parametric equations

$$\left.\begin{aligned} x &= 7 + 3\cos t \\ y &= 4 + 3\sin t \end{aligned}\right\}.$$

$$\cos t = \frac{x-7}{3}\ ; \quad \sin t = \frac{y-4}{3}$$

Pythagorean Identities: *p.* 508

$$\left(\frac{x-7}{3}\right)^2 + \left(\frac{y-4}{3}\right)^2 = 1$$

$$(x-7)^2 + (y-4)^2 = 9$$

which represents a circle with radius 3 and center at point (7, 4) in an orthogonal coordinate system.

- Transform the parametric equations

$$\left.\begin{aligned} x &= \sqrt{5}\,\sin t \\ y &= 3\cos t \\ z &= t \end{aligned}\right\}$$

into a general equation, relating x and y.

We note that $\frac{x}{\sqrt{5}} = \sin t$ and $\frac{y}{3} = \cos t$, so that

Pythagorean Identities: *p.* 508

$$\frac{x^2}{5} + \frac{y^2}{9} = 1,$$

which is the equation of a **right elliptical cylinder**:

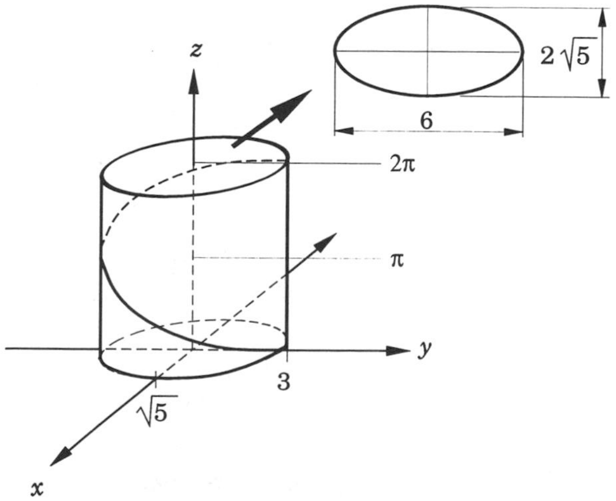

- **Find the corresponding general equation of a parabola represented by the parametric equations**

$$\left.\begin{aligned} &(1) \qquad x = t+3 \\ &(2) \qquad y = \frac{t^2}{4} \end{aligned}\right\}$$

Solve (1) for t:

$$t = x-3\,.$$

Substitute $(x-3)$ for t in (2):

$$y = \frac{(x-3)^2}{4}$$

$$(3) \quad (x-3)^2 = 4y\,,$$

which is the sought equation, representing a parabola with the axis parallel to the y-axis, vertex at (3, 0), and focus at (3, 1).

* * *

– *I say, Doctor, what about making the cut* $y = 2x + 7;\ -5 \le x \le 1$ *?*

Chapter 16

VECTOR ANALYSIS

*** Indicates cross-reference

16.0 Scope and History

Vectors are directed line segments that have both magnitude and direction. They are used to describe physical quantities like velocity and force, and are usually symbolized by arrows.

STEVIN Simon
(1548 - 1620)

The method of calculating the combined action of two or more mechanical forces by adding them according to the parallelogram law had been known ever since the time of Aristotle when, in 1586, the Flemish mathematician **Simon Stevin** analyzed the principle of geometric addition of forces in his treatise *De Beghinselen der Weeghconst* ("Principles of the Art of Weighing"), which caused a major breakthrough in the development of mechanics. But it would take another 200 years for the general concept of vectors to form.

DE
BEGHINSELEN
DER WEEGHCONST
BESCHREVEN DVER
SIMON STEVIN
van Brugghe.

Tot Leyden,
Inde Druckerye van Christoffel Plantijn,
By François van Raphelinghen.
cIↃ. IↃ. LXXXVI.

A chain around a triangular support, as shown here, was at one time thought to be, through the influence of gravitation, capable of perpetual motion. Stevin showed that the device does not work as a perpetual motion machine.

In translation, the inscription reads: "To wonder is no wonder."

WESSEL Caspar
(1745 - 1818)
ARGAND Jean Robert
(1768 - 1822)
WARREN John
(1796 - 1852)
HAMILTON William Rowan
(1788 - 1856)
GAUSS Carl Friedrich
(1777 - 1855)

The germinal ideas of modern vector theory date from around 1800 when **Caspar Wessel** and **Jean Robert Argand** described how a complex number $a + b\,\mathrm{i}$ could be given a geometric interpretation in a coordinate plane. Based on the work of Argand, the English mathematician **John Warren** in 1828 published *A Treatise on the Geometrical Representation of the Square Roots of Negative Numbers*, which inspired the Irish physicist-astronomer-mathematician **W. R. Hamilton** to show, in 1837, that complex numbers can be regarded as ordered pairs of real numbers. Unknown to Hamilton, German mathematician **Carl Friedrich Gauss** had made the same discovery six years earlier, in 1831.

Quaternion

Continuing his investigations, Hamilton found that a hypercomplex number required an ordered set of four real numbers for its representation; $a + b\,\mathrm{i} + c\,\mathrm{j} + d\,\mathrm{k}$, which Hamilton called a **quaternion**, would do if one sacrificed the commutative law of multiplication of traditional algebra:

$$\mathrm{ij} = -\mathrm{ji} = \mathrm{k}; \quad \mathrm{jk} = -\mathrm{kj} = \mathrm{i}; \quad \mathrm{ki} = -\mathrm{ik} = \mathrm{j}$$

and

$$\mathrm{i}^2 = \mathrm{j}^2 = \mathrm{k}^2 = \mathrm{ijk} = -1$$

Just as there are both Euclidean and non-Euclidean geometries, there are also several kinds of algebras that, within their respective realms of validity, are as free of contradictions as conventional algebra.

Latin, *vectus*, perfect participle of *vehere*, "to carry"

vector, "one who carries"

Hamilton was the first to use the term **vector** for a directed line segment in his book *Lectures on Quaternions* (1853), which overflowed in peculiar terms: besides vector, there were vehend, vection, vectum; revector, revehend, revection, revectum; provector ...; transvector ...; *etc.*, which made the book all but unintelligible to the majority of contemporary mathematicians.

GRASSMANN Herman Günther (1809 - 1877)

In 1844, the German mathematician **Hermann Grassmann** – independent of Hamilton – published *Die lineale Ausdehnungslehre* ("Theory of Linear Extension"), in which he treated n-dimensional geometry and hypercomplex systems in a much more general way than did Hamilton.

GIBBS Josiah Willard (1839 - 1903)

From Grassmann's and Hamilton's ideas, American mathematician and physicist **J. W. Gibbs** developed much of vector analysis as we know it today; his treatise *Vector Analysis* appeared in 1881. Yet quaternions in their original form are not forgotten but have an established place in algebra and in quantum physics.

* * *

Many things have more than direction;
The magnitude is also a question.
With acceleration or force,
And many more things, of course,
It's vectors that make the connection.

16.1 Basic Vector Algebra

Scalars and Vectors

Quantities such as mass, length, volume, pressure, temperature, voltage, and time can be characterized by a single real number, scaled to a suitable unit of measurement. Such quantities are called **scalar quantities**, and the real number associated with the magnitude (value) of the quantity is called a **scalar**.

Latin, *scala*, "a ladder"

On the other hand, quantities such as force, velocity, and acceleration require, for their definition, the assignment of not only a *magnitude* but also a *direction*. Such quantities are called **vector quantities** and may be represented mathematically by two or more real numbers.

Latin, *vectus*, "carried"

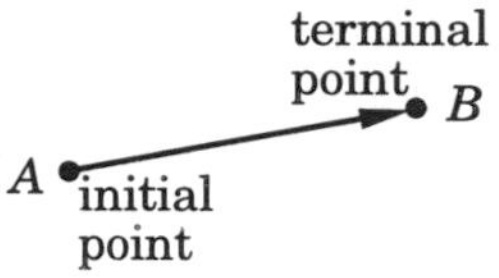

Geometrically, a vector is represented by an arrow (a line segment) with a specific direction, its length corresponding to the magnitude of the vector. A vector representing a displacement from a point A to a point B is said to have the **initial point** or **origin** A and the **terminal point** or **terminus** B.

Several notations are currently in use for vectors and for the magnitude of vectors. In this text we use the following notations:

$\mathbf{a}$	which reads	"Vector **a**"
$\lvert\mathbf{a}\rvert$ or a		"Magnitude of vector **a**"
$\mathbf{0}$		"Null vector"

In handwritten or typewritten text it is customary to underline letters or to use overbars to indicate vector character,

$$\underline{a} \text{ or } \bar{a} \text{ and } \underline{0} \text{ or } \bar{0}$$

In both printed and handwritten texts the notation $\overrightarrow{AB}$ is used for a vector displacement from A to B.

Vectors are **equal** if they have the same direction and the same magnitude. Of the vectors below, **a** equals **b**, and **c** equals **e**:

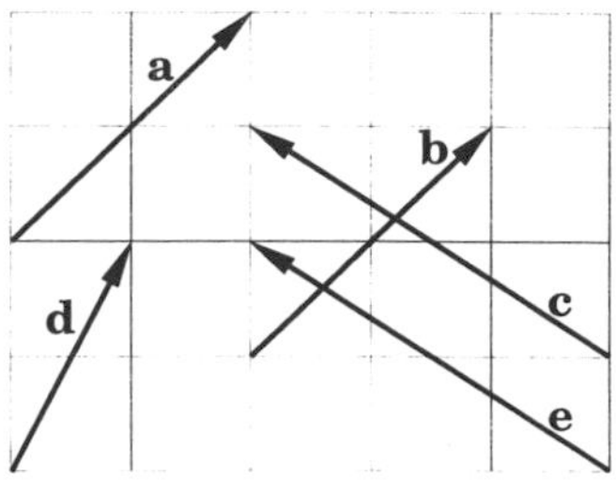

A **scalar multiple** is the product of a scalar and a vector to produce another vector.

If **a** is a vector, then the scalar multiple $\gamma\mathbf{a}$ is a vector with

- the same direction as **a** if $\gamma > 0$;
- the opposite direction of **a** if $\gamma < 0$;

and

- $|\gamma\mathbf{a}| = |\gamma|\,|\mathbf{a}|$.

Below, vector **a** is multiplied by the scalars 0.5, 1.5, and –1:

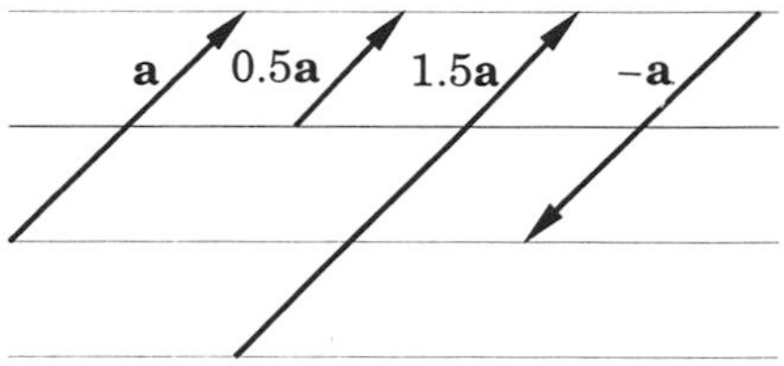

Null Vector

A vector **a** equals zero if its magnitude $|\mathbf{a}|$ is zero; it is called a **null vector** or **zero vector**, and is denoted **0.**

The null vector is the only vector whose direction is indeterminate.

Adding Vectors

Resultant

A vector that is the sum of a given set of vectors is called the **resultant**.

Vectors in Line

Addition of **vectors in one dimension** (vectors on one line) is similar to arithmetical addition. Geometrically, the vectors are positioned along a straight line so that the initial point of one coincides with the terminal point of the other:

Triangle and Parallelogram Laws

To add **vectors in two dimensions** geometrically, we position them so that the initial point of one and the terminal point of another coincide ("head to tail").

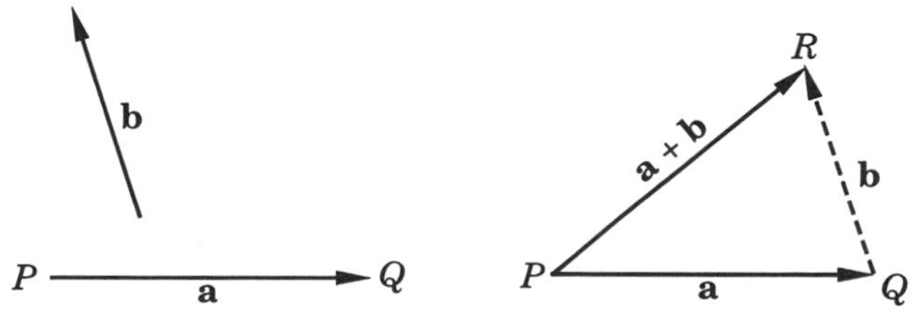

To add vector **b** to vector **a**, we shift **b**, without changing its direction or magnitude, so that its initial point coincides with the terminal point of **a**.

Triangle Law of Addition

Thus, according to the **triangle law of addition** the sum of **a** and **b** equals the displacement of P to R:

$$\mathbf{a}+\mathbf{b} = \overrightarrow{PR}$$

Addition of

$$a+b=b+a$$

is commutative.

To study the additive property of vectors **a** and **b**,

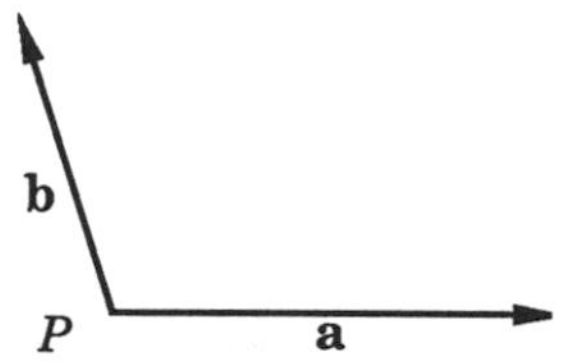

we displace **b** so that its initial point coincides with the terminal point of **a**; the terminal point of **b** will then fall at Q; similarly, if **a** is displaced to make its initial point coincide with the terminal point of **b**, the terminal point of **a** will fall at Q:

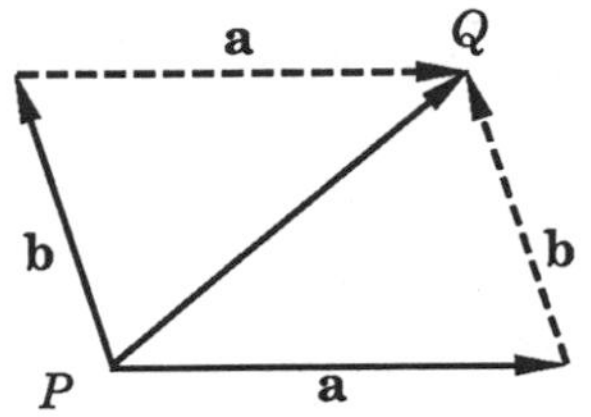

We thus have the commutative property of addition,

$$\mathbf{a}+\mathbf{b} = \mathbf{b}+\mathbf{a};$$

and

$$\mathbf{a}+\mathbf{0} = \mathbf{a}; \quad \mathbf{a}+(-\mathbf{a}) = \mathbf{0}.$$

Parallelogram Law of Addition

The above figure illustrates the **parallelogram law of addition**, another presentation of the triangle law of addition.

The associative property of addition of numbers says that

$$a+(b+c) = (a+b)+c.$$

To show that vectors also satisfy this property, we arrange vectors **a**, **b**, and **c** as the consecutive sides of a polygon:

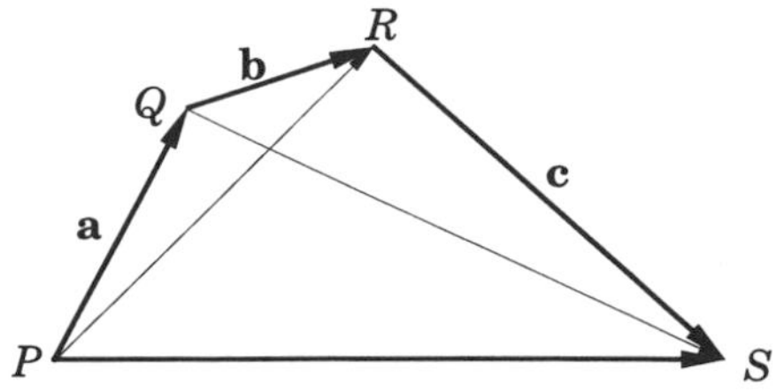

Applying the triangle law of addition, we see that the displacement of point P to point R equals the sum of **a** + **b**, and the displacement of Q to S equals the sum of **b** + **c**.

Consequently, by the associative property of addition:

$$\mathbf{a} + (\mathbf{b} + \mathbf{c}) = (\mathbf{a} + \mathbf{b}) + \mathbf{c}$$

As with addition of scalars, we denote the two equal sums by $\mathbf{a} + \mathbf{b} + \mathbf{c}$.

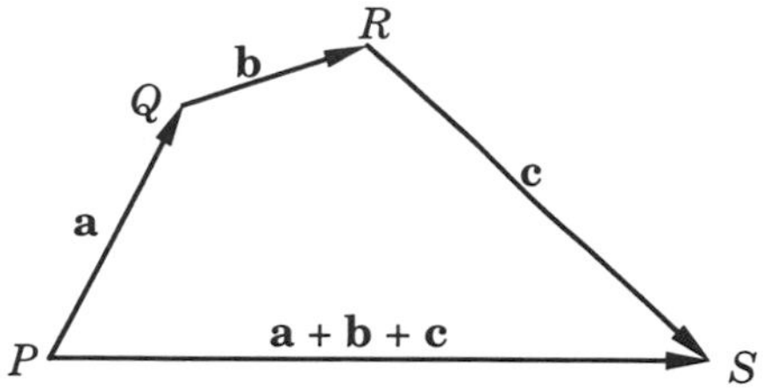

The **resultant** of a larger number of vectors is found by positioning the vectors, successively, "head to tail". The resultant is the straight line joining the initial point of the first vector and the terminal point of the last vector:

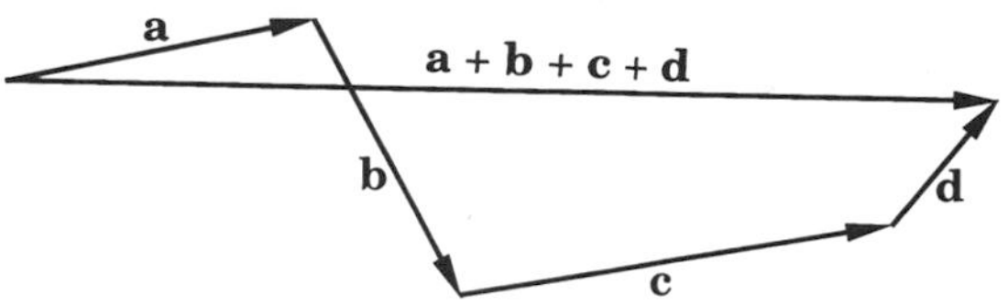

newton (N)

The following example refers to force (F) vectors. The **SI** unit of force is the **newton** (N). 1 N is that force which, when applied to a body having a mass of 1 kg, gives it an acceleration of 1 m/s^2.

- A freight car is being pulled by two cables as shown in the figure below.

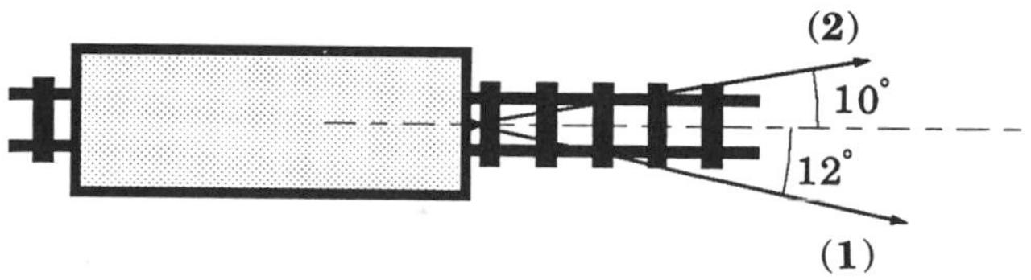

The resultant of the tensile forces has a magnitude of 33.50 kilonewtons.

Find the magnitudes of the tensions in the cables (1) and (2) assuming no side force from the rails.

The freight car moves in the direction of the rails, and the resultant must lie along that axis. Let the tensile forces in cables (1) and (2) be T_1 and T_2, respectively. Construct the force polygon:

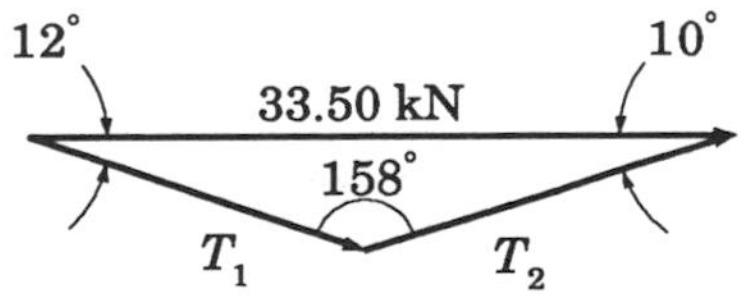

By the law of sines we have:

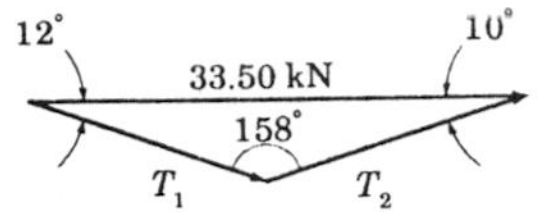

$$\frac{33.50\ \text{kN}}{\sin 158^\circ} = \frac{33.50\ \text{kN}}{\sin 22^\circ} = \frac{T_1}{\sin 10^\circ} = \frac{T_2}{\sin 12^\circ}$$

$$T_1 = 33.50 \cdot \frac{\sin 10^\circ}{\sin 22^\circ} \approx 15.53\ \text{kN}$$

$$T_2 = 33.50 \cdot \frac{\sin 12^\circ}{\sin 22^\circ} \approx 18.59\ \text{kN}$$

Hence, the tensile forces are 15.53 kN in cable (1) and 18.59 kN in cable (2) .

Note: No consideration has been taken to the resultant sideload on the rails.

Subtracting Vectors

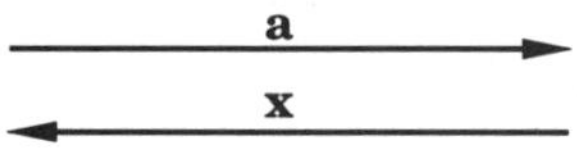

The sum of two parallel vectors of equal length but opposite directions is a null vector:

$$\mathbf{x} = -\mathbf{a}; \quad \mathbf{a} + \mathbf{x} = \mathbf{0}$$

The difference of two vectors **a** and **b** is defined by

$$\mathbf{a} - \mathbf{b} = \mathbf{a} + (-\mathbf{b}).$$

Geometrically the difference between the two vectors **a** and **b** is attained by translating the vector (−**b**) so as to make its initial point coincide with the terminal point of vector **a**, and apply the triangle law of addition.

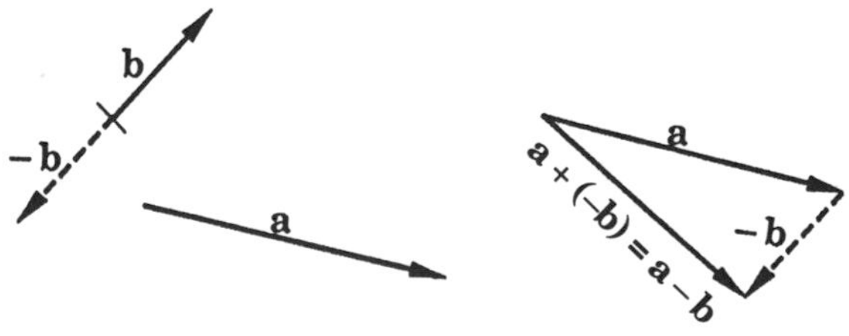

16.2 Scalar and Vector Components

Right-Handed Coordinate Systems

In studying vectors, the use of an orthogonal coordinate system is convenient. In this text the coordinates will be denoted x, y, and z; in other texts the symbols x_1, x_2, and x_3 may be used.

It is customary in vector analysis to consider **right-handed systems** only. When the positive x-axis is rotated counterclockwise into the positive y-axis, a right-handed (standard) screw would advance in the direction of the positive z-axis.

In a right-handed coordinate system, the thumb of the right hand points in the direction of the positive z-axis when the fingers are curled in the direction away from the positive x-axis toward the positive y-axis.

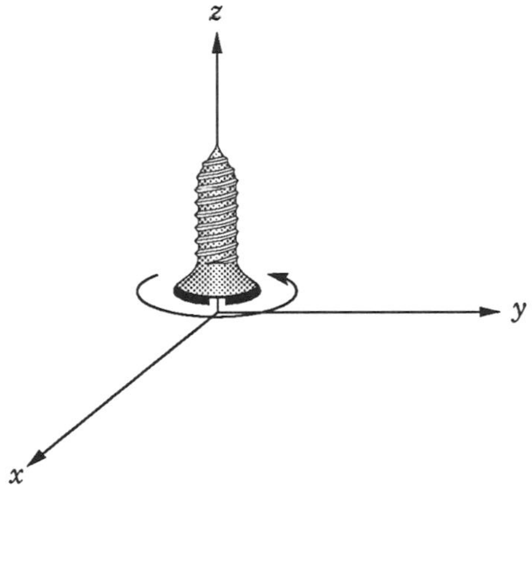

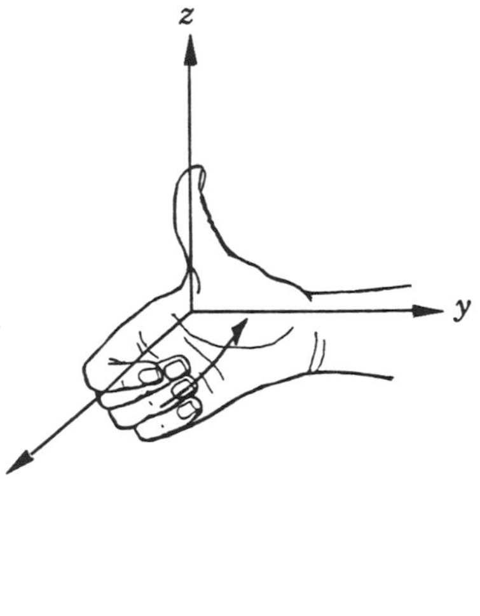

Unit Vectors

A unit vector is a vector whose magnitude is 1.

Unit vectors in the direction of the orthogonal coordinate axes are denoted $\mathbf{e}_x$, $\mathbf{e}_y$, $\mathbf{e}_z$ when the coordinates are denoted x, y, z; they are $\mathbf{e}_1$, $\mathbf{e}_2$, $\mathbf{e}_3$ for coordinates x_1, x_2, x_3. The symbols $\mathbf{i}$, $\mathbf{j}$, and $\mathbf{k}$ may be used irrespective of the notation of the coordinates.

In a two-dimensional orthogonal coordinate system, vectors $\mathbf{i}$ and $\mathbf{j}$ are the unit vectors from the origin to the points (1, 0) and (0, 1), respectively; in a three-dimensional system the unit vectors **i**, **j**, and **k** are (1, 0, 0), (0, 1, 0), and (0, 0, 1), respectively:

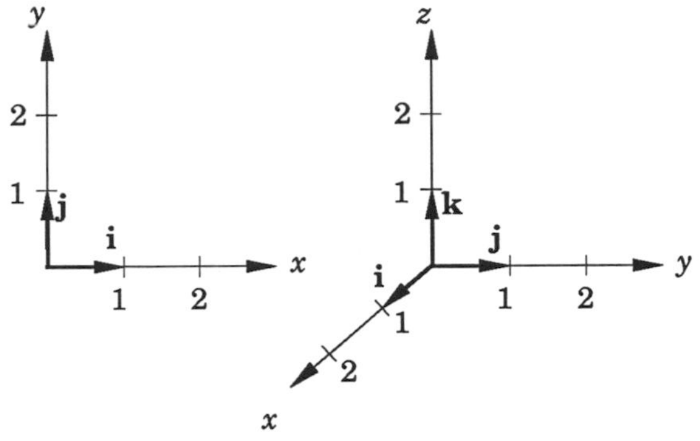

The Position Vector and Its Components

Position Vector

A vector represented by a line segment beginning at the origin of an orthogonal coordinate system is called a **position vector**, and is often denoted **r**.

A vector **r** with its initial point at the origin of an orthogonal coordinate system and its terminal point at (x, y, z) can be written

$$\mathbf{r} = x\,\mathbf{i} + y\,\mathbf{j} + z\,\mathbf{k},$$

Scalar Components
Vector Components

where x, y, z are the **scalar components** of **r**, and $x\,\mathbf{i}$, $y\,\mathbf{j}$, $z\,\mathbf{k}$ are the **vector components**:

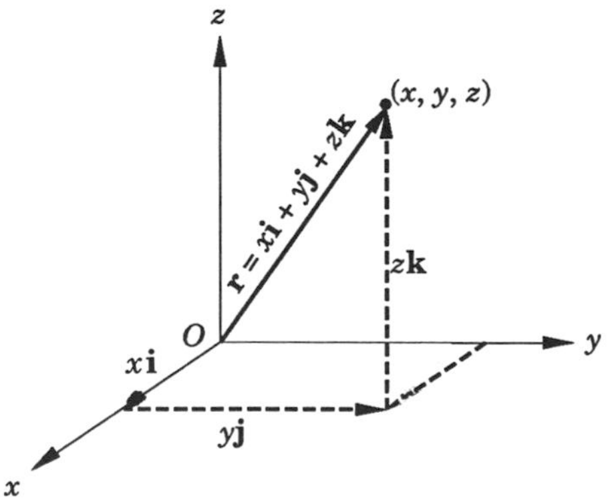

A vector **r**,

$$\mathbf{r} = x\,\mathbf{i} + y\,\mathbf{j} + z\,\mathbf{k},$$

multiplied by the scalar λ,

$$\lambda\mathbf{r} = \lambda(x\,\mathbf{i} + y\,\mathbf{j} + z\,\mathbf{k}) = (\lambda x)\,\mathbf{i} + (\lambda y)\,\mathbf{j} + (\lambda z)\,\mathbf{k},$$

has components λx, λy, and λz.

p. 550

The **magnitude** of a space vector **r** may be calculated by the use of the distance formula:

$$|\mathbf{r}| = \sqrt{x^2 + y^2 + z^2}$$

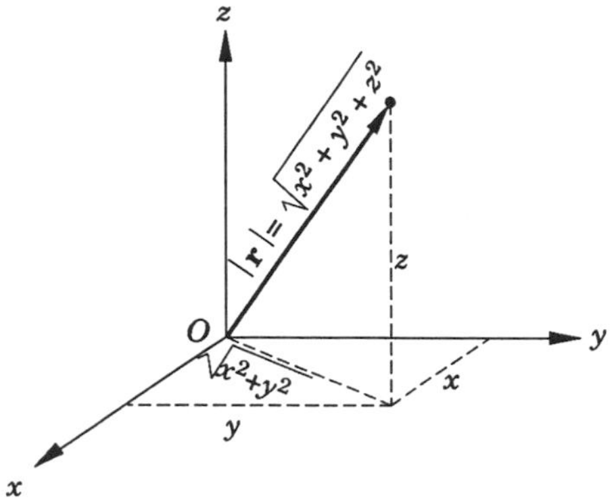

- Given: vectors $\mathbf{a} = 4\,\mathbf{i} - \mathbf{j} - 4\,\mathbf{k}$ and $\mathbf{b} = -\mathbf{i} + 3\,\mathbf{j} + \mathbf{k}$

  Find the vector sum and difference of **a** and **b**, and their magnitudes $|\mathbf{a} + \mathbf{b}|$ and $|\mathbf{a} - \mathbf{b}|$.

  $\mathbf{a} + \mathbf{b} = (4\,\mathbf{i} - \mathbf{j} - 4\,\mathbf{k}) + (-\mathbf{i} + 3\,\mathbf{j} + \mathbf{k}) = 3\,\mathbf{i} + 2\,\mathbf{j} - 3\,\mathbf{k}$;

  $|\mathbf{a} + \mathbf{b}| = \sqrt{3^2 + 2^2 + 3^2} = \sqrt{22}$.

  $\mathbf{a} - \mathbf{b} = (4\,\mathbf{i} - \mathbf{j} - 4\,\mathbf{k}) - (-\mathbf{i} + 3\,\mathbf{j} + \mathbf{k}) = 5\,\mathbf{i} - 4\,\mathbf{j} - 5\,\mathbf{k}$;

  $|\mathbf{a} - \mathbf{b}| = \sqrt{5^2 + 4^2 + 5^2} = \sqrt{66}$.

  The illustration below shows vectors **a**, **b**, **a** + **b**, and **a** – **b**. The translated vectors **b** and –**b** are shown as dashed lines:

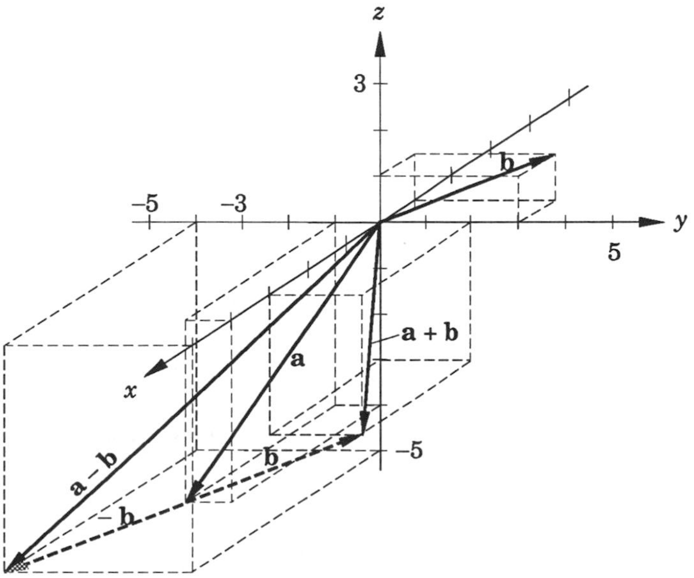

- Three force vectors $\mathbf{F}_1$, $\mathbf{F}_2$, and $\mathbf{F}_3$, with magnitudes 150, 210, and 195 newtons, respectively, have the directions shown in the figure below:

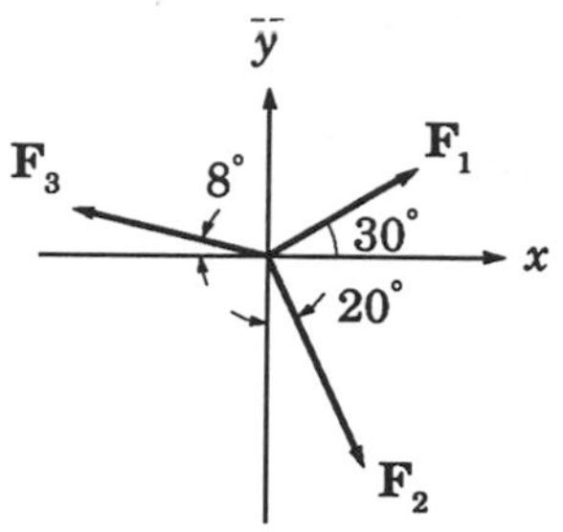

Find the magnitude R and direction θ of the resultant $\mathbf{R}$.

$\mathbf{R} = \mathbf{F}_1 + \mathbf{F}_2 + \mathbf{F}_3$.

If we let F_{1_x}, F_{2_x}, F_{3_x} and F_{1_y}, F_{2_y}, F_{3_y} be the x and y components of $\mathbf{F}_1$, $\mathbf{F}_2$, and $\mathbf{F}_3$, and let R_x and R_y be the x and y components of $\mathbf{R}$, then

$$\mathbf{R} = R_x\,\mathbf{i} + R_y\,\mathbf{j} = (F_{1_x} + F_{2_x} + F_{3_x})\,\mathbf{i} + (F_{1_y} + F_{2_y} + F_{3_y})\,\mathbf{j}\,.$$

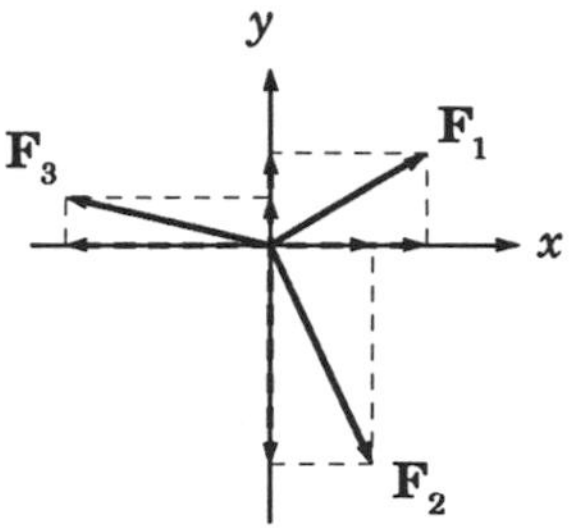

$F_{1_x} = 150 \cos 30^\circ$	$\approx$	129.9 N	$F_{1_y} = 150 \sin 30^\circ$	$\approx$	75.0 N
$F_{2_x} = 210 \sin 20^\circ$	$\approx$	71.8 N	$F_{2_y} = -(210 \cos 20^\circ)$	$\approx$	− 197.3 N
$F_{3_x} = -(195 \cos 8^\circ)$	$\approx$	− 193.1 N	$F_{3_y} = 195 \sin 8^\circ$	$\approx$	27.1 N
$R_x =$		8.6 N	$R_y =$		− 95.2 N

$\mathbf{R} = 8.6\,\mathbf{i}\text{ N} - 95.2\,\mathbf{j}\text{ N}$

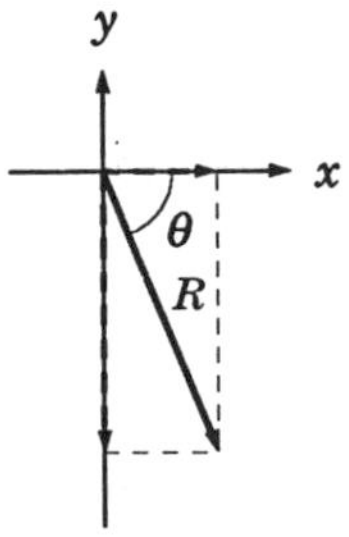

$R = \sqrt{8.6^2 + 95.2^2} \approx 95.6$

$\tan\theta = \dfrac{R_y}{R_x} = \dfrac{-95.2}{8.6}\ ; \quad \theta = -84.8^\circ.$

The resultant is 95.6 newtons at − 84.8° . •

According to Newton, an object moves in the direction of the force acting on it. For the object to be stationary, the resultant of all forces acting on it must be zero; that is, the resultant must be a **null vector**. This state, when all forces acting on an object are self-balancing, is a **state of equilibrium**.

- The three vectors $\mathbf{F}_1$, $\mathbf{F}_2$, and $\mathbf{F}_3$ of the preceding example pull at an object. Determine how $\mathbf{F}_2$ must be changed in order to keep the object motionless.

We have

$$F_{1_x} \approx 129.9\ \text{N} \qquad F_{1_y} \approx 75.0\ \text{N}$$
$$F_{3_x} \approx -193.1\ \text{N} \qquad F_{3_y} \approx 27.1\ \text{N}$$

Equilibrium of the object requires that

$$R_x = 129.9\ \text{N} + F_{2_x} - 193.1\ \text{N} = 0; \quad F_{2_x} = 63.2\ \text{N}$$
$$R_y = 75.0\ \text{N} + F_{2_y} + 27.1\ \text{N} = 0; \quad F_{2_y} = -102.1\ \text{N}$$

and

$$\mathbf{F}_2 = (63.2\,\mathbf{i} - 102.1\,\mathbf{j})\ \text{N}.$$

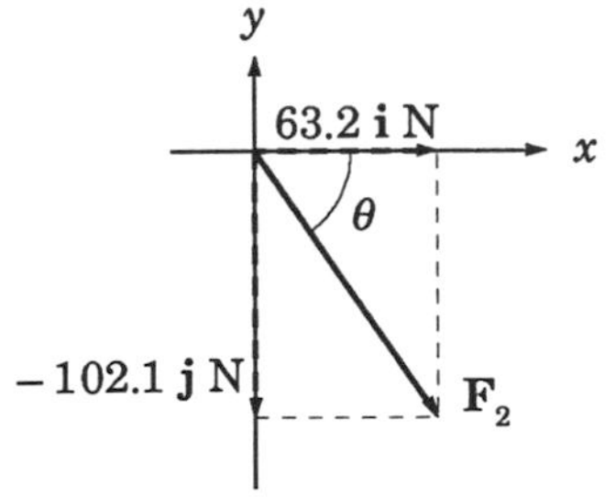

The magnitude, F_2, is

$$F_2 = \sqrt{(63.2)^2 + (-102.1)^2} \approx 120.1\ \text{N}.$$

$$\tan\theta = -\frac{102.1}{63.2}; \quad \theta = -58.2°$$

Hence, to produce equilibrium, the vector $\mathbf{F}_2$ must be 120.1 newtons acting in the direction $-58.2°$. •

Mass

Weight, Force

Acceleration of Free Fall (g)

In the following three-dimensional equilibrium problem reference is made to the quantities mass and weight. The **mass** (m) of a given body is independent of its location on Earth or in the universe. The **weight** of a body is a **force** (F) that equals the product of the mass of the body and the **acceleration of free fall** (g). At sea level, g varies from 9.78 m/s^2 at the equator to 9.83 m/s^2 at the North and South Poles.

- A structural unit is being lifted by means of three cables, AB, AC, and AD. The unit has a mass of 78.5 kilograms (kg). The origin of the coordinates (0, 0, 0) is taken at A, and the coordinates of points B, C, and D are given below (m stands for meter):

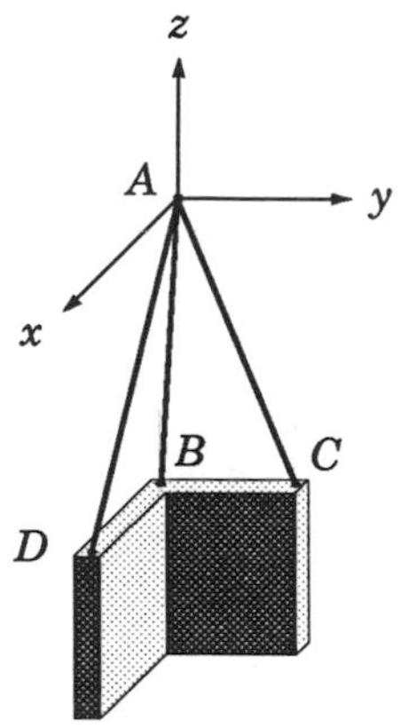

B (– 0.12, – 0.12, – 1.00) m
C (– 0.12, 0.32, – 1.00) m
D (0.32, – 0.12, – 1.00) m
g = 9.81 m/s^2

Find the tensile force in **newtons** in each cable.

The main cable at A must carry the total weight (F) of the structural unit:

$$F = 78.5\ [\text{kg}] \cdot 9.81\ [\text{m/s}^2] = 770\ \text{N}$$

We construct a force diagram at A:

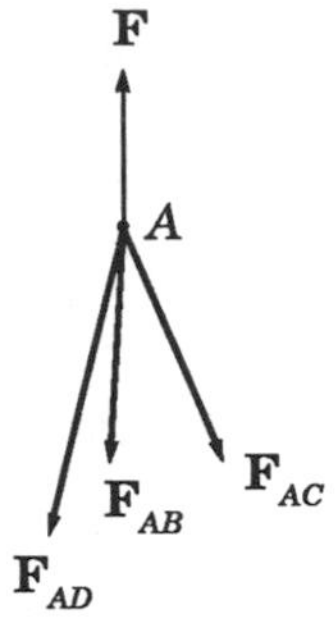

Express the forces $\mathbf{F}_{AB}$, $\mathbf{F}_{AC}$, $\mathbf{F}_{AD}$ as vectors:

$$\left.\begin{aligned} \mathbf{F}_{AB} &= (F_{AB})\,\mathbf{e}_{AB} \\ \mathbf{F}_{AC} &= (F_{AC})\,\mathbf{e}_{AC} \\ \mathbf{F}_{AD} &= (F_{AD})\,\mathbf{e}_{AD} \end{aligned}\right\}$$

where F_{AB} is the magnitude of $\mathbf{F}_{AB}$, and $\mathbf{e}_{AB}$ is a unit vector in the direction of AB, *etc.*

The unit vector $\mathbf{e}_{AB}$ is

$$\mathbf{e}_{AB} = \frac{\mathbf{r}_{AB}}{r_{AB}} = \frac{(x_B - x_A)\,\mathbf{i} + (y_B - y_A)\,\mathbf{j} + (z_B - z_A)\,\mathbf{k}}{\sqrt{(x_B - x_A)^2 + (y_B - y_A)^2 + (z_B - z_A)^2}}\,,\ etc.,$$

where

$$\left.\begin{aligned} \mathbf{r}_{AB} &= -0.12\,\mathbf{i} - 0.12\,\mathbf{j} - 1.00\,\mathbf{k} & r_{AB} &= \sqrt{1.0288} \\ \mathbf{r}_{AC} &= -0.12\,\mathbf{i} + 0.32\,\mathbf{j} - 1.00\,\mathbf{k} & r_{AC} &= \sqrt{1.1168} \\ \mathbf{r}_{AD} &= 0.32\,\mathbf{i} - 0.12\,\mathbf{j} - 1.00\,\mathbf{k} & r_{AD} &= \sqrt{1.1168} \end{aligned}\right\}$$

and thus,

$$\left.\begin{aligned} \mathbf{e}_{AB} &= \frac{-0.12\,\mathbf{i} - 0.12\,\mathbf{j} - 1.00\,\mathbf{k}}{\sqrt{1.0288}} \\ \mathbf{e}_{AC} &= \frac{-0.12\,\mathbf{i} + 0.32\,\mathbf{j} - 1.00\,\mathbf{k}}{\sqrt{1.1168}} \\ \mathbf{e}_{AD} &= \frac{0.32\,\mathbf{i} - 0.12\,\mathbf{j} - 1.00\,\mathbf{k}}{\sqrt{1.1168}} \end{aligned}\right\}$$

and

$$\left.\begin{aligned} \mathbf{F}_{AB} &= (F_{AB})\frac{-0.12\,\mathbf{i} - 0.12\,\mathbf{j} - 1.00\,\mathbf{k}}{\sqrt{1.0288}} \\ \mathbf{F}_{AC} &= (F_{AC})\frac{-0.12\,\mathbf{i} + 0.32\,\mathbf{j} - 1.00\,\mathbf{k}}{\sqrt{1.1168}} \\ \mathbf{F}_{AD} &= (F_{AD})\frac{0.32\,\mathbf{i} - 0.12\,\mathbf{j} - 1.00\,\mathbf{k}}{\sqrt{1.1168}} \end{aligned}\right\}$$

The equilibrium conditions are

$$\begin{aligned}(F_{AB})_x + (F_{AC})_x + (F_{AD})_x &= 0\\ (F_{AB})_y + (F_{AC})_y + (F_{AD})_y &= 0\\ (F_{AB})_z + (F_{AC})_z + (F_{AD})_z &= -770\end{aligned}$$

Substituting the appropriate component values, we have

$$\left.\begin{aligned}-\frac{0.12\,F_{AB}}{\sqrt{1.0288}} - \frac{0.12\,F_{AC}}{\sqrt{1.1168}} + \frac{0.32\,F_{AD}}{\sqrt{1.1168}} &= 0\\ -\frac{0.12\,F_{AB}}{\sqrt{1.0288}} + \frac{0.32\,F_{AC}}{\sqrt{1.1168}} - \frac{0.12\,F_{AD}}{\sqrt{1.1168}} &= 0\\ -\frac{1.00\,F_{AB}}{\sqrt{1.0288}} - \frac{1.00\,F_{AC}}{\sqrt{1.1168}} - \frac{1.00\,F_{AD}}{\sqrt{1.1168}} &= -770\end{aligned}\right\} .$$

Solving the above system of equations, we find

$$\left.\begin{aligned}F_{AB} &= 355\text{ N}\\ F_{AC} &= 222\text{ N}\\ F_{AD} &= 222\text{ N}\end{aligned}\right\} .$$

16.3 Multiplication of Vectors

Products of vectors can be formed in several ways; some of these products are of special interest and importance because of their geometrical interpretations and their uses in describing various phenomena in physics.

We distinguish between **scalar products** and **vector products**.

The Scalar Product

A **scalar product**, $\mathbf{a} \cdot \mathbf{b}$, is the sum of the products of corresponding components of the vectors.

The scalar product is also known as the **inner product**, or the **dot product**, and reads "a dot b".

If

$$\mathbf{a} = x_1\mathbf{i} + y_1\mathbf{j} \quad \text{and} \quad \mathbf{b} = x_2\mathbf{i} + y_2\mathbf{j}$$

then

$$\mathbf{a} \cdot \mathbf{b} = x_1x_2 + y_1y_2 .$$

The scalar product is useful for **determining the angle between two vectors**.

The Scalar Product Theorem

The scalar product of two vectors is equal to the product of the magnitudes of the two vectors and the cosine of their included angle:

$$\mathbf{a} \cdot \mathbf{b} = |\mathbf{a}|\,|\mathbf{b}| \cos\theta$$

$$\cos\theta = \frac{\mathbf{a} \cdot \mathbf{b}}{|\mathbf{a}|\,|\mathbf{b}|}$$

Why?

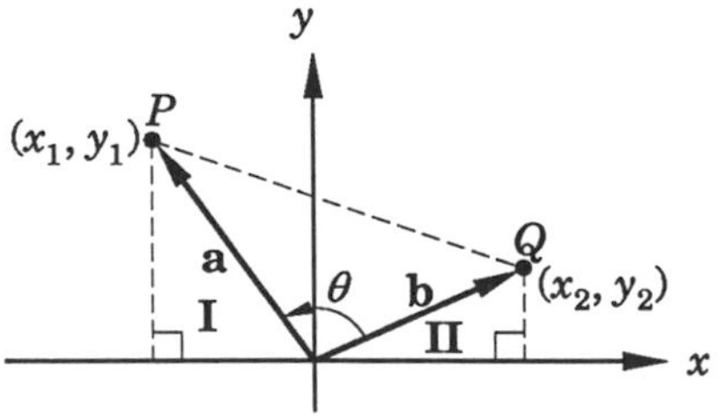

p. 483 The **law of cosines** gives

$$|PQ|^2 = |\mathbf{a}|^2 + |\mathbf{b}|^2 - 2\,|\mathbf{a}|\,|\mathbf{b}| \cos\theta .$$

With

$$\mathbf{a} = x_1\mathbf{i} + y_1\mathbf{j} \quad \text{and} \quad \mathbf{b} = x_2\mathbf{i} + y_2\mathbf{j}$$

p. 550 and

$$|PQ|^2 = (x_2 - x_1)^2 + (y_2 - y_1)^2,$$

we obtain

$$(x_2 - x_1)^2 + (y_2 - y_1)^2 = |\mathbf{a}|^2 + |\mathbf{b}|^2 - 2\,|\mathbf{a}|\,|\mathbf{b}| \cos\theta .$$

Applying the Pythagorean theorem on triangles I and II, we have

$$|\mathbf{a}|^2 = x_1^2 + y_1^2 \quad \text{and} \quad |\mathbf{b}|^2 = x_2^2 + y_2^2$$

and may write

$$(x_2 - x_1)^2 + (y_2 - y_1)^2 = (x_1^2 + y_1^2) + (x_2^2 + y_2^2) - 2\,|\mathbf{a}|\,|\mathbf{b}|\,\cos\theta$$

and

$$x_1 x_2 + y_1 y_2 = |\mathbf{a}|\,|\mathbf{b}|\,\cos\theta.$$

We have already defined the scalar product

$$\mathbf{a}\cdot\mathbf{b} = x_1 x_2 + y_1 y_2,$$

and now have

$$\mathbf{a}\cdot\mathbf{b} = |\mathbf{a}|\,|\mathbf{b}|\,\cos\theta$$

and

$$\cos\theta = \frac{\mathbf{a}\cdot\mathbf{b}}{|\mathbf{a}|\,|\mathbf{b}|}.$$

We note that the scalar product is a scalar and *not* another vector. It is not to be confused with the scalar multiple $\lambda\,\mathbf{a}$ of a vector **a**.

For **three-dimensional vectors** we have, again:

$$\mathbf{a}\cdot\mathbf{b} = |\mathbf{a}|\,|\mathbf{b}|\,\cos\theta$$

Properties of Scalar Products

Scalar Products of Unit Vectors

Since the cosine of 0 is 1, and the cosine of 90° is 0, the scalar products of unit vectors are

$$\mathbf{i}\cdot\mathbf{i} = \mathbf{j}\cdot\mathbf{j} = \mathbf{k}\cdot\mathbf{k} = 1\cdot 1\cos 0 = 1$$

and

$$\mathbf{i}\cdot\mathbf{j} = \mathbf{j}\cdot\mathbf{k} = \mathbf{k}\cdot\mathbf{i} = 1\cdot 1\cos 90^\circ = 0.$$

Commutative Properties
Distributive Properties

Because the coordinates x_1, x_2, y_1, y_2 are real numbers satisfying familiar algebraic properties, it follows that the scalar product of vectors has **commutative** and **distributive** properties:

$$\mathbf{a}\cdot\mathbf{b} = \mathbf{b}\cdot\mathbf{a},$$

$$\mathbf{a}\cdot(\mathbf{b}+\mathbf{c}) = \mathbf{a}\cdot\mathbf{b} + \mathbf{a}\cdot\mathbf{c} = (\mathbf{b}+\mathbf{c})\cdot\mathbf{a}$$

- Determine the angle included between vectors

$$\mathbf{a} = -\mathbf{i} + 3\mathbf{j} + \mathbf{k}; \quad \mathbf{b} = 5\mathbf{i} - 4\mathbf{j} - 5\mathbf{k}.$$

Let the sought angle be θ; then

$$0 \le \theta \le \pi; \quad 0^\circ \le \theta \le 180^\circ.$$

The scalar product is

$$\mathbf{a}\cdot\mathbf{b} = -1\cdot 5 - 3\cdot 4 - 5\cdot 1 = -22.$$

The magnitudes $|\mathbf{a}|$ and $|\mathbf{b}|$ are

$$|\mathbf{a}| = \sqrt{(-1)^2 + 3^2 + 1^2} = \sqrt{11},$$

$$|\mathbf{b}| = \sqrt{(5)^2 + (-4)^2 + (-5)^2} = \sqrt{66}.$$

$$\cos\theta = \frac{\mathbf{a}\cdot\mathbf{b}}{|\mathbf{a}|\,|\mathbf{b}|} = \frac{-22}{\sqrt{11\cdot 66}} = \frac{-22}{11\sqrt{6}} = -\frac{2}{\sqrt{6}};$$

$\theta = 2.526\ldots$ rad; $\theta = 144.735\ldots{}^\circ$.

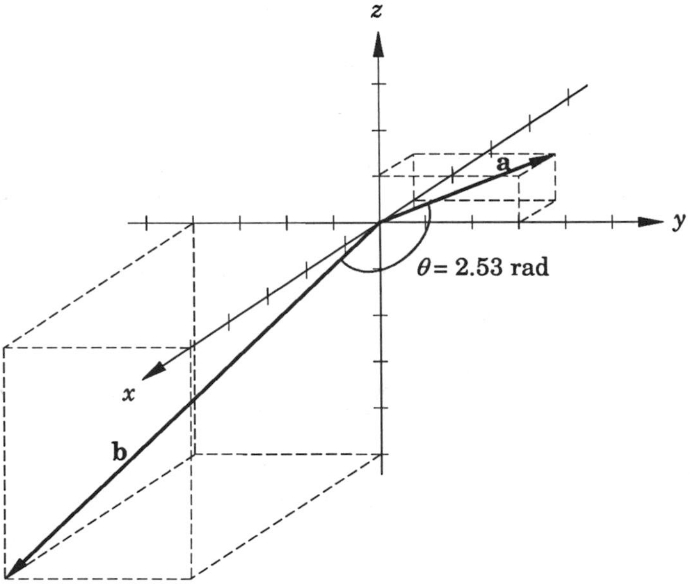

- Return to the example on *pp.* 611 - 12.

  Find the angle θ between cables AB and AC in the suspension point, and the angle ϕ between cable AD and the vertical through point A.

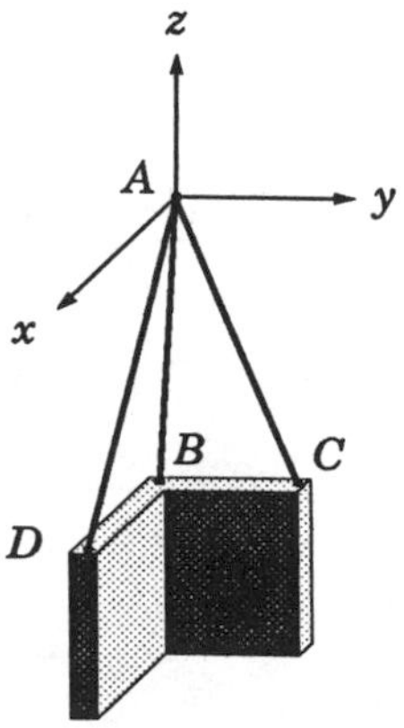

We have

$$\cos\theta = \frac{\mathbf{r}_{AB} \cdot \mathbf{r}_{AC}}{(r_{AB})\,(r_{AC})}$$

where we insert

$$\mathbf{r}_{AB} = -0.12\,\mathbf{i} - 0.12\,\mathbf{j} - 1.00\,\mathbf{k} \qquad r_{AB} = \sqrt{1.0288}$$

$$\mathbf{r}_{AC} = -0.12\,\mathbf{i} + 0.32\,\mathbf{j} - 1.00\,\mathbf{k} \qquad r_{AC} = \sqrt{1.1168}$$

Hence,

$$\cos\theta = \frac{(-0.12)\,(-0.12) + (-0.12)\,(0.32) + (-1.00)\,(-1.00)}{\left(\sqrt{1.0288}\right)\left(\sqrt{1.1168}\right)}$$

$$\theta = 24.42°.$$

The vertical through A is the z-axis. We have

$$\mathbf{r}_{AD} = 0.32\,\mathbf{i} - 0.12\,\mathbf{j} - 1.00\,\mathbf{k} \qquad r_{AD} = \sqrt{1.1168}\,.$$

Hence,

$$\cos\phi = \frac{(0.32)\,0 + (-0.12)\,0 + (-1.00)\,(-1.00)}{1.00 \cdot \sqrt{1.1168}}$$

$$\phi = 18.87°.$$

The Vector Product

The **vector product**, **a** x **b**, of three-dimensional vectors **a** and **b** is a vector orthogonal to the plane that contains vectors **a** and **b**.

The vector product is also known as the **outer product**, or the **cross product**, and reads "a cross b".

We define **a** x **b** as the vector that fulfills the following criteria:

- **a** x **b** is perpendicular to the common plane of vectors **a** and **b**, its **direction** being determined by the right-hand rule.
- The **magnitude** of **a** x **b** is $|\mathbf{a}|\,|\mathbf{b}|\,\sin\theta$, where θ is the included angle between the vectors **a** and **b**:

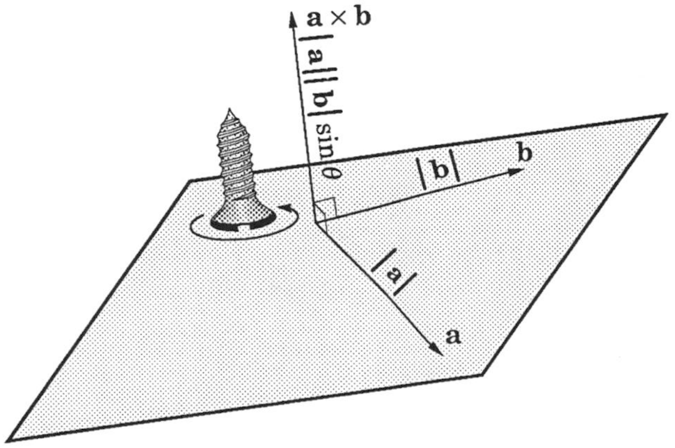

Components of a Vector Product

In a three-dimensional orthogonal coordinate system the unit vectors **i**, **j**, **k** are perpendicular to each other.

The **magnitude** of the vector product $\mathbf{i} \times \mathbf{j}$ is

$$|\mathbf{i}|\,|\mathbf{j}| \sin 90^\circ = 1 \cdot 1 \cdot 1 = 1$$

and its **direction** in a right-handed coordinate system is the direction of the positive z-axis.

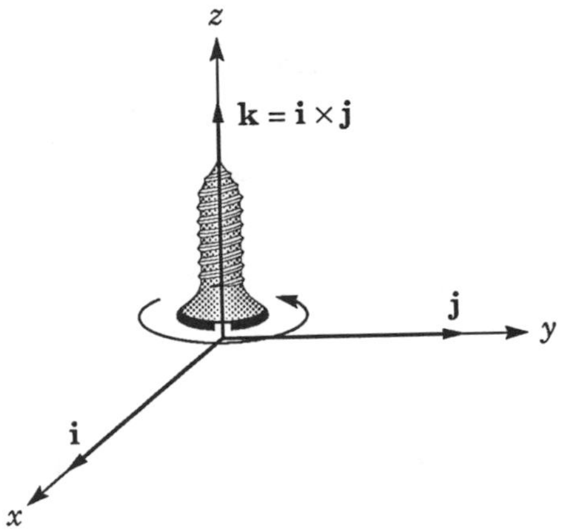

The complete list of unit vector products is:

$$\mathbf{i} \times \mathbf{j} = \mathbf{k} \qquad \mathbf{j} \times \mathbf{k} = \mathbf{i} \qquad \mathbf{k} \times \mathbf{i} = \mathbf{j}$$
$$\mathbf{j} \times \mathbf{i} = -\mathbf{k} \qquad \mathbf{k} \times \mathbf{j} = -\mathbf{i} \qquad \mathbf{i} \times \mathbf{k} = -\mathbf{j}$$
$$\mathbf{i} \times \mathbf{i} = \mathbf{j} \times \mathbf{j} = \mathbf{k} \times \mathbf{k} = \mathbf{0}$$

With

$$\mathbf{a} = x_1 \mathbf{i} + y_1 \mathbf{j} + z_1 \mathbf{k} \quad \text{and} \quad \mathbf{b} = x_2 \mathbf{i} + y_2 \mathbf{j} + z_2 \mathbf{k}$$

the vector product expands as

$$\mathbf{a} \times \mathbf{b} = x_1 \mathbf{i} \times x_2 \mathbf{i} + x_1 \mathbf{i} \times y_2 \mathbf{j} + x_1 \mathbf{i} \times z_2 \mathbf{k} + y_1 \mathbf{j} \times x_2 \mathbf{i} + y_1 \mathbf{j} \times y_2 \mathbf{j}$$
$$+ y_1 \mathbf{j} \times z_2 \mathbf{k} + z_1 \mathbf{k} \times x_2 \mathbf{i} + z_1 \mathbf{k} \times y_2 \mathbf{j} + z_1 \mathbf{k} \times z_2 \mathbf{k},$$

which may be rearranged:

$$\mathbf{a} \times \mathbf{b} = (y_1 z_2 - y_2 z_1)\,\mathbf{i} - (x_1 z_2 - x_2 z_1)\,\mathbf{j} + (x_1 y_2 - x_2 y_1)\,\mathbf{k}.$$

Properties of Vector Products

Since the vector product $\mathbf{a} \times \mathbf{b}$ contains the sine of the included angle, it will be zero when **a** and **b** have the same direction and will have its maximum value $|\mathbf{a}|\,|\mathbf{b}|$ when the vectors are perpendicular to each other.

A scalar may be placed anywhere in the vector product:

$$(\lambda \mathbf{a}) \times \mathbf{b} = \mathbf{a} \times (\lambda \mathbf{b}) = \lambda\,(\mathbf{a} \times \mathbf{b})$$

A vector product is **distributive**:

$$\mathbf{a} \times (\mathbf{b} + \mathbf{c}) = \mathbf{a} \times \mathbf{b} + \mathbf{a} \times \mathbf{c}$$

Anti-commutative

The right-hand rule decrees that the vector products $\mathbf{c}_1 = \mathbf{a} \times \mathbf{b}$ and $\mathbf{c}_2 = \mathbf{b} \times \mathbf{a}$ have opposite directions; that is, the vector product **is not commutative**, but **anti-commutative**,

$$\mathbf{a} \times \mathbf{b} = -(\mathbf{b} \times \mathbf{a})\,.$$

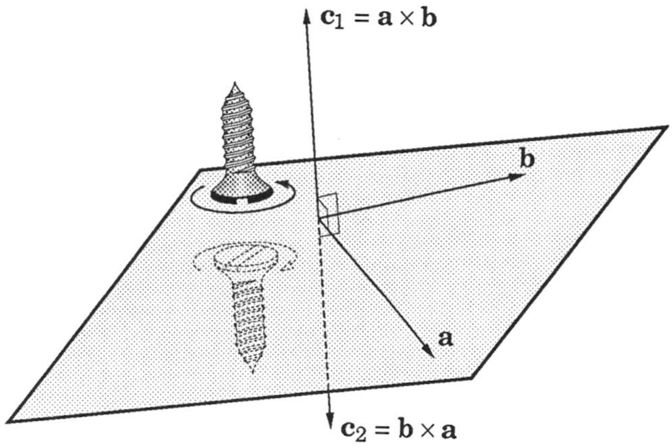

• Given: the vectors $\mathbf{a} = -\mathbf{i} - \mathbf{j} + 3\,\mathbf{k}$ and $\mathbf{b} = 5\,\mathbf{i} + 5\,\mathbf{j} - 4\,\mathbf{k}$

Find the vector products $\mathbf{a} \times \mathbf{b}$ and $\mathbf{b} \times \mathbf{a}$, and their magnitudes.

Solved with determinant: p. 651

$$\begin{aligned}\mathbf{a} \times \mathbf{b} &= (-\mathbf{i} - \mathbf{j} + 3\,\mathbf{k}) \times (5\,\mathbf{i} + 5\,\mathbf{j} - 4\,\mathbf{k})\\ &= -5\,\mathbf{i}\times\mathbf{i} - 5\,\mathbf{i}\times\mathbf{j} + 4\,\mathbf{i}\times\mathbf{k} - 5\,\mathbf{j}\times\mathbf{i} - 5\,\mathbf{j}\times\mathbf{j} + 4\,\mathbf{j}\times\mathbf{k}\\ &\qquad + 15\,\mathbf{k}\times\mathbf{i} + 15\,\mathbf{k}\times\mathbf{j} - 12\,\mathbf{k}\times\mathbf{k}\\ &= -5\times\mathbf{0} - 5\,\mathbf{k} + 4\,(-\mathbf{j}) - 5\,(-\mathbf{k}) - 5\times\mathbf{0} + 4\,\mathbf{i} + 15\,\mathbf{j} + 15\,(-\mathbf{i}) - 12\times\mathbf{0}\\ &= -(11\,\mathbf{i} - 11\,\mathbf{j})\end{aligned}$$

Since $\mathbf{a} \times \mathbf{b} = -(\mathbf{b} \times \mathbf{a})$, we have

$$\mathbf{b} \times \mathbf{a} = 11\,\mathbf{i} - 11\,\mathbf{j}.$$

The results are illustrated in the figure below:

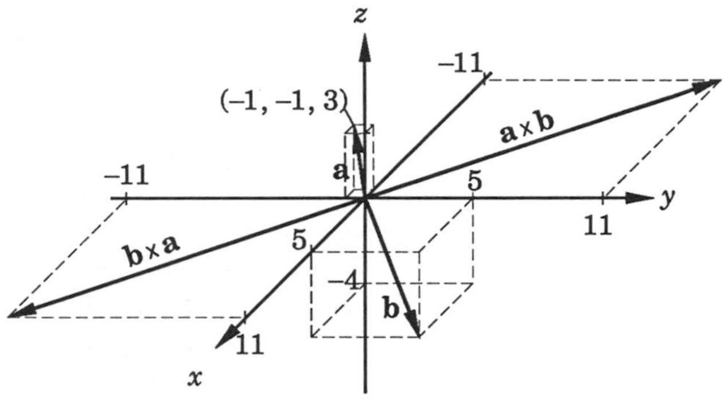

The magnitudes of $\mathbf{a} \times \mathbf{b}$ and $\mathbf{b} \times \mathbf{a}$ are

$$|\mathbf{a} \times \mathbf{b}| = |\mathbf{b} \times \mathbf{a}| = \sqrt{11^2 + 11^2 + 0^2} = 11\sqrt{2}.$$

- Given: $\mathbf{a} = -2\,\mathbf{i} - 2\,\mathbf{j} + \mathbf{k}$; $\mathbf{b} = \mathbf{i} + 2\,\mathbf{j} + \mathbf{k}$

  Find the vector product $\mathbf{a} \times \mathbf{b}$.

Solved with determinant: *p.* 652

$\mathbf{a} \times \mathbf{b} = -4\,\mathbf{i} + 3\,\mathbf{j} - 2\,\mathbf{k}$

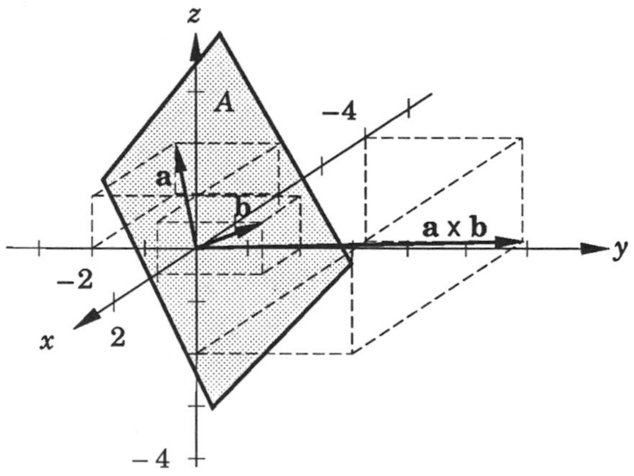

The vector $(-4\,\mathbf{i} + 3\,\mathbf{j} - 2\,\mathbf{k})$ is perpendicular to the plane containing vectors **a** and **b**.

- An architect decides to cut a corner of the top floor of his design as shown in the figure below. The sloping part of the roof is to be covered with sheet glass. For heat transfer calculations, determine the area of the triangular cut OAB.

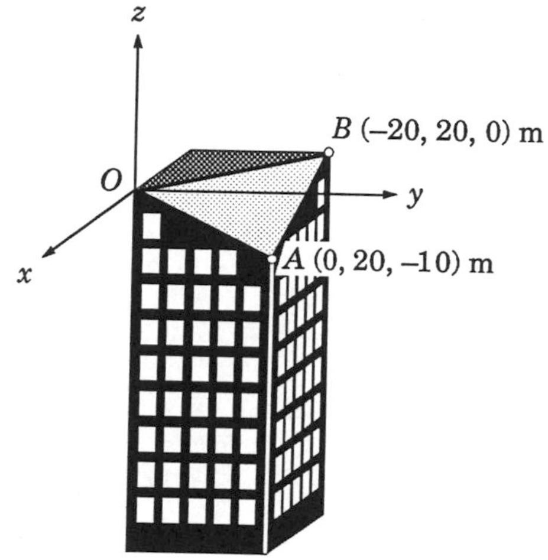

Geometrically, the magnitude of the vector product $\mathbf{a} \times \mathbf{b}$ is the **area** of the parallelogram determined by **a** and **b**.

The magnitude of the vector product of vectors $\mathbf{a} = OA$ and $\mathbf{b} = OB$ is the area of a parallelogram with sides OA and OB,

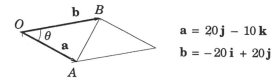

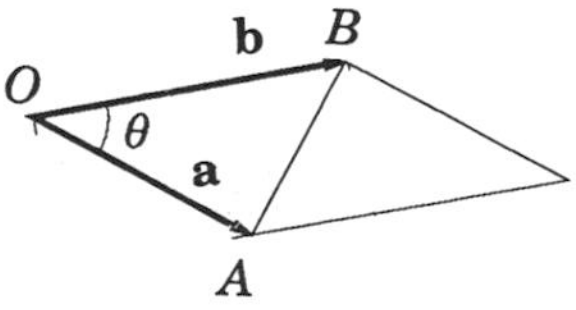

$$\mathbf{a} = 20\,\mathbf{j} - 10\,\mathbf{k}$$

$$\mathbf{b} = -20\,\mathbf{i} + 20\,\mathbf{j}$$

Since AB is the diagonal of the parallelogram, the sought roof area OAB is half the magnitude of the vector product.

Solved with determinant: p. 652

$$\mathbf{a} \times \mathbf{b} = 200\,\mathbf{i} + 200\,\mathbf{j} + 400\,\mathbf{k}$$

$$|\mathbf{a} \times \mathbf{b}| = \sqrt{200^2 + 200^2 + 400^2} = 200\sqrt{6}$$

$$\text{Area}_{OAB} = \frac{1}{2}\,|\mathbf{a} \times \mathbf{b}| = 100\sqrt{6} \approx 245\,\text{m}^2 .$$

There are other ways of calculating the sought surface area: by using the Pythagorean theorem, by trigonometric methods, and by Heron's formula.

Triple Products

Besides the scalar product and the vector product of two vectors, there are several kinds of "mixed" products of the types $\mathbf{a} \cdot (\mathbf{b} \times \mathbf{c})$ and $\mathbf{a} \times (\mathbf{b} \times \mathbf{c})$.

Scalar Triple Product

The combination $\mathbf{a} \cdot (\mathbf{b} \times \mathbf{c})$ is defined as the scalar dot product of a vector $\mathbf{a}$ and the cross product $\mathbf{b} \times \mathbf{c}$ of two vectors $\mathbf{b}$ and $\mathbf{c}$.

If

$$\left.\begin{aligned} \mathbf{a} &= x_1\mathbf{i} + y_1\mathbf{j} + z_1\mathbf{k} \\ \mathbf{b} &= x_2\mathbf{i} + y_2\mathbf{j} + z_2\mathbf{k} \\ \mathbf{c} &= x_3\mathbf{i} + y_3\mathbf{j} + z_3\mathbf{k} \end{aligned}\right\}$$

then

$$\mathbf{b} \times \mathbf{c} = (y_2 z_3 - y_3 z_2)\,\mathbf{i} - (x_2 z_3 - x_3 z_2)\,\mathbf{j} + (x_2 y_3 - x_3 y_2)\,\mathbf{k}$$

Solved with determinant: *p.* 652

and the **scalar triple product**

$$\mathbf{a} \cdot (\mathbf{b} \times \mathbf{c}) = x_1 (y_2 z_3 - y_3 z_2) - y_1 (x_2 z_3 - x_3 z_2) + z_1 (x_2 y_3 - x_3 y_2)\,.$$

The product of the three vectors $\mathbf{a}$, $\mathbf{b}$, $\mathbf{c}$ is positive if they form a right-handed system; otherwise it is negative. Cyclic permutation of the vectors $\mathbf{a}$, $\mathbf{b}$, $\mathbf{c}$ does not change the sign of the product, whereas reversal of any two vectors reverses the sign of the product:

$$\begin{aligned} \mathbf{a} \cdot (\mathbf{b} \times \mathbf{c}) &= \mathbf{b} \cdot (\mathbf{c} \times \mathbf{a}) = \mathbf{c} \cdot (\mathbf{a} \times \mathbf{b}) \\ &= -\mathbf{a} \cdot (\mathbf{c} \times \mathbf{b}) = -\mathbf{b} \cdot (\mathbf{a} \times \mathbf{c}) = -\mathbf{c} \cdot (\mathbf{b} \times \mathbf{a}) \end{aligned}$$

Scalar and vector products of more than two vectors are not defined in a general manner; only two vectors may be multiplied with each other at a time, vector products taking precedence over scalar products; these must be calculated first as they are defined only between vectors.

Geometrically, the absolute value of a scalar triple product is the volume of the parallelepiped with the edges $\mathbf{a}$, $\mathbf{b}$, and $\mathbf{c}$:

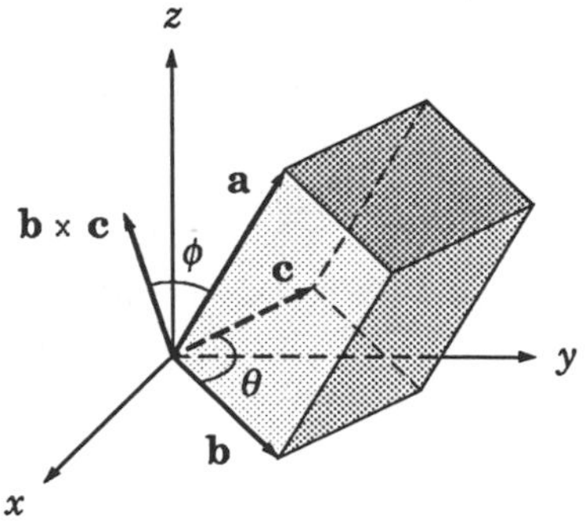

The base of the parallelepiped is a parallelogram with sides $|\mathbf{b}|$, $|\mathbf{c}|$, and the included angle θ. The area of the base is

$$|\,\mathbf{b} \times \mathbf{c}\,| = |\mathbf{b}|\;|\mathbf{c}|\,\sin\theta\,.$$

Since **b** x **c** is a vector product, it is perpendicular to the plane of vectors **b** and **c**. The projection of **a** onto the **b** x **c** plane is the height $|\mathbf{a}|\cos\phi$ of the parallelepiped, whose volume is

$$|\mathbf{b}\times\mathbf{c}|\;|\mathbf{a}|\cos\phi = \mathbf{a}\cdot(\mathbf{b}\times\mathbf{c}),\ \text{if } 0 \le \phi \le 90^\circ .$$

- Find the volume of a parallelepiped defined by vectors

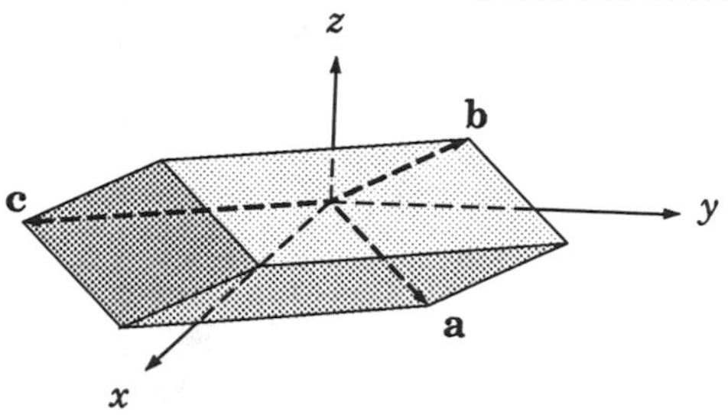

$$\mathbf{a} = \mathbf{i} + 2\mathbf{j} - \mathbf{k}$$
$$\mathbf{b} = 2\mathbf{j} + \mathbf{k}$$
$$\mathbf{c} = \mathbf{i} - 4\mathbf{j} .$$

Solved with determinant: *pp.* 652 - 53

$$\mathbf{a}\cdot(\mathbf{b}\times\mathbf{c}) = 8 \text{ units of volume.}$$

Vector Triple Product

The expression **a** x (**b** x **c**) is known as a **vector triple product**, because the result represents a vector. That the result must be a vector is evident from the fact that vector **a** is "crossed into" the vector formed by **b** x **c**.

The position of the parentheses is crucial,

$$\mathbf{j}\times(\mathbf{j}\times\mathbf{k}) = \mathbf{j}\times\mathbf{i} = -\mathbf{k};$$
$$(\mathbf{j}\times\mathbf{j})\times\mathbf{k} = \mathbf{0}\times\mathbf{k} = \mathbf{0}.$$

Since a vector product is perpendicular to the crossed vectors, **b** x **c** is perpendicular to the plane of **b** and **c**; and **a** x (**b** x **c**) lies in the **bc** plane:

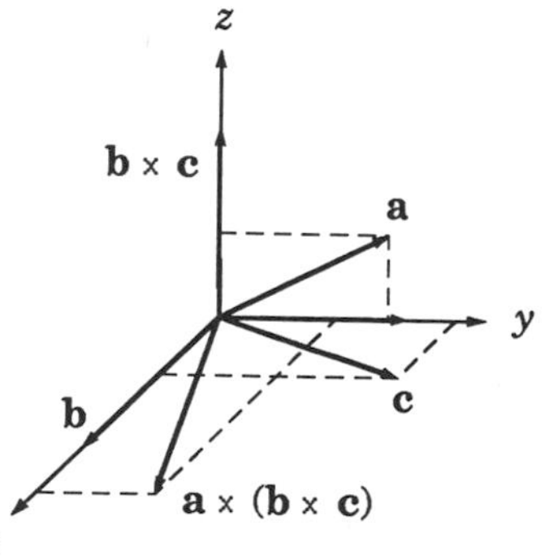

- If $\mathbf{a} = \mathbf{i} + 2\mathbf{j} - \mathbf{k}$; $\mathbf{b} = 2\mathbf{j} + \mathbf{k}$; $\mathbf{c} = \mathbf{i} - 4\mathbf{j}$, find $\mathbf{a}\times(\mathbf{b}\times\mathbf{c})$.

Solved with determinant: *p.* 652

$$\mathbf{b}\times\mathbf{c} = 4\mathbf{i} + \mathbf{j} - 2\mathbf{k};$$

$$\mathbf{a}\times(\mathbf{b}\times\mathbf{c}) = -3\mathbf{i} - 2\mathbf{j} - 7\mathbf{k}.$$

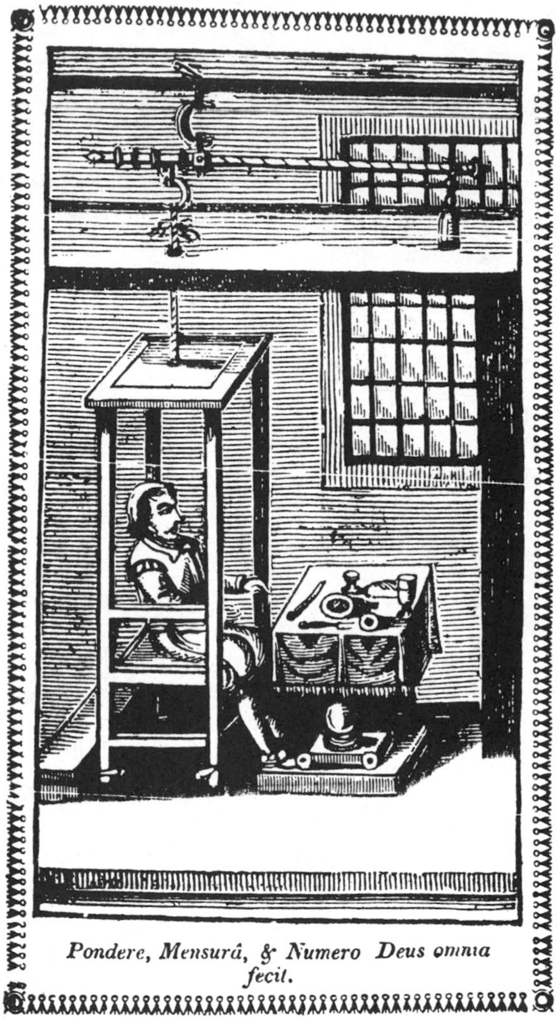
Pondere, Mensurâ, & Numero Deus omnia fecit.

God created everything by weight, measure, and number.

The Apocryphal *Book of Wisdom*

From Sanctorius Sanctorius's *Medicina Statica* (first edition, 1614; above illustration from an English edition printed in 1728).

SANCTORIUS Sanctorius
(1561 - 1636) Italian physician

GALILEI Galileo
(1564 - 1642)

Sanctorius spent long periods of time on a scale, making careful registrations of his own weight and weighing all intake and excretions. Through his observations, Sanctorius made essential conclusions regarding evaporation from the human body. Sanctorius's experimental research was greatly influenced by the methods of his contemporary **Galileo**, who, like Sanctorius, was at the University of Padua.

Chapter 17

FRACTALS

17.0 What Are Fractals?

MANDELBROT Benoit B.
(b. 1924)

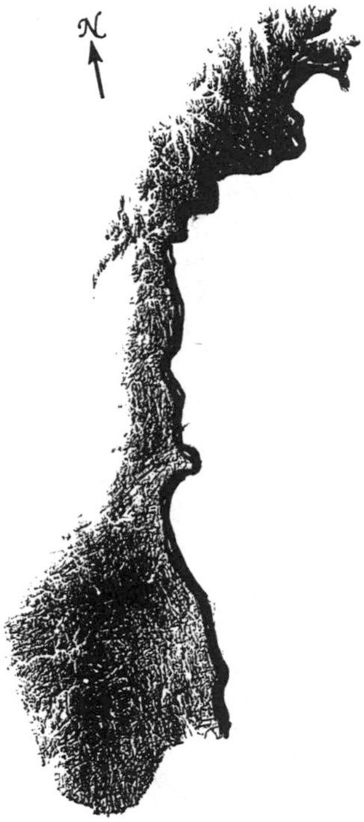

Who can measure Norway's coast line; or any coast line?

Latin, *fractus*, perfect participle of *frangere*, "to break"

"How Long Is the Coast Line of Britain?" is a landmark discourse in *Science*, presented in 1967 by the Polish-born French mathematician **Benoit Mandelbrot**. At first thought, the answer would seem to require measurements of photographs taken from an orbiting satellite or, for greater accuracy, from an aircraft flying along the coast or, for complete accuracy, by walking along the coastline with a measuring tape.

However, the answers obtained will only be successively closer approximations of the length of Britain's coastline. As Mandelbrot pointed out, the length obtained depends on the resolution of measurement, that is, the size of the smallest bend seen on the photograph or measured on site. Consequently, a coastline does not have a determinable length. Analogously, nor does a river have a determinable length.

If we analyze the curve of a coastline or the course of a river, we will find that, when magnified, parts of the curves are identical with the whole, or nearly so; with a certain scale of magnification, the pattern will repeat itself.

Similarly, the more a picture of a cloud is magnified, the more we will be aware of a seemingly endless piling up of forever smaller structures that repeat the general shape of the whole cloud.

The universe is replete with shapes that repeat themselves on different scales within the same object. In Mandelbrot terminology, such objects are said to be **self-similar**.

In the idealized world of mathematics, there are several well-defined figures that are self-similar and an infinite number of such figures may be generated through iteration of functions. These figures have quite unexpected properties, such as a boundless perimeter enclosing a finite area, or a boundless surface area containing a zero volume; the explanation lies in the fact that these figures do not belong within the human experience of a three-dimensional universe. The word **fractal** – coined by Mandelbrot – was intended to describe a dimension that could not be expressed as an integer; today, "fractal" is generally understood to mean a set that is self-similar under magnification.

Unlike mathematical fractals, no object in nature can be magnified an infinite number of times and still present the same shape of every detail in successive magnifications – one reason being the finite size of molecules and atoms. Yet fractal models may provide useful approximations of reality over a finite range of scales.

Mandelbrot and others have applied fractals as explanatory models of natural phenomena involving irregularities on different size scales. This technique is used in graphical analysis in such diverse fields as fluid mechanics, economics, and linguistics and in the study of crystal formation, vascular networks in biological tissue, and population growth.

17.1 The Snowflake Curve

A "snowflake" curve is generated on the perimeter – a broken line curve – of an equilateral triangle by the successive stages of removing the middle third of a line segment and replacing it by the other two sides.

von KOCH Nils Fabian Helge (1870 - 1924)

This curve is also known as **von Koch's curve**, or von Koch's island, named after the Swedish mathematician **Helge von Koch**, who described it in 1904. An aim of von Koch's article was to show that a simple geometric construction can lead to a continuous curve that in every point is void of a tangent and has no well-defined length.

The snowflake is a fractal; the self-similarity relates successive stages: any portion of the curve at stage n will, magnified by a factor of 3, look like a portion of the curve at $(n - 1)$.

The Perimeter

The **perimeter** of a snowflake curve increases indefinitely.

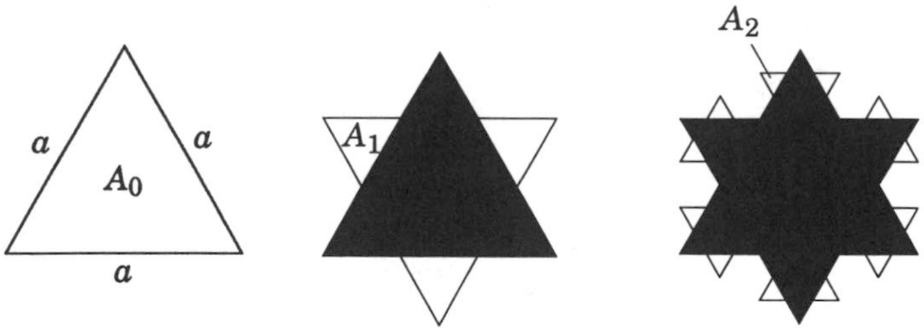

Start with an equilateral triangle of side length a and area A_0. The perimeter of the triangle is $3\,a$.

On the middle third of each side, construct a new equilateral triangle of side length $a/3$ and area A_1. The length of each side of triangle A_0 increases by a factor of 4/3, to $\frac{4}{3}a$.

The fractal is the limit of successive stages; the perimeter of the snowflake increases indefinitely towards infinity:

$$3\,a\left(\frac{4}{3}\right)^n \text{ tends to } \infty \text{ as } n \text{ tends to } \infty .$$

Thus, passing to the limit, the fractal has *infinite perimeter*.

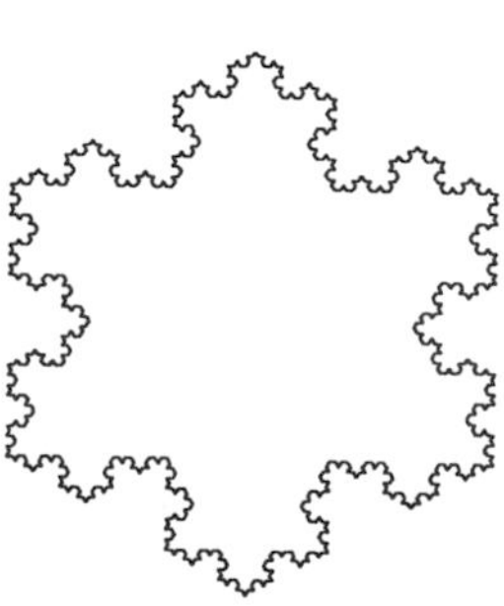

A computer is helpful in the graphic display of fractals, but it can only make a finite number of iterations and will never show the completed fractal. The graph to the left depicts the 4th iteration in the generation of a snowflake curve.

The snowflake curve is continuous, but a unique tangent cannot be drawn anywhere on its periphery. It is continuous everywhere but not smooth anywhere.

The Area

The area enclosed by the perimeter of the snowflake curve is contained within the circumscribed circle of A_0, and thus the snowflake curve encloses a *finite area*. The area approaches an upper limit which is 8/5 the area of the original triangle:

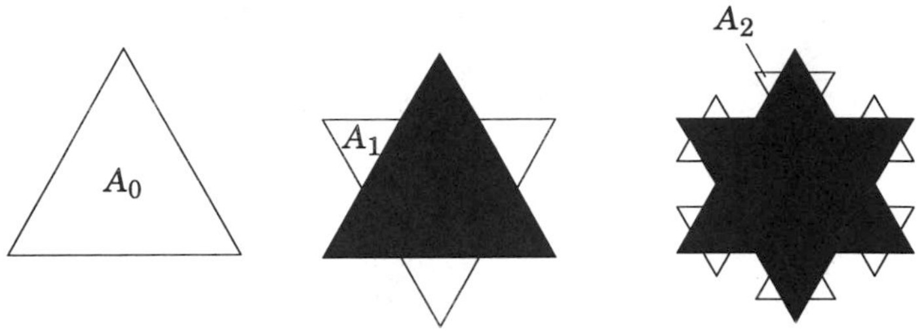

The sides of the original triangle A_0 provide bases for three new triangles A_1, thereby forming a six-pointed star whose twelve sides will serve as bases for the next generation of triangles A_2, *etc.*

Each time a new triangle is added, it will provide four bases for the following generation of triangles, and we have:

$$A_1 = \frac{1}{9} A_0 \qquad \sum A_1 = 3 \cdot \left(\frac{1}{9}\right) \cdot A_0$$

$$A_2 = \left(\frac{1}{9}\right)^2 A_0 \qquad \sum A_2 = 3 \cdot 4 \cdot \left(\frac{1}{9}\right)^2 \cdot A_0$$

$$A_3 = \left(\frac{1}{9}\right)^3 A_0 \qquad \sum A_3 = 3 \cdot 4^2 \cdot \left(\frac{1}{9}\right)^3 \cdot A_0$$

$$\dots \qquad \dots$$

$$A_n = \left(\frac{1}{9}\right)^n A_0 \qquad \sum A_n = 3 \cdot 4^{n-1} \cdot \left(\frac{1}{9}\right)^n \cdot A_0$$

and

$$A = A_0 + \sum_{n=1}^{\infty} 3 \cdot 4^{n-1} \cdot \left(\frac{1}{9}\right)^n \cdot A_0 = A_0 \left[1 + \frac{3}{4} \cdot \sum_{n=1}^{\infty} \left(\frac{4}{9}\right)^n \right]$$

$$= A_0 \left[1 + \frac{3}{4} \cdot \frac{4}{9} \cdot \sum_{n=0}^{\infty} \left(\frac{4}{9}\right)^n \right] = A_0 \left[1 + \frac{1}{3} \cdot \frac{1}{1 - 4/9} \right]$$

$$= A_0 \left[1 + \frac{3}{5} \right] = \frac{8}{5} A_0$$

Thus, the snowflake curve has a perimeter of infinite length enclosing a finite area. This curve and similar geometric constructions were long looked upon mainly as mathematical curiosities, but about 70 years after von Koch's description of the snowflake curve, Mandelbrot used that curve and other simple fractal constructions for his systematic description of natural phenomena. What had been viewed as mathematical oddities now became a means to describe reality!

17.2 Anti-Snowflake and Anti-Square Curves

The Anti-Snowflake Curve

If, unlike the von Koch curve, the triangles are pointed inward, an anti-snowflake curve is generated. The illustration below shows the first four iterations in the course of generating an **anti-snowflake curve**:

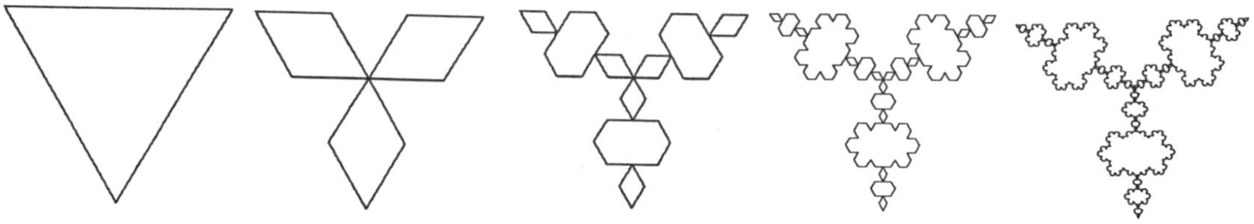

Like the snowflake curve, the perimeter of the anti-snowflake curve increases indefinitely, whereas its total enclosed area approaches a finite limit that is 2/5 of the area of the original triangle.

The proof is similar to that of the snowflake curve but, instead of adding, we now subtract the triangles at each stage.

The Anti-Square Curve

An "**anti-square curve**" may be generated in a manner similar to that of the anti-snowflake curve; the first four iterations show the following pattern:

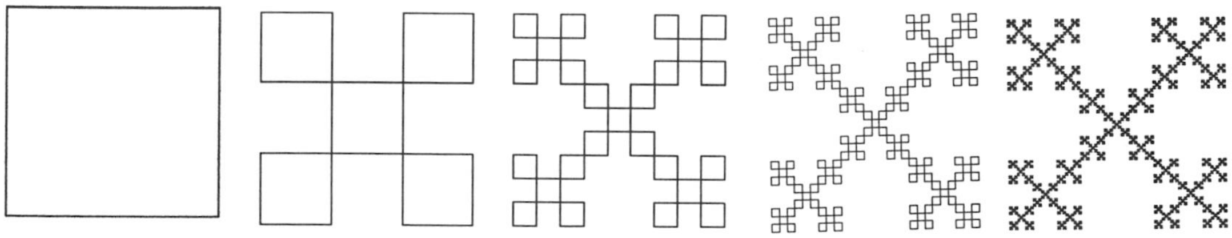

17.3 The Cantor Set

CANTOR Georg: *p.* 257

A Cantor set is generated by removing, in an iterative fashion, a midportion of a true line segment.

The figure below illustrates a Cantor set generated by removing the middle third of the original line segment [0, a] and of every successive set.

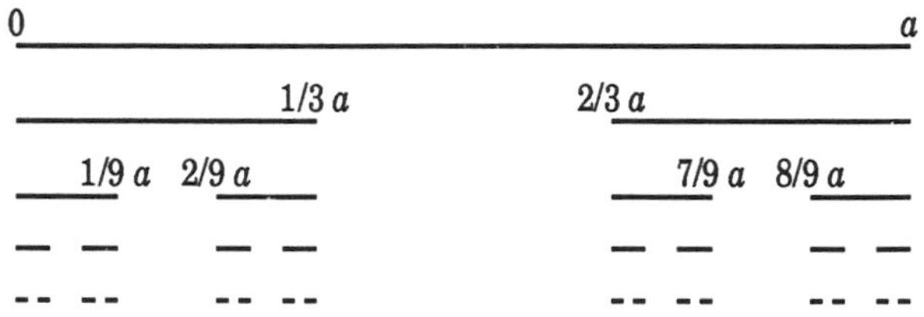

When removing the midportions, we leave the endpoints behind. In a way, a Cantor set may be regarded as the simplest of all fractals; yet it offers several features to ponder.

In the Cantor set, we remove, from a segment of length a, segments whose total length is a; yet we leave behind as many points as we started with.

The whole is made up of two copies of itself at one-third scale. A Cantor set contains no whole line segments, for at some stage its midportion would have been removed, and in the end we have **totally disconnected points** – a "dust".

Once thought to be mathematical curiosities, Cantor sets were first applied by Mandelbrot as mathematical models for solving problems of noise in electronic communication.

Using a solid instead of the one-dimensional line segment as the initial stage, Mandelbrot has extended the concept of the Cantor set to serve as a model of the distribution of matter in the universe and as a model of the distribution of water droplets in a cloud.

Sprinkling flour while rolling pie crust,
He thought about Mandelbrot dust.
Fractals all-round the ledge
Gave the crust a fine edge.
Apple pie done just right was a must.

17.4 Sierpinski Triangle, Carpet, and Sponge

SIERPINSKI Vaclav
(1882 - 1969)

The Sierpinski triangle, carpet, and sponge are named after the Polish mathematician **Vaclav Sierpinski.**

The **Sierpinski triangle**, or **gasket**, starts out as an equilateral triangle where the infinite succession of removals of equilateral triangles takes place *inside* the triangle, the first portion to be removed being a triangle with corners at the midpoints of the sides of the original triangle. The three equilateral triangular areas that remain within the original triangle are each broken up into four equilateral triangles, of which the central one is removed. This process may be iterated *ad infinitum* and a fractal form is generated; the first six iterations are shown here:

The Sierpinski triangle can be regarded as three copies of itself, each at one-half scale.

The final Sierpinski triangle has an unbounded perimeter and a zero area.

The **Sierpinski carpet** starts out as a square which is divided into nine smaller squares, of which the central square is removed. Each of the remaining squares is divided into nine smaller squares from which again the central square is removed. The process is iterated *ad infinitum* and a fractal form is generated.

The Sierpinski carpet can be regarded as the union of eight copies of itself, each at one-third scale.

Like the Sierpinski triangle, the Sierpinski carpet has an unbounded perimeter and zero area.

MENGER Karl
(Austrian-born American mathematician and philosopher; b. 1902)

The **Sierpinski sponge**, or **Menger sponge**, is generated from a cube in a manner analogous to the generation of the Sierpinski carpet from a square. The cube is divided into 27 smaller cubes, and the central cube and those at the center of each face of the original cube are removed. Each of the 20 remaining cubes is divided into 27 yet smaller cubes; the central cube and the cubes facing the center of each face of the preceding cube are removed.

When the process is iterated *ad infinitum*, a fractal form is generated. Each external face of the Sierpinski sponge is a Sierpinski carpet.

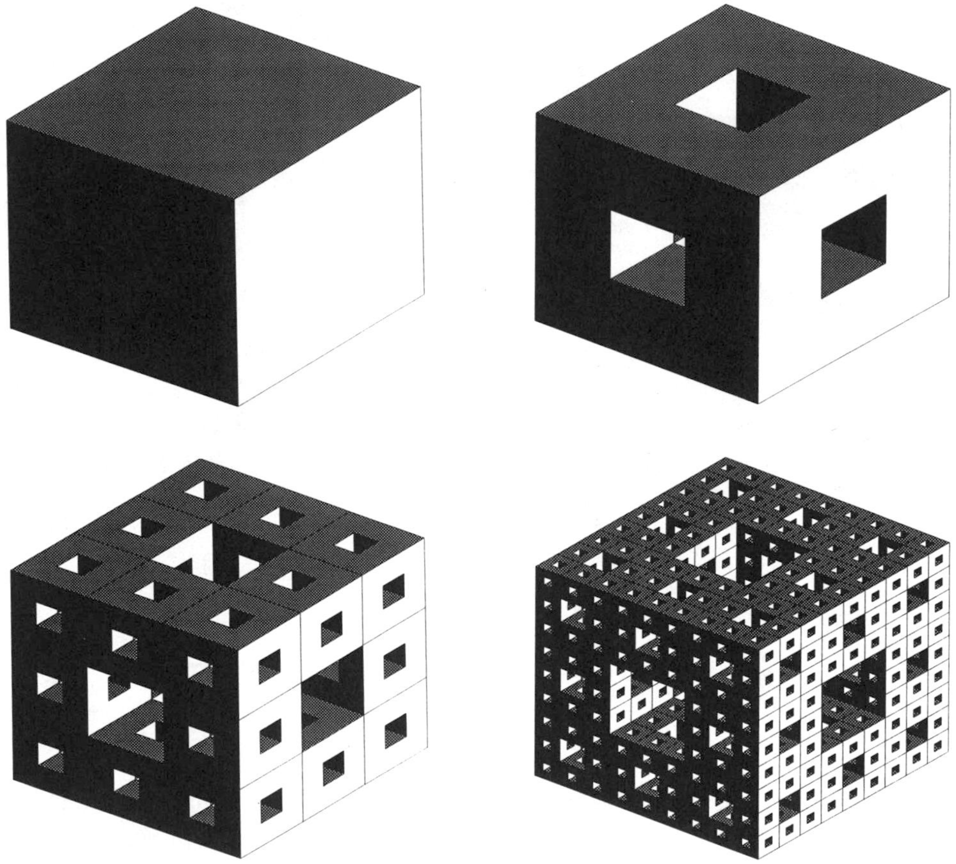

The Sierpinski sponge is made up of 20 copies of itself, each at one-third scale.

The Sierpinski sponge has zero volume, "enclosed" by an unbounded surface area.

17.5 The Mandelbrot Set

The fractal behavior in the complex number plane is demonstrated by iterating a nonlinear function whose variables *include its own result*. If a set of an infinite sequence $f(z)$, $f[f(z)]$, $f\{f[f(z)]\}$..., where z is a complex number, is plotted on a graph, the sequence of iterates may

1. be unbounded; or
2. jump around within a bounded region.

JULIA Gaston Maurice (1892 - 1978) French mathematician

If (2) holds, we say that z lies in the "filled-in **Julia set** for f". In the margin, the Julia set when $f(z) = z^2 + \mathrm{i}$ is illustrated with computer graphics.

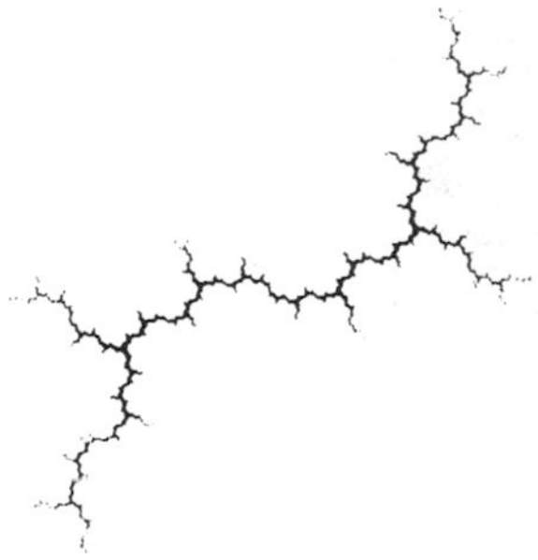

Julia set of $z^2 + \mathrm{i}$.

The **Mandelbrot set** is related to the Julia set, but for it the defining variable is the c in $f(z) = z^2 + c$, where z and c are complex numbers. Starting with $z = 0 + 0\mathrm{i}$, we look for the complex numbers c, such that $0, f(0), f[f(0)]$, ... remain bounded.

If we let $z = 0 + 0\mathrm{i}$ in $f(z) = z^2 + c$, then

$$\begin{aligned} f(z) &= f(0+0\,\mathrm{i}) = (0+0\mathrm{i})^2 + c = c \\ f[f(z)] &= f(c) \qquad\quad\;\; = c^2 + c \\ f\{f[f(z)]\} &= f(c^2+c) \;\; = (c^2+c)^2 + c \end{aligned}$$

and the process may be iterated *ad infinitum*.

If $c = 1 + \mathrm{i}$, the above functions yield

$$\begin{aligned} f(z) &= f(0+0\mathrm{i}) = 1+\mathrm{i} \\ f[f(z)] &= f(1+\mathrm{i}) \;\;= (1+\mathrm{i})^2 + 1 + \mathrm{i} = 1 + 3\,\mathrm{i} \\ f\{f[f(z)]\} &= f(1+3\,\mathrm{i}) = [(1+\mathrm{i})^2 + 1 + \mathrm{i}]^2 + 1 + \mathrm{i} \\ &= -7 + 7\mathrm{i} \text{ and the iterates tend to } \infty. \end{aligned}$$

We are now in a position to define the **Mandelbrot set** as the set of all complex numbers c for which the iterated $f(z) = z^2 + c$ remains bounded. The initial value of z is $0 + 0\mathrm{i}$ and each subsequent value of z is used to find the next one. Using computer graphics, the Mandelbrot set is:

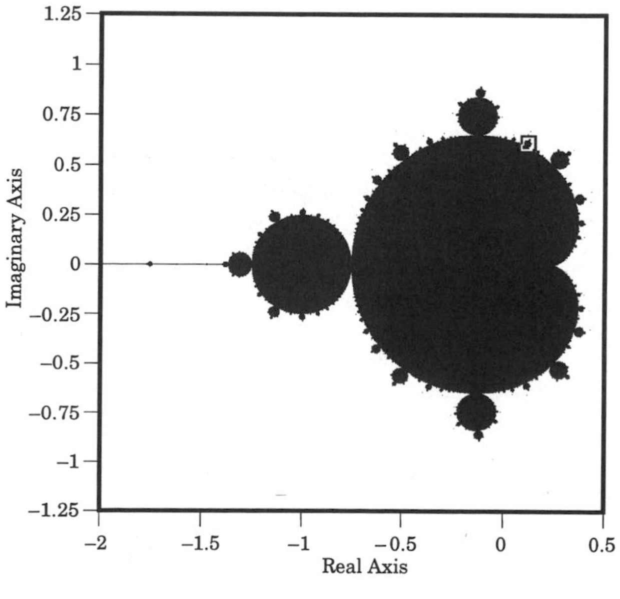

By zooming in on any of the outgrowths of the Mandelbrot set, its self-similar fractal behavior is evident. The picture below is a magnification of the □ area in the picture on the previous page.

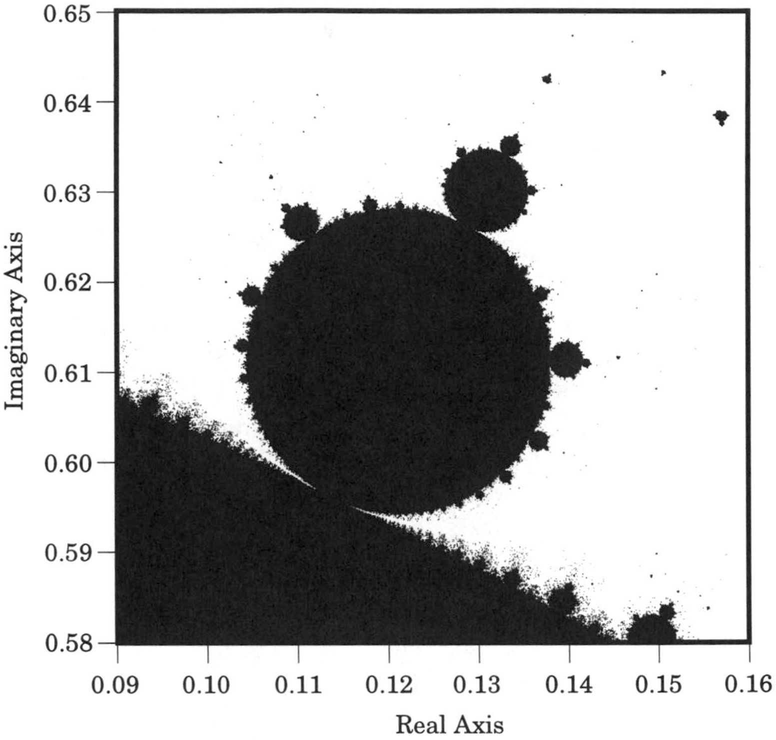

Unlike the snowflake curve and the other fractals discussed above, that of the Mandelbrot set is nonlinear; the difference is that every self-similarity of a magnified Mandelbrot offshoot is more obviously elaborated than its parent; that is, Mandelbrot sets and Julia sets are self-similar in a common-language way, but not in the strict scale-factor way of the Koch snowflake.

DOUADY Adrien (b. 1935)

HUBBARD John (b. 1945)

The geometric link between the Mandelbrot and Julia sets is that the Julia set for $f(z) = z^2 + c$ is disconnected when c lies outside the Mandelbrot set. The Mandelbrot set itself is a connected mass, as was shown by **Adrien Douady** and **John Hubbard**. Thus, the elaborate tendrils of the Mandelbrot set connect what appear here as islands.

17.6 The Dimension Concept

In Euclidean geometry a point has dimension zero; a shape with length alone has dimension one; an area has dimension two; a volume has dimension three.

Peano Curve

PEANO Guiseppe (1858 - 1932)

HILBERT David (1862 - 1943)

The Italian mathematician and logician **Guiseppe Peano** described in 1890 a curve that could pass through every point of a square; the construction of the **Peano curve** was simplified by the German mathematician **David Hilbert**.

Divide a unit square into four sub-squares; join the centers of the sub-squares by a broken line. Then divide every sub-square of a unit square into four "sub-sub"-squares; join the centers of all the sub-sub-squares by a broken line:

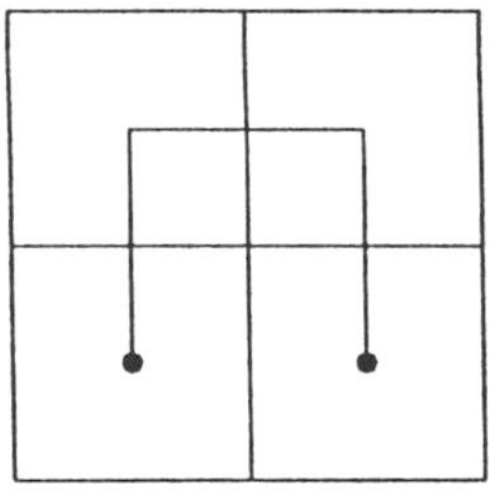

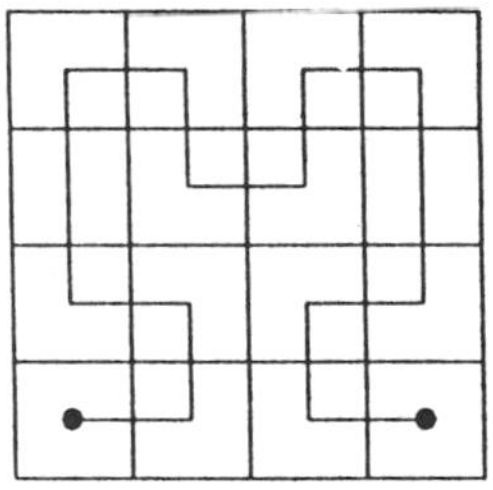

Proceed to the next stage by dividing a unit square into 4 x 4 squares and, again, join every center by a broken line; then fashion 8 x 8 little squares and continue the process of drawing a broken line:

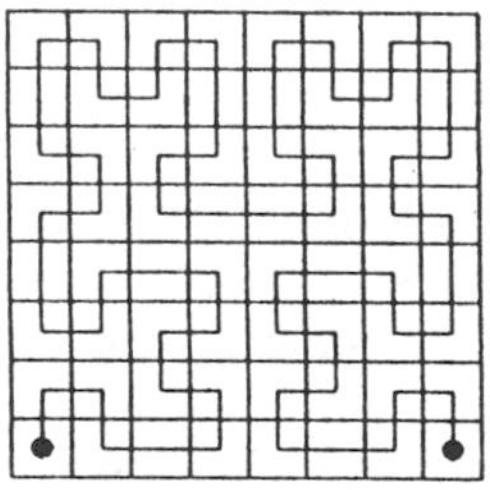

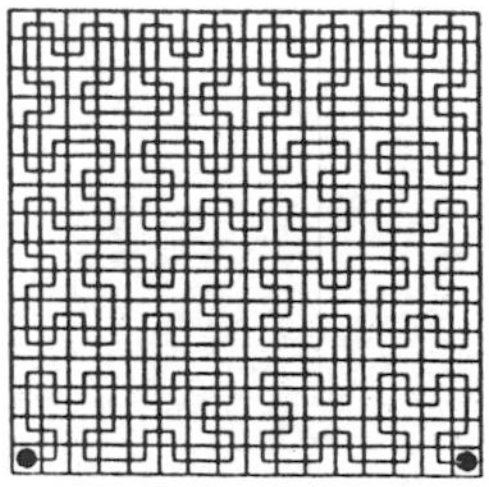

Continuing the division of the square *ad infinitum* and every time joining the centers of all the sub-sub-...sub-squares by a continuous open curve, we approach a stage where the broken line becomes a curve and passes through every point of the plane.

This curve is not a fractal but does indeed present interesting dimensional aspects. Tending to the limit of passing through every point in a plane, the Peano curve may be argued to be of *two dimensions*.

Dimensions of Fractals

If we divide a straight **line segment** into equal parts, each scaled down to length $1/N$-th of the whole, we will need N^1 parts to reassemble the original line segment; the dimension of the object is equal to the exponent.

If we divide a **square** into smaller squares, each scaled down to $1/N$-th of the whole, we will need N^2 parts to reassemble the original square; again, the mathematical dimension is equal to the exponent.

Similarly, if we divide a **cube** into smaller cubes, each scaled to side length $1/N$-th of the original cube, we will need N^3 parts to reassemble the original cube; again, the dimension is equal to the exponent.

In all three cases – the line segment, the square, and the cube – we can piece together N^D pairs, each scaled down by $1/N$ from the original, and rebuild the original.

Similarity Dimension

The **similarity dimension** D is defined as the ratio

$$D = \frac{\log(\text{number of parts})}{\log(1/\text{the linear magnification})}.$$

Thus, for the line segment,

$$D = \frac{\log N^1}{\log N} = \frac{1 \log N}{\log N} = 1;$$

the square,

$$D = \frac{\log N^2}{\log N} = \frac{2 \log N}{\log N} = 2;$$

and the cube,

$$D = \frac{\log N^3}{\log N} = \frac{3 \log N}{\log N} = 3.$$

The similarity dimension is a basic tool in studying fractals.

At every stage of generating the **snowflake curve**, each side of the triangle develops into 4 smaller parts; scaled down to one-third size, the triangles are self-similar. Hence, the dimension

of the snowflake curve is $D = \dfrac{\log 4}{\log 3} = 1.261\ldots;$

of the Sierpinski triangle is $D = \dfrac{\log 3}{\log 2} = 1.584\ldots;$

of the Sierpinski carpet is $D = \dfrac{\log 8}{\log 3} = 1.892\ldots;$

of the Sierpinski sponge is $D = \dfrac{\log 20}{\log 3} = 2.726\ldots;$

of the Cantor set is $D = \dfrac{\log 2}{\log 3} = 0.630\ldots.$

The large dark areas of the **Mandelbrot set** are two-dimensional but the entire set, being not exactly self-similar, does not have a similarity dimension. On the other hand, the similarity dimension is only one of many that yield fractal dimensions. The filigreed boundary of the Mandelbrot set wiggles so violently as to be considered two-dimensional in what is known as the *Hausdorff-Besicovich dimension*.

Chapter 18

MATRICES AND DETERMINANTS

Once a mathematician named Dix
Gardened his lot, just for kicks.
And, as everything grows
In columns and rows,
He cheerfully weeded matrix.

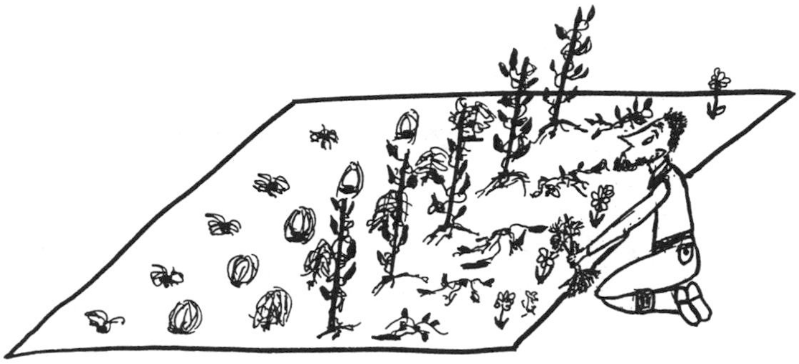

18.0 Scope and History

Scope

A matrix is a rectangular arrangement of *entries*, displayed in rows and columns. The entries may be any kind of numbers, polynomials, or other expressions.

A *square matrix* has a determinant which has a *value* determined by a rule of combination for the entries.

The value of the determinant may be used – either alone or in combination with matrix operations – for solving systems of linear equations, or for solving problems of figures displayed in a coordinate system. Matrices are mainly used to solve large systems of linear equations – discussed in this text – and for transformation of objects in a coordinate system.

History

The ancient Chinese had a method of solving systems of linear equations by representing the coefficients of the unknown quantities by placing bamboo rods on a calculating board in the same pattern as the entries of today's matrix. This led to an early development of the principles of adding and subtracting these rows and columns of the array.

SEKI KOWA (1642 - 1708)

When Chinese science was introduced into Japan, the idea of determinants soon gained acceptance. **Seki Kowa**, greatest of 17th-century Japanese mathematicians, in 1683 wrote *Kai Fukudai no Ho* in which he discussed determinants and methods of their expansion.

von LEIBNIZ Gottfried Wilhelm (1646 - 1716)

CRAMER Gabriel (1704 – 1752)

In 1693, the German **Leibniz** gave a formal description of determinants as applied to the solution of systems of three linear equations in three unknowns. In 1750, the Swiss mathematician **Gabriel Cramer** established the general rule for solving systems of n linear equations in n unknowns.

VANDERMONDE Alexandre-Théophile (1735 - 1796)

Alexandre-Théophile Vandermonde, in 1771, was the first to recognize determinants as independent functions, apart from their use for the solution of systems of linear equations. He also described a number of properties of these functions, and developed a form of notation more complete and appropriate than that suggested by Leibniz. Vandermonde may be regarded as the formal founder of determinant theory.

LAPLACE Pierre Simon de (1749 - 1827)

LAGRANGE Joseph Louis (1736 - 1813)

Laplace in 1772 described a general method of expanding determinants in terms of their minors. **Lagrange**, in 1773, handled second- and third-order determinants, and introduced them for purposes besides the solution of equations.

GAUSS Carl Friedrich (1777 - 1855)

Gauss, in 1801, used determinants in his theory of numbers and suggested the idea of reciprocal determinants; he also came very near the multiplication theorem.

CAUCHY Augustin Louis (1789 - 1857)

The French mathematician **Cauchy**, in 1812, summarized what was known of determinants at the time, and also introduced the word *determinant* to mathematics. He started the theory of determinants as a distinct, independent branch of modern mathematics.

JACOBI Carl Gustav Jakob (1804 - 1857)

The next important contributor to the theory of determinants was **Jacobi**, through whom the term *determinant* received its final acceptance.

HAMILTON William Rowan (1805 - 1865)

Latin, *matrix*, "womb", "mould"

SYLVESTER James Joseph (1814 - 1897)

Thus, determinants were actually well established long before the concept of matrices emerged. Although it had been implicit in earlier writings, particularly by the Irishman **William Rowan Hamilton** in 1839 and 1841, the term **matrix** was not used until 1850 by the English mathematician **James Joseph Sylvester**, who gave it its present meaning of a rectangular array of numbers from which determinants may be formed.

CAYLEY Arthur (1821 -1895)

Credit for understanding and identifying the full significance of the algebraic properties of matrices is generally given to the English mathematician **Arthur Cayley** who presented fundamental work on matrices in 1855.

Because of many years of close collaboration, Cayley and Sylvester are usually considered joint founders of matrix theory, a recognition that should properly be shared by Hamilton.

Matrices, and to a lesser degree determinants, find application in physics and kindred sciences, in engineering, economics, social sciences, and other fields where vast volumes of data are handled.

18.1 Matrices – Presentation

Entries, Elements
Rows Columns

A matrix is an array of real and/or complex numbers, quaternions or any kind of numbers, known as **entries**, or **elements**, disposed in a rectangular pattern of **rows** and **columns**.

The purpose of matrices is to provide a kind of mathematical shorthand to facilitate the study of problems represented by the entries: the matrices may represent transformations (*linear* transformations) of coordinate spaces, or systems of simultaneous linear equations.

Matrices are not numbers and have, in themselves, no specific numerical significance but may be looked upon collectively as a kind of higher-order number system carrying a built-in code of interpretation embodied in the *determinants* of *square matrices*.

In this text, we shall denote matrices by upright boldface capital letters usually from the beginning of the Roman alphabet, and enclose the matrices themselves by square brackets; entries will be in lowercase italics.

Some texts use capital italics or German "Fraktur" characters for matrices, enclosing them in parentheses or, less common, double bars.

$$\begin{bmatrix} \cdots \\ \cdots \\ \cdots \end{bmatrix} \qquad \begin{pmatrix} \cdots \\ \cdots \\ \cdots \end{pmatrix} \qquad \begin{Vmatrix} \cdots \\ \cdots \\ \cdots \end{Vmatrix}$$

Row Index
Column Index

The entries of a matrix carry double subscripts; a_{ij} belongs in the i-th row of the j-th column, i being the **row index** and j the **column index**. A matrix with m rows and n columns,

$$\mathbf{A} = [a_{ij}]_{m\,n} = \begin{bmatrix} a_{11} & a_{12} & a_{13} & \cdots & a_{1n} \\ a_{21} & a_{22} & a_{23} & \cdots & a_{2n} \\ \vdots & \vdots & \vdots & \vdots & \vdots \\ a_{m1} & a_{m2} & a_{m3} & \cdots & a_{mn} \end{bmatrix},$$

is an $m \times n$, or m-by-n, matrix; a_{11} reads a-one-one, not a-eleven; a_{12} is a-one-two; a_{23} is a-two-three, *etc.*

Dimension, Order
Square Matrix

The expression $m \times n$ is the **dimension**, or **order**, of the matrix; matrices with $m = n$ are **square matrices** of the order m, or n, denoted

$$[a_{ij}]_m\,; \quad [a_{ij}]_n\,.$$

Leading Entry, Leading Element

The first non-zero entry in a row of a matrix is called the **leading entry**, or **leading element**, of that row; in the matrix below, 4 is the leading entry in the first row, 7 in the second row:

$$\mathbf{A} = \begin{bmatrix} 0 & 4 & -3 \\ 7 & 2 & 1 \end{bmatrix}$$

Principal Diagonal
Main Diagonal
Diagonal Entries

The entries $a_{11}, a_{22}, a_{33}, \ldots, a_{nn}$ that make up the **principal diagonal**, or **main diagonal** – from upper left to lower right – are **diagonal entries**.

Vector
Algebraic Vector

If either m or n is unity, the matrix is a **vector** or, specifically, an **algebraic vector**, to distinguish it from vectors represented by directed line segments. The two forms of vectors are, actually, different ways of expressing the same data.

Row Vector
Column Vector

A matrix with $m = 1$ consists of a single **row vector**, an n-dimensional row vector; if $n = 1$, we have an m-dimensional **column vector**:

$$\mathbf{X} = [\ a_{11} \quad a_{12} \quad a_{13} \quad \ldots \quad a_{1n}\ ]; \quad \mathbf{Y} = \begin{bmatrix} a_{11} \\ a_{21} \\ \ldots \\ a_{m1} \end{bmatrix}$$

Scalar

If $m = n = 1$, the matrix is reduced to a single entry and becomes a **scalar**.

Lower Triangular Matrix
Upper Triangular Matrix

A matrix in which all entries above the main diagonal are zero is a **lower triangular matrix**; if all entries below the diagonal are zero, the matrix is an **upper triangular matrix**.

When all entries above and below the main diagonal are zero, and all non-zero entries are diagonal entries, the matrix is a diagonal matrix, and is usually denoted **D**.

$$\mathbf{A} = \begin{bmatrix} 3 & 2 & 7 & 2 \\ 0 & 2 & 4 & 9 \\ 0 & 0 & 1 & 1 \\ 0 & 0 & 0 & 8 \end{bmatrix} \quad \mathbf{B} = \begin{bmatrix} 7 & 0 & 0 & 0 \\ 8 & 4 & 0 & 0 \\ 5 & 5 & 4 & 0 \\ 3 & 6 & 3 & 8 \end{bmatrix} \quad \mathbf{D} = \begin{bmatrix} 3 & 0 & 0 \\ 0 & 12 & 0 \\ 0 & 0 & 4 \end{bmatrix}$$

Upper Triangular Matrix   Lower Triangular Matrix   Diagonal Matrix

Trace

The sum of the diagonal entries of a square matrix is the **trace** of the matrix:

$$\text{tr}\mathbf{A} = 14 \qquad \text{tr}\mathbf{B} = 23 \qquad \text{tr}\mathbf{D} = 19$$

Scalar Matrix
Unit Matrix, Identity Matrix
E **I**

A diagonal matrix with $a_{11} = a_{22} = a_{33} = \ldots = a_{nn} = k$, where k is a constant, is a **scalar matrix**. If $k = 1$, the matrix is a **unit matrix**, **E** (for German "Einheit" = unity), or an **identity matrix**, **I**.

$$\mathbf{I} = \begin{bmatrix} 1 & 0 & 0 & 0 \\ 0 & 1 & 0 & 0 \\ 0 & 0 & 1 & 0 \\ 0 & 0 & 0 & 1 \end{bmatrix}$$

Null Matrix

A matrix in which all entries are zero is a **null matrix**, generally denoted **0** or by the Greek letter θ (theta):

$$\theta = \begin{bmatrix} 0 & 0 & 0 & 0 \\ 0 & 0 & 0 & 0 \\ 0 & 0 & 0 & 0 \end{bmatrix} = \mathbf{0}$$

Null Vector

Analogously, a vector with only zero entries is a **null vector**.

18.2 Matrices – Rules of Operation

Addition and Subtraction

Addition of matrices is defined only for matrices of *equal dimensions* which have the same number of rows and the same number of columns, that is, are **conformable for addition**:

$$[a_{ij}]_{mn} + [b_{ij}]_{mn} = [a_{ij} + b_{ij}]_{mn}$$

The actual operation of addition is performed by adding, individually, corresponding entries of the matrices. *Subtraction* is treated as *negative addition*.

$$\mathbf{A} = \begin{bmatrix} 3 & -2 & 7 \\ 0 & 6 & 4 \\ 4 & 0 & 1 \\ 1 & 3 & 0 \end{bmatrix} \qquad \mathbf{B} = \begin{bmatrix} 2 & 4 & 3 \\ 4 & 6 & 0 \\ 1 & 7 & 4 \\ 2 & 3 & 1 \end{bmatrix}$$

$$\mathbf{A}+\mathbf{B} = \begin{bmatrix} 5 & 2 & 10 \\ 4 & 12 & 4 \\ 5 & 7 & 5 \\ 3 & 6 & 1 \end{bmatrix} \qquad \mathbf{A}-\mathbf{B} = \begin{bmatrix} 1 & -6 & 4 \\ -4 & 0 & 4 \\ 3 & -7 & -3 \\ -1 & 0 & -1 \end{bmatrix}$$

The **additive inverse** of a matrix **A** is $-\mathbf{A}$:

$$\mathbf{A} = \begin{bmatrix} 3 & -2 & 7 \\ 0 & 6 & 4 \end{bmatrix} \Rightarrow -\mathbf{A} = \begin{bmatrix} -3 & 2 & -7 \\ 0 & -6 & -4 \end{bmatrix}$$

Multiplication

Multiplication by a Scalar

To multiply a matrix by a real or complex scalar k, every entry a_{ij} of the matrix is to be multiplied separately by k:

$$k\mathbf{A} = k\begin{bmatrix} a_{11} & a_{12} & a_{13} & \cdots & a_{1n} \\ a_{21} & a_{22} & a_{23} & \cdots & a_{2n} \\ \vdots & \vdots & \vdots & \vdots & \vdots \\ a_{m1} & a_{m2} & a_{m3} & \cdots & a_{mn} \end{bmatrix} = \begin{bmatrix} ka_{11} & ka_{12} & ka_{13} & \cdots & ka_{1n} \\ ka_{21} & ka_{22} & ka_{23} & \cdots & ka_{2n} \\ \vdots & \vdots & \vdots & \vdots & \vdots \\ ka_{m1} & ka_{m2} & ka_{m3} & \cdots & ka_{mn} \end{bmatrix}$$

Matrix Products

The product of a row vector $\mathbf{A}_{1m}$ and a column vector $\mathbf{B}_{m1}$ is

$$[a_{11} \quad a_{12} \quad \cdots \quad a_{1m}] \begin{bmatrix} b_{11} \\ b_{21} \\ \vdots \\ b_{m1} \end{bmatrix} = \left[a_{11}b_{11} + a_{12}b_{21} + \ldots + a_{1m}b_{m1}\right].$$

Two matrices **A** and **B** are **conformable for multiplication**, in that order, if the number of columns in **A** is the same as the number of rows in **B**.

The product of two matrices $\mathbf{A}_{mn}$ and $\mathbf{B}_{np}$,

$$\mathbf{A}_{mn} = \begin{bmatrix} a_{11} & a_{12} & \dots & a_{1n} \\ a_{21} & a_{22} & \dots & a_{2n} \\ \vdots & \vdots & \vdots & \vdots \\ a_{m1} & a_{m2} & \dots & a_{mn} \end{bmatrix} \quad \mathbf{B}_{np} = \begin{bmatrix} b_{11} & b_{12} & \dots & b_{1p} \\ b_{21} & b_{22} & \dots & b_{2p} \\ \vdots & \vdots & \vdots & \vdots \\ b_{n1} & b_{n2} & \dots & b_{np} \end{bmatrix},$$

is obtained by multiplying every row of **A** into every column of **B**:

$$\mathbf{A}_{mn} \times \mathbf{B}_{np} = \mathbf{C}_{mp}$$

$$= \begin{bmatrix} a_{11} b_{11} + a_{12} b_{21} + \dots + a_{1n} b_{n1} & a_{11} b_{12} + a_{12} b_{22} + \dots + a_{1n} b_{n2} & \dots & a_{11} b_{1p} + a_{12} b_{2p} + \dots + a_{1n} b_{np} \\ a_{21} b_{11} + a_{22} b_{21} + \dots + a_{2n} b_{n1} & a_{21} b_{12} + a_{22} b_{22} + \dots + a_{2n} b_{n2} & \dots & a_{21} b_{1p} + a_{22} b_{2p} + \dots + a_{2n} b_{np} \\ \vdots & \vdots & \vdots & \vdots \\ a_{m1} b_{11} + a_{m2} b_{21} + \dots + a_{mn} b_{n1} & a_{m1} b_{12} + a_{m2} b_{22} + \dots + a_{mn} b_{n2} & \dots & a_{m1} b_{1p} + a_{m2} b_{2p} + \dots + a_{mn} b_{np} \end{bmatrix}$$

Matrix multiplication is *distributive* and *associative*,

$$\mathbf{A}(\mathbf{B}+\mathbf{C}) = \mathbf{AB}+\mathbf{AC}; \quad (\mathbf{A}+\mathbf{B})\,\mathbf{C} = \mathbf{AC}+\mathbf{BC};$$

$$\mathbf{A}(\mathbf{BC}) = (\mathbf{AB})\,\mathbf{C},$$

but generally *not commutative*: the products **AB** and **BA** are, as a rule, not equal.

• $\mathbf{A} = \begin{bmatrix} 1 & -2 \\ 3 & 4 \end{bmatrix}$; $\mathbf{B} = \begin{bmatrix} 1 & 0 \\ -1 & 1 \end{bmatrix}$.

Find **AB** and **BA**.

$$\mathbf{AB} = \begin{bmatrix} 1 & -2 \\ 3 & 4 \end{bmatrix} \begin{bmatrix} 1 & 0 \\ -1 & 1 \end{bmatrix}$$

$$= \begin{bmatrix} 1 \cdot 1 + (-2)(-1) & 1 \cdot 0 + (-2)\,1 \\ 3 \cdot 1 + 4\,(-1) & 3 \cdot 0 + 4 \cdot 1 \end{bmatrix} = \begin{bmatrix} 3 & -2 \\ -1 & 4 \end{bmatrix}.$$

$$\mathbf{BA} = \begin{bmatrix} 1 & 0 \\ -1 & 1 \end{bmatrix} \begin{bmatrix} 1 & -2 \\ 3 & 4 \end{bmatrix}$$

$$= \begin{bmatrix} 1 \cdot 1 + 0 \cdot 3 & 1\,(-2) + 0 \cdot 4 \\ (-1)\,1 + 1 \cdot 3 & (-1)(-2) + 1 \cdot 4 \end{bmatrix} = \begin{bmatrix} 1 & -2 \\ 2 & 6 \end{bmatrix}$$ •

The commutativity rule does apply, however, if one of the factor matrices is the *identity matrix*:

$$\mathbf{AI} = \mathbf{IA} = \mathbf{A}.$$

Any matrix multiplied by a *null matrix* will give the result zero. There are also matrices that are not null matrices but nevertheless will give zero products:

$$\mathbf{A} = \begin{bmatrix} 0 & 1 & 0 \\ 1 & 0 & 1 \\ 0 & 1 & 0 \end{bmatrix}; \ \mathbf{B} = \begin{bmatrix} 1 & -2 & 1 \\ 0 & 0 & 0 \\ -1 & 2 & -1 \end{bmatrix}; \ \mathbf{AB} = 0; \ \mathbf{BA} \neq 0$$

Premultiply
Postmultiply

In a product **AB** of two matrices **A** and **B**, **B** is said to be **premultiplied** by **A**, and **A** to be **postmultiplied** by **B**.

The following rules apply for square matrices:

- o $\mathbf{A}^m \mathbf{A}^n = \mathbf{A}^{m+n}$
- o $(\mathbf{A}^m)^n = \mathbf{A}^{mn}$

Powers of Diagonal Matrix

A **diagonal matrix** raised to the p-th power remains diagonal. Thus, if

$$\mathbf{D} = \begin{bmatrix} d_{11} & 0_{12} & 0_{13} & \cdots & 0_{1n} \\ 0_{21} & d_{22} & 0_{23} & \cdots & 0_{2n} \\ 0_{31} & 0_{32} & d_{33} & \cdots & 0_{3n} \\ \vdots & & \vdots & \vdots & \vdots \\ 0_{n1} & 0_{n2} & 0_{n3} & \cdots & d_{nn} \end{bmatrix},$$

then

$$(\mathbf{D}_n)^p = \begin{bmatrix} (d_{11})^p & 0_{12} & 0_{13} & \cdots & 0_{1n} \\ 0_{21} & (d_{22})^p & 0_{23} & \cdots & 0_{2n} \\ 0_{31} & 0_{32} & (d_{33})^p & \cdots & 0_{3n} \\ \vdots & \vdots & \vdots & \vdots & \vdots \\ 0_{n1} & 0_{n2} & 0_{n3} & \cdots & (d_{nn})^p \end{bmatrix}.$$

- $\mathbf{D}^4$ of

$$\mathbf{D} = \begin{bmatrix} 3 & 0 & 0 \\ 0 & 2 & 0 \\ 0 & 0 & 7 \end{bmatrix}$$

is

$$\mathbf{D}^4 = \begin{bmatrix} 3^4 & 0 & 0 \\ 0 & 2^4 & 0 \\ 0 & 0 & 7^4 \end{bmatrix} = \begin{bmatrix} 81 & 0 & 0 \\ 0 & 16 & 0 \\ 0 & 0 & 2401 \end{bmatrix}.$$ •

The example below, simple to be sure, illustrates how one can organize data into matrices.

- A baker delivers one kind of bread. He charges \$1.30 for a single-loaf package and \$2.20 for a double-loaf package.

The first week he delivers to a distributor 160 single-loaf packages and 72 double-loaf packages; the second week 190 single-loaf packages and 80 double-loaf packages; the third week 184 single-loaf packages and 100 double-loaf packages.

By the use of matrices, find – for each one of the three weeks – the total number of loaves received by the distributor, and their total value.

In the first row of a matrix **A**, indicate the number of loaves in each package; in the second row, indicate the price per package. Numbers involving double-loaf packages are italicized:

$$\mathbf{A} = \begin{bmatrix} 1 & \mathit{2} \\ 1.30 & \mathit{2.20} \end{bmatrix}$$

In a matrix **B**, indicate in the first row the number of single-loaf packages delivered weekly, and in the second row the number of double-loaf packages delivered weekly,

$$\mathbf{B} = \begin{bmatrix} 160 & 190 & 184 \\ \mathit{72} & \mathit{80} & \mathit{100} \end{bmatrix}.$$

$$\mathbf{AB} = \begin{bmatrix} 1 & \mathit{2} \\ 1.30 & \mathit{2.20} \end{bmatrix} \begin{bmatrix} 160 & 190 & 184 \\ \mathit{72} & \mathit{80} & \mathit{100} \end{bmatrix}$$

$$= \begin{bmatrix} 1 \times 160 + \mathit{2} \times \mathit{72} & 1 \times 190 + \mathit{2} \times \mathit{80} & 1 \times 184 + \mathit{2} \times \mathit{100} \\ 1.30 \times 160 + \mathit{2.20} \times \mathit{72} & 1.30 \times 190 + \mathit{2.20} \times \mathit{80} & 1.30 \times 184 + \mathit{2.20} \times \mathit{100} \end{bmatrix}$$

$$= \begin{bmatrix} 304 & 350 & 384 \\ 366.40 & 423.00 & 459.20 \end{bmatrix}$$

Week	**1**	**2**	**3**
Total Number of Loaves	304	350	384
Total Value $	366.40	423.00	459.20

18.3 Determinants

A determinant is a *value* representing sums and products of a *square matrix*.

The determinant of a matrix **A**, det **A**, is denoted as arrays of numbers or algebraic quantities, called **entries** or **elements**, disposed in horizontal rows and vertical columns and enclosed between single vertical bars:

$$\det \mathbf{A} = \begin{vmatrix} a_{11} & a_{12} & \cdots & a_{1n} \\ \cdots & \cdots & \cdots & \cdots \\ a_{n1} & a_{n2} & \cdots & a_{nn} \end{vmatrix}$$

Order

Unlike matrices, which may be of any rectangular shape, determinants are always square, having an equal number of rows and columns called the **order** of the determinant; thus, only square matrices have determinants.

Main, or Principal, Diagonal

Secondary Diagonal

The diagonal traversing the determinant from upper left to lower right is the **main diagonal**, or **principal diagonal**; the diagonal from lower left to upper right is the **secondary diagonal**.

Every determinant represents a definite numerical value that can be calculated according to specific rules which transform the pattern of numbers into a single number that can be used in calculations.

Determinants with real entries have real values, those with complex entries have complex values.

The numerical value of an n-th order determinant is the algebraic sum of $n!$ terms, each being the product of n different entries taken one each from every row and column of the determinant.

The value of a determinant of the first order is, of course, its single entry; the value of a second-order determinant,

$$\det \mathbf{A}_2 = \begin{vmatrix} a_{11} & a_{12} \\ a_{21} & a_{22} \end{vmatrix},$$

is the sum of its two permutations of two-entry products,

$a_{11} \times a_{22}$ (which reads a-one-one times a-two-two)

$a_{12} \times a_{21}$ (which reads a-one-two times a-two-one).

with their inversion correction factors $(-1)^k$ where k may be measured by the number of inversions of the natural order of the column index j,

$a_{11} a_{22}$	digit order 1 - 2	no inversion	$(-1)^0 = 1$
$a_{12} a_{21}$	digit order 2 - 1	one inversion	$(-1)^1 = -1$,

and thus the determinant

$$\det \mathbf{A}_2 = a_{11} a_{22} - a_{12} a_{21}.$$

A third-order determinant,

$$\det \mathbf{A}_3 = \begin{vmatrix} a_{11} & a_{12} & a_{13} \\ a_{21} & a_{22} & a_{23} \\ a_{31} & a_{32} & a_{33} \end{vmatrix},$$

has $3! = 6$ triple products of entries from every row and column, of which no two triplets may be the same,

$a_{11}\,a_{22}\,a_{33}$	$a_{11}\,a_{23}\,a_{32}$	$a_{12}\,a_{21}\,a_{33}$	$a_{12}\,a_{23}\,a_{31}$	$a_{13}\,a_{21}\,a_{32}$	$a_{13}\,a_{22}\,a_{31}$
1-2-3	1-3-2	2-1-3	2-3-1	3-1-2	3-2-1

no. of inversions:	none	one	one	two	two	three
correction factor:	$+1$	-1	-1	$+1$	$+1$	-1,

and thus the determinant

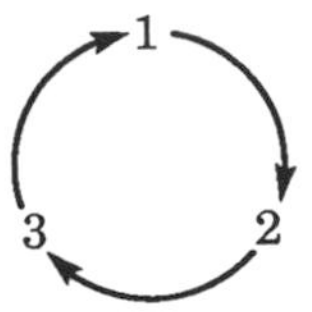

$$\begin{aligned} \det \mathbf{A}_3 = \; & a_{11}\,(a_{22}\,a_{33} - a_{23}\,a_{32}) \\ & - a_{12}\,(a_{21}\,a_{33} - a_{23}\,a_{31}) \\ & + a_{13}\,(a_{21}\,a_{32} - a_{22}\,a_{31})\,. \end{aligned}$$

Example:

$$\det \mathbf{A} = \begin{vmatrix} 2 & 0 & 4 \\ 3 & 1 & 0 \\ 4 & 2 & 2 \end{vmatrix} = 2\,(2-0) - 0\,(6-0) + 4\,(6-4)$$

$$= 4 - 0 + 8 = 12$$

Sarrus's Rule

SARRUS J. P.
(1789 - 1861)

A method of simplifying the development of a third-order determinant – due to the Frenchman **J. P. Sarrus** – extends the development of the determinant toward the right by repeating the first and second columns:

$$\begin{vmatrix} a_{11} & a_{12} & a_{13} \\ a_{21} & a_{22} & a_{23} \\ a_{31} & a_{32} & a_{33} \end{vmatrix} \begin{matrix} a_{11} & a_{12} \\ a_{21} & a_{22} \\ a_{31} & a_{32} \end{matrix} \qquad \begin{vmatrix} a_{11} & a_{12} & a_{13} \\ a_{21} & a_{22} & a_{23} \\ a_{31} & a_{32} & a_{33} \end{vmatrix} \begin{matrix} a_{11} & a_{12} \\ a_{21} & a_{22} \\ a_{31} & a_{32} \end{matrix}$$

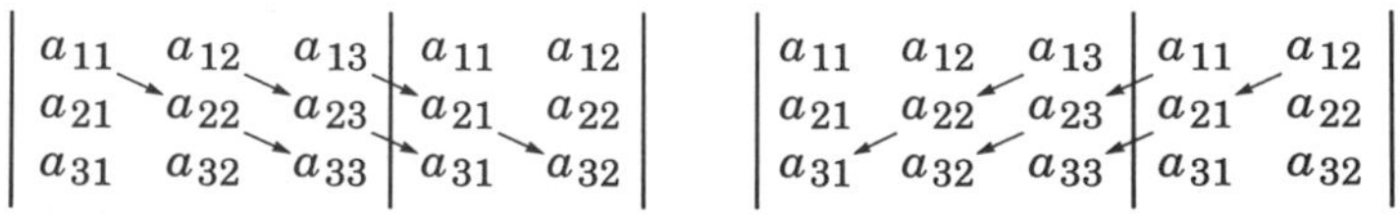

positive products　　　　　　　negative products

The first term of the development is the product $a_{11}\,a_{22}\,a_{33}$ of the main diagonal entries; the other two positive terms are the products of the entries of the parallel diagonals in the extension of the determinant. The three negative terms are products of entries in the secondary diagonal and in the diagonals parallel to it.

The use of Sarrus's rule is restricted to third-order determinants.

The Laplace Expansion

LAPLACE Pierre Simon de (1749 - 1827)

Determinants of order three and higher are usually evaluated by being developed into determinants of lower orders than that of the parent determinant, by **Laplace expansion** into minors, a method named after **Pierre Simon de Laplace**, French mathematician, astronomer, and physicist.

Minor

A **minor** of a determinant is a sub-determinant obtained by removing from the parent determinant an equal number of rows and columns, so as to preserve the square shape.

First Minor
Principal Minor

Deleting the i-th row and the j-th column from an n-th order determinant leaves us with an $(n-1)$-th order **first minor**, or **principal minor**, m_{ij}.

Cofactor
Inversion Correction Factor

A **cofactor** c_{ij} of an entry a_{ij} is a first minor complete with its appropriate *sign* or **inversion correction factor** $(-1)^{i+j}$, which is $(+1)$ when the sum of the row index i and the column index j is even, and (-1) when $(i+j)$ is odd.

The determinant is the sum of all products of entries a_{ij} and their cofactors c_{ij}, complete with inversion correction factors. Since this factor alternates regularly between $(+1)$ and (-1), it is sufficient to establish the sign for the first cofactor of an expansion.

Determinants may be developed by the entries of any row or column by the following procedure:

- Select a row i or column j of the determinant; the more zero entries in it, the easier the calculation.
- Multiply each entry of the chosen row, or column, by the proper cofactor.
- Continue this process until the expansion consists of only second-order determinants.

- Expand the determinant

$$\det \mathbf{A} = \begin{vmatrix} 0 & 4 & -1 & 0 \\ 1 & 0 & 1 & 4 \\ 0 & 6 & 3 & 3 \\ 3 & 5 & 2 & 2 \end{vmatrix}$$

by minors and determine its value.

Since any row or column may be used for the expansion, we select the first row because of its zero entries:

$$\det \mathbf{A} = 0\,(-1)^{(1+1)}\,|m_{11}| + 4\,(-1)^{(1+2)}\,|m_{12}| + (-1)(-1)^{(1+3)}\,|m_{13}| + 0\,(-1)^{(1+4)}\,|m_{14}|$$
$$= -4\,|m_{12}| - |m_{13}|$$

$$|m_{12}| = \begin{vmatrix} 1 & 1 & 4 \\ 0 & 3 & 3 \\ 3 & 2 & 2 \end{vmatrix} = 1\,(-1)^{(1+1)} \begin{vmatrix} 3 & 3 \\ 2 & 2 \end{vmatrix} + 1\,(-1)^{(1+2)} \begin{vmatrix} 0 & 3 \\ 3 & 2 \end{vmatrix} + 4\,(-1)^{(1+3)} \begin{vmatrix} 0 & 3 \\ 3 & 2 \end{vmatrix} = -27$$

$$|m_{13}| = \begin{vmatrix} 1 & 0 & 4 \\ 0 & 6 & 3 \\ 3 & 5 & 2 \end{vmatrix} = 1\,(-1)^{(1+1)} \begin{vmatrix} 6 & 3 \\ 5 & 2 \end{vmatrix} + 0\,(-1)^{(1+2)} \begin{vmatrix} 0 & 3 \\ 3 & 2 \end{vmatrix} + 4\,(-1)^{(1+3)} \begin{vmatrix} 0 & 6 \\ 3 & 5 \end{vmatrix} = -75$$

$$\det \mathbf{A} = -4\,(-27) - (-75) = 108 + 75 = 183$$

Properties of Determinants

The **value of a determinant remains unaffected**:

– If *all* rows and *all* columns are transposed, without changing entry order, so that rows become columns and columns become rows:

$$\det \mathbf{A} = \begin{vmatrix} a_{11} & a_{12} & a_{13} \\ a_{21} & a_{22} & a_{23} \\ a_{31} & a_{32} & a_{33} \end{vmatrix} = a_{11}\begin{vmatrix} a_{22} & a_{23} \\ a_{32} & a_{33} \end{vmatrix} - a_{12}\begin{vmatrix} a_{21} & a_{23} \\ a_{31} & a_{33} \end{vmatrix} + a_{13}\begin{vmatrix} a_{21} & a_{22} \\ a_{31} & a_{32} \end{vmatrix}$$

$$= a_{11}(a_{22}a_{33} - a_{23}a_{32}) - a_{12}(a_{21}a_{33} - a_{23}a_{31}) + a_{13}(a_{21}a_{32} - a_{22}a_{31})$$

$$\det \mathbf{A}^{\mathrm{T}} = \begin{vmatrix} a_{11} & a_{21} & a_{31} \\ a_{12} & a_{22} & a_{32} \\ a_{13} & a_{23} & a_{33} \end{vmatrix} = a_{11}\begin{vmatrix} a_{22} & a_{23} \\ a_{32} & a_{33} \end{vmatrix} - a_{12}\begin{vmatrix} a_{21} & a_{23} \\ a_{31} & a_{33} \end{vmatrix} + a_{13}\begin{vmatrix} a_{21} & a_{22} \\ a_{31} & a_{32} \end{vmatrix}$$

$$= a_{11}(a_{22}a_{33} - a_{32}a_{23}) - a_{12}(a_{21}a_{33} - a_{31}a_{23}) + a_{13}(a_{21}a_{32} - a_{31}a_{22})$$

$$= a_{11}(a_{22}a_{33} - a_{23}a_{32}) - a_{12}(a_{21}a_{33} - a_{23}a_{31}) + a_{13}(a_{21}a_{32} - a_{22}a_{31})$$

Example:

$$\begin{vmatrix} 1 & 2 & 3 \\ 4 & 6 & 7 \\ 5 & 8 & 9 \end{vmatrix} = \begin{vmatrix} 1 & 4 & 5 \\ 2 & 6 & 8 \\ 3 & 7 & 9 \end{vmatrix} = 2$$

Corollary: Theorems stated for rows are also valid for columns.

– If the entries of any one row, column, or a multiple of them are added to, or subtracted from, another row, column, or multiple, then the value of the determinant remains unchanged.

Example:

$$\begin{vmatrix} 1 & 2 & 3 \\ 4 & 6 & 7 \\ 5 & 8 & 9 \end{vmatrix} = \begin{vmatrix} 1 & 2 & 3 \\ 4 + 2 \times 1 & 6 + 2 \times 2 & 7 + 2 \times 3 \\ 5 & 8 & 9 \end{vmatrix} = 2$$

The value of a determinant **retains its numerical value but reverses its sign** if *any two rows, or columns, are interchanged, without changing entry order*:

$$\det \mathbf{A} = \begin{vmatrix} a_{11} & a_{12} & a_{13} \\ a_{21} & a_{22} & a_{23} \\ a_{31} & a_{32} & a_{33} \end{vmatrix} = a_{11}\begin{vmatrix} a_{22} & a_{23} \\ a_{32} & a_{33} \end{vmatrix} - a_{12}\begin{vmatrix} a_{21} & a_{23} \\ a_{31} & a_{33} \end{vmatrix} + a_{13}\begin{vmatrix} a_{21} & a_{22} \\ a_{31} & a_{32} \end{vmatrix}$$

$$= a_{11}(a_{22}a_{33} - a_{23}a_{32}) - a_{12}(a_{21}a_{33} - a_{23}a_{31}) + a_{13}(a_{21}a_{32} - a_{22}a_{31})$$

$$\det \mathbf{B} = \begin{vmatrix} a_{11} & a_{12} & a_{13} \\ a_{31} & a_{32} & a_{33} \\ a_{21} & a_{22} & a_{23} \end{vmatrix} = a_{11}\begin{vmatrix} a_{32} & a_{33} \\ a_{32} & a_{23} \end{vmatrix} - a_{12}\begin{vmatrix} a_{31} & a_{33} \\ a_{21} & a_{23} \end{vmatrix} + a_{13}\begin{vmatrix} a_{31} & a_{32} \\ a_{21} & a_{22} \end{vmatrix}$$

$$= a_{11}(a_{32}a_{23} - a_{33}a_{22}) - a_{12}(a_{31}a_{23} - a_{33}a_{21}) + a_{13}(a_{31}a_{22} - a_{32}a_{21})$$

$$= -[a_{11}(a_{22}a_{33} - a_{23}a_{32}) - a_{12}(a_{21}a_{33} - a_{23}a_{31}) + a_{13}(a_{21}a_{32} - a_{22}a_{31})]$$

Hence,

$$\det \mathbf{B} = -\det \mathbf{A}.$$

Multiplying, or **dividing**, all entries of any one row, or column, *by the same scalar* is equivalent to multiplying, or dividing, the determinant by the scalar.

Corollary: A factor that is common to all entries of a row, or column, may be taken out and placed as a factor in front of the determinant.

Example:

$$\begin{vmatrix} 1 & 2 & 3 \\ 4 & 6 & 7 \\ 3\times 5 & 3\times 8 & 3\times 9 \end{vmatrix} = 3 \begin{vmatrix} 1 & 2 & 3 \\ 4 & 6 & 7 \\ 5 & 8 & 9 \end{vmatrix} = 6$$

If **all entries above or below the principal diagonal** are zero, the value of the determinant is *equal to the product of the entries of the diagonal.*

$$\begin{vmatrix} a_{11} & 0 & 0 \\ a_{21} & a_{22} & 0 \\ a_{31} & a_{32} & a_{33} \end{vmatrix} = a_{11} \begin{vmatrix} a_{22} & 0 \\ a_{32} & a_{33} \end{vmatrix} - 0 \begin{vmatrix} a_{21} & 0 \\ a_{31} & a_{33} \end{vmatrix} + 0 \begin{vmatrix} a_{21} & a_{22} \\ a_{31} & a_{32} \end{vmatrix} = a_{11}\, a_{22}\, a_{33}$$

Example:

$$\begin{vmatrix} 1 & 0 & 0 & 0 & 0 & 0 \\ 4 & -6 & 0 & 0 & 0 & 0 \\ 5 & 8 & 3 & 0 & 0 & 0 \\ 5 & 8 & 9 & 2 & 0 & 0 \\ -1 & 2 & 3 & 1 & -4 & 0 \\ 4 & 6 & 7 & 2 & 6 & 8 \end{vmatrix} = 1\,(-6)\,3 \cdot 2\,(-4)\,8 = 1152$$

Product of Determinants

det (**A B**) = (det **A**) (det **B**)

Example:

For

$$\mathbf{A} = \begin{bmatrix} 1 & -2 \\ 3 & 4 \end{bmatrix}; \ \mathbf{B} = \begin{bmatrix} 1 & 3 \\ -2 & 2 \end{bmatrix}$$

and their products

$$\mathbf{AB} = \begin{bmatrix} 5 & -1 \\ -5 & 17 \end{bmatrix}; \ \mathbf{BA} = \begin{bmatrix} 10 & 10 \\ 4 & 12 \end{bmatrix},$$

det **A** = 10; det **B** = 8; det **AB** = 80; det **BA** = 80.

Though multiplication of matrices is, generally, non-commutative, the products of their determinants are equal.

Predicting Value Zero

The value of a determinant is zero if the entries of any one row, or column, are

– all zero:

$$\begin{vmatrix} a_{11} & a_{12} & a_{13} \\ 0 & 0 & 0 \\ a_{31} & a_{32} & a_{33} \end{vmatrix} = a_{11}\begin{vmatrix} 0 & 0 \\ a_{32} & a_{33} \end{vmatrix} - a_{12}\begin{vmatrix} 0 & 0 \\ a_{31} & a_{33} \end{vmatrix} + a_{13}\begin{vmatrix} 0 & 0 \\ a_{31} & a_{32} \end{vmatrix} = 0$$

– equal, or proportional, to corresponding entries of another row, or column:

$$\begin{vmatrix} a_{11} & a_{12} & a_{13} \\ a_{31} & a_{32} & a_{33} \\ a_{31} & a_{32} & a_{33} \end{vmatrix} = a_{11}\begin{vmatrix} a_{32} & a_{33} \\ a_{32} & a_{33} \end{vmatrix} - a_{12}\begin{vmatrix} a_{31} & a_{33} \\ a_{31} & a_{33} \end{vmatrix} + a_{13}\begin{vmatrix} a_{31} & a_{32} \\ a_{31} & a_{32} \end{vmatrix} = 0;$$

$$\begin{vmatrix} a_{11} & a_{12} & a_{13} \\ p\,a_{31} & p\,a_{32} & p\,a_{33} \\ a_{31} & a_{32} & a_{33} \end{vmatrix} = p \times \begin{vmatrix} a_{11} & a_{12} & a_{13} \\ a_{31} & a_{32} & a_{33} \\ a_{31} & a_{32} & a_{33} \end{vmatrix} = p \times 0 = 0$$

Geometric Significance

The geometric meaning of the absolute value of a second-order determinant,

$$|\det \mathbf{A}| = \left|\begin{vmatrix} a_{11} & a_{12} \\ a_{21} & a_{22} \end{vmatrix}\right|,$$

is that it gives the *area* of a *parallelogram* spanned by vectors $\mathbf{a}_1$ and $\mathbf{a}_2$ whose coordinates are the columns (or rows) of $\mathbf{A}$; the magnitudes of the vectors are taken relative to the unit vectors of the system. Similarly, the absolute value of a third-order determinant gives the *volume* of a *parallelepiped*.

cf. unit vectors: *p.* 607

pp. 617 *et seq.*

The **vector product** of

$$\mathbf{a} = x_1\mathbf{i} + y_1\mathbf{j} + z_1\mathbf{k} \quad \text{and} \quad \mathbf{b} = x_2\mathbf{i} + y_2\mathbf{j} + z_2\mathbf{k}$$

may be expressed as

$$\mathbf{a} \times \mathbf{b} = \begin{vmatrix} \mathbf{i} & \mathbf{j} & \mathbf{k} \\ x_1 & y_1 & z_1 \\ x_2 & y_2 & z_2 \end{vmatrix}.$$

Examples:

cf. p. 620

The products of vectors

$$\mathbf{a} = -\mathbf{i} - \mathbf{j} + 3\mathbf{k} \quad \text{and} \quad \mathbf{b} = 5\mathbf{i} + 5\mathbf{j} - 4\mathbf{k}$$

are

$$\begin{aligned} \mathbf{a} \times \mathbf{b} &= (-\mathbf{i} - \mathbf{j} + 3\mathbf{k}) \times (5\mathbf{i} + 5\mathbf{j} - 4\mathbf{k}) \\ &= -5\,\mathbf{i}\times\mathbf{i} - 5\,\mathbf{i}\times\mathbf{j} + 4\,\mathbf{i}\times\mathbf{k} - 5\,\mathbf{j}\times\mathbf{i} - 5\,\mathbf{j}\times\mathbf{j} + 4\,\mathbf{j}\times\mathbf{k} \\ &\qquad + 15\,\mathbf{k}\times\mathbf{i} + 15\,\mathbf{k}\times\mathbf{j} - 12\,\mathbf{k}\times\mathbf{k} \\ &= -5\times\mathbf{0} - 5\,\mathbf{k} + 4\,(-\mathbf{j}) - 5\,(-\mathbf{k}) - 5\times\mathbf{0} + 4\,\mathbf{i} + 15\,\mathbf{j} + 15\,(-\mathbf{i}) - 12\times\mathbf{0} \\ &= -(11\,\mathbf{i} - 11\,\mathbf{j}); \quad \mathbf{b}\times\mathbf{a} = 11\,\mathbf{i} - 11\,\mathbf{j}. \end{aligned}$$

The products $\mathbf{a}\times\mathbf{b}$ and $\mathbf{b}\times\mathbf{a}$ may, more conveniently, be obtained from the determinants:

$$\mathbf{a}\times\mathbf{b} = \begin{vmatrix} \mathbf{i} & \mathbf{j} & \mathbf{k} \\ -1 & -1 & 3 \\ 5 & 5 & -4 \end{vmatrix} = \mathbf{i}\begin{vmatrix} -1 & 3 \\ 5 & -4 \end{vmatrix} - \mathbf{j}\begin{vmatrix} -1 & 3 \\ 5 & -4 \end{vmatrix} + \mathbf{k}\begin{vmatrix} -1 & -1 \\ 5 & 5 \end{vmatrix}$$

$$= \mathbf{i}\,(4-15) \;\; -\mathbf{j}\,(4-15) \;\; +\mathbf{k}\,(-5+5) = -(11\,\mathbf{i} - 11\,\mathbf{j})$$

and

$$\mathbf{b}\times\mathbf{a} = \begin{vmatrix} \mathbf{i} & \mathbf{j} & \mathbf{k} \\ 5 & 5 & -4 \\ -1 & -1 & 3 \end{vmatrix} = \mathbf{i}\begin{vmatrix} 5 & -4 \\ -1 & 3 \end{vmatrix} - \mathbf{j}\begin{vmatrix} 5 & -4 \\ -1 & 3 \end{vmatrix} + \mathbf{k}\begin{vmatrix} 5 & 5 \\ -1 & -1 \end{vmatrix}$$

$$= \mathbf{i}\,(15-4) \;\; -\mathbf{j}\,(15-4) \;\; +\mathbf{k}\,(5-5) = 11\,\mathbf{i} - 11\,\mathbf{j}$$

cf. p. 620

Similarly, $\mathbf{a}\times\mathbf{b}$ of vectors

$$\mathbf{a} = -2\,\mathbf{i} - 2\,\mathbf{j} + \mathbf{k} \text{ and } \mathbf{b} = \mathbf{i} + 2\,\mathbf{j} + \mathbf{k}$$

is

$$\mathbf{a}\times\mathbf{b} = \begin{vmatrix} \mathbf{i} & \mathbf{j} & \mathbf{k} \\ -2 & -2 & 1 \\ 1 & 2 & 1 \end{vmatrix} = \mathbf{i}\,(-2-2) - \mathbf{j}\,(-2-1) + \mathbf{k}\,(-4+2)$$

$$= -4\,\mathbf{i} + 3\,\mathbf{j} - 2\,\mathbf{k}.$$

p. 621

In the problem on *p.* 621, the vector product of

$$\mathbf{a} = 20\,\mathbf{j} - 10\,\mathbf{k} \text{ and } \mathbf{b} = -20\,\mathbf{i} + 20\,\mathbf{j}$$

may be conveniently determined if written

$$\mathbf{a}\times\mathbf{b} = \begin{vmatrix} \mathbf{i} & \mathbf{j} & \mathbf{k} \\ 0 & 20 & -10 \\ -20 & 20 & 0 \end{vmatrix} = \big\{(0\,\mathbf{i} + 200\,\mathbf{j} + 0\,\mathbf{k}) - [0\,\mathbf{j} + (-200\,\mathbf{i}) + (-400\,\mathbf{k})]\big\}$$

$$= 200\,\mathbf{i} + 200\,\mathbf{j} + 400\,\mathbf{k}.$$

For vectors

$$\left.\begin{aligned} \mathbf{a} &= x_1\mathbf{i} + y_1\mathbf{j} + z_1\mathbf{k} \\ \mathbf{b} &= x_2\mathbf{i} + y_2\mathbf{j} + z_2\mathbf{k} \\ \mathbf{c} &= x_3\mathbf{i} + y_3\mathbf{j} + z_3\mathbf{k} \end{aligned}\right\}$$

p. 622

we have the **scalar triple product**

$$\mathbf{a}\cdot(\mathbf{b}\times\mathbf{c}) = x_1(y_2z_3 - y_3z_2) - y_1(x_2z_3 - x_3z_2) + z_1(x_2y_3 - x_3y_2),$$

which we may write

$$\mathbf{a}\cdot(\mathbf{b}\times\mathbf{c}) = \begin{vmatrix} x_1 & y_1 & z_1 \\ x_2 & y_2 & z_2 \\ x_3 & y_3 & z_3 \end{vmatrix}$$

Examples:

p. 623

The *scalar triple product* $\mathbf{a}\cdot(\mathbf{b}\times\mathbf{c})$ of vectors

$$\mathbf{a} = \mathbf{i} + 2\mathbf{j} - \mathbf{k};\;\; \mathbf{b} = 2\mathbf{j} + \mathbf{k};\;\; \mathbf{c} = \mathbf{i} - 4\mathbf{j}$$

represents the volume of a parallelepiped

$$\mathbf{a} \cdot (\mathbf{b} \times \mathbf{c}) = \begin{vmatrix} 1 & 2 & -1 \\ 0 & 2 & 1 \\ 1 & -4 & 0 \end{vmatrix} = 8 \text{ units of volume.}$$

p. 623

For the *vector triple product* $\mathbf{a} \times (\mathbf{b} \times \mathbf{c})$, we first calculate

$$\mathbf{b} \times \mathbf{c} = \begin{vmatrix} \mathbf{i} & \mathbf{j} & \mathbf{k} \\ 0 & 2 & 1 \\ 1 & -4 & 0 \end{vmatrix} = 4\,\mathbf{i} + \mathbf{j} - 2\,\mathbf{k},$$

then:

$$\mathbf{a} \times (\mathbf{b} \times \mathbf{c}) = \begin{vmatrix} \mathbf{i} & \mathbf{j} & \mathbf{k} \\ 1 & 2 & -1 \\ 4 & 1 & -2 \end{vmatrix} = -3\,\mathbf{i} - 2\,\mathbf{j} - 7\,\mathbf{k}$$

cf. p. 552

The area A of a triangle placed in an orthogonal coordinate system may be conveniently expressed in notations of the absolute value of a **determinant**, where the entries of the third column are unity:

$$A = \left| \frac{1}{2} \begin{vmatrix} x_1 & y_1 & 1 \\ x_2 & y_2 & 1 \\ x_3 & y_3 & 1 \end{vmatrix} \right|$$

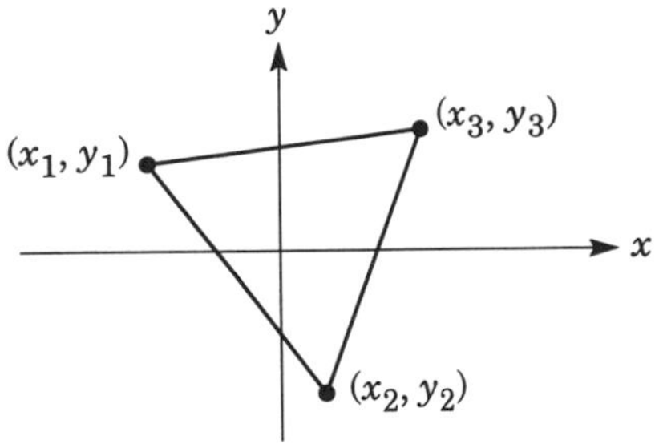

Example:

If the vertices of a triangle are at (3, −2), (−4, 1), (8, 3) in an orthogonal coordinate system, the area A of the triangle is

$$A = \frac{1}{2} \left\| \begin{matrix} 3 & -2 & 1 \\ -4 & 1 & 1 \\ 8 & 3 & 1 \end{matrix} \right\| = \frac{1}{2} \left| (3 - 16 - 12) - (8 + 9 + 8) \right|$$

$$= 25 \text{ square units.}$$

18.4 Special Matrices

The Transpose Matrix

If a matrix **A** is reflected in its main diagonal, so that all rows become columns and all columns become rows without changing their relative order or the order of entries in the rows and columns, the result is a **transpose matrix**, $\mathbf{A}^T$:

$$\mathbf{A} = \begin{bmatrix} a_{11} & a_{12} & a_{13} \\ a_{21} & a_{22} & a_{23} \end{bmatrix} \Rightarrow \mathbf{A}^T = \begin{bmatrix} a_{11} & a_{21} \\ a_{12} & a_{22} \\ a_{13} & a_{23} \end{bmatrix}.$$

A matrix **A** is:

Symmetric Matrix
Skew-Symmetric Matrix

- **symmetric** if $\mathbf{A} = \mathbf{A}^T$
- **skew-symmetric** if $\mathbf{A} = -\mathbf{A}^T$

The matrix $\begin{bmatrix} 4 & -1 & 2 \\ -1 & 1 & 7 \\ 2 & 7 & -1 \end{bmatrix}$ is *symmetric*, since

$$\begin{bmatrix} 4 & -1 & 2 \\ -1 & 1 & 7 \\ 2 & 7 & -1 \end{bmatrix} = \begin{bmatrix} 4 & -1 & 2 \\ -1 & 1 & 7 \\ 2 & 7 & -1 \end{bmatrix}^T;$$

$\begin{bmatrix} 0 & -4 & 7 \\ 4 & 0 & 3 \\ -7 & -3 & 0 \end{bmatrix}$ is *skew-symmetric*, since

$$\begin{bmatrix} 0 & -4 & 7 \\ 4 & 0 & 3 \\ -7 & -3 & 0 \end{bmatrix} = -\begin{bmatrix} 0 & -4 & 7 \\ 4 & 0 & 3 \\ -7 & -3 & 0 \end{bmatrix}^T.$$

The following calculation rules apply for transpose matrices:

- $(\mathbf{A}^T)^T = \mathbf{A}$;
- $(k\mathbf{A})^T = k\mathbf{A}^T$, where k is a scalar quantity;
- $(\mathbf{A} + \mathbf{B})^T = \mathbf{A}^T + \mathbf{B}^T$, if **A** and **B** are conformable for addition.

Example:

$$\mathbf{A} = \begin{bmatrix} 3 & -2 & 7 \\ 0 & 6 & 4 \end{bmatrix} \qquad \mathbf{A}^T = \begin{bmatrix} 3 & 0 \\ -2 & 6 \\ 7 & 4 \end{bmatrix}$$

$$\mathbf{B} = \begin{bmatrix} 2 & 4 & 3 \\ 4 & 6 & -1 \end{bmatrix} \qquad \mathbf{B}^T = \begin{bmatrix} 2 & 4 \\ 4 & 6 \\ 3 & -1 \end{bmatrix}$$

$$\mathbf{A}+\mathbf{B} = \begin{bmatrix} 5 & 2 & 10 \\ 4 & 12 & 3 \end{bmatrix} \qquad (\mathbf{A}+\mathbf{B})^T = \begin{bmatrix} 5 & 4 \\ 2 & 12 \\ 10 & 3 \end{bmatrix} \qquad \mathbf{A}^T+\mathbf{B}^T = \begin{bmatrix} 5 & 4 \\ 2 & 12 \\ 10 & 3 \end{bmatrix}$$

- $(\mathbf{AB})^T = \mathbf{B}^T\mathbf{A}^T$, if **A** and **B** are conformable for multiplication, in this order.

Example:

$$\mathbf{A} = \begin{bmatrix} 3 & -2 \\ 0 & 6 \end{bmatrix} \qquad \mathbf{A}^T = \begin{bmatrix} 3 & 0 \\ -2 & 6 \end{bmatrix}$$

$$\mathbf{B} = \begin{bmatrix} 2 & 4 & 3 \\ 4 & 6 & -1 \end{bmatrix} \qquad \mathbf{B}^T = \begin{bmatrix} 2 & 4 \\ 4 & 6 \\ 3 & -1 \end{bmatrix}$$

$$\mathbf{AB} = \begin{bmatrix} -2 & 0 & 11 \\ 24 & 36 & -6 \end{bmatrix} \qquad (\mathbf{AB})^T = \begin{bmatrix} -2 & 24 \\ 0 & 36 \\ 11 & -6 \end{bmatrix} \qquad \mathbf{B}^T\mathbf{A}^T = \begin{bmatrix} -2 & 24 \\ 0 & 36 \\ 11 & -6 \end{bmatrix}$$

Analogously, $(\mathbf{ABC})^T = \mathbf{C}^T\mathbf{B}^T\mathbf{A}^T$ if **A**, **B**, **C** are conformable for multiplication, and $(\mathbf{ABCD})^T = \mathbf{D}^T\mathbf{C}^T\mathbf{B}^T\mathbf{A}^T$ if **A**, **B**, **C**, **D** are conformable for multiplication, *etc.*

Inverse Matrices

For the inverse $\mathbf{A}^{-1}$ of a matrix **A**, we have

$$\mathbf{A} \cdot \mathbf{A}^{-1} = \mathbf{A}^{-1} \cdot \mathbf{A} = \mathbf{I}.$$

A matrix has an inverse only if its determinant $\neq 0$.

Why? $\det(\mathbf{A} \cdot \mathbf{A}^{-1}) = \det(\mathbf{A}) \cdot \det(\mathbf{A}^{-1})$ and $\det \mathbf{I} = 1$.

If $\det \mathbf{A} = 0$, we have

$$0 \cdot \det(\mathbf{A}^{-1}) \neq 1,$$

and $\mathbf{A} \cdot \mathbf{A}^{-1} = \mathbf{I}$ *does not exist* for **A**.

Only square matrices have inverses.

The product of the square matrix $\mathbf{A}_n$ and its inverse $\mathbf{A}_n^{-1}$,

$$\mathbf{A}_n = \begin{bmatrix} a_{11} & a_{12} & \cdots & a_{1n} \\ a_{21} & a_{22} & \cdots & a_{2n} \\ \vdots & \vdots & \vdots & \vdots \\ a_{n1} & a_{n2} & \cdots & a_{nn} \end{bmatrix} \qquad \mathbf{A}_n^{-1} = \begin{bmatrix} x_{11} & x_{12} & \cdots & x_{1n} \\ x_{21} & x_{22} & \cdots & x_{2n} \\ \vdots & \vdots & \vdots & \vdots \\ x_{n1} & x_{n2} & \cdots & x_{nn} \end{bmatrix},$$

is

$$\begin{bmatrix} a_{11}x_{11} + a_{12}x_{21} + \ldots + a_{1n}x_{n1} & a_{11}x_{12} + a_{12}x_{22} + \ldots + a_{1n}x_{n2} & \cdots & a_{11}x_{1n} + a_{12}x_{2n} + \ldots + a_{1n}x_{nn} \\ a_{21}x_{11} + a_{22}x_{21} + \ldots + a_{2n}x_{n1} & a_{21}x_{12} + a_{22}x_{22} + \ldots + a_{2n}x_{n2} & \cdots & a_{21}x_{1n} + a_{22}x_{2n} + \ldots + a_{2n}x_{nn} \\ \vdots & \vdots & \vdots & \vdots \\ a_{n1}x_{11} + a_{n2}x_{21} + \ldots + a_{nn}x_{n1} & a_{n1}x_{12} + a_{n2}x_{22} + \ldots + a_{nn}x_{n2} & \cdots & a_{n1}x_{1n} + a_{n2}x_{2n} + \ldots + a_{nn}x_{nn} \end{bmatrix}.$$

Since

$$\mathbf{A}_n\ \mathbf{A}_n^{-1} = \mathbf{I} = \begin{bmatrix} 1 & 0 & \dots & 0 \\ 0 & 1 & \dots & 0 \\ \vdots & \vdots & \vdots & \vdots \\ 0 & 0 & \dots & 1 \end{bmatrix},$$

we have

$$\begin{aligned}
a_{11}x_{11} + a_{12}x_{21} + \dots + a_{1n}x_{n1} &= 1 \\
a_{11}x_{12} + a_{12}x_{22} + \dots + a_{1n}x_{n2} &= 0 \\
&\vdots \\
a_{11}x_{1n} + a_{12}x_{2n} + \dots + a_{1n}x_{nn} &= 0 \\[1ex]
a_{21}x_{11} + a_{22}x_{21} + \dots + a_{2n}x_{n1} &= 0 \\
a_{21}x_{12} + a_{22}x_{22} + \dots + a_{2n}x_{n2} &= 1 \\
&\vdots \\
a_{21}x_{1n} + a_{22}x_{2n} + \dots + a_{2n}x_{nn} &= 0 \\[1ex]
a_{n1}x_{11} + a_{n2}x_{21} + \dots + a_{nn}x_{n1} &= 0 \\
a_{n1}x_{12} + a_{n2}x_{22} + \dots + a_{nn}x_{n2} &= 0 \\
&\vdots \\
a_{n1}x_{1n} + a_{n2}x_{2n} + \dots + a_{nn}x_{nn} &= 1 .
\end{aligned}$$

Singular Matrix

Invertible or Nonsingular Matrix

A square matrix for which there exists no inverse – that is, a square matrix whose determinant is 0 – is known as a **singular matrix**; a matrix for which there exists an inverse (determinant ≠ 0) is an **invertible** or **nonsingular** (**irregular**) matrix.

- Find the inverse matrix of $\mathbf{A} = \begin{bmatrix} 1 & 2 & 0 \\ 0 & 3 & 8 \\ 1 & 0 & -5 \end{bmatrix}$.

$$\begin{matrix} (1) \\ (2) \\ (3) \end{matrix} \left[\begin{array}{ccc|ccc} 1 & 2 & 0 & 1 & 0 & 0 \\ 0 & 3 & 8 & 0 & 1 & 0 \\ 1 & 0 & -5 & 0 & 0 & 1 \end{array}\right]$$

$\frac{1}{3} \times (2)$ ——— (2')

$$\begin{matrix} (1) \\ (2') \\ (3) \end{matrix} \left[\begin{array}{ccc|ccc} 1 & 2 & 0 & 1 & 0 & 0 \\ 0 & 1 & 8/3 & 0 & 1/3 & 0 \\ 1 & 0 & -5 & 0 & 0 & 1 \end{array}\right]$$

$(3) - (1)$ ——— (3')

$$\begin{matrix} (1) \\ (2') \\ (3') \end{matrix} \left[\begin{array}{ccc|ccc} 1 & 2 & 0 & 1 & 0 & 0 \\ 0 & 1 & 8/3 & 0 & 1/3 & 0 \\ 0 & -2 & -5 & -1 & 0 & 1 \end{array}\right]$$

$(3') + 2 \times (2')$ ——— (3'')

$$\begin{matrix} (1) \\ (2') \\ (3'') \end{matrix} \left[\begin{array}{ccc|ccc} 1 & 2 & 0 & 1 & 0 & 0 \\ 0 & 1 & 8/3 & 0 & 1/3 & 0 \\ 0 & 0 & 1/3 & -1 & 2/3 & 1 \end{array}\right]$$

$3 \times (3'')$ ——— (3''')

$$\begin{matrix} (1) \\ (2') \\ (3''') \end{matrix} \left[\begin{array}{ccc|ccc} 1 & 2 & 0 & 1 & 0 & 0 \\ 0 & 1 & 8/3 & 0 & 1/3 & 0 \\ 0 & 0 & 1 & -3 & 2 & 3 \end{array}\right]$$

$(2') - \frac{8}{3} \times (3''')$ ——— (2'')

$$\begin{matrix} (1) \\ (2'') \\ (3''') \end{matrix} \left[\begin{array}{ccc|ccc} 1 & 2 & 0 & 1 & 0 & 0 \\ 0 & 1 & 0 & 8 & -5 & -8 \\ 0 & 0 & 1 & -3 & 2 & 3 \end{array}\right]$$

$(1) - 2 \times (2'')$ ——— (1')

$$\begin{matrix} (1') \\ (2'') \\ (3''') \end{matrix} \left[\begin{array}{ccc|ccc} 1 & 0 & 0 & -15 & 10 & 16 \\ 0 & 1 & 0 & 8 & -5 & -8 \\ 0 & 0 & 1 & -3 & 2 & 3 \end{array}\right]$$

$$\mathbf{A}^{-1} = \begin{bmatrix} -15 & 10 & 16 \\ 8 & -5 & -8 \\ -3 & 2 & 3 \end{bmatrix}.$$

- If $\mathbf{B} = \begin{bmatrix} 1 & 2 \\ 3 & 4 \end{bmatrix}$, find the inverse matrix $\mathbf{B}^{-1}$.

With $\mathbf{B}^{-1} = \begin{bmatrix} x_1 & x_2 \\ x_3 & x_4 \end{bmatrix}$, we have, since $\mathbf{B}\mathbf{B}^{-1} = \begin{bmatrix} 1 & 0 \\ 0 & 1 \end{bmatrix}$:

$$\begin{bmatrix} 1 & 2 \\ 3 & 4 \end{bmatrix}\begin{bmatrix} x_1 & x_2 \\ x_3 & x_4 \end{bmatrix} = \begin{bmatrix} x_1 + 2x_3 & x_2 + 2x_4 \\ 3x_1 + 4x_3 & 3x_2 + 4x_4 \end{bmatrix} = \begin{bmatrix} 1 & 0 \\ 0 & 1 \end{bmatrix}$$

$$\left.\begin{aligned} x_1 + 2x_3 &= 1 \\ 3x_1 + 4x_3 &= 0 \end{aligned}\right\} \qquad \left.\begin{aligned} x_2 + 2x_4 &= 0 \\ 3x_2 + 4x_4 &= 1 \end{aligned}\right\}$$

Coefficient matrix and augmented matrix are further described on *p.* 662.

As the coefficient parts of the augmented matrices

$$\begin{bmatrix} 1 & 2 & 1 \\ 3 & 4 & 0 \end{bmatrix} \text{ and } \begin{bmatrix} 1 & 2 & 0 \\ 3 & 4 & 1 \end{bmatrix}$$

are identical, we may augment the parent matrix by the identity matrix and reduce the elimination procedure to only one matrix:

$$\left[\begin{array}{cc|cc} 1 & 2 & 1 & 0 \\ 3 & 4 & 0 & 1 \end{array}\right] \Rightarrow \left[\begin{array}{cc|cc} 1 & 2 & 1 & 0 \\ 0 & -2 & -3 & 1 \end{array}\right] \Rightarrow \left[\begin{array}{cc|cc} 1 & 0 & -2 & 1 \\ 0 & 1 & 3/2 & -1/2 \end{array}\right]$$

$$\mathbf{B}^{-1} = \begin{bmatrix} -2 & 1 \\ 3/2 & -1/2 \end{bmatrix}$$ •

The following **calculation rules** apply if **A** is invertible:

- $(\mathbf{A}^{-1})^{-1} = \mathbf{A}$;
- $(\mathbf{A}^{n})^{-1} = (\mathbf{A}^{-1})^{n}$, where n is a positive integer;
- $(\mathbf{A}^{\mathrm{T}})^{-1} = (\mathbf{A}^{-1})^{\mathrm{T}}$;
- $(\mathbf{A}\,\mathbf{B})^{-1} = \mathbf{B}^{-1}\,\mathbf{A}^{-1}$, if **A** and **B** are invertible and conformable for multiplication;
- $\det(\mathbf{A}_m)^{-1} = 1/\det \mathbf{A}_m$.

Orthogonal Matrices

Orthogonal Matrix

A matrix **A** that is equal to the inverse of its transpose matrix, $(\mathbf{A}^{\mathrm{T}})^{-1}$, is an **orthogonal matrix**.

Example:

$$\mathbf{A} = \begin{bmatrix} \frac{1}{\sqrt{2}} & -\frac{1}{\sqrt{2}} \\ \frac{1}{\sqrt{2}} & \frac{1}{\sqrt{2}} \end{bmatrix}$$ is orthogonal, and therefore

$$\mathbf{A}\,\mathbf{A}^{\mathrm{T}} = \begin{bmatrix} \frac{1}{\sqrt{2}} & -\frac{1}{\sqrt{2}} \\ \frac{1}{\sqrt{2}} & \frac{1}{\sqrt{2}} \end{bmatrix}\begin{bmatrix} \frac{1}{\sqrt{2}} & \frac{1}{\sqrt{2}} \\ -\frac{1}{\sqrt{2}} & \frac{1}{\sqrt{2}} \end{bmatrix} = \begin{bmatrix} 1 & 0 \\ 0 & 1 \end{bmatrix}.$$

If two matrices **A** and **B** are orthogonal and conformable for multiplication, their product **AB** is also an orthogonal matrix.

Complex Matrices

In a **complex matrix**, at least one of the entries is a complex number. If the complex entries of a square matrix **A**,

$$\mathbf{A} = \begin{bmatrix} 2 & 3-i & 4 \\ 2+i & 7 & -i \\ 4 & i & 0 \end{bmatrix},$$

Complex Conjugate Matrix

are replaced by their conjugates, the new matrix is a **complex conjugate matrix**,

$$\bar{\mathbf{A}} = \begin{bmatrix} 2 & 3+i & 4 \\ 2-i & 7 & i \\ 4 & -i & 0 \end{bmatrix},$$

Adjoint Matrix

the transpose of which is an **adjoint matrix**,

$$\mathbf{A}^* = \begin{bmatrix} 2 & 2-i & 4 \\ 3+i & 7 & -i \\ 4 & i & 0 \end{bmatrix},$$

Hermitian Conjugate Matrix

also called an **Hermitian conjugate matrix**.

Normal Matrix

If $\mathbf{A}\,\mathbf{A}^* = \mathbf{A}^*\,\mathbf{A}$, then **A** is known as a **normal matrix**.

Unitary Matrix

A **unitary matrix**, **U**, has the property $\mathbf{U}\,\mathbf{U}^* = \mathbf{I}$, that is, $\mathbf{U}^*$ is $\mathbf{U}^{-1}$. The unitary matrix is normal.

Hermitian Matrix

HERMITE Charles (French mathematician; 1822 - 1901)

A matrix that is its own Hermitian conjugate is an **Hermitian matrix**.

$$\mathbf{A} = \begin{bmatrix} 2 & 3+i & 4 \\ 3-i & 7 & i \\ 4 & -i & 0 \end{bmatrix} \text{ is Hermitian, as } \mathbf{A} = \mathbf{A}^*.$$

The Hermitian matrix is a normal matrix.

Projection Matrix

An Hermitian matrix **P** such that $\mathbf{P}^2 = \mathbf{P}$ is known as a **projection matrix**.

Skew-Hermitian Matrix

A matrix such that $\mathbf{A}^* = -\mathbf{A}$ is a **skew-Hermitian matrix**.

The purpose of mentioning the various subgroups of complex matrices is only for reader awareness; they will not be met again in this text.

From his days at Yale, long past, the skew Hermitian haunted Bill.

18.5 Cofactors and the Inverse of a Matrix

A cofactor of an entry of a matrix is the same as the cofactor of the same entry in the determinant of the matrix and, thus, is defined only for *square matrices*.

p. 648

Example:

Find the cofactor matrix of $\mathbf{A} = \begin{bmatrix} 1 & 2 & 0 \\ 2 & 1 & 4 \\ 4 & 2 & 6 \end{bmatrix}$.

Let

$$\det(\text{cofactor matrix of } \mathbf{A}) = \begin{vmatrix} A_{11} & A_{12} & A_{13} \\ A_{21} & A_{22} & A_{23} \\ A_{31} & A_{32} & A_{33} \end{vmatrix},$$

where

$$A_{11} = +\begin{vmatrix} 1 & 4 \\ 2 & 6 \end{vmatrix} = (1 \times 6 - 4 \times 2) = -2$$

$$A_{12} = -\begin{vmatrix} 2 & 4 \\ 4 & 6 \end{vmatrix} = -(2 \times 6 - 4 \times 4) = 4$$

$$A_{13} = +\begin{vmatrix} 2 & 1 \\ 4 & 2 \end{vmatrix} = (2 \times 2 - 1 \times 4) = 0$$

$$A_{21} = -\begin{vmatrix} 2 & 0 \\ 2 & 6 \end{vmatrix} = -(2 \times 6 - 0 \times 2) = -12$$

$$A_{22} = +\begin{vmatrix} 1 & 0 \\ 4 & 6 \end{vmatrix} = (1 \times 6 - 0 \times 4) = 6$$

$$A_{23} = -\begin{vmatrix} 1 & 2 \\ 4 & 2 \end{vmatrix} = -(1 \times 2 - 2 \times 4) = 6$$

$$A_{31} = +\begin{vmatrix} 2 & 0 \\ 1 & 4 \end{vmatrix} = (2 \times 4 - 0 \times 1) = 8$$

$$A_{32} = -\begin{vmatrix} 1 & 0 \\ 2 & 4 \end{vmatrix} = -(1 \times 4 - 0 \times 2) = -4$$

$$A_{33} = +\begin{vmatrix} 1 & 2 \\ 2 & 1 \end{vmatrix} = (1 \times 1 - 2 \times 2) = -3.$$

Hence,

$$\text{cofactor matrix of } \mathbf{A} = \begin{bmatrix} -2 & 4 & 0 \\ -12 & 6 & 6 \\ 8 & -4 & -3 \end{bmatrix}.$$

The **transpose of the cofactor matrix** may be used to find the inverse of a matrix, if it exists. From cofactor expansion and product properties of matrices, we have

$$\mathbf{A}\,(\text{cofactor matrix of } \mathbf{A})^{\mathrm{T}} = \begin{bmatrix} \det \mathbf{A} & 0 & \dots & 0 \\ 0 & \det \mathbf{A} & \dots & 0 \\ \dots & \dots & \dots & \dots \\ 0 & 0 & \dots & \det \mathbf{A} \end{bmatrix},$$

and by including the factor $1/\det \mathbf{A}$,

$$(1/\det \mathbf{A})\,\mathbf{A}\,(\text{cofactor matrix of } \mathbf{A})^{\mathrm{T}} = \mathbf{I}.$$

By definition $\mathbf{A} \cdot \mathbf{A}^{-1} = \mathbf{I}$, so, if $\mathbf{A}^{-1}$ exists,

$$\frac{1}{\det \mathbf{A}}\,\mathbf{A}\,(\text{cofactor matrix of } \mathbf{A})^{\mathrm{T}} = \mathbf{A} \cdot \mathbf{A}^{-1},$$

and we have a formula for $\mathbf{A}^{-1}$:

$$\mathbf{A}^{-1} = \frac{1}{\det \mathbf{A}}\,(\text{cofactor matrix of } \mathbf{A})^{\mathrm{T}}$$

Example:

To find $\mathbf{A}^{-1}$ of $\mathbf{A} = \begin{bmatrix} 1 & 2 & 0 \\ 2 & 1 & 4 \\ 4 & 2 & 6 \end{bmatrix}$, we have from the above that

$$\text{cofactor matrix of } \mathbf{A} = \begin{bmatrix} -2 & 4 & 0 \\ -12 & 6 & 6 \\ 8 & -4 & -3 \end{bmatrix}$$

$$(\text{cofactor matrix of } \mathbf{A})^{\mathrm{T}} = \begin{bmatrix} -2 & -12 & 8 \\ 4 & 6 & -4 \\ 0 & 6 & -3 \end{bmatrix};$$

$$\det \mathbf{A} = \begin{vmatrix} 1 & 2 & 0 \\ 2 & 1 & 4 \\ 4 & 2 & 6 \end{vmatrix} = 6\,.$$

Hence,

$$\mathbf{A}^{-1} = \frac{1}{6}\begin{bmatrix} -2 & -12 & 8 \\ 4 & 6 & -4 \\ 0 & 6 & -3 \end{bmatrix} = \begin{bmatrix} -1/3 & -2 & 4/3 \\ 2/3 & 1 & -2/3 \\ 0 & 1 & -1/2 \end{bmatrix}.$$

Checking:

$$\mathbf{A}\mathbf{A}^{-1} = \begin{bmatrix} 1 & 2 & 0 \\ 2 & 1 & 4 \\ 4 & 2 & 6 \end{bmatrix}\begin{bmatrix} -1/3 & -2 & 4/3 \\ 2/3 & 1 & -2/3 \\ 0 & 1 & -1/2 \end{bmatrix} = \begin{bmatrix} 1 & 0 & 0 \\ 0 & 1 & 0 \\ 0 & 0 & 1 \end{bmatrix} = \mathbf{I}$$

18.6 Solving Systems of Linear Equations

We shall discuss three methods:

- Elementary row operations
- Use of the inverse of a matrix
- Use of determinants (Cramer's rule)

Elementary Row Operations

By **Gaussian elimination**, or **reduction**, a system of linear equations is transformed into an equivalent system in **echelon form**, which is easy to solve by successive back substitution.

Elementary Operations

The transformation to echelon form is carried out by one or more of the following **elementary operations**:

- Interchange equations;
- Multiply (or divide) an equation by a non-zero quantity;
- Add or subtract any equation, or multiple of it, from another equation of the system.

Consider the system of equations:

$$\left.\begin{array}{lr} (1) & y+4z=6 \\ (2) & y+z=3 \\ (3) & 2x+4y+6z=20 \end{array}\right\}$$

1. Interchange Equations (1) and (3):

$$\left.\begin{array}{lr} (3) & 2x+4y+6z=20 \\ (2) & y+z=3 \\ (1) & y+4z=6 \end{array}\right\}$$

2. Divide Equation (3) by 2 to make the coefficient of x unity:

$$\left.\begin{array}{lr} (3') & x+2y+3z=10 \\ (2) & y+z=3 \\ (1) & y+4z=6 \end{array}\right\}$$

3. Subtract Equation (2) from (1):

$$\left.\begin{array}{lr} (3') & x+2y+3z=10 \\ (2) & y+z=3 \\ (1') & 3z=3 \end{array}\right\}$$

4. Divide Equation (1) by 3 to obtain the **echelon form**:

$$\left.\begin{array}{lr} (3') & x+2y+3z=10 \\ (2) & y+z=3 \\ (1'') & z=1 \end{array}\right\}$$

JORDAN Camille (1838 - 1922)

When the procedure is carried on until the system of equations is completely solved, the systematic method is called **Gauss-Jordan elimination,** or **reduction**.

Continuing from No. 4 above:

5. Multiply (2) by 2 and subtract from (3'):

$$\left.\begin{array}{ll} (3'') & x + z = 4 \\ (2) & y + z = 3 \\ (1'') & z = 1 \end{array}\right\}$$

6. Subtract (1") from (3") and (1") from (2):

$$\left.\begin{array}{ll} (3''') & x = 3 \\ (2') & y = 2 \\ (1'') & z = 1 \end{array}\right\}$$

Systems of linear equations can be advantageously adapted for computer programming by being presented in the form of **matrices containing the coefficients of the variables and the constant terms but leaving out the variables**. This simplified system is permitted since the Gaussian and Gauss-Jordan elimination methods involve only arithmetic operations on coefficients and constant terms.

Instead of using different letters (x, y, z) when handling systems of equations involving several variables, it is more convenient to use one letter with indices $(x_1, x_2, ..., x_n)$.

A system of m linear equations in n unknowns,

$$\left.\begin{array}{c} a_{11}x_1 + a_{12}x_2 + \ldots + a_{1n}x_n = c_1 \\ a_{21}x_1 + a_{22}x_2 + \ldots + a_{2n}x_n = c_2 \\ \vdots \qquad \vdots \quad \vdots \qquad \vdots \\ a_{m1}x_1 + a_{m2}x_2 + \ldots + a_{mn}x_n = c_m \end{array}\right\},$$

Coefficient Matrix

can be expressed by the **coefficient matrix**

$$\begin{bmatrix} a_{11} & a_{12} & \cdots & a_{1n} \\ a_{21} & a_{22} & \cdots & a_{2n} \\ \vdots & \vdots & \vdots & \vdots \\ a_{m1} & a_{m2} & \cdots & a_{mn} \end{bmatrix}$$

Augmented Matrix

and the **augmented matrix**

$$\begin{bmatrix} a_{11} & a_{12} & \cdots & a_{1n} & c_1 \\ a_{21} & a_{22} & \cdots & a_{2n} & c_2 \\ \vdots & \vdots & \vdots & \vdots & \vdots \\ a_{m1} & a_{m2} & \cdots & a_{mn} & c_m \end{bmatrix},$$

which also includes the constant terms.

Every row of the augmented matrix must contain coefficients for all of the unknowns of the equation system; unknowns that are missing in an equation are assigned the coefficient 0.

Elementary Row Operations

The augmented matrix is simplified by a series of **elementary row operations** – a variation of Gauss-Jordan reduction which say that

- rows may be interchanged;
- the entries of any one row may be multiplied by a non-zero constant;
- any multiple of a row may be added to another row.

Using these operations, every square matrix may be transformed into a triangular matrix.

Each step in this reduction process consists in making a selected matrix entry a_{ij} into a "1" and using it to produce zero entries preceding it in the i-th row and above and below it in the j-th column.

Pivot Operation
Pivot Entry
Pivot Row
Target Row

This procedure is known as a **pivot operation**, or **pivoting** for short; a_{ij} is the **pivot entry**, or **pivot**, which must, of course, be non-zero. The row with the pivot entry is called the **pivot row**; a row to be transformed is a **target row**.

Example:

Pivoting about the second entry of the second row (pivot entry in boldface) gives

$$\begin{bmatrix} 1 & 8 & 3 & 2 \\ 0 & \mathbf{2} & 4 & -1 \\ 2 & 4 & 1 & 8 \end{bmatrix} \Rightarrow \begin{bmatrix} 1 & 8 & 3 & 2 \\ 0/2 & \mathbf{2/2} & 4/2 & -1/2 \\ 2 & 4 & 1 & 8 \end{bmatrix}$$

$$\Rightarrow \begin{bmatrix} 1 & 8 & 3 & 2 \\ 0 & \mathbf{1} & 2 & -1/2 \\ 2 & 4 & 1 & 8 \end{bmatrix}$$

$$\Rightarrow \begin{bmatrix} 1-0 & 8-8\cdot\mathbf{1} & 3-8\cdot 2 & 2-8\,(-1/2) \\ 0 & \mathbf{1} & 2 & -1/2 \\ 2-0 & 4-4\cdot\mathbf{1} & 1-4\cdot 2 & 8-4\,(-1/2) \end{bmatrix}$$

$$\Rightarrow \begin{bmatrix} 1 & 0 & -13 & 6 \\ 0 & \mathbf{1} & 2 & -1/2 \\ 2 & 0 & -7 & 10 \end{bmatrix}.$$

Existence of Solutions

Linear independence of two equations, or rows of a matrix, means that there exists no proportionality between the equations or between the rows.

Linearly Independent

For the **linearly independent** binomials

$$x + 2y\,; \quad 2x + y\,; \; x \text{ and } y \text{ not both} = 0$$

we have

$$-(x + 2y) + 2x + y \neq 0\,;$$

Linearly Dependent

for the **linearly dependent**

$$x + 2y\,; \quad 2x + 4y,$$

$$-2\,(x + 2y) + 2x + 4y \equiv 0\,.$$

Rank

The **rank** of a matrix is equal to the dimension of the largest sub-matrix that can be obtained by deleting rows and columns of the parent matrix and that has a non-zero determinant. It is equal to the number of linearly independent rows (or columns) of a matrix in echelon form that do not have exclusively zero entries.

The ranks of the coefficient and augmented matrices may be used to establish the consistency of the equation system and the existence of solutions of the equation.

p. 662

The equations of a system of linear equations are solvable if the matrix formed from the coefficients of the variables, the *coefficient matrix*, has the same rank as the *augmented matrix*.

Finding the Solutions

A matrix reduction may typically be carried out according to the procedure described below.

- Interchange rows, if necessary, so as to make the first row begin with a non-zero entry which is made into a "1" by division.
- Add a suitable multiple of the first row to the second row in order to create a zero in the first position of the second row; this is equivalent to removing one unknown from the corresponding equation.
- Make the leading entry of the transformed second row into a "1" and pivot the matrix about this entry.
- If the intended next pivot entry turns out to be a zero, change this row for another from below it, in order to obtain a non-zero pivot entry.
- Proceed in this manner down the matrix till the last row; all new zero entries correspond to the successive elimination of unknowns in conventional algebra.
- If it proves impossible to obtain a non-zero pivot entry, the reduction procedure is finished, and the matrix is in echelon form.

• Solve

$$\left.\begin{array}{lr} (1) & y+2z = -5 \\ (2) & 5x+y-3z = 15 \\ (3) & 3x+y-2z = 5 \end{array}\right\} .$$

1. Write the system as an augmented matrix.

$$\begin{matrix}(1)\\(2)\\(3)\end{matrix}\begin{bmatrix}0 & 1 & 2 & -5\\ 5 & 1 & -3 & 15\\ 3 & 1 & -2 & 5\end{bmatrix}$$

2. Interchange rows (1) and (2).

$$\begin{matrix}(2)\\(1)\\(3)\end{matrix}\begin{bmatrix}5 & 1 & -3 & 15\\ 0 & 1 & 2 & -5\\ 3 & 1 & -2 & 5\end{bmatrix}$$

3. Pivot about the first entry in row (2).

$$\begin{matrix}(2)\\(1)\\(3)\end{matrix}\begin{bmatrix}1 & 1/5 & -3/5 & 3\\ 0 & 1 & 2 & -5\\ 3-3 & 1-3/5 & -2+9/5 & 5-9\end{bmatrix}$$

$$\begin{matrix}(2)\\(1)\\(3)\end{matrix}\Rightarrow\begin{bmatrix}1 & 1/5 & -3/5 & 3\\ 0 & 1 & 2 & -5\\ 0 & 2/5 & -1/5 & -4\end{bmatrix}$$

4. Pivot the matrix about the second entry in row (2):

$$\begin{matrix}(2)\\(1)\\(3)\end{matrix}\begin{bmatrix}1 & 1/5-1/5 & -3/5-2/5 & 3+1\\ 0 & 1 & 2 & -5\\ 0 & 2/5-2/5 & -1/5-4/5 & -4+2\end{bmatrix}$$

$$\begin{matrix}(2)\\(1)\\(3)\end{matrix}\Rightarrow\begin{bmatrix}1 & 0 & -1 & 4\\ 0 & 1 & 2 & -5\\ 0 & 0 & -1 & -2\end{bmatrix}$$

5. Pivot the matrix about the third entry in row (2).

$$\begin{matrix}(2)\\(1)\\(3)\end{matrix}\begin{bmatrix}1 & 0 & -1+1 & 4+2\\ 0 & 1 & 2-2 & -5-4\\ 0 & 0 & 1 & 2\end{bmatrix}$$

$$\begin{matrix}(2)\\(1)\\(3)\end{matrix}\Rightarrow\begin{bmatrix}1 & 0 & 0 & 6\\ 0 & 1 & 0 & -9\\ 0 & 0 & 1 & 2\end{bmatrix}$$

Hence,

$$x = 6;\ y = -9;\ z = 2.$$

•

This solution follows the strict Gauss-Jordan elimination method, suitable for computer programming. By employing elementary row operations more freely, one might force a pivot entry to be 1, but this can result in cumbersome fractions elsewhere; the strict elimination method is always the most dependable.

Use of the Inverse of a Matrix

When several systems of simultaneous linear equations have the *same coefficient matrix*, the inverse of this matrix suggests itself as an efficient means of solving the systems. Finding the inverse of a matrix is a laborious process, so this method has no advantage over other methods when solving an *isolated* system of equations. If a computer or a scientific calculator with matrix capacity is available, the inverse of a matrix may be readily found and conveniently used for any system of linear equations.

A system of linear equations may be written in matrix form,

$$\mathbf{AX} = \mathbf{C},$$

or, after premultiplication by $\mathbf{A}^{-1}$,

$$\begin{aligned} \mathbf{A}^{-1}\mathbf{AX} &= \mathbf{A}^{-1}\mathbf{C} \\ \mathbf{IX} &= \mathbf{A}^{-1}\mathbf{C} \\ \mathbf{X} &= \mathbf{A}^{-1}\mathbf{C} \end{aligned}$$

where

- **A** is an invertible coefficient matrix;
- $\mathbf{A}^{-1}$ is its inverse matrix;
- **X** is the column vector of unknowns;
- **C** is the column vector of constant terms.

Matrices used here are smaller than generally encountered in practical applications.

- Solve the system

$$\left.\begin{aligned} x_1 + 2x_2 &= 4 \\ 2x_1 + x_2 + 4x_3 &= 3 \\ 4x_1 + 2x_2 + 6x_3 &= 2 \end{aligned}\right\} .$$

$$\mathbf{A} = \begin{bmatrix} 1 & 2 & 0 \\ 2 & 1 & 4 \\ 4 & 2 & 6 \end{bmatrix}; \quad \mathbf{X} = \begin{bmatrix} x_1 \\ x_2 \\ x_3 \end{bmatrix}; \quad \mathbf{C} = \begin{bmatrix} 4 \\ 3 \\ 2 \end{bmatrix}.$$

pp. 655 - 57

$$\mathbf{A}^{-1} = \begin{bmatrix} -1/3 & -2 & 4/3 \\ 2/3 & 1 & -2/3 \\ 0 & 1 & -1/2 \end{bmatrix}$$

$\mathbf{X} = \mathbf{A}^{-1}\mathbf{C}$:

$$\begin{bmatrix} x_1 \\ x_2 \\ x_3 \end{bmatrix} = \begin{bmatrix} -1/3 & -2 & 4/3 \\ 2/3 & 1 & -2/3 \\ 0 & 1 & -1/2 \end{bmatrix} \begin{bmatrix} 4 \\ 3 \\ 2 \end{bmatrix}$$

$$= \begin{bmatrix} -14/3 \\ 13/3 \\ 2 \end{bmatrix}$$

$x_1 = -4\frac{2}{3}$; $x_2 = 4\frac{1}{3}$; $x_3 = 2$ •

- The baker met earlier still charges \$1.30 for a single-loaf package and \$2.20 for a double-loaf package. This time information is available about the total number of loaves delivered during each of three consecutive weeks and the weekly values of the sales:

	Week		
	4	**5**	**6**
Total Number of Loaves	360	416	420
Total Value \$	432.00	492.80	502.00

How many packages of each kind were sold per week?

Let x_{11} be the number of 1-loaf packages sold in the 4th week, x_{12} the number in the 5th week, and x_{13} the number in the 6th week.

Analogously, let x_{21}, x_{22}, and x_{23} be the numbers of the 2-loaf packages sold in the 4th, 5th, and 6th weeks, respectively.

We have the equations

$$\left.\begin{aligned} x_{11} + 2x_{21} &= 360 \\ 1.30\,x_{11} + 2.20\,x_{21} &= 432.00 \end{aligned}\right\};$$

$$\left.\begin{aligned} x_{12} + 2x_{22} &= 416 \\ 1.30\,x_{12} + 2.20\,x_{22} &= 492.80 \end{aligned}\right\};$$

$$\left.\begin{aligned} x_{13} + 2x_{23} &= 420 \\ 1.30\,x_{13} + 2.20\,x_{23} &= 502.00 \end{aligned}\right\}.$$

The coefficient matrix **A**,

$$\mathbf{A} = \begin{bmatrix} 1 & 2 \\ 1.30 & 2.20 \end{bmatrix},$$

displays in the first row the number of loaves of bread and in the second row the cost.

The constant terms matrix **C**,

$$\mathbf{C} = \begin{bmatrix} 360 & 416 & 420 \\ 432.00 & 492.80 & 502.00 \end{bmatrix},$$

displays in the first row the total number of loaves for each week and in the second row the total weekly sales.

Let a matrix **X** contain the sought numbers:

$$\mathbf{X} = \begin{bmatrix} x_{11} & x_{12} & x_{13} \\ x_{21} & x_{22} & x_{23} \end{bmatrix};$$

$$\mathbf{A}^{-1} = \frac{1}{1 \times 2.20 - 2 \times 1.30} \begin{bmatrix} 2.20 & -2 \\ -1.30 & 1 \end{bmatrix}$$

$$= -\frac{1}{0.4} \begin{bmatrix} 2.20 & -2 \\ -1.30 & 1 \end{bmatrix} = \begin{bmatrix} -5.50 & 5.00 \\ 3.25 & -2.50 \end{bmatrix}.$$

Since $\mathbf{A}\,\mathbf{X} = \mathbf{C}$, which gives $\mathbf{X} = \mathbf{A}^{-1}\,\mathbf{C}$, we have

$$\begin{bmatrix} x_{11} & x_{12} & x_{13} \\ x_{21} & x_{22} & x_{23} \end{bmatrix} = -\frac{1}{0.4} \begin{bmatrix} 2.20 & -2 \\ -1.30 & 1 \end{bmatrix} \begin{bmatrix} 360 & 416 & 420 \\ 432.00 & 492.80 & 502.00 \end{bmatrix}$$

$$= \begin{bmatrix} 180 & 176 & 200 \\ 90 & 120 & 110 \end{bmatrix}.$$

	Week 4	Week 5	Week 6
Total Number of 1-Loaf Packages	180	176	200
Total Number of 2-Loaf Packages	90	120	110

Cramer's Rule

CRAMER Gabriel (1704 - 1752)

Cramer's rule – named after the Swiss mathematician and physicist **Gabriel Cramer** – uses determinants to express the solution of a system of linear equations.

For an invertible matrix **A**, we have

p. 660

$$\mathbf{A}^{-1} = \frac{1}{\det \mathbf{A}} \text{ (cofactor matrix of } \mathbf{A})^{\mathrm{T}}$$

p. 666

and, since $\mathbf{X} = \mathbf{A}^{-1}\mathbf{C}$,

$$\mathbf{X} = \frac{1}{\det \mathbf{A}} \text{ (cofactor matrix of } \mathbf{A})^{\mathrm{T}} \mathbf{C} ,$$

that is,

$$\begin{bmatrix} x_1 \\ x_2 \\ \cdots \\ x_m \end{bmatrix} = \frac{1}{\det \mathbf{A}_m} \begin{bmatrix} A_{11} & A_{21} & \cdots & A_{m1} \\ A_{12} & A_{22} & \cdots & A_{m2} \\ \cdots & \cdots & \cdots & \cdots \\ A_{1m} & A_{2m} & \cdots & A_{mm} \end{bmatrix} \begin{bmatrix} c_1 \\ c_2 \\ \cdots \\ c_m \end{bmatrix} .$$

For a system with two linear equations in two unknowns, this reduces to

$$\left.\begin{aligned} a_{11}x_1 + a_{12}x_2 &= c_1 \\ a_{21}x_1 + a_{22}x_2 &= c_2 \end{aligned}\right\} .$$

The coefficient matrix and the transpose of the cofactor matrix of **A** are

$$\mathbf{A} = \begin{bmatrix} a_{11} & a_{12} \\ a_{21} & a_{22} \end{bmatrix}; \text{ (cofactor matrix of } \mathbf{A})^{\mathrm{T}} = \begin{bmatrix} a_{22} & -a_{12} \\ -a_{21} & a_{11} \end{bmatrix},$$

and thus,

$$\begin{bmatrix} x_1 \\ x_2 \end{bmatrix} = \frac{1}{\det \mathbf{A}} \begin{bmatrix} a_{22} & -a_{12} \\ -a_{21} & a_{11} \end{bmatrix} \begin{bmatrix} c_1 \\ c_2 \end{bmatrix},$$

which gives

$$x_1 = \frac{(a_{22}c_1 - a_{12}c_2)}{\det \mathbf{A}}$$

$$x_2 = -\frac{(a_{21}c_1 - a_{11}c_2)}{\det \mathbf{A}} = \frac{(a_{11}c_2 - a_{21}c_1)}{\det \mathbf{A}},$$

whose numerators can be written as determinants

$$\begin{vmatrix} c_1 & a_{12} \\ c_2 & a_{22} \end{vmatrix} \text{ and } \begin{vmatrix} a_{11} & c_1 \\ a_{21} & c_2 \end{vmatrix} .$$

Consequently, we have

$$x_1 = \frac{\begin{vmatrix} c_1 & a_{12} \\ c_2 & a_{22} \end{vmatrix}}{\begin{vmatrix} a_{11} & a_{12} \\ a_{21} & a_{22} \end{vmatrix}} \text{ and } x_2 = \frac{\begin{vmatrix} a_{11} & c_1 \\ a_{21} & c_2 \end{vmatrix}}{\begin{vmatrix} a_{11} & a_{12} \\ a_{21} & a_{22} \end{vmatrix}} .$$

Coefficient Determinant

The denominator of these expressions is the **coefficient determinant**. The numerators are based on the coefficient determinant, with the column representing the sought unknown replaced by the constant-term column.

This method may be extended to apply also for large systems of n linear equations in n variables, provided that the coefficient determinant $\neq 0$, as division by zero is not permissible.

To use **Cramer's rule**, proceed in the following manner.

- Calculate the value of the coefficient determinant:
  - if it is $= 0$, Cramer's rule is not applicable;
  - if it is $\neq 0$, use the determinant as the denominator of the quotients.
- Compute the quotients.

• Solve the system

$$\left.\begin{aligned} 2x_1 + 3x_2 - 2x_3 &= 1 \\ x_1 - 2x_2 - 3x_3 &= -9 \\ 5x_1 + 4x_2 - 4x_3 &= 2 \end{aligned}\right\} .$$

The coefficient determinant is

$$D = \begin{vmatrix} 2 & 3 & -2 \\ 1 & -2 & -3 \\ 5 & 4 & -4 \end{vmatrix} = -21 \neq 0 ;$$

consequently, Cramer's rule is applicable.

To determine x_1, x_2, x_3, replace the first/second/third column of D with the constant-term column of the equations:

$$D_1 = \begin{vmatrix} 1 & 3 & -2 \\ -9 & -2 & -3 \\ 2 & 4 & -4 \end{vmatrix} = -42 ; \quad x_1 = \frac{D_1}{D} = \frac{-42}{-21} = 2$$

$$D_2 = \begin{vmatrix} 2 & 1 & -2 \\ 1 & -9 & -3 \\ 5 & 2 & -4 \end{vmatrix} = -21 ; \quad x_2 = \frac{D_2}{D} = \frac{-21}{-21} = 1$$

$$D_3 = \begin{vmatrix} 2 & 3 & 1 \\ 1 & -2 & -9 \\ 5 & 4 & 2 \end{vmatrix} = -63 ; \quad x_3 = \frac{D_3}{D} = \frac{-63}{-21} = 3$$ •

With an increased number of system equations, Cramer's rule leads to a rapid increase of the number of arithmetic operations. Thus, it has no advantage over Gaussian, Gauss-Jordan, and inverse matrix methods for solving systems of linear equations with *numerical coefficients*, except when the system consists of only two or perhaps three equations.

On the other hand, Cramer's rule is useful for equations with *algebraic coefficients*, such as

$$\left.\begin{aligned} a\,x_1 - b\,x_2 + a\,x_3 &= 2b + a \\ (a+b)\,x_1 + a\,x_2 - a\,x_3 &= 0 \\ a\,x_1 + (a+b)\,x_2 + bx_3 &= b \end{aligned}\right\},$$

where x_2, for instance, can be expressed as

$$x_2 = \frac{\begin{vmatrix} a & 2b+a & a \\ a+b & 0 & -a \\ a & b & b \end{vmatrix}}{\begin{vmatrix} a & -b & a \\ a+b & a & -a \\ b & a+b & b \end{vmatrix}}.$$

An' a'now, folks, a lettle Cramer valtz. Turn on the bubble machine! An' a 1, an' a 2, an' a 3, an' a 1-3-2, an' a 2-1-3, an' a 2-3-1, an' a 3-1-2, an' a 3-2-1 ...

Chapter 19

EMBARKING ON CALCULUS

The scope of calculus is oceanic.

19.1 What Is Calculus?

Calculus (plural: *calculi*) is a diminutive form of Latin *calx*, which means "stone", and originally derives from Greek χαλισ (*chalis*), "limestone" (*cf.* English *chalk*, which we use for writing on the blackboard).

Calculi, in the form of pebbles or beads, were used when reckoning on the countingboard or the abacus – hence the Latin verb *calculare* meaning "to calculate".

The word *calculus* became a common designation to denote all branches of mathematics:

> *We say the Arithmetical Calculus, the Algebraical Calculus, the Differential Calculus, the Exponential Calculus, the Fluxional Calculus, the Integral Calculus, the Literal or Symbolical Calculus . . .*
>
> Charles Hutton , *A Mathematical and Philosophical Dictionary* (1796)

Infinitesimals – quantities smaller than any assignable finite quantity but yet not zero – held a prominent place in the early development of the field of mathematics that concerns differentiation and integration of functions; hence the name "infinitesimal calculus".

Replaced today by the concept of limit values, the word "infinitesimal" is no longer part of modern mathematics terminology and "infinitesimal calculus" has become "the calculus" or just "calculus", encompassing

- **differential calculus**, which deals with derivatives, that is, instantaneous rates of change of continuous functions;
- **integral calculus**, which is the reverse process: finding functions when their derivatives are known;
- **differential equations**, which contain derivatives; and
- **calculus of variations**, which is the study of maxima and minima of functions whose values depend upon a curve or another function; the basic problem is to find a function for which a certain integral assumes a maximum or a minimum value.

Was Calculus invented or discovered
– Was it made or was it uncovered?
Some only deride,
Others try to decide.
Still we aren't sure if it's one or the other!

19.2 History

PYTHAGORAS (b. *c.* 582 B.C.)
ZENON (*c.* 500 B.C.)

Infinitesimals were of vital importance to the followers of **Pythagoras** and to the work of **Zeno** of Elea. The schools of **Plato** and of **Eudoxus** developed the *method of exhaustion*, in which a sought area or volume is confined between two known, or calculable, quantities, one decreasing and one increasing towards the unknown quantity.

Method of Exhaustion

PLATON (429 - 347 B.C.)
EUDOXOS (*c.* 370 B.C.)

If a circle is enclosed between two n-gons, one circum-scribed and one inscribed, with n assuming forever-increasing values, the difference in perimeter and in area of the two polygons becomes successively smaller, until it approaches zero; the perimeter or area of the circle between the polygons can thus be determined with an accuracy that depends only on the magnitude of n.

EVCLEIDIS (*c.* 300 B.C.)
ARCHIMEDES (287 - 212 B.C.)
PAPPOS (A.D. *c.* 400)

Calculation methods of this kind were used by **Euclid**, by **Archimedes**, and by **Pappus** of Alexandria. Archimedes employed them to determine the areas of closed plane figures, the volumes of solids, and the locations of centers of gravity – in a way closely resembling the integral calculus of today.

KEPLER Johannes (1571 - 1630)
GALILEI Galileo (1564 - 1642)
TORRICELLI Evangelista (1608 - 1647)
CAVALIERI Bonaventura (1598 - 1647) *cf. p.* 446

After Pappus, progress in calculus stalled until **Kepler**, **Galileo**, and **Torricelli** applied the results of Archimedes to problems in astronomy and physics. Kepler extended the use of infinitesimals to solving problems of *maxima and minima*, marking the advent of differential and integral calculus.

Bonaventura Cavalieri, a countryman and disciple of Galileo, wrote the first textbook on integration methods, *Geometria indivisibilibus continuorum*, first published in 1635.

GEOMETRIA
INDIVISIBILIBVS
CONTINVORVM
Noua quadam ratione promota.
AVTHORE
P. BONAVENTVRA CAVALERIO
MEDIOLANEN.
Ordinis S.Hieron.Olim in Almo Bononien.Archigyn. Prim. Mathematicarum Profeff.
In hac poftrema editione ab erroribus expurgata.
Ad Illuftrifs. D.D.
MARTIVM VRSINVM
PENNÆ MARCHIONEM &c.
BONONIÆ, M.DC.LIII.

484 GEOMETRIÆ

THEOREMA I. PROPOS. I.

Figuræ planæ quæcunq; in eifdem parallelis conftitutę, in quibus, ductis quibufcunq; eifdem parallelis ęquidiftantibus rectis lineis, conceptæ cuiufcumq; rectæ lineæ portiones funt æquales, etiam inter fe æquales erunt: Et figuræ folidæ quæcumq; in eifdem planis parallelis conftitutæ, in quibus, ductis quibufcunq; planis eifdem planis parallelis æquidiftantibus, conceptæ cuiufcunq; fic ducti plani in ipfis folidis figuræ planæ funt æquales, pariter inter fe æquales erunt. Dicantur autem figuræ æqualiter analogæ, tum planæ, tum ipfæ folidæ inter fe comparatæ, ac etiam iuxta regulas lineas, feu plana parallela, in quibus efse fupponuntur, cum hoc fuerit opus explicare.

LIBER VII.

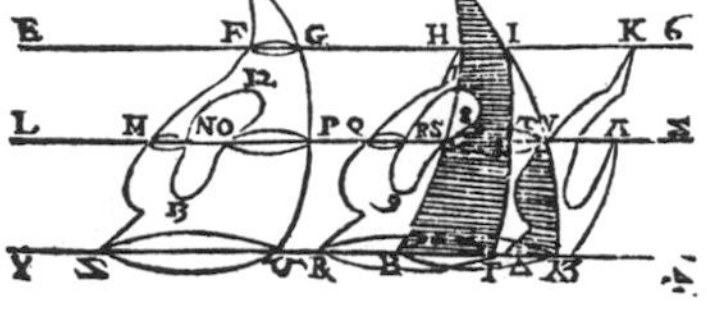

de ROBERVAL Gilles Personne (1602 - 1675)

The French mathematician **Gilles Personne de Roberval** determined the area of the cycloid and of the parabola. Most important was his description of curves as moving points; thus, a tangent was defined by determining the instantaneous direction of the moving point at any position on the curve.

DESCARTES René (1596 - 1650)

A further step in the *tangent problem* was taken by the French mathematician and philosopher **René Descartes**, who applied infinitesimal calculus to find the tangent to a curve at a given point on the curve.

de FERMAT Pierre (1608 - 1665)

The French lawyer **Pierre de Fermat**, "the king of amateurs of mathematics", solved problems for finding maxima and minima with methods at present used in differential calculus, and he used integral calculus to find areas, the lengths of arcs, the locations of centers of gravity, *etc.*, but he did not reduce his procedures to rule-of-thumb methods or point out that integration and differentiation are inverse operations.

WALLIS John (1616 - 1703)

BARROW Isaac (1630 - 1677)

PASCAL Blaise (1623 - 1662)

The new methods were employed to even greater purpose by the English mathematicians **John Wallis** and **Isaac Barrow**, the latter also a classical linguist, and the French mathematician-physicist-philosopher **Blaise Pascal**, all in the mid-17th century. Wallis dealt with infinite series and with products of an infinite number of factors; Pascal introduced multiple integrals; Barrow determined tangents by a method similar to differential calculus and was also the first to recognize that integration and differentiation are inverse operations.

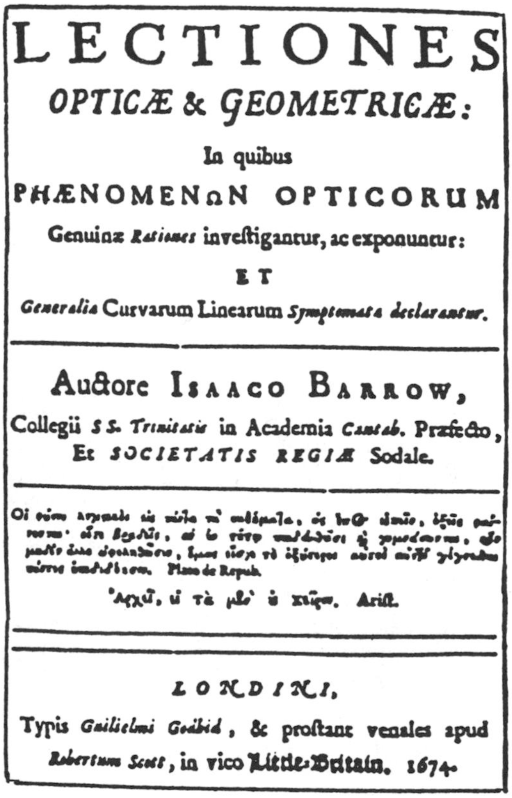

LECTIONES
OPTICÆ & GEOMETRICÆ:
In quibus
PHÆNOMENΩN OPTICORUM
Genuinæ *Rationes* investigantur, ac exponuntur:
ET
Generalia Curvarum Linearum *Symptomata declarantur.*

Auctore ISAACO BARROW,
Collegii *S S. Trinitatis* in Academia *Cantab.* Præfecto,
Et *SOCIETATIS REGIÆ* Sodale.

Plato de Repub.

Arist.

LONDINI,
Typis *Gulielmi Godbid*, & prostant venales apud *Robertum Scott*, in vico Little-Britain. 1674.

By calculations with infinitesimals, mathematicians laid the foundation for calculus in its modern form. Their work included

- in mechanics, notably in statics, the determination of centers of gravity, and in dynamics, the clarification and use of the concepts of velocity, acceleration, *etc*;
- the theory of maxima and minima;
- the calculation of the areas of plane figures, the volumes of solids, and the lengths of arcs;
- the tangent problem; and
- the calculation of infinite series, infinite products, periodic continuous fractions, *etc.*

It only remained for mathematicians to realize the true relationships between these tasks, and to bring order to the methods thus far accessible to only a narrow circle of scientists.

NEWTON Isaac
(1643 - 1727)

von LEIBNIZ Gottfried Wilhelm
(1646 - 1716)

This last important step was taken by the English astronomer, physicist, and mathematician **Isaac Newton** and the German mathematician and philosopher **Gottfried Wilhelm von Leibniz**, who are considered the true founders of today's calculus. Newton had used his "calculus of fluxions" as early as 1665, but did not publish his works until 1704, while Leibniz published his early work on calculus in 1684.

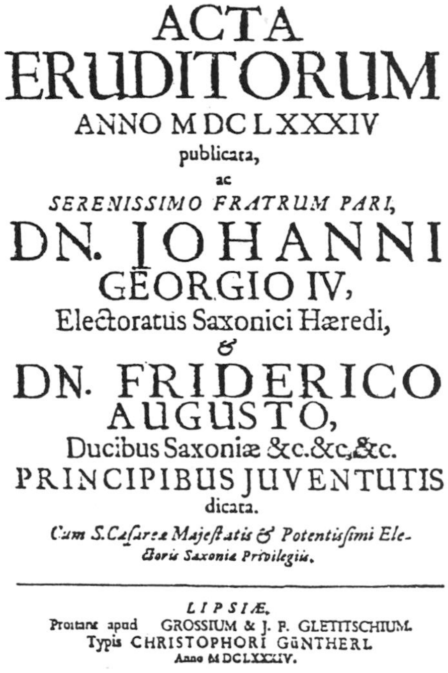

ACTA
ERUDITORUM
ANNO M DC LXXXIV
publicata,
ac
SERENISSIMO FRATRUM PARI,
DN. JOHANNI
GEORGIO IV,
Electoratus Saxonici Hæredi,
&
DN. FRIDERICO
AUGUSTO,
Ducibus Saxoniæ &c.&c.&c.
PRINCIPIBUS JUVENTUTIS
dicata.
Cum S. Cæsareæ Majestatis & Potentissimi Electoris Saxoniæ Privilegiis.

LIPSIÆ,
Prostant apud GROSSIUM & J. F. GLETITSCHIUM.
Typis CHRISTOPHORI GÜNTHERI.
Anno M DC LXXXIV.

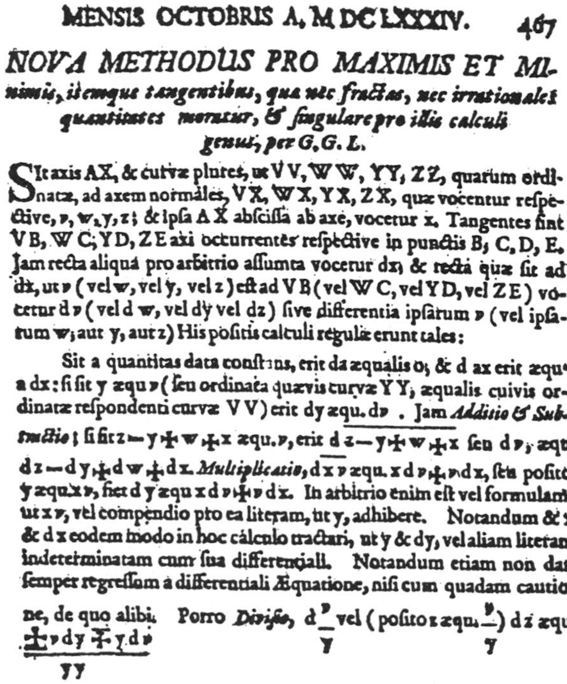

MENSIS OCTOBRIS A. M DC LXXXIV. 467

NOVA METHODUS PRO MAXIMIS ET MInimis, itemque tangentibus, quæ nec fractas, nec irrationales quantitates moratur, & singulare pro illis calculi genus, per G. G. L.

Sit axis AX, & curvæ plures, ut VV, WW, YY, ZZ, quarum ordinatæ, ad axem normales, VX, WX, YX, ZX, quæ vocentur respective, v, w, y, z; & ipsa AX abscissa ab axe, vocetur x. Tangentes sint VB, WC, YD, ZE axi occurrentes respective in punctis B, C, D, E. Jam recta aliqua pro arbitrio assumta vocetur dx, & recta quæ sit ad dx, ut v (vel w, vel y, vel z) est ad VB (vel WC, vel YD, vel ZE) vocetur dv (vel dw, vel dy vel dz) sive differentia ipsarum v (vel ipsarum w, aut y, aut z) His positis calculi regulæ erunt tales:

Sit a quantitas data constans, erit da æqualis 0, & d ax erit æqu. a dx: si sit y æqu v (seu ordinata quævis curvæ YY, æqualis cuivis ordinatæ respondenti curvæ VV) erit dy æqu. dv. Jam *Additio & Subtractio*: si sit z − y + w + x æqu. v, erit d z − y + w + x seu dv, æqu dz − dy + dw + dx. *Multiplicatio*, d x v æqu. x dv + v dx, seu posito y æqu. x v, fiet dy æqu x dv + v dx. In arbitrio enim est vel formulam, ut x v, vel compendio pro ea literam, ut y, adhibere. Notandum & x & dx eodem modo in hoc calculo tractari, ut y & dy, vel aliam literam indeterminatam cum sua differentiali. Notandum etiam non dari semper regressum a differentiali Æquatione, nisi cum quadam cautione, de quo alibi. Porro *Divisio*, $d\frac{v}{y}$ vel (posito z æqu. $\frac{v}{y}$) dz æqu. $\frac{\pm v\,dy \mp y\,dv}{yy}$

"G.G.L." in the above excerpt is the Latin form of Leibniz's initials.

It has been established that Newton had adequately described his calculus as early as 1693, but this work, *Tractatus de quadratura curvarum*, was not published until 1704, when it appeared as an appendix to *Opticks*.

OPTICKS:

OR, A

TREATISE

OF THE

REFLEXIONS, REFRACTIONS, INFLEXIONS and COLOURS

OF

LIGHT.

ALSO

Two TREATISES

OF THE

SPECIES and MAGNITUDE

OF

Curvilinear Figures.

LONDON,
Printed for SAM. SMITH, and BENJ. WALFORD, Printers to the Royal Society, at the *Prince's Arms* in St. *Paul's* Church-yard. MDCCIV.

[170]

TRACTATUS DE Quadratura Curvarum.

QUantitates indeterminatas ut motu perpetuo creſcentes vel decreſcentes, id eſt ut fluentes vel defluentes in ſequentibus conſidero, deſignoq; literis z, y, x, v, & earum fluxiones ſeu celeritates creſcendi noto iiſdem literis punctatis $\dot z$, $\dot y$, $\dot x$, $\dot v$. Sunt & harum fluxionum fluxiones ſeu mutationes magis aut minus celeres quas ipſarum $\dot z$, $\dot y$, $\dot x$, $\dot v$ fluxiones ſecundas nominare licet & ſic dignare $\ddot z$, $\ddot y$, $\ddot x$, $\ddot v$, & harum fluxiones primas ſeu ipſarum $\dot z$, $\dot y$, $\dot x$, $\dot v$ fluxiones tertias ſic $\dddot z$, $\dddot y$, $\dddot x$, $\dddot v$, & quartas ſic $\ddddot z$, $\ddddot y$, $\ddddot x$, $\ddddot v$. Et quemadmodum $\dddot z$, $\dddot y$, $\dddot x$, $\dddot v$ ſunt fluxiones quantitatum $\ddot z$, $\ddot y$, $\ddot x$, $\ddot v$, & hæ ſunt fluxiones quantitatum $\dot z$, $\dot y$, $\dot x$, $\dot v$ & hæ ſunt fluxiones quantitatum primarum z, y, x, v: ſic hæ quantitates conſiderari poſſunt ut fluxiones aliarum quas ſic deſignabo, $\acute z$,

[171]

$\acute z$, $\acute y$, $\acute x$, $\acute v$, & hæ ut fluxiones aliarum $\overset{\prime\prime}{z}$, $\overset{\prime\prime}{y}$, $\overset{\prime\prime}{x}$, $\overset{\prime\prime}{v}$, & hæ ut fluxiones aliarum $\overset{\prime\prime\prime}{z}$, $\overset{\prime\prime\prime}{y}$, $\overset{\prime\prime\prime}{x}$, $\overset{\prime\prime\prime}{v}$. Deſignant igitur $\overset{\prime\prime}{z}$, $\acute z$, z, $\dot z$, $\ddot z$, $\dddot z$, $\ddddot z$ &c. ſeriem quantitatum quarum quælibet poſterior eſt fluxio præcedentis & quælibet prior eſt fluens quantitas fluxionem habens ſubſequentem. Similis eſt ſeries $\overset{\prime\prime}{\overline{\sqrt{az-zz}}}$, $\overset{\prime}{\overline{\sqrt{az-zz}}}$, $\overline{\sqrt{az-zz}}$, $\overset{\cdot}{\overline{\sqrt{az-zz}}}$, $\overset{\cdot\cdot}{\overline{\sqrt{az-zz}}}$, $\overset{\cdot\cdot\cdot}{\overline{\sqrt{az-zz}}}$, ut & ſeries $\frac{\overset{\prime\prime}{az+z^2}}{a-z}$, $\frac{\overset{\prime}{az+z^2}}{a-z}$, $\frac{az+z^2}{a-z}$, $\frac{\overset{\cdot}{az+z^2}}{a-z}$, $\frac{\overset{\cdot\cdot}{az+z^2}}{a-z}$, $\frac{\overset{\cdot\cdot\cdot}{az+z^2}}{a-z}$. Et notandum eſt quod quantitas quælibet prior in his ſeriebus eſt ut area figuræ curvilineæ cujus ordinatim applicata rectangula eſt quantitas poſterior & abſciſſa eſt z: uti $\overset{\prime}{\overline{\sqrt{az-zz}}}$ area curvæ cujus ordinata eſt $\overline{\sqrt{az-zz}}$ & abſciſſa z. Quo autem ſpectant hæc omnia patebit in Propoſitionibus quæ ſequuntur.

TABULA

Curvarum ſimpliciorum quæ quadrari poſſunt.

Curvarum formæ.	Curvarum areæ.
Forma prima.	
$dz^{\eta-1} = y$.	$\frac{d}{\eta} z^{\eta} = t$.
Forma ſecunda.	
$\frac{dz^{\eta-1}}{ee+2efz^{\eta}+ffz^{2\eta}} = y$.	$\frac{dz^{\eta}}{\eta ee+\eta efz^{\eta}} = t$, vel $\frac{-d}{\eta ef+\eta ffz^{\eta}} = t$.
Forma tertia.	
1. $dz^{\eta-1}\sqrt{e+fz^{\eta}} = y$.	$\frac{2d}{3\eta f}R^3 = t$, exiſtente $R = \sqrt{e+fz^{\eta}}$
2. $dz^{2\eta-1}\sqrt{e+fz^{\eta}} = y$.	$\frac{-4e+6fz^{\eta}}{15\eta ff}dR^3 = t$.
3. $dz^{3\eta-1}\sqrt{e+fz^{\eta}} = y$.	$\frac{16ee-24efz^{\eta}+30ffz^{2\eta}}{105\eta f^3}dR^3 = t$.
4. $dz^{4\eta-1}\sqrt{e+fz^{\eta}} = y$.	$\frac{-96e^3+144eefz^{\eta}-180effz^{2\eta}+210f^3z^{3\eta}}{945\eta f^4}dR^3 = t$.
Forma quarta.	
1. $\frac{dz^{\eta-1}}{\sqrt{e+fz^{\eta}}} = y$.	$\frac{2d}{\eta f}R = t$.
2. $\frac{dz^{2\eta-1}}{\sqrt{e+fz^{\eta}}} = y$.	$\frac{-4e+2fz^{\eta}}{3\eta ff}dR = t$.

Expanding on his calculus, Newton in 1671 wrote *Methodus fluxionum et serierum infinitorum*. This work was published posthumously in an English translation (1736); the Latin original did not appear until about 50 years after Newton's death.

THE
METHOD of FLUXIONS
AND
INFINITE SERIES;
WITH ITS
Application to the Geometry of CURVE-LINES.

By the INVENTOR
Sir ISAAC NEWTON, *Kt.*
Late Preſident of the Royal Society.

Tranſlated from the AUTHOR'*s* LATIN ORIGINAL
not yet made publick.

To which is ſubjoin'd,
A PERPETUAL COMMENT upon the whole Work,
Conſiſting of
ANNOTATIONS, ILLUSTRATIONS, and SUPPLEMENTS,
In order to make this Treatiſe
A compleat Inſtitution for the uſe of LEARNERS.

By *JOHN COLSON*, M.A. and F.R.S.
Maſter of Sir *Joſeph Williamſon's* free Mathematical-School at *Rocheſter*.

LONDON:
Printed by HENRY WOODFALL;
And Sold by JOHN NOURSE, at the *Lamb* without *Temple-Bar*.
M.DCC.XXXVI.

PROB. III.

To determine the Maxima *and* Minima *of Quantities.*

1. When a Quantity is the greateſt or the leaſt that it can be, at that moment it neither flows backwards or forwards. For if it flows forwards, or increaſes, that proves it was leſs, and will preſently be greater than it is. And the contrary if it flows backwards, or decreaſes. Wherefore find its Fluxion, by Prob. 1. and ſuppoſe it to be nothing.

2. EXAMP. 1. If in the Equation $x^3 - ax^2 + axy - y^3 = 0$ the greateſt Value of x be required; find the Relation of the Fluxions of x and y, and you will have $3\dot{x}x^2 - 2a\dot{x}x + a\dot{x}y - 3\dot{y}y^2 + a\dot{y}x = 0$. Then making $\dot{x} = 0$, there will remain $-3\dot{y}y^2 + a\dot{y}x = 0$, or $3y^2 = ax$. By the help of this you may exterminate either x or y out of the primary Equation, and by the reſulting Equation you may determine the other, and then both of them by $-3y^2 + ax = 0$.

3. This Operation is the ſame, as if you had multiply'd the Terms of the propoſed Equation by the number of the Dimenſions of the other flowing Quantity y. From whence we may derive the famous

2

and INFINITE SERIES. 45

famous Rule of *Huddenius*, that, in order to obtain the greateſt or leaſt Relate Quantity, the Equation muſt be diſpoſed according to the Dimenſions of the Correlate Quantity, and then the Terms are to be multiply'd by any Arithmetical Progreſſion. But ſince neither this Rule, nor any other that I know yet publiſhed, extends to Equations affected with ſurd Quantities, without a previous Reduction; I ſhall give the following Example for that purpoſe.

4. EXAMP. 2. If the greateſt Quantity y in the Equation $x^3 - ay^2 + \frac{by^3}{a+y} - xx\sqrt{ay+xx} = 0$ be to be determin'd, ſeek the Fluxions of x and y, and there will ariſe the Equation $3\dot{x}x^2 - 2a\dot{y}y + \frac{3ab\dot{y}y^2 + 2b\dot{y}y^3}{a^2 + 2ay + y^2} - \frac{4a\dot{x}xy + 6\dot{x}x^3 + a\dot{y}x^2}{2\sqrt{ay+xx}} = 0$. And ſince by ſuppoſition $\dot{y} = 0$, omit the Terms multiply'd by $\dot{y}$, (which, to ſhorten the labour, might have been done before, in the Operation,) and divide the reſt by xx, and there will remain $3x - \frac{2ay + 3xx}{\sqrt{ay+xx}} = 0$. When the Reduction is made, there will ariſe $4ay + 3xx = 0$, by help of which you may exterminate either of the quantities x or y out of the propos'd Equation, and then from the reſulting Equation, which will be Cubical, you may extract the Value of the other.

Historians have established that Leibniz and Newton both developed the new form of calculus entirely independent of each other – although Leibniz had doubtless heard of Newton's works on calculus during his first visit to London in 1673. During his years as a diplomatic counselor in Paris (1672-76), Leibniz developed many of his mathematical ideas – including calculus – influenced by the Dutch mathematician, astronomer, and physicist **Christiaan Huygens**.

HUYGENS Christiaan
(1629 - 1695)

In any event, we are indebted to Leibniz for presenting the new calculus in a form accessible to a greater number of readers, while Newton's calculus of fluxions remained intelligible only to a select few. The symbols which are still used today in differential and integral calculus derive mainly from Leibniz, who was also the first to realize the importance of the new calculating methods to the entire field of mathematics.

BERNOULLI Jakob (Jacques) (1654 - 1705)

BERNOULLI Johann (Jean) (1667 - 1748)

Subsequent development of calculus was largely the work of the Swiss brothers **Jakob** and **Johann Bernoulli**, champions of Leibniz in his dispute with Newton over precedence to the invention of calculus. The Bernoulli brothers were the first to give public lectures on calculus.

de L'HOSPITAL Guillaume François (1661 - 1704)

The first printed textbook on calculus, *Analyse des infiniment petits*, was published in Paris in 1696. No author is mentioned in the first edition, but the second and following editions give the author as **Guillaume François de L'Hospital**, a French mathematician. In 1691 - 92, L'Hospital was instructed in differential and integral calculus by Johann Bernoulli; correspondence between them suggests that much of the *Analyse* was the work of Bernoulli. The book, which deals only with differential calculus, was largely instrumental in spreading knowledge of differential calculus during the latter half of the 18th century.

ANALYSE
DES
INFINIMENT PETITS,
Pour l'intelligence des lignes courbes.

A PARIS,
DE L'IMPRIMERIE ROYALE.
M. DC. XCVI.

TABLE.

AGNESI Maria Gaetana (1718 - 1799)

The first comprehensive textbook dealing with both differential and integral calculus was the Italian mathematician **Maria Agnesi**'s *Instituzione analitiche* (1748), with a French translation in 1755 and an English translation (*Analytical Institutions*) in 1801. Agnesi's great learning and her proficiency in languages brought forth her remarkable compilation of the works of many authors.

EULER Leonhard (1707 - 1783)
LAGRANGE Joseph Louis (1736 - 1813)
LAPLACE Pierre Simon (1749 - 1827)
CAUCHY Augustin Louis (1789 - 1857)
p. 466

The Swiss mathematician **Leonhard Euler**, the French mathematicians-astronomers-physicists **Joseph Louis Lagrange** (Italian born) and **Pierre Simon Laplace**, and the French mathematician **Augustin Louis Cauchy** were prominent figures in laying the foundation of differential equations.

In *Institutiones calculi differentialis* (Saint Petersburg, 1755) Euler showed how to differentiate the transcendental functions he had described in 1748 in his *Introductio in analysin.*

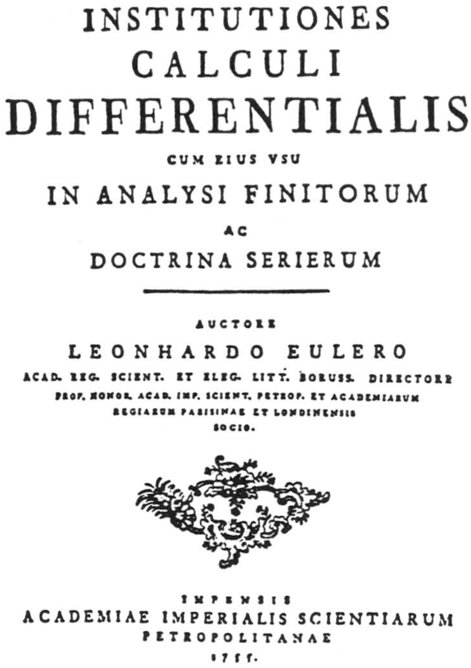
INSTITUTIONES
CALCULI
DIFFERENTIALIS
CUM EIUS VSU
IN ANALYSI FINITORUM
AC
DOCTRINA SERIERUM

AUCTORE
LEONHARDO EULERO
ACAD. REG. SCIENT. ET ELEG. LITT. BORUSS. DIRECTORE
PROF. HONOR. ACAD. IMP. SCIENT. PETROP. ET ACADEMIARUM
REGIARUM PARISINAE ET LONDINENSIS
SOCIO.

IMPENSIS
ACADEMIAE IMPERIALIS SCIENTIARUM
PETROPOLITANAE
1755.

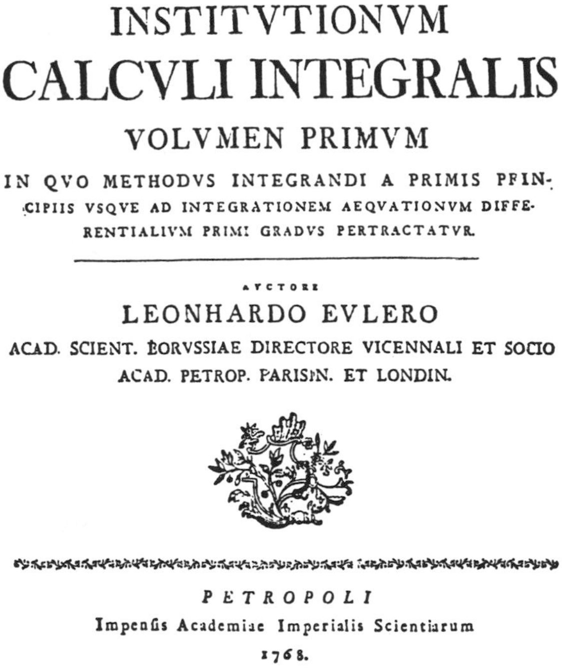
INSTITVTIONVM
CALCVLI INTEGRALIS
VOLVMEN PRIMVM
IN QVO METHODVS INTEGRANDI A PRIMIS PRINCIPIIS VSQVE AD INTEGRATIONEM AEQVATIONVM DIFFERENTIALIVM PRIMI GRADVS PERTRACTATVR.

AVCTORE
LEONHARDO EVLERO
ACAD. SCIENT. BORVSSIAE DIRECTORE VICENNALI ET SOCIO
ACAD. PETROP. PARISIN. ET LONDIN.

PETROPOLI
Impensis Academiae Imperialis Scientiarum
1768.

Euler's *Institutiones calculi integralis* (3 volumes, Saint Petersburg) was published in 1768 - 70. For almost a hundred years the *Introductio*... and the *Institutiones*... were the dominating textbooks in higher mathematical education.

Greek, *isos*, "equal"

ZENODOROS (*c.* 180 B.C.)

Isoperimetric problems deal not only with isoperimetric curves, that is, closed curves with equal perimeters in the plane, but also with closed surfaces in space. The study of such problems emerged in antiquity; **Zenodorus** proposed in *On Isometric Figures* that the circle has the maximum *area* of all isoperimetric figures in a plane, and that the sphere has the maximum *volume* of all bodies with equal surface. Euler developed in *Methodus inviendi* a general method to find a function for which a given integral assumes a maximum or minimum value, thereby inducing isoperimetric problems to become a separate mathematical discipline, later known as **calculus of variations**.

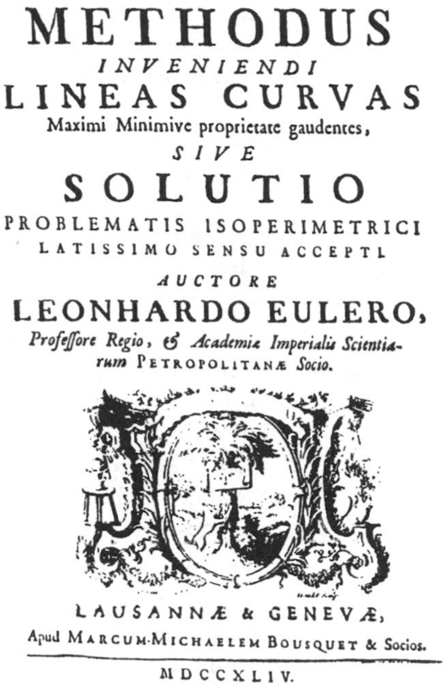
METHODUS
INVENIENDI
LINEAS CURVAS
Maximi Minimive proprietate gaudentes,
SIVE
SOLUTIO
PROBLEMATIS ISOPERIMETRICI
LATISSIMO SENSU ACCEPTI
AUCTORE
LEONHARDO EULERO,
Professore Regio, & Academiæ Imperialis Scientiarum PETROPOLITANÆ *Socio.*

LAUSANNÆ & GENEVÆ,
Apud MARCUM-MICHAELEM BOUSQUET & Socios.
MDCCXLIV.

Leonhard Euler's *A Method for Discovering Curved Lines Having a Maximum or Minimum Property or the Solution of the Isoperimetric Problem taken in its Widest Sense* (1744).

FOURIER Jean Baptiste Joseph (1768 - 1830)

The French engineer-mathematician, Egyptologist, and one of Napoleon's most able administrators **Joseph Fourier** presented, in 1828, in his *Théorie analytique de la chaleur* ("The Analytical Theory of Heat") a method to express any function as the sum of fundamental trigonometric functions. Fourier's discovery is of fundamental importance in both physics and mathematics, where it revolutionized the solving of partial differential equations (that is, differential equations involving more than one independent variable).

GERMAIN Sophie (1776 - 1831)

Although higher education for women was virtually nonexistent, and despite prevailing prejudice and skepticism about feminine pursuits in fields of scientific research, the 18th and 19th centuries brought forth extraordinary achievements by strong-minded, scholarly women. In mathematics, the Frenchwoman **Sophie Germain** – one of the founders of modern applied mathematics – rose to fame in 1816 for a paper on calculus of variations applied to the theory of elasticity.

ABEL Niels Henrik (1802 -1829)

Struggling against poverty and illness, the Norwegian mathematician **Niels Henrik Abel** made significant contributions to several fields of mathematics during his short life. In calculus he will be especially remembered for his contributions to the theory of *integral equations*, that is, equations in which the unknown function belongs to an integral.

CAUCHY Augustin Louis (1789 - 1857)
WEIERSTRASS Karl Theodor Wilhelm (1815 - 1897)
KOVALEVSKI Sonia (1850 - 1891)

In the second half of the 19th century infinitesimals were replaced by the concept of limits, a development due mainly to the work of the French mathematician **A. L. Cauchy** and the German mathematician **Karl Weierstrass**.

A student of Weierstrass's, the Russian mathematician and novelist **Sonia Kovalevski** made important contributions to integral calculus and differential equations. In spite of her academic qualifications and letters of recommendation from the otherwise powerful Karl Weierstrass, she was for a long time unable to obtain an academic position; no university would accept a woman mathematician, however qualified. In 1889 she was appointed professor of mathematics in Stockholm, but many still resented the appointment of a woman, among them the Swedish playwright and novelist August Strindberg.

LIE Marius Sophus (1842 - 1899)
Galois theory: *p.* 301
Lie Groups

The Norwegian mathematician **Sophus Lie** developed a geometric integration theory for certain partial differential equations. Lie is best known for his expansion of group theory, initiated by Galois in the theory of algebraic equations, to include groups of geometric transformation, today known as **Lie groups** and exerting great influence on research in the theory of differential equations and in other fields of mathematics and its applications, especially in quantum physics.

POINCARÉ Jules Henri (1854 - 1912)

The French mathematician, physicist, and astronomer **Henri Poincaré**, "the last mathematical universalist", pioneered work on the theory of differential equations and its applications to dynamics and astronomy. He published over 500 articles and books dealing with practically all the major fields of pure and applied mathematics; his publications on differential equations span from his thesis in 1878 to his last paper, in 1912.

FREDHOLM Erik Ivar (1866 - 1927)

The Swedish mathematician and physicist **Ivar Fredholm** proved the existence of solutions for a general class of integral equations, which is important for the study of boundary problems in mathematical physics. This contribution, sometimes characterized as "long overdue", gave significant inspiration to the development of functional analysis; in a modernized form, Fredholm's ideas are still of fundamental importance in the theory of linear partial differential equations.

SCHWARTZ Laurent (b. 1915)

Before the 1950s the study of partial differential equations was mainly devoted to elliptic, parabolic, and hyperbolic differential equations originating in physics and representing phenomena such as electrostatics, heat conduction, and electromagnetic wave propagation. The introduction of *distribution theory* by the French mathematician **Laurent Schwartz** at the end of the 1940s made it possible to develop a theory of general linear partial differential equations, first with constant coefficients. The extension to equations with variable coefficients led to the development of a *theory of pseudo-differential and Fourier integral operators*, incorporating formal methods used in mathematical physics. This forms the core of *microlocal analysis*, which has also turned out to be an effective tool in the study of nonlinear partial differential equations.

Chapter 20

INTRODUCTION TO DIFFERENTIAL CALCULUS

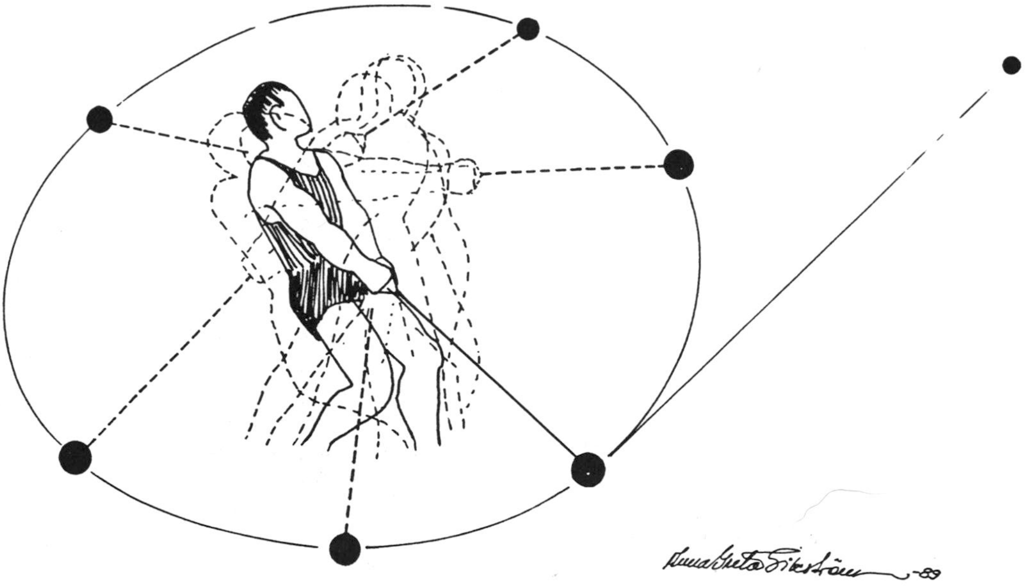

20.1 Derivatives and Differentials

If the independent variable x of a function $f(x)$ is increased by a small amount Δx – the increment "delta x" – it will cause a corresponding change Δf of $f(x)$; the ratio $\Delta f/\Delta x$ is a measure of the rate of change of f with x.

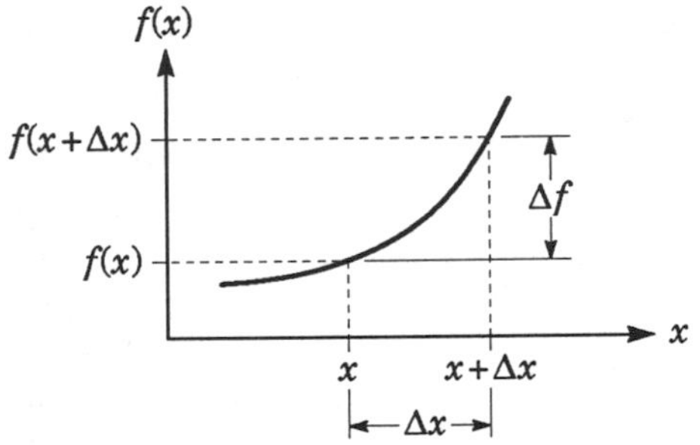

If Δx is made progressively smaller, the limit value

$$\lim_{\Delta x \to 0} \frac{f(x+\Delta x) - f(x)}{\Delta x} = \lim_{\Delta x \to 0} \frac{\Delta f}{\Delta x},$$

when Δx tends to zero (if it exists) is the **first derivative** of $f(x)$ with respect to x,

$$\frac{\mathrm{d}f}{\mathrm{d}x} = f'(x); \quad f' = \text{“}f \text{ prime”},$$

which describes the *instantaneous* change of f at a given point x.

If the limit value $f'(x)$ exists in a specific interval of x, the function $f(x)$ is said to be *differentiable* in this interval.

Differentiation of $f'(x)$ gives the **second derivative** $f''(x)$ – read "f bis" – and, analogously, the derivative of f'' is the **third derivative** $f'''(x)$; higher derivatives are denoted $f^{\mathrm{iv}}(x)$ for the fourth derivative, f^{v} for the fifth derivative, *etc.*

- Find the first and second derivatives of $f(x) = x^2$.

$$\begin{aligned}
f(x) &= x^2 \\
f(x+\Delta x) &= (x+\Delta x)^2 \\
f(x+\Delta x) - f(x) &= x^2 + 2x\,\Delta x + (\Delta x)^2 - x^2 \\
&= 2x\,\Delta x + (\Delta x)^2 \\
\frac{f(x+\Delta x) - f(x)}{\Delta x} &= 2x + \Delta x \\
f'(x) &= \lim_{\Delta x \to 0} (2x + \Delta x) \\
&= 2x.
\end{aligned}$$

$$\begin{aligned}
f'(x) &= 2x \\
f'(x+\Delta x) &= 2(x+\Delta x) \\
f'(x+\Delta x) - f'(x) &= 2x + 2\,\Delta x - 2x \\
\frac{f'(x+\Delta x) - f'(x)}{\Delta x} &= \frac{2x + 2\,\Delta x - 2x}{\Delta x} \\
f''(x) &= \lim_{\Delta x \to 0} 2 \\
&= 2.
\end{aligned}$$

•

The notation

$$\frac{\mathrm{d}^n f(x)}{\mathrm{d}\,x^n}$$

von LEIBNIZ Gottfried Wilhelm (1646 - 1716)

for the n-th derivative of $f(x)$ was introduced by **Leibniz**.

LAGRANGE Joseph Louis (1736 - 1813)

In his treatises *Théorie des fonctions analytiques* (1797) and *Leçons sur le calcul des fonctions* (1800), **Lagrange** used Leibniz's notation, and introduced his own notation

$$f^{(n)}(x).$$

ARBOGAST Louis-François-Antoine (1759 - 1803)

The French mathematician and lawyer **Louis Arbogast** introduced in his *Du calcul des dérivations* (1800) the notation

$$\mathrm{D}^n f(x) \quad \text{or} \quad \mathrm{D}^n f$$

for the n-th derivative of f.

We have so far regarded $\mathrm{d}f/\mathrm{d}x$ and $\mathrm{d}y/\mathrm{d}x$ as inseparable symbols for the derivative of a function $y = f(x)$, but we may consider $\mathrm{d}y/\mathrm{d}x$ as the ratio of two separate quantities $\mathrm{d}y$ and $\mathrm{d}x$, known as **differentials**, which can be handled just as any other algebraic quantities, as we shall see in the chapters on integral calculus and differential equations (Chapters 21 and 34).

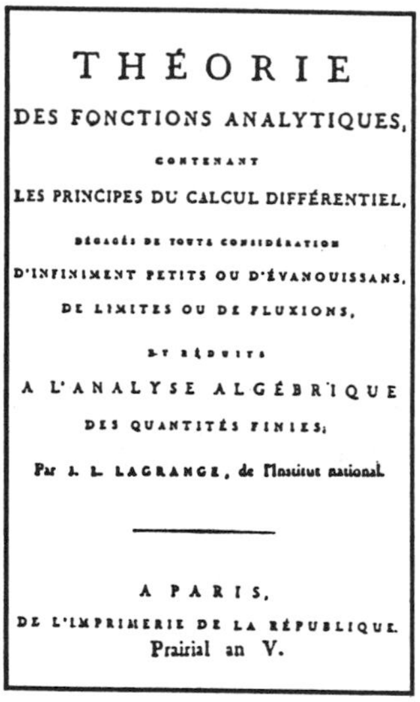
THÉORIE
DES FONCTIONS ANALYTIQUES,
CONTENANT
LES PRINCIPES DU CALCUL DIFFÉRENTIEL,
DÉGAGÉS DE TOUTE CONSIDÉRATION
D'INFINIMENT PETITS OU D'ÉVANOUISSANS,
DE LIMITES OU DE FLUXIONS,
ET RÉDUITS
A L'ANALYSE ALGÉBRIQUE
DES QUANTITÉS FINIES;
Par J. L. LAGRANGE, de l'Institut national.

A PARIS,
DE L'IMPRIMERIE DE LA RÉPUBLIQUE.
Prairial an V.

The publishing date *Prairial an V* refers to the 9th month (20 May - 18 June) of the year five (1797) of the French republican calendar.

Derivatives of Linear Functions

A linear function, that is, a function of degree one, is represented geometrically by a **straight line** with the slope

$$m = \frac{\Delta y}{\Delta x} = \frac{y_2 - y_1}{x_2 - x_1}; \qquad x_1 \neq x_2$$

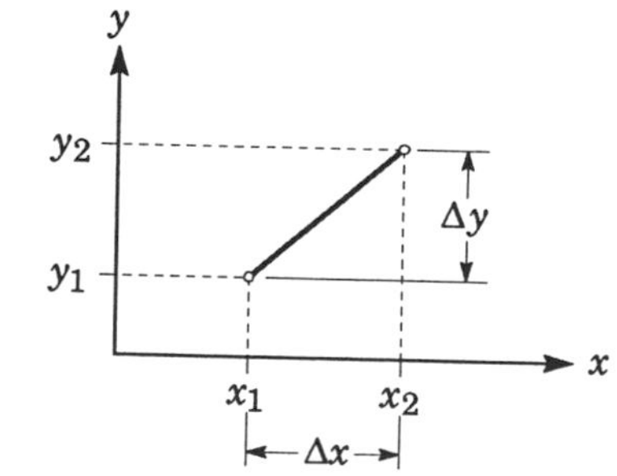

which is also the derivative y' of the linear function $y = f(x)$; all higher derivatives are zero.

In the special case of y = constant, the line is horizontal with the slope = zero; consequently, the **derivative of a constant is zero**.

Derivatives of Nonlinear Functions

Functions of the second and higher degrees are represented geometrically by curves of various descriptions.

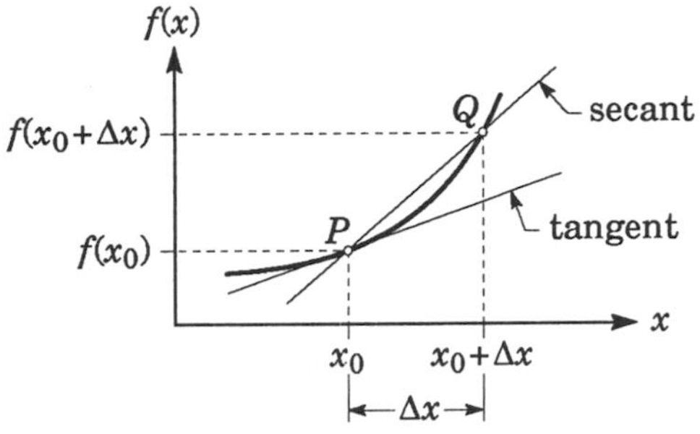

The first derivative of a nonlinear function is the gradient of the curve, that is, the slope of the tangent to the curve.

The slope of the secant PQ is

$$m_{PQ} = \frac{f(x_0 + \Delta x) - f(x_0)}{\Delta x} .$$

As Q moves continuously closer to P, the secant approaches the tangent, with the slope

$$m_P = \lim_{\Delta x \to 0} \frac{f(x_0 + \Delta x) - f(x_0)}{\Delta x} = f'(x),$$

which is the derivative at x_0 of the function $f(x)$.

- Determine the first, second, and third derivatives of the function $f(x) = x^3$ and trace their graphs.

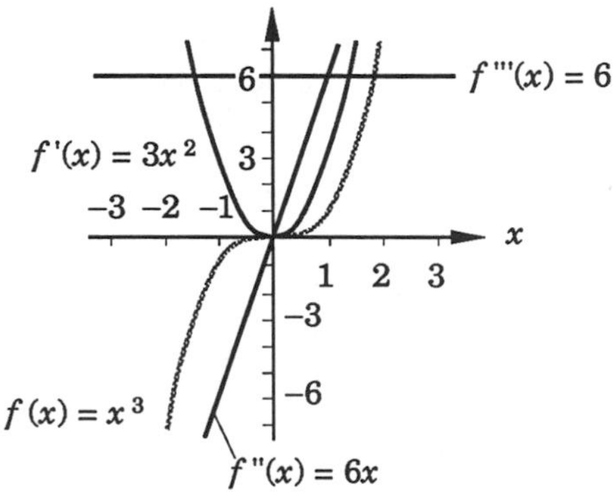

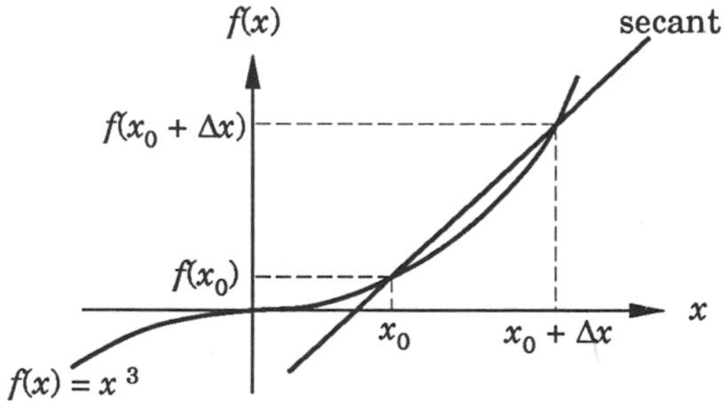

$$f'(x) = \lim_{\Delta x \to 0} \frac{(x + \Delta x)^3 - x^3}{\Delta x} = 3x^2$$

$$f''(x) = \lim_{\Delta x \to 0} \frac{3(x + \Delta x)^2 - 3x^2}{\Delta x} = 6x$$

$$f'''(x) = \lim_{\Delta x \to 0} \frac{6(x + \Delta x) - 6x}{\Delta x} = 6$$ •

The **equation of the tangent** at a specific point on the graph of a function may be determined when the derivative of the function is known.

The derivative of $y = x^2$ is $2x$. At $x = 1.5$; $y = 2.25$, the slope of the tangent is

$$2 \cdot 1.5 = 3$$

and, thus, the equation of the tangent is

$$\frac{y - 2.25}{x - 1.5} = 3$$

$$y = 3x - 2.25$$

and

$$12x - 4y = 9 .$$

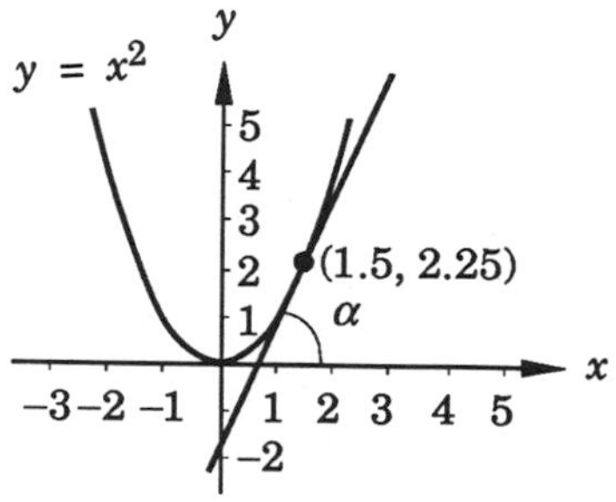

The slope angle α of the tangent is given by

$$\tan\alpha = 3 ; \quad \alpha \approx 72° .$$

- Find the equations of the tangent and the normal of the curve $y = 3x^2$ at (1, 3).

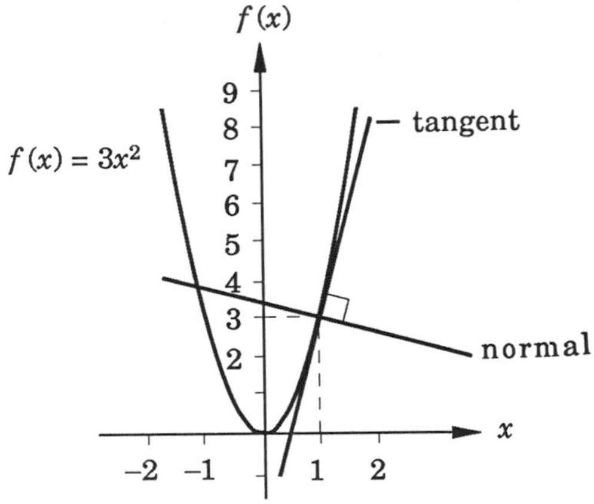

The slope of the tangent is $y' = 6x$, at (1, 3) it is $y'(1) = 6$; the equation of the tangent is then

$$y = 6x + b;$$

since $x = 1, y = 3$, the **equation of the tangent** is

$$6x - y = 3 .$$

The normal is perpendicular to the tangent and thus has the equation

$$y = -\frac{1}{6}x + a ;$$

with (1, 3) inserted, the **equation of the normal** is

$$x + 6y = 19 .$$

20.2 Differentiating Algebraic Functions

The derivative is usually found by proceeding from already known derivatives of the most common functions and applying rules for differentiation of sums, products, and quotients.

The Derivative of a Power of a Variable

$$\frac{d}{dx}[x^n] = \lim_{\Delta x \to 0} \frac{(x + \Delta x)^n - (x)^n}{\Delta x}$$

Binomial Theorem : p. 140

$$= \lim_{\Delta x \to 0} \frac{x^n + n\,x^{n-1}(\Delta x) + \binom{n}{2} x^{n-2}(\Delta x)^2 + \ldots + (\Delta x)^n - x^n}{\Delta x}$$

$$= n\,x^{n-1} + 0 + \ldots + 0$$

$$= n\,x^{n-1},$$

which applies for all real numbers n.

- $f(x) = x^4 \Rightarrow f'(x) = 4x^{4-1} = 4x^3$
  $g(x) = x^{\pi} \Rightarrow g'(x) = \pi\,x^{\pi-1}$
  $h(x) = x^{-1/3} \Rightarrow h'(x) = \frac{-1}{3}x^{-1/3-1} = \frac{-x^{-4/3}}{3} = \frac{-1}{3x^{4/3}}$
  $F(x) = x^{e} \Rightarrow F'(x) = e\,x^{e-1}$
  $G(x) = x^1 \Rightarrow G'(x) = x^{1-1} = x^0 = 1$
  $H(x) = x^0 \Rightarrow H'(x) = 0x^{0-1} = 0$ •

If a function includes a constant factor c, it is reproduced in the derivative,

$$\frac{d}{dx}[c\,f(x)] = c\,\frac{d}{dx}[f(x)]\,.$$

- Find the fourth derivative:
- $g(x) = 2\,x^{1/2}$

$$g'(x) = 2\left(\frac{1}{2}\right)x^{1/2-1} = x^{-1/2} = \frac{1}{x^{1/2}}$$

$$g''(x) = \frac{-1}{2}x^{-1/2-1} = \frac{-x^{-3/2}}{2} = \frac{-1}{2x^{3/2}}$$

$$g'''(x) = \frac{-1}{2}\left(\frac{-3}{2}\right)x^{-3/2-1} = \frac{3}{4}x^{-5/2} = \frac{3}{4x^{5/2}}$$

$$g^{iv}(x) = \frac{3}{4}\left(\frac{-5}{2}\right)x^{-5/2-1} = \frac{-15}{8}x^{-7/2} = \frac{-15}{8x^{7/2}}\,.$$

- $h(x) = 4x^{-1/3}$

$$h'(x) = 4\left(\frac{-1}{3}\right) x^{-1/3-1} = \frac{-4}{3} x^{-4/3} = \frac{-4}{3x^{4/3}}$$

$$h''(x) = \frac{-4}{3}\left(\frac{-4}{3}\right) x^{-4/3-1} = \frac{16}{9} x^{-7/3} = \frac{16}{9x^{7/3}}$$

$$h'''(x) = \frac{16}{9}\left(\frac{-7}{3}\right) x^{-7/3-1} = \frac{-112}{27} x^{-10/3} = \frac{-112}{27x^{10/3}}$$

$$h^{iv}(x) = \frac{-112}{27}\left(\frac{-10}{3}\right) x^{-10/3-1} = \frac{1120}{81} x^{-13/3} = \frac{1120}{81x^{13/3}} .$$

The Derivative of a Sum

The limit value of a sum of functions is the sum of their several limits; thus if

$$h(x) = f(x) + g(x)$$

and

$$f'(x) = \lim_{\Delta x \to 0} \frac{f(x + \Delta x) - f(x)}{\Delta x}$$

$$g'(x) = \lim_{\Delta x \to 0} \frac{g(x + \Delta x) - g(x)}{\Delta x}$$

we have

$$h'(x) = f'(x) + g'(x) .$$

This equality does, of course, apply for a sum of any finite number of functions.

The **subtraction** of a function is equivalent to the addition of its corresponding negative function. The derivative of

$$h(x) = f(x) - g(x)$$

is

$$h'(x) = f'(x) - g'(x) .$$

The Derivative of a Product

The derivative of the product function

$$f(x) \cdot g(x)$$

is

$$h'(x) = \lim_{\Delta x \to 0} \frac{f(x + \Delta x)\, g(x + \Delta x) - f(x)\, g(x)}{\Delta x}$$

where we *add* and *subtract* the expression $f(x + \Delta x)\, g(x)$ in the numerator and obtain, after rearranging,

$$h'(x) = \lim_{\Delta x \to 0} \left[\frac{f(x + \Delta x) - f(x)}{\Delta x} \cdot g(x) + f(x + \Delta x) \cdot \frac{g(x + \Delta x) - g(x)}{\Delta x} \right]$$

$$= f'(x) \cdot g(x) + f(x) \cdot g'(x) .$$

The formula may be generalized to apply to products of any finite number of functions.

If one of the functions is a constant, *e.g.*, f = const, $f' = 0$, the derivative simplifies to

$$h'(x) = f \cdot g'(x) \ .$$

The Derivative of a Quotient

The derivative of the quotient

$$h(x) = \frac{f(x)}{g(x)} \ ; \quad g(x) \neq 0$$

is

$$h'(x) = \lim_{\Delta x \to 0} \frac{\dfrac{f(x + \Delta x)}{g(x + \Delta x)} - \dfrac{f(x)}{g(x)}}{\Delta x}$$

$$= \lim_{\Delta x \to 0} \frac{f(x + \Delta x) \cdot g(x) - f(x)\, g(x + \Delta x)}{\Delta x \cdot g(x) \cdot g(x + \Delta x)} \ ,$$

which we may rewrite

$$h'(x) = \lim_{\Delta x \to 0} \left\{ \frac{\dfrac{f(x + \Delta x) - f(x)}{\Delta x}}{g(x) \cdot g(x + \Delta x)} \cdot g(x) - f(x) \cdot \frac{\dfrac{g(x + \Delta x) - g(x)}{\Delta x}}{g(x) \cdot g(x + \Delta x)} \right\}$$

and obtain, finally,

$$h'(x) = \frac{f'(x) \,.\, g(x) - f(x) \,.\, g'(x)}{\{g(x)\}^2} \ .$$

Derivatives of Composite Functions

In a composite function $y = (f \circ g)(x)$ – "a function of a function" – f is the **external function** and g the **core function**.

The derivative of a composite function is the product of the derivative of the external function f with respect to the core function g multiplied by the derivative of the core function g with respect to the independent variable x,

$$\frac{d}{dx}\left[f\{g(x)\}\right] = \frac{df}{dg} \cdot \frac{dg}{dx} \ ,$$

provided that the functions are differentiable, f with respect to g and g with respect to x.

Chain Rule

The formula is called the **chain rule of differentiation**.

Differentiating a power function of the general form

$$y = u^n ,$$

where u is a differentiable function of a variable x, the chain rule gives

$$y' = \frac{dy}{dx} = \frac{dy}{du} \cdot \frac{du}{dx} = \frac{d}{du}[u^n]\frac{du}{dx}$$

General Power Rule

and thus the **general power rule**,

$$\frac{dy}{dx} = n u^{n-1}\left[\frac{du}{dx}\right] .$$

If one arrives at an answer that is of different algebraic form than a given answer, the two answers may be compared by substituting a constant for the variable. Unequal numerical values definitely signify a wrong answer, whereas equal values indicate the possibility of a correct answer.

There is no stipulation how far a derivative should be simplified, other than that it must be written in a form suitable to its application.

- Find the second derivative of

$$f(x) = (2x + 5)^5 .$$

$$f'(x) = 5(2x+5)^4 \cdot 2 = 10(2x+5)^4$$

$$f''(x) = 4 \cdot 10(2x+5)^3 \cdot 2 = 80(2x+5)^3 .$$

- Find the first derivatives:
  - $g(x) = (4x - 3x^2)^3$;

$$g'(x) = 3(4x - 3x^2)^2(4 - 6x) = 6(2 - 3x)(4x - 3x^2)^2 .$$

  - $f(t) = t\sqrt{1+t^2}$;

$f(t) = t(1+t^2)^{1/2}$, so

$$f'(t) = 1 \cdot (1+t^2)^{1/2} + t \cdot \frac{1}{2}(1+t^2)^{-1/2} \cdot 2t$$

$$= [(1+t^2) + t^2](1+t^2)^{-1/2} = \frac{1+2t^2}{\sqrt{1+t^2}} .$$

  - $F(x) = \sqrt{\frac{x-1}{x+1}}$;

$$F'(x) = \frac{1}{2}\left(\frac{x-1}{x+1}\right)^{-1/2} \frac{d}{dx}\left[\frac{x-1}{x+1}\right]$$

$$= \frac{1}{2}\left(\frac{x+1}{x-1}\right)^{1/2} \frac{(x+1)\cdot 1 - (x-1)\cdot 1}{(x+1)^2}$$

$$= \sqrt{\frac{x+1}{x-1}} \cdot \frac{1}{(x+1)^2} .$$

Differentiability

A function is differentiable only where it is continuous, but the reverse is not necessarily true; a function may be continuous at a certain point without being differentiable at that point.

One-Sided Limits: p. 358

The derivative may have different values

- from the right, $f'(x_0+) = \lim_{\Delta x \to 0+} \dfrac{f(x_0 + \Delta x) - f(x_0)}{\Delta x}$;
- from the left, $f'(x_0-) = \lim_{\Delta x \to 0-} \dfrac{f(x_0 + \Delta x) - f(x_0)}{\Delta x}$ ..

The function is differentiable at x_0 only if the derivatives from the right and from the left are equal.

The absolute-value function $f(x) = |x|$ is a case in point; at $x = 0$, the derivative from the right, for $\Delta x > 0$, is

$$f'(0+) = \lim_{\Delta x \to 0+} \frac{f(0 + \Delta x) - f(0)}{\Delta x} = +1,$$

and the derivative from the left, for $\Delta x < 0$ is

$$f'(0-) = \lim_{\Delta x \to 0-} \frac{f(0 + \Delta x) - f(0)}{\Delta x} = -1.$$

Consequently $f(x) = |x|$ is not differentiable at $x = 0$.

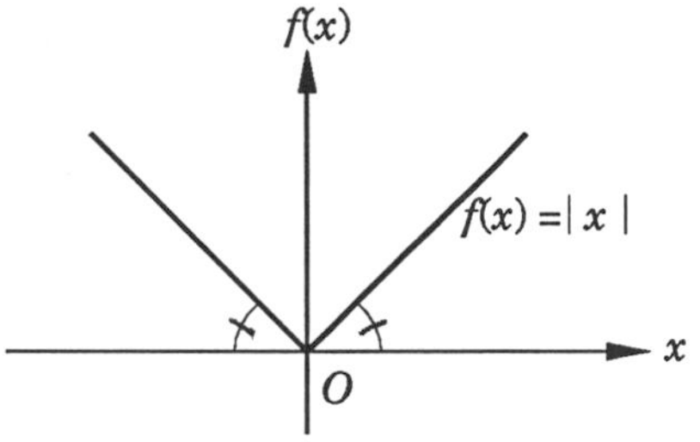

In the general case $f(x) = |x^n|$, we must distinguish two cases when n is odd:

$x \geq 0$; $f(x) = x^n$; $f'(x) = n \cdot x^{n-1}$ [$f'(0) = 0$, only if $n > 1$] ;

$x < 0$; $f(x) = -x^n$; $f'(x) = -n \cdot x^{n-1}$.

A function $f(x)$ may also be continuous but not differentiable at $x = x_0$ if

$$\lim_{\Delta x \to 0} \frac{f(x_0 + \Delta x) - f(x_0)}{\Delta x} = +\infty \text{ or } -\infty ,$$

which means that the tangent of $f(x)$ at x_0 is vertical.

- Is $f(x) = \sqrt[3]{x}$ differentiable at $x = 0$?

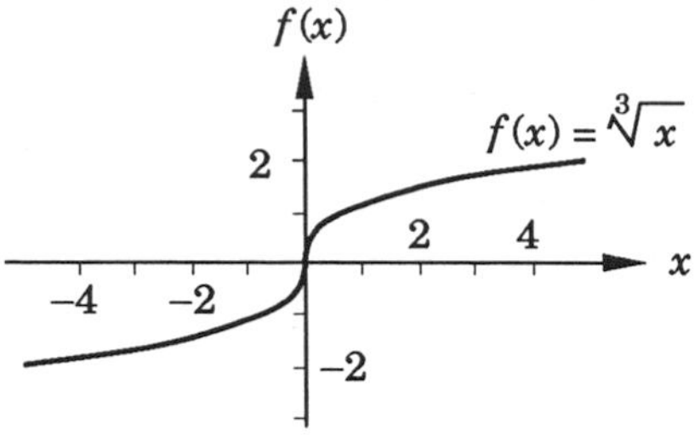

At $x = 0$, we have

$$\underset{x=0}{f'}\left(\sqrt[3]{x}\right) = \lim_{\Delta x \to 0} \frac{\sqrt[3]{0 + \Delta x} - \sqrt[3]{0}}{\Delta x} = \lim_{\Delta x \to 0} \sqrt[3]{\frac{1}{(\Delta x)^2}} = \infty .$$

Although $f(x) = \sqrt[3]{x}$ is continuous at $x = 0$, it is not differentiable as the tangent at $x = 0$ is vertical.

- Find the first derivative of

$$f(x) = |9 - x^2| .$$

$|9 - x^2| = x^2 - 9$ if $(9 - x^2 < 0)$; $\quad f'(x) = 2x$

$|9 - x^2| = 9 - x^2$ if $(9 - x^2 > 0)$; $\quad f'(x) = -2x$

The graph of $f(x) = |9 - x^2|$ shows sharp cusps at $x = 3$ and $x = -3$, where the function is not differentiable:

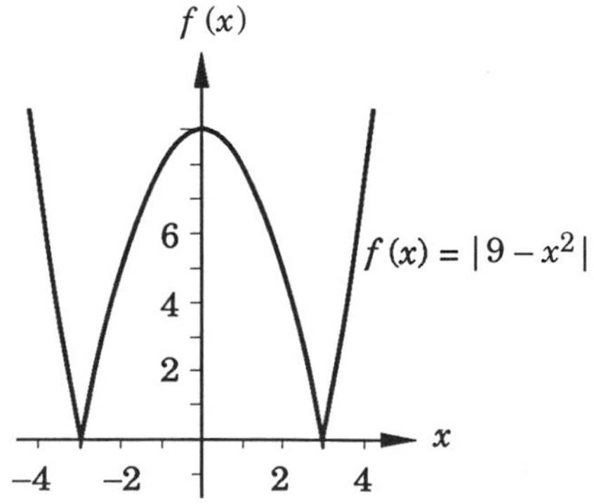

20.3 Differentiating Transcendental Functions

Fundamental Trigonometric Functions

sin x

$$\frac{d}{dx}[\sin x] = \lim_{\Delta x \to 0} \frac{\sin(x + \Delta x) - \sin x}{\Delta x}$$

$$= \lim_{\Delta x \to 0} \frac{\sin x \cos \Delta x + \cos x \sin \Delta x - \sin x}{\Delta x}$$

$$= \cos x \cdot \lim_{\Delta x \to 0} \frac{\sin \Delta x}{\Delta x} - \sin x \cdot \lim_{\Delta x \to 0} \frac{1 - \cos \Delta x}{\Delta x}$$

pp. 529 - 30

$$= 1 \cdot \cos x - 0 \cdot \sin x = \cos x .$$

cos x

$$\frac{d}{dx}[\cos x] = \frac{d}{dx}\left[\sin\left(\frac{\pi}{2} - x\right)\right] = (-1)\sin\left(\frac{\pi}{2} - x\right) = -\sin x .$$

tan x

Using the formula $\frac{d}{dx}\left[\frac{u}{v}\right] = \frac{u'v - uv'}{v^2}$,

$$\frac{d}{dx}[\tan x] = \frac{d}{dx}\left[\frac{\sin x}{\cos x}\right] = \frac{\cos x \cdot \cos x + \sin x \cdot \sin x}{\cos^2 x}$$

$$= \frac{1}{\cos^2 x} = \sec^2 x = 1 + \tan^2 x .$$

cot x

$$\frac{d}{dx}[\cot x] = \frac{d}{dx}\left[\frac{\cos x}{\sin x}\right] = \frac{\sin x \cdot \sin x - \cos x \cdot \cos x}{\sin^2 x}$$

$$= \frac{-1}{\sin^2 x} = -\csc^2 x = -(1 + \cot^2 x) .$$

sec x

Using the formula $\frac{d}{dx}\left[\frac{1}{u}\right] = -\frac{u'}{u^2}$,

$$\frac{d}{dx}[\sec x] = \frac{d}{dx}\left[\frac{1}{\cos x}\right] = -\frac{-\sin x}{\cos^2 x} = \frac{\tan x}{\cos x} = \sec x \cdot \tan x .$$

csc x

$$\frac{d}{dx}[\csc x] = \frac{d}{dx}\left[\frac{1}{\sin x}\right] = -\frac{\cos x}{\sin^2 x} = -\frac{\cot x}{\sin x} = -\csc x \cdot \cot x .$$

- The first derivative of $y = \sin 2x^3$ is

$$y' = 6x^2 \cos 2x^3 .$$

- The first derivative of $y = 5\sqrt{\sin 2x}$ is

$$y' = 5 \cdot \frac{1}{2}(\sin 2x)^{-1/2} \cdot \frac{d}{dx}[\sin 2x]$$

$$= 5 \cdot \frac{1}{2}(\sin 2x)^{-1/2} \cdot 2 \cdot \cos 2x = \frac{5 \cos 2x}{\sqrt{\sin 2x}} .$$

- Find the first derivative of

$$f(x) = \cos x^2; \quad g(x) = x\cos x; \quad h(x) = \cos^2 x .$$

$f'(x) = (-\sin x^2)\cdot 2x = -2x\sin x^2 .$

$g'(x) = x(-\sin x) + (\cos x)\cdot 1 = \cos x - x\sin x .$

p. 512 $h'(x) = -2\sin x\cos x \quad (= -\sin 2x) .$

- Find the second derivative of

$$y = \cos 2x^3 .$$

$$y' = (-\sin 2x^3)(6x^2) = -6x^2\sin 2x^3 ;$$

$$y'' = -6x^2\cdot\cos 2x^3\cdot 6x^2 - 12x\cdot\sin 2x^3$$

$$= -12x(3x^3\cdot\cos 2x^3 + \sin 2x^3) .$$

- Find the second derivative of

$$y = \cos\frac{x}{2} .$$

$$y' = -\frac{1}{2}\sin\frac{x}{2} ;$$

$$y'' = -\frac{1}{4}\cos\frac{x}{2} .$$

- Find the first derivative of

$$y = \frac{\sin x - x\cos x}{\cos x + x\sin x} .$$

$$y' = \frac{(\cos x + x\sin x)\left(\dfrac{d}{dx}[\sin x - x\cos x]\right)}{(\cos x + x\sin x)^2} - \frac{(\sin x - x\cos x)\left(\dfrac{d}{dx}[\cos x + x\sin x]\right)}{(\cos x + x\sin x)^2}$$

$$= \frac{(\cos x + x\sin x)\,x\sin x - (\sin x - x\cos x)\,x\cos x}{(\cos x + x\sin x)^2}$$

p. 508

$$= \frac{x^2(\sin^2 x + \cos^2 x)}{(\cos x + x\sin x)^2} = \frac{x^2}{(\cos x + x\sin x)^2} .$$

- Find the first derivative of

$$f(x) = \cos(\cos x) ; \quad g(x) = \sin(\cos x) ;$$
$$h(x) = \sin(\sin x) ; \quad F(x) = \sin^2(\cos x) .$$

$f'(x) = \{-\sin(\cos x)\}(-\sin x) = \sin x\{\sin(\cos x)\} .$

$g'(x) = \{\cos(\cos x)\}(-\sin x) = -\sin x\{\cos(\cos x)\} .$

$h'(x) = \cos x\{\cos(\sin x)\} .$

$$F'(x) = 2\{\sin(\cos x)\}\left\{\frac{d}{dx}[\sin(\cos x)]\right\}$$
$$= 2\{\sin(\cos x)\}\left\{[\cos(\cos x)](-\sin x)\right\}$$
$$= -2\sin x\{\sin(\cos x)\}\{\cos(\cos x)\}.$$

- Find the first derivative of $y = \tan^3 4t$.

$$y' = 3(\tan 4t)^2\left(\frac{d}{dt}[\tan 4t]\right) = \frac{12\tan^2 4t}{\cos^2 4t}.$$

- Find the second derivative of $f(x) = \tan\frac{x}{2}$.

$$f'(x) = \frac{1}{2}\sec^2\frac{x}{2}.$$
$$f''(x) = \frac{1}{2}\cdot\frac{d}{dx}\left[\sec^2\frac{x}{2}\right] = \frac{1}{2}\cdot 2\cdot\sec\frac{x}{2}\left(\frac{d}{dx}\left[\sec\frac{x}{2}\right]\right)$$
$$= \sec\frac{x}{2}\sec\frac{x}{2}\tan\frac{x}{2}\left(\frac{1}{2}\right) = \frac{1}{2}\sec^2\frac{x}{2}\tan\frac{x}{2}.$$

Inverse Trigonometric Functions

arcsin x If $y = \arcsin x$, then $x = \sin y$; $\frac{dx}{dy} = \cos y$; $dx = \cos y\, dy \Rightarrow \frac{dy}{dx} = \frac{1}{\cos y}$, so

p. 508

$$\frac{d[\arcsin x]}{dx} = \frac{1}{\cos y} = \frac{1}{\sqrt{1-\sin^2 y}} = \frac{1}{\sqrt{1-x^2}}; \quad |x| < 1.$$

arccos x $y = \arccos x$; $x = \cos y$; $dx = -\sin y\ dy$

$$\frac{d[\arccos x]}{dx} = \frac{-1}{\sin y} = \frac{-1}{\sqrt{1-\cos^2 y}} = -\frac{1}{\sqrt{1-x^2}}; \quad |x| < 1.$$

arctan x $y = \arctan x$; $x = \tan y$; $dx = \sec^2 y\ dy$

$$\frac{d[\arctan x]}{dx} = \frac{1}{\sec^2 y} = \frac{1}{1+\tan^2 y} = \frac{1}{1+x^2}.$$

arccot x $y = \operatorname{arccot} x$; $x = \cot y$; $dx = -\csc^2 y\ dy$

$$\frac{d[\text{arcct } x]}{dx} = \frac{-1}{\csc^2 y} = \frac{-1}{1+\cot^2 y} = \frac{-1}{1+x^2}.$$

arcsec x $y = \operatorname{arcsec} x$; $x = \sec y$; $dx = \frac{\tan y}{\cos y}\ dy$

$$\frac{d[\operatorname{arcsec} x]}{dx} = \frac{1}{\sec y\cdot\tan y} = \frac{1}{x\sqrt{x^2-1}}; \quad |x| > 1.$$

arccsc x $y = \operatorname{arccsc} x$; $x = \csc y$; $dx = -\frac{\cot y}{\sin x}\ dy$

$$\frac{d[\operatorname{arccsc} x]}{dx} = \frac{-1}{\csc y\cdot\cot y} = \frac{-1}{x\sqrt{x^2-1}}; \quad |x| > 1.$$

We note that although inverse trigonometric functions are transcendental, their derivatives are **algebraic functions**.

- The first derivative of

$$y = \arctan \frac{x}{3}$$

is

$$y' = \frac{1}{1+\left(\frac{x}{3}\right)^2} \frac{d}{dx}\left[\frac{x}{3}\right] = \frac{1}{1+\frac{x^2}{9}}\left(\frac{1}{3}\right) = \frac{3}{9+x^2}.$$

- The first derivative of

$$y = \arctan(2x^2 - x)$$

is

$$y' = \frac{1}{1+(2x^2-x)^2} \frac{d}{dx}[2x^2 - x] = \frac{4x-1}{4x^4 - 4x^3 + x^2 + 1}.$$

- The first derivative of

$$y = \frac{1+\arctan x}{1-\arctan x}$$

is

$$y' = \frac{(1-\arctan x)\frac{d}{dx}[1+\arctan x] - (1+\arctan x)\frac{d}{dx}[1-\arctan x]}{(1-\arctan x)^2}$$

$$= \frac{(1-\arctan x)\left(\frac{1}{1+x^2}\right) - (1+\arctan x)\left(\frac{-1}{1+x^2}\right)}{(1-\arctan x)^2}$$

$$= \frac{2}{(1+x^2)(1-\arctan x)^2}.$$

At the outset of differentiation
We needed strong dedication.
Now, after a break
And some slices of cake,
We're set for more math gratification.

Logarithmic and Exponential Functions

Natural Logarithms

$$\frac{d}{dx}[\ln x] = \frac{1}{x}$$

Why? $$\frac{d}{dx}[\ln x] = \lim_{\Delta x \to 0} \frac{\ln(x + \Delta x) - \ln(x)}{\Delta x}$$

$$= \lim_{\Delta x \to 0} \left\{ \frac{1}{\Delta x} \cdot \ln\left(\frac{x + \Delta x}{x}\right) \right\},$$

which may be rewritten

$$\frac{d}{dx}[\ln x] = \lim_{\Delta x \to 0} \frac{1}{x} \left\{ \frac{x}{\Delta x} \ln\left(\frac{x + \Delta x}{x}\right) \right\}$$

$$= \frac{1}{x} \left\{ \lim_{\Delta x \to 0} \ln\left(1 + \frac{\Delta x}{x}\right)^{x/\Delta x} \right\},$$

where, with $\frac{x}{\Delta x} = n$ tending to ∞, we recognize the definition of e, and have with $\ln e = 1$,

$$\frac{d}{dx}[\ln x] = \frac{1}{x}.$$

- Find the first derivative:

- $f(x) = \ln(4x^2 - 3)$;

  $f'(x) = \frac{1}{4x^2 - 3} \cdot 8x = \frac{8x}{4x^2 - 3}$.

- $F(x) = \ln(ex)$;

  $F(x) = \ln e + \ln x = 1 + \ln x$;

  $F'(x) = \frac{1}{x}$.

- $g(x) = x^2 \ln x$;

  $g'(x) = x^2 \frac{d}{dx}[\ln x] + \ln x \frac{d}{dx}[x^2] = x(2\ln x + 1)$.

- $G(x) = \frac{\ln x}{x}$;

  $G'(x) = \frac{x \cdot \frac{1}{x} - \ln x}{x^2} = \frac{1 - \ln x}{x^2}$.

- $h(x) = \sin(\ln x)$;

  $h'(x) = \cos(\ln x) \frac{d}{dx}[\ln x] = \frac{\cos(\ln x)}{x}$.

- $H(x) = \ln^3 x^4;$

$$H'(x) = 3(\ln x^4)^2 \frac{d}{dx}[\ln x^4]$$

$$= 3(\ln x^4)^2 \frac{1}{x^4} \cdot 4x^3 = \frac{12(\ln^2 x^4)}{x}.$$

- $y = \ln\{\ln(2x^3)\}$

$$y' = \frac{1}{\ln(2x^3)} \frac{d}{dx}[\ln 2x^3] = \frac{1}{\ln 2x^3} \frac{1}{2x^3} \cdot 6x^2 = \frac{3}{x \ln(2x^3)}.$$

- Find the first derivative of

$$y = \ln(\tan 2x).$$

$$y' = \frac{1}{\tan 2x} \cdot \sec^2 2x \cdot 2$$

p. 510

$$= \frac{\cos 2x}{\sin 2x} \cdot \frac{2}{\cos^2 2x} = \frac{2}{\sin 2x \cdot \cos 2x} = \frac{4}{\sin 4x} = 4 \csc 4x.$$

- Find the first derivative of

$$y = \ln x (\arctan x).$$

$$y' = \ln x \frac{1}{1+x^2} \cdot 1 + \arctan x \frac{1}{x} \cdot 1 = \frac{\ln x}{1+x^2} + \frac{\arctan x}{x}.$$

- Find the equation of the tangent and the normal to the curve $f(x) = x \ln x$ at (e, e).

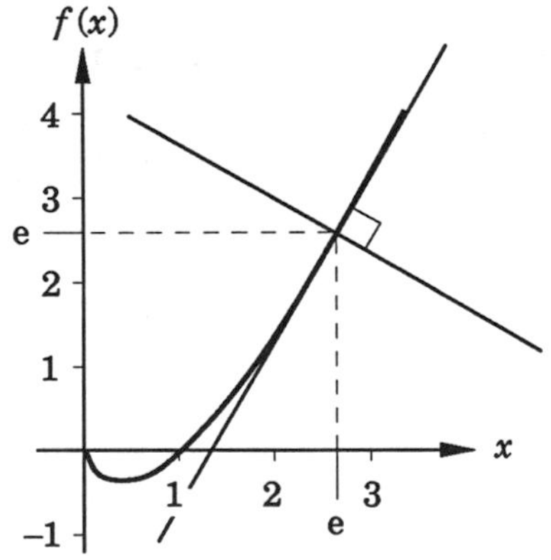

$f'(x) = 1 + \ln x$

The slope of the tangent at (e, e) is $1 + \ln e = 2$; consequently, that of the normal is $-\frac{1}{2}$.

The equation of the tangent is

$$y - e = 2(x - e); \quad y = 2x - e,$$

and that of the normal is

$$y - e = -\frac{1}{2}(x - e); \quad y = -\frac{x}{2} + \frac{3e}{2}.$$

Logarithms to Any Base

$$\frac{d}{dx}[\log_a x] = (\log_a e)\frac{1}{x}$$

p. 156 Why? $y = \log_a x = \dfrac{\ln x}{\ln a}$;

$$\frac{d}{dx}[\log_a x] = \frac{1}{\ln a}\cdot\frac{d}{dx}[\ln x] = \frac{1}{\ln a}\cdot\frac{1}{x} = (\log_a e)\frac{1}{x}.$$

Differentiation is facilitated by choosing e as logarithm base.

Although logarithms are transcendental, their derivatives are **algebraic functions**.

- Find the first derivative of

$$y = \log_{10}\frac{x\sqrt{x-2}}{3}.$$

$$y = \log_{10} x + \frac{1}{2}\log_{10}(x-2) - \log_{10} 3.$$

$$y' = \left(\log_{10} e\right)\frac{1}{x} + \frac{1}{2}\left(\log_{10} e\right)\frac{1}{x-2}$$

$$= \log_{10} e\left\{\frac{1}{x} + \frac{1}{2(x-2)}\right\} = \log_{10} e\left\{\frac{3x-4}{2x(x-2)}\right\}.$$

Exponential Functions

To find the derivative of the exponential function

$$y = e^x,$$

take the logarithm of both sides,

$$\ln y = x; \qquad \frac{dy}{y} = dx; \qquad \frac{dy}{dx} = y = e^x.$$

Thus, we have

$$y = e^x;\; y' = e^x;\; y'' = e^x;\; y''' = e^x,\ etc.,$$

that is, the function e^x is its own derivative *ad infinitum*.

The derivative of the function

$$y = a^{f(x)}$$

is obtained by observing that

$$a^{f(x)} = e^{\ln a \cdot f(x)}$$

$$y' = a^{f(x)}\ln a \cdot \frac{d}{dx}[f(x)].$$

- Find the first derivative:

- $f(x) = 6^{2x}$;

  $f'(x) = 6^{2x} \ln 6 \dfrac{d}{dx}[2x] = 6^{2x}(\ln 6)2 = 2 \cdot \ln 6 \cdot 6^{2x}$.

- $g(x) = e^{-x^3}$;

  $g'(x) = e^{-x^3} \dfrac{d}{dx}[-x^3] = -3x^2 e^{-x^3}$.

- $h(x) = e^{5x^3-2}$;

  $h'(x) = \left(e^{5x^3-2}\right) \dfrac{d}{dx}[5x^3-2] = 15x^2 \cdot e^{5x^3-2}$.

- $F(x) = e^{\ln 7x}$;

  $e^{\ln 7x} = 7x$, so $F'(x) = 7$.

- $G(x) = \ln e^x$;

  $\ln e^x = x$, so $G'(x) = 1$.

Hyperbolic Functions

p. 536

Since hyperbolic functions can be defined in terms of **exponential functions**, their derivatives may be found by applying the rules for differentiating exponential functions.

sinh x

$$\frac{d}{dx}[\sinh x] = \frac{d}{dx}\left[\frac{e^x - e^{-x}}{2}\right] = \frac{e^x + e^{-x}}{2} = \cosh x .$$

cosh x

$$\frac{d}{dx}[\cosh x] = \frac{d}{dx}\left[\frac{e^x + e^{-x}}{2}\right] = \frac{e^x - e^{-x}}{2} = \sinh x .$$

tanh x

$$\frac{d}{dx}[\tanh x] = \frac{d}{dx}\left[\frac{\sinh x}{\cosh x}\right]$$
$$= \frac{\cosh x \cdot \cosh x - \sinh x \cdot \sinh x}{\cosh^2} = \frac{1}{\cosh^2}$$
$$= \operatorname{sech}^2 x .$$

coth x

$$\frac{d}{dx}[\coth x] = \frac{d}{dx}\left[\frac{\cosh x}{\sinh x}\right]$$
$$= \frac{\sinh^2 x - \cosh^2 x}{\sinh^2 x} = \frac{-1}{\sinh^2 x} = -\operatorname{cosech}^2 x .$$

sech x

$$\frac{d}{dx}[\operatorname{sech} x] = \frac{d}{dx}\left[\frac{1}{\cosh x}\right] = \frac{-\sinh x}{\cosh^2 x} = -\operatorname{sech} x \cdot \tanh x .$$

cosech x

$$\frac{d}{dx}[\operatorname{cosech} x] = \frac{d}{dx}\left[\frac{1}{\sinh x}\right] = \frac{-\cosh x}{\sinh^2 x} = -\operatorname{cosech} x \cdot \coth x .$$

- Find the first derivatives:

- $f(x) = \sinh\sqrt{x}$;

$$f'(x) = \left(\cosh\sqrt{x}\right)\frac{d}{dx}\left[\sqrt{x}\right] = \frac{\cosh\sqrt{x}}{2\sqrt{x}} .$$

- $g(x) = \sinh(x^2 + x + 4)$;

$$g'(x) = \cosh(x^2 + x + 4)\frac{d}{dx}[x^2 + x + 4]$$

$$= [\cosh(x^2 + x + 4)](2x + 1) = (2x + 1)\cosh(x^2 + x + 4) .$$

- $h(x) = \ln(\sinh x^2)$;

$$h'(x) = \frac{1}{\sinh x^2}\frac{d}{dx}[\sinh x^2]$$

$$= \frac{\cosh x^2}{\sinh x^2}\frac{d}{dx}[x^2] = \frac{\cosh x^2}{\sinh x^2}\cdot 2x = 2x\coth x^2 .$$

- $F(x) = \coth\frac{1}{x}$;

$$F'(x) = -\left(\operatorname{cosech}^2\frac{1}{x}\right)(-x^{-2}) = \frac{1}{x^2}\operatorname{cosech}^2\frac{1}{x} .$$

- $G(x) = \tanh e^{2x}$;

$$G'(x) = (\operatorname{sech}^2 e^{2x})\, e^{2x}\cdot 2 = 2\,e^{2x}\operatorname{sech} e^{2x} .$$

- $H(x) = \cosh^2(3x^2 + x + 3)$;

$$H'(x) = 2\cosh(3x^2 + x + 3)\frac{d}{dx}[\cosh(3x^2 + x + 3)]$$

$$= 2(6x + 1)\cosh(3x^2 + x + 3)\sinh(3x^2 + x + 3)$$

$$= (6x + 1)\{2\cosh(3x^2 + x + 3)\sinh(3x^2 + x + 3)\} .$$

p. 542 Applying the double-angle identity

$$2\cosh x\sinh x = \sinh 2x ,$$

we have

$$H'(x) = (6x + 1)\sinh 2(3x^2 + x + 3) .$$

- Find the second derivative of $y = \cosh\sqrt{x}$.

$$y' = \left(\sinh\sqrt{x}\right)\frac{d}{dx}\left[\sqrt{x}\right] = \frac{\sinh\sqrt{x}}{2\sqrt{x}} .$$

$$y'' = \frac{d}{dx}\left[\frac{\sinh\sqrt{x}}{2\sqrt{x}}\right] .$$

Differentiating the quotient, we find

$$y'' = \frac{2\sqrt{x}\left(\frac{\cosh\sqrt{x}}{2\sqrt{x}}\right) - \left(\sinh\sqrt{x}\right)\left(\frac{1}{\sqrt{x}}\right)}{4x}$$

$$= \frac{\sqrt{x}\cdot\cosh\sqrt{x} - \sinh\sqrt{x}}{4x\sqrt{x}} .$$

Inverse Hyperbolic Functions

p. 545

Inverse hyperbolic functions are logarithmic and, since the **derivatives** of logarithmic functions are **algebraic functions**, we may expect the derivatives of inverse hyperbolic functions to be algebraic functions.

arsinh x

$$\frac{d}{dx}[\operatorname{arsinh} x] = \frac{d}{dx}\left[\ln\left(x + \sqrt{x^2+1}\right)\right] = \frac{1 + \dfrac{x}{\sqrt{x^2+1}}}{x + \sqrt{x^2+1}}$$

$$= \frac{1}{\sqrt{x^2+1}}.$$

The derivatives of the remaining area hyperbolic functions are found in a similar manner.

arcosh x $\quad \frac{d}{dx}[\operatorname{arcosh} x] = \frac{1}{\sqrt{x^2-1}}; \quad |x| > 1.$

artanh x $\quad \frac{d}{dx}[\operatorname{artanh} x] = \frac{1}{1-x^2}; \quad |x| < 1.$

arcoth x $\quad \frac{d}{dx}[\operatorname{arcoth} x] = \frac{1}{1-x^2}; \quad |x| > 1.$

arsech x $\quad \frac{d}{dx}[\operatorname{arsech} x] = -\frac{1}{x\sqrt{1-x^2}}; \quad 0 < x < 1.$

arcsch x $\quad \frac{d}{dx}[\operatorname{arcosech} x] = -\frac{1}{|x|\sqrt{1+x^2}}; \quad x \neq 0.$

- Find the first derivatives:
  - $f(x) = \operatorname{arsinh} 2x;$

    $$f'(x) = \frac{2}{\sqrt{4x^2+1}}.$$

  - $g(x) = \operatorname{artanh} e^x;$

    $$g'(x) = \frac{1}{1-e^{2x}}e^x = \frac{e^x}{1-e^{2x}}.$$

  - $h(x) = \operatorname{arcosh}\frac{1}{x};$

    $$h'(x) = \frac{1}{\sqrt{\left(\frac{1}{x}\right)^2 - 1}}(-1\,x^{-2}) = \frac{1}{\sqrt{\frac{1}{x^2}-1}}\left(\frac{-1}{x^2}\right)$$

    $\left(\sqrt{x^2} = |x|\right)$

    $$= \frac{1}{\sqrt{\frac{1-x^2}{x^2}}}\left(\frac{-1}{x^2}\right) = \frac{1}{\frac{1}{|x|}\sqrt{1-x^2}}\left(\frac{-1}{x^2}\right) = \frac{-|x|}{x^2\sqrt{1-x^2}}.$$

20.4 Special Techniques of Differentiation

Logarithmic Differentiation

A function is often more easily differentiated if first transformed into logarithmic form; this might considerably reduce the algebraic labor involved in differentiating products and quotients, and in handling composite functions.

Logarithmic differentiation is normal procedure for functions where the variable occurs both as a base and as an exponent.

- $y = x^x$; determine y' by logarithmic differentiation.

Take the natural logarithm of both sides,

$$\ln y = \ln x^x = x \ln x .$$

Differentiate both sides with respect to x:

$$\frac{1}{y} \cdot \frac{dy}{dx} = x \cdot \frac{1}{x} + \ln x = 1 + \ln x$$

$$\frac{dy}{dx} = y(1 + \ln x),$$

where we replace y by x^x,

$$\frac{dy}{dx} = x^x(1 + \ln x) .$$

- $y = x^{x^x}$; determine y' by logarithmic differentiation.

Take the natural logarithm of both sides,

$$\ln y = \ln x^{x^x} = x^x \ln x .$$

Differentiating both sides with respect to x gives

$$\frac{1}{y} \cdot \frac{dy}{dx} = x^x\left(\frac{d}{dx}[\ln x]\right) + \ln x\left(\frac{d}{dx}[x^x]\right)$$

$$= x^x \cdot \frac{1}{x} + \ln x \{x^x(1 + \ln x)\}$$

$$= x^{x-1} + \ln x \{x^x(1 + \ln x)\} .$$

Solve for dy/dx,

$$\frac{dy}{dx} = y\{x^{x-1} + x^x \ln x (1 + \ln x)\},$$

and replace y by x^{x^x},

$$\frac{dy}{dx} = x^{x^x}\{x^{x-1} + x^x \ln x (1 + \ln x)\} .$$

- **Find the first derivative:**

- $y = x^{1/x}$.

$$\ln y = \frac{1}{x} \ln x$$

$$\frac{1}{y} y' = \frac{1}{x} \frac{d}{dx} [\ln x] + \ln x \frac{d}{dx} [x^{-1}] = \frac{1}{x^2} - \frac{\ln x}{x^2}$$

$$y' = \frac{y(1 - \ln x)}{x^2} = \frac{x^{1/x}(1 - \ln x)}{x^2}.$$

- $y = x^{\ln x}$.

$$\ln y = \ln (x^{\ln x}) = \ln x \ln x = (\ln x)^2$$

$$\frac{1}{y} y' = \frac{\ln x}{x} + \frac{\ln x}{x} = (2 \ln x) \cdot \frac{1}{x}$$

$$y' = \frac{2 (\ln x) x^{\ln x}}{x} = 2 (\ln x) x^{\ln x - 1}.$$

- $y = x^{\sqrt{x}}$.

$$\ln y = \sqrt{x} \ln x$$

$$\frac{1}{y} y' = \sqrt{x} \left(\frac{1}{x}\right) + \ln x \left(\frac{1}{2} x^{-1/2}\right)$$

$$= \frac{1}{\sqrt{x}} + \frac{\ln x}{2\sqrt{x}} = \frac{2 + \ln x}{2\sqrt{x}}; \qquad y' = x^{\sqrt{x}} \left(\frac{2 + \ln x}{2\sqrt{x}}\right).$$

- $y = \left(\sqrt{x}\right)^{x^2}$.

$$\ln y = (x^2) \ln \sqrt{x}$$

$$\frac{1}{y} y' = (x^2) \frac{d}{dx} \left[\ln \sqrt{x}\right] + \left(\ln \sqrt{x}\right) 2x$$

$$= (x^2) \frac{1}{\sqrt{x}} \left(\frac{1}{2}\right) x^{-1/2} + 2x \ln \sqrt{x} = \frac{x}{2} + 2x \ln \sqrt{x}$$

$$= \frac{x}{2} + x \ln x;$$

$$y' = y \left(\frac{x}{2} + x \ln x\right) = \left(\sqrt{x}\right)^{x^2} \left(\frac{x}{2} + x \ln x\right).$$

- $y = x^{\cosh x^5}$.

$$\ln y = \ln\left(x^{\cosh x^5}\right) = \cosh x^5 (\ln x).$$

$$\frac{1}{y} y' = \frac{\cosh x^5}{x} + (\ln x)(\sinh x^5) \cdot 5x^4$$

$$y' = y\left\{\frac{\cosh x^5}{x} + 5x^4 (\ln x)(\sinh x^5)\right\}$$

$$= \left(x^{\cosh x^5}\right)\left\{\frac{\cosh x^5}{x} + 5x^4 (\ln x)(\sinh x^5)\right\}.$$ •

Differentiation of functions involving several factors is often more easily handled by considering the natural logarithm of the function.

- $y = e^{-x^2}\sqrt{\dfrac{1-x^2}{1+x^2}}$;

determine y' by logarithmic differentiation.

Take the natural logarithm of both sides,

$$\ln y = \ln\left(e^{-x^2}\sqrt{\frac{1-x^2}{1+x^2}}\right)$$

$$= -x^2 + \frac{1}{2}\left\{\ln(1-x^2) - \ln(1+x^2)\right\}.$$

Differentiate both sides with respect to x,

$$\left(\frac{1}{y}\right)\frac{dy}{dx} = -2x + \left(\frac{1}{2}\right)\frac{-2x}{1-x^2} - \left(\frac{1}{2}\right)\frac{2x}{1+x^2}$$

$$= \frac{2x(x^4-2)}{(1-x^4)}.$$

Solve for dy/dx,

$$\frac{dy}{dx} = y\,\frac{2x(x^4-2)}{(1-x^4)},$$

and replace y by $e^{-x^2}\sqrt{\dfrac{1-x^2}{1+x^2}}$,

$$\frac{dy}{dx} = e^{-x^2}\sqrt{\frac{1-x^2}{1+x^2}}\left\{\frac{2x(x^4-2)}{(1-x^4)}\right\}.$$

- $g(x) = \left(1+\dfrac{1}{x}\right)^{e^x}$; determine $g'(x)$.

$$\ln g(x) = e^x \ln\left(1+\frac{1}{x}\right)$$

$$\frac{1}{y} g'(x) = e^x \frac{d}{dx}\left[\ln\left(1+\frac{1}{x}\right)\right] + \ln\left(1+\frac{1}{x}\right)e^x$$

$$g'(x) = e^x\left\{\ln\left(1+\frac{1}{x}\right) - \frac{1}{x(1+x)}\right\}\left(1+\frac{1}{x}\right)^{e^x}.$$

Implicit Differentiation

If in $f(x, y) = 0$ – a function with variables x and y – the variable y is thought of as dependent on x ("in principle but not in practice, a function within f"), then $f(x, y) = 0$ is said to define y as an implicit function of x.

The process of determining the derivative of one of two variables with respect to the other is known as **implicit differentiation**. In the above $f(x, y) = 0$, the derivative $\mathrm{d}y/\mathrm{d}x$ is determined by treating y as an implied function of x as the entire function is differentiated.

To determine $\mathrm{d}y/\mathrm{d}x$ by implicit differentiation, take the derivative of each term of the equation and remember that y depends on x. Terms involving both variables call for the use of the product rule.

Implicit differentiation of a function given by the equation

$$x^2 + y^2 = 1,$$

where y is thought of as depending on x, thus implying the use of the chain rule when differentiating y^2, gives

$$2x + 2y\left(\frac{\mathrm{d}y}{\mathrm{d}x}\right) = 0$$

$$\left(\frac{\mathrm{d}y}{\mathrm{d}x}\right) = -\frac{x}{y}\ .$$

In this case, it is also possible to differentiate the explicit forms of the equation, and thus pursue the solution by ordinary methods of differentiation of the *explicit functions*

$$y = \sqrt{1-x^2}\,; \quad y = -\sqrt{1-x^2}\,.$$

If, however, the equation of an implicit function cannot be rendered into explicit form, implicit differentiation is indispensable.

- $x^2 + x^2 y + 3xy + y^2 = 0$; determine $\dfrac{\mathrm{d}y}{\mathrm{d}x}$.

Differentiate with respect to x:

$$\frac{\mathrm{d}}{\mathrm{d}x}\,[x^2 + x^2 y + 3xy + y^2] = 0$$

$$2x + \left\{x^2\left(\frac{\mathrm{d}y}{\mathrm{d}x}\right) + y \cdot 2x\right\} + \left\{3x\left(\frac{\mathrm{d}y}{\mathrm{d}x}\right) + 3y\right\} + 2y\left(\frac{\mathrm{d}y}{\mathrm{d}x}\right) = 0$$

Solving for $\mathrm{d}y/\mathrm{d}x$ gives

$$\frac{\mathrm{d}y}{\mathrm{d}x} = -\frac{2x + 3y + 2xy}{x^2 + 3x + 2y}\ .$$

- $x^2 - y^2 = 4$; determine

$$\frac{dy}{dx} \text{ and } \frac{d^2y}{dx^2}$$

by implicit differentiation.

$$\frac{d}{dx}[x^2 - y^2] = 0$$

$$2x - 2y\frac{dy}{dx} = 0$$

$$y' = \frac{x}{y}.$$

$$\frac{d^2y}{dx^2} = \frac{dy'}{dx} = \frac{y - xy'}{y^2} = \frac{y - x\left(\frac{x}{y}\right)}{y^2} = \frac{y^2 - x^2}{y^3}.$$

- $xy = e^y$; determine $\frac{dy}{dx}$ and $\frac{d^2y}{dx^2}$.

The equation cannot be explicitly solved for y, and we use implicit logarithmic differentiation.

$\ln(xy) = \ln e^y$; $\ln x + \ln y = y$.

$$\frac{d}{dx}[\ln x] + \frac{d}{dx}[\ln y] = \frac{dy}{dx}$$

$$\frac{1}{x} + \frac{1}{y}\left(\frac{dy}{dx}\right) = \frac{dy}{dx}$$

$$y' = \frac{y}{x(y-1)}.$$

$$\frac{d^2y}{dx^2} = \frac{dy'}{dx} = \frac{d}{dx}\left[\frac{y}{x(y-1)}\right]$$

$$\ln y' = \ln y - [\ln x + \ln(y-1)]$$
$$= \ln y - \ln x - \ln(y-1).$$

Differentiating with respect to x:

$$\left(\frac{1}{y'}\right)\frac{d^2y}{dx^2} = \frac{1}{y}(y') - \frac{1}{x} - \frac{1}{(y-1)}(y')$$

$$\frac{d^2y}{dx^2} = -\frac{(y')^2}{y(y-1)} - \frac{y'}{x}.$$

Insert $y' = \frac{y}{x(y-1)}$;

thus,

$$\frac{d^2y}{dx^2} = -\frac{2y + y^3 - 2y^2}{x^2(y-1)^3} = -\frac{y(2 + y^2 - 2y)}{x^2(y-1)^3}.$$

20.5 Partial Differentiation

Differentiation with Respect to One Variable

In partial differentiation a function of more than one independent variable, $y = f(x, y, z, t, \ldots)$, is differentiated with respect to one of these variables, the other variables being held constant. The result of the process is a **partial derivative**.

The partial derivative of the function f (of several variables $x, y, \ldots$) with respect to x is denoted

$$\frac{\partial f}{\partial x} \text{ or } \partial f / \partial x \, .$$

The partial n-th derivative of the function f (of several variables $x, y, \ldots$) with respect to x is denoted

$$\frac{\partial^n f}{\partial x^n} \text{ or } \partial^n f / \partial x^n \, .$$

The symbol ∂ ("mirror 6") denotes *partial* and seems to have first been proposed in 1788 by the French mathematician **Lagrange**. It must not be confused with the Greek letter δ (lowercase delta).

For clarity, a subscript may be used to indicate the variable or variables that are held constant, *e.g.*,

$$\left(\frac{\partial f}{\partial x}\right)_{z,\,t} \, .$$

Consider the function

$$w = f(x, z) = x^2 + x z^2 \, .$$

The partial derivative of w with respect to x is obtained when z is held *constant*; thus,

$$\frac{\partial w}{\partial x} = 2x + z^2 \, .$$

In the same way, the partial derivative of w with respect to z is obtained when x is held *constant*; thus,

$$\frac{\partial w}{\partial z} = 2xz \, .$$

The principle of partial differentiation is readily understood if we study the influence of individual variables on any functional relation of several independent variables.

Volume V of a right cylinder varies with its height h and diameter d; thus, volume is a function of height and diameter. To find how the function

$$V(h, d) = \pi \left(\frac{d}{2}\right)^2 h$$

is affected by changes of the independent variables, we consider these variables *one at a time*.

The rate of change of V with respect to d, when h is held constant (the partial derivative of V with respect to d) is

$$\left(\frac{\partial V}{\partial d}\right)_h = \frac{\pi h}{4} \cdot 2 \cdot d = \frac{\pi h}{2} \cdot d \, ;$$

and the rate of change of V with respect to h, when d is held constant (the partial derivative of V with respect to h) is

$$\left(\frac{\partial V}{\partial h}\right)_d = \frac{\pi d^2}{4} \cdot 1 = \frac{\pi d^2}{4} .$$

Partial differentiation follows the basic rules of ordinary differentiation, but it must be kept in mind which variables are held constant.

- $f(x, y) = x^3 + 3x^2y - 2xy^5$; find the first and second partial derivatives of f with respect to x and y.

  To determine $\frac{\partial f}{\partial x}$ and $\frac{\partial^2 f}{\partial x^2}$, the variable y must be held constant.

  $$\frac{\partial}{\partial x}[x^3 + 3x^2y - 2xy^5] = 3x^2 + 3y(2)x - 2y^5(1)$$
  $$= 3x^2 + 6xy - 2y^5 .$$
  $$\frac{\partial^2 f}{\partial x^2} = \frac{\partial}{\partial x}[3x^2 + 6xy - 2y^5] = 6x + 6y(1) - 0$$
  $$= 6x + 6y .$$

  To find $\frac{\partial f}{\partial y}$ and $\frac{\partial^2 f}{\partial y^2}$, x must be held constant.

  $$\frac{\partial}{\partial y}[x^3 + 3x^2y - 2xy^5] = 0 + 3x^2(1) - 2x \cdot 5y^4$$
  $$= 3x^2 - 10xy^4 .$$
  $$\frac{\partial^2 f}{\partial y^2} = \frac{\partial}{\partial y}[3x^2 - 10xy^4] = 0 - 10x \cdot 4y^3 = -40xy^3 .$$

- $g(x, y) = \sin 3x \cos 2y$; find the second partial derivative of g with respect to x, and with respect to y.

  $$\frac{\partial g}{\partial x} = \cos 2y\,(3 \cos 3x) = 3 \cos 3x \cos 2y \, ;$$
  $$\frac{\partial^2 g}{\partial x^2} = 3 \cos 2y\,(-3 \sin 3x) = -9 \sin 3x \cos 2y .$$
  $$\frac{\partial g}{\partial y} = \sin 3x\,(-2 \sin 2y) = -2 \sin 3x \sin 2y \, ;$$
  $$\frac{\partial^2 g}{\partial y^2} = -2 \sin 3x(2 \cos 2y) = -4 \sin 3x \cos 2y .$$

- Determine the first and second partial derivatives of

$$f(x, y) = x\,y\,; \quad g(x, y) = \frac{x}{y}\,; \text{ and } \quad h(x, y) = x^y$$

with respect to x and to y.

$$\frac{\partial f}{\partial x} = y\,; \qquad \frac{\partial^2 f}{\partial x^2} = 0\,; \qquad \frac{\partial f}{\partial y} = x\,; \qquad \frac{\partial^2 f}{\partial y^2} = 0\,.$$

$$\frac{\partial g}{\partial x} = \frac{1}{y}\,; \qquad \frac{\partial^2 g}{\partial x^2} = 0\,;$$

$$\frac{\partial g}{\partial y} = -x\,y^{-2} = -\frac{x}{y^2}\,; \qquad \frac{\partial^2 g}{\partial y^2} = -x\,(-2)\,y^{-3} = \frac{2\,x}{y^3}\,.$$

$$\frac{\partial h}{\partial x} = y\,x^{y-1}\,; \qquad \frac{\partial^2 h}{\partial x^2} = y\,(y-1)\,x^{y-2}\,.$$

p. 701

$$\frac{\partial h}{\partial y} = x^y \ln x\,; \qquad \frac{\partial\, h}{\partial y^2} = x^y \ln^2 x\,.$$

Mixed Partial Derivatives

Second- or higher-order partial derivatives with respect to several *independent* variables are called **mixed partial derivatives**.

The notation

$$\frac{\partial^2 f}{\partial y\ \partial x}$$

implies that the function $f(x, y)$ is first differentiated with respect to x and *then* with respect to y; the notation

$$\frac{\partial^2 f}{\partial x\ \partial y}$$

implies that it is *first* differentiated with respect to y and then with respect to x.

These notations may seem to be written backwards with regard to the order in which the derivatives are determined. To understand the notation, we study the identities

$$\frac{\partial}{\partial y}\left[\frac{\partial f}{\partial x}\right] = \frac{\partial^2 f}{\partial y\ \partial x}\,; \qquad \frac{\partial}{\partial x}\left[\frac{\partial f}{\partial y}\right] = \frac{\partial^2 f}{\partial x\ \partial y}\,.$$

The function

$$w = f(x, y) = x^2 y^3\,,$$

which gives

$$\frac{\partial w}{\partial x} = 2\,x\,y^3 \quad \text{and} \quad \frac{\partial w}{\partial y} = 3\,x^2 y^2$$

and the mixed partial derivatives

$$\frac{\partial^2 w}{\partial y\ \partial x} = \frac{\partial}{\partial y}\,[2\,x\,y^3] = 2\,x \cdot 3\,y^2 = 6\,x\,y^2$$

and

$$\frac{\partial^2 w}{\partial x\ \partial y} = \frac{\partial}{\partial x}\,[3\,x^2 y^2] = y^2 \cdot 3 \cdot 2\,x^1 = 6\,x\,y^2\,.$$

The equality

$$\frac{\partial^2 w}{\partial y\, \partial x} = \frac{\partial^2 w}{\partial x\, \partial y}$$

is no coincidence; it holds good for every function $w = f(x, y)$ whose partial derivatives $\frac{\partial^2 w}{\partial y\, \partial x}$ and $\frac{\partial^2 w}{\partial x\, \partial y}$ are continuous functions.

- The rationale behind the use of mixed partial derivatives is illustrated by this example.

  In an experiment, the volume V of a gas is a function of temperature T and pressure p, according to the equation

$$V(T, p) = \frac{c\,T}{p},$$

where c is a constant.

Sought I: The change in volume with respect to the change in temperature.

$$\frac{\partial V}{\partial T} = \frac{\partial}{\partial T}\left[\frac{c\,T}{p}\right] = \frac{c\,(1)}{p} = \frac{c}{p}\,; \qquad p = \text{constant.}$$

Sought II: The change in volume with respect to the change in pressure.

$$\frac{\partial V}{\partial p} = \frac{\partial}{\partial p}\,[c\,T\,p^{-1}] = c\,T\,(-1)\,p^{-2} = -\frac{c\,T}{p^2}\,; \qquad T = \text{constant.}$$

Sought III: The change in volume with respect to changes in temperature *and* pressure.

The same answer is obtained whether we determine

$$\frac{\partial^2 V}{\partial p\, \partial T} \text{ or } \frac{\partial^2 V}{\partial T\, \partial p}:$$

$$\frac{\partial^2 V}{\partial p\, \partial T} = \frac{\partial}{\partial p}\left[\frac{c}{p}\right] = \frac{\partial}{\partial p}\,[c\,p^{-1}] = c\cdot(-1)\cdot p^{-2} = -\frac{c}{p^2}\,.$$

- $g(x, y) = \sin 3x \cos 2y$; find

$$\frac{\partial^2 g}{\partial x\, \partial y}\,.$$

First differentiate the function with respect to y, regarding x as a constant,

$$\sin 3x\,(-\sin 2y)\,2 = -2 \sin 3x \sin 2y\,;$$

then with respect to x, regarding y as a constant,

$$-2\cdot 3\cos 3x\cdot \sin 2y = -6\cos 3x \sin 2y\,.$$

Hence,

$$\frac{\partial^2 g}{\partial x\, \partial y} = -6\cos 3x \sin 2y\,.$$

- $z = f(x, y) = e^{xy} + x\cos y$

Sought: $\dfrac{\partial^2 z}{\partial y\,\partial x}$.

$$\frac{\partial z}{\partial x} = y\,e^{xy} + \cos y\,; \quad \frac{\partial^2 z}{\partial y\,\partial x} = \frac{\partial}{\partial y}\left[y\,e^{xy} + \cos y\right].$$

$$\frac{\partial}{\partial y}\left[y\,e^{xy}\right] = y\,e^{xy}\left[x\cdot 1 + y\cdot 0\right] + e^{xy}\cdot 1 = x\,y\,e^{xy} + e^{xy}.$$

$$\frac{\partial^2 z}{\partial y\,\partial x} = x\,y\,e^{xy} + e^{xy} - \sin y\,.$$

- $h = f(x, y) = e^x \tan y$

Sought: **a:** $\dfrac{\partial^2 h}{\partial x^2}$ **b:** $\dfrac{\partial^2 h}{\partial y^2}$ **c:** $\dfrac{\partial^2 h}{\partial y\,\partial x}$ **d:** $\dfrac{\partial^2 h}{\partial x\,\partial y}$

a:

y = constant.

e^x is its own derivative.

$$\frac{\partial h}{\partial x} = e^x\cdot 0 + e^x\cdot\tan y = e^x \tan y\,.$$

$$\frac{\partial^2 h}{\partial x^2} = e^x\cdot 0 + e^x\cdot\tan y = e^x \tan y\,.$$

b:

x = constant.

$$\frac{\partial}{\partial y}\left[e^x \tan y\right] = e^x \sec^2 y + 0\cdot\tan y = e^x \sec^2 y\,.$$

$$\frac{\partial^2 h}{\partial y^2} = e^x\, 2\sec y\left(\frac{d}{d\,y}\left[\sec y\right]\right)$$

$$= 2\,e^x \sec^2 y \tan y\,.$$

c:

$$\frac{\partial^2 h}{\partial y\,\partial x} = \frac{\partial}{\partial y}\left[e^x \tan y\right] = e^x \sec^2 y\,.$$

d:

$$\frac{\partial^2 h}{\partial x\,\partial y} = e^x \sec^2 y\,.$$

- $w = f(x, y) = \arctan \dfrac{x}{y}$; find

$$\frac{\partial^2 w}{\partial x \, \partial y}.$$

$$\frac{\partial w}{\partial y} = \frac{1}{1 + \left(\dfrac{x}{y}\right)^2} \left(\frac{\partial}{\partial y}\left[\frac{x}{y}\right]\right) = \frac{y^2}{y^2 + x^2}\left(\frac{\partial}{\partial y}\,[xy^{-1}]\right)$$

$$= \frac{y^2}{y^2 + x^2}\, x\,(-1)\,y^{-2} = -\frac{x}{x^2 + y^2}.$$

Thus,

$$\frac{\partial^2 w}{\partial x \, \partial y} = \frac{\partial}{\partial x}\left[-\frac{x}{x^2 + y^2}\right]$$

and, by the quotient rule,

$$\frac{\partial^2 w}{\partial x \, \partial y} = -\frac{(x^2 + y^2) - x \cdot 2\,x}{(x^2 + y^2)^2} = \frac{x^2 - y^2}{(x^2 + y^2)^2}.$$

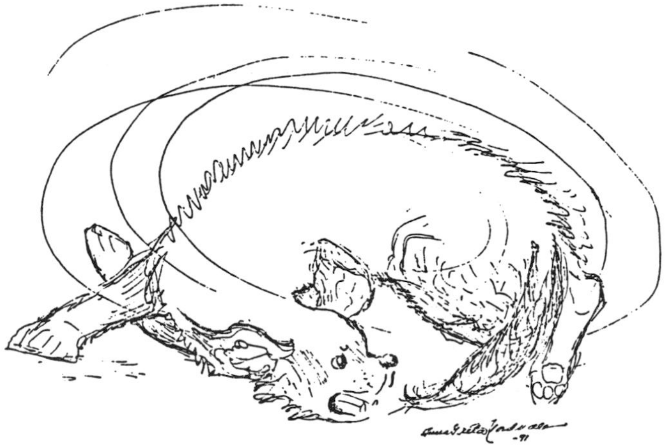

20.6 Mean-Value Theorems

Rolle's Theorem

Suppose f is a continuous function that crosses the x-axis at two points a and b and is differentiable at all points between a and b – that is, it has a tangent at all points on the curve between a and b. Then there is at least one point between a and b where the derivative is 0, and the tangent is parallel to the x-axis.

The function f shown below has four such points $c_1 \dots c_4$ between a and b:

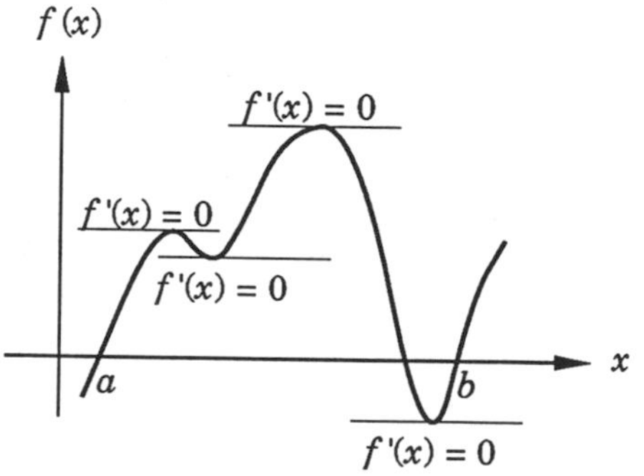

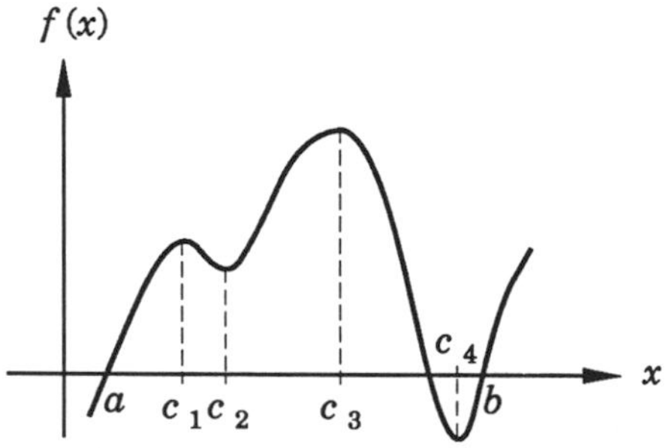

ROLLE Michel
(1652 - 1719)

The theorem is known as **Rolle's theorem** after the French mathematician **Rolle**. In more stringent mathematical language, the theorem can be stated:

> If a function $f(x)$ is continuous in the closed interval $[a, b]$ and differentiable in the open interval $]a, b[$, and if $f(a) = f(b)$, then there exists in the interval at least one value c such that $f'(c) = 0$ $(a < c < b)$.

Lagrange's Mean-Value Theorem

LAGRANGE Joseph Louis
(1736 - 1813)

Rolle's theorem is a special case of the **mean-value theorem**, also called **Lagrange's mean-value theorem** after the Italian-born French mathematician **Lagrange**, who was the first to state it:

> A function $f(x)$ that is continuous in the closed interval $[a, b]$ and differentiable in the open interval $]a, b[$ has in this interval at least one value c such that
>
> $$f'(c) = \frac{f(b) - f(a)}{b - a}.$$

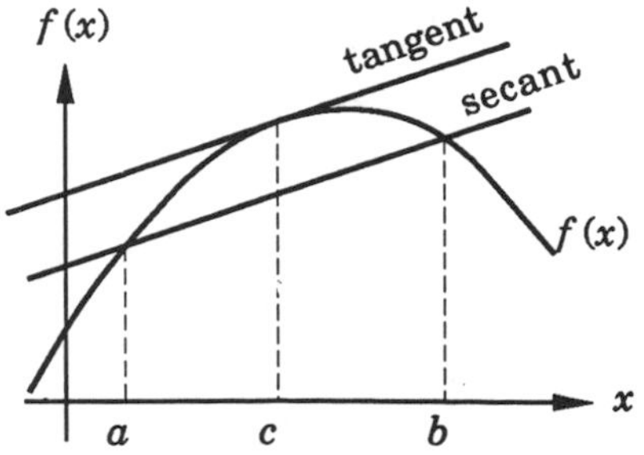

Specifically this means that the slope of a secant which transects the curve of a continuous function at two points a and b is the same as that of a tangent to the curve at some point c between a and b.

To distinguish the Lagrange theorem from the more general Cauchy mean-value theorem, which is concerned with two intersecting curves, the Lagrange theorem is sometimes called the **restricted mean-value theorem**, as one of the curves must be a straight line.

Restricted Mean-Value Theorem

Proof:

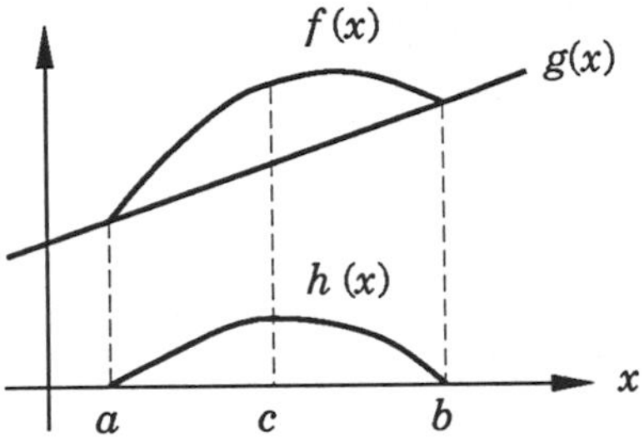

The slope (m) of the secant line $g(x)$ transecting the curve of the differentiable function $f(x)$ at the points $\{a, f(a)\}$ and $\{b, f(b)\}$ is

$$m = \frac{f(b) - f(a)}{b - a} .$$

We have, for all x,

$$m = \frac{g(x) - f(a)}{x - a},$$

which gives the equation of the secant,

$$g(x) = f(a) + \frac{f(b) - f(a)}{b - a}(x - a) .$$

We construct a **support function** $h(x)$,

$$\begin{aligned} h(x) &= f(x) - g(x) \\ &= f(x) - \frac{f(b) - f(a)}{b - a}(x - a) - f(a), \end{aligned}$$

which intersects the x-axis at the points a and b,

$$h(a) = h(b) = 0 .$$

As $h(x)$ is the difference between the differentiable functions $f(x)$ and $g(x)$, it is also differentiable,

$$h'(x) = f'(x) - \frac{f(b) - f(a)}{b - a} .$$

Letting

$$h'(c) = f'(c) - \frac{f(b) - f(a)}{b - a} = 0,$$

we see that there exists a point c such that

$$f'(c) = \frac{f(b) - f(a)}{b - a}; \qquad a < c < b .$$

- A secant to the curve

$$f(x) = -x^2 + 8x + 15$$

is defined by the points (–2, –5) and (7, 22); find the point where the tangent of f is parallel to the secant.

The slope of the secant is

$$m_{\text{sec}} = \frac{22 - (-5)}{7 - (-2)} = \frac{27}{9} = 3 .$$

The derivative of $f(x)$ is $f'(x) = -2x + 8$.

The condition

$$3 = -2x + 8$$

gives

$$x = 2.5 ; \quad f(x) = 28.75 .$$

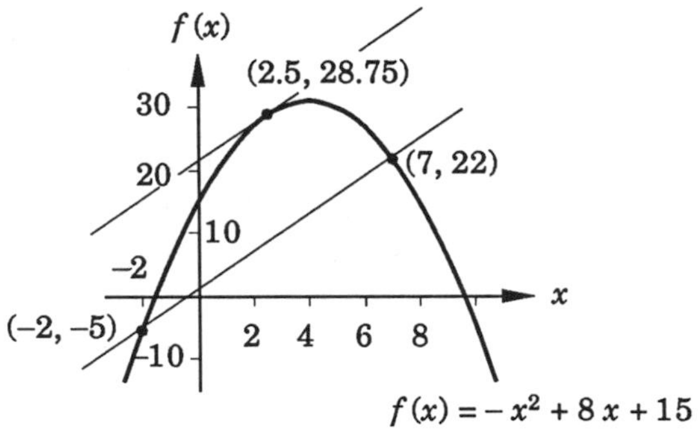

Cauchy's Mean-Value Theorem

The **generalized mean-value theorem** applies to any two continuous and differentiable functions intersecting at two points. It can be stated thus:

> If two functions f and g, which have the same value at $x = a$ and $x = b$, are continuous in the closed interval $[a, b]$ and differentiable in the open interval $]a, b[$, and if $g(b) - g(a) \neq 0$ and $g'(x) \neq 0$ in $]a, b[$, then there exists in $]a, b[$ at least one number c such that

$$\frac{f(b) - f(a)}{g(b) - g(a)} = \frac{f'(c)}{g'(c)} ; \quad a < c < b.$$

The generalized mean-value theorem is also called the **extended mean-value theorem**, or **Cauchy's mean-value theorem**, after the French mathematician **Cauchy**.

CAUCHY Augustin Louis (1789 - 1857)

Proof:

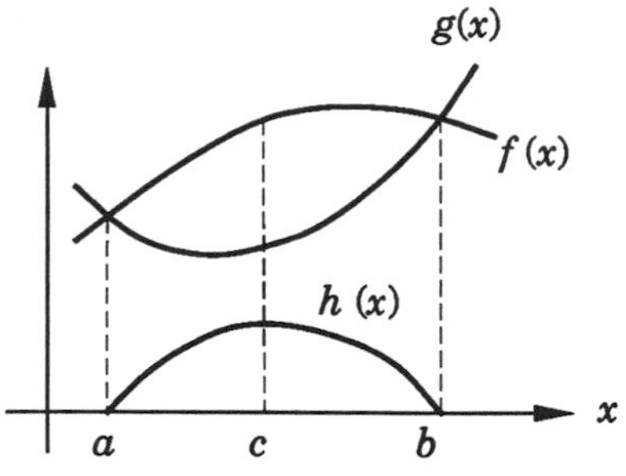

$a < c < b$

By observing that

$$\frac{f(b)-f(a)}{g(b)-g(a)} = 1 = \frac{g(x)-f(a)}{g(x)-g(a)}$$

we find the equation

$$g(x) = f(a) + \frac{f(b)-f(a)}{g(b)-g(a)}\{g(x)-g(a)\} .$$

As before, we construct a **supporting function** $h(x)$,

$$h(x) = f(x) - g(x)$$

$$= f(x) - f(a) - \frac{f(b)-f(a)}{g(b)-g(a)}\{g(x)-g(a)\} .$$

$h(a) = h(b) = 0$, and $h(x)$ is differentiable on $]a, b[$:

$$h'(x) = f'(x) - \frac{f(b)-f(a)}{g(b)-g(a)}\, g'(x) .$$

Letting

$$h'(c) = 0 ,$$

we find a point c in $]a, b[$ where

$$\frac{f(b)-f(a)}{g(b)-g(a)} = \frac{f'(c)}{g'(c)} ; \quad a < c < b .$$

LEÇONS

SUR LE

CALCUL DIFFÉRENTIEL,

PAR M. AUGUSTIN-LOUIS CAUCHY,

INGÉNIEUR EN CHEF DES PONTS ET CHAUSSÉES, PROFESSEUR A L'ÉCOLE ROYALE POLYTECHNIQUE, PROFESSEUR ADJOINT A LA FACULTÉ DES SCIENCES, MEMBRE DE L'ACADÉMIE DES SCIENCES, CHEVALIER DE LA LÉGION D'HONNEUR.

A PARIS,

CHEZ DE BURE FRÈRES, LIBRAIRES DU ROI ET DE LA BIBLIOTHÈQUE DU ROI,
RUE SERPENTE, N.° 7.

1829.

Cauchy gave elementary calculus the character it still bears today.

Chapter 21

INTRODUCTION TO INTEGRAL CALCULUS

*** Indicates cross-reference

21.1 Basic Concepts

In differentiation, relatively simple and straightforward rules enable us to find the derivatives of even quite complicated functions. In integration, which is the reverse process of differentiation, our task is to find a function whose derivative is known. This often involves intelligent guesswork, and it might still not be possible to integrate a given function. When we have arrived at an integration formula, it is always possible, however, to verify it by differentiation.

Integrand
Integral

The function to be integrated is referred to as the **integrand**; the result of an integration is an **integral**.

Latin, *integer*, "complete"; *integralis*, "making up a whole"

von LEIBNIZ Gottfried Wilhelm (1646 - 1716)

The integral sign $\int$, an elongated S denoting sum (Latin: *summa*), was introduced by **Leibniz**, who named integral calculus *calculus summatorius*.

If we differentiate the functions

$$f(x) = x^3 + x\,; \quad g(x) = x^3 + x + 4\,; \quad h(x) = x^3 + x - 7$$

with respect to the independent variable x, we obtain the results

$$f'(x) = g'(x) = h'(x) = 3x^2 + 1$$

in all three cases.

If we reverse the process and integrate the expression $3x^2 + 1$, we obtain $x^3 + x$, but the constant terms of the functions g and h cannot be recovered failing information as to initial or boundary conditions. Therefore, the integral must be written

$$\int (3x^2 + 1)\ \mathrm{d}x = x^3 + x + 4 + C,$$

Arbitrary Constant
Integration Constant

where C is an **arbitrary constant** or **integration constant**.

The notation dx indicates that the integration is to be performed with respect to the variable x; a du would indicate integration with respect to a variable u, *etc.*

A function F is an integral of a function $f(x)$ if $F'(x) = f(x)$ for all x in the domain of f, and, conversely, for every value of a constant C, $F(x) + C$ is also an integral of $f(x)$.

Indefinite Integral
Antiderivative
Primitive Integral

An integral with no restrictions imposed on its independent variable is known as an **indefinite integral**, **antiderivative**, or **primitive integral**,

$$\int f(x)\ \mathrm{d}x,$$

Definite Integral

whereas an integral that is defined by the limit values a and b of the independent variable is a **definite integral**,

$$\int_a^b f(x)\ \mathrm{d}x\ .$$

In Section 21.3 (*p.* 747) we will define the definite integral and see that

$$\int_a^b f(c)\ \mathrm{d}x = F(b) - F(a)$$

for any integral F of f.

Calculating the integral of $(3x^2 + 1)$ from 2 to 3, we find

$$\int_2^3 (3x^2 + 1)\ \mathrm{d}x = \Big|_2^3\ (x^3 + x + C)$$

$$= (27 + 3 + C) - (8 + 2 + C) = 30 - 10 = 20\,,$$

from which we see that a definite integral is independent of the arbitrary constant in the corresponding indefinite integral.

21.2 Methods of Integration

Integrating Power Functions

Differentiating x^n with respect to x gives nx^{n-1}; analogously, x^{n+1} yields $(n+1)\,x^n$, and consequently $\frac{1}{n+1}x^{n+1}$ becomes x^n.

Thus, if $\frac{dy}{dx} = x^n$, we have $y = \frac{1}{n+1}x^{n+1} + C$, that is, the integral of x^n with respect to x is $\frac{x^{n+1}}{n+1}$, or

$$\int x^n \, dx = \frac{x^{n+1}}{n+1} + C,$$

where $n \neq -1$ and C is a constant.

- Integrate:
  - $\int x^{3/2}\, dx = \frac{2}{5}\, x^{5/2} + C$
  - $\int \sqrt{x}\, dx = \int x^{1/2}\, dx = \frac{2}{3}\, x^{3/2} + C$
  - $\int \frac{1}{\sqrt{x}}\, dx = \int x^{-1/2}\, dx = 2\sqrt{x} + C$
  - $\int \sqrt[3]{x}\, dx = \int x^{1/3}\, dx = \frac{3}{4}\, x^{4/3} + C$ •

For x^{-1}, the power rule $\int x^n \, dx = \frac{x^{n+1}}{n+1} + C$ leads to the expression $\frac{x^0}{0} = \frac{1}{0}$. A search through various derivatives reveals that

p. 699 $\frac{d}{dx}[\ln x] = \frac{1}{x}$. Hence, we have

$$\int x^{-1}\, dx = \ln x + C;\ x > 0.$$

More generally, since $\frac{d}{dx}[\ln |x|] = x^{-1}$,

$$\int x^{-1}\, dx = \ln |x| + C;\ x \neq 0.$$

Fundamental Arithmetic Integration Rules

A **constant factor** in the integrand

$$\int \sin x \cdot 5 \cos 2x \, dx = 5 \int \sin x \cdot \cos 2x \, dx$$

or, generally,

$$\int k\, f(x)\, dx = k \int f(x)\, dx, \text{ where } k \text{ is a constant,}$$

can be taken out and placed before the integral sign.

When differentiating a **sum**, we differentiate each term separately; reversing the process, we integrate each term separately,

$$\int [f(x) \pm g(x)]\,dx = \int f(x)\,dx \pm \int g(x)\,dx$$

or

$$\int (3x-5)\,dx = 3\int x\,dx - 5\int dx$$

$$= 3\left(\frac{x^2}{2} + C_1\right) - 5(x + C_2) = \frac{3}{2}x^2 - 5x + 3C_1 - 5C_2,$$

where $(3C_1 - 5C_2)$ is simply another constant:

$$\int (3x-5)\,dx = \frac{3}{2}x^2 - 5x + C.$$

Basic Integration Formulas

Every differential formula, written in reverse, gives an integral:

$$\frac{d}{dx}[\sin x] = \cos x; \quad \int \cos x\,dx = \sin x + C.$$

We can conveniently reverse differentiation formulas to establish a list of basic integration formulas. These, or modifications, form a starting point in the various techniques of integration. Below, t is a variable, a and n are constants.

$$\int t^n\,dt = \frac{t^{n+1}}{n+1} + C; \quad n \neq -1.$$

$$\int t^{-1}\,dt = \int \frac{1}{t}\,dt = \int \left(\frac{\frac{d[t]}{dt}}{t}\right) dt = \ln|t| + C; \quad t \neq 0.$$

$$\int e^t\,dt = e^t + C \quad \text{and} \quad \int e^{at}\,dt = \frac{e^{at}}{a} + C.$$

$$\int a^t\,dt = \frac{a^t}{\ln a} + C; \quad 0 < a \neq 1.$$

$$\int \sin at\,dt = -\frac{1}{a}\cos at + C.$$

$$\int \cos at\,dt = \frac{1}{a}\sin at + C.$$

$$\int \sec at\,dt = \frac{1}{a}\ln|\sec at + \tan at| + C.$$

$$\int \sec^2 at\,dt = \int \frac{1}{\cos^2 at}\,dt = \int (1 + \tan^2 at)\,dt = \frac{1}{a}\tan at + C.$$

$$\int \csc^2 at\,dt = \int \frac{1}{\sin^2 at}\,dt = -\frac{1}{a}\cot at + C.$$

$$\int \frac{1}{\sqrt{a^2 - t^2}}\,dt = \arcsin\frac{t}{a} + C; \quad a > 0.$$

$$\int \frac{1}{a^2+t^2}\,\mathrm{d}t = \frac{1}{a}\arctan\frac{t}{a} + C.$$

$$\int \sinh at\,\mathrm{d}t = \frac{1}{a}\cosh a\,t + C.$$

$$\int \cosh at\,\mathrm{d}t = \frac{1}{a}\sinh a\,t + C.$$

$$\int \frac{1}{\sqrt{t^2 \pm a^2}}\,\mathrm{d}t = \ln\left|t + \sqrt{t^2 \pm a^2}\right| + C.$$

$$\int \frac{1}{\sqrt{t^2+1}}\,\mathrm{d}t = \sinh^{-1} t + C.$$

$$\int \frac{1}{\sqrt{t^2-1}}\,\mathrm{d}t = \cosh^{-1} t + C; \quad |t| > 1.$$

$$\int \frac{1}{a^2-t^2}\,\mathrm{d}t = \frac{1}{a}\tanh^{-1}\frac{t}{a} + C = \frac{1}{2a}\ln\left|\frac{a+t}{a-t}\right| + C.$$

p. 722 The above formulas can all be checked by differentiation.

Tables of Integrals

Tables of some thousand basic integration formulas have been collected over the years. A comprehensive table is often of great help, but no matter how extensive, it will only sporadically give us exactly the needed formula.

Non-Integrability

The existence of a certain elementary function does not necessarily imply that there must exist another elementary function of which the given function is the derivative.

Some functions, such as

$$\int \frac{\sin x}{x}\,\mathrm{d}x \qquad \int \sin(x^2)\,\mathrm{d}x \qquad \int e^{x^3}\,\mathrm{d}x \qquad \int \frac{e^x}{x}\,\mathrm{d}x$$

$$\int \frac{\cos x}{x}\,\mathrm{d}x \qquad \int \sqrt{x}\,\sin x\,\mathrm{d}x \qquad \int e^{-x^2}\,\mathrm{d}x \qquad \int \frac{1}{\ln x}\,\mathrm{d}x,$$

do not possess integrals that can be expressed by a finite number of elementary functions.

Certain radical polynomial expressions (*e.g.*, $\int \sqrt{1-x^2}\,\mathrm{d}x$) can be expressed by a finite number of elementary functions, while others (*e.g.*, $\int \sqrt{1-x^3}\,\mathrm{d}x$) cannot.

Simple Rearrangements

It is often possible to integrate functions directly by using existing tables of integrals. The more comprehensive these tables, the more likely one is to be able to integrate the function directly. Quite simple rearrangements within the integrand will often make it better adapted for integration.

- $$\int \frac{1}{\sqrt[4]{x}}\,dx = \int x^{-1/4}\,dx = \frac{x^{3/4}}{\frac{3}{4}} + C = \frac{4}{3}x^{3/4} + C.$$

p. 725

- $$\int \frac{7}{\sqrt{x^2-4}}\,dx = 7\int \frac{1}{\sqrt{x^2-4}}\,dx = 7\ln\left|x + \sqrt{x^2-4}\right| + C.$$

Adding and Subtracting the Same Quantity

We may add a quantity at one end of the integrand and subtract the same quantity at the other end; at first glance this might seem rather bizarre, but it can serve a purpose.

- Integrate $\int \tan^2 x\,dx$.

$$\int \tan^2 x\,dx = \int (1 + \tan^2 x - 1)\,dx = \int (1 + \tan^2 x)\,dx - \int dx$$

p. 724

$$= \int \sec^2 x\,dx - \int dx = \tan x - x + C.$$

Breaking Up Fractions into Parts

A fraction may often be written as a sum of simpler fractions that can be integrated directly.

- $$\int \frac{3x^3 + 2x^2 - 6 + 9x\sin x}{3x}\,dx$$

$$= \int x^2\,dx + \frac{2}{3}\int x\,dx - 2\int x^{-1}\,dx + 3\int \sin x\,dx$$

$$= \frac{1}{3}x^3 + \frac{1}{3}x^2 - 2\ln|x| - 3\cos x + C.$$

- $$\int \frac{6x^2 - 11x + 3}{2x - 3} = 3\int x\,dx - \int dx = \frac{3}{2}x^2 - x + C.$$

Completing the Square

Several basic integration formulas involve the sum or difference of two squares. By completing the square, we can often extend such formulas to integrands that contain a second-degree polynomial.

Completing the square: *p.* 310

The polynomial $(ax^2 + bx + c)$ is "completed" in the following manner:

- Factor the coefficient a from the rest of the expression:

$$ax^2 + bx + c = a\left(x^2 + \frac{b}{a}x + \frac{c}{a}\right)$$

- Add and subtract the square of half the coefficient $\frac{b}{a}$:

$$ax^2 + bx + c = a\left[x^2 + \frac{b}{a}x + \left(\frac{b}{2a}\right)^2 - \left(\frac{b}{2a}\right)^2 + \frac{c}{a}\right]$$

- Regroup the terms:

$$ax^2 + bx + c = a\left\{\left[x^2 + \frac{b}{a}x + \left(\frac{b}{2a}\right)^2\right] + \left(\frac{c}{a} - \left(\frac{b}{2a}\right)^2\right)\right\}$$

- Depending on positive or negative sign of the radicand, write, respectively

$$ax^2 + bx + c = a\left(x + \frac{b}{2a}\right)^2 + \left(\sqrt{c - \frac{b^2}{4a}}\right)^2$$

or

$$ax^2 + bx + c = a\left(x + \frac{b}{2a}\right)^2 - \left(\sqrt{-c + \frac{b^2}{4a}}\right)^2$$

- To integrate

$$\int \frac{1}{x^2 - 6x + 13}\, dx ,$$

complete the square of the denominator of the integrand,

$$x^2 - 6x + 13 = (x^2 - 6x + 3^2 - 3^2 + 13) = [(x - 3)^2 + 4] .$$

So,

$$\int \frac{1}{x^2 - 6x + 13}\, dx = \int \frac{1}{(x - 3)^2 + 4}\, dx$$

p. 724

$$\int \frac{1}{(x - 3)^2 + 2^2}\, dx = \frac{1}{2} \arctan \frac{x - 3}{2} + C .$$

Use of Trigonometric Identities

Trigonometric identities that transform products into sums or differences are often useful to adapt an integrand for direct integration.

p. 513

- $$\int \sin x \cos 2x\, dx = \int \frac{1}{2}\,[\sin (x + 2x) + \sin (x - 2x)]\, dx$$

$$= \frac{1}{2} \int (\sin 3x - \sin x)\, dx$$

$$= \frac{1}{2} \cos x - \frac{1}{6} \cos 3x + C .$$

p. 511

- $$\int \cos^2 x\, dx = \frac{1}{2} \int (1 + \cos 2x)\, dx = \frac{1}{2}x + \frac{1}{4}\sin 2x + C .$$

Integration by Parts

Integration by parts is the reverse of differentiating a product, where

$$[f(x)\,g(x)]' = f(x)\,g'(x) + g(x)\,f'(x)$$

leads to

$$f(x)\,g(x) = \int f'(x)\,g(x)\,dx + \int f(x)\,g'(x)\,dxs,$$

which may be rearranged

$$\int f(x)\,g'(x)\,dx = f(x)\,g(x) - \int g(x)\,f'(x)\,dx.$$

By introducing $u(x) = f(x)$ and $v(x) = g(x)$, this formula may be rewritten

$$\int u\,v'\,dx = u\,v - \int v\,u'\,dx,$$

where we **choose** u and v so as to make the new integral easier to integrate than the original.

We note the special case

$$\int u\,dx = u\,x - \int x\,u'\,dx\ .$$

The decision as to which part of the integrand to choose for u and for v' is often a matter of trial and error. However, if one part of the integrand is more complex than the other, it is often good practice to choose that part as v'.

- Integrate $\int x \cos x\,dx$.

We decide on

$$\left.\begin{aligned} u &= x \\ v' &= \cos x \end{aligned}\right\} \qquad \left.\begin{aligned} u' &= 1 \\ v &= \sin x + C_1 \end{aligned}\right\}$$

$$\begin{aligned} \int x \cos x\,dx &= x(\sin x + C_1) - \int (\sin x + C_1)(1)\,dx \\ &= x \sin x + C_1 x + \cos x - C_1 x + C \\ &= x \sin x + \cos x + C\ . \end{aligned}$$ •

The constant C_1 will cancel out in the final result; therefore, when integrating by parts, we may drop the first integration constant.

- Integrate $\int x \ln x\,dx$.

Let $u = \ln x$ and $v' = x$; then $u' = \dfrac{1}{x}$ and $v = \dfrac{1}{2}x^2$.

$$\begin{aligned} \int x \ln x\,dx &= \frac{1}{2}x^2 \ln x - \frac{1}{2}\int x^2 \frac{1}{x}\,dx \\ &= \frac{1}{2}x^2 \ln x - \frac{1}{2}\int x\,dx \\ &= \frac{1}{2}x^2 \ln x - \frac{x^2}{4} + C\ . \end{aligned}$$ •

An integrand composed of **only one factor** may be considered to include a factor 1 to permit integration by parts.

- Integrate $\int \ln x \, dx$.

Let $u = \ln x$ and $v' = 1$; then $u' = \frac{1}{x}$ and $v = x$.

$$\int \ln x \, dx = x \ln x - \int x \frac{1}{x} dx = x \ln x - \int dx = x(\ln x - 1) + C.$$

•

The integration of a rather ordinary-appearing expression will sometimes offer unexpected difficulties and surprises.

- Integrate $\int \sqrt{1 - x^2} \, dx$.

Let $u = \sqrt{1 - x^2}$ and $v' = 1$; then $u' = -\frac{x}{\sqrt{1 - x^2}}$ and $v = x$.

$$\int \sqrt{1-x^2} \, dx = x\sqrt{1-x^2} + \int \frac{x^2}{\sqrt{1-x^2}} dx .$$

Rewrite the numerator as a difference $[1 - (1 - x^2)]$ and expand the integral into a difference of two integrals; thus,

$$\int \sqrt{1-x^2} \, dx = x\sqrt{1-x^2} + \int \frac{1-(1-x^2)}{\sqrt{1-x^2}} dx$$

$$= x\sqrt{1-x^2} + \int \frac{1}{\sqrt{1-x^2}} dx - \int \frac{1-x^2}{\sqrt{1-x^2}} dx$$

or

$$= x\sqrt{1-x^2} + \arcsin x - \int \sqrt{1-x^2} \, dx + C_1 .$$

The original integral is now on the right of the equality sign and it might be thought that we have reached a dead end. To the contrary, move it to the left side of the equality sign:

$$2\int \sqrt{1-x^2} \, dx = x\sqrt{1-x^2} + \arcsin x + C_1 ;$$

$$\int \sqrt{1-x^2} \, dx = \frac{1}{2} x\sqrt{1-x^2} + \frac{1}{2} \arcsin x + C .$$

Successive Integration by Parts

Successive integration by parts is sometimes required to complete the integration.

- Integrate $\int e^x \cos x \, dx$.

Let $u = e^x$ and $v' = \cos x$; then $u' = e^x$ and $v = \sin x$.

Integrate by parts:

$$(1) \qquad \int e^x \cos x \, dx = e^x \sin x - \int e^x \sin x \, dx + C_1$$

To integrate by parts, let $u_1 = e^x$ and $v_1' = \sin x$; then $u_1' = e^x$ and $v_1 = -\cos x$:

$$\int e^x \sin x \, dx = -e^x \cos x - \int e^x (-\cos x) \, dx + C_2$$

$$(2) \qquad = -e^x \cos x + \int e^x \cos x \, dx + C_2$$

Adding (1) in (2) and rearranging terms, we find

$$\int e^x \cos x \, dx = \frac{1}{2} e^x (\sin x + \cos x) + C .$$

Reduction Formulas

A **reduction formula** (or **recursion formula**) in integration expresses a given integral as the sum of a function and a known integral.

Several formulas in integration tables normally appear in the form of reduction formulas, some of which are given below.

- Show that

$$\int x^n e^x \, dx = x^n e^x - n \int x^{n-1} e^x \, dx + C ,$$

where n is an integer.

Let $u = x^n$ and $v' = e^x$; then $u' = n x^{n-1}$ and $v = e^x$.

Integrate by parts:

$$\int x^n e^x \, dx = x^n e^x - n \int x^{n-1} e^x \, dx + C$$

- Integrate $\int x^2 e^x \, dx$.

Using the reduction formula, we obtain successively:

$$\int x^2 e^x \, dx = x^2 e^x - 2 \int x^{2-1} e^x \, dx + C_1$$

$$= x^2 e^x - 2 \left[x e^x - \int e^x \, dx \right] + C_1 + C_2$$

$$= x^2 e^x - 2 [x e^x - e^x] + C_1 + C_2 + C_3$$

$$= e^x (x^2 - 2x + 2) + C$$

- Show that

$$\int x^n \sin x \, dx = -x^n \cos x + n \int x^{n-1} \cos x \, dx + C,$$

where n is an integer.

Let $u = x^n$ and $v' = \sin x$; then $u' = n\, x^{n-1}$ and $v = -\cos x$.
Integrate by parts:

$$\int x^n \sin x \, dx = -x^n \cos x - \int n\, x^{n-1} (-\cos x) \, dx + C$$

$$= -x^n \cos x + n \int x^{n-1} \cos x \, dx + C.$$

- Show that

$$\int x^n \cos x \, dx = x^n \sin x - n \int x^{n-1} \sin x \, dx + C,$$

where n is an integer.

Let $u = x^n$ and $v' = \cos x$; then $u' = n\, x^{n-1}$ and $v = \sin x$.
Integrate by parts:

$$\int x^n \cos x \, dx = x^n \sin x - n \int x^{n-1} \sin x \, dx + C .$$

- Show that

$$\int (\ln x)^n \, dx = x (\ln x)^n - n \int (\ln x)^{n-1} \, dx + C,$$

where n is an integer.

$$\int (\ln x)^n \, dx = \int (\ln x)^n \cdot 1 \, dx.$$

Let $u = (\ln x)^n$ and $v' = 1$; then $u' = n\, (\ln x)^{n-1} \left(\frac{1}{x}\right)$ and $v = x$.
Integrate by parts:

$$\int (\ln x)^n \, dx = x (\ln x)^n - \int x\, n\, (\ln x)^{n-1} \left(\frac{1}{x}\right) dx + C$$

$$= x (\ln x)^n - n \int (\ln x)^{n-1} \, dx + C.$$

- Show that

$$\int \sin^n x \, dx = -\frac{\cos x \sin^{n-1} x}{n} + \frac{n-1}{n} \int \sin^{n-2} x \, dx + C,$$

where n is an integer, $\neq 0$.

$$\int \sin^n x \, dx = \int \sin x \sin^{n-1} x \, dx .$$

Let $u = \sin^{n-1} x$ and $v' = \sin x$; then

$$u' = (n-1) \sin^{n-2} x \cos x \text{ and } v = -\cos x .$$

Integrate by parts:

$$\int \sin x \sin^{n-1} x \, dx = -\cos x \sin^{n-1} x - \int (n-1) \sin^{n-2} x \cos x (-\cos x) \, dx + C$$

$$= -\cos x \sin^{n-1} + (n-1) \int \sin^{n-2} x \cos^2 x \, dx + C$$

Pythagorean Identity: p. 508

$$= -\cos x \sin^{n-1} + (n-1) \int \sin^{n-2} x (1 - \sin^2 x) \, dx + C$$

$$= -\cos x \sin^{n-1} + (n-1) \int \sin^{n-2} x \, dx - (n-1) \int \sin^n x \, dx + C.$$

Add $(n-1) \int \sin^n x \, dx$ to both sides of the equation:

$$n \int \sin^n x \, dx = -\cos x \sin^{n-1} + (n-1) \int \sin^{n-2} x \, dx + C.$$

Dividing both sides by n gives

$$\int \sin^n x \, dx = -\frac{\cos x \sin^{n-1}}{n} + \frac{n-1}{n} \int \sin^{n-2} x \, dx + C.$$

•

Analogously, we have

$$\int \cos^n x \, dx = \frac{\sin x \cos^{n-1} x}{n} + \frac{n-1}{n} \int \cos^{n-2} x \, dx + C,$$

where n is an integer, $\neq 0$.

"I'll quit when it stops being fun."

Integration by Substitution

By replacing part of a composite integrand with a new variable, the integrand may be rendered suitable for integration by a basic integration formula. This technique is the reverse of the chain rule for differentiating a composite function.

If F is an integral of f, then the chain rule gives

$$\frac{d}{dx}\,[F\,\{g\,(x)\}] = f\,\{g\,(x)\}\,g'\,(x)$$

or

$$\int f\{g\,(x)\}\,g'\,(x)\;dx = F\{g\,(x)\} + C\,.$$

- To apply the method of integration by substitution, select within the integrand a function of x, say, $g\,(x)$, to be replaced with a new variable, say, u.

  This selection should be made so that the new integrand in u can be integrated on sight.

- Differentiate u with respect to dx, and solve for dx ,

$$\frac{du}{dx} = g'\,(x)\,;\quad dx = \frac{du}{g'(x)}\,.$$

- Make the substitutions $g\,(x) = u$ and $dx = \dfrac{du}{g'(x)}$.
- Place constant factors before the integral sign.
- Eliminate all variables x. If this is not possible, try another substitution or another method of integration.
- Integrate.
- Replace u by $g\,(x)$.

- Integrate $\int x^2\sqrt{x^3-2}\;dx$.

Let $u = x^3 - 2$; then $\dfrac{du}{dx} = 3\,x^2$ and $dx = \dfrac{du}{3\,x^2}$.

We now have

$$\int x^2\sqrt{x^3-2}\;dx = \int x^2\sqrt{u}\;\frac{du}{3\,x^2} = \int x^2\,u^{1/2}\;\frac{du}{3\,x^2}$$

$$= \int u^{1/2}\,\frac{du}{3} = \frac{1}{3}\int u^{1/2}\;du = \frac{1}{3}\cdot\frac{2}{3}\cdot u^{3/2} + C\,.$$

To obtain the **answer in terms of x**, replace u by $(x^3 - 2)$,

$$\int x^2\sqrt{x^3-2}\;dx = \frac{2}{9}(x^3-2)^{3/2} + C.$$

- Integrate $\int x\, e^{-x^2}\, dx$.

Let $u = -x^2$; $dx = -\dfrac{du}{2x}$.

p. 724

$$\int -x\, e^u \frac{du}{2x} = -\frac{1}{2}\int x\, e^u \frac{du}{x} = -\frac{1}{2}\int e^u\, du = -\frac{1}{2} e^u + C.$$

Going back to x gives

$$\int x\, e^{-x^2}\, dx = -\frac{1}{2} e^{-x^2} + C.$$

- Integrate $\int (x^5 + x)^{10} (5x^4 + 1)\, dx$.

Let $u = x^5 + x$; $dx = \dfrac{du}{5x^4 + 1}$.

$$\int u^{10} (5x^4 + 1) \frac{du}{5x^4 + 1} = \int u^{10}\, du = \frac{u^{10+1}}{10+1} + C = \frac{u^{11}}{11} + C.$$

Going back to x gives

$$\int (x^5 + x)^{10} (5x^4 + 1)\, dx = \frac{1}{11} (x^5 + x)^{11} + C.$$

- Integrate $\displaystyle\int \frac{1}{(\ln x)^8 x}\, dx$.

Let $u = \ln x$; $\dfrac{du}{dx} = \dfrac{1}{x}$ and $dx = x\, du$.

$$\int \frac{1}{u^8 x}\, x\, du = \int u^{-8}\, du = \frac{u^{-8+1}}{-8+1} + C = -\frac{1}{7} u^{-7} + C.$$

Going back to x gives

$$\int \frac{1}{(\ln x)^8 x}\, dx = -\frac{1}{7} (\ln x)^{-7} + C.$$

- Integrate $\int \sin x \cos x\, dx$.

Let $u = \sin x$; $dx = \dfrac{du}{\cos x}$.

$$\int u \cos x \frac{du}{\cos x} = \int u\, du = \frac{u^2}{2} + C = \frac{1}{2} \sin^2 x + C.$$

• Integrate $\int x\sqrt{x-1}\ dx$.

Let $u = x - 1$; then $\dfrac{du}{dx} = 1$ and $dx = du$.

$$\int x\sqrt{x-1}\ dx = \int (u+1)\,u^{1/2}\ du = \int u^{3/2} + u^{1/2}\ du$$
$$= 2\left(\frac{u^{5/2}}{5} + \frac{u^{3/2}}{3}\right) + C.$$

Going back to x:

$$\int x\sqrt{x-1}\ dx = 2\left[\frac{\left(\sqrt{x-1}\right)^5}{5} + \frac{\left(\sqrt{x-1}\right)^3}{3}\right] + C. \quad \bullet$$

When integrands contain different roots of the same quantity, it is generally advantageous to use a substitute for the root that is the highest common divisor of the roots involved, and thereby **eliminate all radicals**.

• Integrate $\displaystyle\int \frac{1}{\sqrt[3]{x} + \sqrt[4]{x}}\ dx$.

Let $u = \sqrt[12]{x}$; then $u^{12} = x$, $\sqrt[3]{x} = u^4$, $\sqrt[4]{x} = u^3$, and $dx = 12\,u^{11}\,du$.

We now have

$$\int \frac{1}{\sqrt[3]{x} + \sqrt[4]{x}}\ dx = 12\int \frac{u^8}{u+1}\ du.$$

$$12\int \frac{u^8}{u-1}\ du = 12\int\left(u^7 + u^6 + u^5 + u^4 + u^3 + u^2 + u + 1 + \frac{1}{u-1}\right) du.$$

To evaluate $\displaystyle\int \frac{1}{u-1}\ du$, let $u - 1 = z$; then $du = dz$, and

$$\int \frac{1}{u-1}\ du = \int \frac{1}{z}\ dz = \ln|z| + C_1 = \ln|1-u| + C_1.$$

Hence,

$$\int \frac{1}{\sqrt[3]{x} + \sqrt[4]{x}}\ dx = 12\left(\frac{u^8}{8} + \frac{u^7}{7} + \frac{u^6}{6} + \frac{u^5}{5} + \frac{u^4}{4} + \frac{u^3}{3} + \frac{u^2}{2} + u + \ln|u-1|\right) + C.$$

Going back to x gives

$$\int \frac{1}{\sqrt[3]{x} + \sqrt[4]{x}}\ dx = 12\left(\frac{x^{2/3}}{8} + \frac{x^{7/12}}{7} + \frac{x^{1/2}}{6} + \frac{x^{5/12}}{5} + \frac{x^{1/3}}{4} + \frac{x^{1/4}}{3} + \frac{x^{1/6}}{2} + x^{1/12} + \ln\left|\sqrt[12]{x} - 1\right|\right) + C.$$

Combining Substitution with Integration by Parts

Integration by substitution is often combined with integration by parts.

- Integrate $\int \arcsin x \, dx$.

p. 724 Let $u = \arcsin x$ and $v' = 1$; then $u' = \dfrac{1}{\sqrt{1-x^2}}$ and $v = x$.

$$\int \arcsin dx = x \arcsin x - \int x \frac{1}{\sqrt{1-x^2}} dx$$

$$= x \arcsin x - \int x (1-x^2)^{-1/2} dx; \quad -1 < x < 1.$$

To integrate $\int x (1-x^2)^{-1/2} dx$, substitute:

$$w = 1 - x^2; \ \frac{dw}{dx} = -2x; \ dx = -\frac{dw}{2x}.$$

$$\int x (1-x^2)^{-1/2} dx = \int x w^{-1/2} \left(-\frac{dw}{2x}\right) = -\frac{1}{2}\int w^{-1/2} dw$$

$$= -w^{1/2} + C_1 = -(1-x^2)^{1/2} + C_1$$

$$= -\sqrt{1-x^2} + C_1.$$

Consequently,

$$\int \arcsin x \, dx = x \arcsin x + \sqrt{1-x^2} + C.$$

- Integrate $\int e^{\sqrt{x}} dx$.

Let $z = \sqrt{x}$; then $\dfrac{dz}{dx} = \dfrac{1}{2} x^{-1/2}$ and $dx = 2 x^{1/2} dz$.

Since $z = \sqrt{x} = x^{1/2}$,

$$dx = 2z \, dz.$$

$$\int e^{\sqrt{x}} dx = 2 \int z e^z dz.$$

Let $u = z$ and $v' = e^z$; then $u' = 1$ and $v = e^z$.

$$2 \int e^z z \, dz = 2\left(z e^z - \int e^z dz\right) = 2\left(z e^z - e^z\right) + C.$$

Returning to x gives

$$\int e^{\sqrt{x}} dx = 2\left(\sqrt{x}\, e^{\sqrt{x}} - e^{\sqrt{x}}\right) + C.$$

- Integrate $7 \int x^2 e^{3x} dx$.

Let $u = x^2$ and $v' = e^{3x}$; then $u' = 2x$.

To find v, integrate $\int e^{3x} dx$.

Let $z = 3x$; then $\dfrac{dz}{dx} = 3$ and $dx = \dfrac{dz}{3}$.

$$\int e^{3x}\ dx = \int e^{z}\ \frac{dz}{3} = \frac{1}{3}\int e^{z}\ dz = \frac{1}{3}\ e^{3x} + C_1.$$

With $C_1 = 0$, $v = \frac{1}{3}\ e^{3x}$.

Integrating by parts,

$$7\int x^2\ e^{3x}\ dx = 7\left[x^2\frac{1}{3}\ e^{3x}\right] - \int \frac{1}{3}\ (e^{3x})\ (2\,x)\ dx + C_2$$

$$= \frac{7}{3}\ x^2\ e^{3x} - \frac{14}{3}\int x\ e^{3x}\ dx + C_2.$$

To **repeat integration by parts,**

let $u_1 = x$ and $v_1{}' = e^{3x}$; then $u_1{}' = 1$ and $v_1 = \frac{1}{3}\ e^{3x}$.

$$7\int x^2\ e^{3x}\ dx = \frac{7}{3}x^2\ e^{3x} - \frac{14}{3}\left[x\ \left(\frac{1}{3}e^{3x}\right) - \int \frac{1}{3}\ e^{3x}\ dx\right] + C_3$$

$$= \frac{7}{3}x^2\ e^{3x} - \frac{14}{9}x\,e^{3x} + \frac{14}{27}\ e^{3x} + C.$$

Successive Substitution

It is sometimes advantageous to use successive substitution.

- Integrate $\int \sqrt{1+\sqrt{x}}\ dx$.

Let $u = \sqrt{x}$; then $\frac{du}{dx} = \frac{1}{2}\ x^{-1/2}$ and $dx = 2\,x^{1/2}\,du = 2\,u\ du$.

Thus, $$\int\sqrt{1+\sqrt{x}}\ dx = \int\sqrt{1+u}\ 2\,u\ du,$$

which still does not agree with any basic integration formula.

We make a **new substitution,**

$$z = 1 + u\ ;\ \text{then}\ du = dz\,.$$

The integral now becomes

$$\int\sqrt{1+u}\ 2\,u\,du = \int z^{1/2}\,2\,(z-1)\,dz = 4\left(\frac{z^{5/2}}{5} - \frac{z^{3/2}}{3}\right) + C.$$

Replacing z with $1 + u$ gives

$$\int\sqrt{1+u}\ 2\,u\ du = 4\left(\frac{\left(\sqrt{1+u}\right)^5}{5} - \frac{\left(\sqrt{1+u}\right)^3}{3}\right) + C$$

and replacing u with $\sqrt{x}$ results in

$$\int\sqrt{1+\sqrt{x}}\ dx = 4\left(\frac{\left(\sqrt{1+\sqrt{x}}\right)^5}{5} - \frac{\left(\sqrt{1+\sqrt{x}}\right)^3}{3}\right) + C.$$

Substituting Trigonometric Functions for Algebraic Expressions

Until now, our substitutions have all consisted of letting u represent some function within the integrand. However, for integrands that contain certain radicals, for instance, the square roots $\sqrt{a^2 - x^2}$, $\sqrt{x^2 - a^2}$, and $\sqrt{a^2 + x^2}$, a substitution by a trigonometric function is sometimes more productive.

Consider the triangles

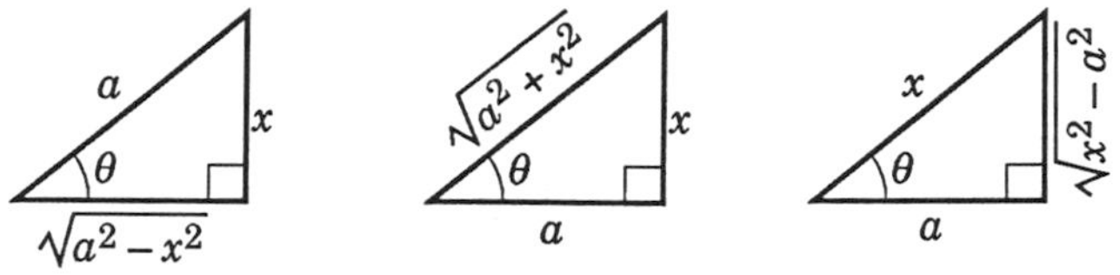

We find that the substitutions

$$x = a \cdot \sin\theta \qquad x = a \cdot \tan\theta \qquad x = a \cdot \sec\theta$$

bring, by definition, the substitutions

$$\sqrt{a^2 - x^2} = a \cdot \cos\theta; \quad \sqrt{a^2 + x^2} = a \cdot \sec\theta; \quad \sqrt{x^2 - a^2} = a \cdot \tan\theta.$$

To ensure a one-to-one relation between the original variable and the substitution, restrictions must be placed on θ. These restrictions are the same as the range of the corresponding inverse trigonometric functions:

pp. 505 - 506

	Substitution	**Range**
$x = a\sin\theta$	$\sqrt{a^2 - x^2} = a\cos\theta$	arcsin θ: $\left[-\frac{\pi}{2}, \frac{\pi}{2}\right]$
$x = a\tan\theta$	$\sqrt{a^2 + x^2} = a\sec\theta$	arctan θ: $\left]-\frac{\pi}{2}, \frac{\pi}{2}\right[$
$x = a\sec\theta$	$\sqrt{x^2 - a^2} = a\tan\theta$	arcsec θ: $\left[-\pi, -\frac{\pi}{2}\right[, \left[0, \frac{\pi}{2}\right[$

- Integrate $\displaystyle\int \frac{\sqrt{4 - x^2}}{x^2}\,dx$.

$$\int \frac{\sqrt{4 - x^2}}{x^2}\,dx = \int \frac{\sqrt{2^2 - x^2}}{x^2}\,dx.$$

Let $x = 2\sin\theta$; $dx = 2\cos\theta\,d\theta$.

$$\int \frac{\sqrt{4 - x^2}}{x^2}\,dx = \int \frac{2\cos\theta}{4\sin^2\theta}\,2\cos\theta\,d\theta = \int \frac{\cos^2\theta}{\sin^2\theta}\,d\theta$$

$$= \int \cot^2\theta\,d\theta = \int(\csc^2\theta - 1)\,d\theta = -\cot\theta - \theta + C.$$

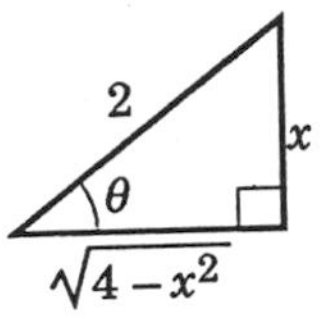

We have

$$\sin\theta = \frac{x}{2}; \quad \cos\theta = \frac{\sqrt{4 - x^2}}{2}; \quad \cot\theta = \frac{\sqrt{4 - x^2}}{x}$$

and obtain

$$\int \frac{\sqrt{4 - x^2}}{x^2}\,dx = -\frac{\sqrt{4 - x^2}}{x} - \arcsin\frac{x}{2} + C. \quad \bullet$$

The example below shows how the integration of a seemingly simple expression may require some ingenuity.

- Integrate $\int \sqrt{x^2+3}\,dx$.

$\sqrt{x^2+3} = \sqrt{x^2+\left(\sqrt{3}\right)^2}$.

Let $x = \sqrt{3}\tan\theta$; $dx = \sqrt{3}\sec^2\theta\, d\theta$.

$\sqrt{x^2+3} = \sqrt{3}\sec\theta$.

We now have

$$\int \sqrt{x^2+3}\,dx = \int \left(\sqrt{3}\sec\theta\right)\cdot\left(\sqrt{3}\sec^2\theta\right)d\theta$$

$$= 3\int \sec^3\theta\, d\theta$$

or

$$3\int \sec\theta\,\sec^2\theta\, d\theta,$$

which we integrate by parts; let $u = \sec\theta$ and $v' = \sec^2\theta$; then $u' = \sec\theta\tan\theta$ and $v = \tan\theta$,

$$\begin{aligned}
\int \sec^3\theta\, d\theta &= \tan\theta\sec\theta - \int \sec\theta\,\tan^2\theta\, d\theta + C_1 \\
&= \tan\theta\sec\theta - \int \sec\theta\,(\sec^2\theta - 1)\, d\theta + C_1 \\
&= \tan\theta\sec\theta - \int \sec^3\theta\, d\theta + \int \sec\theta\, d\theta + C_1 \\
2\int \sec^3\theta\, d\theta &= \tan\theta\sec\theta + \int \sec\theta\, d\theta + C_1 \\
\int \sec^3\theta\, d\theta &= \frac{\tan\theta\sec\theta}{2} + \frac{1}{2}\int \sec\theta\, d\theta + C_1 \\
&= \frac{\tan\theta\sec\theta}{2} + \frac{1}{2}\ln\left|\sec\theta + \tan\theta\right| + C,
\end{aligned}$$

p. 724

and

$$3\int \sec^3\theta\, d\theta = \frac{\left(\sqrt{3}\tan\theta\right)\left(\sqrt{3}\sec\theta\right)}{2} + \frac{3}{2}\ln\left|\sec\theta + \tan\theta\right| + C.$$

Reintroduce $\sqrt{3}\tan\theta = x$ and $\sec\theta = \dfrac{\sqrt{x^2+3}}{\sqrt{3}}$ to obtain

$$\begin{aligned}
\int \sqrt{x^2+3}\,dx &= \frac{x\sqrt{x^2+3}}{2} + \frac{3}{2}\ln\left|\frac{\sqrt{x^2+3}}{\sqrt{3}} + \frac{x}{\sqrt{3}}\right| + C \\
&= \frac{x\sqrt{x^2+3}}{2} + \frac{3}{2}\ln\left|\frac{\sqrt{x^2+3}+x}{\sqrt{3}}\right| + C.
\end{aligned}$$ •

In the above calculations, we have stayed with the basic integration formulas on *pp.* 724 - 25. Judicious use of a more comprehensive table of integrals could simplify the process.

Converting Rational Trigonometric Integrals into Algebraic Integrals

Every rational function of $\sin x$ and $\cos x$ can be converted into a rational algebraic function.

pp. 518 - 19

To solve a trigonometric equation, we converted trigonometric terms to the **tangent of the half-angle**; similarly, the substitution $u = \tan\frac{x}{2}$ to convert rational functions of $\sin x$ and $\cos x$ into rational algebraic functions may prove profitable for integration;

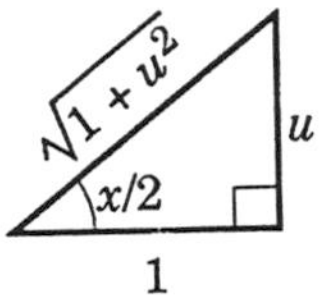

$$u = \tan\frac{x}{2}, \quad \sin\frac{x}{2} = \frac{u}{\sqrt{1+u^2}}, \quad \text{and} \quad \cos\frac{x}{2} = \frac{1}{\sqrt{1+u^2}}.$$

With the identities

pp. 510 - 11

$$\sin\theta = 2\sin\frac{\theta}{2}\cos\frac{\theta}{2}; \quad \cos\theta = \cos^2\frac{\theta}{2} - \sin^2\frac{\theta}{2}$$

and the substitution $u = \tan\frac{x}{2}$, we obtain

$$\sin x = 2\frac{u}{\sqrt{1+u^2}}\left(\frac{1}{\sqrt{1+u^2}}\right) = \frac{2u}{1+u^2};$$

$$\cos x = \left(\frac{1}{\sqrt{1+u^2}}\right)^2 - \left(\frac{u}{\sqrt{1+u^2}}\right)^2 = \frac{1-u^2}{1+u^2},$$

and

$$\frac{du}{dx} = \frac{d}{dx}\left[\tan\frac{x}{2}\right] = \frac{1}{2}\left(\sec^2\frac{x}{2}\right) = \frac{1}{2}\left(1+\tan^2\frac{x}{2}\right) = \frac{1}{2}(1+u^2)$$

$$dx = \frac{2}{1+u^2}\,du.$$

- Integrate $\int \frac{1}{\sin x}\,dx$.

Let $u = \tan\frac{x}{2}$; then $\sin x = \frac{2u}{1+u^2}$ and $dx = \frac{2}{1+u^2}\,du$.

$$\int \frac{1}{\sin x}\,dx = \int \left(\frac{1}{\frac{2u}{1+u^2}}\right)\frac{2}{1+u^2}\,du$$

$$= \int \frac{1}{u}\,du = \ln|u| + C = \ln\left|\tan\frac{x}{2}\right| + C.$$

- Integrate $\int \frac{1}{\cos x}\,dx$.

Let $u = \tan\frac{x}{2}$; then $\cos x = \frac{1-u^2}{1+u^2}$ and $dx = \frac{2}{1+u^2}\,du$.

$$\int \frac{1}{\cos x}\, dx = \int \left(\frac{1}{\frac{1-u^2}{1+u^2}}\right) \frac{2}{1+u^2}\, du = 2\int \frac{1}{1-u^2}\, du$$

p. 725

$$= 2\left(\frac{1}{2}\right) \ln \left|\frac{1+u}{1-u}\right| + C = \ln \left|\frac{1+\tan\frac{x}{2}}{1-\tan\frac{x}{2}}\right| + C.$$

- Integrate $\int \frac{1 - \sin x}{\sin x - \sin x \cos x}\, dx$.

$$\int \frac{1-\sin x}{\sin x - \sin x \cos x}\, dx = \int \frac{1-\sin x}{\sin x\,(1-\cos x)}\, dx.$$

Let $u = \tan\frac{x}{2}$; then $\sin x = \frac{2u}{1+u^2}$, $\cos x = \frac{1-u^2}{1+u^2}$, and $dx = \frac{2}{1+u^2}\, du$.

$$\int \frac{1-\sin x}{\sin x - \sin x \cos x}\, dx = \int \frac{1-\frac{2u}{1+u^2}}{\frac{2u}{1+u^2}\left(1-\frac{1-u^2}{1+u^2}\right)} \left(\frac{2}{1+u^2}\right) du$$

$$= \frac{1}{2}\int (u^{-1} - 2u^{-2} + u^{-3})\, du = \frac{1}{2}\left(\ln|u| + 2u^{-2} - \frac{1}{2}u^{-1}\right) + C$$

$$= \frac{1}{2}\left(\ln\left|\tan\frac{x}{2}\right| + 2\cot\frac{x}{2} - \frac{1}{2}\cot^2\frac{x}{2}\right) + C.$$

Integrating $\int \frac{f'(x)}{f(x)}\, dx$

An integral where the numerator is the derivative of the denominator may be evaluated by the formula

$$\int \frac{f'(x)}{f(x)}\, dx = \ln|f(x)| + C.$$

Why? With $f(x) = u$, $f'(x)\, dx = du$, $dx = \frac{du}{f'(x)}$,

and

$$\int \frac{f'(x)}{f(x)}\, dx = \int \frac{f'(x)}{u} \cdot \frac{du}{f'(x)}\, dx = \int \frac{1}{u}\, du.$$

p. 723 $\int \frac{1}{u}\, du = \ln|u| + C \quad (u \neq 0)$, and

$$\int \frac{f'(x)}{f(x)}\, dx = \ln|f(x)| + C.$$

It is advantageous to watch for integrands whose numerator is the derivative of the denominator, or some multiple thereof, as they are easier to deal with.

- Integrate $\int \frac{2x}{x^2+1}\,dx$.

$\frac{d}{dx}[x^2+1] = 2x$.

Consequently,

$$\int \frac{2x}{x^2+1}\,dx = \ln|x^2+1| + C.$$

Since x^2+1 is always positive, the absolute sign is not needed;

$$\int \frac{2x}{1+x^2}\,dx = \ln(x^2+1) + C.$$

- Integrate $\int \frac{\sec^2 x}{\tan x}\,dx$.

$\frac{d}{dx}[\tan x] = \sec^2 x$;

$$\int \frac{\sec^2 x}{\tan x}\,dx = \ln|\tan x| + C.$$

- Integrate $\int \cot x\,dx$.

$$\int \cot x\,dx = \int \frac{\cos x}{\sin x}\,dx = \ln|\sin x| + C.$$

- Integrate $\int \frac{x-2}{x^2-4x+3}\,dx$.

$$\frac{1}{2}\int \frac{2x-4}{x^2-4x+3}\,dx = \frac{1}{2}\ln|x^2-4x+3| + C.$$

Integrating $\int \frac{f'(x)}{\sqrt{f(x)}}\,dx$

If the denominator of the integrand is the square root of a function of the variable, and the numerator is the derivative of the radicand, then

$$\int \frac{f'(x)}{\sqrt{f(x)}}\,dx = 2\sqrt{f(x)} + C,$$

which we verify by letting $f(x) = u$:

$$\frac{d}{dx}\left[2\sqrt{u} + C\right] = \frac{u'}{\sqrt{u}}.$$

- Integrate $\displaystyle\int \frac{3x^2 + x}{\sqrt{4x^3 + 2x^2}}\, dx$.

Since

$$\frac{d}{dx}\left[4x^3 + 2x^2\right] = 4(3x^2 + x),$$

the integral has the form

$$\frac{1}{4}\int \frac{f'(x)}{\sqrt{f(x)}}\, dx.$$

Hence,

$$\int \frac{3x^2 + x}{\sqrt{4x^3 + 2x^2}}\, dx = \frac{1}{4}\cdot 2\cdot\sqrt{4x^3 + 2x^2} + C = \frac{\sqrt{2}}{2}\sqrt{2x^3 + x^2} + C.$$

•

If the integrand does not immediately lend itself to evaluation by the above formula, it may be rewritten in a manner such that the formula can still be used.

- Integrate $\displaystyle\int \sqrt{\frac{6 - x}{x}}\, dx$.

$$\int \sqrt{\frac{6 - x}{x}}\, dx = \int \frac{6 - x}{\sqrt{6x - x^2}}\, dx = \frac{1}{2}\int \frac{(6 - 2x) + 6}{\sqrt{6x - x^2}}\, dx$$

$$= \frac{1}{2}\int \frac{6 - 2x}{\sqrt{6x - x^2}}\, dx + 3\int \frac{1}{\sqrt{6x - x^2}}\, dx$$

$$= \frac{1}{2}\cdot 2\cdot\sqrt{6x - x^2} + 3\int \frac{1}{\sqrt{3^2 - (x - 3)^2}}\, dx$$

p. 724

$$= \sqrt{6x - x^2} + 3\arcsin\frac{x - 3}{3} + C.$$

•

To evaluate the integral

$$\int \frac{ax + b}{\sqrt{cx^2 + dx + e}}\, dx$$

by the formula $\displaystyle\int \frac{f'(x)}{\sqrt{f(x)}}\, dx = 2\sqrt{f(x)} + C$, we **rewrite the numerator** thus,

$$p\,\frac{d}{dx}\left[cx^2 + dx + e\right] + q,$$

where p and q are constants.

- Integrate $\int \frac{4x-2}{\sqrt{5+4x-x^2}}\,dx$.

$$p\,\frac{d}{dx}\,[5+4x-x^2]+q = 4x-2$$
$$p\,(4-2x)+q = 4x-2,$$

which gives
$$p=-2;\ q=6.$$

Hence,

$$\int \frac{4x-2}{\sqrt{5+4x-x^2}}\,dx = \int \frac{-2\,(4-2x)+6}{\sqrt{5+4x-x^2}}\,dx$$
$$= \int \frac{-2\,(4-2x)}{\sqrt{5+4x-x^2}}\,dx + \int \frac{6}{\sqrt{5+4x-x^2}}\,dx$$
$$= -2\cdot 2\sqrt{5+4x-x^2}+C_1 + 6\int \frac{1}{\sqrt{9-(x-2)^2}}\,dx,$$

p. 724 where
$$\int \frac{1}{\sqrt{3^2-(x-2)^2}}\,dx = \arcsin\frac{x-2}{3}+C_2$$

and, consequently,

$$\int \frac{4x-2}{\sqrt{5+4x-x^2}}\,dx = 6\arcsin\frac{x-2}{3} - 4\sqrt{5+4x-x^2}+C.$$

Integrating $\int f(x)\,f'(x)\,dx$

An integral in two factors where one factor is the derivative of the other is always easy to integrate by the formula

$$\int f(x)\,f'(x)\,dx = \frac{[f(x)]^2}{2}+C.$$

Why? With $f(x)=u$, $f'(x)\,dx = du$, $dx = \frac{du}{f'(x)}$, and

$$\int f(x)\,f'(x)\,dx = \int u\,f'(x)\cdot\frac{du}{f'(x)} = \int u\,du = \frac{u^2}{2}+C = \frac{[f(x)]^2}{2}+C.$$

- Integrate:
  - $\int (x^5+3x^2-7)\,(5x^4+6x)\,dx = \frac{1}{2}\,(x^5+3x^2-7)^2+C.$
  - $\int [\tan x\,(7\sec^2 x)]\,dx = 7\int \tan x\,\sec^2 x\,dx = \frac{7}{2}\tan^2 x + C.$
  - $\int \frac{4\ln x}{3x}\,dx = \frac{4}{3}\int \ln x\left(\frac{1}{x}\right)dx = \frac{2}{3}\cdot\ln^2 x + C.$
  - $\int (\sin x)\,(1-\cos x)\,dx = \frac{1}{2}\,(1-\cos x)^2+C.$

Integration by Partial Fractions

pp. 130 - 31

An algebraic expression containing a polynomial in the denominator, or in the denominator and numerator, can always be integrated by **splitting the function into partial fractions**, each amenable to a known integration formula.

- Integrate

$$\int \frac{x^2 + x + 1}{2x^4 + x^3 + 2x^2 + x}\,dx.$$

$$\int \frac{x^2 + x + 1}{2x^4 + x^3 + 2x^2 + x}\,dx = \int \left(\frac{1}{x} - \frac{6}{5(2x+1)} + \frac{1}{5(x^2+1)} - \frac{2x}{5(x^2+1)}\right) dx$$

$$= \int \frac{1}{x}\,dx - \frac{1}{5}\int \left(\frac{6}{2x+1} - \frac{1}{x^2+1} + \frac{2x}{x^2+1}\right) dx\,.$$

With $u = 2x + 1$, and $du = 2\,dx$ and $dx = \dfrac{du}{2}$, we have

p. 723

$$\int \frac{6}{2x+1}\,dx = 3\int \frac{1}{u}\,du = 3\ln|2x+1| + C_1\,.$$

p. 724

$$\int \frac{1}{x^2+1}\,dx = \arctan x + C_2\,.$$

p. 742

$$\int \frac{2x}{x^2+1}\,dx = 2\int \frac{x}{x^2+1}\,dx = \ln(x^2+1) + C_3\,.$$

$$\int \frac{x^2 + x + 1}{2x^4 + x^3 + 2x^2 + x}\,dx$$

$$= \ln|x| - \frac{1}{5}\,[3\ln|2x+1| - \arctan x + \ln(x^2+1)] + C\,. \bullet$$

If the degree of the numerator is greater than or equal to the degree of the denominator, then the denominator must be divided into the numerator until the degree of the remainder is less than that of the denominator.

- Integrate $\displaystyle\int \frac{x^2+3}{x-3}\,dx\,.$

Division: *pp.* 129 - 30

$$\frac{x^2+3}{x-3} = x + 3 + \frac{12}{x-3}\,.$$

$$\int \frac{x^2+3}{x-3}\,dx = \frac{1}{2}x^2 + 3x + 12\ln|x-3| + C\,.$$

Integration by Power Series Expansion

A function $f(x)$ that can be expanded as a power series in x may always be integrated if $f(x)$ is replaced by the terms of the series.

Integration by power series expansion is used to evaluate integrals that can only be expressed by an infinite number of elementary functions.

The integrals

$$\int e^{-x^2}\,dx \text{ and } \int \frac{\sin x}{x}\,dx$$

cannot be expressed in terms of a finite number of elementary functions.

p. 768

Term-by-term integration of Maclaurin's expansion

$$e^{-x^2} = 1 - \frac{x^2}{1!} + \frac{x^4}{2!} - \dots$$

gives

$$\int e^{-x^2}\,dx = x - \frac{x^3}{3 \cdot 1!} + \frac{x^5}{5 \cdot 2!} - \frac{x^7}{7 \cdot 3!} + \dots + C.$$

Using the expansion

p. 771

$$\sin x = x - \frac{x^3}{3!} + \frac{x^5}{5!} - \frac{x^7}{7!} + \dots$$

we have

$$\frac{\sin x}{x} = 1 - \frac{x^2}{3!} + \frac{x^4}{5!} - \frac{x^6}{7!} + \dots$$

and

$$\int \frac{\sin x}{x}\,dx = x - \frac{x^3}{3 \cdot 3!} + \frac{x^5}{5 \cdot 5!} - \frac{x^7}{7 \cdot 7!} + \dots + C.$$

21.3 The Definite Integral

An integral that is defined between two values a and b of an independent variable is a **definite integral**.

An **indefinite integral** $\int f(x)\,\mathrm{d}x$ is a **function plus an arbitrary constant; a definite integral**

$$\int_a^b f(x)\,\mathrm{d}x$$

is a **quantity.**

Geometrically, a definite integral can be thought of as the area contained between the graph of a function and the x-axis of an orthogonal coordinate system in a closed interval from $x = a$ to $x = b$. Areas located above the x-axis count as positive in integration, areas below the x-axis as negative.

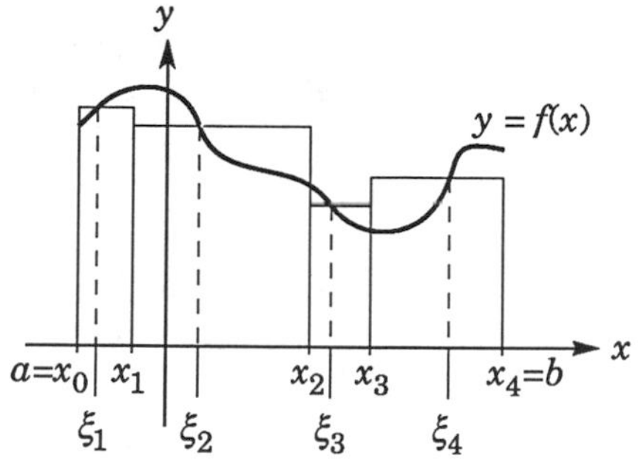

To find an approximation of the area, we divide the interval $[a, b]$ into n sub-intervals at the points

$$a = x_0 < x_1 < x_2 < \ldots < x_i \ldots < x_n = b$$

with

$$\Delta x = \frac{b-a}{n}\,.$$

Within every sub-interval Δa_i we choose an abscissa ξ_i; the total area A under the graph is approximated by the sum of the areas of rectangles, $f(\xi_i) \cdot \Delta x_i$,

$$A \approx \sum_{i=1}^{n} f(\xi_i) \cdot \Delta x_i\,,$$

Riemann Sum

known as a **Riemann sum**.

With an increasing number of subdivisions, and forever smaller sub-intervals Δx_i, the Riemann sum continuously approaches the sought area more closely. That is,

$$A = \lim_{n \to \infty} \sum_{i=1}^{n} f(\xi_i) \cdot \Delta x_i\,.$$

The **definite integral** is defined to be this limit:

$$\int_a^b f(x)\,\mathrm{d}x = \lim_{n \to \infty} \sum_{i=1}^{n} f(\xi_i) \cdot \Delta x_i\,.$$

The limit exists whenever f is a continuous function on the closed interval $[a, b]$.

RIEMANN Georg Friedrich Bernhard (1826 - 1866)

The definite integral is sometimes called a **Riemann integral** after the German mathematician **Bernhard Riemann**, whose work on trigonometric series in 1854 prompted a precise definition of the integral.

The Fundamental Theorem of Calculus

Although some notion of integral calculus is at least 2500 years old, tangent problems and area problems had no unified technique for solution as recently as the latter part of the 17th century – each particular problem involved the employment of special methods.

von LEIBNIZ Gottfried Wilhelm (1646 - 1716)

NEWTON Isaac (1642 - 1727)

A breakthrough in mathematics and physics came in the 1670s when **Leibniz**, in Germany, and **Newton**, in England, recognized the inverse relationship between the tangent problem (differentiation) and the area problem (integration), condensed in the fundamental theorem of calculus. From the plethora of "infinitesimal techniques" developed in the 16th century and the first half of the 17th century, Leibniz and Newton had extracted a powerful system – the infinitesimal calculus or, with present-day terminology, *calculus* – a feat often referred to as their "discovery of calculus".

Let $A(x)$ be the area between the graph of the continuous function $f(x)$, the x-axis, and the verticals through the points $x_0 = a$ and x.

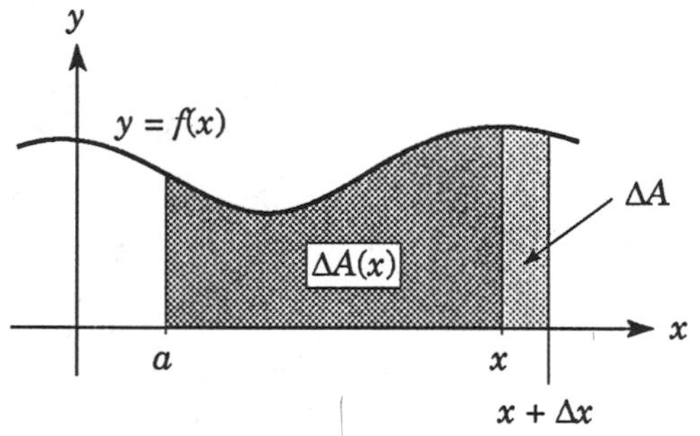

The area increment is

$$\Delta A \approx f(x) \cdot \Delta x \ .$$

When Δx tends to zero, we have the derivative of $A(x)$,

$$\frac{dA}{dx} = \lim_{\Delta x \to 0} \frac{A(x + \Delta x) - A(x)}{\Delta x} = \lim_{\Delta x \to 0} \frac{f(x) \cdot \Delta x}{\Delta x} = f(x),$$

that is, $f(x)$ is the derivative of the area $A(x)$.

If f is a continuous function in the closed interval $[a, b]$, such that

$$F'(x) = f(x) \quad \text{for all } x \text{ in } [a, b] \ ,$$

the **fundamental theorem of calculus** states that the **definite integral**

$$\int_a^b f(x)\ dx = F(b) - F(a) \ .$$

Finding Areas

An important use of definite integrals is the determination of the area between two curves, generally one curve and the x-axis.

Substitution Symbols

For convenience, we use the **substitution symbol** $\Big|_a^b$ or $[\,]_a^b$,

$$\int_a^b f(x)\,dx = \Big|_a^b F(x) = [F(x)]_a^b = F(b) - F(a).$$

A definite integral may be split into two or more integrals of the same function, thus,

$$\int_a^b f(x)\,dx = \int_a^m f(x)\,dx + \int_m^b f(x)\,dx,$$

where $a < m < b$; with the association of definite integrals with area, we have

$$A = A_1 + A_2.$$

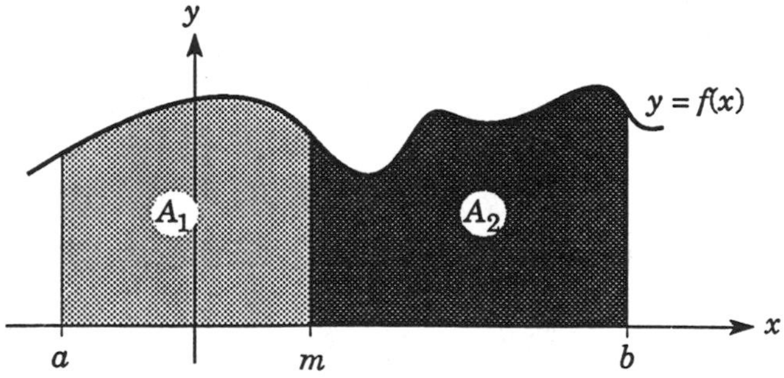

- Determine the area contained between the graph of

$$f(x) = x^3 + 3x^2 - 10x$$

and the x-axis in the interval from $x = -6$ to $x = +4$.

Plotting the function gives the result:

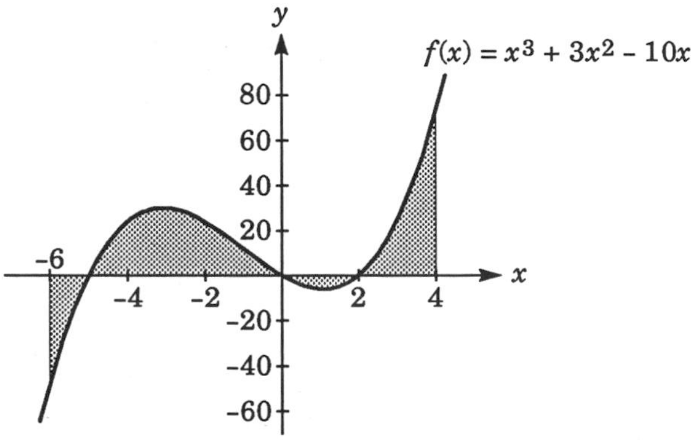

The graph shows that parts of the area are located below the x-axis and, consequently, will appear as negative quantities in integration.

The area between $x = -6$ and $x = +4$ will therefore have to be integrated separately over the intervals

$$x = \quad [-6, -5] \quad [-5, 0] \quad [0, +2] \quad [+2, +4] \, .$$

The indefinite integral is

$$\int (x^3 + 3x^2 - 10x) \, \mathrm{d}x = \frac{x^4}{4} + x^3 - 5x^2 + C ,$$

which gives integral area

$$\Big|_{-6}^{-5} F(x) = \left(\frac{625}{4} - 125 - 125\right) - \left(\frac{1296}{4} - 216 - 180\right)$$

$$= -93\frac{3}{4} + 72 = -21\frac{3}{4} \qquad 21\frac{3}{4}$$

$$\Big|_{-5}^{0} F(x) = 0 + 93\frac{3}{4} \qquad = +93\frac{3}{4} \qquad 93\frac{3}{4}$$

$$\Big|_{0}^{+2} F(x) = \left(\frac{16}{4} + 8 - 20\right) - 0 = -8 - 0 \qquad = -8 \qquad 8$$

$$\Big|_{+2}^{+4} F(x) = \left(\frac{256}{4} + 64 - 80\right) - (-8) = +48 + 8 \qquad = +56 \qquad 56$$

$$120 \qquad 179\frac{1}{2}$$

Changing Limits

An interchange of limits reverses the sign of the definite integral; thus,

$$\int_a^b f(x) \; \mathrm{d}x = -\int_b^a f(x) \; \mathrm{d}x \, .$$

When a definite integral has been evaluated by the method of substitution of a variable, it is generally more convenient to use the limits of the new variable than to convert back to the original variable.

- Find $\displaystyle\int_1^{\sqrt{2}} \frac{\sqrt{4 - x^2}}{x^2} \; \mathrm{d}x$.

$$\int \frac{\sqrt{4 - x^2}}{x^2} \, \mathrm{d}x = \int \frac{\sqrt{2^2 - x^2}}{x^2} \, \mathrm{d}x \, .$$

We have already found by substituting

p. 738

$$x = 2\sin\theta;\ \ dx = 2\cos\theta\, d\theta$$

that

$$\int \frac{\sqrt{4-x^2}}{x^2}\, dx = -\cot\theta - \theta + C\,.$$

Instead of reinstating the original variable x, we determine the limits of the new variable θ.

$$x = 1 \quad \Rightarrow \quad 2\sin\theta = 1\,; \quad \theta = \frac{\pi}{6}$$

$$x = \sqrt{2} \quad \Rightarrow \quad 2\sin\theta = \sqrt{2}\,; \quad \theta = \frac{\pi}{4}$$

Thus,

$$\int_1^{\sqrt{2}} \frac{\sqrt{4-x^2}}{x^2}\, dx = \Big|_{\pi/6}^{\pi/4} (-\cot\theta - \theta) = \left(-1 - \frac{\pi}{4}\right) - \left(-\sqrt{3} - \frac{\pi}{6}\right)$$

$$= \left(\sqrt{3} - 1\right) - \frac{\pi}{12} = 0.470\,25\,\ldots\,.$$

Integrated Mean Values

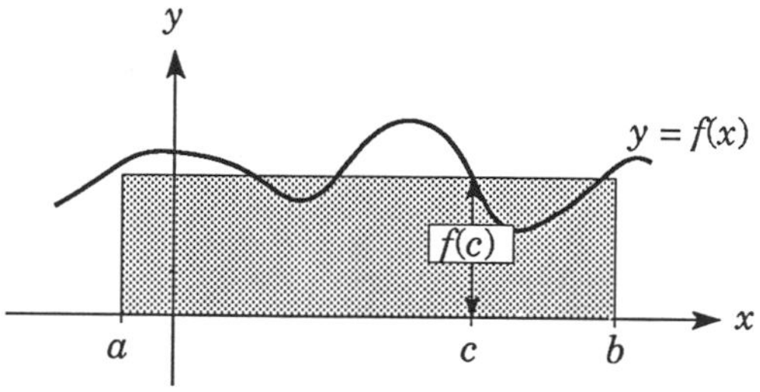

The **mean-value theorem for integrals** states that:

> If f is a continuous function in the interval $[a, b]$, then there exists a number c between a and b such that
>
> $$\int_a^b f(x)\, dx = (b-a)\cdot f(c)\,.$$

The **average value** of a finite quantity of numbers $a_1, a_2 \ldots a_n$ is the arithmetic mean

$$M_n = \frac{a_1 + a_2 + \ldots + a_n}{n}\,.$$

The value $f(c)$, as defined by the mean-value theorem, is referred to as the **average value** of $f(x)$ in $[a, b]$.

To see why $f(c)$ can be thought of as an average value, divide the interval into n equal sub-intervals, each with a length $\Delta x = \dfrac{b-a}{n}$ between the points of division

$$x_1 \; x_2 \; \ldots \; x_n \, ;$$

the average value of the midpoints $f(\xi_1)$, (ξ_2) ... (ξ_n) of the sub-intervals $\Delta x_1, \Delta x_2 \ldots \Delta x_n$ is

$$\frac{f(\xi_1) + f(\xi_2) + \ldots + f(\xi_n)}{n}$$

or, since $n = \dfrac{b-a}{\Delta x}$,

$$\frac{1}{b-a}\,[f(\xi_1) + f(\xi_2) + \ldots + f(\xi_n)]\,\Delta x = \frac{1}{b-a} \sum_{i=1}^{n} f(\xi_i)\,\Delta x .$$

When n tends to infinity, we have

$$\lim_{n \to \infty} \frac{1}{b-a} \sum_{i=1}^{n} f(\xi_i)\,\Delta x = \frac{1}{b-a} \int_a^b f(x)\; dx .$$

To summarize, the **average value** $f(c)$ in $[a, b]$ of a continuous function $f(x)$ is determined by the formula

$$f(c) = \frac{1}{b-a} \int_a^b f(x)\; dx .$$

- Find the average value of

$$f(x) = 1 + 3x - x^2$$

in the interval $[-1, 2]$.

$$f(c) = \frac{1}{2-(-1)} \int_{-1}^{2} (1 + 3x - x^2)\; dx = \frac{1}{3}\left[x + \frac{3}{2}x^2 - \frac{1}{3}x^3\right]_{-1}^{2}$$

$$= \frac{1}{3}\left(\frac{9}{2}\right) = \frac{3}{2} .$$

$$\frac{3}{2} = 1 + 3c - c^2$$

$$c = \frac{1}{2}\left(3 + \sqrt{7}\right), \text{ discarded ;} \qquad c = \frac{1}{2}\left(3 - \sqrt{7}\right) \approx 0.177$$

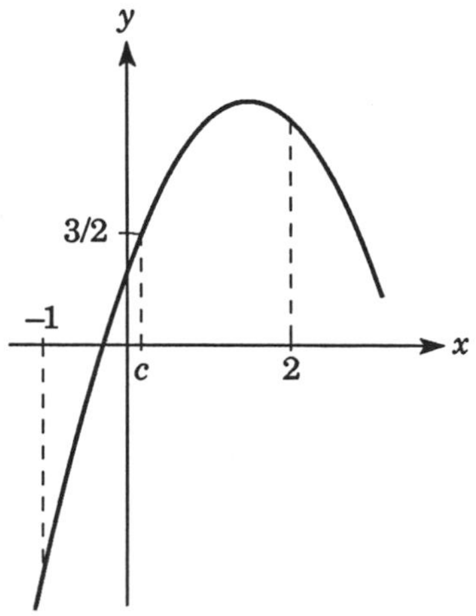

21.4 Multiple Integrals

An integral in which the integrand is integrated twice is a **double integral**,

$$\int\int f(x)\ dx^2;$$

if three times, a **triple integral**,

$$\int\int\int f(x)\ dx^3;$$

etc., to an ***n*-fold iterated integral**,

$$\int\int \ldots \int f(x)\ dx^n.$$

The integration is performed from the inside out.

- Find $\int_{-1}^{3} \int_{0}^{2} 2x\ dx^2$.

$$\int 2x\ dx = x^2 + C_1; \qquad \Big|_0^2\ x^2 = 4.$$

$$\int 4\ dx = 4x + C_2; \qquad \Big|_{-1}^3\ 4x = 16.$$

Hence,

$$\int_{-1}^{3} \int_{0}^{2} 2x\ dx^2 = 16.$$

Functions of **more than one variable** may be integrated with respect to one variable at a time while the other variables are held constant, reversing the process of partial differentiation.

- Find $\int_{0}^{3} \int_{1}^{2} (4y^3 + 2x)\ dx\ dy$.

Hold y constant and integrate with respect to x:

$$\int (4y^3 + 2x)\, dx = 4y^3 x + x^2 + C;$$

$$[4y^3 x + x^2]_{x=1}^{x=2} = (4y^3 \cdot 2 + 2^2) - (4y^3 + 1^2) = 4y^3 + 3.$$

Integrate with respect to y:

$$\int (4y^3 + 3)\, dy = y^4 + 3y + C;$$

$$[y^4 + 3y]_0^3 = 81 + 9 = 90.$$

Hence,

$$\int_{0}^{3} \int_{1}^{2} (4y^3 + 2x)\ dx\ dy = 90.$$

- Find $\int_{-1}^{2}\int_{0}^{1}\int_{1}^{4} (2x + 4y - z)\ dx\ dy\ dz$.

$$\int_{-1}^{2}\int_{0}^{1}\int_{1}^{4} (2x + 4y - z)\ dx\ dy\ dz = \int_{-1}^{2}\int_{0}^{1} \left[x^2 + 4xy - xz\right]_1^4\ dy\ dz$$

$$= \int_{-1}^{2}\int_{0}^{1} (12y - 3z + 15)\ dy\ dz = \int_{-1}^{2} \left[6y^2 - 3yz + 15y\right]_0^1\ dz$$

$$= \int_{-1}^{2} (21 - 3z)\ dz$$

$$= \left[21z - \frac{3}{2}z^2\right]_{-1}^{2} = 58\frac{1}{2}$$ •

A limit of integration may contain a variable of the integrand.

- Find $\int_{1}^{3}\int_{0}^{2y} 3x^2\ dx\ dy$.

$$\int_{1}^{3}\int_{0}^{2y} 3x^2\ dx\ dy = \int_{1}^{3} \left[x^3\right]_{x=0}^{x=2y}\ dy = \int_{1}^{3} 8y^3\ dy = \Big|_1^3\ 2y^4 = 160\ .$$

Geometrical Interpretation of the Double Integral

We are familiar with the geometrical interpretation of the equation $y = f(x)$ as a curve in the two-dimensional x, y-plane, and of the integral $\int_a^b f(x)\ dx$ as an area between the curve and the x-axis.

Similarly, while the equation $z = f(x, y)$ defines a surface in the three-dimensional x, y, z-space, the **double integral** of a continuous function of two variables, $\int_a^b \int_c^d f(x, y)\ dx\ dy$, may be interpreted as a **volume** between the surface $z = f(x, y)$ and the x, y-plane.

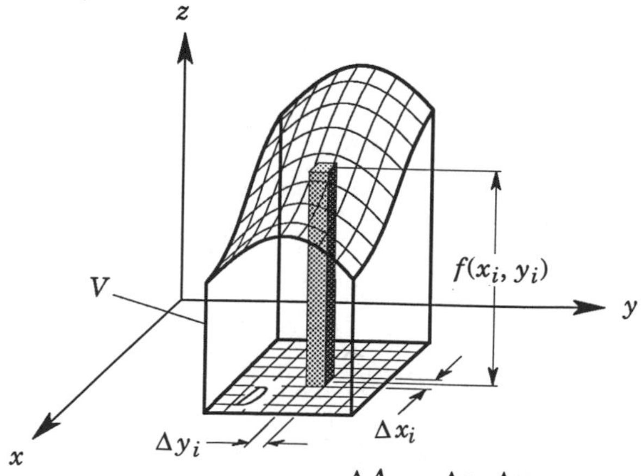

$\Delta A_i = \Delta x_i\ \Delta y_i$

In the graph, the rectangular area ΔA_i in the x, y-plane is projected on the surface $z = f(x, y)$.

The quantity $\Delta A_i = \Delta x_i \; \Delta y_i$ is the area of the bottom surface of a column whose top surface is part of the surface $f(x_i, y_i)$.

The smaller the area ΔA_i, the closer the volume of the column is to that of a parallelepiped measuring $f(x_i, y_i) \, \Delta A_i$.

If the domain D, consisting of (x, y) with $c \le x \le d$, $a \le y \le b$, is divided into an increasingly greater number of rectangles so that ΔA_i tends to 0, then the volume between the surface $z = f(x, y)$ and D equals the sum of all parallelepipeds measuring $f(x_i, y_i) \; \Delta x_i \; \Delta y_i$; thus,

$$V = \lim_{n \to \infty} \sum_{i=1}^{n} f(x_i, y_i) \, \Delta_{x_i} \; \Delta_{y_i} = \int_a^b \int_c^d f(x, y) \; dx \; dy .$$

21.5 Improper or Unrestricted Integrals

Up to this point, we have discussed definite integrals, also referred to as Riemann integrals, which are guaranteed to exist only for continuous functions on closed intervals. However, certain problems create a need for an integrand that is infinite within the range of integration or an integral on an infinite range of integration; such integrals are generally referred to as **improper integrals.**

There is nothing "improper" about so-called improper integrals other than not being covered by the definition of the definite integral, and the term *improper* is rather unfortunate and misleading.

The French *intégrale impropre* is equally infelicitous, less so the German *uneigentliches Integral*, while the Swedish term *generaliserad integral* avoids the negative connotation entirely.

Unrestricted Integral

Since the improper integral has no restriction as to interval or function – other than being unrestricted – we suggest that **unrestricted integral** might be a more suitable term.

Evaluating an unrestricted integral will always involve calculating a definite integral and a limit.

Infinite Integrands

This group of unrestricted (improper) integrals has **integrands that become infinite** in the range of a finite interval of integration.

Using the terminology from infinite series, we say that the **integral converges** if the **limit exists as a finite number**. On the other hand, if the limit does not exist, then the **integral diverges**; a diverging integral may be said not to exist.

We consider the following three typical possibilities for infinite integrands:

1. If f is continuous in $]a, b]$ and has a vertical asymptote at a (that is, f becomes infinite at a), then

$$\int_a^b f(x)\ dx = \lim_{c \to a^+} \int_c^b f(x)\ dx,$$

provided that the limit exists as a finite number.

2. Analogously, if f is continuous in $[a, b[$ and has a vertical asymptote at b (that is, f becomes infinite at b), then

$$\int_a^b f(x)\ dx = \lim_{c \to b^-} \int_a^c f(x)\ dx,$$

provided that the limit exists as a finite number.

3. If f is continuous in $[a, b]$, with the exception of a point c with $a < c < b$, where f has a vertical asymptote (that is, f is infinite at c), then

$$\int_a^b f(x)\,dx = \int_a^c f(x)\,dx + \int_c^b f(x)\,dx$$

provided that $\int_a^c f(x)\,dx$ and $\int_c^b f(x)\,dx$ exist.

In agreement with the use of the term *unrestricted integral*, it is appropriate to say that type **1** is "unrestricted in a", type **2** "unrestricted in b", and type **3** "unrestricted in c".

- Find $\displaystyle\int_0^1 \frac{1}{x^{1/2}}\,dx$.

The integrand becomes infinite at $x = 0$.

We have

$$\int_0^1 \frac{1}{x^{1/2}}\,dx = \lim_{c \to 0^+} \int_c^1 \frac{1}{x^{1/2}}\,dx.$$

Solve in two stages:

1: $\displaystyle\int \frac{1}{x^{1/2}}\,dx = 2x^{1/2} + C$

$= 2\sqrt{x} + C,$

$$\int_c^1 \frac{1}{x^{1/2}}\,dx = \Big|_c^1\, 2\sqrt{x}$$

$= 2 - 2\sqrt{c}$

2: $\displaystyle\lim_{c \to 0^+} \int_c^1 \frac{1}{x^{1/2}}\,dx$

$= \displaystyle\lim_{c \to 0^+} 2 - 2\sqrt{c} = 2 - 0 = 2.$

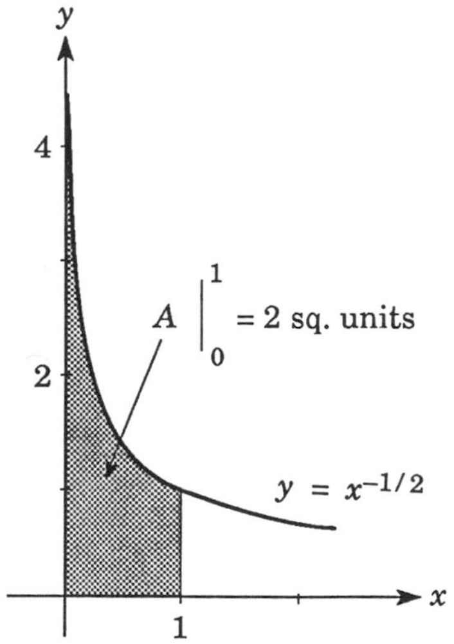

Thus, the integral converges, and

$$\int_0^1 \frac{1}{x^{1/2}}\,dx = 2.$$

The area bounded by the curve of $y = x^{-1/2}$ and the x-axis from $x = 0$ to $x = 1$ is 2 square units.

- Find $\int_0^1 \frac{1}{x^2}\,dx$.

The integrand becomes infinite at $x = 0$.

$$\int_0^1 \frac{1}{x^2}\,dx = \lim_{c \to 0^+} \int_c^1 \frac{1}{x^2}\,dx .$$

Solve in two stages:

1: $\int \frac{1}{x^2}\,dx = -\frac{1}{x} + C,$

$$\int_c^1 \frac{1}{x^2}\,dx = \Big|_c^1 \left(-\frac{1}{x}\right) = \frac{1}{c} - 1 .$$

2: $\lim_{c \to 0^+} \int_c^1 \frac{1}{x^2}\,dx = \lim_{c \to 0^+} \left(\frac{1}{c} - 1\right) = \infty - 1 = \infty .$

Thus, the integral diverges, and the integral $\int_0^1 \frac{1}{x^2}\,dx$ is infinite; and the area bounded by the curve of $y = \frac{1}{x^2}$ and the x-axis from $x = 0$ to $x = 1$ is infinite.

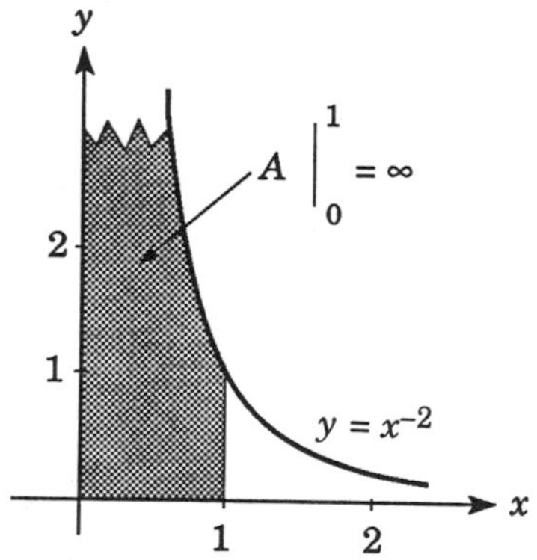

- Find $\int_0^5 \frac{1}{x-1}\,dx$.

This example calls attention to the need for assessing an integrand for asymptotes before evaluating the integral, so that an unrestricted integral is solved in terms of limits and not confused with a definite integral.

The integrand becomes infinite at $x = 1$.

We have

$$\int_0^5 \frac{1}{x-1}\,dx = \int_0^1 \frac{1}{x-1}\,dx + \int_1^5 \frac{1}{x-1}\,dx .$$

Evaluate the components, one at a time, by a two-stage approach.

$$\int_0^1 \frac{1}{x-1}\, dx = \lim_{c \to 1^-} \int_0^c \frac{1}{x-1}\, dx.$$

1: $\int \frac{1}{x-1}\, dx = \ln|x-1| + C;$

$$\int_0^c \frac{1}{x-1}\, dx = \Big|_0^c \ln|x-1| = \ln|c-1|, \text{ for } c < 1.$$

2: $\lim_{c \to 1^-} \int_0^c \frac{1}{x-1}\, dx = \lim_{c \to 1^-} \ln|c-1| = -\infty.$

The nonexistence of a limit at one endpoint is sufficient to manifest the divergence of the integral; as the first component integral diverges, there is no need to evaluate the second, $\int_1^5 \frac{1}{x-1}\, dx.$

Thus, $\int_0^5 \frac{1}{x-1}\, dx$ diverges, and the area bounded by the curve of $y = \frac{1}{x-1}$ and the x-axis from $x = 0$ to $x = 5$ is infinite.

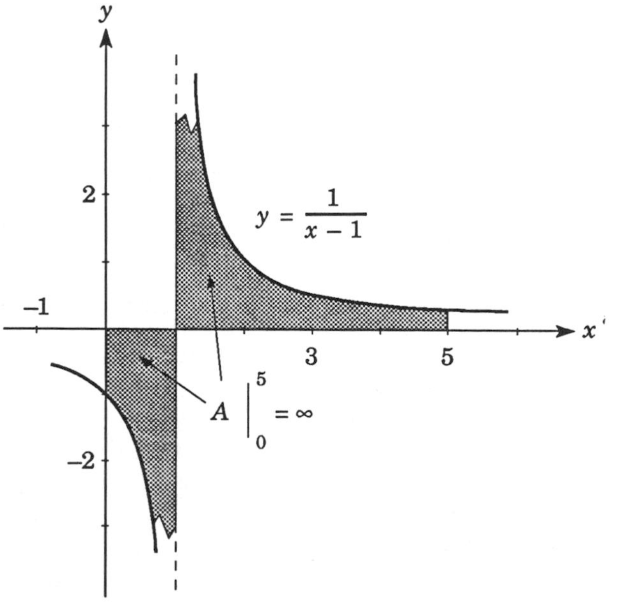

Integrals with Infinite Limits

Assume that f is continuous in the half-open interval $[a, \infty[$; then, for every finite value $b > a$, we have a definite integral

$$F(b) = \int_a^b f(x)\,dx.$$

If b tends to infinity, we have the unrestricted (improper) integral

$$\int_a^\infty f(x)\,dx = \lim_{b \to \infty} \int_a^b f(x)\,dx.$$

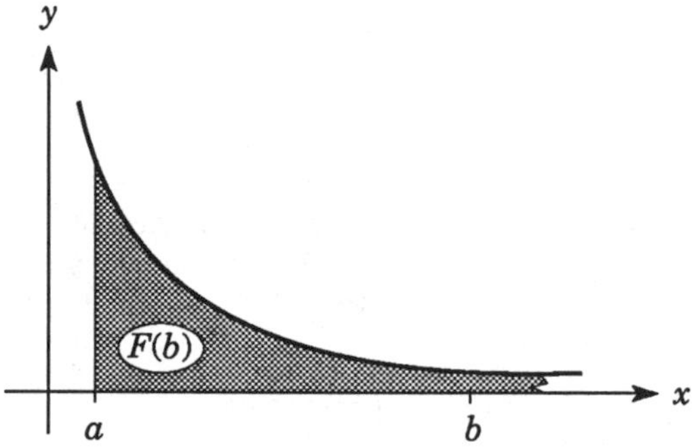

The integral over an infinite interval exists only if the limit exists, and it is then said to **converge**; if the limit does not exist as a finite number, the integral is said to **diverge**.

We have the following possibilities for unrestricted integrals of infinite limits:

1. If f is continuous in $[a, \infty[$, then

$$\int_a^\infty f(x)\,dx = \lim_{b \to \infty} \int_a^b f(x)\,dx,$$

provided that the limit exists as a finite number.

2. If f is continuous in $]-\infty, b]$, then

$$\int_{-\infty}^b f(x)\,dx = \lim_{a \to -\infty} \int_a^b f(x)\,dx,$$

provided that the limit exists as a finite number.

3. If f is continuous in $]-\infty, \infty[$ and c is any real number, then

$$\int_{-\infty}^\infty f(x)\,dx = \int_{-\infty}^c f(x)\,dx + \int_c^\infty f(x)\,dx,$$

provided that both integrals to the right of the equality sign converge, in which case the sum on the right is independent of c.

It is appropriate to say that type **1** is "unrestricted in ∞", type **2** is "unrestricted in $-\infty$", and type **3** is "unrestricted in $-\infty$ and ∞".

- Find $\int_1^{\infty} \frac{1}{x^2}\, dx$.

$$\int_1^{\infty} \frac{1}{x^2}\, dx = \lim_{b \to \infty} \int_1^{b} \frac{1}{x^2}\, dx.$$

1: $\int \frac{1}{x^2}\, dx = -\frac{1}{x} + C;$

$$\int_1^{b} \frac{1}{x^2}\, dx = -\Big|_1^b \frac{1}{x} = 1 - \frac{1}{b}.$$

2: $\lim_{b \to \infty} \int_1^{b} \frac{1}{x^2}\, dx = \lim_{b \to \infty} \left(1 - \frac{1}{b}\right) = 1 - \frac{1}{\infty} = 1.$

The integral converges, and

$$\int_1^{\infty} \frac{1}{x^2}\, dx = 1.$$

The area bounded by the graph of $y = 1/x^2$ and the x-axis from $x = 1$ to $x = \infty$ is 1 square unit.

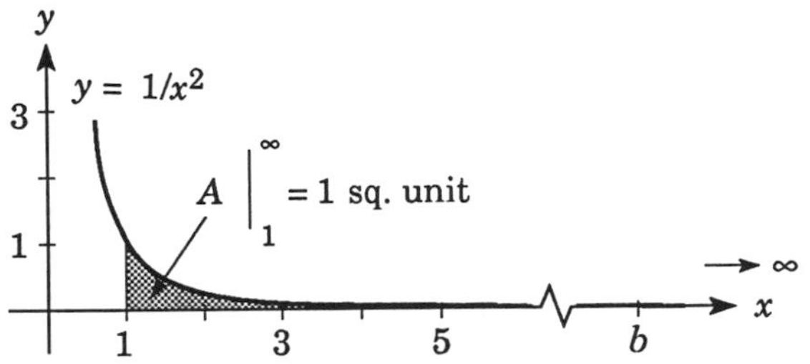

- Find $\int_1^{\infty} \frac{1}{x}\, dx$.

$$\int_1^{\infty} \frac{1}{x}\, dx = \lim_{b \to \infty} \int_1^{b} \frac{1}{x}\, dx.$$

1: $\int \frac{1}{x}\, dx = \ln x + C;$

$$\int_1^{b} \frac{1}{x}\, dx = \Big|_1^b \ln x = \ln b.$$

$$\textbf{2:}\quad \lim_{b\to\infty}\int_1^b \frac{1}{x}\,dx = \lim_{b\to\infty} \ln b = \infty.$$

Thus, the integral diverges, and the area bounded by the curve of $y = \dfrac{1}{x}$ and the x-axis from $x = 1$ to $x = \infty$ is infinite.

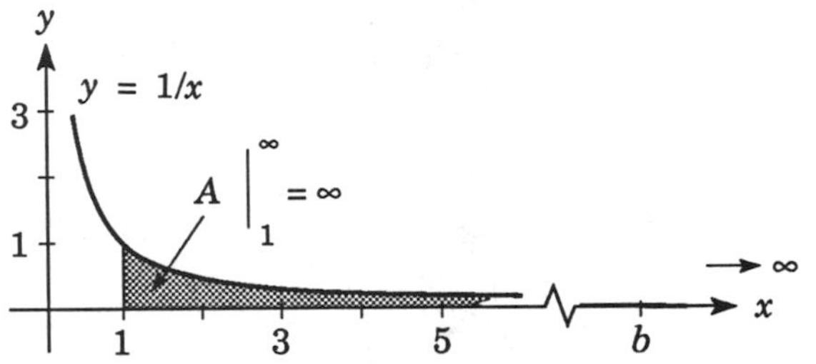

- Find $\displaystyle\int_{-\infty}^{1} \frac{1}{x^2+1}\,dx.$

We have

$$\int_{-\infty}^{1} \frac{1}{x^2+1}\,dx = \lim_{a\to-\infty}\int_a^1 \frac{1}{x^2+1}\,dx.$$

p. 724

$$\textbf{1:}\quad \int \frac{1}{x^2+1}\,dx = \arctan x + C;$$

$$\int_a^1 \frac{1}{x^2+1}\,dx = \Big|_a^1 \arctan x = \left(\frac{\pi}{4}\right) - \arctan a.$$

$$\textbf{2:}\quad \lim_{a\to-\infty}\int_a^1 \frac{1}{x^2+1}\,dx = \frac{\pi}{4} - \lim_{a\to-\infty} \arctan a$$

$$= \frac{\pi}{4} - \left(-\frac{\pi}{2}\right) = \frac{3\pi}{4}.$$

Hence,

$$\int_{-\infty}^{1} \frac{1}{x^2+1}\,dx = \frac{3\pi}{4}.$$

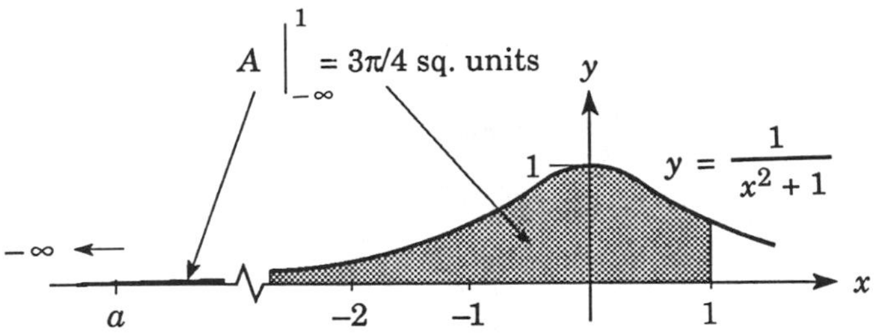

- Find $\int_{-\infty}^{\infty} x\, e^{-x^2}\, dx$.

$$\int_{-\infty}^{\infty} x\, e^{-x^2}\, dx = \int_{-\infty}^{c} x\, e^{-x^2}\, dx + \int_{c}^{\infty} x\, e^{-x^2}\, dx.$$

Choose $c = 0$.

Evaluate the components, one at a time, by the two-stage approach.

$$\int_{-\infty}^{0} x\, e^{-x^2}\, dx = \lim_{a \to -\infty} \int_{a}^{0} x\, e^{-x^2}\, dx\,.$$

p. 734

1: $\int x\, e^{-x^2}\, dx = -\frac{1}{2}\, e^{-x^2} + C\,;$

$$\int_{a}^{0} x\, e^{-x^2}\, dx = -\frac{1}{2}\Big|_{a}^{0} e^{-x^2} = -\frac{1}{2}\left(1 - e^{-a^2}\right).$$

2: $\lim_{a \to -\infty} \int_{a}^{0} x\, e^{-x^2}\, dx = \lim_{a \to -\infty} -\frac{1}{2}\left(1 - e^{-a^2}\right)$

$$= -\frac{1}{2}(1 + 0) = -\frac{1}{2}.$$

Since the first component integral converges, proceed to the next component,

$$\int_{0}^{\infty} x\, e^{-x^2}\, dx = \lim_{b \to \infty} \int_{0}^{b} x\, e^{-x^2}\, dx.$$

1: $\int x\, e^{-x^2}\, dx = -\frac{1}{2}\, e^{-x^2} + C\,;$

$$\int_{0}^{b} x\, e^{-x^2}\, dx = -\frac{1}{2}\Big|_{0}^{b} e^{-x^2} = -\frac{1}{2}\left(e^{-b^2} - 1\right).$$

$$2: \lim_{b \to \infty} \int_0^b x\, e^{-x^2}\, dx = \lim_{b \to \infty} -\frac{1}{2}\left(e^{-b^2} - 1\right)$$

$$= -\frac{1}{2}(0 - 1) = \frac{1}{2}.$$

Thus, both component integrals converge, and

$$\int_{-\infty}^{\infty} x\, e^{-x^2}\, dx = -\frac{1}{2} + \frac{1}{2} = 0.$$

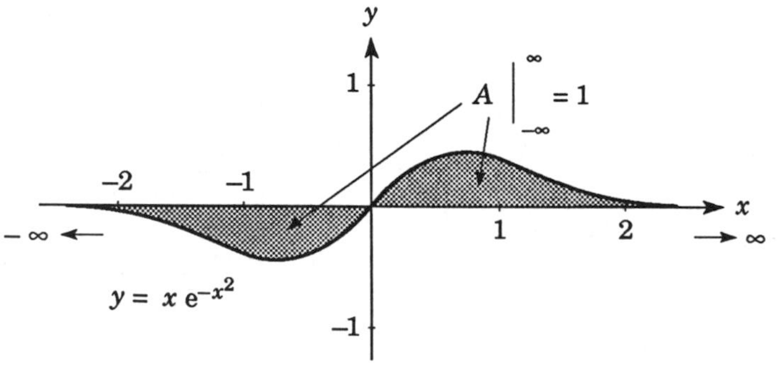

The area bounded by the curve of $y = x\, e^{-x^2}$ and the x-axis is

$$\left|-\frac{1}{2}\right| + \frac{1}{2} = 1 \text{ square unit.}$$

Chapter 22

POWER SERIES

*** Indicates cross-reference

22.1 Convergence

A power series in ascending powers of x may be written

$$\sum_{k=0}^{\infty} a_k x^k = a_0 + a_1 x^1 + a_2 x^2 + a_3 x^3 + \dots,$$

where by convention the 0-th term is taken to be a_0 even if $x = 0$, when x^0 would ordinarily be undefined. More generally, we may seek an expansion of a function $f(x)$ in powers of $(x - c)$,

$$f(x) = \sum_{k=0}^{\infty} a_k (x-c)^k = \lim_{n \to \infty} a_0 + a_1(x-c)^1 + a_2(x-c)^2 + \dots + a_n(x-c)^n;$$

if this limit exists, it defines the sum given. Here c is the **center of convergence**. In the figure below, R denotes the **radius of convergence**; the **interval of convergence** is the interval that consists of all values of x for which the series converges, that is, for which the limit exists:

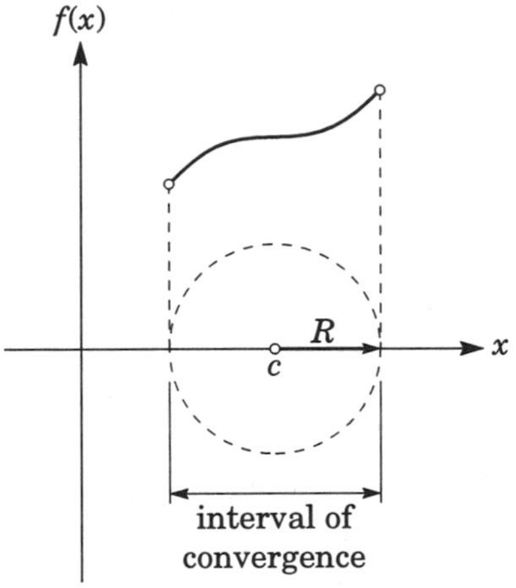

A power series centered at c **converges** for

$$|x - c| < R$$

and **diverges** for

$$|x - c| > R.$$

If the radius of convergence is 0 ($R = 0$), then the series diverges whenever $x \neq c$.

If $R = \infty$, then the series converges for all x.

22.2 Taylor's and Maclaurin's Series

A function f that can be expressed as a **power series centered at c** may be written

$$f(x) = a_0 + a_1(x-c) + a_2(x-c)^2 + a_3(x-c)^3 + \dots .$$

How might the coefficients $a_0, a_1, a_2 \dots$ be determined?

To find a_0, we evaluate $f(c)$, because all terms but the first disappear when $x = c$:

$$f(c) = a_0 + a_1(c-c) + a_2(c-c)^2 + a_3(c-c)^3 + \dots$$

$$a_0 = \frac{f(c)}{0!} = f(c).$$

What about a_1? Assuming that **term-by-term differentiation** is justified, we find the derivative $f'(x)$; as in the previous step, all terms but the first vanish for $x = c$:

$$f'(x) = a_1 + 2a_2(x-c) + 3a_2(x-c)^2 + \dots$$

$$f'(c) = a_1 + 2a_2(c-c) + 3a_2(c-c)^2 + \dots$$

$$f'(c) = a_1 \quad \text{or} \quad a_1 = \frac{f'(c)}{1!} = f'(c).$$

Continuing with f'', we get

$$f''(x) = 2a_2 + 6a_3(x-c) + \dots ,$$

which gives

$$f''(c) = 2a_2 \quad \text{or} \quad a_2 = \frac{f''(c)}{2!}.$$

Continued term-by-term differentiation followed by substitution of x with c, and solving for a_n, gives

$$a_n = \frac{f^{(k)}(c)}{k!},$$

where $k! = 1 \cdot 2 \cdot 3 \cdot \dots \cdot (k-1)\,k$.

Thus, if $f(x)$ is represented by a power series

$$f(x) = a_0 + a_1(x-c) + a_2(x-c)^2 + a_3(x-c)^3 + \dots$$

where c denotes the center of convergence, and term-by-term differentiation is justified, then

$$f(x) = \sum_{k=0}^{\infty} \frac{f^{(k)}(c)}{k!}(x-c)^k$$

$$= f(c) + f'(c)(x-c) + f''(c)\frac{(x-c)^2}{2!} + f'''(c)\frac{(x-c)^3}{3!} + \dots$$

in the interval $(-R+c,\ R+c)$, where R denotes the radius of convergence.

TAYLOR Brook
(1685 - 1731)

GREGORIE James
(1638 - 1675)

This power series is **Taylor's series**, named after the English mathematician **Brook Taylor** and published in his *Methodus incrementorum directa et inversa* (1715).

Although first published by Taylor, this series had been known some 40 years earlier, as evidenced by unpublished papers of the Scottish mathematician and astronomer **James Gregorie**.

METHODUS

Incrementorum

Directa & Inverſa.

AUCTORE

BROOK TAYLOR, LL.D. &

Regiæ Societatis Secretario.

LONDINI

Typis *Pearſonianis*: Proſtant apud *Gul. Innys* ad Inſignia Principis in Cœmeterio *Paulino*. MDCCXV.

(22)

DEMONSTRATIO.

x	$\underset{.}{x}$	$\underset{..}{x}$	$\underset{\vdots}{x}$	$\underset{::}{x}$	&c.
$x+\underset{.}{x}$	$\underset{.}{x}+\underset{..}{x}$	$\underset{..}{x}+\underset{\vdots}{x}$	$\underset{\vdots}{x}+\underset{::}{x}$	&c.	
$x+2\underset{.}{x}+\underset{..}{x}$	$\underset{.}{x}+2\underset{..}{x}+\underset{\vdots}{x}$	$\underset{..}{x}+2\underset{\vdots}{x}+\underset{::}{x}$	&c.		
$x+3\underset{.}{x}+3\underset{..}{x}+\underset{\vdots}{x}$	$\underset{.}{x}+3\underset{..}{x}+3\underset{\vdots}{x}+\underset{::}{x}$	&c.			
$x+4\underset{.}{x}+6\underset{..}{x}+4\underset{\vdots}{x}+\underset{::}{x}$	&c.				
&c.					

Valores ſucceſſivi ipſius x per additionem continuam collecti ſunt x, $x+\underset{.}{x}$, $x+2\underset{.}{x}+\underset{..}{x}$, $x+3\underset{.}{x}+3\underset{..}{x}+\underset{\vdots}{x}$, &c. ut patet per operationem in tabula annexa expreſſam. Sed in his valoribus x coefficientes numerales terminorum x, $\underset{.}{x}$, $\underset{..}{x}$, &c. eodem modo formantur, ac coefficientes terminorum correſpondentium in dignitate binomii.

Et (per Theorema *Newtonianum*) si dignitatis index sit n, coefficientes erunt $1, \frac{n}{1}, \frac{n}{1} \times \frac{n-1}{2}, \frac{n}{1} \times \frac{n-1}{2} \times \frac{n-2}{3}$, &c. Ergò quo tempore z crescendo fit $z + nz$, hoc est $z + v$, fiet x æqualis seriei $x + \frac{n}{1} \dot{x} + \frac{n}{1} \times \frac{n-1}{2} \ddot{x} + \frac{n}{1} \times \frac{n-1}{2} \times \frac{n-2}{3} \dddot{x} +$ &c.

Sed sunt $\frac{n}{1} = \left(\frac{nz}{\dot{z}} = \right) \frac{v}{\dot{z}}$, $\frac{n-1}{2} = \left(\frac{nz - z}{2z} = \right) \frac{v}{2z}$, $\frac{n-2}{3} =$ (nz

(23)

$\left(\frac{nz - 2z}{3z} = \right) \frac{\ddot{v}}{\dot{z}}$, &c. Proinde quo tempore z crescendo fit $z + v$, eodem tempore x crescendo fiet $x + \dot{x} \frac{v}{1\dot{z}} + \ddot{x} \frac{v\dot{v}}{1.2\dot{z}^2} + \dddot{x} \frac{v\dot{v}\ddot{v}}{1.2.3\dot{z}^3} +$ + &c.

C O R O L L. I.

Et ipsis z, $\dot{x}$, $\ddot{x}$, $\dddot{x}$, &c. iisdem manentibus, mutato signo ipsius v, quo tempore z decrescendo fit $z - v$, eodem tempore x decrescendo fiet $x - \dot{x} \frac{v}{1\dot{z}} + \ddot{x} \frac{v\dot{v}}{1.2\dot{z}^2} - \dddot{x} \frac{v\dot{v}\ddot{v}}{1.2.3\dot{z}^3}$ &c. vel juxta notationem nostram $x - \dot{x} \frac{v}{1\dot{z}} + \ddot{x} \frac{v v_{\cdot}}{1.2\dot{z}^2} - \dddot{x} \frac{v v_{\cdot} v_{\cdot\cdot}}{1.2.3\dot{z}^3}$ &c. ipsis $\dot{v}, \ddot{v}$, &c. conversis in $-v_{\cdot}$, $-v_{\cdot\cdot}$, &c.

C O R O L L. II.

Si pro Incrementis evanescentibus scribantur fluxiones ipsis proportionales, factis jam omnibus $\dddot{v}, \ddot{v}, v, v_{\cdot}, v_{\cdot\cdot}$, &c. æqualibus quo tempore z uniformiter fluendo fit $z + v$ fiet x, $x + \dot{x} \frac{v}{1\dot{z}} + \ddot{x} \frac{v^2}{1.2\dot{z}^2} + \dddot{x} \frac{v^3}{1.2.3\dot{z}^3}$ &c. vel mutato signo ipsius v, quo tempore z decrescendo fit $z - v$, x decrescendo fiet $x - \dot{x} \frac{v}{1\dot{z}} + \ddot{x} \frac{v^2}{1.2\dot{z}^2} - \dddot{x} \frac{v^3}{1.2.3\dot{z}^3} +$ &c.

G P R O P.

"Taylor's series" as described (this page and opposite page) in *Methodus incrementorum* ... (1715). Taylor used a system of dots (for increments) and superscript/subscript primes, to which the modern reader is usually unaccustomed.

Maclaurin's Series

If **centered at 0** (that is, $c = 0$), the Taylor series is known as a **Maclaurin series**,

$$f(x) = \sum_{k=0}^{\infty} \frac{f^{(k)}(0)}{k!} x^k$$

$$= f(0) + f'(0)x + \frac{f''(0)}{2!} x^2 + \frac{f'''(0)}{3!} x^3 + \ldots.$$

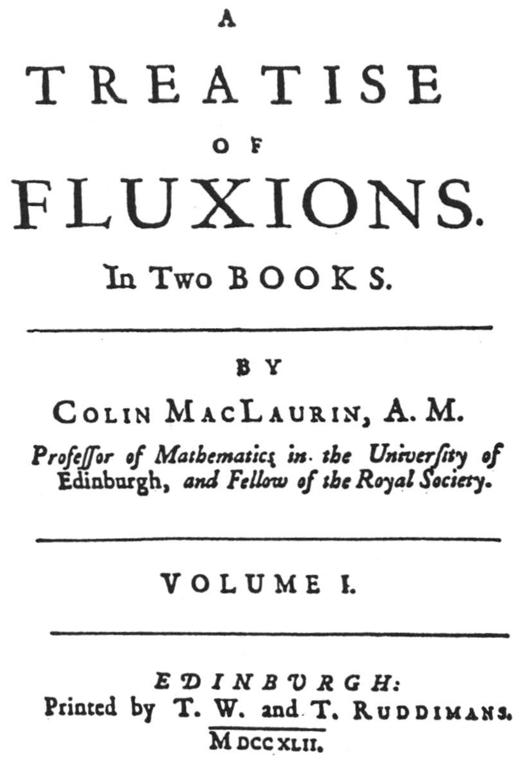

A

TREATISE

OF

FLUXIONS.

In Two BOOKS.

BY

COLIN MACLAURIN, A. M.

Professor of Mathematics in the University of Edinburgh, and Fellow of the Royal Society.

VOLUME I.

EDINBURGH:

Printed by T. W. and T. RUDDIMANS.

MDCCXLII.

MACLAURIN Colin
(1698 - 1746)

STIRLING James
(1692 - 1770)

In his *Treatise of Fluxions* (1742), **Colin Maclaurin** described Taylor's series in a manner still respected in today's textbooks. Maclaurin made no pretense of having discovered the series that now bears his name – on the contrary, he referred to it as a special case of Taylor's series. And even before Maclaurin's treatise, Maclaurin's series had been used by the Scottish mathematician **James Stirling**.

Taylor's Formula with Remainder and Error Estimate

Taylor's formula with remainder approximates functions by polynomials and provides estimates for errors.

If a function $f(x)$ has term-by-term derivatives $f^{(k)}(x)$ in the closed interval $[-R+c,\ R+c]$, where R denotes the radius of convergence, **Taylor's formula with remainder** states that

$$f(x) = f(c) + f'(c)(x-c) + f''(c)\frac{(x-c)^2}{2!}$$

$$+ f'''(c)\frac{(x-c)^3}{3!} \dots + f^{(k-1)}(c)\frac{(x-c)^{k-1}}{(k-1)!} + \textbf{remainder}.$$

The remainder term may be determined, but we leave the discussion of the various methods out of our discussion here.

22.3 Expanding Transcendental Functions

Except for special triangles, we find numerical values of trigonometric functions by trigonometric tables or a calculator. Similarly, tables and the calculator are used to find values of natural logarithms. Power series expansion helps with the calculation of such functions. The values are approximate, but so are those given in tables and by the calculator.

The first term of a converging series is the first approximation, the sum of the first and second terms the next approximation; by adding an increasing number of terms, the degree of accuracy is limited only by our finite capacity of adding terms.

To form polynomials from the terms of a convergent series, Taylor's formula with remainder is the concluding – but not easily applied – tool for approximation.

Some series lend themselves to convenient error estimation. Thus, if for a particular x the terms of the series alternate in sign and decrease in absolute value, the absolute value error introduced when approximating the function at x by the sum of terms from the beginning of the series cannot exceed the absolute value of the first term discarded.

Expanding sine of x

$$\sin x = x - \frac{x^3}{3\,!} + \frac{x^5}{5\,!} - \frac{x^7}{7\,!} + \dots,$$

where x is in radians.

Why? Assuming that $\sin x$ can be expressed as a Maclaurin series, we determine the coefficients of the series:

Let $f(x) = \sin x$, where x is in radians.

$$\sin 0 = 0\,; \quad \cos 0 = 1$$

$$\begin{array}{lllll|lllll}
f(x) & = \sin x & \Rightarrow & f(0) & = 0 & f^{(4)}(x) & = \sin x & \Rightarrow & f^{(4)}(0) & = 0 \\
f'(x) & = \cos x & \Rightarrow & f'(0) & = 1 & f^{(5)}(x) & = \cos x & \Rightarrow & f^{(5)}(0) & = 1 \\
f''(x) & = -\sin x & \Rightarrow & f''(0) & = 0 & f^{(6)}(x) & = -\sin x & \Rightarrow & f^{(6)}(0) & = 0 \\
f'''(x) & = -\cos x & \Rightarrow & f'''(0) & = -1 & f^{(7)}(x) & = -\cos x & \Rightarrow & f^{(7)}(0) & = -1 \\
 & & & & & etc. & & & &
\end{array}$$

Using these results in Maclaurin's series

$$f(x) = f(0) + f'(0)\,x + \frac{f''(0)}{2\,!}\,x^2 + \frac{f'''(0)}{3\,!}\,x^3 + \dots,$$

we obtain

$$f(x) = \sin x = 0 + x + 0 + \left(-\frac{x^3}{3\,!}\right) + 0 + \frac{x^5}{5\,!} + 0 + \left(-\frac{x^7}{7\,!}\right) + \dots$$

$$= x - \frac{x^3}{3\,!} + \frac{x^5}{5\,!} - \frac{x^7}{7\,!} + \dots.$$

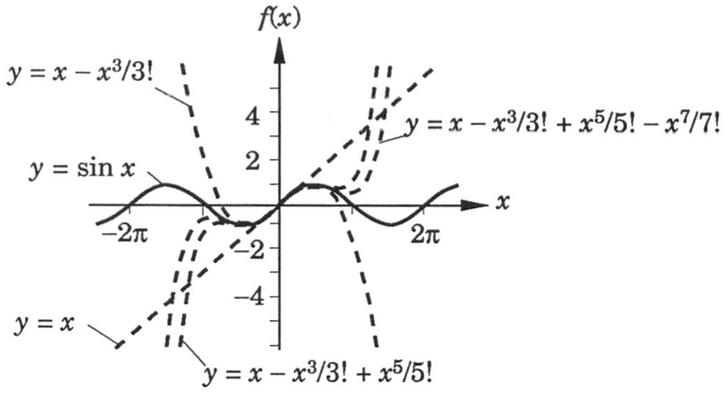

The graph shows improved agreement between the graph of $\sin x$ and graphs of polynomials containing an increasing number of terms obtained by power series expansion of the sine function.

Since the terms of the sine series are of alternating sign and steadily decreasing magnitude for $|x| \le 1$, the error introduced by breaking off the power series at any term will not exceed in magnitude the value of the first term left behind.

- Employ the first eight terms of the Maclaurin series expansion to express the value of

$$\sin 0.2 \text{ rad} .$$

Since

$$\sin x \approx x - \frac{x^3}{3!} + \frac{x^5}{5!} - \frac{x^7}{7!} = x - \frac{x^3}{6} + \frac{x^5}{120} - \frac{x^7}{5040},$$

we obtain

$$\sin 0.2 \text{ rad} \approx 0.2 - \frac{0.2^3}{6} + \frac{0.2^5}{120} - \frac{0.2^7}{5040} .$$

- Find the first four terms of the Maclaurin series of

$$y = \sin(2x + \pi) .$$

$$\begin{aligned}
f(x) &= \sin(2x+\pi) &\Rightarrow\quad f(0) &= \sin\pi = 0\\
f'(x) &= 2\cos(2x+\pi) &\Rightarrow\quad f'(0) &= 2\cos\pi = -2\\
f''(x) &= -4\sin(2x+\pi) &\Rightarrow\quad f''(0) &= -4\sin\pi = 0\\
f'''(x) &= -8\cos(2x+\pi) &\Rightarrow\quad f'''(0) &= -8\cos\pi = 8
\end{aligned}$$

$$f(x) = f(0) + f'(0)\,x + \frac{f''(0)}{2!}\,x^2 + \frac{f'''(0)}{3!}\,x^3 + \ldots$$

$$= 0 + (-2x) + 0 + \frac{8}{3!}x^3 \;\ldots\; \approx \frac{4x^3}{3} - 2x .$$

Expanding cosine of x

$$\cos x \approx 1 - \frac{x^2}{2\,!} + \frac{x^4}{4\,!} - \frac{x^6}{6\,!} + \dots,$$

where x is in radians.

Why? Assuming that $\cos x$ can be expressed as a Maclaurin series, we determine the coefficients of the series:

$\sin 0 = 0$ and $\cos 0 = 1$.

$$\begin{array}{lcllcl}
f(x) &=& \cos x & \Rightarrow & f(0) &= 1 \\
f'(x) &=& -\sin x & \Rightarrow & f'(0) &= 0 \\
f''(x) &=& -\cos x & \Rightarrow & f''(0) &= -1 \\
f'''(x) &=& \sin x & \Rightarrow & f'''(0) &= 0 \\
f^{(4)}(x) &=& \cos x & \Rightarrow & f^{(4)}(0) &= 1 \\
f^{(5)}(x) &=& -\sin x & \Rightarrow & f^{(5)}(0) &= 0 \\
f^{(6)}(x) &=& -\cos x & \Rightarrow & f^{(6)}(0) &= -1
\end{array}$$

etc.

Using these results in Maclaurin's series

$$f(x) = f(0) + f'(0)\,x + \frac{f''(0)}{2\,!}\,x^2 + \frac{f'''(0)}{3\,!}\,x^3 + \dots,$$

we obtain

$$f(x) = \cos x = 1 + 0 - \frac{x^2}{2\,!} + 0 + \frac{x^4}{4\,!} + 0 - \frac{x^6}{6\,!} + \dots$$

$$= 1 - \frac{x^2}{2\,!} + \frac{x^4}{4\,!} - \frac{x^6}{6\,!} + \dots .$$

Expanding Natural Logarithms

$$\ln x = (x-1) - \frac{(x-1)^2}{2} + \frac{(x-1)^3}{3} - \frac{(x-1)^4}{4} +$$

$$\dots + (-1)^k\,\frac{(x-1)^{k-1}}{k-1} + (-1)^{k-1}\,\frac{(x-1)^k}{k} + \dots$$

where x is a real number with $|x - 1| < 1$.

Why? The function $\ln x$ is not defined for $x = 0$, so any Taylor series expansion will have to be centered at some point $c > 0$. The choice $c = 1$ is conducive to convenient differentiation:

Employ Taylor's series; let $c = 1$:

$$f(x) = \ln x = f(1) + f'(1)\,(x-1) + \frac{f''(1)}{2\,!}\,(x-1)^2$$

$$+ \frac{f'''(1)}{3\,!}\,(x-1)^3 + \dots .$$

$$\begin{aligned}
f(x) &= \ln x & \Rightarrow \quad & f(1) = \ln 1 = 0 \quad (e^0 = 1) \\
f'(x) &= \frac{1}{x} & \Rightarrow \quad & f'(1) = 1 \\
f''(x) &= -x^{-2} & \Rightarrow \quad & f''(1) = -1 \\
f'''(x) &= 2x^{-3} & \Rightarrow \quad & f'''(1) = 2 \\
f^{(4)}(x) &= -6x^{-4} & \Rightarrow \quad & f^{(4)}(1) = -6
\end{aligned}$$

etc.

Using the above coefficients,

$$f(x) = \ln x = 0 + (1)(x-1) + \frac{(-1)(x-1)^2}{2!} + \frac{2(x-1)^3}{3!} + \frac{(-6)(x-1)^4}{4!} + \ldots$$

$$= (x-1) - \frac{(x-1)^2}{2} + \frac{(x-1)^3}{3} - \frac{(x-1)^4}{4} + \ldots,$$

which is valid for $0 < x \leq 2$.

- Employ the first four terms of a power series expansion of $\ln x$ to determine the numerical value of $\ln 1.3$.

$$\ln x \approx (x-1) - \frac{(x-1)^2}{2} + \frac{(x-1)^3}{3} - \frac{(x-1)^4}{4}.$$

Discarding the last term of the polynomial, we obtain

$$\ln 1.3 \approx (1.3-1) - \frac{(1.3-1)^2}{2} + \frac{(1.3-1)^3}{3} = 0.264.$$

The terms of the series alternate in sign steadily and decrease in absolute value, so the limit lies trapped between any two successive sums as follows:

$$0.264 < \ln 1.3 < 0.264 + \frac{(1.3-1)^4}{4} = 0.264 + 0.002\,025.$$

So, to two decimal places

$$\ln 1.3 = 0.26\ldots.$$

Expanding e^x

The base of the natural system of logarithms is defined

$$e = \lim_{n \to \infty} \left(1 + \frac{1}{n}\right)^n;$$

it can also be expressed as the sum of an infinite series,

$$e = \lim_{n \to \infty} \left(1 + \frac{1}{1!} + \frac{1}{2!} + \frac{1}{3!} + \ldots + \frac{1}{n!}\right).$$

To expand e^x, employ Maclaurin's series,

$$f(x) = f(0) + \frac{f'(0)\,x}{1!} + \frac{f''(0)}{2!}x^2 + \frac{f'''(0)}{3!}x^3 + \ldots +,$$

and let $f(x) = e^x$.

p. 711 As every derivative of e^x is e^x, and $e^0 = 1$,

$$f(0),\ f\,'(0),\ f\,''(0),\ f\,'''(0)\ \ldots \text{ must all equal } 1;$$

and assuming that e^x can be expressed as a Maclaurin series, we have

$$e^x = \lim_{n \to \infty} \left(1 + \frac{x}{1\,!} + \frac{x^2}{2\,!} + \frac{x^3}{3\,!} + \ldots + \frac{x^n}{n\,!}\right).$$

Alternatively, the series expansion for e^x can be approached using the binomial theorem and $e^x = \lim_{n \to \infty} \left(1 + \frac{x}{n}\right)^n$.

22.4 Binomial Expansion

p. 140 The **binomial theorem**, previously discussed in Chapter 4, says that

$$(x+y)^p = x^p + p\,x^{p-1}y + \frac{p\,(p-1)}{1 \cdot 2} \cdot x^{p-2}\,y^2 + \frac{p\,(p-1)\,(p-2)}{1 \cdot 2 \cdot 3} \cdot x^{p-3}\,y^3 +$$

$$\ldots + \frac{p\,(p-1)\,(p-2) \ldots (p-n+1)}{1 \cdot 2 \cdot 3 \ldots n} \cdot x^{p-n}\,y^n + p\,x\,y^{p-1} + \ldots + y^p,$$

where p is a positive integer, and $n = 0, 1, 2 \ldots p$ may be proved by direct combinatorical arguments.

The general term, the $(n+1)$-th term, is

$$\frac{p\,(p-1)\,(p-2) \ldots (p-n+1)}{n\,!} \cdot x^{p-n}\,y^n,$$

pp. 196 - 97 or

$$\frac{p\,!}{n\,!\,(p-n)\,!} \cdot x^{p-n}\,y^n,$$

or, in notation of the **binomial coefficient**, $\binom{n}{p} \cdot x^{p-n}\,y^n$.

To find the coefficients of the binomial theorem, we may also expand $f(x) = (a+x)^p$ into a **power series**, centered at 0:

$$f(x) = (a+x)^p = f(0) + f\,'(0)\,x + \frac{f\,''(0)}{2\,!}\,x^2 + \frac{f\,'''(0)}{3\,!}\,x^3 + \ldots$$

Determine the coefficients:

$$\begin{array}{lll} f(a) = (a+x)^p & \Rightarrow f(0) = a^p & \\ f\,'(a) = p\,(a+x)^{p-1} & \Rightarrow f\,'(0) = p\,a^{p-1} & \\ f\,''(a) = p\,(p-1)\,(a+x)^{p-2} & \Rightarrow f\,''(0) = p\,(p-1)\,a^{p-2} & \\ \ldots & \ldots & \\ f^{(n)}(a) = p\,(p-1) \ldots (p-n+1)\,(a+x)^{p-n} & \Rightarrow f^{(n)}(0) = p\,(p-1) \ldots (p-n+1)\,(a)^{p-n} & \\ \ldots & \ldots & \\ f^{(p)}(a) = p\,(p-1) \ldots (2)\,(1)\,(a+x)^{p-p} & \Rightarrow f^{(p)}(0) = p\,! & [(a+x)^0 = 1] \\ f^{(n)}(a) = 0;\ n \geq p+1 & & \end{array}$$

For a **positive integer** p, the series is thus reduced to a finite polynomial – and we again have the **binomial theorem**:

$$(x+y)^p = x^p + p\,x^{p-1}y + \frac{p\,(p-1)}{1 \cdot 2} \cdot x^{p-2}\,y^2 + \frac{p\,(p-1)\,(p-2)}{1 \cdot 2 \cdot 3} x^{p-3}\,y^3 +$$

$$\ldots + \frac{p\,(p-1)\,(p-2) \ldots (p-n+1)}{1 \cdot 2 \cdot 3 \ldots n} \cdot x^{p-n}\,y^n + \ldots + p\,x^{p-1} + y^p$$

NEWTON Isaac
(1643 - 1727)

Binomial expansion had been known only for natural number exponents p – the binomial theorem – when **Newton** discovered, in 1664 or 1665, that $(a + x)^p$ expands as an *infinite* power series in x if p is a real number but *not a natural number or zero*. If p is a natural number, the series reduces to the *finite* polynomial of the binomial theorem.

As so often happened in Newton's career, the disclosure came several years after the accomplishment. Only in 1676 did Newton communicate his discovery in two letters directed, via the secretary of the Royal Society, to Leibniz.

The **power series** of

$$f(x) = (1 + x)^p = \sum_{n=0}^{\infty} \binom{p}{n} x^n,$$

where

$$\binom{p}{n} = \frac{p\,(p-1)\,(p-2)\,\ldots\,(p-n+1)}{n\,!}$$

and p is any real number except a positive integer or 0,

converges to $f(x)$ for all real numbers x, such that $-1 < x < 1$ (that is, the radius of convergence = 1).

The proof of the convergence of Newton's binomial series to $(1 + x)^p$ was presented in 1826 – 150 years after Newton's disclosure – by the Norwegian mathematician **Niels Abel**, who investigated the convergence for all real and complex values of the exponent.

ABEL Niels Henrik
(1802 - 1829)

Finding Roots

By expanding a radical expression into a binomial series, we may determine the numerical value through direct calculation; adding an increasing number of terms, the degree of accuracy is limited only by our finite capacity of adding terms.

- To determine a numerical value of

$$\sqrt{1.3}\,,$$

use a seven-term Maclaurin expansion of $\sqrt{a + x}$.

With $f(x) = (a + x)^{1/2}$, expand the binomial into an infinite power series. To do so, first determine the coefficients:

$$f(x) = (a+x)^{1/2} \quad\Rightarrow\quad f(0) = \sqrt{a}$$

$$f'(x) = \frac{1}{2}\,(a+x)^{-1/2} \quad\Rightarrow\quad f'(0) = \frac{1}{2\sqrt{a}}$$

$$f''(x) = -\frac{1}{4}\,(a+x)^{-3/2} \quad\Rightarrow\quad f''(0) = -\frac{1}{4\,a\sqrt{a}}$$

$$f'''(x) = \frac{3}{8}(a+x)^{-5/2} \quad \Rightarrow \quad f'''(0) = \frac{3}{8a^2\sqrt{a}}$$

$$f^{(4)}(x) = -\frac{15}{16}(a+x)^{-7/2} \quad \Rightarrow \quad f^{(4)}(0) = -\frac{15}{16a^3\sqrt{a}}$$

$$f^{(5)}(x) = \frac{105}{32}(a+x)^{-9/2} \quad \Rightarrow \quad f^{(5)}(0) = \frac{105}{32a^4\sqrt{a}}$$

$$f^{(6)}(x) = -\frac{945}{64}(a+x)^{-11/2} \quad \Rightarrow \quad f^{(6)}(0) = -\frac{945}{64a^5\sqrt{a}}$$

$$f(x) = \sqrt{a} + \frac{x}{2\sqrt{a}} - \frac{1}{4a\sqrt{a}}\left(\frac{x^2}{2!}\right) + \frac{3}{8a^2\sqrt{a}}\left(\frac{x^3}{3!}\right)$$
$$- \frac{15}{16a^3\sqrt{a}}\left(\frac{x^4}{4!}\right) + \frac{105}{32a^4\sqrt{a}}\left(\frac{x^5}{5!}\right) - \frac{945}{64a^5\sqrt{a}}\left(\frac{x^6}{6!}\right) + \ldots$$
$$= \sqrt{a} + \frac{x}{2\sqrt{a}} - \frac{x^2}{8a\sqrt{a}} + \frac{x^3}{16a^2\sqrt{a}} - \frac{5x^4}{128a^3\sqrt{a}} + \frac{7x^5}{256a^4\sqrt{a}} - \frac{21x^6}{1024a^5\sqrt{a}} + \ldots.$$

With $a = 1$ and $x = 0.3$, add the first six terms:

$$\sqrt{1.3} = \sqrt{1+0.3} \approx 1 + \frac{0.3}{2} - \frac{0.3^2}{8} + \frac{0.3^3}{16} - \frac{5 \cdot 0.3^4}{128} + \frac{7 \cdot 0.3^5}{256}$$
$$\approx 1.140\,187\,5$$

The value of the seventh term is

$$\frac{21\,(0.3^6)}{1024} = 0.000\,014\ldots$$

Hence,

$$\sqrt{1.3} = 1.1401\ldots.$$

- Determine the numerical value of

$$\sqrt{6}$$

to three reliable decimals.

Use the series expansion of $(a+x)^{1/2}$ as in the preceding example.

Letting $a = 4$ and $x = 2$, we add the first six terms of the series:

$$\sqrt{4+2} = (4+2)^{1/2} \approx 2 + \frac{2}{2 \cdot 2} - \frac{2^2}{8 \cdot 4 \cdot 2} + \frac{2^3}{16 \cdot 4^2 \cdot 2}$$
$$- \frac{5 \cdot 2^4}{128 \cdot 4^3 \cdot 2} + \frac{7 \cdot 2^5}{256 \cdot 4^4 \cdot 2} \approx 2.449\,95$$

The seventh term is

$$-\frac{21\,(2^6)}{1024\,(4^5)\,2} = -0.000\,64\ldots$$

Hence,

$$\sqrt{6} \approx 2.449\,.$$

22.5 The Riemann Zeta Function and Hypothesis

EULER Leonhard
(1707 - 1783)

Euler introduced, in 1740, the zeta function as an infinite series

$$\zeta(s) = 1 + \frac{1}{2^s} + \frac{1}{3^s} + \frac{1}{4^s} + \ldots = \sum_{n=1}^{\infty} \frac{1}{n^s}$$

convergent for all real numbers $s > 1$.

Euler Product

Euler showed that $\zeta(s)$ can also be expressed by a convergent infinite product, now known as the **Euler product**,

$$\zeta(s) = \frac{1}{1-\frac{1}{2^s}} \cdot \frac{1}{1-\frac{1}{3^s}} \cdot \frac{1}{1-\frac{1}{5^s}} \cdot \frac{1}{1-\frac{1}{7^s}} \cdot \frac{1}{1-\frac{1}{11^s}} \cdot \ldots = \prod_{n=1}^{\infty} \left(1 - \frac{1}{p_n{}^s}\right)^{-1},$$

where all values p_n run over *all prime numbers*. We have

$$\zeta(s) = \sum_{n=1}^{\infty} \frac{1}{n^s} = \prod_{n=1}^{\infty} \left(1 - \frac{1}{p_n{}^s}\right)^{-1}.$$

RIEMANN
Georg Friedrich Bernhard
(1826 - 1866)

The German mathematician **Bernhard Riemann**, a pioneer of modern mathematics, treated, in 1859, ζ as a function of a *complex* variable z; for this reason, $\zeta(z)$ is known as the **Riemann zeta function**.

Re(*z*): *p.* 87

The function $\zeta(z)$ has no zeros in $\text{Re}(z) \geq 1$; its only zeros in $\text{Re}(z) \leq 0$ are at $z = -2, -4, -6, \ldots$; it has infinitely many zeros in $0 < \text{Re}(z) < 1$, which are called **nontrivial zeros**.

Riemann conjectured, in 1859, that *all* the nontrivial zeros of $\zeta(z)$ lie on the line $\text{Re}(z) = \frac{1}{2}$, known as the **Riemann hypothesis**.

p. 80

A crucial part of **Hadamard's** and **de la Vallée-Poussin's** proofs (1896) of the **prime number theorem** was to show that $\zeta(z) \neq 0$ for $\text{Re}(z) = 1$.

HARDY Godfrey Harold
(1877 - 1947)

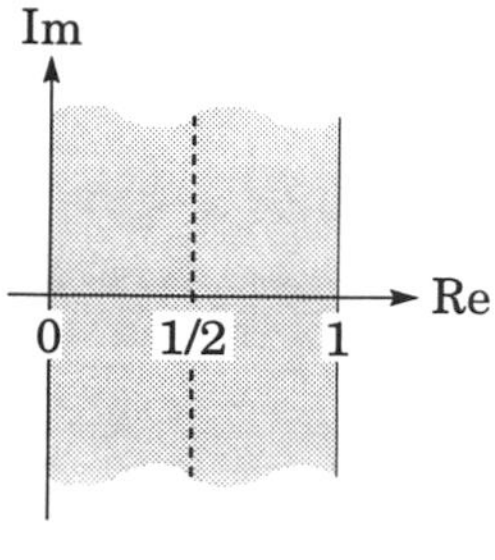

In 1914, the eminent English number theorist **Godfrey Hardy** proved that an infinity of zeros of $\zeta(z)$ lie on $\text{Re}(z) = \frac{1}{2}$, but beyond that, and the fact that it has been shown that the first $1.5 \cdot 10^9$ zeros in $0 < \text{Re}(z) < 1$ are all nontrivial zeros on $\text{Re}(z) = \frac{1}{2}$, the Riemann hypothesis has been neither proved nor disproved.

Since Fermat's last theorem has now been proved, finding a proof of the Riemann hypothesis might be the most sought-after accomplishment in number theory.

Chapter 23

INDETERMINATE LIMITS

Page

PRINCIPIORUM

CALCULI DIFFERENTIALIS

ET

INTEGRALIS

EXPOSITIO ELEMENTARIS

AD NORMAM DISSERTATIONIS AB ACADEMIA SCIENT. REG. PRUSSICA ANNO 1786. PRÆMII HONORE DECORATÆ ELABORATA

AUCTORE

SIMONE L'HUILIER

ACADEMIÆ SCIENT. REGIÆ PRUSSICÆ ET SOCIETATIS REGIÆ LONDINENSIS SOCIO ACADEMIÆ IMPERIALIS PETROPOLITANÆ CORRESPONDENTE.

L'Infini est le gouffre où se perdent nos pensees.

BAILLY Hist. de l'Astron. mod.

TUBINGÆ

APUD JOH. GEORG. COTTAM.

1795

"Infinity is the abyss where our thoughts get lost."

BAILLY Jean Sylvain

(1736 - 1793)

Astronomer, politician. Bailly became mayor of Paris after the destruction of the Bastille in 1789. A royalist, Bailly was guillotined in 1793.

L'HUILIER Simone

(1750 - 1840)

The Swiss mathematician **Simone L'Huilier** made essential contributions to the theory of limit values. The notation "lim" was first used in print by L'Huilier.

23.1 A Retrospect

Limit values have previously been discussed primarily in Chapters 8, 10, 13, and 22; values in the form of sums, differences, products, quotients, or power expressions of zero and infinity require special attention.

p. 361

Determinate Forms

$\infty + \infty = \infty, \quad -\infty - \infty = -\infty$

$\infty \cdot \infty = \infty$

$0^{\infty}, \quad \sqrt[n]{\infty}$

p. 362

Undefined Forms

$\frac{a}{0}$ where a is a non-zero real number, or ∞

p. 360

Indeterminate Forms

$\infty - \infty$

$0 \cdot \infty$

$\frac{0}{0}, \quad \frac{\infty}{\infty}$

$0^0, \quad \infty^0, \quad 1^{\infty}$

The Limit of Nothingness

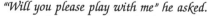

"Will you please play with me" he asked.

"Certainly not," said the lamb. "In the first place, I cannot get into your pen, as I am not old enough to jump over the fence. In the second place, I am not interested in pigs. Pigs mean less than nothing to me."

"What do you mean, less than nothing?" replied Wilbur. "I don't think there is any such thing as less than nothing. Nothing is absolutely the limit of nothingness. It's the lowest you can go. It's the end of the line. How can anything be less than nothing? If there were something that was less than nothing, it would be something – even though it's just a very little bit of something. But if nothing is nothing, then nothing has nothing that is less than it is."

"Oh, be quiet!" said the lamb. "Go play by yourself! I don't play with pigs."

E. B. White, *Charlotte's Web* (1952). Illustrated by Garth Williams.

23.2 L'Hospital's Rule

BERNOULLI Johann (1623 - 1708)

de L'HOSPITAL Guillaume François (1661 - 1704)

Many limits of an indeterminate form can be evaluated by a formula that is probably due to **Johann Bernoulli** but goes by the name of **L'Hospital's rule**; L'Hospital was a pupil of Bernoulli's.

L'Hospital's rule states that:

If f and g are differentiable functions on an open interval that contains a, and

$$\lim_{x \to a} f(x) = 0\,; \quad \lim_{x \to a} g(x) = 0$$

or

$$\lim_{x \to a} f(x) = \pm\infty\,; \quad \lim_{x \to a} g(x) = \pm\infty\,,$$

then

$$\lim_{x \to a} \frac{f(x)}{g(x)} = \lim_{x \to a} \frac{f'(x)}{g'(x)},$$

provided that $g'(x) \neq 0$, if $x \neq a$.

Thus, to apply L'Hospital's rule we **differentiate the numerator and the denominator separately**.

After differentiation, the quotient sometimes still appears in indeterminate form. The rule is then applied – in a repetitive manner – to the quotients of the derivatives.

Even if the quotients of the derivatives continue to yield indeterminate forms, the original quotient may have a limit; it must then be determined by other methods.

One must remember that L'Hospital's rule applies only for indeterminate forms.

Proving L'Hospital's Rule

L'Hospital's rule can be proved by using the **generalized mean-value theorem**. In this text, we base the proof on expansion into **Taylor's series**, assuming that f and g have such expansions. We distinguish two cases, $\frac{0}{0}$ and $\frac{\infty}{\infty}$, which are treated separately.

Case 0/0

Consider

$$\lim_{x \to a} \frac{f(x)}{g(x)}\,; \quad f(a) = 0\,; \quad g(a) = 0\,,$$

and expand the functions f and g in accordance with Taylor's series. Since $f(a) = g(a) = 0$, the first term of both series will be 0:

$$\left.\begin{aligned} f(x) &= 0 + (x-a)f'(a) + \frac{1}{2}(x-a)^2 f''(a) + \ldots \\ g(x) &= 0 + (x-a)g'(a) + \frac{1}{2}(x-a)^2 g''(a) + \ldots \end{aligned}\right\}$$

Thus,

$$\frac{f(x)}{g(x)} = \frac{f(x-a)f'(a) + \frac{1}{2}(x-a)^2 f''(a) + \ldots}{(x-a)g'(a) + \frac{1}{2}(x-a)^2 g''(a) + \ldots},$$

which may be rewritten

$$\frac{f(x)}{g(x)} = \frac{(x-a)\left[f'(a) + \frac{1}{2}(x-a)f''(a) + \ldots\right]}{(x-a)\left[g'(a) + \frac{1}{2}(x-a)g''(a) + \ldots\right]}$$

and, after cancellation,

$$\frac{f(x)}{g(x)} = \frac{f'(a) + \frac{1}{2}(x-a)f''(a) + \ldots}{g'(a) + \frac{1}{2}(x-a)g''(a) + \ldots}.$$

When x tends to a, this becomes

$$\begin{aligned} \lim_{x \to a} \frac{f(x)}{g(x)} &= \lim_{x \to a} \frac{f'(a) + \frac{1}{2}(x-a)f''(a) + \ldots}{g'(a) + \frac{1}{2}(x-a)g''(a) + \ldots} \\ &= \frac{f'(a) + \frac{1}{2}(a-a)f''(a) + \ldots}{g'(a) + \frac{1}{2}(a-a)g''(a) + \ldots} \\ &= \frac{f'(a)}{g'(a)} = \lim_{x \to a} \frac{f'(x)}{g'(x)}, \quad \text{if } g'(a) \neq 0. \end{aligned}$$

If $f'(a) = g'(a) = 0$, this formula develops into

$$\lim_{x \to a} \frac{f(x)}{g(x)} = \lim_{x \to a} \frac{\frac{1}{2}(x-a)f''(a) + \ldots}{\frac{1}{2}(x-a)g''(a) + \ldots} = \lim_{x \to a} \frac{f''(x)}{g''(x)},$$

and so on.

Case ∞/∞

Consider

$$\lim_{x \to a} \frac{f(x)}{g(x)}; \quad \lim_{x \to a} f(x) = \pm\infty; \quad \lim_{x \to a} g(x) = \pm\infty.$$

If the limit $\lim_{x \to a} \frac{f'(x)}{g'(x)}$ exists and f' and g' are defined in an interval about a, though not necessarily at a, and $g'(x) \neq 0$ in some interval about a (though not necessarily at a), then $\lim_{x \to a} \frac{f(x)}{g(x)} = \lim_{x \to a} \frac{f'(x)}{g'(x)}$. We find

$$\lim_{x \to a} \frac{f(x)}{g(x)} = \lim_{x \to a} \frac{\frac{1}{g(x)}}{\frac{1}{f(x)}} \tag{1}$$

and, by the case 0/0,

$$\lim_{x \to a} \frac{\dfrac{1}{g(x)}}{\dfrac{1}{f(x)}} = \lim_{x \to a} \frac{\left(\dfrac{1}{g(x)}\right)'}{\left(\dfrac{1}{f(x)}\right)'} .$$

By the technique of differentiation,

$$\lim_{x \to a} \frac{\left(\dfrac{1}{g(x)}\right)'}{\left(\dfrac{1}{f(x)}\right)'} = \lim_{x \to a} \frac{\dfrac{g'(x)}{[g(x)]^2}}{\dfrac{f'(x)}{[f(x)]^2}} = \frac{1}{\displaystyle\lim_{x \to a} \frac{f'(x)}{g'(x)}} \cdot \left(\lim_{x \to a} \frac{f(x)}{g(x)}\right)^2,$$

and returning to (1), we now have

$$\lim_{x \to a} \frac{f(x)}{g(x)} = \frac{1}{\displaystyle\lim_{x \to a} \frac{f'(x)}{g'(x)}} \cdot \left(\lim_{x \to a} \frac{f(x)}{g(x)}\right)^2$$

and, finally,

$$\lim_{x \to a} \frac{f'(x)}{g'(x)} = \lim_{x \to a} \frac{f(x)}{g(x)} .$$

Applying L'Hospital's Rule

Indeterminate Quotients

- $$\lim_{x \to 0} \frac{1 - \cos x}{x^2} = \lim_{x \to 0} \frac{\sin x}{2x} = \lim_{x \to 0} \frac{\cos x}{2} = \frac{1}{2} .$$

- $$\lim_{x \to 0} \frac{e^x - e^{-x}}{\sin x} = \lim_{x \to 0} \frac{e^x + e^{-x}}{\cos x} = \frac{2}{1} = 2 .$$

- $$\lim_{x \to \infty} \frac{\ln x}{e^x} = \lim_{x \to \infty} \frac{1/x}{e^x} = 0 .$$

- $$\lim_{x \to \infty} \frac{e^{2x}}{x^3} = \lim_{x \to \infty} \frac{2 \cdot e^{2x}}{3x^2} = \lim_{x \to \infty} \frac{4 \cdot e^{2x}}{6x} = \lim_{x \to \infty} \frac{8 \cdot e^{2x}}{6} = \infty .$$

Indeterminate Products ($0 \cdot \infty$)

- $$\lim_{x \to \pi/2} (1 - \sin x) \cdot \tan x = \lim_{x \to \pi/2} \frac{1 - \sin x}{1/\tan x}$$

$$= \lim_{x \to \pi/2} \frac{-\cos x}{-\dfrac{1}{\tan^2 x} \cdot \dfrac{1}{\cos^2 x}} = \lim_{x \to \pi/2} \sin^2 x \cdot \cos x = 0 .$$

Indeterminate Differences ($\infty - \infty$)

- $$\lim_{x \to 0} \left(\frac{1}{x} - \frac{1}{\ln(1+x)} \right) = \lim_{x \to 0} \frac{\ln(1+x) - x}{x \cdot \ln(1+x)}$$

$$= \lim_{x \to 0} \frac{\dfrac{1}{1+x} - 1}{\dfrac{x}{1+x} + \ln(1+x)} = \lim_{x \to 0} \frac{-x}{x + (1+x) \cdot \ln(1+x)}$$

$$= \lim_{x \to 0} \frac{-1}{1 + \dfrac{1+x}{1+x} + \ln(1+x)} = \frac{-1}{1+1+0} = -\frac{1}{2}.$$

Indeterminate Powers (0^0; ∞^0)

- $$\lim_{x \to 0+} x^x = \lim_{x \to 0+} e^{x \cdot \ln x} = e^{\left(\lim\limits_{x \to 0+} x \ln x \right)};$$

$$\lim_{x \to 0+} x \cdot \ln x = \lim_{x \to 0+} \frac{\ln x}{1/x} = \lim_{x \to 0+} \frac{1/x}{-1/x^2}$$

$$= \lim_{x \to 0+} (-x) = 0.$$

Thus,

$$\lim_{x \to 0+} x^x = e^0 = 1.$$

- $$\lim_{x \to 0+} x^{1/\ln x} = \lim_{x \to 0+} e^{(1/\ln x)\ln x} = e^1 = e.$$

- $$\lim_{x \to \infty} \sqrt[x]{x} = \lim_{x \to \infty} e^{(1/x)\ln x} = e^{\left(\lim\limits_{x \to \infty} \frac{1}{x} \ln x \right)};$$

$$\lim_{x \to \infty} \frac{\ln x}{x} = \lim_{x \to \infty} \frac{1/x}{1} = \frac{0}{1} = 0.$$

Thus,

$$\lim_{x \to \infty} \sqrt[x]{x} = e^0 = 1.$$

The concept of limits is often hard to comprehend. However, even an illustration may be enigmatic.

Are the three rings illustrated above joined or not?

Though none of the rings goes through any of the others, they are assembled in such a way that if one is removed, the remaining two fall apart.

We may conclude that taken three at a time the rings are joined, but in pairs they are not.

Borromean Rings

These rings are part of the coat of arms of the noble family Borromeo-Arese of Milan, the source of the name **Borromean rings**. The Borromean family's most illustrious members are Cardinal Carlo Borromeo, canonized in 1610, and Cardinal Federico Borromeo, founder of the Ambrosian Art Gallery in Milan. Their shield bears the inscription *Humilitas*:

Borromeo-Arese

– Milano –

Chapter 24

COMPLEX NUMBERS REVISITED

24.1 Introduction

Chapter 3 p. 87
Chapter 4 p. 131
Chapter 9 pp. 295 *et seq.*
Chapter 17 p. 633

We introduced the imaginary unit and complex numbers in Chapter 3 and described fundamental principles for their handling in Chapter 4; in Chapter 9, we discussed complex roots (solutions) of equations, and in Chapter 17, fractal behavior in the complex number plane. In this chapter we shall explore the use of vectors for calculations with complex numbers, and learn to find their roots and logarithms.

p. 87

If every point in an orthogonal number plane corresponds to a specific complex number, the plane is called a **complex number plane**.

Points on the x-axis represent real numbers, and points on the y-axis, imaginary numbers; the coordinate axes are the **real axis** and the **imaginary axis**.

Cartesian Form of Complex Numbers

In Cartesian coordinates, a complex number

$$z = x + y\mathrm{i}$$

is shown thus:

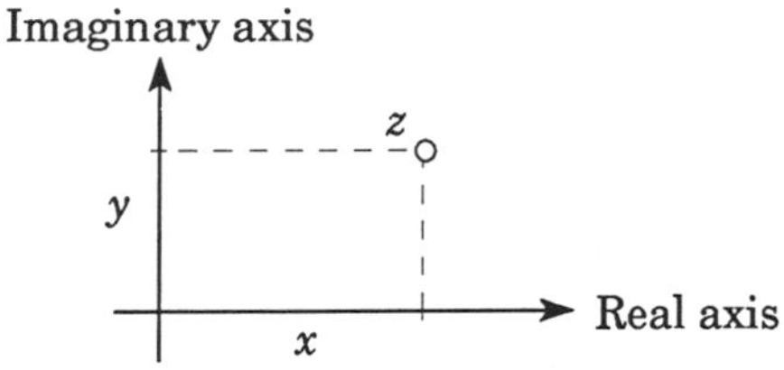

Polar Form of Complex Numbers p. 789

Argument

In a plane polar coordinate system, the complex number z is represented by a radial vector, whose terminal point is $z = x + y\mathrm{i}$ and whose angle θ with the positive real axis is called its **argument**; the argument is determined only up to 2π rad, or multiples of 2π rad.

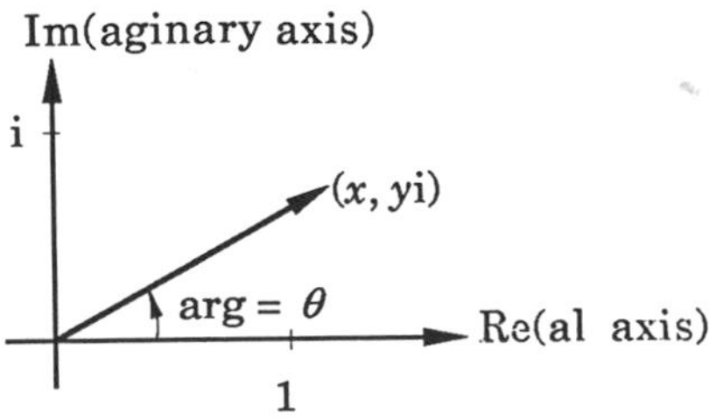

The length of the vector representing the complex number z is denoted r; thus,

$$r = \sqrt{x^2 + y^2},$$

Modulus
Radius Vector

where r is the **modulus** or **radius vector** (or absolute number) of a complex number.

For a complex number z, we have

$$r = |z| = |x + y\mathrm{i}| = \sqrt{x^2 + y^2}.$$

Consider a right triangle whose hypotenuse is r. If θ is the acute angle opposite to side y and adjacent to side x, then

$$\cos\theta = \frac{x}{r}\,; \quad \sin\theta = \frac{y}{r}\,,$$

which gives

$$x = r\cos\theta\,; \quad y = r\sin\theta\,.$$

Polar Form of Complex Numbers

Even if θ is not acute, a complex number $z = x + y\,\mathrm{i}$ in Cartesian form has the **polar form**

$$z = r\cos\theta + (r\sin\theta)\,\mathrm{i}$$

or

$$z = r(\cos\theta + \mathrm{i}\sin\theta)\,.$$

- Find the modulus, argument, and polar form of $z = \sqrt{3} - \mathrm{i}$.

The **modulus** is

$$|z| = |\sqrt{3} + (-\mathrm{i})| = \sqrt{(\sqrt{3})^2 + (-1)^2} = 2\,.$$

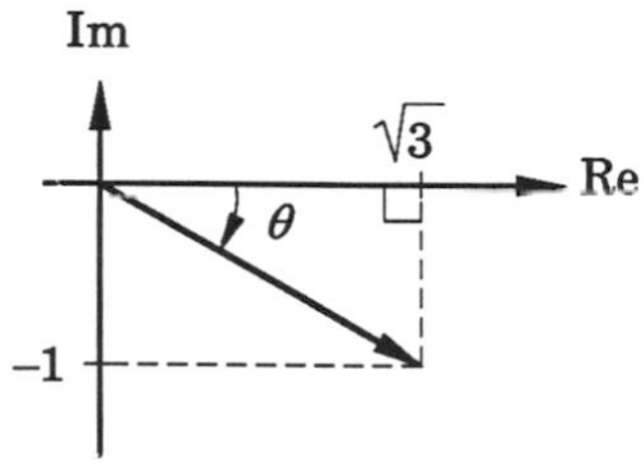

To find the *argument* of a complex number, it is necessary first to locate the proper quadrant in the complex plane; going *clockwise*, the angle θ is *negative*.

$$\tan\theta = -\frac{1}{\sqrt{3}}\,.$$

The **argument** $\theta = \arctan\left(-\frac{1}{\sqrt{3}}\right) = -\pi/6$ rad or $-30°$.

The **polar form** of z is $z = r(\cos\theta + \mathrm{i}\sin\theta)$;

$$z = 2\,[\cos(-\pi/6) + \mathrm{i}\sin(-\pi/6)]\,.$$

24.2 Sums and Differences

Addition and subtraction of complex numbers correspond to the **addition and subtraction of vectors.**

p. 604

Let vectors **a** and **b** represent the complex numbers $2 + 5\mathrm{i}$ and $6 - 3\mathrm{i}$, respectively; then, by the **parallelogram law of addition,** we obtain the sum = $(8 + 2\mathrm{i})$ and the difference = $(-4 + 8\mathrm{i})$:

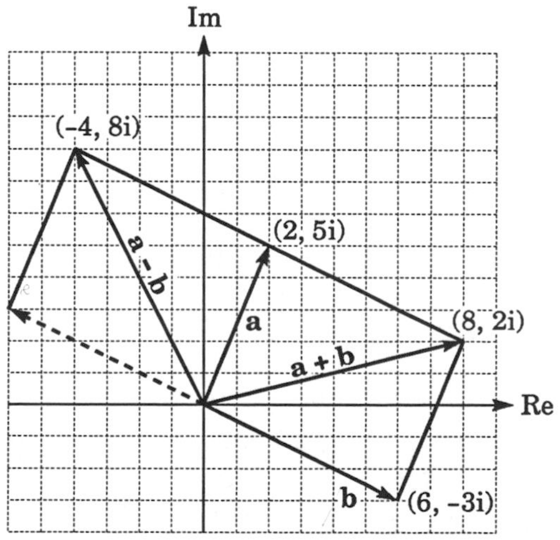

24.3 Products and Quotients

To provide a geometric interpretation of the **product** of two complex numbers, we consider their expressions in polar coordinates,

$$z = r(\cos\theta + i\sin\theta)\,; \quad w = s(\cos\phi + i\sin\phi)\,.$$

We obtain

$$\begin{aligned} zw &= rs(\cos\theta + i\sin\theta)(\cos\phi + i\sin\phi) \\ &= rs(\cos\theta\cos\phi + i\cos\theta\sin\phi + i\sin\theta\cos\phi + i^2\sin\theta\sin\phi) \\ &= rs[(\cos\theta\cos\phi - \sin\theta\sin\phi) + i(\sin\theta\cos\phi + \cos\theta\sin\phi)]\,, \end{aligned}$$

and, finally,

p. 509

$$zw = rs[\cos(\theta+\phi) + i\sin(\theta+\phi)]\,.$$

Compare the preceding product formula with the polar forms of the complex-number factors

$$z = r(\cos\theta + i\sin\theta)\,; \quad w = s(\cos\phi + i\sin\phi)$$

with

$$\text{modulus}(zw) = |zw| = rs$$

$$\arg(zw) = \theta + \phi,$$

where $\theta + \phi$ are determined only up to 2π rad, or multiples of 2π rad.

The **quotient** of the same two complex numbers is

$$\begin{aligned} \frac{z}{w} &= \frac{r(\cos\theta + i\sin\theta)}{s(\cos\phi + i\sin\phi)} \\ &= \frac{r(\cos\theta + i\sin\theta)}{s(\cos\phi + i\sin\phi)} \times \frac{(\cos\phi - i\sin\phi)}{(\cos\phi - i\sin\phi)} \\ &= \frac{r(\cos\theta\cos\phi - i\cos\theta\sin\phi + i\sin\theta\cos\phi - i^2\sin\theta\sin\phi)}{s(\cos^2\phi - i\cos\phi\sin\phi + i\sin\phi\cos\phi - i^2\sin^2\phi)} \\ &= \frac{r(\cos\theta\cos\phi + \sin\theta\sin\phi) + i(\sin\theta\cos\phi - \cos\theta\sin\phi)}{s(\cos^2\phi + \sin^2\phi)} \end{aligned}$$

and, finally,

pp. 508, 509

$$\frac{z}{w} = \frac{r}{s}[\cos(\theta-\phi) + i\sin(\theta-\phi)]$$

with

$$\text{modulus}\left(\frac{z}{w}\right) = \left|\frac{z}{w}\right| = \frac{r}{s}$$

$$\arg\left(\frac{z}{w}\right) = \quad = \theta - \phi\,.$$

- Find the product zw and the quotient z/w of the complex numbers

$$z = 2(\cos 30° + i\sin 30°)\,; \quad w = 3(\cos 170° + i\sin 170°)\,.$$

$|zw| = 2 \times 3 = 6\,.$

$\arg(zw) = 30° + 170°$

$= 200°\,.$

$zw = 6(\cos 200° + i\sin 200°)$

$\approx -5.638 - 2.052\,i\,.$

$|z/w| = \frac{2}{3}\,.$

$\arg(z/w) = 30° - 170°$

$= -140°$, *or* $220°\,.$

$\frac{z}{w} = \frac{2}{3}(\cos 220° + i\sin 220°)$

$\approx -0.511 - 0.429\,i\,.$

24.4 Powers

We have established that the product of two complex numbers

$$z = r(\cos\theta + i\sin\theta)\,; \quad w = s(\cos\phi + i\sin\phi)$$

is

$$z\,w = r\,s\,[\cos(\theta+\phi) + i\sin(\theta+\phi)]\,.$$

With $w = z$ and $\phi = \theta$, we may write

$$z^2 = r^2(\cos\theta + i\sin\theta)^2 = r^2(\cos 2\,\theta + i\sin 2\,\theta)$$

and, similarly,

$$z^3 = r^3(\cos\theta + i\sin\theta)^3 = r^3(\cos 3\,\theta + i\sin 3\,\theta)\,.$$

de Moivre's Formula

Extending to n factors, we obtain **de Moivre's formula**, or theorem,

$$z^n = r^n(\cos n\,\theta + i\sin n\,\theta),$$

de MOIVRE Abraham (1667 - 1754)

which holds good for all rational n. The formula is named for the French mathematician **Abraham de Moivre**.

- Find the value of z^5 when $z = -1 + i\sqrt{3}$.

$$|z| = \sqrt{(-1)^2 + \left(\sqrt{3}\right)^2} = 2\,;$$

$$\theta = \arctan\frac{\sqrt{3}}{-1}\,;\ \theta = \frac{2\pi}{3}\ \text{rad}$$

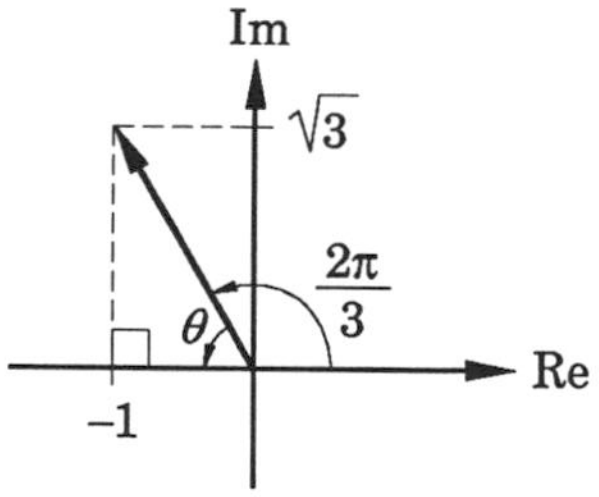

Thus,

$$z = 2\left(\cos\frac{2\,\pi}{3} + i\sin\frac{2\,\pi}{3}\right).$$

Consequently,

$$z^5 = 2^5\left(\cos\frac{2\,\pi}{3} + i\sin\frac{2\,\pi}{3}\right)^5$$

and, according to de Moivre,

$$z^5 = 32\left(\cos\frac{10\,\pi}{3} + i\sin\frac{10\,\pi}{3}\right) = 32\left(\cos\frac{4\,\pi}{3} + i\sin\frac{4\,\pi}{3}\right)$$

$$= 32\left[\cos\left(\frac{\pi}{3} + \pi\right) + i\sin\left(\frac{\pi}{3} + \pi\right)\right]$$

$$= 32\left(-\cos\frac{\pi}{3} - i\sin\frac{\pi}{3}\right) = 32\left(-\frac{1}{2} - i\frac{\sqrt{3}}{2}\right)$$

and, finally,

$$z^5 = -16\left(1 + i\sqrt{3}\right).$$

e^{ix}

Expanding e^{ix} into a power series, we have

$$e^{ix} = 1 + ix - \frac{x^2}{2!} - \frac{ix^3}{3!} + \frac{x^4}{4!} + \frac{ix^5}{5!} + \dots.$$

cf. pp. 774 - 75

Why?

$$\begin{aligned}
f(x) &= e^{ix} & \Rightarrow\ f(0) &= 1 & (e^{i(0)} = 1)\\
f'(x) &= ie^{ix} & \Rightarrow\ f'(0) &= i &\\
f''(x) &= i^2 e^{ix} = -e^{ix} & \Rightarrow\ f''(0) &= -1 & (i^2 = -1)\\
f'''(x) &= -ie^{ix} & \Rightarrow\ f'''(0) &= -i &\\
f^{(4)}(x) &= -i^2 e^{ix} = e^{ix} & \Rightarrow\ f^{(4)}(0) &= 1 &\\
f^{(5)}(x) &= ie^{ix} & \Rightarrow\ f^{(5)}(0) &= i &
\end{aligned}$$

and, after substitutions in

$$f(x) = e^{ix} = f(0) + f'(0)x + \frac{f''(0)}{2!}x^2 + \frac{f'''(0)}{3!}x^3 + \dots,$$

we obtain

$$f(x) = e^{ix} = 1 + ix - \frac{x^2}{2!} - \frac{ix^3}{3!} + \frac{x^4}{4!} + \frac{ix^5}{5!} + \dots.$$

Euler's Formula

$$e^{ix} = \cos x + i \sin x,$$

where i is the imaginary unit.

Why? Since

$$\begin{aligned}
e^{ix} &= 1 + ix - \frac{x^2}{2!} - \frac{ix^3}{3!} + \frac{x^4}{4!} + \frac{ix^5}{5!} - \frac{x^6}{6!} - \frac{ix^7}{7!} + \frac{x^8}{8!} + \frac{ix^9}{9!} \dots\\
&= 1 - \frac{x^2}{2!} + \frac{x^4}{4!} - \frac{x^6}{6!} + \frac{x^8}{8!} + \dots + i\left(x - \frac{x^3}{3!} + \frac{x^5}{5!} - \frac{x^7}{7!} + \frac{x^9}{9!}\right)
\end{aligned}$$

and

p. 773

$$\cos x = 1 - \frac{x^2}{2!} + \frac{x^4}{4!} - \frac{x^6}{6!} + \frac{x^8}{8!} + \dots;$$

p. 771

$$\sin x = x - \frac{x^3}{3!} + \frac{x^5}{5!} - \frac{x^7}{7!} + \frac{x^9}{9!} \dots,$$

we obtain

$$e^{ix} = \cos x + i \sin x$$

and, as $\cos \pi = -1$ and $\sin \pi = 0$, the special case

$$e^{i\pi} = -1.$$

i^i

What one at first might imagine as "the most imaginary number possible" is, after all, a real number; thus:

- The imaginary unit, i, raised to the power i is a real number.

To confirm the thesis, use the special case of Euler's formula,

$$-1 = e^{i\pi},$$

from which we infer that

$$i = e^{i\pi/2};$$

so

$$i^i = (e^{i\pi/2})^i,$$

and we find that

$$i^i = e^{-\pi/2},$$

which is a real number.

24.5 Roots

Let the n-th roots of a complex number

$$w = r(\cos\theta + i\sin\theta)\,;\; -\pi < \theta \le \pi$$

be $z_1, z_2, z_3 \ldots z_n$, all of which satisfy the equation

$$z_p{}^n = w\,.$$

With de Moivre, write

$$z_1 = w^{1/n} = r^{1/n}(\cos\theta + i\sin\theta)^{1/n} = r^{1/n}\left(\cos\frac{\theta}{n} + i\sin\frac{\theta}{n}\right).$$

Adding multiples of 2π to θ, which leaves w unchanged, we find, for $p = 2, 3 \ldots n$, that

- all n values have the **same modulus** ($r^{1/n}$), which is the radius of a circle whose center is the origin of the coordinate system;
- the roots will be **equally spaced** around this circle, that is, the **arguments** of the roots will differ by steps of $2\pi/n$ radians.

Thus, for the complex number w,

$$w = r(\cos\theta + i\sin\theta),$$

de Moivre's theorem gives

$$z_p = r^{1/n}\left[\cos\frac{\theta + 2(p-1)\pi}{n} + i\sin\frac{\theta + 2(p-1)\pi}{n}\right].$$

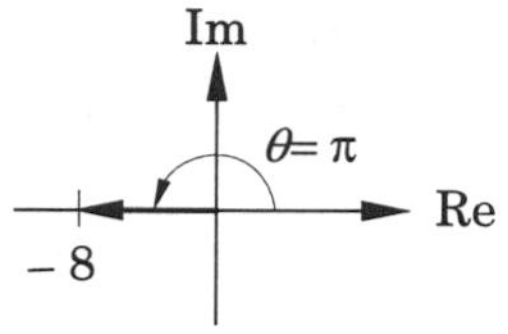

- Find all cubic roots of -8.

With -8 in complex form, $-8+0\,i$,

the modulus is $\sqrt{(-8)^2+0^2} = 8$; the argument (θ) is π, and the roots z_1, z_2, z_3 will be

$$z_1 = 8^{1/3}\left(\cos\frac{\pi}{3} + i\sin\frac{\pi}{3}\right)$$

$$= 2\left(\cos\frac{\pi}{3} + i\sin\frac{\pi}{3}\right) \qquad = 1 + i\sqrt{3}$$

$$z_2 = 2\left(\cos\frac{\pi+2\pi}{3} + i\sin\frac{\pi+2\pi}{3}\right)$$

$$= 2\,(\cos\pi + i\sin\pi) \qquad = -2$$

$$z_3 = 2\left(\cos\frac{\pi+2\times2\pi}{3} + i\sin\frac{\pi+2\times2\pi}{3}\right)$$

$$= 2\left(\cos\frac{5\pi}{3} + i\sin\frac{5\pi}{3}\right) \qquad = 1 - i\sqrt{3}$$

Hence, the three cubic roots of -8 are

$$-2;\; 1 \pm i\sqrt{3},$$

equally spaced around the circumference of a circle whose center is the origin of the coordinate system:

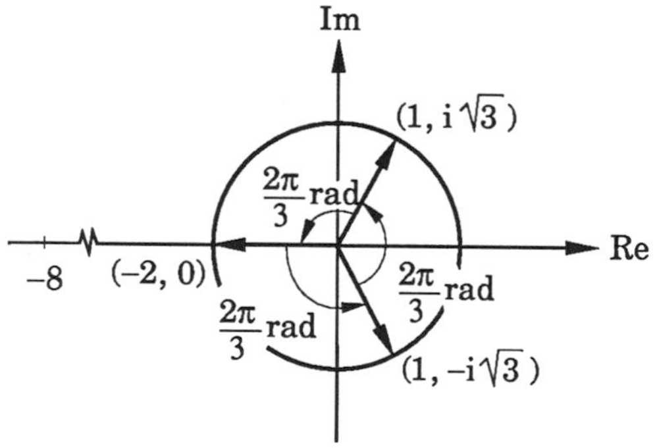

24.6 Logarithms

p. 793 Since $i = e^{i\pi/2}$, we have that

$$\text{principal ln of i is } \frac{\pi}{2}\cdot i\,.$$

A complex number $z = x + i\,y$ may be written in polar form as

$$z = r\,(\cos\theta + i\sin\theta);$$

we obtain for the logarithm

$$\ln z = \ln r + \ln(\cos\theta + i\sin\theta)\,.$$

Since

$$\cos\theta + i\sin\theta = e^{i\theta}; \qquad \ln e^{i\theta} = i\,\theta,$$

we have

$$\ln z = \ln r + i\,\theta,$$

or

$$\ln z = \ln r + i\,(\theta \pm 2\,n\,\pi).$$

Principal Logarithm

Thus, a complex number has infinitely many logarithms, differing by integer multiples of $2\pi i$. Of these, by convention, the **principal logarithm** is the one whose imaginary part is contained in the interval

$$-\pi < \operatorname{Im} z \le +\pi .$$

• Find the principal natural logarithm of $z = 1 + i$.

cf. p. 788 $(z = 1 + 1 \cdot i)$

$|z| = \sqrt{1^2 + 1^2} = \sqrt{2}.$

$\theta = \arcsin \dfrac{1}{\sqrt{2}} = \dfrac{\pi}{4}.$

$z = \sqrt{2}\left(\cos\dfrac{\pi}{4} + i\sin\dfrac{\pi}{4}\right).$

$\ln z = \ln\sqrt{2} + i\left(\dfrac{\pi}{4} \pm 2\,n\,\pi\right)$, where n is any integer.

The principal natural logarithm of z is

$$\ln\sqrt{2} + \frac{i\pi}{4} \approx 0.346 + 0.7854\,i. \qquad \bullet$$

Full of anticipation, Lisa and Stanley had brought their imaginary childhood friends to the lecture, only to find that imaginary numbers aren't imaginary at all, as you can imagine.

From Georg Reisch, *Margarita phylosophica* (first edition, 1496; here reproduced from a 1583 reprint).

In the above illustration, theology watches over the liberal arts. In the Middle Ages, theocratic dogma did not support the introduction of new concepts in learning, such as negative numbers and complex numbers.

The appellation *liberal arts* – for geometry, astronomy, arithmetic, music, grammar, dialectics (logic), and rhetoric – dates from antiquity; *liberal* implies that such study was the privilege only of free citizens.

Chapter 25

EXTREMA AND CRITICAL POINTS

25.1 One Independent Variable

When discussing methods of locating points of relative maximum, points of relative minimum, and points of inflection, the expressions **concave upward** and **concave downward** are less open to misunderstanding than the expressions *concave* and *convex*.

Inflection Point

At the **inflection point** a curve changes from concave upward to concave downward, or vice versa. The slope of the tangent line may have any value.

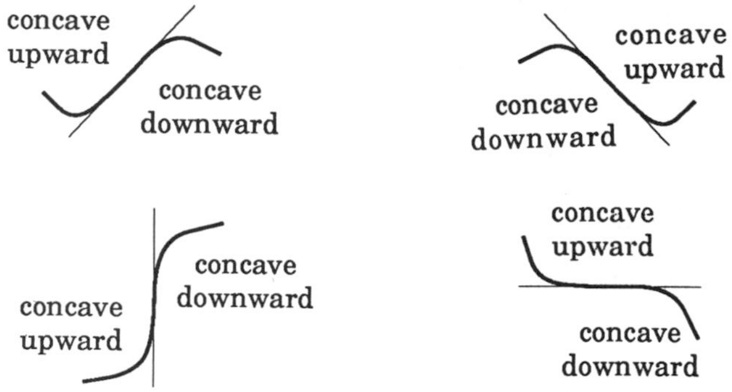

Relative (*or* Local) Maximum
Relative (*or* Local) Minimum
Stationary Points

At points of **relative** (*or* **local**) **maxima**, points of **relative** (*or* **local**) **minima**, and inflection points with horizontal tangent lines, called **stationary points**, the first derivative, if it exists, must be zero; the first derivative is undefined at points where the tangent is vertical.

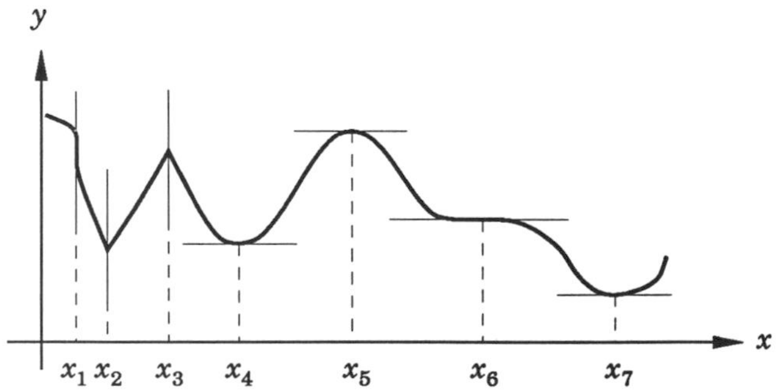

Critical Points

Any point on the curve of a function $y = f(x)$ where the slope of the tangent line of the curve is zero (tangent parallel to x-axis) or undefined (tangent parallel to y-axis) is called a **critical point**. In the above graph, the slope of the tangent line is undefined at the critical points x_1 (inflection point), x_2 (local minimum), and x_3 (local maximum), and zero at the critical points x_4 (local minimum), x_5 (local maximum), x_6 (inflection point), and x_7 (local minimum). The slope of the tangent line at an inflection point may have any value, as exemplified by the inflection points at x_1 and x_6.

Relative Extrema and Critical Points

Relative extrema always occur at critical points of the function:

> If $f(x)$ has a relative extremum at $x = c$, then the first derivative of $f(c)$ is either zero or undefined at c.

Although a relative extremum is always located at a critical point, the opposite does not apply; critical points exist also at other locations.

To distinguish between different kinds of points of interest of a function that is twice differentiable for all values of its domain, we refer to the graph below, which highlights points of relative maximum, relative minimum, and inflection in a function $y = f(x)$ and the corresponding values of $y' = f'(x)$ and $y'' = f''(x)$.

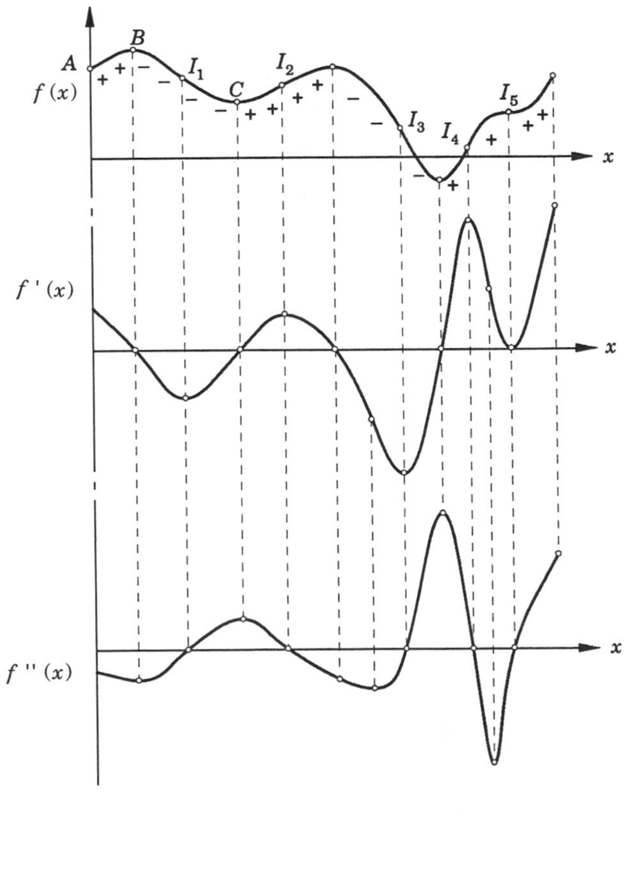

As illustrated in the graph on the preceding page, reproduced here to the left in two parts, the function $f(x)$ and its first and second derivatives have the following particulars:

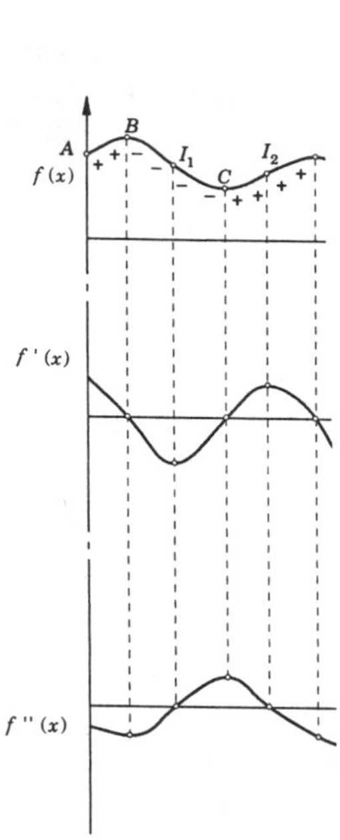

1. Between A and B, $f(x)$ increases as x increases, producing a positive slope of the tangent; the first derivative is positive between A and B.
2. Between B and C, $f(x)$ decreases as x increases, corresponding to a negative slope of the tangent; the first derivative is negative between B and C.
3. At a point of relative maximum, B, the function changes from increasing to decreasing; the first derivative changes from positive to negative, and is zero at the point of maximum.

   Since at a relative maximum the derivative of $f(x)$ changes from positive to negative, it follows that the second derivative is negative at a point of relative maximum.
4. At a point of relative minimum the function changes from decreasing to increasing; the first derivative changes from negative to positive, and is zero at the point of minimum.

   Since at a relative minimum the derivative of $f(x)$ changes from negative to positive, it follows that the second derivative is positive at a point of relative minimum.

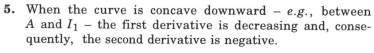

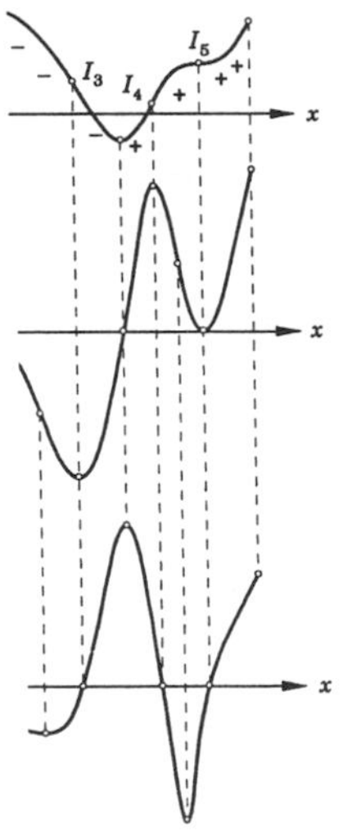

5. When the curve is concave downward – *e.g.*, between A and I_1 – the first derivative is decreasing and, consequently, the second derivative is negative.
6. When the curve is concave upward – *e.g.*, between I_1 and I_2 – the first derivative is increasing and, consequently, the second derivative is positive.
7. At an inflection point the curve changes from concave downward to concave upward (*e.g.*, at I_1) as the first derivative changes from decreasing to increasing, or from concave upward to concave downward (I_2) as the first derivative changes from increasing to decreasing, which implies that the second derivative passes through zero and changes its sign.

   When the tangent line is horizontal at the inflection point (at I_5) and, consequently, the first derivative is zero, the second derivative still changes its sign as it passes through zero, thereby distinguishing the point from an extremum, for which the second derivative would be either positive (a relative minimum) or negative (a relative maximum).

Downward Concavity

Theorems describing downward or upward concavity are cornerstones for the practical use of differentiation.

> If $f(x)$ is a function that has a second derivative $f''(x) < 0$ for all x in the interval $]a, b[$, the graph of $f(x)$ is concave downward in $]a, b[$.

p. 717

Let x_1 and x_2 be any points in $]a, b[$, where $x_1 < x_2$; according to the **restricted mean-value theorem** there will be at least one value c for which

$$f''(c) = \frac{f'(x_2) - f'(x_1)}{x_2 - x_1} \quad ; \quad x_1 < c < x_2.$$

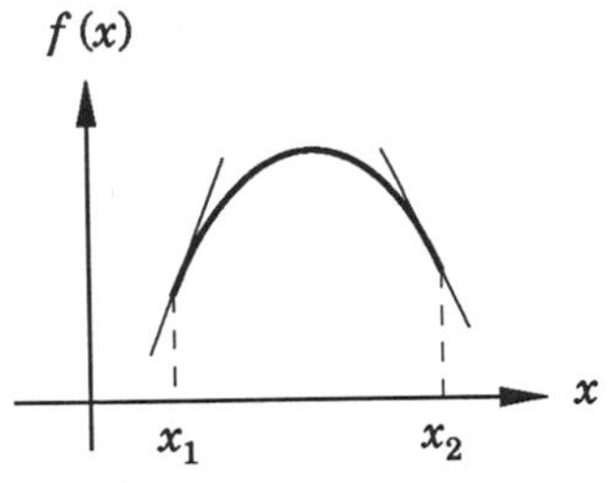

Since the denominator of the above function is positive, the slope of the tangent (the first derivative) of the function is thus decreasing in an interval where the second derivative is < 0.

Upward Concavity

Analogous to the case of downward concavity, we have for upward concavity:

> If $f(x)$ is a function that has a second derivative $f''(x) > 0$ for all x in the interval $]a, b[$, the graph of $f(x)$ is concave upward in $]a, b[$.

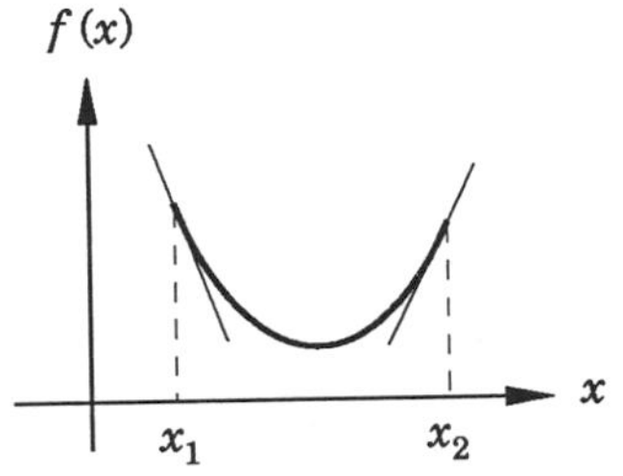

Summary

Consider a function $f(x)$ that is twice differentiable for all x. Let $f'(x) = 0$; solve for x; $x = c$.

We then have:

- If $f''(c) < 0$, c represents a relative maximum of the function.
- If $f''(c) > 0$, then c represents a relative minimum of the original function.

Examples:

$f(x) = -x^2 + 4x$	$g(x) = x^2 + 4x$
$f'(x) = -2x + 4$	$g'(x) = 2x + 4$
$f''(x) = -2$	$g''(x) = 2$
Let $f'(x) = 0$; solve for x.	Let $g'(x) = 0$; solve for x.
Critical point $x = 2$.	Critical point $x = -2$.

Since the second derivative is

negative | positive

the critical point is a

relative maximum. | relative minimum.

If the first derivative is undefined or calculation of the second derivative is cumbersome, it may be more convenient to examine the change of sign of the first derivative at the critical point in order to define the nature of the function: a maximum, a minimum, an inflection point, or some undefined irregularity.

The observations of the sign of the first derivative should be made at points straddling the critical point, and as close to it as possible. The following rules apply:

$f(x)$ has	**a maximum**	**a minimum**	**an inflection point**
if f' changes	$+ + -$	$- + +$	no change

- Locate possible maxima, minima, and inflection points of

$$y = 2x^3 - 3x^2 .$$

$$\frac{dy}{dx} = 6x^2 - 6x ; \qquad \frac{d^2y}{dx^2} = 12x - 6 .$$

Let

$$6x^2 - 6x = 0 ; \; x = 0, \; x = 1,$$

indicating *possible* extrema at $x = 0$ and $x = 1$.

Testing the second derivative for the above values of x gives

$12 \cdot 0 - 6 = -6$, implying a *relative maximum* at $x = 0$

and

$12 \cdot 1 - 6 = 6$, implying a *relative minimum* at $x = 1$.

The corresponding values of y are

$$y = 0 \text{ (relative maximum)}$$

and

$$y = -1 \text{ (relative minimum).}$$

To find possible inflection points, set the second derivative equal to 0:

$$12x - 6 = 0; \; x = \frac{1}{2},$$

indicating a *possible* inflection point at $x = \frac{1}{2}$.

The second derivative is negative for $x < \frac{1}{2}$ and positive for $x > \frac{1}{2}$. Hence, the value of the second derivative changes sign as it passes through 0, and $\left(\frac{1}{2}, -\frac{1}{2}\right)$ is an *inflection point.*

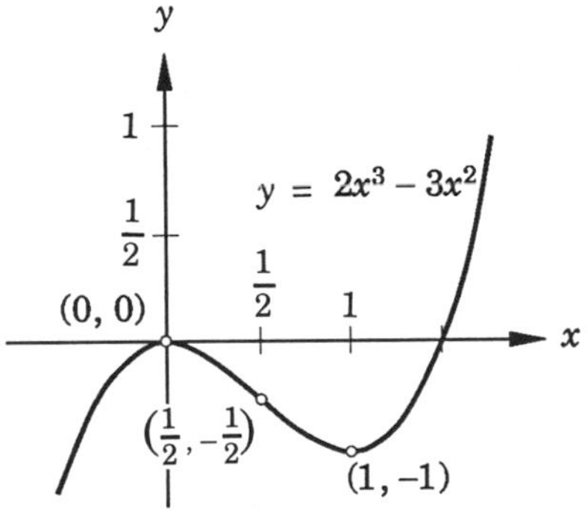

The function y has a relative maximum at (0, 0), an inflection point at $\left(\frac{1}{2}, -\frac{1}{2}\right)$, and a relative minimum at (1, −1). •

The example below typifies how the value of a relative minimum may exceed the value of a relative maximum.

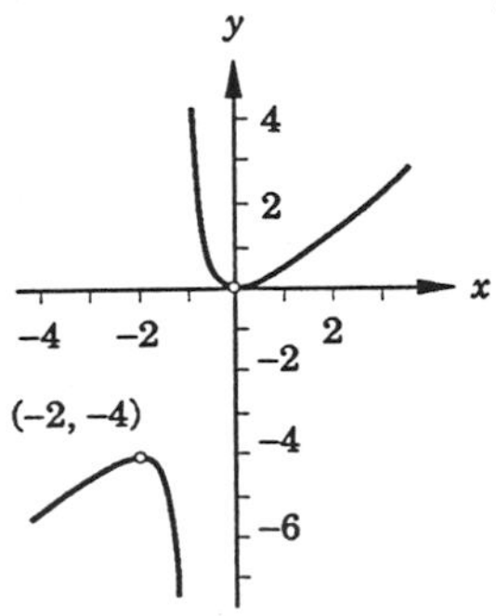

• Locate possible maxima, minima, and inflection points of

$$y = \frac{x^2}{x+1}.$$

$$\frac{dy}{dx} = \frac{x^2 + 2x}{(x+1)^2}; \quad \frac{d^2y}{dx^2} = \frac{2}{(x+1)^3}.$$

$\frac{dy}{dx} = 0$ gives $x = 0$ and $x = -2$.

$$\frac{d^2y}{dx^2} \text{ is } \begin{cases} \text{positive for } x = 0 \\ \text{negative for } x = -2, \end{cases}$$

implying a *relative minimum* at (0, 0) and a *relative maximum* at (− 2, − 4).

The second derivative cannot equal 0; consequently, the curve has no inflection point. •

Endpoint Extremum
Absolute Maximum
Absolute Minimum

When the domain of a function contains one or more closed intervals, points on the curve corresponding to the endpoints of the intervals may represent **endpoint extrema**. **Absolute maxima** and **absolute minima** may be either endpoint extrema or relative extrema.

- Locate absolute extrema of

$$h(x) = x^3 - 6x^2 + 9x; \quad \text{domain: } \left[\frac{1}{2}, 4\frac{1}{5}\right].$$

$h\left(\frac{1}{2}\right) = 3\frac{1}{8}$; $h\left(4\frac{1}{5}\right) = 6\frac{6}{125}$.

The endpoints are

$$\left(\frac{1}{2}, 3\frac{1}{8}\right) \text{ and } \left(4\frac{1}{5}, 6\frac{6}{125}\right).$$

Determine the first and second derivatives,

$$\frac{dy}{dx} = 3x^2 - 12x + 9; \quad \frac{d^2y}{dx^2} = 6x - 12.$$

$x^2 - 4x + 3 = 0$; $(x-3)(x-1) = 0$,

indicating *possible* extrema at $x = 3$ and $x = 1$.

$$\frac{d^2y}{dx^2} \text{ is } \begin{cases} \text{positive for } x = 3 \\ \text{negative for } x = 1, \end{cases}$$

implying a *relative minimum* at (3, 0) and a *relative maximum* at (1, 4).

Since $4 < 6\frac{6}{125}$, the endpoint at $\left(4\frac{1}{5}, 6\frac{6}{125}\right)$ is an *absolute maximum*; and since $0 < 3\frac{1}{8}$, the relative minimum at (3, 0) is also an *absolute minimum*.

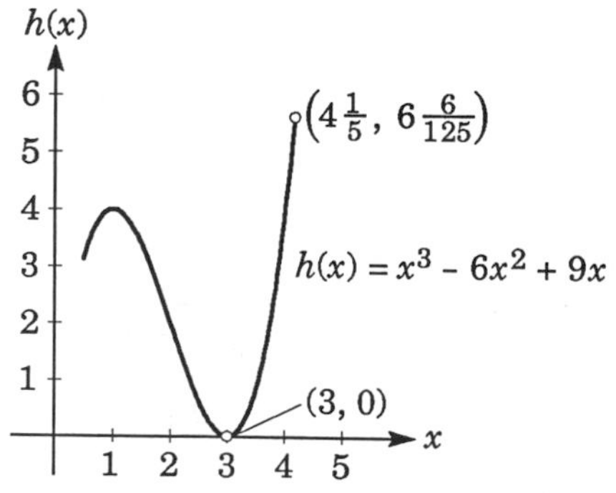

Extrema at Points Where a Function Is Not Differentiable

Continuity of a function at a point does not necessarily imply differentiability at that point and, as illustrated in the example below, it can be expected that **extrema may exist at points where the function has no derivative**.

The function $f(x)$, depicted below, is continuous between a and c. In spite of the fact that the function is not differentiable at x_1 and x_2, a relative minimum does indeed exist at x_1, and a relative maximum exists at x_2. The relative extrema here are also absolute extrema:

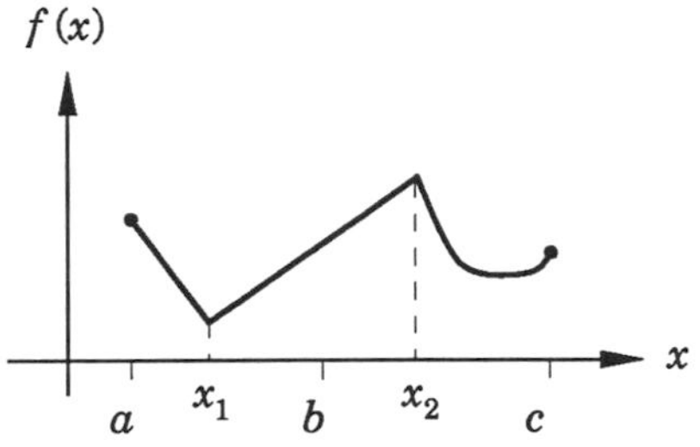

Problems

- Find the values of the relative extrema and sketch the graphs of the functions

$$g(x) = 1 - x^2 \quad \text{and} \quad G(x) = |1 - x^2|.$$

With $y = g(x)$,

$$\frac{dy}{dx} = -2x; \quad \frac{d^2y}{dx^2} = -2.$$

$\frac{dy}{dx} = 0$ gives $x = 0$, suggesting a *possible* extremum.

$\frac{d^2y}{dx^2}$ is negative for $x = 0$, implying a *relative maximum* of $g(x)$ at $(0, 1)$, which is also an absolute maximum.

$g(x)$ intersects the x-axis for $1 - x^2 = 0$; the intersected points are $(-1, 0)$ and $(+1, 0)$.

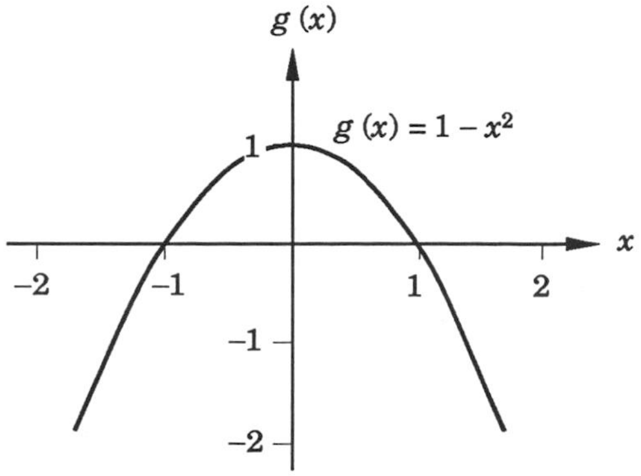

$|1-x^2| = x^2-1 \; ; \; 1-x^2 < 0 \quad \Rightarrow \quad G'(x) = 2x$

$|1-x^2| = 1-x^2 \; ; \; 1-x^2 \geq 0 \quad \Rightarrow \quad G'(x) = -2x$

$G'(x) = 0$ will give $x = 0$.

$G'(x)$ changes sign from positive to negative as it passes through zero. Thus, $G(x)$ has a relative maximum at (0, 1).

$G(x) = 1 - x^2$ meets the x–axis where

$$1-x^2 = 0 ; \; x_1 = 1 , \; x_2 = -1 .$$

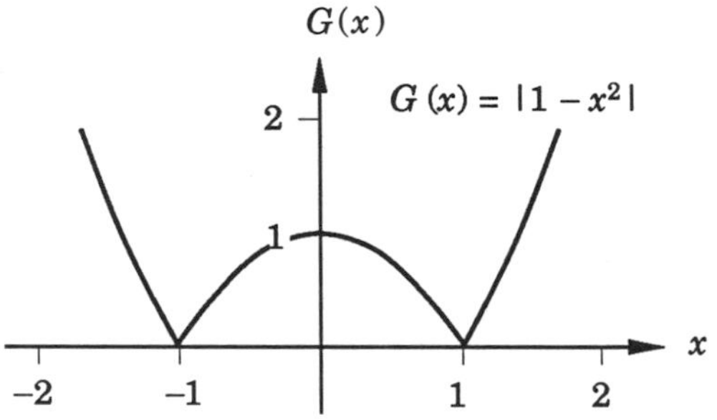

Since $G(x) = |1 - x^2|$ has *only positive values* it returns upward at the relative minima (−1, 0) and (+1, 0), which are also absolute minima.

- Determine extrema and inflection points of

$$f(x) = \frac{1}{x^2+1}$$

and sketch the graph of the function.

$$f'(x) = -\frac{2x}{(x^2+1)^2} \; ; \quad f''(x) = \frac{6x^2-2}{(x^2+1)^3}$$

$$f'(x) = 0 = -\frac{2x}{(x^2+1)^2} = 0 \text{ gives } x = 0.$$

Since $f''(0) = \dfrac{6x^2-2}{(x^2+1)^3} < 0$, a relative maximum (0, 1), which is also an absolute maximum, is located at the critical point.

To determine inflection points, let the second derivative equal 0, solve for x, and test for changes of signs when $f''(x)$ passes through 0:

$$\frac{6x^2-2}{(x^2+1)^3} = 0$$

$$x = \pm \frac{1}{3}\sqrt{3}$$

Test the second derivative for values of x in the intervals between the critical points; for instance:

$$]-\infty, -\tfrac{1}{3}\sqrt{3}[\quad x = -1 \Rightarrow f''(-1) = \frac{6(-1)^2 - 2}{[(-1)^2 + 1]^3} > 0$$

$$]-\tfrac{1}{3}\sqrt{3}, 0[\quad x = -\tfrac{1}{2} \Rightarrow f''\left(-\tfrac{1}{2}\right) = \frac{6\left(-\tfrac{1}{2}\right)^2 - 2}{\left[\left(-\tfrac{1}{2}\right)^2 + 1\right]^3} < 0$$

$$]0, \tfrac{1}{3}\sqrt{3}[\quad x = +\tfrac{1}{2} \Rightarrow f''\left(\tfrac{1}{2}\right) = \frac{6\left(\tfrac{1}{2}\right)^2 - 2}{\left[\left(\tfrac{1}{2}\right)^2 + 1\right]^3} < 0$$

$$]\tfrac{1}{3}\sqrt{3}, \infty[\quad x = +1 \Rightarrow f''(1) = \frac{6(1)^2 - 2}{[(1)^2 + 1]^3} > 0$$

Since the second derivative changes its sign between

$$x = -1 \text{ and } x = -\frac{1}{2}$$

and between

$$x = +\frac{1}{2} \text{ and } x = +1,$$

the curve has inflection points at $\left(-\frac{1}{3}\sqrt{3}, \frac{3}{4}\right)$ and $\left(+\frac{1}{3}\sqrt{3}, \frac{3}{4}\right)$.

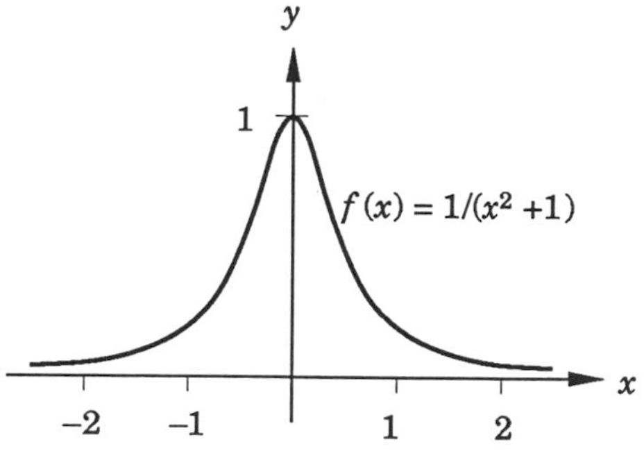

- Find the dimensions (radius and height) that yield the smallest external surface of a 1000-m^3 right cylindrical container.

With the radius r m and the height h m,

$$h = \frac{1000}{\pi r^2} \text{ m.}$$

The total surface area S is a function of r,

$$S(r) = 2\pi r^2 + 2\pi r \frac{1000}{\pi r^2} = 2\pi r^2 + \frac{2000}{r}.$$

$$S'(r) = 4\pi r - 2000 r^{-2}; \quad S''(r) = 4\pi + 4000 r^{-3}.$$

$$4\pi r - 2000 r^{-2} = 0; \quad r = \sqrt[3]{\frac{500}{\pi}} \text{ m.}$$

Since $S''\left(\sqrt[3]{\frac{500}{\pi}}\right) = 4\pi + 4000\left(\sqrt[3]{\frac{500}{\pi}}\right)^{-3} > 0$, the surface area has its minimum value when

$$r = \sqrt[3]{\frac{500}{\pi}} \approx 5.42\,\text{m}\,; \quad h = \frac{1000}{\pi\left(\sqrt[3]{\frac{500}{\pi}}\right)^2} \approx 10.84\,\text{m}.$$

- Find the base angle and the height yielding the maximum volume of a right circular cone whose mantle generatrix is a.

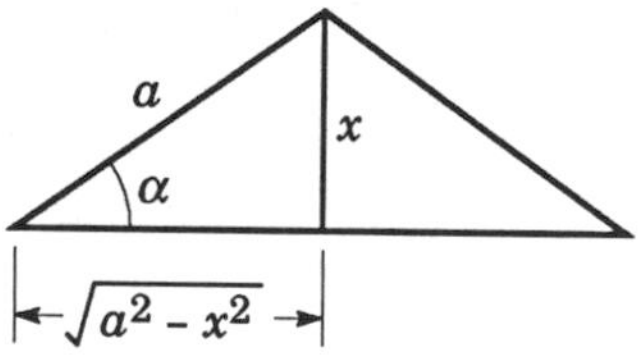

The volume of the cone is

$$V(x) = \frac{1}{3}\pi(a^2 - x^2)\,x = \frac{1}{3}\pi(a^2 x - x^3)\,.$$

$$\frac{dV}{dx} = \frac{1}{3}\pi(a^2 - 3x^2)\,; \quad V'' = -2\pi x < 0\,.$$

$$\frac{1}{3}\pi(a^2 - 3x^2) = 0\,; \quad \text{the } \textbf{height}\ x = \frac{a\sqrt{3}}{3}\,.$$

With the sought **angle** α,

$$\sin\alpha = \frac{h}{a} = \frac{\frac{a\sqrt{3}}{3}}{a} = \frac{\sqrt{3}}{3}\,; \qquad \alpha = 35.26\ldots{}^\circ.$$

- Find the largest possible volume of a right circular cone inscribed in a sphere of radius R

Let the radius of the base of the cone be r, and its height x; the height of the center of the sphere above the center of the base of the cone will be $(x - R)$:

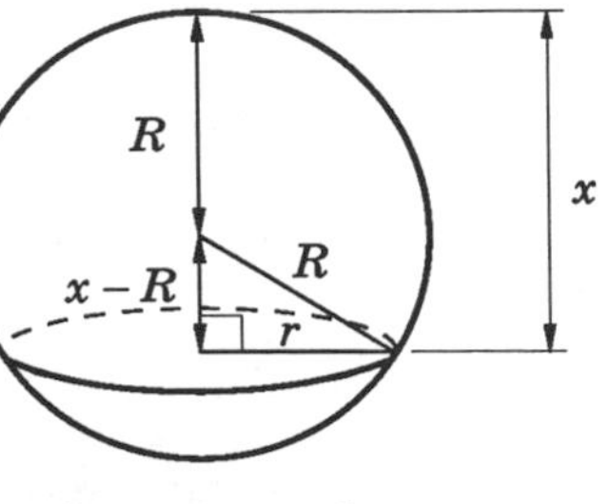

$$R^2 = r^2 + (x - R)^2\,; \quad r^2 = 2Rx - x^2\,.$$

The volume of the cone is

$$V(x) = \frac{1}{3}\pi r^2 x .$$

$$\frac{1}{3}\pi(2Rx - x^2)x = \frac{1}{3}\pi(2Rx^2 - x^3).$$

$$V'(x) = \frac{1}{3}\pi(4Rx - 3x^2); \; V''(x) = \frac{1}{3}\pi(4R - 6x).$$

$$\frac{1}{3}\pi(4Rx - 3x^2) = 0; \quad x_1 = 0 \text{ (discarded)}; \quad x_2 = \frac{4R}{3}.$$

Since

$$V''\left(\frac{4R}{3}\right) = \frac{1}{3}\pi\left[4R - 6\left(\frac{4R}{3}\right)\right] = \frac{1}{3}\pi(-4R) < 0,$$

$x = \dfrac{4R}{3}$ represents a relative (and absolute) maximum.

$$V_{\max} = \frac{1}{3}\pi\left[2R - \left(\frac{4R}{3}\right)\right]\left(\frac{4R}{3}\right)^2 = \frac{32\pi R^3}{81}.$$

- Find the smallest possible volume of a right circular cone circumscribed to a sphere of radius R.

Let the radius of the base of the cone be r, and let the distance between the center of the sphere and the vertex of the cone be h.

The distance between the vertex and the point of contact between the cone and the sphere is $\sqrt{h^2 - R^2}$.

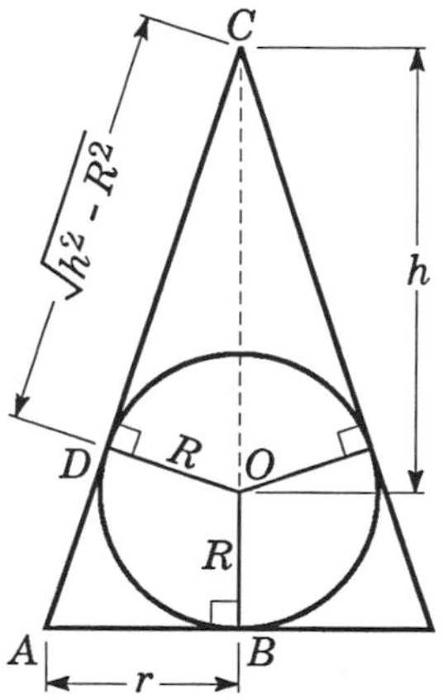

The volume of the cone is

$$V(h) = \frac{1}{3}\pi r^2 (h + R).$$

Similarity of triangles ABC and DOC gives

$$\frac{r}{R} = \frac{h+R}{\sqrt{h^2 - R^2}} = \frac{\sqrt{h+R}\,\sqrt{h+R}}{\sqrt{(h+R)(h-R)}} = \sqrt{\frac{h+R}{h-R}}.$$

$$r = R\sqrt{\frac{h+R}{h-R}}.$$

Substituting for r^2 in $V(h)$ gives us

$$V(h) = \frac{1}{3}\pi R^2 \frac{h+R}{h-R}(h+R) = \frac{\pi R^2}{3}\frac{(h+R)^2}{h-R}.$$

For convenience, let $u = \dfrac{\pi R^2}{3}$.

$$V'(h) = u \cdot \frac{2(h^2-R^2)-(h+R)^2}{(h-R)^2}$$

and

$$V''(h) = u \cdot \frac{(h-R)^2(4h-2h-2R)-[2(h^2-R^2)-(h+R)^2]\,2(h-R)}{(h-R)^4}$$

$V'(h) = 0$ yields

$$h_1 = 3R; \quad h_2 = -R \text{ (discarded).}$$

Since $V''(3R) > 0$, $h = 3R$ represents a relative (and absolute) minimum,

$$V_{min} = \frac{\pi R^2}{3}\frac{(3R+R)^2}{3R-R} = \frac{8\pi R^3}{3}.$$

- What is the largest rectangle that can be inscribed in the ellipse

$$\frac{x^2}{9}+\frac{y^2}{4} = 1?$$

Let the corners of the rectangle be

$$\pm x_0; \ \pm y_0$$

and find the maximum value of the area $A = 4x_0 y_0$.

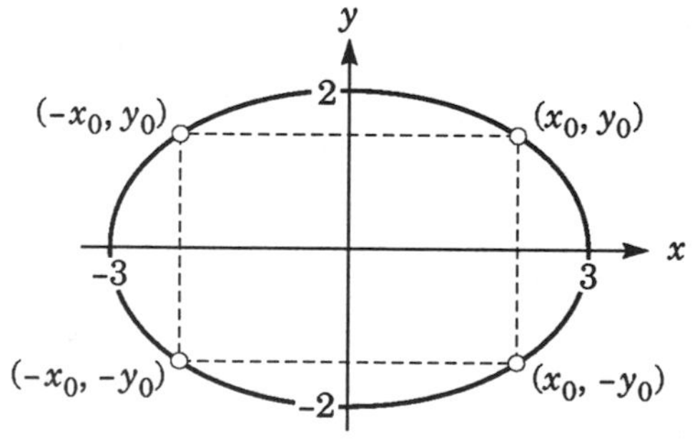

Solving for y yields

$$y = \sqrt{4-\frac{4x^2}{9}}.$$

Let $f(x) = xy$; then

$$f(x) = x\sqrt{4-\frac{4x^2}{9}}.$$

$$f'(x) = x\left(\frac{1}{2}\right)\left(4 - \frac{4x^2}{9}\right)^{-1/2}\left(-\frac{8x}{9}\right) + \sqrt{4 - \frac{4x^2}{9}}$$

$$= -\frac{4x^2}{9\sqrt{4 - \frac{4x^2}{9}}} + \sqrt{4 - \frac{4x^2}{9}}\,.$$

Letting $f'(x) = 0$, and solving for x,

$$x_0 = \frac{3\sqrt{2}}{2} \quad \text{and} \quad y_0 = \sqrt{4 - \frac{4\left(\frac{3\sqrt{2}}{2}\right)^2}{9}} = \sqrt{2}\,.$$

The area of the rectangle is

$$A = 4 \cdot x_0 y_0 = 12 \text{ area units.}$$

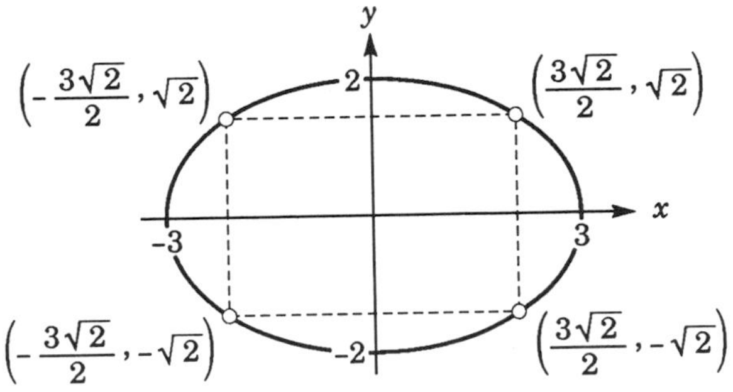

- Find all extrema of

$$f(x) = 2\sin x + \sin 2x$$

in the interval $]0, 2\pi[$.

$$f'(x) = 2\cos x + 2\cos 2x$$

p. 511

$$= 2(\cos x + 2\cos^2 x - 1)$$

$$= 2(2\cos x - 1)(\cos x + 1)$$

$$f''(x) = 2[(2\cos x - 1)(-\sin x) + (\cos x + 1)(-2\sin x)]$$

Let $f'(x) = 0$; solve for x:

$\cos x + 1 = 0$	$2\cos x - 1 = 0$
$\cos x = -1$	$\cos x = \frac{1}{2}$
$x = \pi$	$x = \frac{\pi}{3}\,;\ x = \frac{5\pi}{3}$

Thus, the function has the critical points

$$x = \frac{\pi}{3},\quad x = \pi,\quad x = \frac{5\pi}{3}\,.$$

Examining the second derivative at the critical numbers, we have

$$f''\left(\frac{\pi}{3}\right) = 2\left(2\cos\frac{\pi}{3}+1\right)\left(-\sin\frac{\pi}{3}\right)+\left(\cos\frac{\pi}{3}-1\right)\left(-2\sin\frac{\pi}{3}\right) < 0 \quad \text{(rel. max.)}$$

and

$$f''\left(\frac{5\pi}{3}\right) = 2\left(2\cos\frac{5\pi}{3}+1\right)\left(-\sin\frac{5\pi}{3}\right)+\left(\cos\frac{5\pi}{3}-1\right)\left(-2\sin\frac{5\pi}{3}\right) > 0 \quad \text{(rel. min.)}.$$

On the other hand, we find

$$f''(\pi) = 2\,[(2\cos\pi + 1)(-\sin\pi) + (\cos\pi - 1)(-2\sin\pi)] = 0,$$

giving no information about a local extremum at $x = 0$, but suggesting the possibility of an inflection point.

In the interval $\left]\frac{\pi}{3}, \frac{5\pi}{3}\right[$

$$f'' < 0 \quad \text{for} \quad x < \pi$$

and

$$f'' > 0 \quad \text{for} \quad x > \pi$$

implying that $x = \pi$ represents an inflection point.

Hence,

$$\text{at } x = \frac{\pi}{3} \;\Rightarrow\; f_{max} = 2\sin\frac{\pi}{3} + \sin\left(2\cdot\frac{\pi}{3}\right) \approx 2.6$$

$$\text{at } x = \pi \;\Rightarrow\; f_{infl} = 2\sin\pi + \sin(2\pi) = 0$$

$$\text{at } x = \frac{5\pi}{3} \;\Rightarrow\; f_{min} = 2\sin\frac{5\pi}{3} + \sin\left(2\cdot\frac{5\pi}{3}\right) \approx -2.6\,.$$

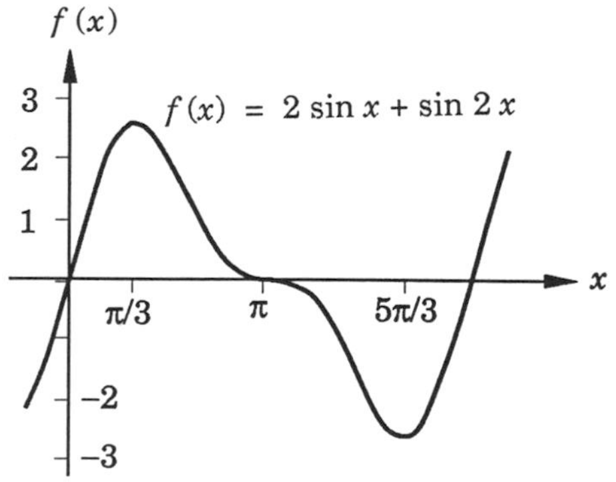

- Are the inequalities $e^x - x \geq 1$ and $e^x \geq e\,x$ true for all real x?

1. If $e^x - x \geq 1$, then $e^x - x - 1 \geq 0$.

   Let $f(x) = e^x - x - 1$.

   Differentiation gives

$$f'(x) = e^x - 1 \quad \text{and} \quad f''(x) = e^x.$$

   Let $f'(x) = 0$,

$$e^x - 1 = 0;\; x = 0 \qquad (e^0 = 1).$$

   $f''(0) = e^0 = 1$.

Since the first derivative is 0 and the second derivative is positive at $x = 0$, $f(x) = e^x - x - 1$ must have a relative minimum at that point, which is also an ***absolute*** *minimum* on $]-\infty, \infty[$

Thus, the inequality $e^x - x - 1 \geq 0$ must be true and, consequently, $e^x - x \geq 1$ for all x.

2. If $e^x \geq e\,x$, then $e^x - e\,x \geq 0$.

Let $g(x) = e^x - e\,x$.

Differentiation gives

$$g'(x) = e^x - e \quad \text{and} \quad g''(x) = e^x .$$

Let $g'(x) = 0$,

$$e^x - e = 0\,; \quad x = 1 \qquad (e^1 = e).$$

$g''(1) = e$.

Since the first derivative is 1 and the second derivative is positive at $x = 1$, $g(x) = e^x - e\,x$ must have a relative minimum at that point, which is also an ***absolute*** *minimum*.

Thus, the inequality $e^x - e\,x \geq 0$ must be true and, consequently, $e^x \geq e\,x$ for all x.

- A company has 2000 units of a given product in stock and can produce 100 more every day.

An order is received for as many units as possible. For delivery on the date of receiving the order, the company will make a profit of \$6 per unit; every consecutive day sees a decline in profit of \$0.20 per item.

How many days of manufacturing should be used for adding to what already is in stock to make the profit the greatest possible?

No units were manufactured on the day the order was received. Delivery will be made after manufacturing hours on the delivery date. There are no other obligations of delivery interfering with the one in question.

Let the number of days of manufacturing be x.

Express the profit as a function $P(x)$,

$$P(x) = (6 - 0.20\,x)(2000 + 100\,x)$$
$$= -20\,x^2 + 200\,x + 12\,000 .$$

$P'(x) = -40\,x + 200$.

$P''(x) = -40$.

Let $P'(x) = 0$,

$$-40\,x + 200 = 0\,; \quad x = 5 .$$

As the second derivative is negative, $x = 5$ must represent a maximum.

5 days of manufacturing should be added.

- A power plant shall supply electric power to a factory situated 1000 m downstream on the opposite bank of a 200-m-wide river. A cable is to be laid for the purpose. The cost of cable and installation is \$50 per meter on land, and \$80 per meter of riverbed. What is the most economical course for the cable?

Let the distance on land be $(1000 - x)$ m; the distance under the river will be

$$d = \sqrt{x^2 + 200^2}\,.$$

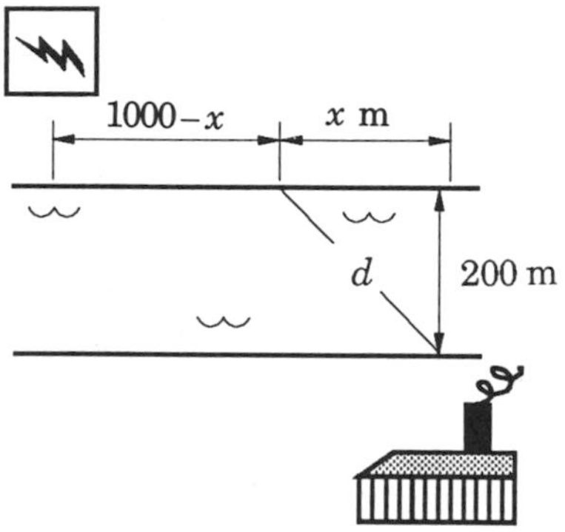

The cost is $C(x)$:

$$C(x) = 50\,(1000 - x) + 80\,\sqrt{x^2 + 200^2}\,.$$

$$C'(x) = -50 + 80 \cdot \frac{\frac{1}{2} \cdot 2x}{\sqrt{x^2 + 200^2}}\,.$$

$C'(x) = 0\,;$

$$5 \cdot \sqrt{x^2 + 200^2} = 8x\,; \quad x \approx 160.1\,.$$

$1000 - x \approx 839.9\,.$

$d \approx \sqrt{x^2 + 200^2} = 256.2\,.$

With the distance 840 m on land, the distance under the river will be 256 m.

* * *

Had he not overlooked the variable of the nearby swamp, his calculations for a maximum of success would have been correct.

25.2 More than One Independent Variable

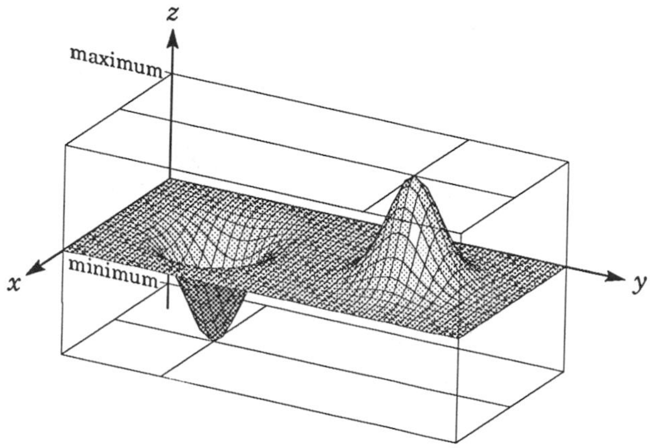

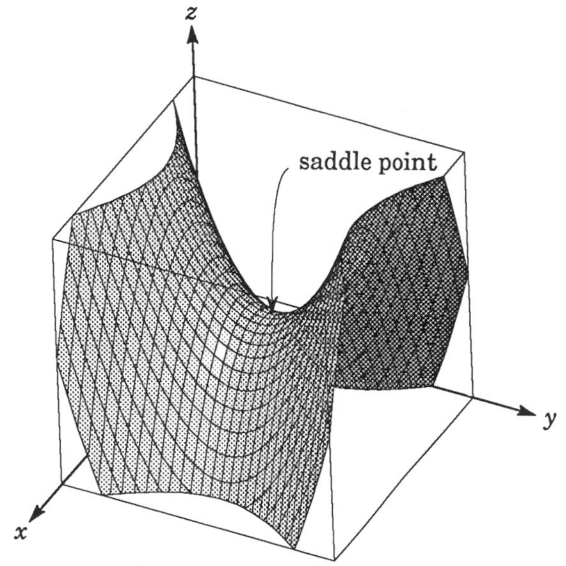

p. 710

Extrema and other critical points of functions of more than one independent variable are traced and identified by **partial differentiation**; such functions are differentiated with respect to all of the independent variables – one at a time, while the others remain unchanged.

Functions of one independent variable, $y = f(x)$, can be depicted in a two-dimensional coordinate system, whereas functions of two independent variables, $z = f(x, y)$, require a three-dimensional system. To display functions of three or more independent variables, colors may be used, although there are no recognized rules for the use of colors in graphic representation.

A differentiable function of one independent variable has a maximum or a minimum where its first derivative is zero; similarly, differentiable functions of two independent variables will have a **critical point** of some description where **both their first-order partial derivatives are zero.**

Saddle Point

This critical feature may be a **local or absolute maximum**, or **minimum**, with a horizontal tangent plane, or a **saddle point**, which is neither a maximum nor a minimum – or, alternatively, it may be considered both, that is, a maximum in some directions and a minimum in other directions, as illustrated by the drawing of a hyperbolic paraboloid on the previous page.

The prime requisite for the existence of an extremum or a saddle point at a point (a, b) is that

$$\frac{\partial f(a,b)}{\partial x} = 0\,; \quad \frac{\partial f(a,b)}{\partial y} = 0\,.$$

The actual nature of the feature is decided by the values of the simple and mixed second-order partial derivatives

$$\frac{\partial^2 f(a,b)}{\partial x^2}; \quad \frac{\partial^2 f(a,b)}{\partial y^2}; \quad \frac{\partial^2 f(a,b)}{\partial x\,\partial y}$$

and of the discriminant

$$g(a,b) = \frac{\partial^2 f(a,b)}{\partial x^2} \cdot \frac{\partial^2 f(a,b)}{\partial y^2} - \left[\frac{\partial^2 f(a,b)}{\partial x\,\partial y}\right]^2 .$$

If $g(a,b) > 0$, in which case the simple partial derivatives must have the same sign, the critical point is

– a **maximum**

if $\dfrac{\partial^2 f(a,b)}{\partial x^2} < 0\,; \quad \dfrac{\partial^2 f(a,b)}{\partial y^2} < 0\,;$

– a **minimum**

if $\dfrac{\partial^2 f(a,b)}{\partial x^2} > 0\,; \quad \dfrac{\partial^2 f(a,b)}{\partial y^2} > 0\,.$

If $g(a,b) < 0$, the critical point is

– a **saddle point.**

If $g(a,b) = 0$, the critical point may or may not be one of the above types.

- Find extrema and saddle points of

$$f(x,y) = x^3 + y^3 + 3xy\,.$$

We have

$$\frac{\partial f}{\partial x} = 3x^2 + 3y\,; \quad \frac{\partial f}{\partial y} = 3y^2 + 3x\,.$$

Let both derivatives be equal to 0 and solve the system of equations

$$\left.\begin{aligned} 3x^2 + 3y &= 0 \\ 3y^2 + 3x &= 0 \end{aligned}\right\} .$$

There are *two critical points*, (0, 0) and (–1 ,– 1).

The values $x = 0, y = 0$ and $x = -1, y = -1$ yield, respectively,

$$f(0,0) = 0,$$
$$f(-1,-1) = 1.$$

The second derivatives are

$$\frac{\partial^2 f}{\partial x^2} = 6x; \qquad \frac{\partial^2 f}{\partial y^2} = 6y; \qquad \frac{\partial^2 f}{\partial x\, \partial y} = 3.$$

For the critical point (0, 0, 0), we find

$$\frac{\partial^2 f(0,0)}{\partial x^2} \cdot \frac{\partial^2 f(0,0)}{\partial y^2} - \left[\frac{\partial^2 f(0,0)}{\partial x\, \partial y}\right]^2 = 0 \cdot 0 - 3^2 < 0.$$

which indicates a *saddle point.*

And for (–1 ,– 1, + 1),

$$\frac{\partial^2 f(-1,-1)}{\partial x^2} \cdot \frac{\partial^2 f(-1,-1)}{\partial y^2} - \left[\frac{\partial^2 f(-1,-1)}{\partial x\, \partial y}\right]^2$$
$$= (-6) \cdot (-6) - 3^2 > 0,$$

which indicates a *relative extremum*, whose nature is determined by the sign of the second partial derivatives of f with respect to x and y. Since

$$\frac{\partial^2 f(-1,-1)}{\partial x^2} = -6 < 0; \frac{\partial^2 f(-1,-1)}{\partial y^2} = -6 < 0,$$

$f(-1,-1)$ represents a *relative maximum.*

Thus, f has a relative maximum at (–1, –1, +1) and a saddle point at (0, 0, 0).

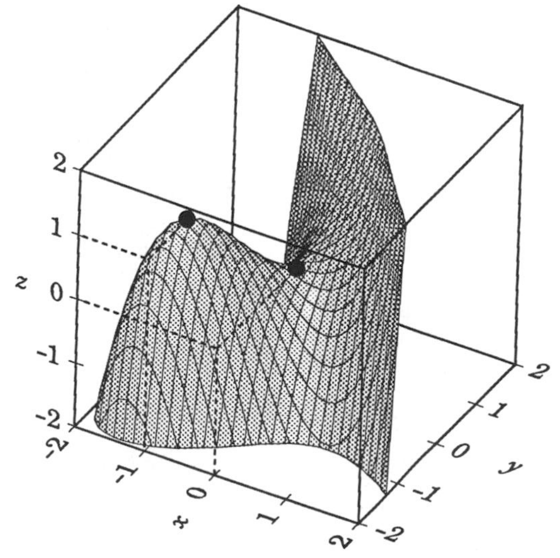

- Identify relative extrema and saddle points of the hyperbolic paraboloid
$$f(x, y) = x^2 - y^2 .$$

$\dfrac{\partial f}{\partial x} = 2x; \quad \dfrac{\partial f}{\partial y} = -2y .$

Equate both derivatives with 0 and solve the equations:

$$\left.\begin{array}{l} x = 0 \\ y = 0 \end{array}\right\} .$$

f has *only one critical point*, (0, 0, 0).

For $x = 0$ and $y = 0$, we find
$$f(0, 0) = 0^3 - 0^3 = 0 .$$

The second-order partial derivatives are
$$\frac{\partial^2 f}{\partial x^2} = 2; \quad \frac{\partial^2 f}{\partial y^2} = -2; \quad \frac{\partial^2 f}{\partial x\, \partial y} = 0 .$$

$$\frac{\partial^2 f(0, 0)}{\partial x^2} \cdot \frac{\partial^2 f(0, 0)}{\partial y^2} - \left[\frac{\partial^2 f(0, 0)}{\partial x\, \partial y}\right]^2 = 2(-2) - 0^2 < 0 ,$$

indicating a *saddle point.*

Thus, f has a saddle point at (0, 0, 0) but no relative extrema.

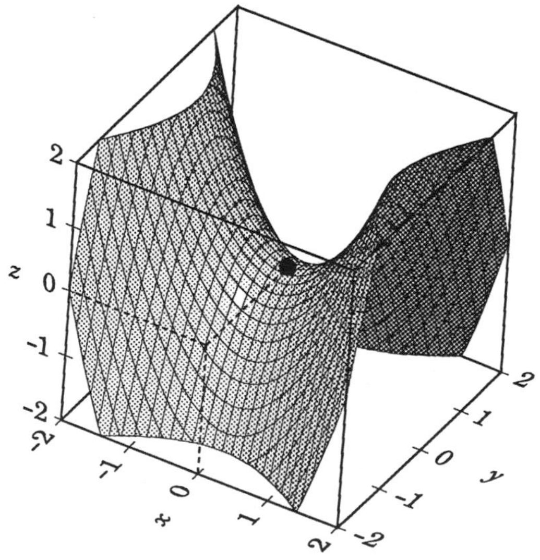

- **Find extrema of**

$$f(x, y) = \sin 3x \cos 2y \,.$$

The maximum and minimum values of $\sin 3x$ and $\cos 2y$ are 1 and -1.

Consequently, f has

– *maxima* for

$$\left.\begin{aligned}\sin 3x &= 1\\ \cos 2y &= 1\end{aligned}\right\} \quad \text{and} \quad \left.\begin{aligned}\sin 3x &= -1\\ \cos 2y &= -1\end{aligned}\right\} ;$$

– *minima* for

$$\left.\begin{aligned}\sin 3x &= -1\\ \cos 2y &= 1\end{aligned}\right\} \quad \text{and} \quad \left.\begin{aligned}\sin 3x &= 1\\ \cos 2y &= -1\end{aligned}\right\} .$$

$$\sin 3x = 1 \quad \Rightarrow \quad 3x = \frac{\pi}{2} + n\,2\pi; \quad x = \frac{\pi}{6} + \frac{n\,2\pi}{3}.$$

$$\sin 3x = -1 \quad \Rightarrow \quad 3x = \frac{3\pi}{2} + n\,2\pi; \quad x = \frac{\pi}{2} + \frac{n\,2\pi}{3}.$$

$$\cos 2y = 1 \quad \Rightarrow \quad 2y = 0 + m\,2\pi; \quad y = m\,\pi.$$

$$\cos 2y = -1 \quad \Rightarrow \quad 2y = \pi + m\,2\pi; \quad y = \frac{\pi}{2} + m\,\pi.$$

Thus, f has iterative maxima at

$$\left(\frac{\pi}{6} + \frac{n\,2\pi}{3},\ m\,\pi,\ +1\right) \quad \text{and} \quad \left(\frac{\pi}{2} + \frac{n\,2\pi}{3},\ \frac{\pi}{2} + m\,\pi,\ +1\right)$$

and iterative minima at

$$\left(\frac{\pi}{6} + \frac{n\,2\pi}{3},\ \frac{\pi}{2} + m\,\pi,\ -1\right) \quad \text{and} \quad \left(\frac{\pi}{2} + \frac{n\,2\pi}{3},\ m\,\pi,\ -1\right).$$

All extrema are absolute extrema.

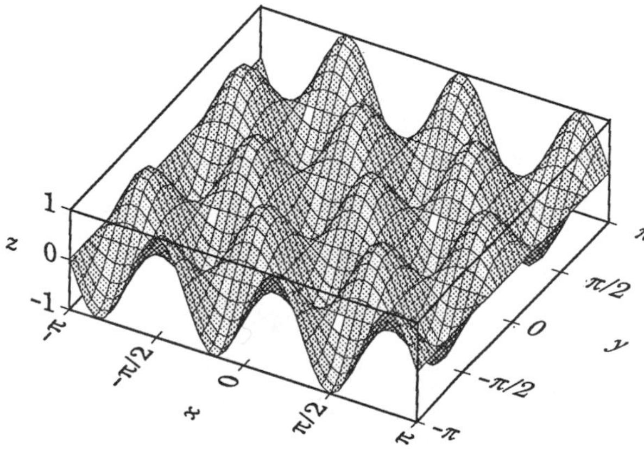

- Find the critical points of the paraboloid of revolution

$$f(x, y) = x^2 + y^2 .$$

$$\frac{\partial f}{\partial x} = 2x ; \quad \frac{\partial f}{\partial y} = 2y .$$

Equate the derivatives with 0 and solve the equations,

$$\left.\begin{array}{l} x = 0 \\ y = 0 \end{array}\right\} .$$

Thus, f has only one critical point, (0, 0, 0).

The second-order partial derivatives are

$$\frac{\partial^2 f}{\partial x^2} = 2 ; \quad \frac{\partial^2 f}{\partial y^2} = 2 ; \quad \frac{\partial^2 f}{\partial x \, \partial y} = 0 .$$

$$\left[\frac{\partial^2 f(0, 0)}{\partial x^2}\right]\left[\frac{\partial^2 f(0, 0)}{\partial y^2}\right] - \left[\frac{\partial^2 f(0, 0)}{\partial x \, \partial y}\right]^2 = 2(2) - 0^2 > 0 ,$$

indicating that f has a relative extremum.

Since $\frac{\partial^2 f(0, 0)}{\partial x^2} = 2 > 0 ;\ \frac{\partial^2 f(0, 0)}{\partial y^2} = 2 > 0$, the point (0, 0, 0) is an absolute minimum, the vertex of the paraboloid.

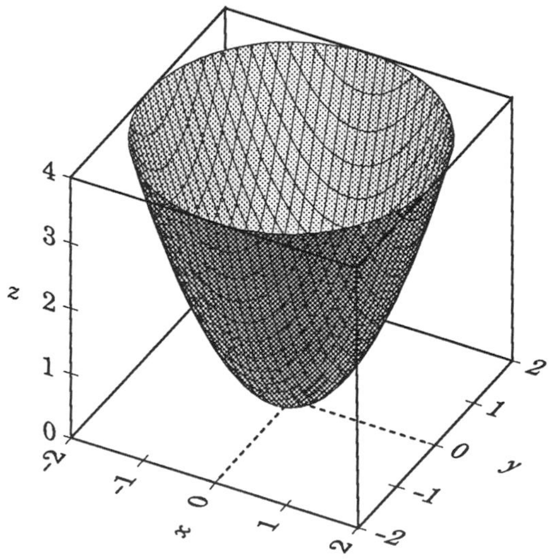

- Find relative extrema and saddle points of

$$g(x, y) = x^2 - y^2 - 4x - 2y + 3 .$$

$$\frac{\partial g}{\partial x} = 2x - 4 ; \quad \frac{\partial g}{\partial y} = -2y - 2 .$$

Equate both derivatives with 0; solve for x and y,

$$\left.\begin{aligned} x &= 2 \\ y &= -1 \end{aligned}\right\} .$$

Thus, there is one critical point, $(2, -1)$, where

$$g(2, -1) = 0 .$$

The partial second derivatives are

$$\frac{\partial^2 g}{\partial x^2} = 2 ; \quad \frac{\partial^2 g}{\partial y^2} = -2 ; \quad \frac{\partial^2 g}{\partial x \, \partial y} = 0 .$$

We have, for the critical point $(2, -1)$,

$$\frac{\partial^2 g(2, -1)}{\partial x^2} \cdot \frac{\partial^2 g(2, -1)}{\partial y^2} - \left[\frac{\partial^2 g(2, -1)}{\partial x \, \partial y}\right]^2 = (2)(-2) - 0^2 < 0 ,$$

indicating a *saddle point.*

Thus, g has a saddle point at $(2, -1, 0)$, but no relative extrema.

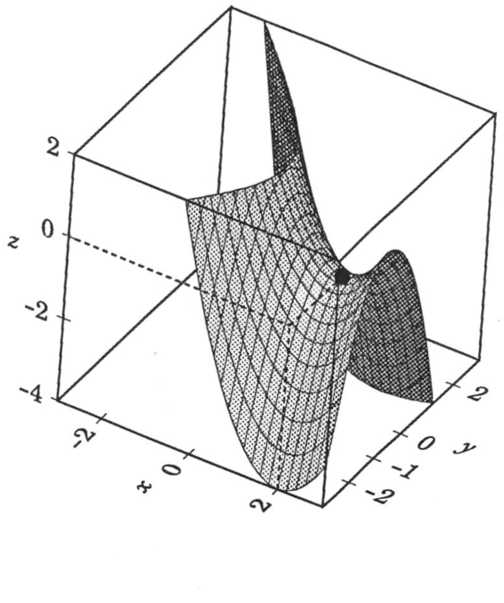

25.3 Functions with Restrictions

Side Condition
Constraint

Extreme values of functions of more than one independent variable can be found only in the case of a **side condition**, or **constraint**, which relates the variables to each other.

Let us study the function

$$z = f(x, y) = 7\left(1 - \frac{1}{4}x - \frac{1}{6}y\right),$$

represented by a plane in space intercepting the coordinate axes at

$$x = 4; \quad y = 6; \quad z = 7,$$

the variables x and y being bound by the constraint

$$g(x, y): \quad (x-2)^2 + (y-3)^2 - 1 = 0,$$

which is a circle of radius 1 centered at (2, 3) in the x, y-plane and a right circular cylinder with the circle as base.

The set of points (x, y, z) that satisfies both the equation $z = f(x, y)$ of the inclined plane and the equation $g(x, y)$ of the upright circular cylinder, is the elliptic section of plane and cylinder, the upper half of which is shown in the figure. We seek the maximum and minimum values of z on the ellipse.

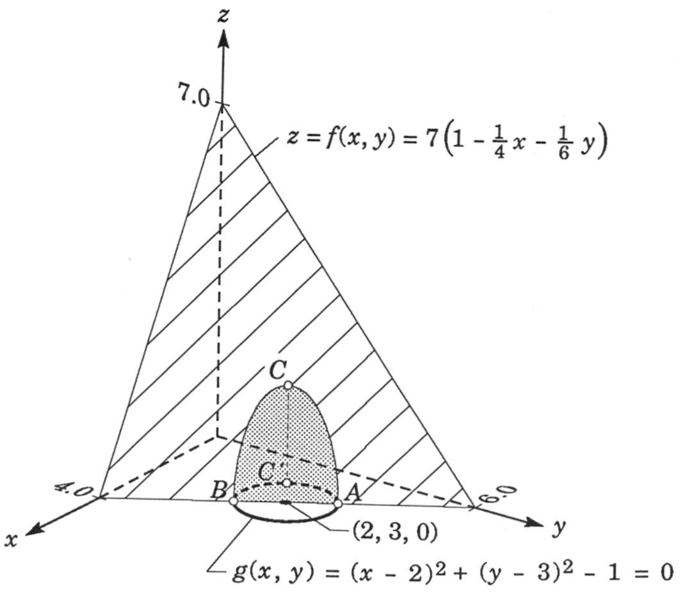

Lagrange's Method of Multipliers

To find extrema of $f(x, y)$ with the constraint $g(x, y) = 0$, we introduce the function

$$F(x, y, \lambda) = f(x, y) + \lambda g(x, y),$$

LAGRANGE Joseph Louis (1736 - 1813)

where λ is **Lagrange's multiplier**, named after the French mathematician and scientist **Joseph Louis Lagrange**.

We differentiate F with respect to x, to y, and to λ,

$$\frac{\partial f}{\partial x} + \lambda \frac{\partial g}{\partial x}; \quad \frac{\partial f}{\partial y} + \lambda \frac{\partial g}{\partial y}; \quad \frac{\partial F}{\partial \lambda},$$

and equate them with zero to form a system of equations, which must be satisfied at any point where a restrained extremum occurs.

Lagrange's method of multipliers can be extended to a function f with any number of independent variables if supplemented by the requisite number of qualifying equations in the same variables.

Thus, keeping within bounds, a function $f(x, y, z, w)$ with the constraining conditions $g(x, y, z, w) = 0$ and $h(x, y, z, w) = 0$ will give the function

$$F(x, y, z, w, \lambda, \mu) = f(x, y, z, w) + \lambda g(x, y, z, w) + \mu h(x, y, z, w),$$

with Lagrange's multipliers λ and μ.

- We return to the function

$$f(x, y) = 7\left(1 - \frac{1}{4}x - \frac{1}{6}y\right)$$

with the constraint

$$g(x, y): \quad (x-2)^2 + (y-3)^2 - 1 = 0;$$

find the extreme values of f.

Define a new function,

$$\begin{aligned} F(x, y, \lambda) &= f(x, y) + \lambda g(x, y) \\ &= 7 - \frac{7}{4}x - \frac{7}{6}y + \lambda(x^2 - 4x + y^2 - 6y + 12). \end{aligned}$$

Partial differentiation of F gives

$$\left(\frac{\partial F}{\partial x}\right) = -\frac{7}{4} + 2\lambda x - 4\lambda; \quad \left(\frac{\partial F}{\partial y}\right) = -\frac{7}{6} + 2\lambda y - 6\lambda; \quad \left(\frac{\partial F}{\partial \lambda}\right) = x^2 - 4x + y^2 - 6y + 12.$$

Letting the first partial derivatives of F equal 0, we arrive at a system of three equations:

$$\left.\begin{aligned} -\frac{7}{4} + 2\lambda x - 4\lambda &= 0 \\ -\frac{7}{6} + 2\lambda y - 6\lambda &= 0 \\ x^2 - 4x + y^2 - 6y + 12 &= 0 \end{aligned}\right\} \quad \left.\begin{aligned} \lambda &= \frac{7}{8x - 16} \\ \lambda &= \frac{7}{12y - 36} \end{aligned}\right\} \quad x = \frac{3}{2}y - \frac{5}{2}.$$

Insert the obtained value of x in $\dfrac{\partial F}{\partial \lambda} = 0$,

$$\left(\frac{3}{2}\,y - 2\right)^2 - 4\left(\frac{3}{2}\,y - 2\right) + y^2 - 6y + 12 = 0$$

$$y^2 - 6y + \frac{113}{13} = 0$$

$$y = 3 \pm \frac{2}{13}\sqrt{13}\,.$$

$$x = 2 \pm \frac{3}{13}\sqrt{13}\,.$$

$$f(x_1, y_1) = 7\left[1 - \frac{1}{4}\left(2 + \frac{3}{13}\sqrt{13}\right) - \frac{1}{6}\left(3 + \frac{2}{13}\sqrt{13}\right)\right] = -\frac{7}{12}\sqrt{13}\,.$$

$$f(x_2, y_2) = 7\left[1 - \frac{1}{4}\left(2 - \frac{3}{13}\sqrt{13}\right) - \frac{1}{6}\left(3 - \frac{2}{13}\sqrt{13}\right)\right] = \frac{7}{12}\sqrt{13}\,.$$

Thus, the maximum value of $f(x, y)$ subject to the constraint $(x-2)^2 + (y-3)^2 = 1$ is

$$f\left(2 - \frac{3}{13}\sqrt{13}\,,\; 3 - \frac{2}{13}\sqrt{13}\right) = \frac{7}{12}\sqrt{13}$$

and the minimum value is

$$f\left(2 + \frac{3}{13}\sqrt{13}\,,\; 3 + \frac{2}{13}\sqrt{13}\right) = -\frac{7}{12}\sqrt{13}\,.$$

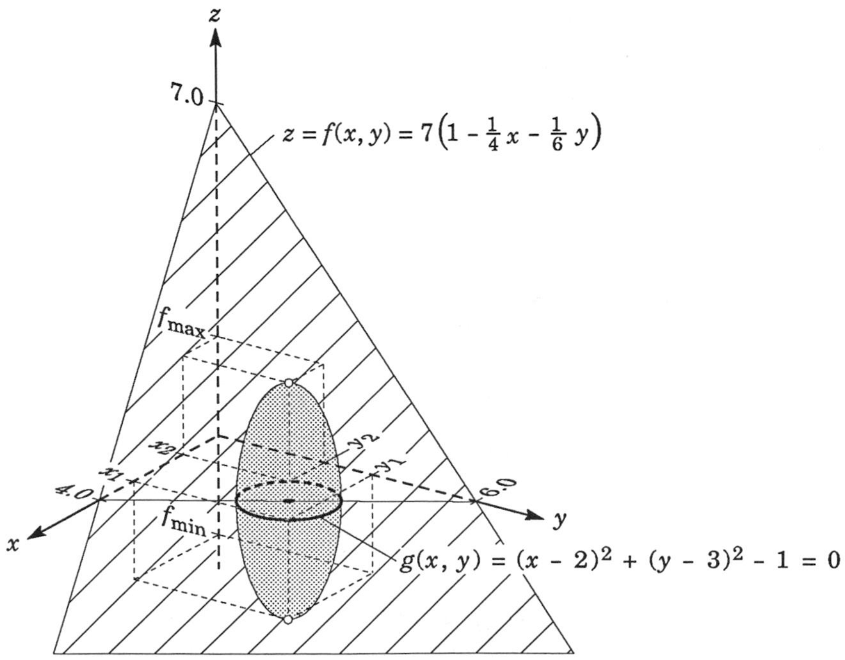

- Find the maximum and minimum values of

$$f(x, y, z) = 5x + y + 2z,$$

subject to the constraints

$$x + y - z - 1 = 0 \text{ and } y^2 + z^2 - 1 = 0.$$

Form the function

$$F(x, y, z, \lambda, \mu) = 5x + y + 2z + \lambda(x + y - z - 1) + \mu(y^2 + z^2 - 1).$$

Equating the derivatives with zero, we have the following system of equations, and solving for x, y, and z:

$$\left.\begin{aligned} 5 + \lambda &= 0 \\ 1 + \lambda + 2\mu y &= 0 \\ 2 - \lambda + 2\mu z &= 0 \\ x + y - z - 1 &= 0 \\ y^2 + z^2 - 1 &= 0 \end{aligned}\right\} \quad \begin{aligned} x_1 &= 1 - \frac{11}{\sqrt{65}} & x_2 &= 1 + \frac{11}{\sqrt{65}} \\ y_1 &= \frac{4}{\sqrt{65}} & y_2 &= -\frac{4}{\sqrt{65}} \\ z_1 &= -\frac{7}{\sqrt{65}} & z_2 &= \frac{7}{\sqrt{65}} \end{aligned}$$

We obtain the extreme values

$$f(x_1, y_1, z_1) = 5\left(1 - \frac{11}{\sqrt{65}}\right) + \frac{4}{\sqrt{65}} - 2 \cdot \frac{7}{\sqrt{65}} = 5 - \sqrt{65}$$

and

$$f(x_2, y_2, z_2) = 5\left(\frac{11}{\sqrt{65}} + 1\right) - \frac{4}{\sqrt{65}} + 2 \cdot \frac{7}{\sqrt{65}} = 5 + \sqrt{65}.$$

Subject to the constraints $(x + y - z - 1 = 0)$ and $(y^2 + z^2 - 1 = 0)$, $f(x, y, z)$ has a maximum

$$f\left(1 + \frac{11}{\sqrt{65}},\ -\frac{4}{\sqrt{65}},\ \frac{7}{\sqrt{65}}\right) = 5 + \sqrt{65}$$

and a minimum

$$f\left(1 - \frac{11}{\sqrt{65}},\ \frac{4}{\sqrt{65}},\ -\frac{7}{\sqrt{65}}\right) = 5 - \sqrt{65}.$$

- Find maxima and minima of

$$f(x, y, z, w) = x\,y\,z\,w$$

with the constraining condition

$$g(x, y, z, w) = 3x^2 + 4y - z + 8w - 84 = 0.$$

We introduce the function

$$\begin{aligned} F(x, y, z, w, \lambda) &= f(x, y, z, w) + \lambda g(x, y, z, w) \\ &= x\,y\,z\,w + \lambda(3x^2 + 4y - z + 8w - 84). \end{aligned}$$

Partial differentiation gives

$$\left(\frac{\partial F}{\partial x}\right) = y\,z\,w + 6\,\lambda\,x \qquad \left(\frac{\partial F}{\partial y}\right) = x\,z\,w + 4\,\lambda$$

$$\left(\frac{\partial F}{\partial z}\right) = x\,y\,w - \lambda \qquad \left(\frac{\partial F}{\partial w}\right) = x\,y\,z + 8\,\lambda$$

$$\left(\frac{\partial F}{\partial \lambda}\right) = 3\,x^2 + 4\,y - z + 8\,w - 84\,.$$

Equating the first derivatives with zero, we have

$$\left.\begin{aligned} y\,z\,w + 6\,\lambda\,x &= 0 \\ x\,z\,w + 4\,\lambda &= 0 \\ x\,y\,w - \lambda &= 0 \\ x\,y\,z + 8\,\lambda &= 0 \\ 3\,x^2 + 4\,y - z + 8\,w - 84 &= 0 \end{aligned}\right\} \quad \left.\begin{aligned} x\,y\,z\,w + 6\,\lambda\,x^2 &= 0 \\ x\,y\,z\,w + 4\,y\,\lambda &= 0 \\ x\,y\,z\,w - z\,\lambda &= 0 \\ x\,y\,z\,w + 8\,w\,\lambda &= 0 \end{aligned}\right\} \quad \begin{aligned} y &= \frac{3}{2}\,x^2 \\ z &= -6\,x^2 \\ w &= \frac{3}{4}\,x^2 \end{aligned}$$

There are also (infinitely many) solutions where $\lambda = 0$. Since $f(x, y, z, w) = 0$ at all of them, these solutions will not affect the final result, and are thus omitted.

Insert the above solutions for y, z, and w in $\frac{\partial F}{\partial \lambda}$ and equate with zero,

$$3\,x^2 + 4 \cdot \frac{3}{2}\,x^2 + 6\,x^2 + 8 \cdot \frac{3}{4}\,x^2 - 84 = 0\,;\; x^2 = 4\,,\;\; x_1 = 2\,,\;\; x_2 = -2\,.$$

$$\left.\begin{aligned} y_1 &= \frac{3}{2}\,x_1^2 = 6 \\ z_1 &= -6\,x_1^2 = -24 \\ w_1 &= \frac{3}{4}\,x_1^2 = 3 \end{aligned}\right\} \qquad \left.\begin{aligned} y_2 &= \frac{3}{2}\,x_2^2 = 6 \\ z_2 &= 6\,x_1^2 = -24 \\ w_2 &= \frac{3}{4}\,x_2^2 = 3 \end{aligned}\right\}$$

The extreme values of $f(x, y, z, w)$ restricted by g are

$$f(2, 6, -24, 3) = 2 \cdot 6 \cdot (-24) \cdot 3 = -864$$

and

$$f(-2, 6, -24, 3) = -2 \cdot 6 \cdot (-24) \cdot 3 = 864\,.$$

Chapter 26

ARC LENGTH

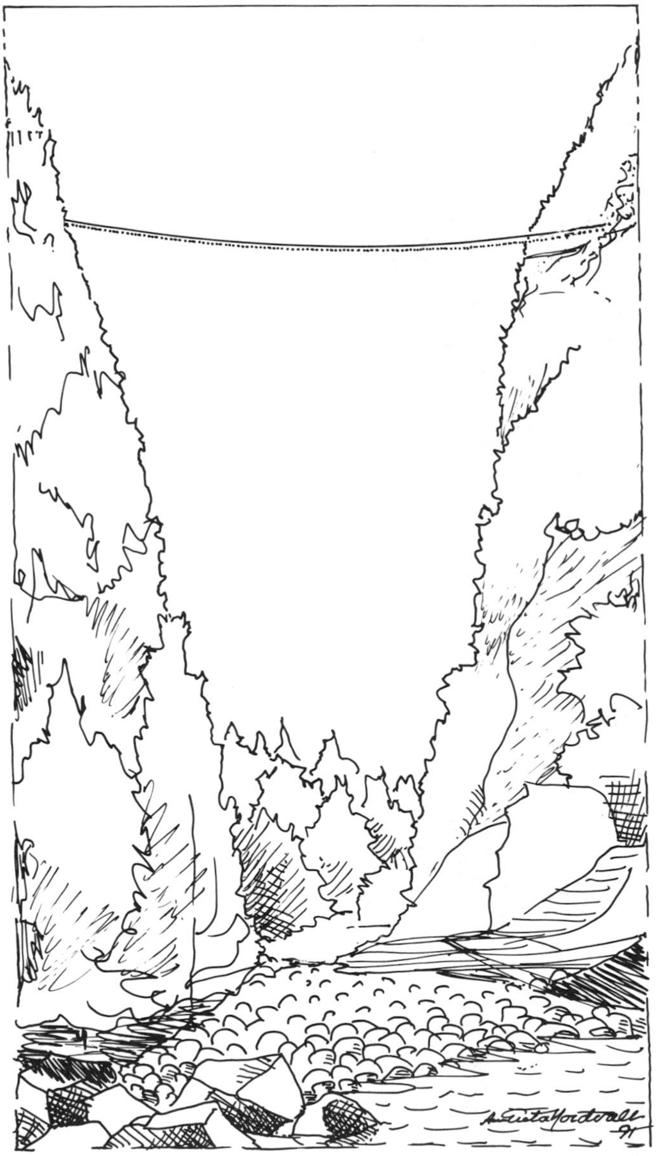

Capilano Suspension Bridge, British Columbia, Canada.

26.1 Basic Principle

The length of an arc of a curve is approximately equal to the sum of the lengths of the straight lines connecting subsequent points on the curve.

Assume that $f(x)$ is a function with a continuous derivative over the interval $[a, b]$ which is divided into n intervals of equal length with endpoints $x_0, x_1 \dots x_n$, so that $[a, b] = [x_0, x_n]$:

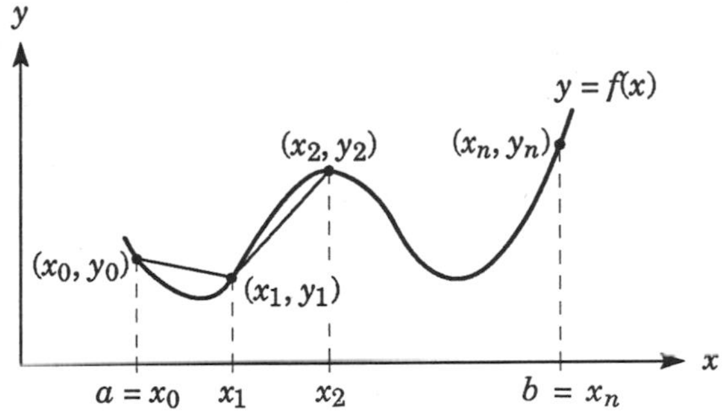

Denoting the arc length by s, we have

p. 550

$$s \approx \sqrt{(x_1 - x_0)^2 + (y_1 - y_0)^2} + \sqrt{(x_2 - x_1)^2 + (y_2 - y_1)^2} + \dots + \sqrt{(x_n - x_{n-1})^2 + (y_n - y_{n-1})^2}$$

or

$$s \approx \sum_{i=1}^{n} \sqrt{(\Delta x_i)^2 + (\Delta y_i)^2} .$$

If we let n tend toward infinity, then

$$s = \lim_{n \to \infty} \sum_{i=1}^{n} \sqrt{(\Delta x_i)^2 + (\Delta y_i)^2} ,$$

which can be rearranged to

$$s = \lim_{n \to \infty} \sum_{i=1}^{n} \sqrt{1 + \left(\frac{\Delta y_i}{\Delta x_i}\right)^2} \Delta x_i .$$

Since $f(x)$ is continuous on each interval $[x_{i-1}, x_i]$, we find by Lagrange's mean-value theorem that there exists a value ξ_i between x_{i-1} and x_i such that

p. 716

$$\frac{f(x_i) - f(x_{i-1})}{x_i - x_{i-1}} = f'(\xi_i) .$$

Since $\Delta y_i = y_i - y_{i-1} = f(x_i) - f(x_{i-1})$ and $\Delta x_i = x_i - x_{i-1}$, we get

$$\frac{\Delta y_i}{\Delta x_i} = f'(\xi_i),$$

and we may write

$$s = \lim_{n \to \infty} \sum_{i=1}^{n} \sqrt{1 + [f'(\xi_i)]^2} \; \Delta x_i ,$$

equivalent to

$$s = \int_a^b \sqrt{1 + [f'(x)]^2} \, dx,$$

which is the formula for the **arc length** of $y = f(x)$ over the interval $[a, b]$ on the x-axis.

Similarly, integration with respect to y gives

$$s = \int_c^d \sqrt{1 + [g'(y)]^2} \, dy.$$

- Find the arc length of the curve

$$f(x) = \sqrt{x^3}$$

from $x = 0$ to $x = 4$.

$$f'(x) = \frac{d}{dx}[x^{3/2}] = \frac{3}{2} x^{1/2} .$$

Let the sought arc length be s;

$$s = \int_0^4 \sqrt{1 + \left(\frac{3}{2} x^{1/2}\right)^2} \, dx = \int_0^4 \sqrt{1 + \frac{9}{4} x} \, dx.$$

Let $1 + \frac{9}{4} x = u$; $dx = \frac{4}{9} du$.

With the new limits

$$x = 0 \quad \Rightarrow \quad u = 1 + \frac{9}{4} \cdot 0 = 1$$

$$x = 4 \quad \Rightarrow \quad u = 1 + \frac{9}{4} \cdot 4 = 10 ,$$

$$s = \int_1^{10} \sqrt{u} \, \frac{4}{9} \, du = \frac{4}{9} \int_1^{10} u^{1/2} \, du = \frac{8}{27} \Big|_1^{10} u^{3/2}$$

$$= \frac{8}{27} \left(\sqrt{1000} - 1\right) = 9.073 \ldots \text{ units.}$$ •

The presence of a radical expression in the integrand of the formula for arc length often precludes the expression of the indefinite integral in terms of an elementary function; therefore, **numerical integration**, described in Chapter 32, is commonly used to find the length of arcs. p. 947

26.2 The Catenary

A catenary is the curve that a uniform, inextensible, and flexible cable or wire, strung freely between two fixed points, assumes under the influence of gravity. The catenary is an application of the **hyperbolic cosine function** and has the equation

p. 538

$$f(x) = a\left(\frac{e^{x/a} + e^{-x/a}}{2}\right) \quad \text{or} \quad f(x) = a\cosh\frac{x}{a}$$

with the vertex at $(0, a)$ in an orthogonal coordinate system.

Let s be the arc of the catenary between $x = 0$ and $x = b$:

$$s = \int_0^b \sqrt{1 + \left(\frac{d}{dx}\, a \cosh\frac{x}{a}\right)^2}\, dx$$

p. 702

$$= \int_0^b \sqrt{1 + \left(\sinh\frac{x}{a}\right)^2}\, dx$$

Pythagorean Identity: *p.* 541

$$= \int_0^b \sqrt{\left(\cosh\frac{x}{a}\right)^2}\, dx$$

$$= \int_0^b \cosh\frac{x}{a}\, dx$$

p. 725

$$= a \sinh\frac{b}{a}.$$

26.3 Arc Length in Parametric Form

A two-dimensional curve may be represented by two **parametric equations**, which give the coordinates x and y in terms of a third variable, the parameter. Similarly, a three-dimensional curve may be represented by three parametric equations, giving the coordinates x, y, and z in terms of the parameter.

A formula for arc length in parametric form can be developed from the formula for arc length,

$$s = \int_a^b \sqrt{1 + [f'(x)]^2}\, dx\,; \quad s = \int_a^b \sqrt{1 + \left(\frac{dy}{dx}\right)^2}\, dx\,.$$

where s is the **length of an arc** of $y = f(x)$, within the interval $[a, b]$ on the x-axis, where $f'(x)$ is required to be continuous.

Consider a plane curve in parametric equations

$$\left.\begin{aligned} x &= x(t) \\ y &= y(t) \end{aligned}\right\}$$

where $a \le t \le b$, and the functions $x(t)$ and $y(t)$ have continuous derivatives in $[a, b]$.

We have

$$ds = \sqrt{1 + \left(\frac{dy}{dx}\right)^2}\, dx\,.$$

Since $\dfrac{dy}{dx} = \dfrac{dy/dt}{dx/dt}$ if $\dfrac{dx}{dt} \ne 0$, we may rewrite this in the variable t; thus,

$$\begin{aligned} ds &= \sqrt{1 + \left(\frac{dy/dt}{dx/dt}\right)^2}\, dx \\ &= \sqrt{\frac{(dx/dt)^2 + (dy/dt)^2}{(dx/dt)^2}}\, dx \\ &= \sqrt{\left(\frac{dx}{dt}\right)^2 + \left(\frac{dy}{dt}\right)^2} \cdot \frac{dt}{dx}\, dx \\ &= \sqrt{\left(\frac{dx}{dt}\right)^2 + \left(\frac{dy}{dt}\right)^2}\, dt\,. \end{aligned}$$

Thus, the formula in **parametric form for arc length s in the plane** is

$$s = \int_a^b \sqrt{\left(\frac{dx}{dt}\right)^2 + \left(\frac{dy}{dt}\right)^2}\, dt$$

for a curve $x = x(t)$, $y = y(t)$, where $a \le t \le b$, and the functions $x(t)$ and $y(t)$ have continuous derivatives in $[a, b]$.

Analogously, the formula in **parametric form for arc length s in space** is

$$s = \int_a^b \sqrt{\left(\frac{dx}{dt}\right)^2 + \left(\frac{dy}{dt}\right)^2 + \left(\frac{dz}{dt}\right)^2}\, dt$$

for a function $x = x(t)$, $y = y(t)$, $z = z(t)$ where $a \le t \le b$, and the functions $x(t)$, $y(t)$, $z(t)$ have continuous derivatives in $[a, b]$.

- Find the length of the perimeter of the astroid

$$\left.\begin{aligned} x &= 2\cos^3 t \\ y &= 2\sin^3 t \end{aligned}\right\} .$$

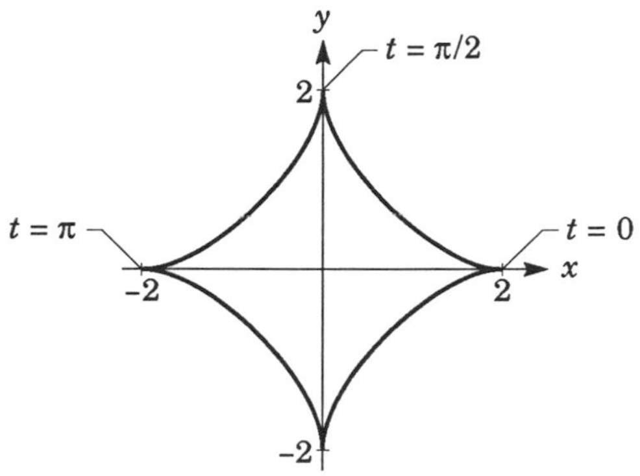

Because of the symmetry of the astroid, the perimeter s is 4 times the length of the arc in one quadrant,

$$s = 4 \int_0^{\pi/2} \sqrt{[x'(t)]^2 + [y'(t)]^2}\, dt .$$

Since

$$\left.\begin{aligned} x'(t) &= 2 \cdot 3\cos^2 t\,(-\sin t) = -6\cos^2 t\,\sin t \\ y'(t) &= 2 \cdot 3\sin^2 t\,(\cos t) = 6\sin^2 t\,\cos t \end{aligned}\right\}$$

we obtain

$$s = 4 \int_0^{\pi/2} \sqrt{(-6\cos^2 t\,\sin t)^2 + (6\sin^2 t\,\cos t)^2}\, dt$$

$$= 24 \int_0^{\pi/2} \sqrt{\cos^4 t\,\sin^2 t + \sin^4 t\,\cos^2 t}\, dt$$

$$s = 24 \int_0^{\pi/2} \sqrt{\sin^2 t \ \cos^2 t \ (\cos^2 t + \sin^2 t)} \ dt$$

$$= 24 \int_0^{\pi/2} \sqrt{\sin^2 t \ \cos^2 t} \ dt$$

$$= 12 \int_0^{\pi/2} 2 \sin t \ \cos t \ dt$$

$$= 12 \int_0^{\pi/2} \sin 2t \ dt$$

$$= -6 \Big|_0^{\pi/2} \cos 2t \ = -6(-1-1) = 12 \text{ length units.}$$

I bake apple pies in threes
For a fellow who likes to keep bees.
We measure arcs in parametric form
While the pies are still warm!
Then we sit eating pie in the trees.

26.4 Arc Length in a Polar Coordinate System

To find the arc length of a curve expressed in polar coordinates, we observe that

$$\left.\begin{aligned} x &= r\cos\theta \\ y &= r\sin\theta \end{aligned}\right\},$$

as illustrated by the graph:

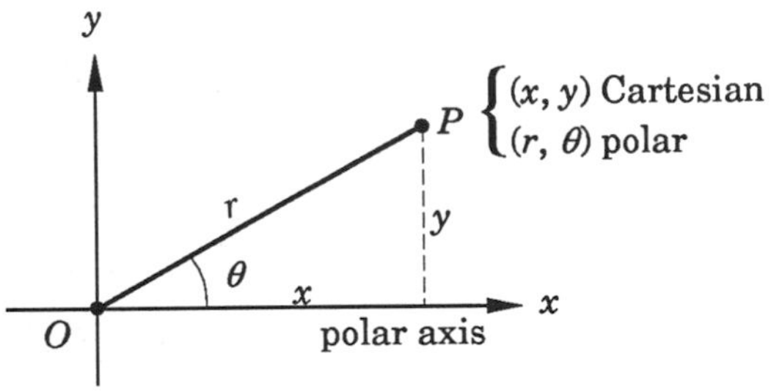

Assuming that $r = f(\theta)$, where f has a continuous derivative, we have the arc length s,

$$s = \int_{\alpha}^{\beta} \sqrt{\left(\frac{dx}{d\theta}\right)^2 + \left(\frac{dy}{d\theta}\right)^2}\, d\theta$$

in which we insert

$$\frac{dx}{d\theta} = \frac{dr}{d\theta}\cos\theta - r\sin\theta; \quad \frac{dy}{d\theta} = \frac{dr}{d\theta}\sin\theta + r\cos\theta$$

and obtain

$$s = \int_{\alpha}^{\beta} \sqrt{\left(\frac{dr}{d\theta}\cos\theta - r\sin\theta\right)^2 + \left(\frac{dr}{d\theta}\sin\theta + r\cos\theta\right)^2}\, d\theta$$

or, after simplification,

$$s = \int_{\alpha}^{\beta} \sqrt{r^2 + \left(\frac{dr}{d\theta}\right)^2}\, d\theta,$$

where s is the length of an arc of $r = f(\theta)$ between $\theta = \alpha$ and $\theta = \beta$.

- Find the length of the arc of the spiral

$$r = 2\,e^{-\theta}$$

between $\theta = 0$ and $\theta = 2\pi$.

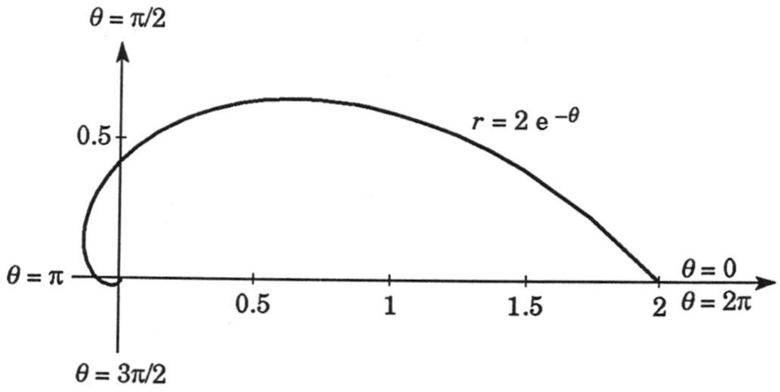

$$s = \int_0^{2\pi} \sqrt{\left(2e^{-\theta}\right)^2 + \left(\frac{d}{d\theta}\,[2\,e^{-\theta}]\right)^2}\; d\theta$$

$$= \int_0^{2\pi} \sqrt{\left(4e^{-2\theta}\right) + \left(4\,e^{-2\theta}\right)}\; d\theta$$

$$= 2\int_0^{2\pi} \sqrt{2e^{-2\theta}}\, d\,\theta = 2\sqrt{2}\int_0^{2\pi} e^{-\theta}\; d\theta$$

$$= 2\sqrt{2}\,\left[-e^{-\theta}\right]_0^{2\pi} = 2\sqrt{2}\,[-e^{-2\pi} + e^{0}]$$

$$= 2\sqrt{2}\,(1 - e^{-2\pi}) \approx 2.823 \text{ length units.}$$

Chapter 27

CENTROIDS

27.1 Mass Point Systems

Suppose that $m_1, m_2 \dots m_n$ are the masses of n objects attached to a rod supported by a knife edge acting as a fulcrum.

Objects to the right of the fulcrum tend to swing the rod clockwise, whereas those on the left move it counterclockwise.

Our task is to find a point on the rod that should be placed directly over the fulcrum in order to obtain exact balance of the rod and the attached objects.

To solve the problem, let the rod – when exactly balanced – coincide with the x-axis of an orthogonal coordinate system, and let the fulcrum be located at a point a:

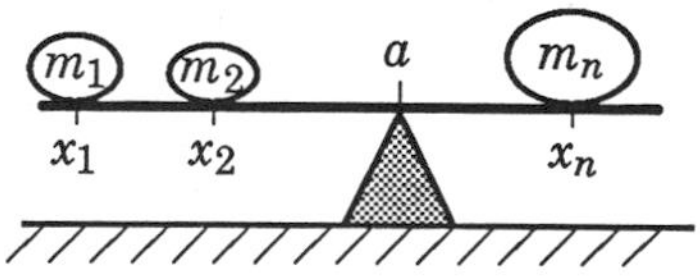

Moment Arm
Moment of Mass

Let d_i be the distance $(x_i - a)$, that is, the **moment arm** of m_i; the product of m_i and the distance d_i is then the **moment of mass** of m_i with respect to a. Assuming the mass of the rod to be negligible in comparison with the masses, the total moment of mass M with respect to a is the sum of all the moments of mass; thus,

$$M = \sum_{i=1}^{n} m_i \, d_i \; .$$

The rod is in exact equilibrium if and only if $M = 0$.

Centroid, Center of Mass,
Center of Gravity, Barycenter
Greek, *barys*, "heavy"

If $M = 0$ around a point a of an object, a is the **centroid** of the object, and commonly denoted $\overline{x}$. Alternative terms are **center of mass**, **center of gravity**, and **barycenter**.

In a state of equilibrium, we have

$$\sum_{i=1}^{n} m_i (x_i - \bar{x}) = 0$$

and

$$\bar{x} = \frac{\sum_{i=1}^{n} m_i x_i}{\sum_{i=1}^{n} m_i} .$$

If m denotes the total sum of the masses of the system,

$$m = \sum_{i=1}^{n} m_i ,$$

and M_0 is the sum of the individual moments of mass,

$$M_0 = \sum_{i=1}^{n} m_i x_i ,$$

we have

$$\bar{x} = \frac{M_0}{m} .$$

- Three objects with masses 2, 3, and 6 are placed on the x-axis at $x = -3$, $x = 4$, and $x = 7$, respectively; find the position of the centroid of the system.

$$\bar{x} = \frac{M_0}{m} = \frac{2 \cdot (-3) + 3 \cdot 4 + 6 \cdot 7}{2 + 3 + 6} = \frac{48}{11}$$

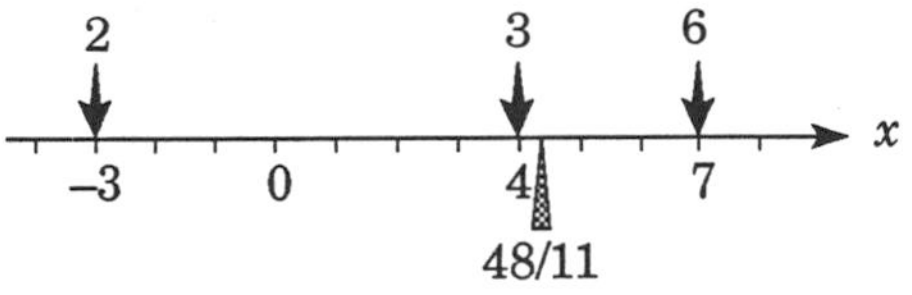

•

The concept of a centroid is applicable also to systems in the plane and in space.

Consider n objects with masses $m_1, m_2 \dots m_n$ located in the x, y-plane at (x_1, y_1), $(x_2, y_2) \dots (x_n, y_n)$, respectively.

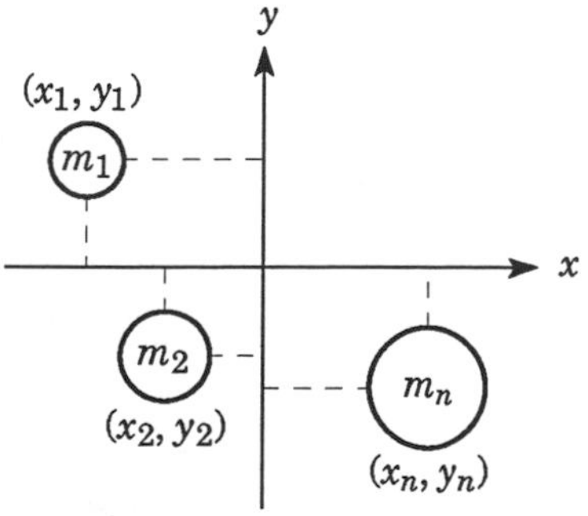

If M_y is the total moment of mass with respect to the y-axis, we have

$$M_y = m_1 x_1 + m_2 x_2 + \ldots + m_n x_n ,$$

and similarly, if M_x is the total moment of mass with respect to the x-axis,

$$M_x = m_1 y_1 + m_2 y_2 + \ldots + m_n y_n .$$

If m is the total mass of the system, the location of the centroid $(\bar{x}, \bar{y})$ is given by

$$\bar{x} = \frac{M_y}{m} \quad \text{and} \quad \bar{y} = \frac{M_x}{m} .$$

- Three masses 2, 3, and 6 are placed at $(-3,-2)$, $(4, 2)$, and $(7, -1)$, respectively, in the x, y-plane; determine the position of the centroid of the system.

$$\bar{x} = \frac{M_y}{m} = \frac{2 \cdot (-3) + 3 \cdot 4 + 6 \cdot 7}{2 + 3 + 6} = \frac{48}{11}$$

$$\bar{y} = \frac{M_x}{m} = \frac{2 \cdot (-2) + 3 \cdot 2 + 6 \cdot (-1)}{2 + 3 + 6} = -\frac{4}{11}$$

The center of mass is at $\left(\frac{48}{11}, -\frac{4}{11}\right)$.

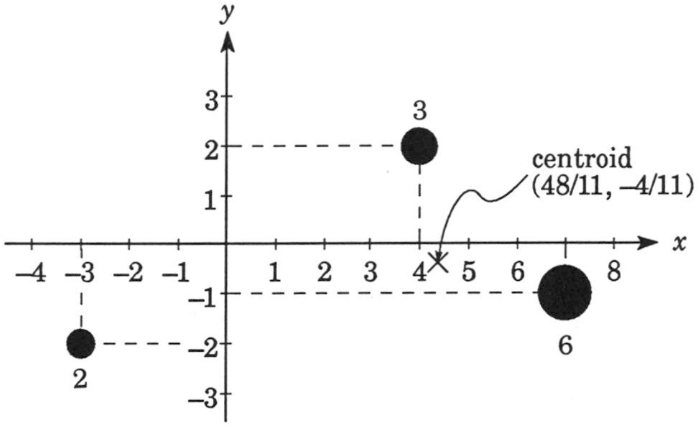

Analogously, for n objects with masses $m_1, m_2 \ldots m_n$ located at (x_1, y_1, z_1), $(x_2, y_2, z_2) \ldots (x_n, y_n, z_n)$ in the x, y, z-space, we have

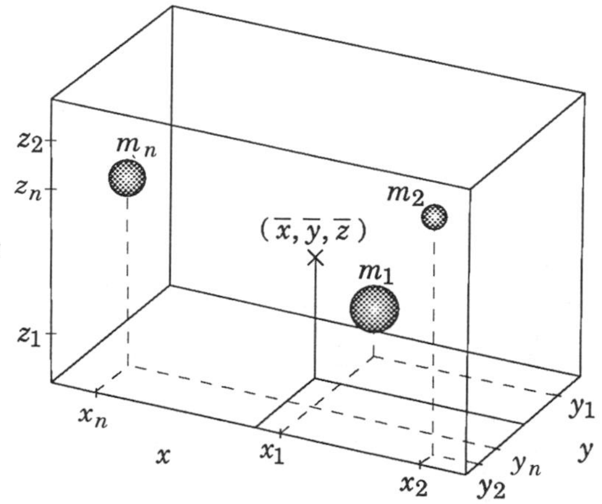

If M_{yz}, M_{zx}, and M_{xy} denote the total moment of mass with respect to the y, z-plane, the z, x-plane, and the x, y-plane, respectively, and m the sum total of the masses $m_1, m_2 \ldots m_n$, the point $(\bar{x}, \bar{y}, \bar{z})$ of the centroid will be

$$(\bar{x}, \bar{y}, \bar{z}) = \left(\frac{M_{yz}}{m}, \frac{M_{zx}}{m}, \frac{M_{xy}}{m}\right).$$

- Three masses 2, 3, and 6 are placed at $(-3, -2, 1)$, $(4, 2, 3)$, and $(7, -1, 0)$, respectively, in an x, y, z-space; find the position of the centroid of the system.

$$\bar{x} = \frac{M_{yz}}{m} = \frac{3 \cdot (-3) + 3 \cdot 4 + 6 \cdot 7}{2 + 3 + 6} = \frac{48}{11}$$

$$\bar{y} = \frac{M_{xz}}{m} = \frac{2 \cdot (-2) + 3 \cdot 2 + 6 \cdot (-1)}{2 + 3 + 6} = -\frac{4}{11}$$

$$\bar{z} = \frac{M_{xy}}{m} = \frac{2 \cdot 1 + 3 \cdot 3 + 6 \cdot 0}{2 + 3 + 6} = 1$$

The centroid is at $\left(\frac{48}{11}, -\frac{4}{11}, 1\right)$.

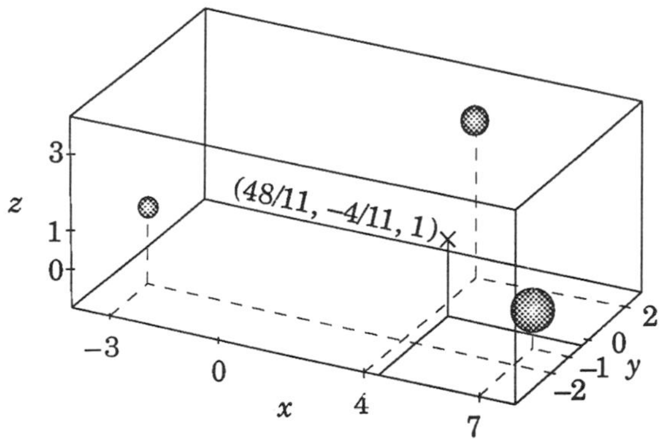

27.2 Plane Figures and Laminas

A thin sheet of uniform thickness and density is known as a lamina. If it has a geometrical center, this point will also be its centroid (or center of mass, center of gravity, or barycenter).

The centroid of a lamina of rectangular shape is the point of intersection of its space diagonals.

- Given a lamina of the shape shown here, find the location of its centroid.

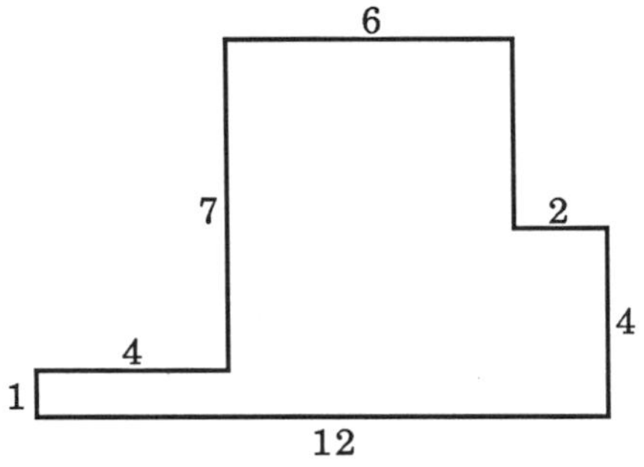

Divide the lamina into rectangular elements as shown by the broken lines and place it on the x-axis of an orthogonal coordinate system, with the y-axis at the midpoint of the bottom line.

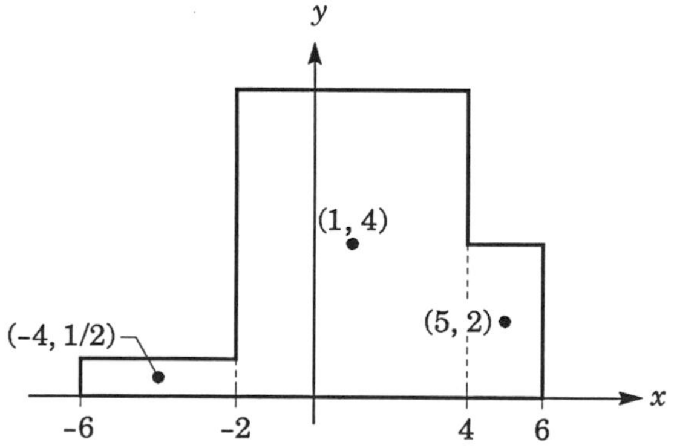

The mass of each section is the product of the area of the rectangle, the thickness δ of the lamina, and the density ρ of the material.

We find

$$\bar{x} = \frac{\left[4 \cdot 1 \cdot (-4) + 6 \cdot 8 \cdot 1 + 2 \cdot 4 \cdot 5\right] \delta\rho}{\left(4 \cdot 1 + 6 \cdot 8 + 2 \cdot 4\right) \delta\rho} = \frac{6}{5} ;$$

$$\bar{y} = \frac{\left(4 \cdot 1 \cdot 0.5 + 6 \cdot 8 \cdot 4 + 2 \cdot 4 \cdot 2\right) \delta\rho}{\left(4 \cdot 1 + 6 \cdot 8 + 2 \cdot 4\right) \delta\rho} = \frac{7}{2} .$$

Thus, the centroid of the lamina is at $\left(1\frac{1}{5}, 3\frac{1}{2}\right)$, if placed on the x-axis (with the midpoint of the bottom line at the origin) of an orthogonal coordinate system. •

To obtain a more general formula for the centroid of a lamina, consider a lamina bounded by the curves of the continuous functions $f(x)$ and $g(x)$ between $x = a$ and $x = b$ $(a < b)$. We divide the interval $[a, b]$ into n sub-intervals with the points of subdivision

$$a = x_0 < x_1 < x_2 < \ldots < x_n = b\,.$$

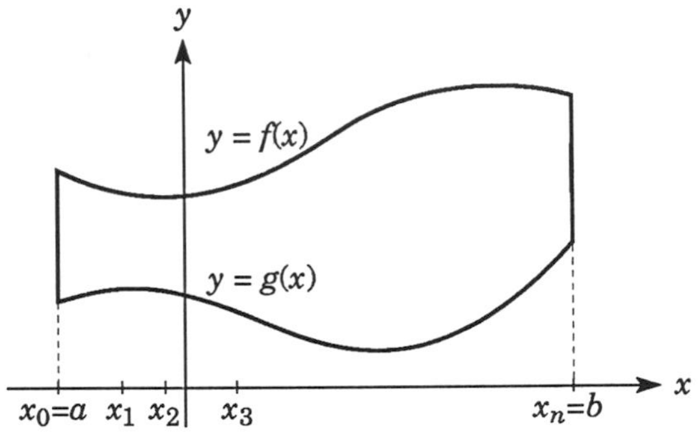

Let x_i be the midpoint of each sub-interval, and construct on each interval a rectangle with sides parallel to the coordinate axes, its breadth Δx_i being equal to the length of the sub-interval and its height equal to $[f(x) - g(x)]$:

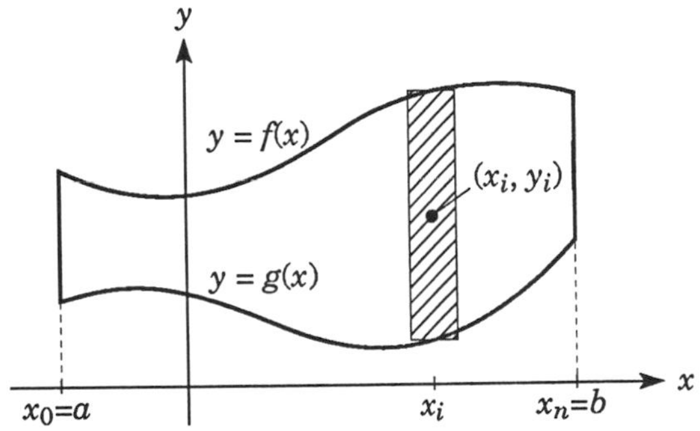

The **geometrical center** of each rectangle is at (x_i, y_i), where

$$y_i = \frac{f(x) + g(x)}{2}\,.$$

The mass m_i of the i-th rectangle is the product of its density ρ, thickness δ, and area $[f(x) - g(x)]\,\Delta x_i$,

$$m_i = \rho\,\delta\,[f(x) - g(x)]\,\Delta x_i\,,$$

and the **total mass** of the lamina can be approximated to

$$m \approx \sum_{i=1}^{n} \rho\,\delta\,[f(x) - g(x)]\,\Delta x_i\,.$$

p. 747

With an increasing number n of forever-diminishing rectangles, this **Riemann sum** becomes a successively better approximation of the total mass,

$$m = \lim_{n \to \infty} \sum_{i=1}^{n} \rho\, \delta\, [f(x) - g(x)]\, \Delta x_i,$$

or

$$m = \rho\, \delta \int_a^b [f(x) - g(x)]\, dx,$$

which is the **formula for the mass of a lamina**, where f and g are continuous functions on the interval $[a, b]$, ρ is the density, and δ is the thickness of the lamina.

Moment of Mass of a Lamina

The moment of mass is the product of mass and moment arm; consequently, for the i-th rectangle the moment of mass about the x-axis is $m_i\, y_i$.

Since $y_i = \dfrac{f(x) + g(x)}{2}$ and $m_i = \rho\, \delta\, [f(x) - g(x)]\, \Delta x_i$, the **total moment of mass about the x-axis** may be approximated by

$$M_x \approx \sum_{i=1}^{n} \rho\, \delta\, \frac{f(x) + g(x)}{2}\, [f(x) - g(x)]\, \Delta x_i$$

$$\approx \sum_{i=1}^{n} \rho\, \delta\, \frac{f^2(x) - g^2(x)}{2}\, \Delta x_i.$$

When the sub-intervals tend to 0, this approximation becomes an equality; that is,

$$M_x = \lim_{n \to \infty} \sum_{i=1}^{n} \rho\, \delta\, \frac{f^2(x) - g^2(x)}{2}\, \Delta x_i$$

or

$$M_x = \frac{\rho\, \delta}{2} \int_a^b [f^2(x) - g^2(x)]\, dx,$$

where, as before, f and g are continuous functions in the interval $[a, b]$, ρ is the density, and δ is the thickness of the lamina.

Analogously, the moment of mass about the y-axis of the i-th rectangle is $m_i\, x_i$, and the formula for the **total moment of mass about the y-axis** is

$$M_y = \rho\, \delta \int_a^b x\, [f(x) - g(x)]\, dx.$$

For the centroid $(\bar{x}, \bar{y})$ of the lamina, we obtain

$$\bar{x} = \frac{M_y}{m} = \frac{\rho\,\delta \int_a^b x\,[f(x) - g(x)]\ dx}{\rho\,\delta \int_a^b [f(x) - g(x)]\ dx} = \frac{\int_a^b x\,[f(x) - g(x)]\ dx}{\int_a^b [f(x) - g(x)]\ dx}$$

$$\bar{y} = \frac{M_x}{m} = \frac{\frac{\rho\,\delta}{2} \int_a^b [f^2(x) - g^2(x)]\ dx}{\rho\,\delta \int_a^b [f(x) - g(x)]\ dx} = \frac{\frac{1}{2} \int_a^b [f^2(x) - g^2(x)]\ dx}{\int_a^b [f(x) - g(x)]\ dx}$$

where the denominator is equal to the area of the lamina.

Thus, the **centroid of the lamina** is

$$\bar{x} = \frac{1}{A} \int_a^b x\,[f(x) - g(x)]\ dx$$

$$\bar{y} = \frac{1}{2A} \int_a^b [f^2(x) - g^2(x)]\ dx,$$

where A is the area of a lamina bounded by $x = a$; $x = b$ and the curves of the continuous functions $f(x)$ and $g(x)$.

In the formula, the factors of density and thickness cancel out, showing, as expected, that the location of the centroid is independent of density and mass.

- Determine the location of the centroid of the region bounded by the parabola $y = x^2$ and the chord $y = 2x + 3$.

The points of intersection are given by

$$x^2 = 2x + 3$$

$$\left.\begin{matrix} x = -1 \\ y = 1 \end{matrix}\right\} \quad \left.\begin{matrix} x = 3 \\ y = 9 \end{matrix}\right\}.$$

[Not to scale]

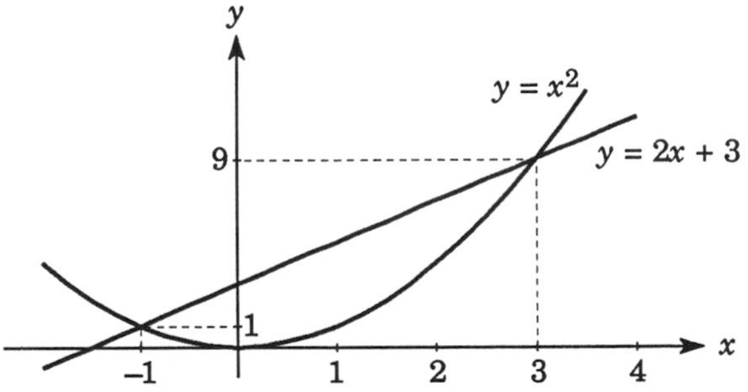

The area A is

$$A = \int_{-1}^{3} (2x + 3 - x^2)\ dx$$

$$= \left[x^2 + 3x - \frac{x^3}{3}\right]_{-1}^{3} = \frac{32}{3}.$$

$$\bar{x} = \frac{1}{A}\int_{-1}^{3} x(2x + 3 - x^2)\ dx$$

$$= \frac{3}{32}\int_{-1}^{3} (2x^2 + 3x - x^3)\ dx$$

$$= \frac{3}{32}\left[\frac{2x^3}{3} + \frac{3x^2}{2} - \frac{x^4}{4}\right]_{-1}^{3}$$

$$= 1$$

$$\bar{y} = \frac{1}{2A}\int_{-1}^{3} \left[(2x + 3)^2 - (x^2)^2\right]\ dx$$

$$= \frac{3}{64}\int_{-1}^{3} (9 + 12x + 4x^2 - x^4)\ dx$$

$$= \frac{3}{64}\left[9x + 6x^2 + \frac{4x^3}{3} - \frac{x^5}{5}\right]_{-1}^{3}$$

$$= \frac{17}{5} = 3.4$$

The coordinates of the centroid are (1, 3.4) .

[Not to scale]

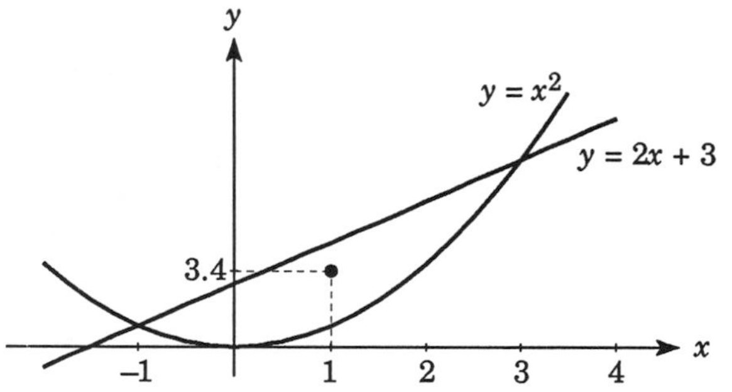

27.3 Center of Mass of Solids of Revolution

While we prefer the term *centroid* for plane figures, the term *center of mass* or *center of gravity* seems more appropriate for solids.

The center of mass of a solid of revolution of uniform density will be situated somewhere on its axis of rotation; the problem is thus reduced to finding out just where on the axis.

We consider a solid of revolution formed when the area bounded by the curve of $f(x)$, the x-axis, and the lines $x = a$ and $x = b$ revolves about the x-axis.

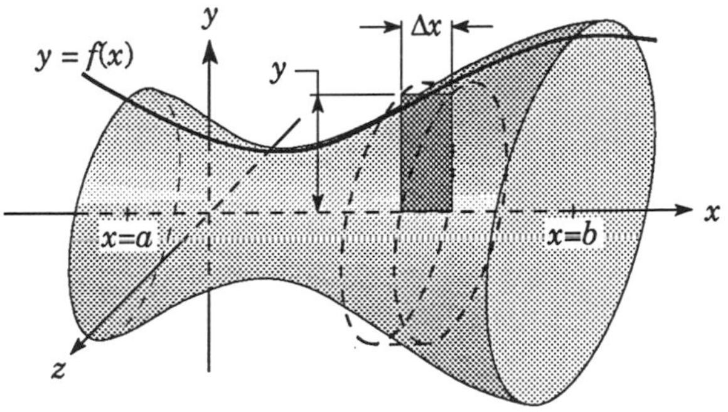

A rectangular section of the area then generates a disk of **volume** $\pi y^2\,\Delta x$ and **mass** $\rho\,\pi y^2\,\Delta x$, where ρ is the density. The center of mass of the disk is also its geometrical center. Since the length of the moment arm is x, the **moment of mass** is $x\,\rho\,\pi y^2\,\Delta x$. Letting Δx tend to 0, we find the **total moment of mass** with respect to the y, z-plane,

$$M_{yz} = \rho\pi \int_a^b x\,[f(x)]^2\,dx,$$

and **the total mass of the solid of rotation**,

$$m = \rho\pi \int_a^b [f(x)]^2\,dx.$$

p. 839 Since $\bar{x} = \dfrac{M_{yz}}{m}$, we obtain the x-coordinate of the **center of mass**,

$$\bar{x} = \frac{\rho\pi \int_a^b x\,[f(x)]^2\,dx}{\rho\pi \int_a^b [f(x)]^2\,dx} = \frac{\int_a^b x\,[f(x)]^2\,dx}{\int_a^b [f(x)]^2\,dx},$$

where $f(x)$ is a continuous function and $a \le x \le b$.

Revolution about the x-axis means that $\bar{y} = \bar{z} = 0$.

By a similar reasoning, the y-coordinate of the **center of mass formed by the area revolving about the y-axis** is

$$\bar{y} = \frac{\int_c^d y\,[g(y)]^2\,dy}{\int_c^d [g(y)]^2\,dy},$$

where $g(y)$ is a continuous function; $c \le y \le d$; and $\bar{x} = \bar{z} = 0$.

- Determine the center of mass of a solid generated by the area rotating about the x-axis and the y-axis, respectively, in the first quadrant of the parabola $y = 9 - x^2$.

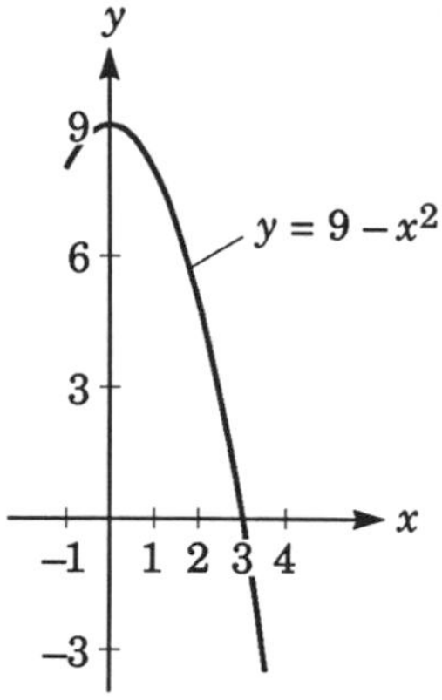

The parabola intersects the coordinate axes at

$$(3, 0) \text{ and } (0, 9).$$

When the curve rotates about the x-axis, the center of mass of the generated solid is on the x-axis; $\bar{y} = \bar{z} = 0$.

$$\bar{x} = \frac{\int_0^3 x\,(9 - x^2)^2\,dx}{\int_0^3 (9 - x^2)^2\,dx}$$

$$= \frac{\int_0^3 (81\,x - 18x^3 + x^5)\,dx}{\int_0^3 (81 - 18\,x^2 + x^4)\,dx}$$

$$= \frac{\left[\frac{81\,x^2}{2} - \frac{9\,x^4}{2} + \frac{x^6}{6}\right]_0^3}{\left[81\,x - 6\,x^3 + \frac{x^5}{5}\right]_0^3} = \frac{15}{16}.$$

The center of mass is $\left(\frac{15}{16}, 0, 0\right)$.

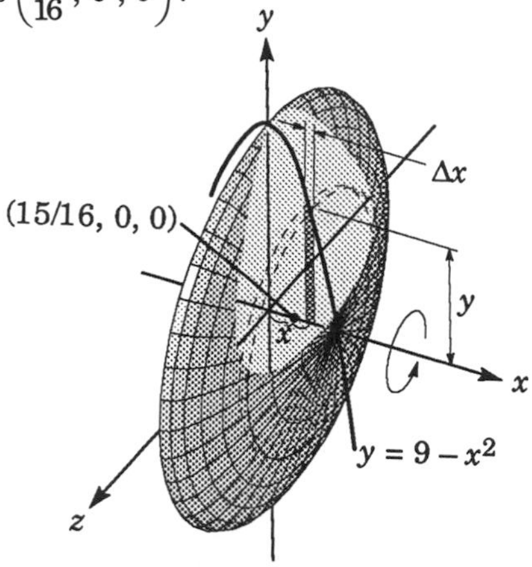

When the curve rotates about the y-axis, the center of mass of the generated solid is on the y-axis; $\bar{x} = \bar{z} = 0$.

$$\bar{y} = \frac{\int_0^9 y\left(\sqrt{9-y}\right)^2 dy}{\int_0^9 \left(\sqrt{9-y}\right)^2 dy} = \frac{\int_0^9 y\,(9-y)\,dy}{\int_0^9 (9-y)\,dy}$$

$$= \frac{\left[\frac{9y^2}{2} - \frac{y^3}{3}\right]_0^9}{\left[9y - \frac{y^2}{2}\right]_0^9} = 3.$$

The center of mass is at $(0, 3, 0)$.

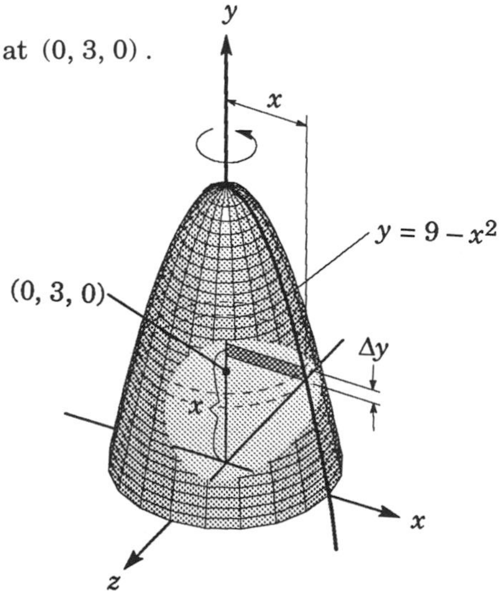

- Find the center of mass of a right circular cone with uniform density and of altitude h and radius r.

The cone may be generated by revolving a right triangle with legs r and h about the leg h. We assume that h is placed along the x-axis of an orthogonal coordinate system with the vertex of the cone at the origin.

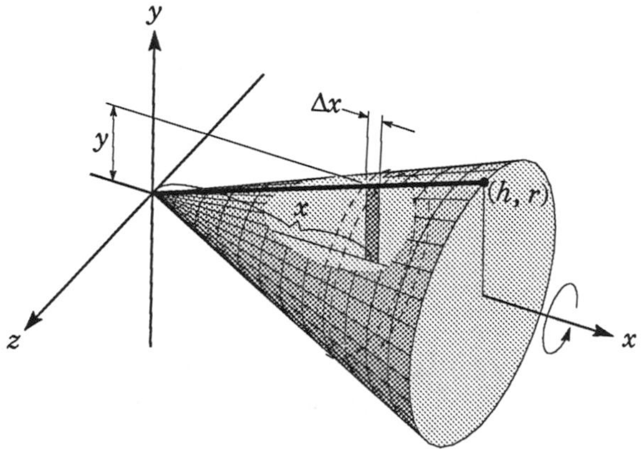

Similar triangles give

$$\frac{r}{h} = \frac{y}{x}\,; \quad y = \frac{r}{h}x\,.$$

$$\overline{x} = \frac{\int_0^h x\left(\frac{r}{h}x\right)^2 \, dx}{\int_0^h \left(\frac{r}{h}x\right)^2 \, dx}$$

$$= \frac{\int_0^h \frac{r^2}{h^2}x^3 \, dx}{\int_0^h \frac{r^2}{h^2}x^2 \, dx} = \Bigg|_0^h \frac{r^2\,x^4}{h^2\,4} \cdot \frac{h^2\,3}{r^2\,x^3} = \frac{3}{4}\cdot h\,.$$

The center of mass of a right circular cone of uniform density is on its axis, at a distance from the vertex of $\frac{3}{4}$ of the height of the cone.

Chapter 28

AREA

28.1 Plane Surfaces

In integration of a function, areas above the x-axis count as positive, those below the x-axis as negative; thus, integration will give the difference between their absolute values.

To find the **actual area**, we must add the absolute values of the integrals, that is, reverse the sign of the area below the x-axis; to be on the safe side, it is good practice to plot a graph of a function whenever in doubt.

Area of an Ellipse

The equation

$$\frac{x^2}{a^2} + \frac{y^2}{b^2} = 1$$

describes an ellipse with the axes $2a$ and $2b$ and with its center at the origin of an orthogonal coordinate system.

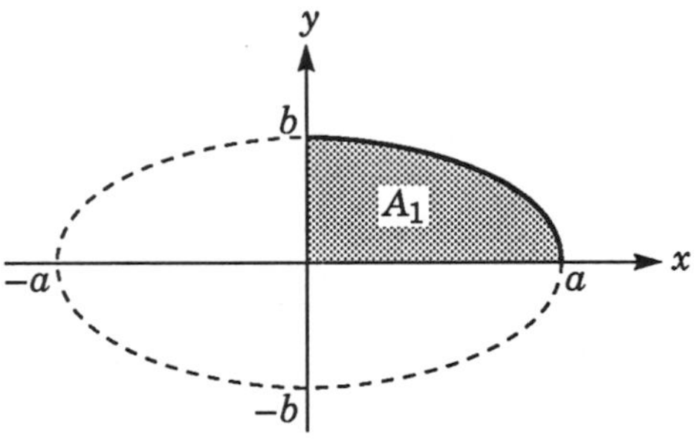

The area of the ellipse is evidently 4 times the area A_1 of the first quadrant. With $y = \frac{b}{a} \cdot \sqrt{a^2 - x^2}$, we have

$$A_1 = \frac{b}{a} \int_0^a \sqrt{a^2 - x^2}\, dx\ .$$

Introducing

p. 738

$$x = a \sin\theta \Rightarrow \sqrt{a^2 - x^2} = a \cos\theta\,;\, dx = a \cos\theta\, d\theta\,,$$

we must find new limit values.

$x = 0:\quad a \sin\theta = 0$	$x = a:\quad a \sin\theta = a$
$\sin\theta = 0$	$\sin\theta = 1$
$\theta = 0$	$\theta = \frac{\pi}{2}$

Thus,

$$A_1 = \frac{b}{a}\int_0^a \sqrt{a^2 - x^2}\,dx = \frac{b}{a}\int_0^{\pi/2} a\cos\theta \cdot a\cos\theta\,d\theta$$

$$= a\,b \int_0^{\pi/2} \cos^2\theta\,d\theta$$

$$= \frac{a\,b}{2}\int_0^{\pi/2} (1 + \cos 2\theta)\,d\theta$$

p. 511

$$= \frac{a\,b}{2}\left[\theta + \frac{1}{2}\sin 2\theta\right]_0^{\pi/2} = \frac{\pi\,a\,b}{4}\,.$$

Hence,

the **area of an ellipse** is $\pi\,a\,b$, where a and b are the semi-axes of the ellipse.

With $a = b = r$,

the **area of a circle** is $\pi\,r^2$, where r is the radius of the circle.

Area between Two Curves

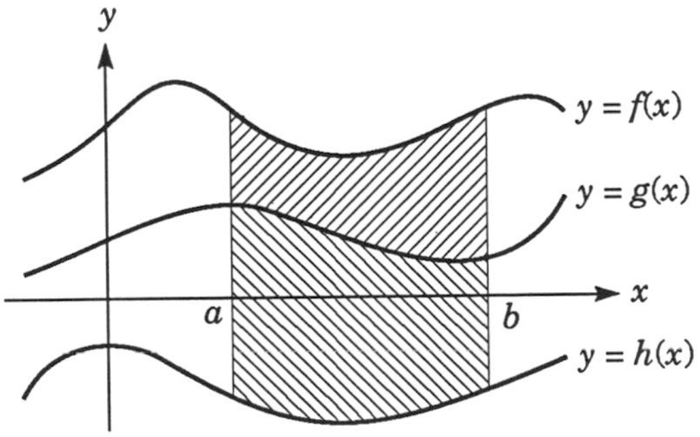

$\int_a^b f(x)\,dx - \int_a^b g(x)\,dx$

The shaded area between the curves of the functions $f(x)$ and $g(x)$ and the lines $x = a$ and $x = b$ may be determined by subtracting the area between $g(x)$ and the x-axis from the area between $f(x)$ and the x-axis.

$\int_a^b g(x)\,dx + \left|\int_a^b h(x)\,dx\right|$

Similarly, the shaded area confined by curves of the functions $g(x)$ and $h(x)$ and the lines $x = a$ and $x = b$ may be determined by adding the area between $g(x)$ and the x-axis to the area between $h(x)$ and the x-axis.

If f and g are continuous functions on $[a, b]$, then the area bounded by the two curves and the lines $x = a$ and $x = b$ is given by

$$\int_a^b [f(x) - g(x)]\,dx$$

if $f(x) \geq g(x)$ for all x in $[a, b]$.

- Find the area between the curves

$$f(x) = x + 1 \quad \text{and} \quad g(x) = x^2 - 2x + 3 .$$

The x-coordinates of the points of intersection are given by

$$x + 1 = x^2 - 2x + 3; \quad x_1 = 2, \quad x_2 = 1 .$$

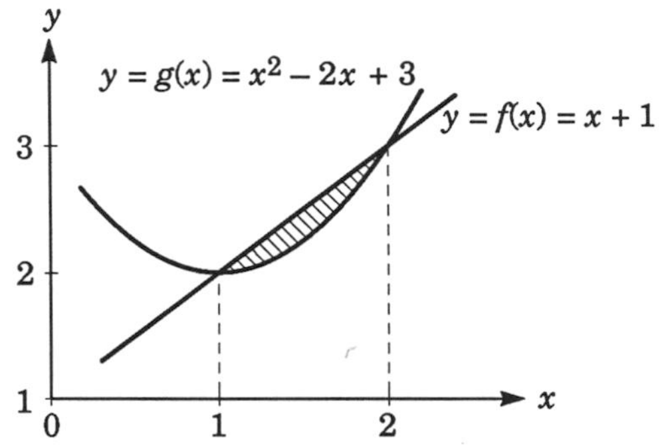

$f(x) \geq g(x)$ in the interval [1, 2] .

We have

$$A = \int_1^2 [f(x) - g(x)]\, dx = \int_1^2 [3x - x^2 - 2]\, dx$$

$$= \left[\frac{3x^2}{2} - \frac{x^3}{3} - 2x\right]_1^2 = \frac{1}{6} .$$

- Find the area between the curves

$$f(x) = x^3 - 4x \quad \text{and} \quad g(x) = 2x^2 - x^3 .$$

The points of intersection are determined by

$$x^3 - 4x = 2x^2 - x^3$$

$$x(x-2)(x+1) = 0$$

$$\left.\begin{aligned} x_1 &= 0 \\ x_2 &= 2 \\ x_3 &= -1 \end{aligned}\right\} .$$

The graph of the two functions shows that $f(x) \geq g(x)$ in the interval [–1, 0], while $g(x) \geq f(x)$ in [0, 2] .

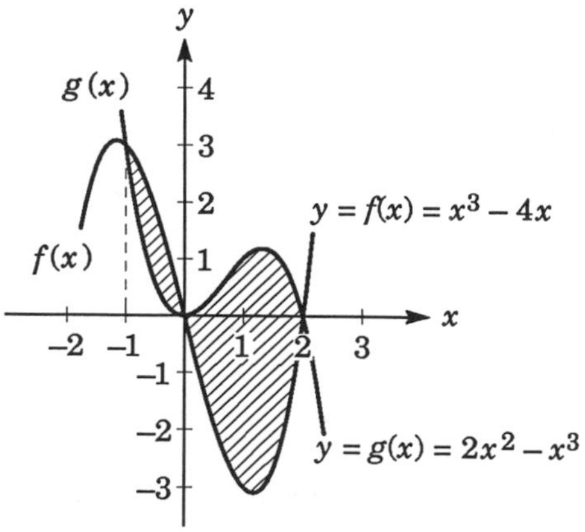

The sought area is

$$A = \int_{-1}^{0} [f(x) - g(x)]\, dx + \int_{0}^{2} [g(x) - f(x)]\, dx$$

$$= \int_{-1}^{0} [(x^3 - 4x) - (2x^2 - x^3)]\, dx + \int_{0}^{2} [(2x^2 - x^3) - (x^3 - 4x)]\, dx$$

$$= \int_{-1}^{0} [2x^3 - 2x^2 - 4x]\, dx + \int_{0}^{2} [2x^2 - 2x^3 + 4x]\, dx$$

$$= \left[\frac{x^4}{2} - \frac{2x^3}{3} - 2x^2\right]_{-1}^{0} + \left[\frac{2x^3}{3} - \frac{x^4}{2} + 2x^2\right]_{0}^{2}$$

$$= \frac{5}{6} + \frac{16}{3} = 6\frac{1}{6} \text{ square units.}$$

- Since the beginning of 1970, the water consumption of a city had grown as

$$f(t) = 1.580 + 0.0020\, t^2,$$

where $f(t)$ is measured in millions of cubic meters and t in years.

To conserve water, better industrial use was encouraged and certain restrictions were imposed in 1986; the next four years (1986-89) saw a reduced growth of consumption described by the function

$$g(t) = 1.580 + 0.0014\, t^2,$$

where $g(t)$ is measured in millions of cubic meters.

Supposing the new trend continues, predict the total volume saved during the ten-year period 1991 through 2000.

Since $t = 0$ at the beginning of 1970, we have

year 1991 $\Rightarrow t = 21$; end of year 2000 $\Rightarrow t = 31$,

and the water saved

$$\int_{21}^{31} [f(t) - g(t)]\, dt = \int_{21}^{31} [(1.580 + 0.0020\, t^2) - (1.580 + 0.0014\, t^2)]\, dt$$

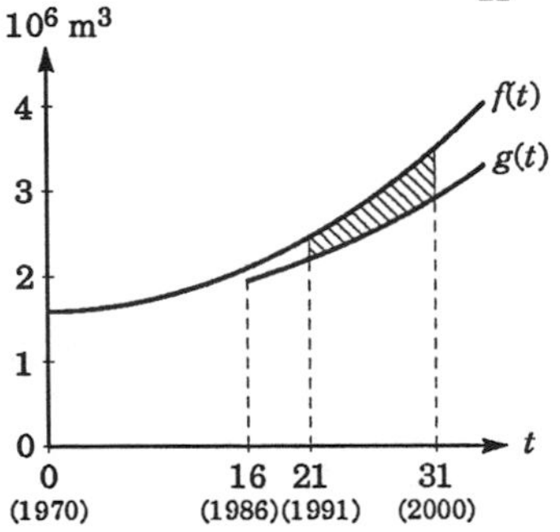

$$= \int_{21}^{31} 0.0006\, t^2\, dt$$

$$= \Big|_{21}^{31}\; 0.0002\, t^3 = 4.106 \cdot 10^6\ \text{m}^3 .$$

Area in Polar Coordinates

Just as rectangles are used to define a Riemann sum in a rectangular coordinate system, sectors of a circle are used to develop a formula for the area of a region whose boundary is given by a polar equation.

Let A be an area bounded by a curve $r = f(\theta)$ and the radius vectors be at $\theta = \alpha$ and $\theta = \beta$. f is assumed to be a continuous, non-negative function in the interval $[\alpha, \beta]$, and $0 \le \alpha, \beta \le 2\pi$. Divide the interval $[\alpha, \beta]$ into n sub-intervals with the points of subdivision

$$\alpha = \theta_0 < \theta_1 < \theta_2 < \ldots < \theta_{n-1} < \theta_n = \beta .$$

For each i let $\Delta\theta_i = \theta_i - \theta_{i-1}$, and choose an arbitrary angle ξ_i within the i-th sub-interval $\Delta\theta_i$.

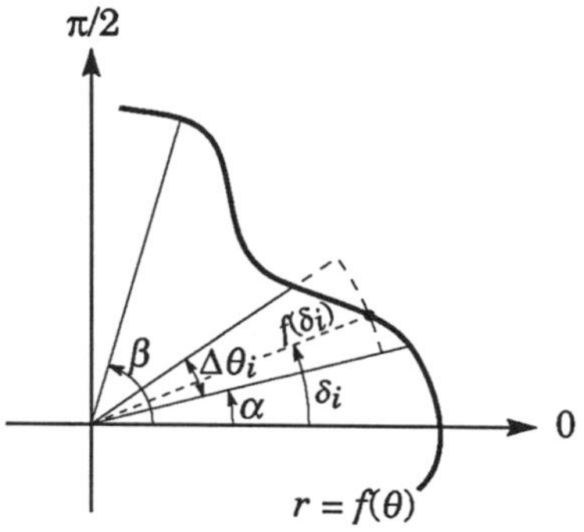

The total area is the sum of the n smaller areas, so

$$A \approx \sum_{i=1}^{n} \frac{1}{2}(\Delta\theta_i)\,[f(\xi_i)]^2 = \frac{1}{2}\sum_{i=1}^{n}[f(\xi_i)]^2\,(\Delta\theta_i) .$$

With an increasing number of points of subdivision, the Riemann sum becomes an increasingly closer approximation of the sought area, so

$$A = \lim_{n\to\infty} \frac{1}{2}\sum_{i=1}^{n}[f(\xi_i)]^2\,(\Delta\theta_i)$$

or

$$A = \frac{1}{2}\int_{\alpha}^{\beta}[f(\theta)]^2\,d\theta .$$

- Find the area enclosed by the curve $r = 8\cos 8\theta$, a 16-petal rhodonea.

To determine the total area of the 16 petals, we calculate the area of a half-petal and multiply the result by 32.

As θ increases from 0, the first zero value of r occurs for $\theta = \frac{\pi}{16}$ rad, where

$$r = 8\cos\frac{8\pi}{16} = 8\cos\frac{\pi}{2} = 0\,.$$

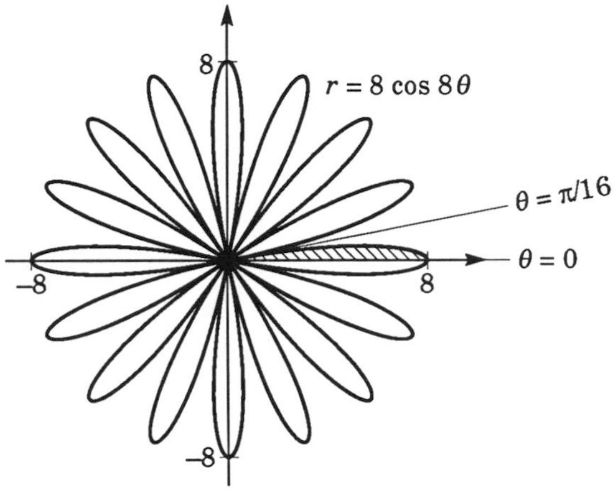

Let A be the area of the half-petal between the angles $\theta = 0$ and $\theta = \frac{\pi}{16}$; then

$$A = \frac{1}{2}\int_0^{\pi/16} (8\cos 8\,\theta)^2\, d\theta = 32\int_0^{\pi/16} \cos^2 8\,\theta\; d\theta.$$

Using the identity $\cos^2\phi = \frac{1}{2}(1 + \cos 2\,\phi)$, we find

$$A = 32\int_0^{\pi/16} \frac{1}{2}(1 + \cos 16\,\theta)\, d\theta$$

$$= 16\int_0^{\pi/16} (1 + \cos 16\,\theta)\, d\theta$$

$$= 16\left[\theta + \frac{1}{16}\sin 16\,\theta\right]_0^{\pi/16} = \pi\,.$$

Thus, the total area of the rhodonea is $32\,\pi$ square units.

As it takes exactly 2π rad to complete the full curve of $r = 8\cos 8\theta$, the area may also be calculated by integrating between 0 and 2π,

$$\int_0^{2\pi} \frac{1}{2}\cdot 8^2\cos^2 8\,\theta\; d\theta = 32\,\pi\,.$$

•

We have assumed until now that $r = f(\theta)$, with $f(\theta) \geq 0$ for all θ. If this requirement is dropped, care must be taken. In the following problem, integration of $f(\theta)$ between 0 and 2π would not give a correct answer.

- Find the area enclosed by a trifolium, $r = \cos 3\theta$.

To determine the area of the three petals, we calculate the area of a half-petal and multiply the result by 6.

We choose the shaded area in the figure below. As θ increases from 0, the first zero value of r occurs for $\theta = \pi/6$ rad, where

$$r = \cos\frac{3\pi}{6} = \cos\frac{\pi}{2} = 0 .$$

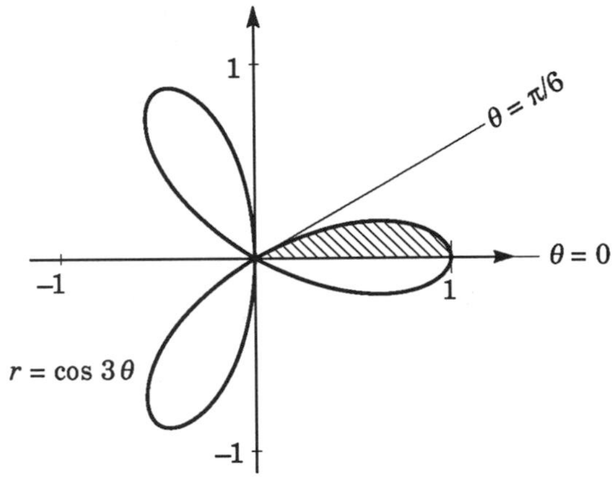

Let A be the area of the half-petal between the two angles $\theta = 0$ and $\theta = \pi/6$,

$$A = \frac{1}{2}\int_0^{\pi/6} (\cos 3\theta)^2 \, d\theta = \frac{1}{2}\int_0^{\pi/6} \cos^2 3\theta \, d\theta .$$

With the identity $\cos^2\theta = \frac{1}{2}(1 + \cos 2\theta)$, we obtain

$$A = \frac{1}{4}\int_0^{\pi/6} (1 + \cos 6\theta)\, d\theta$$

$$= \frac{1}{4}\left[\theta + \frac{1}{6}\sin 6\theta\right]_0^{\pi/6} = \frac{\pi}{24} .$$

Thus, the total area of the trifolium is

$$6\left(\frac{\pi}{24}\right) = \frac{\pi}{4} .$$

As it requires only a change from 0 to π rad to complete the trifolium, the area may also be calculated by integrating between 0 and π,

$$\int_0^{\pi} \frac{1}{2}(\cos 3\theta)^2 \, d\theta = \frac{\pi}{4} .$$

Integrating between 0 and 2π would give twice the correct answer.

cardioid: heart-shaped

When, in a plane, a circle rolls on a fixed circle of equal size, any point of the moving circle outlines a **cardioid**:

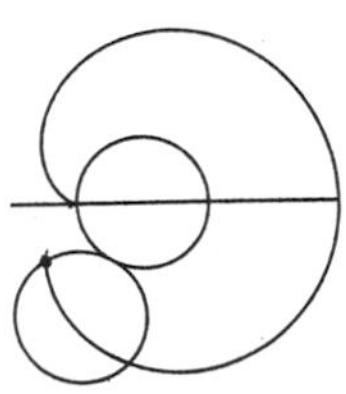

- To enhance signal sensitivity in the forward direction and limit pickup of back signals, a microphone has a characteristic in the shape of a **cardioid**, $r = a\,(1 + \cos\theta)$, where radius vector r describes output signal level as a function of the angle of incidence for an all-round constant sound pressure level.

  Calculate the attenuation of sound emanating from behind the microphone between angles $\theta = \frac{\pi}{2}$ and $\theta = \frac{3\pi}{2}$ when placed in an environment with a constant ambient sound level.

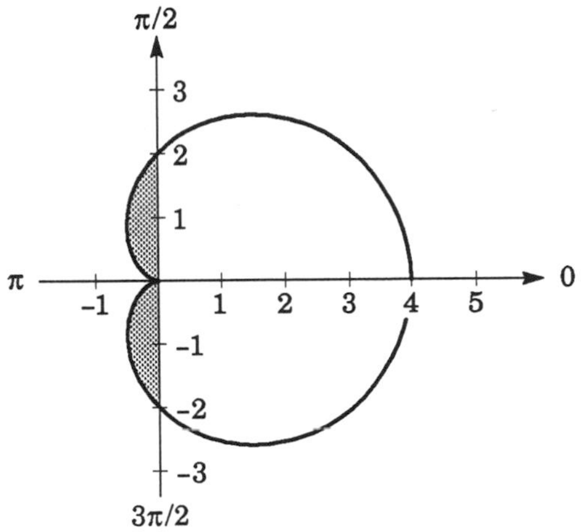

The ratio of back-signal level to total pickup of the microphone is

$$q = \frac{2a^2 \frac{1}{2} \int_{\pi/2}^{\pi} (1 + \cos\theta)^2 \,d\theta}{2a^2 \frac{1}{2} \int_{0}^{\pi} (1 + \cos\theta)^2 \,d\theta} .$$

$$\int (1 + \cos\theta)^2 \,d\theta = \int (1 + 2\cos\theta + \cos^2\theta) \,d\theta$$

$$= \int \left[1 + 2\cos\theta + \frac{1}{2}(1 + \cos 2\theta)\right] d\theta$$

$$= \int \left[\frac{3}{2} + 2\cos\theta + \frac{\cos 2\theta}{2}\right] d\theta = \left[\frac{3}{2}\theta + 2\sin\theta + \frac{1}{4}\sin 2\theta\right] + C .$$

The total energy emission is

$$A_0 = a^2 \left[\frac{3}{2}\theta + 2\sin\theta + \frac{1}{4}\sin 2\theta\right]_0^{\pi} = \frac{3\pi a^2}{2},$$

and the total back radiation is

$$A_b = a^2 \left[\frac{3}{2}\theta + 2\sin\theta + \frac{1}{4}\sin 2\theta\right]_{\pi/2}^{\pi} = \left(\frac{3\pi}{4} - 2\right)a^2 .$$

Thus,

$$q = \frac{\frac{3\pi}{4} - 2}{\frac{3\pi}{2}} = \frac{1}{2} - \frac{4}{3\pi} \approx 0.076 .$$

28.2 Surface of Revolution

A surface of revolution is generated when a plane curve is revolved about an axis in the plane.

Assume that $f(x)$ has a continuous derivative over the interval $[a, b]$, and divide into n intervals of equal length:

$$x_0 = a_1 < x_1 < x_2 < \ldots x_n = b \; .$$

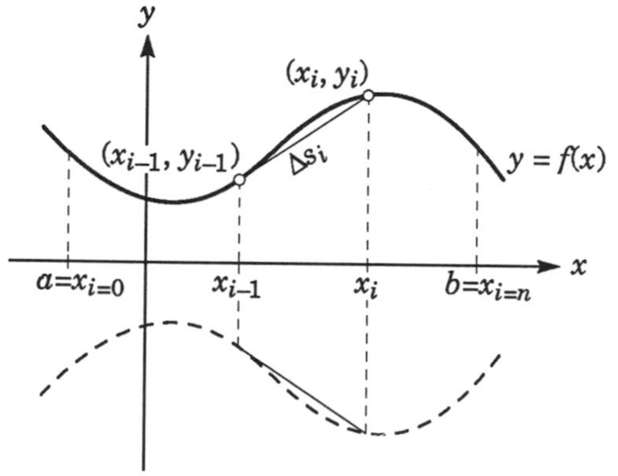

p. 830 Letting $\Delta x_i = x_i - x_{i-1}$, the corresponding **arc length** is approximated by

$$\Delta s_i \approx \sqrt{1 + [f'(\xi_i)]^2} \; \Delta x_t \, ,$$

p. 716 where ξ_i is a point in $[x_{i-1}, x_i]$ by **Lagrange's mean-value theorem**.

If the line segment joining (x_{i-1}, y_{i-1}) and (x_i, y_i) is revolved about the x-axis, it generates a mantle surface of frustum of a right circular cone, whose area approximates the corresponding area, Δs_i, of our surface of revolution.

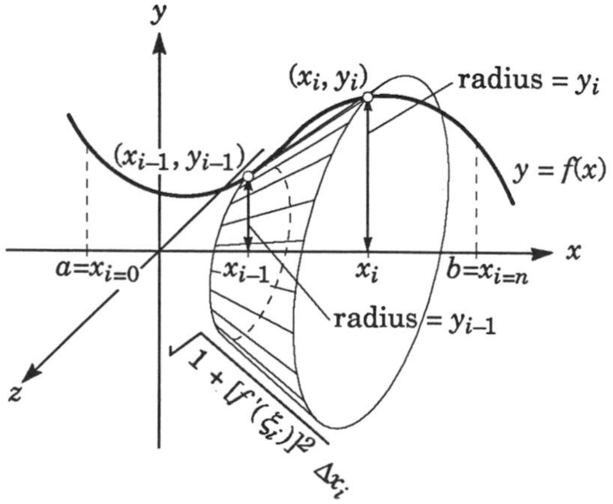

Thus,

$$\Delta S_i \approx 2\pi \, \frac{y_{i-1} + y_i}{2} \sqrt{1 + [f'(\xi_i)]^2} \; \Delta x_i \; .$$

With Δx_i small, we have $y_{i-1} \approx y_i \approx f(\xi_i)$; therefore,

$$\Delta S_i \approx 2\pi f(\xi_i) \sqrt{1 + [f'(\xi_i)]^2} \; \Delta x_i$$

and the total area S generated by revolution of the arc s about the x-axis is given by

$$S = \lim_{n \to \infty} \sum_{i=1}^{n} 2\pi f(\xi_i) \sqrt{1 + [f'(\xi_i)]^2} \; \Delta x_i$$

or

$$S = 2\pi \int_a^b f(x) \sqrt{1 + [f'(x)]^2} \; dx,$$

which is the area of a surface of revolution generated by revolving the curve $y = f(x)$ about the x-axis over the interval $[a, b]$, where f' is continuous.

- Find the surface area generated by revolution of the curve $y = 2\sqrt{x}$ about the x-axis between $x = 0$ and $x = 2$.

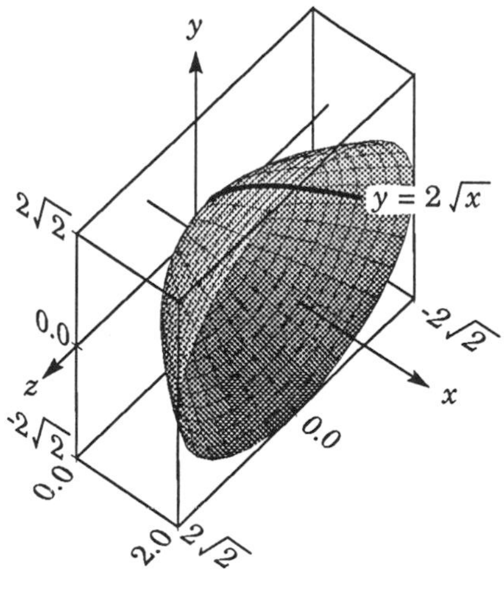

$$S = 2\pi \int_0^2 2\sqrt{x} \sqrt{1 + \left(\frac{d}{dx}\left[2\sqrt{x}\right]\right)^2} \; dx$$

$$= 4\pi \int_0^2 \sqrt{x} \sqrt{1 + (x^{-1/2})^2} \; dx$$

$$= 4\pi \int_0^2 \sqrt{x} \sqrt{1 + x^{-1}} \; dx = 4\pi \int_0^2 \sqrt{x + 1} \; dx.$$

Let $x + 1 = u$; $dx = du$.

The new limits are:

for $x = 0 \Rightarrow u = 1$
for $x = 2 \Rightarrow u = 3$

$$S = 4\pi \int_1^3 \sqrt{u}\ du = \frac{8\pi}{3} \Big|_1^3 u^{3/2}$$

$$= \frac{8\pi}{3}\left(\sqrt{27} - 1\right) = 35.15 \ldots \text{ square units.}$$

- Find the surface area generated by the revolution about the y-axis of the arc between $x = 0$ and $x = 1$ of the curve $y = \sqrt[3]{x}$.

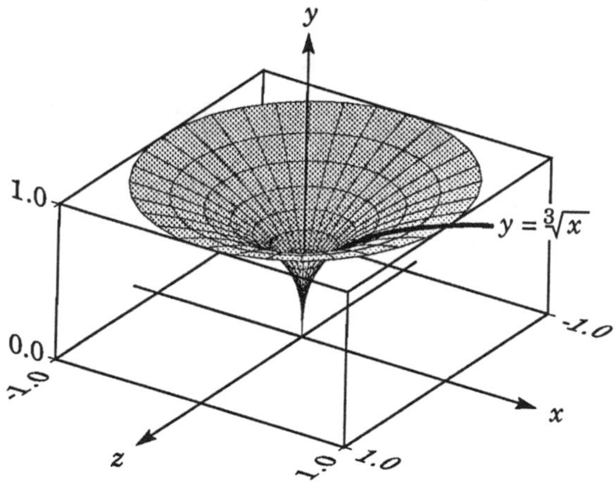

If $y = \sqrt[3]{x}$, then $x = y^3$.

$$\left.\begin{matrix} x = 0 \\ y = 0 \end{matrix}\right\} \qquad \left.\begin{matrix} x = 1 \\ y = 1 \end{matrix}\right\}$$

$$S = 2\pi \int_0^1 y^3 \sqrt{1 + \left(\frac{d}{dx}\,[y^3]\right)^2}\ dy$$

$$= 2\pi \int_0^1 y^3 \sqrt{1 + (3y^2)^2}\ dy = 2\pi \int_0^1 y^3 \sqrt{1 + 9y^4}\ dy.$$

Let $1 + 9y^4 = u$; then $du = 36y^3\ dy$ and $y^3\ dy = \frac{du}{36}$.

The new limits are:

for $y = 0 \Rightarrow u = 1$
for $y = 1 \Rightarrow u = 10$

$$S = 2\pi \int_{1}^{10} \sqrt{u}\, \frac{du}{36} = \frac{\pi}{18} \int_{1}^{10} \sqrt{u}\ du$$

$$= \frac{\pi}{27} \Big|_{1}^{10} u^{3/2} = \frac{\pi}{27} \left(\sqrt{1000} - 1\right) = 3.563\,1\ldots \text{ square units.}$$

Surface Area of a Sphere

Assume a circle of radius r with its center at the origin of an orthogonal coordinate system, $x^2 + y^2 = r^2$, and calculate the area of the surface of revolution generated by the one-fourth of the circumference in the first quadrant. The revolution of this arc through a full circle generates a hemisphere:

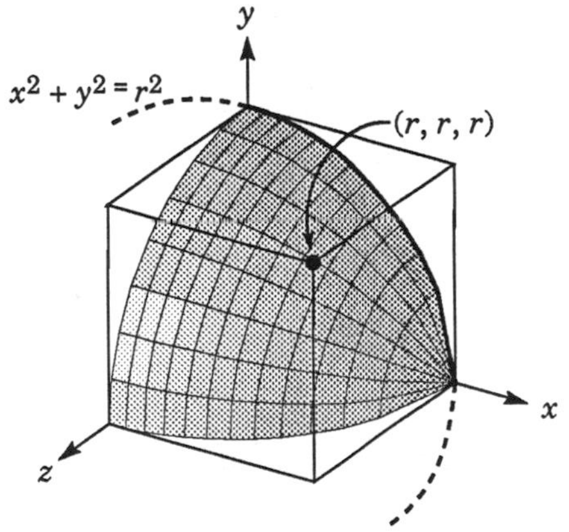

We have $y = \sqrt{r^2 - x^2}$.

$$\frac{S}{2} = 2\pi \int_{0}^{r} \sqrt{r^2 - x^2} \sqrt{1 + \left(\frac{d}{dx}\left[\sqrt{r^2 - x^2}\right]\right)^2}\ dx,$$

which gives

$$S = 4\pi \int_{0}^{r} \sqrt{r^2 - x^2} \sqrt{1 + \left(\frac{-x}{\sqrt{r^2 - x^2}}\right)^2}\ dx$$

$$= 4\pi \int_{0}^{r} \sqrt{r^2 - x^2}\ \frac{r}{\sqrt{r^2 - x^2}}\ dx$$

$$= 4\pi \int_{0}^{r} r\ dx = 4\pi r \Big|_{0}^{r} x$$

and

$$S = 4\pi r^2,$$

which is the well-known formula for the **surface area of a sphere** with radius r.

Parametric Equations and Area of Surface of Revolution

Consider a plane curve given by the parametric equations

$$\left.\begin{aligned} x &= x(t) \\ y &= y(t) \end{aligned}\right\},$$

where $a \le t \le b$ and the functions $x(t)$ and $y(t)$ have continuous derivatives in $[a, b]$.

For a curve described as a graph, we have the surface area S, generated by revolving the arc of $f(x)$ between $x = a$ and $x = b$ about the x-axis,

$$S = 2\pi \int_a^b f(x)\sqrt{1 + [f'(x)]^2}\, dx\,,$$

where the term $\sqrt{1 + [f'(x)]^2}$ represents the arc length.

p. 832 Having established the parametric form of the formula for **arc length**,

$$s = \int_a^b \sqrt{\left(\frac{dx}{dt}\right)^2 + \left(\frac{dy}{dt}\right)^2}\, dt\,,$$

we have as a corollary

$$S = 2\pi \int_a^b y \sqrt{\left(\frac{dx}{dt}\right)^2 + \left(\frac{dy}{dt}\right)^2}\, dt$$

as the **area of a surface of rotation** about the x-axis of a curve $x = x(t)$; $y = y(t)$, where $a \le t \le b$, and the functions $x(t)$ and $y(t)$ have continuous derivatives in $[a, b]$.

- A cycloid with the parametric equations

$$\left.\begin{aligned} x &= t - \sin t \\ y &= 1 - \cos t \end{aligned}\right\}$$

between $t = 0$ and $t = 2\pi$ rotates a full turn around the x-axis; find the generated surface area.

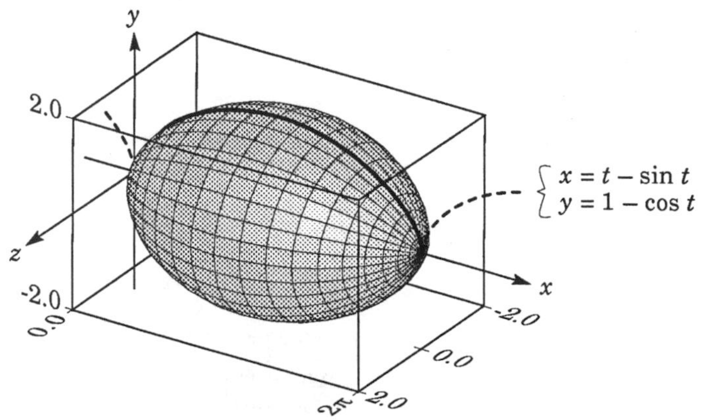

$$S = 2\pi \int_0^{2\pi} (1 - \cos t) \sqrt{\left(\frac{d}{dt}[t - \sin t]\right)^2 + \left(\frac{d}{dt}[1 - \cos t\,]\right)^2}\ dt$$

$$= 2\pi \int_0^{2\pi} (1 - \cos t)\ \sqrt{(1 - \cos t)^2 + \sin^2 t}\ \ dt$$

$$= 2\pi \int_0^{2\pi} (1 - \cos t)\ \sqrt{1 - 2\cos t + \cos^2 t + \sin^2 t}\ \ dt$$

$$= 2\pi \int_0^{2\pi} (1 - \cos t)\ \sqrt{2 - 2\cos t}\ \ dt.$$

Writing

$$S = 2\pi \int_0^{2\pi} 2\left(\frac{1 - \cos t}{2}\right) 2\sqrt{\frac{1 - \cos t}{2}}\ dt$$

p. 512 and using the identity $\sin\frac{t}{2} = \sqrt{\frac{1 - \cos t}{2}}$, we can write

$$S = 8\pi \int_0^{2\pi} \sin^3\frac{t}{2}\ dt\ .$$

p. 731 The reduction formula for $\int \sin^n x\ dx$ yields

$$\int \sin^3\frac{t}{2}\ dt = -\frac{2\cos(t/2)\sin^2(t/2)}{3} + \frac{2}{3}\int \sin(t/2)\ dt\ + C_1$$

$$= -\frac{2\cos(t/2)\sin^2(t/2)}{3} - \frac{2}{3}(2)\cos(t/2) + C_2\ .$$

Hence,

$$S = 8\pi\left[-\frac{2\cos(t/2)\sin^2(t/2)}{3} - \frac{4}{3}\cos(t/2)\right]_0^{2\pi}$$

$$= \frac{64\pi}{3} \approx 67.02\ .$$

Guldin's First Rule

Using *arc length* and *center of mass*, the formula for the surface area of a surface of revolution,

$$S = 2\pi \int_a^b f(x) \sqrt{1 + [f'(x)]^2}\, dx,$$

may be expressed:

Guldin's First Rule

> The surface area of a surface of revolution is equal to the product of the arc length of the generating curve and the length of the path described by its centroid.

KEPLER Johannes (1571 - 1630)

GULDIN Paul (Originally, Habakkuk) (1577 - 1643)

In 1615 **Johannes Kepler** – the eminent German astronomer and mathematician – described and proved this theorem, later known as **Guldin's first rule** after Swiss mathematician **Paul Guldin**, who published it in his four-volume *Centrobaryca* in 1635 - 41.

PAPPOS (c. 300)

The theorem was already known by **Pappus** of Alexandria, whose *Synagoge* ("Collection") was published around A.D. 340 in twelve books, of which the first, part of the second, and the last are lost. The *Synagoge* is considered one of the best sources of information on ancient Greek mathematics.

Surface Area of a Torus

A circle of radius r rotates around an axis in its plane at a distance R from the center of the circle, generating the torus shown here.

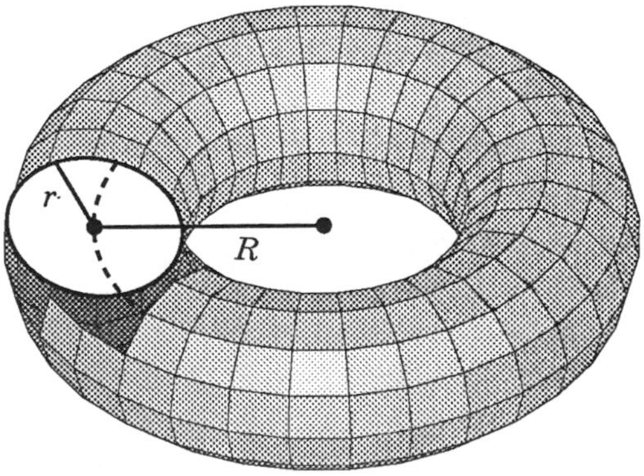

We have:

1. The centroid of the circular area is at its center, at the distance R from the axis of the generated torus; the distance traveled by the centroid is $2\pi R$.
2. The circumference of the circle is $2\pi \cdot r$.

According to Guldin's first rule, the **surface area** of the **torus** is

$$S = 2\pi r \cdot 2\pi R = 4\pi^2 r R.$$

28.3 Work

In natural sciences, statistics, and economics, applications of the determination of area under a curve are legion. As an example we choose to determine **work done by a variable force**.

If an object moves along a straight line, then the force F acting on an object of mass m – in the direction of the movement – will give it an acceleration $a = \dfrac{d^2s}{dt^2}$ (Newton's second law of motion); thus, if distance is s and time is t,

$$F = m\,\frac{d^2s}{dt^2}.$$

In the **SI** (*Système international d'unités*), mass is measured in kilograms (kg), distance in meters (m), time in seconds (s), and the force in newtons (N), and $1\ N = 1\ kg \cdot 1\ m/s^2$.

Thus, acting on a mass of 1 kg, the force of 1 N produces an acceleration of $1\ m/s^2$.

The work (or energy) W is the product of the force F and the distance s that the object moves; thus, $W = F \cdot s$.

JOULE James Prescott (1835 - 1889)

In **SI**, the unit of work is the joule (J), named after the English physicist **James Prescott Joule**. $1\ J = 1\ N \cdot m$ (newton · meter).

Lifting a mass of 3.20 kg straight up 0.50 m against the acceleration of gravity of $9.81\ m \cdot s^{-2}$ requires a force

$$F = 3.20\ [kg] \cdot 9.81\ [m \cdot s^{-2}] = 31.39\ N,$$

and the work produced is

$$W = 3.20\ [kg] \cdot 9.81\ [m \cdot s^{-2}] \cdot 0.50\ [m] = 15.70\ J.$$

Consider **a variable force** $f(x)$, where f is a continuous function, acting on an object, moving it along the x-axis in the positive direction:

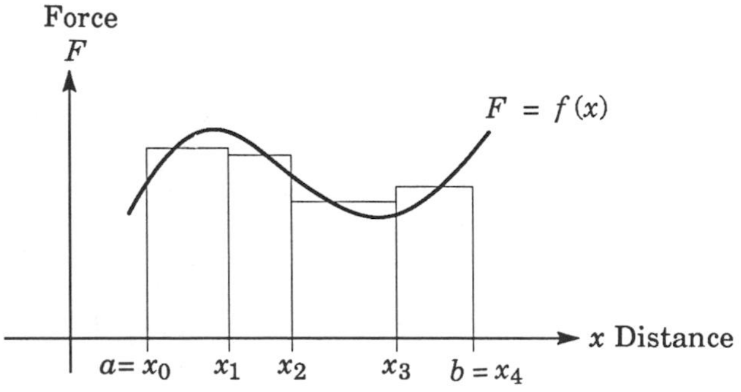

Riemann sum: *p.* 747

As before, we subdivide an interval $[a, b]$ into n sub-intervals of equal length with the points of subdivision

$$a = x_0 < x_1 < x_2 < \ldots < x_n = b$$

and find

$$x_0 = a, \quad x_1 = a + \frac{b-a}{n}, \quad x_2 = a + \frac{2(b-a)}{n}, \quad x_n = a + \frac{n(b-a)}{n} = b.$$

Within the i-th sub-interval, choose an arbitrary point ξ_i .

$$\xi_1 \text{ in } [x_0, x_1], \ \xi_2 \text{ in } [x_1, x_2], \ldots, \xi_n \text{ in } [x_{n-1}, x_n]$$

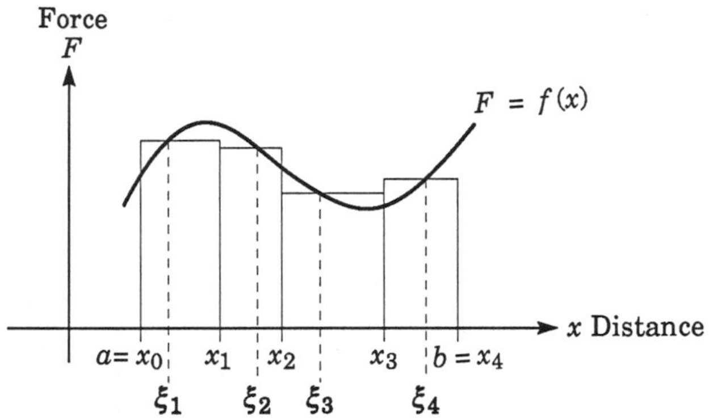

A rectangle whose base is the sub-interval $[x_{i-1}, x_i]$ and whose height (representing force) is $f(\xi_i)$ corresponds to an increment of work W_i; the **total work** done represented by the area A under the curve is the sum of all rectangles W_i,

$$W = A \approx \sum_{i=1}^{n} f(\xi_i)\,(x_i - x_{i-1}),$$

or

$$W \approx \sum_{i=1}^{n} f(\xi_i)\,\Delta x_i \ .$$

With an increasing number of points of subdivision, this sum becomes a forever-closer approximation of the total work produced,

$$W = \lim_{n \to \infty} \sum_{i=1}^{n} f(\xi_i)\,\Delta x_i$$

or

$$W = \int_a^b f(x)\,\mathrm{d}x \ .$$

- A force $F(x) = \dfrac{16}{x+3}$ N, where x denotes the distance in meters from the starting point, moves an object along the x-axis.

  What is the total work, in joules, expended in moving the object from $x = 1$ to $x = 4$?

$$W = \int_1^4 \frac{16}{x+3}\,\mathrm{d}x = 16 \left.\vphantom{\int}\right|_1^4 \ln(x+3) \approx 8.95\ \mathrm{J}.$$

- A spring requires a force of 42 N to stretch it from its natural length of 0.12 m to 0.16 m.

HOOKE Robert
(1635 - 1703)

By Hooke's law – named for the English mathematician and scientist **Robert Hooke** – the force required to stretch (or compress) a spring is proportional to the change in length.

Calculate the work required to stretch the spring from 0.16 m to 0.20 m.

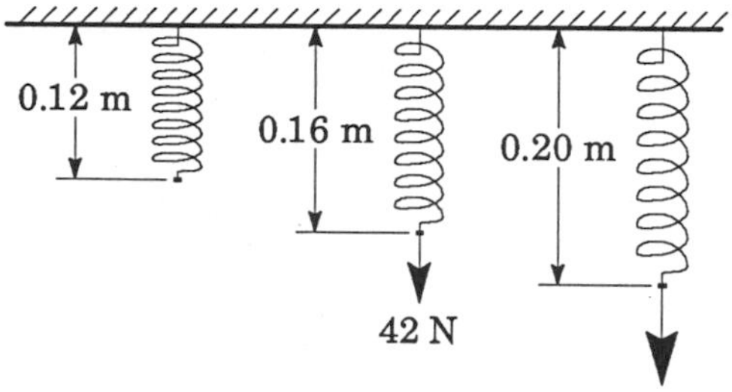

Let $f(x)$ express the force as a function of the distance the spring is stretched from its equilibrium position,

$$f(x) = k \cdot x, \text{ where } k \text{ is the proportionality constant.}$$

$f(0.16 - 0.12) = k \cdot 0.04 = 42\,; \quad k = 1050\,.$

The work required to stretch the spring from 0.16 m to 0.20 m is

$$W = \int_{0.04}^{0.08} 1050\,x \;\; dx = 525 \left|_{0.04}^{0.08} x^2 = 2.52\,\text{J}\,.\right.$$

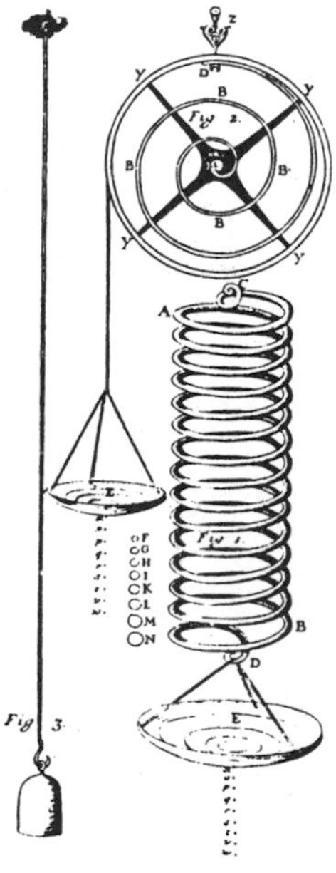

Potentia Reſtitutiva,

OR

SPRING.

The Theory of Springs, though attempted by divers eminent Mathematicians of this Age has hitherto not been Publiſhed by any. It it now about eighteen years ſince I firſt found it out, but deſigning to apply it to ſome particular uſe, I omitted the publiſhing thereof.

About three years ſince His Majeſty was pleaſed to ſee the Experiment that made out this Theory tried at *White-Hall*, as alſo my Spring Watch.

About two years ſince I printed this Theory in an Anagram at the end of my Book of the Deſcriptions of Helioſcopes, *viz.* c e i i i n o s s s t t u u, *id eſt*, *Ut tenſio ſic vis*; That is, The Power of any Spring is in the ſame proportion with the Tenſion thereof: That is, if one power ſtretch or bend it one ſpace, two will bend it two, and three will bend it three, and ſo forward. Now as the Theory is very ſhort, ſo the way of trying it is very eaſie.

In his introduction to *Potentia Restitutiva* (1678), Robert Hooke points out that about two years before, at the end of another publication, he had printed, as an anagram, *ut tensio sic vis* to ensure the priority right of his discovery of the proportionality between force applied and the length of a spring.

From Georg Reisch, *Margarita phylosophica* (first edition, 1496; here reproduced from a 1583 reprint).

The seven categories of liberal arts were, in the Middle Ages, personified by women. Dame Geometry represented "the science of unchangeable shape and form". The illustration points out the practical use of geometry in masonry, carpentry, surveying, and astronomy.

Among the seven liberal arts, the *quadrivium* [geometry, astronomy, arithmetic, and music] constituted a division to which the *trivium* [grammar, dialectics (logic), and rhetoric] was subordinate.

Chapter 29

VOLUME

NOVA

STEREOMETRIA

DOLIORVM VINARIORVM, INPRI-
mis Auſtriaci, figuræ omnium
aptiſsimæ;

ET

USUS IN EO VIRGÆ CUBI-
cæ compendioſiſſimus & plane ſingularis.

Acceſſit

STEREOMETRIÆ ARCHIME-
deæ Supplementum.

Authore

Ioanne Kepplero, Imp. Cæſ. Matthiæ I.
ejuſq; fidd. Ordd. Auſtriæ ſupra Anaſum
Mathematico.

Cum privilegio Cæſareo ad annos XV.

ANNO M. DC. XV.

LINCII

Excudebat JOANNES PLANCVS, ſumptibus Authoris.

Johannes Kepler, "New Solid Geometry of Wine Barrels, Especially Austrian Barrels ..." (1615).

Stocking his wine cellar in 1613, Kepler distrusted the wine merchants' crude methods of volume estimation; he challenged them with computations, which he published in *Nova stereometria....* He computed the volume of solids obtained by the rotation of figures about a line in their plane. Kepler's mathematical trials were of decisive importance for the development of integration methods of calculus.

Kepler's text is also remarkable as one of the first to use methods of differential calculus in the study of maxima and minima. Kepler states: *decrementa habet insitio insensibilia* ("near a maximum the decrements on both sides are in the beginning only imperceptible"). [D.J. Struik, *A Source Book in Mathematics 1200 - 1800*; Princeton University Press, 1986.]

29.0 Introduction

To determine the volume of a **solid of revolution** by integration, one may think of the solid either as consisting of an infinite number of disks with their central axis coinciding with the axis of revolution or as formed by an infinite number of concentric layers; the methods used are called the *disk method* and the *shell method*, respectively. A third method is based on the *area of the generating figure* and the *path of the centroid*.

With **cross-section areas known**, the volume of a solid of any form, whether or not a solid of revolution, can be determined.

29.1 Disk Method

The area under a curve can be determined by dividing it into rectangular strips and summing the areas of all the strips as the number of partitions tends to infinity.

If we want to determine the volume of a solid of revolution formed by revolving the curve of a continuous $f(x)$ about the x-axis, we may divide the solid into a series of disks with thickness Δx, radius $f(x_i)$, and volume $\pi [f(x_i)]^2 \Delta x$, where x_i is the endpoint of a sub-interval on the x-axis and the disks are placed like pineapple slices on the axis of revolution.

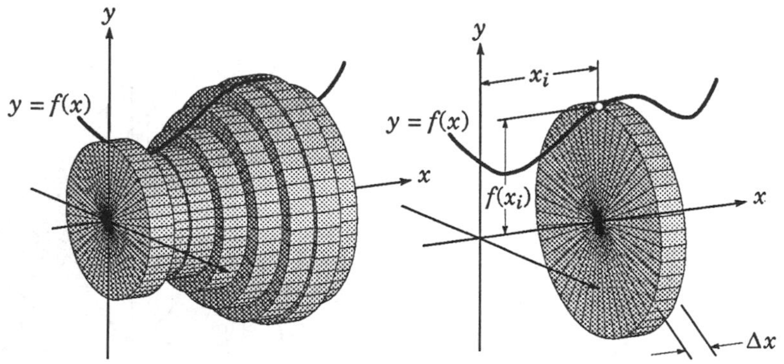

If n is the number of disks on the x-axis and we let n tend to infinity, then the volume of revolution about the x-axis is given by

$$V = \lim_{n \to \infty} \sum_{i=1}^{n} \pi \, [f(x_i)]^2 \, \Delta x$$

or, in integral form,

$$V = \pi \int_a^b [f(x)]^2 \, dx,$$

which is the **volume of a solid of revolution** generated by rotating the region bounded by the graphs of $y = f(x)$, $y = 0$, $x = a$, and $x = b$ **about the x-axis**.

Analogously,

$$V = \pi \int_c^d [g(y)]^2 \, dy$$

is the volume of a solid of revolution generated by rotating the region bounded by the graphs of $x = g(y)$, $x = 0$, $y = c$, and $y = d$ **about the y-axis**.

- Find the volume generated by the part of the graph $y = x^2$ between $x = 1$ and $x = 2$ when rotating about the x-axis.

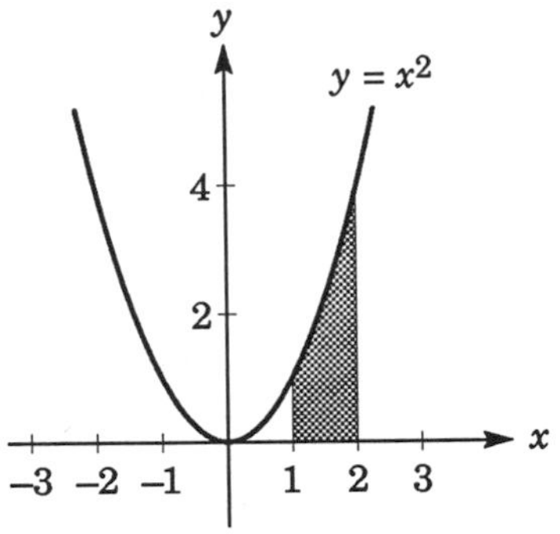

The sought volume is

$$V = \pi \int_1^2 (x^2)^2 \, dx = \frac{\pi}{5} \Big|_1^2 x^5 = \frac{31\pi}{5}.$$

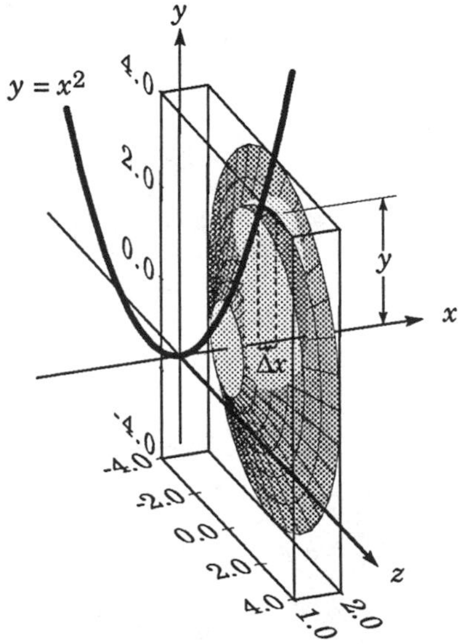

- Find the volume generated by the part of the graph $y = x^2$ between $y = 1$ and $y = 4$ when rotating about the y-axis.

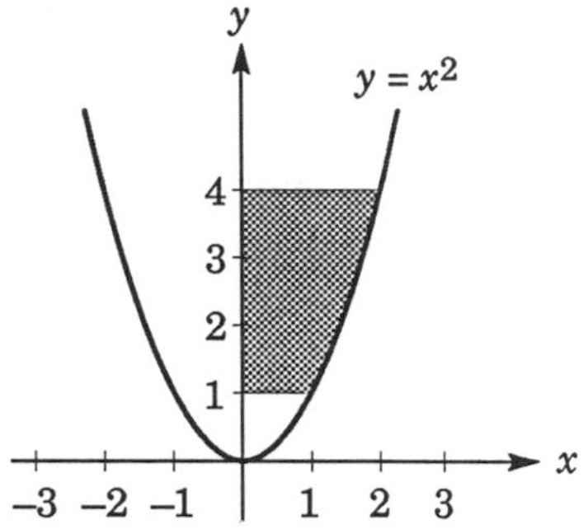

Since the curve revolves about the y-axis, we write the equation $y = x^2$ in the form $x = \sqrt{y}$. The volume is

$$V = \pi \int_1^4 \left(\sqrt{y}\right)^2 dy = \pi \int_1^4 y \; dy = \frac{\pi}{2} \Big|_1^4 y^2 = \frac{15\pi}{2}.$$

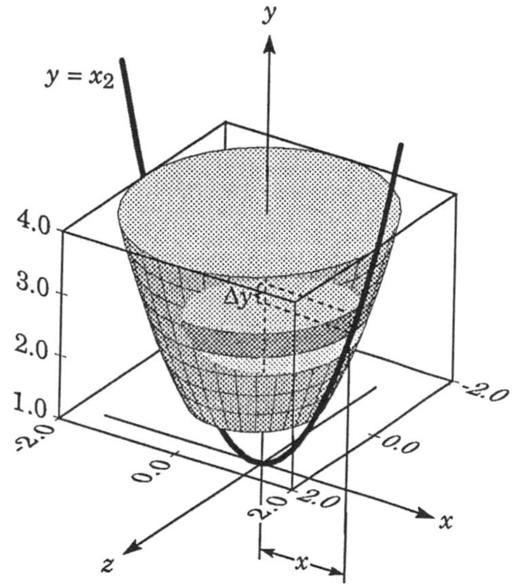

Ellipsoid of Revolution

An ellipsoid of revolution is generated by rotating the ellipse

$$\frac{x^2}{a^2} + \frac{y^2}{b^2} = 1$$

around the x-axis.

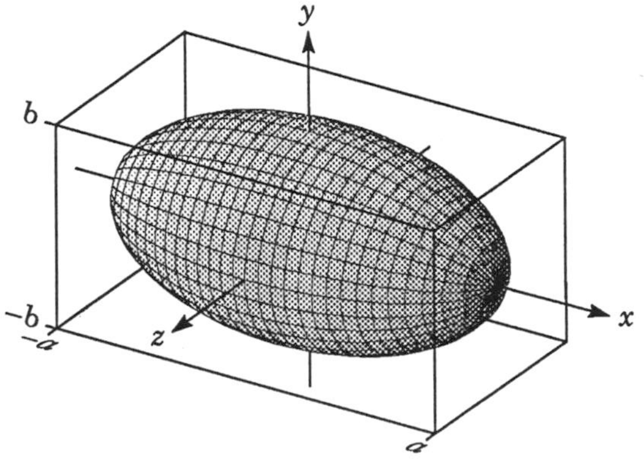

To obtain the volume of the ellipsoid of revolution, we determine the volume generated by the arc; the volume is obtained by integrating the area of the disks from $x = -a$ to $x = a$. With

$$y = \frac{b}{a}\sqrt{a^2 - x^2}$$

we obtain

$$V = \pi \int_{-a}^{a} \left[\frac{b}{a}\sqrt{a^2 - x^2}\right]^2 dx$$

$$= \frac{\pi b^2}{a^2} \int_{-a}^{a} (a^2 - x^2)\ dx$$

$$= \frac{\pi b^2}{a^2} \left[a^2 x - \frac{x^3}{3}\right]_{-a}^{a} = \frac{4\pi a b^2}{3},$$

which is the **volume of an ellipsoid of revolution** generated by an ellipse of axes $2a$ and $2b$ rotating about its major axis $2a$.

If we look upon a sphere of radius r as an ellipsoid of revolution with all axes equal,

$$a = b = r,$$

we obtain the formula for the **volume of a sphere**:

$$V = \frac{4\pi r^3}{3}$$

29.2 Shell Method

In the shell method, the volume of a solid of revolution is calculated as the sum of the volumes of an infinite number of **concentric cylindrical shells** placed concentric with the axis of revolution:

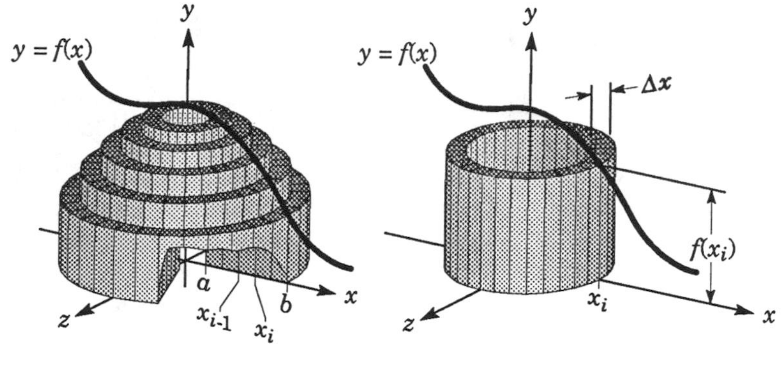

If we let x_i be the endpoint of a subinterval on the x-axis, then the approximate volume of one cylindrical shell is the product of the circumference of the outer edge of the shell, its height $f(x_i)$, and its thickness Δx.

Working on the positive side of the x-axis, divide the solid into a series of cylindrical shells with circumference $2\pi x_i$, height $f(x_i)$, thickness Δx, and volume $2\pi x_i\,[f(x_i)]\,\Delta x$, where x_i is the endpoint of a sub-interval on the x-axis. If n is the number of partitions of the x-axis of the solid of revolution, and we let n tend to infinity, the volume of revolution about the y-axis is

$$V = \lim_{n \to \infty} \sum_{i=1}^{n} 2\pi x_i\,[f(x_i)]\,\Delta x$$

or, in integral form,

$$V = 2\pi \int_a^b x\, f(x)\, dx,$$

which is the **volume of a solid of revolution** generated by rotating the area between the graph $y = f(x)$ and the x-axis and between $x = a$ and $x = b$ **about the y-axis.**

Analogously, working on the positive side of the y-axis, we have

$$V = 2\pi \int_c^d y\, g(y)\, dy,$$

which is the volume of a solid of revolution generated by rotating the area between the graph $x = g(y)$ and the y-axis and between $y = c$ and $y = d$ **about the x-axis.**

- A region bounded by the graphs of $y = x^2$, $y = 0$, $x = 1$, $x = 2$ revolves about the y-axis; find the volume by the shell method.

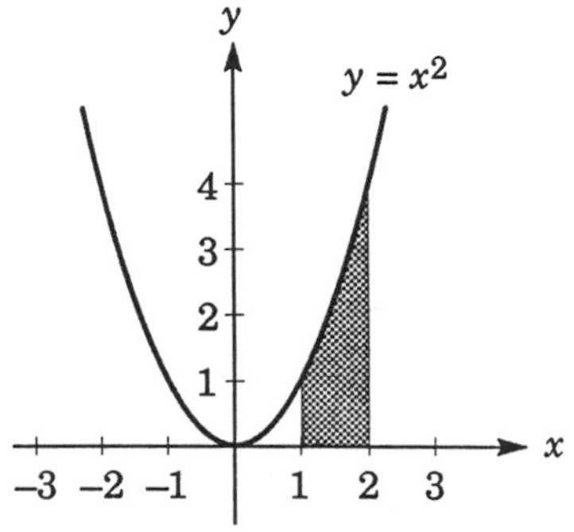

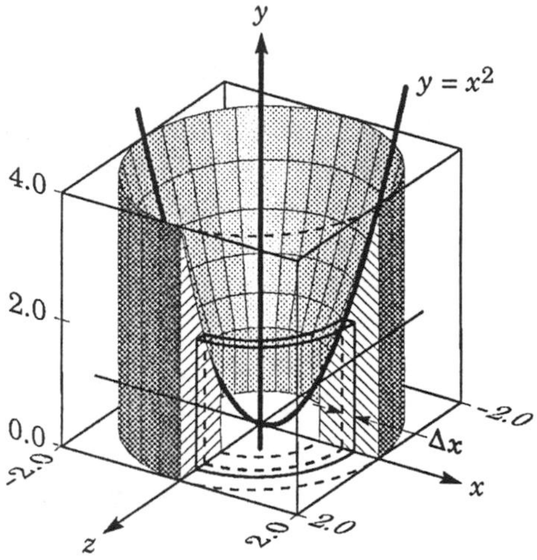

$$V = 2\pi \int_1^2 x\,(x^2)\,dx = 2\pi \int_1^2 x^3\,dx = \frac{\pi}{2}\Big|_1^2 x^4 = \frac{15\pi}{2}.$$

- The region bounded by the graphs of $y = 0$, $x = 0$, $x = \frac{4}{3}$, and $y = 3x^3 - 4x^2 - x + \frac{5}{2}$ revolves about the y-axis; find the volume generated.

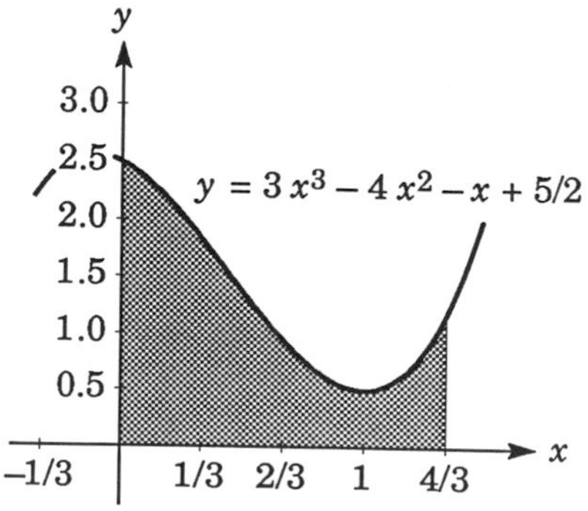

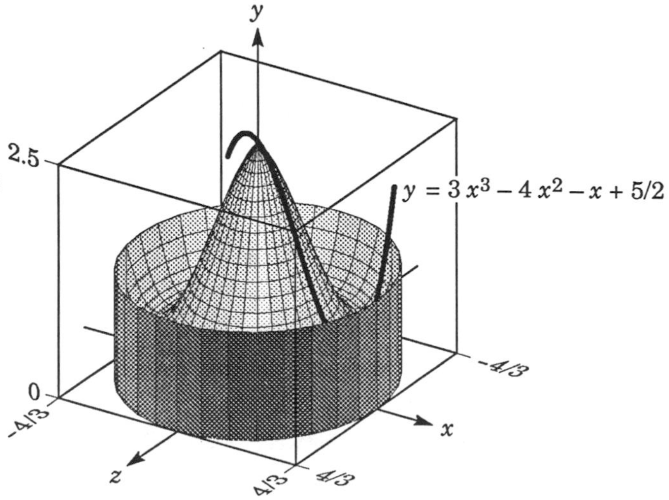

The equation $y = 3x^3 - 4x^2 - x + 5/2$ cannot be rewritten to give x as a function of y and, consequently, is not amenable to the disk method. On the other hand, the equation is suitable for the shell method.

$$V = 2\pi \int_0^{4/3} x\left(3x^3 - 4x^2 - x + \frac{5}{2}\right) dx$$

$$= 2\pi \int_0^{4/3} \left(3x^4 - 4x^3 - x^2 + \frac{5\,x}{2}\right) dx$$

$$= 2\pi \left[\frac{3\,x^5}{5} - x^4 - \frac{x^3}{3} + \frac{5\,x^2}{4}\right]_0^{4/3} = \frac{8\,\pi}{5}.$$

The decision as to which method to use for finding the volume of a solid of revolution depends on the shape and position of the bounded region relative to the axis about which it is to be revolved.

- A region bounded by the graphs of $y = x^2, y = 0$, and $x = 2$ revolves about the y-axis; find the volume generated.

With the shell method,

$$V = 2\pi \int_0^2 x\,(x^2)\ dx = \frac{\pi}{2}\Big|_0^2 x^4 = 8\,\pi\,.$$

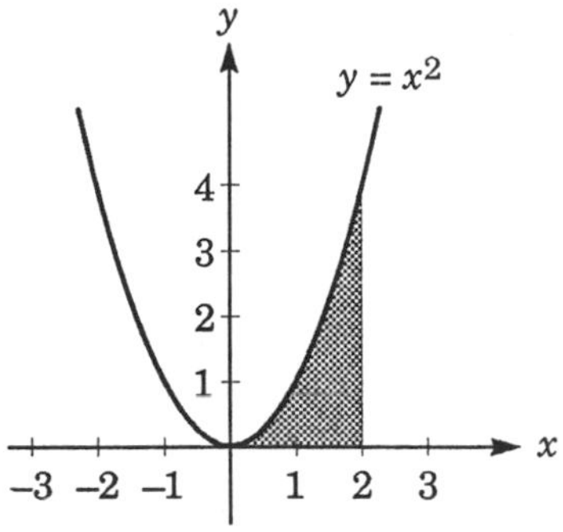

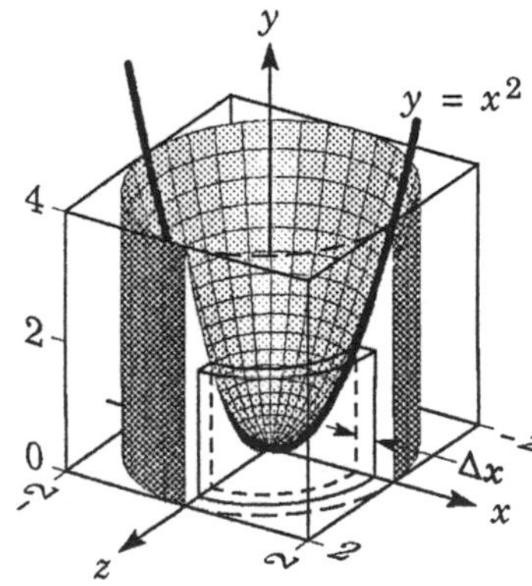

29.3 Solids Generated by Area Bounded by Two Curves

The volume of a solid generated by the revolution of an area bounded by two curves is determined as the difference of the volumes generated by the areas bounded by each of the curves and the axis of revolution.

- A region bounded by the graphs of $y = \sqrt[3]{x}$ and $y = x^2$ revolves about the x-axis; find the volume generated.

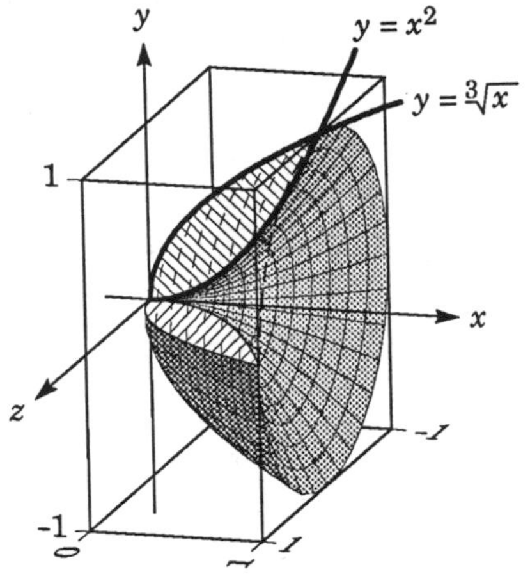

The curves intersect at $x = 0$ and $x = 1$.

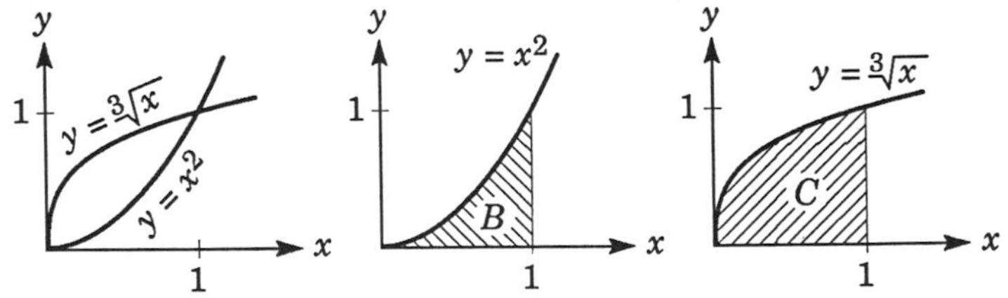

To find the volume of the solid, subtract the volume generated by revolving the area B between the curve $y = x^2$ and the x-axis from the volume generated by revolving the area C between the curve $y = \sqrt[3]{x}$ and the x-axis.

We have, by the disk method,

$$V = \pi \int_0^1 \left(\sqrt[3]{x}\right)^2 \, dx - \pi \int_0^1 (x^2)^2 \, dx$$

$$= \pi \int_0^1 (x^{2/3} - x^4) \, dx = \frac{\pi}{5} \left[3\, x^{5/3} - x^5\right]_0^1 = \frac{2\pi}{5} .$$

- The area between the graphs of

$$y = x^3 + 2\,; \quad y = 3x + 4$$

and the y-axis revolves about the y-axis; find the volume generated.

$$x^3 + 2 = 3x + 4 \Rightarrow x^3 - 3x - 2 = 0 \Rightarrow (x-2)(x+1)^2 = 0\,,$$

and thus $x = 2$ is one solution to the equation.

One point of intersection between the graphs is at $x = 2$. $y = x^3 + 2$ intersects the y-axis at $y = 2$, and $y = 3x + 4$ at $y = 4$.

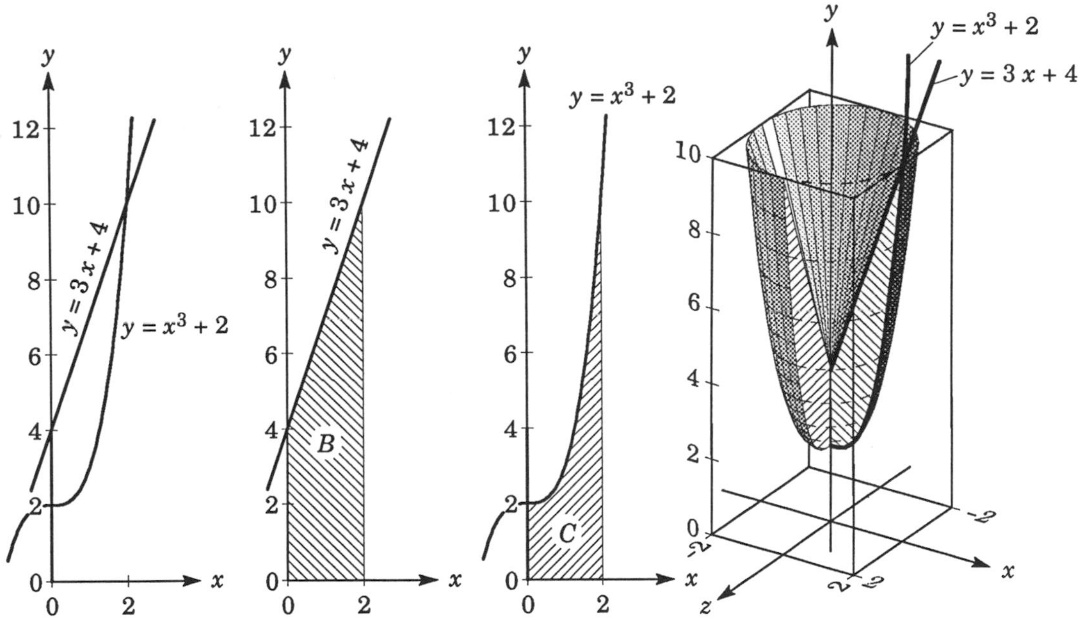

To find the volume of the solid, subtract the volume generated by revolving the area C between the curve $y = x^3 + 2$ and the y-axis from the volume generated by revolving the area B between the curve $y = 3x + 4$ and the y-axis.

Applying the shell method, we have

$$V = 2\pi \int_0^2 x\,(3x + 4)\,dx - 2\pi \int_0^2 x\,(x^3 + 2)\,dx$$

$$= 2\pi \int_0^2 (3x^2 + 2x - x^4)\,dx = 2\pi \left[x^3 + x^2 - \frac{x^5}{5} \right]_0^2 = \frac{56\pi}{5}\,.$$

29.4 Translation of Axes

If a region in the x, y-plane revolves about a line parallel to the x-axis, the volume of the resulting solid of revolution may be determined after translating the region and the line to make the axis of revolution coincide with the x-axis; similarly for an axis of revolution parallel to the y-axis.

- A region bounded by the graphs of $y = -x^2$ and $y = -x$ revolves about

  a: the line $y = -2$; **b:** the line $x = -2$.

Find the volumes generated.

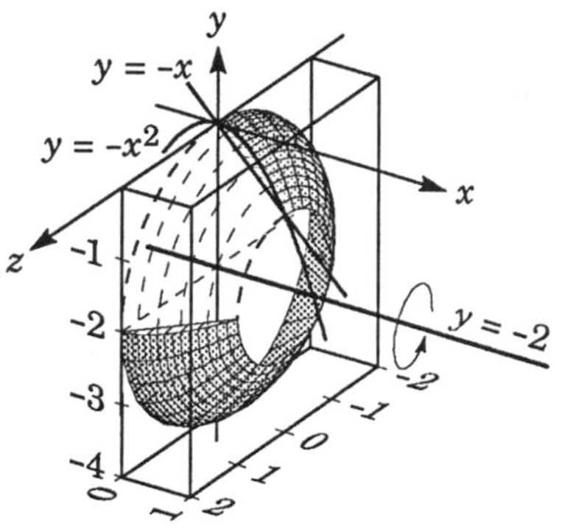

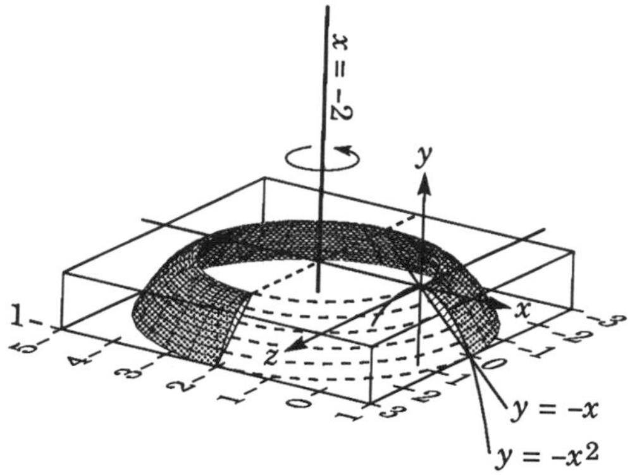

The points of intersection are given by

$$-x^2 = -x\ ; \quad x = 0 \text{ and } x = 1\ .$$

a:

To make the axis of rotation coincide with the x-axis, translate the graphs:

$$y = -x^2 \quad \text{becomes} \quad y = 2 - x^2$$
$$y = -x \quad \text{becomes} \quad y = 2 - x$$

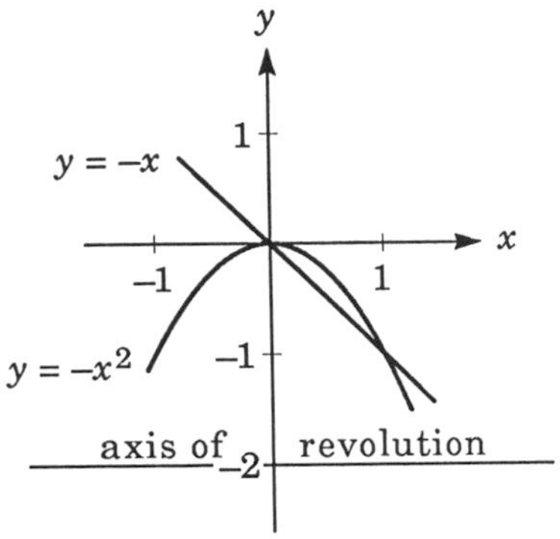

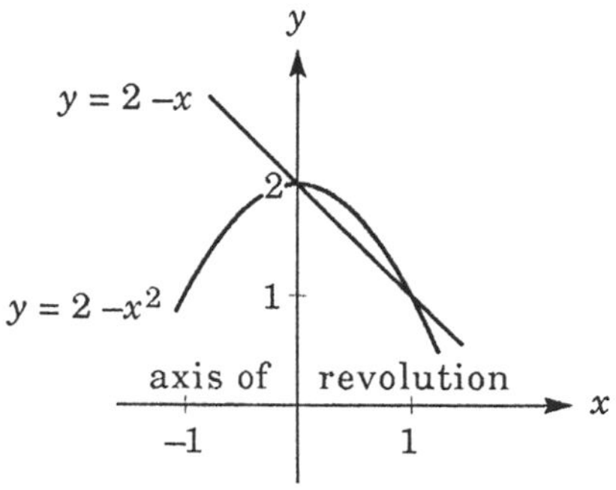

Applying the disk method, we obtain

$$V = \pi \int_0^1 (2 - x^2)^2 \, dx - \pi \int_0^1 (2 - x)^2 \, dx$$

$$= \pi \int_0^1 (x^4 - 5\,x^2 + 4\,x)\, dx = \pi \left[\frac{x^5}{5} - \frac{5\,x^3}{3} + 2\,x^2 \right]_0^1$$

$$= \frac{8\,\pi}{15}\,.$$

b:

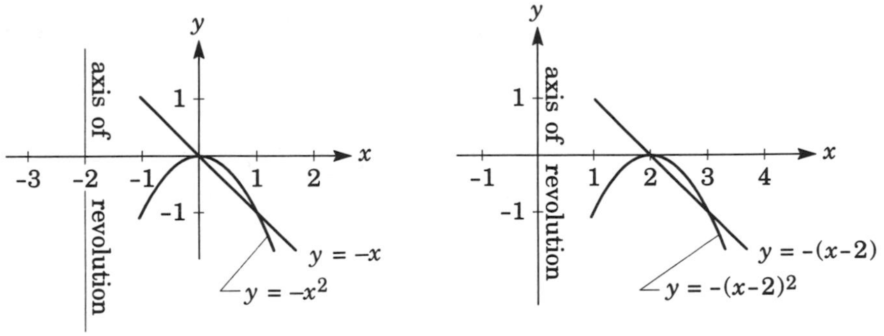

Translate the curves of the graph to the right by two units to change the domain of interest from [0, 1] to [2, 3]:

$$y = -x^2 \quad \text{becomes} \quad y = -(x-2)^2$$
$$y = -x \quad \text{becomes} \quad y = -(x-2)$$

Applying the shell method, we obtain

$$V = \left| 2\,\pi \int_2^3 x\,[-(x-2)]\, dx - 2\,\pi \int_2^3 x\,[-(x-2)^2]\, dx \right|$$

$$= \left| 2\,\pi \int_2^3 (x^3 - 5\,x^2 + 6\,x)\, dx \right| = \left| 2\,\pi \left[\frac{x^4}{4} + \frac{5\,x^3}{3} + 3\,x^2 \right]_2^3 \right| = \frac{5\,\pi}{6}\,.$$

29.5 Guldin's Second Rule

Guldin's Second Rule

> The volume of a solid of revolution is equal to the product of the generating area and the length of the path described by its centroid.

GULDIN Paul (Habakkuk)
(1577 - 1643) p. 866

Pappus's Theorem
PAPPOS
(*c.* 300)

This rule is commonly referred to as **Guldin's second rule**, named for **Paul Guldin** and, like the **first rule**, published in his *Centrobaryca* (1635 - 41). The rule is equally well known as **Pappus's theorem** after **Pappus** of Alexandria. As with Guldin's first rule, Kepler proved Guldin's second rule about 20 years before it was published by Guldin.

We can verify the rule in the case of a solid generated by the rotation around the y-axis of an area contained between the non-intersecting curves $f(x)$ and $g(x)$ and the lines $x = a$ and $x = b$, both on the same side of the y-axis. Using the shell method, we have

$$V = 2\pi \int_a^b x\,[f(x) - g(x)]\,\mathrm{d}x\;; \quad 0 \le a < b \text{ and } f(x) > g(x).$$

If A is the area and $\bar{x}$ the x-coordinate of the centroid, then

p. 845

$$\bar{x} = \frac{1}{A}\int_a^b x\,[f(x) - g(x)]\,\mathrm{d}x\;; \quad \int_a^b x\,[f(x) - g(x)]\,\mathrm{d}x = \bar{x}\,A\;;$$

that is,

$$V = 2\pi\,\bar{x}\,A\,,$$

where $2\pi\bar{x}$ is the length of the path described by the centroid of the area A generating the solid.

Volume of a Torus

A circle of radius r generates a torus by completing one revolution about an axis, located in the plane of the revolving circular area, at a distance R from the center of the circle.

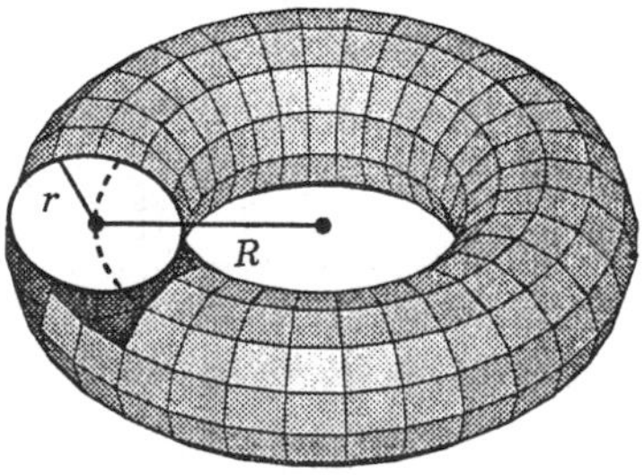

We have:

1. The centroid of the circular area is at its center, at the distance R from the axis of the generated torus; the distance traveled by the centroid is $2\pi R$.
2. The area of the circle is $\pi \cdot r^2$.

According to Guldin's second rule, the **volume of the torus** is

$$V = 2\pi R\,(\pi r^2) = 2\pi^2 r^2 R\,.$$

29.6 Solids with Known Cross-Section Areas

The volumes of solids of any shape whose cross sections are known may be determined by integration,

$$V = \int_a^b A_1(x)\,dx,$$

where V is the volume and $A_1(x)$ is the area of a cross section at x, and

$$V = \int_c^d A_2(y)\,dy,$$

where V is the volume and $A_2(y)$ is the area of a cross section at y.

- Find the volume of the solid whose base is the ellipse

$$\frac{x^2}{16} + \frac{y^2}{9} = 1$$

and whose cross sections – perpendicular to the x-axis – are squares.

The graph of the base of the solid is

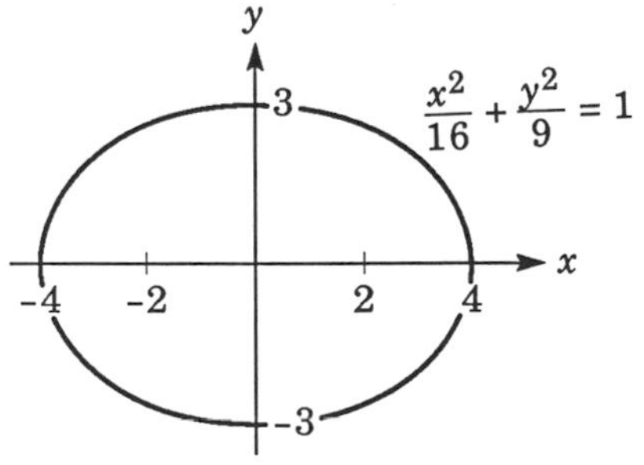

Solving for y,

$$y = 3\sqrt{1 - \frac{x^2}{16}}.$$

Let the height of the solid at x be h; since the cross section is a square,

$$h = 2 \cdot 3\sqrt{1 - \frac{x^2}{16}},$$

and the area of the cross section is h^2; that is,

$$A(x) = \left[6\sqrt{1 - \frac{x^2}{16}}\right]^2.$$

$$V = \int_{-4}^{4} \left[6\sqrt{1 - \frac{x^2}{16}}\right]^2 dx,$$ or, because of symmetry:

$$V = 72 \int_{-4}^{4} \left(1 - \frac{x^2}{16}\right) dx = 72 \left[x - \frac{x^3}{48}\right]_0^4 = 192 .$$

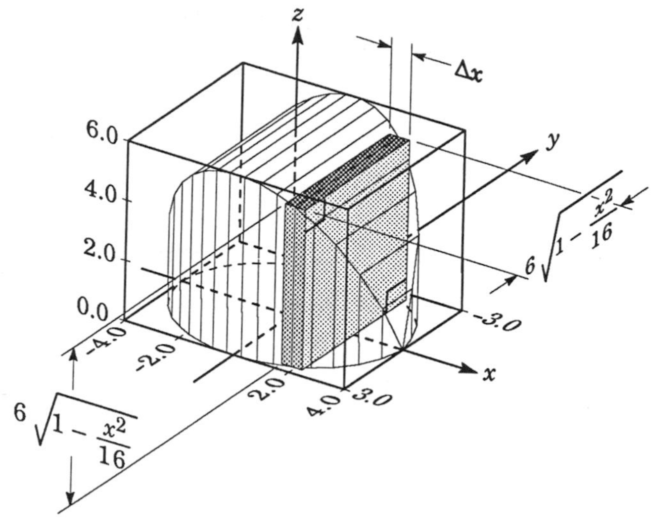

- The horizontal cross section of a solid is an equilateral triangle whose side length is the square of its distance from the topmost vertex of the solid; the height of the solid is 2 length units. Find the volume of the solid.

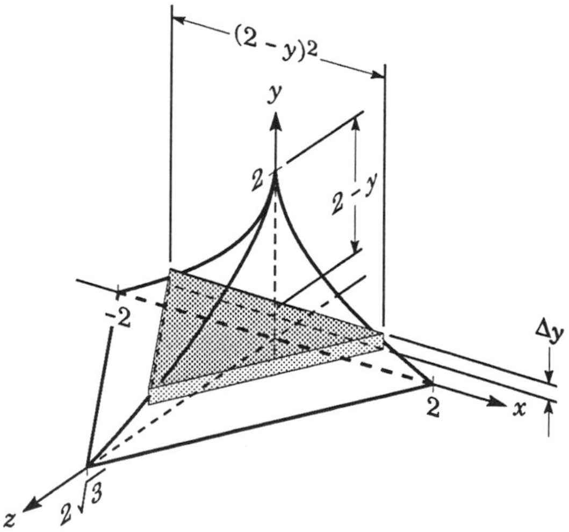

The area A of the equilateral triangle is

$$A = \frac{[(2-y)^2]^2 \sqrt{3}}{4} \text{ square units,}$$

and the volume V of the solid is

$$V = \int_0^2 \frac{(2-y)^4 \sqrt{3}}{4} \, dy = \frac{8}{5} \sqrt{3} \text{ cubic units.}$$

29.7 Transforming Double Integrals from Orthogonal to Polar Coordinates

p. 753

Double integrals over regions such as circles, ellipses, cardioids, and petal curves are generally more conveniently evaluated in polar form than in orthogonal form.

We are familiar with the conversions

p. 583

$$x = r\cos\theta;\ \ y = r\sin\theta;\ \ r = \sqrt{x^2 + y^2}\,.$$

The formula for **transforming double integrals from orthogonal to polar coordinates** is

$$\iint_{R_{xy}} f(x, y)\ \mathrm{d}x\,\mathrm{d}y = \iint_{R_{r\theta}} f(r\cos\theta, r\sin\theta)r\ \mathrm{d}r\,\mathrm{d}\theta,$$

where $f(x, y)$ is a function over the region R_{xy} of the x, y-plane in an orthogonal coordinate system, corresponding to the region $R_{r\theta}$ of the r, θ-plane in a polar coordinate system.

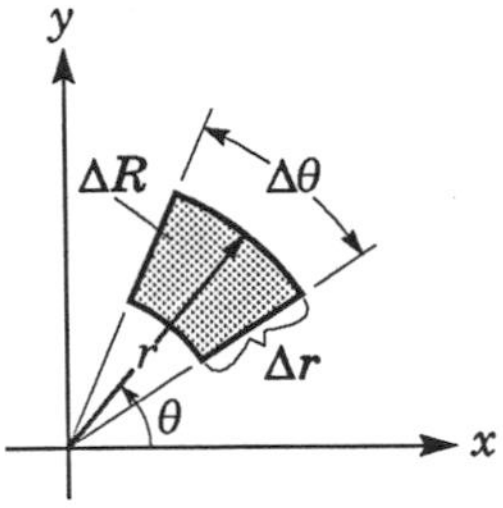

By studying the figure, we can accept the presence of the factor r in the polar form of the integrand. If r is the radius vector, and θ is the central angle in radians, then the length of the larger arc of the sector is $r\,\Delta\,\theta$, and the shaded area $\Delta\,R$ is approximated by

$$\Delta R \approx r\,\Delta\,\theta\,\Delta\,r\,.$$

We shall solve the following problem with and without transformation to polar coordinates. The former method demonstrates a greater simplicity; the latter affords an opportunity of using several interesting integration techniques.

- Find the volume V bounded by the cylinder $x^2 + y^2 = 1$, the x, y-plane, and the graph of the function $f(x, y) = \dfrac{x}{2} + y^3 + 2$.

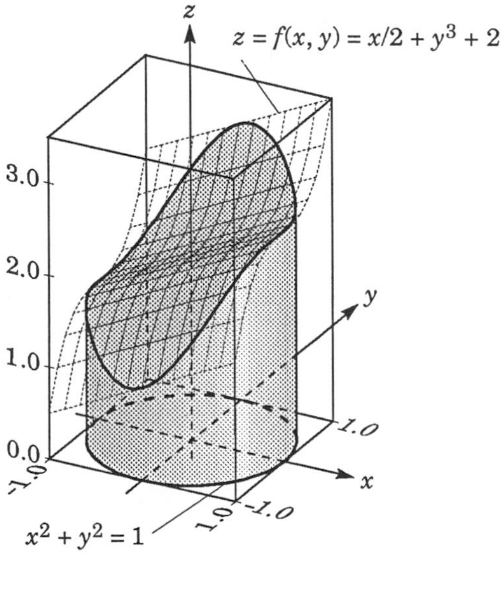

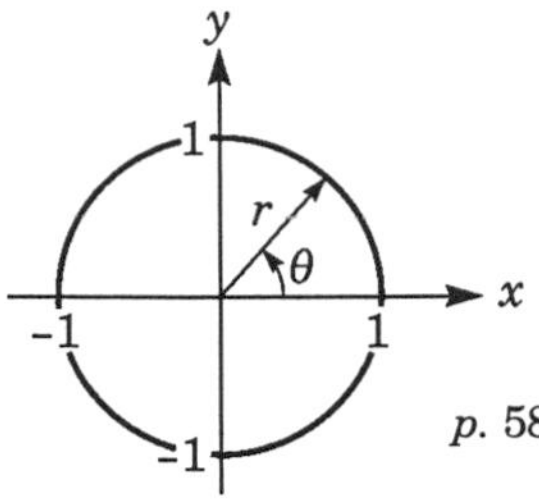

p. 583

1. Solving in Polar Coordinates

The radius r of the given cylinder is 1; thus, in the polar coordinate system the lower limit of r is 0, the upper 1. A complete revolution of the radius vector corresponds to an angle of 2π rad; thus, we can take the lower limit of θ to be 0, the upper to be 2π.

Since $x = r\cos\theta$ and $y = r\sin\theta$, we have

$$V = \int_{\theta=0}^{\theta=2\pi} \int_{r=0}^{r=1} \left[\frac{r\cos\theta}{2} + (r\sin\theta)^3 + 2\right] r \, dr\, d\theta$$

$$= \int_{\theta=0}^{\theta=2\pi} \int_{r=0}^{r=1} \left(\frac{r^2\cos\theta}{2} + r^4\sin^3\theta + 2r\right) dr\, d\theta$$

$$= \int_0^{2\pi} \left[\frac{r^3\cos\theta}{6} + \frac{r^5\sin^3\theta}{5} + r^2\right]_{r=0}^{r=1} d\theta$$

$$= \int_0^{2\pi} \left(\frac{\cos\theta}{6} + \frac{\sin^3\theta}{5} + 1\right) d\theta$$

$\sin^2\theta + \cos^2\theta = 1$
$\sin^2 = 1 - \cos^2\theta$
Thus, $\sin^3 = (1 - \cos^2)\sin\theta$

$$= \int_0^{2\pi} \left(\frac{\cos\theta}{6} + \frac{(1 - \cos^2\theta)\sin\theta}{5} + 1\right) d\theta$$

$$= \int_0^{2\pi} \left(\frac{\cos\theta}{6} + \frac{\sin\theta}{5} - \frac{\cos^2\theta\sin\theta}{5} + 1\right) d\theta$$

$$= \left[\frac{1}{6}\cdot\sin\theta - \frac{1}{5}\cdot\cos\theta + \frac{1}{15}\cdot\cos^3\theta + \theta\right]_0^{2\pi} = 2\pi.$$

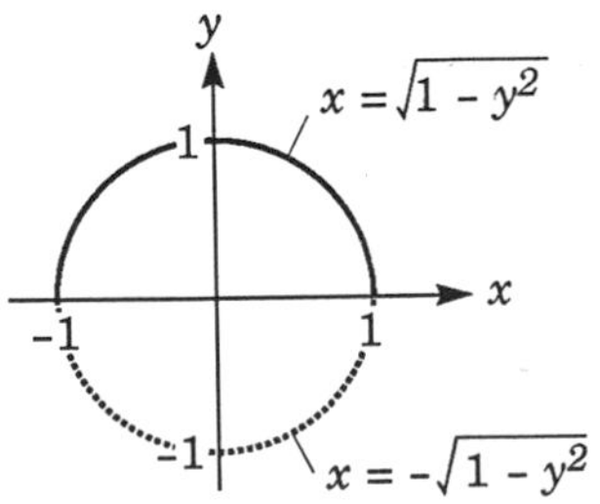

2. Solving in Orthogonal Coordinates

y has the lower limit -1 and the upper limit 1.

Solving $x^2 + y^2 = 1$ for x yields

$$x = \pm\sqrt{1 - y^2}.$$

$$V = \int_{y=-1}^{y=1} \int_{x=-\sqrt{1-y^2}}^{x=\sqrt{1-y^2}} \left(\frac{x}{2} + y^3 + 2\right) dx\, dy$$

$$= \int_{-1}^{1} \left[\frac{x^2}{4} + y^3 x + 2x\right]_{x=-\sqrt{1-y^2}}^{x=\sqrt{1-y^2}} dy$$

$$= \int_{-1}^{1} \left(\left[\frac{1-y^2}{4} + y^3 \sqrt{1-y^2} + 2\sqrt{1-y^2} \right] - \left[\frac{1-y^2}{4} - y^3 \sqrt{1-y^2} - 2\sqrt{1-y^2} \right] \right) dy$$

$$= \int_{-1}^{1} \left(2y^3 \sqrt{1-y^2} + 4\sqrt{1-y^2} \right) dy$$

$$= 2 \int_{-1}^{1} y^3 \sqrt{1-y^2}\, dy + 4 \int_{-1}^{1} \sqrt{1-y^2}\, dy ,$$

which are solved separately.

Substitute $y = \sin\theta$; $\sqrt{1 - y^2} = \cos\theta$, $dy = \cos\theta \; d\theta$, and the new limit values are

$$\text{for } y = -1 \quad \Rightarrow \quad -1 = \sin\theta;\ \theta = -\frac{\pi}{2}$$

$$\text{for } y = 1 \quad \Rightarrow \quad 1 = \sin\theta;\ \theta = \frac{\pi}{2}.$$

$$2 \int_{-1}^{1} y^3 \sqrt{1-y^2}\, dy = 2 \int_{-\pi/2}^{\pi/2} \sin^3\theta \cos\theta \ \cos\theta \ d\theta$$

$$= 2 \int_{-\pi/2}^{\pi/2} \sin^3\theta \cos^2\theta \ d\theta$$

$\sin^2\theta + \cos^2\theta = 1$
$\sin^2 = 1 - \cos^2\theta$
Thus, $\sin^3 = \sin\theta(1 - \cos^2)$

$$= 2 \int_{-\pi/2}^{\pi/2} \sin\theta\,(1 - \cos^2\theta) \cos^2\theta \ d\theta$$

$$= 2 \int_{-\pi/2}^{\pi/2} \sin\theta\,(\cos^2\theta - \cos^4\theta) \ d\theta$$

$$= 2 \left[-\frac{\cos^3\theta}{3} + \frac{\cos^5\theta}{5} \right]_{-\pi/2}^{\pi/2} = 0 .$$

$$4 \int_{-1}^{1} \sqrt{1-y^2}\ dy = 4 \int_{-\pi/2}^{\pi/2} \cos\theta \ \cos\theta \ d\theta$$

$$= 4 \int_{-\pi/2}^{\pi/2} \cos^2\theta \ d\theta = 2 \int_{-\pi/2}^{\pi/2} (1 + \cos 2\theta) \ d\theta$$

$$= \left[2\theta + \sin 2\theta \right]_{-\pi/2}^{\pi/2} = 2\pi .$$

Thus,

$$V = 0 + 2\pi .$$

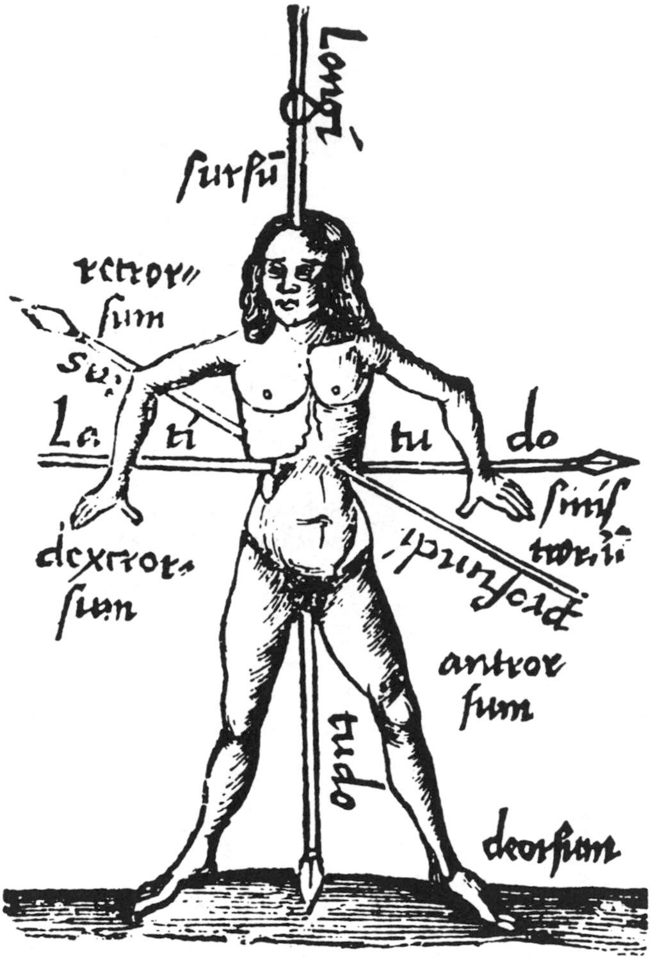

From Georg Reisch, *Margarita phylosophica* (first edition, 1496; here reproduced from a 1583 reprint).

Authors and illustrators of the Middle Ages used poignant scholastic methods. The above illustration is not from a book on warfare or trauma care, but deals with geometry and is accompanied by the following text:

– *What is a solid?*
– *It has length, depth, and breadth. ... Imagine a spear that penetrates the head and reappears at the posterior; it measures the length. Another spear pierces the chest and emerges from the back; it measures the depth. A third one passes through the body from one side to the other; it measures the breadth.*

[Anders Piltz, *Die gelehrte Welt des Mittelalters*, 1982.]

Chapter 30

MOTION

*** Indicates cross-reference

Carl Milles (1875 - 1955), *Pegasus.*

30.1 Laws of Kepler and Newton

ZENON
(*c.* 490 - *c.* 435 B.C.) *cf. p.* 275

To make us ponder the relation between the finite and the infinite, **Zeno** of Elea proposed that a soaring arrow in actual fact stands still. At any specific instant, Zenon argued, the arrow occupies a volume of space precisely equal to that of the arrow; if the arrow indeed moved, it would – in every instant – occupy a volume larger than its own and, equally absurd, every instant would have two portions – a foregoing and a succeeding one.

This paradox and modern physics concur that the arrow is at a specific position at any specific instant, and that there is no inherent difference between an arrow at rest and in motion.

So, where is the fallacy?

> The difference between rest and motion manifests itself only when we look at position at different points in *time*.

If – at different points in time – the arrow is at the same position, it is at rest; if at different positions, in motion.

GALILEI Galileo
(1564 - 1642)

The old belief that the acceleration of a freely falling body is proportional to its mass was refuted by **Galileo Galilei**. In his *Discorsi e Dimostrazioni Matematiche* (1638), Galileo lets "Simplicio" defend the old idea that a body of greater mass falls faster than a body of lesser mass, while "Salvati" develops the doctrine that they have equal velocities.

DISCORSI
E
DIMOSTRAZIONI
MATEMATICHE,
intorno à due nuoue scienze
Attenenti alla
MECANICA & i MOVIMENTI LOCALI,
del Signor
GALILEO GALILEI LINCEO,
Filosofo e Matematico primario del Serenissimo
Grand Duca di Toscana.
Con vna Appendice del centro di grauità d'alcuni Solidi.

IN LEIDA,
Appresso gli Elsevirii. M. D. C. XXXVIII.

Salu. *Mà senz' altre esperienze con breue, e concludente dimostrazione possiamo chiaramente prouare non esser vero, che vn mobile più graue si muoua più velocemente d'un' altro men graue, intendendo di mobili dell' istessa materia; & in somma di quelli de i quali parla Aristotele. Però ditemi S. Simp. se voi ammettete, che di ciascheduno corpo graue cadente sia vna da natura determinata velocità; si che l'accrescergliela, ò diminuirgliela non si possa se non con vsargli violenza, ò opporgli qualche impedimento.*

Simp. *Non si può dubitare, che l'istesso mobile nell' istesso mezzo habbia vna statuita, e da natura determinata velocità, la quale non se gli possa accrescere se non con nuouo impeto conferitogli, ò diminuirgliela saluo che con qualche impedimento che lo ritardi.*

Salu. *Quando dunque noi hauessimo due mobili, le naturali velocità de i quelli fussero ineguali, è manifesto che se noi congiugnessimo il più tardo col più veloce, questo del più tardo sarebbe in parte ritardato, & il tardo in parte velocitato dall' altro più veloce. Non concorrete voi meco in quest' opinione?*

Simp. *Parmi che così debba indubitabilmente seguire.*

Salu. *Mà se questo è, & è insieme vero, che vna pietra grande si muoua per esempio con otto gradi di velocità, & vna minore con quattro, adunque congiugnendole amendue insieme il composto di loro si mouerà con velocità minore di otto gradi; mà le due pietre congiunte insieme fanno vna pietra maggiore, che quella prima che si moueua con otto gradi di velocità, adunque questa maggiore si muoue men velocemente, che la minore; che è contro alla vostra supposizione. Vedete dunque come dal suppor che'l mobile più graue si muoua più velocemente del men graue, io vi concludo il più graue muouersi men velocemente.*

In *Discorsi* ... (previous page), Simplicio accepts that if two bodies of different mass have varying velocities at free fall then – tied together – the body with the greater mass should be delayed, while the one with the lesser mass should be hastened. Salvati declares that he has thus made Simplicio conclude that a body with a greater mass sometimes can fall more slowly than a body with a lesser mass.

PTOLEMAEUS Claudius (*c.* 100 - *c.* 168)

In ancient Greek astronomy and geometry, the circle represented periodic or repetitive motion; planets were believed to move in circular orbits with the Earth as center – the *geocentric system*. Such orbits could, however, not explain why, at times, planets seemed to move backward. **Ptolemy** explained such retrograde motion by superimposing small circles – *epicycles* – on the original circle.

COPERNICUS Nicolaus
KOPERNIK Mikolaj
(1473 - 1543)

The idea of circular orbits and epicycles was retained by **Copernicus**, who introduced the *heliocentric system*, that is, the model of the solar system centered on the Sun, with the Earth and the other planets moving around it.

KEPLER Johannes (1571 - 1630)
BRAHE Tycho (1546 - 1601)

After **Johannes Kepler**'s description of planetary motion, founded primarily on **Tycho Brahe**'s observations, there was no further need to use the concept of epicycles. Kepler's three laws of planetary motion read:

1. All the planets of the solar system describe elliptical orbits, having the Sun as one of the foci.
2. A radius vector joining any planet to the Sun sweeps out equal areas in equal periods of time.
3. The squares of the periods of revolution of the planets about the Sun are directly proportional to the cubes of their mean distances from the Sun (the major semi-axes of the elliptical orbits).

NEWTON Isaac (1643 - 1727)

Newton developed calculus to account for Kepler's laws. Newton's laws of motion describe the effect of external forces on the motion of a body; the same laws of motion apply both to the planets and to motion here on Earth.

Newton's three laws of motion read:

1. A body is at rest or moving at a constant speed in a straight line unless a force compels it to change that state.
2. The force on a body is equal to the product of its mass and acceleration.
3. The actions of two bodies upon each other are always equal in magnitude and opposite in direction.

Four years before Newton was born, the *first law* had been demonstrated experimentally by Galileo. Thus, Galileo understood that force was needed to change motion, not to sustain constant motion. He demonstrated that the Earth's gravity exerted constant downward acceleration, which he measured to be – here in modern units – 9.8 meters per second per second.

The planets are held in their orbital course by the gravitational force produced by the Sun. Newton showed that celestial bodies need not always follow the elliptical courses specified by Kepler's first law but can take paths defined by parabolas or hyperbolas; thus, an object with sufficient energy to surpass the gravitational forces – *e.g.*, a comet – can enter the solar system and depart again.

From *Newton's second law* all of the fundamental equations of dynamics can be derived by methods of calculus.

Kepler never numbered the laws that now bear his name or specially distinguished them from his other discoveries on planetary motion. Neither did Newton number the laws of motion in *Philosophiae naturalis principia mathematica* (1687) in the way we know them. In the facsimile below, "Def. III" is what we now call Newton's first law, or the law of inertia.

PHILOSOPHIÆ
NATURALIS
PRINCIPIA
MATHEMATICA.

Autore *JS.* NEWTON, *Trin. Coll. Cantab. Soc.* Matheseos Professore *Lucasiano*, & Societatis Regalis Sodali.

IMPRIMATUR.
S. PEPYS, *Reg. Soc.* PRÆSES.
Julii 5. 1686.

LONDINI,
Jussu *Societatis Regiæ* ac Typis *Josephi Streater*. Prostat apud plures Bibliopolas. *Anno* MDCLXXXVII.

Def. III.

Materiæ vis insita est potentia resistendi, qua corpus unumquodq;, quantum in se est, perseverat in statu suo vel quiescendi vel movendi uniformiter in directum.

Hæc semper proportionalis est suo corpori, neq; differt quicquam ab inertia Massæ, nisi in modo concipiendi. Per inertiam materiæ fit ut corpus omne de statu suo vel quiescendi vel movendi difficulter deturbetur. Unde etiam vis insita nomine significantissimo vis inertiæ dici possit. Exercet vero corpus hanc vim solummodo in mutatione status sui per vim aliam in se impressam facta, estq; exercitium ejus sub diverso respectu et Resistentia et Impetus: Resistentia quatenus corpus ad conservandum statum suum reluctatur vi impressæ; Impetus quatenus corpus idem, vi resistentis obstaculi difficulter cedendo, conatur statum ejus mutare. Vulgus Resistentiam quiescentibus et Impetum moventibus tribuit; sed motus et quies, uti vulgo concipiuntur, respectu solo distinguuntur ab invicem, neq; semper vere quiescunt quæ vulgo tanquam quiescentia spectantur.

Newton developed and used calculus in his preparation of the *Principia* – the fundamental work for the whole of modern science – but he did not use it in the publication; if the book had been presented with calculus, Newton's contemporaries would have understood the text poorly.

30.2 Differentiating Distance and Velocity

The derivative of a function with respect to an independent variable describes the rate of change of the function when the independent variable changes. An illustration from physics is provided by the motion of an object – which may be **rectilinear**, if the object moves along a straight line, or **curvilinear**, if the path of movement is curved.

Rectilinear Motion
Curvilinear Motion

The simplest form of rectilinear motion is the free fall of a body under the influence of its own weight, which is the product of its mass m and the acceleration of gravity, g. The velocity of the fall increases continually – that is, the movement accelerates – until the body reaches the end of the drop where its kinetic energy is converted into other forms of energy, predominantly heat.

Thus, three fundamental aspects of motion can be distinguished: acceleration, velocity, and distance.

Force
$F = m \cdot a$

Acceleration is proportional to the **force** acting on the body; **mass times acceleration is equal to force**.

When there are no external forces, a body remains at rest or continues to move along a straight line at a constant velocity; thus, its acceleration is zero.

Weight

A body's **weight** is equal to its mass times the value of the gravitational acceleration. While the mass of a body is constant, its weight varies with the local value of the gravitational acceleration.

Units of Measurement

The International System of Units (**SI**) expresses distance in **meters** (m), velocity in **meters per second** (m/s or $m \cdot s^{-1}$), and acceleration in meters per second per second or **meters per second squared** (m/s^2 or $m \cdot s^{-2}$).

Distance

Position

Distance, or **position**, s, is the sum of a body's consecutive displacements along the path of rectilinear motion, that is, the total length of the path traveled during the period of time studied.

Velocity and Speed

Velocity, v, is the rate of change of the position of a body in the direction of movement; that is,

velocity is the first derivative of distance with respect to time,

$$v = \frac{\mathrm{d}s}{\mathrm{d}t} = s'(t) = \dot{s} \ ,$$

where the dot is used in place of the conventional prime to distinguish a derivative with respect to time; Newton used the notation $\dot{s}$ and called it the fluxion of the fluent s.

Velocity has *direction*; if the velocity is considered positive for an object moving in one direction, it is zero when the object reverses its direction and negative when the object moves in the reverse direction.

For $v = \dot{s}$ to be true, $\dot{s}$ must be the *directed distance* from any fixed point (positive in positive direction, negative in negative direction).

Speed is distance passed per unit of time; it is the **absolute value of velocity**.

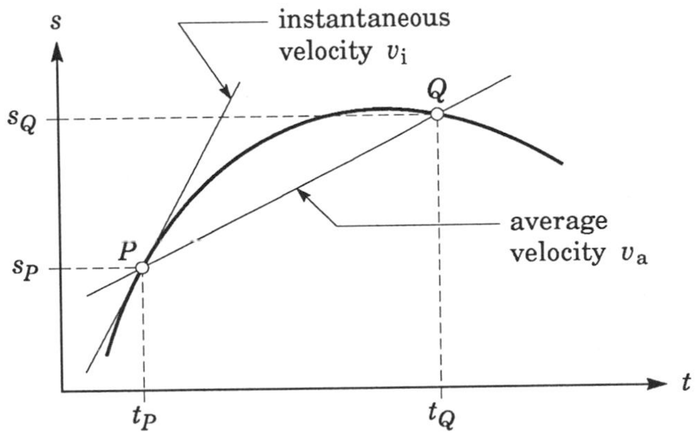

Average Velocity
Instantaneous Velocity

In a distance-time graph – above – the slope of the chord PQ represents the **average velocity** v_a of the moving body over the range from P to Q; the **instantaneous velocity** v_i at the point P, generally called only velocity, is the slope of the tangent at P.

$$v_a = \frac{\text{change in position}}{\text{time}} = \frac{s_Q - s_P}{t_Q - t_P} .$$

Acceleration

Acceleration, a, in its turn, is the rate of increase of velocity, that is,

acceleration is the first derivative of velocity and the second derivative of distance with respect to time,

$$a = \frac{dv}{dt} = \dot{v} = \frac{d^2 s}{dt^2} = \ddot{s} .$$

Deceleration is negative acceleration.

Instantaneous Acceleration
Average Acceleration

As with velocity, we distinguish between **instantaneous acceleration** and **average acceleration**.

The **third derivative of distance** with respect to time is the rate of change of acceleration, which, however, has no practical application.

- An object's motion along a straight line is registered from time 0. At t seconds the position from a fixed point on the line is

$$s(t) = (t^2 - 2t) \text{ meters.}$$

Find: **a:** the acceleration;

b: the initial velocity and speed;

c: the velocity at $t = 2.5$ s;

d: the velocity when its directed distance from the fixed point is 3 m.

With the velocity v, and acceleration a,

$$v(t) = \dot{s} = 2t - 2$$

$$a(t) = \ddot{s} = \dot{v} = 2.$$

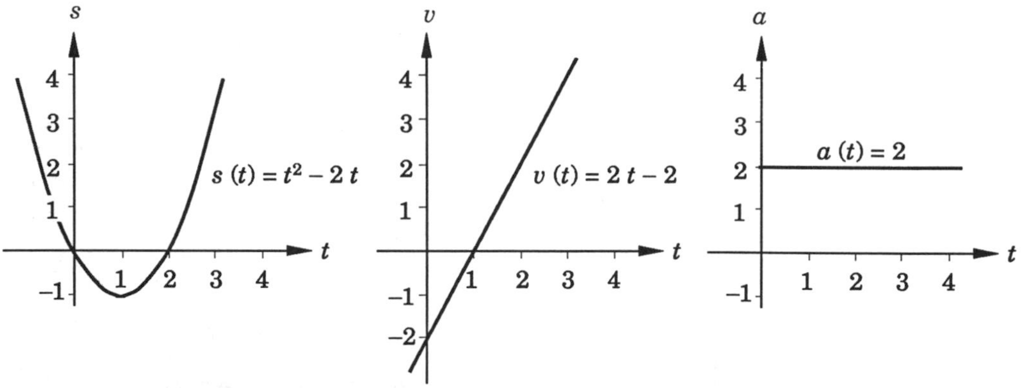

a:

The acceleration is 2 m/s^2.

b:

At time $t = 0$,

$$v(0) = 0 - 2 = -2.$$

The initial velocity is –2 m/s; the initial speed is 2 m/s.

The negative value of the velocity means that the object initially moves in a direction opposite the one determined as "intended", or positive.

c:

$v(2.5) = 2 \cdot 2.5 - 2 = 3$ m/s.

Thus, at 2.5 seconds after start, the velocity is 3 m/s.

d:

Since $s(t) = t^2 - 2t$, we have

$$t^2 - 2t - 3 = 0$$

$$(t+1)(t-3) = 0; \; t = 3\text{ s}; \; t = -1, \text{ discarded.}$$

$v(3) = \dot{s}(3) = 2(3) - 2 = 4$ m/s.

On completion of a distance of 3 meters, the velocity is 4 m/s.

- An experiment shows that from time t, a marble falls freely the distance

$$s(t) = 4.9\,t^2 .$$

Determine when and at what velocity the marble strikes the ground after being dropped from a window 44.1 meters above street level.

$v(t) = \dot{s} = 9.8\,t .$

$44.1 = 4.9\,t^2 ; \quad t = 3\text{ s}.$

$v(3) = 9.8 \cdot 3 = 29.4\text{ m/s}.$

Thus, the marble reaches the ground after 3 s at a velocity of 29.4 m/s.

- An object is thrust upwards and reaches a height

$$s = (73.5\,t - 4.9\,t^2)\text{ meters}$$

at the end of t seconds.

Determine the highest point of the object's ascent.

$v(t) = \dot{s} = 73.5 - 9.8\,t .$

The object reaches its highest point when $v = 0$;

$$0 = 73.5 - 9.8\,t \ ; \ t = 7.5\text{ s}.$$

Thus, the highest point reached is

$$s(7.5) = 73.5 \cdot 7.5 - 4.9 \cdot (7.5)^2 = 275.625 \approx 275.6\text{ m}.$$

Destination, Please?

After a convivial evening in Edinburgh, an Englishman and a Scotsman had embarked on the same train.

Said Jones: "It's a wonderful thing this new railway merger, isn't it."

Answered MacAlstair: "Aye, it is that, man. I'm gaun to Aberdeen an' you're gaun to London an' we're baith on the same train!"

30.3 Integrating Acceleration and Velocity

Since

$$a = \dot{v} \quad \text{and} \quad v = \dot{s},$$

we have

$$v(t) = \int a \cdot dt$$

$$s(t) = \int v \cdot dt.$$

In particular if a is constant, then

$$s(t) = \int \int a \cdot dt^2 = \int a\,t + C_1 \, dt = \frac{1}{2}\,a\,t^2 + C_1 t + C_2 .$$

- A stone is dropped from the roof of a building and strikes the ground after 4.5 seconds. Acceleration of gravity is 9.8 m/s^2.

  Determine the height of the building; disregard air resistance of the stone.

$v(t) = \int 9.8 \; dt = 9.8\,t + C_1;$

at $t = 0$, $v = 0$ so $C_1 = 0$ and $v(t) = 9.8\,t$.

$s(t) = \int 9.8\,t \; dt = 4.9\,t^2 + C_2;$

at $t = 0$, $s = 0$, so $C_2 = 0$ and $s(t) = 4.9\,t^2$.

$s(4.5) = 4.9 \cdot 4.5^2 \approx 99$ m, which is the desired height.

- An aircraft requires 16 seconds and 768 m of airstrip to become airborne. If we assume acceleration to be constant, what is the takeoff velocity?

  After roll-out to start, and with ground brakes engaged, the aircraft races its engines to build up thrust; when the brakes are released, the aircraft starts down the runway with maximum acceleration.

$v(t) = \int a \; dt = a\,t + C_1;$

at $t = 0$, $v = 0$ and $at = 0$, so $C_1 = 0$;

$$v(t) = a\,t.$$

$s(t) = \int v \; dt = \int a\,t \; dt = \dfrac{a\,t^2}{2} + C_2;$

at $t = 0$, $s = 0$ and $\dfrac{a\,t^2}{2} = 0$, so $C_2 = 0$; $s(t) = \dfrac{a\,t^2}{2}$.

At takeoff (t = 16 seconds),

$$768 = \frac{a \cdot 16^2}{2}\,; \quad a = \frac{2 \cdot 768}{16^2} = 6 \text{ m/s}^2 .$$

Hence,

$$v(16) = a \cdot 16 = 6 \cdot 16 = 96 \text{ m/s} \approx 346 \text{ km/h}.$$

30.4 Velocity Vectors and Acceleration Vectors

Velocity has direction as well as magnitude (speed) and therefore is a *vector quantity*, and so is **acceleration**.

To find velocity and acceleration – and their components – for objects moving in curvilinear paths, the use of vector analysis is needed.

p. 608 The derivative of a vector **a** is a vector whose **components** are the derivatives of the components of **a**. Thus, if the vector displacement of a moving particle is

$$\mathbf{r}(t) = x\,\mathbf{i} + y\,\mathbf{j} + z\,\mathbf{k},$$

where *radius vector* **r** is a vector from the origin to the particle at time t in a three-dimensional orthogonal coordinate system, then the velocity vector **v** is

$$\mathbf{v}(t) = \dot{\mathbf{r}} = \frac{dx}{dt}\,\mathbf{i} + \frac{dy}{dt}\,\mathbf{j} + \frac{dz}{dt}\,\mathbf{k},$$

p. 609 with the **magnitude**, or *speed*,

$$|\mathbf{v}(t)| = \sqrt{\left(\frac{dx}{dt}\right)^2 + \left(\frac{dy}{dt}\right)^2 + \left(\frac{dz}{dt}\right)^2};$$

the acceleration vector **a** is

$$\mathbf{a}(t) = \dot{\mathbf{v}} = \frac{d^2\mathbf{r}}{dt^2} = \frac{d^2x}{dt^2}\,\mathbf{i} + \frac{d^2y}{dt^2}\,\mathbf{j} + \frac{d^2z}{dt^2}\,\mathbf{k},$$

with the magnitude

$$|\mathbf{a}(t)| = \sqrt{\left(\frac{d^2x}{dt^2}\right)^2 + \left(\frac{d^2y}{dt^2}\right)^2 + \left(\frac{d^2z}{dt^2}\right)^2}.$$

- A projectile is fired at an angle of elevation θ and with an initial speed of v_0. If air resistance is disregarded, determine the velocity vector **v** and position vector **r** at any time t.

Since the force of gravity acts in a downward direction, the acceleration vector is

$$\mathbf{a} = -g\,\mathbf{j}.$$

$\mathbf{v}(t) = -\int g\,\mathbf{j}\;dt = -\,g\,t\;\mathbf{j} + \mathbf{C}_1,$

where $\mathbf{C}_1$ is a constant vector;

at $t = 0$, $\mathbf{v}(0) = \mathbf{C}_1$, so

$$\mathbf{v}(t) = \mathbf{v}(0) - g\,t\;\mathbf{j}.$$

Further integration gives the position vector

$$\mathbf{r}(t) = \int (\mathbf{v}(0) - g\,t\;\mathbf{j})\;dt = t\,\mathbf{v}(0) - \frac{1}{2}\,g\,t^2\mathbf{j} + \mathbf{C}_2;$$

$\mathbf{r}(0) = \mathbf{0}$, so $\mathbf{C}_2 = \mathbf{0}$, and

(1) $$\mathbf{r}(t) = t\,\mathbf{v}(0) - \frac{1}{2}\,g\,t^2\mathbf{j}.$$

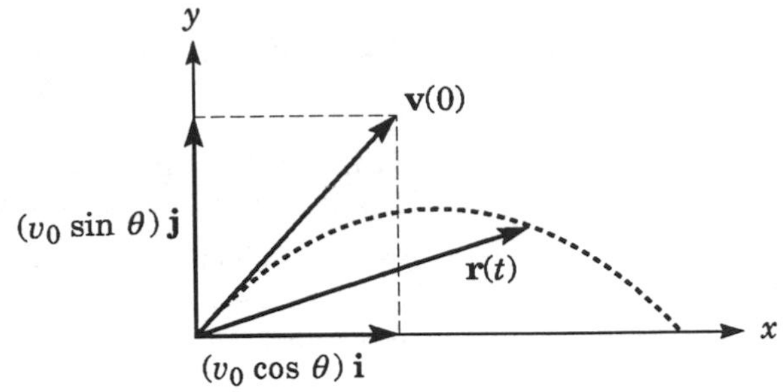

From the graph we have

$$\mathbf{v}(0) = (v_0 \cos\theta)\,\mathbf{i} + (v_0 \sin\theta)\,\mathbf{j}$$

and can rewrite (**1**),

$$\mathbf{r}(t) = t\,[(v_0 \cos\theta)\,\mathbf{i} + (v_0 \sin\theta)\,\mathbf{j}] - \frac{1}{2}\,g\,t^2\,\mathbf{j}$$

$$= t\,(v_0 \cos\theta)\,\mathbf{i} + \left(t\,v_0 \sin\theta - \frac{1}{2}\,g\,t^2\right)\mathbf{j}\,.$$

•

From $\mathbf{r}(t)$ of the above problem, we obtain the parametric equations

$$\left.\begin{aligned} x &= v_0\,t\cos\theta \\ y &= v_0\,t\sin\theta - \frac{1}{2}\,g\,t^2 \end{aligned}\right\}\;.$$

• An object of mass m describes a circular path of radius r; its angular velocity, the rate of change of the angle swept by the radius vector in unit time, is a constant ω.

For time t, find

a: the velocity **v** and the speed of the object;

b: the acceleration **a** and its magnitude;

c: the force vector **F** necessary to produce the motion.

Orienting the axes so that the center of the circle is at the origin and the object is on the positive x-axis when $t = 0$, the polar coordinate θ of its position at time t must be $\omega \cdot t$.

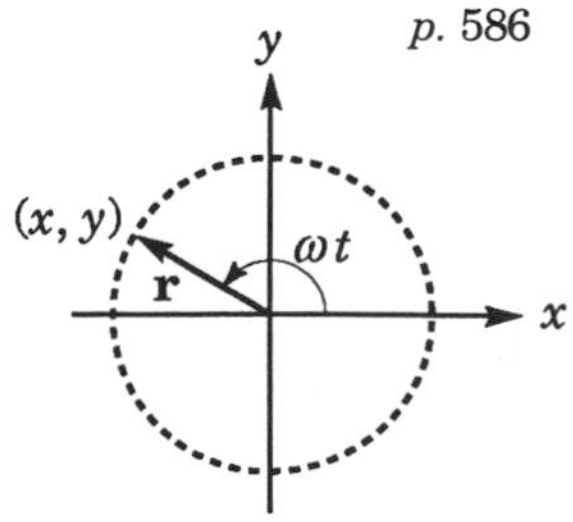

p. 586

The path is thus described by the **parametric form** of the equation for a circle of radius r centered at the origin of an orthogonal coordinate system,

$$\left.\begin{aligned} x &= r\cos\omega t \\ y &= r\sin\omega t \end{aligned}\right\}\;;$$

the position of the object is described by

$$\mathbf{r}(t) = r\,(\mathbf{i}\cos\omega t + \mathbf{j}\sin\omega t)\,.$$

a:

$$\mathbf{v}(t) = \dot{\mathbf{r}} = \omega r\,[\mathbf{i}\,(-\sin\omega t) + \mathbf{j}\cos\omega t\,]\,.$$

p. 550

$$\text{speed} = |\mathbf{v}(t)| = \omega r\sqrt{\sin^2\omega t + \cos^2\omega t} = \omega r\,.$$

b:

$$\mathbf{a}(t) = \dot{\mathbf{v}} = \omega^2 r\,[(-\cos \omega t)\,\mathbf{i} - (\sin \omega t)\,\mathbf{j}]$$
$$= -\omega^2 r\,(\mathbf{i} \cos \omega t + \mathbf{j} \sin \omega t) = -\omega^2 \mathbf{r},$$

which shows that the acceleration vector has a direction opposite to **r** and thus is directed towards the center of the circle.

$$|\mathbf{a}(t)| = \omega^2 r \sqrt{\sin^2 \omega t + \cos^2 \omega t} = \omega^2 r\,.$$

c:

F is the product of mass and vector acceleration,

$$\mathbf{F}(t) = m\,\mathbf{a}(t) = -m\omega^2 r\,(\mathbf{i} \cos \omega t + \mathbf{j} \sin \omega t)\,.$$

The force vector is pointed towards the center of the circle (*centripetal* force, meaning "center-seeking").

30.5 Space-Time; Mass and Energy

Special Theory of Relativity

EINSTEIN Albert (1879 - 1955)

When **Albert Einstein**, in 1905, presented his **special theory of relativity**, he had built on recent findings of the electromagnetic nature of light and was able to break away from classic – Newtonian – physics, where time is viewed independent of space.

Einstein's special theory of relativity is called *special* because it is concerned only with *non-accelerated* relative motion.

General Theory of Relativity

Einstein's **general theory of relativity**, introduced in 1915, addresses, on the other hand, the problem of gravity and *accelerated* motion, and indicates that gravitational forces also affect electromagnetic radiation – which is energy and thereby equivalent to mass. Among other things, the general theory of relativity proposes that gravitational forces bend rays of light, and this has been validated in astronomical observations.

The **special theory of relativity** is based on two postulates:

1. In any two systems that move relative to one another with constant velocity, all physical laws and principles are expressed in the same mathematical form.

   This implies that no experiment can be made to establish whether the system of reference is at rest or moving with constant velocity.

2. The speed of light in vacuum has the same constant value, independent of the velocity of the source or the recipient.

Einstein abolished not only the thought of absolute space but also that of absolute time, concepts that had been accepted but never confirmed in Newtonian physics. Thus, Einstein showed that there is no absolute simultaneity at different locations; the present time – the "right now" – exists only at the location of observation: *now* is only *here*. To understand this, one must get used to viewing time as a coordinate that, like the space coordinates, can be transformed.

Transformations of the coordinates of space and time take place between two systems of reference moving at constant velocity relative to each other; length and time, and also mass, depend on the relative motion.

Lorentz Transformations

LORENTZ Hendrik Antoon (1853 - 1928)

The effects of these space-time transformations, known as **Lorentz transformations** after the Dutch physicist **H. A. Lorentz**, are manifested in the following formulas. The observer is located in a system of reference S that moves at a constant velocity v relative to the system S' of the observed object; the speed of light in vacuum is c.

Contraction of length

1. A rod resting in system S', where its length is l_0, has when remeasured in system S shortened to

$$l = l_0 \cdot \sqrt{1 - \left(\frac{v}{c}\right)^2}.$$

Dilation of time

2. If time measured in system S' is T_0, the observer in system S finds the longer period of time T,

$$T = \frac{T_0}{\sqrt{1 - \left(\frac{v}{c}\right)^2}} .$$

A consequence of the above two formulas and the theory of relativity is that no velocity can be greater than the speed of light in vacuum.

Mass and time

3. An object that, at rest in system S', has the mass m_0 has in system S the mass m,

$$m = \frac{m_0}{\sqrt{1 - \left(\frac{v}{c}\right)^2}} .$$

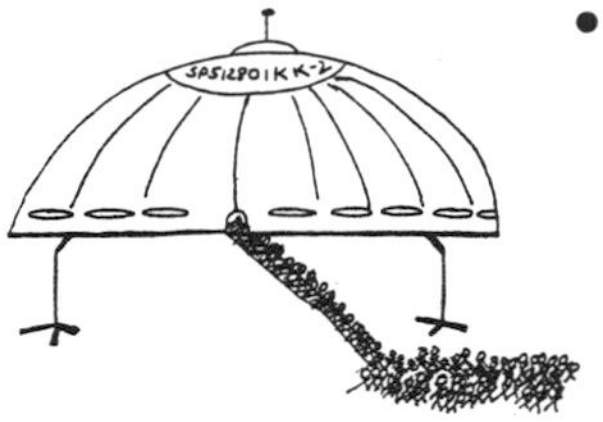

• Launched from planet Ψ-36 at the beginning of the planet's calendar year 1200, a spaceship with a crew of 300 travels at 99.99% warp speed (99.99% of the speed of light) for exactly 3 years by the ship's clock, before returning to the planet. What is the actual Ψ-36 calendar year of return?

$$\frac{3}{\sqrt{1 - \frac{0.9999^2\, c^2}{c^2}}} \approx 212.1 ,$$

indicating 1412 as the year of return. •

Our spaceships attain a mere 0.005% of the speed of light; a 3-year space travel would make our astronauts age only about 12/100ths of a second less than the people remaining on Earth.

Time Warps

– *"Why did you break Tessa's doll?"*
– *"Because then she hit me."*

– *"What's the hurry, Mr. Schein?"*
– *"My past is catching up with me!"*

Einstein's Energy-Mass Relation

Einstein's energy-mass relation

$$E = m \cdot c^2$$

states that an object's total energy is the product of its mass and the speed of light squared.

If an object with the mass m_0 is accelerated to the velocity v, then it has been provided with the kinetic energy E_k,

$$E_k = m \cdot c^2 - m_0 \cdot c^2 = m_0 \cdot c^2 \left[\frac{1}{\sqrt{1 - \left(\frac{v}{c}\right)^2}} - 1 \right].$$

This relativistic expression for kinetic energy can for very small values of v/c be rewritten

$$m_0 \cdot c^2 \left[\frac{1}{\sqrt{1 - \left(\frac{v}{c}\right)^2}} - 1 \right] \approx m_0 \cdot c^2 \left[\sqrt{1 + \left(\frac{v}{c}\right)^2} - 1 \right]$$

$$\approx m_0 \cdot c^2 \left[1 + \frac{1}{2}\left(\frac{v}{c}\right)^2 - 1 \right] = \frac{m_0 \cdot v^2}{2},$$

which is the classic formula for kinetic energy.

• The energy radiated every second by the Sun is $360 \cdot 10^{24}$ J and the speed of light is $300 \cdot 10^6$ m/s; find the Sun's loss of mass per second. [J (joule), m (meter), s (second), and, for mass, kg (kilogram) are coherent units of the International System of Units (SI).]

The relation $E = m \cdot c^2$ means that every energy unit corresponds to the mass $m = E/c^2$, so the Sun's loss of mass per second is

SI multiples: *p.* 54

$$m = \frac{360 \cdot 10^{24}}{(300 \cdot 10^6)^2} = 4 \cdot 10^9 \text{ kg} = 4 \text{ teragrams.}$$ •

Heisenberg's Uncertainty Principle; *Measuring the Elusive*

HEISENBERG Werner Karl (1901 - 1976)

The uncertainty principle, presented in 1927 by the German physicist **Werner Heisenberg**, states that position and velocity of a particle cannot, even in theory, be measured exactly at the same time: the more accurately one knows the position, the less accurately one can know the velocity, and vice versa. We limit our account to the following case history:

There had been a series of nightly break-ins and thefts around where Angus McTaggart had his small croft. One night he heard a foreign noise from his pigsty, so he put on his breeches and went out to have a look-see.

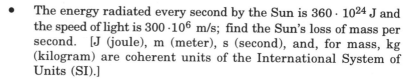

When he came back in, his wife understandably wanted to know if all the piggies were still there, all twenty of them.

Said Angus: "Och, lassie, Ah juist dinna ken. Ah sure counted the nineteen o'them, but Ah spied a wee lil fella wha ran aboot sae perfectly fast Ah couldna richtly charge ma mind to find him a place tae count him."

Chapter 31

HARMONIC ANALYSIS

Page

$x^2\ dx =$
P.G.

31.0 Historical Notes

WIENER Norbert
(1894 - 1964)

American mathematician; founder of the science of **cybernetics**, the study of mathematical structure of control systems and communication systems in living organisms and in machines. Wiener made important contributions to **harmonic analysis**, a field that he widened extensively.

The somewhat snobbish point of view of the purely abstract mathematician would draw but little support from mathematical history. On the other hand, whenever applied mathematics has been merely a technical employment of methods already traditional and jejune, it has been very poor applied mathematics. The desideratum in mathematical as well as physical work is an attitude which is not indifferent to the extremely instructive nature of actual physical situations, yet which is not dominated by these to the dwarfing and paralyzing of its intellectual originality. Viewed as a whole, the theory of harmonic analysis has a very fine record of this sort. It is not a young theory, but neither is it yet in its dotage. There is much more to be learned and much more to be proved.

Norbert Wiener, *The Historical Background of Harmonic Analysis* (1938)

d'ALEMBERT Jean Le Rond
(1717 - 1783)

EULER Leonhard
(1707 - 1783)

BERNOULLI Daniel
(1700 - 1782)

FOURIER Jean Baptiste Joseph
(1768 - 1830)

In the 18th century, the French mathematician and physicist **d'Alembert** and the Swiss mathematician **Euler** described the vibration of strings by means of sums of arbitrary functions, and **Daniel Bernoulli**, a member of the celebrated Swiss family of mathematicians and scientists, used trigonometric functions.

The use of trigonometric functions was further developed by the French mathematician and physicist **Joseph Fourier** in his *Théorie analytique de la chaleur* ("The Analytical Theory of Heat"), published in 1822. Fourier showed that the conduction of heat in solid bodies may be described by infinite series of sines and cosines.

THÉORIE

ANALYTIQUE

DE LA CHALEUR,

PAR M. FOURIER.

A PARIS,

CHEZ FIRMIN DIDOT, PÈRE ET FILS,

LIBRAIRES POUR LES MATHÉMATIQUES, L'ARCHITECTURE HYDRAULIQUE ET LA MARINE, RUE JACOB, N° 24.

1822.

Fourier's work stimulated research in other areas, showing that the type of series used by Fourier is a mathematical prerequisite for the solution of periodic phenomena encountered in practically every branch of science and technology, such as the design of electrical circuits, the prediction of tides and sunspots, and the analysis of sound and electromagnetic waves.

The method of expressing periodic functions as sums of sines and cosines is referred to as **harmonic analysis**.

31.1 Fourier Series

The principal idea of a Fourier series expansion is to represent a function f of period 2π as an infinite series of trigonometric functions, so that

$$f(x) = \frac{a_0}{2} + a_1 \cos x + a_2 \cos 2x + \ldots + b_1 \sin x + b_2 \sin 2x \ldots$$

$$f(x) = \frac{a_0}{2} + \sum_{n=1}^{\infty} (a_n \cos nx + b_n \sin nx),$$

where a_0, a_n, b_n are constants, n is a positive integer, and x takes values between $-\infty$ and $+\infty$. Writing the first term as $\frac{a_0}{2}$ instead of a_0 is, as we shall see, convenient for computation.

The notable quality of a Fourier series expansion is that it can be used for functions that are represented by different expressions in different parts of the interval.

Dirichlet's Conditions

DIRICHLET Peter Gustav Lejeune (1805 - 1859) German mathematician

The theory of Fourier series is complex and is still the object of intense research. For the vast majority of mathematical problems or physical applications the representation above is valid for every x, because functions f arising in practice ordinarily satisfy *Dirichlet's conditions* for convergence of the Fourier series to f. Since sines and cosines both have periods of 2π radians, the above expansion is unchanged by replacing x with $(x + 2k\pi)$, where k is an integer, which corresponds to f having period 2π.

Fourier Coefficients

While derivatives are used to find the coefficients of Taylor's series, the coefficients of the Fourier series for f,

$$\frac{a_0}{2} + \sum_{n=1}^{\infty} (a_n \cos nx + b_n \sin nx),$$

are determined by evaluating certain integrals which exist provided that $f(x)$ is continuous in the interval $-\pi \le x \le \pi$.

a_0 **To find an expression for the coefficient a_0**, we integrate $f(x)$ with respect to x from $x = -\pi$ to $x = \pi$. We assume that $f(x)$ is the sum of the series and that the integral of the sum is the sum of the integrals even though there are infinitely many summands.

$$\int_{-\pi}^{\pi} f(x)\, dx = \frac{a_0}{2}\int_{-\pi}^{\pi} dx + \sum_{n=1}^{\infty}\left(\int_{-\pi}^{\pi} a_n \cos nx\, dx + \int_{-\pi}^{\pi} b_n \sin nx\, dx\right)$$

$$= \frac{a_0}{2}\Big|_{-\pi}^{\pi} x + \sum_{n=1}^{\infty}\left[\frac{a_n}{n}\sin nx - \frac{b_n}{n}\cos nx\right]_{-\pi}^{\pi}$$

$$= \pi\cdot a_0 + \sum_{n=1}^{\infty}\left\{\frac{a_n}{n}[\sin n\pi - \sin(-n\pi)] - \frac{b_n}{n}[\cos n\pi - \cos(-n\pi)]\right\}$$

$$= \pi\cdot a_0 + \sum_{n=1}^{\infty}\left(\frac{2a_n}{n}\cdot \sin n\pi - \frac{2b_n}{n}\cdot 0\right)$$

$$= \pi\cdot a_0 + \sum_{n=1}^{\infty}(0-0) = a_0\pi.$$

$$a_0 = \frac{1}{\pi}\int_{-\pi}^{\pi} f(x)\, dx.$$

a_n **To find an expression for a_n**, $n \geq 1$, multiply $f(x)$ by $\cos mx$, where m is a positive integer, and integrate with respect to x from $x = -\pi$ to $x = \pi$.

$$\int_{-\pi}^{\pi} f(x)\cos mx\, dx = \frac{a_0}{2}\int_{-\pi}^{\pi}\cos mx\, dx + \sum_{n=1}^{\infty}\int_{-\pi}^{\pi} a_n \cos nx\cdot\cos mx\, dx + \sum_{n=1}^{\infty}\int_{-\pi}^{\pi} b_n \sin nx\cdot \cos mx\, dx,$$

which we integrate term by term:

$$\frac{a_0}{2}\int_{-\pi}^{\pi}\cos mx\, dx = \frac{a_0}{2m}\Big|_{-\pi}^{\pi}\sin mx = 0 \qquad \text{(for all } m\text{)}.$$

To evaluate

$$\int_{-\pi}^{\pi} a_n \cos nx\cdot\cos mx\, dx,$$

distinguish two cases, $m \neq n$ and $m = n$.

$\boxed{m \neq n}$

$$a_n \int_{-\pi}^{\pi} \cos nx \cdot \cos mx \ dx$$

$$= \frac{a_n}{2} \int_{-\pi}^{\pi} [\cos (n-m)x + \cos (n+m)x] \ dx$$

$$= \frac{a_n}{2} \left[\frac{\sin (n-m)x}{n-m} + \frac{\sin (n+m)x}{n+m} \right]_{-\pi}^{\pi} = 0 .$$

$\boxed{m = n}$

$$a_n \int_{-\pi}^{\pi} \cos nx \cdot \cos mx \ dx$$

$$= a_n \int_{-\pi}^{\pi} \cos^2 nx \ dx = \frac{a_n}{2} \int_{-\pi}^{\pi} (1 + \cos 2nx) \ dx$$

$$= \frac{a_n}{2} \Big|_{-\pi}^{\pi} x + \frac{a_n}{4n} \Big|_{-\pi}^{\pi} \sin 2nx = \pi \cdot a_n .$$

Thus,

$$\int_{-\pi}^{\pi} a_n \cos nx \cdot \cos mx \ dx = \begin{cases} 0, & \text{if } n \neq m \\ \pi \cdot a_n, & \text{if } n = m . \end{cases}$$

$$b_n \int_{-\pi}^{\pi} \sin nx \cdot \cos mx \ dx = \frac{b_n}{2} \int_{-\pi}^{\pi} [\sin (n+m)x + \sin (n-m)x] \ dx$$

$$= -\frac{b_n}{2} \left[\frac{\cos (n+m)x}{n+m} + \frac{\cos (n-m)x}{n-m} \right]_{-\pi}^{\pi}$$

$$= 0 \ \text{(for all } m \text{ and } n;\ m \neq n) .$$

Thus,

$$\int_{-\pi}^{\pi} f(x) \cos nx \ dx = \begin{cases} 0 & \text{if } m \neq n \\ \pi \cdot a_n & \text{if } m = n . \end{cases}$$

$$a_n = \frac{1}{\pi} \int_{-\pi}^{\pi} f(x) \cdot \cos nx \ dx.$$

$\boldsymbol{b_n}$ **To find an expression for $\boldsymbol{b_n}$**, multiply $f(x)$ by $\sin mx$, where m is a positive integer or zero, and integrate with respect to x from $x = -\pi$ to $x = \pi$.

$$\int_{-\pi}^{\pi} f(x) \sin mx \, dx = \frac{a_0}{2} \int_{-\pi}^{\pi} \sin mx \, dx$$

$$+ \sum_{n=1}^{\infty} \int_{-\pi}^{\pi} a_n \cos nx \sin mx \, dx$$

$$+ \sum_{n=1}^{\infty} \int_{-\pi}^{\pi} b_n \sin nx \sin mx \, dx,$$

which we again integrate term by term:

$$\frac{a_0}{2} \int_{-\pi}^{\pi} a_0 \sin mx \, dx = -\frac{a_0}{2m} \Big|_{-\pi}^{\pi} \cos m\pi = 0 .$$

$$a_n \int_{-\pi}^{\pi} \cos nx \sin mx \, dx = \frac{a_n}{2} \int_{-\pi}^{\pi} [\sin (n+m) x + \sin (n-m) x] \, dx$$

$$= -\frac{a_n}{2} \left[\frac{\cos (n+m)x}{n+m} + \frac{\cos (n-m)x}{n-m} \right]_{-\pi}^{\pi}$$

$$= 0 \text{ (for all } m \text{ and } n;\ m \neq n) .$$

To evaluate

$$b_n \int_{-\pi}^{\pi} \sin n \cdot x \sin mx \, dx,$$

we again distinguish two cases, $m \neq n$ and $m = n$.

$m \neq n$

$$b_n \int_{-\pi}^{\pi} \sin n \cdot x \sin mx \, dx$$

$$= \frac{b_n}{2} \int_{-\pi}^{\pi} [\cos (n-m) x - \cos (n+m) x] \, dx$$

$$= \frac{b_n}{2} \left[\frac{\sin (n-m) x}{n-m} - \frac{\sin (n+m) x}{n+m} \right]_{-\pi}^{\pi}$$

$$= 0 \text{ (for all } m \text{ and } n).$$

$m = n$

$$b_n \int_{-\pi}^{\pi} \sin nx \cdot \sin mx \, dx$$

$$= b_n \int_{-\pi}^{\pi} \sin^2 nx \, dx = \frac{b_n}{2} \int_{-\pi}^{\pi} (1 - \cos 2nx) \, dx$$

$$= \frac{b_n}{2} \left[x - \frac{\sin 2nx}{2n} \right]_{-\pi}^{\pi} = \pi \cdot b_n .$$

Thus,

$$\int_{-\pi}^{\pi} f(x) \sin nx \, dx = \begin{cases} 0 & \text{if } n \neq m \\ \pi \cdot b_n & \text{if } n = m . \end{cases}$$

$$b_n = \frac{1}{\pi} \int_{-\pi}^{\pi} f(x) \sin nx \, dx .$$

Summing up: If a function $f(x)$ is continuous in $-\pi \leq x \leq \pi$, it has the series

$$\frac{a_0}{2} + \sum_{n=1}^{\infty} (a_n \cdot \cos nx + b_n \cdot \sin nx),$$

where

$$a_0 = \frac{1}{\pi} \int_{-\pi}^{\pi} f(x) \, dx ;$$

$$a_n = \frac{1}{\pi} \int_{-\pi}^{\pi} f(x) \cos nx \, dx ;$$

$$b_n = \frac{1}{\pi} \int_{-\pi}^{\pi} f(x) \sin nx \, dx .$$

There exist continuous f such that the Fourier series for f does not converge to $f(x)$ for every x. If, however, f' is also continuous, then we do have for every x that

$$f(x) = \frac{a_0}{2} + \sum_{n=1}^{\infty} (a_n \cos nx + b_n \sin nx) .$$

***

– *Yes?*
– *A remarkable lecture, Professor.*
– *Thank you. What impressed you?*
– *Your splendid integration with the topic and your disintegration of the listeners.*

31.2 Expanding Discontinuous Functions

When using power series expansion, such as Taylor series, we are limited to continuous functions; with Fourier series, we may also expand discontinuous functions.

Hitherto we assumed $f(x)$ to be continuous in $-\pi \leq x \leq \pi$. If $f(x)$ is discontinuous for $x = x_0$ in the interval $-\pi \leq x \leq \pi$ and

$$\text{if } f(x_{0_+}) = \lim_{x \to x_{0_+}} f(x) \text{ and } f(x_{0_-}) = \lim_{x \to x_{0_-}} f(x) \text{ exist,}$$

then the coefficients can be determined as follows:

$$a_0 = \frac{1}{\pi} \int_{-\pi}^{x_0} f(x)\, dx + \frac{1}{\pi} \int_{x_0}^{\pi} f(x)\, dx;$$

$$a_n = \frac{1}{\pi} \int_{-\pi}^{x_0} f(x) \cdot \cos nx \; dx + \frac{1}{\pi} \int_{x_0}^{\pi} f(x) \cdot \cos nx \; dx;$$

$$b_n = \frac{1}{\pi} \int_{-\pi}^{x_0} f(x) \cdot \sin nx \; dx + \frac{1}{\pi} \int_{x_0}^{\pi} f(x) \cdot \sin nx \; dx.$$

The Fourier series will converge to $f(x)$ at every point where $f(x)$ is continuous and to $\frac{1}{2}\,[f(x_+) + f(x_-)]$ at every point where $f(x)$ has a jump of discontinuity.

- Expand the function

$$f(x) = \begin{cases} 0; & -\pi \leq x < 0 \\ 1; & 0 \leq x < \pi \\ f(x + 2k\pi) & \\ \quad k \text{ is an integer} & \end{cases}$$

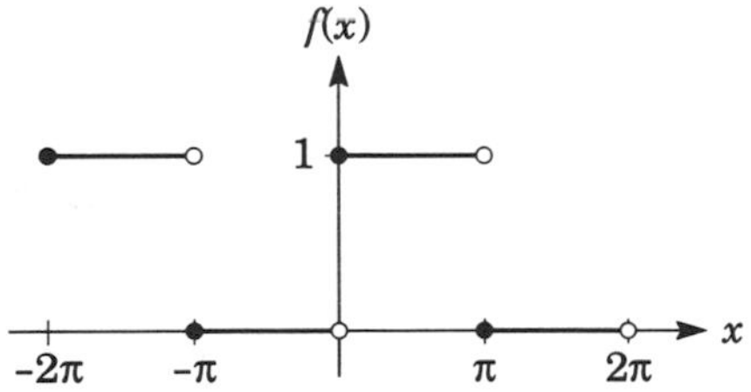

f is a periodic function with the period 2π and may be expanded into a Fourier series.

Our principal concern is to find the coefficients of the Fourier series

$$\frac{a_0}{2} + \sum_{n=1}^{\infty} (a_n \cos nx + b_n \sin nx)\;.$$

We observe that all integrals over the interval $[-\pi, 0]$ which contain $f(x)$ as a factor of the integral are equal to zero. Thus:

$$a_0 = \frac{1}{\pi} \int_0^{\pi} 1 \; dx = 1 \, ;$$

$$a_n = \frac{1}{\pi} \int_0^{\pi} 1 \cdot \cos nx \; dx = 0 \qquad (n = 1, 2, 3, \ldots) \, ;$$

$$b_n = \frac{1}{\pi} \int_0^{\pi} 1 \cdot \sin nx \; dx \qquad (n = 1, 2, 3, \ldots) \, ;$$

$$= \frac{1}{\pi} \left[\frac{-1}{n} \cos nx \right]_0^{\pi} = \frac{1}{n\pi} \Big[\cos nx \Big]_{\pi}^{0}$$

$$= \begin{cases} 0 & \text{for even } n \\ \dfrac{2}{n\pi} & \text{for odd } n \, . \end{cases}$$

The expansion of $f(x)$ contains a constant term and sine terms only.

$$f(x) = \frac{1}{2} + \frac{2}{\pi} \left(\frac{\sin x}{1} + \frac{\sin 3x}{3} + \frac{\sin 5x}{5} + \frac{\sin 7x}{7} + \ldots \right) \text{ if } x \neq k\pi \, .$$

As guaranteed by the theory, if $x = k\pi$, the series is

$$\frac{1}{2} + \frac{2}{\pi} (0 + 0 + \ldots) = \frac{1}{2} \left[f(x_+) + f(x_-) \right] .$$

Graphic Approximation

As shown by the following graphs, just a few terms of the Fourier expansion may suffice to give a fair representation of the graph of $f(x)$:

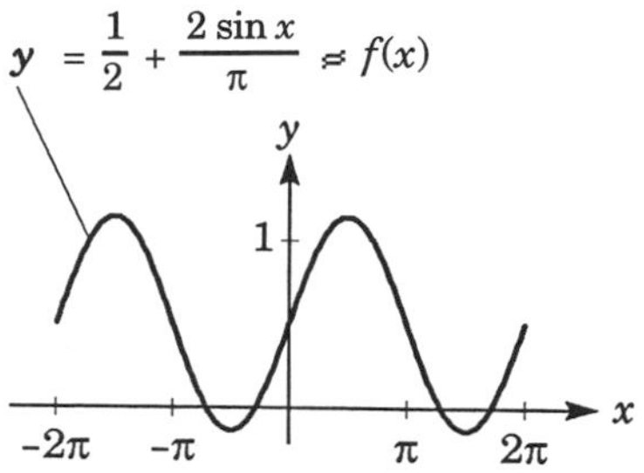

$$y = \frac{1}{2} + \frac{2}{\pi}\left(\frac{\sin x}{1} + \frac{\sin 3x}{3}\right) \approx f(x)$$

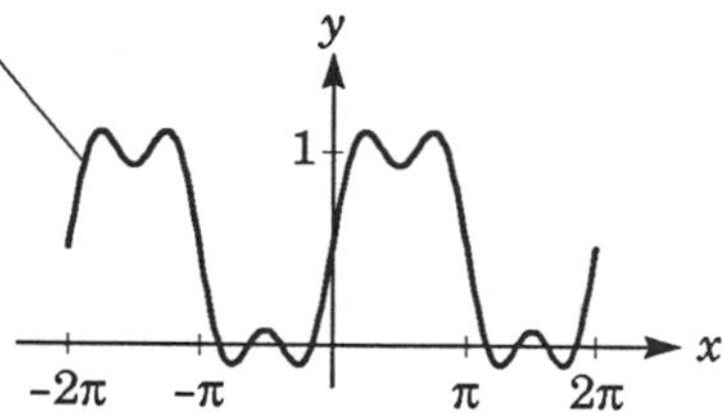

$$y = \frac{1}{2} + \frac{2}{\pi}\left(\frac{\sin x}{1} + \frac{\sin 3x}{3} + \frac{\sin 5x}{5}\right) \approx f(x)$$

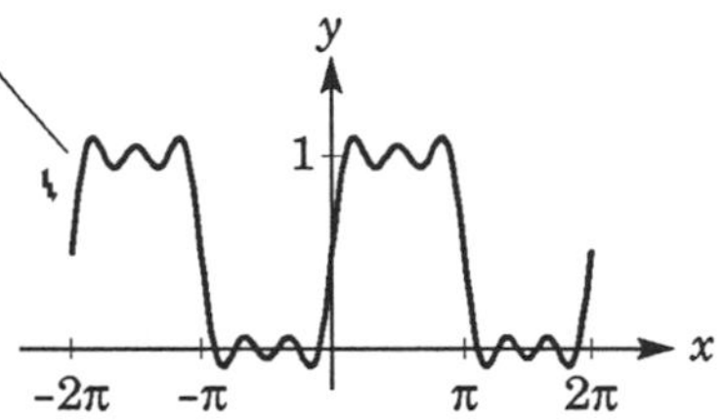

$$y = \frac{1}{2} + \frac{2}{\pi}\left(\frac{\sin x}{1} + \frac{\sin 3x}{3} + \frac{\sin 5x}{5} + \frac{\sin 7x}{7}\right) \approx f(x)$$

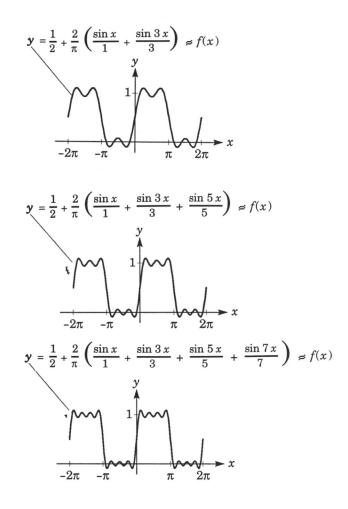

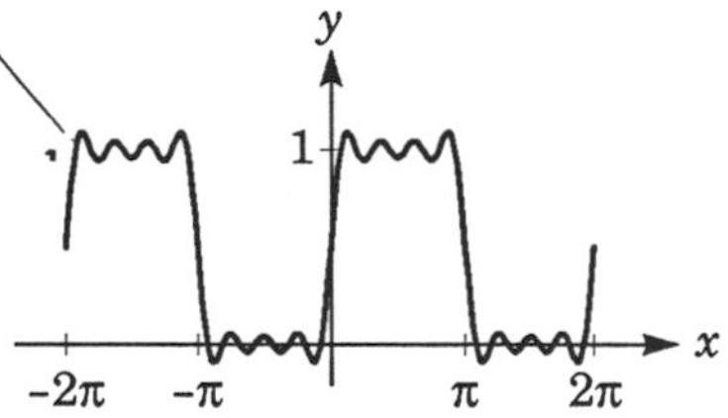

A completed, accurate graphic representation of the Fourier expansion of $f(x)$ requires an infinite number of terms and is therefore unattainable in practice.

31.3 Expanding Even or Odd Functions

cf. Even and Odd Functions: *pp.* 340 and 508

A function f is **even** if $f(-x) = f(x)$ and **odd** if $f(-x) = -f(x)$. A cosine function is an even function; a sine function is odd.

The product of two functions that are both even or both odd is an even function, while the product of one even and one odd function is odd. For the evaluation of Fourier series, it will save work if one can predict whether products of functions are even or odd and realize that an area integral is considered positive above the x-axis, negative below.

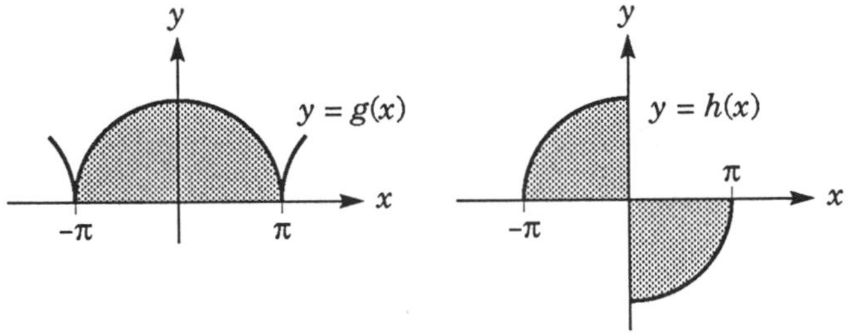

The function $g(x)$ in the above graphs is an even function over the interval $x = -\pi$ to $x = \pi$ and $h(x)$ is odd:

$$\int_{-\pi}^{\pi} g(x)\,dx = 2\int_{0}^{\pi} g(x)\,dx\,; \qquad \int_{-\pi}^{\pi} h(x)\,dx = 0\,.$$

From this, two useful theorems emerge to simplify the evaluation of Fourier series expansions of even or odd functions.

Even Functions

If $f(x)$ is an **even** function defined over $-\pi < x < \pi$, then its Fourier expansion reduces to a cosine series.

In the Fourier expansion for f,

$$\frac{a_0}{2} + \sum_{n=1}^{\infty} (a_n \cos nx + b_n \sin nx),$$

we have, since f is even,

$$a_0 = \frac{1}{\pi}\int_{-\pi}^{\pi} f(x)\,dx = \frac{2}{\pi}\int_{0}^{\pi} f(x)\,dx.$$

The product of two even functions is even; the cosine function is even:

$$a_n = \frac{1}{\pi} \int_{-\pi}^{\pi} f(x) \cdot \cos nx \, dx = \frac{2}{\pi} \int_{0}^{\pi} f(x) \cdot \cos nx \, dx.$$

The sine function is odd; the product of an even function and an odd function is odd:

$$b_n = \frac{1}{\pi} \int_{-\pi}^{\pi} f(x) \cdot \sin nx \, dx = 0 .$$

Thus, for the even function $f(x)$, we have the Fourier series

$$\frac{a_0}{2} + \sum_{n=1}^{\infty} a_n \cos nx ;$$

that is, Fourier expansion of even functions reduces to a cosine series.

Odd Functions

If $f(x)$ is an **odd** function defined over $-\pi \le x \le \pi$, then its Fourier expansion reduces to a sine series.

Since f is odd, we have

$$a_0 = \frac{1}{\pi} \int_{-\pi}^{\pi} f(x) \, dx = 0 .$$

The cosine function is even; the product of an odd function and an even function is odd:

$$a_n = \frac{1}{\pi} \int_{-\pi}^{\pi} f(x) \cdot \cos nx \, dx = 0 .$$

On the other hand, a sine function is odd; the product of two odd functions is even; therefore,

$$b_n = \frac{1}{\pi} \int_{-\pi}^{\pi} f(x) \cdot \sin nx \, dx = 2\frac{1}{\pi} \int_{0}^{\pi} f(x) \cdot \sin nx \, dx.$$

Thus, for the odd function $f(x)$, we have the Fourier series

$$\sum_{n=1}^{\infty} b_n \sin nx ;$$

that is, Fourier expansion of odd functions reduces to a sine series.

- Expand the triangular function

$$f(x) = \begin{cases} |x|; \ -\pi \le x \le \pi \\ f(x + 2k\pi) \\ \quad k \text{ is an integer} \end{cases}$$

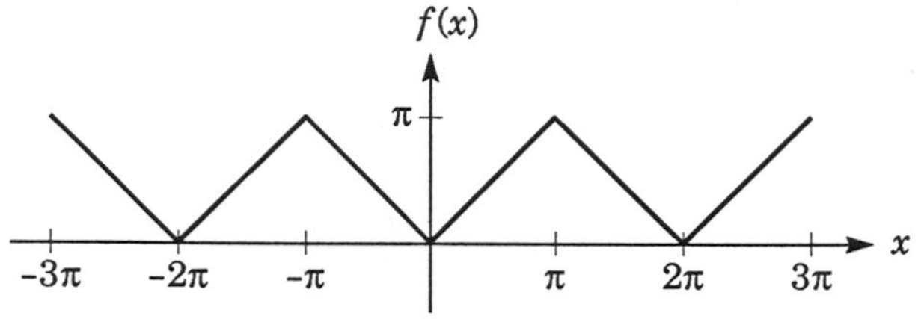

f is a periodic function of period 2π.

Since f is even, $b_n = 0$.

$$f(x) = \frac{a_0}{2} + \sum_{n=1}^{\infty} a_n \cos nx;$$

$$a_0 = \frac{2}{\pi}\int_0^{\pi} |x|\, dx = \frac{2}{\pi} \left. \frac{x^2}{2} \right|_0^{\pi} = \pi;$$

$$a_n = \frac{2}{\pi}\int_0^{\pi} |x| \cos nx \ dx.$$

Integration by parts gives

$$a_n = \frac{2}{\pi}\left[\frac{x \sin nx}{n} + \frac{\cos nx}{n^2}\right]_0^{\pi}$$

$$= 0 + \frac{2}{\pi n^2} \left. \cos nx \right|_0^{\pi} = \begin{cases} 0 & \text{for even } n \\ -\dfrac{4}{\pi n^2} & \text{for odd } n\,. \end{cases}$$

Hence,

$$f(x) = \frac{\pi}{2} - \frac{4}{\pi}\left(\frac{\cos x}{1} + \frac{\cos 3x}{3^2} + \frac{\cos 5x}{5^2} + \frac{\cos 7x}{7^2} + \dots\right) \text{ for all } x.$$

Numerical Value of π

Incidentally, setting $x = 0$, the series may be used to find a numerical value of π; thus,

$$0 = \frac{\pi}{2} - \frac{4}{\pi}\left(\frac{\cos x}{1} + \frac{\cos 3x}{3^2} + \frac{\cos 5x}{5^2} + \frac{\cos 7x}{7^2} + \dots\right)$$

$$= \frac{\pi}{2} - \frac{4}{\pi}\left(1 + \frac{1}{3^2} + \frac{1}{5^2} + \frac{1}{7^2} + \dots\right)$$

or

$$\pi^2 = 8\left(1 + \frac{1}{3^2} + \frac{1}{5^2} + \frac{1}{7^2} + \dots\right).$$

- Expand the sawtooth function

$$f(x) = \begin{cases} x \; ; \; -\pi < x \le \pi \\ f(x + 2k\pi) \\ \qquad k \text{ is an integer} \end{cases}$$

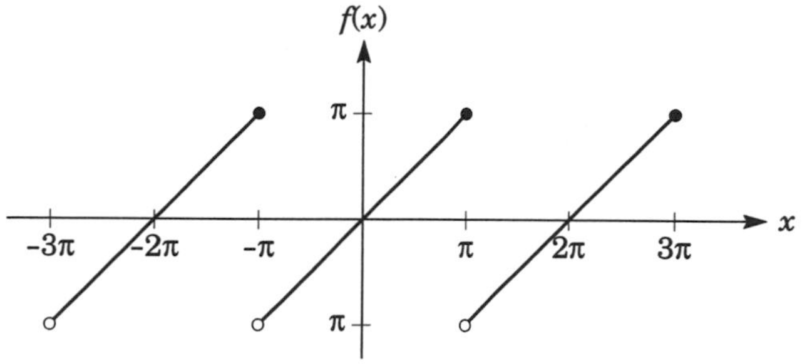

f is a periodic function of period 2π.

Since f is odd, $a_0 = 0$ and $a_n = 0$; $n = 1, 2, \ldots$.

$$f(x) = \sum_{n=1}^{\infty} b_n \sin nx \, ; \quad b_n = \frac{2}{\pi} \int_0^{\pi} f(x) \sin nx \;\; dx.$$

Integration by parts gives

$$b_n = \frac{2}{\pi}\left[\frac{-x \cos nx}{n} + \frac{\sin nx}{n^2}\right]_0^{\pi} ,$$

where the second term is zero, and so

$$b_n = \frac{-2 \cos n\pi}{n} = \left(\frac{2}{n}\right)(-1)^{n-1} .$$

Hence,

$$f(x) = \sum_{n=1}^{\infty} \left(\frac{2}{n}\right)(-1)^{n-1} \sin nx = 2 \sum_{n=1}^{\infty} (-1)^{n-1} \frac{\sin nx}{n}$$

$$= 2\left(\frac{\sin x}{1} - \frac{\sin 2x}{2} + \frac{\sin 3x}{3} - \ldots\right), \text{ if } x \ne (2k+1)\pi .$$

Once again, the series converges to the average

$$0 = \frac{1}{2}\,[f(x_+) + f(x_-)]$$

if $x = (2k+1)\pi$.

From F. Gafurius, *Theorica Musice*, Milan, 1492.
[D. E. Smith, *History of Mathematics*; Dover, 1958.]

Music, and its theory, was for the Pythagoreans (6th to 4th century B.C.) one of the four mathematical categories of science: arithmetic, geometry, music, astronomy.

PYTHAGORAS
(*c.* 580 - 496 B.C.)

Legend supports that **Pythagoras** of Samos discovered that for strings under equal tension, the lengths should be 2 to 1 for the octave of a note, 3 to 2 for the fifth, and 4 to 3 for the fourth. From the knowledge that whole numbers determine musical intervals, the Pythagoreans inferred that the intervals between the heavenly bodies were governed by the laws of musical harmony, and that the same harmony is the true ruler of everything in nature.

From Georg Reisch, *Margarita phylosophica* (first edition, 1496; here reproduced from a 1583 reprint).

Music had a prominent position in schools of the Middle Ages and the Renaissance. Music was one of the seven liberal arts: geometry, astronomy, arithmetic, music, grammar, dialectics (logic), and rhetoric.

Chapter 32

METHODS OF APPROXIMATION

*** Indicates cross-reference

1993
Ann
Gullberg

32.1 Negligible Terms

A quantity

$$A = (a_0 + \Delta a)^p = a_0^p (1 + \varepsilon)^p ,$$

where

$$\frac{\Delta a}{a_0} = \varepsilon << 1,$$

may be calculated with acceptable accuracy by observing that

$$(1 + \varepsilon)^p = 1 + p \cdot \varepsilon + \frac{p\,(p-1)}{2} \cdot \varepsilon^2 + \dots ,$$

where ε^2 and higher terms are negligible.

- $$\sqrt{100.5} = 10\left(1 + \frac{0.5}{100}\right)^{1/2} = 10\,(1 + 0.005)^{1/2} \approx 10\,(1 + 0.0025)$$
  $$= 10.025$$

- $$\sqrt[3]{122} = (125 - 3)^{1/3} = 5\left(1 - \frac{3}{125}\right)^{1/3} \approx 5\,(1 - 0.008) = 4.960$$

- Express $9.9734^{\,9}$ in scientific notation.

  $$9.9734^{\,9} = 10^9\,(1 - 0.0266)^9 \approx 10^8\,(1 - 0.2394) = 9.76 \cdot 10^8$$

- A circular cylindrical container has a nominal diameter $d = 5$ length units and a nominal height $h = 10$ length units; on measurement, we find $d = 5.015$ $h = 9.960$ length units. How do these deviations affect the bottom area and the total volume?

  $$d = 2r = 5.015 = 5\,(1 + 0.003) \qquad h = 9.960 = 10\,(1 - 0.004)$$

  $$A = \pi r^2 \approx A_0 \cdot (1 + 2 \cdot 0.003) \approx 1.006 \cdot A_0$$

  $$V = \pi r^2 h \approx V_0 \cdot (1 + 0.006)\,(1 - 0.004) \approx 1.002 \cdot V_0$$

* * *

All Negligible

IQ Tester: *If I'm 1.78 meters tall, weigh 74 kilograms, commute 97 kilometers a day, and twice the square root of my telephone number is 2431, what is my age?*

Testee: *48, sir.*

IQ Tester: *Right! But how did you figure it out?*

Testee: *I've a cousin who is 24 and he's only half nuts.*

32.2 Interpolation

When the values of a mathematical function corresponding to two arguments (independent variables), called basic points, are known, the value that corresponds to an argument intermediate to the basic points can be found by **interpolation**.

The simplest form of interpolation is **linear interpolation**, which presupposes that the variation of the function value can be described by a straight line passing through the two basic points.

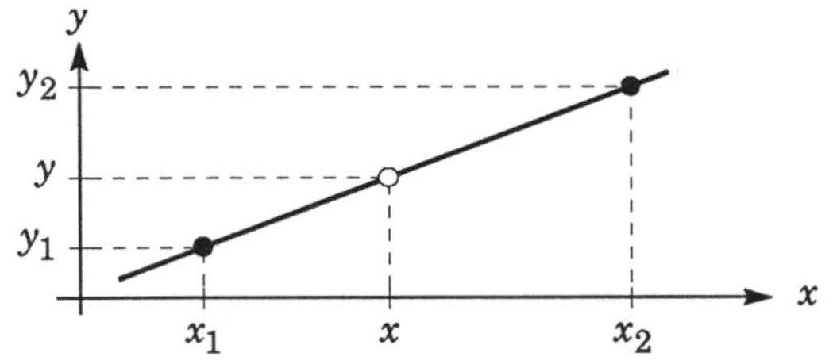

x	lg x	Δ lg x
1.003	0.001 30	
		0.000 43
1.004	0.001 73	
		0.000 44
1.005	0.002 17	
		0.000 43
1.006	0.002 60	
		0.000 43
1.007	0.003 03	
		0.000 43
1.008	0.003 46	
...	...	...
5.03	0.701 57	
		0.000 86
5.04	0.702 43	
		0.000 86
5.05	0.703 29	
		0.000 86
5.06	0.704 15	
		0.000 85
5.07	0.705 00	
		0.000 86
5.08	0.705 86	

If the arguments of the two basic points are x_1 and x_2, and the corresponding function values y_1 and y_2, the value y of the function for an argument x will be

$$y = y_1 + \frac{x - x_1}{x_2 - x_1} \cdot (y_2 - y_1)\,.$$

Referring to a table of common logarithms, we find that the assumption of linearity of variation is permissible, and we may apply the formula to calculate lg 1.0057 and lg 5.057:

$$\lg 1.0057 = \lg 1.005 + \frac{1.005\,7 - 1.005}{1.006 - 1.005} \,.\, (\lg 1.006 - \lg 1.005)$$

$$= 0.002\,17 + 0.7\,(0.002\,60 - 0.002\,17)$$

$$= 0.002\,17 + 0.000\,301 = 0.002\,47(1)\,.$$

$$\lg 5.057 = \lg 5.05 + 0.7 \cdot (\lg 5.06 - \lg 5.05)$$

$$= 0.703\,29 + 0.7 \cdot 0.000\,86$$

$$= 0.703\,29 + 0.000\,602 = 0.703\,89(2)\,.$$

x	e^x	Δe^x
7.1	1212.0	
		127.4
7.2	1339.4	
		140.9
7.3	1480.3	
		155.7
7.4	1636.0	
		172.0
7.5	1808.0	
		190.2
7.6	1998.2	
		210.1
7.7	2208.3	
		232.3

When a table shows that the function does not satisfy the condition of linearity of variation, we may resort to **graphical interpolation** by plotting the function values in a graph and trying to draw a curve passing through the basic points.

For a demonstration, we choose the exponential function e^x.

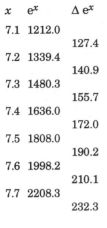

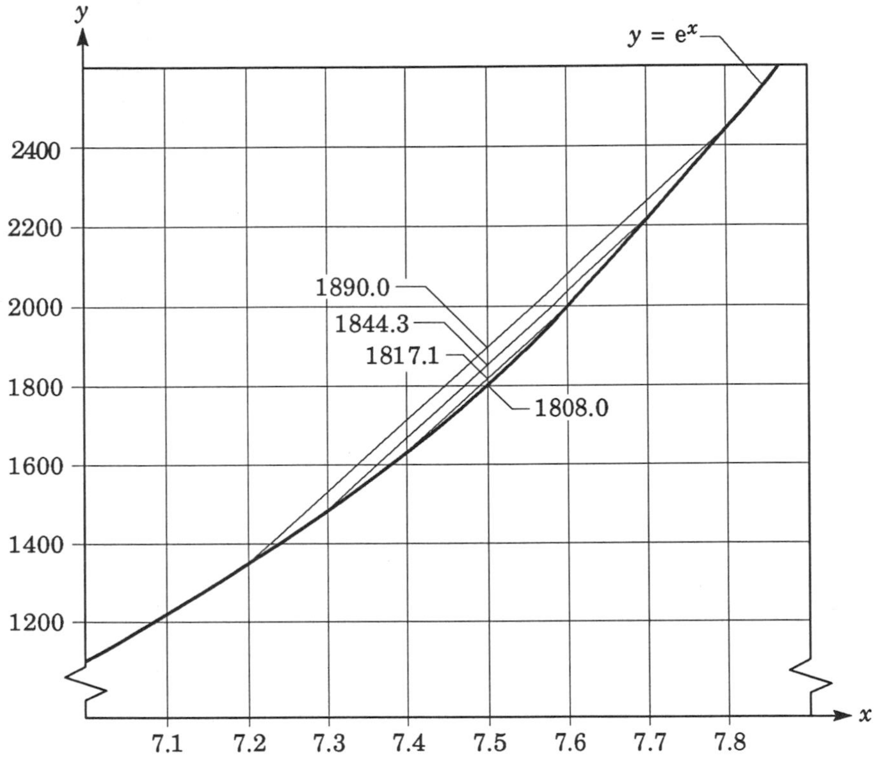

The value $e^{7.5}$ may be calculated by linear interpolation as the arithmetic mean of two adjacent exponential functions:

$$e^{7.5} \approx \begin{cases} \frac{1}{2}(e^{7.4} + e^{7.6}) = \frac{1}{2}(1636.0 + 1998.2) = 1817.1 \\ \frac{1}{2}(e^{7.3} + e^{7.7}) = \frac{1}{2}(1480.3 + 2208.3) = 1844.3 \\ \frac{1}{2}(e^{7.2} + e^{7.8}) = \frac{1}{2}(1339.4 + 2440.6) = 1890.0 \end{cases}$$

The results obtained differ from the table value of 1808.0 by 9.1, 36.3, and 82.0, respectively, corresponding to errors of 0.5%, 2.0%, and 4.5%.

Linear interpolation will generally yield a satisfactory interpolated value for all normal monotonic functions provided that the basic points straddling the sought value are chosen so as to be as close to each other as known values permit.

Graphic interpolation may be a suitable method for functions that rise very steeply with moderate changes of the argument, *e.g.*, the trigonometric tangent function above 80 degrees of angle.

x	$\tan x$	$\lg \tan x$
75°	3.732	0.5719
76°	4.011	0.6032
77°	4.331	0.6366
78°	4.705	0.6725
79°	5.145	0.7113
80°	5.671	0.7537
81°	6.314	0.8003
82°	7.115	0.8522
83°	8.144	0.9109
84°	9.534	0.9784
85°	11.53	1.0580
86°	14.30	1.1554
87°	19.08	1.2806
88°	28.64	1.4569
89°	57.29	1.7581

Linear interpolation is generally acceptable, however, as shown by the graph of the tangent function and its logarithm. The use of the logarithm of a function is often a possible solution for interpolation, followed by conversion of the value found to its antilogarithm.

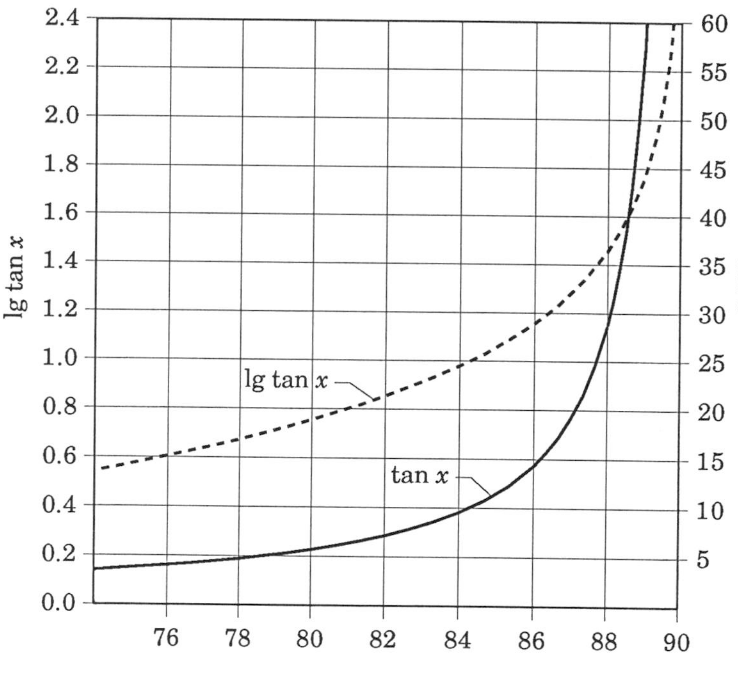

For greater accuracy of interpolation, several methods of replacing the function with polynomial expressions that approximate it to any desired degree of accuracy have been suggested by Lagrange, Newton, Gauss, and other mathematicians of renown.

All mathematical interpolation procedures consist of replacing the original function with simpler functions which provide exact agreement at the known basic points and optimum approximation to the original function in the neighborhood of these points.

Preferred Numbers

A special form of interpolation deals with the dividing up of decade intervals of numbers into sub-intervals of *logarithmically equal width*, so that consecutive points of division form a geometric series with the common ratio

$$q = \sqrt[N]{10},$$

where N may assume the values 5, 10, 20, 40, and, in exceptional cases, 80, according to International Standard (*ISO 3*).

RENARD Charles (1847 - 1905)

In the five series, prefixed by the letter R – after their inventor, the French military officer **Charles Renard** –

	R 5	R 10	R 20	R 40	R 80
step ratio	$\sqrt[5]{10}$	$\sqrt[10]{10}$	$\sqrt[20]{10}$	$\sqrt[40]{10}$	$\sqrt[80]{10}$
approx. value	1.58	1.26	1.12	1.06	1.03

preferred numbers are calculated to five digits and rounded to three digits, as shown here for R 5, R 10, and R 20 (for tables of preferred number series R 40 and R 80, see ISO 3).

Mantissas	Calculated Values	Standard Series		
		R 5	R 10	R 20
000	1.0000	1.00	1.00	1.00
050	1.1220			1.12
100	1.2589		1.25	1.25
150	1.4125			1.40
200	1.5849	1.60	1.60	1.60
250	1.7783			1.80
300	1.9953		2.00	2.00
350	2.2387			2.24
400	2.5119	2.50	2.50	2.50
450	2.8184			2.80
500	3.1623		3.15	3.15
550	3.5481			3.55
600	3.9811	4.00	4.00	4.00
650	4.4668			4.50
700	5.0119		5.00	5.00
750	5.6234			5.60
800	6.3096	6.30	6.30	6.30
850	7.0795			7.10
900	7.9433		8.00	8.00
950	8.9125			9.00
000	10.0000	10.00	10.00	10.00

From the basic tables, further tables of preferred numbers may be created by selecting every second, third, fourth ... n-th term of a series.

The use of preferred numbers benefits industry by restricting the number of machine element sizes, rotational speeds, mechanical and hydraulic pressures, *etc.*, and still satisfying practical industrial requirements with a minimum of components and reduced costs of production and stockkeeping.

In the electronics industry, resistors and capacitors are graded according to another system of preferred numbers based on the 12th root of 10, and rounded off to two figures,

10 - 12 - 15 - 18 - 22 - 26 - 32 - 38 - 47 - 56 - 68 - 82 - 100,

sometimes with the addition of 62, 75, and 91 from the 24th-root series.

With interpolation perforce,
Jack plotted a much better course.
With Jill up the hill,
Oh my! Not a spill,
Skateboarding down from the old water source!

32.3 Graphic and Iterative Methods

Real Zeros

Real numbers of the independent variable (the argument) for which the function is 0 are referred to as the **real zeros** of the function. Thus, if f is a function, then a real root of $f(x) = 0$ is a real zero of f.

p. 304

By the **factor theorem**, every real zero of a polynomial function f corresponds to a first-degree factor of $f(x)$. Consequently, a polynomial function cannot have more real zeros than its degree; *e.g.*, a third-degree polynomial function has at the most three real zeros, a fourth-degree at the most four.

First- and second-degree equations and occasional higher-degree equations can always be solved by using algebraic methods; formulas also exist for solving third- and fourth-degree equations, but they are very complicated. If f is a polynomial of degree five or higher, there is, generally, no algebraic solution. Furthermore, there is no formula by which we can find the exact roots of $\cos x = x$ or other mixed trigonometric equations.

We realize that the number of equations that cannot be solved by algebraic methods many times exceeds the number of equations that are suitable for such methods.

When algebraic methods are not suitable, we can obtain approximative real-number solutions by the use of **graphic methods**, often in combination with **iterative computational methods** by which a result is obtained by replication of a series of operations that successively improves an approximate result.

Graphs

Equations in One Variable

We may solve equations in one variable for **real-number roots** if we change the standard form, $a_0 x^n + a_1 x^{n-1} + \ldots + a_n = 0$, to

$$a_0 x^n + a_1 x^{n-1} + \ldots + a_n = y$$

and graphically determine what points on the curve have $y = 0$.

Graphical solutions are usually time consuming. If the roots are packed closely together, it can be virtually impossible to locate the roots graphically.

• Find the roots of $x^3 - 3x - 1 = 0$.

Write

$$x^3 - 3x - 1 = y$$

and use convenient values of x to find corresponding values of y:

x	−2	−1.5	−1	−0.5	0	0.5	1	1.5	2
y	−3	0.125	1	0.375	−1	−2.375	−3	−2.125	1

Plot the values of x and y on an orthogonal coordinate system and sketch the curve of $x^3 - 3x - 1 = y$:

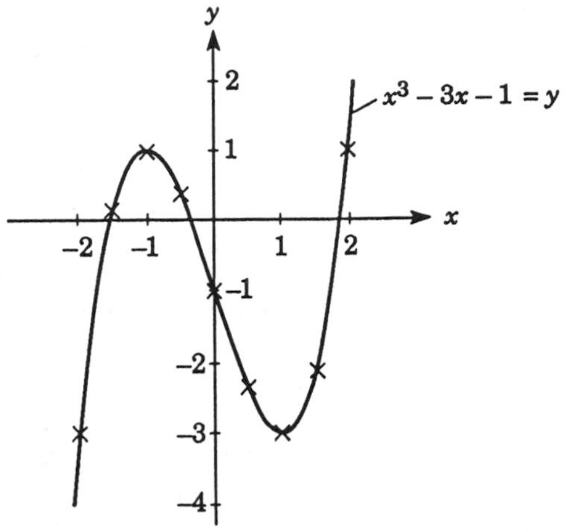

The graph of $x^3 - 3x - 1 = y$ intersects the x-axis in three places; a third-degree equation has three roots and, consequently, all three roots of the equation have real-number values.

- The equation

$$2 \cos x - \sin x = -2 ; \quad 0 \leq x < 15$$

pp. 522 - 23 was solved by algebraic methods in Chapter 13. It may be solved graphically by determining the abscissas of the points of intersection of the curves of

$$y = 2 \cos x ; \quad y = \sin x - 2 .$$

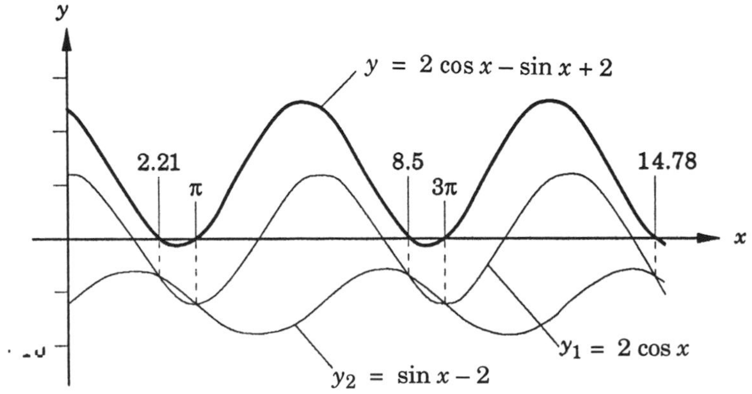

Systems of Equations

- $\left.\begin{array}{l} 2x+y = 5 \\ x^2+y^2 = 25 \end{array}\right\}$

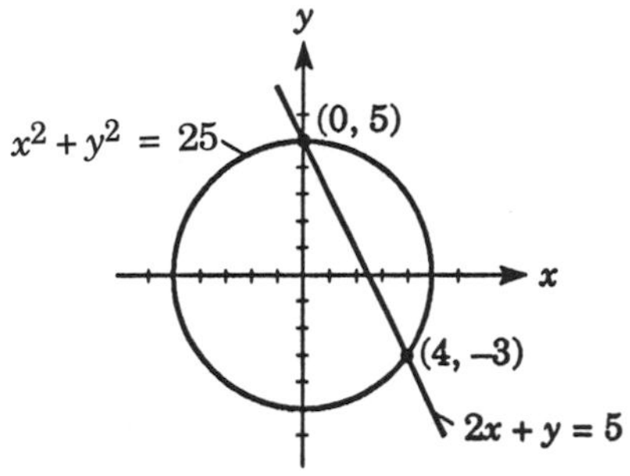

The graphs of the equations intersect at (0, 5) and (4, –3), leading to the solutions

$$\left.\begin{array}{l} x_1 = 0 \\ y_1 = 5 \end{array}\right\}; \qquad \left.\begin{array}{l} x_2 = 4 \\ y_2 = -3 \end{array}\right\}.$$

- $\left.\begin{array}{l} 4x^2+y^2 = 20 \\ x^2 = 2y \end{array}\right\}$

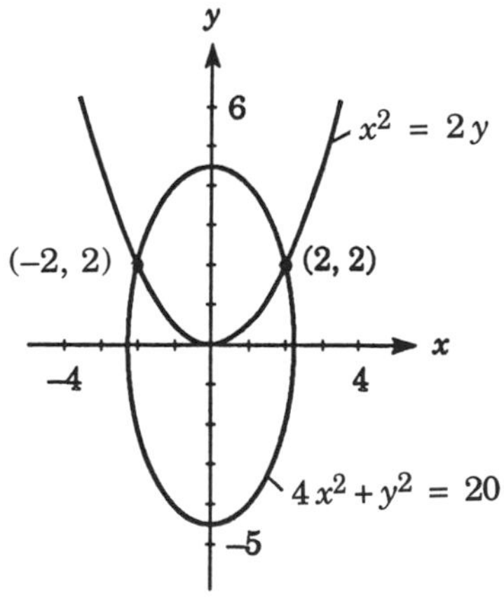

The graphs of the equations intersect at (2, 2) and (–2, 2), leading to the real-number solutions

$$\left.\begin{array}{l} x_1 = 2 \\ y_1 = 2 \end{array}\right\}; \qquad \left.\begin{array}{l} x_2 = -2 \\ y_2 = 2 \end{array}\right\},$$

but the algebraic method gives additional solutions, whose number pairs contain complex-number components,

$$\left.\begin{array}{l} x_3 = \sqrt{-20} \\ y_3 = -10 \end{array}\right\}; \qquad \left.\begin{array}{l} x_4 = -\sqrt{-20} \\ y_4 = -10 \end{array}\right\}.$$

- $$\left.\begin{aligned} x^2 &= 2y \\ x - y &= 1 \end{aligned}\right\}$$

By algebraic means, we find two complex solutions,

$$\left.\begin{aligned} x_1 &= \frac{2+\sqrt{-4}}{2} = 1+\mathrm{i} \\ y_1 &= \frac{\sqrt{-4}}{2} = \mathrm{i} \end{aligned}\right\} ; \qquad \left.\begin{aligned} x_2 &= \frac{2-\sqrt{-4}}{2} = 1-\mathrm{i} \\ y_2 &= \frac{-\sqrt{-4}}{2} = -\mathrm{i} \end{aligned}\right\} ,$$

but there are no real-number solutions; since there are no real-number solutions, the graphs do not intersect:

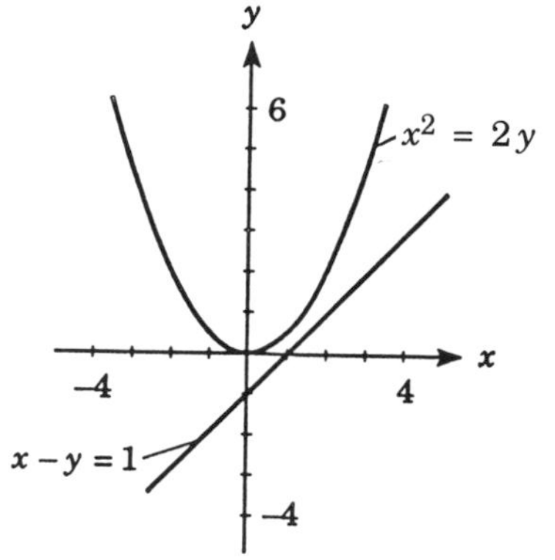

Graphic solutions are approximate, but by determining an increasing number of x, y-points around an intersection of the graph and the x-axis, the accuracy may be improved to any required degree.

Iterative Methods

A real zero of a function f is the same as an x-intercept of the graph of f. There are several numerical procedures for finding real zeros of functions; the two most frequently used methods are the **bisection method** and **Newton's method.**

Bisection Method

Assume that f is a **continuous function** and suppose that f has the real zeros a and b:

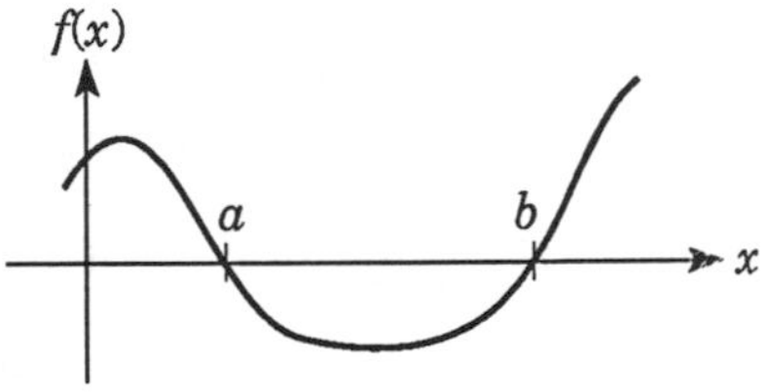

To find a by the bisection method, use initial values that straddle a, say, $x = x_1$ and $x = x_z$, giving $f(x_1)$ and $f(x_z)$ opposite signs. Letting the **midpoint of the interval** $[x_1, x_z]$ be x_{M_1}, we have

p. 551

$$x_{M_1} = \frac{x_1 + x_z}{2}$$

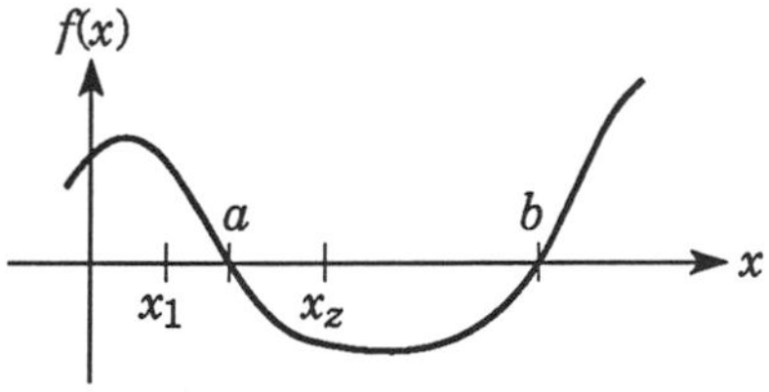

To determine whether the zero of the function is in $[x_1, x_{M_1}]$ or in $[x_{M_1}, x_z]$, we ask:

(**1**) **Is** $\boxed{\begin{array}{c} f(x_1) > 0 \\ \text{and} \\ f(x_{M_1}) < 0 \end{array}}$ **?**

If the answer to question (**1**) is **no**, the resulting question is:

(**2**) **Is** $\boxed{\begin{array}{c} f(x_1) < 0 \\ \text{and} \\ f(x_{M_1}) > 0 \end{array}}$ **?**

If either question (**1**) or question (**2**) has an affirmative answer, then the x-intercept of f is in $[x_1, x_{M_1}]$. On the other hand, if neither question has an affirmative answer, then the x-intercept of f is in $[x_{M_1}, x_z]$.

For our function f, the x intercept is in $[x_1, x_{M_1}]$; zooming in on $[x_1, x_z]$:

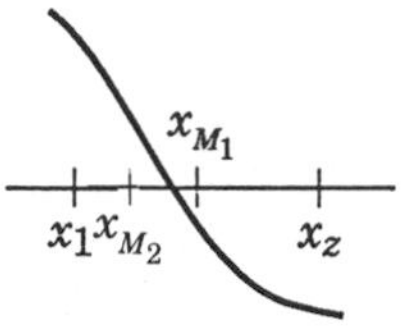

The interval $[x_1, x_{M_1}]$ is now bisected, yielding the intervals $[x_1, x_{M_2}]$ and $[x_{M_2}, x_{M_1}]$.

We now ask:

(3)	Is	$f(x_1) > 0$ and $f(x_{M_2}) < 0$	?

Since the the answer to question (**3**) is **no**, the ensuing question is:

(4)	Is	$f(x_1) < 0$ and $f(x_{M_2}) > 0$	?

Question (**4**) is also answered **no** and, therefore, the x-intercept must be in $[x_{M_2}, x_{M_1}]$:

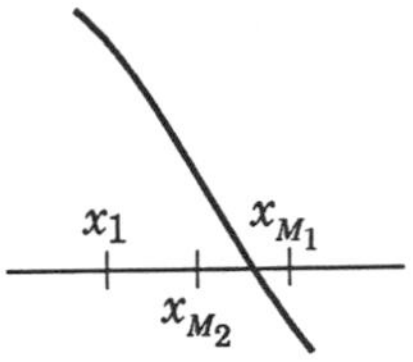

The bisection process is repeated until desired accuracy is obtained.

Finding Initial Values

In order to obtain useful initial values, it is often necessary to make a reasonably detailed graph of the function or, for a function composed of different kinds of elementary functions, to sketch the individual elementary functions.

If given the function $f(x) = \ln x - \sin x$ for $0 < x \leq 4$, we could after some rather lengthy calculations obtain its graph, showing a real zero in the interval [2, 2.5] :

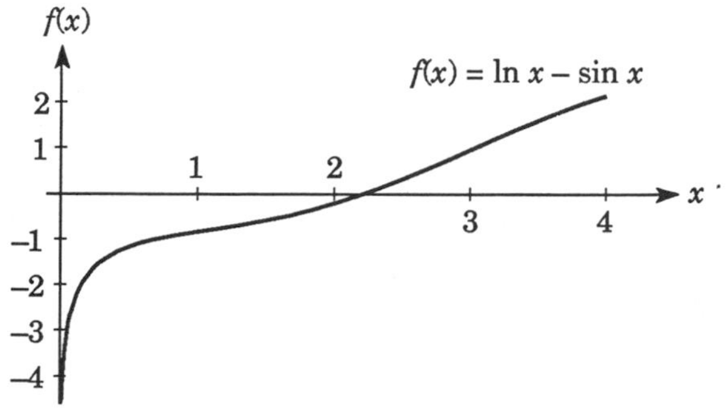

Since $f(x) = \ln x - \sin x = 0$, then $\ln x = \sin x$. The functions $y_1 = \ln x$ and $y_2 = \sin x$ are more easily sketched than f; the point of intersection of y_1 and y_2 is in the interval [2, 2.5], coinciding with the x-intercept of the graph of f:

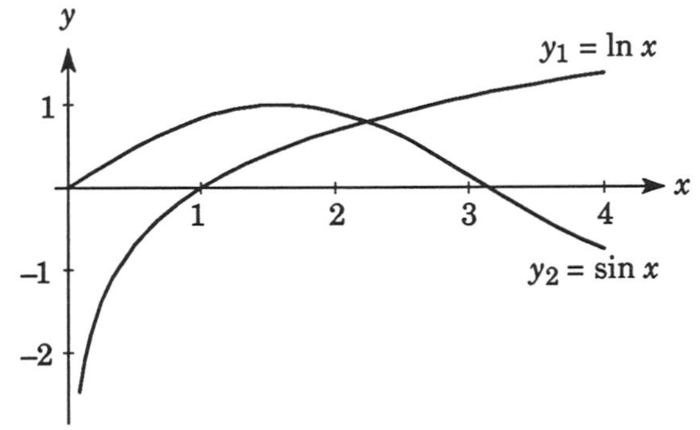

- Solve the equation

$$\cos x = x\ ;$$

round off the result to two decimals.

Let $f(x) = \cos x - x$; sketch $y_1 = \cos x$ and $y_2 = x$:

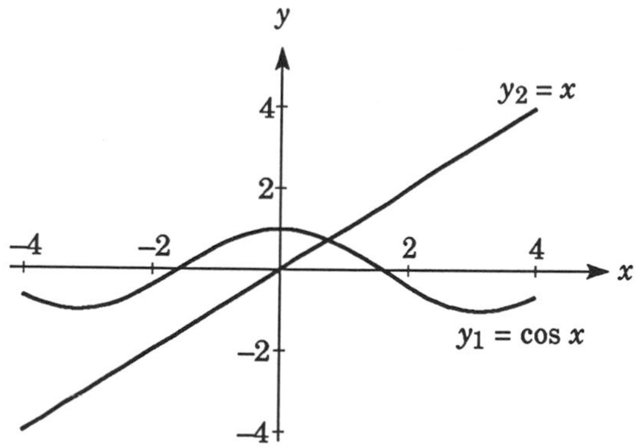

The graphs show a point of intersection between y_1 and y_2. The size chosen for the interval in which the intersection is located depends on the accuracy of the drawing; we choose the initial values 0.6 and 1.

The midpoint M_1 of the interval [0.6, 1] is

$$M_1 = \frac{1 + 0.6}{2} = 0.8\,.$$

Examine the intervals

$$[0.6, 0.8]\ ; \quad [0.8, 1]\,.$$

(1) **Is** $\boxed{\begin{array}{c} f(0.6) > 0 \\ \text{and} \\ f(0.8) < 0 \end{array}}$ **?**

Since $\cos 0.6 - 0.6 > 0$ and $\cos 0.8 - 0.8 < 0$, the answer to (**1**) is **yes**. Consequently, the real zero of f is located in [0.6, 0.8], whose midpoint M_2 is

$$M_2 = \frac{0.6 + 0.8}{2} = 0.7 .$$

Examine the intervals

$$[0.6, 0.7] \; ; \quad [0.7, 0.8] .$$

(**2**) **Is** $\boxed{\begin{matrix} f(0.6) > 0 \\ \text{and} \\ f(0.7) < 0 \end{matrix}}$ **?**

Since $\cos 0.7 - 0.7 \nless 0$, the answer to (**2**) is **no**. Therefore, the next question is:

(**3**) **Is** $\boxed{\begin{matrix} f(0.6) < 0 \\ \text{and} \\ f(0.7) > 0 \end{matrix}}$ **?**

Since $\cos 0.6 - 0.6 \nless 0$, the answer to (**3**) is **no**. Consequently, the real zero of f is located in [0.7, 0.8].

Further iterations of the algorithm produce the intervals

$$[0.7\,,\; 0.75]\,,$$

$$[0.725\,,\; 0.75]\,,$$

$$[0.7375\,,\; 0.75]\,,$$

and

$$[0.7375\,,\; 0.74375]\;.$$

Hence, rounded off to two decimals, the sought root is 0.74 .

Real Zeros Unsuitable for the Bisection Method

The bisection method requires an x-intercept of the function and therefore cannot be employed to find a real zero in an interval where the x-axis is the tangent to the curve. For example, consider

$$f(x) = 2\,x^3 - 3\,x^2 - 36\,x + 81,$$

whose graph for $-5 \le x \le 5$ is:

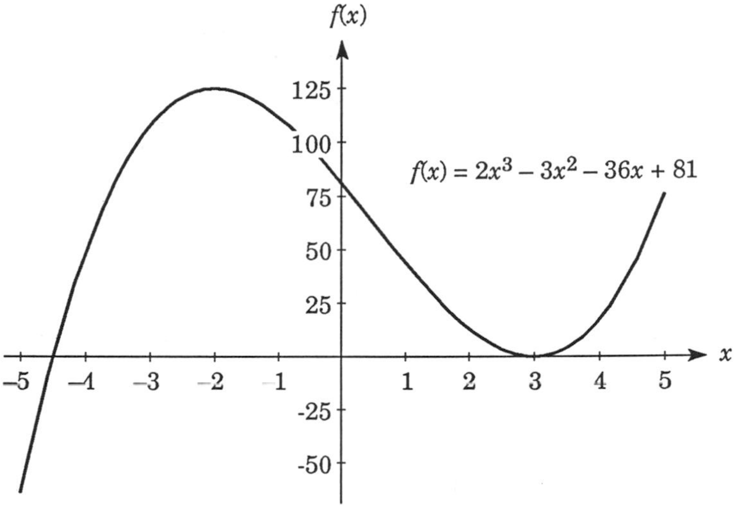

The bisection method is suitable for finding the real zero of f in the interval $[-5, -4]$, but not in $[2, 4]$; the latter interval is p. 941 accessible to examination by **Newton's method**.

Flow Chart for the Bisection Method

Iterative methods are suited for computer programming. In the chart below, A and B are initial values, E is the error tolerance:

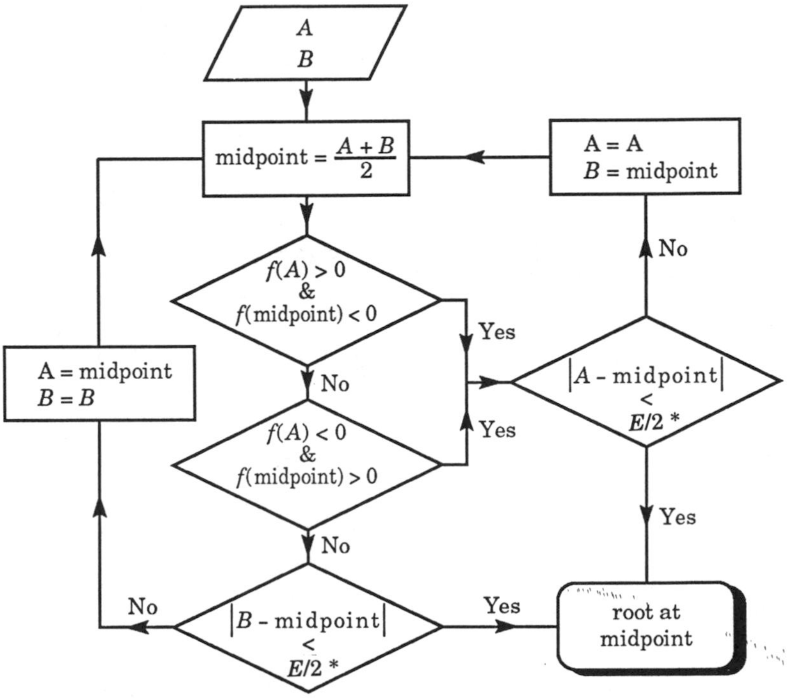

* In order to express the final estimate as a value with an associated error tolerance, one must round or truncate. In either case, using E as the cut-off point in one's calculation rather than E/2, one can easily end up with an estimate ± tolerance (E) which does not include the actual value.

With a computer program, one is less dependent on details of the graph of the function and may without excessive loss of time use initial values that are far apart.

There is always the possibility that the computer program will place the demand on error tolerance too high, causing inordinately long computations or even an infinite loop. It is therefore important to specify not only the error tolerance but also the maximum number of iterations; a real zero to the specified accuracy, or the specified maximum number of iterations (whichever occurs first) will stop the calculations.

Newton's Method

Seed

Newton's method requires a **seed**, that is, an initial rough estimate or guess of the value of the root.

With a suitable seed, the tangent of the curve at the point corresponding to the seed (x_0) will intercept the x-axis. The x-intercept may be located at a point that is at a shorter distance or, as in the graph below, at an intercept (x_1) farther away from the value of the sought root.

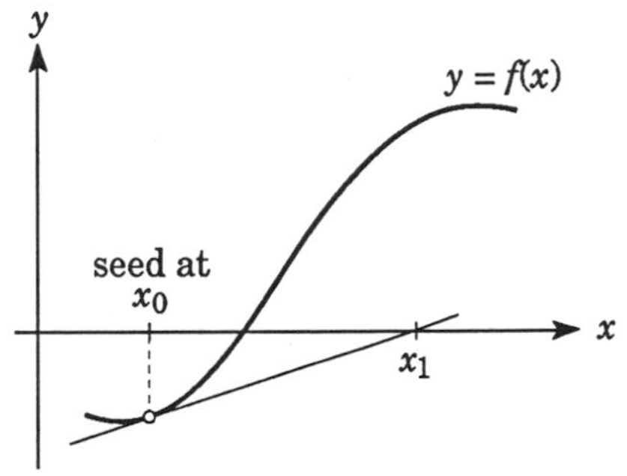

Using the tangent intercept at x_1 as a new seed, we continue the approximation and obtain a new intercept at x_2. In this manner, starting with the approximation $x = x_0$, we obtain a sequence x_1, x_2, x_3 ..., where successive values come ever closer to the sought root.

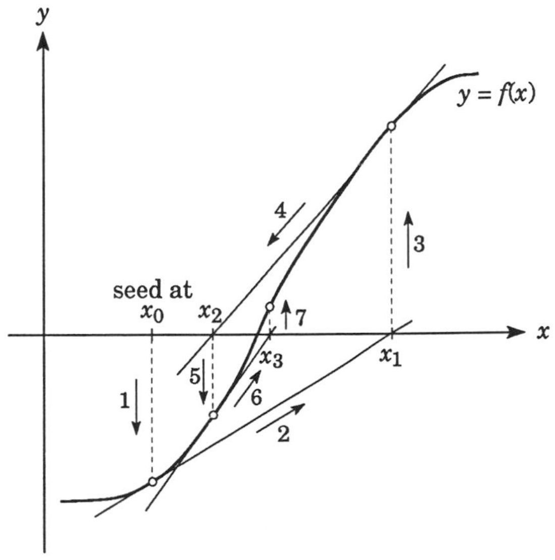

The formula describing the steps of this iterative process is

$$x_{n+1} = x_n - \frac{f(x_n)}{f'(x_n)},$$

where $n = 0, 1, 2, \ldots$ and f and its derivative are continuous.

Why? The slope of the graph of a differentiable function f has, at any point $[x, f(x)]$, the value $f'(x)$.

p. 553 The **point-slope equation** of a tangent line at point $[x_0, f(x_0)]$ is

$$y - f(x_0) = [f'(x_0)](x - x_0) .$$

If this tangent line has the x-intercept x_1, we have the point-slope equation

$$0 - f(x_0) = [f'(x_0)](x_1 - x_0)$$

and can express x_1 in terms of x_0; thus,

$$x_1 = x_0 - \frac{f(x_0)}{f'(x_0)} .$$

Repeating this process for the tangent line at $[x_1, f(x_1)]$ and expressing x_2 in terms of x_1, we find

$$x_2 = x_1 - \frac{f(x_1)}{f'(x_1)},$$

and for $n = 0, 1, 2, \ldots$, the general formula

$$x_{n+1} = x_n - \frac{f(x_n)}{f'(x_n)} .$$

Perils and Fallacies of Newton's Method

Unless one has a rather detailed understanding of the function, Newton's method may give deceptive results; generally, the bisection method is more reliable.

Newton's method may fail if one chooses a seed far away from the sought root, leading to increasingly worse approximations or to convergence towards the wrong root.

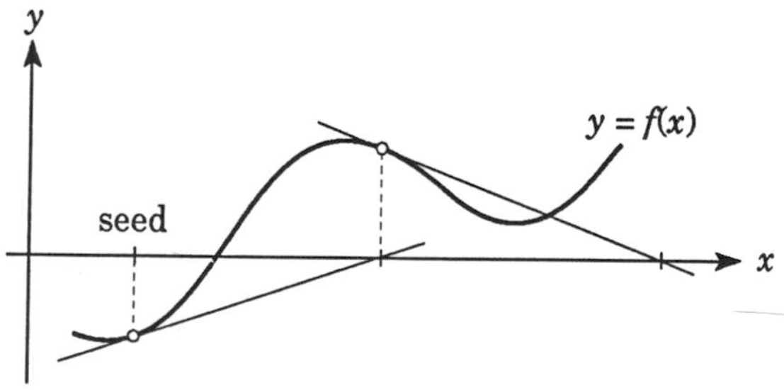

If, in the course of approximations, a tangent line falls on an extremum, then $f'(x_n) = 0$ and, since division by 0 is not possible, x_{n+1} will be undefined. Another choice of seed usually takes care of the problem.

In some instances, the shape of the curve is such that Newton's method will fail for any choice of seed other than the actual x-intercept. It is often helpful to do a few iterations of the bisection method and then switch to Newton's method.

- Determine $\sqrt{2}$ to five decimal places by Newton's method.

To determine $\sqrt{2}$ is equivalent to finding the positive root of the equation

$$x^2 - 2 = 0.$$

With $f(x) = x^2 - 2$,

$$f'(x) = 2x$$

and

$$x_{n+1} = x_n - \frac{x_n^2 - 2}{2x_n}.$$

With $x_0 = 1$ as the seed,

$$x_1 = 1 - \frac{1-2}{2} = 1.5.$$

The iterative process is carried on until successive approximations agree to five decimal places:

$$x_2 = 1.5 - \frac{1.5^2 - 2}{2(1.5)} \approx 1.416\,667$$

$$x_3 = 1.416\,667 - \frac{1.416\,667^2 - 2}{2(1.416\,667)} \approx 1.414\,216$$

$$x_4 = 1.414\,216 - \frac{1.414\,216^2 - 2}{2(1.414\,216)} \approx 1.414\,213$$

Hence, correct to five decimal places,

$$\sqrt{2} = 1.414\,21.$$

- With an error tolerance < 0.0001, find the root of

$$x^3 - x = -1$$

in the interval $[-2, -1]$.

Alternative I: **Newton's Method**

With $f(x) = x^3 - x + 1$,

$$f'(x) = 3x^2 - 1$$

and

$$x_{n+1} = x_n - \frac{x_n^3 - x_n + 1}{3x_n^2 - 1}.$$

With (-1.5) as a seed, $x_0 = -1.5$ and

$$x_1 = -1.5 - \frac{(-1.5)^3 - (-1.5) + 1}{3(-1.5)^2 - 1} = -1.347\,82\ldots.$$

Discontinue the calculations when the difference between two successive values of x is < 0.00005:

$$x_1 = -1.347\,82\ldots$$

$$x_2 = -1.325\,20\ldots$$

$$x_3 = -1.324\,71\ldots$$

$$x_4 = -1.324\,71\ldots.$$

Hence, the sought root is -1.3247 ± 0.0001.

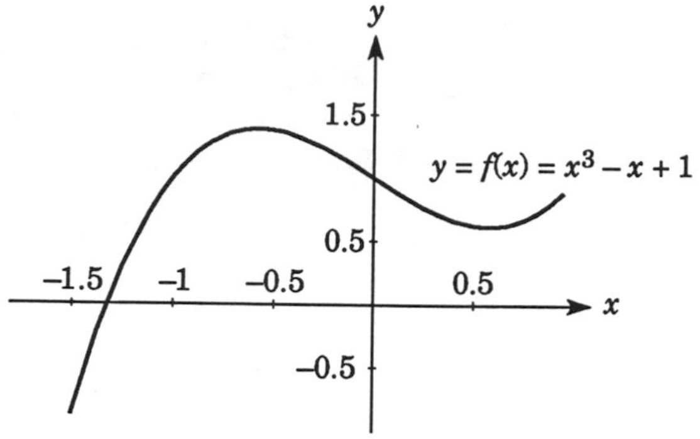

Alternative II: **Bisection Method**

We choose the initial values -2 and -1.

The midpoint M_1 of the interval $[-2, -1]$ is

$$M_1 = \frac{-2 + (-1)}{2} = -\frac{3}{2}.$$

Successive calculations give

$$\begin{aligned}
M_1 &= -1.5 \\
M_2 &= -1.25 \\
M_3 &= -1.375 \\
M_4 &= -1.312\,5 \\
M_5 &= -1.343\,75 \\
M_6 &= -1.328\,125 \\
M_7 &= -1.320\,312\,5 \\
M_8 &= -1.324\,218\,75 \\
M_9 &= -1.326\,171\,875 \\
M_{10} &= -1.325\,195\,312\,5 \\
M_{11} &= -1.324\,707\,031\,25 \\
M_{12} &= -1.324\,951\,171\,875 \\
M_{13} &= -1.324\,829\,101\,562\,5 \\
M_{14} &= -1.324\,768\,066\,406\,25 \\
M_{15} &= -1.324\,737\,548\,828\,125
\end{aligned}$$

The sought root is -1.3247 ± 0.0001. •

When Newton's method fails, the bisection method will usually provide a value of the real zero; similarly, when the bisection method fails, recourse to Newton's method is usually profitable.

• With an error tolerance of < 0.01, find the solution of

$$2x^2 - x^3 = 1\frac{5}{27}$$

in the interval $[1, 2]$.

In the given interval, f shows a maximum on the x-axis; consequently, use of the bisection method would fail to determine the sought root.

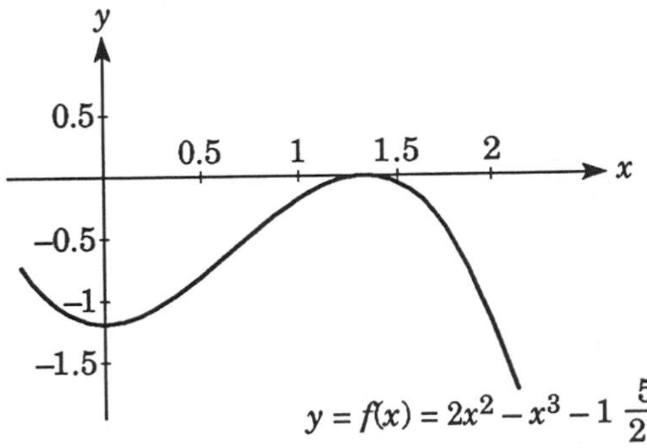

Newton's method takes care of the problem.

With $f(x) = 2x^2 - x^3 - \frac{32}{27}$,

$$f'(x) = 4x - 3x^2$$

and

$$x_{n+1} = x_n - \frac{2x_n^2 - x_n^3 - \frac{32}{27}}{4x_n - 3x_n^2}.$$

With 1 as a seed, $x_0 = 1$ and

$$x_1 = 1 - \frac{2 - 1 - \frac{32}{27}}{4 - 3} = \frac{32}{27}.$$

Discontinue the successive calculations when the difference between two successive values of x is < 0.005:

$$x_1 = 1.185\,18\ldots$$
$$x_2 = 1.262\,34\ldots$$
$$x_3 = 1.298\,50\ldots$$
$$x_4 = 1.316\,07\ldots$$
$$x_5 = 1.324\,74\ldots$$
$$x_6 = 1.329\,04\ldots.$$

Hence, the sought root is $x = 1.33 \pm 0.01$.

Flow Chart for Newton's Method

As with the bisection method, Newton's method is suited for computer programming. In the flow chart below, the seed is x_0 and the error tolerance E.

– *The best part of my new math program is the Pooh book that came with it.*

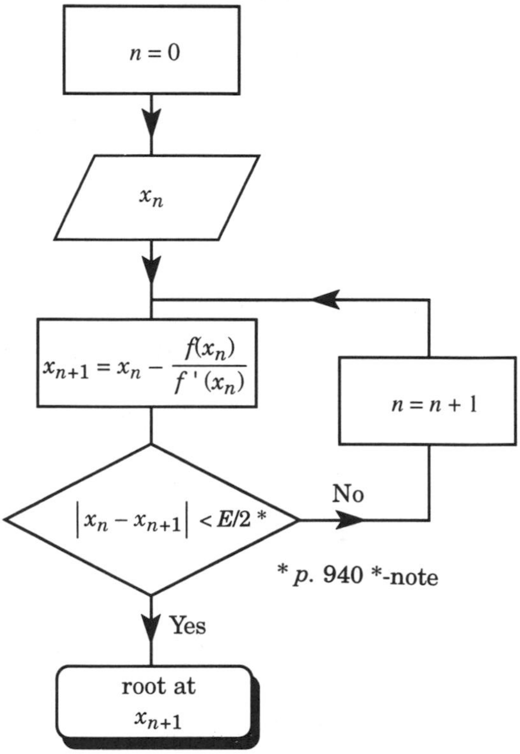

"How do you do Nothing?" asked Pooh, after he had wondered for a long time.

"Well, it is when people call out at you just as you're going off to do it, What are you going to do, Christopher Robin, and you say, Oh, nothing, and then you go and do it."

"Oh, I see," said Pooh.

"This is a nothing sort of thing that we're doing now."

"Oh, I see," said Pooh again.

"It means just going along, listening to all the things you can't hear, and not bothering."

"Oh!" said Pooh.

They walked on, thinking of This and That, and by-and-by they came to an enchanted place on the very top of the Forest called Galleons Lap, which is sixty-something trees in a circle; and Christopher Robin knew that it was enchanted because nobody had ever been able to count whether it was sixty-three or sixty-four, not even when he tied a piece of string round each tree after he had counted it.

A. A. Milne, *The House at Pooh Corner* (1928). Illustrated by Ernest H. Shepard.

32.4 Numerical Integration

To evaluate definite integrals we have, up to this point, used methods requiring that we first find the indefinite integral of the given integrand; by such methods we have obtained exact values.

Numerical integration, on the other hand, is the process of finding an **approximate value** of a definite integral without carrying out the process of evaluating the indefinite integral.

Approximate Integration

The term **approximate integration** is used synonymously with *numerical integration*.

The use of numerical integration to evaluate a definite integral is required

- when the indefinite integral cannot be expressed in terms of an elementary function;
- when the integrand is available only as a table of numerical values or a graph, which is often the case in scientific observations.

Numerical integration is possible only when the limit values of the integral are known. It is often employed to evaluate a complicated integral even if the indefinite integral can be expressed in terms of elementary functions.

Although numerical integration cannot yield an exhaustively defined value, it can be carried on to as many decimal places as desired or as can be achieved in the allotted time.

This does not mean that the method of finding the indefinite integral expressed in terms of elementary functions is superfluous. On the contrary, it is often important to learn what functions constitute the integral.

Numerical integration employs

- series expansion, or
- geometrically oriented methods.

Quadrature Formulas

Formulas of numerical integration are often referred to as **quadrature formulas**, recalling historic attempts to find squares equal to surfaces defined by curved lines.

Infinite Series Expansion

Historically, the investigation of infinite series developed in connection with problems of finding integrals.

If the limits of integration fall within the interval of convergence for power expansion of the integrand, then an approximate value of a definite integral may be obtained by summation of the definite integrals of the early terms of the series.

- Find $\int_0^1 \frac{\sin x}{x}\, dx$.

p. 771

Use the **Maclaurin expansion** of $\sin x$, divide each term by x, and integrate term by term,

$$\int \frac{\sin x}{x}\, dx = x - \frac{x^3}{3 \cdot 3!} + \frac{x^5}{5 \cdot 5!} - \frac{x^7}{7 \cdot 7!} + \ldots$$

$$\int_0^1 \frac{\sin x}{x}\, dx = 1 - \frac{1}{3 \cdot 3!} + \frac{1}{5 \cdot 5!} - \frac{1}{7 \cdot 7!} + \ldots .$$

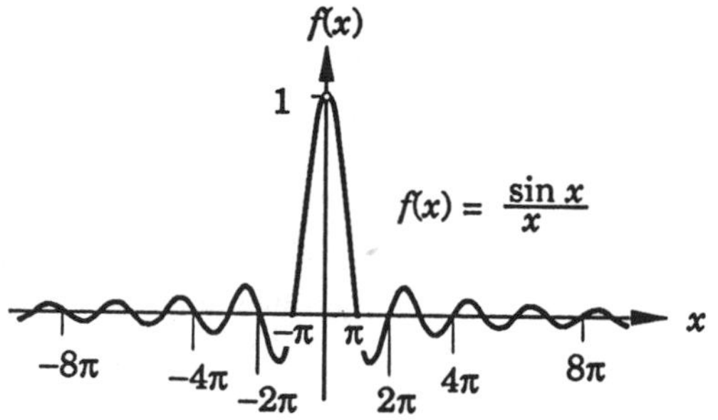

By including an increasing number of terms in the sum, we achieve ever-greater accuracy. Since the expansion produces an alternating series, the error is bounded by the last term used in the expansion.

- Find $\int_{-1}^1 e^{-x^2}\, dx$.

$$\int e^{-x^2}\, dx = x - \frac{x^3}{3 \cdot 1!} + \frac{x^5}{5 \cdot 2!} - \ldots .$$

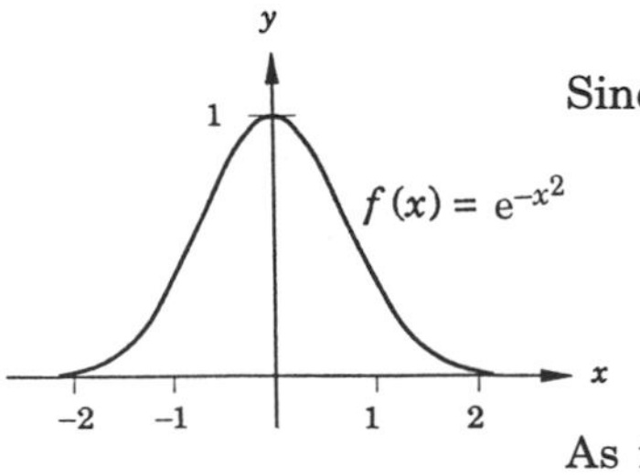

Since $\int_{-1}^1 e^{-x^2}\, dx = 2 \int_0^1 e^{-x^2}\, dx$, we have

$$\int_{-1}^1 e^{-x^2}\, dx = 2\left(1 - \frac{1}{3 \cdot 1!} + \frac{1}{5 \cdot 2!} - \ldots\right).$$

As in the previous example, the error is bounded by the last term used in the expansion.

The Trapezoidal Rule

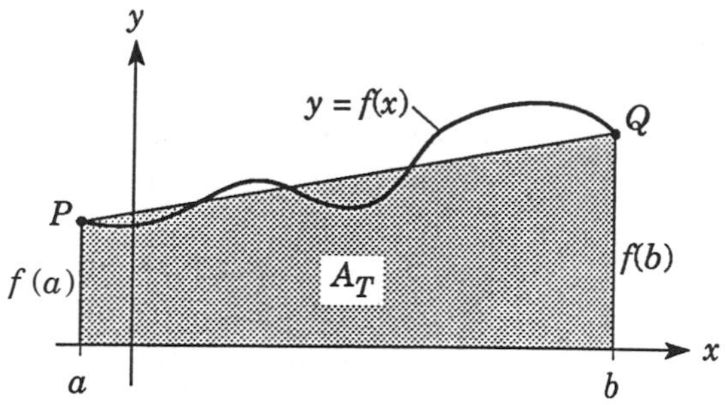

The simplest approximation of a curve $y = f(x)$ extending between two points P and Q is a chord connecting the two points.

Letting the domain of f be $[a, b]$, the area A_T between the chord and the x-axis is a trapezoid with the length $(b - a)$ and the parallel sides $f(a)$ and $f(b)$; thus

$$A_T = \frac{1}{2}\,(b-a)\,[f(a) + f(b)]\ .$$

The smaller the intervals that are replaced by chords, the better the approximation of the area will be. In the graph below, $[a, b]$ is divided into n sub-intervals of equal length, $\dfrac{b-a}{n}$:

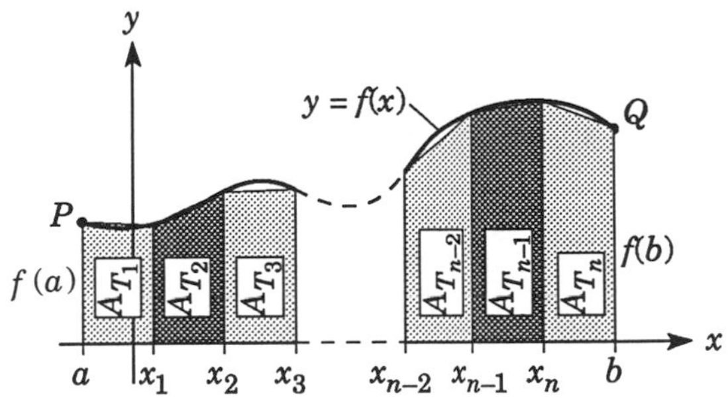

By adding the areas of the sub-intervals, we have

$$\begin{aligned} A_{T_1} + A_{T_2} + \ldots + A_{T_n} &= \frac{1}{2}\,\frac{b-a}{n}\,[f(a) + f(x_1)] + \frac{1}{2}\,\frac{b-a}{n}\,[f(x_1) + f(x_2)] + \ldots \\ &\quad + \frac{1}{2}\,\frac{b-a}{n}\,[f(x_{n-2}) + f(x_{n-1})] + \frac{1}{2}\,\frac{b-a}{n}\,[f(x_{n-1}) + f(b)] \\ &= \frac{b-a}{2n}\,[f(a) + 2f(x_1) + 2f(x_2) + \ldots + 2f(x_{n-1}) + f(b)]\ , \end{aligned}$$

and the **trapezoidal rule**,

$$\int_a^b f(x)\ dx \approx \frac{b-a}{2n}\,[f(a) + 2f(x_1) + 2f(x_2) + \ldots + 2f(x_{n-1}) + f(b)]\ ,$$

where n is a positive integer.

- Approximate the values of the integral

$$\int_0^1 e^{-x^2}\, dx$$

by the trapezoidal rule for $n = 1$, $n = 2$, $n = 4$. Truncate the answers after the fifth decimal.

$\boxed{n = 1}$

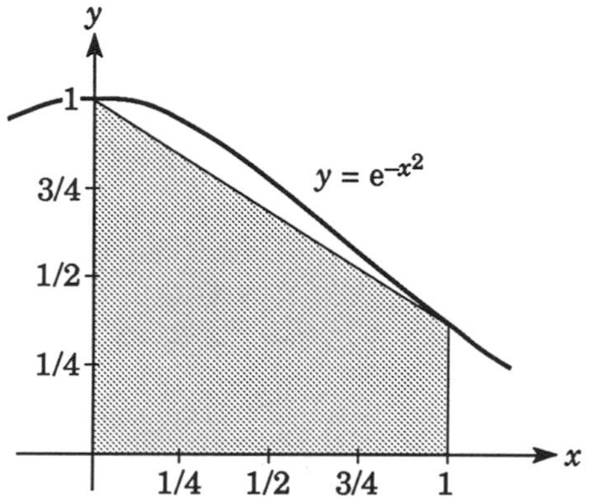

$$\int_0^1 e^{-x^2}\, dx \approx \frac{1-0}{2 \cdot 1}\,(e^{-0} + e^{-1})$$

$$\approx \frac{1}{2}\left(1 + \frac{1}{e}\right) \approx 0.683\,93\ldots .$$

$\boxed{n = 2}$

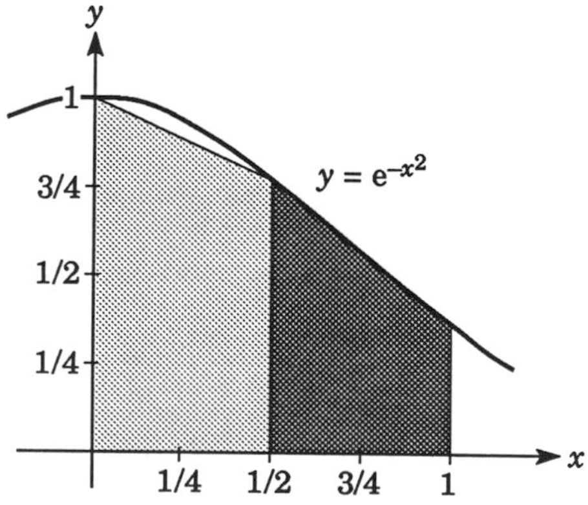

$$\int_0^1 e^{-x^2}\, dx \approx \frac{1-0}{2 \cdot 2}\,(e^{-0} + 2\, e^{-1/4} + e^{-1})$$

$$\approx \frac{1}{4}\left[1 + 2\left(\frac{1}{e^{0.25}}\right) + \frac{1}{e}\right] \approx 0.731\,37\ldots .$$

$\boxed{n = 4}$

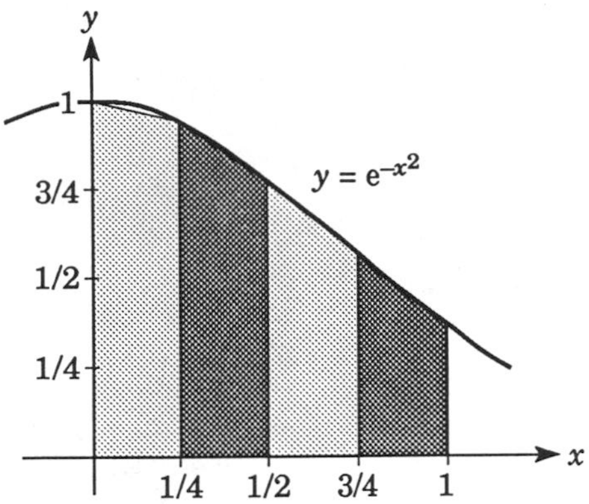

$$\int_0^1 e^{-x^2}\,dx \approx \frac{1-0}{2\cdot 4}\left(e^{-0} + 2\,e^{-1/16} + 2\,e^{-1/4} + 2\,e^{-9/16} + e^{-1}\right)$$

$$\approx \frac{1}{8}\left[1 + 2\left(\frac{1}{e^{0.0625}}\right) + 2\left(\frac{1}{e^{0.25}}\right) + 2\left(\frac{1}{e^{0.3625}}\right) + \frac{1}{e}\right]$$

$$\approx 0.742\,98\ldots.$$

•

Although we do not know a function, we can still evaluate its integral if we have sets of points relating x and y.

• Given:

x	3	4	5	6	7	8
y	4.66	5.72	6.30	6.33	5.85	5.23

Use the trapezoidal rule to approximate the value of the integral of the function f, defined by the above points.

$\boxed{n = 5}$

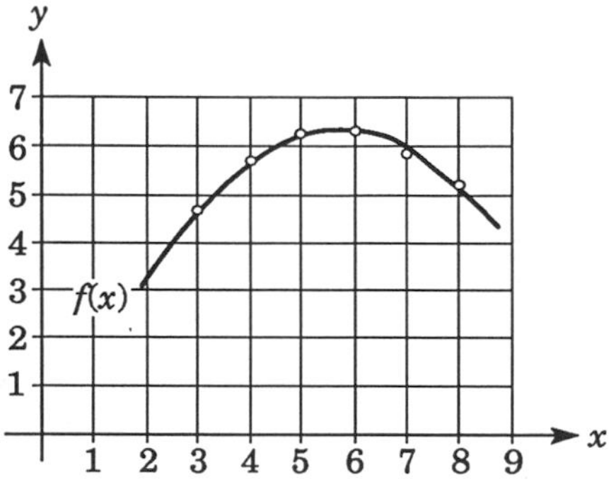

$$\int_3^8 f(x)\,dx \approx \frac{8-3}{2\cdot 5}\left(4.66 + 2\cdot 5.72 + 2\cdot 6.30 + 2\cdot 6.33 + 2\cdot 5.85 + 5.23\right)$$

$$\approx 29.145\,.$$

Boundary Estimation and Convergence

The error R of an approximate integral T may be defined as

$$R = T - I,$$

where I is the true value of the integral.

If $T < I$, R is negative;

if $T > I$, R is positive.

In an interval where the curve of a function f is concave downward and, consequently, its second derivative is negative, the error is always negative irrespective of the location of the curve above or below the x-axis:

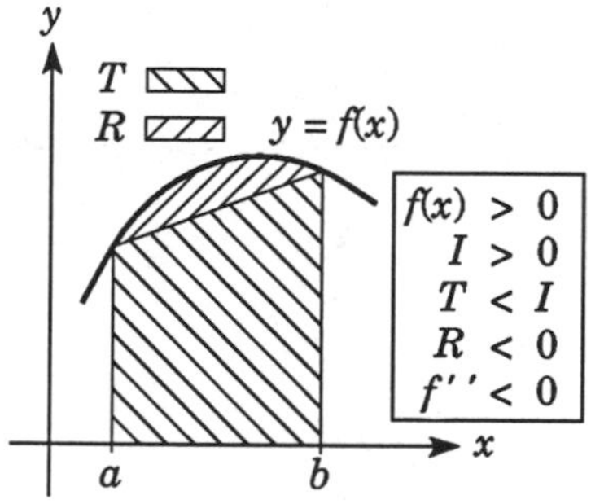

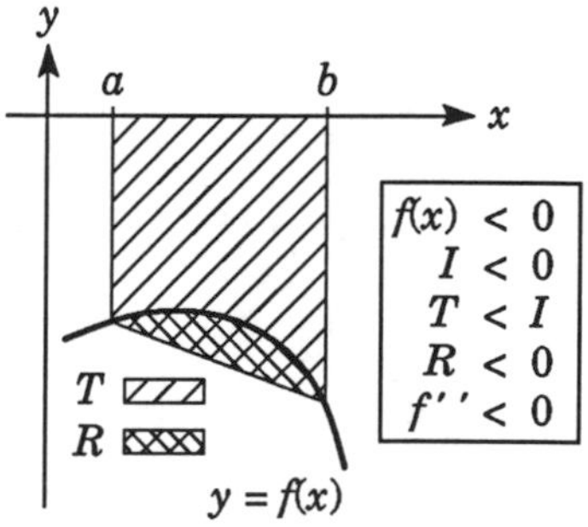

Analogously, when the curve is concave upward the error is positive; **the error always has the same sign as the second derivative of the function**.

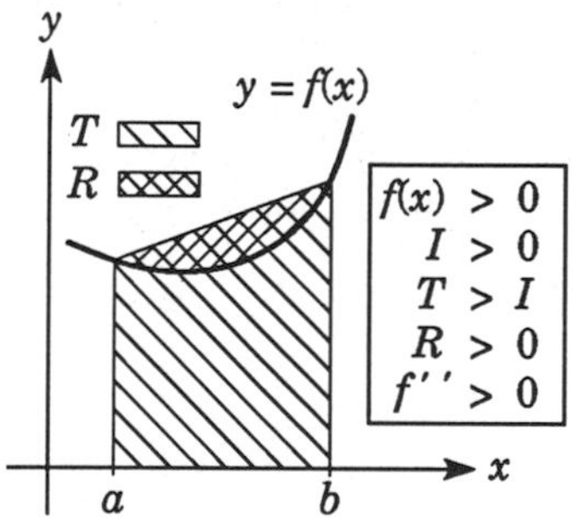

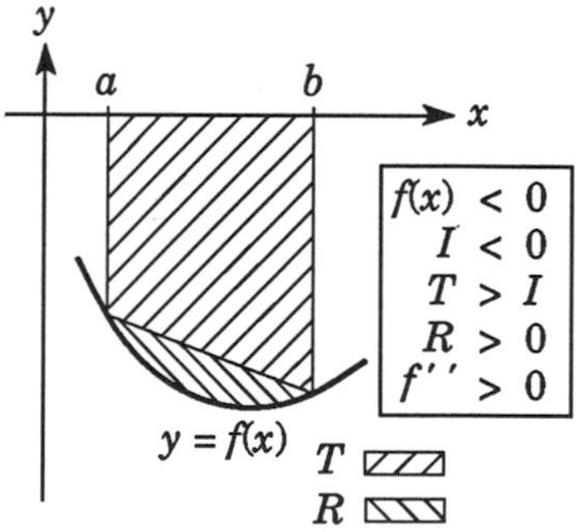

The closer $y = f(x)$ agrees with a straight line, the smaller $f''(x)$ becomes; **the magnitude of the error** can be expected to be **proportional to the magnitude of $f''(x)$**. Therefore, the lower boundary of T must be a factor of the minimum value of $f''(x)$ and the upper boundary a factor of the maximum value of $f''(x)$, which leads to the **boundary formula of the trapezoidal rule**,

$$L_T \frac{(b-a)^3}{12\,n^2} \le T_n - \int_a^b f(x)\,dx \le U_T \frac{(b-a)^3}{12\,n^2},$$

where L_T is the lower boundary value and U_T is the upper boundary value of $f''(x)$ on the interval $[a, b]$.

- The trapezoidal rule for $n = 4$ has given

$$\int_0^1 e^{-x^2}\, dx \approx 0.742\,98\ldots;$$

determine the number of reliable decimals.

Let $f(x) = \int e^{-x^2}\, dx$; then
$$f'(x) = -2x\, e^{-x^2};$$
$$f''(x) = (4x^2 - 2)\, e^{-x^2};$$
$$f'''(x) = 4x\,(3 - 2x^2)\cdot e^{-x^2}.$$

In [0, 1] the minimum value of $f''(x)$ is $f''(0) = -2$; the maximum value is $f''(1) = 2\,e^{-1}$; $f'''(x) \geq 0$, that is, the curve of $y = f'''(x)$ is located above the x-axis. Consequently, $f''(x)$ is increasing and $f''(0) = -2$ and $f''(1) = 2\,e^{-1}$ must be the upper and lower boundary values, respectively.

Use the boundary formula of the trapezoidal rule:

$$(-2)\,\frac{(1-0)^3}{12\cdot 4^2} \leq 0.742\,98 - \int_0^1 e^{-x^2}\, dx \leq 2\,e^{-1}\cdot\frac{(1-0)^3}{12\cdot 4^2}$$

$$-\frac{1}{96} \leq 0.742\,98 - \int_0^1 e^{-x^2}\, dx \leq \frac{1}{96\,e}.$$

Multiplying by (-1) gives

$$-\frac{1}{96\,e} \leq \int_0^1 e^{-x^2}\, dx - 0.742\,98 \leq \frac{1}{96},$$

which we rearrange to

$$-\frac{1}{96\,e} + 0.742\,98 \leq \int_0^1 e^{-x^2}\, dx \leq \frac{1}{96} + 0.742\,98,$$

$$0.739\,14 \leq \int_0^1 e^{-x^2}\, dx \leq 0.753\,40.$$

Since

$$0.742\,98 - 0.739\,14 = 0.003\,84$$

and

$$0.753\,40 - 0.742\,98 = 0.010\,42$$

and the absolute error must not exceed a half-unit of the digit, the approximation 0.742 98 is reliable only to the first decimal place.

Hence,

$$\int_0^1 e^{-x^2}\, dx \approx 0.7.$$

The trapezoidal rule may be programmed for computer use, which makes investigation for convergence simple. When satisfactory convergence can be demonstrated, there is usually no need to determine the boundary.

It is helpful to use a program that successively doubles the number of sub-intervals until the results repeat themselves; a computer programmed to truncate the result at 8 significant digits and to cease operation when two successive values had identical digits gave the following results:

Integral	Number of Intervals	Approximate Value
$\int_0^1 e^{-x^2}\,dx$	1	0.683 939 72...
	2	0.731 370 25...
	4	0.742 984 10...
	8	0.745 865 61...
	16	0.746 584 60...
	32	0.746 764 25...
	64	0.746 809 16...
	128	0.746 820 39...
	256	0.746 823 20...
	512	0.746 823 90...
	1024	0.746 824 07...
	2048	0.746 824 12...
	4096	0.746 824 13...
	8192	0.746 824 13...

– *My trapezoidal act is done and, as a final demonstration of the simplicity of complexities in numerical calisthenics, I give you Mr. Simpson for his parabolic roller coaster.*

Simpson's Rule

While the trapezoidal rule uses the principle that a chord is the easiest approximation of a curve extending between two specified points, Simpson's rule is based on the idea that the simplest curve which can be made to agree with **three arbitrary points** is a parabola. Thus, instead of sums of areas of trapezoids, Simpson's method uses sums of areas under parabolas to determine approximate values of definite integrals. Since a parabola generally approximates better than a straight line to the shape of a curve, we can expect better results by Simpson's rule than by the trapezoidal rule.

SIMPSON Thomas (1710 - 1761)

The rule is named after the English mathematician **Thomas Simpson**.

Deriving the Definite Integral of a Quadratic Polynomial

A function

$$p(x) = A\,x^2 + B\,x + C,$$

where A, B, and C are constants and $A \neq 0$, represents a parabola:

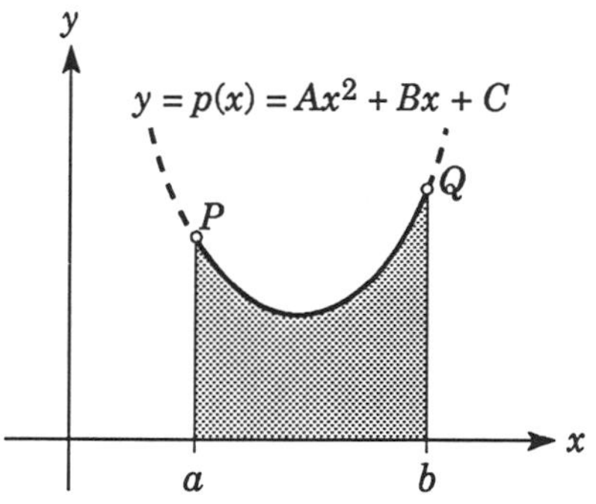

The area contained between the arc PQ, the x-axis, and the ordinates at a and b is given by

$$\int_a^b (A\,x^2 + B\,x + C)\,\mathrm{d}x.$$

To fit a parabola to the curve $y = f(x)$ between a and b, we choose the point

$$c = \frac{a+b}{2}.$$

Give the parabola a coordinate system of its own whose origin is the point $[c, p\,(c)]$. t is the independent variable of the new coordinate system,

$$t = x - c.$$

We must now define limit values for the new coordinate system.

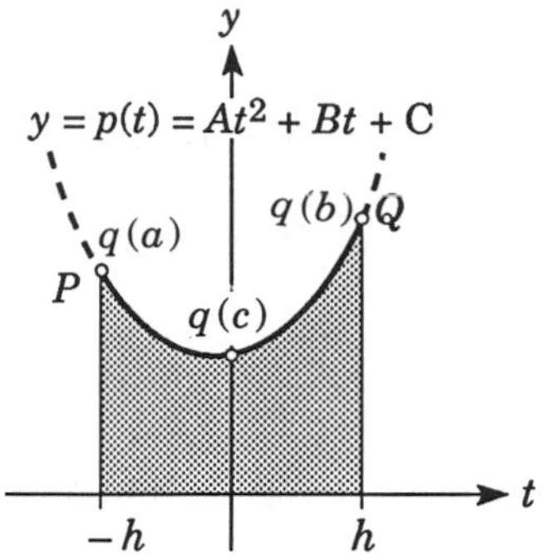

If $(-h)$ and h are the lower and upper limits, we have

$$-h = a-c\,; \quad h = b-c\,.$$

We have the equation

$$p\,(x-c) = p\,(t) = A\,t^2 + B\,t + C$$

and the relations

$$\left.\begin{array}{lll} p(-h) &= A\,(-h)^2 + B\,(-h) + C = A\,h^2 - B\,h + C &= q(a) \\ p(0) &= A\cdot 0^2 + B\cdot 0 + C = C &= q(c) \\ p(h) &= A\,h^2 + B\,h + C &= q(b) \end{array}\right\}\,,$$

which give

$$\left.\begin{array}{ll} q\,(a) + q\,(b) &= 2A\,h^2 + 2\,C \\ q\,(c) &= C \end{array}\right\}$$

to be inserted in the equation

$$\int_{-h}^{h} (A\,t^2 + B\,t + C)\,\mathrm{d}\,t = \left[\frac{A\,t^3}{3} + \frac{B\,t^2}{2} + C\,t\right]_{-h}^{h}$$

$$= \frac{1}{3}\,h\,[(2\,A\,h^2 + 2\,C) + 4\,C]$$

$$= \frac{1}{3}\,h\,[q\,(a) + q\,(b) + 4\;q\,(b)]\,.$$

Since

$$\int_{-h}^{h} (A\,(t+c)^2 + B\,(t+c) + C)\,\mathrm{d}t = \int_{a}^{b} (A\,x^2 + B\,x + C)\,\mathrm{d}x$$

and

$$h = \frac{b-a}{2}\,; \quad c = \frac{a+b}{2}$$

we can write

$$\int_{a}^{b} (A\,x^2 + B\,x + C)\,\mathrm{d}x = \frac{1}{6}\,(b-a)\left[q\,(a) + q\,(b) + 4\,q\left(\frac{a+b}{2}\right)\right],$$

which we may use to find an approximate value of any definite integral.

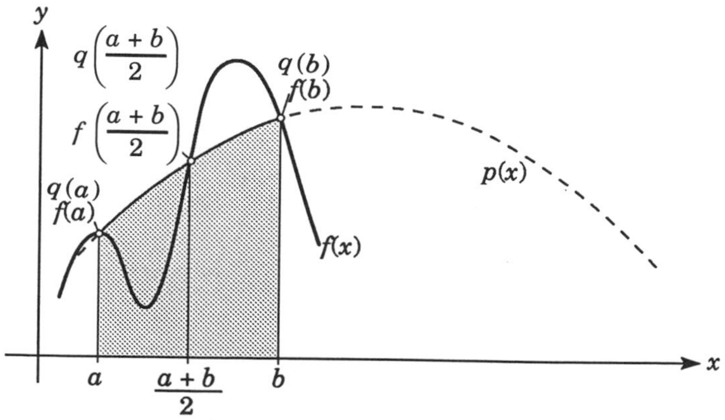

$$\int_a^b f(x)\,dx \approx \frac{b-a}{6}\left[p(a)+p(b)+4p\left(\frac{a+b}{2}\right)\right]$$
$$= \frac{b-a}{6}\left[f(a)+f(b)+4f\left(\frac{a+b}{2}\right)\right].$$

In the graph below, $[a, b]$ is divided into an **even number** n of sub-intervals with equal length, $\frac{b-a}{n}$. As long as the arcs of the function corresponding to each of the sub-intervals are consistently concave downward *or* concave upward, we may expect the approximation of the curve to an arc of the parabola, represented in the graph by the quadratic functions p_1 to $p_{n/2}$, to be better the shorter the intervals are.

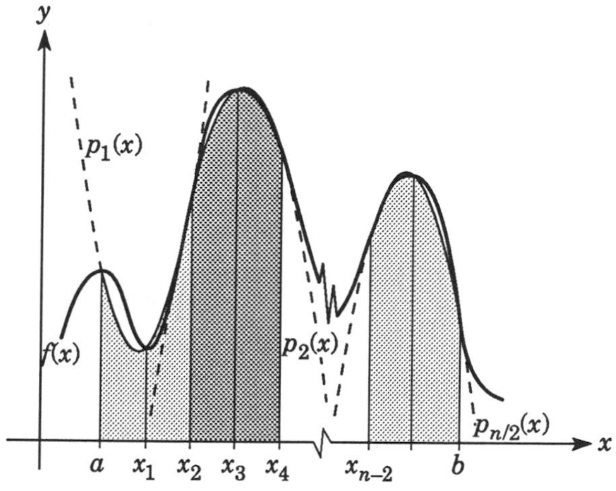

Since an integral is equal to the sum of its parts,

$$\int_a^b f(x)\,dx = \int_a^{x_2} f(x)\,dx + \int_{x_2}^{x_4} f(x)\,dx + \ldots + \int_{x_{n-2}}^{b} f(x)\,dx,$$

we obtain

$$\int_a^b f(x)\,d\,x \approx S_1 + S_2 + \ldots + S_n$$
$$= \frac{1}{6}(x_2 - a)\,[f(a) + f(x_2) + 4f(x_1)]$$
$$+ \frac{1}{6}(x_4 - x_2)\,[f(x_2) + f(x_4) + 4f(x_3)] + \ldots$$
$$+ \frac{1}{6}(b - x_{n-2})\,[f(x_{n-2}) + f(b) + 4f(x_{n-1})],$$

which gives us **Simpson's rule**,

$$\int_a^b f(x)\,dx \approx \frac{b-a}{3n}\,[f(a) + 4f(x_1) + 2f(x_2) + 4f(x_3)$$
$$+ 2f(x_4) + \ldots + 2f(x_{n-2}) + 4f(x_{n-1}) + f(b)],$$

where n is a positive even integer, indicating n equal sub-intervals on $[a, b]$.

- Determine an approximate value of the integral

$$\int_0^2 e^{-x^2}\, dx$$

by Simpson's rule for $n = 2$ and $n = 4$.

It is understood that no guarantee can be given regarding reliable digits.

$n = 2$

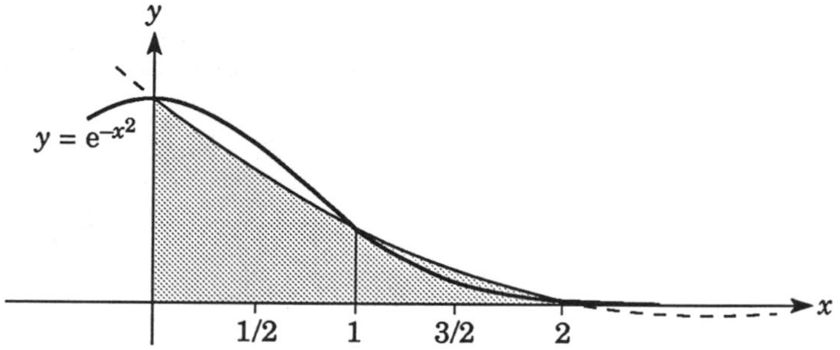

$$\int_0^2 e^{-x^2}\, dx \approx \frac{2-0}{3\cdot 2}\left(e^{-0} + 4\,e^{-1} + e^{-4}\right)$$

$$= \frac{1}{3}\left(1 + \frac{4}{e} + \frac{1}{e^4}\right) = 0.829\,94\ldots .$$

$n = 4$

$$\int_0^2 e^{-x^2}\, dx \approx \frac{2-0}{3\cdot 4}\left(e^{-0} + 4\,e^{-1/4} + 2\,e^{-1} + 4\,e^{-9/4} + e^{-4}\right)$$

$$= 0.881\,81\ldots .$$

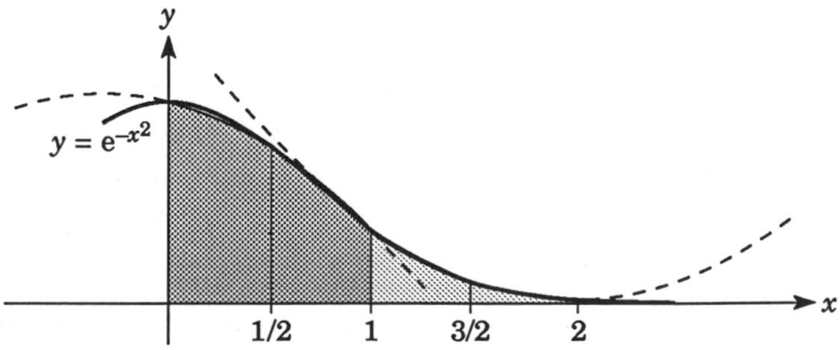

•

When convergence can be satisfactorily demonstrated for Simpson's rule, there is usually no need to determine the boundary values. The table below was obtained with a computer programmed to truncate the result at 8 significant digits until two successive values are identical:

Integral	**Number of Intervals**	**Approximate Value**
$\int_0^2 e^{-x^2}\, dx$	2	0.829 944 47...
	4	0.881 812 43...
	8	0.882 065 51...
	16	0.882 080 40...
	32	0.882 081 33...
	64	0.882 081 39...
	128	0.882 081 39...

The steady convergence towards 0.882 081 39... is a strong indication that the value is correct to at least the 7th place of decimals.

Boundary Estimation and Convergence

Generally, Simpson's rule leads to a satisfactory value faster than the trapezoidal rule. On the other hand, error estimation is a much more formidable task with Simpson's rule than with the trapezoidal rule.

With Simpson's rule the magnitude of the error is proportional to the magnitude of the 4th derivative of the integrand, which leads to the boundary formula of Simpson's rule,

$$L_S \frac{(b-a)^3}{180\, n^4} \le T_n - \int_a^b f(x)\, dx \le U_S \frac{(b-a)^3}{180\, n^4},$$

where L_S is the lower boundary value and U_S is the upper boundary value of $f^{(4)}$ in the interval $[a, b]$, and T_n is the approximate value of the integral evaluated by Simpson's rule over n sub-intervals.

We try to avoid the use of the boundary formula of Simpson's rule, for we must find not only the 4th derivative of f but – in order to determine upper and lower boundary values of $f^{(4)}$ – also the 5th derivative of f, which can be a formidable task.

Like $\int e^{-x^2} x$, the indefinite integral $\int \sin(x^2)\,dx$ cannot be expressed by elementary functions. The table below compares computer results obtained with Simpson's rule and with the trapezoidal rule.

Integral	**Number of Intervals**	**Approximate Value**	
		Simpson's Rule	**Trapezoidal Rule**
$\int_0^{10} \sin(x^2)\,dx$	1		−2.531 828 21...
	2	−1.726 287 74...	−1.927 672 85...
	4	−1.733 979 05...	−1.782 402 50...
	8	5.896 281 91...	3.976 610 81...
	16	1.270 177 02...	1.946 785 47...
	32	0.515 778 43...	0.873 530 19...
	64	0.543 626 35...	0.626 102 31...
	128	0.581 723 28...	0.592 818 04...
	256	0.583 575 75...	0.585 886 32...
	512	0.583 665 27...	0.584 220 53...
	1 024	0.583 670 55...	0.583 808 05...
	2 048	0.583 670 88...	0.583 705 17...
	4 096	0.583 670 90...	0.583 679 47...
	8 192	0.583 670 90...	0.583 673 04...
	16 384		0.583 671 44...
	32 768		0.583 671 03...
	65 536		0.583 670 93...
	131 072		0.583 670 91...
	262 144		0.583 670 90...
	524 288		0.583 670 90...

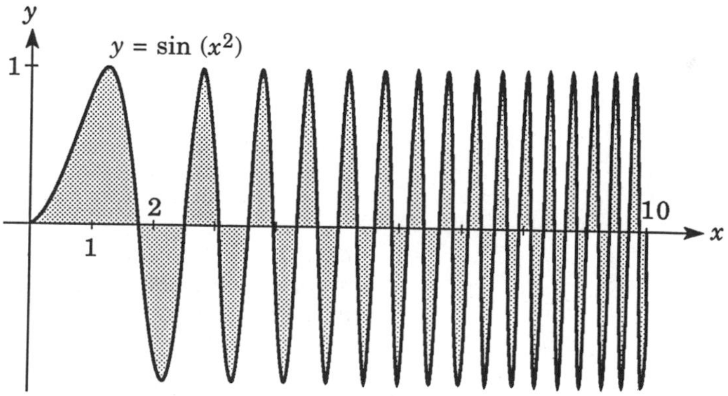

Chapter 33

PROBABILITY THEORY

33.0 Introduction

Probability theory is a branch of **statistics**, a science that employs mathematical methods of collection, organization, and interpretation of data, with applications in practically all scientific areas.

Greek, *stochazesthai*, "to guess at", "to aim at"

When working with probability theory, we analyze random – or **stochastic** – phenomena and assess the *likelihood* that an event will occur.

In this chapter we discuss some fundamental aspects of probability theory and explore its use of integral calculus.

Wheel of Fortune

Gregor Reisch, *Margarita phylosophica* (first edition, 1496; 1503 reprint).

33.1 History

CARDANO Gerolamo
(1501 - 1576)

One of the earliest mathematical studies on probability was *Liber de ludo aleae* ("On Casting the Die"), written by the 16th-century Italian mathematician and physician **Gerolamo Cardano**; it was not published until 1663, 87 years after Cardano's death. Cardano introduced concepts of combinatorics into calculations of probability and defined probability as "the number of favorable outcomes divided by the number of possible outcomes". It is likely that Cardano would be known as "the father of the theory of probability" had the publication not been delayed.

PASCAL Blaise
(1623 - 1662)
FERMAT Pierre de
(1601 - 1665)
HUYGENS Christiaan
(1629 - 1695)

In the 17th century, questions about the probability of events occurring in games of chance were discussed in correspondence between the French mathematicians **Blaise Pascal** and **Pierre de Fermat**. Building on their results, the Dutch physicist-astronomer-mathematician **Christiaan Huygens** published, in 1656, *De ratiociniis in ludo aleae* ("On Reasoning in Games of Chance").

BERNOULLI Jakob
(1654 - 1705)

The Swiss mathematician **Jakob Bernoulli** was an early advocate of the use of probability theory in medicine and meteorology in his work *Ars conjectandi* ("The Art of Conjecture"), published posthumously in 1713.

JACOBI BERNOULLI,
Profeſſ. Baſil. & utriuſque Societ. Reg. Scientiar.
Gall. & Pruſſ. Sodal.
MATHEMATICI CELEBERRIMI,

ARS CONJECTANDI,
OPUS POSTHUMUM.
Accedit
TRACTATUS
DE SERIEBUS INFINITIS,
Et EPISTOLA Gallicè ſcripta
DE LUDO PILÆ
RETICULARIS.

BASILEÆ,
Impenſis THURNISIORUM, Fratrum.
cIↄ Iↄcc xiii.

In the late 18th century, it became increasingly evident that analogies exist between games of chance and random phenomena in physical, biological, and social sciences.

LAGRANGE Joseph Louis (1736 - 1813)

LAPLACE Pierre Simon de (1749 - 1827)

GAUSS Carl Friedrich (1777 - 1855)

POISSON Siméon-Denis (1781 - 1840)

de MOIVRE Abraham (1667 - 1754)

Contributions of fundamental importance to probability theory were made in the latter half of the 18th century and the beginning of the 19th century by the French mathematicians, astronomers, and physicists **Joseph Louis Lagrange** and **Pierre Simon Laplace**, the omnipresent German mathematician and astronomer **Carl Friedrich Gauss**, and the French mathematician **Siméon-Denis Poisson**. The most important publication on probability theory in this era is Laplace's *Théorie analytique des probabilités* (1812), which discussed practical applications of the theory and developed the concepts of normal distribution, first discovered by **Abraham de Moivre**. In the 1837 *Recherches sur la probabilité des jugements...* ("Researches on the Probability of Opinions...") Poisson introduced what we now know as the Poisson distribution, or Poisson law of large numbers, an approximate method used to describe probable occurrence of unlikely events in a large number of unconnected trials.

CHEBYSHEV Pafnuty Lvovich (1821 - 1894)

MARKOV Andrei Andreevich (1856 - 1922)

LYAPUNOV Alexandr Mikhailovich (1857 - 1918)

About 1850 - 1900, the Russian (Petersburg) school of probability theory, emphasizing stringent mathematical methods, dominated its development. Prominent figures of this school were **Pafnuty Chebyshev** and, originally disciples of Chebyshev's, **Andrei Markov** and **Alexandr Lyapunov**.

EINSTEIN Albert (1879 - 1955)

RUTHERFORD Ernest (1871 - 1937)

CHARLIER Carl Vilhelm Ludvig (1864 - 1934)

In the beginning of the 20th century, the need for applications of probability theory increased in physics, economics, insurance, and telephone comunication. Important impulses were given by **Albert Einstein**, the New Zealand-born English physicist **Ernest Rutherford**, and the Swedish astronomer **C. V. L. Charlier**. Applications often precipitated new probability problems which had to be tackled within the field of theoretical probability, and thus a fruitful interplay between the sciences was created.

PEARSON Karl (1857 - 1936)

The English mathematician **Karl Pearson** is the founder of modern hypothesis testing – he developed the chi-square test of statistical significance. Besides making major contributions in mathematics and probability theory, Pearson also practiced law, was active in politics, published literary works, and wrote *The Grammar of Science* (1892), a classic in the philosophy of science.

FISHER Ronald Aylmer (1890 - 1962)

One of the most eminent scientists of the 20th century, the English geneticist and statistician **R. A. Fisher**, professor of genetics at Cambridge from 1943 to 1957, developed methods of *multivariate analysis* – analysis of problems involving more than one variable – and used them in his investigations of the linkage of genes to various traits. He also introduced the idea of *likelihood* in statistical inference, that is, how to draw conclusions on the basis of the relative probability of different events.

Fisher's *Statistical Methods for Research Workers* (1925) was used extensively as a textbook and a reference book and remained in print for more than 50 years.

Statistical Methods for Research Workers

BY

R. A. FISHER, M.A.

Fellow of Gonville and Caius College, Cambridge
Chief Statistician, Rothamsted Experiment Station

OLIVER AND BOYD
EDINBURGH: TWEEDDALE COURT
LONDON: 33 PATERNOSTER ROW, E.C.
1925

I

INTRODUCTORY

1. The Scope of Statistics

THE science of statistics is essentially a branch of Applied Mathematics and may be regarded as mathematics applied to observational data. As in other mathematical studies the same formula is equally relevant to widely different groups of subject matter. Consequently the unity of the different applications has usually been overlooked, the more naturally because the development of the underlying mathematical theory has been much neglected. We shall therefore consider the subject matter of statistics under three different aspects, and then show in more mathematical language that the same types of problems arise in every case. Statistics may be regarded as (i.) the study of populations, (ii.) as the study of variation, (iii.) as the study of methods of the reduction of data.

PEARSON Egon (1895 - 1980)

NEYMAN Jerzy (1894 - 1981)

Along with **Egon Pearson** (son of Karl Pearson) and Fisher, **Jerzy Neyman** was one of the principal founders of modern statistical analysis. Neyman lived in Poland until he was forty, then in England, and finally settled in the United States. In 1955, he became professor of statistics at the University of California at Berkeley, where his department became a world center for the development of mathematical statistics. Egon Pearson and Neyman founded what is now known as the school of statistical inference.

KHINCHIN Alexandr Yakovlevich (1894 - 1959)

KOLMOGOROV Andrey Nikolayevich (1903 - 1987)

The Russian mathematicians **Alexandr Khinchin** and **A. N. Kolmogorov** are the founders of the Moscow school of probability theory, one of the most influential in the 20th century. One may say that the present golden age of probability theory started in 1933 with Kolmogorov's *Grundbegriffe der Wahrscheinlichkeitsrechnung* ("Foundations of Probability Theory"). Kolmogorov introduced several fundamental postulates in statistics and probability theory; he showed that probability theory may be founded on the concepts of set theory and mathematical measure theory.

Theory of Games

von NEUMANN John (Johann) (1903 - 1957) p. 301

The **theory of games** was founded by **John von Neumann**, preeminent 20th-century innovator in many fields of pure and applied mathematics. He created a mathematical model for games of chance, such as poker and bridge, that involve free choices – *strategy* – for the players. His first paper on this subject was presented in 1926; von Neumann's theories were further developed in his major work *Theory of Games and Economic Behavior* (1944), co-authored with the economist **Oskar Morgenstern**. The theory of games is now a mathematical discipline of its own, with far-reaching applications to economics and social sciences.

MORGENSTERN Oskar (1902 - 1977)

33.2 The Basics

A stopped clock shows the correct time twice every 24 hours.

Defining Probability

The classical definition of **mathematical probability** is the ratio of the number of favorable outcomes to the total number of possible outcomes.

The **total probability of favorable and unfavorable outcomes** is 1; if there is no probability of favorable outcomes, the probability is 0. If there are 9 marbles in a box,

2 red, 3 green, 4 yellow,

the probability of drawing a red, green, ***or*** yellow marble from the box is 1, whereas the probability of white is 0.

The first time, the probability of drawing

a red marble is $\frac{2}{9}$,

a green marble is $\frac{3}{9} = \frac{1}{3}$,

a yellow marble is $\frac{4}{9}$;

the probability of drawing a red ***or*** a green marble is

$$\frac{2}{9} + \frac{1}{3} = \frac{5}{9} .$$

In two successive draws, one marble at a time and *putting the marble back after the first draw*, the probability of drawing two red marbles is

$$\frac{2}{9} \times \frac{2}{9} = \frac{4}{81} = 0.049 \ldots .$$

If the first draw is not replaced, the probability of drawing two red marbles is

$$\frac{2}{9} \times \frac{1}{8} = \frac{1}{36} = 0.027 \ldots .$$

Conditional Probability

Draw two cards from a deck of cards; what is the probability that the second card will be the seven of spades?

There are several answers to this question:

- If we have not looked at the first card drawn but simply laid it aside, the probability for the second card will be the same as for the first card, that is, 1/52.
- If we looked at the first card, there are two possibilities:

  If the first card *was not* the seven of spades, the chance that the second card will be is 1/51.

  If the first card *was* the seven of spades, the probability for the second card is *nil*.

This simple example shows that "probability" should always be viewed in the light of what *information* there is at hand.

– *I'll see you. What've you got?*
– *Five aces – and you?*
– *Two revolvers.*
– *All right, you win.*

The Law of Large Numbers

When the number of observations of an experiment repeated under identical conditions is sufficiently large, then – by the **law of large numbers** – the proportion of some specific outcome will be close to the underlying probability of that outcome; the greater the number of observations, the closer the agreement.

Conclusions can never be drawn with absolute certainty from statistical data; they only provide evidence about the likelihood that a conclusion is correct. Suppose a die is cast 100 times and a 6 turns up every time; if the die is well balanced, the probability of this happening is

$$\left(\frac{1}{6}\right)^{100};$$

or rather something has happened which would be *very* unlikely if the die is *not* weighted.

Random Numbers

In a set of numbers, *e.g.*, 0, 1, 2 ... 9, each random number has an equal chance of being selected, and each selection is independent of all prior selections. To select numbers at random, we may write each number of a set on a piece of paper which we then place in an urn; each time a number has been drawn, we return it to the urn and mix. The sequence

$$6, 7, 8, 8, 6, 4, 7, 1, 6, 8, 6, 8, 3, 2, 1$$

is equally as likely as the sequence

$$1, 2, 3, 4, 5, 6, 7, 8, 9, 0, 1, 2, 3, 4, 5.$$

There are tables of random numbers; computers may be programmed to generate such numbers.

Monte Carlo Methods

ULAM Stanislav Marcin (1909 - 1984)

von NEUMANN John (Johann) (1903 - 1957) p. 301

Processes that are too complicated to allow exact analysis may often be solved by probabilistic methods that employ the law of large numbers. Named after the famous gambling casino, they are known as Monte Carlo methods. They are used anywhere from estimating the strength of a hand in bridge to modeling the statistics of a nuclear chain reaction. Key inventors of the Monte Carlo methods were **John von Neumann** and the Polish mathematician **Stanislav Ulam**.

A Composite Area Problem

To determine the approximate area of a plane figure of irregular or complex outline, we may

- o enclose that area by a figure of known area;
- o place a given number of random points within the enclosing figure;
- o find the number of points that hit the sought area;
- o then, with p being the probability that a random point will hit the sought area (or areas),

$$p = \frac{\text{sought area}}{\text{known area}} \approx \frac{\text{points inside sought area}}{\text{points inside known area}} .$$

Example:

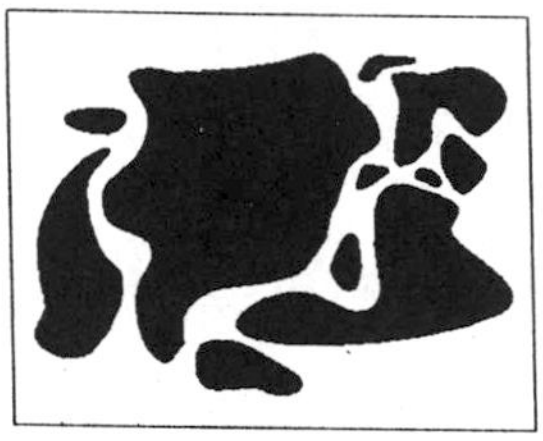

An oil spill breaks up over the sea in smaller and larger sections. An aerial photograph is taken of the area and it is estimated that the sections of oil and open water can be enclosed by a 4 x 5-kilometer rectangle.

– Find the total surface area of the oil sections.

The photograph may be digitized and fed into a computer, programmed to give the number of random points that hit the sought area. The greater the number of total points, the more reliable the estimate.

The area enclosed is $4 \times 5 = 20 \text{ km}^2$.

We might obtain:

Number of points: total	in the oil	sought area (km²)	
10	6	$\approx \frac{6}{10} \cdot 20$	$\approx 12 \text{ km}^2$
100	61	$\approx \frac{61}{100} \cdot 20$	$\approx 12.2 \text{ km}^2$
1 000	441	$\approx \frac{441}{1\,000} \cdot 20$	$\approx 8.8 \text{ km}^2$
10 000	4 654	$\approx \frac{4\,654}{10\,000} \cdot 20$	$\approx 9.3 \text{ km}^2$
100 000	48 581	$\approx \frac{48\,581}{100\,000} \cdot 20$	$\approx 9.7 \text{ km}^2$

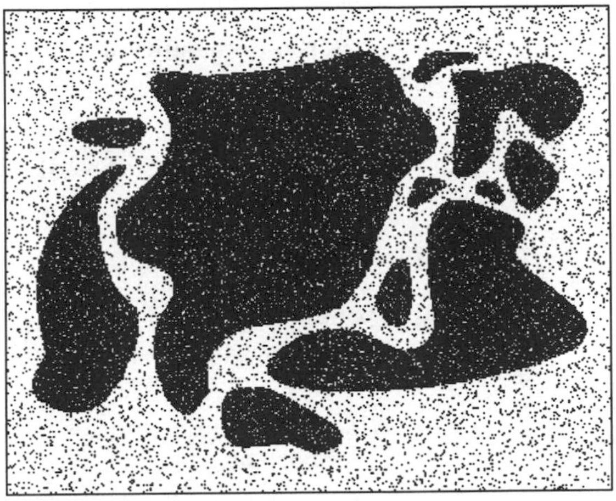

Total: 10 000 dots.

Binomial Distribution

If there are two possible outcomes of an event (*e.g.*, success or failure), and the possibilities of the outcomes are independent and constant, the distribution of probabilities is called a **binomial distribution**.

Examples of binomial distribution are probabilities regarding the number of heads or tails when tossing a coin and of red or black when cutting and reshuffling a deck of cards.

p. 196

The notations ${}_nC_k$ and $\binom{n}{k}$, denoting **combinations**, give the number of ways one can select k objects among n, disregarding order.

Binomial Theorem: *pp.* 140, 775

The probabilities of the binomial distribution are given by the terms in the expansion of $(p + q)^n$:

$$(p+q)^n = \binom{n}{0} p^n + \binom{n}{1} p^{n-1} q + \binom{n}{2} p^{n-2} q^2 + \ldots + \binom{n}{n-1} p\, q^{n-1} + \binom{n}{n} q^n \,.$$

In a binomial distribution, where the probability of success is p and that of failure is $q = 1 - p$, the probability of k successes is

$${}_nC_k \, p^k q^{n-k} = \frac{n\,!}{(n-k)\,!\,k\,!}\, p^k q^{n-k} \,.$$

- What is the probability of a run of 4 boys in a family with 4 children?

$${}_4C_4 \, 0.5^4 \, (1-0.5)^{4-4} = \frac{4\,!}{(4-4)\,!\,4\,!}\, 0.5^4 \, (1-0.5)^{4-4} = 0.062\,5 \,.$$

- What is the probability of a run of 4 boys and 4 girls in a family with eight children?

Comparing with the previous problem, we find that the probability is

$$0.0625\,(0.0625) = 0.003\,906\,25 \,.$$

- What is the probability of a run of 8 girls in a family with 8 children?

$$ {}_8C_8\, 0.5^8\, (1-0.5)^{8-8} = \frac{8\,!}{(8-8)\,!\;8\,!}\, 0.5^8\, (1-0.5)^{8-8} $$
$$ = 0.003\,906\,25\,; $$

  thus, the same probability as in the previous example.

- In a population, 32% of people have blood group A and 68% have one of the other blood groups, B, AB, and 0. Five people are available for an emergency blood donation.
  - What are the probabilities that 0, 1, 2, 3, 4, 5 of the possible donors are blood group A?

Number of donors with blood group A	**Probability**
0	$\frac{5\,!}{(5-0)\,!\;0\,!}\, 0.32^0\, (0.68^{5-0}) \approx 0.1454$
1	$\frac{5\,!}{(5-1)\,!\;1\,!}\, 0.32^1\, (0.68^{5-1}) \approx 0.3421$
2	$\frac{5\,!}{(5-2)\,!\;2\,!}\, 0.32^2\, (0.68^{5-2}) \approx 0.3220$
3	$\frac{5\,!}{(5-3)\,!\;3\,!}\, 0.32^3\, (0.68^{5-3}) \approx 0.1515$
4	$\frac{5\,!}{(5-4)\,!\;4\,!}\, 0.32^4\, (0.68^{5-4}) \approx 0.0356$
5	$\frac{5\,!}{(5-5)\,!\;5\,!}\, 0.32^5\, (0.68^{5-5}) \approx 0.0034$

The Null Hypothesis

A hypothesis that is being tested for rejection is known as a **null hypothesis** – an assumption that observations are due to chance alone.

Type I Error
Type II Error

The rejection of a null hypothesis when in fact true is known as a **type I error**; if accepted despite being false, a **type II error** is committed.

Significance Level

The **significance level** of a test is the probability of committing a type I error. The lower the significance level, the lower the probability of a type I error and the stronger the evidence against the hypothesis. If we can reject the null hypothesis at the decided-upon significance level, we say that our results are significant, and our hypothesis is rejected.

Beyond Doubt

– *Has it been proved that women live longer than men?*

– *Yes, at least for widows.*

$\int x\,dx = \frac{1}{2}x^2$

$\int x^2\,dx = \frac{1}{3}x^3$

33.3 The Probability Density Function

Random Variable

A **random variable** is a numeric quantity that can be measured in a random experiment. More precisely, it is a function of the possible outcomes of the experiment.

The probability distribution of a continuous random variable x is a function of the numerical values of x; such a function is referred to as a **probability density function**, and it is so constructed that the area between its curve and the x-axis when calculated over the entire range of x for which $f(x)$ is defined equals 1.

If $f(x)$ is a probability density function of a continuous random variable x that can assume values between $x = a$ and $x = b$, then

$$\int_a^b f(x)\, dx = 1 .$$

The graph of a probability density function may take several forms; *e.g.*,

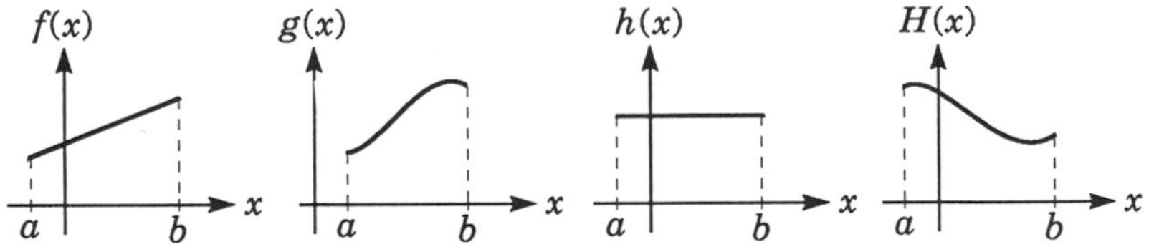

$$f(x) = 4x^3 \quad \text{for} \quad 0 \leq x \leq 1$$

is a probability density function, because $f(x)$ is positive for all real values of x, and

$$\int_0^1 f(x)\, dx = \int_0^1 4x^3\, dx = 4\left(\frac{x^4}{4}\right)\Bigg|_0^1 = 1 .$$

- With a probability density function

$$f(x) = 4x(1 - x^2) \quad \text{for} \quad 0 \leq x \leq 1 ,$$

determine the probability that the random variable falls in [0.2, 0.4] .

$$\int_{0.2}^{0.4} 4x(1 - x^2)\, dx = \Bigg|_{0.2}^{0.4} 4\left(\frac{x^2}{2} - \frac{x^4}{4}\right)$$

$$= 4\left[\left(\frac{0.16}{2} - \frac{0.0256}{4}\right) - \left(\frac{0.04}{2} - \frac{0.0016}{4}\right)\right]$$

$$= 0.216 .$$

The sought probability is 0.216, corresponding to the shaded area of the graph:

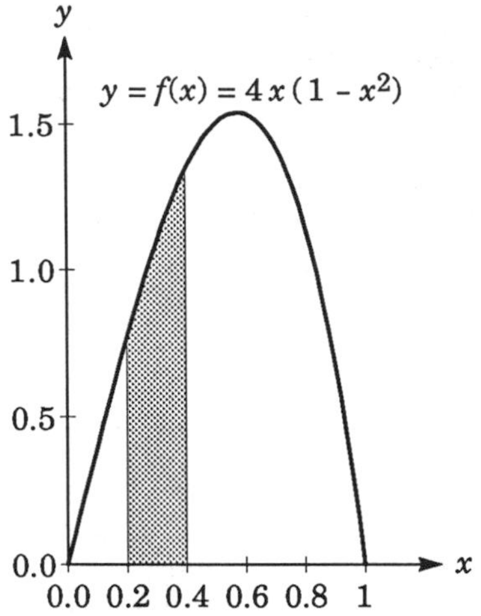

- The time for a certain type of surgical suture to be absorbed by the tissue of the human body varies according to the probability density function

$$f(x) = 0.001\, e^{-0.001h}\ ; \quad h \geq 0\,,$$

where h denotes hours.

What is the probability that a suture will remain in the tissue

a: for no more than 600 hours;

b: beyond 1000 hours.

a:

$$\int_0^{600} 0.001\, e^{-0.001h}\, dh = \Big|_0^{600} (-e^{-0.001h}) = 1 - e^{-0.6} \approx 0.45\,.$$

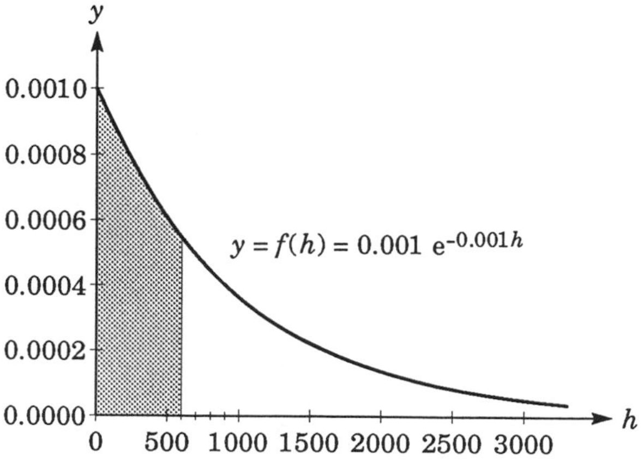

The sought probability for case **a** is 0.45 .

b:

$$1-\int_0^{1000} 0.001\,e^{-0.001h}\,dh = 1-\left[-e^{-0.001h}\right]_0^{1000}$$

$$= 1-(1-e^{-0.1}) \approx 0.37\,.$$

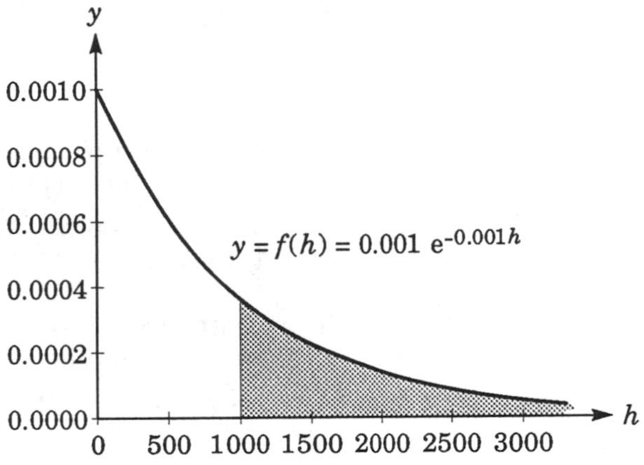

The sought probability for case **b** is 0.37.

33.4 Central Tendency

A measure of central tendency is a summary of a whole distribution of events or measurements. The three most useful measurements are the mode, the median, and the mean.

Mode

The value of events or experimental data that occur most frequently is the mode of the data.

The mode of the values 1, 3, 3, 3, 5, 5, 7, 9, 9 is 3; the modes of 2, 2, 2, 3, 3, 9, 4, 9, 9, 1, 2, 9, 5 are 2 and 9.

- Playing with two dice, the sum of the throws may vary from 2 to 12. What would be the expected mode after a large number of throws of the dice? What is the probability, each time, of throwing a seven, a two, or a twelve?

1 + 1;	2;	3;	4;	5;	6	=	2;	3;	4;	5;	6; 7
2 + 1;	2;	3;	4;	5;	6	=	3;	4;	5;	6;	7; 8
3 + 1;	2;	3;	4;	5;	6	=	4;	5;	6;	7;	8; 9
4 + 1;	2;	3;	4;	5;	6	=	5;	6;	7;	8;	9; 10
5 + 1;	2;	3;	4;	5;	6	=	6;	7;	8;	9;	10; 11
6 + 1;	2;	3;	4;	5;	6	=	7;	8;	9;	10;	11; 12

Sums of the dice: 1 two; 2 threes; 3 fours; 4 fives; 5 sixes; 6 sevens; 5 eights; 4 nines; 3 tens; 2 elevens; 1 twelve.

Sum total: 36

The mode is seven.

The probability of throwing a seven is 6/36 ≈ 0.167; that of throwing a two is 1/36 ≈ 0.03, as is that of throwing a twelve.

Median

The median is the middle value of a sequence of observations arranged in order of magnitude; the number of values greater than the median equals the number of values smaller than the median. If the number of observations is even, the median is the arithmetic mean of the two middle observations.

For instance, the median of 7, 9, 72, 82, 100, 101, 119 is 82; the median of 3, 9, 11, 14, 16, 19, 23, 27 is 15.

For a continuous random variable x with probability density function f over $[a, b]$, the median of x is the number m for which

$$\int_a^m f(x)\,\mathrm{d}x = \int_m^b f(x)\,\mathrm{d}x = \frac{1}{2}\,.$$

Mean

When simply referring to the **mean**, or **average**, we usually have in mind the arithmetic mean:

The **arithmetic mean** μ of $x_1, x_2 \ldots x_n$ is

$$\mu = \frac{1}{n} \sum_{i=1}^{n} x_i = \frac{1}{n} (x_1 + x_2 + \ldots + x_n).$$

- If in 120 observations the numbers

0, 1, 2, 3, 4, and 5

occur, respectively,

0, 8, 16, 24, 32, and 40 times,

then

$$\mu = \frac{0 \cdot 0 + 8 \cdot 1 + 16 \cdot 2 + 24 \cdot 3 + 32 \cdot 4 + 5 \cdot 40}{120} = \frac{11}{3} \approx 3.67 .$$

The probabilities of the finite values of observations can be expressed using the function

$$f(x_i) = \begin{cases} 0/120 & \text{when } x_i = 0 \\ 8/120 & x_i = 1 \\ 16/120 & x_i = 2 \\ 24/120 & x_i = 3 \\ 32/120 & x_i = 4 \\ 40/120 & x_i = 5 \end{cases}$$

and the arithmetic mean as

$$\mu = \sum x_i \ f(x_i)$$
$$= 0\,(0/120) + 1\,(8/120) + 2\,(16/120) + 3\,(24/120) + 4\,(32/120) + 5\,(40/120)$$
$$= \frac{11}{3} \approx 3.67 .$$

Weighted Means

Weighted means give mark to the importance of different data.

- If entrance to an educational program weighs English as 4, Mathematics as 3.5, Physics and Chemistry as 3, Biology as 2, and all other subjects as 0, then a student with a 4 in English, 3 in Mathematics, 2 in Physics, and 3 in Chemistry and Biology has the weighted arithmetic mean μ_w:

$$\mu_w = \frac{4 \cdot 4 + 3.5 \cdot 3 + 3 \cdot 2 + 3 \cdot 3 + 2 \cdot 3}{4 + 3.5 + 3 + 3 + 2} = 3.06$$

Mean of a Continuous Random Variable

For a **continuous** random variable x with probability density function f over $[a, b]$, the arithmetic mean μ of x is

$$\mu = \int_a^b x \ f(x) \ dx.$$

33.5 Dispersion

Range

The simplest measure of dispersion, or spread, of data is the **range**, that is, the difference, or distance, between the highest and lowest observed values. When dealing with the extreme values, the range is a rather rough value of the dispersion.

Variance and Standard Deviation

Taking into account the deviation of every observed value from the mean, the variance and the standard deviation are sensitive indicators of the degree of variability of statistical observations. The standard deviation is simply the square root of the variance.

If μ stands for the mean, we have for a data set:

- the **variance** σ^2 (lowercase *sigma*),

$$\sigma^2 = \frac{\sum(x-\mu)^2}{n-1} ;$$

- the **standard deviation** σ,

$$\sigma = \sqrt{\frac{\sum(x-\mu)^2}{n-1}} .$$

The reason for specifying $(n-1)$ and not n in the denominator is due to the fact that there are only $(n-1)$ independent pieces of information besides the mean. For great values of n, it is usually of no significance whether n or $(n-1)$ is used.

- Determine the standard deviation of the observations

$$3, 4, 5, 6, 7, 9, 10, 12 .$$

$$\mu = \frac{1}{8}(3+4+5+6+7+9+10+12) = 7 .$$

Calculate the deviation of each of the 8 observations from the mean; square the results:

$$\begin{aligned}
(3-7)^2 &= 16 \\
(4-7)^2 &= 9 \\
(5-7)^2 &= 4 \\
(6-7)^2 &= 1 \\
(7-7)^2 &= 0 \\
(9-7)^2 &= 4 \\
(10-7)^2 &= 9 \\
(12-7)^2 &= 25
\end{aligned}$$

The variance is

$$\sigma^2 = \frac{1}{8-1}(16+9+4+1+0+4+9+25) = \frac{68}{7},$$

and the standard deviation is

$$\sigma = \sqrt{\frac{68}{7}} = 3.1167\ldots.$$

Chebyshev's Theorem

If a distribution of measurements has a small standard deviation, most measurements are likely to be closely grouped around the arithmetic mean; on the other hand, a large value of the standard deviation indicates a greater variability, and therefore the measurements are likely to be more spread out from the mean.

This idea is formally expressed in **Chebyshev's theorem**, or **Chebyshev's inequality**, given here without proof:

CHEBYSHEV Pafnuty Lvovich (1821 - 1894)

If a probability distribution has the arithmetic mean μ and the standard deviation σ, then, for $k > 0$, *at least* $1 - \frac{1}{k^2}$ of the measurements of any set of observations will have a value within $k\sigma$ of the mean of the measurements.

p. 967

The **law of large numbers** is an immediate consequence of Chebyshev's theorem.

- Consider a random sample of 400 components, manufactured to a nominal diameter of 18.62 mm, with a mean of 18.64 mm and a standard deviation of 0.03 mm. To determine the interval of 300 measurements, we have

$$400\left(1 - \frac{1}{k^2}\right) = 300$$

$$k = \pm 2$$

and find

$$18.64 \pm 2\,(0.03) = 18.64 \pm 0.06 \text{ mm}.$$

From the result, we may infer that *at least* 300 of the 400 machined components have a diameter 18.64 ± 0.06 mm, and that *no more* than 100 components have a diameter below 18.58 mm or above 18.70 mm.

for μ and σ, then $k > 0$

z Score

To compare, or rank, observations from two independent sets of observations or populations, we may convert the individual observations into standard units, referred to as z scores or z values.

In a population with the arithmetic mean μ and the standard deviation σ, the z score is given by

$$z = \frac{x - \mu}{\sigma}.$$

- For jobs as city guides for foreigners, a group of linguistically inclined candidates was given tests in French, Spanish, Italian, and German. One of them, Madelaine, who felt that French was her strong point, was particularly eager to make use of her French, but would also accept a job as a guide for Italian tourists. She knew that her German was rather rusty, but she still had to take the German test.

Results:

Language	Mean	Standard deviation	Madelaine's score
French	550	40	530
Spanish	380	40	400
Italian	480	60	500
German	300	25	350

Madelaine's z scores were:

$$\text{French:}\quad z = \frac{530 - 550}{40} = -\frac{1}{2}$$

$$\text{Spanish:}\quad z = \frac{400 - 380}{40} = \frac{1}{2}$$

$$\text{Italian:}\quad z = \frac{500 - 480}{60} = \frac{1}{3}$$

$$\text{German:}\quad z = \frac{350 - 300}{25} = 2$$

We see that although Madelaine got her highest score in French, this score was $\frac{1}{2}$ of the standard deviation below the mean of the competitors' French scores. As she expected, she got her lowest score in German, but this score was nevertheless 2 standard deviations above the mean of the German scores. She was accordingly offered a job as a guide for German tourists but did not take it.

33.6 Normal Distribution

Historical Notes

de MOIVRE Abraham (1667 - 1754)
LAPLACE Pierre Simon de (1749 - 1827)
POISSON Siméon-Denis (1781 - 1840)
GAUSS Carl Friedrich (1777 - 1855)

In the early 18th century it became apparent to scientists who studied the distribution of errors in repeated measurements that observations of a great number of different measurements often tend to show a similar form of distribution, now called **normal distribution** or **Gaussian distribution**. In 1733, its mathematical equation was formulated by the mathematician and statistician **Abraham de Moivre**, born in France, educated in Belgium, and finally resident in England. The mathematical properties of the normal distribution were studied and explained by **Pierre Simon Laplace, Siméon-Denis Poisson,** and **Carl Friedrich Gauss**.

QUETELET Adolphe (1796 - 1874)

The Belgian astronomer, mathematician, and statistician **Adolphe Quetelet** had studied astronomy and probability with Laplace in Paris and was the first to apply normal distribution to the study of sociology. Quetelet presented his concept of the "average human being" (*l'homme moyen*) around whom measurements of human traits were grouped in normal probability distributions. His observations of the numerical consistency of what had been supposed to be voluntary acts of crime provoked extensive discussions about free will versus social determinism; such studies are still an important subject for research on social behavior and criminology. Noncontroversial, however, was Quetelet's development of statistical methods for collecting and analyzing large numbers of simultaneous astronomical, meteorological, and geodetic observations made at points in Europe chosen at random.

GALTON Sir Francis (1822 - 1911)

Francis Galton, English explorer and anthropologist, and an early authority on fingerprinting, developed Quetelet's observation of the distribution of certain measurable human characteristics and became a pioneer in using the normal distribution in work on heredity.

Normal Distribution Functions

The probability density function of a continuous random variable with a normal, or Gaussian, distribution is

$$f(x) = \frac{1}{\sigma\sqrt{2\pi}}\, e^{-\frac{1}{2}\left(\frac{x-\mu}{\sigma}\right)^2} \quad ; \; -\infty < x < \infty ,$$

where σ and μ denote the standard deviation and its arithmetic mean, respectively.

$f(x)$ is of bell-shaped form:

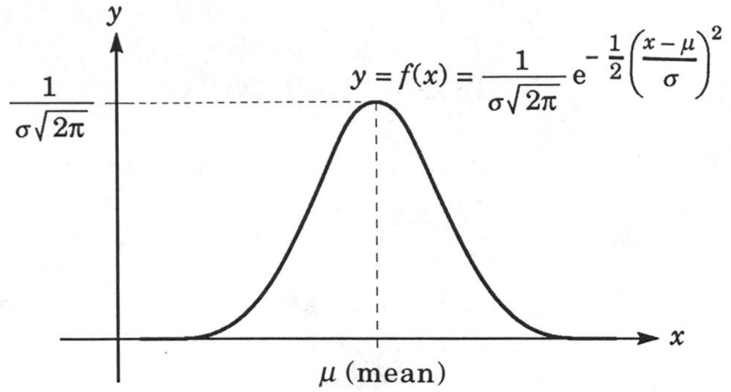

For $f(x)$ to be a probability density function, its integral over the interval $-\infty < x < \infty$ must equal 1.

$$\frac{1}{\sigma\sqrt{2\pi}} \int_{-\infty}^{\infty} e^{-\frac{1}{2}\left(\frac{x-\mu}{\sigma}\right)^2} dx = 1 .$$

Why? Let $z = \frac{x-\mu}{\sigma}$; then

$$\frac{dz}{dx} = \frac{1}{\sigma} \quad \text{and} \quad dx = \sigma\, dz .$$

Therefore, if

$$\frac{1}{\sigma\sqrt{2\pi}} \int_{x_1}^{x_2} e^{-\frac{1}{2}\left(\frac{x-\mu}{\sigma}\right)^2} dx = 1$$

we have

$$\frac{1}{\sqrt{2\pi}} \int_{-\infty}^{\infty} e^{-z^2/2}\, dz = 1$$

or

$$\int_{-\infty}^{\infty} e^{-z^2/2}\, dz = \sqrt{2\pi} .$$

p. 887 To confirm this result, use the method of **transforming double integrals from orthogonal to polar coordinates**.

Let $I = \int_{-\infty}^{\infty} e^{-z^2/2}\, dz$.

To obtain a double integral, write

$$I^2 = \left(\int_{-\infty}^{\infty} e^{-z^2/2}\, dz \right)^2 .$$

For $z = x$ and $z = y$, we have

$$I^2 = \int_{-\infty}^{\infty} e^{-x^2/2}\, dx \int_{-\infty}^{\infty} e^{-y^2/2}\, dy$$

$$= \int_{-\infty}^{\infty}\int_{-\infty}^{\infty} e^{-(x^2+y^2)/2}\, dx\ dy\ .$$

To evaluate this double integral, we transform it to polar coordinates. To achieve a better understanding of this procedure, we can think of the above expression as a volume bounded by the function $f(x, y) = e^{-(x^2+y^2)/2}$ and the infinite x, y-plane.

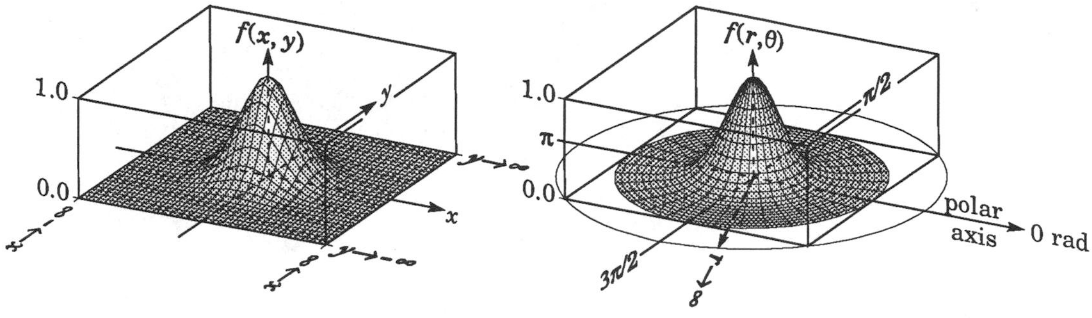

We note that the range of x and y from $-\infty$ to ∞ in rectangular coordinates is equivalent to an infinite radius sweeping one complete revolution (2π rad) about the origin in polar coordinates.

p. 583 Since $r = \sqrt{x^2 + y^2}$, we have

$$I^2 = \int_{-\infty}^{\infty}\int_{-\infty}^{\infty} e^{-(x^2+y^2)/2}\, dx\ dy$$

$$= \int_{0}^{2\pi}\int_{0}^{\infty} e^{-r^2/2}\, r\ d\theta\ dr = 2\pi \int_{0}^{\infty} e^{-r^2/2}\, r\ dr\ .$$

With $t = -r^2/2$,

$$\frac{dt}{dr} = -r, \text{ or } dr = -\frac{dt}{r}.$$

$$I^2 = 2\pi \int_{0}^{-\infty} e^{t}\, r \left(-\frac{1}{r}\right) dt = -2\pi \int_{0}^{-\infty} e^{t}\, dt = 2\pi \int_{0}^{\infty} e^{-t}\, dt$$

$$= 2\pi\ .$$

Thus,

$$I = \int_{-\infty}^{\infty} e^{-z^2/2}\, dz = \sqrt{2\pi}$$

and

$$\frac{1}{\sqrt{2\pi}} \int_{-\infty}^{\infty} e^{-z^2/2}\, dz = 1.$$

Using the substitution $z = \dfrac{x-\mu}{\sigma}$ to reinstate the variable x, we find

$$\frac{1}{\sqrt{2\pi}} \int_{-\infty}^{\infty} e^{-\frac{1}{2}\left(\frac{x-\mu}{\sigma}\right)^2} \left(\frac{1}{\sigma}\right) dx = 1$$

and

$$\frac{1}{\sigma\sqrt{2\pi}} \int_{-\infty}^{\infty} e^{-\frac{1}{2}\left(\frac{x-\mu}{\sigma}\right)^2} dx = 1.$$

Use of Numerical Integration and Standard Normal Distribution

Exact numerical values may be determined of

$$f(x) = \frac{1}{\sigma\sqrt{2\pi}}\, e^{-\frac{1}{2}\left(\frac{x-\mu}{\sigma}\right)^2}$$

for the intervals $]-\infty, \infty[$ and $[0, \infty[$, but rigorous mathematical integration for intervals with a finite number of intervals is not possible.

One way of handling such cases is **numerical integration**; another way is to use the substitution $z = \dfrac{x-\mu}{\sigma}$ and transform $f(x)$ into a **standard normal distribution**

$$\phi(z) = \frac{1}{\sqrt{2\pi}}\, e^{-z^2/2}$$

whose graph is symmetrical about the arithmetic mean, located at $z = 0$.

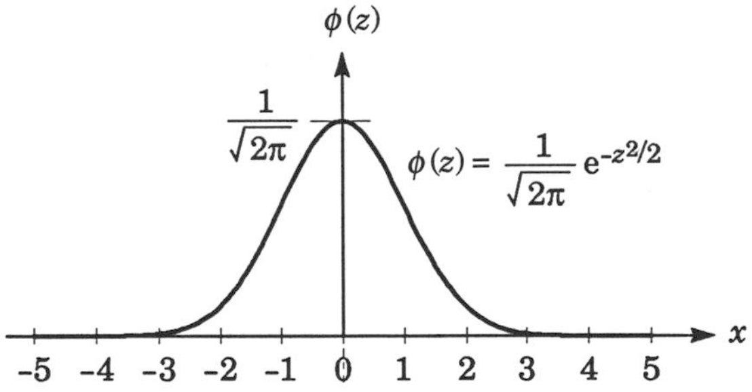

Tables, such as the one on *p.* 985, are available for determining area under the standard normal distribution curve. The use of such tables dispenses with the need for laborious numerical integration.

- Given a normal distribution with arithmetic mean 9.4 and standard deviation 3, what is the probability of a random variable assuming a value between 7 and 12.4?

We have

$$z_1 = \frac{x_1-\mu}{\sigma} = \frac{7-9.4}{3} = -0.8 \qquad\qquad z_2 = \frac{x_2-\mu}{\sigma} = \frac{12.4-9.4}{3} = 1.$$

Thus, if the probability is p,

$$p\,(7 < x < 12.4) = p\,(-0.8 < z < 1)\,.$$

Alternative I

Using the table for standard normal distribution, we find

$$\phi(1) = 0.8413$$

and

$$\phi(-0.8) = 1 - 0.7881 = 0.2119 .$$

Thus

$$p(-0.8 < z < 1) = 0.8413 - 0.2119 = 0.629 .$$

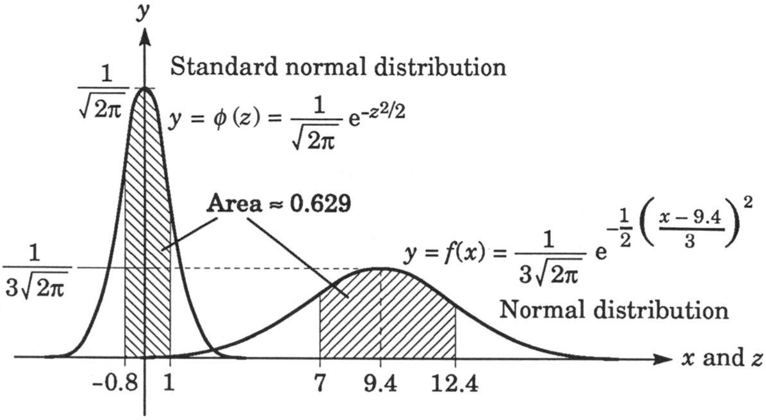

Transformation to standard normal distribution

Original distribution

Alternative II

$$p(-0.8 < z < 1) = \frac{1}{\sqrt{2\pi}} \int_{-0.8}^{1} e^{-z^2/2}\, dz$$

Using a computer programmed for numerical integration, by the **trapezoidal rule**, we find p. 949

Integral	Number of Intervals	Approximate Value
$\frac{1}{\sqrt{2\pi}} \int_{-0.8}^{1} e^{-z^2/2}\, dz$	1	0.478 496 05
	2	0.596 505 32
	4	0.621 435 55
	8	0.627 487 14
	16	0.628 989 49
	32	0.629 364 43
	64	0.629 458 12
	128	0.629 481 54
	256	0.629 487 40

Standard Normal Distribution

$$\phi(z_i) = \frac{1}{\sqrt{2\pi}} \int_{-\infty}^{z} e^{-z^2/2}\, dz$$

z_i	0.00	0.01	0.02	0.03	0.04	0.05	0.06	0.07	0.08	0.09
0.0	0.5000	0.5040	0.5080	0.5120	0.5160	0.5199	0.5239	0.5279	0.5319	0.5359
0.1	0.5398	0.5438	0.5478	0.5517	0.5557	0.5596	0.5636	0.5675	0.5714	0.5753
0.2	0.5793	0.5832	0.5871	0.5910	0.5948	0.5987	0.6026	0.6064	0.6103	0.6141
0.3	0.6179	0.6217	0.6255	0.6293	0.6331	0.6368	0.6406	0.6443	0.6480	0.6517
0.4	0.6554	0.6591	0.6628	0.6664	0.6700	0.6736	0.6772	0.6808	0.6844	0.6879
0.5	0.6915	0.6950	0.6985	0.7019	0.7054	0.7088	0.7123	0.7157	0.7190	0.7224
0.6	0.7257	0.7291	0.7324	0.7357	0.7389	0.7422	0.7454	0.7486	0.7517	0.7549
0.7	0.7580	0.7611	0.7642	0.7673	0.7704	0.7734	0.7764	0.7794	0.7823	0.7852
0.8	0.7881	0.7910	0.7939	0.7967	0.7995	0.8023	0.8051	0.8078	0.8106	0.8133
0.9	0.8159	0.8186	0.8212	0.8238	0.8264	0.8289	0.8315	0.8340	0.8365	0.8389
1.0	0.8413	0.8438	0.8461	0.8485	0.8508	0.8531	0.8554	0.8577	0.8599	0.8621
1.1	0.8643	0.8665	0.8686	0.8708	0.8729	0.8749	0.8770	0.8790	0.8810	0.8830
1.2	0.8849	0.8869	0.8888	0.8907	0.8925	0.8944	0.8962	0.8980	0.8997	0.9015
1.3	0.9032	0.9049	0.9066	0.9082	0.9099	0.9115	0.9131	0.9147	0.9162	0.9177
1.4	0.9192	0.9207	0.9222	0.9236	0.9251	0.9265	0.9279	0.9292	0.9306	0.9319
1.5	0.9332	0.9345	0.9357	0.9370	0.9382	0.9394	0.9406	0.9418	0.9429	0.9441
1.6	0.9452	0.9463	0.9474	0.9484	0.9495	0.9505	0.9515	0.9525	0.9535	0.9545
1.7	0.9554	0.9564	0.9573	0.9582	0.9591	0.9599	0.9608	0.9616	0.9625	0.9633
1.8	0.9641	0.9649	0.9656	0.9664	0.9671	0.9678	0.9686	0.9693	0.9699	0.9706
1.9	0.9713	0.9719	0.9726	0.9732	0.9738	0.9744	0.9750	0.9756	0.9761	0.9767
2.0	0.9772	0.9778	0.9783	0.0788	0.9793	0.9798	0.9803	0.9808	0.9812	0.9817
2.1	0.9821	0.9826	0.9830	0.9834	0.9838	0.9842	0.9846	0.9850	0.9854	0.9857
2.2	0.9861	0.9864	0.9868	0.9871	0.9875	0.9878	0.9881	0.9884	0.9887	0.9890
2.3	0.9893	0.9896	0.9898	0.9901	0.9904	0.9906	0.9909	0.9911	0.9913	0.9916
2.4	0.9918	0.9920	0.9922	0.9925	0.9927	0.9929	0.9931	0.9932	0.9934	0.9936
2.5	0.9938	0.9940	0.9941	0.9943	0.9945	0.9946	0.9948	0.9949	0.9951	0.9952
2.6	0.9953	0.9955	0.9956	0.9957	0.9959	0.9960	0.9961	0.9962	0.9963	0.9964
2.7	0.9965	0.9966	0.9967	0.9968	0.9969	0.9970	0.9971	0.9972	0.9973	0.9974
2.8	0.9974	0.9975	0.9976	0.9977	0.9977	0.9978	0.9979	0.9979	0.9980	0.9981
2.9	0.9981	0.9982	0.9983	0.9983	0.9984	0.9984	0.9985	0.9985	0.9986	0.9986
3.0	0.9987	0.9987	0.9987	0.9988	0.9988	0.9989	0.9989	0.9989	0.9990	0.9990
3.1	0.9990	0.9991	0.9991	0.9991	0.9992	0.9992	0.9992	0.9992	0.9993	0.9993
3.2	0.9993	0.9993	0.9994	0.9994	0.9994	0.9994	0.9994	0.9995	0.9995	0.9995
3.3	0.9995	0.9995	0.9995	0.9996	0.9996	0.9996	0.9996	0.9996	0.9996	0.9997
3.4	0.9997	0.9997	0.9997	0.9997	0.9997	0.9997	0.9997	0.9997	0.9997	0.9998
3.5	0.99977									
3.6	0.99984									
3.7	0.99989									
3.8	0.99993									
3.9	0.99995									
4.0	0.99997									

Since the function is symmetric around 0, only positive values of z need be given in the table; then one can use $\phi(-z) = 1 - \phi(z)$.

For instance, for $z = -1.4$ we find $\phi(-1.4) = 1 - \phi(1.4) = 1 - 0.9192 = 0.0808$.

Alternative III

To integrate with respect to x between $x_1 = 7$ and $x_2 = 12.4$, we may also use the normal density function

$$f(x) = \frac{1}{\sigma\sqrt{2\,\pi}}\; e^{-\frac{1}{2}\left(\frac{x-\mu}{\sigma}\right)^2} .$$

p. 949 Using the trapezoidal rule, we find

Integral	Number of Intervals	Approximate Value
$\frac{1}{3\sqrt{2\,\pi}} \int_{7}^{12.4} e^{-\frac{1}{2}\left(\frac{x-9.4}{3}\right)^2} dx$	1	0.478 496 05
	2	0.596 505 32
	4	0.621 435 55
	8	0.627 487 14
	16	0.628 989 49
	32	0.629 364 43
	64	0.629 458 12
	128	0.629 481 54
	256	0.629 487 40

The sought probability is 0.629 .

- A type of battery has an average useful service life of 3300 hours, with a standard deviation of 500 hours.

  What is the probability that a given battery will last less than 2500 hours?

$$z = \frac{2500 - 3300}{500} = -1.6 .$$

The sought probability p is

$$p(x < 2500) = p(z < -1.6) = \frac{1}{\sqrt{2\,\pi}} \int_{-\infty}^{-1.6} e^{-z^2/2}\, dz,$$

which we solve by using the table on p. 985,

$$\Phi(-1.6) = 1 - 0.9452 = 0.0548 .$$

The sought probability $p = 0.0548$ corresponds to the shaded area in the graph.

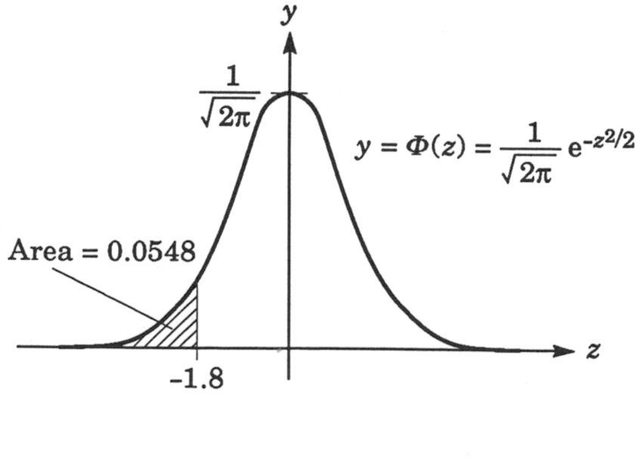

- A batch of 4000 screws has a nominal length of 30.00 mm with a normal distribution of the lengths. A sample of 300 screws was found to have a mean length of 29.99 mm with a standard deviation of 0.02 mm.

Find: **a:** How many screws are likely to be 30.00 mm or more in length?

b: Between what lengths are 99.994% of the screws likely to be?

a:

The probability of a length of at least 30.00 mm is defined by

$$z = \frac{30.00 - 29.99}{0.02} = 0.50 .$$

The sought probability equals the definite integral

$$\frac{1}{\sqrt{2\pi}} \int_{0.50}^{\infty} e^{-z^2/2}\, dz,$$

which may be solved with numerical integration or using the table on *p.* 985. Using the latter method and denoting the sought probability p,

$$\begin{aligned} p\,(x \geq 30.00) &= p\,(z \geq 0.50) \\ &= 1 - p\,(z < 0.50) \\ &= 1 - 0.6915 = 0.3085 . \end{aligned}$$

This is equivalent to

$$\begin{aligned} p\,(x \geq 30.00) &= p\,(z \geq 0.50) \\ &= 1 - \frac{1}{\sqrt{2\pi}} \int_{-\infty}^{0.50} e^{-z^2/2}\, dz, \end{aligned}$$

that is, the total area under the curve less the shaded area in the graph below.

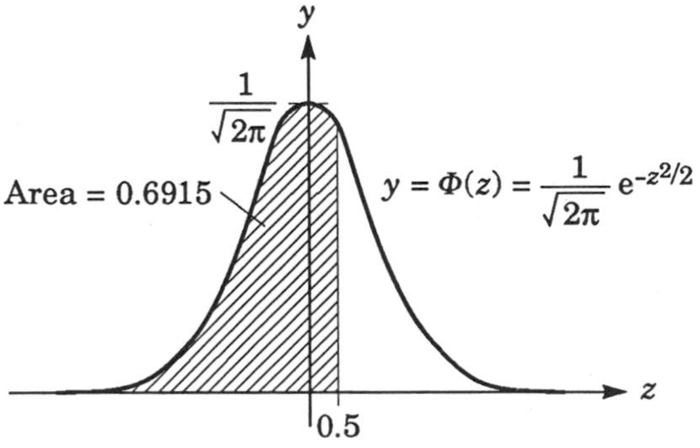

Hence,

$$0.3085 \cdot 4000 = 1234,$$

suggesting that about 1200 screws are likely to have a length $\geq$ 30.00 mm.

b:

The table on *p*. 985 shows that the area under the standard normal distribution curve, between $z = -4$ and $z = 4$, is

$$0.99997 - (1 - 0.99997) = 0.99994,$$

equivalent to

$$\frac{1}{\sqrt{2\pi}} \int_{-4}^{4} e^{-z^2/2}\, dz \approx 0.99994 .$$

Therefore, if the sought boundaries are x_1 and x_2, then

$$-4 = \frac{x_1 - 29.99}{0.02}$$

$$x_1 = 29.91 \text{ mm}$$

and

$$4 = \frac{x_2 - 29.99}{0.02}$$

$$x_2 = 30.07 \text{ mm.}$$

The result may be verified by

$$\frac{1}{0.02\sqrt{2\pi}} \int_{29.91}^{30.07} e^{-\frac{1}{2}\left(\frac{x - 29.99}{0.02}\right)^2} dx \approx 0.99994 .$$

Hence, 99.994% of the screws will have a length between 29.91 mm and 30.07 mm.

* * *

Taking No Chance

Ten yellow and ten blue socks are in a drawer in a dark bedroom. How many socks should one pick to be certain to get:

A. One pair of matching socks?

B. One pair of a specific color?

Answer: **A**. Three socks. **B**. Twelve socks.

Not Even Half a Chance!

Betty is at the fair with 32 dollars in cash. She bets on the toss of a coin, six times. Betty wagers half her cash every time; her opponent matches her bets. Betty wins half the tosses. How much cash did she have in the end?

Answer: $13.50, irrespective of the order of the losses and winnings!

Chapter 34

DIFFERENTIAL EQUATIONS

34.1 Fundamental Concepts

A differential equation contains one or more terms involving derivatives of one variable (the dependent variable, y) with respect to another variable (the independent variable, x), such as

$$\frac{dy}{dx} = 2x.$$

Unlike algebraic equations, the solutions of differential equations are functions and not just numbers. The solution of the above equation is the integral

$$y = \int 2x \, dx = x^2 + C,$$

where C is an arbitrary constant.

In physics, chemistry, biology, and other areas of natural science, as well as areas outside natural science such as engineering and economics, we often encounter the task of solving the **relationships between rates of change of continuously varying quantities**, which is exactly what the terms of a differential equation represent; thus, differential equations are essential to all scientific investigation.

Ordinary and Partial Differential Equations

A differential equation that involves a function of a **single variable** and some of its derivatives is an **ordinary differential equation**. Differentiating

$$f(x) = y = x^3 + 5x^2 + 3x + 2$$

gives the ordinary differential equation

$$\frac{dy}{dx} = 3x^2 + 10x + 3.$$

If the differential equation involves functions of **two or more variables** and some of their partial derivatives, the equation is a **partial differential equation**. If we differentiate

$$f(x, y) = z = x^3 + 3x^2y + 3x - 2xy^5$$

with respect to x and to y, we get the two partial differential equations

$$\frac{\partial z}{\partial x} = 3x^2 + 6xy + 3 - 2y^5 \quad \text{and} \quad \frac{\partial z}{\partial y} = 3x^2 - 10xy^4.$$

Order and Degree

The **order of a differential equation** is the order of the highest derivative that appears in the equation; consequently,

$$\frac{dy}{dx} = 3x^2 + 10x + 3; \quad \frac{\partial z}{\partial x} = 3x^2 + 6xy + 3 - 2y^5; \quad \frac{\partial z}{\partial y} = 3x^2 - 10xy^4$$

are all **first-order** differential equations.

The equation

$$\frac{d^2y}{dx^2} = \frac{d}{dx}\,[3x^2 + 10x + 3] = 6x + 10$$

is a **second-order** differential equation; the equations

$$y''' = 6 \quad \text{and} \quad y^{(4)} = 0$$

are **third- and fourth-order** differential equations, respectively.

The equations

$$\frac{\partial^2 z}{\partial x^2} = 6x + 6y\,; \qquad \frac{\partial^2 z}{\partial x\,\partial y} = 6x - 10y^4\,; \qquad \frac{\partial^2 z}{\partial y^2} = -40\,x\,y^3$$

are all second-order differential equations.

Degree

The **degree** of a differential equation whose terms are polynomials in the derivatives is defined as the highest power of the highest-order derivative; consequently, the differential equations

$$\frac{dy}{dx} + 5xy = x^2; \qquad x\,\frac{d^2y}{dx^2} + \frac{dy}{dx} = 3; \qquad x\,\frac{d^2y}{dx^2} + \left(\frac{dy}{dx}\right)^3 = 15$$

are first-degree differential equations, and

$$\left(\frac{dy}{dx}\right)^2 + 5xy = 2x^2; \quad x\left(\frac{d^2y}{dx^2}\right)^2 + \frac{dy}{dx} = 7; \quad x\left(\frac{d^2y}{dx^2}\right)^2 + \left(\frac{dy}{dx}\right)^3 = 32$$

are second-degree differential equations.

General and Particular Solutions

A solution

$$y = \int 2x\,dx = x^2 + C$$

is the **general solution** of the differential equation $\frac{dy}{dx} = 2x$ and represents an infinite number of solutions, where the arbitrary constant C may be any real or complex number.

Family of Curves

Plotting the solution for different values of C, we obtain a **family of curves**, as shown in the graph below for some integer values of C.

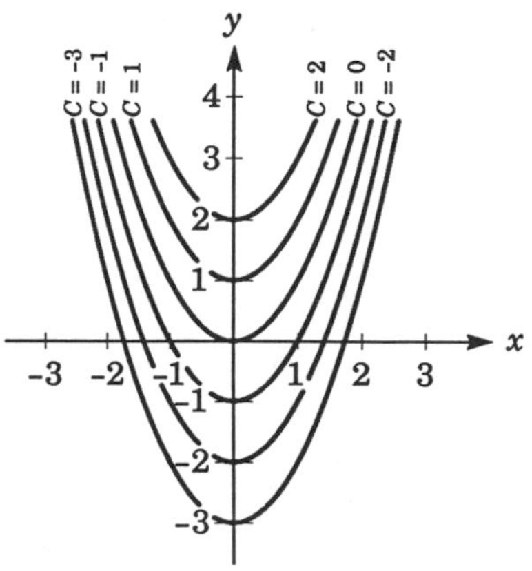

Particular Solution

The correspondence of a particular value of the independent variable x with a certain numerical value of the dependent variable y will give the constant C a specific value, and this solution is referred to as a **particular solution**. If $x = 1$ corresponds to $y = 3$, we insert these values in the general solution $y = x^2 + C$ and obtain $C = 2$.

Thus,

$$y = x^3 + 5x^2 + 3x + 2$$

is a **particular solution** of all the differential equations

$$y' = 3x^2 + 10x + 3; \quad y'' = 6x + 10; \quad y''' = 6,$$

whose **general solutions** are, respectively,

$$y = \int (3x^2 + 10x + 3)\ dx = x^3 + 5x^2 + 3x + C_1;$$

$$y = \int\int (6x + 10)\ dx^2$$

$$= \int (3x^2 + 10x + C_1)\ dx = x^3 + 5x^2 + C_1 x + C_2;$$

$$y = \int\int\int 6\ dx^3 = \int\int (6x + C_1)\ dx^2$$

$$= \int (3x^2 + C_1 x + C_2)\ dx = x^3 + \frac{1}{2} C_1 x^2 + C_2 x + C_3.$$

Verifying Solutions

Ordinarily, when we refer to the solution of a differential equation, we mean its general solution.

The solution of a differential equation contains no derivatives; together with its derivatives, the solution must satisfy the original differential equation.

To verify a solution of a differential equation, we simply substitute the solution and its derivatives into the given differential equation; a correct solution will result in an identity.

- Verify that $y = (C_1 x^2 + C_2 x)$, where C_1 and C_2 are arbitrary constants, is a solution of

$$x^2 y'' - 2xy' + 2y = 0.$$

$$y = C_1 x^2 + C_2 x$$

$$y' = 2C_1 x + C_2$$

$$y'' = 2C_1$$

$2y$	$=$	$2C_1 \cdot x^2$	$+ 2C_2 \cdot x$
$-2xy'$	$=$	$-4C_1 \cdot x^2$	$- 2C_2 \cdot x$
$x^2 y''$	$=$	$2C_1 \cdot x^2$	
Σ	$=$	$0 \cdot x^2$	$+ 0 \cdot x$

34.2 First-Order Ordinary Differential Equations

To solve a differential equation, some method must be found of rearranging or transforming the equation so that we can *integrate its terms*.

Directly Integrable Equations

A differential equation of the type

$$\frac{dy}{dx} = f(x)$$

may be solved by integrating both sides directly.

- Solve the equations

$$\frac{dy}{dx} = 2x+5; \qquad \frac{dy}{dx} = 6e^{3x} + 2/x .$$

Direct integration gives

$$y = \int (2x+5)\,dx = x^2 + 5x + C_1 \qquad\Big|\qquad y = \int \left(6e^{3x} + \frac{2}{x}\right) dx = 2e^{3x} + 2\ln x + C_2$$

where C_1 and C_2 are arbitrary constants.

Separation of Variables

A differential equation

$$f(y)\,\frac{dy}{dx} = g(x),$$

where f is a function of y only and g is a function of x only, may be solved by **separation of the variables**.

We recall that a derivative $\frac{dy}{dx}$ can always be written in differential form; thus,

$$f(y)\,dy = g(x)\,dx.$$

If f and g are continuous functions, then

$$\int f(y)\,dy = \int g(x)\,dx .$$

- Solve the equation $\frac{dy}{dx} = x y^2$.

Writing in differential form, $dy = x y^2\,dx$, and separating the variables,

$$\int \frac{1}{y^2}\,dy = \int x\,dx; \quad -\frac{1}{y} = \frac{x^2}{2} + C.$$

- Find the general solution of $\frac{dy}{dx} = y \sin x$, and the particular solution if $y = 1$ for $x = 0$.

$$\int \frac{1}{y}\, dy = \int \sin x \;\, dx$$

$$\ln y = -\cos x + C; \quad \ln 1 = -\cos 0 + C; \quad C = 1,$$

$$\ln y = -\cos x + 1.$$

- Solve the equation $\frac{dy}{dx} = 4x(y-2)$.

$$\int \frac{1}{y-2}\, dy = \int 4x \;\, dx; \quad \ln(y-2) = 2x^2 + C_1$$

$$y - 2 = e^{C_1} \cdot e^{2x^2},$$

and with $e^{C_1} = C$, $y = 2 + C e^{2x^2}$.

Substitution of Variables

First-order differential equations are not always separable. Homogeneous differential equations, however, may be transformed into separable equations by the substitution of a variable.

Homogeneous Differential Equations

The degree of a term is the sum of the exponents of the variables in the term. An expression is said to be **homogeneous** if all terms have the same degree;

$$x^2 y + 3xy^2 - 4y^3$$

is a homogeneous equation in x and y, all terms being of the third degree.

Similarly, $f(x, y) = \sqrt{x^2 + xy}$ is a homogeneous function of the first degree, but $f(x, y) = \sqrt{x^2 y + xy}$ is not.

Thus,

$$2x \;\, dx + 3\sqrt{x^2 + y^2}\; dy = 0$$

is a **homogeneous differential equation**, since $2x$ and $3\sqrt{x^2 + y^2}$ are both functions of the first degree.

The term *homogeneous* is also used to indicate that the right-hand member of a linear differential equation is 0.

In the equation

$$\frac{dy}{dx} = \frac{x + 2y}{3x},$$

the variables are not separable in the manner described above, but they can be made so by introducing the substitution

$$y = vx,$$

where v is a function of x.

Differentiating $y = v\,x$ with respect to x, we obtain

$$\frac{dy}{dx} = v + x\frac{dv}{dx} .$$

We can now separate the variables; thus,

$$v + x\frac{dv}{dx} = \frac{x + 2\,v\,x}{3\,x}\ ;\ x\frac{dv}{dx} = \frac{1+2\,v}{3} - v\ ;\ x\ dv = \frac{1-v}{3}\ dx ,$$

and

$$\int \frac{1}{1-v}\ dv = \int \frac{1}{3\,x}\ dx.$$

$$-\ln(1-v) = \frac{1}{3}\ln x + C_1\ ;\quad \ln(1-v)^{-1} = \ln\sqrt[3]{x} + \ln C_2 ,$$

$$\frac{1}{1-v} = C_2\sqrt[3]{x}\ ;\quad y = x - C\,x^{2/3} .$$

- Solve the homogeneous equation

$$2\,x\,y\frac{dy}{dx} = x^2 + 2\,y^2\ ;\ \ y(1) = 0 .$$

Rewrite as

$$\frac{dy}{dx} = \frac{x^2 + 2\,y^2}{2\,x\,y} .$$

Substitute $y = v\,x$;

$$v + x\frac{dv}{dx} = \frac{x^2 + 2\,(v\,x)^2}{2\,x\,v\,x}\ ;\quad x\frac{dv}{dx} = \frac{1}{2\,v}\ ;\quad 2\,v\ dv = \frac{dx}{x} .$$

$$\int 2\,v\ dv = \int \frac{1}{x}\ dx\ ;\ \ v^2 = \ln x + C_1 .$$

Inserting $v = \frac{y}{x}$ gives the general solution

$$\frac{y^2}{x^2} = \ln x + C;\ \ y^2 = x^2\ln x + C\,x^2 .$$

$y(1) = 0$ gives $C = 0$.

Hence, the particular solution is

$$y^2 = x^2\ln x .$$

Integrating Factors

A linear differential equation of order n is one that can be written

$$a_n(x)\frac{d^n y}{dx^n} + a_{n-1}(x)\frac{d^{n-1}y}{dx^{n-1}} + \ldots + a_1(x)\frac{dy}{dx} + a_0(x)\,y = f(x),$$

where $a_0(x)$, $a_1(x)$, ..., $a_{n-1}(x)$, $a_n(x)$, and $f(x)$ are given functions of x, or constants.

The linear differential equation of the first order,

$$\frac{dy}{dx} + P(x)y = Q(x) ,$$

arises in many applications of differential equations. Its solution is simple and straightforward but differs from the methods described above.

Euler's Multiplier

To solve this type of first-order linear differential equation, we use an **integrating factor** or **Euler's multiplier**.

With the integrating factor $e^{\int P(x)\,dx}$, we have

$$\frac{dy}{dx}\left(e^{\int P(x)\,dx}\right) + P(x)\left(e^{\int P(x)\,dx}\right)y = Q(x)\left(e^{\int P(x)\,dx}\right),$$

where the left-hand member is the derivative of $y\,e^{\int P(x)\,dx}$; therefore,

$$\frac{d}{dx}\left[y\,e^{\int P(x)\,dx}\right] = Q(x)\left(e^{\int P(x)\,dx}\right)$$

and

$$y\,e^{\int P(x)\,dx} = \int Q(x)\,e^{\int P(x)\,dx}\,dx .$$

Thus, the **general solution** of the first-order linear differential equation

$$\frac{dy}{dx} + P(x)y = Q(x)$$

becomes

$$y = \frac{1}{e^{\int P(x)\,dx}}\int Q(x)\,e^{\int P(x)\,dx}\,dx .$$

- Solve the linear differential equation

$$\frac{dy}{dx} + 2y = 3 .$$

As an integrating factor, use $e^{\int 2\,dx} = e^{2x}$,

$$y = \frac{1}{e^{2x}}\int 3\,e^{2x}\,dx = e^{-2x}\left(\frac{3}{2}\,e^{2x} + C\right) = \frac{3}{2} + Ce^{-2x}. \quad \bullet$$

A given differential equation must sometimes be rearranged to the standard form.

- Solve the equation $x\,\frac{dy}{dx} - y = x^3$.

Rearranging the equation, we have

$$\frac{dy}{dx} - \frac{1}{x}\,y = x^2 .$$

Using

$$e^{\int(-x^{-1})\,dx} = e^{-\int x^{-1}\,dx} = e^{-\ln x} = x^{-1}$$

as an integrating factor, we obtain

$$y = \frac{1}{x^{-1}}\int x^2\,x^{-1}\,dx = x\left(\frac{x^2}{2} + C\right) = \frac{x^3}{2} + Cx .$$

- Find the particular solution to $x^2 y' + 2xy = \cos x$ if $y\left(\frac{\pi}{2}\right) = 0$.

Rewrite the equation in standard form,

$$y' + \frac{2}{x} y = \frac{\cos x}{x^2},$$

and introduce the integrating factor

$$e^{\int P(x)\,dx} = e^{\int (2/x)\,dx} = e^{2 \ln x} = x^2$$

and insert, with $Q(x) = \frac{\cos x}{x^2}$, in the general solution:

$$y = \frac{1}{x^2} \int \frac{\cos x}{x^2} \cdot x^2\,dx = \frac{1}{x^2} \int \cos x\,dx = \frac{1}{x^2} (\sin x + C).$$

$$y\left(\frac{\pi}{2}\right) = 0 \text{ gives } C = -1. \qquad y = \frac{1}{x^2} (\sin x - 1).$$

The Bernoulli Equation

$$\frac{dy}{dx} + P(x)y = Q(x)\,y^n$$

BERNOULLI Jakob (1654 - 1705)

BERNOULLI Johann (1667 - 1748)

– named after the brothers **Jakob** and **Johann Bernoulli**, second generation of the celebrated but rivalry-filled family (fathers against sons, brothers against brothers) of Swiss mathematicians – differs from the equation $\frac{dy}{dx} + P(x)y = Q(x)$ only by the factor y^n of the right member.

Convert the Bernoulli equation to linear form by dividing each term by y^n:

$$y^{-n} \frac{dy}{dx} + P(x)\,y^{1-n} = Q(x).$$

Introducing the new function $z(x) = y^{1-n}$, we have

$$\frac{dz}{dx} = (1-n)\,y^{-n} \frac{dy}{dx}, \quad \text{or} \quad y^{-n} \frac{dy}{dx} = \frac{1}{1-n} \frac{dz}{dx},$$

and thus

$$\frac{1}{1-n} \frac{dz}{dx} + P(x)z = Q(x).$$

Multiplying each term by $(1 - n)$, the original equation becomes

$$\frac{dz}{dx} + (1-n)\,P(x)z = (1-n)\,Q(x),$$

giving the **general solution of a Bernoulli equation,**

$$y^{1-n} = \frac{1}{e^{\int (1-n)\,P(x)\,dx}} \int (1-n)\,Q(x)\,e^{\int (1-n)\,P(x)\,dx}\,dx.$$

- Solve the Bernoulli equation

$$y' + \frac{1}{x}\, y = x\, y^2 .$$

We introduce

$$P = \frac{1}{x}\,;\quad Q = x\,;\quad n = 2\,;\quad (1-n) = -\,1$$

and have

$$\int (1-n)\cdot P \;\mathrm{d}x = -\int \frac{1}{x}\,\mathrm{d}x = -\ln x\,;\qquad \frac{1}{\mathrm{e}^{-\ln x}} = x$$

$$\int (-\,1)\cdot Q\cdot \frac{1}{x}\,\mathrm{d}x = -\int \mathrm{d}x = -x + C$$

$$y^{-1} = x\,(C-x) = C\,x - x^2$$

$$y = \frac{1}{C\,x - x^2}\,.$$

* * *

Great Bernoulli family Swiss
To math never gave a miss.
And 'twas always a fight
As to who had the right
To publish which, what, and this.

34.3 Formulating Differential Equations

Phenomena and processes in science and technology may sometimes develop by discrete steps but do so more often in a continuous manner, and then require differential equations for their description and mathematical analysis. The formulation of these differential equations requires physical observations to be put into mathematical form, which is generally not a very formidable undertaking.

These differential equations are of varied kinds and so are the ways of solving them. Some equations, however, will resist algebraic solution by purely mathematical procedures, but may always be solved approximately by numerical or graphical methods.

Population Growth

In biology, population is defined as the number of individuals or organisms living in a certain area or, especially in botany, the total mass of the species being considered.

A culture of bacteria of a given species "multiplies by division" at a known rate of growth that is proportional to the number P of bacteria in the culture, which is a function of time,

$$\frac{dP}{dt} = q \cdot P \,; \quad \int \frac{dP}{P} = \int q \; dt \,; \quad \ln P = C + q\,t \,;$$

$$P = P_0 \cdot e^{q\,t} \quad (\text{where } P_0 = e^C = \text{initial population}),$$

which is the **general solution** of the equation of growth.

The coefficient q can be determined experimentally by photographing the culture through a microscope and counting the number of bacteria that divide in a given space of time.

• At the beginning of a study, a culture contained 1000 bacteria of a species which were found to split at an average rate of 1.2 times an hour. Assuming unchanged conditions, determine the number of bacteria in the culture after 3 hours and after 24 hours.

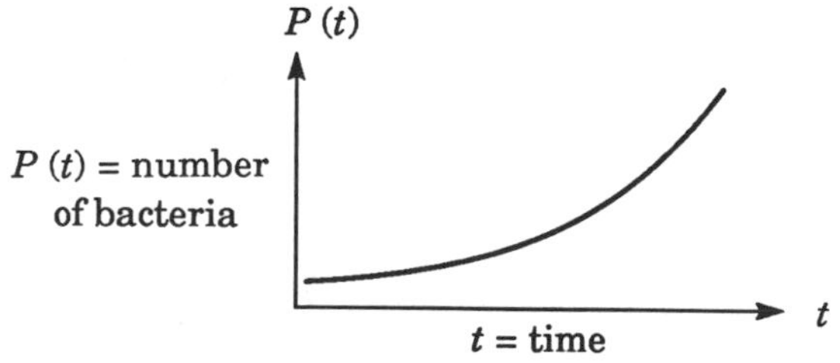

We have

$$P(3) = 1000 \cdot e^{1.2 \cdot 3} \approx 36\,600$$

$$P(24) = 1000 \cdot e^{1.2 \cdot 24} \approx 3.2 \cdot 10^{15}\,. \quad •$$

There are several limiting growth-restricting factors – more difficult to assess and to account for in a strictly mathematical way – such as natural death and an accelerated demise of bacteria through increased population density.

Human population increase and forest growth are other examples of the laws of exponential growth, which apply equally for its inverse – exponential decay – as exemplified by the dating of fossils by radioactive decay, deforestation by acid rain, and the near-extinction of whole species of birds and small mammals by the importune use of pesticides.

- From 1890 onward, an area has experienced a steady average annual population growth of 0.3%, and in 1990 numbered 500 780 inhabitants.
  - What was the population in 1890?
  - Assuming a continued growth at the same rate, what will the population be in 2090?

With 1990 as zero year, the population in the year t is $P(t)$, increasing at the annual rate of $q = 0.3\%$,

$$\frac{dP}{dt} = q \cdot P; \quad P = P_0 \cdot e^{0.003\,t},$$

where P_0 is the population in 1990, that is, $P_0 = 500\,780$.

$$P(1890) = 500\,780 \cdot e^{-0.003 \cdot 100} \approx 371\,000$$

$$P(2090) = 500\,780 \cdot e^{0.003 \cdot 100} \approx 676\,000.$$

Human Population Explosion

In 1600 the Earth's human population was about $0.5 \cdot 10^9$; in the second half of the 19th century it passed $1 \cdot 10^9$; in 1930 it reached about $2 \cdot 10^9$, a figure that in 1995 was nearly threefold, or $5.7 \cdot 10^9$.

In the above problems, we assumed unchanged conditions. The reality of a continued human population growth is that the Earth ultimately will reach a stage of insufficient resources to support even more people, which would put a halt to the population growth; persistent high birth rates would result in generally shortened lives. The alternative would be humankind's willingness to decrease reproduction to no more children than to replace the parents; fewer births would result in increased longevity.

Iteration and Chaos: *p.* 348

Predictions of population growths may suffer from *chaotic changes* that result from *sensitive dependence on initial conditions* used in the mathematical formula.

Radioactive Decay

Radioactive isotopes decay exponentially at a rate characterized by their half-life, that is, the time required for the intensity of the radioactive emission to diminish by half.

Radioactive decay is described mathematically by the same differential equation as exponential growth, with the difference that the proportionality factor is negative instead of positive.

- Measurements of the radiation from a given isotope have established that it diminishes by 2% in a one-year period; calculate the half-life of the isotope.

If p is the decay constant, and $R(t)$ the radiation at time t, we have

$$\frac{dR}{dt} = pR,$$

whose general solution is

$$R = Ce^{pt},$$

where C is the arbitrary constant.

To find C, use the information that at $t = 0$ the amount of the radioactive element is $R(0)$,

$$R(0) = Ce^0; \quad C = R(0).$$

To find p, note that when $t = 1$ year, there is 98% left of the original amount $R(0)$,

$$0.98R(0) = R(0)e^p$$

$$\ln e^p = \ln 0.98$$

$$p \ln e = \ln 0.98; \quad p = \ln 0.98.$$

With the **half-life** x years, we find

$$0.5\,R(0) = R(0)\,e^{px}$$

and, inserting the value of p,

$$0.5\,R(0) = R(0)\,e^{x \ln 0.98}$$

$$\ln 0.5 = x \ln 0.98 \ln e; \quad x \approx 34.3 \text{ years.}$$ •

Radiocarbon Dating

The radioactive carbon isotope ^{14}C is continually formed in the Earth's atmosphere. Present in carbon dioxide, ^{14}C is absorbed from the air by plants and passed on to animals through the food chain, resulting in the same proportion of ^{14}C in the carbon contents of *living* organisms as in the carbon reservoir of the atmosphere. When an organism dies, it ceases to absorb ^{14}C, whose proportion of the total amount of carbon then steadily decreases. The half-life of ^{14}C is 5730 ± 40 years.

Developed by the American physicist **Willard F. Libby** about 1946, the **radiocarbon method** is widely used to date 500- to 50 000-year-old fossils and archaeological finds. By measuring the amount of residual ^{14}C of a dead organism and comparing the amount with the content of a living organism, the death date can be estimated, or the age determined of plant and animal products (papyrus, paper, cloth, charcoal, hides, *etc.*).

- Estimate the age of a quiver which has 77.7% of its original ^{14}C, whose half-life is 5730 ± 40 years.

If p is the decay constant, and $R(t)$ the radiation at time t, we have

$$\frac{dR}{dt} = pR; \quad R = Ce^{pt},$$

where C is the arbitrary constant.

$R(0) = C e^{p \cdot 0}; \; C = R(0)$.

With the half-life 5730 ± 40 years,

$$0.5\,R(0) = R(0)\, e^{p\,(5730 \pm 40)}$$

$$p = \frac{\ln 0.5}{5730 \pm 40} .$$

When $t = x$ years, there is 77.7% left of the original amount $R(0)$,

$$0.777 R(0) = R(0)\, e^{px}$$

$$p\,x \ln e = \ln 0.777 ; \quad x = \frac{\ln 0.777}{p} .$$

$$x = \frac{(5730 \pm 40) \cdot \ln 0.777}{\ln 0.5} = 2086 \pm 15 \text{ years.}$$

Compound Interest

Capital invested with interest may be compounded at any agreed interval, *e.g.*, annually, quarterly, daily, or continuously.

• $100 is deposited at a 6% annual interest rate, compounded continuously; find the total value of the investment at the end of 10 years.

The principal S grows at the rate given by

$$\frac{dS}{dt} = 0.06\,S ;$$

$$\int \frac{1}{S}\, dS = \int 0.06 \; dt$$

1000(e^{.08·10} + e^{.08·9}

$$\ln S = 0.06\,t + C_1$$

$$S = e^{0.06\,t + C_1} = e^{0.06\,t}\,[e^{C_1}] ,$$

and, as e^{C_1} may be written C, the general solution is

$$S = C e^{0.06 t} .$$

The initial condition gives

$$S(0) = 100 = C e^{0.06 \cdot 0} ; \; C = 100 .$$

At the end of 10 years, the principal will be

$$S(10) = 100\, e^{0.06 \cdot 10} = \$182.21 .$$ •

If compounded *annually* at 6% annual interest rate, the principal of the deposited $100 would be

$$S = 100\,(1 + 0.06)^{10} = \$179.08;$$

quarterly,

$$S = 100 \left(1 + \frac{0.06}{4}\right)^{4 \cdot 10} = \$181.40;$$

thus – to the surprise of many a disgruntled saver – there is no great difference in the principal obtained with continuously compounded interest.

Continuous Dilution

- A tank contains 10 kg of sodium chloride (NaCl) in 1000 liters of water, which is being continuously diluted by 5 l/min of fresh water being added to it during active mixing by means of a rotary impeller, at the same time forcing an equal amount of the solution out of the tank.

  The NaCl concentration of the solution is originally 10 g/l, that is, 10 kg in 1000 liters; how many minutes will be required for the concentration to drop to 3.2 g/l ?

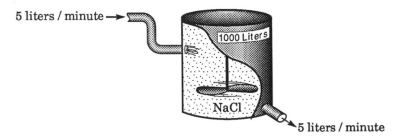

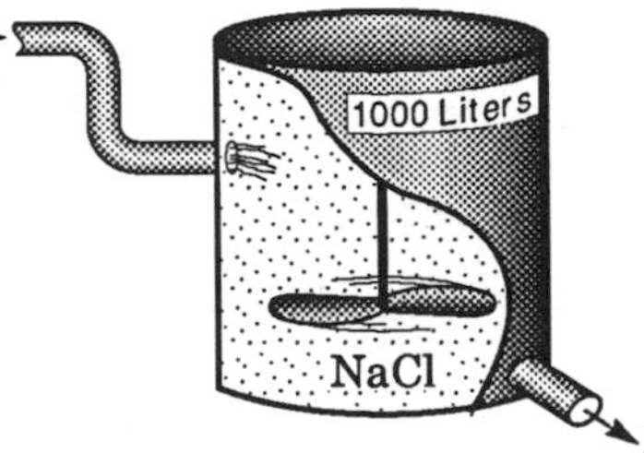

The drainpipe removes 5 liters of solution every minute, corresponding to 0.005 of the total volume. Thus the concentration obeys the law

$$c(t) = 10 \cdot e^{-0.005\,t} ;$$

for the sought time T, we have

$$3.2 = 10 \cdot e^{-0.005\,T} ;$$

$$T = \left| \frac{\ln 0.32}{0.005} \right| = 200 \left| \ln 0.32 \right| \approx 228 \text{ min} = 3^{\text{h}}\,48^{\text{min}} .$$

- A 6-m^3 tank contains 4.20% CO_2 (that is, 4.20% of the total number of molecules in the gas mixture is CO_2). Air with a CO_2 content of 0.03% is led into the tank at the rate of 2 m^3/minute and thoroughly mixed with the gas in the tank. The mixture leaves the tank at the same rate.

  In a mixture of gases, each gas expands to fill the entire volume available to the mixture; thus, every individual gas occupies a volume equal to the entire mixture.

  What is the concentration of CO_2 in the tank after 10 minutes?

Let each m^3 of the gas mixture contain Q molecules.

The number of CO_2 molecules in the tank at t minutes is $N(t)$; at $t(0)$,

$$N(0) = 6 \cdot 0.0420 \cdot Q = 0.2520 \cdot Q .$$

CO_2 enters the tank at the rate of

$$2 \cdot 0.0003 \cdot Q = 0.0006 \cdot Q \text{ molecules/min,}$$

and gas mixture leaves the tank at the rate of

$$2 \cdot Q \cdot \frac{N(t)}{6 \cdot Q} = \frac{N(t)}{3} \text{ molecules/min.}$$

The concentration of CO_2 in the tank changes at the rate $\dfrac{dN}{dt}$,

$$\frac{dN}{dt} = 0.0006 \cdot Q - \frac{N}{3}; \quad \frac{dN}{dt} + \frac{N}{3} = 0.0006 \cdot Q,$$

with the general solution

p. 997

$$N = \frac{0.0006 \cdot Q}{e^{\int dt/3}} \cdot \int e^{\int dt/3}\, dt = 0.0018 \cdot Q + C \cdot e^{-t/3}.$$

Since $N(0) = 0.2520 \cdot Q$, we have

$$0.2520 \cdot Q = 0.0018 \cdot Q + C \cdot e^0; \quad C = 0.2502 \cdot Q,$$

and at $t = 10$,

$$N(10) = 0.0018 \cdot Q + 0.2502 \cdot Q \cdot e^{-10/3}$$
$$\approx 0.0107 \cdot Q \text{ molecules of } CO_2$$

and the concentration

$$\frac{0.0107\, Q}{6\, Q} \cdot 100 \approx 0.18\%\ CO_2.$$

Cooling and Heating

The rate of change of temperature of an object is proportional to the temperature difference between the object and the ambient medium. To enhance the concept of differential equations, all other physical laws are ignored in the following two problems.

- A turkey is taken from the refrigerator at 2° C and placed in an oven preheated to 200° C and kept at that temperature; after 30 minutes the internal temperature of the turkey has risen to 16° C. The fowl is ready to be taken out when its internal temperature reaches 88° C.

Determine the cooking time required.

The temperature θ of the turkey increases at the rate $d\theta/dt$,

$$\frac{d\theta}{dt} = c\,(200 - \theta); \quad \int \frac{d\theta}{200 - \theta} = c \int dt,$$

with the **general solution**

$$c \cdot t = C_1 - \ln(200 - \theta),$$

where we insert the given initial and intermediate conditions:

t (min.)	θ (° C)	$200 - \theta$ (° C)	
0	2	198	$C_1 = \ln 198$
30	16	184	$c = \dfrac{1}{30}(\ln 198 - \ln 184)$
T	88	112	

and obtain

$$T = 30 \cdot \frac{\ln 198 - \ln 112}{\ln 198 - \ln 184} = 233 \text{ min} = 3^{h}\,53^{min} \approx 4 \text{ hours}.$$

- A slice is cut from a loaf of rye bread fresh from the oven at 180° C and placed in a room with a constant temperature of 20° C. After 1 minute, the temperature of the slice is 140° C.

  When has the slice of bread cooled to 32° C?

The temperature θ of the bread decreases at a rate proportional to the temperature difference between it and the room air, that is,

$$-\frac{d\theta}{dt} = c\,(\theta-20) \;\Rightarrow\; \frac{d\theta}{dt} = -c\,(\theta-20)\,; \quad \int \frac{1}{\theta-20}\,d\theta = -c\int dt\,,$$

with the general solution

$$c \cdot t = C_1 - \ln(\theta-20)\,,$$

where we insert the given and intermediate conditions:

t (min.)	θ (° C)	$\theta-20$ (° C)	
0	180	160	$C_1 = \ln 160$
1	140	120	$c = \ln 160 - \ln 120$
T	32	12	

The cooling time is

$$T = \frac{\ln 160 - \ln 12}{\ln 160 - \ln 120} \approx 9 \text{ min.}$$

The Proverbial Bathtub

For a long time, maybe since the days of Archimedes, bathtub problems have appeared in mathematical textbooks. Here is ours.

- Mr. Crackpot's bathtub leaks and loses water at a rate that is proportional to the volume of the water in the tub. With the tap open all the way, he continuously adds 5 l/min into the tub, which will just balance the loss through the drain when the tub contains 105 liters of water.

  One morning, arising late but still wanting his bath, he could only allow the water to run for 15 minutes before he got into the tub. Determine the water volume at the beginning of the bath.

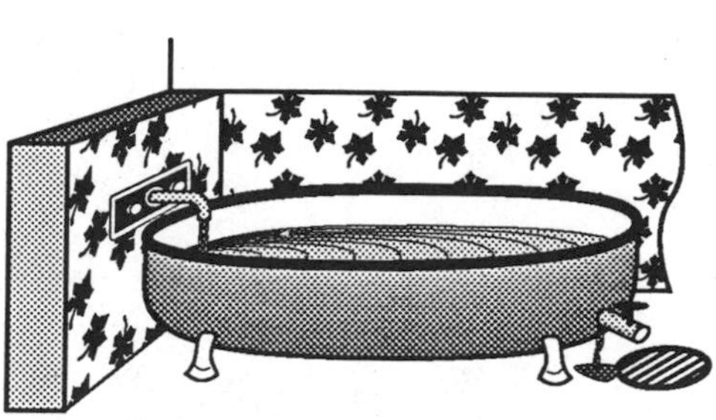

Solution:

Let the volume in the tub be $v(t)$, the water supply rate q_1, and the drain rate q_2,

$$q_2 = c \cdot v(t),$$

where c is determined by the equilibrium conditions,

$$q_1 = q_2; \quad 5 = c \cdot 105; \quad c = \frac{1}{21}.$$

We then have the differential equation

$$q_1 - q_2 = \frac{dv}{dt} = 5 - \frac{v}{21},$$

and the **general solution**

p. 997

$$v = 5 \cdot \frac{\int e^{\int dt/21}\, dt}{e^{\int dt/21}} = 5 \cdot \frac{21 \cdot e^{t/21} + C_1}{e^{t/21}} = 105 + C_2 \cdot e^{-t/21},$$

where the initial condition $v(0) = 0$ gives $C_2 = -105$, and

$$v(15) = 105\,(1 - e^{-15/21}) \approx 54 \text{ [liters]}.$$

Fall through a Resisting Medium

The acceleration a of a moving body is directly proportional to the force F that acts on it and inversely proportional to its mass m,

$$F = k\,m\,a,$$

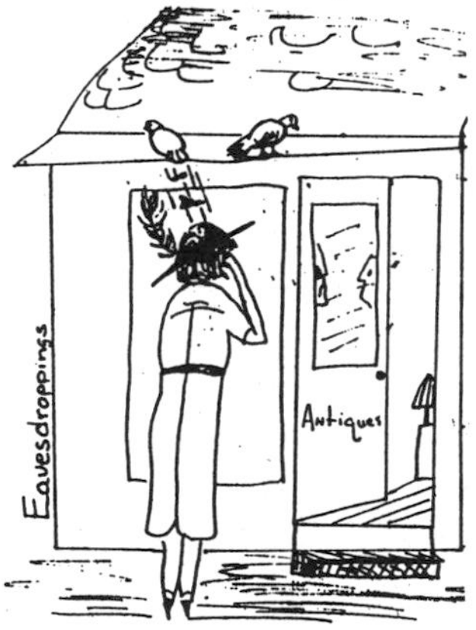

where the magnitude of the coefficient k depends on the units used. With **SI** units (mass in kilograms, acceleration in meters per second squared, and force in newtons), the coefficient k becomes 1,

$$1\,\text{N} = 1\,\text{kg} \cdot 1\,\text{m} \cdot \text{s}^{-2}.$$

The forces acting on a body in free fall through the atmosphere are its weight, which is the product of mass m and the acceleration of gravity $g = 9.81$ m$\cdot$ s^{-2}, and the air resistance, which may be considered to be proportional to the square of the velocity of the fall.

As acceleration is the first derivative of velocity, we can set up the equation

$$F = m\,\frac{dv}{dt} = m \cdot g - k \cdot v^2,$$

where the coefficient k can be approximated by 0.3 with SI units and in the relevant velocity region.

- An object of mass 12 kg falls freely through the atmosphere. Acceleration of gravity is 9.8 m/s^2.
  - If the initial velocity is zero, what is the velocity after 15 seconds?
  - What is the maximum velocity attainable?

Solution:

The equation describing the fall is

$$12 \cdot \frac{dv}{dt} = 12 \cdot 9.8 - 0.3\,v^2$$

or

$$\frac{dv}{dt} + 0.025\,v^2 = 9.8\,.$$

With the integrating factor $e^{\int 0.025 dt} = e^{0.025\,t}$, we obtain

$$v^2 = \frac{1}{e^{0.025\,t}} \int 9.8\,e^{0.025t}\,dt$$

$$= Ke^{-0.025\,t} + 392\,.$$

The initial condition $t = 0$; $v = 0$ gives $K = -392$;

At $t = 15$, we have

$$v^2 = -392\,e^{-0.025 \cdot 15} + 392 \approx 123\ [\text{m/s}]^2\,; \quad v \approx 11.07\text{ m/s}\,.$$

At maximum attainable velocity, the acceleration is 0; thus, $\frac{dv}{dt} = 0$. Solving for v in

p. 1007

$$F = 12 \cdot \frac{dv}{dt} = 0 = 12 \cdot 9.8 - 0.3\,v^2$$

we find the maximum velocity $\sqrt{392}$ m/s $\approx$ 19.80 m/s.

Oscillating Motion

Natural Oscillations
Free Oscillations

Natural oscillations – or **free oscillations** – develop in a mechanical system whose equilibrium has been disturbed, when the system returns to normal without interference from outside forces.

A simple illustration of free oscillations is offered by a mechanical spring at whose free end is placed a concentrated mass m which causes an extension d of the spring beyond its natural length l to a point of equilibrium p, where the weight of the mass is balanced by the tension force in the spring.

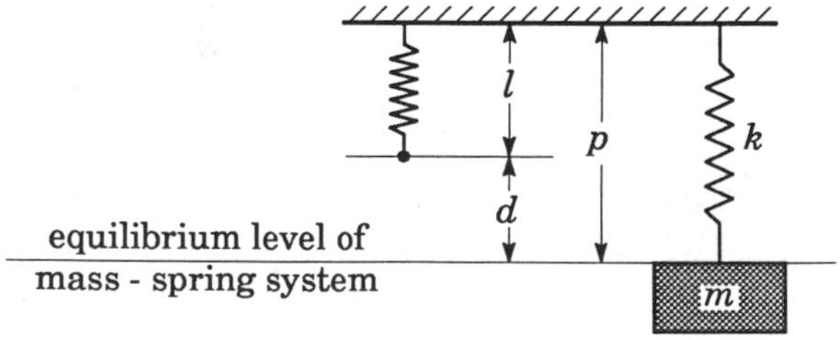

If the mass is pulled down a distance s_0 below the equilibrium level, the additional force set up in the spring is proportional by a factor k, the **spring constant**, to the elongation s_0 of the spring.

Harmonic Oscillations

When the mass is released, the system will attempt to return to its equilibrium position by **harmonic oscillations** with the amplitude s_0.

Since force is proportional to the product of mass and acceleration, which is the second derivative of distance with respect to time, we have

$$m\,\ddot{s} = -k\,s\,,$$

the negative sign signifying a reaction force; the "dotted" derivative $\ddot{s}$ denotes the second derivative of distance with respect to time.

By rearrangement, we have the **fundamental equation of an undamped harmonic oscillation,**

$$\ddot{s} + \frac{k}{m} \cdot s = 0,$$

where $\ddot{s}$ denotes the second derivative of distance with respect to time.

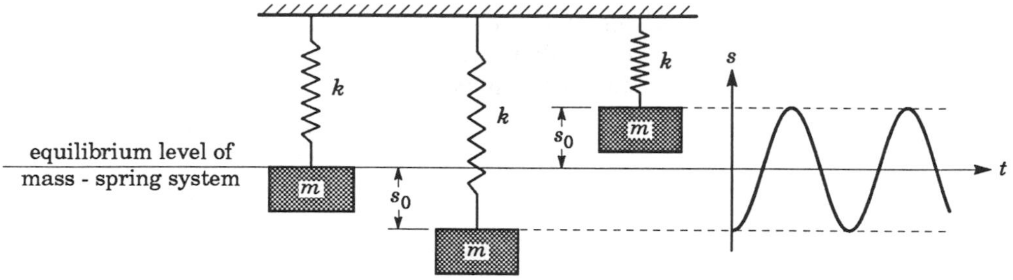

If the surrounding medium presents resistance to the motion, the equation must be supplemented by a term representing the attenuating force, which is often proportional to the velocity of the oscillating body, that is, to the first time derivative of distance.

We then have the **fundamental equation of a damped harmonic oscillation,**

$$\ddot{s} + \frac{c}{m} \cdot \dot{s} + \frac{k}{m} \cdot s = 0\,.$$

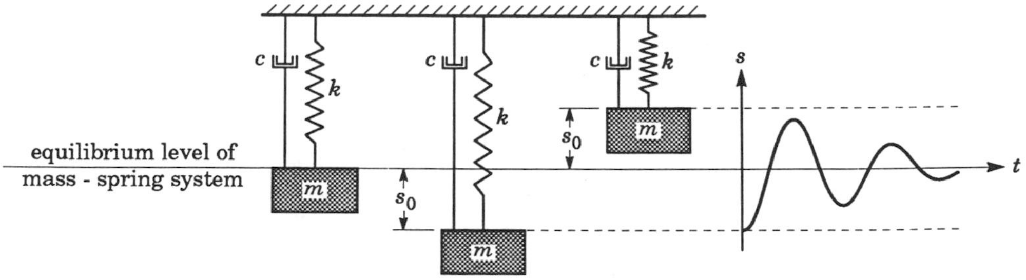

- A body whose mass is 1/2 kg pulls a spring down 1/5 m and gains equilibrium. The body is then pushed 1/10 m above the equilibrium position and released. A damping force numerically equal to the velocity is present. Acceleration of gravity is 10 m/s^2.

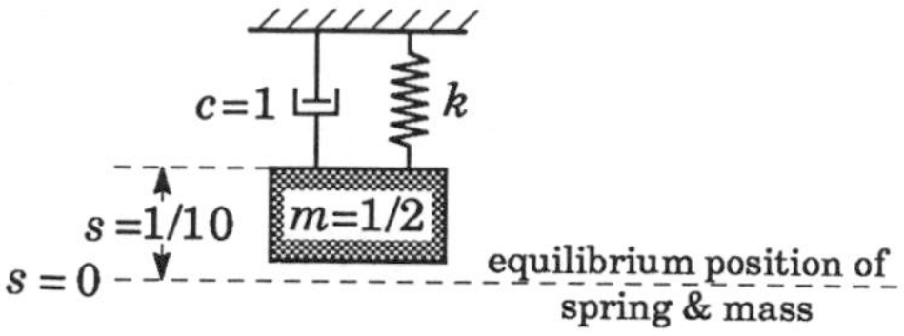

Find the differential equation describing the position of the body as a function of time; the mass of the body is assumed to be concentrated at the end of the spring.

Solution:

To determine the spring constant k, we have

$$\frac{1}{2}\text{ (kg)} \times 10\text{ (m/s}^2) = \frac{1}{5}\,k\text{ (m)}; \quad k = 25\text{ (N/m)}.$$

Using the value of k and the information that the damping force is numerically equal to the velocity, we may form the **differential equation**

$$0.5\,\ddot{s} + \dot{s} + 25\,s = 0$$

or

Solved on *p.* 1021

$$\ddot{s} + 2\,\dot{s} + 50\,s = 0\,.$$

The particular solution is obtained from the initial values

$$\text{distance } s = -\frac{1}{10} \text{ meter at } t=0$$

and

$$\text{velocity } \dot{s} = 0 \text{ at } t=0\,.$$

•

Damped oscillations eventually die out; forced oscillations continue for as long as the external activating force remains coupled to the system.

Forced Oscillations

Forced oscillations develop in a system regularly activated by external forces (*e.g.*, the pendulum of a clock, a swing pushed on the downswing).

The **fundamental equation of a forced oscillation** has the form

$$\ddot{s} + \frac{c}{m}\cdot\dot{s} + \frac{k}{m}\cdot s = F(t),$$

where $F(t)$ may be any periodic function.

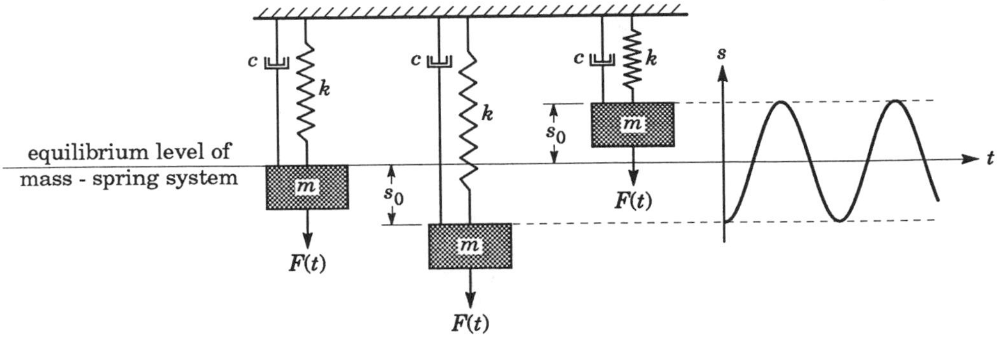

The differential equation

$$\frac{1}{3}\ddot{s} + 4\,\dot{s} + 30\,s = 3\cos 5\,t\,,$$

with initial values $s(0) = \frac{1}{2}$ and $\dot{s}(0) = 0$, may represent a system composed of a mass of 1/3 unit, which is attached to a spring whose spring constant is 30 units, with a damping force numerically equal to 4 times the velocity and an external force equal to $3\cos 5t$ units; the external force $3\cos 5t$ is, of course, periodic. The initial values indicate that the mass is pulled down 1/2 length unit from the equilibrium position and released with an initial velocity of 0:

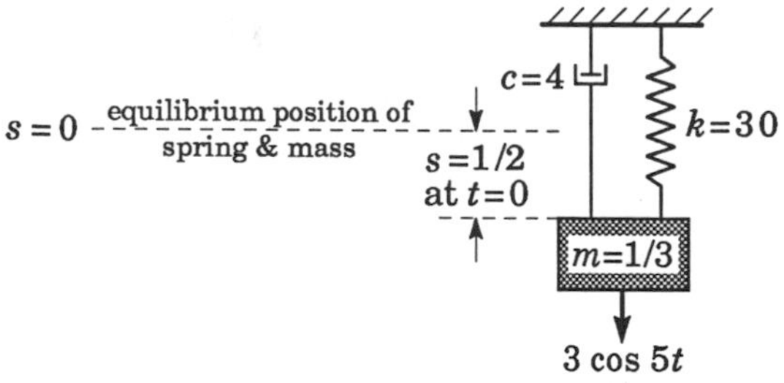

Generality of a Mathematical Model

Many physical and technical phenomena or processes may be described by the same mathematical model. Thus, equivalent mathematical descriptions exist between units of mechanics and, *e.g.*, units of acoustics, hydraulics, gas flow, and fluid flow.

As an example, we note the following analogies between mechanical and electrical quantities:

Mechanical Quantity		**Electrical Quantity**
mass (m)	$\Rightarrow$	inductance (L)
position (y)	$\Rightarrow$	electric charge (Q)
velocity $\left(v = \frac{dy}{dt}\right)$	$\Rightarrow$	electric current (I)
friction coefficient (μ) [damping constant (c)]	$\Rightarrow$	resistance (R)
spring constant (k)	$\Rightarrow$	1/capacitance ($1/C$) [capacitance (C)]
external force [$F(t)$]	$\Rightarrow$	electromotive force (E)

p. 1010 By the differential equation for forced oscillatory motion,

$$m \frac{d^2y}{dt^2} + c\frac{dy}{dt} + k\,y = F(t),$$

and the above analogies, we have the differential equation

$$L\frac{d^2Q}{dt^2} + R\frac{dQ}{dt} + \frac{Q}{C} = E(t)$$

or

$$L\frac{dI}{dt} + RI + \frac{Q}{C} = E(t)$$

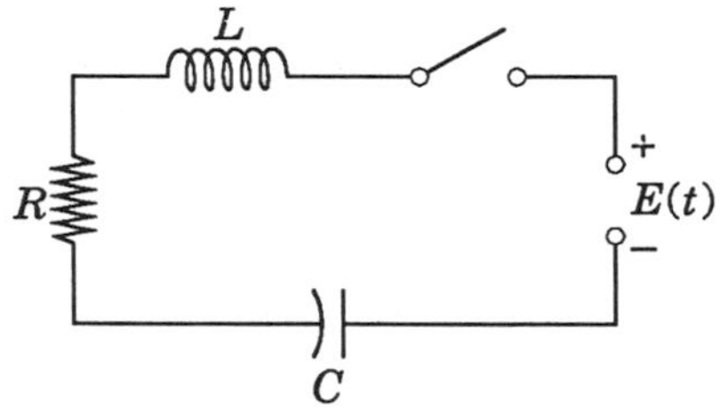

Orthogonal Trajectories

A curve which intersects all curves of a given family at the same angles is referred to as a **trajectory**; if the intersection is at a right angle, we have an **orthogonal trajectory**.

The concept of orthogonal trajectories is used in many branches of physics, *e.g.*, electrostatics (lines of force due to an electric charge and lines of constant potential); thermodynamics (isothermals and heat flow lines); hydrodynamics (streamlines and lines of constant velocity).

The task of finding equations of orthogonal trajectories involves interesting applications of fundamental principles of analytic geometry, differential and integral calculus, and differential equations.

The family of circles, $x^2 + y^2 = r^2$, and the family of straight lines, $y = m\,x$, are examples of families of curves that are orthogonal trajectories of each other:

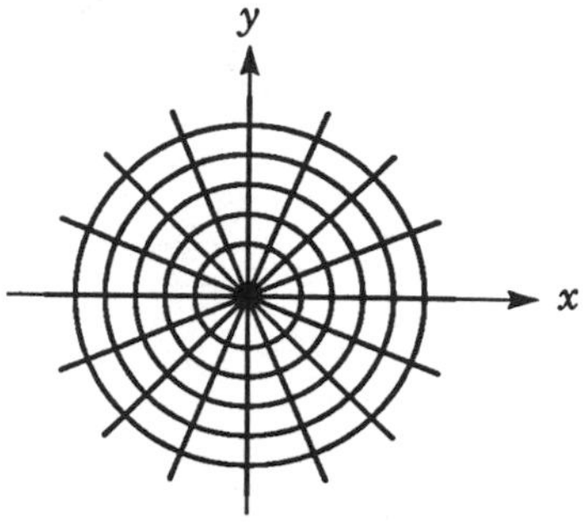

The slope of the family of straight lines is defined by

(1) $$m = \frac{y}{x} .$$

Since the derivative may be interpreted as the slope of a curve, we also have

(2) $$\frac{dy}{dx} = m .$$

Combining (**1**) and (**2**), we have

(3) $$\frac{dy}{dx} = \frac{y}{x} .$$

p. 555 The slopes of orthogonal lines are negative reciprocals. Consequently, the slope of the orthogonal trajectory line of (**1**) is the negative reciprocal of (**3**),

(4) $$\frac{dy}{dx} = -\frac{x}{y} .$$

Solving (**4**), by separation of variables, we obtain

$$\int y \ dy = -\int x \ dx$$

$$\frac{1}{2} y^2 = -\frac{1}{2} x^2 + C_1 ,$$

which we rearrange as

$$\frac{y^2 + x^2}{2} = C_1$$

or

$$y^2 + x^2 = C .$$

Replacing C with r^2 gives

$$y^2 + x^2 = r^2,$$

which, as expected, is a family of circles with centers at the origin of the orthogonal coordinate system.

The orthogonal trajectories of the following example are not so easy to predict as the one in the introductory example.

- Describe the family of orthogonal trajectories of

$$y = p\, x^2 .$$

To find the slope of the curves represented by the given equation, we differentiate

$$\frac{dy}{dx} = 2 p x .$$

Eliminate p by inserting $p = \dfrac{y}{x^2}$.

Thus,

$$\frac{dy}{dx} = 2 \frac{y}{x^2} x = \frac{2y}{x} .$$

Introducing the negative reciprocal of the right member of the above equation, we find

$$\frac{dy}{dx} = -\frac{x}{2y},$$

which represents the slope of the orthogonal trajectory of the given curve.

Solving the differential equation, we obtain

$$\int 2y \ dy = -\int x \ dx$$

$$y^2 = -\frac{1}{2}x^2 + C,$$

where C is an arbitrary constant.

Replace C by p and rearrange,

$$y^2 + \frac{x^2}{2} = p; \quad \frac{y^2}{p} + \frac{x^2}{2p} = 1,$$

which is a family of ellipses with centers at the origin of the orthogonal coordinate system; they are orthogonal trajectories of the family of parabolas $y = p\,x^2$.

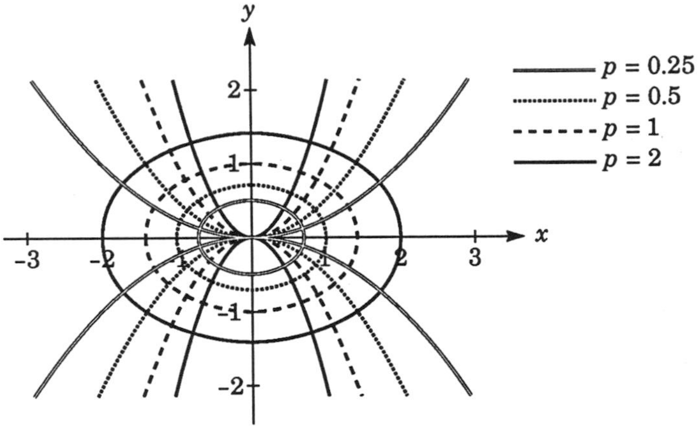

34.4 Second-Order Ordinary Differential Equations

Directly Integrable Equations

A differential equation of the type

$$\frac{d^2y}{dx^2} = f(x)$$

can be solved by direct integration of both sides.

- We have the equation

$$\frac{d^2y}{dx^2} = \cos x .$$

By integration, we find, first,

$$\frac{dy}{dx} = \int \cos x \, dx = \sin x + C_1 ,$$

and then

$$y = \int (\sin x + C_1) \, dx = -\cos x + C_1 x + C_2 ,$$

where C_1 and C_2 are arbitrary constants.

- Find the particular solution of

$$x \frac{d^2y}{dx^2} - x^2 = 1$$

if $y(1) = 2$ and $y'(1) = 1$.

Rewrite the equation $\frac{d^2y}{dx^2} = x + x^{-1}$ and integrate.

The first integration gives

$$\frac{dy}{dx} = \int (x + x^{-1}) \, dx = \frac{x^2}{2} + \ln x + C_1 ,$$

where C_1 is determined by the condition $y'(1) = 1$,

$$y'(1) = 1 = \frac{1^2}{2} + \ln 1 + C_1 ; \quad C_1 = \frac{1}{2} .$$

The second integration gives

$$y = \int \left(\frac{x^2}{2} + \ln x + \frac{1}{2} \right) dx = \frac{x^3}{6} + x \ln x - \frac{x}{2} + C_2 ;$$

C_2 is determined by the condition

$$y(1) = \frac{1^3}{6} + 1 \ln 1 - \frac{1}{2} + C_2 = 2 ; \quad C_2 = \frac{7}{3} .$$

The particular solution is

$$y = \frac{1}{6}(x^3 - 3x + 14) + x \ln x .$$

Linear Differential Equations

The **general linear differential equation** of order n is

$$a_n(x)\frac{d^n y}{dx^n} + a_{n-1}(x)\frac{d^{n-1}y}{dx^{n-1}} + \ldots + a_1(x)\frac{dy}{dx} + a_0(x)y = f(x) ,$$

where $a_n(x)$, $a_{n-1}(x) \ldots a_0(x)$, and $f(x)$ are given functions of x, or constants.

In terms of the **differential operator**

$$D = \frac{d}{dx} ,$$

ARBOGAST Louis (1759 - 1803)

introduced by the French lawyer and mathematician **Louis Arbogast** in his *Du calcul des dérivations* (1800), the equation becomes

$$a_n(x)D^n y + a_{n-1}(x)D^{n-1}y + \ldots + a_1(x)Dy + a_0(x)y = f(x)$$

or, simplified,

$$\sum_{i=0}^{n} a_i(x)\, D^i y = f(x) .$$

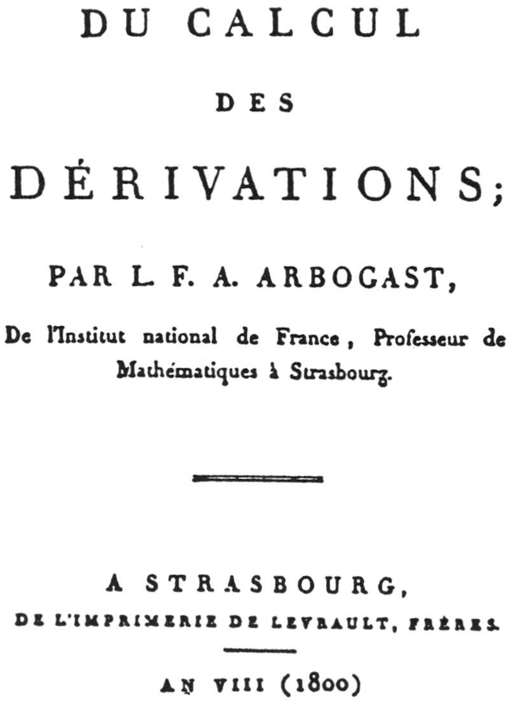
DU CALCUL

DES

DÉRIVATIONS;

PAR L. F. A. ARBOGAST,

De l'Institut national de France, Professeur de Mathématiques à Strasbourg.

A STRASBOURG,

DE L'IMPRIMERIE DE LEVRAULT, FRÈRES.

AN VIII (1800)

Homogeneous
Nonhomogeneous

If $f(x) = 0$, the linear differential equation is known as **homogeneous**; if $f(x) \neq 0$, it is **nonhomogeneous** (or inhomogeneous).

This use of the term *homogeneous* has nothing in common with its use to denote differential equations consisting exclusively of homogeneous functions of the same degree in x and y. To avoid unnecessary confusion, we will specify **linear** when referring to the types of equation considered here.

Linear Dependence
Linear Independence

Linear dependence and **independence** are concepts of importance in the study of homogeneous and nonhomogeneous linear differential equations. A collection of functions is linearly dependent if one of them can be expressed as a sum of constant multiples of the others; if this is not possible, the collection is said to be linearly independent.

$y_1(x) = 3x$ and $y_2(x) = x$ are linearly dependent, one being a multiple of the other; $y_1(x) = 3x$ and $y_3(x) = x^2$, on the other hand, are linearly independent.

Homogeneous Linear Equations

If $y_1, y_2 \dots y_n$ are linearly independent solutions of a **homogeneous linear differential equation** of order n,

$$a_n(x)\frac{d^n y}{dx^n} + a_{n-1}(x)\frac{d^{n-1}y}{dx^{n-1}} + \dots + a_1(x)\frac{dy}{dx} + a_0(x)y = 0$$

or

$$\sum_{i=0}^{n} a_i(x)\,D^i y = 0\ ,$$

then the **general solution** is

$$y = \sum_{i=0}^{n} C_i y_i = C_1 y_1 + C_2 y_2 + \dots + C_n y_n\ ,$$

where all $C_1, C_2, \dots, C_n$ are arbitrary constants.

Nonhomogeneous Linear Equations

A nonhomogeneous linear differential equation of order n,

$$\sum_{i=0}^{n} a_i(x)\,D^i y = f(x)\ ,$$

Reduced Equation

differs from its homogeneous counterpart, often called the **reduced equation** in that the right-hand member of the equation is $\neq 0$.

The **general solution of a nonhomogeneous linear differential equation** is the sum of

any of its particular solutions, and
a complementary solution,

the latter of which is the

general solution of the reduced equation.

Homogeneous Constant-Coefficient Linear Differential Equations

The Auxiliary Equation

The method of solution to be described is applicable to higher-order homogeneous linear differential equations of the type

$$\sum_{i=0}^{n} a_i(x)\, \mathrm{D}^i y = 0\,,$$

whose coefficients are constants, but we will deal here specifically with second-order equations,

$$a_2 \frac{\mathrm{d}^2 y}{\mathrm{d}x^2} + a_1 \frac{\mathrm{d}y}{\mathrm{d}x} + a_0 y = 0\,.$$

Attempting $y = \mathrm{e}^{rx}$ as a general solution, we have

$$(a_2 r^2 + a_1 r + a_0)\, \mathrm{e}^{rx} = 0\,;$$

as e^{rx} cannot be 0, we see that $y = \mathrm{e}^{rx}$ is a solution of the given differential equation, if

$$a_2 r^2 + a_1 r + a_0 = 0\,.$$

Auxiliary Equation
Characteristic Equation

This **auxiliary equation**, or **characteristic equation**, has the roots

$$r_1 = \frac{-a_1 + \sqrt{a_1{}^2 - 4\,a_0\,a_2}}{2\,a_2}\,;\quad r_2 = \frac{-a_1 - \sqrt{a_1{}^2 - 4\,a_0\,a_2}}{2\,a_2}\,;$$

$y_1 = \mathrm{e}^{r_1 x}$ and $y_2 = \mathrm{e}^{r_2 x}$ will then be **particular solutions** of the differential equation.

p. 311

Depending upon the value of the **discriminant**

$$D = a_1{}^2 - 4\,a_0\,a_2\,,$$

we distinguish three cases:

$D > 0$:	r_1, r_2	real and distinct (unequal)
$D = 0$:	r_1, r_2	real and equal
$D < 0$:	r_1, r_2	complex conjugates

Real and Distinct r_1 and r_2

If the discriminant is positive, the roots r_1 and r_2 of the auxiliary equation

$$a_2 r^2 + a_1 r + a_0 = 0$$

are **real and distinct** and the general solution of the homogeneous linear differential equation,

$$a_2\, y'' + a_1\, y' + a_0\, y = 0,$$

is

$$y = C_1 \mathrm{e}^{r_1 x} + C_2 \mathrm{e}^{r_2 x}\,,$$

where C_1 and C_2 are arbitrary constants.

Real and Equal r_1 and r_2

If the discriminant is 0, the roots of the auxiliary equation are real and equal,

$$r_1 = r_2 = r\,;$$

the formula above gives

$$y_1 = C_1 e^{rx} + C_2 e^{rx} = C_3 e^{rx}\,,$$

which contains only one arbitrary constant and therefore does not qualify as a general solution. To get around this difficulty, we consider

$$y_2 = x\, e^{rx}$$

as a possible second solution, and verify that it is.

$$a_2 \cdot \frac{d^2}{dx^2}\,[x \cdot e^{rx}] + a_1 \cdot \frac{d}{dx}\,[x \cdot e^{rx}] + a_0 \cdot x \cdot e^{rx}$$

$$= a_2\,(2\,r\,e^{rx} + r^2\,x \cdot e^{rx}) + a_1\,(e^{rx} + r\,x\,e^{rx}) + a_0 \cdot x \cdot e^{rx}$$

$$= e^{rx}\,(2\,a_2\,r + a_1) + x\,e^{rx}\,(a_2\,r^2 + a_1\,r + a_0)\,.$$

Both expressions $(2\,a_2\,r + a_1)$ and $(a_2\,r^2 + a_1\,r + a_0)$ are = 0 when $r = -\dfrac{a_1}{2\,a_2}$, that is, when

$$a_1{}^2 - 4\,a_0\,a_2 = 0\,.$$

Since $y_1 = e^{rx}$ and $y_2 = x\,e^{rx}$ are linearly independent, the general solution of the homogeneous linear differential equation

$$a_2\,y'' + a_1\,y' + a_0\,y = 0$$

is

$$y = C_1 e^{rx} + C_2\,x\,e^{rx}\,,$$

where C_1 and C_2 are arbitrary constants.

Complex Conjugate r_1 and r_2

If the discriminant is negative, the roots of the auxiliary equation $(a_2\,r^2 + a_1\,r + a_0 = 0)$ are the **conjugate complex numbers**,

$$r_1\,r_2 = \frac{-a_1}{2\,a_2} \pm i\,\frac{\sqrt{\left|a_1{}^2 - 4\,a_0\,a_2\right|}}{2\,a_2}$$

or

$$r_1 = a + b\,i\,; \quad r_2 = a - b\,i\,,$$

where

$$a = -\frac{a_1}{2\,a_2}\,; \quad b = \frac{1}{2\,a_2}\sqrt{\left|a_1{}^2 - 4\,a_0\,a_2\right|}\,.$$

The general solution becomes

$$y = C_1 e^{(a+b\,i)\,x} + C_2 e^{(a-b\,i)\,x}$$

$$= e^{a\,x}\,(C_1 e^{i\,b\,x} + C_2 e^{-i\,b\,x})\,.$$

Equations having solutions of this type occur primarily in oscillation problems.

p. 792

To obtain a solution without imaginary or complex number exponents, we use **Euler's formula**

$$e^{ix} = \cos x + i \sin x ,$$

and the corollaries

$$e^{ibx} = \cos bx + i \sin bx ; \quad e^{-ibx} = \cos bx - i \sin bx ,$$

which give

$$y = e^{ax}(A_1 \cos bx + A_2 \sin bx) ,$$

where

$$A_1 = C_1 + C_2 ; \quad A_2 = i(C_1 - C_2).$$

- Find the general solution of

$$y'' + y' - 2y = 0$$

and the particular solution for $y(0) = 1$; $y'(0) = 0$.

The auxiliary equation

$$r^2 + r - 2 = 0 ; \quad (r-1)(r+2) = 0$$

has the roots

$$r_1 = 1 ; \quad r_2 = -2 .$$

Inserting those values in $y = C_1 e^{r_1 x} + C_2 e^{r_2 x}$, we obtain the **general solution**

$$y = C_1 e^x + C_2 e^{-2x} ,$$

where C_1 and C_2 are arbitrary constants.

Differentiating y gives

$$y' = C_1 e^x - 2C_2 e^{-2x} .$$

$y(0) = 1$ and $y'(0) = 0$ require that

$$\left.\begin{aligned} C_1 e^0 + C_2 e^{-0} &= 1 \\ C_1 e^0 - 2C_2 e^{-0} &= 0 \end{aligned}\right\} \qquad C_1 = \frac{2}{3} ; \; C_2 = \frac{1}{3},$$

and the particular solution is

$$y = \frac{2}{3} \cdot e^x + \frac{1}{3} \cdot e^{-2x} .$$

- Solve the equation $\quad y'' + y' + y = 0 .$

The auxiliary equation $r^2 + r + 1 = 0$ has the roots

$$r_1 r_2 = -\frac{1}{2} \pm i\frac{\sqrt{3}}{2} .$$

Replacing in

$$y = e^{ax}(A_1 \cos bx + A_2 \sin bx) ,$$

we find the solution

$$y = e^{-x/2}\left(A_1 \cos \frac{\sqrt{3}}{2}x + A_2 \sin \frac{\sqrt{3}}{2}x\right),$$

where A_1 and A_2 are arbitrary constants.

- $y'' + 2y' + 50y = 0$ is the equation of a damped oscillation.
  Find the particular solution for $y(0) = -\frac{1}{10}$ and $y'(0) = 0$.

The auxiliary equation

$$r^2 + 2r + 50 = 0$$

has the roots

$$r_1, r_2 = -1 \pm 7\,\mathrm{i}.$$

The general solution is

$$y = \mathrm{e}^{-t}(A_1 \cos 7t + A_2 \sin 7t).$$

Since $y(0) = -\frac{1}{10}$, we have $A_1 = -\frac{1}{10}$;

$$y = \mathrm{e}^{-t}\left(A_2 \sin 7t - \frac{1}{10}\cos 7t\right)$$

$$y' = \left[\left(7A_2 + \frac{1}{10}\right)\cos 7t - \left(A_2 - \frac{7}{10}\right)\sin 7t\right]\mathrm{e}^{-t}.$$

Since $y'(0) = 0$, we have

$$\left[\left(7A_2 + \frac{1}{10}\right)\cdot 1 - \left(A_2 + \frac{7}{10}\right)\cdot 0\right]\cdot 1 = 0;\quad A_2 = -\frac{1}{70}.$$

Hence, the particular solution is

$$y(t) = \mathrm{e}^{-t}\left(-\frac{1}{70}\sin 7t - \frac{1}{10}\cos 7t\right).$$

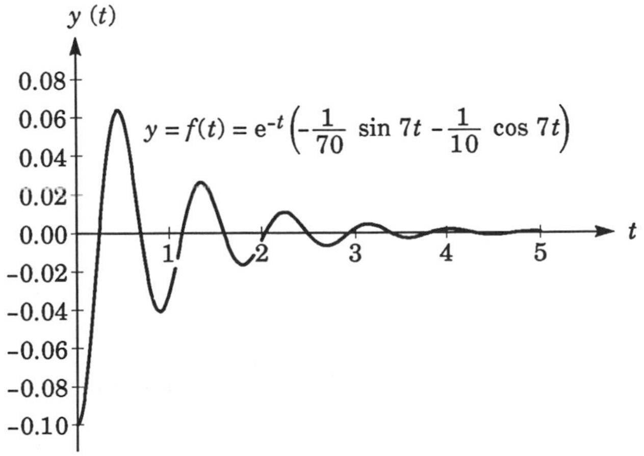

Harmonic Differential Equations

Equations of the form

$$\frac{d^2y}{dx^2} + n^2 y = 0,$$

where n is any number, are called **simple harmonic differential equations**, and occur in problems involving periodic motion described by sine or cosine functions of time.

Its auxiliary equation, $r^2 + 0\,r + n^2 = 0$, has the roots $r = 0 \pm \mathrm{i}\,n$.

We find the solution

$$y = e^{0x}(A_1 \cos n\,x + A_2 \sin n\,x) = A_1 \cos n\,x + A_2 \sin n\,x,$$

where A_1 and A_2 are arbitrary constants.

Similarly, for $\dfrac{d^2y}{dx^2} - n^2 y = 0$, we find $r = \pm\, n$.

Consequently, $y_1 = e^{nx}$ and $y_2 = e^{-nx}$ are two linearly independent solutions, and we obtain the solution

$$y = C_1 e^{nx} + C_2 e^{-nx}.$$

p. 536 The above solution may be expressed in terms of **hyperbolic functions**:

$$\left.\begin{aligned} \cosh n\,x &= \frac{e^{nx} + e^{-nx}}{2} \\ \sinh n\,x &= \frac{e^{nx} - e^{-nx}}{2} \end{aligned}\right\} \qquad \left.\begin{aligned} e^{nx} &= \cosh n\,x + \sinh n\,x \\ e^{-nx} &= \cosh n\,x - \sinh n\,x \end{aligned}\right\}.$$

We may now write

$$y = (C_1 + C_2)\cosh n\,x + (C_1 - C_2)\sinh n\,x,$$

or

$$y = A_1 \cosh n\,x + A_2 \sinh n\,x.$$

- Find the particular solution of

$$\frac{d^2y}{dx^2} - 4y = 0$$

for $y(0) = 1$ and $y'(0) = 5$, expressed in exponential functions and in hyperbolic functions.

Rewriting the equation

$$\frac{d^2y}{dx^2} - 2^2 y = 0,$$

the general solution is

$$y = C_1 e^{2x} + C_2 e^{-2x}$$

with

$$y' = 2C_1 e^{2x} - 2C_2 e^{-2x}.$$

$y(0) = 1$ and $y'(0) = 5$ give

$$\left.\begin{aligned} C_1 e^0 + C_2 e^0 &= 1 \\ 2C_1 e^0 - 2C_2 e^0 &= 5 \end{aligned}\right\} \quad \left.\begin{aligned} C_1 + C_2 &= 1 \\ C_1 - C_2 &= 5/2 \end{aligned}\right\} \quad \left.\begin{aligned} C_1 &= 7/4 \\ C_2 &= -3/4 \end{aligned}\right\}.$$

Thus, the particular solution in exponential functions is

$$y = \frac{7}{4} e^{2x} - \frac{3}{4} e^{-2x}.$$

We have

$$e^{2x} = \cosh 2x + \sinh 2x$$

$$e^{-2x} = \cosh 2x - \sinh 2x;$$

$$y = \frac{7}{4}\cosh 2x + \frac{7}{4}\sinh 2x - \frac{3}{4}\cosh 2x + \frac{3}{4}\sinh 2x$$

$$y = \cosh 2x + \frac{5}{2}\sinh 2x.$$

- In the differential equation

$$\frac{d^2y}{dt^2} + \frac{k}{m} y = 0,$$

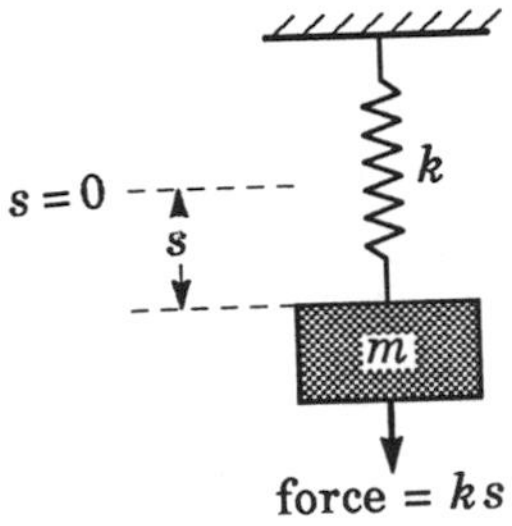

describing the undamped oscillatory motion of a spring-mass system, m denotes the mass and k is the spring constant.

Find the general solution and the particular solution for $k = 18$ N/m, $m = 2$ kg and $y(0) = -0.05$ m, $y'(0) = 0$ m/s.

We obtain the general solution

$$y = A_1 \cos\sqrt{\frac{k}{m}}\, t + A_2 \sin\sqrt{\frac{k}{m}}\, t.$$

Inserting the values of k and m in the general solution,

$$y = A_1 \cos\sqrt{\frac{18}{2}}\, t + A_2 \sin\sqrt{\frac{18}{2}}\, t$$

$$= A_1 \cos 3t + A_2 \sin 3t.$$

$$y(0) = A_1 \cos 0 + A_2 \sin 0; \quad A_1 = -0.05.$$

So

$$y = -0.05 \cos 3t + A_2 \sin 3t;$$

$$y' = 0.15 \sin 3t + A_2 3 \cos 3t;$$

$$y'(0) = 0.15 \sin 0 + A_2 3 \cos 0 = 0; \quad A_2 = 0,$$

and the particular solution is

$$y = -0.05 \cos 3t.$$

Nonhomogeneous Constant-Coefficient Linear Differential Equations

Introduction

The **general solution** of a nonhomogeneous linear differential equation of order n is

$$y = y_p + C_1 y_1 + C_2 y_2 + \ldots + C_n y_n ,$$

where y_p is some **particular solution** of the equation and $C_1 y_1 + C_2 y_2 + \ldots + C_n y_n$ is the **complementary solution**, that is, the general solution of the corresponding homogeneous linear differential equation.

We shall discuss a method of solving higher-order nonhomogeneous linear differential equations with **constant coefficients**,

$$a_n \frac{d^n y}{dx^n} + a_{n-1} \frac{d^{n-1} y}{dx^{n-1}} + \ldots + a_1 \frac{dy}{dx} + a_0 y = f(x) ,$$

but restrict ourselves here specifically to second-order equations,

$$a_2 \frac{d^2 y}{dx^2} + a_1 \frac{dy}{dx} + a_0 y = f(x) .$$

A simple equation of this type, $\frac{d^2 y}{dx^2} + y = x$, has the **reduced equation**

$$\frac{d^2 y}{dx^2} + y = 0,$$

Complementary Solution

with the solution – known as the **complementary solution** –

$$y = A_1 \cos x + A_2 \sin x ,$$

where A_1 and A_2 are arbitrary constants.

Examination shows that $y_p = x$ is a **particular solution**, and we obtain the general solution

$$y = A_1 \cos x + A_2 \sin x + x .$$

It is only rarely that a particular solution of a nonhomogeneous constant-coefficient linear differential equation can be found by mere inspection, as in the above case. We shall therefore describe both the more general method of **variation of constants** and the simpler method of **undetermined coefficients**, which is adequate for special cases of nonhomogeneous constant-coefficient linear differential equations, commonly encountered in physics.

Variation of Constants

LAGRANGE Joseph Louis (1736 - 1813)

The method of variation of constants, also called **variation of parameters**, is due to **Lagrange**, French mathematician and physicist.

If the differential equation

(1) $$a_2 \frac{d^2y}{dx^2} + a_1 \frac{dy}{dx} + a_0 y = f(x)$$

has the **complementary solution**

(2) $$y_c = C_1 y_1 + C_2 y_2$$

then a **particular solution** of the form

(3) $$y_p = u(x) y_1 + v(x) y_2$$

can be obtained by **restricting** the functions $u(x)$ and $v(x)$ to satisfying the condition

(4) $$u'(x)\, y_1 + v'(x)\, y_2 = 0 .$$

Differentiating (**3**), we obtain

$$\frac{d}{dx}[y_p] = u(x) y'_1 + y_1 u'(x) + v(x) y'_2 + y_2 v'(x) ,$$

which by virtue of (**4**) may be reduced to

(5) $$\frac{d}{dx}[y_p] = u(x) y'_1 + v(x) y'_2 .$$

Differentiating (**5**), we find

(6) $$\frac{d^2}{dx^2} y_p = u y''_1 + y'_1 u' + v y''_2 + y'_2 v' .$$

Inserting (**3**), (**5**), and (**6**) in (**1**) leads to

$$a_2 (u y''_1 + y'_1 u' + v y''_2 + y'_2 v') + a_1 (u y'_1 + v y'_2) + a_0 (u y_1 + v y_2) = f(x)$$

or, rearranged,

(7) $$u\,(a_2 y''_1 + a_1 y'_1 + a_0 y_1) + v\,(a_2 y''_2 + a_1 y'_2 + a_0 y_2) + a_2 (y'_1 u' + y'_2 v') = f(x) .$$

Since, by definition, y_1 and y_2 are solutions of the homogeneous equation, we have

$$a_2 y''_1 + a_1 y'_1 + a_0 y_1 = 0 ; \quad a_2 y''_2 + a_1 y'_2 + a_0 y_2 = 0$$

and (**7**) becomes

(8) $$y'_1 u' + y'_2 v' = \frac{f(x)}{a_2} ; \quad a_2 \neq 0 .$$

We now have the simultaneous equations (**4**) and (**8**)

$$\left.\begin{aligned} y_1 u' + y_2 v' &= 0 \\ y'_1 u' + y'_2 v' &= \frac{f(x)}{a_2} \end{aligned}\right\} ,$$

which give

$$\left.\begin{aligned} u' &= -\frac{y_2 f(x)}{a_2 (y_1 y'_2 - y'_1 y_2)} \\ v' &= \frac{y_1 f(x)}{a_2 (y_1 y'_2 - y'_1 y_2)} \end{aligned}\right\} .$$

By integration,

(9) $$u = -\frac{1}{a_2} \int \frac{y_2 f(x)}{y_1 y'_2 - y'_1 y_2} \, dx$$

(10) $$v = \frac{1}{a_2} \int \frac{y_1 f(x)}{y_1 y'_2 - y'_1 y_2} \, dx ,$$

which we insert in (**3**) and obtain

$$y_p = y_2 \frac{1}{a_2} \int \frac{y_1 f(x)}{y_1 y'_2 - y'_1 y_2} \, dx - y_1 \frac{1}{a_2} \int \frac{y_2 f(x)}{y_1 y'_2 - y'_1 y_2} \, dx ,$$

which is a **particular solution** of

$$a_2 \frac{d^2 y}{dx^2} + a_1 \frac{dy}{dx} + a_0 y = f(x) ,$$

whose complementary solution is $y_c = C_1 y_1 + C_2 y_2$.

As may be expected, solving a differential equation by the method of variation of constants is often quite a formidable task.

- Solve the equation $y'' + 4y = \csc 2x$.

The homogeneous equation $y'' + 4y = 0$ gives the complementary solution

$$y_c = A_1 \cos 2x + A_2 \sin 2x .$$

The particular solution is (since $a_2 = 1$)

$$y_p = y_2 \int \frac{y_1 \cdot f(x)}{y_1 y'_2 - y'_1 y_2} \, dx - y_1 \int \frac{y_2 \cdot f(x)}{y_1 y'_2 - y'_1 y_2} \, dx,$$

where $f(x) = \csc 2x$, and

$$\left.\begin{aligned} y_1 &= \sin 2x \\ y_2 &= \cos 2x \end{aligned}\right\} \quad \left.\begin{aligned} y_1' &= 2 \cdot \cos 2x \\ y_2' &= -2 \cdot \sin 2x \end{aligned}\right\} \quad \left.\begin{aligned} y_1 \cdot \csc 2x &= 1 \\ y_2 \cdot \csc 2x &= \frac{\cos 2x}{\sin 2x} \end{aligned}\right\} ;$$

$$y_1 \cdot y_2' - y_1' \cdot y_2 = -2 \cdot \sin^2 2x - 2 \cdot \cos^2 2x = -2 ;$$

thus,

$$y_p = \frac{\sin 2x}{2} \int \frac{\cos x}{\sin x} \, dx - \frac{\cos 2x}{2} \int dx$$

$$= \frac{1}{4} \cdot \sin 2x \cdot \ln |\sin x| - \frac{1}{2} x \cdot \cos 2x .$$

The general solution is

$$y = A_1 \sin 2x + A_2 \cos 2x + \frac{1}{4} \sin 2x \cdot \ln |\sin x| - \frac{1}{2} \cdot x \cdot \cos 2x .$$

Undetermined Coefficients

Perturbation Functions

By identifying some special cases of **perturbation functions** $f(x)$ in nonhomogeneous constant-coefficient linear differential equations,

$$a_2 \frac{d^2y}{dx^2} + a_1 \frac{dy}{dx} + a_0 y = f(x) ,$$

we may dispense with the often cumbersome method of variation of constants in favor of the **method of undetermined coefficients**.

The basic principle of this method is to *assume* a particular solution y_p that is similar to the perturbation function except that its coefficients are *so far* undetermined, and to solve the system for the coefficients.

Amenable to treatment by the method of undetermined coefficients are cases where the perturbation function is

- a polynomial;
- an exponential function;
- a function of trigonometric sines or cosines;
- a function of hyperbolic sines or cosines;
- a product or sum of such functions.

Polynomial Perturbation Functions

If the perturbation function $f(x)$ is a polynomial of degree n, the first step is to attempt an n-th degree polynomial particular solution y_p,

$$y_p = A_n x^n + A_{n-1} x^{n-1} + \ldots + A_1 x + A_0 ,$$

where A_n, A_{n-1}, …, A_0 are undetermined coefficients.

The next step is to insert y_p and its relevant derivatives in the given differential equation, then equate the coefficients of both members of the equation and solve for A_n, A_{n-1}, …, A_0.

Exponential Perturbation Functions

If the perturbation function is an exponential function $f(x) = b\,e^{cx}$, a first attempt at a particular solution y_p should be

$$y_p = A \cdot x^k \cdot e^{cx} ,$$

where A is a constant to be determined by the integrating conditions and k is the lowest power of x that will prevent duplication of terms that are already part of the complementary solution – the terms must be linearly independent.

If no roots of the auxiliary equation are equal to c, then $k = 0$, and we have

$$y_p = A \cdot e^{cx} .$$

The constant A is determined by inserting y_p and its derivatives in the given differential equation.

Trigonometric and Hyperbolic Perturbation Functions

If $f(x)$ is equal to $(b \cdot \sin cx)$ or $(b \cdot \cos cx)$, where b and c are constants, we assume a particular solution

$$y_p = x^k\,(A \cdot \cos c\,x + B \cdot \sin c\,x)\,;$$

if $f(x)$ is equal to $(b \cdot \sinh cx)$ or $(b \cdot \cosh cx)$, we choose a particular solution

$$y_p = x^k\,(A \cdot \cosh c\,x + B \cdot \sinh c\,x)\,,$$

where A and B are constants to be determined and k is again the lowest power of x that prevents duplication of terms in the complementary solution.

The chosen y_p and its derivatives are inserted in the given differential equation and the coefficients of the equation members are equated.

Product-Type Perturbation Functions

If the perturbation function is a product of several functions of the kinds that allow solving the equation by the method of undetermined coefficients, a particular solution may be obtained as a product of factors of particular solutions.

Sums of Functions

If

$$a_2 \cdot \frac{d^2y}{dx^2} + a_1 \cdot \frac{dy}{dx} + a_0\,y = f(x) = g_1(x) + g_2(x)$$

and y_{p_1} and y_{p_2} are particular solutions of

$$a_2\,y'' + a_1\,y' + a_0\,y = g_1(x) \quad \text{and} \quad a_2\,y'' + a_1\,y' + a_0\,y = g_2(x)\,,$$

respectively, then

$$y_p = y_{p_1} + y_{p_2}$$

is a particular solution of the given differential equation.

Problems

- Find a particular solution of

$$y'' + 2\,y' - 4\,y = 4\,x^2 - 8\,x + 12\,.$$

Since the right-hand member of the equation is a second-degree polynomial, the particular solution is of the form

$$y_p = A_2\,x^2 + A_1\,x + A_0\,.$$

Differentiation gives

$$y_p{}' = 2\,A_2\,x + A_1\,; \qquad y_p{}'' = 2\,A_2\,.$$

Inserting y_p and its derivatives, we obtain

$$-4\,A_2\,x^2 + (4\,A_2 - 4\,A_1)\,x + (2\,A_2 + 2\,A_1 - 4\,A_0) = 4\,x^2 - 8\,x + 12\,.$$

Equating the coefficients gives

$$\left.\begin{aligned} -4A_2 &= 4 \\ 4A_2 - 4A_1 &= -8 \\ 2A_2 + 2A_1 - 4A_0 &= 12 \end{aligned}\right\} \qquad \left.\begin{aligned} A_0 &= -3 \\ A_1 &= 1 \\ A_2 &= -1 \end{aligned}\right\}.$$

Hence, a particular solution is

$$y_p = -x^2 + x - 3 .$$

- Find the particular solution of

$$y'' - 3y' + 2y = 4x$$

for $y(0) = 7$; $y'(0) = 2$.

The auxiliary equation $r^2 - 3r + 2 = 0$ has the roots $r_1 = 2$ and $r_2 = 1$; consequently, the complementary solution y_c of the given equation is

$$y_c = C_1 e^{2x} + C_2 e^x$$

and the particular solution is of the form

$$y_p = A_1 x + A_0$$

with

$$y_p' = A_1 \text{ and } y_p'' = 0 .$$

Inserting y_p and its derivatives, we have

$$\left.\begin{aligned} -3A_1 + 2(A_1 x + A_0) &= 4x + 0 \\ 2A_1 x + (2A_0 - 3A_1) &= 4x + 0 \end{aligned}\right\} \qquad \left.\begin{aligned} A_0 &= 3 \\ A_1 &= 2 \end{aligned}\right\}$$

and

$$y_p = 2x + 3 .$$

The general solution of the given equation is

$$y = C_1 e^{2x} + C_2 e^x + 2x + 3,$$

where C_1 and C_2 are to be determined.

$y(0) = 7 = C_1 e^0 + C_2 e^0 + 2 \cdot 0 + 3; \quad C_1 + C_2 = 4$

Differentiation gives

$y' = 2C_1 e^{2x} + C_2 e^x + 2;$

$y'(0) = 2 = 2C_1 e^0 + C_2 e^0 + 2; \quad 2C_1 + C_2 = 0$

$$\left.\begin{aligned} C_1 + C_2 &= 4 \\ 2C_1 + C_2 &= 0 \end{aligned}\right\} \qquad \begin{aligned} C_1 &= -4 \\ C_2 &= 8 \end{aligned}$$

Hence, the sought particular solution is

$$y = -4e^{2x} + 8e^x + 2x + 3 .$$

- Find the general solution of

$$y'' + 4y' + 4y = 5e^{3x} .$$

Begin by finding the complementary solution y_c. The auxiliary equation $r^2 + 4r + 4 = 0$ has the roots

$$r_1 = r_2 = -2 ;$$

thus the complementary solution is

$$y_c = C_1 e^{-2x} + C_2 x e^{-2x}$$

with the arbitrary constants C_1 and C_2.

To find a particular solution y_p, assume

$$y_p = A x^k e^{3x}.$$

In the expression $y_p = A x^k e^{cx}$, c is not a root of the auxiliary equation. Therefore, $k = 0$, and

$$y_p = A e^{3x}.$$

Differentiation gives

$$y_p' = 3A e^{3x}\ ;\ \ y_p'' = 9A e^{3x}.$$

Inserting y_p and its derivatives, we obtain

$$9A\ e^{3x} + 4(3A e^{3x}) + 4A e^{3x} = 5e^{3x}\ ;\ A = \frac{1}{5},$$

and

$$y_p = \frac{1}{5} e^{3x}.$$

The general solution of the given equation is $y = y_c + y_p$, or

$$y = C_1 e^{-2x} + C_2 x e^{-2x} + \frac{e^{3x}}{5}.$$

- Find the general solution of

$$y'' + 4y' + 4y = 5e^{-2x}.$$

The complementary solution is

$$y_c = C_1 e^{-2x} + C_2 x e^{-2x},$$

where C_1 and C_2 are arbitrary constants.

To find a particular solution y_p, assume

$$y_p = A x^k e^{-2x}.$$

The terms of the complementary solution have the factors e^{-2x} and $x\ e^{-2x}$. By choosing $k = 2$, duplication of terms that already appear in the complementary solution is prevented. Therefore,

$$y_p = A x^2 e^{-2x}.$$

Differentiation gives

$$\left.\begin{aligned} y_p' &= 2A(x - x^2)e^{-2x} \\ y_p'' &= 2A(1 - 4x + 2x^2)e^{-2x} \end{aligned}\right\}.$$

Inserting these in the given equation, we find $A = \frac{5}{2}$,

and thus

$$y_p = \frac{5}{2} x^2 e^{-2x}$$

and the general solution is

$$y = y_c + y_p = C_1 e^{-2x} + C_2 x e^{-2x} + \frac{5x^2 e^{-2x}}{2}.$$

- Find the general solution of

$$y'' + 4y = \sin 2x .$$

Solving the homogeneous equation $y'' + 4y = 0$, we obtain the complementary solution

$$y_c = A_1 \cos 2x + A_2 \sin 2x ,$$

where A_1 and A_2 are arbitrary constants.

To find a particular solution y_p, assume

$$y_p = x^k \,(A \cos 2x + B \sin 2x) .$$

In order to prevent duplication of terms let $k = 1$, which gives

$$y_p = x\,(A \cos 2x + B \sin 2x) .$$

Differentiation gives

$$\left.\begin{aligned} y_p' &= -2Ax \sin 2x + A \cos 2x + 2Bx \cos 2x + B \sin 2x \\ y_p'' &= -4Ax \cos 2x - 4A \sin 2x - 4Bx \sin 2x + 4B \cos 2x \end{aligned}\right\} .$$

Inserting y_p and y_p'', we have

$$4B \cos 2x - 4A \sin 2x = 0 \cos 2x + \sin 2x ,$$

and by equating the coefficients,

$$A = -\frac{1}{4} \; ; \; B = 0 .$$

We now have

$$y_p = -\frac{1}{4}\, x \cos 2x$$

and the general solution is

$$y = y_c + y_p = A_1 \cos 2x + A_2 \sin 2x - \frac{1}{4}\, x \cos 2x .$$

- Find the general solution of

$$y'' - y = 3 \sin x .$$

The auxiliary equation $r^2 - 1 = 0$ has the roots $r_1, r_2 = \pm 1$, giving the complementary solution

$$y_c = C_1 e^x + C_2\, e^{-x}$$

with the arbitrary constants C_1 and C_2.

To find a particular solution y_p, assume

$$y_p = x^k \,(A \cos x + B \sin x) .$$

There is no duplication of terms from the complementary solution; therefore, $k = 0$, and we have

$$\left.\begin{aligned} y_p &= A \cos x + B \sin x \\ y_p' &= -A \sin x + B \cos x \\ y_p'' &= -A \cos x - B \sin x \end{aligned}\right\} .$$

Inserting y_p and y_p'' in the given equation, we find $A = 0$ and $B = -3/2$. Thus,

$$y_p = 0 \cos x - \frac{3}{2} \sin x = -\frac{3}{2} \sin x ,$$

and the general solution is

$$y = C_1 e^x + C_2\, e^{-x} - \frac{3}{2} \sin x .$$

- Find the general solution of

$$y'' + y = (x-1)\sin x\,.$$

The auxiliary equation $r^2 + 1 = 0$ of the reduced equation $y'' + y = 0$ gives $r = \pm i$.

The complementary solution y_c is

$$y_c = A_1 \cos x + A_2 \sin x\,,$$

where A_1 and A_2 are arbitrary constants.

Assume a particular solution y_p of a form similar to the right member of the given equation, except that the coefficients are undetermined,

$$y_p = (A\,x + B)\,x^k \cos x + (C\,x + D)\,x^k \sin x\,,$$

where we let $k = 1$ to prevent duplication; thus,

$$\left.\begin{aligned} y_p &= (A\,x^2 + B\,x)\cos x + (C\,x^2 + D\,x)\sin x \\ y_p{}' &= \cos x\,(2\,A\,x + B + C\,x^2 + D\,x) \\ &\quad + \sin x\,(-A\,x^2 - B\,x + 2\,C\,x + D) \\ y_p{}'' &= \cos x\,[-(A\,x^2 + B\,x) + 4\,C\,x + 2\,(A + D)] \\ &\quad + \sin x\,[-(C\,x^2 + D\,x) - 4\,A\,x - 2\,(B - C)] \end{aligned}\right\}\;;$$

$$\begin{aligned} y_p{}'' + y_p &= x \cdot \sin x - \sin x \\ &= 4\,C\,x \cdot \cos x + 2\,(A + D)\cos x \\ &\quad - 4\,A\,x \cdot \sin x - 2\,(B - C)\sin x \end{aligned}$$

and, by equating coefficients,

$$A = -\frac{1}{4}\,;\; B = \frac{1}{2}\,;\; C = 0\,;\; D = \frac{1}{4}\,.$$

We now have

$$\begin{aligned} y_p &= \left(-\frac{1}{4}x^2 + \frac{1}{2}x\right)\cos x + \left(0\,x^2 + \frac{1}{4}x\right)\sin x \\ &= \frac{x \sin x}{4} + \frac{x \cos x}{2} - \frac{x^2 \cos x}{4}\,. \end{aligned}$$

The general solution of the given equation is $y = y_c + y_p$, or

$$y = A_1 \cos x + A_2 \sin x + \frac{x \cos x}{2} - \frac{x^2 \cos x}{4} + \frac{x \sin x}{4}\,.$$

- Find the particular solution of

$$y'' + 2\,y' - 3\,y = x\;e^{-3\,x}$$

for $y(0) = 1$ and $y'(0) = 2$.

$$y'' + 2y' - 3y = 0\,;\quad r_1 = 1\,;\; r_2 = -3\,;$$

the complementary solution y_c is

$$y_c = C_1\,e^x + C_2\,e^{-3\,x}\,,$$

where C_1 and C_2 are arbitrary constants.

The right-hand member of the given equation may be written $(x + 0)\,e^{-3\,x}$ and we therefore assume

$$y_p = (A\,x + B)\,x^k\;e^{-3\,x}\,.$$

To prevent duplication, let $k = 1$,

$$\left.\begin{aligned} y_p &= e^{-3x}(A x^2 + B x) \\ y_p' &= e^{-3x}(-3 A x^2 + 2 A x - 3 B x + B) \\ y_p'' &= e^{-3x}(9 A x^2 - 12 A x + 9 B x + 2A - 6B) \end{aligned}\right\} ;$$

$$\begin{aligned} y_p'' + 2 y_p' - 3 y_p = {} & e^{-3x}\,(9 A x^2 - 12 A x + 9 B x + 2A - 6B \\ & - 6 A x^2 + 4 A x - 6 B x + 2B \\ & - 3 A x^2 - 3 B x) \\ = {} & e^{-3x}(- 8 A x + 2A - 4B) \end{aligned}$$

$$A = -\frac{1}{8}\,; \quad B = -\frac{1}{16}\,.$$

We now have

$$y_p = -\left(\frac{x^2}{8} + \frac{x}{16}\right) \cdot e^{-3x}$$

and the general solution of the given equation is $y = y_c + y_p$,

$$\left.\begin{aligned} y &= C_1 \cdot e^x + C_2 \cdot e^{-3x} - \left(\frac{x^2}{8} + \frac{x}{16}\right) \cdot e^{-3x} \\ y' &= C_1 \cdot e^x - 3 C_2 \cdot e^{-3x} + \left(\frac{3x^2}{8} - \frac{x}{16} - \frac{1}{16}\right) \cdot e^{-3x} \end{aligned}\right\} .$$

$$\begin{aligned} y(0) = 1 &\Rightarrow & C_1 + C_2 &= 1 \\ y'(0) = 2 &\Rightarrow & C_1 - 3C_2 - \frac{1}{16} &= 2 \end{aligned} \quad \Bigg\} \quad C_1 = \frac{81}{64}, \quad C_2 = -\frac{17}{64}\,.$$

Thus, the particular solution is

$$y = \frac{81}{64} e^x - \frac{17}{64} e^{-3x} - \frac{1}{8} x + \frac{1}{16} x\, e^{-3x}\,.$$

- Solve the equation

$$y'' + 4y' + 4y = 5\,e^{3x} + 5\,e^{-2x}\,.$$

Particular solutions of

$$y'' + 4y' + 4y = 5\,e^{3x}\,; \quad y'' + 4y' + 4y = 5\,e^{-2x}$$

are, respectively,

p. 1030

$$y_{p_1} = \frac{e^{3x}}{5}\,; \qquad y_{p_2} = \frac{5\,x^2\,e^{-2x}}{2}\,.$$

A particular solution y_p of the given equation is

$$y_p = \frac{e^{3x}}{5} + \frac{5\,x^2\,e^{-2x}}{2}\,;$$

its complementary solution is

$$y_c = C_1\,e^{-2x} + C_2\,x \cdot e^{-2x}\,,$$

and the general solution is $y = y_c + y_p$, or

$$y = C_1\,e^{-2x} + C_2\,x \cdot e^{-2x} + \frac{e^{3x}}{5} + \frac{5\,x^2\,e^{-2x}}{2}\,,$$

where C_1 and C_2 are arbitrary constants.

Series Solution of Second-Order Ordinary Differential Equations

Up till now, we have only discussed methods for solving second- or higher-order linear differential equations which are suitable for equations with constant coefficients.

Solutions for second-order linear differential equations of the form

$$\frac{d^2y}{dx^2} + P(x)\,\frac{dy}{dx} + Q(x)\,y = 0\,,$$

that is, linear homogeneous differential equations whose coefficients P and Q are functions of x, can be found by power expansion methods.

Assume the solution

$$y = f(x) = C_0 x^0 + C_1 x^1 + C_2 x^2 + C_3 x^3 + \ldots + C_n x^n = \sum_{n=0}^{\infty} C_n x^n$$

and substitute power series expansions of y and its derivatives for the terms of the given equation.

The result will be a power series formula, which is useful provided that the series meets the proper convergence criteria.

- Find the general solution and the particular solution of

$$y'' + x\,y' + y = 0$$

for the initial values $y(0) = 5$ and $y'(0) = -3$.

Assume the series solution

$$y = \sum_{n=0}^{\infty} C_n x^n\,.$$

$$y' = \sum_{n=0}^{\infty} n\,C_n x^{n-1}\,;\quad y'' = \sum_{n=0}^{\infty} n\,(n-1)\,C_n x^{n-2}\,;$$

$$x\,y' = x\sum_{n=0}^{\infty} n\,C_n x^{n-1} = \sum_{n=0}^{\infty} n\,C_n x^n\,.$$

Insert in the given equation,

$$\sum_{n=0}^{\infty} n\,(n-1)\,C_n x^{n-2} + \sum_{n=0}^{\infty} n\,C_n x^n + \sum_{n=0}^{\infty} C_n x^n = 0\,,$$

and rearrange,

$$\sum_{n=0}^{\infty} n\,(n-1)\,C_n x^{n-2} = -\sum_{n=0}^{\infty} (n+1)\,C_n x^n\,.$$

The next step is to **obtain equal powers** of x in the series expressions, so that we can **equate the coefficients**.

Replacing n by $(n + 2)$ in the series expression for y'', we obtain

$$\sum_{n=0}^{\infty} n(n-1) C_n x^{n-2} = \sum_{n+2=0}^{\infty} (n+2)[(n+2)-1] C_{n+2} x^{[(n+2)-2]}$$
$$= \sum_{n=-2}^{\infty} (n+2)(n+1) C_{n+2} x^n .$$

The rearranged equation becomes

$$\sum_{n=-2}^{\infty} (n+2)(n+1) C_{n+2} x^n = -\sum_{n=0}^{\infty} (n+1) C_n x^n ,$$

that is,

$$(n+2)(n+1) C_{n+2} = -(n+1) C_n ,$$

from which

$$C_{n+2} = -\frac{C_n}{n+2}, \text{ where } n = 0, 1, 2, 3, \ldots .$$

Thus, we find the coefficients

$C_2 = -\frac{C_0}{2}$	$C_3 = -\frac{C_1}{3}$
$C_4 = -\frac{C_2}{4} = \frac{C_0}{2 \cdot 4}$	$C_5 = -\frac{C_3}{5} = \frac{C_1}{3 \cdot 5}$
$C_6 = -\frac{C_4}{6} = \frac{-C_0}{2 \cdot 4 \cdot 6}$	$C_7 = \frac{C_5}{7} = \frac{-C_1}{3 \cdot 5 \cdot 7}$
$\vdots$	$\vdots$
$C_{2n} = \frac{(-1)^n C_0}{2 \cdot 4 \cdot 6 \ldots (2n)}$	$C_{2n+1} = \frac{(-1)^n C_1}{3 \cdot 5 \cdot 7 \ldots (2n+1)}$

With the assumption $y = \sum_{n=0}^{\infty} C_n x^n$, we now have the **general solution**

$$y = C_0 + C_1 x - \frac{C_0}{2} x^2 - \frac{C_1}{3} x^3 + \frac{C_0}{2 \cdot 4} x^4 + \frac{C_1}{3 \cdot 5} x^5 - \ldots$$

or, by gathering terms with like coefficients C_0 and C_1, respectively,

$$y = C_0\left(1 - \frac{x^2}{2} + \frac{x^4}{2 \cdot 4} - \frac{x^6}{2 \cdot 4 \cdot 6} + \ldots\right)$$
$$+ C_1\left(x - \frac{x^3}{3} + \frac{x^5}{3 \cdot 5} - \frac{x^7}{3 \cdot 5 \cdot 7} + \ldots\right).$$

$y(0) = 5$ gives $C_0 = 5$.

Differentiating y yields

$$y' = C_0\left(0 - \frac{2x}{2} + \frac{4x^3}{2 \cdot 4} - \frac{6x^5}{2 \cdot 4 \cdot 6} + \dots\right)$$

$$+ C_1\left(1 - \frac{3x^2}{3} + \frac{5x^4}{3 \cdot 5} - \frac{7x^6}{3 \cdot 5 \cdot 7} + \dots\right).$$

$y'(0) = -3$ gives $C_1 = -3$.

Inserting C_0 and C_1, we obtain the **particular solution**

$$y = 5\left(1 - \frac{x^2}{2} + \frac{x^4}{2 \cdot 4} - \frac{x^6}{2 \cdot 4 \cdot 6} + \dots\right)$$

$$-3\left(x - \frac{x^3}{3} + \frac{x^5}{3 \cdot 5} - \frac{x^7}{3 \cdot 5 \cdot 7} + \dots\right).$$

p. 277 The series converges for $|x| \le 1$ by the *ratio test.*

Alternative Answers:

Sometimes a series may be recognized as the expansion of an elementary function. Indeed, for one of the series obtained in the last example, we have

$$1 - \frac{x^2}{2} + \frac{x^4}{2 \cdot 4} - \frac{x^6}{2 \cdot 4 \cdot 6} + \dots$$

$$= 1 + \left(-\frac{x^2}{2}\right) + \frac{1}{2!}\left(-\frac{x^2}{2}\right)^2 + \frac{1}{3!}\left(-\frac{x^2}{2}\right)^3 + \dots = e^{-x^2/2}.$$

The second series that is part of the general solution is not an elementary function; and the particular solution for the initial values $y(0) = 5$ and $y'(0) = -3$ may be written

$$y = 5\,e^{-x^2/2} - 3\left(x - \frac{x^3}{3} + \frac{x^5}{3 \cdot 5} - \frac{x^7}{3 \cdot 5 \cdot 7} + \dots\right).$$

If the initial values are specified, as here, for $x = 0$, the solution $y = \sum_{n=0}^{\infty} C_n x^n$ is a suitable assumption. However, if the initial values of x are prescribed at points $x \neq 0$, we must choose another series expansion of y. The following example is a case in point.

- Find the particular solution of

$$y'' - 2(x-1)y' + y = 0$$

for the initial values $y(1) = 3$ and $y'(1) = -2$.

Since the initial conditions are specified at $x = 1$, it is advisable to find a series solution in powers of $(x - 1)$ instead of powers of x. The advantage will be evident when solving for the constants of the particular solution.

With $X = (x - 1)$, the given equation becomes

$$y'' - 2Xy' + y = 0 .$$

Assume the solution

$$y = \sum_{n=0}^{\infty} C_n X^n$$

with

$$y' = \sum_{n=0}^{\infty} n\, C_n X^{n-1}; \quad y'' = \sum_{n=0}^{\infty} n\,(n-1)\, C_n X^{n-2}$$

and

$$Xy' = X \sum_{n=0}^{\infty} n\, C_n X^{n-1} = \sum_{n=0}^{\infty} n\, C_n X^n .$$

To obtain equal powers of x in both series expressions replace n with $(n + 2)$ in the expansion of y''; thus, as in the previous example,

$$y'' = \sum_{n=0}^{\infty} n\,(n-1)\, C_n X^{n-2} = \sum_{n=-2}^{\infty} (n+2)\,(n+1)\, C_{n+2} X^n .$$

The given equation may now be written

$$\sum_{n=-2}^{\infty} (n+2)\,(n+1)\, C_{n+2} X^n - 2 \sum_{n=0}^{\infty} n\, C_n X^n + \sum_{n=0}^{\infty} C_n X^n = 0$$

or

$$\sum_{n=-2}^{\infty} (n+2)\,(n+1)\, C_{n+2} X^n - \sum_{n=0}^{\infty} (2n-1)\, C_n X^n = 0$$

or

$$\sum_{n=0}^{\infty} (n+2)\,(n+1)\, C_{n+2} X^n = \sum_{n=0}^{\infty} (2n-1)\, C_n X^n .$$

Equating the coefficients, we have

$$(n+2)\,(n+1)\, C_{n+2} = (2n-1)\, C_n$$

and

$$C_{n+2} = \frac{(2n-1)}{(n+2)\,(n+1)}\, C_n ; \qquad n = 0, 1, 2, 3 \ldots .$$

We find the coefficients

$$C_{0+2} = C_2 = -\frac{C_0}{2 \cdot 1}$$

$$C_4 = \frac{3\,C_2}{4 \cdot 3} = \frac{-3\,C_0}{4 \cdot 3 \cdot 2 \cdot 1}$$

$$C_6 = \frac{7\,C_4}{6 \cdot 5} = \frac{-7 \cdot 3C_0}{6 \cdot 5 \cdot 4 \cdot 3 \cdot 2 \cdot 1}$$

$$C_8 = \frac{11\,C_4}{8 \cdot 7} = \frac{-11 \cdot 7 \cdot 3C_0}{8 \cdot 7 \cdot 6 \cdot 5 \cdot 4 \cdot 3 \cdot 2 \cdot 1}$$

etc.

$$C_{1+2} = C_3 = \frac{C_1}{3 \cdot 2}$$

$$C_5 = \frac{5\,C_3}{5 \cdot 4} = \frac{5\,C_1}{5 \cdot 4 \cdot 3 \cdot 2}$$

$$C_7 = \frac{9\,C_5}{7 \cdot 6} = \frac{9 \cdot 5C_1}{7 \cdot 6 \cdot 5 \cdot 4 \cdot 3 \cdot 2}$$

$$C_9 = \frac{13\,C_5}{9 \cdot 8} = \frac{13 \cdot 9 \cdot 5C_1}{9 \cdot 8 \cdot 7 \cdot 6 \cdot 5 \cdot 4 \cdot 3 \cdot 2}$$

etc.

We can distinguish a pattern; thus:

$C_2 = -\dfrac{C_0}{2\,!}$	$C_3 = \dfrac{C_1}{3\,!}$
$C_4 = -\dfrac{3\,C_0}{4\,!}$	$C_5 = \dfrac{5\,C_1}{5\,!}$
$C_6 = -\dfrac{3 \cdot 7\,C_0}{6\,!}$	$C_7 = \dfrac{5 \cdot 9\,C_1}{7\,!}$
$C_8 = -\dfrac{3 \cdot 7 \cdot 11\,C_0}{8\,!}$	$C_9 = \dfrac{5 \cdot 9 \cdot 13\,C_1}{9\,!}$
$\vdots$	$\vdots$
$C_{2n} = -\dfrac{3 \cdot 7 \cdot 11 \ldots (4n-5)}{(2n)\,!}\,C_0$	$C_{2n+1} = \dfrac{5 \cdot 9 \ldots (4n-3)}{(2n+1)\,!}\,C_1$

$n = 1, 2, 3 \ldots .$

$y = \sum_{n=0}^{\infty} C_n X^n$ gives the **general solution**

$$y = C_0\left(1 - \frac{1}{2\,!}X^2 - \frac{3}{4\,!}X^4 - \frac{3 \cdot 7}{6\,!}X^6 - \ldots\right)$$

$$+C_1\left(X + \frac{1}{3\,!}X^3 + \frac{5}{5\,!}X^5 + \frac{5 \cdot 9}{7\,!}X^7 + \ldots\right)$$

or, with $X = x - 1$,

$$y = C_0\left[1 - \frac{1}{2\,!}(x-1)^2 - \frac{3}{4\,!}(x-1)^4 - \frac{3 \cdot 7}{6\,!}(x-1)^6 - \ldots\right]$$

$$+C_1\left[(x-1) + \frac{1}{3\,!}(x-1)^3 + \frac{5}{5\,!}(x-1)^5 + \frac{5 \cdot 9}{7\,!}(x-1)^7 + \ldots\right],$$

which for the given initial value $y(1) = 3$ gives $C_0 = 3$.

Differentiating y yields

$$y' = C_0\left[0 - \frac{2}{2\,!}(x-1) - \frac{3 \cdot 4}{4\,!}(x-1)^3 - \ldots\right]$$

$$+C_1\left[1 + \frac{3}{3\,!}(x-1)^2 + \frac{5 \cdot 5}{5\,!}(x-1)^4 + \ldots\right],$$

which for $y'(1) = -2$ gives $C_1 = -2$.

Inserting these numerical values into the general solution, we now obtain the **particular solution**

$$y = 3\left[1 - \frac{1}{2\,!}(x-1)^2 - \frac{3}{4\,!}(x-1)^4 - \frac{3 \cdot 7}{6\,!}(x-1)^6 - \ldots\right]$$

$$-2\left[(x-1) + \frac{1}{3\,!}(x-1)^3 + \frac{5}{5\,!}(x-1)^5 + \frac{5 \cdot 9}{7\,!}(x-1)^7 + \ldots\right].$$

"I still don't have all the answers, but I'm beginning to ask the right questions."

The reach of differential equations is boundless and much more can be said about methods for their solution. But not here. With the foundation laid in this book, more advanced texts on differential equations and other branches of mathematics may now be the reader's oyster.

Bibliography

The list represents essential works that have inspired me and provided me with knowledge, but it could never be made complete. It may serve as a recommendation for further reading, yet one must hunt down the things one is interested in and not rely on book lists of others.

The list represents editions consulted by me; some works now appear in later editions. Some of the non-English-language works are available in English translation.

Aabe, A.: *Episodes from the Early History of Mathematics*. New York: Random House, 1964.

Adams, Colin C.: The Knot Book. New York: W.H. Freeman and Company, 1994.

Anderson, Robert F. V.: *Introduction to Linear Algebra*. New York: Holt, Reinhart and Winston, 1988.

Andrews, Larry C.: *Ordinary Differential Equations*. Glenview, IL: Scott, Forseman, 1982.

Appel, Kenneth, and Wolfgang Haken: "The Solution of the Four-Color Map Problem". *Scientific American*, May 1977.

Ardalan, Neder, and Laleh Bakhtiar: *The Sense of Unity. The Sufi Tradition in Persian Architecture*. Chicago: University of Chicago Press, 1973.

Arfken, George: *Mathematical Methods for Physicists*. Orlando: Academic Press, 1985.

Armstrong, B. A.: *Basic Topology*. New York: Springer, 1983.

Ashurst, F. Gareth: *Founders of Mathematics*. London: Frederick Muller, 1982.

Aspray, William, and Philip Kitcher (editors): *History and Philosophy of Modern Mathematics*. Minneapolis: University of Minnesota Press, 1988.

Barr, Stephen: *Experiments in Topology* (1964). New York: Dover, 1989.

Baugh, Albert C.: *A History of the English Language*. New York: Appleton, 1935.

Beckmann, Petr: *A History of π*, 3rd ed. New York: St. Martin's Press, 1971.

Bell, Eric Temple: *Mathematics, Queen and Servant of Science* (1951). Redmond, WA: Tempus, 1989.

Bergendal, Gunnar, Mats Håstad, and Lennart Råde: *Statistik och sannolikhetslära*. Stockholm: Biblioteksförlaget, 1967.

Bittinger, Marvin L.: *Logic, Proof, and Sets*, 2nd ed. Reading, MA: Addison-Wesley, 1982.

Björner, Anders, Lars Brandell, and Olof Schönbeck: *Matematik för naturvetare*. 3 vol. Stockholm: Matematiska institutionen, Stockholms universitet, 1978.

Blomfield, Leonard: *Language*. London, 1935.

Boas, Mary L.: *Mathematical Methods in the Physical Sciences*, 3rd ed. New York: John Wiley & Sons, 1983.

Bold, Benjamin: *Famous Problems of Geometry and How to Solve Them*. New York: Dover, 1969.

Bombieri, Enrico: "Prime Territory: Exploring the Infinite Landscape at the Base of the Number System". *The Sciences*, Vol. 32, No. 5, Sept./Oct. 1992.

Boyer, Carl B.: "The Invention of Analytic Geometry". *Scientific American*, Jan. 1949.

– : *A History of Mathematics.* New York: John Wiley & Sons, 1968.

– : *A History of Mathematics*; revision by Uta C. Merzbach. New York: John Wiley & Sons, 1991.

Bronshtein, I. N., and K. A. Semendyayev: *Handbook of Mathematics.* Translated from the German. New York: Verlag Harri Deutsch/VNR, 1985.

Brun, Viggo: *Alt er tall.* Norway: Universitetsforlaget, 1964.

Buck, Carl Darling: *A Dictionary of Selected Synonyms in the Principal Indo-European Languages.* Chicago: University of Chicago Press, 1949.

Bunt, Lucas N. H., Philip S. Jones, and Jack D. Bedient: *The Historical Roots of Elementary Mathematics.* New York: Dover, 1976.

Cajori, F.: *A History of Mathematical Notations*, 2 vol. La Salle, IL: Open Court, 1928, 1929. New York: Dover, 1993.

Cambell, A.: *Old English Grammar.* Oxford: Claredon Press (1959). New York: Dover, 1993.

Cardano, Girolamo: *Ars Magna or the Rules of Algebra.* Translated by T. Richard Witmer. The Massachusetts Institute of Technology (1968). New York: Dover, 1993

Carnap, Rudolf: "What Is Probability?" *Scientific American*, Sept. 1953.

Closs, Michael P. (editor): *Native American Mathematics.* Austin, TX: University of Texas Press, 1986.

Coe, Michael D.: *Breaking the Maya Code.* New York: Thames and Hudson, 1992.

Cohen, Joel E.: *How Many People Can the Earth Support?* New York: W. W. Norton & Co., 1995.

Cohen, Paul J., and Reuben Hersh: "Non-Cantorian Set Theory". *Scientific American*, Dec. 1967.

Compact Edition of the Oxford English Dictionary, The. Oxford: Oxford University Press, 1971.

Courant, R.: *Differential and Integral Calculus*, 2nd ed., 2 vol. (1937). Translated from the German. New York: Interscience/Wiley Classics Library, 1988.

Courant, Richard: "Mathematics in the Modern World". *Scientific American*, Sept. 1964.

Courant, Richard, and Herbert Robbins: *What Is Mathematics? An Elementary Approach to Ideas and Methods.* Oxford: Oxford University Press, 1941.

Dantzig, Tobias: *Number: The Language of Science*, 4th ed. (1954). New York: Free Press/Macmillan, 1954.

Dauben, Joseph W.: "Georg Cantor and the Origins of Transfinite Set Theory". *Scientific American*, June 1983.

Davis, Harry F., and Arthur David Snider: *Introduction to Vector Analysis*, 5th ed. Dubuque, IA: Wm. C. Brown, 1987.

Davis, Philip J.: "Number". *Scientific American*, Sept. 1964.

Davis, Philip J., and Reuben Hersch: *The Mathematical Experiment.* Boston: Houghton Mifflin, 1981.

Dedron, P., and J. Itard: *Mathematics and Mathematicians*, 2 vol. Translated from the French. Milton Keynes, England: Open University Press/ Richard Sadler, 1973.

Devaney, Robert L.: *Chaos, Fractals, and Dynamics: Computer Experiments in Mathematics.* Menlo Park, CA: Addison-Wesley, 1990.

Devlin, Keith: *The Joy of Sets*, 2nd ed. New York: Springer-Verlag, 1993.

Dewdney, A. K.: "Computer Recreations. A Computer Microscope Zooms in for a Look at the Most Complex Object in Mathematics". *Scientific American*, Aug. 1985.

– : "Computer Recreations. Beauty and Profundity: The Mandelbrot Set and a Flock of Its Cousins Called Julia". *Scientific American*, Nov. 1987.

Dilke, O. A. W.: *Mathematics and Measurement.* Berkeley: University of California Press/British Museum, 1987.

Dobinson, John: *Mathematics for Technology*, 2 vol. London: Penguin, 1969, 1972.

Dunham, William: *Journey through Genius.* New York: John Wiley & Sons, 1990.

– : *The Mathematical Universe.* New York: John Wiley & Sons, 1994.

Dunkels, Andrejs, Håkan Ekblom, Anders Grennberg, Torbjörn Hedberg, Eilif Hensvold, Henry Kallioniemi, and Reinhold Näslund: *Derivator, integraler och sånt .... Analys med numeriska metoder för tekniska högskolor*, 2 vol. Sweden: Centek, 1985, 1986.

Dyson, Freeman J.: "Mathematics in the Physical Sciences". *Scientific American*, Sept. 1964.

Ebbinghaus, H.-D., H. Hermes, F. Hirzebruch, M. Koecher, K. Mainzer, J. Neukirch, A. Prestel, and R. Remmert: *Numbers.* New York: Springer, 1990.

Einstein, Albert: *Über die spezielle und allgemeine Relativitätsteorie.* Braunschweig, 1972.

Ekeland, Ivar: *Le calcul, l'Imprévu: Les figures du temps de Kepler á Thom.* Paris: Editions du Seuil, 1984.

Encyclopædia Britannica, 15th ed. Chicago: Encyclopædia Britannica, 1986.

Englefield, M. J.: *Mathematical Methods for Engineering and Science Students.* London: Edward Arnold, 1987.

Euclid: *Euclid's Elements.* From the text of Heiberg (Heiberg's text published 1883 - 88) with introduction and comments by Thomas L. Heath (original second edition 1928), 3 vol. New York: Dover, 1956.

Eves, Howard: *An Introduction to the History of Mathematics*, 6th ed. Philadelphia: Saunders, 1990.

– : *Great Moments in Mathematics before 1650.* Dolciani Mathematical Expositions; No. 5. The Mathematical Association of America, 1983.

– : *Great Moments in Mathematics after 1650.* Dolciani Mathematical Expositions; No. 7. The Mathematical Association of America, 1983.

Fauvel, John, and Jeremy Gray (editors): *The History of Mathematics: A Reader.* London: Macmillan Education Ltd., The Open University, 1987.

Fibonacci Association: *The Fibonacci Quarterly*, Vol. 32, No. 1, Feb. 1994.

Finkbeiner, Daniel Talbot: *Introduction to Matrices and Linear Transformations*, 3rd ed. San Francisco: W. H. Freeman and Company, 1978.

Flato, Moshé: *Le pouvoir des mathématiques.* Paris: Hachette, 1990.

Fletcher, Peter: *Foundations of Higher Mathematics.* Boston: PWS-KENT, 1988.

Folkerts, Menso, Knobloch, E., and Reich, K.: *Maß, Zahl und Gewicht: Mathematik als Schlüssel zu Weltverständnis und Weltbeherrschung.* Wolfenbüttel, Germany: VCH, 1989.

Friberg, Jöran: "Numbers and Measures in the Earliest Written Records". *Scientific American*, Feb. 1984.

Friedrichs, K. O.: *From Pythagoras to Einstein.* The Mathematical Association of America, 1965.

Gårding, Lars: *Encounter with Mathematics.* New York: Springer, 1977.

Gask, H.: *Ordinära differentialekvationer*, 8th ed. Lund, Sweden: Lunds matematiska sällskap, 1978.

Gellert, W., H. Küstner, M. Hellwich, and H. Kästner (editors): *The VNR Concise Encyclopedia of Mathematics*. New York: Van Nostrand Reinhold, 1977.

Gillispie, Charles Coulston (editor), The American Council of Learned Societies: *Dictionary of Scientific Biography,* 15 vol., 1 suppl. New York: Charles Scribner's Sons, 1970 - 80.

Godman, A., and J. F. Talbert: *Additional Mathematics, Pure and Applied,* 2nd ed. London: Longman, 1973

Goldfeld, Dorian: "Beyond the Last Theorem". *The Sciences*; March/April, 1995.

Gradshteyn, I. S., and I. M. Ryzhik: *Table of Integrals, Series, and Products*. New York: Academic Press, 1980.

Greenberg, Michael D.: *Advanced Engineering Mathematics*. Englewood Cliffs: Prentice Hall, 1988.

Griffiths, H. B., and P. J. Hilton: *Classical Mathematics*. London: Van Nostrand Reinhold, 1970.

Gullberg, Jan: *Vätska-Gas-Energi*, 2nd ed. Stockholm: Nordiska Bokhandeln, 1980

Hadley, James: *Greek Grammar for Schools and Colleges*. New York: Appleton, 1884.

Hall, Tord: *Matematikens utveckling*. Lund, Sweden: Gleerups, 1970.

Hamilton, A. G.: *A First Course in Linear Algebra*. Cambridge: Cambridge University Press, 1987.

Harkness, Albert: *A Complete Latin Grammar*. New York: American Book Co., 1898.

Hay, G. E.: *Vector and Tensor Analysis*. NY: Dover, 1953.

Heath, T. L.: *A History of Greek Mathematics*, 2 vol. (1921). New York: Dover, 1981.

– : See above under Euclid.

Hendricks, John R.: "The Magic Hexagram". *Journal of Recreational Mathematics*, Vol. 25, No.1, 1993.

Hilbert, David: *Grundlagen der Geometrie*. Stuttgart, Germany: B.G. Teuben, 1898 - 99. English translation from the tenth edition: *Foundations of Geometry*. La Salle, IL: Open Court (1971). New York: Dover, 1994.

Hilton, Peter, Friedrich Hirzebruch, and Reinhold Remmert (editors): *Miscellanea Mathematica*. Berlin: Springer-Verlag, 1991.

Hobson, E. W.: *"Squaring the Circle": A History of the Problem*. New York: Chelsea Publishing Company, 1969.

Hoffman, Banesh: *About Vectors* (1966). NY: Dover, 1975.

Hogben, Lancelot: *Mathematics for the Million* (1937). New York: W. W. Norton, 1993.

Ifrah, Georges: *Histoire universelle des chiffres*. Paris: Editions Seghers, 1981.

International Organization for Standardization: *ISO Standards Handbook 2: Units of Measurement*, 2nd ed. Switzerland: ISO, 1982 and later supplements.

James/James: *Mathematics Dictionary*. New York: Van Nostrand Reinhold, 1976 (4th ed.) and 1992 (5th ed.).

Jeffrey, Alan: *Handbook of Mathematical Formulas and Integrals*. San Diego: Academic Press, 1995.

Johnson, Richard E., and Fred L. Kiokemeister: *Calculus with Analytic Geometry*, 2nd ed. Boston: Allyn and Bacon, 1960.

Kac, Mark: "Probability". *Scientific American*, Sept. 1964.

Kasner, Edward, and James R. Newman: *Mathematics and the Imagination* (1st ed. 1940). Redmond, WA: Tempus, 1989.

Kiepert, Ludwig: *Grundriß der Differential- und Integral-Rechnung.* Hannover, Germany: Helwingsche Verlagsbuchhandlung, 1910.

Kiyosi, Itó (editor), Mathematics Society of Japan: *Encyclopedic Dictionary of Mathematics*, 2 vol., 2nd ed. Cambridge, MA: MIT Press, 1993.

Kline, Morris: "Geometry ". *Scientific American*, Sept. 1964.

– : *Mathematical Thought from Ancient to Modern Times*, 3 vol. New York: Oxford University Press, 1972.

– : *Mathematics and the Search for Knowledge.* New York: Oxford University Press, 1985.

– : *Mathematics in Western Culture.* New York: Oxford University Press, 1953.

Kneale, William, and Martha Kneale: *The Development of Logic.* New York: Oxford University Press, 1962.

Kramer, Edna E.: *The Nature and Growth of Modern Mathematics.* Princeton: Princeton University Press, 1982.

Larson, Roland E., and Robert P. Hostetler: *Calculus with Analytic Geometry*, 3rd ed. Lexington, MA: Heath, 1986.

Lewis, Charlton T.: *An Elementary Latin Dictionary.* New York: American Book Co., 1890.

Lindblom, Verner: *Klassiska klurigheter.* Västerås, Sweden: ICA Bokförlag AG, 1992.

MacDuffee, Cyrus Colton: *The Theory of Matrices.* New York: Chelsea Publishing Company, 1956.

Mac Lane, Saunders: *Mathematics: Form and Function.* New York: Springer, 1986.

Mandelbrot, B.: "How Long Is the Coast of Britain? Statistical Self-Similarity and Fractional Dimension". *Science*, Vol. 156, 1967 *pp.* 636 - 38.

Martensen, Johan C.: *Från Bologna till Stockholm.* Stockholm: Distribution by Nordiska Bokhandeln, 1989.

Matthews, W. H.: *Mazes and Labyrinths: Their History and Development* (1922). New York: Dover, 1970.

Menninger, Karl.: *Zahlwort ond Ziffer: Eine Kulturgeschichte der Zahlen*, 2 vol. Göttingen, Germany: Vadenhoek & Ruprech, 1957 - 58.

Munhe, Mustafa A., and David J. Foulis: *Algebra and Trigonometry with Applications.* New York: Worth, 1982.

Nagel, Ernest, and James R. Newman: "Gödel's Proof". *Scientific American*, June 1956.

National Council of Teachers of Mathematics: *Historical Topics for the Mathematics Classroom.* Washington, DC: National Council of Teachers of Mathematics, 1969.

Needham, Joseph: *Science and Civilization in China: Mathematics and the Sciences of Heavens and the Earth.* Cambridge: Cambridge University Press, 1968.

Neugebauer, Otto: *The Exact Sciences in Antiquity.* Princeton: Princeton University Press, 1952.

Neuwirth, Lee: "The Theory of Knots". *Scientific American*, June 1979.

Newman, James R.: "The Rhind Papyrus". *Scientific American*, Aug. 1952.

– : *The World of Mathematics*, 4 vol. (1956). Redmond, WA: Tempus, 1988.

Osen, Lynn M.: *Women in Mathematics.* Cambridge, MA: MIT Press, 1974.

Paulos, John Allen: *Beyond Numeracy.* New York: Knopf, 1991.

Pearson, Carl E. (editor): *Handbook of Applied Mathematics*, 2nd ed. New York: Van Nostrand Reinhold, 1983.

Pedoe, Dan: *The Gentle Art of Mathematics*. London: English UP, 1958.

Peterson, Ivars: *The Mathematical Tourist: Snapshots of Modern Mathematics*. New York: W. H. Freeman, 1989.

– : *Islands of Truth: A Mathematical Mystery Cruise*. New York: W. H. Freeman, 1990.

Philips, Anthony: "Turning a Surface Inside Out". *Scientific American*, May 1966.

Phillips, H. B.: *Differential Equations*, 3rd ed. New York: John Wiley & Sons, 1934.

Porkess, Roger: *Dictionary of Statistics*. London: Collins, 1988.

Quine, W. V.: "Foundations of Mathematics". *Scientific American*, Sept. 1964.

Råde, Lennart, and Bertil Westergren: *Beta: Mathematics Handbook*. Lund, Sweden: Studentlitteratur/Chartwell-Bratt, 1988.

Rasmussen, Chr.: *Grønlandsk Sproglære*. Copenhagen, 1888.

Resnikoff, H. L., and R. O. Wells, Jr.: *Mathematics in Civilization*. New York: Dover, 1984.

Ronan, Colin A.: *The Cambridge Illustrated History of the World's Science*. Cambridge: Cambridge University Press, Newnes Books, 1983.

Roslund, Curt: "Orientation and Geometry of Ale's Stones". *Archeology*, Vol 3, No. 4, 1980.

– : "EDM Technique Applied to the Prehistoric Monument Ale's Stones". *Archeology and Natural Science*, Vol 1, 1993.

Ruhlen, Merritt: *The Origin of Language*. New York: John Wiley & Sons, 1994.

Russell, David M.: *MathView™ Professional*. Computer software. Calabasos, CA: BrainPower Inc., 1987.

Singer, Charles: *A Short History of Scientific Ideas to 1900*. Oxford: Oxford University Press, 1959.

Skeat, Walter W.: *A Concise Etymological Dictionary of the English Language*. New York: Putnam, 1980.

Smart, James R.: *Modern Geometries*, 3rd ed. Pacific Grove, CA: Brooks/Cole, 1988.

Smith, D. E.: *History of Mathematics*, 2 vol. New York: Dover, 1958.

– : *A Source Book in Mathematics*. New York: McGraw-Hill, 1929.

Steen, Lynn Arthur (editor): *Mathematics Today: Twelve Informal Essays*. New York: Springer, 1978.

Stephenson, G.: *Mathematical Methods for Science Students*, 2nd ed. London: Longman, 1973.

Stewart, Ian: "Gauss". *Scientific American*, July 1977.

Stewart, James: *Calculus*. Monterey, CA: Brooks/Cole, 1987.

Stillwell, John: *Mathematics and Its History*. New York: Springer, 1989.

Stoll, Robert R.: *Set Theory and Logic* (1963). New York: Dover, 1979.

Stone, Richard: "Mathematics in the Social Sciences". *Scientific American*, Sept. 1964.

Strang, Gilbert: *Introduction to Applied Mathematics*. Wellesley, MA: Wellesley-Cambridge, 1986.

Stroud, K. A.: *Engineering Mathematics*, 2nd ed. New York: Springer, 1982.

– : *Further Engineering Mathematics*, 2nd ed. New York: Springer, 1986.

Struik, D. J.: *A Concise History of Mathematics*, 3rd ed. New York: Dover, 1987.

– : *A Source Book in Mathematics 1200 - 1800* (1969). Princeton University Press, 1986.

Swetz, Frank J. (editor): *From Five Fingers to Infinity*. Chicago and La Salle, IL: Open Court, 1994.

Tarski, Alfred: "Truth and Proof". *Scientific American*, June 1969.

Thomas, Ivor: *Greek Mathematical Works. I: Thales to Euclid* (1939). Cambridge, MA: Harvard University Press, 1980.

Thompson, J. E.: *Algebra for the Practical Worker*, 4th ed. New York: Van Nostrand Reinhold, 1982.

– : *Calculus for the Practical Worker*, 4th ed. New York: Van Nostrand Reinhold, 1982.

– : *Geometry for the Practical Worker*, 4th ed. New York: Van Nostrand Reinhold, 1982.

Thompson, Silvanus P.: *Calculus Made Easy* (1st ed. 1910), 3rd ed. New York: St. Martin's Press, 1946.

Titchmarsh, E. C.: *Mathematics for the General Reader* (1959). New York: Dover, 1981.

Vygodsky, M.: *Mathematical Handbook: Elementary Mathematics.* Translated from the Russian. Moscow: MIR, 1979.

Wagner, Ulrich: *Das Bamberger Rechenbuch von 1483.* Berlin: Akademie-Verlag, 1988.

Walpole, Ronald E.: *Elementary Statistical Concepts.* New York: Collier Macmillan, 1976.

Washington, Allyn J.: *Basic Technical Mathematics with Calculus.* Menlo Park, CA: Cummings, 1970.

Weaver, Warren: "Probability". *Scientific American*, Oct. 1950.

Whitehead, Alfred North: *An Introduction to Mathematics.* London: Oxford University Press, 1948.

Wolfram, Stephen: *Mathematica™: A System for Doing Mathematics by Computer.* Computer software. Champaign, IL: Wolfram Research, 1988 - 91.

– : *Mathematica™: A System for Doing Mathematics by Computer*, 2nd ed. Redwood City, CA: Addison-Wesley, 1991.

Woodcock, Jim, and Martin Loomes: *Software Engineering Mathematics.* Reading, MA: Addison-Wesley, 1988.

Yoshino, Y.: *The Japanese Abacus Explained* (1937). New York: Dover, 1963.

Works Cited

The strength of those who have gone before should not be gleaned solely from histories or random quotations. True inspiration and power lie in the original works.

Numbers in bold face at the end of the entries refer to page number in this book.

∗ before a page number indicates display of a reproduction of the work.

Abel, Niels Henrik: Article in *Journal für die reine und angewandte Mathematik* (*"Crelle's Journal"*, 1826); **300**

Agnesei, Maria: *Instituzione analitiche* (1748); **680**

– : Analytical Institutions (1801); **680**

Apollonius of Perga: *Conics* (*c.* 200 B.C.); **408**

Arbogast, L.-F.-A.: *Du calcul des dérivations* (1800); **686, ∗1016**

Archimedes: *On the Measurement of the Circle*; **90**

– : *The Sand Reckoner*; **29**

Argand, Jean Robert : *Les quantitiés imaginaires* (1806); **∗88**

Bamberger Rechenbuch (1483); **∗74, ∗118**

Barrow, Isaac: *Lectiones Opticae & Geometricae* (1674); **675**

– : *Euclid's Elements* (travel edition, 1686); **∗416**

Bashkara: *Lilavati* (*c.* 1150); **299**

– : *Vija-ganita* (*c.* 1150); **299**

Bernoulli, Jacob: *Ars Conjectandi* (1713); **30, ∗963**

Bible, 1 Kings 7.23; **93**

Bible, Mark 5.9; **14**

Bible (1562 misprint), Matt. 5.9; **46**

Bombelli, Raffaele: *L'Algebra* (1572); **87**

Boole, G.: *Mathematical Analysis of Logic* (1847); **217**

– : *An Investigation into the Laws of Thought, on which are founded the Mathematical Theories of Logic and Probabilities* (1854); **217**

Briggs, H.: *Logarithmorum chilias prima* (1617); **152**

– : *Arithmetica logarithmica* (1624); **152, 155**

Bürgi, J.: *Aritmetische und Geometrische Progress Tabulen* (1620); **153**

Buteo, Johannes: *Opera geometrica* (1559); **108**

Cantor, G.: Publications in *Acta Mathematica*; **262**

– : *Über eine Eigenschaft des Inbegriffes aller reellen Zahlen* (1874); **257**

– : "Beiträge zur Begründung der transfeniten Mengenlehre", *Matematische Annalen* (1895 - 97); **257**

– : *Contributions to the Founding of the Theory of Transfinite Numbers* (1915) (this is a translation from the German by P. E. B. Jourdain of the above "Beiträge ..."); **257**

Cardano, G.: *Artis magna sive de regulis algebraicis liber unus* (1545); **∗73, ∗299, 316, 322**

– : *Liber de ludo aleae* (1663); **963**

Carroll, Lewis: *Through the Looking-Glass* (1872); **227**

Cauchy, A.-L.: *Cours d'analyse* (1821); ∗**345**
– : *Leçons sur le calcul différentiel* (1829); ∗**719**
Cavalieri, Bonaventura: *Geometria indivisibilibus continuorum*; ∗**674**
Ceulen, Ludolph van: *Van den Circkel* (1596); ∗**92**
– : *Aritmetische en Geometrische fondamenten* (1615); **92**
Chu Shih-Chieh: *Precious Mirror of the Four Elements* (1303); ∗**141**
Codex Mendoza; **60**
Copernicus: *De revolutionibus*; **464**
Corpus Iuris Civilis; ∗**40**
Cotes, Roger: *Harmonia mensurarum* (1722); **581**
Cremona, Gerard of: *The Almagest* (translation, *c.* 1175); **461, 463**
Decker, Ezechiel de, and Adriaan Vlacq: *Het tweede deel vand de Nieuwe telkonst* (1627); **152**
– : *Arithmetica logarithmica* (1628); **152**
Dedekind, R.: *Stetigkeit und irrationale Zahlen* (1872); **258**
– : *Was sind und was sollen die Zahlen* (1887); **70**
Defoe, Daniel: *The Life and Strange Surprising Adventures of Robinson Crusoe of York, Marincr* (1719); **32**
Delamain, R.: *Grammologia or the Mathematical Ring* (1630); **176**
– : *The Making, Description and Use of a Small Portable Instrument for the Pocket ... Called a Horizontal Quadrant* (1632); **177**
Desargues, Girard: *Broullion projet ...* (1639); **434**
Descartes, R.: *La géométrie* (1673); **367**, ∗**368, 548**
Diophantus: *Arithmetica* (*c.* A.D. 250); **259**
Eco, Umberto: *Foucault's Pendulum* (1989); **6**
– : *The Name of the Rose* (198); **7, 219**
Euclid: *Elements* (*Stoicheia*) and revised publications; ∗**365, 366, 367**, ∗**381, 386**, ∗**413**, ∗**416**, ∗**425**
Euler, L.: *Mechanica* (1736); **85**, ∗**86**
– : *Methodus inviendi lineas curvas maximi minimi ...* (1744); **681**
– : *Introductio in analysin infinitorum* (1748); ∗**465, 680**
– : *Institutionis calculi differentialis* (1755); ∗**680**
– : *Institutionum calculi integralis* (1768 - 70); ∗**680**
Faulhaber, J.: *Arithmetischer Cubiccossischer Lustgarten* (1604) ; **212**
Fermat, Pierre de: *Cogita physico-mathematica* (1644); **79**
– : *Ad locos planos et solidos isagoge* (1679); **548**
Fibonacci, L. de (Leonardo of Pisa): *Liber abaci* (1202); **102, 286, 299**
– : *Practica geometriae* (1220); **91, 464**
Fibonacci Association: *The Fibonacci Quarterly*; **288**
Fincke, Thomas: *Geometriae rotundi* (1583); **465, 486**
Fisher, R. A.: *Statistical Methods for Research Workers* (1925); ∗**964**
Fourier, J.: *Théorie analytique de la chaleur* (1822); **336**, ∗**909**
Frege, G.: *Begriffsschrift: Eine der arithmetischen nachgebildete Formelschprache des reinen Denkens* (1879); **217**
– : *Grundsetze der Arithmetik* (1893, 1903); **217**
Gafurius, F.: *Theorica Musice* (1492); ∗**922**
Galilei, G.: *Dialogue Concerning the Two Chief World Systems* (1632); **257**
– : *Discorsi e Dimostrazione Matematiche* (1638); ∗**893**

Gauss, C. F.: *Demonstratio nova theorematis omnem functionem algebraicum* ... (1799); ∗**300**
– : *Disquisitiones arithmeticae* (1801); ∗**417**
Gibbs, J. W.: *Vector Analysis* (1881); **601**
Grassmann, G.: *Die lineale Ausdehnungslehre* (1844); **601**
Guldin, P.: *Centrobaryca* (1635 - 41); **866, 884**
Hamilton, W. R.: *Lectures on Quaternions* (1853); **601**
Harriot, T.: *Artis analyticae praxis* (1631); **104**
Hasper, W.: *Handbuch der Buchdruckerkunst* (1835); ∗**39**
Haÿ, R.-J.,: *Traité de christallographie* (1822); ∗**399**
Hawking, Stephen W.: A Brief History of Time; From the Big Bang to Black Holes (1988); **294**
Hilbert, David: *Grundlagen der Geometrie* (1899); **369**
Hobbes, T.: *Leviathan*; **3**
Høeg, Peter: *Smilla's Sense of Snow* (1993); **70, 370**
Hooke, R.: *Potentia Restitutiva* (1678); ∗**869**
Horace: *Ars poetica*; **17**
Hospital, G. F. de L': *Analyse des infiniment petits* (1696); ∗**679**
Huilier, S. L': *Principorum calculi differentialis er integralis* (1795); **780**
Hulsius, L.: *Erster Tractat Der Mechanischen Instrumenten* (1604); ∗**466**
Hutton, Charles: *A Mathematical and Philosophical Dictionary* (1796); **673**
Huygens, C.: *De ratiociniis in ludo aleae* (1656); **963**
– : *Horlogium Oscillatorium* (1673); ∗**591**
I Ching (*c.* 2000 B.C.); **184**
Isidore of Seville: *Etymologiae* (*c.* A.D. 600); **5**
Jode, C. de: *De quadrante geometrico in quo quidquid* (1594); ∗**490**
Jones, William: *Synopsis palmariorum mathesos* (1706); **85**
al-Kashi, J. M.: *Risala al-muhitiyya* ("Treatise on the Circumference") (1424); **91**
Kepler, J.: *Mysterium Cosmographicum* (1596); ∗**398**
– : *Astronomia Nova* (1609); ∗**398**
– : *Nova stereometria doliorum vinariorum* (1613); ∗**872**
– : *Harmonices mundi* (1619); ∗**399**
– : *Tabulae Rudolphinae* (1627); **152**
al-Khowarizmi: *Liber algorismi de numero indorum* (825; 1120); **48-49**
– : *Hisab al-jabr w'al-musqabalah* (825); **100, 298**
Kolmogorov, A. N.: *Grundbegriffe der Wahrscheinlichkeitsrechnung*; **965**
Lagrange, J. L.: *Théorie des fonctions analytiques* (1797); ∗**686**
– : *Leçons sur le calcul des fonctions* (1800); **686**
Lambert, J. H.: *Theorie der Parallellinien* (1786); **369**
Laplace, P. S.: *Théorie analytique des probabilités* (1812); **964**
Leibniz, G. W. von: *Dissertatio de arte combinatoria* (1666); ∗**217**
– : *Acta Eruditorem* (1694); **336**
– : "Nova methodus pro maximis et minimis" in *Acta Eruditorem* (1694); ∗**676**
L'Hospital, G. F. de, see Hospital, G. F. L'
L'Huilier, S., see Huilier, S. L'
Lincoln, A.: *Gettysburg Address* (1863); **21**
Linné, C. von: *Critica botanica* (1737); **5**
Listing, J. B.: *Vorstudien zur Topologie* (1848); **369**

Luneschloss, J. de: *Thesaurus mathematatum reservatus per algebram novam*; **296**
Maclaurin, Colin: *Treatise of Fluxions* (1742); ***770**
Mandelbrot, B.: "How Long is the Coast of Britain? Statistical Self-Similarity and Fractional Dimension". *Science*, Vol. 156, 1967, *pp.* 636 - 38; **626**
Menelaus: *Sphaerica* (c. A.D. 100); **463**
Milne, A. A.: *The House at Pooh Corner* (1928); ***303, 305, *946**
Monge, Gaspard: *Géométrie descriptive* (1794); **367**
Napier, J.: *Mirifici logarithmorum canonis descriptio* (1614); **152**
– : *A Description of the Admirable Table of Logarithms* (1616); ***151**
– : *Rabdologiae* (1617); **174**
– : *Mirifici logarithmorum canonis constructio* (1619); **152**
Nehemiah: *Textbook on Geometry* (c. A.D. 150); **95**
Neumann, John von, and Oskar Morgenstern: *Theory of Games and Economic Behavior* (1944); **965**
Newton, Isaac: *Methodus fluxionum et serierum infinitorum* (1671); **678**
– : *Philosophiae naturalis principia mathematica* (1687),*"Principia"*; **548, *895**
– : "Tractatus de quadratura curvarum" in *Opticks* (1704); ***677**
– : *The Method of Fluxions and Infinite Series* (1736); ***678**
Otho, Valentinus: *Opus palatinum de triangulis* (1596); **465**
Oughtred, William: *Clavis mathematicae* (1631, 1647); **104**,
– : *Clavis mathematicae* (3rd ed., 1652); ***109, 110**
– : *The Circle of Proportion and the Horizontal Instrument* (1632); **177**
: *Canones sinuum* (1657); **110**
Pacioli, Luca: *De divina proportioni* (1509); ***399, 418**
Pao Chhi-Shou: *Pi Nai Shan Fang Chi* ("Pi Nai Mountain Hut Records"); **214**
Pappus: *Synagoge* (c. A.D. 340); **866**
Pascal, B.: *Essay pour les coniques* (1640); **434**
– : *Traité du triangle arithméthique* (1665); ***141**
Peano, Giuseppe: *Arthmetics principia, nova methodo exposita* (1899); **157**
Pearson, K.: *The Grammar of Science* (1892); **964**
Peet, T. E.: *The Rhind Mathematical Papyrus* (1923); **99**
Penrose, L.S. and R. Penrose: "Impossible Objects: A Special Type of Visual Illusions"; *British Journal of Psychology*, Vol. 49 (1958); **374**
Piltz, A.: *Die gelehrte Welt des Mittelalters*. Cologne: Vöhlan Verlag, 1982; **218, 890**
Pitiscus, B.: *Trigonometria; sive de solutione triangularum tractatus brevis et perspicuus* (1595); **458, 465**
– : *Trigonometria sive de dimensione triangulae* (1600); **458, *459, 465**
– : *A Canon of Triangles* (1614); **465**
Poincaré, J. H.: *Analysis situs* (1895); **369**
Poisson, S.-D.: *Recherches sur la probabilité de jugements* (1837); **964**
Ptolemy: *Almagest* (c. A.D. 150); **367, 461, 463**
: *Syntaxis mathematica* (c. A.D. 150); **463**
Rahn, Johann Heinrich: *Teutsche Algebra* (1659); **105**
Recorde, R.: *Ground of Artes* (1542), 1558 ed.; ***103, 118, 170**
– : *Whetstone of Witte* (1567); ***107**
Regiomontanus, J.: *De triangulis omnimodis* (c. 1464, in print 1533); ***464**
Reisch, Gregor: *Margarita phylosophica* (1446, 1503, 1583); ***115, 116, *218, *463 *870, *890, *922, *962**

Rhæticus, G. J: *De lateribus et angulis triangulorum* (1542); **464**

– : *Canon doctrinae triangulorum* (1551); **464**

Rhind Papyrus (*c.* 1650 B.C.); ✳**98, 134, 297**

Riemann, Bernhard: *Über die Hypothesen, welche der Geometrie zu Grunde liegen* (1854); **383**

Riese, Adam: *Rechenbuch* (1520, 1574); ✳**113**

Robert of Chester: *Liber algorismi de numero indorum* (1120); **49**

Roller, H.: *Perspectiva* (1546); **372**

Rudolff, C.: *Die Coss* (1525); **135**

Ruffini, Paola: *Della insolubilità delle equazioni algebraiche generali di grado superiore al quarto* (1803 - 13); **290**

Russell, B.: *Principles of Mathematics* (1903); **218**

Russell, B., and A. N. Whitehead: *Principia mathematica* (1910, 1912, 1913); **218**

Sanctorius, S.: *Medicina statica* (1614, 1728); ✳**624**

Schedel, Hartmann: *Buch der Cronicken* (1493); ✳**41**

Seki, Kova: *Kai Fukudai no Ho* (1683); **639**

Stevin, Simon: *De Beghinselen der Weegcoonst* (1586); ✳**600**

Stifel, Michael: *Arithmetica integra* (1544); ✳**134, 150**

– : *Die Coss* (1553); ✳**299**

Stör, L.: *Geometria et perspectiva* (1556); ✳**373**, ✳**396**

Tartaglia, Niccolo: *General tractato di numeri et misure*; ✳**114**

Taylor, Brook: *Methodus incrementorum directa et inversa* (1715); ✳**768**, ✳**769**

Taylor, Richard and Andrew Wiles: Article in *Annals of Mathematics* (May 1995); **334**

Trenchant, Jean: *Arithmetique* (1557); **146**

Veblen, O., and J. H. C. Whitehead: *Foundations of Differential Geometry* (1932); **370**

Vesalius, A.: *De humani corporis fabrica* (1543); **96**

Viète, François: *In artem analyticem isagoge* (1591); **299**

– : *Variorum de rebus mathematicis responsorum liber VIII* (1593); **93**

– : *De aequationum recognitione et emendatione* (1615); **319**

Vlacq, Adrian, and E. de Decker: *Het tweede deel van de Nieuwe telkonst* (1627); **152**

– : *Arithmetica logarithmica* (1628); **104, 152**

Wallis, J.: *Arithmetica infinitorum* (1655); **30, 93**

– : *De algebra tractatus* (1685); **87**

Walther, Hans: *Proverbia sententiaeque latinitatis* ; (Göttingen, 1963); **17**

Warren, John: *A Treatise on the Geometrical Representation of the Square Roots of Negative Numbers* (1828); **600**

Wessel, Caspar: *Om Directionens analytiske Betegning* (1797); **88**

White, E. B.: *Charlotte's Web*. HarperCollins Publishers (1952); ✳**781**

White, E. E.: *A Complete Artithmetic*. Cincinnati: Van Antwerp, Brigg & Co. (1870); ✳**124**

Whitehead, J. H. C. and O. Veblen: *Foundations of Differential Geometry* (1932); **370**

Widmann, J.: *Behennde unnd hüpsche Rechnug auff allen Kauffmannschaften* (1489); ✳**103**

Wiener, N.: "The Historical Background of Harmonic Analysis" (1938); extract cited from Felix E. Browder: "Mathematics and the Sciences", in Aspray, William, and Philip Kitcher (editors): *History and Philosophy of Modern Mathematics*. Minneapolis: University of Minnesota Press (1988); **909**

Wiles, Andrew and Richard Taylor: Article in *Annals of Mathematics* (May 1995); **334**

Xylander, G.: *Arithmetica* (1575); **108**

Zenoderus: *On Isometric Figures* (*c.* 180 B.C.); **670**

Name Index

al-KHOWARIZMI, 48, 100, 298

Subject Index

An indexed page number might be the first page of a subject and further information could be found on following pages.

Symbols in Common Use

Notations and symbols are listed in the order of their presentation in the text; only those in current use are included.

Symbol	Page
∞	30
$\frac{a}{b}$; a/b	74
. or , (decimal marker)	75
e	85
π	85
i ; j	87
Re (z)	87
Im (z)	87
$+$	103
$-$	103
$\pm$; $\mp$	104
Δ	104
$\times$ or $\cdot$	104
: or $\div$	105
Σ	105
Π	105
!	106
$=$	108
$\not\approx$	108
$\equiv$	108
$\stackrel{\text{def}}{=}$	108
$\neq$	108
$>$	109
$<$	109
$\geq$	109
$\leq$	109
$>>$	109
$<<$	109
$\widehat{=}$	110
$\sim$ or $\propto$	111
(,); [,]; [, [; (,]; [,)	111-12
( ); [ ]; { }	122
mod	123
z^*, $\bar{z}$	131
a^p	134
$\sqrt{\ }$; $\sqrt[n]{\ }$	135

Symbol	Page
$\log_a$	154
lg	154
ln	154
lb	154
QED	159
${}_nP_p$; ${}^nP_p$; P_p^n ; $P(n, p)$	188
${}_nC_p$; $\binom{n}{p}$	196
$\neg$ or $\sim$	220
$\forall$	220
$\in$	221
$\exists$	221
$\wedge$	221
$\vee$	222
$\Rightarrow$	222
$\Leftrightarrow$	224
$\{\ ,\ldots,\ \}$	233
$\in$	233
$\notin$	233
$\ni$	233
$\not\ni$	233
$\emptyset$	234
$\mathbf{N}$	234
$\mathbf{N}^*$	234
$\mathbf{Z}$	234
$\mathbf{Z}_+$	234
$\mathbf{Z}_-$	234
$\mathbf{Z}^*$	234
$\mathbf{Q}$	234
$\mathbf{Q}_+$	234
$\mathbf{Q}_-$	234
$\mathbf{Q}^*$	234

Symbol	Page
$\mathbf{R}$	234
$\mathbf{R}_+$	234
$\mathbf{R}_-$	234
$\mathbf{R}^*$	234
$\mathbf{C}$	234
$\mathbf{C}^*$	234
$\{\mid\}$; $\{:\}$; $\{;\}$; $\{.\}$	235
$\subseteq$, $\nsubseteq$	236
$\subset$	236
$\supseteq$, $\nsupseteq$	236
$\supset$	236
$\subsetneq$, $\subset$; $\supsetneq$, $\supset$	237
$\mathscr{P}$; P ; P	238
$\cup$	238
$\bigcup$	238
$\cap$	239
$\bigcap$	239
$\setminus$ or $-$	239
$\complement$; $\sim$; $-$	240
(,); (, … ,)	241
$\times$	241
$'$ or $^-$	252
$\oplus$ or $+$	252
$\otimes$ or $\cdot$ or $\times$	252
$\aleph_0, \aleph_1, \aleph_2, \aleph_3$ *etc.*	258
c	262
$\lim_{n \to a}$	270
$f(x)$	336
$f \circ g$; $(f \circ g)(x)$	345
f^{-1}	346

Symbol	Page
rad	410
sr (Ω)	412
$\cong$	427
$\sim$	427
sin	470
cos	470
tan	470
cot	470
sec	470
csc	470
arcsin	477
arccos	477
arctan	477
arccot	477
arcsec	477
arccsc	477
sinh	534
cosh	534
tanh	534
coth	534
sech	534
cosech	534
arsinh	539
arcosh	539
artanh	539
arcoth	539
arsech	539
arcsch	539
$\mathbf{a}$	602
$\lvert\mathbf{a}\rvert$ or a	602
$\mathbf{e}_x, \mathbf{e}_y, \mathbf{e}_z;\ \mathbf{e}_1, \mathbf{e}_2, \mathbf{e}_3$	607
$\mathbf{i}, \mathbf{j}, \mathbf{k}$	607
$\mathbf{a} \cdot \mathbf{b}$	614
$\mathbf{a} \times \mathbf{b}$	617
$\begin{bmatrix} a_{11} & \cdots & a_{1n} \\ \vdots & \vdots & \vdots \\ a_{m1} & \cdots & a_{mn} \end{bmatrix}$	640
$\mathbf{I};\ \mathbf{E}$	641
tr $\mathbf{A}$	641

Symbol	Page
$\begin{vmatrix} a_{11} & \cdots & a_{1n} \\ \vdots & & \vdots \\ a_{n1} & \cdots & a_{nn} \end{vmatrix}$	646
det $\mathbf{A}$	646
$\mathbf{A}^{\mathrm{T}}$	654
$\mathbf{A}^{-1}$	655
$\bar{\mathbf{A}}$	658
$\mathbf{A}^*$	658
$\mathbf{U}$	658
$\frac{\mathrm{d}f(x)}{\mathrm{d}x}$	685
$f'(x), f''(x), f'''(x)$	685
$\frac{\mathrm{d}^n f(x)}{\mathrm{d}x^n}$	686
$\mathrm{D}f(x)$	686
$\mathrm{D}^n f(x)$	686
$f^{(n)}(x)$	686
$\frac{\partial f}{\partial x}$	710
$\frac{\partial^n f}{\partial x^n}$	710
$\frac{\partial^2 f}{\partial y\ \partial x}$	712
$\int f(x)\ \mathrm{d}x$	722
$\int_a^b f(x)\ \mathrm{d}x$	747
$\Big\vert_a^b$	749
$[\ ]_a^b$	749
$\dot{s}$	896-97
$\ddot{s}$	897